New Video Resources

Author in Action videos are available for every learning objective of every section. The authors are all active teachers, and these Camtasia Studio® videos present the concepts as if they were explaining them to their own students.

- **Short, manageable video clips** cover all the important points of each learning objective, and are perfect for studying or reviewing on the students' schedule.

- **Icons** in the text and eText alert students to when a video is available.

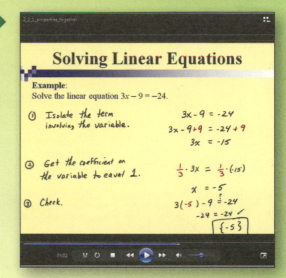

NEW! The **Video Notebook** is a note-taking tool that students use in conjunction with the "Author in Action" videos.

- A **Video Guide** for every section organizes the content by learning objective.

- **Ample space** is provided to allow students to write down important definitions and procedures, as well as show their work on examples, while watching the video.

- The **unbound, three-hole-punched format** allows students to insert additional class notes or homework, helping them build a course notebook and develop good study skills for future classes.

Course: Name:

Instructor: Section:

Section 2.2 Video Guide
Linear Equations: Using the Properties Together

Objectives:
1. Use the Addition and Multiplication Properties of Equality to Solve Linear Equations
2. Combine Like Terms and Use the Distributive Property to Solve Linear Equations
3. Solve a Linear Equation with the Variable on Both Sides of the Equation
4. Use Linear Equations to Solve Problems

Section 2.2 – Objective 1: Use the Addition and Multiplication Properties of Equality to Solve Linear Equations
Video Length – 6:27

We will now solve linear equations where we need to use both the Addition Property of Equality and the Multiplication Property of Equality.

1. **Example:** Solve the linear equation $3x - 9 = -24$.

Write the steps in words	Show the steps with math
Step 1	
Step 2	
Step 3	

Final answer: _____

2. **Example:** Solve the linear equation $\frac{5}{4}x + 2 = 17$.

Final answer: _____

MyMathLab® = Your Resource for Success

In the lab, at home,...

- Access videos, PowerPoint® slides, and animations.
- Complete assigned homework and quizzes.
- Learn from your own personalized Study Plan.
- Print out the Video Notebook for additional practice.
- Explore even more tools for success.

...and on the go.

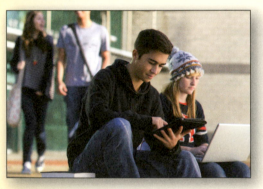

 Download the free MyDashBoard App to see instructor announcements and check your results on your Apple® or Android™ device. MyMathLab log-in required.

 Download the free Pearson eText App to access the full eText on your Apple® or Android™ device. MyMathLab log-in required.

 Use your Chapter Test as a study tool! Chapter Test Prep Videos show step-by-step solutions to all Chapter Test exercises. Access these videos in MyMathLab or by scanning the code.

Scan the code or go to: www.youtube.com/SullivanElementary3e

Don't Miss Out! Log In Today.

MyMathLab delivers proven results in helping individual students succeed. It provides engaging experiences that personalize, stimulate, and measure learning for each student. And, it comes from a trusted partner with educational expertise and an eye on the future.

To learn more about how MyMathLab combines proven learning applications with powerful assessment, visit **www.mymathlab.com**

VIDEOS • POWERPOINT SLIDES • ANIMATIONS • HOMEWORK •
QUIZZES • PERSONALIZED STUDY PLAN • TOOLS FOR SUCCESS

Elementary Algebra

Elementary Algebra

Third Edition

Michael Sullivan, III
Joliet Junior College

Katherine R. Struve
Columbus State Community College

Janet Mazzarella
Southwestern College

PEARSON

Boston   Columbus   Indianapolis   New York   San Francisco   Upper Saddle River
Amsterdam   Cape Town   Dubai   London   Madrid   Milan   Munich   Paris   Montréal   Toronto
Delhi   Mexico City   São Paulo   Sydney   Hong Kong   Seoul   Singapore   Taipei   Tokyo

Editorial Director, Mathematics: Christine Hoag
Editor-in-Chief: Michael Hirsch
Acquisitions Editor: Mary Beckwith
VP/Executive Director of Development & Market Development: Carol Trueheart
Development Editor: Lenore Parens
Executive Content Editor: Kari Heen
Senior Content Editor: Lauren Morse
Editorial Assistant: Matthew Summers
Senior Managing Editor: Karen Wernholm
Senior Project Manager: Patty Bergin
Digital Assets Manager: Marianne Groth
Supplements Production Coordinator: Katherine Roz
Media Producer: Audra Walsh
Content Development Manager: Rebecca Williams
Senior Content Developer: Mary Durnwald
Executive Marketing Manager: Michelle Renda
Marketing Assistant: Susan Mai
Senior Author Support/Technology Specialist: Joe Vetere
Rights and Permissions Advisor: Cheryl Besenjak
Image Manager: Rachel Youdelman
Procurement Manager/Boston: Evelyn Beaton
Procurement Specialist: Linda Cox
Media Procurement Specialist: Ginny Michaud
Associate Director of Design, USHE EMSS/HSC/EDU: Andrea Nix
Senior Design Specialist: Heather Scott
Text Design: Tamara Newnam
Production Management, Composition, and Illustrations: Cenveo Publisher Services
Cover Design: Tamara Newnam
Cover Images: Blue cap—Angelo Gilardelli/Shutterstock; Wood: Kool99/Shutterstock; Hook: Elenamiv/Shutterstock

For permission to use copyrighted material, grateful acknowledgment is made to the copyright holders on page PC-1, which is hereby made part of this copyright page.

Many of the designations used by manufacturers and sellers to distinguish their products are claimed as trademarks. Where those designations appear in this book, and Pearson Education was aware of a trademark claim, the designations have been printed in initial caps or all caps.
Library of Congress Cataloging-in-Publication Data on File

2 3 4 5 6 7 8 9 10—V011—16 15 14 13

www.pearsonhighered.com

ISBN-13: 978-0-321-88015-4 (Student Edition)
ISBN-10: 0-321-88015-3

To my father, Michael Sullivan, an unbelievable mentor,
to the memory of my Mother, who is missed dearly,
and to Stefan and Zofia Wilk, who helped
in so many ways.
—Michael Sullivan

To my husband, Dan Struve, for his
encouragement and support.

—Katherine R. Struve

To my students, who have inspired me more than
I can say; and to the two of whom I am most proud,
my children, Kellen and Jillian.

—Janet Mazzarella

About the Authors

With training in mathematics, statistics, and economics, Michael Sullivan, III has a varied teaching background that includes 23 years of instruction in both high school and college-level mathematics. He is currently a full-time professor of mathematics at Joliet Junior College. Michael has numerous textbooks in publication, including an Introductory Statistics series and a Precalculus series, which he writes with his father, Michael Sullivan.

Michael believes that his experiences writing texts for college-level math and statistics courses give him a unique perspective as to where students are headed once they leave the developmental mathematics tract. This experience is reflected in the philosophy and presentation of his developmental text series. When not in the classroom or writing, Michael enjoys spending time with his three children, Michael, Kevin, and Marissa, and playing golf. Now that his two sons are getting older, he has the opportunity to do both at the same time!

Kathy Struve has been a classroom teacher for nearly 35 years, first at the high school level and, for the past 20 years, at Columbus State Community College. Kathy embraces classroom diversity: diversity of age, learning styles, and previous learning success. She is aware of the challenges of teaching mathematics at a large, urban community college, where students have varied mathematics backgrounds and may enter college with a high level of mathematics anxiety.

Kathy served as Lead Instructor of the Developmental Algebra sequence at Columbus State, where she developed curriculum, conducted workshops, and provided leadership to adjunct faculty in the mathematics department. She embraces the use of technology in instruction, and teaches web and hybrid classes in addition to traditional face-to-face classes. She is always looking for ways to more fully involve students in the learning process. In her spare time Kathy enjoys spending time with her two adult daughters, her two granddaughters, and biking, hiking, and traveling with her husband.

Born and raised in San Diego county, Janet Mazzarella spent her career teaching in culturally and economically diverse high schools before taking a position at Southwestern College 22 years ago. Janet has taught a wide range of mathematics courses from arithmetic through calculus for math/science/engineering majors and has training in mathematics, education, engineering, and accounting.

Janet has worked to incorporate technology into the curriculum by participating in the development of Interactive Math and Math Pro. At Southwestern College, she helped develop the self-paced developmental mathematics program and spent two years serving as its director. In addition, she has been the Chair of the Mathematics Department, the faculty union president, and the faculty coordinator for Intermediate Algebra. In the past, free time consisted of racing motorcycles off-road in the Baja 500 and rock climbing, but recently she has given up the adrenaline rush of these activities for the thrill of traveling in Europe.

Contents

Preface

We would like to thank the reviewers, class testers, and users of the previous edition of *Elementary Algebra* who helped to make the book an overwhelming success. Their thoughtful comments and suggestions provided strong guidance for improvements to the third edition that we believe will enhance this solid, student-friendly text.

The Elementary Algebra course serves a diverse group of students. Some of them are new to algebra, while others were introduced to the material but have not yet grasped all the concepts. Still other students realized success in the course in the past but need a refresher. Not only do the backgrounds of students vary with respect to their mathematical abilities, but students' motivation, reading, and study skills also range considerably.

This diversity makes teaching Elementary Algebra challenging. It is imperative that texts recognize the diversity of the classroom and address the array of needs of the many students.

Elementary Algebra introduces students to the logic and precision of mathematics. We expect students to leave the course with an appreciation of this precision as well as of the power of mathematics. Our students need to understand that the concepts we teach in this course form the basis for future mathematics courses. Once they have a conceptual understanding of algebra, students recognize that the material is not merely a series of unconnected topics. Instead, they see a story in which each new chapter builds on concepts learned in previous chapters.

To reinforce this idea, we remind our students of a helpful fact—mathematics is about taking a problem and reducing it to another problem that we have already seen. Taking a problem and reducing it to parts that are easier to solve help students to see the forest for the trees (and, to carry the metaphor further, prevent them from feeling that they are lost in the woods).

In short, to address the many needs of today's Elementary Algebra students, we established the following as our goals for this text:

- Provide students with a strong conceptual foundation in mathematics through a clear, distinct, thorough presentation of concepts.
- Offer comprehensive exercise sets that build students' skills, show various intriguing applications of mathematics, begin to build mathematical thinking skills, and reinforce mathematical concepts.
- Provide students with ample opportunity to see the connections among the various topics learned in the course.
- Present a variety of study aids and tips so students quickly come to view the text as a useful and reliable tool that can increase success in the course.

New to the Third Edition

- The popular Author in Action videos have been edited so that the length of each video is under 12 minutes. Therefore, some objectives will have more than one lecture video. Students will be alerted to the availability of a video with the ▶ icon. The videos are available in MyMathLab, the Multimedia Textbook (in MyMathLab), or on DVD.
- A new video notebook will be available with the third edition—ideal for online, emporium/redesign courses, or inverted classrooms. This video notebook assists students in taking thorough, organized, and understandable notes as they watch the Author in Action videos by asking students to complete definitions, procedures, and examples based on the content of the videos.
- The Quick Check exercises now have more fill-in-the-blank and True/False questions to assess the student's understanding of vocabulary and formulas. The end-of-section exercises went through a thorough review. The goal of the review was to

be sure that all the various problem types are covered within the exercise set. The result is a comprehensive set of exercises that slowly build in level of difficulty.

- The text went through an extensive review of the exposition with the goal of reducing word count and making the reading more accessible to students. Now, the text provides concise explanations with a page design that is friendly and inviting.

Develop an Effective Text for Use In and Out of the Classroom

Given the hectic lives led by most students, coupled with the anxiety and trepidation with which they approach this course, an outstanding developmental mathematics text must provide pedagogical support that makes the text valuable to students as they study and do assignments. Pedagogy must be presented within a framework that teaches students how to study math; pedagogical devices must also address what students see as the "mystery" of mathematics—and solve that mystery.

To encourage students and to clarify the material, we developed a set of pedagogical features that help students develop good study skills, garner an understanding of the connections between topics, and work smarter in the process. The pedagogy used is based upon the more than 70 years of classroom teaching experience that the authors bring to this text.

Examples are often the determining factor in how valuable a textbook is to a student. Students look to examples to provide them with guidance and instruction when they need it most—the times when they are away from the instructor and the classroom. We have developed two example formats in an attempt to provide superior guidance and instruction for the students.

Innovative Sullivan/Struve Examples

The innovative *Sullivan/Struve Example* has a two-column format in which annotations are provided to the **left** of the algebra, rather than the right, as is the practice in most texts. Because we read from **left to right,** placing the annotation on the left will make more sense to the student. It becomes clear that the annotation describes what we are about to do instead of what was just done. The annotations may be thought of as the teacher's voice offering clarification immediately before writing the next step in the solution on the board. Consider the following:

EXAMPLE 3   **Combining Like Terms to Solve a Linear Equation**

Solve the equation: $2x - 6 + 3x = 14$

Solution

$$2x - 6 + 3x = 14$$

Combine like terms: $5x - 6 = 14$

Add 6 to each side of the equation: $5x - 6 + 6 = 14 + 6$

$$5x = 20$$

Divide both sides by 5: $\dfrac{5x}{5} = \dfrac{20}{5}$

$$x = 4$$

Check

$$2x - 6 + 3x = 14$$

Substitute 4 for x in the original equation: $2(4) - 6 + 3(4) \overset{?}{=} 14$

$$8 - 6 + 12 \overset{?}{=} 14$$

$$14 = 14 \quad \text{True}$$

The solution of the equation is 4, or the solution set is $\{4\}$.   ●

Quick ✓

In Problems 6–9, solve each equation.

6. $7b - 3b + 3 = 11$ **7.** $-3a + 4 + 4a = 13 - 27$

8. $6c - 2 + 2c = 18$ **9.** $-12 = 5x - 3x + 4$

Showcase Examples

Showcase Examples are used strategically to introduce key topics or important problem-solving techniques. These examples provide "how-to" instruction by offering a guided, step-by-step approach to solving a problem. Students can then immediately see how each of the steps is employed. We remind students that the *Showcase Example* is meant to provide "how-to" instruction by including the words "how to" in the example title. The *Showcase Example* has a three-column format in which the left column describes a step, the middle column provides a brief annotation, as needed, to explain the step, and the right column presents the algebra. With this format, students can see each step in the problem-solving process in context so that the steps make more sense. This approach is more effective than simply stating each step in the text.

EXAMPLE 3 **How to Solve an Inequality Using the Addition Property of Inequality**

Solve the linear inequality $4y - 5 \leq 3y - 2$. Express the solution set using set-builder notation and interval notation. Graph the solution set.

Step-by-Step Solution

Step 1: Get the terms containing variables on the left side of the inequality.	Subtract 3y from both sides:	$4y - 5 \leq 3y - 2$ $4y - 5 - 3y \leq 3y - 2 - 3y$ $y - 5 \leq -2$
Step 2: Isolate the variable y on the left side.	Add 5 to both sides:	$y - 5 + 5 \leq -2 + 5$ $y \leq 3$

Figure 14

The solution set using set-builder notation is $\{y \mid y \leq 3\}$. The solution set using interval notation is $(-\infty, 3]$. The solution set is graphed in Figure 14. ●

Quick ✓

12. To _____ an inequality means to find the set of all values of the variable for which the statement is true.

13. Write the inequality that results from adding 7 to each side of $3 < 8$. What property of inequalities does this illustrate?

In Problems 14–17, find the solution of the linear inequality. Express the solution set in set-builder notation and interval notation. Graph the solution set.

14. $n - 2 > 1$

15. $-2x + 3 < 7 - 3x$

16. $5n + 8 \leq 4n + 4$

17. $3(4x - 8) + 12 > 11x - 13$

Quick Check Exercises

Placed at the conclusion of most examples, the *Quick Check* exercises provide students with an opportunity for immediate reinforcement. By working the problems that mirror the example just presented, students get instant feedback and gain confidence in their understanding of the concept. All *Quick Check* exercise answers are provided in the back of the text. **The *Quick Check* exercises should be assigned as homework to encourage students to read, consult, and use the text regularly.**

Superior Exercise Sets: Paired with Purpose

Students learn algebra by doing algebra. The superior end-of-section exercise sets in this text provide students with ample practice of both procedures and concepts. The exercises are paired and present problem types with every possible derivative. The exercises also present a gradual increase in difficulty level. The early, basic exercises keep the student's focus on as few "levels of understanding" as possible. The later or higher-numbered exercises are "multi-task" (or Mixed Practice) exercises where students are required to utilize multiple skills, concepts, or problem-solving techniques.

Throughout the textbook, the exercise sets are grouped into eight categories—some of which appear only as needed:

1. **Are You Ready? . . .** problems are located at the opening of the section. They are problems that deal with prerequisite material for the section along with page references so students may remediate, if necessary. Answers to the Ready? . . . problems appear as a footnote on the page.

2. **Quick Check** exercises, which provide the impetus to get students into the text, follow most examples and are numbered sequentially as the first problems in each section exercise set. By doing these problems as homework and the first exercises attempted, the student is directed into the material in the section. If a student gets stuck, he or she will learn that the example immediately preceding the Quick Check exercise illustrates the concepts needed to solve the problem.

3. **Building Skills** exercises are drill problems that develop the student's understanding of the procedures and skills in working with the methods presented in the section. These exercises can be linked back to a single learning objective in the section. Notice that the Building Skills problems begin the numbering scheme where the Quick Checks leave off. For example, if the last Quick Check exercise is Problem 20, then we begin the Building Skills exercises with Problem 21. This serves as a reminder that Quick Check exercises should be assigned as homework.

4. **Mixed Practice** exercises are also drill problems, but they offer a comprehensive assessment of the skills learned in the section by asking problems that relate to more than one concept or objective. In addition, we may present problems from previous sections so students must first recognize the type of problem and then employ the appropriate technique to solve the problem.

5. **Applying the Concepts** exercises are problems that allow students to see the relevance of the material learned within the section. Problems in this category either are situational problems that use material learned in the section to solve "real-world" problems or are problems that ask a series of questions to enhance a student's conceptual understanding of the mathematics presented in the section.

6. **Extending the Concepts** exercises can be thought of as problems that go beyond the basics. Within this block of exercises an instructor will find a variety of problems to sharpen students' critical-thinking skills.

7. **Explaining the Concepts** problems require students to think about the big picture concepts of the section and express these ideas in their own words. It is our belief that students need to improve their ability to communicate complicated ideas both orally and in writing. When they are able to explain mathematical methods or concepts to another individual, they have truly mastered the ideas. These problems can serve as a basis for classroom discussion or can be used as writing assignments.

8. Finally, we include coverage of the **graphing calculator.** Instructors' philosophies about the use of graphing devices vary considerably. Because instructors disagree about the value of this tool, we have made an effort to make graphing technology

entirely optional. When appropriate, technology exercises are included at the close of a section's exercise set.

Problem Icons In addition to the carefully structured categories of exercises, selected problems are flagged with icons to denote that:

- Problems whose number is green have complete worked-out solutions found in MyMathLab.
- △ These problems focus on geometry concepts.
- ⌨ A calculator will be useful in working the problem.

Hallmark Features

Quick Check Exercises: Encourage Study Skills that Lead to Independent Learning

What is one of the overarching goals of an education? We believe it is to learn to solve problems independently. In particular, we would like to see students develop the ability to pick up a text or manual and teach themselves the skills they need. In our mathematics classes, however, we are often frustrated because students rarely read the text and often struggle to understand the concepts independently.

To encourage students to use the text more effectively and to help them achieve greater success in the course, we have structured the exercises in the third edition of our text differently from other mathematics textbooks. The aim of this structure is to get students "into the text" in order to increase their ability and confidence to work any math problem—particularly when they are away from the classroom and an instructor who can help.

Each section's exercise set begins with the *Quick Check* exercises. The *Quick Checks* are consecutively numbered. The end-of-section exercises begin their numbering scheme based on where the *Quick Checks* end. For example:

- Section 1.2: *Quick Checks* end at Problem 48, so the end-of-section exercise set starts with Problem 49 (see page 16).
- Section 1.3: *Quick Checks* end at Problem 24, so the end-of-section exercise set starts with Problem 25 (see page 25).

The *Quick Checks* follow most examples and provide the platform for students to get "into the text." By integrating these exercises into the exercise set, we direct students to the instructional material in that section. Our hope is that students will then become more aware of the instructional value of the text and will be more likely to succeed when studying away from the classroom and the instructor.

Answer annotations to Quick Checks and exercises have been placed directly next to each problem in the Annotated Instructor's Edition to make it easier for instructors to create assignments.

We have used the same background color for the Quick Checks and the exercise sets to reinforce the connection between them visually. The colored background will also make the Quick Checks easier to find on the page.

Answers to Selected Exercises at the back of the text integrate the answers to *every* Quick Check exercise with the answers to *every odd* problem from the section exercise sets.

Study Skills and Student Success

We have included study skills and student success as regular themes throughout this text starting with *Section 1.1, Success in Mathematics*. In addition to this dedicated section that covers many of the basics that are essential to success in any math course, we have included several recurring study aids that appear in the margin. These features are designed to anticipate the student's needs and to provide immediate help—as if the teacher were looking over his or her shoulder. These margin features include *In Words; Work Smart;* and *Work Smart: Study Skills.*

Section 1.1: *Success in Mathematics* focuses the student on basic study skills, including what to do during the first week of the semester; what to do before, during, and after class; how to use the text effectively; and how to prepare for an exam.

In Words helps to address the difficulty that students have in reading mathematically precise definitions and theorems by explaining them in plain English.

Work Smart provides "tricks of the trade" hints, tips, reminders, and alerts. It also identifies some common errors to avoid and helps students work more efficiently.

Work Smart: Study Skills reminds students of study skills that will help them to succeed at various points in the course. Attention to these practices will help them to become better, more proficient learners.

Test Preparation and Student Success

The Chapter Tests in this text and the companion Chapter Test Prep Videos have been designed to help students make the most of their valuable study time.

Chapter Test In preparation for their classroom test, students should take the practice test to make sure they understand the key topics in the chapter. The exercises in the Chapter Tests have been crafted to reflect the level and types of exercises a student is likely to see on a classroom test.

Chapter Test Prep Videos The Chapter Test Prep Videos provide students with help at the critical juncture when they are studying for a test. The videos present step-by-step solutions to the exact exercises found in each of the book's Chapter Tests. Easy video navigation allows students instant access to the worked-out solutions to the exercises they want to study or review. These videos are available in MyMathLab or the video lecture series DVD.

Seeing the Connections: The Big Picture

Another important role of the pedagogy in this text is to help students see and understand the connection between the mathematical topics being presented. Several section-opening and margin features help to reinforce connections:

The Big Picture: Putting It Together (Chapter Opener) This feature is based on how we start each chapter in the classroom—with a quick sketch of what we plan to cover. Before tackling a chapter, we tie concepts and techniques together by summarizing material covered previously and then relate these ideas to material we are about to discuss. It is important for students to understand that content truly builds from one chapter to the next. We find that students need to be reminded that the familiar operations of addition, subtraction, multiplication, and division are being applied to different or more complex objects.

Are You Ready for This Section? As part of this building process, we think it is important to remind students of specific skills that they will need from earlier in the course to be successful within a given section. The *Are You Ready? . . .* feature that begins each section not only provides a list of prerequisite skills that a student should understand before tackling the content of a new section but also acts as a short quiz to test students' preparedness. Answers to the quiz are provided in a footnote on the same page, and a cross-reference to the material in the text is provided so that the student can remediate when necessary.

Mixed Practice These problems exist within each end-of-section exercise set and draw upon material learned from multiple objectives. Sometimes, these problems simply represent a mixture of problems presented within the section, but they also may include a mixture of problems from various sections. For example, students may need to distinguish between linear and quadratic equations or students may need to distinguish between the direction *simplify* versus *solve*.

Putting the Concepts Together (**Mid-Chapter Review**) Each chapter has a group of exercises at the appropriate point in the chapter, entitled *Putting the Concepts Together*. These exercises serve as a review—synthesizing material introduced up to that point in the chapter. The exercises in these mid-chapter reviews are carefully chosen to assist students in seeing the "big picture."

Cumulative Review Learning algebra is a building process, and building involves considerable reinforcement. The Cumulative Review exercises at the end of each odd-numbered chapter, starting with Chapter 3, help students to reinforce and solidify their knowledge by revisiting concepts and using them in context. This way, studying for the final exam should be fairly easy.

In Closing

When we started writing this textbook, we discussed what improvement we could make in coverage; in staples such as examples and problems; and in any pedagogical features that we found truly useful. After writing and rewriting, and reading many thoughtful reviews from instructors, we focused on the following features of the text to set it apart.

- The **innovative *Sullivan/Struve Examples*** and ***Showcase Examples*** provide students with superior guidance and instruction when they need it most—when they are away from the instructor and the classroom. Each of the margin features ***In Words, Work Smart,*** and ***Work Smart: Study Skills*** are designed to improve study skills, make the textbook easier to navigate, and increase student success.

- **Exercise Sets: Paired with Purpose**—The exercise sets are structured to assess student understanding of vocabulary, concepts, drill, problem solving, and applications. The exercise sets are graded in difficulty level to build confidence and to enhance students' mathematical thinking. The ***Quick Check*** exercises provide students with immediate reinforcement and instant feedback to determine their understanding of the concepts presented in the examples.

- **The Big Picture**—Each section opens with a *Preparing for This Section quiz* that allows students to review material learned earlier in the course that is needed in the upcoming section. **Mixed Practice** problems require students to utilize material learned from multiple objectives to solve a problem. Often, these problems require students to first determine the correct approach to solving the problem prior to actually solving it. ***Putting the Concepts Together*** helps students see the big picture and provide a structure for learning each new concept and skill in the course.

Student and Instructor Resources

STUDENT RESOURCES

Available for purchase at MyPearsonStore.com

Student's Solutions Manual
(ISBN: 0321879910/9780321879912)
Complete worked solutions to the odd-numbered problems in the end-of-section exercises and all of the Quick Checks and end-of-chapter exercises.

***Do the Math* Workbook**
(ISBN: 032188003X/9780321880031)
A collection of 5-Minute Warm-Up exercises; Guided Practice exercises; and *Do the Math* exercises for each section in the text. These exercises are designed for students to show their work in homework, during class, or in a lab setting.

Author in Action Videos on DVD for *Elementary Algebra*, 3/e
(ISBN: 0321880021/9780321880024)
Keyed to each section in the text, objective specific mini-lectures provide a 12-minute review of the key concepts. Videos include optional subtitles in English and Spanish. The Chapter Test Prep videos are included in the package.

Video Notebook
(ISBN: 0321881389/9780321881380)
Ideal for online, emporium based/redesign courses or inverted classrooms. Students learn to take organized notes as they watch the Author in Action videos. Students are asked to complete definitions, procedures, and examples based on the videos.

INSTRUCTOR RESOURCES

Available through your Pearson representative

Annotated Instructor's Edition
(ISBN: 0321879961/9780321879967)

Instructor's Solutions Manual
(Available for download from the IRC)

Instructor's Resource Manual
(Available for download from the IRC)
Includes Mini-Lectures (one-page lesson plans at-a-glance) for each section in the text that feature key examples and Teaching Tips on how students respond to the material. Printed Test Forms (free response and multiple choice), Additional Exercises.

PowerPoint Lecture Slides
(Available for download from the IRC)

TestGen® for *Elementary Algebra*, 3/e
(Available for download from the IRC)

Online Resources

MyMathLab® (access code required)

MathXL® (access code required)

Acknowledgments

Textbooks are written by authors but evolve through the efforts of many people. We would like to extend our thanks to the following individuals for their important contributions to the project. From Pearson: Mary Beckwith, Michael Hirsch, Lauren Morse, Melissa Parkin, Michelle Renda, Chris Hoag, Patty Bergin, Rose Kernan, Heather Scott and, finally, the Pearson Arts & Sciences sales team for their confidence and support of our books.

We would also like to thank George Seki, Cindy Trimble and John Bialas for their attention to details and consistency in accuracy checking the text and answer sections and Val Villegas for his work on the video notebook. We offer many thanks to all the instructors from across the country who participated in reviewer conferences and focus groups, reviewed or class-tested some aspect of the manuscript, and taught from the previous editions. Their insights and ideas form the backbone of this text. Hundreds of instructors contributed their time, energy, and ideas to help us shape this text. We will attempt to thank them all here. We apologize for any omissions.

The following individuals, many of whom reviewed or class-tested the previous edition, provided direction and guidance in shaping the third edition.

Marwan Abu–Sawwa, *Florida Community College—Jacksonville*
Darla Aguilar, *Pima State University*
Grant Alexander, *Joliet Junior College*
Philip Anderson, *South Plains College*
MaryAnne Anthony, *Santa Ana College*
Mary Lou Baker, *Columbia State Community College*
Bill Bales, *Rogers State*
Tony Barcellos, *American River College*
John Beachy, *Northern Illinois University*
Donna Beatty, *Ventura College*
David Bell, *Florida Community College—Jacksonville*
Sandy Berry, *Hinds Community College*
John Bialas, *Joliet Junior College*
Linda Blanco, *Joliet Junior College*
Kevin Bodden, *Lewis and Clark College*
Rebecca Bonk, *Joliet Junior College*
Cherie Bowers, *Santa Ana College*
Becky Bradshaw, *Lake Superior College*
Lori Braselton, *Georgia Southern University*
Tim Britt, *Jackson State Community College*
Beverly Broomell, *Suffolk Community College*
Joanne Brunner, *Joliet Junior College*
Hien Bui, *Hillsborough Community College—Dale Mabry*
Connie Buller, *Metropolitan Community College*
Annette Burden, *Youngstown State University*
James Butterbach, *Joliet Junior College*
Marc Campbell, *Daytona Beach Community College*
Elena Catoiu, *Joliet Junior College*

Nancy Chell, *Anne Arundel Community College*
John F. Close, *Salt Lake Community College*
Bobbi Cook, *Indian River Community College*
Carlos Corona, *San Antonio College*
Faye Dang, *Joliet Junior College*
Shirley Davis, *South Plains College*
Vivian Dennis-Monzingo, *Eastfield College*
Alvio Dominguez, *Miami Dade College—Wolfson*
Karen Driskell, *South Plains College*
Thomas Drucker, *University of Wisconsin—Whitewater*
Brenda Dugas, *McNeese State University*
Doug Dunbar, *Okaloosa-Walton Junior College*
Laura Dyer, *Southwestern Illinois State University*
Bill Echols, *Houston Community College—Northwest*
Erica Egizio, *Lewis University*
Laura Egner, *Joliet Junior College*
Jason Eltrevoog, *Joliet Junior College*
Nancy Eschen, *Florida College Jacksonville*
Mike Everett, *Santa Ana College*
Phil Everett, *Ohio State University*
Scott Fallstrom, *Shoreline Community College*
Betsy Farber, *Bucks County Community College*
Fitzroy Farquharson, *Valencia Community College—West*
Jacqueline Fowler, *South Plains College*
Dorothy French, *Community College of Philadelphia*

Randy Gallaher, *Lewis and Clark College*
Sanford Geraci, *Broward Community College*
Donna Gerken, *Miami Dade College—Kendall*
Adrienne Goldstein, *Miami Dade College—Kendall*
Marion Graziano, *Montgomery County Community College*
Susan Grody, *Broward College*
Tom Grogan, *Cincinnati State University*
Barbara Grover, *Salt Lake Community College*
Shawna Haider, *Salt Lake Community College*
Margaret Harris, *Milwaukee Area Technical College*
Teresa Hasenauer, *Indian River College*
Mary Henderson, *Okaloosa-Walton Junior College*
Celeste Hernandez, *Richland College*
Paul Hernandez, *Palo Alto College*
Pete Herrera, *Southwestern College*
Bob Hervey, *Hillsborough College—Dale Mabry*
Teresa Hodge, *Broward College*
Sandee House, *Georgia Perimeter College*
Becky Hubiak, *Tidewater Community College—Virginia Beach*
Sally Jackman, *Richland College*
John Jarvis, *Utah Valley State College*
Nancy Johnson, *Broward College*
Steven Kahn, *Anne Arundel Community College*
Linda Kass, *Bergen Community College*
Donna Katula, *Joliet Junior College*
Mohammed Kazemi, *University of North Carolina—Charlotte*

Doreen Kelly, *Mesa Community College*
Mike Kirby, *Tidewater Community College—Virginia Beach*
Keith Kuchar, *College of Dupage*
Carla Kulinsky, *Salt Lake Community College*
Julie Labbiento, *Leigh Carbon Community College*
Kathy Lavelle, *Westchester Community College*
Deanna Li, *North Seattle Community College*
Brian Macon, *Valencia Community College—West*
Lynn Marecek, *Santa Ana College*
Jim Matovina, *Community College of Southern Nevada*
Jean McArthur, *Joliet Junior College*
Michael McComas, *Marshall University*
Mikal McDowell, *Cedar Valley College*
Lee McEwen, *Ohio State University*
David McGuire, *Joliet Junior College*
Angela McNulty, *Joliet Junior College*
Debbie McQueen, *Fullerton College*
Judy Meckley, *Joliet Junior College*
Lynette Meslinsky, *Erie Community College—City Campus*
Kausha Miller, *Lexington Community College*
Chris Mizell, *Okaloosa Walton Junior College*
Jim Moore, *Madison Area Technical College*
Ronald Moore, *Florida College Jacksonville*
Elizabeth Morrison, *Valencia College—West*
Roya Namavar, *Rogers State University*
Hossein Navid-Tabrizi, *Houston Community College*
Carol Nessmith, *Georgia Southern University*

Kim Neuburger, *Portland Community College*
Larry Newberry, *Glendale Community College*
Elsie Newman, *Owens Community College*
Charlotte Newsome, *Tidewater Community College*
Charles Odion, *Houston Community College*
Viann Olson, *Rochester Community and Technical College*
Linda Padilla, *Joliet Junior College*
Carol Perry, *Marshall Community and Technical College*
Faith Peters, *Miami Dade College—Wolfson*
Dr. Eugenia Peterson, *Richard J. Daley College*
Jean Pierre-Victor, *Richard J. Daley College*
Philip Pina, *Florida Atlantic University*
Carol Poos, *Southwestern Illinois University*
Elise Price, *Tarrant County College*
R.B. Pruitt, *South Plains College*
William Radulovich, *Florida College Jacksonville*
Pavlov Rameau, *Miami Dade College—Wolfson*
David Ray, *University of Tennessee—Martin*
Nancy Ressler, *Oakton Community College*
Michael Reynolds, *Valencia College—West*
George Rhys, *College of the Canyons*
Jorge Romero, *Hillsborough College—Dale Mabry*
David Ruffato, *Joliet Junior College*
Carol Rychly, *Augusta State University*
David Santos, *Community College of Philadelphia*

Togba Sapolucia, *Houston Community College*
Doug Smith, *Tarrant County College*
Catherine J.W. Snyder, *Alfred State College*
Gisela Spieler-Persad, *Rio Hondo College*
Raju Sriram, *Okaloosa-Walton Junior College*
Patrick Stevens, *Joliet Junior College*
Bryan Stewart, *Tarrant County College*
Jennifer Strehler, *Oakton Community College*
Elizabeth Suco, *Miami Dade College—Wolfson*
Katalin Szucs, *East Carolina University*
KD Taylor, *Utah Valley State College*
Mary Ann Teel, *University of North Texas*
Suzanne Topp, *Salt Lake Community College*
Suzanne Trabucco, *Nassau Community College*
Jo Tucker, *Tarrant County College*
Bob Tuskey, *Joliet Junior College*
Mary Vachon, *San Joaquin Delta College*
Carol Walker, *Hinds Community College*
Kim Ward, *Eastern Connecticut State University*
Richard Watkins, *Tidewater Community College*
Natalie Weaver, *Daytona Beach College*
Darren Wiberg, *Utah Valley State College*
Rachel Wieland, *Bergen Community College*
Christine Wilson, *Western Virginia University*
Brad Wind, *Miami Dade College—North*
Roberta Yellott, *McNeese State University*
Steve Zuro, *Joliet Junior College*

Additional Acknowledgments

We also would like to extend thanks to our colleagues at Joliet Junior College, Columbus State Community College, and Southwestern College, who provided encouragement, support, and the teaching environment where the ideas and teaching philosophies in this text were developed.

Michael Sullivan, III
Katherine R. Struve
Janet Mazzarella

Elementary Algebra

CHAPTER

1 Operations on Real Numbers and Algebraic Expressions

In the year 1202, the Italian mathematician Leonardo Fibonacci posed this problem: A certain man put a pair of rabbits in a place surrounded on all sides by a wall. How many pairs of rabbits can be produced from that pair in a year if every month each pair begets a new pair that is productive from the second month on?

The answer to Fibonacci's puzzle leads to a sequence of numbers called the *Fibonacci sequence*. See Problem 155 in Section 1.4.

The Big Picture: Putting It Together

Welcome to algebra! This course is taken by a diverse group of individuals. Some of you may never have taken an algebra course, while others may have taken algebra at some time in the past. In any case, we have written this text with both groups in mind.

The first chapter of the text reviews arithmetic. The material is presented with an eye on the future, which is algebra. This means that we will slowly build our discussion so that the shift from arithmetic to algebra is painless. Carefully study the methods used in this section, because these same methods will be used again in later chapters.

1.1 Success in Mathematics

Objectives

1 What to Do the First Week of the Semester

2 What to Do Before, During, and After Class

3 How to Use the Text Effectively

4 How to Prepare for an Exam

Let's start by having a discussion about the "big picture" goals of the course and how this text can help you to be successful at mathematics. Our first "big picture" goal is to develop algebraic skills and gain an appreciation for the power of algebra and mathematics. But there is also a second "big picture" goal. By studying mathematics, we develop a sense of logic and exercise the part of our brains that deals with logical thinking. The examples and problems in this text are like the crunches we do in a gym to exercise our bodies. The goal of running or walking is to get from point A to point B, so doing fifty crunches on a mat does not accomplish that goal, but crunches do make our upper bodies, backs, and hearts stronger when we need to run or walk.

Logical thinking can assist us in solving difficult everyday problems, and solving algebra problems "builds the muscles" in the part of our brain that performs logical thinking. So, when you are studying algebra and getting frustrated with the amount of work that needs to be done, and you say, "My brain hurts," remember the phrase we all use in the gym, "No pain, no gain."

Another phrase to keep in mind is "Success breeds success." Mathematics is everywhere. You already are successful at doing some everyday mathematics. With practice, you can take your initial successes and become even more successful. Have you ever done any of the following everyday activities?

- Compare the price per ounce of different sizes of jars of peanut butter or jam.
- Leave a tip at a restaurant.
- Figure out how many calories your bowl of breakfast cereal provides.
- Take an opinion survey along with many other people.
- Measure the distances between cities as you plan a vacation.
- Order the appropriate number of gallons of paint to cover the walls of a room.
- Buy a car and take out a car loan with interest.
- Double a cookie recipe.
- Exchange American dollars for Canadian dollars.
- Fill up a basketball or soccer ball with air (balls are spheres, after all).
- Coach a Little League team (scores, statistics, catching, and throwing all involve math).
- Check the percentages of saturated and unsaturated fats in a chocolate bar.

You may do five or ten mathematical activities in a single day! The everyday mathematics that you already know is the foundation for your success in this course.

1 What to Do the First Week of the Semester

The first week of the semester gives you the opportunity to prepare for a successful course. Here are the things you should do:

1. **Pick a good seat.** Choose a seat that gives you a good view of the room. Sit close enough to the front so you can easily see the board and hear the professor.

2. **Read the syllabus to learn about your instructor and the course.** Take note of your instructor's name, office location, e-mail address, telephone number, and office hours. Pay attention to any additional help available, such as tutoring centers, videos in the library, software, online tutorials, and so on. Be sure you fully understand all of the instructor's policies for the class, including the policy on absences, missed exams or quizzes, and homework. Ask questions.

3. **Learn the names of some of your classmates and exchange contact information.** One of the best ways to learn math is through group study sessions. Try to create time each week to study with your classmates. Knowing how to get in contact with classmates is also useful if you ever miss class, because you can obtain the assignment for the day.

4. Budget your time. Most students have a tendency to "bite off more than they can chew." To help with time management, consider the following general rule: Plan on studying *at least* two hours outside of class for each hour in class. Thus, if you enrolled in a four-hour math class, you should set aside at least eight hours each week to study for the course. You will also need to set aside time for other courses. Consider your work schedule and personal life when creating your time budget.

❷ What to Do Before, During, and After Class

Now that the semester is under way, we present the following ideas for what to do before, during, and after each class meeting. These suggestions may sound overwhelming, but we guarantee that by following them, you will be successful in mathematics (and other courses). Also, you will find that studying for exams becomes much easier by following this plan.

Before Class Begins

1. Read the section or sections that will be covered in the upcoming class meeting. Watch the video lectures that accompany the text.

2. Based on your reading, write down a list of questions. Your questions will probably be answered through the lecture. You can then ask any questions that are not answered completely.

3. Make sure you are mentally prepared for class. Your mind should be alert and ready to concentrate for the entire class. (Invest in a cup of coffee and eat lots of protein for breakfast!)

During Class

1. Arrive early enough to prepare your mind and material for the lecture.

2. Stay alert. Do not doze off or daydream during class. If you do so, understanding the lecture will be very difficult when you "return to class."

3. Take thorough notes. It is normal not to understand certain topics the first time you hear them in a lecture. However, this does not mean that you throw your hands up in despair. Rather, continue to take class notes.

4. Do not be afraid to ask questions. In fact, instructors love questions, for two reasons. First, if one student has a question, other students probably have the same question. Second, by asking questions, you teach the teacher what topics cause difficulty.

After Class

1. Reread (and possibly rewrite) your class notes. In our experience as students, we were amazed how often our confusion during class disappeared after we studied our in-class notes after class.

2. Reread the section. This is an especially important step. Once you have heard the lecture, the section will make more sense and you will understand much more.

3. Do your homework. **Homework is not optional.** There is an old Chinese proverb that says,

> I hear … and I forget
>
> I see … and I remember
>
> I do … and I understand

This proverb applies to any situation in life in which you want to succeed. Would a pianist expect to be the best if she didn't practice? The only way you are going to learn algebra is by doing algebra.

4. When you get a problem wrong, try to figure out why you got the problem wrong. If you can't discover your error, be sure to ask for help.

5. If you have questions, visit your professor during office hours. You can also ask someone in your study group or go to the tutoring center on campus, if available.

Learning Is a Building Process

Learning is the art of making connections between thousands of neurons (specialized cells) in the brain. Memory is the ability to reactivate these neural networks—it is a conversation among neurons.

Math isn't a mystery. You already know some math. But you do have to practice what you know and expand your knowledge. Why? The brain contains thousands of neurons. Through repeated practice, signals in the brain travel faster. The cells "fire" more quickly, and connections are made faster and with less effort. Practice allows us to retrieve concepts and facts at test time. Remember those crunches, which are a way of making your body more robust and nimble—learning does the same to your brain.

Have We Mentioned Asking Questions?

To move information from short-term memory to long-term memory, we need to think about the information, comprehend its meaning, and ask questions about it.

❸ How to Use the Text Effectively

When we sat down to write this text, we knew from our teaching experience that students typically do not read their math books. We decided to accept the steps students usually go through:

1. Attend the lecture and watch the instructor do some problems on the board. Perhaps work some problems in class.

2. Go home and work on the homework assignment.

3. After each problem, check the answer in the back of the text. If you were right, move on, but if wrong, go back and see where the solution went wrong.

4. If the mistake cannot be identified, go to your class notes or try to find a similar example in the text. With a little luck, the student can determine where the solution went wrong in the problem.

5. If Step 4 fails, mark the problem and ask about it in the next class meeting, which leads us back to Step 1.

With this model in mind, we started to develop this text so that there is more than one way to extract the information you need from it.

All of the features in the text are here to help you succeed. These features are based on techniques we use in class. The features that appear, an explanation of the purpose of each feature, and how each can be used to help you succeed in this course are outlined in the following paragraphs.

Are You Ready for This Section?: Warming Up

Each section (after Section 1.2) begins with a short "readiness quiz." This quiz asks questions about material that was presented earlier in the course and is needed for the upcoming section. Take the readiness quiz to be sure you understand the material that the new section is based on. Answers to the quiz appear in a footnote on the page of the quiz. Check your answers. If you get a problem wrong or don't know how to do a problem, go back to the section listed and review the material.

Objectives: A "Road Map" through the Course

To the left of the readiness quiz is a list of objectives to be covered in the section. If you follow the objectives, you will get a good idea of the section's "big picture"—the important concepts, techniques, and procedures.

The objectives are numbered. (See the numbered headline at the beginning of this section.) When we begin discussing a particular objective within the section, the objective number appears along with the stated objective.

Examples: Where to Look for Information

Examples are meant to provide you with guidance and instruction when you are away from the instructor and the classroom. With this in mind, we have developed two special example formats.

Step-by-Step Examples have a three-column format where the left column describes a step, the middle column briefly explains the step, and the right column presents the algebra. Thus the left and middle columns can be thought of as your instructor's voice during a lecture. *Step-by-Step Examples* introduce key topics or important problem-solving strategies. They provide easy-to-understand, practical instructions by including the words "how to" in the examples' title.

Annotated Examples have a two-column format with explanations to the left of the algebra. The explanation clearly describes what we are about to do in the order in which we will do it. Again, annotations are like your instructor's voice as he or she writes each step of the solution on the board.

Authors in Action: Lecture Videos to Help You Learn

Every objective has one or more classroom lecture videos of the authors teaching their students. These "live" classroom lectures can be used to supplement your instructor's presentations and your reading of the text. These videos can be found in the Multimedia Library of MyMathLab or on a DVD video series and are marked with an ▶ icon in the text.

In Words: Math in Everyday Language

Have you ever been given a math definition in class and said, "What in the world does that mean?" We have heard that from our students. So we added the "In Words" feature, which restates mathematical definitions in everyday language. This margin feature will help you understand the language of mathematics better. See page 11.

Work Smart

These "tricks of the trade" that appear in the margin can help you solve problems. They also show alternative problem-solving approaches. There is more than one way to solve a math problem! See page 9.

Work Smart: Study Skills

These margin notes highlight the study skills required for success in this and other mathematics courses. See page 7.

Exercises: A Unique Numbering Scheme

As teachers, we know that students typically jump right to the exercises after attending class. This means they tend to skip all of the examples and explanations of concepts in the section. To help you use the text most effectively to learn the math, we have structured the exercises differently from other texts you have used. Our structure is designed to encourage the reading of the text, while increasing your confidence and ability to work any mathematical problem. For this reason, the exercises in each section are broken into as many as seven parts. Each exercise set will have some, or all, of the following exercise types.

1. Quick Checks
2. Building Skills
3. Mixed Practice

4. Applying the Concepts

5. Extending the Concepts

6. Explaining the Concepts

7. The Graphing Calculator

1. **Quick Checks: Learning to Ride a Bicycle with Training Wheels**   Do you remember when you were first learning to ride a bicycle? Training wheels were placed on the bicycle to assist you in learning balance. The Quick Checks are like exercises with training wheels. These exercises appear right after the example or examples that illustrate the concept being taught. So, if you get stuck on a Quick Check problem, you can simply consult the example immediately preceding it, rather than searching through the text. The Quick Check exercises also verify your understanding of new vocabulary. See Quick ✓ on page 9 in Section 1.2.

2. **Building Skills: Learning to Ride a Bicycle with Assistance**   Once you felt ready to ride without training wheels, you probably had an adult follow closely behind you, holding the bicycle for balance and building your confidence. The Building Skills problems serve a similar purpose. They are keyed to the objectives within the section, so the directions for the problem indicate which objective is being developed. As a result, you know exactly which objective (but not exactly which example) to consult if you get stuck. See page 16 in Section 1.2.

3. **Mixed Practice: Now You Are Ready to Ride!**   After mastering training wheels and learning to balance with assistance, you are ready to ride alone. This stage corresponds to the Mixed Practice exercises. These exercises include problems that develop your ability to see the big picture of mathematics. They are not keyed to a particular objective and require you to determine the appropriate approach to solving a problem on your own. See page 17 in Section 1.2.

4. **Applying the Concepts: Where Will I Ever Use This Stuff?**   The Applying the Concepts exercises not only illustrate the application of mathematics in your life but also provide problems that test your conceptual understanding of the mathematics. See pages 17–18 in Section 1.2.

5. **Extending the Concepts: Stretching Your Mind**   Sometimes we need to be challenged. These exercises extend your skills to a new level and provide further insight into where mathematics can be used. See page 18 in Section 1.2.

6. **Explaining the Concepts: Verbalize Your Understanding**   These problems require you to express the section's big-picture concepts in your own words. Students need to improve their ability to communicate complicated ideas (both oral and written). If you truly understand the material in the section, you should be able to articulate the concepts clearly. See page 27 in Section 1.3.

7. **The Graphing Calculator**   The graphing calculator is a great tool for verifying answers and for helping to visualize results. These exercises illustrate how the graphing calculator can be incorporated into the material of the section. See page 259 in Section 4.1.

Chapter Review

The chapter review is arranged section by section. For each section, we state key concepts, key terms, and objectives. We also list the examples and page numbers from the text that illustrate each objective. Also, for each objective, we list the problems in the review exercises that test your understanding. If you get a problem wrong, use this feature to determine where to look in the book to help you to work the problem.

Chapter Test

We have included a chapter test. Once you think you are prepared for the exam, take the chapter test. If you do well on the chapter test, chances are you will do well on your in-class exam. Be sure to take the test under the conditions you will face in class. If you are unsure how to solve a problem in the chapter test, watch the Chapter Test Prep Videos, which shows an instructor solving each chapter test problem.

Cumulative Review: Reinforcing Your Knowledge

The building process of learning algebra involves a lot of reinforcement. Thus we provide cumulative reviews at the end of every odd-numbered chapter starting with Chapter 3. Do these cumulative reviews after each chapter test, so that you are always refreshing your memory—making those neurons do their calisthenics. This way, studying for the final exam should be fairly easy.

④ How to Prepare for an Exam

The following steps are time-tested suggestions to help you prepare for an exam.

Step 1: Revisit your homework and the chapter review problems About one week before your exam, start to redo your homework assignments. If you don't understand a topic, seek out help. Work the problems in the chapter review as well. The problems are keyed to the section objectives. If you get a problem wrong, identify the objective and examples that illustrate the objective. Then review this material and try the problem in the chapter review again. If you get the problem wrong again, seek out help.

Step 2: Test yourself A day or two before the exam, take the chapter test under test conditions. Be sure to check your answers. If you got any problems wrong, determine why you got them wrong and remedy the situation.

Step 3: View the Chapter Test Prep Videos These videos show step-by-step solutions to the problems found in each of the book's chapter tests. Follow the worked-out solutions to any of the exercises on the chapter test that you want to study or review.

Work Smart: Study Skills

Do not "cram" for an exam by pulling an "all-nighter."

Step 4: Follow these rules as you train Be sure to arrive early at the location of the exam. Prepare your mind for the exam. Be sure you are well rested. Don't try to pull "all-nighters." If you need to study all night for an exam, then your time management is poor, and you should rethink how you are using your time or whether you have enough time set aside for the course.

1.1 Exercises   PRACTICE

1. Why do you want to be successful in mathematics? Are your goals positive or negative? If you stated your goal negatively ("Just get me out of this course!"), can you restate it positively?

2. Name three activities in your daily life that involve the use of math (for instance, playing cards, operating your computer, or reading a credit-card bill).

3. What is your instructor's name?

4. What are your instructor's office hours? Where is your instructor's office?

5. What is your instructor's e-mail address?

6. Does your class have a website? Do you know how to access it? What information is located on the website?

7. Are there tutors available for this course? If so, where are they located? When are they available?

8. Name two other students in your class. What is their contact information? When can you meet with them to study?

9. List some of the things that you should do before class begins.

10. List some of the things that you should do during class.

11. List some of the things that you should do after class.

12. What is the point of the Chinese proverb quoted in this section?

13. What is the "readiness quiz"? How should it be used?

14. Name three features that appear in the margins. What is the purpose of each of them?

15. Name the categories of exercises that appear in this text.

16. How should the chapter review material be used?

17. How should the chapter test be used? What are the Chapter Test Prep Videos?

18. How should the cumulative review be used?

19. List the four steps that should be followed when preparing for an exam. Can you think of other methods of preparing for an exam that have worked for you?

20. Use the chart below to help manage your time. Be sure to fill in time allocated to various activities in your life, including school, work, and leisure.

	Monday	Tuesday	Wednesday	Thursday	Friday	Saturday	Sunday
7 A.M.							
8 A.M.							
9 A.M.							
10 A.M.							
11 A.M.							
Noon							
1 P.M.							
2 P.M.							
3 P.M.							
4 P.M.							
5 P.M.							
6 P.M.							
7 P.M.							
8 P.M.							
9 P.M.							

1.2 Fractions, Decimals, and Percents

Objectives

1 Factor a Number as a Product of Prime Factors

2 Find the Least Common Multiple of Two or More Numbers

3 Write Equivalent Fractions

4 Write a Fraction in Lowest Terms

5 Round Decimals

6 Convert Between Fractions and Decimals

7 Convert Between Percents and Decimals

Work Smart

The first six primes are 2, 3, 5, 7, 11, and 13.

Work Smart: Study Skills

The icon ▶ means a video is available for this content. See page 5 for a description.

We base our discussion in this section on *natural numbers*. **Natural numbers** are the numbers 1, 2, 3, 4, and so on.

▶ **1** Factor a Number as a Product of Prime Factors

When we multiply, the numbers that are multiplied together are the **factors** and the answer is the **product.**

$$\underset{\text{factor}}{7} \cdot \underset{\text{factor}}{5} = \underset{\text{product}}{35}$$

When we write a number as a product, we say that we **factor** the number. For example, when we write 20 as the product $10 \cdot 2$, we say that we have factored 20.

Some natural numbers are *prime* numbers and others are *composite*.

> **Definition**
>
> A natural number is **prime** if its only factors are 1 and itself. Natural numbers that are not prime are called **composite**. The number 1 is neither prime nor composite.

Examples of prime numbers are 2, 3, 5, 7, 11, and 13. When we write a composite number as the product of prime numbers, we say that we are writing the **prime** factorization of the number. We can use a *factor tree* to find the prime factorization of a number. The process begins with finding two factors of the given number. Continue to factor until all factors are prime.

EXAMPLE 1

Finding the Prime Factorization

Write the prime factorization of 24.

Solution

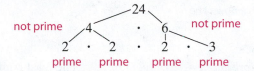

Work Smart

We could have begun the factorization of 24 with the factors 8 and 3 instead of 4 and 6. Try this for yourself.

All the numbers in the last row are prime, so we are done. The prime factorization of 24 is $2 \cdot 2 \cdot 2 \cdot 3$. Order is not important in multiplying factors. The product could also be written as $3 \cdot 2 \cdot 2 \cdot 2$ or $2 \cdot 3 \cdot 2 \cdot 2$.

> **Quick ✓**
>
> **1.** A natural number is _____ if its only factors are 1 and itself.
> **2.** In the statement $6 \cdot 8 = 48$, 6 and 8 are called _____ and 48 is called the _____.
>
> *In Problems 3–6, find the prime factorization of each number. If a number is prime, state so.*
> **3.** 12   **4.** 120
> **5.** 31   **6.** 117

▶ ② Find the Least Common Multiple of Two or More Numbers

A **multiple** of a number is the product of that number and any natural number. For example, the multiples of 2 are

$$2 \cdot 1 = 2, \quad 2 \cdot 2 = 4, \quad 2 \cdot 3 = 6, \quad 2 \cdot 4 = 8, \quad 2 \cdot 5 = 10, \quad 2 \cdot 6 = 12, \quad \text{and so on.}$$

Multiples of 3 are

$$3 \cdot 1 = 3, \quad 3 \cdot 2 = 6, \quad 3 \cdot 3 = 9, \quad 3 \cdot 4 = 12, \quad 3 \cdot 5 = 15, \quad 3 \cdot 6 = 18, \quad \text{and so on.}$$

Notice that the numbers 2 and 3 have 6 and 12 as common multiples. The smallest common multiple, called the *least common multiple*, of 2 and 3 is 6.

> **Definition**
>
> The **least common multiple (LCM)** of two or more natural numbers is the smallest number that is a multiple of each of the numbers.

For example, to find the least common multiple of 6 and 15, we could list the multiples of each number until we find the smallest common multiple, as follows:

Multiples of 6:	6, 12, 18, 24, 30, 36, 42, . . .
Multiples of 15:	15, 30, 45, 60, . . .

The least common multiple is 30. This approach works just fine for numbers, but it does not work for algebra. For this reason, we recommend that you follow the steps used in Example 2 to find the least common multiple so that you will be better prepared when we discuss the LCM again later in the course.

EXAMPLE 2 **How to Find the Least Common Multiple**

Find the least common multiple of 6 and 15.

Step-by-Step Solution

Step 1: Write each number as the product of prime factors, aligning common factors vertically.

Arrange the common factor of 3 in its own column:
$$6 = 2 \cdot 3$$
$$15 = 3 \cdot 5$$

Step 2: Write down the common factor(s), if any. Then write down the remaining factors.

The common factor is 3.
The remaining factors are 2 and 5.

Step 3: Multiply the factors listed in Step 2. The product is the least common multiple (LCM).

The least common multiple of 6 and 15 is $2 \cdot 3 \cdot 5 = 30$

●

EXAMPLE 3 **Finding the Least Common Multiple**

Find the least common multiple of 18 and 15.

Solution

We first write each number as the product of prime factors.

Write the prime factors in each column that the numbers share, if any. Then write down the remaining factors. Find the product of the factors.

$$18 = 2 \cdot 3 \cdot 3$$
$$15 = 3 \cdot 5$$
$$2 \cdot 3 \cdot 3 \cdot 5$$

The LCM is $2 \cdot 3 \cdot 3 \cdot 5 = 90$.

●

Quick ✔

7. The ____ _____ _____ of two or more natural numbers is the smallest number that is a multiple of each of the numbers.

In Problems 8–11, find the LCM of the numbers.

8. 6 and 8

9. 45 and 72

10. 14 and 9

11. 12, 18, and 30

▶ ❸ **Write Equivalent Fractions**

Figure 1

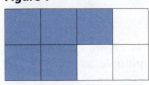

A fraction represents a part of a whole. For example, the fraction $\frac{5}{8}$ means "5 parts out of 8 parts." A fraction also indicates division: $\frac{5}{8}$ means "five divided by eight" and may be written as $8\overline{)5}$. Figure 1 shows the fraction $\frac{5}{8}$ visually.

In the fraction $\frac{5}{8}$, the number 5 is the **numerator** and the number 8 is the **denominator.**

The denominator tells the number of equal parts that the whole is divided into, and the numerator tells the number of equal parts that are shaded. For example, in Figure 1 the box is divided into 8 equal parts, 5 of which are shaded.

We use **whole numbers**, the natural numbers plus 0, for the numerator of a fraction, and we use natural numbers for the denominator.

Fractions without common denominators can be rewritten in equivalent forms so they have the same denominator.

Work Smart

The denominator of a number such as 7 is 1 because $7 = \frac{7}{1}$.

Figure 2

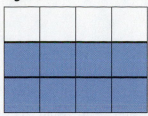

Definition
Equivalent fractions are fractions that represent the same part of a whole.

For example, $\frac{2}{3}$ and $\frac{8}{12}$ are equivalent fractions. To understand why, consider Figure 2. Break the whole into 12 parts and shade 8 of these parts, so that the shaded region represents $\frac{8}{12}$ of the rectangle. If we consider only the 3 parts separated by the thick black lines, we can see that 2 parts are shaded for a fraction of $\frac{2}{3}$. In each case, the same portion of the rectangle is shaded, so $\frac{2}{3}$ and $\frac{8}{12}$ are equivalent fractions.

How do we obtain equivalent fractions? The answer lies in the following property.

If a, b, and c are whole numbers, then

$$\frac{a}{b} = \frac{a \cdot c}{b \cdot c} \quad \text{if } b \neq 0, c \neq 0$$

EXAMPLE 4 **Writing an Equivalent Fraction**

Write the fraction $\frac{3}{4}$ as an equivalent fraction with a denominator of 20.

Solution

We want to know $\frac{3}{4}$ equals "what" over 20, or $\frac{3}{4} = \frac{?}{20}$. To write $\frac{3}{4}$ with a denominator of 20, we multiply the numerator and denominator of $\frac{3}{4}$ by 5. Do you see why?

$$\frac{3}{4} = \frac{3 \cdot 5}{4 \cdot 5}$$

$$= \frac{15}{20}$$

Quick ✓

12. In the fraction $\frac{7}{12}$, 7 is called the _____ and 12 is called the _____.

13. Fractions that represent the same portion of a whole are called _____ _____.

In Problems 14 and 15, rewrite each fraction with the denominator indicated.

14. $\frac{1}{2}$; 10

15. $\frac{5}{8}$; 48

We sometimes need to rewrite two or more fractions so that they both have the same denominator. For example, we could rewrite the fractions $\frac{5}{6}$ and $\frac{3}{8}$ with a common denominator of 24, 48, 96 and so on because these are common multiples of the denominators 6 and 8. Notice that 24 is the least common multiple of 6 and 8.

Definition
The **least common denominator (LCD)** is the least common multiple of the denominators of a group of fractions.

EXAMPLE 5 **How to Write Two Fractions as Equivalent Fractions with the LCD**

Write $\dfrac{5}{8}$ and $\dfrac{9}{20}$ as equivalent fractions with the least common denominator.

Step-by-Step Solution

Step 1: Find the least common denominator of the fractions.

The denominators of $\dfrac{5}{8}$ and $\dfrac{9}{20}$ are 8 and 20.

Write each denominator as the product of prime factors:
$$8 = 2 \cdot 2 \cdot 2$$
$$20 = 2 \cdot 2 \cdot \cdot 5$$

Write the common factors; then write the uncommon factors:
$$\text{LCD} = 2 \cdot 2 \cdot 2 \cdot 5$$
$$= 40$$

Step 2: Rewrite each fraction with the least common denominator.

Multiply the numerator and denominator of $\dfrac{5}{8}$ by 5:
$$\frac{5}{8} = \frac{5 \cdot 5}{8 \cdot 5}$$
$$= \frac{25}{40}$$

Multiply the numerator and denominator of $\dfrac{9}{20}$ by 2:
$$\frac{9}{20} = \frac{9 \cdot 2}{20 \cdot 2}$$
$$= \frac{18}{40}$$

●

> **Quick ✓**
>
> **16.** The ____ _____ _____ is the least common multiple of the denominators of a group of fractions.
>
> *In Problems 17 and 18, write the equivalent fractions with the least common denominator.*
>
> **17.** $\dfrac{1}{4}$ and $\dfrac{5}{6}$ **18.** $\dfrac{9}{20}$ and $\dfrac{11}{16}$

▶ ❹ Write a Fraction in Lowest Terms

> **Definition**
>
> A fraction is written in **lowest terms** if the numerator and the denominator share no common factor other than 1.

In Words

To write a fraction in lowest terms, find any common factors between the numerator and denominator, and divide out the common factors.

We can write fractions in lowest terms using the fact that

$$\frac{a \cdot c}{b \cdot c} = \frac{a}{b}$$

Thus, to write a fraction in lowest terms, we write the numerator and the denominator as a product of primes and then divide out common factors.

EXAMPLE 6 **Writing a Fraction in Lowest Terms**

Write $\dfrac{24}{40}$ in lowest terms.

Solution

Write the numerator and the denominator as the product of primes and divide out common factors.

Work Smart

Use different slash marks to keep track of factors that have divided out. Also, we may use nonprime factors when writing a fraction in lowest terms. In Example 6 we could write

$$\frac{24}{40} = \frac{8 \cdot 3}{8 \cdot 5} = \frac{3}{5}$$

$$\frac{24}{40} = \frac{2 \cdot 2 \cdot 2 \cdot 3}{2 \cdot 2 \cdot 2 \cdot 5}$$

Divide out common factors:

$$= \frac{2 \cdot 2 \cdot 2 \cdot 3}{2 \cdot 2 \cdot 2 \cdot 5}$$

$$= \frac{3}{5}$$

Quick ✓

19. A fraction is written in _____ _____ if the numerator and the denominator share no common factor other than 1.

In Problems 20–22, write each fraction in lowest terms.

20. $\dfrac{45}{80}$ **21.** $\dfrac{4}{9}$ **22.** $\dfrac{16}{56}$

▶ ⑤ Round Decimals

Decimals and percentages commonly occur in everyday life. You receive a 92% on your test, there is a 10% discount on jeans, we pay 7.75% in sales tax, 45% of the people polled support a proposition. Before we discuss decimals and percents, we consider place value.

Figure 3 shows how we interpret the place value of each digit in the number 9186.347. For example, the 7 is in the thousandths position, 3 is in the tenths position, and the 8 is in the tens position.

Figure 3

The number 9186.347 is read "nine thousand, one hundred eighty-six and three hundred forty-seven thousandths."

Quick ✓

In Problems 23–26, tell the place value of the digit in the given number.

23. 235.71; the 1   **24.** 56,701.28; the 2

25. 278,403.95; the 8   **26.** 0.189; the 9

We round decimals in the same way we round whole numbers. First, identify the specified place value in the decimal. If the digit to the right is 5 or more, add 1 to the digit; if the digit to the right is 4 or less, leave the digit as it is. Then drop the digits to the right of the specified place value.

EXAMPLE 7   **Rounding a Decimal Number**

(a) Round 8.726 to the nearest hundredth.

(b) Round 0.9451 to the nearest thousandth.

Solution

(a) To round to the nearest hundredth, we determine that 2 is in the hundredths place: 8.726. The number to the right of 2 is 6. Since 6 is greater than 5, we round 8.726 to 8.73.

(b) To round 0.9451 to the nearest thousandth, we see that 5 is in the thousandths place: 0.9451. The number to the right of 5 is 1. Since 1 is less than 5, we round 0.9451 to 0.945.

Quick ✔

In Problems 27–31, round each number to the given decimal place.

27. 0.173 to the nearest tenth

28. 0.932 to the nearest hundredth

29. 1.396 to the nearest hundredth

30. 690.004 to the nearest hundredth

31. 59.98 to the nearest tenth

⑥ Convert Between Fractions and Decimals

▶ Convert a Fraction to a Decimal

To convert a fraction to a decimal, divide the numerator of the fraction by the denominator of the fraction until the remainder is 0 or the remainder repeats.

EXAMPLE 8 **Converting a Fraction to a Decimal**

Convert each number to a decimal.

(a) $\dfrac{9}{20}$ **(b)** $\dfrac{2}{3}$

Solution

(a)

$$\frac{9}{20} = 20\overline{)9.00} \quad \begin{array}{r} 0.45 \\ \hline \end{array}$$
$$\begin{array}{r} 8\,0 \\ \hline 100 \\ 100 \\ \hline 0 \end{array}$$

Therefore, $\dfrac{9}{20} = 0.45$.

(b)

$$\frac{2}{3} = 3\overline{)2.000} \quad \begin{array}{r} 0.666 \\ \hline \end{array}$$
$$\begin{array}{r} 1\,8 \\ \hline 20 \\ 18 \\ \hline 20 \\ 18 \\ \hline 2 \end{array}$$

Notice that the remainder, 2, repeats. So $\dfrac{2}{3} = 0.666\ldots$. ●

In Example 8(a), the decimal 0.45 is called a **terminating decimal** because the decimal stops after the 5. In Example 8(b), the number 0.666 . . . is called a **repeating decimal** because the 6 repeats indefinitely. The decimal 0.666 . . . can also be written as $0.\overline{6}$. The bar over the 6 means the 6 repeats.

Quick ✔

In Problems 32–35, write the fraction as a decimal.

32. $\dfrac{2}{5}$ **33.** $\dfrac{5}{6}$

34. $\dfrac{11}{8}$ **35.** $\dfrac{3}{7}$

Based on Examples 8(a) and (b) and Quick Check Problems 32–35, you should notice that **every fraction has a decimal representation that either terminates or repeats.**

▶ Convert a Decimal to a Fraction

To convert a decimal to a fraction, identify the place value of the last digit in the decimal. Write the decimal as a fraction using the place value of the last digit as the denominator, and write in lowest terms.

EXAMPLE 9 **Writing a Decimal as a Fraction**

Convert each decimal to a fraction and write in lowest terms.

(a) 0.8   (b) 0.77   (c) 4.237

Solution

(a) 0.8 is equivalent to 8 tenths, or $\dfrac{8}{10}$. Because $\dfrac{8}{10} = \dfrac{4 \cdot 2}{5 \cdot 2} = \dfrac{4}{5}$, we write $0.8 = \dfrac{4}{5}$.

(b) 0.77 is equivalent to 77 hundredths, or $\dfrac{77}{100}$.

(c) 4.237 is equivalent to 4237 thousandths, or $\dfrac{4237}{1000}$.

> **Quick ✓**
> *In Problems 36–38, write the decimal as a fraction and write in lowest terms.*
>
> **36.** 0.6   **37.** 0.17   **38.** 0.625

❼ Convert Between Percents and Decimals

When computing with percents, it is convenient to write percents as decimals. How is a percent converted to a decimal? Let's see.

▶ Convert a Percent to a Decimal

> **Definition**
> The word **percent** means **parts per hundred** or **parts out of one hundred.**

So 25% means 25 parts out of 100 parts. Therefore, $25\% = \dfrac{25}{100} = \dfrac{1 \cdot 25}{4 \cdot 25} = \dfrac{1}{4}$.

Since the word percent means "parts per hundred," 100% means "100 parts per 100," so 100% = 1. Therefore, to convert from a percent to a decimal, multiply the percent by $\dfrac{1}{100\%}$.

EXAMPLE 10 **Writing a Percent as a Decimal**

Write the following percents as decimals:

(a) 27%   (b) 150%

Solution

Work Smart

To convert from a percent to a decimal, move the decimal point two places to the left and drop the % symbol.

(a) $27\% = 27\% \cdot \dfrac{1}{100\%}$

$= \dfrac{27}{100}$

$= 0.27$

(b) $150\% = 150\% \cdot \dfrac{1}{100\%}$

$= \dfrac{150}{100}$

$= 1.5$

Quick ✔

39. The word percent means parts per _____, so 35% means __ parts out of 100 parts or $\dfrac{\ \ \ }{100}$.

In Problems 40–43, write the percent as a decimal.

40. 23%   **41.** 1%

42. 72.4%   **43.** 127%

▶ Convert a Decimal to a Percent

Because $100\% = 1$, to convert a decimal to a percent, multiply the decimal by $\dfrac{100\%}{1}$.

EXAMPLE 11 **Writing a Decimal as a Percent**

Write the following decimals in percent form:

(a) 0.445   **(b)** 1.42

Solution

(a) $0.445 = 0.445 \cdot \dfrac{100\%}{1}$

$\quad\quad = 44.5\%$

(b) $1.42 = 1.42 \cdot \dfrac{100\%}{1}$

$\quad\quad = 142\%$

Work Smart

To convert from a decimal to a percent, move the decimal point two places to the right and add the % symbol.

Quick ✔

44. To convert a decimal to a percent, multiply the decimal by _____.

In Problems 45–48, write the decimal as a percent.

45. 0.15   **46.** 0.8

47. 1.372   **48.** 0.004

1.2 Exercises   Exercise numbers in green have complete video solutions in MyMathLab.

*Problems **1–48** are the Quick ✔s that follow the EXAMPLES.*

Building Skills

In Problems 49–62, find the prime factorization of each number. If a number is prime, state so. See Objective 1.

49. 25   **50.** 9   **51.** 28

52. 100   **53.** 21   **54.** 35

55. 36   **56.** 54   **57.** 50

58. 70   **59.** 53   **60.** 79

61. 252   **62.** 315

In Problems 63–74, find the LCM of each set of numbers. See Objective 2.

63. 6 and 21   **64.** 10 and 14

65. 18 and 63   **66.** 42 and 18

67. 15 and 14   **68.** 55 and 6

69. 30 and 45   **70.** 8 and 60

71. 5, 6, and 12   **72.** 9, 15, and 20

73. 3, 8, and 9   **74.** 4, 18, and 20

In Problems 75–80, write each fraction with the given denominator. See Objective 3.

75. Write $\dfrac{2}{3}$ with denominator 12.

76. Write $\dfrac{4}{5}$ with denominator 15.

77. Write $\dfrac{3}{4}$ with denominator 24.

78. Write $\dfrac{5}{14}$ with denominator 28.

79. Write 7 with denominator 3.

80. Write 4 with denominator 10.

In Problems 81–88, write the equivalent fractions with the least common denominator. See Objective 3.

81. $\dfrac{1}{2}$ and $\dfrac{3}{8}$ **82.** $\dfrac{3}{4}$ and $\dfrac{5}{12}$

83. $\dfrac{3}{5}$ and $\dfrac{2}{3}$ **84.** $\dfrac{1}{4}$ and $\dfrac{2}{9}$

85. $\dfrac{1}{12}$ and $\dfrac{5}{18}$ **86.** $\dfrac{5}{12}$ and $\dfrac{7}{15}$

87. $\dfrac{2}{9}$ and $\dfrac{7}{18}$ and $\dfrac{7}{30}$

88. $\dfrac{7}{10}$ and $\dfrac{1}{4}$ and $\dfrac{5}{6}$

In Problems 89–96, write each fraction in lowest terms. See Objective 4.

89. $\dfrac{14}{21}$ **90.** $\dfrac{9}{15}$ **91.** $\dfrac{38}{18}$

92. $\dfrac{81}{36}$ **93.** $\dfrac{22}{66}$ **94.** $\dfrac{9}{27}$

95. $\dfrac{18}{3}$ **96.** $\dfrac{36}{4}$

In Problems 97–102, tell the place value of the indicated digit in the given number. See Objective 5.

97. 3465.902; the 0

98. 549,813.0267; the 8

99. 357.469; the 5

100. 9124.786; the 7

101. 2018.3764; the 6

102. 539.016; the 9

In Problems 103–110, round each number to the given place. See Objective 5.

103. 578.206 to the nearest tenth

104. 7298.0845 to the nearest hundredth

105. 354.678 to the nearest hundredth

106. 543.56 to the nearest whole number

107. 3682.0098 to the nearest thousandth

108. 683.098 to the nearest hundredth

109. 29.96 to the nearest whole number

110. 37.999 to the nearest tenth

In Problems 111–120, convert each fraction to a decimal. See Objective 6.

111. $\dfrac{5}{8}$ **112.** $\dfrac{3}{4}$

113. $\dfrac{2}{7}$ **114.** $\dfrac{2}{9}$

115. $\dfrac{5}{16}$ **116.** $\dfrac{11}{32}$

117. $\dfrac{3}{13}$ **118.** $\dfrac{6}{13}$

119. $\dfrac{29}{25}$ **120.** $\dfrac{57}{50}$

In Problems 121–126, write each decimal as a fraction in lowest terms. See Objective 6.

121. 0.75 **122.** 0.25 **123.** 0.5

124. 0.4 **125.** 0.982 **126.** 0.358

In Problems 127–132, write each percent as a decimal. See Objective 7.

127. 37% **128.** 59%

129. 6.02% **130.** 8.25%

131. 0.1% **132.** 0.5%

In Problems 133–138, write each decimal as a percent. See Objective 7.

133. 0.2 **134.** 0.5

135. 0.275 **136.** 0.349

137. 2 **138.** 1

Mixed Practice

In Problems 139–144, write each fraction as a decimal, rounded to the indicated place.

139. $\dfrac{13}{6}$ to the nearest tenth

140. $\dfrac{15}{8}$ to the nearest tenth

141. $\dfrac{8}{3}$ to the nearest hundredth

142. $\dfrac{9}{7}$ to the nearest hundredth

143. $\dfrac{14}{27}$ to the nearest thousandth

144. $\dfrac{18}{31}$ to the nearest thousandth

Applying the Concepts

145. Planets in Our Solar System At a certain point, Mercury, Venus, and Earth lie on a straight line. If it takes these planets 3, 7, and 12 months, respectively, to revolve around the sun, what is the fewest number of months until they align this way again?

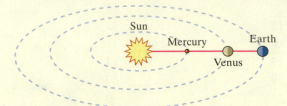

146. Talladega Raceway At Talladega, one of the crew chiefs discovered that in a given time interval, Jeff Gordon completed 21 laps, Dale Earnhardt Jr. completed 18 laps, and Robby Gordon completed 15 laps. Suppose all three drivers begin at the same time. How many laps would need to be completed so that all three drivers were at the finish line at exactly the same time?

147. Sam's Medication Bob gives his dog Sam one type of medication every 4 days, and a second type of medication every 10 days. How often does Bob give Sam both medications on the same day?

148. Visiting Columbus Pamela and Geoff both visit Columbus on business. Pamela flies to Columbus from Atlanta every 14 days, and Geoff takes the train to Columbus from Cincinnati every 20 days. How often are both Pamela and Geoff in Columbus on business?

149. Survey Data In a survey of 500 students, 325 stated that they work at least 25 hours per week. Express the fraction of students that work at least 25 hours per week as a fraction in lowest terms.

150. Survey Data In a survey of 750 students, 450 stated that they are enrolled in 15 or more semester hours. Express the fraction of students who are enrolled in 15 or more semester hours as a fraction in lowest terms.

In Problems 151–158, express answers to the nearest hundredth of a percent, if necessary.

151. Eating Healthy? In a poll conducted by Zogby International of 1200 adult Americans, 840 stated that they believe that they eat healthy foods. What percentage of adult Americans believe that they eat healthy foods?

152. Ghosts In a survey of 1100 adult women conducted by Harris Interactive, it was determined that 640 believe in ghosts. What percentage of adult women believe in ghosts?

153. Test Score A student earns 85 points out of a total of 110 points on an exam. Express this score as a percent.

154. Test Score A student earns 80 points out of a total of 115 points on an exam. Express this score as a percent.

155. Time Utilization In a 24-hour day, Jackson sleeps for 8 hours, works for 4 hours, and goes to school and studies for 6 hours.

(a) What percent of the time does Jackson sleep?

(b) What percent of the time does Jackson work?

(c) What percent of the time does Jackson go to school and study?

156. Tree Inventory An arborist counted the number of ash trees in Whetstone Park. He found that there were 48 white ash trees, 51 green ash trees, and 2 blue ash trees.

(a) What percent of the ash trees in Whetstone Park were white ashes?

(b) What percent of the ash trees in Whetstone Park were green ashes?

(c) What percent of the ash trees in Whetstone Park were blue ashes?

157. Cashews A single serving of cashews contains 14 grams of fat. Of this, 3 grams is saturated fat. What percentage of fat grams is saturated fat in a single serving of cashews?

158. Cheese Pizza A single serving of cheese pizza contains 11 grams of fat. Of this, 5 grams are saturated fat. What percentage of fat grams is saturated fat in a single serving of cheese pizza?

Extending the Concepts

159. The Sieve of Eratosthenes Eratosthenes (276 B.C.– 194 B.C.) was born in Cyrene, which is now in Libya in North Africa. He devised an algorithm (a series of steps that are followed to solve a problem) for identifying prime numbers. The algorithm works as follows.

Step 1: List all the natural numbers that are greater than or equal to 2.

Step 2: The first number in the list, 2, is prime. Cross out all multiples of 2. For example, cross out 2, 4, 6, ...

Step 3: Identify the next number in the list after the most recently identified prime number. For example, we already know 2 is a prime number, so the next number in the list, 3, is also prime. Cross out all multiples of this number.

Step 4: Repeat Step 3.

Use the algorithm to find all the prime numbers less than 100.

1.3 The Number Systems and the Real Number Line

Objectives

1. Classify Numbers
2. Plot Points on a Real Number Line
3. Use Inequalities to Order Real Numbers
4. Compute the Absolute Value of a Real Number

Are You Ready for This Section?

Before getting started, take the following readiness quiz. If you get a problem wrong, go back to the section cited and review the material.

R1. Write $\dfrac{5}{8}$ as a decimal.   [Section 1.2, p. 14]

R2. Write $\dfrac{9}{11}$ as a decimal.   [Section 1.2, p. 14]

Work Smart

The use of the word "real" to describe numbers leads us to question, "Are there 'nonreal' numbers?" The answer is yes. We use the word "imaginary" to describe nonreal numbers. Imaginary does not mean that these numbers are made up, however. We will discuss imaginary numbers later in the text.

This section discusses the *real number system*. We use the real numbers every day, so it is an idea that you are already familiar with. In short, real numbers are numbers that we use to count or measure things, such as 25 students in your class, 18.4 miles per gallon, or a $130 debt.

As we proceed through the text, we will deal with various types of numbers that are organized in *sets*. A **set** is a well-defined collection of objects. For example, we can identify the students enrolled in Elementary Algebra at your college as a set. The collection of numbers 0, 1, 2, 3, 4, 5, 6, 7, 8, and 9 may also be identified as a set. If we let A represent this set of numbers, then we can write

$$A = \{0, 1, 2, 3, 4, 5, 6, 7, 8, 9\}$$

In this notation, braces $\{\ \}$ are used to enclose the objects, or **elements,** in the set. A set with no elements in it is called an **empty set.** Empty sets are denoted by the symbol $\varnothing$ or $\{\ \}$.

EXAMPLE 1

Writing a Set

Write the set that represents the vowels.

Solution

The vowels are *a, e, i, o,* and *u.* If we let *V* represent this set, then

$$V = \{a, e, i, o, u\}$$

Quick ✓

1. Write the set that represents the first four positive, odd numbers.
2. Write the set that represents the states in the United States with names that begin with the letter A.
3. Write the set that represents the states in the United States with names that begin with the letter Z.

1 Classify Numbers

We will develop the real number system by looking at the history of numbers. The first types of numbers that humans worked with are the *natural numbers* or *counting numbers*. We introduced natural numbers in Section 1.2. We now present a formal definition using a set.

Definition

The **natural numbers**, or **counting numbers**, are the numbers in the set $\{1, 2, 3, \ldots\}$.

The three dots in this definition are called an *ellipsis* and indicate that the pattern continues indefinitely.

Figure 4
The natural numbers.

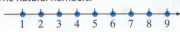

The natural numbers are often used to count things. For example, we can count the number of cars waiting at a Wendy's drive-thru. We can represent the counting numbers graphically using a number line. See Figure 4. The arrow on the right indicates the direction in which the numbers increase.

Since we do not count the number of cars waiting in the drive-thru by saying, "zero, one, two, three...," zero is not a natural, or counting, number. When we add the number 0 to the set of counting numbers, we get the set of *whole numbers*.

> **Definition**
>
> The **whole numbers** are the numbers in the set $\{0, 1, 2, 3, \ldots\}$.

Figure 5
The whole numbers.

Figure 5 represents the whole numbers on the number line. Notice that the set of natural numbers is included in the set of whole numbers.

By expanding the numbers to the left of zero on the number line, we obtain the set called *integers*.

> **Definition**
>
> The **integers** are the numbers in the set $\{\ldots, -3, -2, -1, 0, 1, 2, 3, \ldots\}$.

Figure 6
The integers.

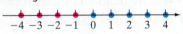

Figure 6 represents the integers on the number line. Notice that the whole numbers and natural numbers are included in the set of integers.

Integers are useful in many situations. For example, we could not discuss temperatures above 0°F (positive counting numbers) or below 0°F (negative counting numbers) without integers. A debt of 300 dollars can be represented as an integer by -300 dollars.

How can we represent a part of a whole, such as a part of last night's leftover pizza or part of a dollar? To address this problem, we enlarge our number system to include *rational numbers*.

In Words

A rational number is a number that can be expressed as a fraction where the numerator is any integer and the denominator is any nonzero integer.

> **Definition**
>
> A **rational number** is a number that can be written in the form $\dfrac{p}{q}$, where p and q are integers. However, q cannot equal zero.

Work Smart

Remember, all integers are also rational numbers. For example,
$$42 = \frac{42}{1}.$$

Examples of rational numbers are $\dfrac{2}{5}, \dfrac{5}{2}, \dfrac{0}{8}, -\dfrac{7}{9}$, and $\dfrac{31}{4}$. Because $\dfrac{p}{1} = p$ for any integer p, it follows that all integers are also rational numbers. For example, 7 is an integer, but it is also a rational number because it can be written as $\dfrac{7}{1}$. We illustrate this idea below.

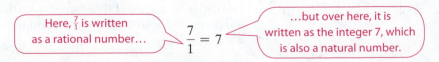

In addition to representing rational numbers as fractions, we can also represent rational numbers in decimal form as either repeating decimals or terminating decimals. Table 1 shows various rational numbers in fraction form and decimal form.

Table 1

Fraction Form of Rational Number	Decimal Form of Rational Number	Terminating or Repeating Decimal
$\frac{7}{2}$	3.5	Terminating
$\frac{1}{3}$	$0.333\ldots = 0.\overline{3}$	Repeating
$-\frac{3}{8}$	-0.375	Terminating
$-\frac{15}{11}$	$-1.3636\ldots = -1.\overline{36}$	Repeating

The repeating decimal $0.\overline{3}$ and the terminating decimal -0.375 are rational numbers because they represent fractions (see Section 1.2).

Decimals that neither terminate nor repeat are called *irrational numbers.*

In Words

Numbers that cannot be written as the ratio of two integers are irrational.

Definition

An **irrational number** is a number that has a decimal representation that neither terminates nor repeats. Therefore, irrational numbers cannot be written as the quotient (ratio) of two integers.

An example of an irrational number is 1.343343334... because the decimal neither terminates nor repeats. Other examples of irrational numbers are the symbols $\sqrt{2}$, whose value is approximately 1.41421, and π, whose value is approximately 3.141593.

Now we are ready for a formal definition of the set of *real numbers.*

Definition

The set of rational numbers combined with the set of irrational numbers is called the set of **real numbers.**

Figure 7 shows the relationships among the various types of numbers. Note that the oval that represents the whole numbers surrounds the oval that represents the natural numbers. This means the set of whole numbers includes all the natural numbers.

Figure 7

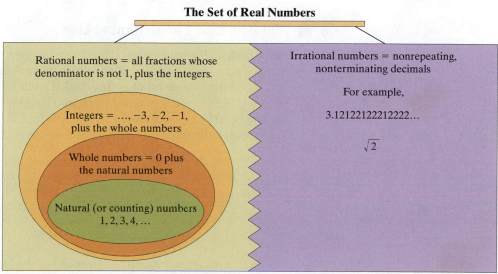

The set of real numbers is composed of the set of rational numbers and the set of irrational numbers

EXAMPLE 2 **Classifying Numbers in a Set**

List the numbers in the set

$$\left\{ 9, -\frac{2}{7}, -4, 0, -4.010010001\ldots, 3.\overline{632}, 18.3737\ldots \right\}$$

that are

(a) Natural numbers   (b) Whole numbers

(c) Integers   (d) Rational numbers

(e) Irrational numbers   (f) Real numbers

Solution

(a) 9 is the only natural number.

(b) 0 and 9 are the whole numbers.

(c) 9, −4, and 0 are the integers.

(d) $9, -\dfrac{2}{7}, -4, 0, 3.\overline{632}$, and 18.3737... are the rational numbers.

(e) −4.010010001... is the only irrational number because the decimal does not repeat, nor does it terminate.

(f) All the numbers listed are real numbers. Real numbers consist of rational numbers together with irrational numbers. ●

Quick ✔

4. *True or False* Every integer is a rational number.

5. Real numbers that can be represented with a terminating or repeating decimal are called _____ numbers.

In Problems 6–11, list the numbers in the set $\left\{ \dfrac{11}{5}, -5, 12, 2.\overline{76}, 0, 2.737737773\ldots, \dfrac{18}{4} \right\}$

that are

6. Natural numbers   7. Whole numbers

8. Integers   9. Rational numbers

10. Irrational numbers   11. Real numbers

▶ **② Plot Points on a Real Number Line**

Look back at Figure 6 (page 20). Notice the gaps between the integers plotted on the number line. These gaps are filled in with the real numbers that are not integers.

To construct a **real number line,** pick a point on a line somewhere in the center, and label it 0. This point is called the **origin.** The point 1 unit to the right of 0 corresponds to the real number 1. The distance between 0 and 1 determines the **scale** of the number line. For example, the point representing 2 is twice as far from 0 as 1 is. See Figure 8.

Figure 8
The real number line.

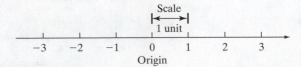

Notice that an arrowhead on the right end of the line indicates the direction in which the numbers increase. Points to the left of 0 correspond to the real numbers −1, −2, and so on.

> **Definition**
>
> The real number associated with a point P is called the **coordinate** of P.

EXAMPLE 3 **Plotting Points on a Real Number Line**

On a real number line, label the points with coordinates $0, 6, -2, 2.5, -\dfrac{1}{2}$.

Solution

We draw a real number line and then plot the points. See Figure 9. Notice that 2.5 is midway between 2 and 3. Also notice that $-\dfrac{1}{2}$ is midway between -1 and 0.

Figure 9

> **Quick** ✓
>
> **12.** The point on the real number line whose coordinate is 0 is called the _____.
>
> **13.** On a real number line, label the points with coordinates $0, 3, -2, \dfrac{1}{2}$, and 3.5.

The real number line consists of three classes (or categories) of real numbers, as shown in Figure 10.

Figure 10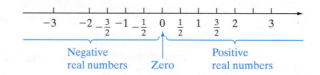

- The **negative real numbers** are the coordinates of points to the left of 0.
- The real number 0 is the coordinate of the origin.
- The **positive real numbers** are the coordinates of points to the right 0.

The **sign** of a number refers to whether the number is a positive or a negative real number. For example, the sign of -4 is negative and the sign of 100 is positive.

Figure 11

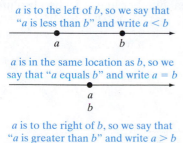

a is to the left of b, so we say that "a is less than b" and write $a < b$

a is in the same location as b, so we say that "a equals b" and write $a = b$

a is to the right of b, so we say that "a is greater than b" and write $a > b$

▶ ③ Use Inequalities to Order Real Numbers

Given two numbers (points) a and b, a must be to the left of b (denoted $a < b$) or the same as b (denoted $a = b$) or to the right of b (denoted $a > b$). See Figure 11.

If a is less than or equal to b, we write $a \le b$. Similarly, $a \ge b$ means that a is greater than or equal to b. Collectively, the symbols $<, >, \le$, and $\ge$ are called **inequality symbols.** The "arrowhead" in an inequality always points to the smaller number. For $3 < 5$, the "arrowhead" points to 3.

Note that $a < b$ and $b > a$ mean the same thing. For example, $2 < 3$ and $3 > 2$ mean the same thing. Do you see why?

EXAMPLE 4 **Using Inequality Symbols**

(a) We know that \$3 is less than \$7 and that 3 apples is fewer than 7 apples. Using the real number line, we say $3 < 7$ because the point whose coordinate is 3 lies to the left of the point whose coordinate is 7 on a real number line.

(b) Being $2 in debt is not as bad as being $5 in debt, so $-2 > -5$. Using the real number line, $-2 > -5$ because the point whose coordinate is -2 lies to the right of the point whose coordinate is -5 on a real number line.

(c) $2.7 > \dfrac{5}{2}$ because $\dfrac{5}{2} = 2.5$ and $2.7 > 2.5$.

(d) $\dfrac{5}{6} > \dfrac{4}{5}$ because $\dfrac{5}{6} = \dfrac{25}{30}$ and $\dfrac{4}{5} = \dfrac{24}{30}$, and 25 out of 30 parts is more than 24 out of 30 parts. We could also write $\dfrac{5}{6} = 0.8\overline{3}$ and $\dfrac{4}{5} = 0.8$. Since $0.8\overline{3}$ is greater than 0.80, $\dfrac{5}{6} > \dfrac{4}{5}$.

Work Smart

Write fractions with a common denominator or change fractions to decimals to compare the location of the numbers on the number line.

Quick ✓

14. The symbols $<, >, \le, \ge$ are called _____ symbols.

In Problems 15–20, replace the question mark by $<, >,$ or $=$, whichever is correct.

15. $2 \; ? \; 9$

16. $-5 \; ? \; -3$

17. $\dfrac{4}{5} \; ? \; \dfrac{1}{2}$

18. $\dfrac{4}{7} \; ? \; 0.5$

19. $\dfrac{4}{3} \; ? \; \dfrac{20}{15}$

20. $-\dfrac{4}{3} \; ? \; -\dfrac{5}{4}$

Based upon the discussion so far, we conclude that

$$a > 0 \qquad \text{is equivalent to} \qquad a \text{ is positive}$$
$$a < 0 \qquad \text{is equivalent to} \qquad a \text{ is negative}$$

We sometimes read $a > 0$ as "a is positive." If $a \ge 0$, then $a > 0$ or $a = 0$, so we may read this as "a is nonnegative" or "a is greater than or equal to zero."

▶ ➍ Compute the Absolute Value of a Real Number

The real number line can be used to describe the concept of *absolute value*.

In Words

Think of absolute value as the number of units you must count to get from 0 to a number. The absolute value of a number can never be negative because it represents a distance.

Definition

The **absolute value** of a number a, written $|a|$, is the distance from 0 to a on a real number line.

For example, because the distance from 0 to 3 on a real number line is 3, the absolute value of 3, $|3|$, is 3. Because the distance from 0 to -3 on a real number line is 3, $|-3| = 3$. See Figure 12.

Figure 12

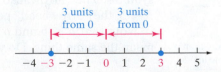

EXAMPLE 5 **Computing Absolute Value**

Evaluate each of the following:

(a) $|6|$ **(b)** $|-7|$ **(c)** $|0|$ **(d)** $-|-1.5|$

Solution

(a) $|6| = 6$ because the distance from 0 to 6 on a real number line is 6.

(b) $|-7| = 7$ because the distance from 0 to -7 on a real number line is 7.

(c) $|0| = 0$ because the distance from 0 to 0 on a real number line is 0.

(d) $-|-1.5| = -1.5$

Quick ✔

21. The distance from zero to a point on a real number line whose coordinate is a is called the _____ _____ of a.

In Problems 22–24, evaluate each expression.

22. $|-15|$

23. $\left|\dfrac{3}{4}\right|$

24. $-|-4|$

1.3 Exercises MyMathLab® Exercise numbers in green have complete video solutions in MyMathLab

Problems **1–24** *are the* Quick ✔*s that follow the* EXAMPLES.

Building Skills

In Problems 25–30, write each set. See Objective 1.

25. A is the set of *whole* numbers less than 5.

26. B is the set of *natural* numbers less than 25.

27. D is the set of *natural* numbers less than 5.

28. C is the set of *integers* between -6 and 4, not including -6 or 4.

29. E is the set of even *natural* numbers less than 1.

30. F is the set of *whole* numbers less than 0.

In Problems 31–36, list the elements in the set $\left\{-4, 3, -\dfrac{13}{2}, 0, 2.303003000\ldots\right\}$ *that are described. See Objective 1.*

31. natural numbers

32. whole numbers

33. integers

34. rational numbers

35. irrational numbers

36. real numbers

In Problems 37–42, list the elements in the set $\left\{-4.2, 3.\overline{5}, \pi, \dfrac{5}{5}\right\}$ *that are described. See Objective 1.*

37. real numbers

38. rational numbers

39. irrational numbers

40. integers

41. whole numbers

42. natural numbers

In Problems 43 and 44, plot the points in each set on a real number line. See Objective 2.

43. $\left\{0, \dfrac{3}{3}, -1.5, -2, \dfrac{4}{3}\right\}$

44. $\left\{\dfrac{3}{4}, \dfrac{0}{2}, -\dfrac{5}{4}, -0.5, 1.5\right\}$

In Problems 45–52, determine whether the statement is True or False. See Objective 3.

45. $-2 > -3$

46. $0 < -5$

47. $-6 \le -6$

48. $-3 > -5$

49. $\dfrac{3}{2} = 1.5$

50. $4.7 = 4.\overline{7}$

51. $\pi = 3.14$

52. $\dfrac{1}{3} = 0.33$

In Problems 53–60, replace the ? with the correct symbol: $>, <, =$. *See Objective 3.*

53. $-1 \ ? \ 0$

54. $-8 \ ? \ -8.5$

55. $\dfrac{5}{8} \ ? \ \dfrac{6}{11}$

56. $\dfrac{5}{12} \ ? \ \dfrac{2}{3}$

57. $\dfrac{2}{9} \ ? \ 0.22$

58. $\dfrac{5}{11} \ ? \ 0.\overline{45}$

59. $\dfrac{42}{6} \ ? \ 7$

60. $\dfrac{3}{4} \ ? \ \dfrac{3}{5}$

In Problems 61–68, evaluate each expression. See Objective 4.

61. $|-12|$

62. $|-8|$

63. $|4|$

64. $|7|$

65. $\left|-\dfrac{3}{8}\right|$

66. $\left|-\dfrac{13}{9}\right|$

67. $-|-2.1|$

68. $-|-3.2|$

Mixed Practice

In Problems 69 and 70, (a) plot the points on a real number line, (b) write the numbers in ascending order, and (c) list the numbers that are (i) integers, (ii) rational numbers.

69. $\left\{\dfrac{3}{5}, -1, -\dfrac{1}{2}, 1, 3.5, |-7|, -4.5\right\}$

70. $\left\{8, -2, |-4|, -1.5, -\dfrac{4}{3}, 0, -\dfrac{15}{3}\right\}$

Applying the Concepts

In Problems 71–78, place a ✓ in the box if the given number belongs to that set.

		Natural	Whole	Integers	Rational	Irrational	Real
71.	-100						
72.	0						
73.	-10.5						
74.	$\sqrt{2}$						
75.	$\dfrac{75}{25}$						
76.	4						
77.	$7.56556555\ldots$						
78.	$6.\overline{45}$						

In Problems 79–88, determine whether the statement is True or False.

79. Every *whole* number is also an *integer*.

80. Every *decimal* number is a *rational* number.

81. There are numbers that are both *rational* and *irrational*.

82. 0 is a positive number.

83. Every *natural* number is also a *whole* number.

84. Every *integer* is also a *real* number.

85. Every terminating decimal is a *rational* number.

86. Some numbers in the form $\dfrac{p}{q}$, $q \neq 0$ are *integers*.

87. 0 is a nonnegative *integer*.

88. -1 is a nonpositive *integer*.

In Problems 89–94, name the set, or give the elements of the set, that matches each description.

89. nonterminating and nonrepeating decimals

90. nonnegative integers

91. the set of rational numbers combined with the set of irrational numbers

92. terminating or repeating decimals

93. numbers that are both nonnegative and nonpositive

94. numbers that are both negative and positive

Extending the Concepts

*If every element of set A is also an element of set B, we say A is a **subset** of B and we write $A \subseteq B$. In Problems 95–98, use this definition and the following sets to answer True or False to each statement.*

$$X = \{a, b, c, d, e\} \quad Y = \{c, e\} \quad Z = \{c, e, f\}$$

95. $Y \subseteq X$

96. $Z \subseteq Y$

97. $Y \subseteq Z$

98. $Z \subseteq X$

The **intersection** of two sets is the set that contains the elements common to both A and B and is written $A \cap B$. The **union** of two sets is the set of all elements that are in either A or B and is written

A ∪ B. In Problems 99–104, write the elements of each set, using sets A, B, and C below.

$A = \{7, 8, 9, 10, 11, 12\}$ $B = \{10, 11, 12, 13, 14, 15\}$
$C = \{11, 12, 13, 14, 15\}$

99. $A \cup B$

100. $A \cap B$

101. $B \cup C$

102. $B \cap C$

103. $A \cap C$

104. $A \cup C$

105. If $A = \{$even integers$\}$ and $B = \{$whole numbers less than 11$\}$, find $A \cap B$.

106. If $X = \{48, 49, 50, \ldots\}$ and $Y = \{60, 62, 64, \ldots, 80\}$, find $X \cap Y$.

107. When writing subsets, it is important to be orderly when creating the list. Think of a pattern and then answer the following:

(a) List all possible subsets of set Z where $Z = \{1, 2, 3, 4\}$. *Hint:* The empty set is a subset of every set.

(b) How many subsets did you find?

108. Use the set $M = \{a, b, c\}$ to answer the following:

(a) List all possible subsets of M. *Hint:* The empty set is a subset of every set.

(b) How many subsets did you find?

(c) Determine a rule for finding the number of subsets of a set that has n elements.

Explaining the Concepts

109. Write a definition of "rational number" in your own words. Describe the characteristics to look for when deciding whether a number is in this set.

110. Write a definition of "irrational number" in your own words. Describe the characteristics to look for when deciding whether a number is in this set.

1.4 Adding, Subtracting, Multiplying, and Dividing Integers

Objectives

1. Add Integers
2. Determine the Additive Inverse of a Number
3. Subtract Integers
4. Multiply Integers
5. Divide Integers

Are You Ready for This Section?

Before getting started, take this readiness quiz. If you get a problem wrong, go back to the section cited and review the material.

R1. Write $\dfrac{16}{36}$ as a fraction in lowest terms.   [Section 1.2, pp. 12–13]

R2. $|-5| =$ ___.   [Section 1.3, pp. 24–25]

In this section, we perform addition, subtraction, multiplication, and division, called **operations,** on integers. The symbols used in algebra for these operations are $+$, $-$, $\cdot$, and $/$, respectively. The results of these four operations are called the **sum, difference, product,** and **quotient,** respectively. Table 2 summarizes these ideas.

Table 2

Operation	Symbols	Words
Addition	$a + b$	Sum: a plus b
Subtraction	$a - b$	Difference: a minus b
Multiplication	$a \cdot b, (a) \cdot b, a \cdot (b), (a) \cdot (b),$ $ab, (a)b, a(b), (a)(b)$	Product: a times b
Division	a/b or $\dfrac{a}{b}$	Quotient: a divided by b

Ready?...Answers   **R1.** $\dfrac{4}{9}$ **R2.** 5

In algebra, we avoid using the multiplication sign $\times$ used in arithmetic. Instead, we multiply two expressions that are placed next to each other without an operation symbol, as in ab or that are in parentheses, as in $(a)(b)$, or we use $\cdot$ as in $a \cdot b$.

A **mixed number** is a whole number followed by a fraction. We do not use mixed numbers in algebra. When you see a mixed number, rewrite it as a fraction. Recall that to write $3\dfrac{2}{5}$ as a fraction, we multiply the whole number 3 by the denominator 5, obtaining 15, and then add this result to the numerator 2 to get 17. This result is the numerator of the fraction. The denominator remains 5. Thus

$$3\frac{2}{5} = \frac{17}{5} \quad \leftarrow 3\cdot5 + 2$$

In algebra, mixed numbers are confusing because the lack of an operation symbol between two terms means multiplication. To avoid confusion, write $3\dfrac{2}{5}$ as 3.4 or as $\dfrac{17}{5}$.

Work Smart

Do not use mixed numbers in algebra.

▶ ① Add Integers

Adding Integers with the Same Sign Using a Number Line

We will use a real number line to discover a pattern for adding integers. When we add a positive integer, we move to the right on the number line, and when we add a negative integer, we move to the left on the number line.

Remember, the *sign* of a number indicates whether the number is positive or negative. For example, the sign of 4 is positive, while the sign of -12 is negative. We will first consider adding integers with the same sign.

EXAMPLE 1 **Adding Two Positive Integers Using a Number Line**

Find the sum: $5 + 3$

Solution

We begin at 5 on the number line and move 3 spaces to the right, so $5 + 3 = 8$. See Figure 13.

Figure 13

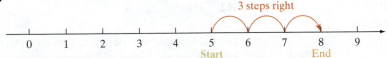

EXAMPLE 2 **Adding Two Negative Integers Using a Number Line**

Find the sum: $-7 + (-4)$

Solution

We begin at -7 on the number line and move 4 spaces to the left, so $-7 + (-4) = -11$. See Figure 14.

Figure 14

Quick ✔

1. The answer to an addition problem is called the ___

In Problems 2 and 3, use a number line to find each sum.

2. $8 + 6$ **3.** $-3 + (-5)$

Adding Integers with Different Signs Using a Number Line

We now consider the sum of two integers with different signs.

EXAMPLE 3   **Adding Integers with Different Signs Using a Number Line**

Find the sum: $-5 + 3$

Solution

We begin at -5 and then move 3 units to the right. From Figure 15 we see that $-5 + 3 = -2$.

Figure 15

●

EXAMPLE 4   **Adding Integers with Different Signs Using a Number Line**

Find the sum: $7 + (-4)$

Solution

We begin at 7 and move 4 spaces to the left. We see that $7 + (-4) = 3$. See Figure 16.

Figure 16

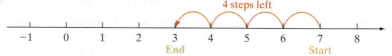

●

Quick ✔

In Problems 4–7, use a number line to find each sum.

4. $-1 + 4$ **5.** $3 + (-4)$ **6.** $-8 + 4$ **7.** $17 + (-3)$

Adding Integers Using Absolute Value

Did you discover a pattern for adding integers from Examples 1–4? To add integers with the same sign (both positive or both negative), add the absolute values of the integers and attach the common sign. To add integers with different signs (one positive and one negative), subtract the smaller absolute value from the larger absolute value and attach the sign of the integer having the larger absolute value.

EXAMPLE 5   **How to Add Integers with the Same Sign Using Absolute Value**

Find $-16 + (-24)$ using absolute value.

Step-by-Step Solution

Step 1: Add the absolute values of the two integers.

We have $|-16| = 16$ and $|-24| = 24$. So
$$16 + 24 = 40$$

Step 2: Attach the common sign, either positive or negative.

Both integers are negative in the original problem, so
$$-16 + (-24) = -40$$

●

EXAMPLE 6 **How to Add Integers with Different Signs Using Absolute Value**

Find $-31 + 16$ using absolute value.

Step-by-Step Solution

Step 1: Subtract the smaller absolute value from the larger absolute value.

We have $|-31| = 31$ and $|16| = 16$. The smaller absolute value is 16, so we compute

$$31 - 16 = 15$$

Step 2: Attach the sign of the integer with the larger absolute value.

The larger absolute value is 31, which was a negative number in the original problem, so the sum is negative. Therefore,

$$-31 + 16 = -15$$

Adding Two Integers Using Absolute Value

To add integers with the same sign (both positive or both negative),

Step 1: Add the absolute values of the two integers.

Step 2: Attach the common sign.

To add integers with different signs (one positive and one negative),

Step 1: Subtract the smaller absolute value from the larger absolute value.

Step 2: Attach the sign of the integer with the larger absolute value.

Quick ✔

8. The sum of two negative integers will be _____.

In Problems 9–12, use absolute value to find each sum.

9. $-11 + 7$ **10.** $5 + (-8)$

11. $-8 + (-16)$ **12.** $-94 + 38$

▶ ❷ **Determine the Additive Inverse of a Number**

Work Smart

The additive inverse of a, $-a$ should not be called the *negative* of a because it suggests that the opposite is a negative number, which may not be true! For example, the additive inverse of -11 is 11, a positive number.

What is $3 + (-3)$? What is $10 + (-10)$? What is $-143 + 143$? The answer to all three of these questions is 0. These results are true in general.

Additive Inverse Property

For any real number a other than 0, there is a real number $-a$, called the **additive inverse**, or **opposite**, of a, having the following property.

$$a + (-a) = -a + a = 0$$

Any two numbers whose sum is zero are additive inverses, or opposites, of each other.

EXAMPLE 7 **Finding an Additive Inverse**

(a) The additive inverse of 9 is -9 because $9 + (-9) = 0$.

(b) The additive inverse of -12 is $-(-12) = 12$ because $-12 + 12 = 0$.

Notice from Example 7(b) that $-(-a) = a$ **for any real number** a.

▶ ❸ Subtract Integers

To subtract integers, we rewrite the subtraction problem as an addition problem and use our addition rules.

From arithmetic, we write $10 - 6 = 4$. Using the additive inverse, we can write $10 - 6 = 4$ as the addition problem $10 + (-6) = 4$.

> **In Words**
> To subtract b from a, add the opposite of b to a.

Definition

The **difference** $a - b$, read "a minus b" or "a less b," is defined as

$$a - b = a + (-b)$$

EXAMPLE 8 **How to Subtract Integers**

Compute the difference: $-18 - (-40)$

Step-by-Step Solution

Step 1: Change the subtraction problem to an equivalent addition problem. $-18 - (-40) = -18 + 40$

Step 2: Find the sum. $= 22$ ●

> **In Words**
> The problem in Example 8 is read "negative eighteen minus negative 40" not "negative 18 minus minus 40."

Subtracting Nonzero Integers

1. Change the subtraction problem to an equivalent addition problem using $a - b = a + (-b)$.
2. Find the sum.

For the remainder of this text, to **evaluate** will mean to find the numerical value of an expression. To evaluate an expression that has both addition and subtraction, change all subtraction to addition. Then add from left to right.

EXAMPLE 9 **Evaluating an Expression with Three Integers**

Evaluate: $10 - 18 + 25$

Solution

> **Work Smart**
> When adding and subtracting more than two numbers, add in order from left to right.

$$10 - 18 + 25 = 10 + (-18) + 25$$
Add from left to right: $= -8 + 25$
$$= 17$$ ●

EXAMPLE 10 **Evaluating an Expression with Four Integers**

Evaluate: $162 - (-46) + 80 - 274$

Solution

$$162 - (-46) + 80 - 274 = 162 + 46 + 80 + (-274)$$
$$\text{Add from left to right:} = 208 + 80 + (-274)$$
$$= 288 + (-274)$$
$$= 14$$

Quick ✓

In Problems 23–26, evaluate each expression.

23. $8 - 13 + 5$ **24.** $-27 - 49 + 18$

25. $3 - (-14) - 8 + 3$ **26.** $-825 + 375 - (-735) + 265$

▶ ④ Multiply Integers

In the statement $9 \cdot 2 = 18$, 9 and 2 are the **factors** and 18 is the **product.**

$$\underbrace{9}_{\text{factor}} \cdot \underbrace{2}_{\text{factor}} = \underbrace{18}_{\text{product}}$$

Recall that we can think of multiplication as repeated addition.

$$3 \cdot 4 = \underbrace{4 + 4 + 4}_{\text{Add 4 three times}} = 12$$

Notice that the product of two positive factors is positive. We knew that from arithmetic! But what is $3 \cdot (-4)$?

$$3 \cdot (-4) = \underbrace{-4 + (-4) + (-4)}_{\text{Add } -4 \text{ three times}} = -12$$

We conclude that the product of two real numbers with *different signs* is negative.

What about the product of two negative numbers? Consider the pattern below.

$$-4 \cdot 3 = -12$$
$$-4 \cdot 2 = -8$$
$$-4 \cdot 1 = -4$$
$$-4 \cdot 0 = 0$$

Each time the second factor decreases by 1, the product increases by 4. Assuming this pattern continues, we would have

$$-4 \cdot -1 = 4$$
$$-4 \cdot -2 = 8$$

The pattern suggests that the product of two negative numbers is a positive number.

In Words

The product is *positive* if the signs of the two factors are the same and *negative* if the signs of the two factors are different.

Rules of Signs for Multiplying Two Integers

1. The product of two positive integers is positive.

2. The product of one positive integer and one negative integer is negative.

3. The product of two negative integers is positive.

EXAMPLE 11 **Multiplying Integers**

(a) $2(-4) = -8$ (b) $-6(5) = -30$

(c) $(-7)(-8) = 56$ (d) $-25(18) = -450$

Quick ✔

27. The product of two integers with the same sign is _____.

In Problems 28–32, find the product.

28. $-3(7)$ **29.** $13(-4)$ **30.** $5 \cdot 16$

31. $-9(-12)$ **32.** $(-13)(-25)$

Find the Product of Several Integers

To find the product of more than two integers, multiply in order, from left to right.

EXAMPLE 12 **Multiplying Three or More Integers**

Find the product:

(a) $3 \cdot (-4) \cdot (-7)$ (b) $-8 \cdot (-1) \cdot 4 \cdot (-5)$

Solution

To find each product, multiply from left to right.

(a) $\begin{aligned} 3 \cdot (-4) \cdot (-7) &= -12 \cdot (-7) \\ &= 84 \end{aligned}$ (b) $\begin{aligned} -8 \cdot (-1) \cdot 4 \cdot (-5) &= 8 \cdot 4 \cdot (-5) \\ &= 32 \cdot (-5) \\ &= -160 \end{aligned}$

Work Smart

If we multiply an *even* number of negative factors, the product is *positive*.

If we multiply an *odd* number of negative factors, the product is *negative*.

Quick ✔

33. *True or False* The product of thirteen negative factors is negative.

In Problems 34 and 35, find the product.

34. $-3 \cdot 9 \cdot (-4)$ **35.** $3 \cdot (-4) \cdot (-5) \cdot (-6)$

▶ ⑤ Divide Integers

When we divide, the numerator is the **dividend,** the denominator is the **divisor,** and the answer is the **quotient.**

$$\text{dividend} \rightarrow \frac{28}{7} = 4 \leftarrow \text{quotient} \quad \text{or} \quad \underbrace{28}_{\text{dividend}} \div \underbrace{7}_{\text{divisor}} = \underbrace{4}_{\text{quotient}}$$

To discuss division of integers, we need to introduce the idea of a *multiplicative inverse*.

In Words

Any two numbers whose product is 1 are called multiplicative inverses, or reciprocals, of each other.

Multiplicative Inverse (Reciprocal) Property

For each *nonzero* real number a, there is a real number $\frac{1}{a}$, called the **multiplicative inverse,** or **reciprocal,** of a, having the following property:

$$a \cdot \frac{1}{a} = \frac{1}{a} \cdot a = 1 \quad \text{where } a \neq 0$$

EXAMPLE 13 **Finding the Multiplicative Inverse (or Reciprocal) of an Integer**

(a) The multiplicative inverse, or reciprocal, of 5 is $\dfrac{1}{5}$.

(b) The multiplicative inverse, or reciprocal, of -8 is $-\dfrac{1}{8}$.

Quick ✓

In Problems 36 and 37, find the multiplicative inverse, or reciprocal, of each integer.

36. 6   **37.** -2

Now we can define division of integers.

> **Definition**
>
> If b is a nonzero integer, the **quotient** $\dfrac{a}{b}$, read as "a divided by b" or "the **ratio** of a to b," is defined as
>
> $$\frac{a}{b} = a \cdot \frac{1}{b} \quad \text{if} \quad b \neq 0$$

For example, $\dfrac{40}{8} = 40 \cdot \dfrac{1}{8}$ and $\dfrac{12}{7} = 12 \cdot \dfrac{1}{7}$. Because division can be represented as multiplication, the same rules of signs that apply to multiplication also apply to division.

> **Rules of Signs for Dividing Two Integers**
>
> **1.** The quotient of two positive integers is positive. That is, $\dfrac{+a}{+b} = \dfrac{a}{b}$.
>
> **2.** The quotient of one positive integer and one negative integer is negative. That is, $\dfrac{-a}{b} = \dfrac{a}{-b} = -\dfrac{a}{b}$.
>
> **3.** The quotient of two negative integers is positive. That is, $\dfrac{-a}{-b} = \dfrac{a}{b}$.

In Words

These are the same rules we saw in multiplying integers. A positive divided by a positive is positive, a positive divided by a negative is negative, and so on.

Finding the quotient of two integers is the same as writing a fraction in lowest terms.

EXAMPLE 14 **Finding the Quotient of Two Integers**

Find each quotient:

(a) $\dfrac{-90}{20}$ (b) $200 \div (-5)$

Solution

(a)
$$\frac{-90}{20} = \frac{-9 \cdot 10}{2 \cdot 10}$$

Divide out the common factor:
$$= \frac{-9 \cdot \cancel{10}}{2 \cdot \cancel{10}}$$

$$= \frac{-9}{2}$$

$\dfrac{-a}{b} = -\dfrac{a}{b}:$
$$= -\frac{9}{2}$$

(b)
$$200 \div (-5) = \frac{200}{-5}$$
$$= \frac{40 \cdot 5}{-1 \cdot 5}$$

Divide out the common factor: $= \dfrac{40 \cdot \cancel{5}}{-1 \cdot \cancel{5}}$

$$= -40$$

Work Smart: Study Skills

Selected problems in the exercise sets are identified by green color. For extra help, worked solutions to these problems are available in MyMathLab.

Quick ✔

38. In the division problem $\dfrac{18}{3} = 6$, 18 is called the _____, 3 is called the _____, and 6 is called the _____.

39. *True or False:* The quotient of two negative numbers is positive.

In Problems 40–42, find the quotient.

40. $\dfrac{20}{-4}$

41. $\dfrac{-72}{20}$

42. $-63 \div -7$

1.4 Exercises

MyMathLab® PRACTICE

Exercise numbers in green have complete video solutions in MyMathLab.

*Problems **1–42** are the Quick ✔s that follow the EXAMPLES.*

Building Skills

In Problems 43–58, find the sum. See Objective 1.

43. $8 + 7$ **44.** $6 + 4$

45. $-5 + 9$ **46.** $-4 + 12$

47. $8 + (-12)$ **48.** $7 + (-13)$

49. $-11 + (-8)$ **50.** $-13 + (-5)$

51. $-16 + 37$ **52.** $-32 + 49$

53. $-119 + (-209)$ **54.** $-145 + (-68)$

55. $-14 + 21 + (-18)$

56. $(-13) + 37 + (-22)$

57. $74 + (-13) + (-23) + 5$

58. $-34 + 46 + (-12) + 72$

In Problems 59–62, determine the additive inverse of each real number. See Objective 2.

59. -325 **60.** -34

61. 125 **62.** 7

In Problems 63–78, find the difference. See Objective 3.

63. $23 - 12$ **64.** $35 - 23$

65. $9 - 17$ **66.** $12 - 19$

67. $-20 - 8$ **68.** $-15 - 9$

69. $13 - (-41)$ **70.** $14 - (-18)$

71. $-36 - (-36)$ **72.** $-15 - (-15)$

73. $0 - 41$ **74.** $0 - 18$

75. $-93 - (-62)$ **76.** $46 - (-25)$

77. $86 - (-86)$ **78.** $49 - (-49)$

In Problems 79–94, find the product. See Objective 4.

79. $5 \cdot 8$ **80.** $7 \cdot 9$

81. $8(-7)$ **82.** $9(-7)$

83. $(0)(-21)$ **84.** $-21 \cdot 0$

85. $(-48)(-3)$ **86.** $(-22)(-5)$

87. $(-42)3$ **88.** $(-128)7$

89. $-5 \cdot 6 \cdot 3$ **90.** $-6 \cdot 4 \cdot 8$

91. $-10(3)(-7)$ **92.** $-8(2)(-9)$

93. $(-2)(4)(-1)(3)(5)$

94. $(-3)(-4)(6)(-1)$

In Problems 95–100, find the multiplicative inverse (or reciprocal) of each number. See Objective 5.

95. 8 **96.** 10 **97.** -4

98. -3 **99.** 1 **100.** 2

In Problems 101–112, find the quotient. See Objective 5.

101. $10 \div 2$ **102.** $36 \div 9$ **103.** $\dfrac{-56}{-8}$

104. $\dfrac{-63}{-7}$ **105.** $\dfrac{-45}{3}$ **106.** $\dfrac{-144}{6}$

107. $\dfrac{35}{10}$ **108.** $\dfrac{20}{16}$ **109.** $\dfrac{60}{-42}$

110. $\dfrac{120}{-66}$ **111.** $\dfrac{-105}{-12}$ **112.** $\dfrac{-80}{-12}$

Mixed Practice

In Problems 113–128, evaluate the expression.

113. $-4 \cdot 18$ **114.** $7 \cdot (-15)$

115. $-16 - (-76)$ **116.** $87 - 19$

117. $-9 \cdot (-19)$ **118.** $7 \cdot 209$

119. $\dfrac{120}{-8}$ **120.** $\dfrac{-156}{-26}$

121. $-98 + 56$ **122.** $103 + (-66)$

123. $\dfrac{75}{|-20|}$ **124.** $\dfrac{|-42|}{12}$

125. $|-14| + |-26|$ **126.** $|-10| + (-62)$

127. $|-389| - 627$ **128.** $|-193| - (-20)$

In Problems 129–136, write each expression using mathematical symbols. Then evaluate the expression.

129. the sum of 28 and -21

130. the sum of 32 and -64

131. -21 minus 47

132. -85 subtracted from -16

133. -12 multiplied by 18

134. 32 multiplied by -8

135. -36 divided by -108 **136.** -40 divided by 100

Applying the Concepts

In Problems 137–142, write the positive or negative number for each amount or measurement.

137. Stock The price of IBM stock fell by 3.25 dollars.

138. Temperature The current temperature in Juneau, Alaska, is 14° below zero.

139. Chargers Football The Chargers lost 6 yards on the play.

140. Profit Mark's auto dealership showed a profit of $125,000 this quarter.

141. Checking Account Leila's checking account is now overdrawn by $48.

142. Census The number of people in Brian's hometown grew by 12,368.

143. Hiking Loren and Richard went on a hiking trip. They walked 8 miles to the base of Snow Creek Falls, where they set up camp. They went another 3 miles to see the Falls and then returned to their campsite. How many miles did they walk that day?

144. Football The Cary High School Eagles took over possession of the ball on their own 15-yard line. The following plays occurred: QB sack, loss of 7 yards; Jon Anderson ran for 14 yards; Juan Ramirez caught a pass for 26 yards. What yard marker is the ball on now?

145. Bank Balance When Martha balanced her checkbook, she had $563 in the account. Then the following transactions occurred: She wrote a check to Home Depot for $46, deposited $233, wrote a check to Vons for $63, and wrote a check to Petco for $32. What is Martha's new balance?

146. Checkbook Balance Josie began the month with $399 in her bank account. She deposited her paycheck of $839. She paid her $69 telephone bill, the electric bill for $78, and rent of $739. How much does she have left for spending money?

147. Warehouse Inventory The warehouse began the month with 725 cases of soda. During the month the following transactions occurred: 120 cases were shipped out, 590 cases were shipped out, and 310 cases were delivered to the warehouse. Does the warehouse have enough stock on hand to fill an order for 450 cases of soda? What is the difference between what it has and what has been requested?

148. Altitude of an Airplane A pilot leveled off his airplane at 35,000 feet at the beginning of the flight. The following adjustments were made during the trip: gained 4290 feet, dropped 10,400 feet, and then dropped 2605 feet. At what altitude is the plane currently flying?

149. Distance An airplane flying at 25,350 feet is directly over a submarine that is 375 feet below sea level. What is the distance between a person in the plane and a person in the submarine?

150. Elevation The highest point in California is Mt. Whitney at an elevation of 14,495 feet, and the lowest point is in Death Valley at 280 feet below sea level. What is the maximum difference in elevation between the two points in California?

Extending the Concepts

151. Find two integers whose sum is -8 and whose product is 15.

152. Find two integers whose sum is 2 and whose product is -24.

153. Find two integers whose sum is -10 and whose product is -24.

154. Find two integers whose sum is -18 and whose product is 45.

155. The Fibonacci Sequence
The numbers in the Fibonacci sequence are 1, 1, 2, 3, 5, 8, 13, 21, 34, 55, . . ., where each term after the second term is the sum of the two preceding terms. This famous sequence of numbers can be used to model many phenomena in nature.

(a) Form fractions of consecutive terms in the sequence. Find the decimal approximations to $\dfrac{1}{1}, \dfrac{2}{1}, \dfrac{3}{2}, \dfrac{5}{3}$, and so on.

(b) What number do the ratios get close to? This number is called the **golden ratio** and has application in many different areas.

(c) Research Fibonacci numbers and cite three different applications.

Explaining the Concepts

156. Write a sentence or two that justifies the fact that the product of a positive number and a negative number is a negative number. You may use an example.

157. Explain how $42 \div 4$ may be written as a multiplication problem.

1.5 Adding, Subtracting, Multiplying, and Dividing Rational Numbers

Objectives

1. Multiply Rational Numbers in Fractional Form
2. Divide Rational Numbers in Fractional Form
3. Add or Subtract Rational Numbers in Fractional Form
4. Add, Subtract, Multiply, or Divide Rational Numbers in Decimal Form

In Words
To find the product of two or more fractions, multiply the numerators together. Then multiply the denominators together. Write the fraction in lowest terms, if necessary.

Are You Ready for This Section?

Before getting started, take this readiness quiz. If you get a problem wrong, go back to the section cited and review the material.

R1. Find the least common denominator of $\dfrac{5}{12}$ and $\dfrac{3}{16}$. [Section 1.2, pp. 11–12]

R2. Rewrite $\dfrac{4}{5}$ as an equivalent fraction with a denominator of 30. [Section 1.2, pp. 10–11]

R3. Write each rational number in lowest terms. [Section 1.4, pp. 34–35]
(a) $-\dfrac{18}{30}$ **(b)** $-\dfrac{24}{4}$

Now that we are comfortable with operations on integers, we can perform operations on rational numbers. We begin with operations on rational numbers expressed as fractions and end the section with the operations on rational numbers in decimal form.

All of the properties included for integers in Section 1.4 apply to rational numbers as well. In fact, these properties apply to all real numbers.

▶ ❶ Multiply Rational Numbers in Fractional Form

We use the following property to multiply two rational numbers in fractional form:

> **Multiplying Fractions**
>
> $$\frac{a}{b} \cdot \frac{c}{d} = \frac{a \cdot c}{b \cdot d} \quad \text{where } b \text{ and } d \neq 0$$

The rules of signs that apply to integers also apply to rational numbers: The product of two positive rational numbers is positive; the product of a positive rational number and a negative rational number is negative; and the product of two negative rational numbers is positive.

Ready?...Answers **R1.** LCD $= 48$
R2. $\frac{4}{5} = \frac{24}{30}$ **R3.** a. $-\frac{3}{5}$ b. -6

EXAMPLE 1 **Multiplying Rational Numbers (Fractions)**

Find the product: $\dfrac{2}{9} \cdot \left(-\dfrac{15}{19}\right)$

Solution

We begin by rewriting the rational number $-\dfrac{15}{19}$ as $\dfrac{-15}{19}$. Then we multiply the numerators and multiply the denominators.

$$\frac{2}{9} \cdot \left(\frac{-15}{19}\right) = \frac{2 \cdot (-15)}{9 \cdot 19}$$

Work Smart

Notice that we do not multiply in the numerator or denominator until we divide out common factors.

Write the numerator and the denominator as products of prime factors: $= \dfrac{2 \cdot 3 \cdot (-5)}{3 \cdot 3 \cdot 19}$

Divide out common factors: $= \dfrac{2 \cdot \cancel{3} \cdot (-5)}{\cancel{3} \cdot 3 \cdot 19}$

$= \dfrac{2 \cdot (-5)}{3 \cdot 19}$

Multiply: $= -\dfrac{10}{57}$

●

> **Quick ✓**
>
> *In Problems 1–5, find each product, and write in lowest terms.*
>
> **1.** $\dfrac{3}{4} \cdot \dfrac{9}{8}$ **2.** $\dfrac{-5}{7} \cdot \dfrac{56}{15}$ **3.** $\dfrac{12}{45} \cdot \left(-\dfrac{18}{20}\right)$
>
> **4.** $-\dfrac{25}{75} \cdot \left(-\dfrac{9}{4}\right)$ **5.** $\dfrac{7}{3} \cdot \dfrac{1}{14} \cdot \left(-\dfrac{9}{11}\right)$

▶ ② **Divide Rational Numbers in Fractional Form**

To divide rational numbers, we must know how to find the reciprocal of a rational number. In Section 1.4, we saw that two numbers are *reciprocals*, or multiplicative inverses, if their product is 1. This definition applies to any nonzero real number. Thus, $\dfrac{3}{2}$ and $\dfrac{2}{3}$ are reciprocals because $\dfrac{3}{2} \cdot \dfrac{2}{3} = 1$; 9 and $\dfrac{1}{9}$ are reciprocals because $9 \cdot \dfrac{1}{9} = 1$; $-\dfrac{4}{7}$ and $-\dfrac{7}{4}$ are reciprocals because $-\dfrac{4}{7} \cdot \left(-\dfrac{7}{4}\right) = 1$.

> **Quick ✓**
>
> **6.** Two numbers are called multiplicative inverses, or reciprocals, if their product is equal to _.
>
> *In Problems 7–10, find the reciprocal of each number.*
>
> **7.** 12   **8.** $\dfrac{7}{5}$ **9.** $-\dfrac{1}{4}$ **10.** $-\dfrac{31}{20}$

We divide rational numbers by rewriting the division as an equivalent multiplication problem.

> **Dividing Rational Numbers Expressed as Fractions**
>
> $$\frac{a}{b} \div \frac{c}{d} = \frac{a}{b} \cdot \frac{d}{c} = \frac{a \cdot d}{b \cdot c} \quad \text{where } b, c, d \neq 0$$

EXAMPLE 2   **How to Divide Rational Numbers (Fractions)**

Find the quotient: $\dfrac{3}{10} \div \dfrac{12}{25}$

Step-by-Step Solution

Step 1: Write the equivalent multiplication problem.

$$\frac{3}{10} \div \frac{12}{25} = \frac{3}{10} \cdot \frac{25}{12}$$

Step 2: Write the product in factored form and divide out common factors.

$$= \frac{3 \cdot 25}{10 \cdot 12}$$

$$= \frac{3 \cdot 5 \cdot 5}{5 \cdot 2 \cdot 4 \cdot 3}$$

$$= \frac{\cancel{3} \cdot \cancel{5} \cdot 5}{\cancel{5} \cdot 2 \cdot 4 \cdot \cancel{3}}$$

$$= \frac{5}{2 \cdot 4}$$

Step 3: Multiply the remaining factors.

$$= \frac{5}{8}$$

Quick ✔

In Problems 11–14, find the quotient.

11. $\dfrac{5}{7} \div \dfrac{7}{10}$

12. $-\dfrac{9}{12} \div \dfrac{14}{7}$

13. $\dfrac{8}{35} \div \left(-\dfrac{1}{10}\right)$

14. $-\dfrac{18}{63} \div \left(-\dfrac{54}{35}\right)$

▶ ❸ **Add or Subtract Rational Numbers in Fractional Form**

Add or Subtract Rational Numbers (Fractions) with Like Denominators

Figure 17

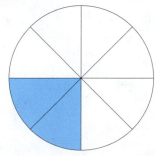

In Figure 17, two of the eight equal regions are shaded. Each of the regions represents the fraction $\dfrac{1}{8}$. Together, the two shaded regions make up $\dfrac{1}{4}$ of the circle. This implies that

$$\frac{1}{8} + \frac{1}{8} = \frac{1 + 1}{8}$$

$$= \frac{2}{8}$$

$$= \frac{1}{4}$$

Or, suppose Bobby has \$0.25 and his grandma gives him \$0.50. He now has \$0.25 + \$0.50 = \$0.75, or $\dfrac{3}{4}$ of a dollar. Because $0.25 = \dfrac{1}{4}$ and $0.50 = \dfrac{1}{2}$, we can determine Bobby's good fortune using fractions:

$$\frac{1}{4} + \frac{1}{2} = \frac{1}{4} + \frac{2}{4}$$

$$= \frac{3}{4}$$

Based on these results, we might conclude that to add fractions with the same denominators, we add the numerators and write the result over the common denominator.

This conclusion is correct. Also, because we can write any subtraction problem as an equivalent addition problem, we have the following methods for adding or subtracting rational numbers.

In Words
To add or subtract fractions with a common denominator, add or subtract the numerators and retain the denominator.

Adding or Subtracting Rational Numbers (Fractions) with the Same Denominator

$$\frac{a}{c} + \frac{b}{c} = \frac{a+b}{c} \quad \text{where } c \neq 0 \qquad \frac{a}{c} - \frac{b}{c} = \frac{a-b}{c} = \frac{a+(-b)}{c} \quad \text{where } c \neq 0$$

EXAMPLE 3 **Adding Rational Numbers (Fractions) with the Same Denominator**

Find the sum and write in lowest terms: $-\dfrac{1}{8} + \dfrac{3}{8}$

Solution

$$-\frac{1}{8} + \frac{3}{8} = \frac{-1}{8} + \frac{3}{8}$$

Write the numerators as a sum over the common denominator:
$$= \frac{-1+3}{8}$$

Add the numerators:
$$= \frac{2}{8}$$

Factor 8 and divide out the 2s:
$$= \frac{1 \cdot \cancel{2}}{4 \cdot \cancel{2}}$$

$$= \frac{1}{4}$$

EXAMPLE 4 **Subtracting Rational Numbers (Fractions) with the Same Denominator**

Find the difference and write in lowest terms: $\dfrac{9}{16} - \dfrac{3}{16}$

Solution

$$\frac{9}{16} - \frac{3}{16} = \frac{9-3}{16}$$

$$= \frac{6}{16}$$

Factor 6, factor 16, and divide out the 2s:
$$= \frac{3 \cdot \cancel{2}}{8 \cdot \cancel{2}}$$

$$= \frac{3}{8}$$

Quick ✓

15. $-\dfrac{5}{7} + \dfrac{3}{7} = \dfrac{\underline{} + 3}{\underline{}}$

In Problems 16–19, find the sum or difference, and write in lowest terms.

16. $\dfrac{8}{11} + \dfrac{2}{11}$

17. $-\dfrac{18}{35} + \dfrac{3}{35}$

18. $\dfrac{19}{63} - \dfrac{10}{63}$

19. $-\dfrac{9}{10} - \dfrac{3}{10}$

Work Smart: Study Skills

Note the title of Example 5: "How to Add Rational Numbers with Unlike Denominators." This three-column example provides a guided, step-by-step approach to solving a problem so you can see each of the steps. Cover the third column and try to work the example yourself. Then look at the entries in the third column and check your solution. Was it correct?

▶ **Adding or Subtracting Rational Numbers (Fractions) with Unlike Denominators**

How do we add rational numbers with different denominators? We must find the least common denominator of the two rational numbers. Recall from Section 1.2 that the **least common denominator (LCD)** is the smallest number that each denominator has as a common multiple.

EXAMPLE 5 **How to Add Rational Numbers (Fractions) with Unlike Denominators**

Find the sum: $\dfrac{5}{6} + \dfrac{3}{8}$

Step-by-Step Solution

Step 1: Find the least common denominator of the fractions.

Write each denominator as the product of prime factors, arranging like factors vertically:

$$6 = 2 \qquad \cdot 3$$
$$8 = 2 \cdot 2 \cdot 2$$

Find the product of each of the prime factors the greatest number of times they appear in any factorization:

$$LCD = 2 \cdot 2 \cdot 2 \cdot 3$$
$$= 24$$

Step 2: Write each rational number with the denominator found in Step 1.

Use $1 = \dfrac{4}{4}$ to change the denominator 6 to 24, and use $1 = \dfrac{3}{3}$ to change the denominator 8 to 24:

$$\frac{5}{6} + \frac{3}{8} = \frac{5}{6} \cdot \frac{4}{4} + \frac{3}{8} \cdot \frac{3}{3}$$
$$= \frac{20}{24} + \frac{9}{24}$$

Step 3: Add the numerators and write the result over the common denominator.

$$= \frac{20 + 9}{24}$$
$$= \frac{29}{24}$$

Step 4: Write in lowest terms.

The rational number is in lowest terms, so $\dfrac{5}{6} + \dfrac{3}{8} = \dfrac{29}{24}$. ●

EXAMPLE 6 **How to Subtract Rational Numbers (Fractions) with Unlike Denominators**

Find the difference: $-\dfrac{9}{14} - \dfrac{1}{6}$

Step-by-Step Solution

Step 1: Find the least common denominator of the fractions.

Write each denominator as the product of prime factors, aligning like factors vertically:

$$14 = 2 \cdot 7$$
$$6 = 2 \qquad \cdot 3$$

Find the product of each of the prime factors the greatest number of times it appears in any factorization:

$$LCD = 2 \cdot 7 \cdot 3$$
$$= 42$$

Step 2: Write each rational number with the denominator found in Step 1.

Use $1 = \dfrac{3}{3}$ to change the denominator 14 to 42, and use $1 = \dfrac{7}{7}$ to change the denominator 6 to 42:

$$-\frac{9}{14} - \frac{1}{6} = \frac{-9}{14} \cdot \frac{3}{3} - \frac{1}{6} \cdot \frac{7}{7}$$
$$= \frac{-27}{42} - \frac{7}{42}$$

(continued)

Step 3: Subtract the numerators
and write the result over the
common denominator.

$$= \frac{-27 - 7}{42}$$

$$= \frac{-34}{42}$$

Step 4: Write in lowest terms.

Factor -34 and 42 and
divide out like factors:

$$= \frac{2 \cdot (-17)}{2 \cdot 21}$$

$$= -\frac{17}{21}$$

Adding or Subtracting Rational Numbers (Fractions) with Unlike Denominators

Step 1: Find the LCD of the rational numbers.

Step 2: Write each rational number with the LCD.

Step 3: Add or subtract the numerators and write the result over the common denominator.

Step 4: Write the result in lowest terms.

Quick ✔

In Problems 20–23, find each sum or difference, and write in lowest terms.

20. $\dfrac{3}{14} + \dfrac{10}{21}$

21. $\dfrac{5}{12} - \dfrac{5}{18}$

22. $-\dfrac{23}{6} + \dfrac{7}{12}$

23. $-\dfrac{1}{15} - \dfrac{1}{12}$

Remember, the direction "evaluate" means to find the value of the expression.

EXAMPLE 7 **Evaluating an Expression Containing Rational Numbers**

Evaluate and write in lowest terms: $4 - \dfrac{2}{3}$

Solution

The key is to remember that $4 = \dfrac{4}{1}$.

$$4 - \frac{2}{3} = \frac{4}{1} - \frac{2}{3}$$

Rewrite each fraction with LCD $= 3$:

$$= \frac{4}{1} \cdot \frac{3}{3} - \frac{2}{3}$$

$$= \frac{12}{3} - \frac{2}{3}$$

$$= \frac{10}{3}$$

Quick ✔

In Problem 24 and 25, evaluate and write in lowest terms.

24. $-2 + \dfrac{7}{16}$

25. $6 - \dfrac{9}{4}$

4 Add, Subtract, Multiply, or Divide Rational Numbers in Decimal Form

▶ **Adding or Subtracting Decimals**

To add or subtract decimals, arrange the numbers in a column with the decimal points aligned. Then add or subtract the digits in the like place values. Put the decimal point in the answer directly below the decimal points in the problem.

EXAMPLE 8 **Adding or Subtracting Decimals That Have the Same Sign**

Evaluate each expression:

(a) $2.93 + 7.2 + 3.026$ (b) $76.4 - 4.95$

Solution

(a) Use zeros as placeholders:

$$
\begin{array}{r}
2.930 \\
7.200 \\
+3.026 \\
\hline
13.156
\end{array}
$$

Work Smart

A whole number has an implied decimal point. For example,

$74 = 74.000$

(b) Use zero as a placeholder:

$$
\begin{array}{r}
\overset{5\ \ 13\ 10}{76.40} \\
-4.95 \\
\hline
71.45
\end{array}
$$

EXAMPLE 9 **Adding or Subtracting Decimals That Have Different Signs**

Evaluate each expression:

(a) $100.32 - (-32.015)$ (b) $-23.03 + 18.49$

Solution

(a) Remember, $a - (-b) = a + b$, so $100.32 - (-32.015) = 100.32 + 32.015$.

$$
\begin{array}{r}
100.320 \\
+32.015 \\
\hline
132.335
\end{array}
$$

(b) Recall that to add real numbers with different signs, subtract the smaller absolute value from the larger absolute value and attach the sign of the larger absolute value. Because $|-23.03| = 23.03$ and $|18.49| = 18.49$, we compute $23.03 - 18.49$ and attach a negative sign to the difference.

$$
\begin{array}{r}
23.03 \\
-18.49 \\
\hline
4.54
\end{array}
$$

So, $-23.03 + 18.49 = -4.54$.

Quick ✓

In Problems 26–29, find the sum or difference.

26. $9.67 - (-11.344)$ **27.** $-17.39 + 81.96$

28. $-74.28 + 14.832$ **29.** $-180.782 + 100.3 + 9.07$

▶ Multiplying Decimals

The rules for multiplying decimals come from the rules for multiplying rational numbers in fractional form. For example,

$$\underbrace{-0.7}_{\substack{1 \text{ decimal} \\ \text{place}}} \cdot \underbrace{0.03}_{\substack{2 \text{ decimal} \\ \text{places}}} = \frac{-7}{10} \cdot \frac{3}{100} = \frac{-21}{1000} = \underbrace{-0.021}_{\substack{3 \text{ decimal} \\ \text{places}}}$$

Notice the number of digits to the right of the decimal point in the answer is equal to the sum of the numbers of digits to the right of each decimal point in the factors.

The rules of signs that we learned in Section 1.4 apply to decimals as well.

EXAMPLE 10   **Multiplying Decimals**

Find the product:

(a) $3.43 \cdot 2.6$ **(b)** $-3.17 \cdot 0.02$

Solution

(a)

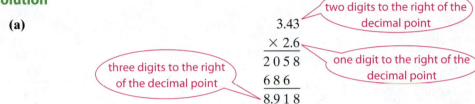

(b)

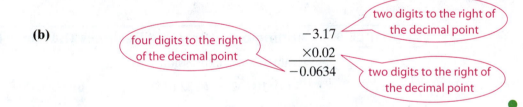

Work Smart

The number of digits to the right of the decimal point in the product is the *sum* of the numbers of digits to the right of each decimal point in the factors.

> **Multiplying Decimals**
>
> **Step 1:** Multiply the factors as if they were whole numbers.
>
> **Step 2:** Place the decimal point so the number of digits to the right of the decimal point in the product equals the *sum* of the numbers of digits to the right of each decimal point in the factors.

Quick ✔

In Problems 30–33, find the product.

30. $23.9 \cdot 0.2$ **31.** $257 \cdot (-3.5)$

32. $-3.45 \cdot 0.03$ **33.** $-0.03 \cdot (-0.45)$

▶ Dividing Decimals

In the division problem $2\overline{)7.94}$ with 3.97, the number 2 is the *divisor*, 7.94 is the *dividend*, and 3.97 is the *quotient*. Notice that when we divide by a whole number, we line up the decimal points in the quotient and the dividend. In algebra, we typically write this division problem as $\dfrac{7.94}{2} = 3.97$.

To divide decimals, we want the divisor to be a whole number, so we multiply the dividend and the divisor by a power of 10 that will make the divisor a whole number. Then we divide as described above.

EXAMPLE 11 **Dividing Decimals**

(a) Divide: $\dfrac{22.26}{15.9}$ 　　　　　(b) Divide: $\dfrac{0.03724}{-0.38}$

Solution

(a) Since the divisor 15.9 is fifteen and nine-tenths, we multiply $\dfrac{22.26}{15.9}$ by $\dfrac{10}{10}$ to make the divisor a whole number and obtain $\dfrac{22.26}{15.9} \cdot \dfrac{10}{10} = \dfrac{222.6}{159}$. Now we divide.

$$
\begin{array}{r}
1.4 \\
159\overline{)222.6} \\
\underline{159} \\
63\ 6 \\
\underline{63\ 6} \\
0
\end{array}
$$

So $\dfrac{22.26}{15.9} = 1.4$.

In Words

To divide decimals, change the divisor to a whole number and divide.

(b) Because we have a positive number divided by a negative number, the quotient will be negative. The divisor 0.38 is thirty-eight hundredths, so we multiply $\dfrac{0.03724}{0.38}$ by $\dfrac{100}{100}$ to make the divisor a whole number and obtain

$\dfrac{0.03724}{0.38} \cdot \dfrac{100}{100} = \dfrac{3.724}{38}$. Now we divide.

$$
\begin{array}{r}
0.098 \\
38\overline{)3.724} \\
\underline{3\ 42} \\
304 \\
\underline{304} \\
0
\end{array}
$$

Therefore, $\dfrac{0.03724}{-0.38} = -0.098$.

Dividing Decimals

Step 1: Multiply the dividend and divisor by a power of 10 that will make the divisor a whole number.

Step 2: Divide as though working with whole numbers.

Work Smart: Study Skills

Selected problems in the exercise sets are identified by green color. For extra help, worked solutions to these problems are in MyMathLab.

Quick ✓

In Problems 34–36, find the quotient.

34. $\dfrac{18.25}{73}$ 　　　　**35.** $\dfrac{1.0032}{0.12}$ 　　　　**36.** $\dfrac{-4.2958}{45.7}$

1.5 Exercises MyMathLab® Math XL PRACTICE

Exercise numbers in green have complete video solutions in MyMathLab.

Problems 1–36 are the Quick ✓s that follow the EXAMPLES.

Building Skills

In Problems 37–42, write each rational number in lowest terms.

37. $\dfrac{14}{21}$ **38.** $\dfrac{9}{15}$ **39.** $-\dfrac{38}{18}$

40. $-\dfrac{81}{36}$ **41.** $-\dfrac{22}{44}$ **42.** $-\dfrac{24}{27}$

In Problems 43–52, find the product and write in lowest terms. See Objective 1.

43. $\dfrac{6}{5}\cdot\dfrac{2}{5}$ **44.** $\dfrac{7}{8}\cdot\dfrac{10}{21}$ **45.** $-\dfrac{5}{2}\cdot 10$

46. $-\dfrac{3}{7}\cdot 63$ **47.** $-\dfrac{3}{2}\cdot\dfrac{4}{9}$ **48.** $-\dfrac{5}{2}\cdot\dfrac{16}{25}$

49. $-\dfrac{22}{3}\cdot\left(-\dfrac{12}{11}\right)$ **50.** $-\dfrac{60}{75}\cdot\left(-\dfrac{25}{4}\right)$

51. $\dfrac{3}{16}\cdot\dfrac{8}{9}$ **52.** $\dfrac{4}{27}\cdot\dfrac{9}{16}$

In Problems 53–56, find the reciprocal of each number. See Objective 2.

53. $\dfrac{3}{5}$ **54.** $\dfrac{9}{4}$

55. -5 **56.** -8

In Problems 57–66, find the quotient and write in lowest terms. See Objective 2.

57. $\dfrac{4}{9}\div\dfrac{8}{15}$ **58.** $\dfrac{3}{2}\div\dfrac{9}{8}$

59. $-\dfrac{1}{3}\div 3$ **60.** $-\dfrac{1}{4}\div 4$

61. $\dfrac{5}{6}\div\left(-\dfrac{5}{4}\right)$ **62.** $\dfrac{4}{3}\div\left(-\dfrac{9}{10}\right)$

63. $\dfrac{2}{5}\div\dfrac{22}{5}$ **64.** $\dfrac{44}{63}\div\dfrac{88}{21}$

65. $-8\div\left(-\dfrac{1}{4}\right)$ **66.** $-3\div\left(-\dfrac{1}{6}\right)$

In Problems 67–84, find the sum or difference and write in lowest terms. See Objective 3.

67. $\dfrac{3}{4}+\dfrac{3}{4}$ **68.** $\dfrac{6}{11}+\dfrac{16}{11}$

69. $\dfrac{9}{8}-\dfrac{5}{8}$ **70.** $\dfrac{12}{5}-\dfrac{2}{5}$

71. $\dfrac{6}{7}-\left(-\dfrac{8}{7}\right)$ **72.** $\dfrac{2}{3}-\left(-\dfrac{7}{3}\right)$

73. $-\dfrac{5}{3}+2$ **74.** $-\dfrac{7}{8}+4$

75. $6-\dfrac{7}{2}$ **76.** $3-\dfrac{5}{3}$

77. $-\dfrac{4}{3}+\dfrac{1}{4}$ **78.** $-\dfrac{2}{5}+\left(-\dfrac{2}{3}\right)$

79. $\dfrac{7}{5}+\left(-\dfrac{23}{20}\right)$ **80.** $\dfrac{7}{15}-\left(-\dfrac{4}{3}\right)$

81. $\dfrac{8}{15}-\dfrac{7}{10}$ **82.** $\dfrac{17}{6}-\dfrac{13}{9}$

83. $-\dfrac{33}{10}-\left(-\dfrac{33}{8}\right)$ **84.** $-\dfrac{29}{6}-\left(-\dfrac{29}{20}\right)$

In Problems 85–102, perform the indicated operation(s). See Objective 4.

85. $-10.5+4$ **86.** $-13.2+7$

87. $-(-3.5)+4.9$ **88.** $-(-32.9)+10.3$

89. $39.1-(-16.82)$ **90.** $29.23-(-12.98)$

91. $-5.21-(-6.7)$ **92.** $-4.94-(-3.87)$

93. $45-2.45$ **94.** $32-5.68$

95. $4.3\cdot 5.8$ **96.** $3.1\cdot 10.9$

97. $0.075\cdot(-120)$ **98.** $0.065\cdot(-340)$

99. $\dfrac{136.08}{5.6}$ **100.** $\dfrac{332.59}{7.9}$

101. $\dfrac{-25.48}{0.052}$ **102.** $\dfrac{-48}{0.03}$

Mixed Practice

In Problems 103–134, evaluate and write in lowest terms.

103. $-\dfrac{5}{6}+\dfrac{7}{15}$ **104.** $-\dfrac{8}{9}+\left(-\dfrac{16}{21}\right)$

105. $-\dfrac{10}{21}\cdot\dfrac{14}{5}$ **106.** $\dfrac{24}{5}\cdot\left(-\dfrac{35}{4}\right)$

107. $\dfrac{3}{64}\div\left(-\dfrac{9}{16}\right)$ **108.** $-\dfrac{12}{7}\div\left(-\dfrac{4}{21}\right)$

109. $-\dfrac{5}{12}+\dfrac{2}{12}$ **110.** $-\dfrac{4}{9}+\dfrac{1}{9}$

111. $-\dfrac{2}{7}-\dfrac{17}{5}$ **112.** $-\dfrac{3}{4}-\dfrac{1}{5}$

113. $-8.7-(-10.3)$ **114.** $-4.63-(-12.9)$

115. $\dfrac{1}{12}+\left(-\dfrac{5}{28}\right)$ **116.** $\dfrac{3}{16}+\left(-\dfrac{7}{40}\right)$

117. $-12.03\cdot 4.2$ **118.** $34.2\cdot(-8.43)$

119. $36\cdot\left(-\dfrac{4}{9}\right)$ **120.** $-\dfrac{8}{3}\cdot 15$

121. $-27\div\dfrac{9}{5}$ **122.** $-24\div\dfrac{8}{7}$

123. $3.62-10.2$ **124.** $4.75-6.2$

125. $\dfrac{-145.518}{18.42}$

126. $\dfrac{-297.078}{22.17}$

127. $\dfrac{12}{7} - \dfrac{17}{14} - \dfrac{48}{21}$

128. $\dfrac{9}{4} - \dfrac{21}{6} - \dfrac{11}{8}$

129. $54.2 - 18.78 - (-2.5) + 20.47$

130. $90.3 - 100.9 - (-34.26) + 32.95$

131. $(400)(-25.8)(0.003)$

132. $(500)(-12.4)(-0.02)$

133. $-\dfrac{11}{12} - \left(-\dfrac{1}{6}\right) + \dfrac{7}{8}$

134. $\dfrac{8}{15} - \left(-\dfrac{7}{9}\right) + \dfrac{2}{3}$

Applying the Concepts

135. Watching TV If Rachel spends $\dfrac{1}{8}$ of her life watching TV, how many hours of TV does she watch in one week?

136. Halloween Candy Henry decided to make $\dfrac{2}{3}$-oz bags of candy for treats at Halloween. If he bought 16 oz of candy, how many bags will he have to give away?

137. Biology Class Susan's biology class begins with 36 students. If $\dfrac{2}{3}$ will finish the course and $\dfrac{3}{4}$ of those get a passing grade, how many students will pass Susan's biology class this term?

138. Pizza Time Joyce and Ramie bought a pizza. Joyce ate $\dfrac{2}{5}$ of the pizza and Ramie ate $\dfrac{1}{9}$ of what was left. What fraction of the pizza remains uneaten?

139. Hourly Pay Last week, Jonathon received a paycheck for $442.80. The withholding for federal, state, and FICA (social security and Medicare) taxes was $97.20. If Jonathon worked 30 hours last week, what is his hourly pay rate?

140. Average Revenue Aqsa runs an online store selling beauty products. Last year, the total revenue of the company was $29,409.12. What was the average revenue per month?

141. Stock Prices The price per share of Intel stock has been up and down lately. On Monday it rose 2.75; on Tuesday it rose 0.87; on Wednesday it dropped 1.12; on Thursday it rose 0.52; and on Friday it fell 0.62. What was the net change in Intel's stock price per share for the week?

142. Bank Balance Henry started the month with $43.68 in his checking account. During the month the following transactions occurred: He deposited his paycheck of $929.30; and he wrote checks for rent $650, phone $33.49, credit card $229.50, cable service $75.50, and groceries $159.30. How much does he have in his account now?

Extending the Concepts

Problems 143–146 use the following definition.

If P and Q are two points on a real number line with coordinates a and b, respectively, then the **distance between P and Q**, denoted by $d(P, Q)$, is

$$d(P, Q) = |b - a|$$

143. Find the distance between the points P and Q on the real number line if $P = -9.7$ and $Q = 3.5$.

144. Find the distance between the points P and Q on the real number line if $P = -12.5$ and $Q = 2.6$.

145. Find the distance between the points P and Q on the real number line if $P = -\dfrac{13}{3}$ and $Q = \dfrac{7}{5}$.

146. Find the distance between the points P and Q on the real number line if $P = -\dfrac{5}{6}$ and $Q = 4$.

Explaining the Concepts

147. We know that 6 divided by 2 is 3. Explain why 6 divided by $\dfrac{1}{2}$ is 12.

148. Use a figure like Figure 17 on page 39 to explain why $\dfrac{1}{6} + \dfrac{2}{6} = \dfrac{1}{2}$.

Putting the Concepts Together (Sections 1.2–1.5)

We designed these problems so that you can review Sections 1.2–1.5 and show your mastery of the concepts. Take time to work these problems before proceeding with the next section. The answers to these problems are located at the back of the text on page AN-2.

1. Write $\dfrac{7}{8}$ and $\dfrac{9}{20}$ as equivalent fractions with the least common denominator.

2. Write $\dfrac{21}{63}$ as a fraction in lowest terms.

3. Convert $\dfrac{2}{7}$ to a decimal.

4. Write 0.375 as a fraction in lowest terms.

5. Write 12.3% as a decimal.

6. Write 0.0625 as a percent.

7. Use the set $\left\{ -12, -\dfrac{14}{7}, -1.25, 0, \sqrt{2}, 3, 11.2 \right\}$ to list all of the elements that are:

   (a) integers
   (b) rational numbers
   (c) irrational numbers
   (d) real numbers

8. Replace the ? with the correct symbol $>, <, =$: $\dfrac{1}{8}$? 0.5

In Problems 9–30, perform the indicated operation and write in lowest terms.

9. $17 + (-28)$

10. $-23 + (-42)$

11. $18 - 45$

12. $3 - (-24)$

13. $-18 - (-12.5)$

14. $(-5)(2)$

15. $25(-4)$

16. $(-8)(-9)$

17. $\dfrac{-35}{7}$

18. $\dfrac{-32}{-2}$

19. $27 \div -3$

20. $-\dfrac{4}{5} - \dfrac{11}{5}$

21. $7 - \dfrac{4}{5}$

22. $\dfrac{7}{12} + \dfrac{5}{18}$

23. $-\dfrac{5}{12} - \dfrac{1}{18}$

24. $\dfrac{6}{25} \cdot 15 \cdot \dfrac{1}{2}$

25. $\dfrac{2}{7} \div (-8)$

26. $\dfrac{0}{-8}$

27. $3.56 - (-7.2)$

28. $18.946 - 11.3$

29. $62.488 \div 42.8$

30. $(7.94)(2.8)$

1.6 Properties of Real Numbers

Objectives

1 Use the Identity Properties of Addition and Multiplication

2 Use the Commutative Properties of Addition and Multiplication

3 Use the Associative Properties of Addition and Multiplication

4 Understand the Multiplication and Division Properties of 0

Are You Ready for This Section?

Before getting started, take this readiness quiz. If you get a problem wrong, go back to the section cited and review the material.

R1. Find the sum: $12 + 3 + (-12)$ [Section 1.4, pp. 28–30]

R2. Find the product: $\dfrac{3}{4} \cdot 11 \cdot \dfrac{4}{3}$ [Section 1.5, pp. 37–38]

This section presents properties of real numbers. A property in mathematics is a rule that is always true. These properties will be used throughout this text and in future math courses, so it is extremely important that you understand these properties and know how to use them.

▶ **1** **Use the Identity Properties of Addition and Multiplication**

The real number 0 is the only number that when added to any real number a results in the same real number a.

Identity Property of Addition

For any real number a,

$$0 + a = a + 0 = a$$

That is, the sum of any number and 0 is that number. We call 0 the **additive identity.**

Recall that $3 \cdot 5$ is equivalent to adding 5 three times, so $3 \cdot 5 = 5 + 5 + 5$. Therefore, $1 \cdot 5$ means to add 5 once, so $1 \cdot 5 = 5$. This property about the real number 1 is true in general.

Identity Property of Multiplication

For any real number a,

$$a \cdot 1 = 1 \cdot a = a$$

That is, the product of any number and 1 is that number. We call 1 the **multiplicative identity.**

The multiplicative identity lets us create expressions that are equivalent to other expressions. For example, the expressions $\frac{4}{5}$ and $\frac{4}{5} \cdot \frac{3}{3}$ are equivalent because $\frac{3}{3} = 1$.

Conversion

One use of the multiplicative identity is *conversion*. **Conversion** is changing the units of measure (such as inches or pounds). For example, we might change a length from inches to feet or a weight from pounds to ounces.

EXAMPLE 1 **Converting from Inches to Feet**

Janice measures her family room and finds that its length is 184 inches. How many feet long is Janice's family room? (*Note*: 12 inches = 1 foot)

Solution

When doing conversions, make sure the units of measure you are trying to remove get divided out and the new units of measure remain. In this problem, we want the inches to divide out and feet to remain. Because 12 inches equals 1 foot, multiplying 184 inches by $\dfrac{1 \text{ foot}}{12 \text{ inches}}$ is the same as multiplying by 1.

$$184 \text{ inches} = 184 \text{ inches} \cdot \overset{\text{Multiplying by 1}}{\frac{1 \text{ foot}}{12 \text{ inches}}}$$

$$= \frac{184}{12} \text{ feet}$$

$184 = 2 \cdot 2 \cdot 2 \cdot 23; \; 12 = 2 \cdot 2 \cdot 3;$
Divide out common factors:
$$= \frac{2 \cdot 2 \cdot 2 \cdot 23}{2 \cdot 2 \cdot 3} \text{ feet}$$

$$= \frac{46}{3} \text{ feet}$$

So 184 inches equals $\dfrac{46}{3}$ feet. Because 46 divided by 3 is 15 with a remainder of 1 (Why? See the Work Smart), $\dfrac{46}{3}$ feet is equivalent to $15\dfrac{1}{3}$ feet. Since $\dfrac{1}{3}$ foot $\cdot \dfrac{12 \text{ inches}}{1 \text{ foot}} = 4$ inches, we have that $\dfrac{46}{3}$ feet is equivalent to 15 feet, 4 inches.

Quick ✓

1. Because $a \cdot 1 = 1 \cdot a = a$ for any real number a, we call 1 the _____ _____.

In Problems 2–4, convert each measurement to the indicated unit of measurement.

2. 96 inches = ? feet [1 foot = 12 inches]
3. 500 minutes = ? hours [60 minutes = 1 hour]
4. 88 ounces = ? pounds [16 ounces = 1 pound]

▶ ❷ Use the Commutative Properties of Addition and Multiplication

---

EXAMPLE 2 **Illustrating the Commutative Properties**

(a) $4 + 7 = 11$ and
    $7 + 4 = 11$ so
    $4 + 7 = 7 + 4$

(b) $3 \cdot 8 = 24$ and
    $8 \cdot 3 = 24$ so
    $3 \cdot 8 = 8 \cdot 3$

The results of Example 2 are true in general.

Commutative Properties of Addition and Multiplication

If a and b are real numbers, then
$$a + b = b + a \quad \text{and} \quad a \cdot b = b \cdot a$$

In Words

The Commutative Property of real numbers states that the order in which we add or multiply real numbers does not affect the final result.

We add real numbers from left to right. We multiply real numbers from left to right. The **Commutative Property** allows us to write $3 + 5$ as $5 + 3$ and write $3 \cdot 5$ as $5 \cdot 3$ without affecting the value of the expression. Why is this important? Rearranging addition or multiplication problems makes some expressions easier to evaluate.

EXAMPLE 3 **Using the Commutative Property of Addition**

Evaluate the expression: $18 + 3 + (-18)$

Solution

$$\begin{aligned} 18 + 3 + (-18) &= 18 + (-18) + 3 \\ &= 0 + 3 \\ &= 3 \end{aligned}$$

If we add from left to right, we get $18 + 3 + (-18) = 21 + (-18) = 3$. Rearranging the numbers made the problem easier!

Does subtraction obey the Commutative Property? In other words, does $3 - 14 = 14 - 3$? Because $3 - 14 = -11$, but $14 - 3 = 11$, we see that **subtraction is not commutative.**

Quick ✓

5. The Commutative Property of Addition states that for any real numbers a and b, $a + b =$ _ + _.

6. The Commutative Property of Multiplication states that for any real numbers a and b, _ · _ = _ · _.

In Problems 7–9, use the Commutative Property of Addition and the Additive Inverse Property to find the sum of the real numbers.

7. $(-8) + 22 + 8$

8. $\dfrac{8}{15} + \dfrac{3}{20} + \left(-\dfrac{8}{15}\right)$

9. $2.1 + 11.98 + (-2.1)$

EXAMPLE 4 **Using the Commutative Property of Multiplication**

Find each product.

(a) $-27 \cdot 7 \cdot \left(-\dfrac{2}{9}\right)$

(b) $100 \cdot 307.5 \cdot 0.01$

Solution

(a) $-27 \cdot 7 \cdot \left(-\dfrac{2}{9}\right) = -27 \cdot \left(-\dfrac{2}{9}\right) \cdot 7$

$$= -\overset{3}{\cancel{27}} \cdot \left(-\dfrac{2}{\underset{1}{\cancel{9}}}\right) \cdot 7$$

$$= -3 \cdot (-2) \cdot 7$$

$$= 6 \cdot 7$$

$$= 42$$

(b) $100 \cdot 307.5 \cdot 0.01 = 100 \cdot 0.01 \cdot 307.5$

$$= 1 \cdot 307.5$$

$$= 307.5$$

Quick ✓

In Problems 10–12, find the product of the real numbers.

10. $\left(-\dfrac{4}{3}\right) \cdot (-13) \cdot \left(-\dfrac{3}{4}\right)$

11. $\dfrac{5}{22} \cdot \dfrac{18}{331} \cdot \left(-\dfrac{44}{5}\right)$

12. $100{,}000 \cdot 349 \cdot 0.00001$

Work Smart

Neither division nor subtraction is commutative.

Does division obey the Commutative Property? That is, does $a \div b = b \div a$? For example, does $8 \div 2 = 2 \div 8$? Because $8 \div 2 = 4$, but $2 \div 8 = \dfrac{2}{8} = \dfrac{1}{4}$, we conclude that **division is not commutative**.

▶ ❸ **Use the Associative Properties of Addition and Multiplication**

Sometimes **grouping symbols** such as parentheses ( ), brackets [ ], or braces { } are used to indicate that the operation within the grouping symbols is to be performed first. For example, $5 \cdot (8 + 3)$ means that we should first add 8 and 3 and then multiply this sum by 5.

Earlier we mentioned that addition is performed from left to right. We also stated that multiplication is performed from left to right. But does the order in which we add (or multiply) three or more numbers matter? Let's see.

EXAMPLE 5   **Illustrating the Associative Properties**

(a) $2 + (8 + 6) = 2 + 14 = 16$ and $(2 + 8) + 6 = 10 + 6 = 16$

so
$$2 + (8 + 6) = (2 + 8) + 6$$

(b) $-4 \cdot (9 \cdot 2) = -4 \cdot (18) = -72$ and $(-4 \cdot 9) \cdot 2 = -36 \cdot 2 = -72$

so
$$-4 \cdot (9 \cdot 2) = (-4 \cdot 9) \cdot 2$$

Example 5 illustrates the **Associative Properties of Addition and Multiplication**.

> **Associative Properties of Addition and Multiplication**
>
> If a, b, and c are real numbers, then
> $$a + (b + c) = (a + b) + c = a + b + c$$
> $$a \cdot (b \cdot c) = (a \cdot b) \cdot c = a \cdot b \cdot c$$

EXAMPLE 6   **Using the Associative Property of Addition**

Use the Associative Property of Addition to evaluate $23 + 453 + (-453)$.

Solution

Because 453 and -453 are additive inverses, we use the Associative Property of Addition to insert parentheses around these numbers.

$$23 + 453 + (-453) = 23 + (453 + (-453))$$
$$= 23 + 0$$
$$= 23$$

EXAMPLE 7   **Using the Associative Property of Multiplication**

Use the Associative Property of Multiplication to evaluate $-\dfrac{3}{11} \cdot \dfrac{9}{4} \cdot \dfrac{8}{3}$.

Solution

Because the second and third factors have common factors that can be divided out, use the Associative Property of Multiplication to insert parentheses around $\dfrac{9}{4} \cdot \dfrac{8}{3}$ and perform this operation first.

$$-\frac{3}{11} \cdot \frac{9}{4} \cdot \frac{8}{3} = -\frac{3}{11} \cdot \left(\frac{9}{4} \cdot \frac{8}{3}\right)$$
$$= -\frac{3}{11} \cdot \left(\frac{\overset{3}{\cancel{9}}}{\underset{1}{\cancel{4}}} \cdot \frac{\overset{2}{\cancel{8}}}{\underset{1}{\cancel{3}}}\right)$$
$$= -\frac{3}{11} \cdot (3 \cdot 2)$$
$$= -\frac{3}{11} \cdot 6$$
$$= -\frac{18}{11}$$

Quick ✓

In Problems 13–16, use an Associative Property to evaluate each expression.

13. $14 + 101 + (-101)$

14. $14 \cdot \frac{1}{5} \cdot 5$

15. $-34.2 + 12.6 + (-2.6)$

16. $\frac{19}{2} \cdot \frac{4}{38} \cdot \frac{50}{13}$

▶ ④ Understand the Multiplication and Division Properties of 0

Now let's look at multiplication by zero.

> **Multiplication Property of Zero**
>
> For any real number a, the product of a and 0 is always 0; that is,
>
> $$a \cdot 0 = 0 \cdot a = 0$$

We now introduce some division properties of zero.

Work Smart

Division by zero is not allowed. That is, 0 cannot be used as a divisor.

> **Division Properties of Zero**
>
> For any nonzero real number a,
>
> **1.** The quotient of 0 and a is 0. That is, $\frac{0}{a} = 0$.
>
> **2.** The quotient of a and 0 is **undefined**. That is, $\frac{a}{0}$ is undefined.

Why are these statements true? When we divide, we can check the quotient by multiplication. For example, $\frac{12}{4} = 3$ because $4 \cdot 3 = 12$. In the same way, $\frac{0}{4} = 0$ because $4 \cdot 0 = 0$. But what is the value of $\frac{12}{0}$? To determine this quotient, we should be able to determine a real number such that $0 \cdot \square = 12$. But since the product of 0 and any real number is 0, there is no value for $\square$.

EXAMPLE 8 **Using Zero as a Divisor and a Dividend**

Find the quotient:

(a) $\frac{23}{0}$

(b) $\frac{0}{17}$

Solution

(a) $\frac{23}{0}$ is undefined because 0 is the divisor.

(b) $\frac{0}{17} = 0$ because 0 is the dividend. ●

Quick ✓

In Problems 17 and 18, tell whether the quotient is zero or undefined.

17. $\frac{0}{22}$

18. $\frac{-11}{0}$

We now summarize the properties of addition, multiplication, and division.

Work Smart

The Commutative Property changes **order** and the Associative Property changes **grouping**.

Summary Properties of Addition

Identity Property of Addition   For any real number a, $0 + a = a + 0 = a$.

Commutative Property of Addition   If a and b are real numbers, then $a + b = b + a$.

Additive Inverse Property   For any real number a, $a + (-a) = -a + a = 0$.

Associative Property of Addition   If a, b, and c are real numbers, then $a + (b + c) = (a + b) + c$.

Summary Properties of Multiplication and Division

Identity Property of Multiplication $a \cdot 1 = 1 \cdot a = a$ for any real number a.

Commutative Property of Multiplication   If a and b are real numbers, then $a \cdot b = b \cdot a$.

Multiplicative Inverse Property $a \cdot \dfrac{1}{a} = \dfrac{1}{a} \cdot a = 1$ provided that $a \neq 0$.

Work Smart: Study Skills

Selected problems in the exercise sets are identified by a green color. For extra help, worked solutions to these problems are in MyMathLab.

Associative Property of Multiplication   If a, b, and c are real numbers, then $a \cdot (b \cdot c) = (a \cdot b) \cdot c$.

Multiplication Property of Zero   For any real number a, $a \cdot 0 = 0 \cdot a = 0$.

Division Properties of Zero   For any nonzero number a, $\dfrac{0}{a} = 0$ and $\dfrac{a}{0}$ is undefined.

1.6 Exercises   MyMathLab®   PRACTICE

Exercise numbers in green have complete video solutions in MyMathLab.

Problems 1–18 are the Quick ✔ s that follow the EXAMPLES.

Building Skills

In Problems 19–28, convert each measurement to the indicated unit of measurement. See Objective 1. Use the following conversions:

1 foot = 12 inches	3 feet = 1 yard
1 gallon = 4 quarts	100 centimeters = 1 meter
16 ounces = 1 pound	

19. 13 feet to inches

20. 13 yards to feet

21. 4500 centimeters to meters

22. 5900 centimeters to meters

23. 42 quarts to gallons

24. 58 quarts to gallons

25. 180 ounces to pounds

26. 120 ounces to pounds

27. 16,200 seconds to hours

28. 22,500 seconds to hours

In Problems 29–44, state the property of real numbers that is being illustrated. See Objectives 2, 3, and 4.

29. $16 + (-16) = 0$

30. $4 \cdot 63 \cdot \dfrac{1}{4} = 4 \cdot \dfrac{1}{4} \cdot 63$

31. $\dfrac{3}{4}$ is equivalent to $\dfrac{3}{4} \cdot \dfrac{5}{5}$

32. $4 + 5 + (-4)$ is equivalent to $4 + (-4) + 5$

33. $12 \cdot \dfrac{1}{12} = 1$

34. $-236 + 236 = 0$

35. $34.2 + (-34.2) = 0$

36. $(4 \cdot 5) \cdot 7 = 4 \cdot (5 \cdot 7)$

37. $\dfrac{0}{17}$

38. $\dfrac{-8}{0}$

39. $\dfrac{2}{3} \cdot \left(-\dfrac{12}{43}\right) \cdot \dfrac{3}{2} = \dfrac{2}{3} \cdot \dfrac{3}{2} \cdot \left(-\dfrac{12}{43}\right)$

40. $\dfrac{5}{12} \cdot \dfrac{12}{5} = 1$

41. $5.23 + 4.98 + (-5.23) = 5.23 + (-5.23) + 4.98$

42. $16.4 \cdot 0 = 0$

43. $\dfrac{21}{0}$

44. $\dfrac{0}{106}$

Mixed Practice

In Problems 45–64, evaluate each expression by using the properties of real numbers.

45. $54 + 29 + (-54)$

46. $46 + 59 + (-46)$

47. $\dfrac{9}{5} \cdot \dfrac{5}{9} \cdot 18$

48. $\dfrac{4}{9} \cdot \dfrac{9}{4} \cdot 28$

49. $-25 \cdot 13 \cdot \dfrac{1}{5}$

50. $36 \cdot (-12) \cdot \dfrac{1}{6}$

51. $347 + 456 + (-456)$

52. $593 + 306 + (-306)$

53. $\dfrac{9}{2} \cdot \left(-\dfrac{10}{3}\right) \cdot 6$

54. $\dfrac{13}{2} \cdot \dfrac{8}{39} \cdot \dfrac{39}{4}$

55. $\dfrac{7}{0}$

56. $\dfrac{0}{100}$

57. $100(-34)(0.01)$

58. $4000(0.5)(0.001)$

59. $569.003 \cdot 0$

60. $104 \cdot \dfrac{1}{104}$

61. $\dfrac{45}{3902} + \left(-\dfrac{45}{3902}\right)$

62. $30 \cdot \dfrac{4}{4}$

63. $-\dfrac{5}{44} \cdot \dfrac{80}{3} \cdot \dfrac{11}{5}$

64. $\dfrac{7}{48} \cdot \left(-\dfrac{21}{4}\right) \cdot \dfrac{12}{7}$

Applying the Concepts

65. Balancing the Checkbook Alberto's checking account balance at the start of the month was \$321.03. During the month, he wrote checks for \$32.84, \$85.03, and \$120.56. He also deposited a check for \$120.56. What is Alberto's balance at the end of the month?

66. Stock Price Before the opening bell on Monday, a certain stock was priced at \$32.04. On Monday the stock was up \$0.54, on Tuesday it was down \$0.32, and on Wednesday it was down \$0.54. What was the closing price of the stock on Wednesday?

Extending the Concepts

In Problems 67–70, insert parentheses to make the statement true.

67. $-3 - 4 - 10 = 3$

68. $-6 - 4 + 10 = -20$

69. $-15 + 10 - 4 - 8 = -1$

70. $25 - 6 - 10 - 1 = 28$

71. Convert 30 miles per hour to feet per second. (*Note*: 1 mile $=$ 5280 feet)

72. Convert 40 miles per hour to feet per second.

(*Note*: 1 mile $=$ 5280 feet)

Explaining the Concepts

73. In your own words, explain why 0 does not have a multiplicative inverse.

74. Why does $2(4 \cdot 5)$ not equal $(2 \cdot 4) \cdot (2 \cdot 5)$?

75. Why does $\dfrac{0}{4} = 0$? Why is $\dfrac{4}{0}$ undefined?

76. How is the Identity Property of Addition related to the Additive Inverse Property?

77. How is the Identity Property of Multiplication related to the Multiplicative Inverse Property?

78. Let a be a real number such that $a > 0$, and let b represent a real number such that $b < 0$. Indicate whether each of the following statements are true or false, and explain your decision using actual values for a and b.

(a) $a > -a$

(b) $b > -b$

(c) $a + (-a) = 0$

(d) $-b + b = 0$

(e) $a - (-b) > 0$

(f) $ab < 0$

(g) $-ab > 0$

(h) $\dfrac{-a}{-b} > 0$

1.7 Exponents and the Order of Operations

Objectives

1 Evaluate Exponential Expressions

2 Apply the Rules for Order of Operations

Are You Ready for This Section?

Before getting started, take this readiness quiz. If you get a problem wrong, go back to the section cited and review the material.

R1. Find the sum: $9 + (-19)$ [Section 1.4, pp. 28–30]

R2. Find the difference: $28 - (-7)$ [Section 1.4, pp. 31–32]

R3. Find the product: $-7 \cdot \dfrac{8}{3} \cdot 36$ [Section 1.5, pp. 37–38]

R4. Find the quotient: $\dfrac{100}{-15}$ [Section 1.4, pp. 34–35]

▶ **1** Evaluate Exponential Expressions

If we wanted to multiply 2 eight times, we would write $2 \cdot 2 \cdot 2 \cdot 2 \cdot 2 \cdot 2 \cdot 2 \cdot 2$. That's a lot of writing! To reduce the amount of writing needed to show repeated multiplication, we use **exponential notation**, where we write $2 \cdot 2 \cdot 2 \cdot 2 \cdot 2 \cdot 2 \cdot 2 \cdot 2$ as 2^8. In 2^8, 2 is called the **base** and 8 is called the **exponent**.

> ### Exponential Notation
>
> If n is a natural number and a is a real number, then
>
> $$a^n = \underbrace{a \cdot a \cdot a \cdot \,\cdots\, \cdot a}_{n \text{ factors}}$$
>
> where a is the **base** and n is the **exponent** or **power.** The exponent tells the number of times the base is used as a factor.

An expression written in the form a^n is said to be in **exponential form.** The expression 6^2 is read "six squared," 8^3 is read "eight cubed," and the expression 11^4 is read "eleven to the fourth power." In general, we read a^n as "a to the nth power."

EXAMPLE 1   **Writing a Numerical Expression in Exponential Form**

Write each expression in exponential form.

    (a) $5 \cdot 5 \cdot 5$ **(b)** $(-4) \cdot (-4) \cdot (-4) \cdot (-4) \cdot (-4) \cdot (-4)$

Solution

    (a) The expression $5 \cdot 5 \cdot 5$ contains three factors of 5, so $5 \cdot 5 \cdot 5 = 5^3$.

    (b) The expression $(-4) \cdot (-4) \cdot (-4) \cdot (-4) \cdot (-4) \cdot (-4)$ contains six factors of -4, so $(-4) \cdot (-4) \cdot (-4) \cdot (-4) \cdot (-4) \cdot (-4) = (-4)^6$.

Quick ✓

1. In the expression 3^6, 3 is the _____ and 6 is the _____ or _____.

In Problems 2 and 3, write each expression in exponential form.

2. $11 \cdot 11 \cdot 11 \cdot 11 \cdot 11$ **3.** $(-7) \cdot (-7) \cdot (-7) \cdot (-7)$

To evaluate an exponential expression, write the expression in **expanded form**. For example, 2^8 in expanded form is $2 \cdot 2 \cdot 2 \cdot 2 \cdot 2 \cdot 2 \cdot 2 \cdot 2$.

EXAMPLE 2 **Evaluating an Exponential Expression**

Evaluate each exponential expression:

(a) 6^4 (b) $\left(\dfrac{5}{3}\right)^5$

Solution

(a) $6^4 = 6 \cdot 6 \cdot 6 \cdot 6$
$= 1296$

(b) $\left(\dfrac{5}{3}\right)^5 = \left(\dfrac{5}{3}\right)\left(\dfrac{5}{3}\right)\left(\dfrac{5}{3}\right)\left(\dfrac{5}{3}\right)\left(\dfrac{5}{3}\right)$

$= \dfrac{5 \cdot 5 \cdot 5 \cdot 5 \cdot 5}{3 \cdot 3 \cdot 3 \cdot 3 \cdot 3}$

$= \dfrac{3125}{243}$

EXAMPLE 3 **Evaluating an Exponential Expression—Odd Exponent**

Evaluate each exponential expression:

(a) $(-5)^3$ (b) -5^3

Solution

(a) $(-5)^3 = (-5) \cdot (-5) \cdot (-5)$
$= -125$

(b) $-5^3 = -(5 \cdot 5 \cdot 5)$
$= -125$

EXAMPLE 4 **Evaluating an Exponential Expression—Even Exponent**

Evaluate each exponential expression:

(a) $(-5)^4$ (b) -5^4

Solution

(a) $(-5)^4 = (-5) \cdot (-5) \cdot (-5) \cdot (-5)$
$= 625$

(b) $-5^4 = -(5 \cdot 5 \cdot 5 \cdot 5)$
$= -625$

Work Smart

There is a difference between $(-5)^4$ and -5^4.

The parentheses in $(-5)^4$ tell us to use four factors of -5. However, in the expression -5^4, we use 5 as a factor four times and then multiply the result by -1. We could also read -5^4 as "Take the opposite of the quantity 5^4."

Quick ✔

In Problems 4–9, evaluate each exponential expression.

4. 2^4

5. $(-7)^2$

6. $\left(-\dfrac{1}{6}\right)^3$

7. $(0.9)^2$

8. -2^4

9. $(-2)^4$

▶ ② **Apply the Rules for Order of Operations**

To evaluate $3 \cdot 5 + 4$, do you multiply first and then add to get $15 + 4 = 19$ *or* do you add first and then multiply to get $3 \cdot 9 = 27$?

Because $3 \cdot 5$ is equivalent to $5 + 5 + 5$, we have

$$3 \cdot 5 + 4 = 5 + 5 + 5 + 4$$
$$= 19$$

In Words

Multiply first then add.

Based on this, **whenever addition and multiplication appear in the same expression, always multiply first and then add.**

Because any division problem can be written as a multiplication problem, we divide before adding as well. Also, because any subtraction problem can be written as an addition problem, we always multiply and divide before adding and subtracting.

EXAMPLE 5 **Evaluating an Expression Containing Multiplication, Division, and Addition**

Evaluate each expression:

(a) $11 + 2 \cdot (-6)$ (b) $7 + 12 \div 3 \cdot 5$

Solution

(a)   Multiply first: $11 + 2 \cdot (-6) = 11 + (-12)$

                       Add: $= -1$

(b) Multiply/divide left to right: $7 + 12 \div 3 \cdot 5 = 7 + 4 \cdot 5$

                      Multiply: $= 7 + 20$

                                            $= 27$

Quick ✓

In Problems 10–13, evaluate each expression.

10. $1 + 7 \cdot 2$ **11.** $-3 \div \left(-\dfrac{1}{2}\right) + 18$

12. $9 \cdot 4 \div 2 + 5$ **13.** $\dfrac{15}{2} \div (-5) \cdot 8 - 7$

▶ **Parentheses**

If we want to add two numbers first and then multiply, we use parentheses and write $(3 + 5) \cdot 4$. In other words, **always evaluate the expression in parentheses first.**

EXAMPLE 6 **Finding the Value of an Expression Containing Parentheses**

Evaluate each expression:

(a) $(5 + 3) \cdot 2$ (b) $\left(\dfrac{3}{2} - \dfrac{5}{2}\right)\left(\dfrac{7}{3} + \dfrac{2}{3}\right)$

Solution

(a) $(5 + 3) \cdot 2 = 8 \cdot 2$ (b) $\left(\dfrac{3}{2} - \dfrac{5}{2}\right)\left(\dfrac{7}{3} + \dfrac{2}{3}\right) = \left(-\dfrac{2}{2}\right)\left(\dfrac{9}{3}\right)$

            $= 16$ $= (-1)(3)$

                                                       $= -3$

Quick ✓
In Problems 14–16, evaluate each expression.

14. $8(2 + 3)$ **15.** $(2 - 9) \cdot (5 + 4)$ **16.** $\left(\dfrac{6}{7} + \dfrac{8}{7}\right) \cdot \left(\dfrac{11}{8} + \dfrac{5}{8}\right)$

The Division Bar

If an expression contains a division bar, we treat the terms above and below the division bar as if they were in parentheses. For example,

$$\frac{3+5}{9+7} = \frac{(3+5)}{(9+7)} = \frac{8}{16} = \frac{\cancel{8} \cdot 1}{\cancel{8} \cdot 2} = \frac{1}{2}$$

EXAMPLE 7 **Finding the Value of an Expression That Contains a Division Bar**

Evaluate each expression:

(a) $\dfrac{7 \cdot 3}{3 + 9 \cdot 2}$

(b) $\dfrac{1 + 7 \div \dfrac{1}{5}}{-6 \cdot 2 + 8}$

Solution

(a) Multiply: $\dfrac{7 \cdot 3}{3 + 9 \cdot 2} = \dfrac{21}{3 + 18}$

Add: $= \dfrac{21}{21}$

$= 1$

(b) Write division as multiplication: $\dfrac{1 + 7 \div \dfrac{1}{5}}{-6 \cdot 2 + 8} = \dfrac{1 + 7 \cdot 5}{-6 \cdot 2 + 8}$

Multiply: $= \dfrac{1 + 35}{-12 + 8}$

Add: $= \dfrac{36}{-4}$

$= \dfrac{9 \cdot \cancel{4}}{-1 \cdot \cancel{4}}$

$= -9$

Quick ✔

In Problems 17–19, evaluate each expression.

17. $\dfrac{2 + 5 \cdot 6}{-3 \cdot 8 - 4}$

18. $\dfrac{(12 + 14) \cdot 2}{13 \cdot 2 + 13 \cdot 5}$

19. $\dfrac{4 + 3 \div \dfrac{1}{7}}{2 \cdot 9 - 3}$

▶ Multiple Grouping Symbols

Grouping symbols include parentheses ( ), brackets [ ], braces { }, and absolute value symbols, | |. Operations within the grouping symbols are performed first. **When multiple grouping symbols occur, evaluate the expression in the innermost grouping symbols first and work outward.**

EXAMPLE 8 **Finding the Value of an Expression Containing Grouping Symbols**

Evaluate each expression:

(a) $2 \cdot [3 \cdot (6 + 3) - 7]$

(b) $\left[4 + \left(\dfrac{2}{3} \cdot (-9) \right) \right] \cdot 3$

Solution

(a) Perform the operation in parentheses first: $2 \cdot [3 \cdot (6 + 3) - 7] = 2 \cdot [3 \cdot 9 - 7]$

Perform the operations in brackets, multiply first: $= 2 \cdot [27 - 7]$

$= 2 \cdot [20]$

$= 40$

(b) Perform the operation in parentheses first: $\left[4 + \left(\dfrac{2}{3} \cdot (-9) \right) \right] \cdot 3 = [4 + (-6)] \cdot 3$

Perform the operation in brackets: $= -2 \cdot 3$

$= -6$ ●

> **Quick** ✓
>
> *In Problems 20 and 21, evaluate each expression.*
>
> **20.** $4 \cdot [2 \cdot (3 + 7) - 15]$ **21.** $2 \cdot \{4 \cdot [26 - (9 + 7)] - 15\} - 10$

⊙ When do we evaluate exponents in the order of operations? In $2 \cdot 4^3$, do we multiply first and then evaluate the exponent to obtain $2 \cdot 4^3 = 8^3 = 512$, or do we evaluate the exponent first and then multiply to obtain $2 \cdot 4^3 = 2 \cdot 64 = 128$? Because $2 \cdot 4^3 = 2 \cdot 4 \cdot 4 \cdot 4 = 128$, we **evaluate exponents before multiplication.** Don't forget to evaluate the expression in the grouping symbol first.

EXAMPLE 9 **Finding the Value of an Expression Containing Exponents**

Evaluate each of the following:

(a) $2 + 7(-4)^2$ (b) $\dfrac{2 \cdot 3^2 + 4}{3(2 - 6)}$

Solution

Evaluate the exponent
↓

(a) $2 + 7(-4)^2 = 2 + 7 \cdot 16$

Multiply: $= 2 + 112$

Add: $= 114$

Parentheses first
↓

(b) $\dfrac{2 \cdot 3^2 + 4}{3(2 - 6)} = \dfrac{2 \cdot 3^2 + 4}{3(-4)}$

Evaluate the exponent: $= \dfrac{2 \cdot 9 + 4}{3(-4)}$

Find products: $= \dfrac{18 + 4}{-12}$

Add terms in numerator: $= \dfrac{22}{-12}$

Write in lowest terms: $= \dfrac{2 \cdot 11}{2 \cdot -6}$

$= -\dfrac{11}{6}$ ●

> **Quick** ✓
>
> *In Problems 22–24, evaluate each of the following:*
>
> **22.** $\dfrac{7 - 5^2}{2}$ **23.** $3(7 - 3)^2$ **24.** $\dfrac{(-3)^2 + 7(1 - 3)}{3 \cdot 2^3 + 6}$

> ## Order of Operations
>
> **Step 1:** Perform all operations within *grouping symbols* first. When an expression has more than one set of grouping symbols, begin within the innermost grouping symbols and work outward.
>
> **Step 2:** Evaluate expressions containing exponents.
>
> **Step 3:** Perform *multiplication and division,* working from *left to right.*
>
> **Step 4:** Perform *addition and subtraction,* working from *left to right.*

EXAMPLE 10 **How to Evaluate an Expression Using Order of Operations**

Evaluate: $18 + 7(2^3 - 26) + 5^2$

Step-by-Step Solution

Steps 1 and 2: Evaluate the expression in the parentheses first. In the parentheses, evaluate the expression containing the exponent first.	$18 + 7(2^3 - 26) + 5^2 = 18 + 7(8 - 26) + 25$ Evaluate $8-26$ in the parentheses: $\quad = 18 + 7(-18) + 25$
Step 3: Perform multiplication and division, working from left to right.	Multiply $7(-18)$: $\quad = 18 - 126 + 25$
Step 4: Perform addition and subtraction, working from left to right.	$= -108 + 25$ $= -83$ ●

EXAMPLE 11 **Evaluating a Numerical Expression Using Order of Operations**

Evaluate: $\left(\dfrac{2^3 - 6}{10 - 2 \cdot 3}\right)^2$

Solution

Evaluate the exponential expression inside the parentheses:
$$\left(\frac{2^3 - 6}{10 - 2 \cdot 3}\right)^2 = \left(\frac{8 - 6}{10 - 2 \cdot 3}\right)^2$$

Multiply inside the parentheses:
$$= \left(\frac{8 - 6}{10 - 6}\right)^2$$

Add/subtract inside the parentheses:
$$= \left(\frac{2}{4}\right)^2$$

Simplify:
$$= \left(\frac{1}{2}\right)^2$$

Evaluate the exponential expression:
$$= \frac{1}{4} \quad ●$$

> ## Quick ✓
>
> *In Problems 25–28, evaluate each expression.*
>
> **25.** $\dfrac{(4 - 10)^2}{2^3 - 5}$
>
> **26.** $-3[(-4)^2 - 5(8 - 6)]^2$
>
> **27.** $\dfrac{(2.9 + 7.1)^2}{5^2 - 15}$
>
> **28.** $\left(\dfrac{4^2 - 4(-3)(1)}{7 \cdot 2}\right)^2$

Work Smart: Study Skills

Selected problems in the exercise sets are in green. For extra help, worked solutions to these problems are in MyMathLab.

1.7 Exercises MyMathLab® PRACTICE

Exercise numbers in green have complete video solutions in MyMathLab.

Problems 1–28 are the Quick ✔s that follow the EXAMPLES.

Building Skills

In Problems 29–32, write in exponential form. See Objective 1.

29. $5 \cdot 5$

30. $4 \cdot 4 \cdot 4 \cdot 4 \cdot 4$

31. $\left(-\dfrac{3}{5}\right) \cdot \left(-\dfrac{3}{5}\right) \cdot \left(-\dfrac{3}{5}\right)$

32. $(-8)(-8)(-8)$

In Problems 33–54, evaluate each exponential expression. See Objective 1.

33. 8^2

34. 4^2

35. $(-8)^2$

36. $(-4)^2$

37. 10^3

38. 2^5

39. -10^3

40. -2^5

41. $(-10)^3$

42. $(-2)^5$

43. $(1.5)^2$

44. $(0.04)^2$

45. -8^2

46. -4^2

47. -1^{20}

48. $(-1)^{19}$

49. 0^4

50. 1^6

51. $\left(-\dfrac{1}{2}\right)^6$

52. $\left(-\dfrac{3}{2}\right)^5$

53. $\left(-\dfrac{1}{3}\right)^3$

54. $\left(-\dfrac{3}{4}\right)^2$

In Problems 55–86, evaluate each expression. See Objective 2.

55. $2 + 3 \cdot 4$

56. $12 + 8 \cdot 3$

57. $-5 \cdot 3 + 12$

58. $-3 \cdot 12 + 9$

59. $100 \div 2 \cdot 50$

60. $50 \div 5 \cdot 4$

61. $156 - 3 \cdot 2 + 10$

62. $86 - 4 \cdot 3 + 6$

63. $(2 + 3) \cdot 4$

64. $(7 - 5) \cdot \dfrac{5}{2}$

65. $8 \div 4 \cdot 2$

66. $4 \div 7 \cdot 21$

67. $\dfrac{4 + 2}{2 + 8}$

68. $\dfrac{5 + 3}{3 + 15}$

69. $\dfrac{14 - 6}{6 - 14}$

70. $\dfrac{15 - 7}{7 - 15}$

71. $13 - [3 + (-8)4]$

72. $12 - [7 + (-6)3]$

73. $(-8.75 - 1.25) \div (-2)$

74. $(-11.8 - 15.2) \div (-2)$

75. $4 - 2^3$

76. $10 - 4^2$

77. $15 + 4 \cdot 5^2$

78. $10 + 3 \cdot 2^4$

79. $-2^3 + 3^2 \div (2^2 - 1)$

80. $-5^2 + 3^2 \div (3^2 + 9)$

81. $\left(\dfrac{4^2 - 3}{12 - 2 \cdot 5}\right)^2$

82. $\left(\dfrac{7 - 5^2}{8 + 4 \cdot 2}\right)^2$

83. $-2 \cdot [5 \cdot (9 - 3) - 3 \cdot 6]$

84. $3 \cdot [6 \cdot (5 - 2) - 2 \cdot 5]$

85. $\left(\dfrac{4}{3} + \dfrac{5}{6}\right)\left(\dfrac{2}{5} - \dfrac{9}{10}\right)$

86. $\left(\dfrac{3}{4} + \dfrac{1}{2}\right)\left(\dfrac{2}{3} - \dfrac{1}{2}\right)$

Mixed Practice

In Problems 87–110, evaluate each expression.

87. $4^2 - 3 \cdot 4 + 7$

88. $(-2)^2 + 4 \cdot (-2) + 11$

89. $4 + 2 \cdot (6 - 2)$

90. $3 + 6 \cdot (9 - 5)$

91. $\dfrac{12 - 16 \div 4 + (-24)}{16 \cdot 2 - 4 \cdot 0}$

92. $\dfrac{6 + 15 \div 3 + 16}{6 + 10 \cdot 0}$

93. $\left(\dfrac{2 - (-4)^3}{5^2 - 7 \cdot 2}\right)^2$

94. $\left(\dfrac{9 \cdot 2 - (-2)^3}{4^2 + 3(-1)^5}\right)^2$

95. $\dfrac{5^2 - 10}{3^2 + 6}$

96. $\dfrac{12(2)^3}{4^2 + 4 \cdot 5}$

97. $\left|6 \cdot (5 - 3^2)\right|$

98. $-6 \cdot (2 + |2 \cdot 3 - 4^2|)$

99. $\dfrac{4 - (-6)}{1 - (-1)}$

100. $\dfrac{-2 - 12}{3 - (-4)}$

101. $-\dfrac{7}{20} + \dfrac{3}{8} \div \dfrac{1}{2}$

102. $-\dfrac{4}{5} + \dfrac{3}{10} \div \dfrac{2}{9}$

103. $\dfrac{21 - 3^2}{1 + 3}$

104. $\dfrac{5 + 3^2}{2 + 5}$

105. $\dfrac{3}{4} \cdot \left[\dfrac{5}{4} \div \left(\dfrac{3}{8} - \dfrac{1}{8}\right) - 3\right]$

106. $\left[\dfrac{9}{10} \div \left(\dfrac{2}{5} + \dfrac{1}{5}\right) + \dfrac{7}{2}\right] \cdot \dfrac{1}{10}$

107. $\left(\dfrac{4}{3}\right)^3 - \left(\dfrac{1}{2}\right)^2 \cdot \left(\dfrac{8}{3}\right) + 2 \div 3$

108. $\dfrac{1}{18} \cdot \dfrac{46}{5} - \left(\dfrac{2}{3}\right)^2$

109. $\dfrac{5^2 - 3^3}{|4 - 4^2|}$

110. $\dfrac{3 \cdot 2^3 - 2^2 \cdot 12}{3 + 3^2}$

Applying the Concepts

In Problems 111–114, express each number as the product of prime factors. Write the answer in exponential form.

111. 72   **112.** 675

113. 48   **114.** 200

In Problems 115–120, insert grouping symbols so that the expression has the desired value.

115. $4 \cdot 3 + 6 \cdot 2$ results in 36

116. $4 \cdot 7 - 4^2$ results in -36

117. $4 + 3 \cdot 4 + 2$ results in 42

118. $6 - 4 + 3 - 1$ results in 0

119. $6 - 4 + 3 - 1$ results in 4

120. $4 + 3 \cdot 2 - 1 \cdot 6$ results in 42

121. Cost of a TV The total amount paid for a flat-screen television that costs $479, plus sales tax of 7.5%, is found by evaluating the expression $479 + 0.075(479)$. Evaluate this expression rounded to the nearest cent.

122. Manufacturing Cost Evaluate the expression

$$3000 + 6(100) - \frac{100^2}{1000}$$ to find the weekly production

cost of manufacturing 100 calculators.

123. Surface Area The surface area of a right circular cylinder whose radius is 6 inches and height is 10 inches is given approximately by $2 \cdot 3.1416 \cdot 6^2 + 2 \cdot 3.1416 \cdot 6 \cdot 10$. Evaluate this expression. Round the answer to two decimal places.

124. Volume of a Cone The volume of a cone whose radius is 3 centimeters and whose height is 12 centimeters is given approximately by $\frac{1}{3} \cdot 3.1416 \cdot 3^2 \cdot 12$. Evaluate this expression. Round the answer to two decimal places.

125. Investing If $1000 is invested at 3% annual interest and remains untouched for 2 years, the amount of money that is in the account after 2 years is given by the expression $1000(1 + 0.03)^2$. Evaluate this expression, rounded to the nearest cent.

126. Investing If $5000 is invested at 4.5% annual interest and remains untouched for 5 years, the amount of money that is in the account after 5 years is given by the expression $5000(1 + 0.045)^5$. Evaluate this expression, rounded to the nearest cent.

Extending the Concepts

The Angle Addition Postulate from geometry states that the measure of an angle is equal to the sum of the measures of its parts. Refer to the figure. Use the Angle Addition Postulate to answer Problems 127 and 128.

127. If the measure of $\angle XYQ = 46.5°$ and the measure of $\angle QYZ = 69.25°$, find the measure of the measure of $\angle XYZ$.

128. If the measure of $\angle QYZ = 18°$ and the measure of $\angle XYZ = 57°$, find the measure of $\angle XYQ$.

Explaining the Concepts

129. Explain the difference between -3^2 and $(-3)^2$. Identify the distinguishing characteristics of the two problems, and explain how to evaluate each expression.

1.8 Simplifying Algebraic Expressions

Objectives

1. Evaluate Algebraic Expressions
2. Identify Like Terms and Unlike Terms
3. Use the Distributive Property
4. Simplify Algebraic Expressions by Combining Like Terms

Are You Ready for This Section?

Before getting started, take this readiness quiz. If you get a problem wrong, go back to the section cited and review the material.

R1. Find the sum: $-3 + 8$ [Section 1.4, pp. 28–30]

R2. Find the difference: $-7 - 8$ [Section 1.4, pp. 31–32]

R3. Find the product: $-\frac{4}{3}(27)$ [Section 1.5, pp. 37–38]

Ready?...Answers R1. 5
R2. -15 **R3.** -36

What is algebra? The word "algebra" is derived from the Arabic word *al-jabr*, which means "restoration." Today algebra means more. **Algebra** uses symbols to represent quantities and to express general relationships that hold for all members of the set. In this course, the set of numbers referred to in the definition will usually be the set of real numbers.

▶ ① Evaluate Algebraic Expressions

In arithmetic, we work with numbers. In algebra, we use letters such as x, y, a, b, and c to represent numbers.

> **Definition**
>
> When a letter represents any number from a set of numbers, it is called a **variable**.

For most of the text, we use the set of real numbers.

> **Definition**
>
> A **constant** is either a fixed number, such as 5, or a letter or symbol that represents a fixed number.

For example, in Einstein's Theory of Relativity, $E = mc^2$, the E and m are variables that represent total energy and mass, respectively, and c is a constant that represents the speed of light (299,792,458 meters per second).

> **Definition**
>
> An **algebraic expression** is any combination of variables, constants, grouping symbols, and mathematical operations such as addition, subtraction, multiplication, division, and exponents.

Some examples of algebraic expressions are

$$x - 5 \qquad \frac{1}{2}x \qquad 2y - 7 \qquad z^2 + 3 \qquad \text{and} \qquad \frac{b - 1}{b + 1}$$

Recall that a variable represents any number from a set of numbers. One of the procedures we perform on algebraic expressions is *evaluating an algebraic expression*.

> **Definition**
>
> To **evaluate an algebraic expression**, substitute a numerical value for each variable into the expression and simplify the result.

EXAMPLE 1   **Evaluating an Algebraic Expression**

Evaluate each expression for the given value of the variable.

   (a) $2x + 5$ for $x = 8$ **(b)** $a^2 - 2a + 4$ for $a = -3$

Solution

   (a) We substitute 8 for x in the expression $2x + 5$:

$$2(8) + 5 = 16 + 5$$
$$= 21$$

   (b) We substitute -3 for a in $a^2 - 2a + 4$:

$$(-3)^2 - 2(-3) + 4 = 9 + 6 + 4$$
$$= 19$$

EXAMPLE 2   **An Algebraic Expression for Revenue**

The expression $4.50x + 2.50y$ represents the total amount of money, in dollars, received at a school play, where x represents the number of adult tickets sold and y represents the number of student tickets sold. Evaluate $4.50x + 2.50y$ for $x = 50$ and $y = 82$. Interpret the result.

Solution

We substitute 50 for x and 82 for y in the expression $4.50x + 2.50y$.

$$4.50(50) + 2.50(82) = 225 + 205 = 430$$

So $430 was collected by selling 50 adult tickets and 82 student tickets.

Quick ✓

1. When a letter represents any number from a set of numbers, it is called a _____.

2. To _____ an algebraic expression, substitute a numerical value for each variable into the expression and simplify the result.

In Problems 3 and 4, evaluate each expression for the given value of the variable.

3. $-3k + 5$ for $k = 4$ **4.** $-2y^2 - y + 8$ for $y = -2$

5. The Amadeus Coffee Shop creates a breakfast blend of two types of coffee. They mix x pounds of a mild coffee that sells for $7.00 per pound with y pounds of a robust coffee that sells for $10.00 per pound. An algebraic expression that represents the value of the breakfast blend, in dollars, is $7x + 10y$. Evaluate this expression for $x = 8$ pounds and $y = 16$ pounds

---

▶ ❷ **Identify Like Terms and Unlike Terms**

Algebraic expressions consist of *terms*.

Definition

A **term** is a constant or the product of a constant and one or more variables raised to a power.

In algebraic expressions, the terms are separated by addition signs.

EXAMPLE 3   **Identifying the Terms in an Algebraic Expression**

Identify the terms in the following algebraic expressions.

   (a) $4a^3 + 5b^2 - 8c + 12$ **(b)** $\dfrac{x}{4} - 7y + 8z$

Solution

   (a) Rewrite $4a^3 + 5b^2 - 8c + 12$ so it contains only addition signs.

$$4a^3 + 5b^2 + (-8c) + 12$$

   The four terms are $4a^3$, $5b^2$, $-8c$, and 12.

   (b) The algebraic expression $\dfrac{x}{4} - 7y + 8z$ has three terms: $\dfrac{x}{4}$, $-7y$, and $8z$.

Quick ✔

6. *True or False* A constant by itself can be a term.

In Problems 7–9, identify the terms in each algebraic expression.

7. $5x^2 + 3xy$

8. $9ab - 3bc + 5ac - ac^2$

9. $\dfrac{2mn}{5} - \dfrac{3n}{7}$

Definition

The **coefficient** of a term is the numerical factor of the term.

For example, the coefficient of $7x$ is 7; the coefficient of $-2x^2y$ is -2. Terms that have no number as a factor, such as mn, have a coefficient of 1 since $mn = 1 \cdot mn$. The coefficient of $-y$ is -1 since $-y = -1 \cdot y$. If a term consists of just a constant, the coefficient is the number itself. For example, the coefficient of 14 is 14.

EXAMPLE 4 **Determining the Coefficient of a Term**

Determine the coefficient of each term:

(a) $\dfrac{1}{2}xy^2$ **(b)** $-\dfrac{t}{12}$ **(c)** ab^3 **(d)** 12

Solution

(a) The coefficient of $\dfrac{1}{2}xy^2$ is $\dfrac{1}{2}$.

(b) The coefficient of $-\dfrac{t}{12}$ is $-\dfrac{1}{12}$ because $-\dfrac{t}{12}$ can be written as $-\dfrac{1}{12} \cdot t$.

(c) The coefficient of ab^3 is 1 because ab^3 can be written as $1 \cdot ab^3$.

(d) The coefficient of 12 is 12 because the coefficient of a constant is the number itself.

Quick ✔

In Problems 10–14, determine the coefficient of each term.

10. $2z^2$ **11.** xy **12.** $-b$

13. 5 **14.** $-\dfrac{2}{3}z$

Sometimes we can simplify algebraic expressions by combining *like terms*.

Work Smart

Like terms can have different coefficients, but they cannot have different variables or different exponents on those variables.

Definition

Terms that have the same variable factor(s) with the same exponent(s) are called **like terms**.

For example, $3x^2$ and $-7x^2$ are like terms because both contain x^2, but $3x^2$ and $-7x^3$ are not like terms because the variable x is raised to different powers. Constant terms such as -9 and 6 are like terms.

EXAMPLE 5   **Classifying Terms as Like or Unlike**

Classify the following pairs of terms as *like* or *unlike*.

(a) $2p^3$ and $-5p^3$ (b) $7kr$ and $\frac{1}{4}k^2r$ (c) 5 and 8

Solution

(a) $2p^3$ and $-5p^3$ are *like* terms because both contain p^3.

(b) $7kr$ and $\frac{1}{4}k^2r$ are *unlike* terms because k is raised to different powers.

(c) 5 and 8 are *like* terms because both are constants.   ●

Quick ✓

15. *True or False* Like terms can have different coefficients or different exponents on the same variable.

In Problems 16–20, tell whether the terms are like or unlike.

16. $-\frac{2}{3}p^2$ and $\frac{4}{5}p^2$ 17. $\frac{m}{6}$ and $4m$ 18. $3a^2b$ and $-2ab^2$

19. $8a$ and 11   20. -7 and 12

▶ ❸ **Use the Distributive Property**

The *Distributive Property* will be used throughout this course and in future courses.

The Distributive Property

If a, b, and c are real numbers, then

$$a \cdot (b + c) = a \cdot b + a \cdot c$$
$$(a + b) \cdot c = a \cdot c + b \cdot c$$

That is, multiply each of the terms inside the parentheses by the factor on the outside.

Because $b - c = b + (-c)$, it is also true that $a(b - c) = a \cdot b - a \cdot c$.

EXAMPLE 6   **Using the Distributive Property to Remove Parentheses**

Use the Distributive Property to remove the parentheses.

(a) $3(x + 5)$ (b) $-\frac{1}{3}(6x - 12)$

Solution

(a) To use the Distributive Property, multiply each term in the parentheses by 3:

$$3(x + 5) = 3 \cdot x + 3 \cdot 5$$
$$= 3x + 15$$

Work Smart

The long name for the Distributive Property is the Distributive Property of Multiplication over Addition. This name helps to remind us that we do not distribute across multiplication. For example,

$$6x(5xy) \neq 6x \cdot 5x \cdot 6x \cdot y$$

(b) Multiply each term in the parentheses by $-\frac{1}{3}$:

$$-\frac{1}{3}(6x - 12) = -\frac{1}{3} \cdot 6x - \left(-\frac{1}{3}\right) \cdot 12$$
$$= -2x + 4$$

> **Quick ✔**
>
> **21.** $a \cdot (b + c) = a \cdot \_ + a \cdot \_$
>
> *In Problems 22–25, use the Distributive Property to remove the parentheses.*
>
> **22.** $6(x + 2)$ **23.** $-5(x + 2)$
>
> **24.** $-2(k - 7)$ **25.** $(8x + 12)\dfrac{3}{4}$

▶ ④ Simplify Algebraic Expressions by Combining Like Terms

An algebraic expression that contains the sum or difference of like terms may be simplified using the Distributive Property "in reverse." When we use the Distributive Property to add coefficients of like terms, we say that we are **combining like terms**.

EXAMPLE 7 **Using the Distributive Property to Combine Like Terms**

Combine like terms:

(a) $2x + 7x$ **(b)** $x^2 - 5x^2$

Solution

(a) $2x + 7x = (2 + 7)x$ **(b)** $x^2 - 5x^2 = (1 - 5)x^2$
   $= 9x$ $= -4x^2$

Look carefully at the results of Example 7. Notice that when we combine like terms, we add the coefficients of the like terms and keep the variables and exponents the same.

> **Quick ✔**
>
> **26.** $4x^2 + 9x^2 = (\_ + \_)x^2$
>
> *In Problems 27–29, combine like terms.*
>
> **27.** $3x - 8x$ **28.** $-5x^2 + x^2$ **29.** $-7x - x + 6 - 3$

Sometimes we must rearrange terms using the Commutative Property of Addition to combine like terms.

EXAMPLE 8 **Combining Like Terms Using the Commutative Property**

Combine like terms: $4x + 5y + 12x - 7y$

Solution

Use the Commutative Property to rearrange the terms.

$$4x + 5y + 12x - 7y = 4x + 12x + 5y - 7y$$

Use the Distributive Property "in reverse": $= (4 + 12)x + (5 - 7)y$
$= 16x + (-2y)$

Write the answer in simplest form: $= 16x - 2y$

Quick ✔

In Problems 30–33, combine like terms.

30. $3a + 2b - 5a + 7b - 4$

31. $5ac + 2b + 7ac - 5a - b$

32. $5ab^2 + 7a^2b + 3ab^2 - 8a^2b$

33. $\frac{4}{3}rs - \frac{3}{2}r^2 + \frac{2}{3}rs - 5$

Often, we need to remove parentheses by using the Distributive Property before we can combine like terms. Recall that the rules for order of operations on real numbers place multiplication before addition or subtraction. In this section, the direction **simplify** will mean to remove all parentheses and combine like terms.

EXAMPLE 9 **Combining Like Terms Using the Distributive Property**

Simplify the algebraic expression: $3 - 4(2x + 3) - (5x + 1)$

Solution

First, we use the Distributive Property to remove parentheses.

$$3 - 4(2x + 3) - (5x + 1) = 3 - 8x - 12 - 5x - 1$$

Rearrange terms using the Commutative Property of Addition: $= -8x - 5x + 3 - 12 - 1$

Combine like terms: $= -13x - 10$ ●

Quick ✔

34. Explain what it means to simplify an algebraic expression.

In Problems 35–38, simplify each expression.

35. $3x + 2(x - 1) - 7x + 1$

36. $m + 2n - 3(m + 2n) - (7 - 3n)$

37. $2(a - 4b) - (a + 4b) + b$

38. $\frac{1}{2}(6x + 4) - \frac{1}{3}(12 - 9x)$

Work Smart

Remember that multiplication comes before subtraction. In the first step of Example 9, do not compute $3 - 4$ first to obtain $-1(2x + 3)$.

Work Smart: Study Skills

Selected problems in the exercise sets are in green. For extra help, worked solutions to these problems are in MyMathLab.

Simplifying an Algebraic Expression

Step 1: Remove any parentheses using the Distributive Property.

Step 2: Combine any like terms.

1.8 Exercises

 MyMathLab® MathXP PRACTICE

Exercise numbers in **green** have complete video solutions in MyMathLab.

Problems 1–38 are the **Quick** ✔*s that follow the* **EXAMPLES.**

Building Skills

In Problems 39–50, evaluate each expression using the given values of the variables. See Objective 1.

39. $2x + 5$ for $x = 4$

40. $3x + 7$ for $x = 2$

41. $x^2 + 3x - 1$ for $x = 3$

42. $n^2 - 4n + 3$ for $n = 2$

43. $4 - k^2$ for $k = -5$

44. $-2p^2 + 5p + 1$ for $p = -3$

45. $\dfrac{9x - 5y}{x + y}$ for $x = 3, y = 5$

46. $\dfrac{3y + 2z}{y - z}$ for $y = 4, z = -2$

47. $(x + 3y)^2$ for $x = 3, y = -4$

48. $(a - 2b)^2$ for $a = 1, b = 2$

49. $b^2 - 4ac$ for $a = 1, b = 4, c = 3$

50. $\dfrac{a^2 - 4}{a^2 + 5a - 14}$ for $a = -3$

In Problems 51–54, for each expression, identify the terms and then name the coefficient of each term. See Objective 2.

51. $2x^3 + 3x^2 - x + 6$ **52.** $3m^4 - m^3n^2 + 4n - 1$

53. $z^2 + \dfrac{2y}{3}$ **54.** $t^3 - \dfrac{t}{4}$

In Problems 55–62, determine whether the terms are like or unlike. See Objective 2.

55. $8x$ and 8 **56.** $11p$ and 11

57. 54 and -21 **58.** -13 and 38

59. $12b$ and $-b$ **60.** $6a^2$ and $-3a^2$

61. r^2s and rs^2 **62.** x^2y^3 and y^2x^3

In Problems 63–70, use the Distributive Property to remove the parentheses. See Objective 3.

63. $3(m + 2)$ **64.** $3(4s + 2)$

65. $(3n^2 + 2n - 1)6$ **66.** $(6a^4 - 4a^2 + 2)3$

67. $-(x - y)$ **68.** $-5(k - n)$

69. $(8x - 6y)\left(-\dfrac{1}{2}\right)$ **70.** $(20a - 15b)\left(-\dfrac{2}{5}\right)$

In Problems 71–98, simplify each expression by using the Distributive Property to remove parentheses and combining like terms. See Objective 4.

71. $5x - 2x$ **72.** $14k - 11k$

73. $4z - 6z + 8z$ **74.** $9m - 8m + 2m$

75. $2m + 3n + 8m + 7n$ **76.** $x + 2y + 5x + 7y$

77. $0.3x^7 + x^7 + 0.9x^7$ **78.** $1.7n^4 - n^2 + 2.1n^4$

79. $-3y^6 + 13y^6$ **80.** $-7p^5 + 2p^5$

81. $-(6w + 12y - 13z)$ **82.** $-(-6m + 9n - 8p)$

83. $5(k + 3) - 8k$ **84.** $3(7 - z) - z$

85. $7n - (3n + 8)$ **86.** $18m - (6 + 9m)$

87. $(7 - 2x) - (x + 4)$ **88.** $(3k + 1) - (4 - k)$

89. $(7n - 8) - (3n - 6)$ **90.** $(5y - 6) - (11y + 8)$

91. $-6(n - 3) + 2(n + 1)$

92. $-9(7r - 6) + 9(10r + 3)$

93. $\dfrac{2}{3}x + \dfrac{1}{6}x$ **94.** $\dfrac{3}{5}y + \dfrac{7}{10}y$

95. $\dfrac{1}{2}(8x + 5) - \dfrac{2}{3}(6x + 12)$

96. $\dfrac{1}{5}(60 - 15x) + \dfrac{3}{4}(12 - 4x)$

97. $2(0.5x + 9) - 3(1.5x + 8)$

98. $3(0.2x + 6) - 5(1.6x + 1)$

Mixed Practice

In Problems 99–110, (a) evaluate the expression for the given value(s) of the variable(s) before combining like terms, (b) simplify the expression by combining like terms and then evaluate the expression for the given value(s) of the variable(s). Compare your results.

99. $5x + 3x; x = 4$ **100.** $8y + 2y; y = -3$

101. $-2a^2 + 5a^2; a = -3$ **102.** $4b^2 - 7b^2; b = 5$

103. $4z - 3(z + 2); z = 6$

104. $8p - 3(p - 4); p = 3$

105. $5y^2 + 6y - 2y^2 + 5y - 3; y = -2$

106. $3x^2 + 8x - x^2 - 6x; x = 5$

107. $\dfrac{1}{2}(4x - 2) - \dfrac{2}{3}(3x + 9); x = 3$

108. $\dfrac{1}{5}(5x - 10) - \dfrac{1}{6}(6x + 12); x = -2$

109. $3a + 4b - 7a + 3(a - 2b); a = 2, b = 5$

110. $-4x - y + 2(x - 3y); x = 3, y = -2$

Applying the Concepts

In Problems 111–116, evaluate each expression using the given values of the variables.

111. $\dfrac{1}{2}h(b + B); h = 4, b = 5, B = 17$

112. $\dfrac{1}{2}h(b + B); h = 9, b = 3, B = 12$

113. $\dfrac{a - b}{c - d}; a = 6, b = 3, c = -4, d = -2$

114. $\dfrac{a - b}{c - d}; a = -5, b = -2, c = 7, d = 1$

115. $b^2 - 4ac; a = 7, b = 8, c = 1$

116. $b^2 - 4ac; a = 2, b = 5, c = 3$

117. Renting a Truck The cost of renting a truck from Hamilton Truck Rental is $59.95 per day plus $0.15 per mile. The expression $59.95 + 0.15m$ represents the cost of renting a truck for one day and driving it m miles. Evaluate $59.95 + 0.15m$ for $m = 125$.

118. Renting a Car The cost of renting a compact car for one day from CMH Auto is $29.95 plus $0.17 per mile. The expression $29.95 + 0.17m$ represents the total daily cost. Evaluate the expression $29.95 + 0.17m$ for $m = 245$.

119. Ticket Sales The Center for Science and Industry sells adult tickets for $12 and children's tickets for $7. The expression $12a + 7c$ represents the total revenue from selling a adult tickets and c children's tickets. Evaluate the algebraic expression $12a + 7c$ for $a = 156$ and $c = 421$.

120. Ticket Sales A community college theatre group sold tickets to a recent production. Student tickets cost $5, and nonstudent tickets cost $8. The algebraic expression $5s + 8n$ represents the total revenue from selling s student tickets and n nonstudent tickets. Evaluate $5s + 8n$ for $s = 76$ and $n = 63$.

△ **121. Rectangle** The width of a rectangle is w yards, and the length of the rectangle is $(3w - 4)$ yards. The perimeter of the rectangle is given by the algebraic expression $2w + 2(3w - 4)$.

  (a) Simplify the algebraic expression $2w + 2(3w - 4)$.

  (b) Determine the perimeter of a rectangle whose width w is 5 yards.

△ **122. Rectangle** The length of a rectangle is l meters, and the width of the rectangle is $(l - 11)$ meters. The perimeter of the rectangle is given by the algebraic expression $2l + 2(l - 11)$.

  (a) Simplify the expression $2l + 2(l - 11)$.

  (b) Determine the perimeter of a rectangle whose length l is 15 meters.

123. Finance Novella invested some money in two investment funds. She placed s dollars in stocks that yield 5.5% annual interest and b dollars in bonds that yield 3.25% annual interest. Evaluate the expression $0.055s + 0.0325b$ for $s = \$2950$ and $b = \$2050$. Round your answer to the nearest cent.

124. Finance Jonathan received an inheritance from his grandparents. He invested x dollars in a Certificate of Deposit that pays 2.95% and y dollars in an off-shore oil drilling venture that is expected to pay 12.8%. Evaluate the algebraic expression $0.0295x + 0.128y$ for $x = \$2500$ and $y = \$1000$.

Extending the Concepts

125. Simplify the algebraic expression (cleverly!!) using the Distributive Property—in reverse!
$$2.75(-3x^2 + 7x - 3) - 1.75(-3x^2 + 7x - 3)$$

126. Simplify the algebraic expression using the Distributive Property in reverse.
$$11.23(7.695x + 81.34) + 8.77(7.695x + 81.34)$$

Explaining the Concepts

127. Explain why the sum $2x^2 + 4x^2$ is *not* equivalent to $6x^4$. What is the correct answer?

128. Use $x = 4$ and $y = 5$ to answer parts (a), (b), and (c).

  (a) Evaluate $x^2 + y^2$.

  (b) Evaluate $(x + y)^2$.

  (c) Are the results the same? Is $(x + y)^2$ equal to $x^2 + y^2$? Explain your response.

Chapter 1 Activity: The Math Game

Focus: Applying order of operations and simplifying expressions
Time: 10 minutes
Group size: 2–4

- The instructor will announce when the groups may begin solving the problems to the right.
- When your group has completed all of the problems, ask the instructor to check the answers. The instructor will tell you how many answers are correct, but not which ones.
- The first group to complete all of the problems correctly will win a prize, as determined by the instructor.

1. Evaluate: $-8 \div 2^2 \cdot 6 + (-2)^3$

2. Evaluate: $\dfrac{6(-3) + 4^2}{25 + 4(-9 + 4)}$

3. Evaluate: $x^3 - x^2$ for $x = -3$

4. Evaluate: $\dfrac{(x + 2y)^2}{xy}$ for $x = 1, y = -2$

5. Simplify: $-2(4x + 3) - (5x - 1)$

6. Simplify: $\dfrac{3}{4}(8x^2 + 16) - 2x^2 + 3x$

Chapter 1 Review

Section 1.2 Fractions, Decimals, and Percents

KEY CONCEPTS

- To find the least common multiple (LCM) of two numbers, (1) factor each number as the product of prime factors; (2) write the factor(s) that the numbers share, if any; (3) write down the remaining factors the greatest number of times that the factors appear in any number. The product of the factors is the LCM.

- To write a fraction in lowest terms, find the common factors between the numerator and denominator, and use the fact that $\frac{a \cdot c}{b \cdot c} = \frac{a}{b}$ to divide out the common factors.

- To round a decimal, identify the specified place value in the decimal. If the digit to the right is 5 or more, add 1 to the digit; if the digit to the right is 4 or less, leave the digit as it is. Then drop the digits to the right of the specified place value.

- To convert a fraction to a decimal, divide the numerator of the fraction by the denominator of the fraction until the remainder is 0 or the remainder repeats.

- To convert a decimal to a fraction, identify the place value of the denominator, write the decimal as a fraction using the given denominator, and write in lowest terms.

- To convert a percent to a decimal, move the decimal point two places to the left and drop the % symbol.

- To convert a decimal to a percent, move the decimal point two places to the right and add the % symbol.

KEY TERMS

Factor
Product
Prime
Composite
Prime factorization
Multiple
Least Common Multiple (LCM)
Numerator
Denominator
Equivalent fractions
Least Common Denominator (LCD)
Terminating decimal
Repeating decimal
Percent

You Should Be Able To...	EXAMPLE	Review Exercises
❶ Factor a number as a product of prime factors (p. 8)	Example 1	1–4
❷ Find the least common multiple of two or more numbers (p. 9)	Examples 2 and 3	5, 6
❸ Write equivalent fractions (p. 10)	Examples 4 and 5	7–10
❹ Write a fraction in lowest terms (p. 12)	Example 6	11–13
❺ Round decimals (p. 13)	Example 7	14, 15
❻ Convert between fractions and decimals (p. 14)	Examples 8 and 9	16–22
❼ Convert between percents and decimals (p. 15)	Examples 10 and 11	23–31

In Problems 1–4. factor each number as a product of primes, if possible.

1. 75
2. 87
3. 81
4. 17

In Problems 5 and 6, find the LCM of the numbers.

5. 18 and 24
6. 4, 8, and 18

In Problems 7 and 8, write each number with the given denominator.

7. Write $\frac{7}{15}$ with the denominator 30.

8. Write 3 with the denominator 4.

In Problems 9 and 10, write equivalent fractions with the least common denominator.

9. $\frac{1}{6}$ and $\frac{3}{8}$ **10.** $\frac{9}{16}$ and $\frac{7}{24}$

In Problems 11–13, write each fraction in lowest terms.

11. $\frac{25}{60}$ **12.** $\frac{125}{250}$ **13.** $\frac{96}{120}$

In Problems 14 and 15, round each number to the given place.

14. 21.7648 to the neared hundredth

15. 14.91 to the nearest one (unit)

16. Write $\frac{8}{9}$ as a repeating decimal.

17. Write $\frac{9}{32}$ as a terminating decimal.

In Problems 18 and 19, write each fraction as a decimal rounded to the indicated place.

18. $\frac{11}{6}$ to the nearest hundredth.

19. $\frac{19}{8}$ to the nearest tenth.

In Problems 20–22, write each decimal as a fraction in lowest terms.

20. 0.6   **21.** 0.375   **22.** 0.864

In Problems 23–26, write each percent as a decimal.

23. 41%   **24.** 760%

25. 9.03%   **26.** 0.35%

In Problems 27–30, write each decimal as a percent.

27. 0.23   **28.** 1.17

29. 0.045   **30.** 3

31. A student earns 12 points out of a total of 20 points on a quiz.

(a) Express this score as a fraction in lowest terms.

(b) Express this score as a percentage.

Section 1.3 The Number Systems and the Real Number Line

KEY CONCEPTS

- $a < b$ means a is to the left of b on a real number line.
- $a = b$ means a and b are in the same position on a real number line.
- $a > b$ means a is to the right of b on a real number line.
- $|a|$ is the distance from 0 to a on a real number line.

KEY TERMS

Set	Real number line
Elements	Origin
Empty set	Scale
Natural numbers	Coordinate
Counting numbers	Negative real numbers
Whole numbers	Zero
Integers	Positive real numbers
Rational number	Sign
Irrational number	Inequality symbols
Real numbers	Absolute value

You Should Be Able To...	EXAMPLE	Review Exercises
❶ Classify numbers (p. 19)	Example 2	32–41, 57, 58
❷ Plot points on a real number line (p. 22)	Example 3	42
❸ Use inequalities to order real numbers (p. 23)	Example 4	43–47, 51–56
❹ Compute the absolute value of a real number (p. 24)	Example 5	48–50, 55, 56

In Problems 32–35, write each set.

32. A is the set of whole numbers less than 7.

33. B is the set of natural numbers less than or equal to 3.

34. C is the set of integers greater than -3 and less than or equal to 5

35. D is the set of whole numbers less than 0.

In Problems 36–41, use the set
$$\left\{-6, -3.25, 0, 5.030030003\ldots, \frac{9}{3}, 11, \frac{5}{7}\right\}.$$

List all the elements that are

36. natural numbers

37. whole numbers

38. integers

39. rational numbers

40. irrational numbers

41. real numbers

42. Plot the points $\{-3, -\frac{4}{3}, 0, 2, 3.5\}$ on a real number line.

In Problems 43–47, determine whether the statement is True or False.

43. $-3 > -1$

44. $5 \le 5$

45. $-5 \le -3$

46. $\frac{1}{2} = 0.5$

47. $\frac{2}{3} > \frac{5}{6}$

In Problems 48–50, evaluate each expression.

48. $-\left|\frac{1}{2}\right|$

49. $|-7|$

50. $-|-6|$

In Problems 51–56, replace the ? with the correct symbol: $>$, $<$, $=$.

51. $\frac{1}{4}$? 0.25

52. -6 ? 0

53. 0.83 ? $\frac{3}{4}$

54. $\frac{-2}{|-2|}$? $-|-1|$

55. $|-4|$? $|-3|$

56. $\frac{4}{5}$? $\left|-\frac{5}{6}\right|$

57. Explain the difference between a rational number and an irrational number. Be sure that your explanation includes a discussion of terminating decimals and nonterminating decimals.

58. What do we call the set of positive integers?

Section 1.4 Adding, Subtracting, Multiplying, and Dividing Integers

KEY CONCEPTS

- **Rules of Signs for Multiplying Two Integers**
  1. The product of two positive integers is positive.
  2. The product of one positive integer and one negative integer is negative.
  3. The product of two negative integers is positive.

- **Rules of Signs for Dividing Two Integers**
  1. The quotient of two positive integers is positive. That is, $\dfrac{+a}{+b} = \dfrac{a}{b}$.
  2. The quotient of one positive integer and one negative integer is negative. That is, $\dfrac{-a}{b} = \dfrac{a}{-b} = -\dfrac{a}{b}$.
  3. The quotient of two negative integers is positive. That is, $\dfrac{-a}{-b} = \dfrac{a}{b}$.

KEY TERMS

Operations
Sum
Difference
Product
Quotient
Mixed number
Additive inverse
Opposite
Evaluate
Factors
Dividend
Divisor
Multiplicative inverse
Reciprocal
Ratio
Golden ratio

You Should Be Able To...	EXAMPLE	Review Exercises
❶ Add integers (p. 28)	Examples 1 through 6	59–66, 71, 72, 89, 90, 93, 99, 100, 102
❷ Determine the additive inverse of a number (p. 30)	Example 7	87, 88
❸ Subtract integers (p. 31)	Examples 8 through 10	67–72, 91, 92, 94, 99–101
❹ Multiply integers (p. 32)	Examples 11 and 12	73–78, 95, 96, 102
❺ Divide integers (p. 33)	Example 14	79–86, 97, 98

In Problems 59–86, perform the indicated operation.

59. $-2 + 9$

60. $6 + (-10)$

61. $-23 + (-11)$

62. $-120 + 25$

63. $-|-2 + 6|$

64. $-|-15| + |-62|$

65. $-110 + 50 + (-18) + 25$

66. $-28 + (-35) + (-52)$

67. $-10 - 12$

68. $18 - 25$

69. $-11 - (-32)$

70. $0 - (-67)$

71. $34 - 18 + 10$

72. $-49 - 8 + 21$

73. $-6(-2)$

74. $4(-10)$

75. $13(-86)$

76. $-19(423)$

77. $(11)(13)(-5)$

78. $(-53)(-21)(-10)$

79. $\dfrac{-20}{-4}$

80. $\dfrac{60}{-5}$

81. $\dfrac{|-55|}{11}$

82. $-\left|\dfrac{-100}{4}\right|$

83. $\dfrac{120}{-15}$

84. $\dfrac{64}{-20}$

85. $\dfrac{-180}{54}$

86. $\dfrac{-450}{105}$

In Problems 87 and 88, determine the additive inverse of each number.

87. 13

88. -45

In Problems 89–98, write the expression using mathematical symbols, and then evaluate the expression.

89. -43 plus 101

90. 45 plus -28

91. -10 minus -116

92. 74 minus 56

93. the sum of 13 and -8

94. the difference between -60 and -10

95. -21 multiplied by -3

96. 54 multiplied by -18

97. -34 divided by -2

98. -49 divided by 14

99. Football Matt Forte had three possessions of the football within the first few minutes of the game. On his first possession he gained 20 yards, on his second possession he lost 6 yards, and on his third possession he gained 12 yards. What was his total yardage?

100. Temperature On a winter day in Detroit, Michigan, the temperature was 10°F in the morning. The temperature rose 12°F in the afternoon and then fell 25°F by midnight. What was the temperature at midnight in Detroit?

101. Temperature One day in Bismarck, North Dakota, the high temperature was 6°F above zero and the low temperature was 18°F below zero. What was the difference between the high and low temperatures on that day in Bismarck?

102. Test Score Ms. Rosen awards 5 points for each correct multiple-choice question and awards 8 points for each correct free-response question. On one of Ms. Rosen's tests, Sarah got 11 multiple-choice questions correct and 4 free-response questions correct. What was Sarah's test score?

Section 1.5 Adding, Subtracting, Multiplying, and Dividing Rational Numbers

KEY CONCEPTS

- **Multiplying Fractions**

$$\frac{a}{b} \cdot \frac{c}{d} = \frac{a \cdot c}{b \cdot d} \quad \text{where } b \text{ and } d \neq 0$$

- **Dividing Fractions**

$$\frac{a}{b} \div \frac{c}{d} = \frac{a}{b} \cdot \frac{d}{c} = \frac{a \cdot d}{b \cdot c} \quad \text{where } b, c, d \neq 0$$

- **Adding or Subtracting Fractions with the Same Denominator**

$$\frac{a}{c} + \frac{b}{c} = \frac{a + b}{c} \quad \text{where } c \neq 0$$

$$\frac{a}{c} - \frac{b}{c} = \frac{a - b}{c} = \frac{a + (-b)}{c} \quad \text{where } c \neq 0$$

- **Adding or Subtracting Fractions with Unlike Denominators**

  Step 1: Find the LCD of the fractions.

  Step 2: Find equivalent fractions with the LCD by multiplying by a fraction equivalent to 1.

  Step 3: Add or subtract the numerators, and write the result over the common denominator.

  Step 4: Simplify the result.

KEY TERMS

Least common denominator

You Should Be Able To...	EXAMPLE	Review Exercise
❶ Multiply rational numbers in fractional form (p. 37)	Example 1	103–106, 136
❷ Divide rational numbers in fractional form (p. 38)	Example 2	107–110
❸ Add or subtract rational numbers in fractional form (p. 39)	Examples 3 through 7	111–122, 137
❹ Add, subtract, multiply, or divide rational numbers in decimal form (p. 43)	Examples 8 through 11	123–135, 138

In Problems 103–122, perform the indicated operation. Write in lowest terms.

103. $\dfrac{2}{3} \cdot \dfrac{15}{8}$

104. $-\dfrac{3}{8} \cdot \dfrac{10}{21}$

105. $\dfrac{5}{8} \cdot \left(-\dfrac{2}{25}\right)$

106. $5 \cdot \left(-\dfrac{3}{10}\right)$

107. $\dfrac{24}{17} \div \dfrac{18}{3}$

108. $-\dfrac{5}{12} \div \dfrac{10}{16}$

109. $-\dfrac{27}{10} \div 9$

110. $20 \div \left(-\dfrac{5}{8}\right)$

111. $\dfrac{2}{9} + \dfrac{1}{9}$

112. $-\dfrac{6}{5} + \dfrac{4}{5}$

113. $\dfrac{5}{7} - \dfrac{2}{7}$

114. $\dfrac{7}{5} - \left(-\dfrac{8}{5}\right)$

115. $\dfrac{3}{10} + \dfrac{1}{20}$

116. $\dfrac{5}{12} + \dfrac{4}{9}$

117. $-\dfrac{7}{35} - \dfrac{2}{49}$

118. $\dfrac{5}{6} - \left(-\dfrac{1}{4}\right)$

119. $-2 - \left(-\dfrac{5}{12}\right)$

120. $-5 + \dfrac{9}{4}$

121. $-\dfrac{1}{10} + \left(-\dfrac{2}{5}\right) + \dfrac{1}{2}$

122. $-\dfrac{5}{6} - \dfrac{1}{4} + \dfrac{3}{24}$

In Problems 123–134, perform the indicated operation.

123. $30.3 + 18.2$

124. $-43.02 + 18.36$

125. $201.37 - 118.39$

126. $-35.1 - 18.64$

127. $(-0.04)(-2.01)$

128. $(87.3)(-2.98)$

129. $\dfrac{69.92}{3.8}$

130. $-\dfrac{1.08318}{0.042}$

131. $12.5 - 18.6 + 8.4$

132. $-13.5 + 10.8 - 20.2$

133. $(12.9)(1.4)(-0.3)$

134. $(2.4)(6.1)(-0.05)$

135. Checking Account Lee had a balance of $256.75 in her checking account. Lee wrote a check for $175.68 on Wednesday and wrote a check for $180.00 on Thursday. What is her checking account balance now? Is Lee's account overdrawn?

136. Super Bowl Party Jarred had 36 friends at his Super Bowl party. Two-thirds of his friends wanted the NFC team to win. How many of Jarred's friends wanted the NFC team to win the Super Bowl?

137. Ribbon Cutting Tara has a piece of ribbon that is 15 inches long. If she cuts off a $3\frac{1}{2}$-inch piece of the ribbon, what is the length of the piece that remains?

138. Buying Clothes While shopping at her favorite store, Sierra bought 5 sweaters. If the sweaters cost $35 each and sales tax is 6.75% of the net price (net price = price × quantity), how much did Sierra spend on the clothes?

Section 1.6 Properties of Real Numbers

KEY CONCEPTS

- **Identity Property of Addition**
  For any real number a, $0 + a = a + 0 = a$.
- **Commutative Property of Addition**
  If a and b are real numbers, then $a + b = b + a$.
- **Additive Inverse Property**
  For any real number a, $a + (-a) = -a + a = 0$.
- **Associative Property of Addition**
  If a, b, and c are real numbers, then $a + (b + c) = (a + b) + c$.
- **Commutative Property of Multiplication**
  If a and b are real numbers, then $a \cdot b = b \cdot a$.
- **Multiplication Property of Zero**
  For any real number a, the product of a and 0 is always 0; that is, $a \cdot 0 = 0 \cdot a = 0$.
- **Multiplicative Identity**
  $a \cdot 1 = 1 \cdot a = a$ for any real number a.
- **Associative Property of Multiplication**
  If a, b, and c are real numbers, then $a \cdot (b \cdot c) = (a \cdot b) \cdot c$.
- **Multiplicative Inverse Property**
  $a \cdot \frac{1}{a} = \frac{1}{a} \cdot a = 1$ provided that $a \neq 0$
- **Division Properties of Zero**
  For any nonzero number a,
  1. The quotient of 0 and a is 0. That is, $\frac{0}{a} = 0$.
  2. The quotient of a and 0 is undefined. That is, $\frac{a}{0}$ is undefined.

KEY TERMS

Additive Identity
Multiplicative Identity
Conversion
Commutative Property
Grouping Symbols
Associative Property
Undefined

You Should Be Able To...	EXAMPLE	Review Exercises
❶ Use the Identity Properties of Addition and Multiplication (p. 48)	Example 1	140, 141, 144–146, 148, 151–156, 164–166
❷ Use the Commutative Properties of Addition and Multiplication (p. 50)	Examples 2 through 4	142, 143, 147, 151–154, 157, 158, 161, 162, 167, 168
❸ Use the Associative Properties of Addition and Multiplication (p. 51)	Examples 5 through 7	139, 150, 155, 156
❹ Understand the Multiplication and Division Properties of 0 (p. 53)	Example 8	149, 159, 160, 163

In Problems 139–150, state the property of real numbers that is being illustrated.

139. $(5 \cdot 12) \cdot 10 = 5 \cdot (12 \cdot 10)$

140. $20 \cdot \dfrac{1}{20} = 1$

141. $\dfrac{8}{3} \cdot \dfrac{3}{8} = 1$

142. $\dfrac{5}{3} \cdot \left(-\dfrac{18}{61}\right) \cdot \dfrac{3}{5} = \dfrac{5}{3} \cdot \dfrac{3}{5} \cdot \left(-\dfrac{18}{61}\right)$

143. $9 \cdot 73 \cdot \dfrac{1}{9} = 9 \cdot \dfrac{1}{9} \cdot 73$

144. $23.9 + (-23.9) = 0$

145. $36 + 0 = 36$

146. $-49 + 0 = -49$

147. $23 + 5 + (-23)$ is equivalent to $23 + (-23) + 5$

148. $\dfrac{7}{8}$ is equivalent to $\dfrac{7}{8} \cdot \dfrac{3}{3}$

149. $14 \cdot 0 = 0$

150. $-5.3 + (5.3 + 2.8) = (-5.3 + 5.3) + 2.8$

In Problems 151–168, evaluate each expression, if possible, by using the properties of real numbers.

151. $144 + 29 + (-144)$

152. $76 + 99 + (-76)$

153. $\dfrac{19}{3} \cdot 18 \cdot \dfrac{3}{19}$

154. $\dfrac{14}{9} \cdot 121 \cdot \dfrac{9}{14}$

155. $3.4 + 42.56 + (-42.56)$

156. $5.3 + 3.6 + (-3.6)$

157. $\dfrac{9}{7} \cdot \left(-\dfrac{11}{3}\right) \cdot 7$

158. $\dfrac{13}{5} \cdot \dfrac{18}{39} \cdot 5$

159. $\dfrac{7}{0}$

160. $\dfrac{0}{100}$

161. $1000(-334)(0.001)$

162. $400(0.5)(0.01)$

163. $43{,}569{,}003 \cdot 0$

164. $154 \cdot \dfrac{1}{154}$

165. $\dfrac{3445}{302} + \left(-\dfrac{3445}{302}\right)$

166. $130 \cdot \dfrac{42}{42}$

167. $-\dfrac{7}{48} \cdot \dfrac{20}{3} \cdot \dfrac{12}{7}$

168. $\dfrac{9}{8} \cdot \left(-\dfrac{25}{13}\right) \cdot \dfrac{48}{9}$

Section 1.7 Exponents and the Order of Operations

KEY CONCEPTS

- **Exponential Notation**

  If n is a natural number and a is a real number, then

  $$a^n = \underbrace{a \cdot a \cdot a \cdot \ldots \cdot a}_{n \text{ factors}}$$

  where a is called the base and the natural number n is called the exponent or power.

- **Rules for Order of Operations**

  Step 1: Perform all operations within *grouping symbols* first. When an expression has multiple grouping symbols, begin with the innermost pair of grouping symbols and work outward.

  Step 2: Evaluate expressions containing *exponents*.

  Step 3: Perform *multiplication and division* in the order in which they occur, working from *left to right*.

  Step 4: Perform *addition and subtraction* in the order in which they occur, working from *left to right*.

KEY TERMS

Exponential notation
Base
Exponent
Power
Exponential form
Expanded form

You Should Be Able To...	EXAMPLE	Review Exercises
❶ Evaluate exponential expressions (p. 56)	Examples 1 through 4	169–178
❷ Apply the rules for order of operations (p. 57)	Examples 5 through 11	179–186

In Problems 169–172, write in exponential form.

169. $3 \cdot 3 \cdot 3 \cdot 3$

170. $\dfrac{2}{3} \cdot \dfrac{2}{3} \cdot \dfrac{2}{3}$

171. $(-4)(-4)$

172. $(-3)(-3)(-3)$

In Problems 173–178, evaluate each expression.

173. 5^3

174. -5^3

175. $(-3)^4$

176. $(-5)^3$

177. -3^4

178. $\left(\dfrac{1}{2}\right)^6$

In Problems 179–186, evaluate each expression.

179. $-2 + 16 \div 4 \cdot 2 - 10$

180. $-4 + 3[2^3 + 4(2 - 10)]$

181. $(12 - 7)^3 + (19 - 10)^2$

182. $5 - (-12 \div 2 \cdot 3) + (-3)^2$

183. $\dfrac{2 \cdot (4 + 8)}{3 + 3^2}$

184. $\dfrac{3 \cdot (5 + 2^2)}{2 \cdot 3^3}$

185. $\dfrac{6 \cdot [\,12 - 3 \cdot (5 - 2)\,]}{5 \cdot [\,21 - 2 \cdot (4 + 5)\,]}$

186. $\dfrac{4 \cdot [3 + 2 \cdot (8 - 6)]}{5 \cdot [14 - 2 \cdot (2 + 3)]}$

Section 1.8 Simplifying Algebraic Expressions

KEY CONCEPT

- **Distributive Property**
  If a, b, and c are real numbers, then $a \cdot (b + c) = a \cdot b + a \cdot c$ and $(a + b) \cdot c = a \cdot c + b \cdot c$

KEY TERMS

Algebra	Term
Variable	Coefficient
Constant	Like terms
Algebraic expression	Combining like terms
Evaluate an algebraic expression	Simplify

You Should Be Able To...	EXAMPLE	Review Exercises
❶ Evaluate algebraic expressions (p. 64)	Examples 1 and 2	187–190, 205
❷ Identify like terms and unlike terms (p. 65)	Examples 3 through 5	191–196
❸ Use the Distributive Property (p. 67)	Example 6	200–204
❹ Simplify algebraic expressions by combining like terms (p. 68)	Examples 7 through 9	197–204

In Problems 187–190, evaluate each expression using the given values of the variables.

187. $x^2 - y^2$ for $x = 5$, $y = -2$

188. $x^2 - 3y^2$ for $x = 3$, $y = -3$

189. $(x + 2y)^3$ for $x = -1$, $y = -4$

190. $\dfrac{a - b}{x - y}$ for $a = 5$, $b = -10$, $x = -3$, $y = 2$

In Problems 191 and 192, identify the terms and then name the coefficient of each term.

191. $3x^2 - x + 6$

192. $2x^2y^3 - \dfrac{y}{5}$

In Problems 193–196, determine whether the terms are like or unlike.

193. $4xy^2$, $-6xy^2$

194. $-3x$, $4x^2$

195. $-6y$, -6

196. -10, 4

In Problems 197–204, simplify each algebraic expression.

197. $4x - 6x - x$

198. $6x - 10 - 10x - 5$

199. $0.2x^4 + 0.3x^3 - 4.3x^4$

200. $-3(x^4 - 2x^2 - 4)$

201. $20 - (x + 2)$

202. $-6(2x + 5) + 4(4x + 3)$

203. $5 - (3x - 1) + 2(6x - 5)$

204. $\frac{1}{6}(12x + 18) - \frac{2}{5}(5x + 10)$

205. Moving Van The cost of renting a moving van for one day is $19.95 plus $0.25 per mile. The expression $19.95 + 0.25m$ represents the total cost of renting the truck for one day and driving m miles. Evaluate the expression $19.95 + 0.25m$ for $m = 315$.

Chapter 1 Test

Step-by-step test solutions are found on the Chapter Test Prep Videos available in MyMathLab® or on YouTube.

1. Find the LCM of 2, 6, and 14.

2. Write $\frac{21}{66}$ in lowest terms.

3. Write $\frac{13}{9}$ as a decimal rounded to the nearest hundredth.

4. Write 0.425 as a fraction in lowest terms.

5. Write 0.6% as a decimal.

6. Write 0.183 as a percent.

In Problems 7–15, perform the indicated operation. Write in lowest terms.

7. $\frac{4}{15} - \left(-\frac{2}{30}\right)$

8. $\frac{21}{4} \cdot \frac{3}{7}$

9. $-16 \div \frac{3}{20}$

10. $14 - 110 - (-15) + (-21)$

11. $-14.5 + 2.34$

12. $(-4)(-1)(-5)$

13. $16 \div 0$

14. -6 subtracted from -20

15. -110 divided by -2

16. Use the set $\left\{-2, -\frac{1}{2}, 0, 2.5, 6\right\}$. List all of the elements that are:

   (a) natural numbers

   (b) whole numbers

   (c) integers

   (d) rational numbers

   (e) irrational numbers

   (f) real numbers

In Problems 17 and 18, replace the ? with the correct symbol >, <, or =.

17. $-|-14|$? -12

18. $\left|-\frac{2}{5}\right|$? 0.4

In Problems 19–21, evaluate each expression.

19. $-16 \div 2^2 \cdot 4 + (-3)^2$

20. $\dfrac{4(-9) - 3^2}{25 + 4(-6 - 1)}$

21. $8 - 10[6^2 - 5(2 + 3)]$

22. Evaluate $(x - 2y)^3$ for $x = -1$ and $y = 3$.

In Problems 23 and 24, simplify each algebraic expression.

23. $-6(2x + 5) - (4x - 2)$

24. $\frac{1}{2}(4x^2 + 8) - 6x^2 + 5x$

25. Bank Account Latoya started with $675.15 in her bank account. She wrote a check for $175.50, withdrew $78.00 in cash, and made a deposit of $110.20. How much money does Latoya have in her bank account now?

26. Perimeter The length of a rectangle is 5 feet more than its width. The algebraic expression $2(x + 5) + 2x$ represents the perimeter of the rectangle. Simplify the expression $2(x + 5) + 2x$.

2 Equations and Inequalities in One Variable

You and your friends want to buy pizza to eat while watching the Super Bowl. Your local grocery store sells medium (12″) frozen pizzas for $9.99 and small (8″) frozen pizzas that are on special for $4.49 each. Which should you buy to get the best deal?

Understanding mathematical formulas and models allows you to solve everyday problems such as this one. See Problem 83 in Section 2.4.

The Big Picture: Putting It Together

In Chapter 1, we reviewed the arithmetic skills needed throughout the course. We also introduced algebraic expressions and discussed how to simplify and evaluate them.

In this chapter, we begin the study of algebra. The word "algebra" comes from the Arabic word *al-jabr*. The word *al-jabr* means "restoration." This is a reference to the fact that if a number is added to one side of an equation, then it must also be added to the other side in order to "restore" the equality. While algebra now means a whole lot more than "restoration," we will concentrate on the "restoration" part of algebra in this chapter.

Outline

2.1 Linear Equations: The Addition and Multiplication Properties of Equality

Objectives

❶ Determine Whether a Number Is a Solution of an Equation

❷ Use the Addition Property of Equality to Solve Linear Equations

❸ Use the Multiplication Property of Equality to Solve Linear Equations

Are You Ready for This Section?

Before getting started, take this readiness quiz. If you get a problem wrong, go back to the section cited and review the material.

R1. Determine the additive inverse of 3. What is the sum of a real number and its additive inverse? [Section 1.4, p. 30]

R2. Determine the multiplicative inverse of $-\frac{4}{3}$. What is the product of a nonzero number and its multiplicative inverse? [Section 1.4, pp. 33–34]

R3. Evaluate: $\frac{2}{3}\left(\frac{3}{2}\right)$ [Section 1.5, pp. 37–38]

R4. Use the Distributive Property to simplify: $-4(2x + 3)$ [Section 1.8, pp. 67–68]

R5. Simplify: $11 - (x + 6)$ [Section 1.8, pp. 68–69]

Work Smart: Study Skills

Remember, the green icon ▶ means there is a lecture video on this material. See page 5 for a complete description.

▶ ❶ **Determine Whether a Number Is a Solution of an Equation**

We begin with a definition.

> **Definition**
>
> A **linear equation in one variable** is an equation that can be written in the form $ax + b = c$, where a, b, and c are real numbers and a does not equal 0.

Examples of linear equations (in the variable x) are

$$x - 5 = 8 \qquad \frac{1}{2}x - 7 = \frac{3}{2}x + 8 \qquad 0.2(x + 5) - 1.5 = 4.25 - (x + 3)$$

The algebraic expressions in the equation are called the **sides** of the equation. For example, in the equation $\frac{1}{2}x - 7 = \frac{3}{2}x + 8$, the algebraic expression $\frac{1}{2}x - 7$ is the **left side** of the equation, and $\frac{3}{2}x + 8$ is the **right side** of the equation.

An equation may be true or false. For example, the equation $x - 5 = 8$ is true if the variable x is replaced by 13, but false if x is replaced by 2.

Because replacing x by 13 in the equation $x - 5 = 8$ results in a true statement, we say that 13 is a *solution* of the equation $x - 5 = 8$. We could also say that $x = 13$ *satisfies* the equation.

> **Definition**
>
> The **solution** of a linear equation is the value or values of the variable that make the equation a true statement. The set of all solutions of an equation is called the **solution set.** We sometimes say that the solution **satisfies** the equation.

We use set notation to indicate the solution set of an equation. For example, the solution set of $x - 5 = 8$ is $\{13\}$.

To determine whether a number satisfies an equation, we replace the variable with the number and see whether the left side of the equation equals the right side of the equation. If it does, we have a true statement, and the replacement value is a solution of the equation.

EXAMPLE 1   **Determine Whether a Number is a Solution of an Equation**

Determine whether the given value of the variable is a solution of the equation $4x + 7 = 19$.

(a) $x = -2$ (b) $x = 3$

Solution

(a)
$$4x + 7 = 19$$
Replace x with -2: $\quad 4(-2) + 7 \stackrel{?}{=} 19$
Simplify: $\quad -8 + 7 \stackrel{?}{=} 19$
$$-1 = 19 \quad \text{False}$$

Since the left and right sides of the equation are not equal when we replace x by -2, $x = -2$ is *not* a solution of the equation.

(b)
$$4x + 7 = 19$$
Replace x with 3: $\quad 4(3) + 7 \stackrel{?}{=} 19$
Simplify: $\quad 12 + 7 \stackrel{?}{=} 19$
$$19 = 19 \quad \text{True}$$

Since both sides of the equation are equal when we replace x by 3, $x = 3$ is a solution. ●

Quick ✓

1. The _____ of a linear equation is the value or values of the variable that make the equation a true statement.

In Problems 2–5, determine whether the given number is a solution of the equation.

2. $a - 4 = -7; a = -3$ 3. $2x + 1 = 11; x = \dfrac{21}{2}$

4. $3x - (x + 4) = 8; x = 6$ 5. $-9b + 3 + 7b = -3b + 8; b = -3$

▶ ❷ **Use the Addition Property of Equality to Solve Linear Equations**

We solve linear equations by writing a series of *equivalent equations* that result in the equation

$$x = a\ number$$

We form **equivalent equations** using mathematical properties that transform the original equation into a new equation that has the same solution. The first property is called the *Addition Property of Equality*.

In Words
The Addition Property of Equality says that whatever you add to one side of the equation, you must also add to the other side.

Addition Property of Equality

The **Addition Property of Equality** states that for real numbers a, b, and c,

$$\text{if } a = b, \text{ then } a + c = b + c$$

Remember, since $a - b$ is equivalent to $a + (-b)$, the Addition Property of Equality can also be used to subtract a real number from each side of the equation.

Because the goal in solving a linear equation is to get the variable by itself with a coefficient of 1, we say that we want to **isolate the variable.**

EXAMPLE 2 **How to Use the Addition Property of Equality to Solve a Linear Equation**

Solve the linear equation: $x - 6 = 11$

Step-by-Step Solution

Since the coefficient of the variable x in this equation is 1, we just need to "get x by itself."

Step 1: Isolate the variable x on the left side of the equation.

$$x - 6 = 11$$

Add 6 to each side of the equation: $\qquad x - 6 + 6 = 11 + 6$

Step 2: Simplify the left and right sides of the equation.

Use the Additive Inverse Property, $a + (-a) = 0$: $\qquad x + 0 = 17$

Use the Additive Identity Property, $a + 0 = a$: $\qquad x = 17$

Step 3: Check Verify that $x = 17$ is the solution.

$$x - 6 = 11$$

Replace x by 17 in the original equation to see whether a true statement results:

$$17 - 6 \stackrel{?}{=} 11$$
$$11 = 11 \quad \text{True}$$

Because $x = 17$ satisfies the original equations, the solution is 17, or the solution set is $\{17\}$. ●

EXAMPLE 3 **Using the Addition Property of Equality**

Solve the linear equation: $x + \dfrac{5}{2} = \dfrac{1}{4}$

Solution

$$x + \frac{5}{2} = \frac{1}{4}$$

Subtract $\dfrac{5}{2}$ from each side of the equation: $\qquad x + \dfrac{5}{2} - \dfrac{5}{2} = \dfrac{1}{4} - \dfrac{5}{2}$

$a + (-a) = 0$; LCD = 4: $\qquad x + 0 = \dfrac{1}{4} - \dfrac{5}{2} \cdot \dfrac{2}{2}$

$a + 0 = a$: $\qquad x = \dfrac{1}{4} - \dfrac{10}{4}$

$$x = -\frac{9}{4}$$

Check Verify that $x = -\dfrac{9}{4}$ is the solution.

$$x + \frac{5}{2} = \frac{1}{4}$$
$$-\frac{9}{4} + \frac{5}{2} \stackrel{?}{=} \frac{1}{4}$$
$$-\frac{9}{4} + \frac{10}{4} \stackrel{?}{=} \frac{1}{4}$$
$$\frac{1}{4} = \frac{1}{4} \quad \text{True}$$

The solution is $-\dfrac{9}{4}$, or the solution set is $\left\{-\dfrac{9}{4}\right\}$.

6. *True or False* The Addition Property of Equality says that whatever you add to one side of the equation, you must also add to the other side.

In Problems 7–12, solve each equation using the Addition Property of Equality.

7. $x - 11 = 21$ 8. $y + 7 = 21$ 9. $-8 + a = 4$

10. $-3 = 12 + c$ 11. $z - \dfrac{2}{3} = \dfrac{5}{3}$ 12. $\dfrac{5}{4} + x = \dfrac{1}{6}$

EXAMPLE 4   **How Much Is the MP3 Player?**

The total cost to buy an MP3 player, including \$3.60 in sales tax, was \$63.59. To find the price p of the player before tax, solve $p + 3.60 = 63.59$ for p.

Solution

$$p + 3.60 = 63.59$$

Subtract 3.60 from each side: $p + 3.60 - 3.60 = 63.59 - 3.60$

Use the Additive Inverse Property, $a + (-a) = 0$: $p + 0 = 59.99$

Use the Additive Identity Property, $a + 0 = a$: $p = 59.99$

Since $p = 59.99$, the MP3 player costs \$59.99 before tax.   ●

13. The total cost for a used car, including tax, title, and dealer preparation charges of \$1472.25, is \$13,927.25. To find the price p of the car before the extra charges, solve the equation $p + 1472.25 = 13,927.25$ for p.

▶ ❸ **Use the Multiplication Property of Equality to Solve Linear Equations**

A second property that allows us to create equivalent equations is called the *Multiplication Property of Equality*.

In Words

The Multiplication Property of Equality says that when you multiply one side of an equation by a nonzero quantity, you must also multiply the other side by the same nonzero quantity.

Multiplication Property of Equality

The **Multiplication Property of Equality** states that for real numbers a, b, and c, where c does not equal 0,

$$\text{if } a = b, \quad \text{then} \quad ac = bc$$

EXAMPLE 5   **How to Solve a Linear Equation Using the Multiplication Property of Equality**

Solve the equation: $5x = 30$

Step-by-Step Solution

Step 1: *Get the coefficient of the variable x to be 1.* $5x = 30$

Multiply each side of the equation by $\dfrac{1}{5}$: $\dfrac{1}{5}(5x) = \dfrac{1}{5}(30)$

Step 2: Simplify the left and right sides of the equation.

Use the Associative Property of Multiplication: $\left(\dfrac{1}{5} \cdot 5\right)x = \dfrac{1}{5}(30)$

Use the Multiplicative Inverse Property, $a \cdot \dfrac{1}{a} = 1$ $\qquad 1 \cdot x = 6$

Use the Multiplicative Identity Property, $1 \cdot a = a$: $\qquad x = 6$

Step 3: Check Verify the solution.

$$5x = 30$$

Replace x by 6 in the original equation: $\quad 5(6) \overset{?}{=} 30$

$$30 = 30 \quad \text{True}$$

The solution is 6, or the solution set is $\{6\}$.

In Step 1 of Example 5, we multiplied both sides of the equation by $\dfrac{1}{5}$ to make the coefficient of x to equal 1. We could also divide both sides of the equation by 5, because dividing by 5 is the same as multiplying by the reciprocal of 5, $\dfrac{1}{5}$.

$$5x = 30$$

Divide each side of the equation by 5: $\quad \dfrac{5x}{5} = \dfrac{30}{5}$

Simplify: $\qquad x = 6$

Which approach do you prefer?

EXAMPLE 6 **Using the Multiplication Property of Equality**

Solve the equation: $-4n = 18$

Solution

$$-4n = 18$$

Multiply each side of the equation by $-\dfrac{1}{4}$: $\quad -\dfrac{1}{4}(-4n) = -\dfrac{1}{4} \cdot 18$

Use the Associative Property of Multiplication: $\quad \left(-\dfrac{1}{4} \cdot (-4)\right)n = -\dfrac{18}{4}$

Use the Multiplicative Inverse Property; Factor: $\quad 1 \cdot n = -\dfrac{9 \cdot 2}{2 \cdot 2}$

Use the Multiplicative Identity; Divide out common factors: $\quad n = -\dfrac{9}{2}$

Check Verify the solution by replacing $n = -\dfrac{9}{2}$ in the original equation to see whether a true statement results.

$$-4n = 18$$

$$-4\left(-\dfrac{9}{2}\right) \overset{?}{=} 18$$

Divide out the common factor: $\quad \overset{-2}{\cancel{-4}}\left(-\dfrac{9}{\cancel{2}_{1}}\right) \overset{?}{=} 18$

$$18 = 18 \quad \text{True}$$

The solution is $-\dfrac{9}{2}$, or the solution set is $\left\{-\dfrac{9}{2}\right\}$.

14. *The* _____ *Property of Equality states that for real numbers* $a, b,$ *and* $c, c \neq 0,$ *if* $a = b,$ *then* $ac = bc.$

In Problems 15–18, solve each equation using the Multiplication Property of Equality.

15. $8p = 16$ **16.** $-7n = 14$ **17.** $6z = 15$ **18.** $-12b = 28$

EXAMPLE 7   **Solving a Linear Equation with a Fraction as a Coefficient**

Solve the equation: $\dfrac{x}{3} = -7$

Solution

The left side of the equation, $\dfrac{x}{3}$, is equivalent to $\dfrac{1}{3}x$. To eliminate the fraction, we will multiply both sides of the equation by 3, the reciprocal of $\dfrac{1}{3}$.

$$\frac{x}{3} = -7$$

$$\frac{1}{3} \cdot x = -7$$

Multiply both sides of the equation by 3, the reciprocal of $\dfrac{1}{3}$: $3 \cdot \left(\dfrac{1}{3}x\right) = 3 \cdot -7$

Use the Associative Property of Multiplication: $\left(3 \cdot \dfrac{1}{3}\right)x = -21$

Use the Multiplicative Inverse Property, $a \cdot \dfrac{1}{a} = 1$: $1 \cdot x = -21$

Use the Multiplicative Identity, $1 \cdot a = a$: $x = -21$

Check   Let $x = -21$ in the original equation to see whether a true statement results.

$$\frac{x}{3} = -7$$

Let $x = -21$: $\dfrac{-21}{3} \stackrel{?}{=} -7$

$$-7 = -7 \quad \text{True}$$

The solution is -21, or the solution set is $\{-21\}$.   ●

EXAMPLE 8   **Solving a Linear Equation with a Fraction as a Coefficient**

Solve the equation: $12 = \dfrac{2}{3}x$

Solution

$$12 = \frac{2}{3}x$$

Multiply both sides of the equation by $\dfrac{3}{2}$, the reciprocal of $\dfrac{2}{3}$: $\dfrac{3}{2}(12) = \dfrac{3}{2}\left(\dfrac{2}{3}x\right)$

Use the Associative Property of Multiplication: $18 = \left(\dfrac{3}{2} \cdot \dfrac{2}{3}\right)x$

Work Smart

When the variable is on the right side of an equation, we isolate the variable in the same way we did when it was on the left side of the equation.

Use the Multiplicative Inverse Property, $a \cdot \dfrac{1}{a} = 1$: $18 = 1 \cdot x$

Use the Multiplicative Identity Property, $1 \cdot a = a$: $18 = x$

If $b = a$, then $a = b$: $x = 18$

Check Let $x = 18$ in the original equation to see whether a true statement results.

$$12 = \frac{2}{3}x$$

Let $x = 18$: $12 \overset{?}{=} \dfrac{2}{3}(18)$

$$12 \overset{?}{=} \frac{2}{\overset{}{\underset{1}{3}}}(\overset{6}{18})$$

$$12 = 12 \quad \text{True}$$

The solution is 18, or the solution set is $\{18\}$.

> **Quick ✓**
>
> *In Problems 19–21, solve each equation using the Multiplication Property of Equality.*
>
> **19.** $\dfrac{4}{3}n = 12$ **20.** $-21 = \dfrac{7}{3}k$ **21.** $15 = -\dfrac{z}{2}$

EXAMPLE 9 **Solving a Linear Equation with Fractions**

Solve the equation: $\dfrac{4}{5} = -\dfrac{2}{15}p$

Solution

$$\frac{4}{5} = -\frac{2}{15}p$$

Multiply both sides by $-\dfrac{15}{2}$: $-\dfrac{15}{2} \cdot \dfrac{4}{5} = -\dfrac{15}{2} \cdot \left(-\dfrac{2}{15}p\right)$

Simplify; use the Associative Property of Multiplication: $-\dfrac{\overset{3}{15}}{\underset{1}{2}} \cdot \dfrac{\overset{2}{4}}{\underset{1}{5}} = \left(-\dfrac{15}{2} \cdot \left(-\dfrac{2}{15}\right)\right)p$

$$-6 = p$$

Check

$$\frac{4}{5} = -\frac{2}{15}p$$

Let $p = -6$ in the original equation: $\dfrac{4}{5} \overset{?}{=} -\dfrac{2}{15}(-6)$

Simplify: $\dfrac{4}{5} \overset{?}{=} -\dfrac{2}{\underset{5}{15}}(\overset{-2}{-6})$

$$\frac{4}{5} = \frac{4}{5} \quad \text{True}$$

The solution is -6, or the solution set is $\{-6\}$.

Working with fractional coefficients can be tricky. The three equations $\dfrac{-x}{3} = 7$, $-\dfrac{1}{3}x = 7$, and $\dfrac{x}{-3} = 7$ are all equivalent. Do you see why?

Work Smart

To solve an equation of the form $ax = b$, where a is an integer, either multiply by the reciprocal of a or divide by a.

To solve an equation of the form $ax = b$, where a is a fraction, multiply by the reciprocal of a.

Multiply by the reciprocal of a

$$2x = 7 \qquad 5x = -\frac{15}{2}$$

$$\frac{1}{2}(2x) = \frac{1}{2} \cdot 7 \qquad \frac{1}{5}(5x) = \frac{1}{5}\left(-\frac{15}{2}\right)$$

$$x = \frac{7}{2} \qquad x = -\frac{3}{2}$$

Divide by a

$$-3x = 72$$

$$\frac{-3x}{-3} = \frac{72}{-3}$$

$$x = -24$$

Multiply by the reciprocal of a

$$\frac{4}{5}x = -16$$

$$\frac{5}{4}\left(\frac{4}{5}x\right) = \frac{5}{4}(-16)$$

$$x = \frac{5}{4} \cdot \frac{\overset{-4}{\cancel{-16}}}{1}$$

$$x = -20$$

Work Smart: Study Skills

Selected problems in the exercise sets are identified by a green color. For extra help, worked solutions to these problems are in MyMathLab.

Quick ✓

In Problems 22–24, solve each equation using the Multiplication Property of Equality.

22. $\dfrac{3}{8}b = \dfrac{9}{4}$

23. $-\dfrac{4}{9} = \dfrac{-t}{6}$

24. $\dfrac{1}{4} = -\dfrac{7}{10}m$

2.1 Exercises

MyMathLab® Math XP PRACTICE

Exercise numbers in green have complete video solutions in MyMathLab

Problems 1–24 are the Quick ✓s that follow the EXAMPLES.

Building Skills

In Problems 25–32, determine whether the given value is a solution to the equation. Answer Yes or No. See Objective 1.

25. $3x - 1 = 5; x = 2$

26. $4t + 2 = 16; t = 3$

27. $4 - (m + 2) = 3(2m - 1); m = 1$

28. $3(x + 1) - x = 5x - 9; x = -3$

29. $8k - 2 = 4; k = \dfrac{3}{4}$

30. $-15 = 3x - 16; x = \dfrac{1}{3}$

31. $r + 1.6 = 2r + 1; r = 0.6$

32. $3s - 6 = 6s - 3.4; s = -1.2$

In Problems 33–48, solve the equation using the Addition Property of Equality. Be sure to check your solution. See Objective 2.

33. $x - 9 = 11$

34. $y - 8 = 2$

35. $x + 4 = -8$

36. $r + 3 = -1$

37. $12 = n - 7$

38. $13 = u - 6$

39. $-8 = x + 5$

40. $-2 = y + 13$

41. $x - \dfrac{2}{3} = \dfrac{4}{3}$

42. $x - \dfrac{1}{8} = \dfrac{3}{8}$

43. $z + \dfrac{1}{2} = \dfrac{3}{4}$

44. $n + \dfrac{3}{5} = \dfrac{7}{10}$

45. $\dfrac{5}{12} = x - \dfrac{3}{8}$

46. $\dfrac{3}{8} = y - \dfrac{1}{6}$

47. $w + 3.5 = -2.6$

48. $z + 4.9 = -2.6$

In Problems 49–72, solve the equation using the Multiplication Property of Equality. Be sure to check your solution. See Objective 3.

49. $5c = 25$

50. $8b = 48$

51. $-7n = 28$

52. $-8s = 40$

53. $4k = 14$

54. $4z = 30$

55. $-6w = 15$

56. $-8p = 20$

57. $\dfrac{5}{3}a = 35$

58. $\dfrac{4}{3}b = 16$

59. $-\dfrac{3}{11}p = -33$

60. $-\dfrac{6}{5}n = -36$

61. $\dfrac{n}{5} = 8$

62. $-\dfrac{y}{6} = 3$

63. $\dfrac{6}{5} = 2x$

64. $\dfrac{9}{2} = 3b$

65. $5y = -\dfrac{5}{3}$

66. $4r = -\dfrac{12}{5}$

67. $\dfrac{1}{2}m = \dfrac{9}{2}$

68. $\dfrac{1}{4}w = \dfrac{7}{2}$

69. $-\dfrac{3}{8}t = \dfrac{1}{6}$

70. $\dfrac{3}{10}q = -\dfrac{1}{6}$

71. $\dfrac{5}{24} = -\dfrac{y}{8}$

72. $\dfrac{11}{36} = -\dfrac{t}{9}$

Mixed Practice

In Problems 73–100, solve the equation. Be sure to check your solution.

73. $n - 4 = -2$

74. $m - 6 = -9$

75. $b + 12 = 9$

76. $c + 4 = 1$

77. $2 = 3x$

78. $9 = 5y$

79. $-4q = 24$

80. $-6m = 54$

81. $-39 = x - 58$

82. $-637 = c - 142$

83. $-18 = -301 + x$

84. $-46 = -51 + q$

85. $-\dfrac{x}{5} = -10$

86. $\dfrac{z}{3} = -12$

87. $m - 56.3 = -15.2$

88. $p - 26.4 = -471.3$

89. $-40 = -6c$

90. $-45 = 12x$

91. $14 = -\dfrac{7}{2}c$

92. $12 = -\dfrac{3}{2}n$

93. $\dfrac{3}{4} = -\dfrac{x}{16}$

94. $\dfrac{5}{9} = -\dfrac{h}{36}$

95. $x - \dfrac{5}{16} = \dfrac{3}{16}$

96. $w - \dfrac{7}{20} = \dfrac{9}{20}$

97. $-\dfrac{3}{16} = -\dfrac{3}{8} + z$

98. $\dfrac{5}{2} = -\dfrac{13}{4} + y$

99. $\dfrac{5}{6} = -\dfrac{2}{3}z$

100. $-\dfrac{4}{9} = \dfrac{8}{3}b$

Applying the Concepts

101. New Car The total cost for a new car, including tax, title, and dealer preparation charges of $2722.50, is $27,685.77. To find the price of the car without the extra charges, solve the equation $y + 2722.50 = 27,685.77$, where y represents the price of the car without the extra charges.

102. New Kayak The total cost for a new kayak is $862.92, including sales tax of $63.92. To find the cost of the kayak without tax, solve the equation $k + 63.92 = 862.92$, where k represents the cost of the kayak.

103. Discount The cost of a sleeping bag has been discounted by $17, so that the sale price of the bag is $51. Find the original price of the sleeping bag, p, by solving the equation $p - 17 = 51$.

104. Discount The cost of a computer has been discounted by $239, so that the sale price is $1230. Find the original price of the computer, c, by solving the equation $c - 239 = 1230$.

105. Eating Out Rebecca purchased several "Happy Meals" at McDonald's for her child's play group. Each Happy Meal costs $4 and she spent a total of $48 on the food. To determine the number of Happy Meals, h, she purchased, solve the equation $4h = 48$.

106. Paperback Books Anne bought 3 paperback books to read on the flight to Europe. She paid $36 for the books (without sales tax). To find the average price, p, of each book, solve the equation $3p = 36$.

107. Interest Suppose you have a credit card debt of $3,000. Last month, the bank charged you $45 interest on the debt. The solution to the equation $45 = \dfrac{3000}{12} \cdot r$ represents the annual interest rate, r, on the credit card. Find the annual interest rate on the credit card.

108. Interest Suppose you have a credit card debt of $4,000. Last month, the bank charged you $40 interest on the debt. The solution to the equation $40 = \dfrac{4000}{12} \cdot r$ represents the annual interest rate, r, on the credit card. Find the annual interest rate on the credit card.

Extending the Concepts

109. Let λ represent some real number. Solve the equation $x + \lambda = 48$ for x.

110. Let β represent some real number. Solve the equation $x - \beta = 25$ for x.

111. Let θ represent some real number except 0. Solve the equation $14 = \theta x$ for x.

112. Let ψ represent some real number except 0. Solve the equation $\dfrac{2}{5} = \psi x$ for x.

113. Find the value of λ in the equation $x + \lambda = \dfrac{16}{3}$ so that the solution is $x = -\dfrac{2}{9}$.

114. Find the value of β in the equation $x - \beta = -13.6$ so that the solution is $x = -4.79$.

115. Find the value of θ in the equation $-\dfrac{3}{4} = \theta x$ so that the solution is $x = \dfrac{7}{8}$.

116. Find the value of ψ in the equation $-11.92 = \psi x$ so that the solution is $x = 2.98$.

Explaining the Concepts

117. Explain what is meant by finding the *solution* of an equation.

118. Consider the equation $3z = \dfrac{15}{4}$. You could either multiply both sides of the equation by $\dfrac{1}{3}$ or divide both sides by 3. Explain which operation you would choose and why you think it is easier.

119. Explain the difference between an algebraic expression and an algebraic equation. Write an equation involving the algebraic expression $x - 10$. Then solve the equation.

120. A classmate suggests that to solve the equation $4x = 2$, you must divide by 4 and the result will be $x = 2$. Explain why this reasoning is or is not correct.

121. Explain the Addition Property of Equality. How is it used to solve $x - 5 = 12$? What role does the Identity Property of Addition play in solving this equation?

2.2 Linear Equations: Using the Properties Together

Objectives

❶ Use the Addition and the Multiplication Properties of Equality to Solve Linear Equations

❷ Combine Like Terms and Use the Distributive Property to Solve Linear Equations

❸ Solve a Linear Equation with the Variable on Both Sides of the Equation

❹ Use Linear Equations to Solve Problems

Ready?...Answers **R1.** $10 - 3x$
R2. -3

Are You Ready for This Section?

Before getting started, take this readiness quiz. If you get a problem wrong, go back to the section cited and review the material.

R1. Simplify by combining like terms: $6 - (4 + 3x) + 8$ [Section 1.8, pp. 68–69]

R2. Evaluate the expression $2(3x + 4) - 5$ for $x = -1$.   [Section 1.8, pp. 64–65]

❶ Use the Addition and the Multiplication Properties of Equality to Solve Linear Equations

In the last section, we solved equations such as $x + 3 = 7$ and $\dfrac{1}{2}z = 8$ using the Addition Property or the Multiplication Property of Equality, but not both. We now look at equations that require using both the Addition and Multiplication Properties of Equality to solve for the variable. For example, the equation $2x + 3 = 7$ can be read as "Two *times* a number x *plus* three equals seven." We must "undo" both the multiplication and the addition to find the solution of the equation.

EXAMPLE 1   **How to Solve a Linear Equation Using the Addition and Multiplication Properties of Equality**

Solve the equation: $2z - 3 = 9$

Step-by-Step Solution

Step 1: Isolate the term containing the variable.   Use the Addition Property of Equality by adding 3 to each side of the equation:

$$2z - 3 = 9$$
$$2z - 3 + 3 = 9 + 3$$
$$2z = 12$$

Step 2: Get the coefficient of the variable to be 1.   Use the Multiplication Property of Equality by dividing both sides of the equation by 2 $\left(\text{or multiply both sides by } \dfrac{1}{2}\right)$:

$$\frac{2z}{2} = \frac{12}{2}$$
$$z = 6$$

(continued)

Step 3: Check Verify that $z = 6$ is the solution of the equation.

$$2z - 3 = 9$$

Substitute 6 for z into the original equation: $2(6) - 3 \stackrel{?}{=} 9$

$$12 - 3 \stackrel{?}{=} 9$$

$$9 = 9 \quad \text{True}$$

Because $z = 6$ satisfies the equation, the solution is 6, or the solution set is $\{6\}$. ●

EXAMPLE 2 **Solving a Linear Equation Using the Addition and Multiplication Properties of Equality**

Solve the equation: $\dfrac{3}{2}p + 3 = 12$

Solution

$$\frac{3}{2}p + 3 = 12$$

Subtract 3 from both sides of the equation: $\dfrac{3}{2}p + 3 - 3 = 12 - 3$

Simplify: $\dfrac{3}{2}p = 9$

Multiply both sides of the equation by $\dfrac{2}{3}$: $\dfrac{2}{3}\left(\dfrac{3}{2}p\right) = \dfrac{2}{3}(9)$

Simplify: $p = 6$

Check

$$\frac{3}{2}p + 3 = 12$$

Let $p = 6$ in the original equation: $\dfrac{3}{2}(6) + 3 \stackrel{?}{=} 12$

Simplify: $9 + 3 \stackrel{?}{=} 12$

$$12 = 12 \quad \text{True}$$

The solution is $p = 6$, or the solution set is $\{6\}$. ●

Quick ✓

1. *True or False* To solve the equation $2x - 11 = 40$, the first step is to add 11 to each side of the equation.

In Problems 2–5, solve each equation.

2. $5x - 4 = 11$

3. $8 = \dfrac{2}{3}k - 4$

4. $8 - 5r = -2$

5. $-\dfrac{3}{2}n + 2 = -\dfrac{1}{4}$

▶ ❷ **Combine Like Terms and Use the Distributive Property to Solve Linear Equations**

Often, we must combine like terms before we can use the Addition or Multiplication Property of Equality.

EXAMPLE 3 **Combining Like Terms to Solve a Linear Equation**

Solve the equation: $2x - 6 + 3x = 14$

Solution

$$2x - 6 + 3x = 14$$

Combine like terms: $5x - 6 = 14$

Add 6 to each side of the equation: $5x - 6 + 6 = 14 + 6$

$$5x = 20$$

Divide both sides by 5: $\dfrac{5x}{5} = \dfrac{20}{5}$

$$x = 4$$

Check

$$2x - 6 + 3x = 14$$

Substitute 4 for x in the original equation: $2(4) - 6 + 3(4) \overset{?}{=} 14$

$$8 - 6 + 12 \overset{?}{=} 14$$

$$14 = 14 \quad \text{True}$$

The solution of the equation is 4, or the solution set is $\{4\}$.

Quick ✔

In Problems 6–9, solve each equation.

6. $7b - 3b + 3 = 11$

7. $-3a + 4 + 4a = 13 - 27$

8. $6c - 2 + 2c = 18$

9. $-12 = 5x - 3x + 4$

When an equation contains parentheses, we use the Distributive Property to remove the parentheses before we use the Addition or Multiplication Property of Equality.

EXAMPLE 4 **Solving a Linear Equation Using the Distributive Property**

Solve the equation: $4(2x + 3) - 7 = -11$

Solution

$$4(2x + 3) - 7 = -11$$

Use the Distributive Property to remove parentheses: $8x + 12 - 7 = -11$

Combine like terms: $8x + 5 = -11$

Subtract 5 from each side of the equation: $8x + 5 - 5 = -11 - 5$

$$8x = -16$$

Divide both sides by 8: $\dfrac{8x}{8} = \dfrac{-16}{8}$

$$x = -2$$

Check

$$4(2x + 3) - 7 = -11$$

Substitute -2 for x in the original equation: $4[2(-2) + 3] - 7 \overset{?}{=} -11$

$$4(-4 + 3) - 7 \overset{?}{=} -11$$

$$4(-1) - 7 \overset{?}{=} -11$$

$$-4 - 7 \overset{?}{=} -11$$

$$-11 = -11 \quad \text{True}$$

The solution of the equation is -2, or the solution set is $\{-2\}$.

Quick ✔

In Problems 10–14, solve each equation.

10. $2(y + 5) - 3 = 11$

11. $\frac{1}{2}(4 - 6x) + 5 = 3$

12. $4 - (6 - x) = 11$

13. $8 + \frac{2}{3}(2n - 9) = 10$

14. $\frac{1}{3}(2x + 9) + \frac{x}{3} = 5$

▶ ❸ Solve a Linear Equation with the Variable on Both Sides of the Equation

When solving a linear equation, our goal is to get the terms that contain the variable on one side of the equation and the constant terms on the other side.

EXAMPLE 5 **Solving a Linear Equation with the Variable on Both Sides of the Equation**

Solve the equation: $9y - 5 = 5y + 9$

Solution

$$9y - 5 = 5y + 9$$

Subtract 5y from each side of the equation: $9y - 5 - 5y = 5y + 9 - 5y$

$$4y - 5 = 9$$

Add 5 to each side of the equation: $4y - 5 + 5 = 9 + 5$

$$4y = 14$$

Divide both sides by 4: $\dfrac{4y}{4} = \dfrac{14}{4}$

Simplify: $y = \dfrac{7}{2}$

Check

$$9y - 5 = 5y + 9$$

Substitute $\frac{7}{2}$ for y in the original equation: $9\left(\dfrac{7}{2}\right) - 5 \overset{?}{=} 5\left(\dfrac{7}{2}\right) + 9$

$$\dfrac{63}{2} - 5 \overset{?}{=} \dfrac{35}{2} + 9$$

$$\dfrac{63}{2} - \dfrac{10}{2} \overset{?}{=} \dfrac{35}{2} + \dfrac{18}{2}$$

$$\dfrac{53}{2} = \dfrac{53}{2} \qquad \text{True}$$

The solution of the equation is $\dfrac{7}{2}$, or the solution set is $\left\{\dfrac{7}{2}\right\}$.

●

Quick ✔

In Problems 15 and 16, solve each equation.

15. $3x + 4 = 5x - 8$

10. $10m + 3 = 6m - 11$

EXAMPLE 6 **How to Solve a Linear Equation in One Variable**

Solve the equation: $2(z - 4) + 3z = 4 - (z + 2)$

Step-by-Step Solution

Step 1: Remove any parentheses using the Distributive Property.

$$2(z - 4) + 3z = 4 - (z + 2)$$
$$2z - 8 + 3z = 4 - z - 2$$

Step 2: Combine like terms on each side of the equation.

$$5z - 8 = 2 - z$$

Step 3: Use the Addition Property of Equality to get the terms with the variable on one side of the equation and the constants on the other side.

Add z to both sides of the equation: $\quad 5z - 8 + z = 2 - z + z$

Simplify: $\quad 6z - 8 = 2$

Add 8 to both sides of the equation: $\quad 6z - 8 + 8 = 2 + 8$

Simplify: $\quad 6z = 10$

Step 4: Use the Multiplication Property of Equality to get the coefficient of the variable to be 1.

Divide both sides of the equation by 6: $\quad \dfrac{6z}{6} = \dfrac{10}{6}$

Simplify: $\quad z = \dfrac{5}{3}$

Step 5: Check the solution to verify that it satisfies the original equation.

We leave the check to you.

The solution of the equation is $\dfrac{5}{3}$, or the solution set is $\left\{ \dfrac{5}{3} \right\}$.

Quick ✓

17. *True or False* To solve the equation $13 - 2(7x + 1) + 8x = 12$, the first step is to subtract 2 from 13 and get $11(7x + 1) + 8x = 12$.

In Problems 18 and 19, solve each equation.

18. $-9x + 3(2x - 3) = -10 - 2x$ $\qquad$ **19.** $3 - 4(p + 5) = 5(p + 2) - 12$

We now summarize the steps that you should follow to solve an equation in one variable. Not all the steps may be necessary to solve every equation, but you should use this summary as a guide.

Solving an Equation in One Variable

Step 1: Remove any parentheses using the Distributive Property.

Step 2: Combine like terms on each side of the equation.

Step 3: Use the Addition Property of Equality to get the terms with the variable on one side of the equation and the constants on the other side.

Step 4: Use the Multiplication Property of Equality to get the coefficient of the variable to be 1.

Step 5: Check the solution to verify that it satisfies the original equation.

4 Use Linear Equations to Solve Problems

We continue to solve problems that are modeled by algebraic equations.

EXAMPLE 7 **How Much Does Alejandro Make in an Hour?**

Alejandro, a hospital engineer, earns double time on all hours worked in excess of 40 hours each week. One week, Alejandro worked 46 hours and earned $1326 before taxes. To determine Alejandro's hourly wage, w, solve the equation $40w + 6(2w) = 1326$.

Solution

$$40w + 6(2w) = 1326$$

Simplify: $\quad 40w + 12w = 1326$

Combine like terms: $\quad 52w = 1326$

Divide both sides of the equation by 52: $\quad \dfrac{52w}{52} = \dfrac{1326}{52}$

Simplify: $\quad w = 25.5$

Alejandro makes $25.50 per hour.

Work Smart: Study Skills

Selected problems in the exercise sets are identified by a green color. For extra help, worked solutions to these problems are in MyMathLab.

Quick ✓

20. Marcella works at a clothing store. Whenever she works more than 40 hours in a week, she gets paid twice her regular hourly wage of $10 per hour. One week, Marcella earned $640 before taxes. To determine how many hours, h, Marcella worked, solve the equation $400 + 20(h - 40) = 640$ for h.

2.2 Exercises MyMathLab® MathXL PRACTICE

Exercise numbers in green have complete video solutions in MyMathLab

Problems 1–20 are the **Quick ✓**s that follow the **EXAMPLES**.

Building Skills

In Problems 21–34, solve the equation. Check your solution. See Objective 1.

21. $3x + 4 = 7$

22. $5t + 1 = 11$

23. $2y - 1 = -5$

24. $6z - 2 = -8$

25. $-3p + 1 = 10$

26. $-4x + 3 = 15$

27. $8y + 3 = 15$

28. $6z - 7 = 3$

29. $5 - 2z = 11$

30. $1 - 3k = 4$

31. $\dfrac{2}{3}x + 1 = 9$

32. $\dfrac{5}{4}a + 3 = 13$

33. $\dfrac{7}{2}y - 1 = 13$

34. $\dfrac{1}{5}p - 3 = 2$

In Problems 35–46, solve the equation. Check your solution. See Objective 2.

35. $3x - 7 + 2x = -17$

36. $5r + 2 - 3r = -14$

37. $2k - 7k - 8 = 17$

38. $2b + 5 - 8b = 23$

39. $2(x + 1) + 5 = -9$

40. $3(t - 4) + 7 = -11$

41. $-3(2 + r) = 9$

42. $-5(6 + z) = -20$

43. $17 = 2 - (n + 6)$

44. $21 = 5 - (2a - 1)$

45. $-8 = 5 - (7 - z)$

46. $-9 = 3 - (6 + 4y)$

In Problems 47–60, solve the equation. Check your solution. See Objective 3.

47. $2x + 9 = x + 1$

48. $7z + 13 = 6z + 8$

49. $2t - 6 = 3 - t$

50. $3 + 8x = 21 - x$

51. $14 - 2n = -4n + 7$

52. $6 - 12m = -3m + 3$

53. $-3(5 - 3k) = 6k + 6$

54. $-4(10 - 7x) = 3x + 10$

55. $2(2x + 3) = 3(x - 4)$

56. $3(5 + x) = 2(2x + 11)$

57. $13 + 3(x - 1) = 9x$

58. $-8 + 4(p + 6) = 10p$

59. $9(6 + a) + 33a = 10a$

60. $5(12 - 3w) + 25w = 2w$

Mixed Practice

In Problems 61–76, solve the equation. Check your solution.

61. $-5x + 11 = 1$ **62.** $-6n + 14 = -10$

63. $4m + 5 = 2$ **64.** $7x + 1 = -9$

65. $-2(3n - 2) = 2$ **66.** $-5(2n - 3) = 10$

67. $4k - (3 + k) = -(2k + 3)$

68. $11w - (2 - 4w) = 13 + 2(w - 1)$

69. $2y + 36 = 6 + 6y$

70. $7a - 26 = 13a + 2$

71. $\frac{1}{2}(-4k + 28) = 6 + 14k$

72. $\frac{2}{3}(9a - 12) = -6a - 11$

73. $-\frac{5}{2}(x + 6) + \frac{3}{2}x = -8$

74. $\frac{4}{3}a - \left(\frac{7}{3}a + 6\right) = -15$

75. $-3(2y + 3) - 1 = -4(y + 6) + 2y$

76. $-5(b + 2) + 3b = -2(1 + 5b) + 6$

Applying the Concepts

77. Burger King A Burger King Tendercrisp Chicken garden salad contains 6 more grams of fat than a McDonald's Southwestern salad with crispy chicken. If there are 38 grams of fat in the two salads, find the number of grams of fat in each salad by solving the equation $x + (x + 6) = 38$, where x represents the number of fat grams in the McDonald's Southwestern salad and $x + 6$ represents the number of fat grams in the Burger King Tendercrisp salad. (SOURCE: *Burger King and McDonald's websites*)

78. Wendy's A Wendy's Mandarin Chicken salad contains 10 grams of fat less than a Taco Bell Zesty Chicken Border Bowl. If there are 60 grams of fat in the two salads, find the number of grams of fat in the Taco Bell Zesty Chicken Border Bowl by solving the equation $x + (x - 10) = 60$, where x represents the number of grams of fat in a Taco Bell Border Bowl and $x - 10$ represents the number of grams of fat in the Wendy's Mandarin Chicken salad. (SOURCE: *Wendy's and Taco Bell websites*)

△**79. Dog Run** The length of a rectangular dog run is 2 feet more than twice the width, w, and the perimeter of the dog run is 30 feet. Solve the equation $2w + 2(2w + 2) = 30$ to find the width,

w, of the dog run. Then find the length, $2w + 2$, of the dog run.

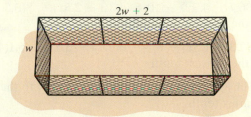

△**80. Garden** The width of a rectangular garden is 1 yard more than one-half the length, L. The perimeter of the garden is 26 yards. Find the length, L, of the garden by solving the equation $2L + 2\left(\frac{1}{2}L + 1\right) = 26$.

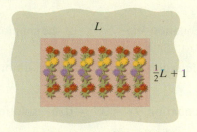

81. Overtime Pay Jennifer worked 44 hours last week, including 4 hours of overtime, and earned $552. She is paid at a rate of 1.5 times her regular hourly rate for overtime hours. Solve the equation $40x + 4(1.5x) = 552$ to find her regular hourly pay rate, x.

82. Overtime Pay Juan worked a total of 50 hours last week and earned $855. He earned 1.5 times his regular hourly rate for 6 hours, and double his hourly rate for 4 holiday hours. Solve the equation $40x + 6(1.5x) + 4(2x) = \$855$, where x is Juan's regular hourly rate.

△**83. Hanging Wallpaper** Becky purchased a remnant of 42 feet of a wallpaper border to hang in a rectangular bedroom. She knows that one wall of the bedroom is 5 feet longer than the other wall. Let x represent the length of the shorter wall. Assuming the length of the shorter wall is 8 feet, does Becky have enough wallpaper to hang the border? Use the equation $2x + 2(x + 5) = 42$ to answer the question.

△**84. Perimeter of a Triangle** The perimeter of a triangle is 210 inches. If the sides are made up of 3 consecutive even integers, find the lengths of all 3 sides by solving the equation $x + (x + 2) + (x + 4) = 210$, where x represents the length of the shortest side.

Extending the Concepts

In Problems 85 and 86, solve the equation. Check your solution.

85. $8[4 - 6(x - 1)] + 5[(2x + 3) - 5] = 18x - 338$

86. $3[10 - 4(x - 3)] + 2[(3x + 6) - 2] = 2x + 360$

In Problems 87–90, use a calculator to solve the equation, and round your answer to the indicated place.

87. $3(36.7 - 4.3x) - 10 = 4(10 - 2.5x) - 8(3.5 - 4.1x)$ to the nearest hundredth

88. $12(2.3 - 1.5x) - 6 = -3(18.4 - 3.5x) - 6.1(4x + 3)$ to the nearest tenth

89. $3.5\{4 - [6 - (2x + 3)] + 5\} = -18.4$ to the nearest tenth

90. $9\{3 - [4(2.3z - 1)] + 6.5\} = -406.3$ to the nearest hundredth

In Problems 91–94, determine the value of d to make the statement true.

91. In the equation $3d + 2x = 12$, the solution is -4.

92. In the equation $5d - 2x = -2$, the solution is -6.

93. In the equation $\frac{2}{3}x - d = 1$, the solution is $-\frac{3}{8}$.

94. In the equation $\frac{2}{5}x + 3d = 0$, the solution is $\frac{15}{8}$.

Explaining the Concepts

95. Explain the difference between $6x - 2(x + 1)$ and $6x - 2(x + 1) = 6$. In general, what is the difference between an algebraic expression and an algebraic equation?

96. Explain the Addition and Multiplication Properties of Equality. In the Multiplication Property of Equality, why do you think we cannot multiply both sides of the equation by 0?

97. A classmate begins to solve the equation $7x + 3 - 2x = 9x - 5$ by adding $2x$ in the following manner:

$$7x + 3 - 2x = 9x - 5$$
$$\underline{+ 2x \qquad\qquad + 2x}$$

Will this lead to the correct solution? Why or why not? Write an explanation telling the steps you would use to solve this equation.

98. A *corollary* is a rule or theorem that is closely related to a previous rule. Write a corollary to the Addition Property of Equality and title it the Subtraction Property of Equality. Write a corollary to the Multiplication Property of Equality, called the Division Property of Equality. What restrictions would you place on the Division Property of Equality? Why do you think these properties were not included in the text?

2.3 Solving Linear Equations Involving Fractions and Decimals; Classifying Equations

Objectives

1. Use the Least Common Denominator to Solve a Linear Equation Containing Fractions
2. Solve a Linear Equation Containing Decimals
3. Classify a Linear Equation as an Identity, a Conditional Equation, or a Contradiction
4. Use Linear Equations to Solve Problems

Work Smart

Although removing fractions is not required to solve an equation, it frequently makes the arithmetic easier.

Are You Ready for This Section?

Before getting started, take this readiness quiz. If you get a problem wrong, go back to the section cited and review the material.

R1. Find the LCD of $\frac{3}{5}$ and $\frac{3}{4}$. [Section 1.2, pp. 10–12]

R2. Find the LCD of $\frac{3}{8}$ and $-\frac{7}{12}$. [Section 1.2, pp. 10–12]

1 Use the Least Common Denominator to Solve a Linear Equation Containing Fractions

Sometimes equations are easier to solve if they do not contain fractions. To solve a linear equation containing fractions, we can multiply each side of the equation by the least common denominator (LCD) to rewrite the equation without fractions. Recall that the LCD is the smallest number that each denominator has as a common multiple. The Multiplication Property of Equality allows us to multiply both sides of the equation by the LCD.

Ready?...Answers **R1.** 20 **R2.** 24

EXAMPLE 1 **How to Solve a Linear Equation That Contains Fractions**

Solve the linear equation: $\dfrac{1}{2}x + \dfrac{2}{3}x = \dfrac{14}{3}$

Step-by-Step Solution

Before following the steps in Section 2.2 on page 95, we can rewrite the equation as an equivalent equation by multiplying both sides of the equation by 6, the LCD of $\dfrac{1}{2}, \dfrac{2}{3},$ and $\dfrac{14}{3}$.

$$6\left(\dfrac{1}{2}x + \dfrac{2}{3}x\right) = 6\left(\dfrac{14}{3}\right)$$

Now we can follow Steps 1–5 from the summary in Section 2.2 to solve the equation.

Step 1: Apply the Distributive Property to remove parentheses.

$$6\left(\dfrac{1}{2}x + \dfrac{2}{3}x\right) = 6\left(\dfrac{14}{3}\right)$$

Use the Distributive Property: $6\left(\dfrac{1}{2}x\right) + 6\left(\dfrac{2}{3}x\right) = 6\left(\dfrac{14}{3}\right)$

$$3x + 4x = 28$$

Step 2: Combine like terms.

$$7x = 28$$

Step 3: Use the Addition Property of Equality to get the terms with the variable on one side of the equation and the constants on the other side.

This step is not necessary because in this problem, the only variable term is already on one side of the equation, and the constant is on the other side of the equation.

Step 4: Get the coefficient of the variable to be 1. Divide both sides of the equation by 7:

$$\dfrac{7x}{7} = \dfrac{28}{7}$$
$$x = 4$$

Step 5: Check Verify that $x = 4$ is the solution of the equation.

$$\dfrac{1}{2}x + \dfrac{2}{3}x = \dfrac{14}{3}$$

Substitute 4 for x in the original equation:

$$\dfrac{1}{2}(4) + \dfrac{2}{3}(4) \overset{?}{=} \dfrac{14}{3}$$
$$2 + \dfrac{8}{3} \overset{?}{=} \dfrac{14}{3}$$
$$\dfrac{6}{3} + \dfrac{8}{3} \overset{?}{=} \dfrac{14}{3}$$
$$\dfrac{14}{3} = \dfrac{14}{3} \quad \text{True}$$

The solution of the equation is 4, or the solution set is $\{4\}$.

Quick ✔

1. To solve a linear equation containing fractions, we can multiply each side of the equation by the ___ _____ _____ to rewrite the equation without fractions.

In Problems 2 and 3, solve each equation by multiplying by the LCD.

2. $\dfrac{2}{5}x - \dfrac{1}{4}x = \dfrac{3}{2}$

3. $\dfrac{5}{6}x + \dfrac{2}{9} = -\dfrac{1}{3}x - \dfrac{5}{18}$

EXAMPLE 2 **Solving a Linear Equation Using the LCD**

Solve the linear equation: $\dfrac{7n + 5}{8} = 2 + \dfrac{3n + 15}{10}$

Solution

Because the equation contains fractions, we multiply both sides of the equation by 40, the LCD.

$$\frac{7n + 5}{8} = 2 + \frac{3n + 15}{10}$$

Multiply both sides by the LCD, 40:
$$40\left(\frac{7n + 5}{8}\right) = 40\left(2 + \frac{3n + 15}{10}\right)$$

Use the Distributive Property:
$$40\left(\frac{7n + 5}{8}\right) = 40(2) + 40\left(\frac{3n + 15}{10}\right)$$

Divide out common factors:
$$\overset{5}{40}\left(\frac{7n + 5}{\underset{1}{8}}\right) = 40(2) + \overset{4}{40}\left(\frac{3n + 15}{\underset{1}{10}}\right)$$

Use Distributive Property to remove parentheses:
$$5(7n + 5) = 40(2) + 4(3n + 15)$$
$$35n + 25 = 80 + 12n + 60$$

Combine like terms:
$$35n + 25 = 140 + 12n$$

Isolate terms containing n:
$$35n + 25 - 12n = 140 + 12n - 12n$$
$$23n + 25 = 140$$

Isolate the constants:
$$23n + 25 - 25 = 140 - 25$$
$$23n = 115$$

Divide both sides by 23:
$$\frac{23n}{23} = \frac{115}{23}$$
$$n = 5$$

Check

$$\frac{7n + 5}{8} = 2 + \frac{3n + 15}{10}$$

Substitute 5 for n in the original equation:
$$\frac{7(5) + 5}{8} \overset{?}{=} 2 + \frac{3(5) + 15}{10}$$
$$\frac{35 + 5}{8} \overset{?}{=} 2 + \frac{15 + 15}{10}$$
$$\frac{40}{8} \overset{?}{=} 2 + \frac{30}{10}$$
$$5 \overset{?}{=} 2 + 3$$
$$5 = 5 \quad \text{True}$$

The solution is 5, or the solution set is $\{5\}$. ●

Quick ✓

4. The result of multiplying the equation $\dfrac{1}{5}x + 7 = \dfrac{3}{10}$ by 10, the LCD of $\dfrac{1}{5}$ and $\dfrac{3}{10}$, is $2x + \underline{} = 3$.

In Problems 5 and 6, solve each equation by multiplying by the LCD.

5. $\dfrac{2a + 3}{4} + 3 = \dfrac{3a + 19}{8}$

6. $\dfrac{3x - 2}{4} - 1 = \dfrac{6}{5}x$

Section 2.3 Solving Linear Equations Involving Fractions and Decimals; Classifying Equations **101**

▶ ❷ Solve a Linear Equation Containing Decimals

When decimals occur in a linear equation, we can rewrite the equation as an equivalent equation without decimals by multiplying both sides of the equation by a power of 10 so that the decimal is cleared. For example, $0.8x$ is equivalent to $\dfrac{8}{10}x$, so multiplying by 10 would clear the decimal from the term. Because 0.34 is equivalent to $\dfrac{34}{100}$, multiplying by 100 would clear the decimal from the term.

EXAMPLE 3 | **Solving a Linear Equation with a Decimal Coefficient**

Solve the equation: $0.3x - 4 = 11$

Solution

To rewrite the equation as an equivalent equation without decimals, we can multiply both sides of the equation by 10. Do you see why? Multiplying $0.3x = \dfrac{3}{10}x$ by 10 will clear the decimal.

$$0.3x - 4 = 11$$

Multiply both sides of the equation by 10: $\quad 10 \cdot (0.3x - 4) = 10 \cdot 11$

Distribute: $\quad 10 \cdot 0.3x - 10 \cdot 4 = 110$

$$3x - 40 = 110$$

Add 40 to both sides of the equation: $\quad 3x = 150$

Divide both sides of the equation by 3: $\quad x = 50$

Check

$$0.3x - 4 = 11$$

Let $x = 50$ in the original equation: $\quad 0.3(50) - 4 \stackrel{?}{=} 11$

$$15 - 4 \stackrel{?}{=} 11$$

$$11 = 11 \quad \text{True}$$

The solution is 50, or the solution set is $\{50\}$. ●

It's not a requirement to clear fractions or decimals before solving an equation. Here is another way to solve $0.3x - 4 = 11$.

$$0.3x - 4 = 11$$

Add 4 to each side: $\quad 0.3x - 4 + 4 = 11 + 4$

$$0.3x = 15$$

Divide each side by 0.3: $\quad \dfrac{0.3x}{0.3} = \dfrac{15}{0.3}$

$$x = 50$$

Which method do you prefer?

Quick ✓

7. To clear the decimals in the equation $0.25x + 5 = 7 - 0.3x$, multiply both sides of the equation by ___.

In Problems 8 and 9, solve each equation.

8. $0.2z = 20$ **9.** $0.15p - 2.5 = 5$

EXAMPLE 4 **Solving a Linear Equation Containing Decimals**

Solve the equation: $p + 0.08p = 129.6$

Solution

Because $0.08 = \dfrac{8}{100}$ and $129.6 = \dfrac{1296}{10}$, multiplying by the LCD, 100, will clear the decimals. However, we will first combine like terms on the left-hand side of the equation.

$$p + 0.08p = 129.6$$

$p = 1 \cdot p:\quad 1p + 0.08p = 129.6$

Combine like terms:$\quad 1.08p = 129.6$

Multiply both sides of the equation by 100:$\quad 100 \cdot 1.08p = 100 \cdot 129.6$

Simplify:$\quad 108p = 12{,}960$

Divide both sides of the equation by 108:$\quad \dfrac{108p}{108} = \dfrac{12{,}960}{108}$

Simplify:$\quad p = 120$

We leave the check to you. The solution is 120, or the solution set is $\{120\}$. ●

> **Quick ✓**
>
> **10.** Simplify: $n + 0.25n = $ ___ n.
>
> *In Problems 11 and 12, solve each equation.*
>
> **11.** $p + 0.05p = 52.5$ **12.** $c - 0.25c = 120$

EXAMPLE 5 **Solving a Linear Equation That Contains Decimals**

Solve the equation: $0.05x + 0.08(10{,}000 - x) = 680$

Solution

Before we clear the equation of decimals, we will distribute the 0.08 and combine like terms.

$$0.05x + 0.08(10{,}000 - x) = 680$$

$$0.05x + 800 - 0.08x = 680$$

Combine like terms:$\quad -0.03x + 800 = 680$

Subtract 800 from both sides:$\quad -0.03x = -120$

Multiply both sides of the equation by 100 to clear the decimal:$\quad 100(-0.03x) = 100(-120)$

$$-3x = -12{,}000$$

Divide both sides by -3:$\quad x = 4000$

We leave the check to you. The solution is 4000, or the solution set is $\{4000\}$. ●

> **Quick ✓**
>
> *In Problems 13 and 14, solve each equation.*
>
> **13.** $0.36y - 0.5 = 0.16y + 0.3$ **14.** $0.12x + 0.05(5000 - x) = 460$

▶ ❸ **Classify a Linear Equation as an Identity, a Conditional Equation, or a Contradiction**

All of the linear equations we have solved so far have had a single solution. All other values of the variable make the equation false. These types of equations have a special name.

> **Definition**
>
> A **conditional equation** is an equation that is true for some values of the variable and false for other values of the variable.

For example, $x + 5 = 11$ is a conditional equation because it is true when $x = 6$ and false for every other real number x. The solution set of $x + 5 = 11$ is {6}. All the equations we have studied to this point have been conditional equations.

Some equations are false for all values of the variable.

> **Definition**
>
> A **contradiction** is an equation that is false for every value of the variable.

For example, the equation $2x + 3 = 7 + 2x$ is a contradiction because it is false for all values of x. Contradictions are identified by creating equivalent equations. For example, subtracting $2x$ from both sides of $2x + 3 = 7 + 2x$ gives $3 = 7$, which is clearly false. **Contradictions have no solution, so the solution set is the empty set, written as { } or ∅.**

Work Smart

Do not write the empty set as {∅}.

EXAMPLE 6 **Solving a Linear Equation That Is a Contradiction**

Solve the equation: $3y - (5y + 4) = 12y - 7(2y - 1)$

Solution

$$3y - (5y + 4) = 12y - 7(2y - 1)$$

Use the Distributive Property to remove parentheses:
$$3y - 5y - 4 = 12y - 14y + 7$$

Combine like terms:
$$-2y - 4 = -2y + 7$$

Isolate terms containing y:
$$-2y + 2y - 4 = -2y + 2y + 7$$
$$-4 = 7 \quad \text{False}$$

The statement $-4 = 7$ is false, so the equation is a contradiction. The solution set is ∅ or { }.

Some equations are true for all real numbers for which the equation is defined.

> **Definition**
>
> An **identity** is an equation that is satisfied for all values of the variable for which both sides of the equation are defined.

For example, the equation $2(x - 5) + 1 = 5x - (9 + 3x)$ is an identity because any real number x makes the equation true. Just as with contradictions, identities are found by forming equivalent equations. For example, if we use the Distributive Property and combine like terms in the equation $2(x - 5) + 1 = 5x - (9 + 3x)$, we obtain $2x - 9 = 2x - 9$. After subtracting $2x$ from both sides we obtain $-9 = -9$, which is true. **The solution set of a linear equation that is an identity is the set of all real numbers.**

EXAMPLE 7 **Solving a Linear Equation That Is an Identity**

Solve the equation: $2(x + 5) = 4x - (2x - 10)$

Solution

$$2(x + 5) = 4x - (2x - 10)$$

Use the Distributive Property to remove parentheses:
$$2x + 10 = 4x - 2x + 10$$

Combine like terms:
$$2x + 10 = 2x + 10$$

Isolate terms containing x:
$$2x - 2x + 10 = 2x - 2x + 10$$
$$10 = 10$$

The statement $10 = 10$ is true for all real numbers x. Therefore, the solution set is the set of all real numbers, and the equation is an identity.

Quick ✓

15. A _____ _____ is an equation that is true for some values of the variable and false for other values of the variable.

16. A(n) _____ is an equation that is false for every value of the variable. A(n) _____ is an equation that is satisfied for all values of the variable for which both sides of the equation are defined.

17. *True or False* The solution of the equation $4x + 1 = 4x + 7$ is the empty set.

In Problems 18–21, solve each equation and state the solution set.

18. $3(x + 4) = 4 + 3x + 18$

19. $\frac{1}{3}(6x - 9) - 1 = 6x - [4x - (-4)]$

20. $-5 - (9x + 8) + 23 = 7 + x - (10x - 3)$

21. $\frac{3}{2}x - 8 = x + 7 + \frac{1}{2}x$

Summary

- A *conditional* equation is true for some values of the variable and false for others.
- A *contradiction* is false for all values of the variable.
- An *identity* is true for all of the permitted values of the variable.

EXAMPLE 8 **Classifying a Linear Equation**

Solve the equation $2(2 + a) - 12a = -a + 5 - 9a$. State whether the equation is a contradiction, an identity, or a conditional equation.

Solution

$$2(2 + a) - 12a = -a + 5 - 9a$$

Use the Distributive Property to remove parentheses: $\quad 4 + 2a - 12a = -a + 5 - 9a$

Combine like terms: $\qquad\qquad 4 - 10a = 5 - 10a$

Isolate terms containing a: $\quad 4 - 10a + 10a = 5 - 10a + 10a$

$$4 = 5$$

Since $4 = 5$ is false, the equation is a contradiction. The solution set is $\varnothing$ or $\{\ \}$. ●

Quick ✓

In Problems 22–25, solve each equation and state whether each equation is a contradiction, an identity, or a conditional equation.

22. $2(x - 7) + 8 = 6x - (4x + 2) - 4$

23. $\dfrac{4(7 - x)}{3} = x$

24. $4(5x - 4) + 1 = 20x$

25. $\dfrac{1}{2}(4x - 6) = 6\left(\dfrac{1}{3}x - \dfrac{1}{2}\right) + 4$

---

④ Use Linear Equations to Solve Problems

We continue to solve problems that are modeled by algebraic equations.

EXAMPLE 9   **Solving a Problem from Finance**

You invest part of $5000 in a certificate of deposit (CD) that earns 2% simple interest compounded annually, and the rest in bonds that earn 4% simple interest compounded annually. To determine the amount, c, you should invest in the CD to earn $170 interest at the end of one year, solve the equation $0.02c + 0.04(5000 - c) = 170$.

Solution

$$0.02c + 0.04(5000 - c) = 170$$

Use the Distributive Property $0.02c + 200 - 0.04c = 170$

Combine like terms: $-0.02c + 200 = 170$

Isolate term containing c: $-0.02c + 200 - 200 = 170 - 200$

$$-0.02c = -30$$

Multiply both sides by 100 to eliminate the decimal: $-2c = -3000$

Divide both sides by -2: $c = 1500$

Because $c = 1500$, you must invest $1500 in the CD to earn $170 in interest at the end of one year.

Quick ✓

26. Janet Majors invested part of her lottery winnings in a savings account that pays 4% annual interest, and $250 more than that in a mutual fund that pays 6% annual interest. Her total interest was $65. To determine the amount, s, that she invested in the savings account, solve the equation $0.04s + 0.06(s + 250) = 65$.

2.3 Exercises   MathXL® PRACTICE

Exercise numbers in green have complete video solutions in MyMathLab

Problems 1–26 are the Quick ✓s that follow the EXAMPLES.

Building Skills

In Problems 27–42, solve the equation. Check your solution. See Objective 1.

27. $\dfrac{1}{5}x + \dfrac{3}{2} = \dfrac{3}{10}$

28. $\dfrac{3}{2}n - \dfrac{4}{11} = \dfrac{91}{22}$

29. $\dfrac{a}{4} - \dfrac{a}{3} = -\dfrac{1}{2}$

30. $\dfrac{3}{2}b - \dfrac{4}{5}b = \dfrac{28}{5}$

31. $\dfrac{3x + 2}{4} = \dfrac{x}{2}$

32. $\dfrac{2x - 3}{5} = \dfrac{3}{10}x$

33. $\dfrac{2k - 1}{4} = 2$

34. $\dfrac{3a + 2}{5} = -1$

35. $-\dfrac{2x}{3} + 1 = \dfrac{5}{9}$

36. $\dfrac{4}{3}m - 1 = \dfrac{1}{9}$

37. $\dfrac{2}{3}(x + 1) = \dfrac{1}{9}(x + 4)$

38. $\dfrac{2}{3}(6 - x) = \dfrac{5}{6}x$

39. $\dfrac{y}{10} + 3 = \dfrac{y}{4} + 6$

40. $\dfrac{p}{8} - 1 = \dfrac{7p}{6} + 2$

41. $\dfrac{4x - 9}{3} + \dfrac{x}{6} = \dfrac{x}{2} - 2$

42. $\dfrac{3x + 2}{4} - \dfrac{x}{12} = \dfrac{x}{3} - 1$

In Problems 43–62, solve the equation. Check your solution. See Objective 2.

43. $0.4w = 12$

44. $0.3z = 6$

45. $-1.3c = 5.2$

46. $-1.7q = -8.5$

47. $1.05p = 52.5$

48. $1.06z = 31.8$

49. $p + 1.5p = 12$

50. $2.5a + a = 7$

51. $p + 0.05p = 157.5$

52. $p + 0.04p = 260$

53. $0.3x + 2.3 = 0.2x + 1.1$

54. $0.7y - 4.6 = 0.4y - 2.2$

55. $0.65x + 0.3x = x - 3$

56. $0.5n - 0.35n = 2.5n + 9.4$

57. $3 + 1.5(z + 2) = 3.5z - 4$

58. $5 - 0.2(m - 2) = 3.6m + 1.6$

59. $0.02(2c - 24) = -0.4(c - 1)$

60. $0.3(6a - 4) = -0.10(2a - 8)$

61. $0.15x + 0.10(250 - x) = 28.75$

62. $0.03t + 0.025(1000 - t) = 27.25$

In Problems 63–74, solve the equation. State whether the equation is a contradiction, an identity, or a conditional equation. See Objective 3.

63. $3(z + 1) = 2(z - 3) + z$

64. $4(y - 2) = 6(y + 1) - 2y$

65. $6q - (q - 3) = 2q + 3(q + 1)$

66. $-3x + 2 + 5x = 2(x + 1)$

67. $9a - 5(a + 1) = 2(a - 3)$

68. $7b + 2(b - 4) = 8b - (3b + 2)$

69. $\dfrac{4x - 9}{6} - \dfrac{x}{2} = \dfrac{x}{6} + 3$

70. $\dfrac{2m + 1}{4} - \dfrac{m}{6} = \dfrac{m}{3} - 1$

71. $\dfrac{5z + 1}{5} = \dfrac{2z - 3}{2}$

72. $\dfrac{2y - 7}{4} = \dfrac{3y - 13}{6}$

73. $\dfrac{q}{3} + \dfrac{4}{5} = \dfrac{5q + 12}{15}$

74. $\dfrac{2x}{3} + \dfrac{x + 3}{12} = \dfrac{3x + 1}{4}$

Mixed Practice

In Problems 75–100, solve the equation.

75. $-3(2n + 4) = 10n$

76. $-15(z - 3) = 25z$

77. $-2x + 5x = 4(x + 2) - (x + 8)$

78. $5m - 3(m + 1) = 2(m + 1) - 5$

79. $-6(x - 2) + 8x = -x + 10 - 3x$

80. $3 - (x + 10) = 3x + 7$

81. $\dfrac{3}{4}x = \dfrac{1}{2}x - 5$

82. $\dfrac{1}{3}x = 2 + \dfrac{5}{6}x$

83. $\dfrac{1}{2}x + 2 = \dfrac{4x + 1}{4}$

84. $\dfrac{x}{2} + 4 = \dfrac{x + 7}{3}$

85. $0.3p + 2 = 0.1(p + 5) + 0.2(p + 1)$

86. $1.6z - 4 = 2(z - 1) - 0.4z$

87. $-0.7x = 1.4$

88. $0.2a = -6$

89. $\dfrac{3(2y - 1)}{5} = 2y - 3$

90. $\dfrac{4(2n + 1)}{3} = 2n - 6$

91. $0.6x - 0.2(x - 4) = 0.4(x - 2)$

92. $0.3x - 1 = 0.5(x + 2) - 0.2x$

93. $\dfrac{3x - 2}{4} = \dfrac{5x - 1}{6}$

94. $\dfrac{x - 1}{4} = \dfrac{x - 4}{6}$

95. $0.3x + 2.6x = 5.7 - 1.8 + 2.8x$

96. $0.3(z - 10) - 0.5z = -6$

97. $\dfrac{3}{2}x - 6 = \dfrac{2(x - 9)}{3} + \dfrac{1}{6}x$

98. $\dfrac{2x - 3}{4} + 5 = \dfrac{3(x + 3)}{4} - \dfrac{x}{2} + 2$

99. $\dfrac{2}{3}\left[4 - \left(\dfrac{x}{2} + 6\right) - 2x\right] + 3 = \dfrac{5x}{6}$

100. $\dfrac{1}{2}\left[3 - \left(\dfrac{2x}{3} - 1\right) + 3x\right] = \dfrac{-4x + 1}{3} + 1$

In Problems 101–104, solve the equation and round to the indicated place.

101. $2.8x + 13.754 = 4 - 2.95x$ to the nearest hundredth

102. $-4.88x - 5.7 = 2(-3.41x) + 1.2$ to the nearest whole number

103. $x - \{1.5x - 2[x - 3.1(x + 10)]\} = 0$ to the nearest tenth

104. $-3x - 2\{4 + 3[x - (1 + x)]\} = 12$ to the nearest hundredth

Applying the Concepts

105. Sales Tax The price of a pair of jeans, including sales tax of 6%, is $53. To find the price, p, of the jeans, solve the equation $1.06p = 53$.

106. Gardening Bob Adams rented a rototiller at a cost of $7.50 per hour. Bob paid a bill of $37.50. Find the number of hours, h, he rented the tiller by solving the equation $7.50h = 37.50$.

107. Purchasing a Car The total cost (including 6% sales tax) for the purchase of an automobile was $19,080. To determine the cost of the auto before the sales tax was added, solve the equation $x + 0.06x = 19,080$, where x represents the cost of the car before taxes.

108. Purchasing a Kayak The total cost, including 5.5% sales tax, for the purchase of a kayak was $1266. Find the price of the kayak, k, before sales tax, by solving the equation $k + 0.055k = 1266$.

109. Hourly Pay Bob recently received a 4% pay increase. His hourly wage is now $8.84. To determine his hourly wage, w, before the 4% pay raise, solve the equation $w + 0.04w = 8.84$.

110. Hourly Pay A union representing airline employees recently agreed to a 6% cut in hourly wage in order to help the airline avoid filing for bankruptcy. A baggage handler will now earn $26.32 per hour. To find the baggage handler's hourly wage, w, before the pay cut, solve the equation $w - 0.06w = 26.32$.

111. MP3 Player Tamara purchased an MP3 player at a "25% off" sale for $60.00. To determine the original price, p, of the MP3 player, solve the equation $p - 0.25p = 60$.

112. Team Sweatshirt Your favorite college has logo sweatshirts on sale for $33.60. The sweatshirt has been marked down by 30%. To find the original price of the sweatshirt, x, solve the equation $x - 0.30x = 33.60$.

113. Piggy Bank Celeste saves dimes and quarters in a piggy bank. She opened the bank and discovered that she had $7.05 and that the number of dimes was 3 more than twice the number of quarters. To find the number of quarters, q, solve the equation $0.25q + 0.10(2q + 3) = 7.05$.

114. Clean Car Pablo cleaned out his car and found nickels and quarters in the car seats. He found $4.25 in change and noticed that the number of quarters was 5 less than twice the number of nickels. Solve the equation $0.05n + 0.25(2n - 5) = 4.25$ to find n, the number of nickels Pablo found.

△ **115. Comparing Perimeters** The two rectangles shown in the figure have the same perimeter. Solve the equation

$$2x + 2(x + 3) = 2\left(\frac{1}{2}x\right) + 2(x + 6)$$ for x, the

width of the first rectangle

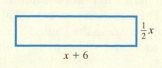

△ **116. Comparing Perimeters** The square and the rectangle shown have the same perimeter. Solve the equation

$$4x = 2\left(\frac{1}{2}x\right) + 2(x + 5)$$ for x, the length of the

side of the square.

117. Paying Your Taxes You are single and just determined that you paid $1372.50 in federal income taxes for 2012. The solution to the equation $1372.50 = 0.15(x - 8700) + 870$ represents your adjusted gross income in 2012. Determine your adjusted gross income in 2012. (SOURCE: *Internal Revenue Service*)

118. Paying Your Taxes You are married and just determined that you paid $13,310 in federal income taxes in 2012. The solution to the equation $13,310 = 0.25(x - 70,700) + 9735$ represents the adjusted gross income of you and your spouse in 2012. Determine the adjusted gross income of you and your spouse in 2012. (SOURCE: *Internal Revenue Service*)

Explaining the Concepts

119. Make up a linear equation that has one solution. Make up a linear equation that has no solution. Make up a linear equation that is an identity. Discuss the differences and similarities in making up each equation.

120. A student solved the equation $3(x + 8) = \frac{1}{2}(6x + 4)$ and wrote the answer $24 = 2$. The instructor did not give full credit for this answer. Explain the student's error and determine the correct solution.

121. When solving the equation $\frac{2}{3}x - 5 = \frac{1}{2}x$, a student decided to multiply both sides by the LCD and wrote $6 \cdot \frac{2}{3}x - 5 = \frac{1}{2}x \cdot 6$. This resulted in the next line $4x - 5 = 3x$. Is this correct? Explain the steps necessary to finish using this technique and then suggest another list of steps that would arrive at the correct solution.

122. Explain how you can tell from the last step of the solution process whether the solution of an equation is the set of all real numbers or the empty set. For example, if the last line of the solution of an equation is $-6 = 8$, is the solution the set of all real numbers or the empty set? If the last line of the solution of an equation is $9 = 9$, is the solution the set of all real numbers or the empty set?

2.4 Evaluating Formulas and Solving Formulas for a Variable

Objectives

1. Evaluate a Formula
2. Solve a Formula for a Variable

Are You Ready for This Section?

Before getting started, take this readiness quiz. If you get a problem wrong, go back to the section cited and review the material.

R1. Evaluate the expression $2L + 2W$ for $L = 7$ and $W = 5$. [Section 1.8, pp. 64–65]

R1. Round the expression 0.5873 to the hundredths place. [Section 1.2, pp. 13–14]

▶ ① Evaluate a Formula

In this section, we use *formulas* to solve mathematical problems.

> **Definition**
>
> A mathematical **formula** is an equation that describes how two or more variables are related.

For example, a formula for the area of a rectangle is *area = length · width*, or $A = l \cdot w$. You use mathematical formulas every day. For instance, to paint your bedroom, you must find the surface area of the walls to compute the amount of paint you will need. To determine the number of gallons of gasoline you can afford to put in your car, you may also use a formula, as shown in the next example.

EXAMPLE 1 **Evaluating a Formula**

The number of gallons of gasoline that you put in your car is found by using the formula $\dfrac{C}{p} = n$, where C is the total cost, p is the price per gallon, and n is the number of gallons. How many gallons of gas can you purchase for $51.00 if gas costs $4.25 per gallon?

Solution

$$\frac{C}{p} = n$$

Replace *C* with cost, $51, and *p* with price per gallon, $4.25:

$$\frac{\$51}{\$4.25} = n$$

Evaluate the numerical expression:

$$12 = n$$

You can purchase 12 gallons for $51. ●

> **Quick ✓**
>
> 1. A _____ is an equation that describes how two or more variables are related.
> 2. Your best friend, who is on spring break in Europe, e-mailed you that the temperature in Paris today is 15° Celsius. Use the formula $F = \dfrac{9}{5}C + 32$, where F is degrees Fahrenheit and C is degrees Celsius, to find the approximate Paris temperature in degrees Fahrenheit.
> 3. The size of a dress purchased in Europe is different from that of one purchased in the United States. The equation $c = a + 30$ gives the European (Continental) dress size c in terms of the size in the United States, a. Find the Continental dress size that corresponds to a size 10 dress in the United States.

EXAMPLE 2 **Evaluating a Formula Containing a Percent**

The formula $S = P - 0.25P$ gives the sale price S of an item originally costing P dollars that was reduced by 25%. Find the sale price of a pair of jeans that originally cost $40.00.

Solution

$$S = P - 0.25P$$

Replace P with the original price of the jeans, $40: $\quad S = 40 - 0.25(40)$

Evaluate the numerical expression: $\quad S = 40 - 10$

$$S = 30$$

The sale price of the jeans is $30.

Quick ✓

4. The formula $E = 250 + 0.05S$ is a formula for the earnings, E, of a salesman who receives $250 per week plus 5% commission on all weekly sales, S. Find the earnings of a salesman who had weekly sales of $1250.

5. The formula $N = p + 0.06p$ models the new population, N, of a town with current population p if the town is expecting population growth of 6% next year. Find the new population of a town whose current population is 5600 persons.

The Simple Interest Formula

Interest is money paid for the use of money. The total amount borrowed is the **principal.** The principal can be in the form of a loan (an individual borrows from the bank) or a deposit (the bank borrows from the individual). The **rate of interest,** expressed as a percent, is the amount charged for the use of the principal for a given period of time. The time frame is usually a year or some fraction of a year (such as 1 month, or $\frac{1}{12}$ of a year).

Simple Interest Formula

If an amount of money, P, called the **principal** is invested for a period of t years at an annual interest rate r, expressed as a decimal, the amount of interest I earned is

$$I = Prt$$

Interest earned according to this formula is called **simple interest.**

EXAMPLE 3 **Evaluating Using the Simple Interest Formula**

Janice invested $500 in a two-year certificate of deposit (CD) at the simple interest rate of 4% per year. Find the amount of interest Janice will earn in 6 months and the total amount of money she will have at the end of 6 months.

Solution

The amount of money Janice invested, P, is $500. The interest rate r is 4% $= 0.04$.

Because 6 months is half a year, we have that $t = \dfrac{1}{2}$.

Work Smart

Be sure to express t in years and r as a decimal when using the simple interest formula.

$$I = Prt$$

Let $p = \$500$, $r = 0.04$, and $t = \dfrac{1}{2}$: $\quad I = 500 \cdot 0.04 \cdot \dfrac{1}{2}$

Evaluate the numerical expression: $\quad I = 10$

At the end of 6 months Janice earned $10 and will have a total of $500 + $10 = $510.

Quick ✓

6. The total amount borrowed in a loan is called _____. _____ is money paid for the use of money.

7. Bill invested his $2500 Virginia Lottery winnings in an 8-month certificate of deposit that earns 3% simple interest per annum. Find the amount of interest Bill's investment will earn at the end of 8 months. Also find the total amount of money Bill will have at the end of 8 months.

▶ Geometry Formulas

Let's review a few common terms from geometry.

Definitions

The **perimeter** is the sum of the lengths of all the sides of a figure.

The **area** is the amount of space enclosed by a two-dimensional figure, measured in square units.

The **surface area** of a solid is the sum of the areas of the surfaces of a three-dimensional figure.

The **volume** is the amount of space occupied by a three-dimensional figure, measured in units cubed.

The **radius** r of a circle is any line segment that extends from the center of the circle to any point on the circle.

The **diameter** of a circle is any line segment that extends from one point on the circle through the center to a second point on the circle. The length of a diameter is two times the length of the radius, $d = 2r$.

In circles, we use the term **circumference** to mean the perimeter.

Formulas from geometry are useful in solving many types of problems. We list some of these formulas in Table 1.

Table 1

Plane Figures	Formulas	Plane Figures	Formulas
Square	**Area:** $A = s^2$ **Perimeter:** $P = 4s$	Trapezoid	**Area:** $A = \dfrac{1}{2}h(B + b)$ **Perimeter:** $P = a + b + c + B$
Rectangle	**Area:** $A = lw$ **Perimeter:** $P = 2l + 2w$	Parallelogram	**Area:** $A = bh$ **Perimeter:** $P = 2a + 2b$
Triangle	**Area:** $A = \dfrac{1}{2}bh$ **Perimeter:** $P = a + b + c$	Circle	**Area:** $A = \pi r^2$ **Circumference:** $C = 2\pi r$ $= \pi d$

Table 1 (*continued*)

Solids	Formulas	Solids	Formulas
Cube	**Volume:** $V = s^3$ **Surface Area:** $S = 6s^2$	Right Circular Cylinder	**Volume:** $V = \pi r^2 h$ **Surface Area:** $S = 2\pi r^2 + 2\pi rh$
Rectangular Solid	**Volume:** $V = lwh$ **Surface Area:** $S = 2lw + 2lh + 2wh$	Cone	**Volume:** $V = \frac{1}{3}\pi r^2 h$
Sphere	**Volume:** $V = \frac{4}{3}\pi r^3$ **Surface Area:** $S = 4\pi r^2$		

EXAMPLE 4 **Evaluating the Formula for the Perimeter of a Rectangle**

Find the number of yards of fencing needed to enclose a garden that is 10.5 yards long and 4.25 yards wide, as illustrated in Figure 1.

Solution

The garden is rectangular, so we use the formula for the perimeter of a rectangle, $P = 2l + 2w$, where $l = 10.5$ yards and $w = 4.25$ yards.

$$P = 2l + 2w$$

Replace *l* with 10.5 and *w* with 4.25: $P = 2(10.5) + 2(4.25)$

Evaluate the numerical expression: $P = 21 + 8.5$

$$P = 29.5$$

Figure 1

10.5 yards

4.25 yards

A total of 29.5 yards of fencing is needed to enclose the garden.

Quick ✔

8. The _____ is the sum of the lengths of all the sides of a figure.

9. The _____ is the amount of space enclosed by a two-dimensional figure, measured in square units.

10. The _____ is the amount of space occupied by a three-dimensional figure, measured in units cubed.

11. The _____ of a circle is any line segment that extends from the center of the circle to any point on the circle.

12. *True or False* The area A of a circle whose radius is r is found using the formula $A = \pi r^2$.

13. Find the area of a trapezoid with $h = 4.5$ inches, $B = 9$ inches, and $b = 7$ inches.

Sometimes we need to use more than one geometry formula to solve a problem.

EXAMPLE 5 **Finding the Area of a Lawn and the Cost of Sod**

A circular swimming pool with a 30-foot diameter is to be installed in a 100-foot by 60-foot rectangular yard. Grass will then be installed on the remaining land. See Figure 2 on the following page.

(a) Determine the area of land that is to receive grass.

(b) If sod costs $0.25 per square foot installed, what will the cost of the lawn be?

Figure 2

60 feet

30 feet

100 feet

Solution

(a) The area of a rectangle is $A = lw$ and the area of a circle is $A = \pi r^2$. We find the area of the lawn by subtracting the area of the circular swimming pool from the area of the rectangular yard.

$$\text{Area of lawn} = \text{Area of rectangle} - \text{Area of circle}$$
$$= \quad lw \quad - \quad \pi r^2$$

The length l of the lawn is 100 feet and the width is 60 feet. The radius of the circle is one-half its diameter, so the radius is 15 feet. Substituting into the formulas, we obtain

$$\text{Area of remaining lawn} = (100 \text{ feet})(60 \text{ feet}) - \pi (15 \text{ feet})^2$$
$$= 6000 \text{ feet}^2 - 225\pi \text{ feet}^2$$
$\pi \approx 3.14159\text{:} \qquad \approx 5293 \text{ feet}^2$

Approximately 5293 square feet of sod is needed.

(b) The cost of the sod is the product of the square footage and the cost per square foot.

$$\text{Cost for sod} = 5293 \text{ square feet} \cdot \frac{\$0.25}{1 \text{ square foot}}$$
$$= \$1323.25$$

The sod will cost $1323.25.

In Example 5(b), notice that the units "square feet" divide out, giving an answer in dollars. Getting the appropriate units in the answer gives us more confidence that the answer is correct.

> **In Words**
> The word "per" means for each one. For example, the sod costs $0.25 for each square foot installed. When you see the word "per," think fraction. Thus $0.25 per square foot is represented as
> $$\frac{\$0.25}{1 \text{ square foot}}$$

Quick ✓

14. A circular water feature whose diameter is 4 feet is to be installed in a rectangular garden that is 20 feet by 10 feet. Once the water feature is installed, sod is to be installed on the remaining land.

(a) Determine the area, to the nearest square foot, of land that is to receive sod.

(b) If sod costs $0.25 per square foot installed, what will be the cost of the lawn?

Geometry formulas can even be used to make us savvy consumers.

EXAMPLE 6   **Determining the Better Buy**

At Mamma da Vinci's Pizza Parlor, a 16-inch pizza costs $14.99 and a 12-inch pizza costs $9.99. Which is the "better" buy?

Solution

The "better" buy is the pizza that costs less per square inch. Our plan is: (1) find the area of each pizza and (2) find the price per square inch of each pizza. Keep in mind that a 16″ pizza has a 16″ *diameter* and an 8″ *radius*. A 12″ pizza has a 6″ radius.

1. **Find the area of each pizza.** Since pizzas are circular, use the formula for the area of a circle.

	Area of 16″ pizza	Area of 12″ pizza
	$A = \pi r^2$	$A = \pi r^2$
Replace r with its given value:	$A = \pi \cdot 8^2$	$A = \pi \cdot 6^2$
Evaluate the numerical expression:	$A \approx 201.06 \text{ in.}^2$	$A \approx 113.10 \text{ in.}^2$

2. **Find the price per square inch of each pizza.** The price per square inch is found by dividing the price of the pizza by the number of square inches of area.

price per square inch of 16″ pizza: $\dfrac{\$14.99}{201.06 \text{ in.}^2} \approx \$0.075$ per square inch

price per square inch of 12″ pizza: $\dfrac{\$9.99}{113.10 \text{ in.}^2} \approx \$0.088$ per square inch

The price per square inch of the 16-inch pizza is about 7.5¢. The price per square inch of the 12-inch pizza is about 8.8¢. The 16-inch pizza is the "better" buy.

Quick ✓

15. A homeowner plans to construct a circular brick pad for his barbeque grill. The diameter of the brick pad is to be 6 feet. Find, to the nearest hundredth of a square foot, the area of the barbeque pad.

16. An extra large (18″) pizza at Dante's Pizza costs $16.99 and a small (9″) pizza costs $8.99. Which is the better buy?

▶ ❷ Solve a Formula for a Variable

To "solve for a variable" means to isolate the variable with a coefficient of 1 on one side of the equation and all other variables and constants, if any, on the other side by forming equivalent equations. For example, the formula for the area of a rectangle, $A = lw$, is solved for A because A is by itself with a coefficient of 1 on one side of the equation while all other variables are on the other side.

The steps we follow to solve formulas for a certain variable are identical to those we followed when solving equations.

Solve for x: $\quad\quad 15 = \dfrac{3}{2}x$	**Solve for h:** $\quad\quad A = \dfrac{1}{2}bh$
Multiply both sides of the equation by 2 to clear the fraction. $\quad 2(15) = 2\left(\dfrac{3}{2}x\right)$	Multiply both sides of the equation by 2 to clear the fraction. $\quad 2(A) = 2\left(\dfrac{1}{2}bh\right)$
$30 = 3x$	$2A = bh$
Divide by 3 to isolate the variable x. $\quad \dfrac{30}{3} = \dfrac{3x}{3}$	Divide by b to isolate the variable h. $\quad \dfrac{2A}{b} = \dfrac{bh}{b}$
$10 = x$	$\dfrac{2A}{b} = h$

EXAMPLE 7 **Solve a Formula for a Variable**

Solve the simple interest formula $I = Prt$ for t.

Solution

To solve for t, we isolate t on one side of the equation.

$$I = Prt$$

Divide both sides of the equation by Pr: $\quad \dfrac{I}{Pr} = \dfrac{Prt}{Pr}$

Simplify: $\quad \dfrac{I}{Pr} = t$

Use the Symmetric Property: $\quad t = \dfrac{I}{Pr}$

Work Smart

The Symmetric Property states that if $a = b$, then $b = a$.

EXAMPLE 8 **Solve a Formula for a Variable**

Solve the perimeter of a rectangle formula $P = 2l + 2w$ for w.

Solution

$$P = 2l + 2w$$

Subtract 2*l* from both sides of the equation: $\quad P \qquad = 2l + 2w$

Simplify: $\quad P - 2l = 2w$

Divide both sides of the equation by 2: $\quad \dfrac{P - 2l}{2} = \dfrac{2w}{2}$

$$\dfrac{P - 2l}{2} = w$$

Use the Symmetric Property: $\qquad w = \dfrac{P - 2l}{2}$

The next example illustrates a skill that is extremely important in the study of graphing lines.

EXAMPLE 9 **Solving for a Variable in a Formula**

Solve the equation $3x + 2y = 6$ for y.

Solution

To solve for y, we isolate y on one side of the equation and the variable x and constants on the other side.

$$3x + 2y = 6$$

Subtract 3*x* from both sides of the equation to isolate the term containing *y*: $\quad 3x + 2y - 3x = 6 - 3x$

Simplify: $\quad 2y = 6 - 3x$

Divide both sides by 2: $\quad \dfrac{2y}{2} = \dfrac{6 - 3x}{2}$

Simplify: $\quad y = \dfrac{6 - 3x}{2}$

Work Smart

$\dfrac{A + B}{C} = \dfrac{A}{C} + \dfrac{B}{C}$, so

$\dfrac{6 - 3x}{2} = \dfrac{6}{2} - \dfrac{3x}{2}$.

The equation $y = \dfrac{6 - 3x}{2}$ may also be written as $y = 3 - \dfrac{3}{2}x$, or $y = -\dfrac{3}{2}x + 3$, by dividing each term in the numerator by 2. Do you see why these three equations are equivalent? See the Work Smart.

Quick ✔

In Problems 21 and 22, solve each equation for the indicated variable.

21. $6x - 3y = 18$; for y **22.** $4x + 6y = 15$; for y

2.4 Exercises   MyMathLab®   Exercise numbers in green
are complete video solutions
in MyMathLab

PRACTICE

*Problems **1–22** are the Quick ✔s that follow the **EXAMPLES**.*

Building Skills

In Problems 23–32, substitute the given values into the formula and then evaluate to find the unknown quantity. Label units in the answer. If the answer is not exact, round your answer to the nearest hundredth. See Objective 1.

23. How Far Can I Go? The Chevrolet Volt is an electric car that also has a gas engine. The distance D, in miles, the car can travel with a fully charged battery on g gallons of gasoline is given by the formula $D = 36 + 35g$. Determine the distance the Volt can travel using 4 gallons of gasoline.

24. Maximum Heart Rate Maximum heart rate, in beats per minute, is the highest heart rate a person can safely achieve while exercising. One formula for the maximum heart rate H of an individual who is a years of age is $H = 208 - 0.7a$. Determine the maximum heart rate of an individual who is 40 years old.

25. Buying a Digital Media Player The formula $S = P - 0.20P$ gives the sale price S of an item whose original cost P dollars was reduced by 20%. Find the sale price of a digital media player that originally cost $130.00.

26. Buying a Computer The formula $S = P - 0.35P$ gives the sale price S of an item whose original cost P dollars was reduced by 35%. Find the sale price of a computer that originally cost $950.00.

27. Salesperson's Earnings The formula $E = 500 + 0.15S$ is a formula for the earnings, E, of a salesperson who receives $500 per week plus 15% commission on all sales, S. Find the earnings of a salesperson who had weekly sales of $1000.

28. Salesperson's Earnings The formula $E = 750 + 0.07S$ is a formula for the earnings, E, of a salesperson who receives $750 per week plus 7% commission on all sales, S. Find the earnings of a salesperson who had weekly sales of $1200.

29. Planning a Trip A businessperson is planning a trip from Tokyo to San Diego, California, in March. She learns that the average high temperature in San Diego in March is 68° Fahrenheit. Use the formula $C = \dfrac{5}{9}(F - 32)$ to convert 68° Fahrenheit to degrees Celsius.

30. Planning a Trip An English literature professor plans to take twenty students on a study-abroad trip to London in March. He learns that the average daytime temperature in London in March is 10 degrees Celsius. Use the formula $F = \dfrac{9}{5}C + 32$ to convert 10 degrees Celsius to degrees Fahrenheit.

31. Lottery Earnings Therese invested her $200 West Virginia Lottery winnings in a 6-month certificate of deposit (CD) that earns 3% simple interest per annum. Find the amount of interest Therese's investment will earn.

32. Investing an Inheritance Christopher invested his $5000 inheritance from his grandmother in a 9-month certificate of deposit that earns 4% simple interest per annum. Find the amount of interest Christopher's investment will earn.

△ **33.** Find

   (a) the perimeter and

   (b) the area of the rectangle.

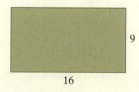

16   9

△ **34.** Find

   (a) the perimeter and

   (b) the area of the rectangle.

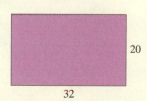

32   20

△ **35.** Find

   (a) the perimeter and

   (b) the area of the rectangle.

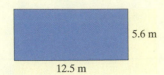

5.6 m

12.5 m

△ **36.** Find

   (a) the perimeter and

   (b) the area of the rectangle.

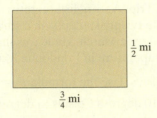

$\frac{1}{2}$ mi

$\frac{3}{4}$ mi

△ **37.** Find

   (a) the perimeter and

   (b) the area of the square.

9

△ **38.** Find

   (a) the perimeter and

   (b) the area of the square.

3.5

△ **39.** Find

   (a) the circumference and

   (b) the area of the circle. Use $\pi \approx 3.14$.

$r = 5$ cm

△ **40.** Find

   (a) the circumference and

   (b) the area of the circle. Use $\pi \approx 3.14$.

$r = 2.8$ yards

△ **41. Area of a Circle** Find the area of a circle A when $\pi \approx \dfrac{22}{7}$ and $r = \dfrac{14}{3}$ inches.

△ **42. Area of a Circle** Find the area of a circle A when $\pi \approx 3.14$ and $r = 2.5$ km.

In Problems 43–56, solve each formula for the stated variable. See Objective 2.

43. $d = rt$; solve for r **44.** $A = lw$; solve for w

45. $C = \pi d$; solve for d **46.** $F = mv^2$; solve for m

47. $I = Prt$; solve for t **48.** $V = LWH$; solve for W

49. $A = \dfrac{1}{2}bh$; solve for b **50.** $V = \dfrac{1}{3}Bh$; solve for B

51. $P = a + b + c$; solve for a **52.** $S = a + b + c$; solve for b

53. $A = P + Prt$; solve for r

54. $P = 2l + 2w$; solve for l

55. $A = \dfrac{1}{2}h(B + b)$; solve for b

56. $S = 2\pi r(r + h)$; solve for h

In Problems 57–64, solve for y. See Objective 2.

57. $3x + y = 12$ **58.** $-2x + y = 18$

59. $10x - 5y = 25$ **60.** $12x - 6y = 18$

61. $4x + 3y = 13$ **62.** $5x + 6y = 18$

63. $\dfrac{1}{2}x - \dfrac{1}{6}y = 2$ **64.** $\dfrac{2}{3}x - \dfrac{5}{2}y = 5$

Applying the Concepts

In Problems 65–78, (a) solve for the indicated variable, and then (b) find the value of the unknown quantity. When units are given, label units in the answer.

65. Profit = Revenue − Cost: $P = R - C$

   (a) Solve for C.

   (b) Find C when $P = \$1200$ and $R = \$1650$.

66. Profit = Revenue − Cost: $P = R - C$

   (a) Solve for R.

   (b) Find R when $P = \$4525$ and $C = \$1475$.

67. Simple Interest: $I = Prt$

(a) Solve for r.

(b) Find r when $I = \$225$, $P = \$5000$, and $t = 1.5$ years.

68. Simple Interest: $I = Prt$

(a) Solve for t.

(b) Find t when $I = \$42$, $P = \$525$, and $r = 4\%$.

69. Statistics Formula: $Z = \dfrac{x - \mu}{\sigma}$

(a) Solve for x.

(b) Find x when $Z = 2$, $\mu = 100$, and $\sigma = 15$.

70. Physics Formula: $P = mgh$

(a) Solve for m.

(b) Find m when $P = 8192$, $g = 32$, and $h = 1$.

71. Algebra: $y = mx + 5$

(a) Solve for m.

(b) Find m when $x = 3$ and $y = -1$.

72. Fahrenheit/Celsius Temperature Conversion:
$F = \dfrac{9}{5}C + 32$

(a) Solve for C.

(b) Find C when $F = 59°$.

73. Finance: $A = P + Prt$

(a) Solve for r.

(b) Find r when $A = \$540$, $P = \$500$, and $t = 2$.

74. Finance: $A = P + Prt$

(a) Solve for t.

(b) Find t when $A = \$249$, $P = \$240$, and $r = 2.5\% = 0.025$.

△ **75. Volume of a Right Circular Cylinder:** $V = \pi r^2 h$

(a) Solve for h.

(b) Find h when $V = 320\pi$ mm^3 and $r = 8$ mm.

△ **76. Volume of a Cone:** $V = \dfrac{1}{3}\pi r^2 h$

(a) Solve for h.

(b) Find h when $V = 75\pi$ in.^{3} and $r = 5$ in.

△ **77. Area of a Triangle:** $A = \dfrac{1}{2}bh$

(a) Solve for b.

(b) Find b when $A = 45$ ft and $h = 5$ ft.

△ **78. Area of a Trapezoid:** $A = \dfrac{1}{2}h(b + B)$

(a) Solve for h.

(b) Find h when $A = 99$ cm^2, $b = 19$ cm, and $B = 3$ cm.

79. Energy Expenditure Basal energy expenditure (E) is the amount of energy required to maintain the body's normal metabolic activity such as respiration, maintenance of body temperature, and so on. For males, the basal energy expenditure is given by the formula

$$E = 66.67 + 13.75W + 5H - 6.76A$$

where W is the weight of the male (in kilograms), H is the height of the male (in centimeters), and A is the age of the male. Determine the basal energy expenditure of a 37-year-old male who is 178 cm (5 feet, 10 inches) tall and weighs 82 kg (180 pounds).

80. Energy Expenditure See Problem 79. The basal energy expenditure for females is given by

$$E = 665.1 + 9.56W + 1.85H - 4.68A$$

Compute the basal energy expenditure of a 40-year-old female who is 168 cm tall (5 feet, 6 inches) and weighs 57 kg (125 pounds).

△ **81. Cylinders** The volume V of a right circular cylinder is given by the formula $V = \pi r^2 h$, where r is the radius and h is the height.

(a) Solve the formula for h.

(b) Find the height of a right circular cylinder whose volume is 90π cubic inches and whose radius is 3 inches.

△ **82. Soup Can** The formula $S = 2\pi rh + 2\pi r^2$ gives the surface area S of a right circular cylinder whose radius is r and height is h.

(a) Solve the formula for h.

(b) Find the height of a right circular cylinder whose surface area is 8.25π square inches and radius is 1.5 inches.

△ **83. Grocery Store** You are standing at the freezer case at your local grocery store trying to decide which is the "better" buy: a medium (12″) pizza for $9.99 or 2 small (8″) pizzas that are on special for $4.49 each. Which should you choose to get the best deal?

△ **84. Pizza for Dinner** Mama Mimi's Take and Bake Pizzeria is running a special on southwestern-style pizzas: a large 16″ pizza for $13.99 or two small 8″ pizzas for $12.99. Which should you choose to get the best deal?

85. Taking a Trip Jason drives a truck as an independent contractor. He bills himself out at $28 per hour. Suppose Jason has a contract that calls for him to leave a dock at 9:00 A.M. and travel 600 miles to a warehouse. Jason has driven this route many times and figures that he can travel at an average speed of 50 miles per hour.

(a) Using the formula $d = rt$, where d is the distance traveled, r is the average speed, and t is the time spent traveling, determine how long Jason expects the trip to take.

(b) How much money does Jason expect to earn from this contract?

86. Taking a Trip Messai drives a truck as an independent contractor. He bills himself out at \$32 per hour. Messai has a contract that calls for him to leave a dock at 8:00 A.M. and travel 145 miles to a warehouse. At the warehouse, he will wait while the truck is loaded (this takes 2 hours) and then return to his original dock. Messai has driven this route many times and figures that he can travel at an average speed of 58 miles per hour.

(a) Using the formula $d = rt$, where d is the distance traveled, r is the average speed, and t is the time spent traveling, determine how long Messai expects the round-trip to take. Exclude the time Messai waits for the truck to be loaded.

(b) How much money does Messai expect to earn from this contract driving his truck?

△ **87. Area of a Region** Find the area of the figure below.

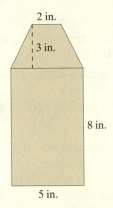

2 in.

3 in.

8 in.

5 in.

△ **88. Area of a Region** Find the area of the figure below.

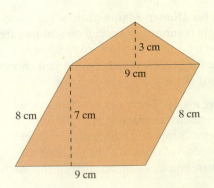

3 cm

9 cm

8 cm   7 cm   8 cm

9 cm

△ **89. Ice Cream Cone** Find the amount of ice cream in a cone if the radius of the cone is 4 cm and its height is 10 cm. The ice cream fully fills the cone, and the hemisphere of ice cream on the top has a radius of 4 cm.

△ **90. Window** Find the area of the window given that the upper portion is a semicircle:

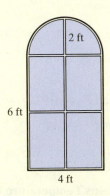

2 ft

6 ft

4 ft

91. Federal Income Taxes For a single filer with an annual adjusted income in 2011 over \$35,350, but less than \$85,650, the federal income tax T for an annual adjusted income I is found using the formula $T = 0.25(I - 35,350) + 4867.5$.

(a) Solve the formula for I.

(b) Determine the adjusted income of a single filer whose tax bill is \$14,780.

92. Computing a Grade Jose's art history instructor uses the formula $G = \dfrac{a + b + 2c + 2d}{6}$ to compute her students' semester grade. The variables a and b represent the grades on two tests, c represents the grade on a research paper, d represents the final exam grade, and G represents the student's average.

(a) Solve the equation for d, Jose's final exam grade.

(b) A final average of 84 will earn Jose a B in the course. Compute the grade Jose must make on his final exam in order to earn a B for the semester, if he scored 78 and 74 on his tests, and 84 on his research paper.

△ **93. Remodel a Bathroom** You plan to remodel your bathroom and you've chosen 1-foot-by-1-foot ceramic tiles for the floor. The bathroom is 7 feet 6 inches long and 8 feet 2 inches wide.

(a) How many tiles do you need to cover the floor of your bathroom?

(b) Each tile costs \$6. How much will it cost to tile your floor?

(c) The store from which you purchase the tile offers a discount of 10% on orders over $350. Does your order qualify for the discount?

△ **94. Painting a Room** A gallon of paint can cover about 500 square feet. Find the number of gallon containers of paint that must be purchased to paint two coats on each wall of a rectangular room measuring 8 feet by 12 feet, with a 10-foot ceiling. *Note:* You cannot purchase a partial can of paint!

△ **95. Landscaping a Back Yard** A circular swimming pool whose diameter is 24 feet is to be installed in a rectangular yard that is 60 feet by 90 feet. Once the pool is installed, sod is to be installed on the remaining land.

(a) Determine the area of land that is to receive sod. Round your answer to the nearest foot. Use $\pi \approx 3.14159$.

(b) If sod costs $0.25 per square foot installed, what will be the cost of the lawn?

△ **96. Landscaping a Back Yard** A rectangular swimming pool whose dimensions are 12 feet by 24 feet is to be installed in a rectangular yard that is 80 feet by 40 feet. Once the pool is installed, sod is to be installed on the remaining land.

(a) Determine the area of land that is to receive sod.

(b) A pallet of sod covers 500 square feet. How many pallets of sod are required?

(c) Each pallet of sod costs $96. What is the cost of the sod?

Extending the Concepts

△ **97. Conversion** A rectangle has length 5 feet and width 18 inches.

(a) What is the area in square inches?

(b) What is the area in square feet?

△ **98. Conversion** A rectangle has length 9 yards and width 8 feet.

(a) What is the area in square feet?

(b) What is the area in square yards?

△ **99. Conversion** Determine a formula for converting square inches to square feet.

△ **100. Conversion** Determine a formula for converting square yards to square feet.

Explaining the Concepts

101. A student solved the equation $x + 2y = 6$ for y and obtained the result $y = \dfrac{-x + 6}{2}$. Another student solved the same equation for y and obtained the result $y = -\dfrac{1}{2}x + 3$. Are both solutions correct? Explain why or why not.

102. Make up an example of a linear equation whose coefficients are integers. Make up a similar example where the coefficients are constants (letters). Solve both and then explain which steps in the solution are alike and which are different.

Putting the Concepts Together (Sections 2.1–2.4)

We designed these problems so that you can review Sections 2.1–2.4 and show your mastery of the concepts. Take time to work these problems before proceeding with the next section. The answers are at the back of the text on page AN-4.

1. Determine whether the given value of the variable is a solution of the equation

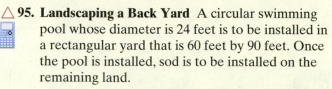

(a) $x = \dfrac{3}{2}$ **(b)** $x = -\dfrac{5}{2}$

2. Determine whether the given value of the variable is a solution of the equation

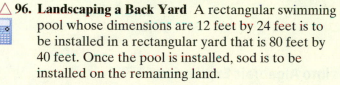

(a) $x = -4$ **(b)** $x = 1$

In Problems 3–14, solve the equation and check the solution.

3. $x + \dfrac{1}{2} = -\dfrac{1}{6}$ **4.** $-0.4m = 16$

5. $14 = -\dfrac{7}{3}p$ **6.** $8n - 11 = 13$

7. $\dfrac{5}{2}n - 4 = -19$ **8.** $-(5 - x) = 2(5x + 8)$

9. $7(x + 6) = 2x + 3x - 15$

10. $-7a + 5 + 8a = 2a + 8 - 28$

11. $-\dfrac{1}{2}(x - 6) + \dfrac{1}{6}(x + 6) = 2$

12. $0.3x - 1.4 = -0.2x + 6$

13. $5 + 3(2x + 1) = 5x + x - 10$

14. $3 - 2(x + 5) = -2(x + 2) - 3$

15. Investment You have $7500 to invest, and your financial advisor suggests that you put part of the money in a certificate of deposit (CD) that earns 2.4% simple interest and the remainder in bonds that earn 4% simple interest. Determine the amount you should invest in the CD to earn $220 interest at the end of one year, by solving the equation $0.024x + 0.04(7500 - x) = 220$, where x represents the amount of money invested in CDs.

16. Area of a trapezoid: $A = \frac{1}{2}h(B + b)$

   (a) Solve for b.

   (b) Find b when $A = 76$ in.^{2}, $h = 8$ in., and $B = 13$ in.

17. Volume of a right circular cylinder: $V = \pi r^2 h$

   (a) Solve for h.

   (b) Find h when $V = 117\pi$ in.^{2} and $r = 3$ in.

18. Solve the equation $3x + 2y = 14$ for y.

2.5 Problem Solving: Direct Translation

Objectives

1 Translate English Phrases into Algebraic Expressions

2 Translate English Sentences into Equations

3 Build Models for Solving Direct Translation Problems

Are You Ready for This Section?

Before getting started, take this readiness quiz. If you get a problem wrong, go back to the section cited and review the material.

R1. Solve the equation: $x + 34.95 = 60.03$ [Section 2.1, pp. 83–85]

R2. Solve the equation: $x + 0.25x = 60$ [Section 2.3, pp. 102]

▶ **1 Translate English Phrases into Algebraic Expressions**

One of the neat features of mathematics is that its symbols let us express English phrases briefly and consistently as algebraic expressions. For example, the English phrase "5 more than a number x" is represented algebraically as $x + 5$.

Some words or phrases easily translate into mathematical symbols, as shown in Table 2.

Table 2 Math Symbols and the Words They Represent

Add (+)	Subtract (−)	Multiply (·)	Divide (/)
sum	difference	product	quotient
plus	minus	times	divided by
greater than	subtracted from	of	per
more than	less	twice	ratio
exceeds by	less than	double	
in excess of	decreased by	half	
added to	fewer		
increased by			
combined			
altogether			

EXAMPLE 1 **Writing English Phrases Using Math Symbols**

Express each English phrase using mathematical symbols.

   (a) The sum of 2 and 5

   (b) The difference of 12 and 7

   (c) The product of −3 and 8

    (d) The quotient of 10 and 2

    (e) 9 less than 15

    (f) A number z decreased by 11

    (g) Three times the sum of a number x and 8

Solution

    (a) The word sum tells us to use the $+$ symbol, so "The sum of 2 and 5" is represented mathematically as $2 + 5$.

    (b) The word difference tells us to use the $-$ symbol, so "The difference of 12 and 7" is represented mathematically as $12 - 7$.

    (c) "The product of -3 and 8" is represented as $-3 \cdot 8$.

    (d) "The quotient of 10 and 2" is represented mathematically as $\dfrac{10}{2}$.

    (e) "9 less than 15" is represented mathematically as $15 - 9$.

    (f) "A number z decreased by 11" is represented algebraically as $z - 11$.

    (g) "Three times the sum of a number x and 8" is represented algebraically as $3(x + 8)$.

Work Smart

In Example 1(e), think of money when translating. Bob owes Mary $15. Mary says, "No, you owe me $9 less than $15", or $6.

Work Smart

When translating from English to math, try some specific examples. For example, to translate "a number z decreased by 11," pick specific values of z, as in "16 decreased by 11," which would be $16 - 11$, or 5. So "z decreased by 11" is $z - 11$.

In Example 1(g), the mathematical representation is $3(x + 8)$ rather than $3x + 8$ because the phrase "three times the sum" means to multiply the sum of the two numbers by 3. The phrase that would result in $3x + 8$ might be "the sum of three times a number and 8." Do you see the difference?

Quick ✓

In Problems 1–6, express each phrase using mathematical symbols.

1. The sum of 5 and 17

2. The difference of 7 and 4

3. The quotient of 25 and 3

4. The product of -2 and 6

5. Twice a less 2

6. Five times the difference of m and 6.

EXAMPLE 2   **Translating an English Phrase into an Algebraic Expression**

Write an algebraic expression for each problem.

    (a) The Raiders scored p points in a football game. The Packers scored 12 more points than the Raiders. Write an algebraic expression for the number of points the Packers scored.

    (b) A lumberman cuts a 50-foot log into two pieces. One piece is t feet long. Express the length of the second piece as an algebraic expression in t.

    (c) The number of quarters in a vending machine is two less than the number of dimes, d, in the machine. Write an expression for the number of quarters as an algebraic expression in d.

Solution

    (a) The phrase "more than" implies addition. The Packers scored $p + 12$ points in the football game.

    (b) The log is 50 feet long. If the lumberman cuts one piece 20 feet long, then the other piece must be $50 - 20 = 30$ feet. In general, if the lumberman cuts one piece t feet long, then the remaining piece must be $(50 - t)$ feet.

    (c) The phrase "less than" implies subtraction. Two less than the number of dimes is represented algebraically as $d - 2$.

Quick ✓

In Problems 7–9, translate each phrase into an algebraic expression.

7. Terry earned z dollars last week. Anne earned $50 more than Terry last week. Write an algebraic expression for Anne's earnings in terms of z.

8. Melissa paid x dollars for her college math book and $15 less than that for her college sociology book. Express the cost of her sociology book in terms of x.

9. Tim raided his piggy bank and found he had 75 dimes and quarters. Tim has d dimes. Express the number of quarters in his piggy bank as an algebraic expression in d.

EXAMPLE 3

Translating an English Phrase into an Algebraic Expression

Write an algebraic expression for each problem.

(a) The number of student tickets sold for a play is five fewer than four times the number n of nonstudent tickets sold. Write an algebraic expression for the number of student tickets sold in terms of n.

(b) The height of a full-grown maple tree is ten feet more than three times the height h of a sapling. Write an algebraic expression for the height of the full-grown tree in terms of h.

(c) Como believes he owes Abigail d dollars. Abigail says Como owes her 5 dollars less than three times the amount Como thought was owed. Write an algebraic expression for the amount Abigail says Como owes her in terms of d.

Solution

(a) The phrase "fewer than" implies subtraction. The number of student tickets sold is five fewer than four times the number of nonstudent tickets sold. So the algebraic expression representing the number of student tickets sold is 4 times the number of nonstudent tickets sold minus 5, or $4n - 5$.

(b) The full-grown tree is ten feet more than three times the height of the sapling, so $(3h + 10)$ feet represents the height of the full-grown tree.

(c) The phrase "less than" implies subtraction. Como owes Abigail 5 dollars less than 3 times d, so he owes her $3d - 5$ dollars.

●

Quick ✓

In Problems 10–12, translate each phrase into an algebraic expression.

10. The width of a platform is 2 feet less than three times the length, l. Express the width of the platform as an algebraic expression in terms of l.

11. T.J. has quarters and dimes in his piggy bank. The number of dimes is three more than twice the number of quarters. T.J. has q quarters. Express the number of dimes as an algebraic expression in terms of q.

12. The number of red M&Ms in a bowl is five less than three times the number of brown M&Ms, b, in the bowl. Express the number of red M&Ms in the bowl as an expression in terms of b.

▶ ❷ Translate English Sentences into Equations

Nearly every word problem that we do in algebra requires some type of translation. Learning to speak the language of math is the same as learning to speak any language. Now that we know how to translate English phrases to algebraic expressions, we can learn to translate English sentences to algebraic equations.

In Words
English phrase is to algebraic expression as English sentence is to algebraic equation.

In English, a complete sentence must contain a subject and a verb, so expressions or "phrases" are not complete sentences. For example, the expression "Beats me!" does not contain a subject, and the expression "5 more than a number x" does not contain a verb. Therefore neither expression is a complete sentence. The statement "5 more than a number x is 18" is a complete sentence because it contains a subject and a verb, so we can translate it into a mathematical statement. In mathematics, statements can be represented symbolically as equations.

Work Smart
We learned in Section 2.1 that an equation is a statement in which two algebraic expressions are equal.

In English, statements can be true or false. For example, "The moon is made of green cheese" is a false statement, while "The sky is blue" is a true statement. Mathematical statements can be true or false as well—we call them conditional equations.

Table 3 lists words that typically translate into an equal sign.

Table 3 Words That Translate into an Equal Sign

is	yields	are	equals
was	gives	results in	is equal to
is equivalent to			

Notice that these words are verbs. The equal sign in an equation acts like a verb in a sentence.

EXAMPLE 4 **Translating English Sentences into Equations**

Translate each sentence into an equation. Do not solve the equation.

(a) Five more than a number x is 20.

(b) Four times the sum of a number z and 3 is 15.

(c) The difference of x and 5 equals the quotient of x and 2.

Solution

(a) 5 more than a number x is   20
$$x + 5 \qquad = \qquad 20$$

(b) We first need to determine the sum and then multiply this result by 4.

Four times the sum of a number z and 3   is   15
$$4(z + 3) \qquad = \qquad 15$$

(c) The difference of x and 5   equals   the quotient of x and 2
$$x - 5 \qquad = \qquad \frac{x}{2}$$

Work Smart
5 more than a number x can be written as $5 + x$ or $x + 5$. Be careful! $5 - x$ is not the same as $x - 5$. Can you explain why?

Work Smart
The English sentence "The sum of four times a number z and 3 is 15" would be expressed mathematically as $4z + 3 = 15$. Do you see how this differs from Example 4(b)?

Quick ✓

13. In mathematics, English statements can be represented symbolically as _____.

In Problems 14–17, translate each English statement into an equation. Do not solve the equation.

14. The product of 3 and y is equal to 21.

15. The difference of x and 10 equals the quotient of x and 2.

16. Three times the sum of n and 2 is 15.

17. The sum of three times n and 2 is 15.

▶ **An Introduction to Problem Solving and Mathematical Models**

Every day we encounter various types of problems that must be solved. **Problem solving** is the ability to use information, tools, and our own skills to achieve a goal. For example, suppose 4-year-old Kevin wants a glass of water, but he is too short to reach the sink.

Kevin has a problem. To solve the problem, he finds a step stool and pulls it over to the sink. He uses the step stool to climb on the counter, opens the kitchen cabinet, and pulls out a cup. He then crawls along the counter top, turns on the faucet, fills the cup, and proceeds to drink the water. Problem solved!

Of course, Kevin could solve the problem other ways. Just as there are various ways to solve life's everyday problems, there are many ways to solve problems using mathematics. However, regardless of the approach, there are always some common aspects in solving any problem. For example, regardless of how Kevin ultimately ends up with his cup of water, someone must get a cup from the cabinet and someone must turn on the faucet.

One of the purposes of learning algebra is to be able to solve certain types of problems. To solve these problems, we will translate the verbal descriptions in the problems into equations we can solve. The process of turning a verbal description into a mathematical equation is known as **mathematical modeling**. We call the equation that is developed the **mathematical model**.

Not all models are mathematical. In general, a **model** is a way of using graphs, pictures, equations, or even verbal descriptions to represent a real-life situation. Because the world is a complex place, we need to simplify information when we develop a model. For example, a map is a model of our road system. Maps don't show all the details of the system (such as trees, buildings, or potholes), but they do a good job of describing how to get from point *A* to point *B*. Mathematical models are similar in that we often make assumptions regarding our world in order to make the mathematics less complicated.

Every problem is unique in some way. However, because many problems are similar, we can categorize problems. In this text, we will solve five categories of problems.

Five Categories of Problems

1. **Direct Translation**—problems in which we translate from English directly to an equation by using key words in the verbal description.
2. **Geometry**—problems in which the unknown quantities are related through geometric formulas.
3. **Mixtures**—problems in which two or more quantities are combined in some fashion.
4. **Uniform Motion**—problems in which an object travels at a constant speed.
5. **Work Problems**—problems in which two or more entities join forces to complete a job.

In this section, we will concentrate on direct translation problems. However, the following guidelines will help you solve any category of problem.

Solving Problems with Mathematical Models

Step 1: Identify What You Are Looking For Read the problem very carefully, perhaps two or three times. Identify the type of problem and the information that you wish to find. It is fairly typical that the last sentence in the problem indicates what it is we wish to solve for in the problem.

Step 2: Give Names to the Unknowns Assign variables to the unknown quantities. Choose a variable that is representative of the unknown quantity. For example, use *t* for time.

Step 3: Translate the Problem into Mathematics Read the problem again. This time, after each sentence is read, determine whether the sentence can be translated into a mathematical statement or expression in terms of the variables identified in Step 2. It is often helpful to create a table, chart, or figure. If necessary, combine the mathematical statements or expressions into an equation that can be solved.

Step 4: Solve the Equation(s) Found in Step 3 Solve the equation for the variable.

Step 5: Check the Reasonableness of Your Answer Check your answer to be sure that it makes sense. If it does not, go back and try again.

Step 6: Answer the Question Write your answer in a complete sentence.

Let's review each of these steps, one at a time.

- **Identify** Carefully read the problem. Reading a verbal description of a problem is not like reading a spy novel. You may not know how to solve the problem while reading it, but you should get a sense of which of the five categories the problem falls into, what information you are given, and what you are being asked to do.
- **Name** Assign variables to the unknowns. Write down the name of each variable and what it represents. Use this to check your final answer.
- **Translate** In this step, you develop a model (an equation) that mathematically describes the problem.
- **Solve the equation** This is generally the easy part. Most students say, "I could solve the problem if I could find the right equation."
- **Check** Checking your answer can be difficult because you can make two types of errors: 1. You correctly translate the problem into a model but then make an error solving the equation. 2. You misinterpret the problem and develop an incorrect model. Your solution will satisfy your model, but it probably will not be the solution to the original problem. We can check for this type of error by determining whether the solution is reasonable.
- **Answer the question** Always be sure to answer the question being asked.

> ### Quick ✓
>
> **18.** Letting variables represent unknown quantities and then expressing relationships among the variables in the form of equations is called _____ _____.

❸ Build Models for Solving Direct Translation Problems

Let's look at a "direct translation" problem. Remember, these problems can be set up by reading the problem and translating the verbal description into an equation.

EXAMPLE 5) **Solving a Direct Translation Problem**

The price of a "premium" ticket for a Broadway show in New York is $69 more than twice the price of a premium ticket for the same show in Chicago. If you buy a ticket for this show in Chicago, and your friend buys a ticket for the same show in New York, and the total cost of the tickets is $436.50, what does each ticket cost?

Solution

Step 1: Identify This is a direct translation problem. We want to know the price of a premium ticket for a Broadway show in New York and for the same show in Chicago.

Step 2: Name We know the price of a ticket in New York is $69 more than twice the price of a ticket for that show in Chicago. We will let c represent the ticket price in Chicago. Then $2c + 69$ represents the ticket price in New York.

Step 3: Translate Because we know that the total cost for the tickets to both performances is $436.50, we use the equation

price of ticket in Chicago plus price of ticket in New York equals total cost

$$c \quad + \quad 2c + 69 \quad = \quad 436.50 \quad \text{The Model}$$

Step 4: Solve We now solve the equation

$$c + (2c + 69) = 436.50$$

Combine like terms: $\quad 3c + 69 = 436.50$

Subtract 69 from each side of the equation: $\quad 3c + 69 - 69 = 436.50 - 69$

$$3c = 367.50$$

Divide each side by 3: $\quad \dfrac{3c}{3} = \dfrac{367.50}{3}$

$$c = 122.50$$

Work Smart

Remember that you can use any letter to represent the unknown(s) when you make your model. Choose a letter that reminds you what it represents. For example, use t for time.

Work Smart

It is helpful to assign the variable to the quantity that you know the **least** about.

So, the price of a ticket for the show in Chicago is $c = \$122.50$. The price of a ticket for the same Broadway show in New York is $2c + 69 = 2(122.50) + 69 = 314$. It costs $314 to see the show in New York.

Step 5: Check Is the total cost of both tickets $436.50? Since $\$122.50 + \$314 = \$436.50$, our answers are correct.

Step 6: Answer The price of a premium ticket to a Broadway show in Chicago is $122.50, and the price of a premium ticket to the same Broadway show in New York is $314.

> **Quick ✓**
>
> *Translate the problem into an algebraic equation and solve the equation for the unknowns.*
>
> **19.** Sean and Connor decide to buy a pizza. The pizza costs $15 and they decide to split the cost based upon how much pizza each eats. Connor eats two-thirds of the amount that Sean eats, so Connor pays two-thirds of the amount that Sean pays. How much does each pay?

Work Smart

Examples of consecutive even integers are

$$16, 18, 20$$
$$-10, -8, -6$$
$$752, 754, 756$$

Examples of consecutive odd integers are

$$9, 11, 13$$
$$-35, -33, -31$$
$$623, 625, 627$$

▶ Consecutive Integer Problems

Recall that an *integer* is a member of the set $\{\ldots, -3, -2, -1, 0, 1, 2, 3, \ldots\}$. An *even integer* is an integer that is divisible by 2, such as, 16 and 24. An *odd integer* is an integer that is not even, such as 9 and 17. Consecutive integers differ by 1, so if n represents the first integer, then $n + 1$ represents the second integer, $n + 2$ represents the third integer, and so on. Consecutive *even* integers, such as 14 and 16, differ by 2, so if n represents the first even integer, then $n + 2$ represents the second even integer and $n + 4$ represents the third even integer. Consecutive odd integers also differ by 2 (41 and 43, for example), so if n represents the first odd integer, then $n + 2$ represents the second odd integer, and $n + 4$ represents the third odd integer. See Figure 3.

Figure 3

Consecutive integers
$n \quad n+1 \quad n+2$

Consecutive **even** integers, if n is even
$n \quad n+2 \quad n+4$

Consecutive **odd** integers, if n is odd
$n \quad n+2 \quad n+4$

EXAMPLE 6 **Solving a Direct Translation Problem: Consecutive Integers**

The sum of three consecutive even integers is 324. Find the integers.

Solution

Step 1: Identify This is a direct translation problem. We need three consecutive even integers whose sum is 324.

Step 2: Name We will let n represent the first even integer, so $n + 2$ is the second even integer, and $n + 4$ is the third even integer.

Step 3: Translate The sum of the three consecutive even integers is 324, so our equation is

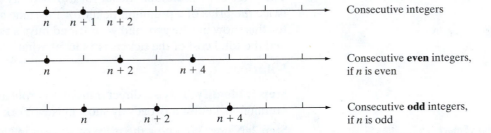

first even integer   second even integer   third even integer   sum

$$n \quad + \quad (n+2) \quad + \quad (n+4) \quad = 324 \qquad \text{The Model}$$

Step 4: Solve We solve the equation.

$$n + (n + 2) + (n + 4) = 324$$

Combine like terms: $\qquad 3n + 6 = 324$

Subtract 6 from each side of the equation: $\qquad 3n + 6 - 6 = 324 - 6$

$$3n = 318$$

Divide each side by 3: $\qquad \dfrac{3n}{3} = \dfrac{318}{3}$

$$n = 106$$

Since n, the first even integer, is 106, the remaining even integers are 108 and 110.

Step 5: Check The numbers are all even integers and $106 + 108 + 110 = 324$. The answer is correct.

Step 6: Answer The three consecutive even integers are 106, 108, and 110. ●

> **Quick ✓**
>
> **20.** *True or False* If n represents the first of three consecutive odd integers, then $n + 1$ and $n + 3$ represent the next two odd integers.
>
> *In Problems 21 and 22, translate the problem into an algebraic equation and solve the equation for the unknowns.*
>
> **21.** The sum of three consecutive even integers is 270. Find the integers.
>
> **22.** The sum of 4 consecutive odd integers is 72. Find the integers.

EXAMPLE 7 **Solving a Direct Translation Problem: Piece Lengths**

A carpenter is building a shelving system and cuts a 14-foot length of cherry shelving into three pieces. The second piece is twice as long as the first, and the third piece is 2 feet longer than the first. Find the length of each piece of cherry shelving.

Solution

Step 1: Identify This is a direct translation problem. We are looking for the length of each piece of shelving. We know that the board is 14 feet long.

Step 2: Name Because the lengths of the second and third pieces are described in terms of the length of the first piece, we will let x represent the length of the first piece. The second piece is $2x$ because it is twice as long as the first piece. The third piece is $x + 2$ because it is 2 feet longer than the first piece. See Figure 4.

Figure 4

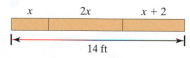

Step 3: Translate The lengths of the three pieces must sum to 14 feet, so our equation is

length of first piece		length of second piece		length of third piece		total length	
x	$+$	$2x$	$+$	$(x + 2)$	$=$	14	The Model

Step 4: Solve We solve the equation.

$$x + 2x + (x + 2) = 14$$

Combine like terms: $\qquad 4x + 2 = 14$

Subtract 2 from each side of the equation: $\qquad 4x + 2 - 2 = 14 - 2$

$$4x = 12$$

Divide each side by 4: $\qquad \dfrac{4x}{4} = \dfrac{12}{4}$

$$x = 3$$

Work Smart

Sometimes these problems are called "the whole equals the sum of the parts" problems because the values of the "parts" must sum to the value of the "whole."

Since the length of the first piece is 3 feet, the second piece is $2x = 2 \cdot 3 = 6$ feet and the third piece is $x + 2 = 3 + 2 = 5$ feet.

Step 5: Check The sum of the three pieces of cherry shelving is $3 + 6 + 5 = 14$, so we have the correct answer.

Step 6: Answer The three pieces of shelving are 3 feet, 6 feet, and 5 feet long. ●

Quick ✔

In Problem 23, translate the problem into an algebraic equation and solve.

23. A 76-inch length of ribbon is to be cut into three pieces. The longest piece is to be 24 inches longer than the shortest piece, and the third piece is to be half the length of the longest piece. Find the length of each piece of ribbon.

⊚ **EXAMPLE 8** **Investment Decisions**

A total of $25,000 is to be invested in bonds and certificates of deposit (CDs). The amount in CDs is to be $8000 less than the amount in bonds. How much is to be invested in each type of investment?

Solution

Step 1: Identify We want to know the amount invested in CDs and the amount invested in bonds.

Step 2: Name Let b represent the amount invested in bonds.

Step 3: Translate Suppose we invested $18,000 in bonds, then the amount in CDs will be $8000 less than this amount, or $10,000. In general, if b is the amount invested in bonds, then $b - 8000$ represents the amount invested in CDs. Also, our total investment in bonds and CDs is $25,000, so

$$\underbrace{b}_{\text{Amount invested in bonds}} + \underbrace{(b - 8000)}_{\text{Amount invested in CDs}} = \underbrace{25{,}000}_{\text{Total investment}} \quad \text{The Model}$$

Step 4: Solve
$$b + (b - 8000) = 25{,}000$$

Combine like terms: $\quad 2b - 8000 = 25{,}000$

Add 8000 to each side of the equation: $\quad 2b = 33{,}000$

Divide both sides by 2: $\quad b = 16{,}500$

Step 5: Check If we invest $16,500 in bonds, then the amount invested in CDs should be $8000 less than this amount, or $8500. The total investment is $16{,}500 + \$8500 = \$25{,}000$, so the answer checks.

Step 6: Answer Invest $16,500 in bonds and $8500 in CDs. ●

Quick ✔

In Problem 24, translate the problem into an algebraic equation and solve.

24. A total of $18,000 is to be invested in stocks and bonds. If the amount invested in bonds is twice that invested in stocks, how much is invested in each category?

⊚ **EXAMPLE 9** **Choosing a Long-Distance Carrier**

MCI has a long-distance phone plan that charges $6.99 a month plus $0.04 per minute of usage. Sprint has a long-distance phone plan that charges $5.95 a month plus $0.05 per minute of usage. For how many minutes of long-distance calls will the monthly costs for the two plans be the same? (SOURCE: *MCI and Sprint*)

Solution

Step 1: Identify This is a direct translation problem. We are looking for the number of minutes for which the two plans cost the same.

Step 2: Name Let m represent the number of long-distance minutes used in the month.

Step 3: Translate The monthly fee for MCI is $6.99 plus $0.04 for each minute used. So if one minute is used, the fee is $6.99 + 0.04(1) = 7.03$ dollars. For two minutes, the fee is $6.99 + 0.04(2)$ dollars. In general, for m minutes, the monthly fee is $(6.99 + 0.04m)$ dollars. Similar logic results in the monthly fee for Sprint being $(5.95 + 0.05m)$ dollars. To find the number of minutes for which the two plans cost the same, we need to solve

$$\underset{\text{Cost for MCI}}{6.99 + 0.04m} = \underset{\text{Cost for Sprint}}{5.95 + 0.05m} \quad \text{The Model}$$

Step 4: Solve

$$6.99 + 0.04m = 5.95 + 0.05m$$

Subtract 5.95 from both sides: $\quad 1.04 + 0.04m = 0.05m$

Subtract $0.04m$ from both sides: $\quad 1.04 = 0.01m$

Multiply both sides by 100: $\quad 104 = m$

Step 5: Check The cost of MCI's plan will be $6.99 + 0.04(104) = \$11.15$. The cost of Sprint's plan will be $5.95 + 0.05(104) = \$11.15$. They are the same!

Step 6: Answer the Question The two plans will cost the same if 104 minutes are used. ●

Quick ✓

25. You need to rent a moving truck. EZ-Rental charges $30 per day plus $0.15 per mile. Do It Yourself Rental charges $15 per day plus $0.25 per mile. For how many miles will the daily cost of renting from these companies be the same?

26. Your grandmother wants a cell phone for emergency use only. You have found that Company A charges $15 per month plus $0.05 per minute, while Company B charges $0.20 per minute with no monthly charge. For how many minutes will the monthly cost be the same?

2.5 Exercises   Exercise numbers in **green** have complete video solutions in MyMathLab

Problems 1–26 are the Quick ✓s that follow the EXAMPLES.

Building Skills

In Problems 27–44, translate each phrase into an algebraic expression. Let x represent the unknown number. See Objective 1.

27. the sum of 5 and a number

28. a number increased by 32.3

29. the product of a number and $\dfrac{2}{3}$

30. the product of -2 and a number

31. half of a number

32. double a number

33. a number less -25

34. 8 less than a number

35. the quotient of a number and 3

36. the quotient of -14 and a number

37. $\dfrac{1}{2}$ more than a number

38. $\dfrac{4}{5}$ of a number

39. 9 more than 6 times a number

40. 21 more than 4 times a number

41. twice the sum of 13.7 and a number

42. 50 less than half of a number

43. the sum of twice a number and 31

44. the sum of twice a number and 45

In Problems 45–52, choose a variable to represent one quantity. State what that quantity represents and then express the second quantity in terms of the first. See Objective 1.

45. The Columbus Clippers scored 5 more runs than the Richmond Braves.

46. The Toronto Blue Jays scored 3 fewer runs than the Cleveland Indians.

47. Jan has $0.55 more in her piggy bank than Bill.

48. Beryl has $0.25 more than 3 times the amount Ralph has.

49. Janet and Kathy will share the $200 grant.

50. Juan and Emilio will share the $1500 lottery winnings.

51. There were 1433 visitors to the Arts Center Spring show. Some were adults and some were children.

52. There were 12,765 fans at a recent NBA game. Some held paid admission tickets, and some held special promotion tickets.

In Problems 53–60, translate each statement into an equation.
Let x represent the unknown number. DO NOT SOLVE.
See Objective 2.

53. The sum of a number and 15 is -34.

54. The sum of 43 and a number is -72.

55. 35 is 7 less than triple a number.

56. 49 is 3 less than twice a number.

57. The quotient of a number and -4, increased by 5, is 36.

58. The quotient of a number and -6, decreased by 15, is 30.

59. Twice the sum of a number and 6 is the same as 3 more than the number.

60. Twice the sum of a number and 5 is the same as 7 more than the number.

Applying the Concepts

61. **Number Sense** The sum of a number and -12 is 71. Find the number.

62. **Number Sense** The difference between a number and 13 is -29. Find the number.

63. **Number Sense** 25 less than twice a number is -53. Find the number.

64. **Number Sense** The sum of 13 and twice a number is -19. Find the number.

65. **Consecutive Integers** The sum of three consecutive integers is 165. Find the numbers.

66. **Consecutive Integers** The sum of three consecutive odd integers is 81. Find the numbers.

67. **Bridges** The longest bridge in the United States is the Verrazano-Narrows Bridge. The second-longest bridge in the United States is the Golden Gate Bridge, which is 60 feet shorter than the Verrazano-Narrows Bridge. The combined length of the two bridges is 8460 feet. Find the length of each bridge.

68. **Towers** The tallest buildings in the world (those having the most floors) are Burj Khalifa in Dubai, UAE, and the Abraj Al-Bait Tower in Mecca, Saudi Arabia. The Abraj Al-Bait Tower has 43 fewer floors than Burj Khalifa. The two buildings together have 283 floors. Find the number of floors in each building.

69. **Buying a Motorcycle** The total price for a new motorcycle is $11,894.79. The tax, title, and dealer preparation charges amount to $679.79. Find the price of the motorcycle before the extra charges.

70. **Buying a Desk** The total price for a new desk is $285.14, including sales tax of $16.14. Find the original cost of the desk.

71. **Finance** A total of $20,000 is to be invested, some in bonds and some in certificates of deposit (CDs). The amount invested in bonds is to be $3000 greater than the amount invested in CDs. How much is to be invested in each type of investment?

72. **Finance** A total of $10,000 is to be divided between Sean and George. George is to receive $3000 less than Sean. How much will each receive?

73. **Investments** Suppose that your Aunt May has left you an unexpected inheritance of $32,000. You have decided to invest the money rather than blow it on frivolous purchases. Your financial advisor has recommended that you diversify by placing some of the money in stocks and some in bonds. Based upon current market conditions, she has recommended that the amount in bonds should equal three-fifths of the amount invested in stocks. How much should be invested in stocks? How much should be invested in bonds?

74. **Investments** Jack and Diane have $40,000 to invest. Their financial advisor has recommended that they diversify by placing some of the money in stocks and some in bonds. Based upon current market conditions, he has recommended that the amount in bonds should equal two-thirds of the amount invested in stocks. How much should be invested in stocks? How much should be invested in bonds?

75. **Cereal** A serving of Kashi Go Lean Crunch cereal contains 4 times as much fiber as a serving of Kellogg's Smart Start Cereal. If you eat a serving of each cereal, you will consume 10 g of fiber. Find the amount of fiber in each cereal.

76. **Books** A paperback edition of a book costs $12.50 less than the hardback edition of the book. If you purchase one of each version, you will pay $37.40. Find the cost of the paperback edition of the book.

77. Income On a joint income tax return, Elizabeth Morrell's adjusted gross income was $2549 more than her husband Dan's adjusted gross income. Their combined adjusted gross income was $55,731. Find Elizabeth Morrell's adjusted gross income.

78. Spring Break Allison went shopping to prepare for her Spring Break trip. Her bathing suit cost $8 more than a pair of shorts, and a T-shirt cost $2 less than the shorts. Find the cost of the bathing suit if Allison spent $60 on the items, before sales tax.

79. Truck Rentals You need to rent a moving truck. You have identified two companies that rent trucks. EZ-Rental charges $35 per day plus $0.15 per mile. Do It Yourself Rental charges $20 per day plus $0.25 per mile. For how many miles will the daily cost of renting be the same?

80. Cellular Telephones You need a new cell phone for emergencies only. Company A charges $12 per month plus $0.10 per minute, while Company B charges $0.15 per minute with no monthly service charge. For how many minutes will the monthly cost be the same?

81. Comparing Printers Samuel is trying to decide between two laser printers, one manufactured by Hewlett-Packard, the other by Brother. Both have similar features and warranties, so price is the determining factor. The Hewlett-Packard costs $200, and printing costs are approximately $0.03 per page. The Brother costs $240, and printing costs are approximately $0.01 per page. How many pages need to be printed for the cost of the two printers to be the same?

82. Comparing Job Offers Hans has just been offered two sales jobs selling vacuums. The first job offer is a base monthly salary of $2000 plus a commission of $50 for each vacuum sold. The second job offer is a base monthly salary of $1200 plus a commission of $60 for each vacuum sold. How many vacuums must be sold for the two jobs to pay the same salary?

83. Adjusted Gross Income On a joint income tax return, Jensen Beck's adjusted gross income was $249 more than his wife Maureen's adjusted gross income. Their combined adjusted gross income was $72,193. Find Jensen and Maureen Beck's adjusted gross income.

84. Camping Trip Jaime Juarez purchased some new camping equipment. He spent $199 on a cookware set, a lantern, and a cook stove. The cookware set cost $30 more than the lantern, and the cook stove cost $34 more than the lantern. Find the cost of each item.

85. Computing Grades Going into the final, which will count as three tests, Monica has test scores of 76, 90, 94, and 88. What score does Monica need on the final exam in order to have an average score of 90?

86. Computing Grades Going into the final exam, which will count as two tests, Brooke has test scores of 80, 83, 71, 61, and 95. What score does Brooke need on the final exam in order to have an average score of 80?

Extending the Concepts

In Problems 87–90, write a problem that would translate into the given equation.

87. $10x = 370$

88. $5n + 10 = 170$

89. $\dfrac{x + 74}{2} = 80$

90. $n + n + 2 = 98$

△ **91. Angles** The sum of the measures of the three angles in any triangle is 180 degrees. The measure of the smallest angle of a certain triangle is half the measure of the second angle. The measure of the largest angle is 40° more than 4 times the measure of the smallest. Find the measure of each angle.

△ **92. Angles** The sum of the measures of the three angles in any triangle is 180 degrees. The measure of one angle of a certain triangle is 1° more than three times the measure of the smallest angle. The measure of the third angle is 13° less than twice the measure of the second angle. Find the measure of each angle.

Explaining the Concepts

93. How is mathematical modeling related to problem solving? Why do we make assumptions when creating mathematical models?

94. What is the difference between an algebraic expression and an equation? How is each related to English phrases and English statements?

95. Two students write an equation to solve a word problem with consecutive odd integers. One student assigns the variables as $n - 1, n + 1, n + 3$, where n is an even integer. A second student uses $n, n + 2, n + 4$, where n is an odd integer. Which student is correct? Will the value for n be the same for both? Make up a problem that can be solved in more than one way, and explain how the variables were assigned.

96. Using the algebraic expression $3x + 5$, make up a problem that uses the direction "evaluate." Using the same algebraic expression, make up a problem that uses the direction "solve."

2.6 Problem Solving: Problems Involving Percent

Objectives

1. Solve Problems Involving Percent
2. Solve Business Problems That Involve Percent

Are You Ready for This Section?

Before getting started, take the following readiness quiz. If you get a problem wrong, go back to the section cited and review the material.

R1. Write 45% as a decimal. [Section 1.2, p. 15]

R2. Write 0.2875 as a percent. [Section 1.2, p. 16]

R3. Simplify: (a) $0.4 \cdot 50$ (b) $\dfrac{15}{0.3}$ [Section 1.5, pp. 44–45]

R4. Simplify: $p + 0.05p$ [Section 2, 3, p. 102]

▶ 1 Solve Problems Involving Percent

Percent means "divided by 100" or "per hundred." The symbol % denotes percent, so 45% means 45 out of 100 or $\dfrac{45}{100}$ or 0.45. In applications involving percents, we often see the word "of," as in 20% of 60. The word "of" translates into multiplication in mathematics, so 20% of 60 means $0.20 \cdot 60$.

EXAMPLE 1 **Solving an Equation Involving Percent**

A number is 35% of 40. Find the number.

Solution

Step 1: Identify We want to know the unknown number.

Step 2: Name Let n represent the number.

Step 3: Translate

$$\underset{n}{\underbrace{\text{a number}}} \quad \underset{=}{\underbrace{\text{is}}} \quad \underset{0.35}{\underbrace{35\%}} \quad \underset{\cdot}{\underbrace{\text{of}}} \quad \underset{40}{\underbrace{40}}$$

Step 4: Solve

$$n = 0.35\,(40)$$

Multiply: $n = 14$

Step 5: Check Check the multiplication: $0.35\,(40) = 14$.

Step 6: Answer 14 is 35% of 40.

Quick ✓

1. Percent means "divided by ___."

2. *True or False* 40% of 120 means $0.4 \cdot 120$.

3. A number is 60% of 90. Find the number.

4. A number is 3% of 80. Find the number.

5. A number is 150% of 24. Find the number.

6. A number is $8\dfrac{3}{4}\%$ of 40. Find the number.

EXAMPLE 2 **Solving an Equation Involving Percent**

The number 240 is what percent of 800?

Solution

Step 1: Identify We want to know the percent.

Step 2: Name Let x represent the percent.

Step 3: Translate

$$\underset{240}{\underline{240}} \quad \underset{=}{\underline{\text{is}}} \quad \underset{x}{\underline{\text{what percent}}} \quad \underset{\cdot}{\underline{\text{of}}} \quad \underset{800}{\underline{800?}}$$

Step 4: Solve

$$240 = 800x$$

Divide each side by 800: $\quad \dfrac{240}{800} = \dfrac{800x}{800}$

$$0.3 = x$$

Since we are finding a percent and our answer is a decimal, we must change 0.3 to a percent by moving the decimal point two places to the right: 0.30 = 30%.

Step 5: Check Is 240 equal to 30% of 800? Does $(0.30)(800) = 240$? Yes!

Step 6: Answer The number 240 is 30% of 800. ●

> **Quick ✔**
>
> **7.** The number 8 is what percent of 20?
>
> **8.** The number 15 is what percent of 40?
>
> **9.** The number 44 is what percent of 40?

EXAMPLE 3 **Solving an Equation Involving Percent**

42 is 35% of what number?

Solution

Step 1: Identify We want to know the number.

Step 2: Name Let x represent the number.

Step 3: Translate

$$\underset{42}{\underline{42}} \quad \underset{=}{\underline{\text{is}}} \quad \underset{0.35}{\underline{35\%}} \quad \underset{\cdot}{\underline{\text{of}}} \quad \underset{x}{\underline{\text{what number?}}}$$

Step 4: Solve

$$42 = 0.35x$$

Divide each side by 0.35: $\quad \dfrac{42}{0.35} = \dfrac{0.35x}{0.35}$

$$120 = x$$

Step 5: Check Is 35% of 120 equal to 42? Because $(0.35)(120) = 42$, the answer is correct.

Step 6: Answer 42 is 35% of 120. ●

> **Quick ✔**
>
> **10.** 14 is 28% of what number?   **11.** 14.8 is 18.5% of what number?
>
> **12.** 102 is 136% of what number?

EXAMPLE 4 **Educational Attainment of U.S. Residents**

In 2010, the number of U.S. residents aged 25 years or older was approximately 200,000,000. If 31% these U.S. residents are high school graduates, determine the number of U.S. residents aged 25 years or older in 2010 who were high school graduates. (SOURCE: *U.S. Census Bureau*)

Solution

Step 1: Identify We want to know the number of U.S. residents aged 25 years or older in 2010 who are high school graduates.

Step 2: Name Let h represent the number of high school graduates.

Step 3: Translate We know that 31% of U.S. residents aged 25 years or older are high school graduates. We also know that in the year 2010, there were approximately 200,000,000 U.S. residents aged 25 years or older. We translate the words of the problem:

$$\underbrace{31\%}\ \underbrace{of}\ \underbrace{U.S.\ residents}\ \underbrace{are}\ \underbrace{high\ school\ graduates}$$
$$0.31\ \cdot\ 200{,}000{,}000\ =\ h \qquad \text{The Model}$$

Step 4: Solve We solve the equation.

$$(0.31)(200{,}000{,}000) = h$$
$$\text{Multiply:} \qquad 62{,}000{,}000 = h$$

Step 5: Check We can recheck our arithmetic. Because $0.31 \cdot 200{,}000{,}000 = 61{,}000{,}000$, the answer is correct.

Step 6: Answer In the year 2010, the number of U.S. residents aged 25 years or older who were high school graduates was 62,000,000.

Quick ✓

13. In 2010, the number of U.S. residents aged 25 years or older was approximately 200,000,000. If 20% of all U.S. residents aged 25 years or older have bachelor's degrees, determine the number of U.S. residents aged 25 years or older in 2010 who have bachelor's degrees.

▶ ② Solve Business Problems That Involve Percent

Now let's look at percent problems from business. Typically, these problems involve discounts or mark-ups that businesses use in determining their prices. They may also include finding the cost of an item excluding sales tax, as shown in Example 5.

EXAMPLE 5 **Finding the Cost of an Item Excluding Sales Tax**

You just purchased a new pair of jeans for $41.34, including 6% sales tax. How much did the jeans cost before sales tax?

Solution

Step 1: Identify We want to know the price of the jeans before sales tax.

Step 2: Name Let p represent the price of the jeans before sales tax.

Step 3: Translate The total cost is the sum of the price of the jeans, p, and the amount of tax. The amount of tax is 6% of the price of the jeans. Thus the total price is

Work Smart

To change a percent to a decimal, move the decimal point two places to the left.

$$\underbrace{original\ price}\ \ \underbrace{amount\ of\ tax}\ \ \underbrace{total\ price}$$
$$p\ \ +\ \ 0.06p\ \ =\ 41.34 \qquad \text{The Model}$$

Step 4: Solve

$$p + 0.06p = 41.34$$

Combine like terms:

$$1.06p = 41.34$$

Divide each side of the equation by 1.06:

$$\frac{1.06p}{1.06} = \frac{41.34}{1.06}$$

$$p = 39$$

Step 5: Check If the jeans cost $39, then the jeans plus the 6% sales tax on $39 equals $39 + (0.06)($39) = $39 + $2.34 = $41.34, so our answer is correct.

Step 6: Answer The jeans cost $39 before the sales tax.   ●

Quick ✔

14. Suppose you just purchased a used car. The price of the car, including 7% sales tax, was $7811. What was the price of the car before sales tax?

15. As a reward for being named "Teacher of the Year," Janet will receive a 2.5% pay raise. If Janet's current salary is $59,000, determine her new salary.

Another type of percent problem involves the discounts or mark-ups that businesses use in determining their prices. For these problems, it is helpful to remember the following:

Original Price − Discount = Sale Price

Wholesale Price + Markup = Selling Price

EXAMPLE 6   **Markdown**

A local clothing store is going out of business and all merchandise is marked down by 40%. The sale price of a jacket is $108. What was the original price?

Solution

Step 1: Identify In this direct translation problem, we want the original price of a jacket that was marked down by 40%, and we know that its sale price is $108.

Step 2: Name Let p represent the original price of the jacket.

Step 3: Translate We know that the original price minus the discount is the sale price, which is $108, so

$$p - \text{discount} = 108$$

"Marked down by 40%" means that the discount is 40% of the original price, so the discount is represented by $0.40p$. Substitute this into the equation $p - \text{discount} = 108$:

$$p - 0.40p = 108 \quad \text{The Model}$$

Step 4: Solve

$$p - 0.40p = 108$$
$$1p - 0.40p = 108$$

Combine like terms:

$$0.60p = 108$$

Divide each side by 0.60:

$$\frac{0.60p}{0.60} = \frac{108}{0.60}$$

$$p = 180$$

Step 5: Check If p, the original price of the jacket, was $180, then the discount would be $0.40($180$) = 72. Subtracting $72 from the original price of $180 gives $108, the sale price. Also notice that our answer is reasonable—the sale price is less than the original price.

Step 6: Answer The original price of the jacket was $180. ●

Quick ✓

16. Suppose a gas station marks its gasoline up 10%. If the gas station charges $4.29 per gallon of 87 octane gasoline, what does it pay for the gasoline?

17. A furniture store marks recliners down by 25%. The sale price, excluding the sales tax, is $494.25. Find the original price of each recliner.

18. Albert's house lost 2% of its value last year. The value of the house is now $148,000. To the nearest dollar, what was the value of the house one year ago?

2.6 Exercises

Exercise numbers in green have complete video solutions in MyMathLab

Problems 1–18 are the Quick ✓s that follow the EXAMPLES.

Building Skills
In Problems 19–36, find the unknown in each percent question. See Objective 1.

19. What is 50% of 160?

20. What is 80% of 50?

21. 7% of 200 is what number?

22. 75% of 20 is what number?

23. What number is 16% of 30?

24. What number is 150% of 9?

25. 18 is 15% of what number?

26. 40% of what number is 122?

27. 60% of 120 is what number?

28. 45% of what number is 900?

29. 24 is 120% of what number?

30. 11 is 5.5% of what number?

31. What percent of 60 is 24?

32. 15 is what percent of 75?

33. 1.5 is what percent of 20?

34. 4 is what percent of 25?

35. What percent of 300 is 600?

36. What percent of 16 is 12?

Applying the Concepts

37. **Sales Tax** The sales tax in Delaware County, Ohio, is 6%. The total cost of purchasing a tennis racket, including tax, is $57.24. Find the cost of the tennis racket before sales tax.

38. **Sales Tax** The sales tax in Franklin County, Ohio, is 5.75%. The total cost of purchasing a used Honda Civic, including sales tax, is $8460. Find the cost of the car before sales tax.

39. **Pay Cut** Todd works for a computer firm, and last year he earned a salary of $120,000. Recently, Todd was required to take a 15% pay cut. Find Todd's salary after the pay cut.

40. **Pay Raise** MaryBeth works from home as a graphic designer. Recently she raised her hourly rate by 5% to cover increased costs. Her new hourly rate is $29.40. Find MaryBeth's previous hourly rate.

41. **Bad Investment** After Mrs. Fisher lost 9% of her investment, she had $22,750. What was Mrs. Fisher's original investment?

42. **Good Investment** Perry just learned that his house increased in value by 4% over the past year. The value of the house is now $208,000. What was the value of the home one year ago?

43. **Marking Up Furniture** Darvin Furniture marks up the price of a dining room set 60%. What will be the selling price of a dining room set that Darvin buys for $1400?

44. **Hailstorm** Toyota Town had a 15%-off sale on vehicles that had been damaged in a hailstorm. The original price of a truck was $28,000. What is the sale price?

45. **Discount Pricing** A wool suit that was originally priced at $700 is on the 30%-off clearance rack. What is the sale price of the suit?

46. Business: Marking Up the Price of Books A college bookstore marks up the price that it pays the publisher for a book by 30%. If the selling price of a book is $117, how much did the bookstore pay for the book?

47. Furniture Sale A furniture store discounted a dining room table by 40%. The discounted price of the dining room table was $240. Determine the original price of the dining room table.

48. Vacation Package The Liberty Travel Agency advertised a 5-night vacation package in Jamaica for 30% off the regular price. The sale price of the package is $1139. To the nearest dollar, how much was the vacation package before the 30%-off sale? (SOURCE: *New York Times*, 2/10/08)

49. Voting In an election for school president, the loser received 60% of the number of votes that the winner received. If 848 votes were cast, how many did each receive?

50. Voting On a committee consisting of Republicans and Democrats, there are twice as many Republicans as Democrats. If 30% of the Republicans and 20% of the Democrats voted in favor of a bill and there were 8 yes votes, how many people are on the committee?

51. Commission Melanie receives a 3% commission on every house she sells. If she received a commission of $8571, what was the value of the house she sold?

52. Commission Mario collects a commission for bringing in advertisers to his magazine company. He receives 8% on $450 full-page ads and 5% for $300 half-page ads. If he sells twice as many half-page as full-page ads and his commission was $5610, how many of each type did he bring in?

53. Job Growth According to the Bureau of Labor Statistics, the demand for physical therapists is expected to grow 30% over the next 10 years. If there are currently 1,400,000 physical therapists in the United States, how many are there expected to be in 10 years?

54. Grades If 15% of Grant's astronomy class received an A, how many students were in his astronomy class if 6 students earned A's this term?

55. Bachelors Based on data obtained from the U.S. Census Bureau, 30% of the 111 million males aged 18 years or older have never married. How many males aged 18 years or older have never married?

56. Census Data Based on data obtained from the U.S. Census Bureau, 24% of the 118 million females aged 18 years or older have never married. How many females aged 18 years or older have never married?

According to the U.S. Census Bureau, the level of education of the U.S. population is changing. The table below shows the number of persons (in thousands) aged 25 years and older and their educational attainment in 2004 and 2010. In Problems 57 and 58, use the information in the table to answer each question.

	2004		2010	
	Number	Percent	Number	Percent
Population 25 years and older	186,534	100.0	199,928	100.0
Associate's degree	13,244		18,259	
Bachelor's degree	32,084		38,784	

SOURCE: *United States Census Bureau.*

57. Find, to the nearest tenth of a percent, the percent of the U.S. population aged 25 years and older that held a bachelor's degree in 2004 and in 2010.

58. Find, to the nearest tenth of a percent, the percent of the U.S. population aged 25 years and older that held an associate's degree in 2004 and in 2010.

Extending the Concepts

59. Discount Pricing Suppose that you are the manager of a clothing store and have just purchased 100 shirts for $12 each. After 1 month of selling the shirts at the regular price, you plan to have a sale giving 25% off the original selling price. However, you still want to make a profit of $6 on each shirt at the sale price. What should you price the shirts at initially to ensure this?

In Problems 60–65, find the percent increase or decrease. The percent increase or percent decrease is defined as $\dfrac{\text{amount of change}}{\text{original amount}} \times 100\%.$

60. Population Growth The population in a small fishing town grew from 2500 to 2825. Find the percent increase.

61. Gas Mileage The gas mileage on your van decreases due to the heavy weight of extra passengers and a boat on a trailer. With the extra weight your van gets 15 mpg, and without the weight it gets 21 mpg.

To the nearest tenth, what is your percent decrease in gas mileage with extra weight?

62. Salaries Your current consulting position pays you $77,000 each year. A competing firm has offered you $86,000 to join it. To the nearest tenth, what percent increase in salary would you get if you joined the competing firm?

63. Car Depreciation A new car decreases in value by 25% each year. At *www.bmwusa.com* Misha priced his 335i convertible at $55,670.

   (a) After 2 years, what will the car be worth?

   (b) To the nearest tenth of a percent, after 2 years, what is the overall decrease in the value of the car?

64. Stock Prices One day Buffalo Wild Wings stock price went from $82.98 to $84.64. To the nearest hundredth, what was the percent increase?

65. Stock Prices One day Cisco Systems stock went from $20.43 to $20.00. To the nearest tenth, what is the percent decrease in the price?

Explaining the Concepts

66. The sales tax rate is 6%. Explain why $1.06p$ will correctly calculate the total purchase price of any item that sells for p dollars.

67. A problem on a quiz stated: Write an equation to solve the following problem. "Jack received a 5% raise in hourly wage, effective on his birthday. Jack's new hourly wage will be $12.81. Find Jack's current hourly wage." You wrote the equation $x + 0.05 = 12.81$, and your instructor counted your answer wrong. Explain why your equation is incorrect, and then write the correct equation and solve it for Jack's current hourly wage.

68. An item is reduced by 10% and then this is reduced by another 20%. Is this the same as reducing the item by 30%? Explain why or why not.

2.7 Problem Solving: Geometry and Uniform Motion

Objectives

1. Set Up and Solve Complementary and Supplementary Angle Problems
2. Set Up and Solve "Angles of a Triangle" Problems
3. Use Geometry Formulas to Solve Problems
4. Set Up and Solve Uniform Motion Problems

Figure 5

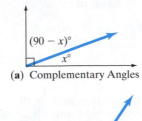

(a) Complementary Angles

(b) Supplementary Angles

Ready?...Answers   R1. {70}   R2. {4}

Are You Ready for This Section?

Before getting started, take the following readiness quiz. If you get a problem wrong, go back to the section cited and review the material.

R1. Solve: $q + 2q - 30 = 180$ [Section 2.2, pp. 92–93]

R2. Solve: $30w + 20(w + 5) = 300$ [Section 2.2, p. 93]

In this section, we continue using the six-step method introduced in Section 2.5.

1 Set Up and Solve Complementary and Supplementary Angle Problems

We begin by defining *complementary* and *supplementary angles*.

> **Definitions**
>
> Two angles whose measures sum to 90° are called **complementary angles.** Each angle is called the *complement* of the other. Two angles whose measures sum to 180° are called **supplementary angles.** Each angle is called the *supplement* of the other.

For example, the angles shown in Figure 5(a) are complements because their sum is 90°. Notice the use of the symbol ⌐ to show the 90° angle. The angles shown in Figure 5(b) are supplementary because their sum is 180°.

EXAMPLE 1   **Solving a Complementary Angle Problem**

Find the measure of two complementary angles if the measure of the larger angle is 6° greater than twice the measure of the smaller angle.

Solution

Step 1: Identify We are looking for the measure of two complementary angles, so their sum must be 90°.

Step 2: Name We know least about the measure of the smaller angle, so we call it x.

Step 3: Translate We let the measure of the larger angle be $(2x + 6)$ because it is 6° more than twice the measure of the smaller angle.

The angles are complementary, so the sum of the measures of the two angles must equal 90°.

measure of smaller angle   measure of larger angle

$$x \quad + \quad (2x + 6) \quad = 90 \quad \text{The Model}$$

Step 4: Solve

$$x + (2x + 6) = 90$$
$$\text{Combine like terms:} \quad 3x + 6 = 90$$
$$\text{Subtract 6 from each side of the equation:} \quad 3x = 84$$
$$\text{Divide each side by 3:} \quad x = 28$$

Step 5: Check The measure of the smaller angle, x, is 28° and the measure of the larger angle is $(2x + 6)$ degrees $= 2(28) + 6 = 56 + 6 = 62°$. Because the sum of the measures of these angles is $28° + 62° = 90°$, our answer is correct.

Step 6: Answer The two complementary angles measure 28° and 62°.   ●

> **Quick ✓**
>
> 1. Complementary angles are angles whose measures sum to __ degrees.
>
> 2. Find two complementary angles if the measure of the large angle is 12° more than the measure of the smaller angle.
>
> 3. Find two supplementary angles if the measure of the larger angle is 30° less than twice the measure of the smaller angle.

Figure 6

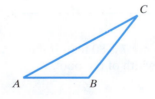

▶ ❷ Set Up and Solve "Angles of a Triangle" Problems

The sum of the measures of the interior angles of a triangle is 180°. In the triangle in Figure 6, the sum of the measures of angles A, B, and C must equal 180°, which we write as

$$m \angle A + m \angle B + m \angle C = 180°.$$

EXAMPLE 2   **Solving an "Angles of a Triangle" Problem**

The measure of the largest angle of a triangle is 20° more than twice the measure of the smallest angle, and the measure of the middle angle is 10° more than twice the measure of the smallest angle. Find the measure of each angle.

Solution

Step 1: Identify This is an "angles of a triangle" problem, so we know that the sum of the angles is 180°.

Step 2: Name We know least about the measure of the smallest angle, so we call its measure x and call the angle A.

Step 3: Translate The largest angle measures 20° more than twice the smallest angle, so $(2x + 20)$ is its measure. The middle angle measures 10° more than twice the smallest angle so $(2x + 10)$ is its measure. We will call the largest angle C and the middle angle B. Since $m \angle A + m \angle B + m \angle C = 180°$; we have

$$\underbrace{m \angle A}_{x} + \underbrace{m \angle B}_{(2x + 10)} + \underbrace{m \angle C}_{(2x + 20)} = 180 \quad \text{The Model}$$

Step 4: Solve

$$x + (2x + 10) + (2x + 20) = 180$$

$$\text{Combine like terms:} \quad 5x + 30 = 180$$

$$\text{Subtract 30 from each side of the equation:} \quad 5x = 150$$

$$\text{Divide each side by 5:} \quad x = 30$$

Step 5: Check Since $x = 30$ degrees $= m \angle A$, we know that $m \angle C = (2x + 20) = 2(30) + 20 = 60 + 20 = 80$ degrees and that $m \angle B = (2x + 10) = 2(30) + 10 = 60 + 10 = 70$ degrees.

The sum of the measures of these angles is $30° + 70° + 80° = 180°$, so our answer is correct.

Step 6: Answer The measures of the angles of the triangle are 30°, 70°, and 80°. ●

> **Quick ✓**
>
> **4.** The sum of the measures of the angles of a triangle is ___ degrees.
>
> **5.** The measure of the smallest angle of a triangle is one-third the measure of the largest angle. The measure of the middle angle is 65° less than the measure of the largest angle. Find the measure of the angles of the triangle.

▶ ❸ Use Geometry Formulas to Solve Problems

Recall from Section 2.4 that the perimeter of a figure is the sum of the lengths of its sides.

EXAMPLE 3 **Solving a Perimeter Problem**

The perimeter of the rectangular swimming pool shown in Figure 7 is 80 feet. If the length is 10 feet more than the width, find the length and the width of the pool.

Figure 7

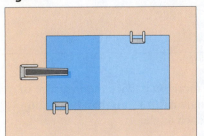

Solution

Step 1: Identify This is a perimeter problem. We want to find the length and the width of the pool, given the perimeter.

Step 2: Name Because we know the length in terms of the width, we let w represent the width of the pool.

Step 3: Translate The length of the pool is 10 feet more than the width, so the length is $w + 10$. The formula for perimeter of a rectangle is $P = 2l + 2w$, where l is the length and w is the width. So we have

$$\underbrace{2l}_{2(w + 10)} + \underbrace{2w}_{2w} = \underbrace{P}_{80} \quad \text{The Model}$$

Step 4: Solve

$$2(w + 10) + 2w = 80$$

Use the Distributive Property: $2w + 20 + 2w = 80$

Combine like terms: $4w + 20 = 80$

Subtract 20 from each side of the equation: $4w = 60$

Divide each side by 4: $w = 15$

Step 5: Check The width of the pool is $w = 15$ feet, so the length is $w + 10 = 15 + 10 = 25$ feet. The perimeter is $15 + 15 + 25 + 25 = 80$ feet, so our answer is correct.

Step 6: Answer The length of the pool is 25 feet, and the width of the pool is 15 feet. ●

Quick ✓

6. *True or False* The perimeter of a square is $4s$, where s is the length of a side of the square.

7. *True or False* The perimeter of a rectangle can be found by multiplying the length of a rectangle by the width of the rectangle.

8. The perimeter of a small rectangular garden is 9 feet. If the length is twice the width, find the width and length of the garden.

Recall that the area of a plane (that is, a two-dimensional) figure is the number of square units the figure contains, such as square feet, square inches, or square centimeters.

EXAMPLE 4 **Solving an Area Problem**

A garden in the shape of a trapezoid has an area of 18 square feet. The height is 3 feet, and the length of the shorter base is 2 feet less than the length of the longer base. Find the length of each base of the trapezoid. See Figure 8.

Figure 8

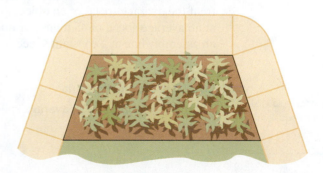

Solution

Step 1: Identify This problem is about the area of a trapezoid. The formula for the area of a trapezoid is $A = \frac{1}{2}h(B + b)$, where h is the height, B is the length of the longer base, and b is the length of the shorter base. We are given the area and the height of the trapezoid.

Step 2: Name We know that one base is 2 feet shorter than the other. Let B represent the length of the longer base.

Step 3: Translate Since one base is 2 feet shorter than the other base, and B represents the length of the longer base, $B - 2$ represents the length of the shorter base. Using the area of a trapezoid formula, we replace the values we know for A, h, B, and b.

$$A = \frac{1}{2}h(B + b)$$

$$\underbrace{18}_{\text{area}} = \frac{1}{2} \cdot \underbrace{3}_{\text{height}} \; (\underbrace{B}_{B} + \underbrace{B - 2}_{b}) \qquad \text{The Model}$$

Step 4: Solve

$$18 = \frac{1}{2} \cdot 3(B + B - 2)$$

Combine like terms in the parentheses: $\qquad 18 = \frac{1}{2} \cdot 3(2B - 2)$

Multiply by 2 to clear fractions: $\qquad 2[18] = 2\left[\frac{1}{2} \cdot 3(2B - 2)\right]$

$$36 = 3(2B - 2)$$

Use the Distributive Property: $\qquad 36 = 6B - 6$

Add 6 to each side of the equation: $\qquad 42 = 6B$

Divide each side by 6: $\qquad 7 = B$

Work Smart

Instead of distributing the 3 in

$$36 = 3(2B - 2)$$

we could divide both sides by 3. Try it! Which approach do you prefer?

Step 5: Check The longer base, B, is 7 feet long. The smaller base is $B - 2 = 7 - 2 = 5$ feet long. The area of the trapezoidal garden is $\frac{1}{2} \cdot 3(7 + 5) = \frac{1}{2} \cdot 3(12) = 18$ square feet. The answers 5 feet and 7 feet are correct.

Step 6: Answer The lengths of the two bases are 5 feet and 7 feet. ●

Quick ✓

9. The surface area of a rectangular box is 62 square feet. If the length of the box is 3 feet and the width is 2 feet, find the height of the box.

▶ ❹ Set Up and Solve Uniform Motion Problems

Objects that move at a constant velocity (speed) are said to be in **uniform motion.** We treat the average speed of an object as its constant velocity. For example, a car traveling at an average speed of 45 miles per hour is in uniform motion. An object traveling down an assembly line at a constant speed is also in uniform motion.

In Words

The uniform motion formula states that distance equals rate times time.

Uniform Motion Formula

If an object moves at an average speed r, the distance d covered in time t is given by the formula

$$d = r \cdot t$$

We use r in the uniform motion formula because "rate" is another term for speed. When solving uniform motion problems, we often write $d = rt$ as $rt = d$. We will use a chart to set up uniform motion problems as shown in Table 4.

Table 4

	Rate	·	Time	=	Distance
Object 1	rate 1		time 1		distance 1
Object 2	rate 2		time 2		distance 2

Rate, time, and distance must be expressed in corresponding units. For example, if rate (speed) is stated in miles per hour, then distance must be in miles, and time must be in hours. If rate is measured in kilometers per minute, then distance is in kilometers and time is in minutes.

EXAMPLE 5 **Solving a Uniform Motion Problem for Time**

Bob and Karen drove from Atlanta, Georgia, to Durham, North Carolina, a distance of 390 miles, to attend a family reunion. Their average speed for the first part of the trip was 60 miles per hour (mph). Due to road construction, their average speed for the remainder of the trip was 45 mph. How long did they travel at 45 mph if they drove 3 hours longer at 60 mph than at 45 mph?

Solution

Step 1: Identify This is a uniform motion problem. We wish to know the number of hours Bob and Karen drove at 45 mph.

Step 2: Name Let t represent the number of hours Bob and Karen drove at 45 mph. Since they traveled 3 hours longer at 60 mph, $t + 3$ represents the number of hours driven at 60 mph.

Step 3: Translate We set up Table 5. For the first part of the trip, the rate is 60 mph and the time is $t + 3$ hours, so the distance is $60(t + 3)$. For the second part of the trip, the rate is 45 mph and the time is t, so the distance is $45t$.

Table 5

	Rate (in mph)	·	Time	=	Distance
First part of trip	60		$t + 3$		
Second part of trip	45		t		
Total					390

The total distance that Bob and Karen traveled is 390 miles, so

$$\underset{\substack{\text{distance traveled} \\ \text{at 60 mph}}}{60(t + 3)} + \underset{\substack{\text{distance traveled} \\ \text{at 45 mph}}}{45t} = \underset{\substack{\text{total} \\ \text{distance}}}{390} \qquad \text{The Model}$$

Step 4: Solve

$$60(t + 3) + 45t = 390$$

Use the Distributive Property: $\quad 60t + 180 + 45t = 390$

Combine like terms: $\quad 105t + 180 = 390$

Subtract 180 from both sides: $\quad 105t + 180 - 180 = 390 - 180$

$$105t = 210$$

Divide both sides by 105: $\quad \dfrac{105t}{105} = \dfrac{210}{105}$

$$t = 2$$

Step 5: Check Bob and Karen drove for $t = 2$ hours at 45 mph. So they drove $t + 3 = 2 + 3 = 5$ hours at 60 mph. Does 2 hours driven at 45 mph plus 5 hours at 60 mph equal a distance of 390 miles? Because 2 hours · 45 mi/hr + 5 hours · 60 mi/hr = 90 miles + 300 miles = 390 miles, our answer checks.

Step 6: Answer the Question Bob and Karen drove for 2 hours at 45 mph. ●

EXAMPLE 6 **Solving a Uniform Motion Problem for Rate**

Two groups of friends took a canoe trip down Big Darby Creek. The first group left at 12 noon. One-half hour later, the second group left the same location, paddling at an average speed that was 0.75 mph faster than the first group. At 2:30 P.M. the second group caught up to the first group. How fast was each group paddling?

Solution

Step 1: Identify This is a uniform motion problem. We want to find the rate at which each group paddled.

Step 2: Name Let r represent the rate at which the first group paddled. The second group's paddling rate is $r + 0.75$, because its rate is 0.75 mph greater than that of the first group.

Step 3: Translate We have a specific time that the first group left, 12 noon. The second group left $\frac{1}{2}$ hour later. In our formula time is measured in hours. So the first group traveled for two-and-a-half (or 2.5) hours (12 noon until 2:30 P.M.), and the second group traveled for 2 hours. We're now ready to fill in Table 6.

Table 6

	Rate (in mph) $\cdot$	Time $=$	Distance
First group	r	2.5	
Second group	$r + 0.75$	2	

Because the second group caught up to the first group, the distances the two groups traveled were the same.

$$\underset{\text{distance traveled by first group}}{2.5r} = \underset{\text{distance traveled by second group}}{2(r + 0.75)} \qquad \text{The Model}$$

Step 4: Solve
$$2.5r = 2(r + 0.75)$$

Use the Distributive Property:
$$2.5r = 2r + 1.5$$

Subtract $2r$ from both sides:
$$2.5r - 2r = 2r - 2r + 1.5$$
$$0.5r = 1.5$$

Divide both sides by 0.5:
$$\frac{0.5r}{0.5} = \frac{1.5}{0.5}$$
$$r = 3$$

Work Smart

In solving $0.5r = 1.5$, we could also multiply both sides by 2. Why?

$$2(0.5\,r)\ 2 \cdot \frac{1}{2}r = r$$

Step 5: Check The first group paddled at $r = 3$ mph, and the second group paddled at $r + 0.75 = 3 + 0.75 = 3.75$ mph. Did both groups travel the same distance? Because $(3)(2.5) = 7.5$ miles and $(2)(3.75)$ also equals 7.5 miles, our answers are correct.

Step 6: Answer the Question The first group paddled at 3 miles per hour and the second group paddled at 3.75 miles per hour. ●

Quick ✓

10. *True or False* When using $d = rt$ to calculate the distance traveled, it is not necessary to assume that the object travels at a constant speed.

11. Two bikers, José and Luis, start at the same point at the same time and travel in opposite directions. José's average speed is 5 miles per hour more than that of Luis, and after 3 hours the bikers are 63 miles apart. Find the average speed of each biker.

12. Tanya, a long-distance runner, runs at an average speed of 8 miles per hour. Two hours after Tanya leaves your house, you leave in your car and follow the same route. If your average speed is 40 miles per hour, how long will it be before you catch up to Tanya? How far will each of you be from your home?

Problems **1–12** *are the Quick ✔s that follow the* **EXAMPLES**.

Building Skills

In Problems 13–16, identify the measure of each of the angles. See Objective 1.

△ **13.** Find two supplementary angles if the measure of the first angle is three times the measure of the second.

△ **14.** Find two supplementary angles if the measure of the first angle is four times the measure of the second.

△ **15.** The measures of two complementary angles are consecutive even integers. Find the measure of each angle.

△ **16.** Find two complementary angles if the measure of the first angle is 25° less than the measure of the second.

In Problems 17–20, find the measures of the angles of the triangle. See Objective 2.

△ **17.**

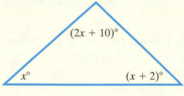

△ **18.**

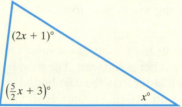

△ **19.** The measures of the angles of a triangle are consecutive even integers. Find the measure of each angle.

△ **20.** In a triangle, the second angle measures four times the first. The measure of the third angle is 18° more than the measure of the second. Find the measures of the three angles.

In Problems 21–26, use a formula from geometry to solve for the unknown quantity. See Objective 3.

△ **21.** The length of a rectangle is 8 feet longer than the width. If the perimeter is 88 feet, find the length and width of the rectangle.

△ **22.** The width of a rectangle is 10 meters less than the length. If the perimeter is 56 meters, find the length and width of the rectangle.

△ **23.** A rectangular field is divided into 2 squares of the same size and shape. If it takes 294 yards of fencing

to enclose the field and divide the field into the two parcels, find the dimensions of the field. See the figure.

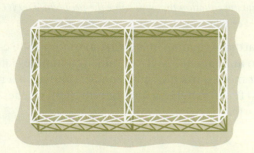

△ **24.** A rectangular plot has been divided into two fields so that the length of one of the fields is twice the length of the other. The smaller field is a square, and the larger field is a rectangle. If it takes 279 meters of fencing to enclose entire plot and to divide it into two fields, find the dimensions of the original plot.

△ **25.** A trapezoid has an area of 900 square meters. The height of the trapezoid is 40 meters, and the length of the longer base is twice that of the shorter base. Find the length of each base of the trapezoid.

△ **26.** A parallelogram has a perimeter of 120 inches. If one side of the parallelogram is 10 inches longer than the other, find the dimensions of the figure.

In Problems 27 and 28, set up the model to solve the uniform motion problems by answering parts (a)–(d). See Objective 4.

27. Two cars leave Chicago, one traveling north and the other south. The car going north is traveling at 62 mph, and the car going south is traveling at 68 mph. How long before they are 585 miles apart?

 (a) Write an algebraic expression for the distance traveled by the car going north.

 (b) Write an algebraic expression for the distance traveled by the car going south.

 (c) Write an algebraic expression for the total distance traveled by the two cars.

 (d) Write an equation to answer the question.

28. Two trains leave Albuquerque, traveling in the same direction on parallel tracks. One train is traveling at 72 mph, and the other is traveling at 66 mph. How long before they are 45 miles apart?

 (a) Write an algebraic expression for the distance traveled by the faster train.

(b) Write an algebraic expression for the distance traveled by the slower train.

(c) Write an algebraic expression for the difference in distance between the two.

(d) Write an equation to answer the question.

In Problems 29 and 30, fill in the table from the information given. Then write the equation that will solve the problem. DO NOT SOLVE. See Objective 4.

29. Martha is running in her first marathon. She can run at a rate of 528 ft per min. Ten minutes later her mom starts the same course, running at a rate of 880 ft per min. How long before Mom catches up to Martha?

	Rate	·	Time	=	Distance
Martha	?		?		?
Mom	?		?		?

30. A 580-mile trip in a small plane took a total of 5 hours. The first two hours were flown at one rate, and then the plane encountered a headwind and was slowed by 10 mph. Find the rate for each portion of the trip.

	Rate	·	Time	=	Distance
Beginning of trip	?		?		?
Rest of the trip	?		?		?
Total			?		?

Applying the Concepts

△ **31. Isosceles Triangle** An isosceles triangle has exactly two sides that are equal in length (*congruent*). If the base (the third side) measures 45 inches and the perimeter is 98 inches, find the length of the two congruent sides, called *legs*.

△ **32. Isosceles Triangle** See Problem 31. An isosceles triangle has a base of 17 cm. If the perimeter is 95 cm, find the length of each of the legs.

△ **33. Billboard** A billboard along a highway has a perimeter of 110 feet. Find the length of the billboard if its height is 15 feet.

△ **34. Buying Wallpaper** Erika is buying wallpaper for her bedroom. She remembers that the perimeter of the room is 54 ft and that the room is twice as long as it is wide.

(a) Find the dimensions of the room.

(b) If the walls are 8 ft high, how many square feet of wallpaper does she need to buy?

(c) Erika arrives at the decorating store and finds that wallpaper is sold by the square yard. How many square yards of wallpaper does Erika need to buy?

△ **35. Back Yard** Bob's backyard is in the shape of a trapezoid with height of 60 feet. The shorter base is 8 feet shorter than the longer base, and the area of the backyard is 2160 square feet. Find the length of each base of the trapezoidal yard.

△ **36. Buying Fertilizer** Melinda has to buy fertilizer for a flower garden in the shape of a right triangle. If the area of the garden is 54 square feet and the base of the garden measures 9 feet, find the height of the triangular garden.

△ **37. Garden** The perimeter of a rectangular garden is 60 yards. The width of the garden is three yards less than twice the length.

(a) Find the length and width of the garden.

(b) What is the area of the garden?

△ **38. Table** The Jacksons are having a custom rectangular table made. The length of the table is 18 inches more than the width, and the perimeter is 180 inches. Find the length and the width of the table.

△ **39. Boats** Two boats leave a port at the same time, one going north and the other traveling south. The northbound boat travels 16 mph faster than the southbound boat. If the southbound boat is traveling at 47 mph, how long will it be before they are 1430 miles apart?

40. Cyclists Two cyclists leave a city at the same time, one going east and the other going west. The west-bound cyclist bikes 4 mph faster than the east-bound cyclist. After 5 hours they are 200 miles apart. How fast is the east-bound cyclist riding?

41. Road Trip Two cars leave a city on the same road, one driving 12 mph faster than the other. After 4 hours, the car traveling faster stops for lunch. After 4 hours and 30 minutes, the car traveling slower stops for lunch. Assuming that the person in the faster car is still eating lunch, the cars are now 24 miles apart. How fast is each car driving?

42. Passenger and Freight Trains Two trains leave a city on parallel tracks, traveling the same direction. The passenger train is going twice as fast as the freight train. After 45 minutes, the trains are 30 miles apart. Find the speed of each train.

43. Down the Highway A 360-mile trip began on a freeway in a car traveling at 62 mph. Once the road became a 2-lane highway, the car slowed to 54 mph. If the total trip took 6 hours, find the time spent on each type of road.

44. River Trip Max lives on a river, 30 miles from town. Max travels downstream (with the current) at 20 mph. Returning upstream (against the current), his rate is 12 mph. If the total trip to town and back took 4 hours, how long did his upstream trip take?

45. Walking and Jogging Carol can complete her neighborhood jog in 10 minutes. It takes 30 minutes to cover the same distance when she walks. If her jogging rate is 4 mph faster than her walking rate, find the speed at which she jogs.

46. Trip to School Dien drives to school at 40 mph. Five minutes $\left(\dfrac{1}{12}\text{hour}\right)$ after he leaves home, his mother sees that he forgot his homework and leaves to take it to him, driving 48 mph. If they arrive at school at the same time, how far away is the school?

△ **47. Isosceles Triangle** In an isosceles triangle, the base angles (angles opposite the two congruent legs) are equal in measure (*congruent*). Find the measures of the angles of an isosceles triangle in which the third angle (called the *vertex angle*) has a measure that is 16° less than twice the measure of a base angle.

△ **48. Isosceles Triangle** See Problem 47. In an isosceles triangle, the measure of the vertex angle is 4 degrees less than twice the measure of each base angle. Find the measure of each of the angles of the triangle.

Extending the Concepts

Parallel lines are lines in the same plane that never intersect (think of railroad tracks going infinitely far out into space). A line that cuts two parallel lines is called a *transversal*. The transversal forms eight different angles that are related in following ways:

Corresponding angles are equal in measure.

Alternate interior angles are equal in measure.

Interior angles on the same side of the transversal are supplementary.

In the figure shown, lines L_1 and L_2 are parallel $(L_1 \| L_2)$ and the transversal is labeled t. In this figure, there are four pairs of corresponding angles:

$\angle 1$ and $\angle 5$, $\angle 2$ and $\angle 6$, $\angle 3$ and $\angle 7$, $\angle 4$ and $\angle 8$

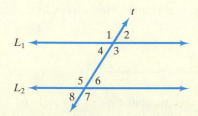

There are two pairs of alternate interior angles:
$\angle 3$ and $\angle 5$, $\angle 4$ and $\angle 6$.

There are two pairs of interior angles on the same side of the transversal: $\angle 3$ and $\angle 6$, $\angle 4$ and $\angle 5$.

In Problems 49–54, given $L_1 \| L_2$, use the appropriate properties from geometry to solve for x.

△ **49.**
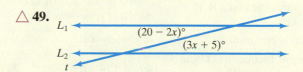

△ **50.**
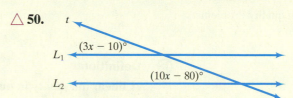

△ **51.** △ **52.**
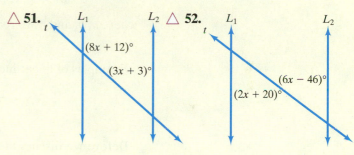

△ **53.**
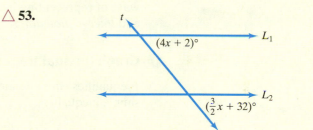

△ **54.**
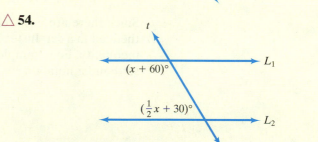

Explaining the Concepts

55. Explain the difference between complementary angles and supplementary angles.

56. Explain the difference between the area of a rectangle and the perimeter of a rectangle.

57. When setting up a uniform motion problem, you wrote $65t + 40t = 115$. Your classmate wrote $65t - 40t = 115$. Write a word problem for each of these equations, and explain the keys to recognizing the difference between the two types.

2.8 Solving Linear Inequalities in One Variable

Objectives

1. Graph Inequalities on a Real Number Line
2. Use Interval Notation
3. Solve Linear Inequalities Using Properties of Inequalities
4. Model Inequality Problems

Are You Ready for This Section?

Before getting started, take this readiness quiz. If you get a problem wrong, go back to the section cited and review the material.

In Problems R1–R4, replace the question mark with $<$, $>$, or $=$ to make the statement true.

R1. $4 ? 19$ **R2.** $-11 ? -24$ **R3.** $\frac{1}{4} ? 0.25$ **R4.** $\frac{5}{6} ? \frac{4}{5}$ [Section 1.3, pp. 23–24]

▶ **Definition**

A **linear inequality in one variable** is an inequality that can be written in the form

$$ax + b < c \text{ or } ax + b \le c \text{ or } ax + b > c \text{ or } ax + b \ge c$$

where a, b, and c are real numbers and $a \ne 0$.

Examples of linear inequalities in one variable, x, are

$$x + 2 > 7 \quad \frac{1}{2}x + 4 \le 9 \quad 5x > 0 \quad 3x + 7 \ge 8x - 3$$

Before we discuss methods for solving linear inequalities, we will present three ways of representing inequalities: *set-builder notation*, graphing on a real number line, and *interval notation*.

1 Graph Inequalities on a Real Number Line

Inequalities with one inequality symbol are called **simple inequalities**. For example, the simple inequality

$$x > 2 \quad \text{means} \quad \text{the set of all real numbers } x \text{ greater than 2}$$

Since there are infinitely many real numbers that are greater than 2, we cannot list them all in a set. Instead we use **set-builder notation** to express the set formed by this inequality. For example, the set of all real numbers greater than 2 is represented in set-builder notation is

$$\{ \quad x \quad | \quad x > 2\}$$

The set of all x such that x is greater than 2

Representing an inequality on a number line is called graphing the inequality, and the picture is called the **graph of the inequality**. For example, we graph $\{x | x > 2\}$ on a real number line by shading the portion of the number line that is to the right of 2. The number 2 is called an **endpoint**. We use a parenthesis to indicate that 2 is *not* included in the solution set. See Figure 9.

Figure 9
$x > 2$

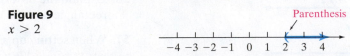

To graph the inequality $x \ge 2$, also shade the number line to the right of 2, but use a bracket on the endpoint (to indicate that 2 is included in the solution set). See Figure 10.

Figure 10
$x \ge 2$

In a similar way, we graph the inequality $x < 4$ in Figure 11(a) and $x \le 4$ in Figure 11(b).

Figure 11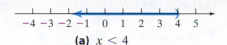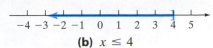

(a) $x < 4$ (b) $x \le 4$

EXAMPLE 1

Graphing Simple Inequalities on a Real Number Line

Graph each inequality on a real number line.

(a) $x > 5$ (b) $x \le -4$

Solution

(a) The inequality $x > 5$ represents all real numbers greater than five. Because 5 is not included in this set, we place a parenthesis at 5 and shade to the right. See Figure 12.

Figure 12
$x > 5$

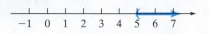

Work Smart

The inequalities $x > 5$ and $5 < x$ have the same graph. Do you see why?

(b) The inequality $x \le -4$ represents all real numbers less than or equal to negative four. Because -4 is included in the set, we place a bracket at -4 and shade to the left. See Figure 13.

Figure 13
$x \le -4$

> **Quick ✓**
>
> 1. *True or False* To graph an inequality that contains $>$ or $<$, use a parenthesis.
>
> *In Problem 2–5, graph each inequality on a real number line.*
>
> 2. $n \ge 8$ 3. $a < -6$
>
> 4. $x > -1$ 5. $p \le 0$

▶ ❷ **Use Interval Notation**

A third way to represent inequalities uses *interval notation*. To use interval notation, we need a new symbol, ∞.

The symbol ∞ (which is read as "infinity") is not a real number; it is used to indicate that there is no upper bound on an inequality. The symbol $-\infty$ (which is read as "minus infinity" or "negative infinity") also is not a real number and means that there is no lower bound on an inequality.

Work Smart

We use a parenthesis when either $-\infty$ or ∞ is an endpoint because these symbols are not real numbers.

Intervals Involving ∞

$[a, \infty)$	consists of all real numbers x for which $x \ge a$
(a, ∞)	consists of all real numbers x for which $x > a$
$(-\infty, a]$	consists of all real numbers x for which $x \le a$
$(-\infty, a)$	consists of all real numbers x for which $x < a$
$(-\infty, \infty)$	consists of all real numbers x (or $-\infty < x < \infty$)

We now have three ways to represent all real numbers greater than 2.

SET-BUILDER NOTATION	NUMBER LINE GRAPH	INTERVAL NOTATION
$\{x \mid x > 2\}$	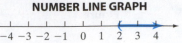	$(2, \infty)$

Table 7 summarizes set builder notation, interval notation, and the graphs.

Work Smart

Intervals are always read from left to right. For example $(5, \infty)$ represents all real numbers greater than 5 because 5 is the left endpoint.

Table 7

Set-Builder Notation	Interval Notation	Graph
$\{x \mid x \geq a\}$	$[a, \infty)$	⊢──────→ a
$\{x \mid x > a\}$	(a, ∞)	(──────→ a
$\{x \mid x \leq a\}$	$(-\infty, a]$	←──────┤ a
$\{x \mid x < a\}$	$(-\infty, a)$	←──────) a
$\{x \mid x \text{ is a real number}\}$	$(-\infty, \infty)$	←──────→

EXAMPLE 2 · **Writing an Inequality in Interval Notation**

Write each inequality using interval notation.

(a) $x > -4$ **(b)** $x \leq 8$

(c) ⊢─[──→ 1.5 1.75 2 2.25 **(d)** ←────) 16 17 18 19

Solution

Remember, if the inequality contains $<$ or $>$, use a parenthesis, (or). If the inequality contains $\leq$ or $\geq$, use a bracket, [or].

(a) $(-4, \infty)$ **(b)** $(-\infty, 8]$ **(c)** $[1.75, \infty)$ **(d)** $(-\infty, 19)$ ●

Quick ✓

6. *True or False* The inequality $x \geq 3$ is written in interval notation as $[3, \infty]$.

7. *True or False* The inequality $x < -4$ is written in interval notation as $(-4, -\infty)$.

In Problems 8–11, write the inequality in interval notation.

8. $x \geq -3$ 9. $x < 12$

10. ←────┤ 1.0 1.5 2.0 2.5 3.0 11. ⊢──(──→ 115 120 125 130 135

▶ ❸ **Solve Linear Inequalities Using Properties of Inequalities**

To **solve an inequality** means to find all values of the variable for which the statement is true. These values are the **solutions** of the inequality. The set of all solutions is the **solution set**. As with equations, one method for solving a linear inequality is to replace it by a series of *equivalent inequalities* until an inequality with an obvious solution, such as $x > 2$, is obtained.

Two inequalities having the same solution set are called **equivalent inequalities.** We obtain equivalent inequalities by using some of the same operations used to find equivalent equations.

Consider the inequality $3 < 8$. If we add 2 to both sides of the inequality, the left side becomes 5 and the right side becomes 10. Since $5 < 10$, we see that adding the same quantity to both sides of an inequality does not change the sense, or direction, of the inequality. This result is called the **Addition Property of Inequality**.

In Words
The Addition Property of Inequality states that the direction of the inequality does not change when the same quantity is added to each side of the inequality.

Addition Property of Inequality

For real numbers a, b, and c,

$$\text{If} \quad a < b, \quad \text{then} \quad a + c < b + c$$
$$\text{If} \quad a > b, \quad \text{then} \quad a + c > b + c$$

In Words
Subtracting a quantity from both sides of an inequality also does not change the direction of the inequality.

The Addition Property of Inequality also holds true for subtracting a real number from both sides of an inequality, since $a - b$ is equivalent to $a + (-b)$.

EXAMPLE 3 **How to Solve an Inequality Using the Addition Property of Inequality**

Solve the linear inequality $4y - 5 \le 3y - 2$. Express the solution set using set-builder notation and interval notation. Graph the solution set.

Step-by-Step Solution

Step 1: Get the terms containing variables on the left side of the inequality.

Subtract 3y from both sides:

$$4y - 5 \le 3y - 2$$
$$4y - 5 - 3y \le 3y - 2 - 3y$$
$$y - 5 \le -2$$

Step 2: Isolate the variable y on the left side.

Add 5 to both sides:

$$y - 5 + 5 \le -2 + 5$$
$$y \le 3$$

Figure 14

The solution set using set-builder notation is $\{y \mid y \le 3\}$. The solution set using interval notation is $(-\infty, 3]$. The solution set is graphed in Figure 14. ●

Quick ✓

12. To _____ an inequality means to find the set of all values of the variable for which the statement is true.

13. Write the inequality that results from adding 7 to each side of $3 < 8$. What property of inequalities does this illustrate?

In Problems 14–17, find the solution of the linear inequality. Express the solution set in set-builder notation and interval notation. Graph the solution set.

14. $n - 2 > 1$

15. $-2x + 3 < 7 - 3x$

16. $5n + 8 \le 4n + 4$

17. $3(4x - 8) + 12 > 11x - 13$

We've seen what happens when we add a real number to both sides of an inequality. Let's look at two examples from arithmetic to see if we can figure out what happens when we multiply or divide both sides of an inequality by a nonzero constant.

EXAMPLE 4 **Multiplying or Dividing an Inequality by a Positive Number**

(a) Write the inequality that results from multiplying both sides of the inequality $-2 < 5$ by 3.

(b) Write the inequality that results from dividing both sides of the inequality $18 > 14$ by 2.

Solution

(a) Multiplying both sides of $-2 < 5$ by 3 results in the numbers -6 and 15 on each side of the inequality, so we have $-6 < 15$.

(b) Dividing both sides of $18 > 14$ by 2 results in the numbers 9 and 7 on each side of the inequality, so we have $9 > 7$.   ●

From Example 4 it would seem that multiplying (or dividing) both sides of an inequality by a positive real number does not change the direction of the inequality.

EXAMPLE 5 **Multiplying or Dividing an Inequality by a Negative Number**

(a) Write the inequality that results from multiplying both sides of the inequality $-2 < 5$ by -3.

(b) Write the inequality that results from dividing both sides of the inequality $18 > 14$ by -2.

Solution

(a) Multiplying both sides of $-2 < 5$ by -3 results in the numbers 6 and -15 on each side of the inequality, so we have $6 > -15$.

(b) Dividing both sides of $18 > 14$ by -2 results in the numbers -9 and -7 on each side of the inequality, so we have $-9 < -7$.   ●

Example 5 suggests that multiplying (or dividing) both sides of an inequality by a negative real number results in an inequality that changes the direction of the original inequality. The results of Examples 4 and 5 lead us to the **Multiplication Properties of Inequality.**

In Words

The Multiplication Properties of Inequality state that if you multiply (or divide) both sides of an inequality by a positive number, the inequality symbol remains the same, but if you multiply (or divide) both sides by a negative number, the inequality symbol reverses.

> **Multiplication Properties of Inequality**
>
> Let a, b, and c be real numbers.
>
> If $a < b$ and c is positive, then $ac < bc$.
> If $a > b$ and c is positive, then $ac > bc$.
> If $a < b$ and c is negative, then $ac > bc$.
> If $a > b$ and c is negative, then $ac < bc$.

The Multiplication Property of Inequality also holds for dividing both sides of an inequality by a nonzero real number.

EXAMPLE 6 **Solving a Linear Inequality Using the Multiplication Properties of Inequality**

Solve each linear inequality. Express the solution set using set-builder notation and interval notation. Graph the solution set.

(a) $\dfrac{1}{3}x > -2$ 　　　　　　　(b) $-4x \geq 24$

Solution

(a) $$\frac{1}{3}x > -2$$

Multiply both sides of the inequality by 3: $3 \cdot \frac{1}{3}x > 3 \cdot (-2)$

$$x > -6$$

The solution set using set-builder notation is $\{x \mid x > -6\}$. The solution set using interval notation is $(-6, \infty)$. The graph of the solution set is shown in Figure 15.

(b) $$-4x \geq 24$$

Divide both sides of the inequality by -4. $\dfrac{-4x}{-4} \leq \dfrac{24}{-4}$
Remember to reverse the inequality symbol!

$$x \leq -6$$

The solution set using set-builder notation is $\{x \mid x \leq -6\}$. The solution set using interval notation is $(-\infty, -6]$. The graph of the solution set is shown in Figure 16.

Figure 15   **Figure 16**

▶ We can now solve inequalities using both the Addition and Multiplication Properties of Inequality. In each solution, we isolate the variable on the left side of the inequality, so the inequality is easier to read. If the variable ends up on the right side, remember that

$$a < x \text{ is equivalent to } x > a \quad \text{and}$$
$$a > x \text{ is equivalent to } x < a$$

EXAMPLE 7   **How to Solve an Inequality Using Both the Addition and Multiplication Properties of Inequality**

Solve the inequality $4(x + 1) - 2 < 8x - 26$. Express the solution set using set-builder notation and interval notation. Graph the solution set.

Step-by-Step Solution

Step 1: Remove parentheses. $4(x + 1) - 2 < 8x - 26$

Use the Distributive Property: $4x + 4 - 2 < 8x - 26$

(continued)

Step 2: Combine like terms on each side of the inequality.

$$4x + 2 < 8x - 26$$

Step 3: Get the terms containing variables on the left side of the inequality and the constants on the right side.

Subtract 8x from both sides:

$$4x + 2 - 8x < 8x - 26 - 8x$$
$$-4x + 2 < -26$$

Subtract 2 from both sides:

$$-4x + 2 - 2 < -26 - 2$$
$$-4x < -28$$

Step 4: Get the coefficient of the variable term to be 1.

Divide both sides by -4:
Remember to reverse the inequality symbol!

$$\frac{-4x}{-4} > \frac{-28}{-4}$$
$$x > 7$$

Figure 17

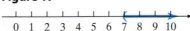

The solution set using set-builder notation is $\{x | x > 7\}$, or, using interval notation, $(7, \infty)$. The graph of the solution is shown in Figure 17. ●

After solving an inequality, we can substitute a value for the variable that is in the solution set into the original inequality and see whether we obtain a true statement. If we obtain a true statement, then we have some evidence that our solution is correct. However, this does *not* prove that our solution is correct. The check for Example 7 is shown below.

Check Since the solution of the inequality $4(x + 1) - 2 < 8x - 26$ is any real number greater than 7, let's replace x by 10.

$$4(x + 1) - 2 < 8x - 26$$

Replace x with 10:

$$4(10 + 1) - 2 \overset{?}{<} 8(10) - 26$$
$$4(11) - 2 \overset{?}{<} 80 - 26$$

Perform the arithmetic:

$$42 < 54 \qquad \text{True}$$

$42 < 54$ is a true statement, so $x = 10$ is in the solution set. Also, if we substitute 6 for x, which is less than 7, we obtain $26 < 22$, which is false. We have some evidence that our solution set, $(7, \infty)$, is correct.

> **Quick** ✓
>
> *In Problems 24–27, find the solution of the linear inequality. Express the solution set using set-builder notation and interval notation. Graph the solution set.*
>
> **24.** $3x - 7 > 14$ **25.** $-4n - 3 < 9$
>
> **26.** $2x - 6 < 3(x + 1) - 5$ **27.** $-4(x + 6) + 18 \geq -2x + 6$

EXAMPLE 8   **Solving a Linear Inequality Containing Fractions**

Solve the inequality $\frac{1}{2}(x - 4) \geq \frac{3}{4}(2x + 1)$. Express the solution using set-builder notation and interval notation. Graph the solution set.

Solution

Clear the fractions by multiplying both sides of the inequality by 4, the least common denominator of $\frac{1}{2}$ and $\frac{3}{4}$.

$$\frac{1}{2}(x - 4) \geq \frac{3}{4}(2x + 1)$$

$$4 \cdot \frac{1}{2}(x - 4) \geq 4 \cdot \frac{3}{4}(2x + 1)$$

$$2(x - 4) \geq 3(2x + 1)$$

Use the Distributive Property: $\quad 2x - 8 \geq 6x + 3$

Subtract $6x$ from both sides: $\quad 2x - 8 - 6x \geq 6x + 3 - 6x$

$$-4x - 8 \geq 3$$

Add 8 to both sides: $-4x - 8 + 8 \geq 3 + 8$

$$-4x \geq 11$$

Divide both sides by -4 and remember
to reverse the inequality symbol: $\quad \dfrac{-4x}{-4} \leq \dfrac{11}{-4}$

$$x \leq -\frac{11}{4}$$

The solution is $\left\{ x \,\middle|\, x \leq -\dfrac{11}{4} \right\}$, or, using interval notation, $\left(-\infty, -\dfrac{11}{4} \right]$. The graph of
the solution is given in Figure 18.

Figure 18

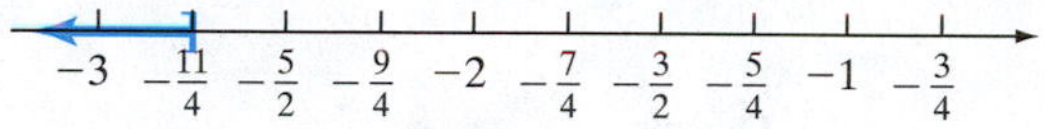

Number line from -3 to $-\frac{3}{4}$ with shaded ray extending left from $-\frac{11}{4}$; marks at -3, $-\frac{11}{4}$, $-\frac{5}{2}$, $-\frac{9}{4}$, -2, $-\frac{7}{4}$, $-\frac{3}{2}$, $-\frac{5}{4}$, -1, $-\frac{3}{4}$.

Quick ✓

In Problems 28 and 29, find the solution of the linear inequality and express the solution set using set-builder notation and interval notation. Graph the solution set.

28. $\dfrac{1}{2}(x + 2) > \dfrac{1}{5}(x + 17)$ **29.** $\dfrac{4}{3}x - \dfrac{2}{3} \leq \dfrac{4}{5}x + \dfrac{3}{5}$

Some inequalities are true for all values of the variable, and some are false for all values
of the variable. We present these special cases now.

EXAMPLE 9 **Solving an Inequality Whose Solution Set Is All Real Numbers**

Solve the inequality $3(x + 4) - 5 > 7x - (4x + 2)$. Express the solution using
set-builder notation and interval notation, and graph the solution set.

Solution

$$3(x + 4) - 5 > 7x - (4x + 2)$$

Use the Distributive Property: $\quad 3x + 12 - 5 > 7x - 4x - 2$

$$3x + 7 > 3x - 2$$

Subtract $3x$ from each side: $\quad 3x + 7 - 3x > 3x - 2 - 3x$

$$7 > -2$$

Figure 19

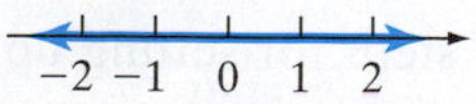

Number line marked at -2, -1, 0, 1, 2 with entire line shaded.

The statement $7 > -2$ is true, so the solution is all real numbers. The solution set is
$\{ x \,|\, x \text{ is any real number} \}$, or $(-\infty, \infty)$ in interval notation. Figure 19 shows the graph
of the solution set.

EXAMPLE 10 | **Solving an Inequality Whose Solution Set Is the Empty Set**

Solve the inequality $8\left(\frac{1}{2}x - 1\right) + 2x \leq 6x - 10$. Express the solution using set-builder notation and interval notation, if possible, and graph the solution set.

Solution

$$8\left(\frac{1}{2}x - 1\right) + 2x \leq 6x - 10$$

Use the Distributive Property: $\quad 4x - 8 + 2x \leq 6x - 10$

$$6x - 8 \leq 6x - 10$$

Subtract $6x$ from each side: $\quad 6x - 8 - 6x \leq 6x - 10 - 6x$

$$-8 \leq -10$$

The statement $-8 \leq -10$ is false, so this inequality has no solution. The solution set is the empty set, $\varnothing$ or $\{\ \}$. Figure 20 shows the graph of the solution set on a real number line.

Figure 20

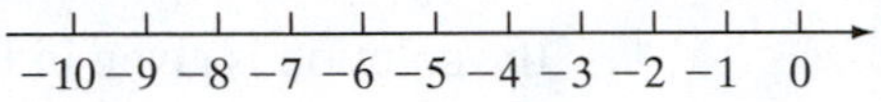

Quick ✓

30. *True or False* When you are solving an inequality, if the variable is eliminated and the result is a true statement, the solution is all real numbers.

31. *True or False* When you are solving an inequality, if the variable is eliminated and the result is a false statement, the solution is the empty set or $\varnothing$.

In Problems 32–35, find the solution of the linear inequality. Express the solution set using set-builder notation and interval notation, if possible. Graph the solution set.

32. $-2x + 7(x + 5) \leq 6x + 32$ **33.** $-x + 7 - 8x \geq 2(8 - 5x) + x$

34. $\frac{3}{2}x + 5 - \frac{5}{2}x < 4x - 3(x + 1)$ **35.** $x + 3(x + 4) \geq 2x + 5 + 3x - x$

▶ ❹ Model Inequality Problems

When solving word problems modeled by inequalities, look for key words that indicate the type of inequality symbol to use. Some key words and phrases are listed in Table 8.

Table 8

Word or Phrase	Inequality Symbol	Word or Phrase	Inequality Symbol
at least	$\geq$	at most	$\leq$
no less than	$\geq$	no more than	$\leq$
more than	$>$	fewer than	$<$
greater than	$>$	less than	$<$

Work Smart

The context of the words "less than" is important! "3 **less than** x" is $x - 3$, but "3 is **less than** x" is $3 < x$. Do you see the difference?

To solve applications involving linear inequalities, use the same steps for setting up applied problems that we introduced in Section 2.5, on page 124.

EXAMPLE 11 **A Handyman's Fee**

A handyman charges a flat fee of $60 plus $22 per hour for a job. How many hours does this handyman need to work to make at least $500?

Solution

Step 1: Identify We want to know the number of hours the handyman must work to make at least $500.

Step 2: Name Let h represent the number of hours that the handyman must work.

Step 3: Translate We know that the sum of the flat fee that the handyman charges and the hourly charge must exceed $500. So

flat fee	plus	hourly wage	times	no. hours worked	at least	500	
60	+	22	$\cdot$	h	$\geq$	500	**The Model**

Step 4: Solve

$$60 + 22h \geq 500$$

Subtract 60 from each side: $60 - 60 + 22h \geq 500 - 60$

Simplify: $22h \geq 440$

Divide both sides by 22: $\dfrac{22h}{22} \geq \dfrac{440}{22}$

$$h \geq 20$$

Step 5: Check If the handyman works 23 hours (for example), will he earn more than $500? Since $60 + 22(23) = 566$ is greater than 500, we have evidence that our answer is correct.

Step 6: Answer The handyman must work no less than 20 hours to earn at least $500.

Quick ✓

36. A worker in a large apartment complex uses an elevator to move supplies. The elevator has a weight limit of 2000 pounds. The worker weighs 180 pounds, and each box of supplies weighs 91 pounds. Find the maximum number of boxes of supplies the worker can move on one trip in the elevator.

2.8 Exercises MyMathLab® PRACTICE

Exercise numbers in **green** have complete video solutions in MyMathLab

*Problems **1–36** are the **Quick ✓**s that follow the **EXAMPLES**.*

Building Skills

In Problems 37–44, graph each inequality on a number line, and write each inequality in interval notation. See Objectives 1 and 2.

37. $x > 4$ **38.** $n > 8$

39. $x \leq -1$ **40.** $x \leq 6$

41. $z \geq -3$ **42.** $x \geq -2$

43. $x < 4$ **44.** $y < -3$

In Problems 45–50, use interval notation to express the inequality shown in each graph. See Objective 2.

45. ← 0 1 2 3

46. −2 −1 0 1 2 3

47. −2 −1 0 1 2 3

48. −2 −1 0 1 2 3

49. −2 −1 0 1 2

50. −2 −1 0 1 2 3

In Problems 51–58, fill in the blank with the correct symbol. State which property of inequality is being used. See Objective 3.

51. If $x - 7 < 11$, then x ____ 18

52. If $3x - 2 > 7$, then $3x$ ____ 9

53. If $\dfrac{1}{3}x > -2$, then x ____ -6

54. If $\dfrac{5}{3}x \geq -10$, then x ____ -6

55. If $3x + 2 \le 13$, then $3x$___11

56. If $\frac{3}{4}x + 1 \le 9$, then $\frac{3}{4}x$___8

57. If $-3x \ge 15$, then x___-5

58. If $-4x < 28$, then x___-7

In Problems 59–88, solve the inequality. Express the solution set in set-builder notation and interval notation. Graph the solution set on a real number line. See Objective 3.

59. $x + 1 < 5$

60. $x + 4 \le 3$

61. $x - 6 \ge -4$

62. $x - 2 < 1$

63. $3x \le 15$

64. $4x > 12$

65. $-5x < 35$

66. $-7x \ge 28$

67. $3x - 7 > 2$

68. $2x + 5 > 1$

69. $3x - 1 \ge 3 + x$

70. $2x - 2 \ge 3 + x$

71. $1 - 2x \le 3$

72. $2 - 3x \le 5$

73. $-2(x + 3) < 8$

74. $-3(1 - x) > x + 8$

75. $4 - 3(1 - x) \le 1$

76. $8 - 4(2 - x) \le -2x$

77. $\frac{1}{2}(x - 4) > x + 8$

78. $3x + 4 > \frac{1}{3}(x - 2)$

79. $4(x - 1) > 3(x - 1) + x$

80. $2y - 5 + y < 3(y - 2)$

81. $5(n + 2) - 2n \le 3(n + 4)$

82. $3(p + 1) - p \ge 2(p + 1)$

83. $2n - 3(n - 2) < n - 4$

84. $4x - 5(x + 1) \le x - 3$

85. $4(2w - 1) \ge 3(w + 2) + 5(w - 2)$

86. $3q - (q + 2) > 2(q - 1)$

87. $3y - (5y + 2) > 4(y + 1) - 2y$

88. $8x - 3(x - 2) \ge x + 4(x + 1)$

In Problems 89–98, write the given statement using inequality symbols. Let x represent the unknown quantity. See Objective 4.

89. Karen's salary this year will be at least $16,000.

90. Bob's salary this year will be at most $120,000.

91. There will be at most 20,000 fans at the Cleveland Indians game today.

92. The cost of a new lawnmower is at least $250.

93. The cost to remodel a kitchen is more than $12,000.

94. There are no more than 25 students in your math class on any given day.

95. x is a positive number

96. x is a nonnegative number

97. x is a nonpositive number

98. x is a negative number

Mixed Practice

In Problems 99–116, solve the inequality and express the solution set in set-builder notation and interval notation, if possible. Graph the solution set on a real number line.

99. $-1 < x - 5$

100. $6 \ge x + 15$

101. $-\frac{3}{4}x > -\frac{9}{16}$

102. $-\frac{5}{8}x > \frac{25}{48}$

103. $3(x + 1) > 2(x + 1) + x$

104. $5(x - 2) < 3(x + 1) + 2x$

105. $-4a + 1 > 9 + 3(2a + 1) + a$

106. $-5b + 2(b - 1) \le 6 - (3b - 1) + 2b$

107. $n + 3(2n + 3) > 7n - 3$

108. $2k - (k - 4) \ge 3k + 10 - 2k$

109. $\frac{x}{2} \ge 1 - \frac{x}{4}$

110. $\frac{x}{3} \ge 2 + \frac{x}{6}$

111. $\frac{x + 5}{2} + 4 > \frac{2x + 1}{3} + 2$

112. $\frac{3z - 1}{4} + 1 \le \frac{6z + 5}{2} + 2$

113. $-5z - (3 + 2z) > 3 - 7z$

114. $2(4a - 3) \leq 5a - (2 - 3a)$

115. $1.3x + 3.1 < 4.5x - 15.9$

116. $4.9 + 2.6x < 4.2x - 4.7$

Applying the Concepts

117. Auto Rental A car can be rented from Certified Auto Rental for $55 per week plus $0.18 per mile. How many miles can you drive if you have at most $280 to spend for weekly transportation?

118. Truck Rental A truck can be rented from Acme Truck Rental for $80 per week plus $0.28 per mile. How many miles can you drive if you have at most $100 to spend on truck rental?

119. Final Grade Yvette has scores of 72, 78, 66, and 81 on her algebra tests. What score must she make on the final exam in order to pass the course with at least 360 points? The final exam counts as two test grades.

120. Final Grade To earn an A in Mrs. Smith's elementary statistics class, Elizabeth must earn at least 540 points. Thus far, Elizabeth has earned scores of 85, 83, 90, and 96. The final exam counts as two test grades. How many points does Elizabeth have to score on the final exam to earn an A?

121. Calling Plan Imperial Telephone has a long-distance calling plan that has a monthly fee of $10 and a charge of $0.03 per minute. Mayflower Communications has a monthly fee of $6 and a fee of $0.04 per minute. For how many minutes is Imperial Telephone the cheaper plan?

122. Commission A recent college graduate had an offer of a sales position that pays $15,000 per year plus 1% of all sales.

  (a) Write an expression for the total annual salary based on sales of S dollars.

  (b) For what total sales amount will the college graduate earn in excess of $150,000 annually?

123. Borrowing Money The amount of money that a lending institution will allow you to borrow mainly depends on the interest rate and your annual income. The equation $L = 2.98I - 76.11$ describes the amount of money, L, that a bank will lend at an interest rate of 7.5% for 30 years, based upon annual income, I. For what annual income, I, will a bank lend at least $150,000? (SOURCE: *Information Please Almanac*)

124. Advertising A marketing firm found that the equation $S = 2.1A + 224$ describes how the amount

of sales S of a product depends on A, the amount spent on advertising the product. Both S and A are measured in thousands of dollars. For what amount, A, is the sales of a product at least 350 thousand dollars?

Extending the Concepts

125. Grades In your Economics 101 class, you have scores of 68, 82, 87, and 89 on the first four of five tests. To earn a grade of B or higher, the average of the first five test scores must be greater than or equal to 80. Find the minimum score that you can make on the last test and earn a B.

126. Delivery Service A messenger service charges $10 to make a delivery to an address. In addition, each letter delivered costs $3 and each package delivered costs $8. If there are 15 more letters than packages delivered to this address, what is the maximum number of items that can be delivered for $85?

Compound inequalities are solved using the properties of inequality. Study the example below, and then solve Problems 127–134. State the solution in set-builder notation.

$$-15 \leq 2x - 1 < 37$$
$$-15 + 1 \leq 2x - 1 + 1 < 37 + 1$$
$$\frac{-14}{2} \leq \frac{2x}{2} < \frac{38}{2}$$
$$-7 \leq x < 19$$

127. $-3 < x + 30 < 16$

128. $-41 \leq x - 37 \leq 26$

129. $-6 \leq \frac{3x}{2} \leq 9$

130. $-4 < \frac{8x}{9} < 12$

131. $-7 \leq 2x - 3 < 15$

132. $4 < 3x + 7 \leq 28$

133. $4 < 6 - \frac{x}{2} \leq 10$

134. $-1 \leq 5 - \frac{x}{3} < 1$

Explaining the Concepts

135. In graphing an inequality, when is a left parenthesis used? When is a left bracket used?

136. Explain the circumstances in which the direction of the inequality symbol is reversed when one is solving a simple inequality.

137. Explain how you recognize when the solution of an inequality is all real numbers. Explain how you recognize when the solution of an inequality is the empty set.

138. Why is the interval notation $[-7, -\infty)$ for $x > -7$ incorrect? What is the correct notation?

Chapter 2 Activity: Pass to the Right

Focus: Solving linear equations and inequalities as a group

Time: 20–30 minutes

Group size: 3–4

1. Each member of the group should choose one of the following four equations and write it down on a piece of paper. Do not begin to solve.

(a) $\dfrac{x + 2}{2} - \dfrac{5x - 12}{6} = 1$

(b) $-\dfrac{1}{9}(x + 27) + \dfrac{1}{3}(x + 3) = x + 6$

(c) $\dfrac{x + 3}{3} - \dfrac{2x - 12}{9} = 1$

(d) $-\dfrac{1}{2}(x + 6) + \dfrac{1}{7}(x + 7) = x + 3$

2. Each member of the group should pass her or his equation to the person on the right. This member should perform the first step in solving the equation

and, when finished with that step, should pass the paper to the next group member on her or his right.

3. Upon receipt of the equation, each group member should check the previous member's work and then perform the next step. If an error is found, discuss and correct the error.

4. Continue passing the problems until all equations have been solved.

5. As a group, discuss the results.

6. If time permits, repeat this activity except choose one of the following inequalities. Express your final answer using interval notation.

(a) $\dfrac{1}{4}(2x + 12) > \dfrac{3}{8}(x - 1)$

(b) $\dfrac{3x + 1}{10} - \dfrac{1 + 6x}{5} \leq -\dfrac{1}{2}$

(c) $\dfrac{1}{2}(2x + 14) > \dfrac{3}{4}(x - 1)$

(d) $\dfrac{3x + 1}{21} - \dfrac{1 + 4x}{7} \leq -\dfrac{1}{3}$

Chapter 2 Review

Section 2.1 Linear Equations: The Addition and Multiplication Properties of Equality

KEY CONCEPTS

- **Linear Equation in One Variable**
  An equation equivalent to one of the form $ax + b = c$, where a, b, and c are real numbers and $a \neq 0$.

- **Addition Property of Equality**
  For real numbers a, b, and c, if $a = b$, then $a + c = b + c$.

- **Multiplication Property of Equality**
  For real numbers a, b, and c, where $c \neq 0$, if $a = b$, then $ac = bc$.

KEY TERMS

Linear equation
Sides
Left side
Right side
Solution
Solution set
Satisfies
Equivalent equations

You Should Be Able To...	EXAMPLE	Review Exercises
❶ Determine whether a number is a solution of an equation (p. 82)	Example 1	1–4
❷ Use the Addition Property of Equality to solve linear equations (p. 83)	Examples 2 through 4	5–10, 15, 16, 19
❸ Use the Multiplication Property of Equality to solve linear equations (p. 85)	Examples 5 through 9	11–14, 17, 18, 20

In Problems 1–4, determine whether the given value is a solution to the equation. Answer Yes or No.

1. $3x + 2 = 7$; $x = 5$

2. $5m - 1 = 17$; $m = 4$

3. $6x + 6 = 12$; $x = \dfrac{1}{2}$

4. $9k + 3 = 9$; $k = \dfrac{2}{3}$

In Problems 5–18, solve the equation. Check your solution.

5. $n - 6 = 10$

6. $n - 8 = 12$

7. $x + 6 = -10$

8. $x + 2 = -5$

9. $-100 = m - 5$

10. $-26 = m - 76$

11. $\frac{2}{3}y = 16$ **12.** $\frac{x}{4} = 20$

13. $-6x = 36$ **14.** $-4x = -20$

15. $z + \frac{5}{6} = \frac{1}{2}$ **16.** $m - \frac{1}{8} = \frac{1}{4}$

17. $1.6x = 6.4$ **18.** $1.8m = 9$

19. Discount The cost of a used Honda Accord has been discounted by $1200, so that the sale price is $18,900. Find the original price of the Honda Accord, p, by solving the equation $p - 1200 = 18{,}900$.

20. Coffee While studying for an exam, Randi drank coffee at a local coffeehouse. She bought 3 cups of coffee for a total of $7.65. Find the cost of each cup of coffee, c, by solving the equation $3c = 7.65$.

Section 2.2 Linear Equations: Using the Properties Together

KEY CONCEPT

- The steps for solving an equation in one variable are given on page 95.

KEY TERM

Distributive Property

You Should Be Able To...	EXAMPLE	Review Exercises
1 Use the Addition and Multiplication Properties of Equality to solve linear equations (p. 91)	Examples 1 and 2	21–24
2 Combine like terms and use the Distributive Property to solve linear equations (p. 92)	Examples 3 and 4	25–30
3 Solve a linear equation with the variable on both sides of the equation (p. 94)	Examples 5 and 6	31–34
4 Use linear equations to solve problems (p. 96)	Example 7	35, 36

In Problems 21–34, solve the equation. Check your solution.

21. $5x - 1 = -21$ **22.** $-3x + 7 = -5$

23. $\frac{2}{3}x + 5 = 11$ **24.** $\frac{5}{7}x - 2 = -17$

25. $-2x + 5 + 6x = -11$ **26.** $3x - 5x + 6 = 18$

27. $2m + 0.5m = 10$ **28.** $1.4m + m = -12$

29. $-2(x + 5) = -22$ **30.** $3(2x + 5) = -21$

31. $5x + 4 = -7x + 20$ **32.** $-3x + 5 = x - 15$

33. $4(x - 5) = -3x + 5x - 16$

34. $4(m + 1) = m + 5m - 10$

35. Ages Skye is 4 years older than Beth. The sum of their ages is 24. Find Skye's age by solving the equation $x + x + 4 = 24$, where x represents Beth's age and $x + 4$ represents Skye's age.

36. Parking Lot The length of a rectangular parking lot is 10 yards longer than the width, w, and the perimeter of the parking lot is 96 yards. Solve the equation $2w + 2(w + 10) = 96$ to find the width, w, of the parking lot. Then find the length of the parking lot, $w + 10$.

Section 2.3 Solving Linear Equations Involving Fractions and Decimals; Classifying Equations

KEY CONCEPTS

- A **conditional equation** is an equation that is true for some values of the variable and false for other values of the variable.

- An equation that is false for every replacement value of the variable is called a **contradiction**.

- An equation that is satisfied for every choice of the variable for which both sides of the equation are defined is called an **identity**.

KEY TERMS

Least common denominator (LCD)

Identity

Conditional equation

Contradiction

You Should Be Able To...	EXAMPLE	Review Exercises
❶ Use the least common denominator to solve a linear equation containing fractions (p. 98)	Examples 1 and 2	37–40, 45, 46
❷ Solve a linear equation containing decimals (p. 101)	Examples 3 through 5	41–44, 47, 48
❸ Classify a linear equation as an identity, a conditional equation, or a contradiction (p. 102)	Examples 6 through 8	49–54
❹ Use linear equations to solve problems (p. 104)	Example 9	55, 56

In Problems 37–48, solve the equation. Check your solution.

37. $\dfrac{6}{7}x + 3 = \dfrac{1}{2}$

38. $\dfrac{1}{4}x + 6 = \dfrac{5}{6}$

39. $\dfrac{n}{2} + \dfrac{2}{3} = \dfrac{n}{6}$

40. $\dfrac{m}{8} + \dfrac{m}{2} = \dfrac{3}{4}$

41. $1.2r = -1 + 2.8$

42. $0.2x + 0.5x = 2.1$

43. $1.2m - 3.2 = 0.8m - 1.6$

44. $0.3m + 0.8 = 0.5m + 1$

45. $\dfrac{1}{2}(x + 5) = \dfrac{3}{4}$

46. $-\dfrac{1}{6}(x - 1) = \dfrac{2}{3}$

47. $0.1(x + 80) = -0.2 + 14$

48. $0.35(x + 6) = 0.45(x + 7)$

In Problems 49–54, solve each equation. State whether the equation is a contradiction, an identity, or a conditional equation.

49. $4x + 2x - 10 = 6x + 5$

50. $-2(x + 5) = -5x + 3x + 2$

51. $-5(2n + 10) = 6n - 50$

52. $8m + 10 = -2(7m - 5)$

53. $10x - 2x + 18 = 2(4x + 9)$

54. $-3(2x - 8) = -3x - 3x + 24$

55. T-Shirt Al purchased a tie-dyed T-shirt at a "20% off" sale for $12.60. What was the original price, p, of the shirt? Solve the equation $p - 0.20p = 12.60$.

56. Couch Cushions Juanita was cleaning under her couch cushions and found some nickels and dimes. She found $0.55 in change, with the number of nickels one less than twice the number of dimes. Solve the equation $0.10d + 0.05(2d - 1) = 0.55$ to find d, the number of dimes Juanita found.

Section 2.4 Evaluating Formulas and Solving Formulas for a Variable

KEY CONCEPTS

- **Simple interest formula**
  $I = Prt$, I represents the amount of interest, P is the principal, r is the rate of interest (expressed as a decimal), and t is time (expressed in years).

- **Geometry Formulas (pages 110–111)**

KEY TERMS

Formula
Interest
Principal
Rate of interest

You Should Be Able To...	EXAMPLE	Review Exercises
❶ Evaluate a formula (p. 108)	Examples 1 through 6	57–60, 67–70
❷ Solve a formula for a variable (p. 113)	Examples 7 through 9	61–66, 67, 68

In Problems 57–60, substitute the given values into the formula and then simplify to find the unknown quantity. Include units in the answer.

57. area of a rectangle: $A = lw$; Find A when $l = 8$ inches and $w = 6$ inches.

58. perimeter of a square: $P = 4s$; Find P when $s = 16$ cm.

59. perimeter of a rectangle: $P = 2l + 2w$; Find w when $P = 16$ yards and $l = \dfrac{13}{2}$ yards.

60. circumference of a circle: $C = \pi d$; Find C when $d = \dfrac{15}{\pi}$ mm.

In Problems 61–66, solve each formula for the stated variable.

61. $V = LWH$; solve for H.

62. $I = Prt$; solve for P.

63. $S = 2LW + 2LH + 2WH$; solve for W.

64. $\rho = mv + MV$; solve for M.

65. $2x + 3y = 10$; solve for y.

66. $6x - 3y = 15$; solve for y.

 67. Finance The formula $A = P(1 + r)^t$ can be used to find the future value A of a deposit of P dollars in an account that earns an annual interest rate r (expressed as a decimal) after t years.

(a) Solve the formula for P.

(b) How much would you have to deposit today in order to have $3000 in 6 years in a bank account that pays 5% annual interest? Round your answer to the nearest cent.

68. Cylinders The surface area A of a right circular cylinder is given by the formula $A = 2\pi rh + 2\pi r^2$, where r is the radius and h is the height.

(a) Solve the formula for h.

(b) Determine the height of a right circular cylinder whose surface area is 72π square centimeters and whose radius is 4 centimeters.

69. Christmas Bonus Samuel invested his $500 Christmas bonus in a 9-month certificate of deposit that earns 3% simple interest. Use the formula $I = Prt$ to find the amount of interest Samuel's investment will earn.

70. Coffee Table Find the area of a circular coffee table top with a diameter of 3 feet.

Section 2.5 Problem Solving: Direct Translation

KEY CONCEPTS

- **Six steps for solving problems with mathematical models**
  - **Identify** what you are looking for
  - **Name** the unknown(s)
  - **Translate** to a mathematical equation
  - **Solve** the equation
  - **Check** the answer
  - **Answer** the question in a complete sentence

KEY TERMS

Problem solving
Mathematical modeling
Mathematical model
Model

You Should Be Able To...	EXAMPLE	Review Exercises
1 Translate English phrases into algebraic expressions (p. 120)	Examples 1 through 3	71–76; 83–86
2 Translate English sentences into equations (p. 122)	Example 4	77–82
3 Build models for solving direct translation problems (p. 125)	Examples 5 through 9	87–90

In Problems 71–76, translate each phrase into an algebraic expression. Let x represent the unknown number.

71. the difference between a number and 6

72. eight subtracted from a number

73. the product of -8 and a number

74. the quotient of a number and 10

75. twice the sum of 6 and a number

76. four times the difference of 5 and a number

In Problems 77–82, translate each statement into an equation. Let x represent the unknown number. DO NOT SOLVE.

77. The sum of 6 and a number is equal to twice the number increased by 5.

78. The product of 6 and a number decreased by 10 is one more than double the number.

79. Eight less than a number is the same as half of the number.

80. The ratio of 6 to a number is the same as the number added to 10.

81. Four times the sum of twice a number and 8 is 16.

82. Five times the difference between double a number and 8 is -24.

In Problems 83–86, choose a variable to represent one quantity. State what the quantity represents, and then express the second quantity in terms of the first.

83. Jacob is seven years older than Sarah.

84. José runs twice as fast as Consuelo.

85. Irene has $6 less than Max.

86. Victor and Larry will share $350.

87. Losing Weight Over the past year Lee Lai lost 28 pounds. Find Lee Lai's weight one year ago if her current weight is 125 pounds.

88. Consecutive Integers The sum of three consecutive integers is 39. Find the integers.

89. Finance A total of $20,000 is to be divided between Roberto and Juan, with Roberto to receive $2000 less than Juan. How much will each receive?

90. Truck Rentals You need to rent a moving truck. You identified two companies that rent trucks. ABC-Rental charges $30 per day plus $0.15 per mile. U-Do-It Rental charges $15 per day plus $0.30 per mile. For how many miles will the daily cost of renting be the same?

Section 2.6   Problem Solving: Problems Involving Percent

KEY CONCEPTS

- Percent means divided by 100 or per hundred.
- Total Cost = Original Cost + Sales Tax
- Original Price − Discount = Sale Price
- Wholesale Price + Markup = Selling Price

You Should be Able To...	EXAMPLE	Review Exercises
1 Solve problems involving percent (p. 132)	Examples 1 through 4	91–94
2 Solve business problems that involve percent (p. 134)	Examples 5 and 6	95–100

In Problems 91–94, find the unknown in each percent question.

91. What is 6.5% of 80?

92. 18 is 30% of what number?

93. 15.6 is what percent of 120?

94. 110% of what number is 55?

95. Sales Tax The sales tax in Florida is 6%. The total cost of purchasing a leotard, including tax, was $19.61. Find the cost of the leotard before sales tax.

96. Tutoring Mei Ling is a private tutor. She raised her hourly fee by 8.5% to cover her traveling expenses. Her new hourly fee is $32.55. Find Mei Ling's previous hourly fee.

97. Business: Discount Pricing A sweater, discounted by 70% for an end-of-the-year clearance sale, has a price tag of $12. What was the sweater's original price?

98. Business: Mark-Up A clothing store marks up the price that it pays for a suit 80%. If the selling price of a suit is $360, how much did the store pay for the suit?

99. Salary Tanya earns $500 each week plus 2% of the value of the computers she sells each week. If Tanya wishes to earn $3000 this week, what must be the total value of the computers she sells?

100. Voting In an election for school president, the loser received 80% of the winner's votes. If 900 votes were cast, how many did each receive?

Section 2.7   Problem Solving: Geometry and Uniform Motion

KEY CONCEPTS

- Complementary angles are two angles whose measures sum to 90°.
- Supplementary angles are two angles whose measures sum to 180°.
- The sum of the measures of the interior angles of a triangle is 180°.

- **Uniform Motion**
  If an object moves at an average speed r, the distance d covered in time t is given by the formula $d = rt$.

KEY TERMS

Complementary angles
Supplementary angles
Uniform motion

You Should be Able To...	EXAMPLE	Review Exercises
❶ Set up and solve complementary and supplementary angle problems (p. 138)	Example 1	101, 102
❷ Set up and solve "angles of a triangle" problems (p. 139)	Example 2	103, 104
❸ Use geometry formulas to solve problems (p. 140)	Examples 3 and 4	105–108
❹ Set up and solve uniform motion problems (p. 142)	Examples 5 and 6	109, 110

△ **101. Complementary Angles** Find two complementary angles such that the measure of the first angle is 20° more than six times the second.

△ **102. Supplementary Angles** Find two supplementary angles such that the measure of the first angle is 60° less than twice the second.

△ **103. Triangles** In a triangle, the measure of the second angle is twice the first. The measure of the third angle is 30° more than the second. Find the measures of the three angles.

△ **104. Triangles** In a triangle, the measure of the second angle is 5° less than the first. The measure of the third angle is 5° less than twice the second. Find the measures of the three angles.

△ **105. Rectangle** The length of a rectangle is 15 inches longer than twice its width. If the perimeter is 78 inches, find the length and width of the rectangle.

△ **106. Rectangle** The length of a rectangle is four times its width. If the perimeter is 70 cm, find the length and width of the rectangle.

△ **107. Garden** The perimeter of a rectangular garden is 120 feet. The width of the garden is twice the length.

(a) Find the length and width of the garden.

(b) What is the area of the garden?

△ **108. Back Yard** Yvonne's back yard is in the shape of a trapezoid with height of 80 feet. The shorter base is 10 feet shorter than the longer base, and the area of the back yard is 3600 square feet. Find the length of each base of the trapezoidal yard.

109. Boating Two motorboats leave the same dock at the same time, traveling in the same direction. One boat travels at 18 miles per hour, and the other travels at 25 miles per hour. In how many hours will the motorboats be 35 miles apart?

110. Train Station Two trains leave a station at the same time. One train is traveling east at 10 miles per hour faster than the other train, which is traveling west. After 6 hours, the two trains are 720 miles apart. At what speed did the faster train travel?

Section 2.8 Solving Linear Inequalities in One Variable

KEY CONCEPTS

- A linear inequality is of the form $ax + b < c$, $ax + b \leq c$, $ax + b > c$, or $ax + b \geq c$, where a, b, and c are real numbers and $a \neq 0$.

- **Addition Property of Inequality**
  For real numbers a, b, and c

  If $a < b$, then $a + c < b + c$.

  If $a > b$, then $a + c > b + c$.

- **Multiplication Properties of Inequality**
  For real numbers a, b, and c

  If $a < b$, and if $c > 0$, then $ac < bc$.

  If $a > b$, and if $c > 0$, then $ac > bc$.

  If $a < b$, and if $c < 0$, then $ac > bc$.

  If $a > b$, and if $c < 0$, then $ac < bc$.

KEY TERMS

Linear inequality
Simple inequality
Set-builder notation
Graph of an inequality
Endpoint
Interval
Solve an inequality
Equivalent inequalities

SET-BUILDER NOTATION	INTERVAL NOTATION	GRAPH
$\{x \mid x \geq a\}$	$[a, \infty)$	
$\{x \mid x > a\}$	(a, ∞)	
$\{x \mid x \leq a\}$	$(-\infty, a]$	
$\{x \mid x < a\}$	$(-\infty, a)$	
$\{x \mid x \text{ is any real number}\}$	$(-\infty, \infty)$	

You Should be Able To...	EXAMPLE	Review Exercises
❶ Graph inequalities on a real number line (p. 148)	Example 1	111–116
❷ Use interval notation (p. 149)	Example 2	117–120
❸ Solve linear inequalities using properties of inequalities (p. 150)	Examples 3 through 10	121–128
❹ Model inequality problems (p. 156)	Example 11	129, 130

In Problems 111–116, graph each inequality on a number line.

111. $x \leq -3$

112. $x > 4$

113. $m < 2$

114. $m \geq -5$

115. $0 < n$

116. $-3 \leq n$

In Problems 117–120, write the inequality in interval notation.

117. $x < -4$

118. $x \geq 7$

119.

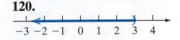

120.

In Problems 121–128, solve the inequality and express the solution in set-builder notation and interval notation if possible. Graph the solution set on a real number line.

121. $4x + 3 < 2x - 10$

122. $3x - 5 \geq -12$

123. $-4(x - 1) \leq x + 8$

124. $6x - 10 < 7x + 2$

125. $-3(x + 7) > -x - 2x$

126. $4x + 10 \leq 2(2x + 7)$

127. $\frac{1}{2}(3x - 1) > \frac{2}{3}(x + 3)$

128. $\frac{5}{4}x + 2 < \frac{5}{6}x - \frac{7}{6}$

129. Moving The Rent-A-Moving-Van Company charges a flat rate of $19.95 per day plus $0.20 for each mile driven. How many miles can be driven by a customer who can afford to spend at most $32.95 for one day?

130. Bowling Travis is bowling three games in a tournament. In the first game, his score was 148. In the second game, his score was 155. What score must Travis get in the third game for his tournament average to be greater than 151?

Chapter 2 Test

Remember to use your Chapter Test Prep Video to see fully worked-out solutions to any of these problems you would like to review.

In Problems 1–8, solve the equation. Check your solution.

1. $x + 3 = -14$

2. $-\frac{2}{3}m = \frac{8}{27}$

3. $5(2x - 4) = 5x$

4. $-2(x - 5) = 5(-3x + 4)$

5. $-\frac{2}{3}x + \frac{3}{4} = \frac{1}{3}$

6. $-0.6 + 0.4y = 1.4$

7. $8x + 3(2 - x) = 5(x + 2)$

8. $2(x + 7) = 2x - 2 + 16$

In Problems 9 and 10, (a) solve for the indicated variable, and then (b) find the value of the unknown quantity. Include units in the answer.

9. Volume of a rectangular solid: $V = lwh$

   (a) Solve for l.

   (b) Find l when $V = 540$ in.^{3}, $w = 6$ in., and $h = 10$ in.

10. Equation of a line: $2x + 3y = 12$

   (a) Solve for y.

   (b) Find y when $x = 8$.

11. Translate the following statement into an equation: Six times the difference between a number and 8 is equal to 5 less than twice the number. DO NOT SOLVE.

12. 18 is 30% of what number?

13. **Consecutive Integers** The sum of three consecutive integers is 48. Find the integers.

14. **Trading Spaces** On the show *Trading Spaces*, designer Vern Yip constructs a triangular art piece for a bedroom. The longest side is two inches longer than the middle side. The shortest side is 14 inches shorter than the middle side. If the perimeter of the art piece is 60 inches, what is the length of each side?

15. **Buses** Kimberly and Clay leave a concert hall at the same time, traveling in buses going in opposite directions. Kimberly's bus travels at 40 mph and Clay's bus travels at 60 mph. In how many hours will Kimberly and Clay be 350 miles apart?

16. **Construction** A carpenter cuts an oak board 21 feet long into two pieces. The longer board is 1 foot longer than three times the shorter one. Find the length of each board.

17. **New Backpack** Sherry purchased a new backpack for her daughter. The backpack was on sale for 20% off the regular price. If Sherry paid $28.80 for the backpack without sales tax, what was the original price?

In Problems 18 and 19, solve the inequality and express the solution in set-builder notation and interval notation. Graph the solution set on a real number line.

18. $3(2x - 5) \leq x + 15$

19. $-6x - 4 < 2(x - 7)$

20. **Cell Phone** Danielle's cell phone plan has a $30 monthly fee and an extra charge of $0.35 a minute. For how many minutes can Danielle use her cell phone so that the monthly bill is at most $100?

CHAPTER

3 Introduction to Graphing and Equations of Lines

Your legs feel like they are on fire. Exactly how steep is this hill? We can use mathematics to describe the steepness of a hill. See Example 7 and Problems 71 and 72 in Section 3.3.

The Big Picture: Putting It Together

It is now time to switch gears. In Chapter 2, we solved linear equations and inequalities involving one unknown. We are now going to focus our attention on linear equations and inequalities involving two unknowns.

When we deal with a single unknown, we can represent solutions to equations or inequalities graphically using real number lines. In that case, we need only one dimension to represent the solution. When we have two unknowns, we need to work in two dimensions. This is accomplished through the *rectangular coordinate system*.

3.1 The Rectangular Coordinate System and Equations in Two Variables

Objectives

1. Plot Points in the Rectangular Coordinate System
2. Determine Whether an Ordered Pair Satisfies an Equation
3. Create a Table of Values That Satisfy an Equation

Are You Ready for This Section?

Before getting started, take the following readiness quiz. If you get a problem wrong, go back to the section cited and review the material.

R1. Plot the following points on a real number line: [Section 1.3, pp. 22–23]

$$4, -3, \frac{1}{2}, 5.5$$

R2. Evaluate: **(a)** $|3|$ **(b)** $|-5|$ [Section 1.3, pp. 24–25]

R3. Evaluate $3x + 5$ for **(a)** $x = 4$ **(b)** $x = -1$ [Section 1.8, pp. 64–65]

R4. Evaluate $2x - 5y$ for
(a) $x = 3, y = 2$ **(b)** $x = 1, y = -4$ [Section 1.8, pp. 64–65]

R5. Solve: $3x + 5 = 14$ [Section 2.2, pp. 91–92]

R6. Solve: $5(x - 3) - 2x = 3x + 2$ [Section 2.3, pp. 102–104]

▶ ❶ Plot Points in the Rectangular Coordinate System

Figure 1

Because pictures allow individuals to visualize ideas, they are typically more powerful than other forms of printed communication. Figure 1, which shows the results of the Manhattan Project from the test conducted on July 16, 1945, illustrates the power of the atom in a way that words never could.

Although the pictures that we use in mathematics might not deliver as powerful a message as Figure 1, they are powerful nonetheless. To draw pictures of mathematical relationships, we need a "canvas." The "canvas" that we use in this chapter is the *rectangular coordinate system*.

In Section 1.3, we learned how to plot points on a real number line. We can think of plotting points on the real number line as plotting in one dimension. In this chapter, we use the *rectangular coordinate system* to plot points in two dimensions.

We begin by drawing two real number lines, one horizontal and one vertical, that intersect at right (90°) angles. We call the horizontal real number line the **x-axis** and the vertical real number line the **y-axis.** The point where the x-axis and y-axis intersect is called the **origin, O.** See Figure 2.

Figure 2
The rectangular coordinate system.

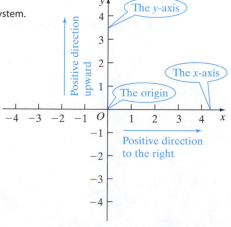

The origin O has a value of 0 on each axis. Points on the x-axis to the right of O represent positive real numbers, and points on the x-axis to the left of O represent negative real numbers. On the y-axis, points above O represent positive real numbers, and points below O represent negative real numbers. Notice in Figure 2 that we label the x-axis "x" and the y-axis "y." An arrow at the end of each axis shows the positive direction.

The coordinate system in Figure 2 is called a **rectangular** or **Cartesian coordinate system,** named after René Descartes (1596–1650), a French mathematician, philosopher, and theologian. The plane formed by the x-axis and y-axis is often called the **xy-plane,** and the x-axis and y-axis are called the **coordinate axes.**

We can represent any point P in the rectangular coordinate system by using an **ordered pair (x, y)** of real numbers. We say that x represents the distance that P is from the y-axis. If $x > 0$ (that is, if x is positive), then P is x units to the right of the y-axis. If $x < 0$, then P is $|x|$ units to the left of the y-axis. We say that y represents the distance that P is from the x-axis. If $y > 0$, then P is y units above the x-axis. If $y < 0$, then P is $|y|$ units below the x-axis. The ordered pair (x, y) is also called the **coordinates** of P. To plot the point with coordinates $(2, 5)$, begin at the origin, move 2 units to the right, and then move 5 units up, as shown in Figure 3(a).

Work Smart

Be careful: The order in which numbers appear in the ordered pairs matters. For example, (3, 2) represents a different point from (2, 3).

Figure 3

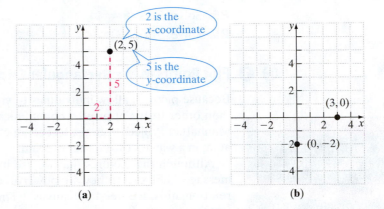

(a)   (b)

The origin O has coordinates $(0, 0)$. Any point on the x-axis has coordinates of the form $(x, 0)$, and any point on the y-axis has coordinates of the form $(0, y)$. For example, the points with coordinates $(3, 0)$ and $(0, -2)$, respectively, are shown in Figure 3(b).

If point P has coordinates (x, y), then x is called the **x-coordinate** of P, and y is called the **y-coordinate** of P.

The x- and y-axes divide the plane into four separate regions or **quadrants.** In quadrant I, both the x-coordinate and the y-coordinate are positive. In quadrant II, x is negative and y is positive. In quadrant III, both x and y are negative. In quadrant IV, x is positive and y is negative. Points on the coordinate axes do not belong to a quadrant. See Figure 4.

Figure 4

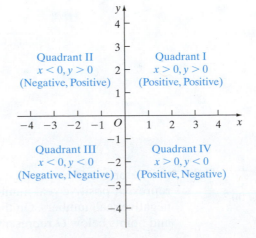

Let's summarize what we've learned so far.

> ## The Rectangular Coordinate System
>
> - Composed of two real number lines—one horizontal (the x-axis) and one vertical (the y-axis). The x- and y-axes intersect at the origin.
> - Also called the Cartesian coordinate system or xy-plane.
> - Points in the rectangular coordinate system are denoted (x, y) and are called the coordinates of the point. We call x the x-coordinate and y the y-coordinate.
> - If both x and y are positive, the point lies in quadrant I; if x is negative and y is positive, the point lies in quadrant II; if x is negative and y is negative, the point lies in quadrant III; if x is positive and y is negative, the point lies in quadrant IV.
> - Points on the x-axis have a y-coordinate of 0; points on the y-axis have an x-coordinate of 0. Points on the x- or y-axis do not lie in a quadrant.

EXAMPLE 1 Plotting Points in the Rectangular Coordinate System

Plot the following ordered pairs in the rectangular coordinate system. Tell which quadrant each point lies in or state that the point lies on the x- or y-axis.

(a) $A(3, 1)$ **(b)** $B(-4, 2)$ **(c)** $C(3, -5)$

(d) $D(4, 0)$ **(e)** $E(0, -3)$ **(f)** $F\left(-\dfrac{5}{2}, -\dfrac{7}{2}\right)$

Solution

Before we can plot the points, we draw a rectangular or Cartesian coordinate system. See Figure 5(a). We now plot the points.

Figure 5

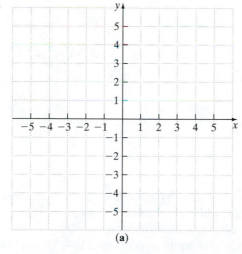

(a)

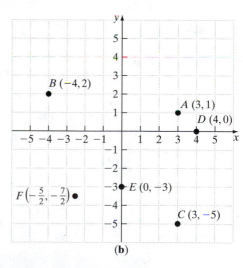

(b)

(a) To plot the point whose coordinates are $A(3, 1)$, begin at the origin O and travel 3 units to the right and then 1 unit up. Label the point A. Point A lies in quadrant I because both x and y are positive. See Figure 5(b).

(b) To plot $B(-4, 2)$, begin at the origin O and travel 4 units to the left and 2 units up. Label the point B. See Figure 5(b). Point B lies in quadrant II.

(c) See Figure 5(b). Point C lies in quadrant IV.

(d) See Figure 5(b). Point D lies on the x-axis, not in a quadrant.

(e) See Figure 5(b). Point E lies on the y-axis, not in a quadrant.

(f) It helps to convert the fractions into decimals, so $F\left(-\dfrac{5}{2}, -\dfrac{7}{2}\right) = F(-2.5, -3.5)$.

The x-coordinate of the point is halfway between -3 and -2; the y-coordinate is halfway between -4 and -3. See Figure 5(b). Point F lies in quadrant III. ●

Quick ✓

1. In the rectangular coordinate system, we call the horizontal real number line the _____ and we call the vertical real number line the _____. The point where these two axes intersect is called the _____.

2. If (x, y) are the coordinates of a point P, then x is called the _____ of P and y is called the _____ of P.

3. *True or False* The point whose ordered pair $(-2, 4)$ is located in quadrant IV.

4. *True or False* The ordered pairs $(7, 4)$ and $(4, 7)$ represent the same point in the Cartesian plane.

In Problems 5 and 6, plot the ordered pairs in the rectangular coordinate system. Tell which quadrant each point lies in or state that the point lies on the x-axis or y-axis.

5. (a) $(5, 2)$ (b) $(-4, -3)$ (c) $(1, -3)$ (d) $(-2, 0)$ (e) $(0, 6)$ (f) $\left(-\dfrac{3}{2}, \dfrac{5}{2}\right)$

6. (a) $(-6, 2)$ (b) $(1, 7)$ (c) $(-3, -2)$ (d) $(4, 0)$ (e) $(0, -1)$ (f) $\left(\dfrac{3}{2}, -\dfrac{7}{2}\right)$

EXAMPLE 2   **Identifying Points in the Rectangular Coordinate System**

Identify the coordinates of each point labeled in Figure 6.

Figure 6

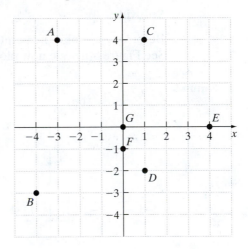

Solution

In an ordered pair (x, y), remember that x represents the position of the point left or right of the y-axis, while y represents the position of the point above or below the x-axis.

Point A is 3 units left of the y-axis so it has an x-coordinate of -3; point A is 4 units above the x-axis, so its y-coordinate is 4. The ordered pair $(-3, 4)$ corresponds to point A. We find the remaining coordinates in a similar fashion.

Point	Position	Ordered Pair
A	3 units left of the y-axis, 4 units above the x-axis	$(-3, 4)$
B	4 units left of the y-axis, 3 units below the x-axis	$(-4, -3)$
C	1 unit right of the y-axis, 4 units above the x-axis	$(1, 4)$
D	1 unit right of the y-axis, 2 units below the x-axis	$(1, -2)$
E	4 units right of the y-axis, on the x-axis	$(4, 0)$
F	On the y-axis, 1 unit below the x-axis	$(0, -1)$
G	On the x-axis, on the y-axis	$(0, 0)$

Quick ✓

7. Identify the coordinates of each point labeled in the figure below.

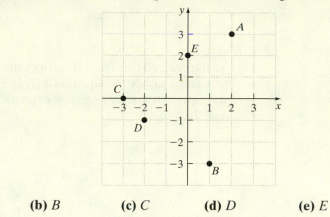

| (a) A | (b) B | (c) C | (d) D | (e) E |

▶ ❷ Determine Whether an Ordered Pair Satisfies an Equation

Recall from Sections 2.1–2.3, that the solution of a linear equation in one variable is either a single value of the variable (conditional equation), the empty set (contradiction), or all real numbers (identity). See Table 1.

Table 1 Categories of Linear Equations in One Variable

Conditional Equation	Contradiction	Identity
$3x + 2 = 11$	$4(x - 2) - x = 2(x + 1) + x$	$-2(x + 3) + 4x = 2(x - 3)$
$3x = 9$	$4x - 8 - x = 2x + 2 + x$	$-2x - 6 + 4x = 2x - 6$
$x = 3$	$3x - 8 = 3x + 2$	$2x - 6 = 2x - 6$
	$-8 = 2$	$-6 = -6$
	no solution	solution is all real numbers

We will now look at equations in two variables.

Definition

An **equation in two variables,** x and y, is a statement in which the algebraic expressions involving x and y are equal. The expressions are called **sides** of the equation.

For example, the following statements are all equations in two variables.

$$x + 2y = 6 \qquad y = -3x + 7 \qquad y = x^2 + 3$$

Since an equation is a statement, it may be true or false, depending on the values of the variables. Any values of the variables that make the equation a true statement **satisfy** the equation.

The first equation, $x + 2y = 6$, is satisfied when $x = 4$ and $y = 1$.

$$x + 2y = 6$$

Substitute $x = 4$ and $y = 1$: $4 + 2(1) \stackrel{?}{=} 6$

$$6 = 6 \quad \text{True}$$

We can also say that the ordered pair $(4, 1)$ satisfies the equation. Does $x = 8$ and $y = -1$ satisfy the equation $x + 2y = 6$?

$$x + 2y = 6$$
$$x = 8, y = -1: \quad 8 + 2(-1) \overset{?}{=} 6$$
$$6 = 6 \quad \text{True}$$

So the ordered pair $(8, -1)$ satisfies the equation as well. In fact, infinitely many choices of x and y satisfy the equation $x + 2y = 6$, but some choices of x and y do not satisfy the equation $x + 2y = 6$. For example, $x = 3$ and $y = 4$ do not satisfy the equation $x + 2y = 6$.

$$x + 2y = 6$$
$$x = 3, y = 4: \quad 3 + 2(4) \overset{?}{=} 6$$
$$11 = 6 \quad \text{False}$$

EXAMPLE 3 **Determining Whether an Ordered Pair Satisfies an Equation**

Determine whether the following ordered pairs satisfy the equation $x + 2y = 8$.

(a) $(2, 3)$ **(b)** $(6, -1)$ **(c)** $(-2, 5)$

Solution

(a) Check to see whether $x = 2, y = 3$ satisfies the equation $x + 2y = 8$.

$$x + 2y = 8$$
$$\text{Let } x = 2, y = 3: \quad 2 + 2(3) \overset{?}{=} 8$$
$$2 + 6 \overset{?}{=} 8$$
$$8 = 8 \quad \text{True}$$

The statement is true, so $(2, 3)$ satisfies the equation $x + 2y = 8$.

(b) For the ordered pair $(6, -1)$, we have

$$x + 2y = 8$$
$$\text{Let } x = 6, y = -1: \quad 6 + 2(-1) \overset{?}{=} 8$$
$$6 - 2 \overset{?}{=} 8$$
$$4 = 8 \quad \text{False}$$

The statement $4 = 8$ is false, so $(6, -1)$ does not satisfy the equation.

(c) For the ordered pair $(-2, 5)$, we have

$$x + 2y = 8$$
$$\text{Let } x = -2, y = 5: \quad -2 + 2(5) \overset{?}{=} 8$$
$$-2 + 10 \overset{?}{=} 8$$
$$8 = 8 \quad \text{True}$$

The statement is true, so $(-2, 5)$ satisfies the equation. ●

Quick ✓

8. *True or False* An equation in two variables can have more than one solution.

9. Determine whether the following ordered pairs satisfy the equation $x + 4y = 12$.

     (a) $(4, 2)$ **(b)** $(-2, 4)$ **(c)** $(1, 8)$

10. Determine whether the following ordered pairs satisfy the equation $y = 4x + 3$.

     (a) $(1, 3)$ **(b)** $(-2, -5)$ **(c)** $\left(-\dfrac{3}{2}, -3\right)$

▶ ❸ Create a Table of Values That Satisfy an Equation

In Example 3 we learned how to determine whether a given ordered pair satisfies an equation. Now, we learn how to find an ordered pair that satisfies an equation.

EXAMPLE 4 **How to Find an Ordered Pair That Satisfies an Equation**

Find an ordered pair that satisfies the equation $3x + y = 5$.

Step-by-Step Solution

Step 1: *Choose any value for one of the variables in the equation.*

Choose any value of x or y that you wish. We will let $x = 2$.

Step 2: *Substitute the chosen value of the variable into the equation. Solve for the remaining variable.*

Substitute 2 for x in $3x + y = 5$ and then solve for y.

$$3x + y = 5$$
$$\text{Let } x = 2: \quad 3(2) + y = 5$$
$$\text{Simplify:} \quad 6 + y = 5$$
$$\text{Subtract 6 from both sides:} \quad 6 - 6 + y = 5 - 6$$
$$y = -1$$

One ordered pair that satisfies the equation is $(2, -1)$. ●

> **Quick ✓**
>
> *In Problems 11–13, determine an ordered pair that satisfies the given equation by substituting the given value of the variable into the equation.*
>
> **11.** $2x + y = 10; x = 3$ **12.** $-3x + 2y = 11; y = 1$ **13.** $4x + 3y = 0; x = \dfrac{1}{2}$

Look again at Example 3. Did you notice that two different ordered pairs satisfy the equation? In fact, an infinite number of ordered pairs satisfy the equation $x + 2y = 8$ because for any real number y, we can find a value of x that makes the equation a true statement. One way to find some of the solutions of an equation in two variables is to create a table of values that satisfy the equation. The table is created by choosing values of x and using the equation to find the corresponding value of y (as we did in Example 4) or by choosing a value of y and using the equation to find the corresponding value of x.

EXAMPLE 5 **Creating a Table of Values That Satisfy an Equation**

Use the equation $y = -2x + 5$ to complete Table 2, and then use the table to list some of the ordered pairs that satisfy the equation.

Solution

The first entry in the table is $x = -2$. We substitute -2 for x and use the equation $y = -2x + 5$ to find y.

$$y = -2x + 5$$
$$x = -2: \quad y = -2(-2) + 5$$
$$y = 4 + 5$$
$$y = 9$$

Table 2

x	y	(x, y)
−2		
0		
1		

Now substitute 0 for x in the equation $y = -2x + 5$.

$$y = -2x + 5$$
$$x = 0: \quad y = -2(0) + 5$$
$$y = 0 + 5$$
$$y = 5$$

Finally, substitute 1 for x in the equation $y = -2x + 5$.

$$y = -2x + 5$$
$$x = 1: \quad y = -2(1) + 5$$
$$y = -2 + 5$$
$$y = 3$$

The completed table is shown as Table 3. Three ordered pairs that satisfy the equation are $(-2, 9)$, $(0, 5)$, and $(1, 3)$.

Table 3

x	y	(x, y)
−2	9	(−2, 9)
0	5	(0, 5)
1	3	(1, 3)

Quick ✓

In Problems 14 and 15, use the equation to complete the table. Use the table to list some of the ordered pairs that satisfy the equation.

14. $y = 5x - 2$

x	y	(x, y)
−2		
0		
1		

15. $y = -3x + 4$

x	y	(x, y)
−1		
2		
5		

EXAMPLE 6	**Creating a Table of Values That Satisfy an Equation**

Use the equation $2x - 3y = 12$ to complete Table 4, and then use the table to list some of the ordered pairs that satisfy the equation.

Table 4

x	y	(x, y)
−3		
	−4	
6		

Solution

The first entry in the table is $x = -3$. Substitute -3 for x and use the equation $2x - 3y = 12$ to find y.

$$2x - 3y = 12$$
$$x = -3: \quad 2(-3) - 3y = 12$$
$$-6 - 3y = 12$$

Add 6 to both sides: $\quad -3y = 18$

Divide both sides by -3: $\quad y = -6$

Substitute -4 for y in the equation $2x - 3y = 12$.

$$2x - 3y = 12$$
$$y = -4: \quad 2x - 3(-4) = 12$$
$$2x + 12 = 12$$

Subtract 12 from both sides: $\quad 2x = 0$

Divide both sides by 2: $\quad x = 0$

Substitute 6 for x in the equation $2x - 3y = 12$.

$$2x - 3y = 12$$
$$x = 6: \quad 2(6) - 3y = 12$$
$$12 - 3y = 12$$

Subtract 12 from both sides: $\quad -3y = 0$

Divide both sides by -3: $\quad y = 0$

Table 5

x	y	(x, y)
−3	−6	(−3, −6)
0	−4	(0, −4)
6	0	(6, 0)

Table 5 shows that three ordered pairs that satisfy the equation are $(-3, -6)$, $(0, -4)$, and $(6, 0)$.

Quick ✓

In Problems 16 and 17, use the equation to complete the table. Use the table to list some of the ordered pairs that satisfy the equation.

16. $2x + y = -8$

x	y	(x, y)
−5		
	−4	
2		

17. $2x - 5y = 18$

x	y	(x, y)
−6		
	−4	
2		

When modeling a situation, we can use variables other than x and y. For example, we might use the equation $C = 1.20m + 3$ to represent the cost of taking a taxi where C represents the cost (in dollars) and m represents the number of miles driven.

Recall from Section 1.3 that the scale of a number line refers to the distance between tick marks on the number line. In Figures 2 through 6, we used a scale of 1 on both the x-axis and the y-axis. In applications, a different scale is often used on each axis.

EXAMPLE 7 **An Electric Bill**

In North Carolina, Duke Power determines that the monthly electric bill for a household will be C dollars for using x kilowatt-hours (kWh) of electricity using the formula

$$C = 9.90 + 0.092896x$$

SOURCE: *Duke Power*

(a) Complete Table 6, and use the table to list ordered pairs that satisfy the equation. Express answers rounded to the nearest penny.

Table 6

x (kWh)	50 kWh	100 kWh	200 kWh
C ($)			

(b) Plot in a rectangular coordinate system the ordered pairs (x, C) found in part (a).

Solution

(a) The first entry in the table is $x = 50$. Substitute 50 for x into the equation $C = 9.90 + 0.092896x$ to find C. Round C to two decimal places because C represents the cost, so our answer is rounded to the nearest penny.

$$C = 9.90 + 0.092896x$$

$x = 50$: $\quad C = 9.90 + 0.092896(50)$

Use a calculator: $\quad C = 14.54$

Now substitute 100 for x into the equation to find C.

$$C = 9.90 + 0.092896x$$

$x = 100$: $\quad C = 9.90 + 0.092896(100)$

Use a calculator: $\quad C = 19.19$

Figure 7

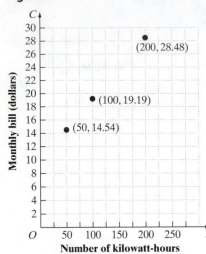

Now substitute 200 for x into the equation to find C.

$$C = 9.90 + 0.092896x$$

$x = 200$: $C = 9.90 + 0.092896\,(200)$

Use a calculator: $C = 28.48$

Table 7 shows the completed table. Three ordered pairs that satisfy the equation are (50, 14.54), (100, 19.19), and (200, 28.48).

Table 7

x (kWh)	50 kWh	100 kWh	200 kWh
C ($)	$14.54	$19.19	$28.48

(b) Since x represents the number of kilowatt-hours used, we label the horizontal axis x. Because C represents the bill, we label the vertical axis C. The ordered pairs found in part (a) are plotted in Figure 7. Note that we used different scales on the horizontal and vertical axes.   ●

Notice in Figure 7 that we labeled the horizontal and vertical axes so that it is clear what they represent. Labeling the axes is a good practice to follow whenever you draw a graph.

Quick ✓

18. Piedmont Natural Gas charges its customers in North Carolina a monthly fee of C dollars for using x therms of natural gas using the formula

$$C = 10 + 0.86676x$$

SOURCE: *Piedmont Natural Gas*

(a) Complete the table, and use the results to list ordered pairs (x, C) that satisfy the equation. Express answers rounded to the nearest penny.

x (therms)	50 therms	100 therms	150 therms
C ($)			

(b) Plot in a rectangular coordinate system the ordered pairs found in part (a).

3.1 Exercises   MyMathLab®   PRACTICE   Exercise numbers in green have complete video solutions in MyMathLab.

*Problems **1–18** are the Quick ✓ s that follow the EXAMPLES.*

Building Skills

In Problems 19–22, plot the following ordered pairs in the rectangular coordinate system. Tell which quadrant each point lies in or state that the point lies on the x-axis or y-axis. See Objective 1.

19. $A(-3, 2)$; $B(4, 1)$; $C(-2, -4)$; $D(5, -4)$; $E(-1, 3)$; $F(2, -4)$

20. $P(-3, -2)$; $Q(2, -4)$; $R(4, 3)$; $S(-1, 4)$; $T(-2, -4)$; $U(3, -3)$

21. $A\left(\dfrac{1}{2}, 0\right)$; $B\left(\dfrac{3}{2}, -\dfrac{1}{2}\right)$; $C\left(4, \dfrac{7}{2}\right)$; $D\left(0, -\dfrac{5}{2}\right)$; $E\left(\dfrac{9}{2}, 2\right)$; $F\left(-\dfrac{5}{2}, -\dfrac{3}{2}\right)$; $G(0, 0)$

22. $P\left(\dfrac{3}{2}, -2\right)$; $Q\left(0, \dfrac{5}{2}\right)$; $R\left(-\dfrac{9}{2}, 0\right)$; $S(0, 0)$; $T\left(-\dfrac{3}{2}, -\dfrac{9}{2}\right)$; $U\left(3, \dfrac{1}{2}\right)$; $V\left(\dfrac{5}{2}, -\dfrac{7}{2}\right)$

In Problems 23 and 24, plot the following ordered pairs in the rectangular coordinate system. Tell the location of each point: positive x-axis, negative x-axis, positive y-axis, or negative y-axis. See Objective 1.

23. $A(3,0)$; $B(0,-1)$; $C(0,3)$; $D(-4,0)$

24. $P(0,-1)$; $Q(-2,0)$; $R(0,3)$; $S(1,0)$

In Problems 25 and 26, identify the coordinates of each point labeled in the figure. Name the quadrant in which each point lies or state that the point lies on the x- or y-axis. See Objective 1.

25.

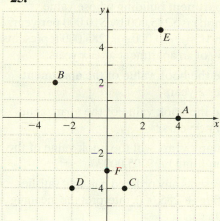

26.

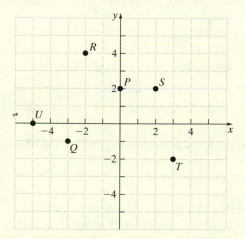

In Problems 27–32, determine whether or not the ordered pair satisfies the equation. See Objective 2.

27. $y = -3x + 5$; $A(-2,-1)$ $B(2,-1)$ $C\left(\frac{1}{3},4\right)$

28. $y = 2x - 3$; $A(-1,-5)$ $B(4,-5)$ $C(-2,-7)$

29. $3x + 2y = 4$; $A(0,2)$ $B(1,0)$ $C(4,-4)$

30. $5x - y = 12$; $A(2,0)$ $B(0,12)$ $C(-2,-22)$

31. $\frac{4}{3}x + y - 1 = 0$; $A(3,-3)$ $B(-6,-9)$ $C\left(\frac{3}{4},0\right)$

32. $\frac{3}{4}x + 2y = 0$; $A\left(-4,\frac{3}{2}\right)$ $B(0,0)$ $C\left(1,-\frac{3}{2}\right)$

For Problems 33–38, see Objective 3.

33. Find an ordered pair that satisfies the equation $x + y = 5$ by letting $x = 4$.

34. Find an ordered pair that satisfies the equation $x + y = 7$ by letting $x = 2$.

35. Find an ordered pair that satisfies the equation $2x + y = 9$ by letting $y = -1$.

36. Find an ordered pair that satisfies the equation $-4x - y = 5$ by letting $y = 7$.

37. Find an ordered pair that satisfies the equation $-3x + 2y = 15$ by letting $x = -3$.

38. Find an ordered pair that satisfies the equation $5x - 3y = 11$ by letting $y = 3$.

In Problems 39–52, use the equation to complete the table. Use the table to list some of the ordered pairs that satisfy the equation. See Objective 3.

39. $y = -x$

x	y	(x, y)
-3		
0		
1		

40. $y = x$

x	y	(x, y)
-4		
0		
2		

41. $y = -3x + 1$

x	y	(x, y)
-2		
-1		
4		

42. $y = 4x - 5$

x	y	(x, y)
-3		
1		
2		

43. $2x + y = 6$

x	y	(x, y)
-1		
2		
3		

44. $3x + 4y = 2$

x	y	(x, y)
-2		
2		
4		

45. $y = 6$

x	y	(x, y)
-4		
1		
12		

46. $x = 2$

x	y	(x, y)
	-4	
	0	
	8	

47. $x - 2y + 6 = 0$

x	y	(x, y)
1		
	1	
-2		

48. $2x + y - 4 = 0$

x	y	(x, y)
-1		
	-4	
	2	

49. $y = 5 + \dfrac{1}{2}x$

x	y	(x, y)
	7	
-4		
	2	

50. $y = 8 - \dfrac{1}{3}x$

x	y	(x, y)
	10	
	9	
	27	

51. $\dfrac{x}{2} + \dfrac{y}{3} = -1$

x	y	(x, y)
0		
	0	
	-6	

52. $\dfrac{x}{5} - \dfrac{y}{2} = 1$

x	y	(x, y)
	0	
0		
	-5	

In Problem 53–64, for each equation find the missing value in the ordered pair. See Objective 3.

53. $y = -3x - 10$ $A(\underline{\quad}, -16)$ $B(-3, \underline{\quad})$

   $C\left(\underline{\quad}, -9\right)$

54. $y = 5x - 4$ $A(-1, \underline{\quad})$ $B(\underline{\quad}, 31)$

   $C\left(-\dfrac{2}{5}, \underline{\quad}\right)$

55. $x = -\dfrac{1}{3}y$ $A(2, \underline{\quad})$ $B(\underline{\quad}, 0)$ $C\left(\underline{\quad}, -\dfrac{1}{2}\right)$

56. $y = \dfrac{2}{3}x$ $A\left(\underline{\quad}, -\dfrac{8}{3}\right)$ $B(0, \underline{\quad})$ $C\left(-\dfrac{5}{6}, \underline{\quad}\right)$

57. $x = 4$ $A(\underline{\quad}, -8)$ $B(\underline{\quad}, -19)$ $C(\underline{\quad}, 5)$

58. $y = -1$ $A(6, \underline{\quad})$ $B(-1, \underline{\quad})$ $C(0, \underline{\quad})$

59. $y = \dfrac{2}{3}x + 2$ $A(\underline{\quad}, 4)$ $B(-6, \underline{\quad})$ $C\left(\dfrac{1}{2}, \underline{\quad}\right)$

60. $y = -\dfrac{5}{4}x - 1$ $A(\underline{\quad}, -6)$ $B(-8, \underline{\quad})$

   $C\left(\underline{\quad}, -\dfrac{11}{6}\right)$

61. $\dfrac{1}{2}x - 3y = 2$ $A\left(-4, \underline{\quad}\right)$ $B(\underline{\quad}, -1)$

   $C\left(-\dfrac{2}{3}, \underline{\quad}\right)$

62. $\dfrac{1}{3}x + 2y = -1$ $A\left(-4, \underline{\quad}\right)$ $B\left(\underline{\quad}, -\dfrac{3}{4}\right)$

   $C\left(0, \underline{\quad}\right)$

63. $0.5x - 0.3y = 3.1$ $A(20, \underline{\quad})$ $B(\underline{\quad}, -17)$

   $C(2.6, \underline{\quad})$

64. $-1.7x + 0.2y = -5$ $A(\underline{\quad}, -110)$ $B(40, \underline{\quad})$

   $C(2.4, \underline{\quad})$

Applying the Concepts

65. Book Value Residential investment property such as apartment buildings may be depreciated over 28.5 years, according to the Internal Revenue Service (IRS). The IRS allows an investor to depreciate an apartment building whose value (excluding the land) is $285,000 by $10,000 per year.
   The equation $V = -10,000x + 285,000$ represents the book value, V, of an apartment after x years.

   (a) What will be the book value of the apartment building after 2 years?

   (b) What will be the book value of the apartment building after 5 years?

   (c) After how many years will the book value of the apartment building be $205,000?

   (d) If (x, V) represents any ordered pair that satisfies $V = -10,000x + 285,000$, interpret the meaning of $(3, 255000)$.

66. Taxi Ride The cost to take a taxi is $1.70 plus $2.00 per mile for each mile driven. The total cost, C, is given by the equation $C = 1.7 + 2m$, where m represents the total miles driven.

   (a) How much will it cost to take a taxi 5 miles?

   (b) How much will it cost to take a taxi 20 miles?

   (c) If you spent $32.70 on cab fare, how far was your trip?

   (d) If (m, C) represents any ordered pair that satisfies $C = 1.7 + 2m$, interpret the meaning of $(14, 29.7)$ in the context of this problem.

67. College Graduates An equation to approximate the percentage of the U.S. population 25 years of age or older with a bachelor's degree can be given by the model $P = 0.441n + 25.6$, where n is the number of years after 2000.

   (a) According to the model, what percentage of the U.S. population 25 years of age or older had a bachelor's degree in 2000 ($n = 0$)?

   (b) According to the model, what percentage of the U.S. population 25 years of age or older had a bachelor's degree in 2010 ($n = 10$)?

   (c) According to the model, what percentage of the U.S. population 25 years of age or older will have a bachelor's degree in 2020 ($n = 20$)?

   (d) In which year will 50% of the U.S. population 25 years of age or older have a bachelor's degree? Round your answer to the nearest year.

(e) According to the model, 100% of the U.S. population 25 years of age or older will have a bachelor's degree in 2169. Do you think this is reasonable? Why or why not?

68. Life Expectancy The model $A = 0.183n + 67.895$ is used to estimate the life expectancy A of residents of the United States born n years after 1950.

(a) According to the model, what is the life expectancy for a person born in 1950?

(b) According to the model, what is the life expectancy for a person born in 1980 ($n = 30$)?

(c) If the model holds true for future generations, what is the life expectancy for a person born in 2020?

(d) If a person has a life expectancy of 77 years, to the nearest year, when was the person born?

(e) Do you think life expectancy will continue to increase in the future? What could happen that would change this model?

77. Find the value of k for which $\left(-8, -\dfrac{5}{2}\right)$ satisfies $kx - 4y = 6$.

78. Find the value of k for which $\left(-9, \dfrac{1}{2}\right)$ satisfies $kx + 2y = -2$.

In Problems 79 and 80, use the equation to complete the table. Choose any value for x and solve the resulting equation to find the corresponding value for y. Then plot these ordered pairs in a rectangular coordinate system. Connect the points and describe the figure.

79. $3x - 2y = -6$

x	y	(x, y)

80. $-x + y = 4$

x	y	(x, y)

In Problems 69–72, use the equation to complete the table. Use the table to list some of the ordered pairs that satisfy the equation.

69. $4a + 2b = -8$

a	b	(a, b)
2		
	-4	
	6	

70. $2r - 3s = 3$

r	s	(r, s)
	3	
	-1	
-3		

71. $\dfrac{2p}{5} + \dfrac{3q}{10} = 1$

p	q	(p, q)
0		
	0	
-10		

72. $\dfrac{4a}{3} + \dfrac{2b}{5} = -1$

a	b	(a, b)
0		
	0	
	10	

In Problems 81–84, use the equation to complete the table. Then plot the points in a rectangular coordinate system.

81. $y = x^2 - 4$

x	y	(x, y)
-2		
-1		
0		
1		
2		

82. $y = -x^2 + 3$

x	y	(x, y)
-2		
-1		
0		
1		
2		

83. $y = -x^3 + 2$

x	y	(x, y)
-2		
-1		
0		
1		
2		

84. $y = 2x^3 - 1$

x	y	(x, y)
-2		
-1		
0		
1		
2		

Extending the Concepts

In Problems 73–78, determine the value of k so that the given ordered pair is a solution to the equation.

73. Find the value of k for which $(1, 2)$ satisfies $y = -2x + k$.

74. Find the value of k for which $(-1, 10)$ satisfies $y = 3x + k$.

75. Find the value of k for which $(2, 9)$ satisfies $7x - ky = -4$.

76. Find the value of k for which $(3, -1)$ satisfies $4x + ky = 9$.

Explaining the Concepts

85. Describe how the quadrants in the rectangular coordinate system are labeled and how you can determine the quadrant in which a point lies. Describe the characteristics of a point that lies on either the x- or y-axis.

86. Describe how to plot the point whose coordinates are $(3, -5)$.

The Graphing Calculator

Graphing calculators can also create tables of values that satisfy an equation. To do this, we first solve the equation for y. For example, to obtain a table of values that satisfy the equation $2x - 3y = 12$ (Example 6), we solve for y as follows:

$$2x - 3y = 12$$

Subtract 2x from both sides: $\qquad -3y = -2x + 12$

Divide both sides by -3: $\qquad y = \dfrac{-2x + 12}{-3}$

Divide -3 into each term in the numerator: $\qquad y = \dfrac{-2x}{-3} + \dfrac{12}{-3}$

Simplify: $\qquad y = \dfrac{2}{3}x - 4$

We now enter the equation $y = \dfrac{2}{3}x - 4$ into the calculator and create the table shown.

X	Y1
-3	-6
-2	-5.333
-1	-4.667
0	-4
1	-3.333
2	-2.667
3	-2

$Y_1 \blacksquare (2/3)X-4$

In Problems 87–94, use a graphing calculator to create a table of values that satisfy each equation. Have the table begin at -3 and increase by 1.

87. $y = 2x - 9$

88. $y = -3x + 8$

89. $y = -x + 8$

90. $y = 2x - 4$

91. $y + 2x = 13$

92. $y - x = -15$

93. $y = -6x^2 + 1$

94. $y = -x^2 + 3x$

3.2 Graphing Equations in Two Variables

Objectives

1. Graph a Line by Plotting Points
2. Graph a Line Using Intercepts
3. Graph Vertical and Horizontal Lines

Are You Ready For This Section?

Before getting started, take the following readiness quiz. If you get a problem wrong, go back to the section cited and review the material.

R1. Solve: $4x = 24$ [Section 2.1, pp. 85–87]

R2. Solve: $-3y = 18$ [Section 2.1, pp. 85–87]

R3. Solve: $2x + 5 = 13$ [Section 2.2, pp. 91–92]

▶ ① Graph a Line by Plotting Points

In the previous section, we found values of x and y that satisfy an equation. What does this mean? Well, it means that the ordered pair (x, y) is a point on the graph of the equation.

In Words

The graph of an equation is a geometric way of representing the set of all ordered pairs that make the equation a true statement. Think of the graph as a picture of the solution set.

Definition

The **graph of an equation in two variables** x and y is the set of points whose coordinates, (x, y), in the xy-plane satisfy the equation.

But how do we obtain the graph of an equation? One method for graphing an equation is the **point-plotting method.**

EXAMPLE 1 **How to Graph an Equation Using the Point-Plotting Method**

Graph the equation $y = 2x - 3$ using the point-plotting method.

Step-by-Step Solution

Step 1: We find ordered pairs that satisfy the equation by choosing some values of x and using the equation to find the corresponding values of y. See Table 8.

Table 8

x	y	(x, y)
	$y = 2(-2) - 3$ $= -4 - 3$ $= -7$	$(-2, -7)$
	$y = 2(-1) - 3$ $= -5$	$(-1, -5)$
	$y = 2(0) - 3$ $= -3$	$(0, -3)$
	$y = 2(1) - 3$ $= -1$	$(1, -1)$
	$y = 2(2) - 3$ $= 1$	$(2, 1)$

Step 2: Plot in a rectangular coordinate system the points whose coordinates, (x, y), were found in Step 1. See Figure 8.

Figure 8

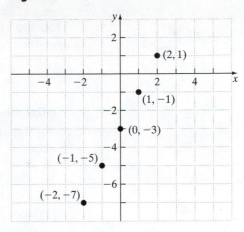

Step 3: Connect the points with a straight line. See Figure 9.

Figure 9
$y = 2x - 3$

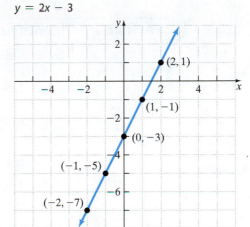

Work Smart

Remember, all coordinates of the points on the graph shown in Figure 9 satisfy the equation $y = 2x - 3$.

The graph of the equation in Figure 9 does not show all the points that satisfy the equation. For example, the point $(5, 7)$ is a part of the graph of $y = 2x - 3$ but is not shown in Figure 9. Since the graph of $y = 2x - 3$ could be extended as far as we please, we use arrows to indicate that the pattern shown continues. It is important to show enough of the graph so that anyone can "see" how it continues. This is called a **complete graph.**

For the remainder of the text, we will say "the point (x, y)" for short rather than "the point whose coordinates are (x, y)."

Graphing an Equation Using the Point-Plotting Method

Step 1: Find several ordered pairs that satisfy the equation.

Step 2: Plot the points found in Step 1 in a rectangular coordinate system.

Step 3: Connect the points with a smooth curve or line.

Quick ✔

1. The _____ of an equation in two variables is the set of points whose coordinates, (x, y), in the xy-plane satisfy the equation.

In Problems 2 and 3, draw a complete graph of each equation using point plotting.

2. $y = 3x - 2$

3. $y = -4x + 8$

A question you may be asking yourself is "How many points do I need to find before I can be sure that I have a complete graph?" It depends on the type of equation you are graphing. The equation that we graphed in Example 1 is called a *linear equation*.

> **Definition**
>
> A **linear equation in two variables** is an equation that can be written in the form
>
> $$Ax + By = C$$
>
> where A, B, and C are real numbers. A and B cannot both be 0. A linear equation written in the form $Ax + By = C$ is said to be in **standard form.** *

EXAMPLE 2   **Identifying Linear Equations in Two Variables**

Determine whether the equation is a linear equation in two variables.

(a) $3x - 4y = 9$ **(b)** $\frac{1}{2}x + \frac{2}{3}y = 4$

(c) $x^2 + 5y = 10$ **(d)** $-2y = 5$

Solution

(a) The equation $3x - 4y = 9$ is a linear equation in two variables because it is written in the form $Ax + By = C$ with $A = 3$, $B = -4$, and $C = 9$.

(b) The equation $\frac{1}{2}x + \frac{2}{3}y = 4$ is a linear equation in two variables because it is written in the form $Ax + By = C$ with $A = \frac{1}{2}$, $B = \frac{2}{3}$, and $C = 4$.

(c) The equation $x^2 + 5y = 10$ is not a linear equation because x is squared.

(d) The equation $-2y = 5$ is a linear equation in two variables because it is written in the form $Ax + By = C$ with $A = 0$, $B = -2$, and $C = 5$.   ●

Quick ✓

4. A(n) _____ equation is an equation that can be written in the form $Ax + By = C$, where A, B, and C are real numbers, and A and B are not both zero. Equations written in this form are said to be in _____ ____.

In Problems 5–7, determine whether or not the equation is a linear equation in two variables.

5. $4x - y = 12$ **6.** $5x - y^2 = 10$

7. $5x = 20$

Work Smart

When graphing a line, be sure to find three points—just to be safe!

For the remainder of the text, we will refer to linear equations in two variables as **linear equations.** The graph of a linear equation is a **line.** To graph a linear equation requires only two points; however, we recommend finding a third point as a check.

*Some texts call $Ax + By = C$ the **general form** of a line.

EXAMPLE 3

Graphing a Linear Equation Using the Point-Plotting Method

Graph the linear equation $2x + y = 4$.

Solution

Because the coefficient of y is 1, it is easier to choose values of x and find the corresponding values of y. We will determine the value of y for $x = -2, 0$, and 2. There is nothing magical about these choices. Any three different values of x will give us the results we want.

Work Smart

Choose values of x (or y) that make the algebra easy.

$x = -2$:	$x = 0$:	$x = 2$:
$2x + y = 4$	$2x + y = 4$	$2x + y = 4$
Let $x = -2$: $\quad 2(-2) + y = 4$	Let $x = 0$: $\quad 2(0) + y = 4$	Let $x = 2$: $\quad 2(2) + y = 4$
$-4 + y = 4$	$y = 4$	$4 + y = 4$
$y = 8$		$y = 0$

We plot the three points from Table 9 and connect them with a straight line. See Figure 10.

Table 9

x	y	(x, y)
-2	8	$(-2, 8)$
0	4	$(0, 4)$
2	0	$(2, 0)$

Figure 10

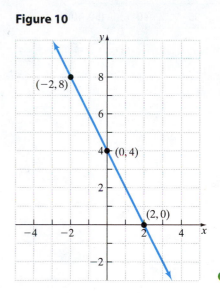

Quick ✓

8. The graph of a linear equation is a ___.

In Problems 9 and 10, graph each linear equation using the point-plotting method.

9. $-3x + y = -6$

10. $2x + 3y = 12$

EXAMPLE 4

Cost of Renting a Car

A car-rental agency quotes you the cost of renting a car in Washington, D.C., as $30 per day plus $0.20 per mile. The linear equation $C = 0.20m + 30$ models the cost, where C represents total cost and m represents the number of miles that were traveled in one day.

(a) Complete Table 10 to find ordered pairs that satisfy the equation. Round answers to the nearest penny.

(b) Graph the linear equation $C = 0.20m + 30$ using the points obtained in part (a).

Table 10

m	C	(m, C)
0		
50		
100		

Solution

(a) The first entry in the table is $m = 0$.
We substitute 0 for m in the equation
$C = 0.20m + 30$ to find C.

$$C = 0.20m + 30$$
$m = 0$: $\quad C = 0.20(0) + 30$
$$C = 30$$

Now substitute 50 for m in the
equation $C = 0.20m + 30$ to find C.

$$C = 0.20m + 30$$
$m = 50$: $\quad C = 0.20(50) + 30$
$$C = 40$$

Now substitute 100 for m in the
equation to find C.

$$C = 0.20m + 30$$
$m = 100$: $\quad C = 0.20(100) + 30$
$$C = 50$$

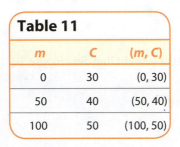

Table 11

m	C	(m, C)
0	30	(0, 30)
50	40	(50, 40)
100	50	(100, 50)

Table 11 shows the completed table. The ordered pair (50, 40) means that if a car is driven 50 miles in a day, the rental cost will be $40.

(b) Since m represents the number of miles driven, we label the horizontal axis m. We label the vertical axis C, the cost of renting the car. We let the scale of the horizontal axis be 10, so each tick mark represents 10 miles. We scale the vertical axis to 5, so each tick mark represents 5 dollars. Scaling in this way makes it easier to plot the ordered pairs. In Figure 11, we plot the points found in part (a) and then draw the line.

Figure 11

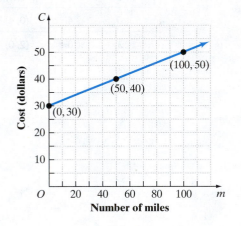

Quick ✓

11. Michelle sells computers. Her monthly salary is $3000 plus 8% of total sales. The linear equation $S = 0.08x + 3000$ models Michelle's monthly salary, S, where x represents her total sales in the month.

(a) Complete the table and use the results to list ordered pairs that satisfy the equation.

x	S	(x, S)
0		
10,000		
25,000		

(b) Graph the linear equation $S = 0.08x + 3000$ using the points obtained in part (a).

Figure 12

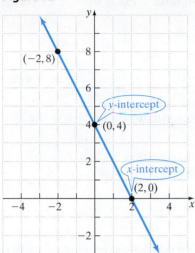

▶ ❷ Graph a Line Using Intercepts

Intercepts should always be displayed in a complete graph.

> **Definitions**
>
> The **intercepts** are the points, if any, where a graph crosses or touches a coordinate axis. A point at which the graph crosses or touches the *x*-axis is an ***x*-intercept**, and a point at which the graph crosses or touches the *y*-axis is a ***y*-intercept.**

See Figure 12 for an illustration. The graph in Figure 12 is the graph obtained in Example 3.

EXAMPLE 5 | **Finding Intercepts from a Graph**

Find the intercepts of the graphs shown in Figures 13(a) and 13(b). What are the *x*-intercepts? What are the *y*-intercepts?

In Words
An *x*-intercept exists when $y = 0$.
A *y*-intercept exists when $x = 0$.

Figure 13

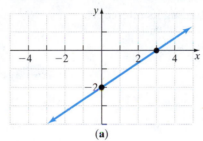
(a)

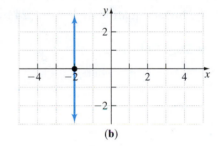
(b)

Solution

(a) The intercepts of the graph in Figure 13(a) are the points $(0, -2)$ and $(3, 0)$. The *x*-intercept is $(3, 0)$. The *y*-intercept is $(0, -2)$.

(b) The intercept of the graph in Figure 13(b) is the point $(-2, 0)$. The *x*-intercept is $(-2, 0)$. There are no *y*-intercepts. ●

> **Quick ✔**
>
> **12.** The _____ are the points, if any, where a graph crosses or touches a coordinate axis.
>
> *In Problems 13 and 14, find the intercepts of the graphs shown in the figures. What are the x-intercepts? What are the y-intercepts?*
>
> **13.**
>
>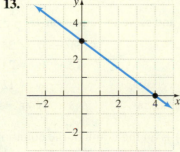
>
> **14.**
>
>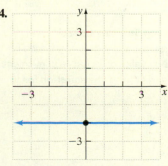

Let's see how to find intercepts algebraically. From Figure 12, it should be clear that an x-intercept exists when the value of y is 0 and that a y-intercept exists when the value of x is 0. This leads to the following procedure.

Procedure for Finding Intercepts

1. To find the x-intercept(s), if any, of the graph of an equation, let $y = 0$ in the equation and solve for x.

2. To find the y-intercept(s), if any, of the graph of an equation, let $x = 0$ in the equation and solve for y.

Work Smart

Every point on the x-axis has a y-coordinate of 0. That's why we set $y = 0$ to find an x-intercept. Likewise, every point on the y-axis has an x-coordinate of 0. That's why we set $x = 0$ to find a y-intercept.

We can use the intercepts to graph a line. Because the intercepts represent only two points, we find a third point so that we can check our work.

EXAMPLE 6 **How to Graph a Linear Equation by Finding Its Intercepts**

Graph the linear equation $4x - 3y = 24$ by finding its intercepts.

Step-by-Step Solution

Step 1: Find the y-intercept by letting $x = 0$ and solving the equation for y.

$$4x - 3y = 24$$
$$\text{Let } x = 0: \quad 4(0) - 3y = 24$$
$$0 - 3y = 24$$
$$-3y = 24$$
$$\text{Divide both sides by } -3: \quad y = -8$$

The y-intercept is $(0, -8)$.

Step 2: Find the x-intercept by letting $y = 0$ and solving the equation for x.

$$4x - 3y = 24$$
$$\text{Let } y = 0: \quad 4x - 3(0) = 24$$
$$4x - 0 = 24$$
$$4x = 24$$
$$\text{Divide both sides by } 4: \quad x = 6$$

The x-intercept is $(6, 0)$.

Step 3: Find one additional point on the graph by choosing any value of x that is convenient and solving the equation for y.

We will let $x = 3$ and solve the equation $4x - 3y = 24$ for y.

$$\text{Let } x = 3: \quad 4(3) - 3y = 24$$
$$12 - 3y = 24$$
$$\text{Subtract 12 from both sides:} \quad -3y = 12$$
$$\text{Divide both sides by } -3: \quad y = -4$$

The point $(3, -4)$ is on the graph of the equation.

Step 4: Plot the points found in Steps 1–3 and draw in the line.

We plot the points $(0, -8)$, $(6, 0)$, and $(3, -4)$. Connect the points with a straight line and obtain the graph in Figure 14.

Figure 14
$4x - 3y = 24$

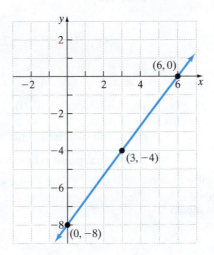

EXAMPLE 7   **Graphing a Linear Equation by Finding Its Intercepts**

Graph the linear equation $\dfrac{1}{2}x - 2y = 3$ by finding its intercepts.

Solution

Figure 15
$\dfrac{1}{2}x - 2y = 3$

x-intercept:

$$\frac{1}{2}x - 2y = 3$$

Let $y = 0$: $\quad \dfrac{1}{2}x - 2(0) = 3$

$$\frac{1}{2}x = 3$$

$$x = 6$$

y -intercept:

$$\frac{1}{2}x - 2y = 3$$

Let $x = 0$: $\quad \dfrac{1}{2}(0) - 2y = 3$

$$-2y = 3$$

$$y = -\frac{3}{2}$$

Additional point (choose $x = 2$):

$$\frac{1}{2}x - 2y = 3$$

Let $x = 2$: $\quad \dfrac{1}{2}(2) - 2y = 3$

$$1 - 2y = 3$$

$$-2y = 2$$

$$y = -1$$

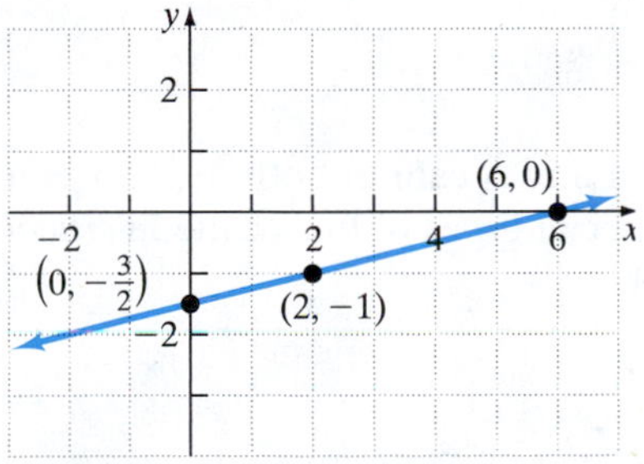

Plot the points $(6, 0)$, $\left(0, -\dfrac{3}{2}\right)$, and $(2, -1)$ and connect them with a straight line. See Figure 15.

Quick ✓

15. *True or False* To find the *y*-intercept(s), if any, of the graph of an equation, let $y = 0$ in the equation and solve for *x*.

In Problems 16–18, graph each linear equation by finding its intercepts.

16. $x + y = 3$

17. $2x - 5y = 20$

18. $\dfrac{3}{2}x - 2y = 9$

EXAMPLE 8 **Graphing a Linear Equation of the Form *Ax* + *By* = 0**

Graph the linear equation $2x + 3y = 0$ by finding its intercepts.

Solution

x-intercept:

$$2x + 3y = 0$$
Let $y = 0$: $2x + 3(0) = 0$
$$2x + 0 = 0$$
Divide both sides by 2: $x = 0$

The *x*-intercept is $(0, 0)$.

y-intercept:

$$2x + 3y = 0$$
Let $x = 0$: $2(0) + 3y = 0$
$$3y = 0$$
Divide both sides by 3: $y = 0$

The *y*-intercept is $(0, 0)$.

Because both the *x*- and *y*-intercepts are $(0, 0)$, we find *two* additional points.

Additional point (choose $x = 3$):

$$2x + 3y = 0$$
Let $x = 3$: $2(3) + 3y = 0$
$$6 + 3y = 0$$
Subtract 6 from both sides: $3y = -6$
Divide both sides by 3: $y = -2$

Additional point (choose $x = -3$):

$$2x + 3y = 0$$
Let $x = -3$: $2(-3) + 3y = 0$
$$-6 + 3y = 0$$
Add 6 to both sides: $3y = 6$
Divide both sides by 2: $y = 2$

Work Smart

Linear equations of the form $Ax + By = 0$, where $A \neq 0$ and $B \neq 0$, have only one intercept at $(0, 0)$, so two additional points should be plotted to obtain the graph.

Figure 16
$2x + 3y = 0$

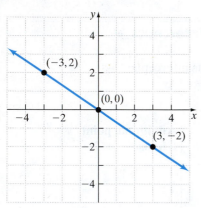

The points $(3, -2)$ and $(-3, 2)$ are on the graph of the equation. We plot the points $(0, 0)$, $(-3, 2)$, and $(3, -2)$ and connect them with a straight line. See Figure 16. ●

Quick ✓

In Problems 19 and 20, graph the equation by finding its intercepts.

19. $y = \dfrac{1}{2}x$

20. $4x + y = 0$

❸ Graph Vertical and Horizontal Lines

In the equation of a line, $Ax + By = C$, we said that A and B cannot both be zero. But what if $A = 0$ or $B = 0$? We find that this leads to special types of lines called *vertical lines* (when $B = 0$) and *horizontal lines* (when $A = 0$).

EXAMPLE 9 **Graphing a Vertical Line**

Graph the equation $x = 3$ using the point-plotting method.

Solution

Because the equation $x = 3$ can be written as $1x + 0y = 3$, we know the graph is a line. For the equation $x = 3$, any value of y has a corresponding *x*-value of 3. For example, if $y = -1$, then

$$1x + 0(-1) = 3$$
$$x = 3$$

See Table 12 for other choices for y. The points $(3, -2)$, $(3, -1)$, $(3, 0)$, $(3, 1)$, and $(3, 2)$ are all on the line. See Figure 17.

Table 12

x	y	(x, y)
3	−2	(3, −2)
3	−1	(3, −1)
3	0	(3, 0)
3	1	(3, 1)
3	2	(3, 2)

Figure 17
$x = 3$

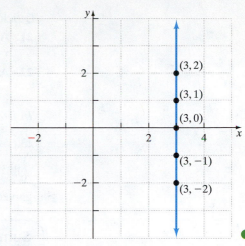

The results of Example 9 lead to a definition of a *vertical line*.

> **Definition Equation of A Vertical Line**
>
> A **vertical line** is given by an equation of the form
>
> $$x = a$$
>
> where $(a, 0)$ is the *x*-intercept.

EXAMPLE 10

Graphing a Horizontal Line

Graph the equation $y = -2$ using the point-plotting method.

Solution

Because the equation $y = -2$ can be written as $0x + 1y = -2$, we know the graph is a line. For $y = -2$, any value of *x* has a corresponding *y*-value of -2. For example, if $x = -2$, then

$$0(-2) + 1y = -2$$
$$y = -2$$

See Table 13 for other choices of *x*. The points $(-2, -2)$, $(-1, -2)$, $(0, -2)$, $(1, -2)$, and $(2, -2)$ are all on the line. See Figure 18.

Table 13

x	y	(x, y)
−2	−2	(−2, −2)
−1	−2	(−1, −2)
0	−2	(0, −2)
1	−2	(1, −2)
2	−2	(2, −2)

Figure 18
$y = -2$

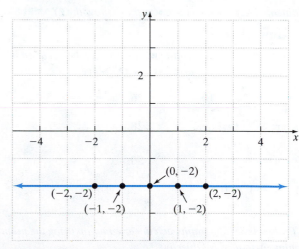

The results of Example 10 lead to a definition of a *horizontal line*.

> **Definition Equation of a Horizontal Line**
>
> A **horizontal line** is given by an equation of the form
>
> $$y = b$$
>
> where $(0, b)$ is the *y*-intercept.

Quick ✔

21. A _____ line is given by an equation of the from $x = a$, where _____ is the x-intercept.

22. A _____ line is given by an equation of the from $y = b$, where _____ is the y-intercept.

In Problems 23 and 24, graph each equation.

23. $x = -5$ 24. $y = -4$

We covered a lot of material in this section. We present a summary below to help you organize the information presented.

Summary Intercepts and Equations of Lines

Topic	Comments
Intercepts: Points where the graph crosses or touches a coordinate axis.	Intercepts need to be shown for a graph to be complete.
x-intercept: Point where the graph crosses or touches the x-axis. Found by letting $y = 0$ in the equation.	
y-intercept: Point where the graph crosses or touches the y-axis. Found by letting $x = 0$ in the equation.	
Standard Form of an Equation of a Line: $Ax + By = C$, where A and B are not both zero	Can be graphed using point-plotting or intercepts.
Equation of a Vertical Line: $x = a$	Graph is a vertical line whose x-intercept is $(a, 0)$.
Equation of a Horizontal Line: $y = b$	Graph is a horizontal line whose y-intercept is $(0, b)$.

3.2 Exercises   MyMathLab®   MathXL PRACTICE

Exercise numbers in **green** have complete video solutions in MyMathLab.

Problems 1–24 are the Quick ✔s that follow the EXAMPLES.

Building Skills

In Problems 25–32, determine whether or not the equation is a linear equation in two variables. See Objective 1.

25. $2x - 5y = 10$ 26. $y^2 = 2x + 3$

27. $x^2 + y = 1$ 28. $y - 2x = 9$

29. $y = \dfrac{4}{x}$ 30. $x - 8 = 0$

31. $y - 1 = 0$ 32. $y = -\dfrac{2}{x}$

In Problems 33–50, graph each linear equation using the point-plotting method. See Objective 1.

33. $y = 2x$ 34. $y = 3x$

35. $y = 4x - 2$ 36. $y = -3x - 1$

37. $y = -2x + 5$ 38. $y = x - 6$

39. $x + y = 5$ 40. $x - y = 6$

41. $-2x + y = 6$ 42. $5x - 2y = -10$

43. $4x - 2y = -8$ 44. $x + 3y = 6$

45. $x = -4y$ 46. $x = \dfrac{1}{2}y$

47. $y + 7 = 0$ 48. $x - 6 = 0$

49. $y - 2 = 3(x + 1)$ 50. $y + 3 = -2(x - 2)$

In Problems 51–58, find the intercepts of each graph. See Objective 2.

51.

52.

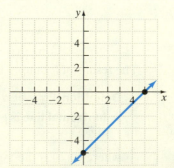

53.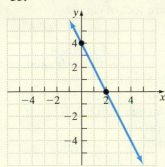

54.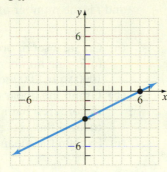

55.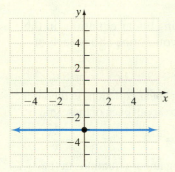

56.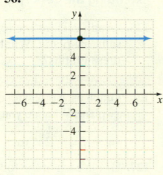

57.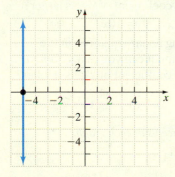

58.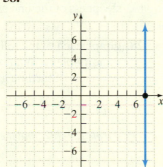

In Problems 59–70, find the intercepts of each equation. See Objective 2.

59. $2x + 3y = -12$

60. $3x - 5y = 30$

61. $x = -6y$

62. $y = 10x$

63. $y = x - 5$

64. $y = -x + 7$

65. $\dfrac{x}{6} + \dfrac{y}{8} = 1$

66. $\dfrac{x}{2} - \dfrac{y}{8} = 1$

67. $x = 4$

68. $y = 6$

69. $y + 2 = 0$

70. $x + 8 = 0$

In Problems 71–86, graph each linear equation by finding its intercepts. See Objective 2.

71. $3x + 6y = 18$

72. $3x - 5y = 15$

73. $-x + 5y = 15$

74. $-2x + y = 14$

75. $\dfrac{1}{2}x = y + 3$

76. $\dfrac{4}{3}x = -y + 1$

77. $9x - 2y = 0$

78. $\dfrac{1}{3}x - y = 0$

79. $y = -\dfrac{1}{2}x + 3$

80. $y = \dfrac{2}{3}x - 3$

81. $\dfrac{1}{3}y + 2 = 2x$

82. $\dfrac{1}{2}x - 3 = 3y$

83. $\dfrac{x}{2} + \dfrac{y}{3} = 1$

84. $\dfrac{y}{4} - \dfrac{x}{3} = 1$

85. $4y - 2x + 1 = 0$

86. $2y - 3x + 2 = 0$

In Problems 87–94, graph each horizontal or vertical line. See Objective 3.

87. $x = 5$

88. $x = -7$

89. $y = -6$

90. $y = 2$

91. $y - 12 = 0$

92. $y + 3 = 0$

93. $3x - 5 = 0$

94. $2x - 7 = 0$

Mixed Practice

In Problems 95–106, graph each linear equation by the point-plotting method or by finding intercepts.

95. $y = 2x - 5$

96. $y = -3x + 2$

97. $y = -5$

98. $x = 2$

99. $2x + 5y = -20$

100. $3x - 4y = 12$

101. $2x = -6y + 4$

102. $5x = 3y - 10$

103. $x - 3 = 0$

104. $y + 4 = 0$

105. $3y - 12 = 0$

106. $-4x + 8 = 0$

107. If $(3, y)$ is a point on the graph of $4x + 3y = 18$, find y.

108. If $(-4, y)$ is a point on the graph of $3x - 2y = 10$, find y.

109. If $(x, -2)$ is a point on the graph of $3x + 5y = 11$, find x.

110. If $(x, -3)$ is a point on the graph of $4x - 7y = 19$, find x.

Applying the Concepts

111. Plot the points $(3, 5)$ and $(-2, 5)$ and draw a line through the points. What is the equation of this line?

112. Plot the points $(-1, 2)$ and $(5, 2)$ and draw a line through the points. What is the equation of this line?

113. Plot the points $(-2, -4)$ and $(-2, 1)$ and draw a line through the points. What is the equation of this line?

114. Plot the points $(3, -1)$ and $(3, 2)$ and draw a line through the points. What is the equation of this line?

In Problems 115–118, find the equation of each line.

115.

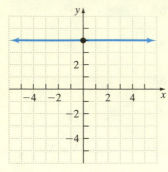

116.

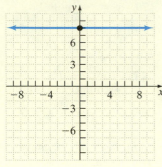

117.

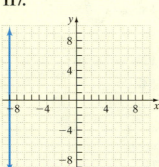

118.

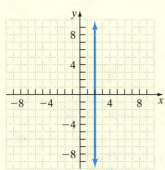

119. Create a set of ordered pairs in which the *x*-coordinates are twice the *y*-coordinates. What is the equation of this line?

120. Create a set of ordered pairs in which the *y*-coordinates are twice the *x*-coordinates. What is the equation of this line?

121. Create a set of ordered pairs in which the *y*-coordinates are 2 more than the *x*-coordinates. What is the equation of this line?

122. Create a set of ordered pairs in which the *x*-coordinates are 3 less than the *y*-coordinates. What is the equation of this line?

123. Calculating Wages Marta earns $500 per week plus $100 in commission for every car she sells. The linear equation that calculates her weekly earnings is $E = 100n + 500$, where E represents her weekly earnings in dollars and n represents the number of cars she sold during the week.

(a) Create a set of ordered pairs (n, E) if, in three consecutive weeks, she sold 0 cars, 4 cars, and 10 cars.

(b) Graph the linear equation $E = 100n + 500$ using the ordered pairs obtained in part (a). Be sure to label the axes appropriately.

(c) Explain the meaning of the *E*-intercept.

124. Carpet Cleaning Harry's Carpet Cleaning charges a $50 service charge plus $0.10 for each square foot of carpeting to be cleaned. The linear equation that calculates the total cost to clean a carpet is

$C = 0.1f + 50$, where C is the total cost in dollars and f is the number of square feet of carpet.

(a) Create a set of ordered pairs (f, C) for the following number of square feet to be cleaned: 1000 sq ft, 2000 sq ft, 2500 sq ft.

(b) Graph the linear equation $C = 0.1f + 50$ using the ordered pairs obtained in part (a). Be sure to label the axes appropriately.

(c) Explain the meaning of the *C*-intercept.

Extending the Concepts

125. Graph each of the following linear equations in the same *xy*-plane. What do you notice about each of these graphs?
$$y = 2x - 1 \quad y = 2x + 3 \quad 2x - y = 5$$

126. Graph each of the following linear equations in the same *xy*-plane. What do you notice about each of these graphs?
$$y = 3x + 2 \quad 6x - 2y = -4 \quad x = \frac{1}{3}y - \frac{2}{3}$$

127. Graph each of the following linear equations in the same *xy*-plane. What statement can you make about the steepness of the line as the coefficient of *x* gets larger?
$$y = x \quad y = 2x \quad y = 10x$$

128. Graph each of the following linear equations in the same *xy*-plane. What statement can you make about the steepness of the line as the coefficient of *x* gets smaller?
$$y = x + 2 \quad y = \frac{1}{2}x + 2 \quad y = \frac{1}{8}x + 2$$

In Problems 129–132, find the intercepts of each graph.

129.

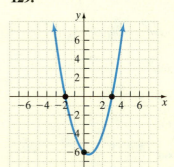

130.

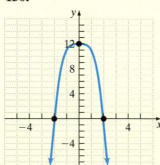

131.

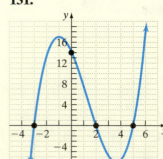

132.

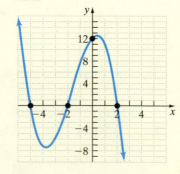

Explaining the Concepts

133. Explain what the graph of an equation represents.

134. What is meant by a complete graph?

135. How many points are required to graph a line? Explain your reasoning and why you might include additional point(s) when graphing a line.

136. Explain how to use the intercepts to graph the equation $Ax + By = C$, where A, B, and C are not equal to zero. Explain how to graph the same equation when C is equal to zero. Can you use the same techniques for both equations? Why or why not?

The Graphing Calculator

Graphing calculators can graph equations. In fact, graphing calculators also use the point-plotting method to obtain the graph by choosing 95 values of x and using the equation to find the corresponding values of y. As with creating tables, we first solve the equation for y. For example, to obtain a table of values that satisfy the equation $2x - 3y = 12$

(Example 6 from Section 3.1), we must solve for y, and we obtain $y = \dfrac{2}{3}x - 4$. We enter the equation $y = \dfrac{2}{3}x - 4$ into the calculator and create the graph shown below.

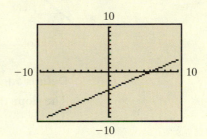

In Problems 137–142, use a graphing calculator to graph each equation.

137. $y = 2x - 9$ **138.** $y = -3x + 8$

139. $y + 2x = 13$ **140.** $y - x = -15$

141. $y = -6x^2 + 1$ **142.** $y = -x^2 + 3x$

3.3 Slope

Objectives

1 Find the Slope of a Line Given Two Points

2 Find the Slope of Vertical and Horizontal Lines

3 Graph a Line Using Its Slope and a Point on the Line

4 Work with Applications of Slope

Are You Ready for This Section?

Before getting started, take the following readiness quiz. If you get a problem wrong, go back to the section cited and review the material.

R1. Evaluate: $\dfrac{5 - 2}{8 - 7}$ [Section 1.7, p. 59]

R2. Evaluate: $\dfrac{3 - 7}{9 - 3}$ [Section 1.7, p. 59]

R3. Evaluate: $\dfrac{-3 - 4}{6 - (-1)}$ [Section 1.7, p. 59]

Figure 19

(a)   (b)

Pretend you are on snow skis for the first time in your life. The ski resort that you are visiting has two "beginner" hills, as shown in Figure 19. Which hill would you prefer to go down? Why?

It is clear that the hill in Figure 19(a) is not as steep as the hill in Figure 19(b). Mathematicians like to numerically describe situations such as the steepness of a hill. Measuring the steepness of each hill allows for an easy comparison. The numerical measure that describes the steepness of a hill is its *slope*.

▶ **1 Find the Slope of a Line Given Two Points**

Consider the staircase drawn in Figure 20(a) on the next page. If we draw a line through the top of each riser on the staircase (in blue), we can see that each step contains exactly the same horizontal change (or **run**) and the same vertical change (or **rise**). We define *slope* in terms of the rise and run.

Ready?...Answers **R1.** 3   **R2.** $-\dfrac{2}{3}$

R3. -1

Figure 20

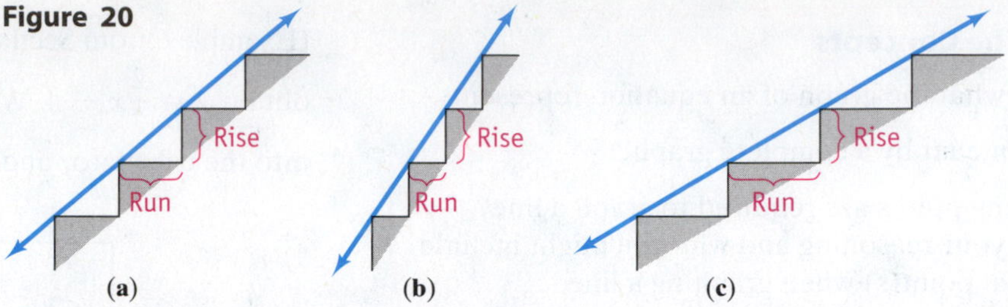

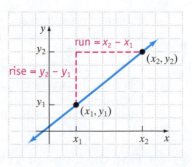

(a)   (b)   (c)

Definition

The **slope** of a line, denoted by the letter m, is the ratio of the rise to the run. That is,

$$\text{Slope} = m = \frac{\text{rise}}{\text{run}}$$

The symbol for the slope of a line is m, which comes from the French word *monter*, which means "to go up, ascend, or climb." Slope is a numerical measure of the steepness of the line. For example, if the run is decreased and the rise remains the same, then the staircase becomes steeper. See Figure 20(b). If the run is increased and the rise remains the same, then the staircase becomes less steep. See Figure 20(c).

If the staircase in Figure 20(a) has a rise of 6 inches and a run of 6 inches, then the slope of the line is

$$m = \frac{\text{rise}}{\text{run}} = \frac{6 \text{ inches}}{6 \text{ inches}} = 1$$

If the rise of the staircase is increased to 9 inches, then the slope of the line is

$$m = \frac{\text{rise}}{\text{run}} = \frac{9 \text{ inches}}{6 \text{ inches}} = \frac{3}{2}$$

The main idea is the steeper the line, the larger the slope. We can define the slope of a line using rectangular coordinates.

Work Smart

The subscripts 1 and 2 on $x_1, x_2, y_1,$ and y_2 do not represent a computation (as superscripts do in x^2). Instead, they are used to indicate that the values of the variable x_1 may be different from x_2, and y_1 may be different from y_2.

Definition

If $x_1 \neq x_2$, the **slope m** of the line containing the points (x_1, y_1) and (x_2, y_2) is defined by the formula

$$m = \frac{y_2 - y_1}{x_2 - x_1} \qquad x_1 \neq x_2$$

Figure 21 illustrates the slope of a line.

Figure 21

Notice that the slope m of a line may be viewed as

$$m = \frac{\text{rise}}{\text{run}} = \frac{y_2 - y_1}{x_2 - x_1}$$

We can also write the slope m of a line as

$$m = \frac{y_2 - y_1}{x_2 - x_1} = \frac{\text{change in } y}{\text{change in } x} = \frac{\Delta y}{\Delta x}$$

The symbol Δ is the Greek letter delta. In mathematics, we read the symbol Δ as "change in." Thus we read $\dfrac{\Delta y}{\Delta x}$ as "change in y divided by change in x." The symbol Δ comes from the first letter of the Greek word for "difference," *diaphora*.

EXAMPLE 1 **How to Find the Slope of a Line**

Find the slope of the line drawn in Figure 22.

Figure 22

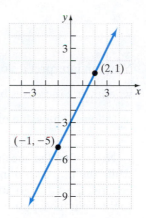

Step-by-Step Solution

Step 1: Let one of the points be (x_1, y_1) and the other point be (x_2, y_2).

Let's say that $(x_1, y_1) = (-1, -5)$ and $(x_2, y_2) = (2, 1)$.

Step 2: Find the slope by evaluating

$$m = \frac{y_2 - y_1}{x_2 - x_1}$$

$$m = \frac{y_2 - y_1}{x_2 - x_1}$$

$$x_1 = -1, y_1 = -5;\ \ x_2 = 2, y_2 = 1: \quad = \frac{1 - (-5)}{2 - (-1)}$$

$$= \frac{6}{3}$$

$$m = 2$$

Figure 23

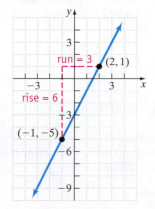

Work Smart

It doesn't matter which point is called (x_1, y_1) and which is called (x_2, y_2). The answer will be the same. In Example 1, if we let $(x_1, y_1) = (2, 1)$ and $(x_2, y_2) = (-1, -5)$, we get

$$m = \frac{y_2 - y_1}{x_2 - x_1} = \frac{-5 - 1}{-1 - 2} = \frac{-6}{-3} = 2$$

Remember that slope can be thought of as "rise divided by run." This description of the slope of a line is illustrated in Figure 23. We interpret the slope of the line drawn in Figure 23 as follows: "The value of y will increase by 6 units whenever x increases by 3 units." Or, because $\dfrac{6}{3} = 2 = \dfrac{2}{1}$, "the value of y will increase by 2 units whenever x increases by 1 unit." Both interpretations are acceptable.

EXAMPLE 2 **Finding and Interpreting the Slope of a Line**

Plot the points $(-1, 3)$ and $(2, -2)$ and draw a line through the two points. Find and interpret the slope of the line.

Solution

We plot the points $(x_1, y_1) = (-1, 3)$ and $(x_2, y_2) = (2, -2)$ in the rectangular coordinate system and draw a line through the two points. See Figure 24. The slope of the line drawn in Figure 24 is

$$m = \frac{y_2 - y_1}{x_2 - x_1} = \frac{-2 - 3}{2 - (-1)}$$

$$= \frac{-5}{3}$$

$$= -\frac{5}{3}$$

Figure 24

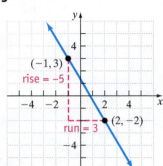

You can interpret a slope of $-\frac{5}{3} = \frac{-5}{3}$ this way: The value of y will go down 5 units whenever x increases by 3 units. Because $-\frac{5}{3} = \frac{5}{-3} = \frac{\text{rise}}{\text{run}}$, a second interpretation is as follows: The value of y will increase by 5 units whenever x decreases by 3 units. ●

Notice that the line drawn in Figure 23 goes up and to the right and the slope is positive, while the line drawn in Figure 24 goes down and to the right and slope is negative. In general, lines that go down and to the right have negative slope, and lines that go up and to the right have positive slope. See Figure 25.

Figure 25

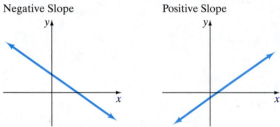

Negative Slope   Positive Slope

Line goes down and to the right   Line goes up and to the right

Quick ✓

1. If the run of a line is 10 and its rise is 6, then its slope is __.

2. *True or False* If $P = (x_1, y_1)$ and $Q = (x_2, y_2)$ are two distinct points with $y_1 \neq y_2$, the slope m of the line that contains points P and Q is defined by the formula $m = \dfrac{x_2 - x_1}{y_2 - y_1}$.

3. *True or False* If the slope of a line is $\dfrac{3}{2}$, then y will increase by 3 units when x increases by 2 units.

4. If the graph of a line goes up as you move to the right, then the slope of this line must be _____.

In Problems 5 and 6, plot the points in a rectangular coordinate system. Then draw a line through the two points. Find and interpret the slope of the line containing the points.

5. $(0, 2)$ and $(4, 10)$ 6. $(-2, 2)$ and $(3, -7)$

▶ ❷ Find the Slope of Vertical and Horizontal Lines

Did you notice that in the definition of slope, $m = \dfrac{y_2 - y_1}{x_2 - x_1}$, we have the restriction that $x_1 \neq x_2$? This means the formula does not apply if the x-coordinates of the two points are the same. Why? See Example 3.

EXAMPLE 3　**The Slope of a Vertical Line**

Plot the points $(2, -1)$ and $(2, 3)$ in a rectangular coordinate system. Then draw a line through the two points. Find and interpret the slope of the line.

Solution

We plot the points $(x_1, y_1) = (2, -1)$ and $(x_2, y_2) = (2, 3)$ in the rectangular coordinate system and draw a line through the two points. See Figure 26. The slope of the line drawn in Figure 26 is

$$m = \frac{y_2 - y_1}{x_2 - x_1} = \frac{3 - (-1)}{2 - 2}$$

$$= \frac{4}{0}$$

Figure 26

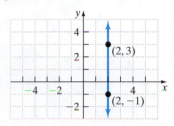

Because division by 0 is undefined, we say that the slope of the line is undefined. When y increases by 4, there is no change in x.

Let's generalize the results of Example 3. Let (x_1, y_1) and (x_2, y_2) be two distinct points. If $x_1 = x_2$, then we have a **vertical line** whose slope m is **undefined** (since this results in division by 0). See Figure 27.

Figure 27
Vertical line

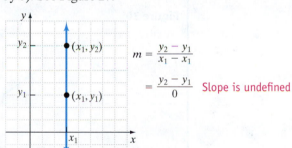

$$m = \frac{y_2 - y_1}{x_1 - x_1}$$

$$= \frac{y_2 - y_1}{0} \quad \text{Slope is undefined}$$

Okay, but what if $y_1 = y_2$?

EXAMPLE 4　**The Slope of a Horizontal Line**

Plot the points $(-2, 4)$ and $(3, 4)$ in a rectangular coordinate system. Then draw a line through the two points. Find and interpret the slope of the line.

Solution

We plot the points $(x_1, y_1) = (-2, 4)$ and $(x_2, y_2) = (3, 4)$ in the rectangular coordinate system and draw a line through the two points. See Figure 28. The slope of the line drawn in Figure 28 is

$$m = \frac{y_2 - y_1}{x_2 - x_1} = \frac{4 - 4}{3 - (-2)}$$

$$= \frac{0}{5}$$

$$= 0$$

Figure 28

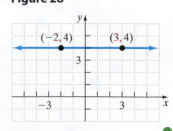

The slope of the line is 0. A slope of 0 can be interpreted as: There is no change in y when x increases by 1 unit.

Let's generalize the results of Example 4. Let (x_1, y_1) and (x_2, y_2) be two distinct points. If $y_1 = y_2$, then we have a **horizontal line** whose slope m is 0. See Figure 29.

Figure 29
Horizontal line

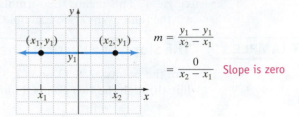

$$m = \frac{y_1 - y_1}{x_2 - x_1}$$

$$= \frac{0}{x_2 - x_1} \quad \text{Slope is zero}$$

Quick ✓

7. The slope of a horizontal line is _____, while the slope of a vertical line is _____.

In Problems 8 and 9, plot the given points in a rectangular coordinate system. Then draw a line through the two points. Find and interpret the slope of the line.

8. $(2, 5)$ and $(2, -1)$ **9.** $(2, 5)$ and $(6, 5)$

Summary **The Slope of a Line**

Figure 30 illustrates the four possibilities for the slope of a line. Remember, just as we read a text from left to right, we also read graphs from left to right.

Figure 30

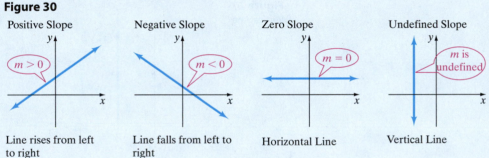

Positive Slope — $m > 0$ — Line rises from left to right

Negative Slope — $m < 0$ — Line falls from left to right

Zero Slope — $m = 0$ — Horizontal Line

Undefined Slope — m is undefined — Vertical Line

▶ ❸ **Graph a Line Using Its Slope and a Point on the Line**

We now illustrate how to use slope to graph lines.

EXAMPLE 5   **Graphing a Line Given a Point and Its Slope**

Draw a graph of the line that contains the point $(1, 3)$ and has a slope of 2.

Solution

We know that $m = 2 = \frac{2}{1} = \frac{\text{rise}}{\text{run}}$. This means

that y will increase by 2 units (the rise), when x increases by 1 unit (the run). So if we start at $(1, 3)$ and move 2 units up and then 1 unit to the right, we end up at the point $(2, 5)$. We then draw a line through the points $(1, 3)$ and $(2, 5)$ to obtain the graph of the line. See Figure 31.

Figure 31

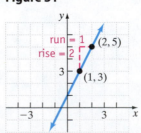

EXAMPLE 6 **Graphing a Line Given a Point and Its Slope**

Draw a graph of the line that contains the point $(1, 3)$ and has a slope of $-\dfrac{2}{3}$.

Solution

Figure 32

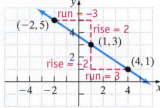

Because slope $= -\dfrac{2}{3} = \dfrac{-2}{3} = \dfrac{\text{rise}}{\text{run}}$, y will decrease by 2 units when x increases by 3 units. If we start at $(1, 3)$ and move 2 units down and then 3 units to the right, we end up at the point $(4, 1)$. We then draw a line through the points $(1, 3)$ and $(4, 1)$ to obtain the graph of the line. See Figure 32.

It is perfectly acceptable to set $\dfrac{\text{rise}}{\text{run}} = -\dfrac{2}{3} = \dfrac{2}{-3}$ so that we move 2 units up from $(1, 3)$ and then 3 units to the left. We would then end up at $(-2, 5)$, which is also on the graph of the line, as indicated in Figure 32.

> **Quick ✓**
>
> **10.** Draw a graph of the line that contains the point $(1, 2)$ and has a slope of
>
> **(a)** $\dfrac{1}{2}$ **(b)** -3 **(c)** 0

▶ ❹ **Work with Applications of Slope**

In its simplest form, slope is a ratio of rise over run. For example, a hill whose grade is $5\% \left(= 0.05 = \dfrac{5}{100} \right)$ goes up 5 feet (the rise) for every 100 feet it goes horizontally (the run). See Figure 33.

Work Smart

If the "rise" is positive, we go up. If the "rise" is negative, we go down. Similarly, if the "run" is positive, then we go to the right. If the "run" is negative, then we go to the left.

Figure 33

100 feet

5 feet

Work Smart

The pitch of a roof or grade of a road is always represented as a positive number.

Consider the pitch of a roof. If a roof's pitch is $\dfrac{5}{12}$, then every 5-foot measurement upward results in a horizontal measurement of 12 feet. See Figure 34.

Figure 34

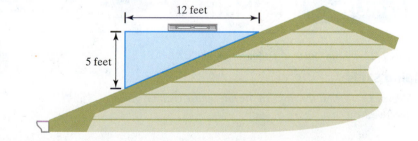

12 feet

5 feet

EXAMPLE 7 **Finding the Grade of a Road**

In Heckman Pass, British Columbia, there is a road that rises 9 feet for every 50 feet of horizontal distance covered. What is the grade of the road?

Solution

The grade of the road is given by $\dfrac{\text{rise}}{\text{run}}$. Since a rise of 9 feet is accompanied by a run of 50 feet, the grade of the road is $\dfrac{9 \text{ feet}}{50 \text{ feet}} = 0.18 = 18\%$.

The slope *m* of a line measures the amount that *y* changes as *x* changes from x_1 to x_2. The slope of a line is also called the **average rate of change** of *y* with respect to *x*.

In applications, we are often interested in knowing how the change in one variable might affect some other variable. For example, if your income increases by $1000, how much will your spending (on average) change? Or, if the speed of your car increases by 10 miles per hour, how much (on average) will your car's gas mileage change?

EXAMPLE 8   **Slope as an Average Rate of Change**

In Naples, Florida, the price of a new three-bedroom house that is 2100 square feet is $239,000. The price of a new three-bedroom house that is 2500 square feet is $295,000. Find and interpret the slope of the line joining the points (2100, 239,000) and (2500, 295,000). (SOURCE: *www.fiddlerscreek.com*)

Solution

Let *x* represent the square footage of the house and *y* represent the price. Let $(x_1, y_1) = (2100, 239{,}000)$ and $(x_2, y_2) = (2500, 295{,}000)$ and compute the slope as

$$m = \frac{y_2 - y_1}{x_2 - x_1} = \frac{295{,}000 - 239{,}000}{2500 - 2100}$$

$$= \frac{56{,}000}{400}$$

$$= 140$$

The unit of measure of *y* is dollars, and the unit of measure for *x* is square feet. So, the slope can be interpreted as follows: Between 2100 and 2500 square feet, the price increases by $140 per square foot, on average.   ●

> **Quick** ✓
>
> **11.** A road rises 4 feet for every 50 feet of horizontal distance covered. What is the grade of the road?
>
> **12.** The average annual cost of operating a Chevy Cobalt is $1370 when it is driven 10,000 miles. The average annual cost of operating a Chevy Cobalt is $1850 when it is driven 14,000 miles. Find and interpret the slope of the line joining (10,000, 1370) and (14,000, 1850).

3.3 Exercises   MyMathLab® Math XL PRACTICE   Exercise numbers in green have complete video solutions in MyMathLab.

Problems 1–12 are the **Quick** ✓*s that follow the* **EXAMPLES**.

Building Skills

In Problems 13–18, find the slope of the line whose graph is given.
See Objective 1.

13.

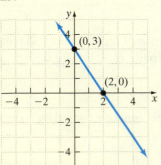

14.

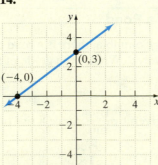

15.

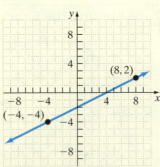

16.

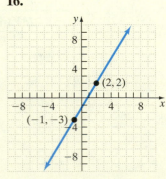

17.

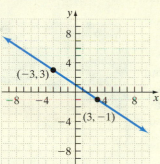

18.

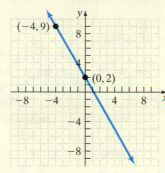

39.

40.

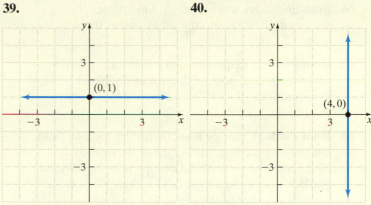

In Problems 19–22, (a) plot the points in a rectangular coordinate system, (b) draw a line through the points, (c) and find and interpret the slope of the line. See Objective 1.

19. $(-3, 2)$ and $(3, 5)$ **20.** $(2, 6)$ and $(-2, -4)$

21. $(2, -9)$ and $(-2, -1)$ **22.** $(4, -5)$ and $(-2, -4)$

In Problems 23–36, find and interpret the slope of the line containing the given points. See Objective 1.

23. $(10, 4)$ and $(6, 12)$ **24.** $(7, 3)$ and $(0, -11)$

25. $(4, -4)$ and $(12, -12)$ **26.** $(-3, 2)$ and $(2, -3)$

27. $(7, -2)$ and $(4, 3)$ **28.** $(-8, -1)$ and $(2, 3)$

29. $(0, 6)$ and $(-4, 0)$ **30.** $(-5, 0)$ and $(0, 3)$

31. $(-4, -1)$ and $(2, 3)$ **32.** $(5, 1)$ and $(-1, -1)$

33. $\left(\dfrac{1}{2}, \dfrac{3}{4}\right)$ and $\left(-\dfrac{5}{2}, -\dfrac{1}{4}\right)$ **34.** $\left(-\dfrac{1}{3}, \dfrac{2}{5}\right)$ and $\left(\dfrac{2}{3}, -\dfrac{3}{5}\right)$

35. $\left(\dfrac{1}{2}, \dfrac{1}{3}\right)$ and $\left(\dfrac{3}{4}, \dfrac{5}{6}\right)$ **36.** $\left(\dfrac{1}{4}, -\dfrac{4}{3}\right)$ and $\left(-\dfrac{5}{4}, \dfrac{1}{3}\right)$

In Problems 37–40, find the slope of the line whose graph is given. See Objective 2.

37.

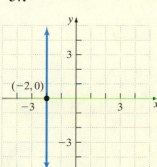

38.

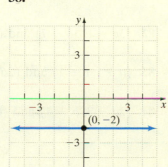

In Problems 41–44, find and interpret the slope of the line containing the given points. See Objective 2.

41. $(4, -6)$ and $(-1, -6)$ **42.** $(-1, -3)$ and $(-1, 2)$

43. $(3, 9)$ and $(3, -2)$ **44.** $(5, 1)$ and $(-2, 1)$

In Problems 45–62, draw a graph of the line that contains the given point and has the given slope. See Objective 3.

45. $(4, 2)$; $m = 1$ **46.** $(3, -1)$; $m = -1$

47. $(0, 6)$; $m = -2$ **48.** $(-1, 3)$; $m = 3$

49. $(-1, 0)$; $m = \dfrac{1}{4}$ **50.** $(5, 2)$; $m = -\dfrac{1}{2}$

51. $(2, -3)$; $m = 0$ **52.** $(1, 4)$; m is undefined

53. $(2, 1)$; $m = \dfrac{2}{3}$ **54.** $(-2, -3)$; $m = \dfrac{5}{2}$

55. $(-1, 4)$; $m = -\dfrac{5}{4}$ **56.** $(0, -2)$; $m = -\dfrac{3}{2}$

57. $(0, 0)$; m is undefined   **58.** $(3, -1)$; $m = 0$

59. $(0, 2)$; $m = -4$ **60.** $(0, 0)$; $m = \dfrac{1}{5}$

61. $(2, -3)$; $m = \dfrac{3}{4}$ **62.** $(-3, 0)$; $m = -3$

Applying the Concepts

In Problems 63–66, draw the graph of the two lines with the given properties on the same rectangular coordinate system.

63. Both lines pass through the point $(2, -1)$. One has slope of 2 and the other has slope of $-\dfrac{1}{2}$.

64. Both lines pass through the point $(3, 0)$. One has slope of $\dfrac{2}{3}$ and the other has slope of $-\dfrac{3}{2}$.

65. Both lines have a slope of $\dfrac{3}{4}$. One passes through the point $(-1, -2)$, and the other passes through the point $(2, 1)$.

66. Both lines have a slope of -1. One passes through the point $(0, -3)$, and the other passes through the point $(2, -1)$.

67. Roof Pitch A carpenter who was installing a new roof on a garage noticed that for every 1-foot horizontal run, the roof was elevated by 4 inches. What is the pitch of this roof?

68. Roof Pitch A canopy is set up on the football field. On the 45-yard line, the height of the canopy is 68 inches. The peak of the canopy is at the 50-yard marker, where the height is 84 inches. What is the pitch of the roof of the canopy?

69. Building a Roof To build a shed in his backyard, Moises has decided to use a pitch of $\dfrac{2}{5}$ for his roof. The shed measures 30 in. from the side to the center. How much height should he add to the roof to get the desired pitch?

30 inches

70. Building a Roof The design for the bedroom of a house requires a roof pitch of $\dfrac{7}{20}$. If the room measures 5 feet from the wall to the center, how high above the ceiling is the peak of the roof?

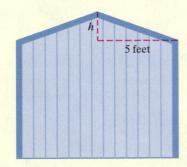

5 feet

71. Road Grade Fall River Road was completed in 1920 and was the first road built through the Rocky Mountains in Colorado. It was so steep that sometimes the early model cars had to drive up the hill in reverse to maximize the output of their weak engines and fuel systems. If the road rises 200 feet

for every 1250 feet of horizontal change, what is the grade of this road in percent?

72. Road Grade Barbara decided to take a bicycle trip up to the observatory on Mauna Kea on the island of Hawaii. The road has a vertical rise of 120 feet for every 800 feet of horizontal change. In percent, what is the grade of this road?

73. Population Growth The population of the United States was 123,202,624 in 1930 and 309,330,219 in 2010. Use the ordered pairs (0, 123 million) and (80, 309 million) to find and interpret the slope of the line representing the average rate of change in the population of the United States.

74. Earning Potential On average, a person who graduates from high school can expect to have lifetime earnings of 1.2 million dollars. It takes four years to earn a bachelor's degree, but the lifetime earnings will increase to 2.1 million dollars. Use the ordered pairs (0, 1.2 million) and (4, 2.1 million) to find and interpret the slope of the line representing the increase in earnings due to finishing college.

Extending the Concepts

In Problems 75–78, find any two ordered pairs that lie on the given line. Graph the line and then determine the slope of the line.

75. $3x + y = -5$

76. $2x + 5y = 12$

77. $y = 3x + 4$

78. $y = -x - 6$

In Problems 79–84, find the slope of the line containing the given points.

79. $(2a, a)$ and $(3a, -a)$

80. $(4p, 2p)$ and $(-2p, 5p)$

81. $(2p + 1, q - 4)$ and $(3p + 1, 2q - 4)$

82. $(3p + 1, 4q - 7)$ and $(5p + 1, 2q - 7)$

83. $(a + 1, b - 1)$ and $(2a - 5, b + 5)$

84. $(2a - 3, b + 4)$ and $(4a + 7, 5b - 1)$

In economics, **marginal revenue** is a rate of change defined as the change in total revenue divided by the change in output. If Q_1 represents the number of units sold, then the total revenue from selling these goods is represented by R_1. If Q_2 represents a different number of units sold, then the total revenue from this sale is represented by R_2.

We compute marginal revenue as $MR = \dfrac{R_2 - R_1}{Q_2 - Q_1}$.

So marginal revenue is a rate of change, or slope. Marginal revenue is important in economics because it is used to determine the level of output that maximizes profits for a company. Use the marginal revenue formula to solve Problems 85 and 86.

85. Determine and interpret marginal revenue if total revenue is \$1000 when 400 hot dogs are sold at a baseball game and total revenue is \$1200 when 500 hot dogs are sold.

86. Determine and interpret marginal revenue if total revenue is \$300 when 30 compact disks are sold and total revenue is \$400 when 50 compact disks are sold.

Explaining the Concepts

87. Describe a line that has one x-intercept but no y-intercept. Give two ordered pairs that could lie on this line and then describe how to find its slope.

88. Describe a line that has one y-intercept but no x-intercept. Give two ordered pairs that could lie on this line and then describe how to find its slope.

3.4 Slope-Intercept Form of a Line

Objectives

1 Use the Slope-Intercept Form to Identify the Slope and y-Intercept of a Line

2 Graph a Line Whose Equation Is in Slope-Intercept Form

3 Graph a Line Whose Equation Is in the Form $Ax + By = C$

4 Find the Equation of a Line Given Its Slope and y-Intercept

5 Work with Linear Models in Slope-Intercept Form

Are You Ready for This Section?

Before getting started, take the following readiness quiz. If you get a problem wrong, go back to the section cited and review the material.

R1. Solve $4x + 2y = 10$ for y.　　　　　[Section 2.4, pp. 114–115]

R2. Solve: $10 = 2x - 8$　　　　　　　　[Section 2.2, pp. 91–92]

▶ 1 Use the Slope-Intercept Form to Identify the Slope and y-Intercept of a Line

In this section, we use the slope and y-intercept to graph a line. This method for graphing will be more efficient than plotting points. Why? Well, suppose we wish to graph the equation $-2x + y = 5$ by plotting points. To do this, we might first solve the equation for y (get y by itself) by adding $2x$ to both sides of the equation.

$$-2x + y = 5$$
Add 2x to both sides: 　　$$y = 2x + 5$$

Table 14

x	$y = 2x + 5$	(x, y)
-2	$2(-2) + 5 = 1$	$(-2, 1)$
-1	$2(-1) + 5 = 3$	$(-1, 3)$
0	$2(0) + 5 = 5$	$(0, 5)$

We now create Table 14, which gives us points on the graph of the equation. Figure 35 shows the graph of the line.

Notice two things about the line in Figure 35. First, the slope is $m = 2$. Second, the y-intercept is $(0, 5)$. If you look at the equation after we solved for y, $y = 2x + 5$, you should notice that the coefficient of the variable x is 2 and the constant is 5. This is no coincidence!

Figure 35
$-2x + y = 5$

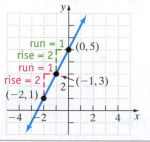

Slope-Intercept Form of an Equation of a Line

An equation of a line with slope m and y-intercept $(0, b)$ is

$$y = mx + b$$

Ready?...Answers **R1.** $y = -2x + 5$
R2. $\{9\}$

EXAMPLE 1 **Finding the Slope and *y*-Intercept of a Line**

Find the slope and *y*-intercept of the line whose equation is $y = -3x + 1$.

Solution

We compare the equation $y = -3x + 1$ to the slope-intercept form of a line $y = mx + b$. The coefficient of x, -3, is the slope, and the constant is 1, so the *y*-intercept is $(0, 1)$.

●

EXAMPLE 2 **Finding the Slope and *y*-Intercept of a Line Whose Equation Is in Standard Form**

Find the slope and *y*-intercept of the line whose equation is $3x + 2y = 6$.

Solution

Rewrite the equation $3x + 2y = 6$ so that it is in the form $y = mx + b$.

$$3x + 2y = 6$$

Subtract $3x$ from both sides: $\qquad 2y = -3x + 6$

Divide both sides by 2: $\qquad y = \dfrac{-3x + 6}{2}$

$\dfrac{a + b}{c} = \dfrac{a}{c} + \dfrac{b}{c}$: $\qquad y = \dfrac{-3}{2}x + \dfrac{6}{2}$

Simplify: $\qquad y = -\dfrac{3}{2}x + 3$

> **Work Smart**
>
> Notice that after we subtracted $3x$ from both sides, we wrote the equation as $2y = -3x + 6$ rather than $2y = 6 - 3x$. This is because we want to get the equation in the form $y = mx + b$, so the term involving x should be first.

Now we compare the equation $y = -\dfrac{3}{2}x + 3$ to the slope-intercept form of a line $y = mx + b$. The coefficient of x, $-\dfrac{3}{2}$, is the slope, and the constant is 3, so the *y*-intercept is $(0, 3)$.

●

Any equation of the form $y = b$ can be written $y = 0x + b$. Thus the slope of the line whose equation is $y = b$ is 0, and the *y*-intercept is $(0, b)$.

Any equation of the form $x = a$ cannot be written in the form $y = mx + b$. The line whose equation is $x = a$ has an undefined slope and no *y*-intercept.

> **Quick ✓**
>
> **1.** An equation of a line with slope m and *y*-intercept $(0, b)$ is _____
>
> *In Problems 2–6, find the slope and y-intercept of the line whose equation is given.*
>
> **2.** $y = 4x - 3$ $\qquad$ **3.** $3x + y = 7$ $\qquad$ **4.** $2x + 5y = 15$
>
> **5.** $y = 8$ $\qquad$ **6.** $x = 3$

▶ ② **Graph a Line Whose Equation Is in Slope-Intercept Form**

If an equation is in slope-intercept form, we can graph the line by plotting the *y*-intercept and using the slope to find another point on the line.

EXAMPLE 3   **How to Graph a Line Whose Equation Is in Slope-Intercept Form**

Graph the line $y = 3x - 1$ using the slope and y-intercept.

Step-by-Step Solution

Step 1: Identify the slope and y-intercept of the line.

$$y = 3x - 1$$
$$y = 3x + (-1)$$

$m = 3$ $b = -1$

The slope is $m = 3$ and the y-intercept is $(0, -1)$.

Step 2: Plot the y-intercept and then use the slope to find a second point on the graph. Draw a line through the points.

Plot the y-intercept at $(0, -1)$. Use the slope $m = \dfrac{3}{1} = \dfrac{\text{rise}}{\text{run}}$

to find a second point on the graph. See Figure 36.

Figure 36
$y = 3x - 1$

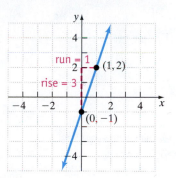

Quick ✓

In Problems 7 and 8, graph the line using the slope and y-intercept.

7. $y = 2x - 5$ **8.** $y = \dfrac{1}{2}x - 5$

EXAMPLE 4   **Graphing a Line Whose Equation Is in Slope-Intercept Form**

Graph the line $y = -\dfrac{4}{3}x + 2$ using the slope and y-intercept.

Figure 37
$y = -\dfrac{4}{3}x + 2$

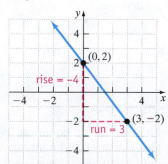

Solution

First, we determine the slope and y-intercept.

$$y = -\dfrac{4}{3}x + 2$$

$m = -\dfrac{4}{3}$ $b = 2$

Plot the y-intercept at $(0, 2)$. Now use the slope $m = -\dfrac{4}{3} = \dfrac{-4}{3} = \dfrac{\text{rise}}{\text{run}}$ to find a second point on the graph. Then draw a line through these two points. See Figure 37.

Quick ✓

In Problems 9 and 10, graph the line using the slope and y-intercept.

9. $y = -3x + 1$ **10.** $y = -\dfrac{3}{2}x + 4$

▶ ❸ **Graph a Line Whose Equation Is in the Form** $Ax + By = C$

If a linear equation is written in standard form $Ax + By = C$, we can still use the slope and y-intercept to obtain the graph of the equation. Let's see how.

EXAMPLE 5   **How to Graph a Line Whose Equation Is in the Form** $Ax + By = C$

Graph the line $8x + 2y = 10$ using the slope and y-intercept.

Step-by-Step Solution

Step 1: Solve the equation for y to put it in the form $y = mx + b$.

$$8x + 2y = 10$$

Subtract 8x from both sides:
$$2y = -8x + 10$$

Divide both sides by 2:
$$y = \frac{-8x + 10}{2}$$

Simplify:
$$y = -4x + 5$$

Step 2: Identify the slope and y-intercept of the line.

The slope is $m = -4$ and the y-intercept is $(0, 5)$.

Step 3: Plot the y-intercept and then use the slope to find a second point on the graph. Draw a line through the points.

Plot the point $(0, 5)$ and use the slope $m = -4 = \dfrac{-4}{1} = \dfrac{\text{rise}}{\text{run}}$ to find a second point on the graph. See Figure 38.

Figure 38
$8x + 2y = 10$

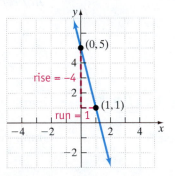

Work Smart

An alternative to graphing the equation in Example 5 using the slope and y-intercept would be to graph the line using intercepts.

Quick ✓

In Problems 11–13, graph each line using the slope and y-intercept.

11. $-2x + y = -3$ **12.** $6x - 2y = 2$ **13.** $3x + 5y = 0$

14. List three techniques that can be used to graph a line.

▶ ❹ **Find the Equation of a Line Given Its Slope and y-Intercept**

Up to now, we have identified the slope and y-intercept of a line from an equation. We will now reverse the process and find the equation of a line given its slope and y-intercept. This is a straightforward process—replace m with the slope and b with the y-intercept.

EXAMPLE 6   **Finding the Equation of a Line Given Its Slope and y-Intercept**

Find the equation of a line whose slope is $\frac{3}{8}$ and whose y-intercept is $(0, -4)$. Graph the line.

Solution

The slope is $m = \frac{3}{8}$ and the y-intercept is $(0, b) = (0, -4)$. Substitute $\frac{3}{8}$ for m and -4 for b in the slope-intercept form of a line, $y = mx + b$, to obtain

$$y = \frac{3}{8}x - 4$$

Figure 39 shows the graph of the equation.

Figure 39

$y = \frac{3}{8}x - 4.$

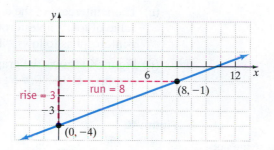

Quick ✓

In Problems 15–18, find the equation of the line whose slope and y-intercept are given. Graph the line

15. $m = 3$, $(0, -2)$ **16.** $m = -\frac{1}{4}$, $(0, 3)$ **17.** $m = 0$, $(0, -1)$ **18.** $m = 1$, $(0, 0)$

▶ ⑤ **Work with Linear Models in Slope-Intercept Form**

In many situations, we can use a linear equation to describe the relationship between two variables. For example, your long-distance phone bill depends linearly on the number of minutes used, or the cost of renting a moving truck depends linearly on the number of miles driven.

EXAMPLE 7   **A Model for Total Cholesterol**

When you have a physical exam, your doctor draws blood for your cholesterol test. Your total cholesterol count is measured in milligrams per deciliter (mg/dL). It is the sum of low-density lipoprotein cholesterol (LDL)—sometimes called "bad cholesterol"—and high-density lipoprotein cholesterol (HDL)—sometimes called "good cholesterol." Based on data from the National Center for Health Statistics, a woman's total cholesterol y is related to her age x by the following linear equation:

$$y = 1.1x + 157$$

(a) Use the equation to predict the total cholesterol of a 40-year-old woman.

(b) Determine and interpret the slope of the equation.

(c) Determine and interpret the y-intercept of the equation.

(d) Graph the equation in a rectangular coordinate system.

Solution

(a) We substitute 40 for x, the women's age, in the equation $y = 1.1x + 157$ to find the total cholesterol y.

$$y = 1.1x + 157$$
$$x = 40: \quad y = 1.1(40) + 157$$
$$= 201$$

We predict that a 40-year-old woman will have a total cholesterol of 201 mg/dL.

(b) The slope of the equation $y = 1.1x + 157$ is 1.1. Because slope equals
$$\frac{\text{rise}}{\text{run}} = \frac{1.1 \text{ mg/dL}}{1 \text{ year}},$$ we interpret the slope as follows: "The total cholesterol of a female increases by 1.1 mg/dL as age increases by 1 year."

(c) The y-intercept of the equation $y = 1.1x + 157$ is (0, 157). The y-intercept is the value of total cholesterol, y, when $x = 0$. Since x represents age, we interpret the y-intercept as follows: "The total cholesterol of a newborn girl is 157 mg/dL."

(d) Figure 40 shows the graph of the equation. Because it does not make sense for x to be less than 0, we graph the equation only in quadrant I.

Work Smart

Notice that we did not use the slope to obtain an additional point on the graph of the equation. It would be difficult to find an additional point with a slope of 1.1. For example, from the y-intercept, we would go up 1.1 mg/dL and right 1 year and end up at (1, 158.1). It would be hard to draw the line through these two points!

Figure 40
$y = 1.1x + 157$

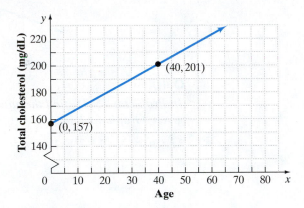

In Figure 40, notice the "broken line" (⌇) on the y-axis near the origin. This symbol indicates that a portion of the graph has been removed. This is done to avoid a lot of white space in the graph. Whenever you are reading a graph, always look carefully at how the axes are labeled and at the units on each axis.

Quick ✓

19. Based on data obtained from the National Center for Health Statistics, the birth weight y of a baby, measured in grams, is linearly related to gestation period x (in weeks) according to the equation

$$y = 143x - 2215$$

(a) Use the equation to predict the birth weight of a baby if the gestation period is 30 weeks.

(b) Use the equation to predict the birth weight of a baby if the gestation period is 36 weeks.

(c) Determine and interpret the slope of the equation.

(d) Explain why it does not make sense to interpret the y-intercept of the equation.

(e) Graph the equation in a rectangular coordinate system for $28 \leq x \leq 43$.

We know that slope can be interpreted as a rate of change. For this reason, when information in a problem is given as a rate of change (as in miles per gallon or dollars per pound), the rate of change will represent the slope in a linear model.

EXAMPLE 8 **Cost of Owning and Operating a Car**

Some costs involved in owning a car are affected by the number of miles driven (gas and maintenance), while others are not (comprehensive insurance, license plates, depreciation). Suppose the annual cost of operating a Chevy Cobalt is $0.25 per mile plus $3000.

(a) Write a linear equation that relates the annual cost of operating the car y to the number of miles driven in a year x.

(b) What is the annual cost of driving 11,000 miles?

(c) Graph the equation in a rectangular coordinate system.

Solution

(a) The rate of change in the problem is $0.25 per mile. We can express this as $\dfrac{\$0.25}{1 \text{ mile}}$, which is the slope m of the linear equation. The cost of $3000 is a cost that does not change with the number of miles driven. Put another way, if we drive 0 miles, the cost will be $3000, so this value represents the y-intercept, $(0, b)$. The linear equation that relates cost y to number of miles driven x is

$$y = 0.25x + 3000$$

(b) Let $x = 11{,}000$ in the equation $y = 0.25x + 3000$.

$$y = 0.25(11{,}000) + 3000$$
$$= 2750 + 3000$$
$$= \$5750$$

The cost of driving 11,000 miles in a year is $5750. This cost includes gas, insurance, maintenance, and depreciation in the value of the vehicle.

(c) See Figure 41.

Figure 41
$y = 0.25x + 3000$

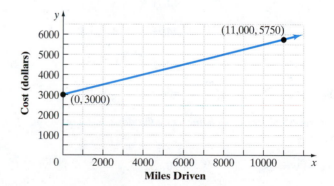

Quick ✓

20. The daily cost, y, of renting a 16-foot moving truck for a day is $50 plus $0.38 per mile driven, x.

(a) Write a linear equation relating the daily cost y to the number of miles driven, x.

(b) Determine the cost of renting the truck if the truck is driven 75 miles.

(c) If the cost of renting the truck is $84.20, how many miles were driven?

(d) Graph the linear equation.

3.4 Exercises

 MyMathLab® **MathⓍL** PRACTICE

Exercise numbers in green have complete video solutions in MyMathLab.

*Problems **1–20** are the Quick ✓ s that follow the **EXAMPLES**.*

Building Skills

In Problems 21–40, find the slope and y-intercept of the line whose equation is given. See Objective 1.

21. $y = 5x + 2$ **22.** $y = 7x + 1$

23. $y = x - 9$ **24.** $y = x - 7$

25. $y = -10x + 7$ **26.** $y = -6x + 2$

27. $y = -x - 9$ **28.** $y = -x - 12$

29. $2x + y = 4$ **30.** $3x + y = 9$

31. $2x + 3y = 24$ **32.** $6x - 8y = -24$

33. $5x - 3y = 9$ **34.** $10x + 6y = 24$

35. $x - 2y = 5$ **36.** $-x - 5y = 3$

37. $y = -5$ **38.** $y = 3$

39. $x = 6$ **40.** $x = -2$

In Problems 41–48, use the slope and y-intercept to graph each line whose equation is given. See Objective 2.

41. $y = x + 3$ **42.** $y = x + 4$

43. $y = -2x - 3$ **44.** $y = -4x - 1$

45. $y = \dfrac{2}{3}x + 2$ **46.** $y = \dfrac{4}{3}x - 3$

47. $y = -\dfrac{5}{2}x - 2$ **48.** $y = -\dfrac{2}{5}x + 3$

In Problems 49–56, graph each line using the slope and y-intercept. See Objective 3.

49. $4x + y = 5$ **50.** $3x + y = 2$

51. $x + 2y = -6$ **52.** $x - 2y = -4$

53. $3x - 2y = 10$ **54.** $4x + 3y = -6$

55. $6x + 3y = -15$ **56.** $5x - 2y = 6$

In Problems 57–68, find the equation of the line with the given slope and intercept. See Objective 4.

57. slope is -1; y-intercept is $(0, 8)$

58. slope is 1; y-intercept is $(0, 10)$

59. slope is $\dfrac{6}{7}$; y-intercept is $(0, -6)$

60. slope is $\dfrac{4}{7}$; y-intercept is $(0, -9)$

61. slope is $-\dfrac{1}{3}$; y-intercept is $\left(0, \dfrac{2}{3}\right)$

62. slope is $\dfrac{1}{4}$; y-intercept is $\left(0, \dfrac{3}{8}\right)$

63. slope is undefined; x-intercept is $(-5, 0)$

64. slope is 0; y-intercept is $(0, -2)$

65. slope is 0; y-intercept is $(0, 3)$

66. slope is undefined; x-intercept is $(4, 0)$

67. slope is 5; y-intercept is $(0, 0)$

68. slope is -3; y-intercept is $(0, 0)$

Mixed Practice

In Problems 69–92, graph each equation using any method you wish.

69. $y = 2x - 7$ **70.** $y = -4x + 1$

71. $3x - 2y = 24$ **72.** $2x + 5y = 30$

73. $y = -5$ **74.** $y = 4$

75. $x = -6$ **76.** $x = -3$

77. $6x - 4y = 0$ **78.** $3x + 8y = 0$

79. $y = -\dfrac{5}{3}x + 6$ **80.** $y = -\dfrac{3}{5}x + 4$

81. $2y = x + 4$ **82.** $3y = x - 9$

83. $y = \dfrac{x}{3}$ **84.** $y = -\dfrac{x}{4}$

85. $2x = -8y$ **86.** $-3x = 5y$

87. $y = -\dfrac{2}{3}x + 1$ **88.** $y = \dfrac{3}{2}x - 4$

89. $x + 2 = -7$ **90.** $y - 4 = -1$

91. $5x + y + 1 = 0$ **92.** $2x - y + 4 = 0$

Applying the Concepts

93. Weekly Salary Dien is paid a salary of $400 per week plus an 8% commission on all sales he makes during the week.

(a) Write a linear equation that calculates his weekly income, where y represents his income and x represents the amount of sales.

(b) What is Dien's weekly income if he sold $1200 worth of merchandise?

(c) Graph the equation in a rectangular coordinate system. Label the axes appropriately.

94. Car Rental To rent a car for a day, Gloria pays $75 plus $0.10 per mile.

(a) Write a linear equation that calculates the daily cost, y, to rent a car that will be driven x miles.

(b) What is the cost to drive this car for 200 miles?

(c) If Gloria paid $87.50, how many miles did she drive?

(d) Graph the equation in a rectangular coordinate system. Label the axes appropriately.

95. Cell Phone Costs The cost per minute of talk time for cell phone users has gone down over the years. In 1995, cell phone users paid, on the average, 56¢ per minute. In 2011, they paid 5¢ per minute. Assuming that the rate of decline of the cost per minute was constant, the cost per minute can be calculated by the equation $y = -3.1875x + 56$, where x represents the number of years after 1995 and y represents the cost per minute of cell phone usage in cents. (SOURCE: *CTIA, The Wireless Association*)

(a) What was the cost per minute for a cell phone user in 1999?

(b) In which year did a cell phone user pay 17.75¢ per minute?

(c) Interpret the slope of $y = -3.1875x + 56$.

(d) Can this trend continue indefinitely?

(e) Graph the equation in a rectangular coordinate system. Label the axes appropriately.

96. Counting Calories According to a National Academy of Sciences Report, the recommended daily intake of calories for males between the ages of 7 and 15 can be calculated by the equation $y = 125x + 1125$, where x represents the boy's age and y represents the recommended caloric intake.

(a) What is the recommended caloric intake for a 12-year-old boy?

(b) What is the age of a boy whose recommended caloric intake is 2250 calories?

(c) Interpret the slope of $y = 125x + 1125$.

(d) Why would this equation not be accurate for a 3-year-old male?

(e) Graph the equation in a rectangular coordinate system. Label the axes appropriately.

Extending the Concepts

In Problems 97–102, find the value of the missing coefficient so that the line will have the given property.

97. $2x + By = 12$; slope is $\dfrac{1}{2}$

98. $Ax + 2y = 5$; slope is $\dfrac{3}{2}$

99. $Ax - 2y = 10$; slope is -2

100. $12x + By = -1$; slope is -4

101. $x + By = \dfrac{1}{2}$; y-intercept is $-\dfrac{1}{6}$

102. $4x + By = \dfrac{4}{3}$; y-intercept is $\dfrac{2}{3}$

In Problems 103 and 104, use the following information. In business, a cost equation relates the total cost of producing a product or good such as a refrigerator, rug, or blender to the number of goods produced. The simplest cost model is the linear cost model. In the linear cost model, the slope of the linear equation represents the cost of producing one additional unit of a good. Variable cost is reported as a rate of change, such as $40 per calculator. Examples of variable costs include labor costs and materials. The y-intercept of the linear equation represents the fixed costs of production—these are costs that exist regardless of the level of production. Fixed costs would include the cost of the manufacturing facility and insurance.

103. Cost Equations Suppose the variable cost of manufacturing a graphing calculator is $40 per calculator, and the daily fixed cost is $4000.

(a) Write a linear equation that relates cost y to the number of calculators manufactured x.

(b) What is the daily cost of manufacturing 500 calculators?

(c) One day, the total cost was $19,000. How many calculators were manufactured?

(d) Graph the equation relating cost and number of calculators manufactured.

104. Cost Equations Suppose the variable cost of manufacturing a cellular telephone is $35 per phone, and the daily fixed cost is $3600.

(a) Write a linear equation that relates the daily cost y to the number of cellular telephones manufactured x.

(b) What is the daily cost of manufacturing 400 cellular phones?

(c) One day, the total cost was $13,225. How many cellular telephones were manufactured?

(d) Graph the equation relating cost and number of cellular telephones manufactured.

(g) $-5x + 2y = 12$

(h) $x - y = -3$

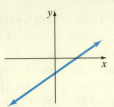

Explaining the Concepts

105. Describe the line whose graph is shown. Which of the following equations could have the graph that is shown?

(a) $y = 3x - 2$ **(b)** $y = -2x + 5$

(c) $y = 3$ **(d)** $2x + 3y = 6$

(e) $3x - 2y = 8$ **(f)** $4x - y = -4$

106. Without graphing, describe the orientation of each line (rises from left to right, and so on). Explain how you came to this conclusion.

(a) $y = 4x - 3$ **(b)** $y = -2x + 5$

(c) $y = x$ **(d)** $y = 4$

3.5 Point-Slope Form of a Line

Objectives

1 Find the Equation of a Line Given a Point and a Slope

2 Find the Equation of a Line Given Two Points

3 Build Linear Models Using the Point-Slope Form of a Line

Are You Ready For This Section?

Before getting started, take the following readiness quiz. If you get a problem wrong, go back to the section cited and review the material.

R1. Solve $y - 3 = 2(x + 1)$ for y.   [Section 2.4, pp. 114–115]

R2. Evaluate: $\dfrac{7 - 3}{4 - 2}$ [Section 1.7, p. 59]

▶ **1** Find the Equation of a Line Given a Point and a Slope

Figure 42

We have discussed two forms for the equation of a line: the standard form of a line, $Ax + By = C$, where A and B are not both zero, and the slope-intercept form of a line, $y = mx + b$, where m is the slope and b is the y-intercept. We now introduce another form for the equation of a line.

Suppose we have a nonvertical line with slope m containing the point (x_1, y_1). See Figure 42. For any other point (x, y) on the line, the slope of the line is

$$m = \frac{y - y_1}{x - x_1}$$

Multiplying both sides by $x - x_1$ gives us

$$m(x - x_1) = y - y_1 \quad \text{or} \quad y - y_1 = m(x - x_1)$$

> **Point-Slope Form of an Equation of a Line**
>
> An equation of a nonvertical line with slope m containing the point (x_1, y_1) is
>
> Slope
> ↓
> $$y - y_1 = m(x - x_1)$$
> ↑ Given point ↑

The point-slope form of a line can be used to write an equation in either slope-intercept form ($y = mx + b$) or standard form ($Ax + By = C$).

EXAMPLE 1 | **Using the Point-Slope Form of an Equation of a Line—Positive Slope**

Find the equation of a line that has a slope of 3 and contains the point $(-1, 4)$. Write the equation in slope-intercept form. Graph the line.

Figure 43
$y = 3x + 7$

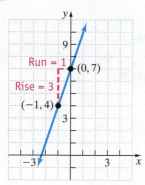

Solution

Because we are given the slope and a point on the line, we use the point-slope form of a line with $m = 3$ and $(x_1, y_1) = (-1, 4)$.

$$y - y_1 = m(x - x_1)$$

$m = 3, x_1 = -1, y_1 = 4:$ $\quad y - 4 = 3(x - (-1))$

$$y - 4 = 3(x + 1)$$

To put the equation in slope-intercept form, $y = mx + b$, we solve the equation for y.

Distribute: $\quad y - 4 = 3x + 3$

Add 4 to both sides: $\quad y = 3x + 7$

See Figure 43 for a graph of the line.

EXAMPLE 2 | **Using the Point-Slope Form of an Equation of a Line—Negative Slope**

Find the equation of a line that has a slope of $-\dfrac{3}{4}$ and contains the point $(-4, 3)$. Write the equation in slope-intercept form. Graph the line.

Solution

Because we are given the slope and a point on the line, we use the point-slope form of a line with $m = -\dfrac{3}{4}$ and $(x_1, y_1) = (-4, 3)$.

$$y - y_1 = m(x - x_1)$$

$m = -\dfrac{3}{4}, x_1 = -4, y_1 = 3:$ $\quad y - 3 = -\dfrac{3}{4}(x - (-4))$

Simplify: $\quad y - 3 = -\dfrac{3}{4}(x + 4)$

Distribute: $\quad y - 3 = -\dfrac{3}{4}x - 3$

Add 3 to both sides: $\quad y = -\dfrac{3}{4}x$

Figure 44
$y = -\dfrac{3}{4}x$

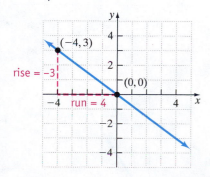

See Figure 44 for a graph of the line.

Quick ✓

1. The point-slope form of a nonvertical line that has a slope of m and contains the point (x_1, y_1) is _____.

2. *True or False* The slope of the line $y - 3 = 4(x - 1)$ is 4.

In Problems 3–6, find the equation of the line with the given properties. Write the equation in slope-intercept form. Graph the line.

3. $m = 3$ containing $(x_1, y_1) = (2, 1)$

4. $m = \dfrac{1}{3}$ containing $(x_1, y_1) = (3, -4)$

5. $m = -4$ containing $(x_1, y_1) = (-2, 5)$

6. $m = -\dfrac{5}{2}$ containing $(x_1, y_1) = (-4, 5)$

EXAMPLE 3 **Finding the Equation of a Horizontal Line**

Find the equation of a horizontal line that contains the point $(-4, 2)$. Write the equation in slope-intercept form. Graph the line.

Solution

The line is horizontal, so its slope is 0. Because we know the slope and a point on the line, we use the point-slope form with $m = 0$, $x_1 = -4$, and $y_1 = 2$.

Work Smart

When the slope of a line is 0, the equation of the line will always be in the form "y = some number."

$$y - y_1 = m(x - x_1)$$
$$m = 0, x_1 = -4, y_1 = 2: \quad y - 2 = 0(x - (-4))$$
$$y - 2 = 0$$

Add 2 to both sides: $\quad y = 2 \quad$ Slope-intercept form, $y = 0x + 2$

See Figure 45 for a graph of the line.

Figure 45
$y = 2$

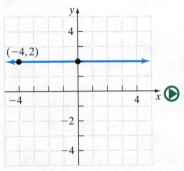

Quick ✓

7. Find the equation of a horizontal line that contains the point $(-2, 3)$. Write the equation of the line in slope-intercept form. Graph the line.

▶ ❷ Find the Equation of a Line Given Two Points

From Section 3.2, we know that we need only two points to graph a line. Given two points, we can find the equation of the line through the points by first finding the slope of the line and then using the point-slope form of a line.

EXAMPLE 4 **How to Find the Equation of a Line from Two Points**

Find the equation of a line through the points $(1, 3)$ and $(4, 9)$. Write the equation in slope-intercept form. Graph the line.

Step-by-Step Solution

Step 1: Find the slope of the line containing the points.

Let $(x_1, y_1) = (1, 3)$ and $(x_2, y_2) = (4, 9)$. Substitute these values into the formula for the slope of a line.

$$m = \frac{y_2 - y_1}{x_2 - x_1} = \frac{9 - 3}{4 - 1} = \frac{6}{3} = 2$$

Step 2: Substitute the slope found in Step 1 and either point into the point-slope form of a line to find the equation.

$$y - y_1 = m(x - x_1)$$
Use $m = 2, x_1 = 1, y_1 = 3: \quad y - 3 = 2(x - 1)$

Step 3: Solve the equation for y.

Distribute the 2: $\quad y - 3 = 2x - 2$
Add 3 to both sides: $\quad y = 2x + 1$

The slope-intercept form of the equation is $y = 2x + 1$. The slope of the line is 2, and the y-intercept is $(0, 1)$. See Figure 46 for the graph.

Figure 46
$y = 2x + 1$

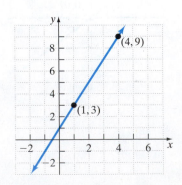

Quick ✓

In Problems 8 and 9, find the equation of the line containing the given points.
Write the equation in slope-intercept form. Graph the line.

8. $(0, 2); (3, 5)$ **9.** $(-1, 4); (1, -2)$

Work Smart: Study Skills

To write the equation of a nonvertical line, we must know either the slope of the line and a point on the line or two points on the line.

- If the **slope** and the **y-intercept** are known, use the slope-intercept form, $y = mx + b$.
- If the **slope** and a **point** that is not the *y*-intercept are known, use the **point-slope** form, $y - y_1 = m(x - x_1)$.
- If **two points** are known, first find the **slope** and then use that slope and one of the **points** in the **point-slope** formula, $y - y_1 = m(x - x_1)$.

EXAMPLE 5 **Finding the Equation of a Vertical Line from Two Points**

Find the equation of a line through the points $(-3, 2)$ and $(-3, -4)$. Write the equation in slope-intercept form, if possible. Graph the line.

Solution

Let $(x_1, y_1) = (-3, 2)$ and $(x_2, y_2) = (-3, -4)$ in the formula for the slope of a line.

$$m = \frac{y_2 - y_1}{x_2 - x_1} = \frac{-4 - 2}{-3 - (-3)} = \frac{-6}{0}$$

Work Smart

The equation of a vertical line cannot be written in slope-intercept form.

The slope is undefined, so the line is vertical. No matter what value of *y* we choose, the *x*-coordinate of the point on the line will be -3. The equation of the line is $x = -3$, which cannot be written in slope-intercept form. See Figure 47 for the graph.

Figure 47
$x = -3$

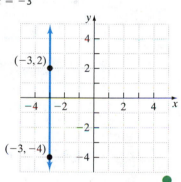

Quick ✓

10. Find the equation of the line containing the points $(3, 2)$ and $(3, -4)$. Write the equation in slope-intercept form, if possible. Graph the line.

Summary Equations of Lines

Form of Line	Formula	Comments
Horizontal line	$y = b$	Graph is a horizontal line (slope is 0) with *y*-intercept $(0, b)$.
Vertical line	$x = a$	Graph is a vertical line (undefined slope) with *x*-intercept $(a, 0)$.
Point-slope	$y - y_1 = m(x - x_1)$	Useful for finding the equation of a line, given a point and the slope, or two points.
Slope-intercept	$y = mx + b$	Useful for finding the equation of a line, given the slope and *y*-intercept, or for quickly determining the slope and *y*-intercept of the line, given the equation of the line.
Standard form	$Ax + By = C$	Useful for finding the *x*- and *y*-intercepts.

Quick ✓

11. List the five forms of the equation of a line: Horizontal line: _____; Vertical line: _____; Point-slope: _____; Slope-intercept: _____; Standard form: _____.

❸ Build Linear Models Using the Point-Slope Form of a Line

We can use the point-slope form of a line to build linear models from data.

EXAMPLE 6 **Building a Linear Model from Data**

Health care costs are skyrocketing. For individuals 20 years of age or older, the percentage of total income y that an individual spends on health care increases linearly with age x. According to data obtained from the Bureau of Labor Statistics, a 35-year-old spends about 4.0% of income on health care, while a 65-year-old spends about 11.2% of income on health care.

(a) Plot the points $(35, 4.0)$ and $(65, 11.2)$ in a rectangular coordinate system and graph the line. Find the linear equation in slope-intercept form that relates the percent of income spent on health care y to age x.

(b) Use the equation found in part (a) to predict the percentage of income that a 50-year-old spends on health care.

(c) Interpret the slope.

Solution

(a) We plot the ordered pairs $(35, 4.0)$ and $(65, 11.2)$ and draw a line through the points. Be careful to start the graph at $x = 20$. Do you see why? See Figure 48.

Figure 48

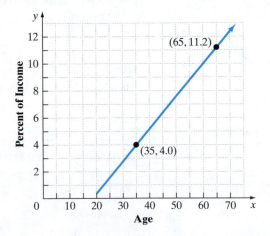

Because we know two points on the line, we will find the slope of the line and then use the point-slope form of a line to find the equation of the line.

$$m = \frac{y_2 - y_1}{x_2 - x_1} = \frac{11.2 - 4.0}{65 - 35}$$

$$= \frac{7.2}{30}$$

$$= 0.24$$

We use the point-slope form of a line with $m = 0.24$, $x_1 = 35$, and $y_1 = 4.0$:

$$y - y_1 = m(x - x_1)$$

$m = 0.24, x_1 = 35,$ and $y_1 = 4.0$: $\quad y - 4.0 = 0.24(x - 35)$

$$y - 4.0 = 0.24x - 8.4$$

Add 4.0 to both sides of the equation: $\quad y = 0.24x - 4.4$

The equation $y = 0.24x - 4.4$ relates the percent of income spent on health care y to the age x.

(b) We substitute 50 for x in the equation found in part (a), so the predicted percentage of income spent on health care for a 50-year old is

$$y = 0.24x - 4.4$$

Let $x = 50$: $\qquad = 0.24(50) - 4.4$

$$= 7.6$$

We predict that 7.6% of a 50-year-old's income is spent on health care.

(c) The slope is 0.24. The percentage of income spent on health care for the individual increases by 0.24% as the individual ages by one year. ●

Quick ✓

12. Armando owns a gas station. He has found that when the price of regular unleaded gasoline is \$3.90, he sells 400 gallons of gasoline between the hours of 7:00 A.M. and 8:00 A.M. When the price of regular unleaded gasoline is \$4.10, he sells 380 gallons of gasoline between the hours of 7:00 A.M. and 8:00 A.M. Suppose that the relation between quantity of gasoline sold and price is linear.

(a) Plot the points in a rectangular coordinate system and graph the line. Find the linear equation in slope-intercept form that relates quantity of gasoline sold y to price x.

(b) Use the equation found in part (a) to predict the number of gallons of gasoline sold if the price is \$4.00.

(c) Interpret the slope.

3.5 Exercises

MyMathLab® PRACTICE

Exercise numbers in green have complete video solutions in MyMathLab.

Problems 1–12 are the Quick ✓ *s that follow the* **EXAMPLES**.

Building Skills

In Problems 13–28, find the equation of the line that contains the given point and has the given slope. Write the equation in slope-intercept form and graph the line. See Objective 1.

13. $(2, 5)$; slope $= 3$

14. $(4, 1)$; slope $= 6$

15. $(-1, 2)$; slope $= -2$

16. $(6, -3)$; slope $= -5$

17. $(8, -1)$; slope $= \dfrac{1}{4}$

18. $(-8, 2)$; slope $= -\dfrac{1}{2}$

19. $(0, 13)$; slope $= -6$

20. $(0, -4)$; slope $= 9$

21. $(5, -7)$; slope $= 0$

22. $(3, 12)$; undefined slope

23. $(-4, 5)$; undefined slope

24. $(-7, -1)$; slope $= 0$

25. $(-3, 0)$; slope $= \dfrac{2}{3}$ **26.** $(-10, 0)$; slope $= -\dfrac{4}{5}$

27. $(-8, 6)$; slope $= -\dfrac{3}{4}$ **28.** $(-4, -6)$; slope $= \dfrac{3}{2}$

In Problems 29–36, find the equation of the line that contains the given point and satisfies the given information. Write the equation in slope-intercept form, if possible. See Objectives 1 and 2.

29. Vertical line that contains $(-3, 10)$

30. Horizontal line that contains $(-6, -1)$

31. Horizontal line that contains $(-1, -5)$

32. Vertical line that contains $(4, -3)$

33. Horizontal line that contains $(0.2, -4.3)$

34. Vertical line that contains $(3.5, 2.4)$

35. Vertical line that contains $\left(\dfrac{1}{2}, \dfrac{7}{4}\right)$

36. Horizontal line that contains $\left(\dfrac{3}{2}, \dfrac{9}{4}\right)$

In Problems 37–52, find the equation of the line that contains the given points. Write the equation in slope-intercept form, if possible. See Objective 2.

37. $(0, 4)$ and $(-2, 0)$ **38.** $(0, 3)$ and $(6, 0)$

39. $(1, 2)$ and $(0, 6)$ **40.** $(2, 4)$ and $(0, 8)$

41. $(-3, 2)$ and $(1, -4)$ **42.** $(-2, 4)$ and $(2, -2)$

43. $(-3, -11)$ and $(2, -1)$ **44.** $(4, 18)$ and $(-1, 3)$

45. $(4, -3)$ and $(-3, -3)$ **46.** $(-6, 5)$ and $(7, 5)$

47. $(2, -1)$ and $(2, -9)$ **48.** $(-3, 8)$ and $(-3, 1)$

49. $(0.1, 0.6)$ and $(0.5, 0.7)$ **50.** $(0.7, 0.8)$ and $(0.2, 0.4)$

51. $\left(\dfrac{1}{2}, -\dfrac{9}{4}\right)$ and $\left(\dfrac{5}{2}, -\dfrac{1}{4}\right)$ **52.** $\left(\dfrac{1}{3}, \dfrac{12}{5}\right)$ and $\left(\dfrac{4}{3}, \dfrac{2}{5}\right)$

Mixed Practice

In Problems 53–70, find the equation of the line described. Write the equation in slope-intercept form, if possible. Graph the line.

53. Contains $(4, -2)$ with slope $= 5$

54. Contains $(3, 2)$ with slope $= 4$

55. Horizontal line that contains $(-3, 5)$

56. Vertical line that contains $(-4, 2)$

57. Contains $(1, 3)$ and $(-4, -2)$

58. Contains $(-2, -8)$ and $(2, -6)$

59. Contains $(-2, 3)$ with slope $= \dfrac{1}{2}$

60. Contains $(-8, 3)$ with slope $= \dfrac{1}{4}$

61. Vertical line that contains $(5, 2)$

62. Horizontal line that contains $(-2, -6)$

63. Contains $(3, -19)$ and $(-1, 9)$

64. Contains $(-3, 13)$ and $(4, -22)$

65. Contains $(6, 3)$ with slope $= -\dfrac{2}{3}$

66. Contains $(-6, 3)$ with slope $= -\dfrac{2}{3}$

67. Contains $(-2, 3)$ and $(4, -6)$

68. Contains $(5, -3)$ and $(-3, 3)$

69. x-intercept: $(5, 0)$; y-intercept: $(0, -2)$

70. x-intercept: $(-6, 0)$; y-intercept: $(0, 4)$

Applying the Concepts

71. Shipping Packages The shipping department for a warehouse has noted that if 60 packages are shipped during a month, the total expenses for the department are \$1635. If 120 packages are shipped during a month, the total expenses for the shipping department are \$1770. Let x represent the number of packages and y represent the total expenses for the shipping department.

(a) Interpret the meaning of the point $(60, 1635)$ in the context of this problem.

(b) Plot the ordered pairs $(60, 1635)$ and $(120, 1770)$ in a rectangular coordinate system and graph the line through the points.

(c) Find the linear equation, in slope-intercept form, that relates the total expenses for the shipping department, y, to the number of packages sent, x.

(d) Use the equation found in part (c) to find the total expenses during a month when 200 packages were sent.

(e) Interpret the slope.

72. Retirement Plans Based on the retirement plan available by his employer, Kei knows that if he retires after 20 years, his monthly retirement income will be $3150. If he retires after 30 years, his monthly income increases to $3600. Let x represent the number of years of service and y represent the monthly retirement income.

(a) Interpret the meaning of the point $(30, 3600)$ in the context of this problem.

(b) Plot the ordered pairs $(20, 3150)$ and $(30, 3600)$ in a rectangular coordinate system and graph the line through the points.

(c) Find the linear equation, in slope-intercept form, that relates the monthly retirement income, y, to the number of years of service, x.

(d) Use the equation found in part (c) to find the monthly income for 15 years of service.

(e) Interpret the slope.

73. Credit Scores Your Fair Isaacs Corporation (FICO) credit score is used to determine your ability to get credit (such as a car loan or a credit card). FICO scores have a range of 300 to 850, with a higher score indicating a better credit history. Suppose a bank offers a person with a credit score of 600 a 15% interest rate on a 3-year car loan, while a person with a credit score of 750 is offered a 6% interest rate. Let x represent a person's credit score and y represent the interest rate.

(a) Fill in the ordered pairs: $(\_\_, 15); (\_\_, 6)$

(b) Plot the ordered pairs from part (a) in a rectangular coordinate system, and graph the line through the points.

(c) Find the linear equation, in slope-intercept form, that relates the interest rate (in percent), y, to the credit score, x.

(d) Use the equation found in part (c) to predict the interest rate for a person with a credit score of 700.

(e) Interpret the slope.

74. U.S. Traffic Fatalities Nationwide, the statistics for traffic fatalities show a decline. In 1999, the United States had 41,717 fatal crashes, and in 2009 the number dropped to 33,808. Let y represent the number of traffic fatalities and x represent the number of years since 1999.

(SOURCE: *National Highway Traffic Safety Administration*)

(a) Fill in the ordered pairs: $(\_, 41{,}717); (\_, 33{,}808)$.

(b) Plot the ordered pairs from part (a) in a rectangular coordinate system and graph the line through the points.

(c) Find the linear equation, in slope-intercept form, that relates the number of traffic fatalities, y, to the number of years since 1999, x.

(d) Use the equation found in part (c) to find the number of traffic fatalities in 2019.

(e) Interpret the slope.

Extending the Concepts

Up to this point, when we knew the slope of a line and a point on the line, we found the equation of the line using point-slope form. We could also use the slope-intercept form to find this equation.

For example, suppose that we were asked to find the equation of the line that has slope 6 and passes through the point $(2, -5)$. Let's use the slope-intercept form, $y = mx + b$, to write the equation of this line. We know $m = 6$, so we have $y = 6x + b$. We also know that $y = -5$ when $x = 2$. So we substitute 2 for x and -5 for y into the equation and solve for b.

$$y = 6x + b$$

Let $x = 2$ and $y = -5$: $\quad -5 = 6(2) + b$

Multiply: $\quad -5 = 12 + b$

Subtract 12 from each side: $\quad -5 - 12 = 12 - 12 + b$

$$-17 = b$$

We now know that $b = -17$, and because we also know that $m = 6$, the equation of the line is $y = 6x - 17$. Use this technique to find the slope-intercept form of the line for Problems 75–86.

75. $(-4, 2)$; slope $= 3$

76. $(5, -2)$; slope $= 4$

77. $(3, -8)$; slope $= -2$

78. $(-1, 7)$; slope $= -5$

79. $\left(\dfrac{2}{3}, \dfrac{1}{2}\right)$; slope $= 6$

80. $\left(\dfrac{4}{3}, -\dfrac{3}{2}\right)$; slope $= -9$

81. $(6, -13)$ and $(-2, -5)$

82. $(-10, -5)$ and $(2, 7)$

83. $(5, -1)$ and $(-10, -4)$

84. $(-6, -3)$ and $(9, 2)$

85. $(-4, 8)$ and $(2, -1)$

86. $(4, -9)$ and $(8, -19)$

Explaining the Concepts

87. You are asked to write the equation of the line through the points $(3, 1)$ and $(4, 7)$ in slope-intercept form. After calculating the slope of the line, you choose the point-slope form and assign $x_1 = 3$ and $y_1 = 1$. Your friend lets $x_1 = 4$ and $y_1 = 7$. Will you and your friend obtain the same answer? Explain why or why not.

88. You are asked to write the equation of the line through $(-1, 3)$ and $(0, 4)$. Which form of a line would you choose to find the equation? Explain why you chose this form. Could you also use one of the other forms?

3.6 Parallel and Perpendicular Lines

Objectives

1. Determine Whether Two Lines Are Parallel
2. Find the Equation of a Line Parallel to a Given Line
3. Determine Whether Two Lines Are Perpendicular
4. Find the Equation of a Line Perpendicular to a Given Line

Are You ready for This Section?

Before getting started, take the following readiness quiz. If you get a problem wrong, go back to the section cited and review the material.

R1. Determine the reciprocal of 3.   [Section 1.4, pp. 33–34]

R2. Determine the reciprocal of $-\dfrac{3}{5}$.   [Section 1.4, pp. 33–34]

▶ ① Determine Whether Two Lines Are Parallel

Two lines in the rectangular coordinate system that do not intersect (that is, have no points in common) are said to be *parallel*. The equations of lines can be used to determine whether the lines are parallel.

> ### Definition
> Two nonvertical lines are **parallel** *if and only if* their slopes are equal and they have different *y*-intercepts. Vertical lines are parallel if they have different *x*-intercepts.

Work Smart

The use of the words "if and only if" in the definition of parallel lines means that there are two statements being made:

1. If two nonvertical lines are parallel, then their slopes are equal and they have different *y*-intercepts.
2. If two nonvertical lines have equal slopes and different *y*-intercepts, then they are parallel.

Figure 49(a) shows nonvertical parallel lines. Figure 49(b) shows vertical parallel lines.

Figure 49
Parallel lines.

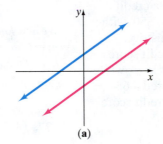

(a)　　　　　　(b)

To determine whether two lines are parallel, we find the slope and *y*-intercept of each line by putting the equations of the lines in slope-intercept form. If the slopes are the same but the *y*-intercepts are different, then the lines are parallel.

EXAMPLE 1

Determining Whether Two Lines Are Parallel

Determine whether the line $y = 4x - 5$ is parallel to $y = 3x - 2$. Graph the lines to confirm your results.

Solution

The line $y = 4x - 5$ has slope 4 and the *y*-intercept $(0, -5)$. The line $y = 3x - 2$ has slope 3 and *y*-intercept $(0, -2)$. Because the lines have different slopes, they are not parallel. Figure 50 shows that the lines intersect at $(3, 7)$. Therefore, the lines are not parallel.

Figure 50

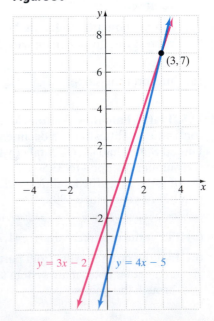

EXAMPLE 2 **Determining Whether Two Lines Are Parallel**

Determine whether the line $-3x + y = 5$ is parallel to $6x - 2y = -2$. Graph the lines to confirm your results.

Solution

We solve each equation for y so that each is in slope-intercept form.

$$-3x + y = 5$$

Add $3x$ to both sides: $\quad\quad y = 3x + 5$

The slope of the line $-3x + y = 5$ is 3 and the y-intercept is $(0, 5)$.

$$6x - 2y = -2$$

Subtract $6x$ from both sides: $\quad -2y = -6x - 2$

Divide both sides by -2: $\quad y = \dfrac{-6x - 2}{-2}$

Divide each term in the numerator by -2: $\quad y = 3x + 1$

The slope of $6x - 2y = -2$ is 3 and the y-intercept is $(0, 1)$.

The lines have the same slope, 3, but different y-intercepts, so they are parallel. Figure 51 shows the graph of the two lines. ●

Work Smart

Make sure both criteria for parallel lines are satisfied.

1. Same slope

2. Different y-intercepts

Lines with the same slope and same y-intercept are called *coincident lines*.

Figure 51

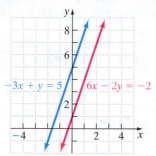

$-3x + y = 5$ $6x - 2y = -2$

Quick ✓

1. Two nonvertical lines are parallel if and only if their _____ are equal and they have different _____. Vertical lines are parallel if they have different _____.

In Problems 2–4, determine whether the two lines are parallel. Graph the lines to confirm your results.

2. $y = 2x + 1$
$\quad y = -2x - 3$

3. $6x + 3y = 3$
$\quad 10x + 5y = 10$

4. $4x + 5y = 10$
$\quad 8x + 10y = 20$

▶ ❷ Find the Equation of a Line Parallel to a Given Line

Now that we know how to identify parallel lines, we can find the equation of a line that is parallel to a given line.

EXAMPLE 3 **How to Find the Equation of a Line Parallel to a Given Line**

Find the equation for the line that is parallel to $2x + y = 5$ and contains the point $(-1, 3)$. Write the equation of the line in slope-intercept form. Graph the lines.

Step-by-Step Solution

Step 1: Find the slope of the given line.

$$2x + y = 5$$

Subtract $2x$ from both sides: $\quad y = -2x + 5$

The slope of the line is -2, so the slope of the parallel line is also -2.

Step 2: Use the point-slope form of a line with the given point and the slope found in Step 1 to find the equation of the parallel line.

$$y - y_1 = m(x - x_1)$$

$m = -2, x_1 = -1, y_1 = 3:$ $\quad y - 3 = -2(x - (-1))$

Step 3: Put the equation in slope-intercept form
by solving for y.

$$y - 3 = -2(x + 1)$$
Distribute the -2: $\quad y - 3 = -2x - 2$
Add 3 to both sides: $\quad\quad y = -2x + 1$

The equation of the line parallel to
$2x + y = 5$ is $y = -2x + 1$. Figure 52
shows the graph of the parallel lines.

Figure 52

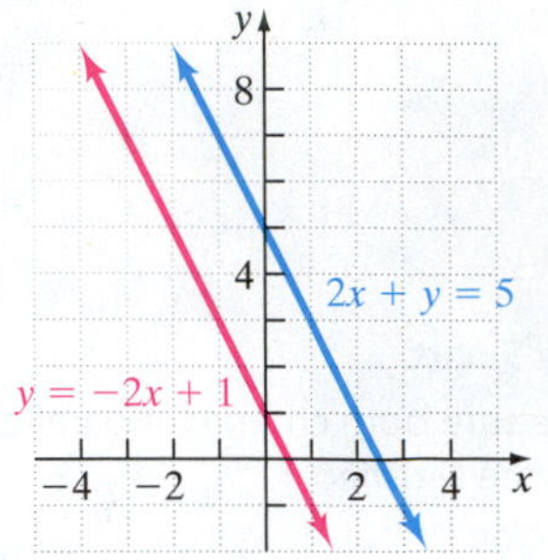

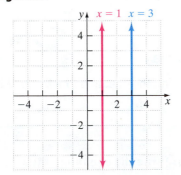

Quick ✓

In Problems 5 and 6, find the equation of the line that contains the given point and is parallel to the given line. Write the line in slope-intercept form. Graph the lines.

5. $y = 2x + 1$ containing $(2, 3)$

6. $3x + 2y = 4$ containing $(-2, 3)$

EXAMPLE 4 **Finding the Equation of a Line Parallel to a Given Line**

Find the equation for the line that is parallel to $x = 3$ and contains the point $(1, 5)$.
Graph the lines.

Solution

Figure 53

The equation of the given line is $x = 3$. This is a vertical line, so the line parallel to it
will also be vertical. Vertical lines have equations of the form $x = a$. The line parallel
to $x = 3$ that contains the point $(1, 5)$ is $x = 1$. Figure 53 shows the graph of the lines
$x = 3$ and $x = 1$.

Quick ✓

*In Problems 7 and 8, find the equation of the line that contains the given point and is
parallel to the given line. Write the line in slope-intercept form, if possible. Graph the
lines.*

7. $x = -2$ containing $(3, 1)$

8. $y + 3 = 0$ containing $(-2, 5)$

▶ ❸ **Determine Whether Two Lines Are Perpendicular**

Two lines that intersect at a right (90°) angle are
perpendicular. See Figure 54.

   Just as slopes can tell us whether two lines are
parallel, slopes can also tell us whether two lines
are perpendicular.

Figure 54
Perpendicular lines.

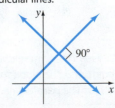

> **Definition**
>
> Two nonvertical lines are **perpendicular** if and only if the product of their slopes is -1. Put another way, two nonvertical lines are perpendicular if their slopes are negative reciprocals of each other. Any vertical line is perpendicular to any horizontal line.

EXAMPLE 5 **Finding the Slope of a Line Perpendicular to a Given Line**

Find the slope of a line perpendicular to a line whose slope is (a) 5 (b) $-\dfrac{2}{3}$.

Work Smart

If m_1 and m_2 are negative reciprocals of each other, then $m_1 = \dfrac{-1}{m_2}$. For example, the numbers 4 and $-\dfrac{1}{4}$ are negative reciprocals. Watch out, though, because the use of the word "*negative*" does not mean that the slope of the perpendicular line must be negative—it means that the nonvertical lines have slopes that are opposite in sign. One is positive and one is negative.

Solution

(a) To find the slope of a line perpendicular to a given line, we determine the negative reciprocal of the slope of the given line. The negative reciprocal of 5 is $\dfrac{-1}{5} = -\dfrac{1}{5}$. Any line whose slope is $-\dfrac{1}{5}$ will be perpendicular to the line whose slope is 5.

(b) The negative reciprocal of $-\dfrac{2}{3}$ is $\dfrac{-1}{-\frac{2}{3}} = -1 \cdot \left(-\dfrac{3}{2}\right) = \dfrac{3}{2}$. Any line whose slope is $\dfrac{3}{2}$ will be perpendicular to the line whose slope is $-\dfrac{2}{3}$.

> **Quick** ✓
>
> **9.** Given any two nonvertical lines, if the product of their slopes is -1, then the lines are _____.
>
> **10.** *True or False* Two different lines L_1 and L_2 have slopes $m_1 = 4$ and $m_2 = -4$, so L_1 is perpendicular to L_2.
>
> *In Problems 11–13, find the slope of a line perpendicular to the line whose slope is given.*
>
> **11.** -4 **12.** $\dfrac{5}{4}$ **13.** $-\dfrac{1}{5}$

EXAMPLE 6 **Determining Whether Two Lines Are Perpendicular**

Determine whether the line $y = 3x - 2$ is perpendicular to $y = \dfrac{1}{3}x + 1$. Graph the lines to confirm your results.

Solution

We first need to find the slope of each line. If the product of the slopes is -1 (or the slopes are negative reciprocals of each other), then the lines are perpendicular. The slope of $y = 3x - 2$ is $m_1 = 3$.

The slope of $y = \dfrac{1}{3}x + 1$ is $m_2 = \dfrac{1}{3}$.

Because the product of the slopes,

$$m_1 \cdot m_2 = 3 \cdot \left(\dfrac{1}{3}\right) = 1 \neq -1,$$ the lines

are not perpendicular. Notice that the slopes are reciprocals of each other, but not *negative* reciprocals of each other. Figure 55 shows the graph of the two lines.

Figure 55

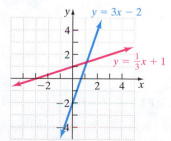

EXAMPLE 7 **Determining Whether Two Lines Are Perpendicular**

Determine whether the line $2x + 3y = -6$ is perpendicular to $3x - 2y = 2$. Graph the lines to confirm your results.

Solution

Write each equation in slope-intercept form to find the slopes of the two lines.

$$2x + 3y = -6$$

Subtract 2x from both sides: $\quad 3y = -2x - 6$

Divide both sides by 3: $\quad y = \dfrac{-2x - 6}{3}$

Divide 3 into each term in the numerator: $\quad y = -\dfrac{2}{3}x - 2$

The slope of $2x + 3y = -6$ is $m_1 = -\dfrac{2}{3}$.

$$3x - 2y = 2$$

Subtract 3x from both sides: $\quad -2y = -3x + 2$

Divide both sides by -2: $\quad y = \dfrac{-3x + 2}{-2}$

Divide -2 into each term in the numerator: $\quad y = \dfrac{3}{2}x - 1$

The slope of $3x - 2y = 2$ is $m_2 = \dfrac{3}{2}$.

The product of the slopes is $m_1 \cdot m_2 = -\dfrac{2}{3} \cdot \dfrac{3}{2} = -1$, so the lines are perpendicular.

Put another way, because the slopes are negative reciprocals of each other, the lines are perpendicular. See Figure 56 for the graph of the two lines. ●

Figure 56

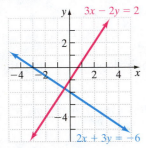

$3x - 2y = 2$

$2x + 3y = -6$

Quick ✓

In Problems 14–16, determine whether the given lines are perpendicular. Graph the lines.

14. $y = 4x - 3$
$\quad\ \ y = -\dfrac{1}{4}x - 4$

15. $2x - y = 3$
$\quad\ \ x - 2y = 2$

16. $5x + 2y = 8$
$\quad\ \ 2x - 5y = 10$

▶ ❹ **Find the Equation of a Line Perpendicular to a Given Line**

Now that we know how to find the slope of a line perpendicular to a second line, we can find the equation of a line that is perpendicular to a given line.

EXAMPLE 8 **How to Find the Equation of a Line Perpendicular to a Given Line**

Find the equation of the line that is perpendicular to the line $y = 3x - 2$ and contains the point $(3, -1)$. Write the equation of the line in slope-intercept form. Graph the two lines.

Step-by-Step Solution

Step 1: Find the slope of the given line. $\qquad$ The slope of the line $y = 3x - 2$ is 3.

Step 2: Find the slope of the perpendicular line.

The slope of the perpendicular line is $\dfrac{-1}{3} = -\dfrac{1}{3}$.

Step 3: Use the point-slope form of a line with the given point and the slope found in Step 2 to find the equation of the perpendicular line.

$$y - y_1 = m(x - x_1)$$

$m = -\dfrac{1}{3}, x_1 = 3, y_1 = -1:$ $y - (-1) = -\dfrac{1}{3}(x - 3)$

Step 4: Put the equation in slope-intercept form by solving for y.

$$y + 1 = -\dfrac{1}{3}(x - 3)$$

Distribute the $-\dfrac{1}{3}$: $y + 1 = -\dfrac{1}{3}x + 1$

Subtract 1 from both sides: $y = -\dfrac{1}{3}x$

The equation of the line perpendicular to $y = 3x - 2$ through $(3, -1)$ is $y = -\dfrac{1}{3}x$. Figure 57 shows the graph of the two lines.   ●

Figure 57

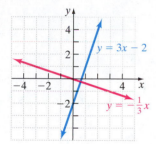

Quick ✓

In Problems 17 and 18, find the equation of the line that contains the given point and is perpendicular to the given line. Write the line in slope-intercept form. Graph the lines.

17. $(-4, 2); y = 2x + 1$ **18.** $(-2, -1); 2x + 3y = 3$

(**EXAMPLE 9**)   **Finding the Equation of a Line Perpendicular to a Given Line**

Find the equation of the line that is perpendicular to the line $x = 2$ and contains the point $(-4, 3)$. Write the equation of the line in slope-intercept form, if possible. Graph the two lines.

Solution

The equation $x = 2$ is the equation of a vertical line. Therefore, the line perpendicular to $x = 2$ will be horizontal. Horizontal lines have slopes equal to 0. To find the equation of the perpendicular line, we use the point-slope formula.

$$y - y_1 = m(x - x_1)$$

$m = 0, x_1 = -4, y_1 = 3:$ $y - 3 = 0(x - (-4))$

$$y - 3 = 0$$

Add 3 to both sides: $y = 3$

Work Smart

The line perpendicular to a vertical line is horizontal, and vice versa.

The equation of the line perpendicular to $x = 2$ through the point $(-4, 3)$ is $y = 3$. See Figure 58 for the graph of the two lines.   ●

Figure 58

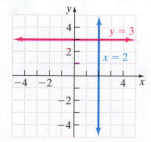

Quick ✓

In Problems 19 and 20, find the equation of the line that contains the given point and is perpendicular to the given line. Write the line in slope-intercept form, if possible. Graph the lines.

19. $x = -4$ containing $(-1, -5)$ **20.** $y + 2 = 0$ containing $(3, -2)$

*Problems **1–20** are the Quick ✓s that follow the* **EXAMPLES**.

Building Skills

In Problems 21–28, fill in the chart with the missing slopes. See Objectives 1 and 3.

	Slope of the Given Line	Slope of a Line Parallel to the Given Line	Slope of a Line Perpendicular to the Given Line
21.	$m = -3$		
23.	$m = \dfrac{1}{2}$		
25.	$m = -\dfrac{4}{9}$		
27.	$m = 0$		

	Slope of the Given Line	Slope of a Line Parallel to the Given Line	Slope of a Line Perpendicular to the Given Line
22.	$m = 4$		
24.	$m = -\dfrac{1}{8}$		
26.	$m = \dfrac{5}{2}$		
28.	$m = $ undefined		

In Problems 29–42, determine whether the lines are parallel, perpendicular, or neither. See Objectives 1 and 3.

29. $L_1: y = 2x - 3$
$L_2: y = -\dfrac{1}{2}x + 1$

30. $L_1: y = -4x + 3$
$L_2: y = 4x - 1$

31. $L_1: y = \dfrac{3}{4}x + 2$
$L_2: y = 0.75x - 1$

32. $L_1: y = 0.8x + 6$
$L_2: y = \dfrac{4}{5}x + \dfrac{19}{3}$

33. $L_1: y = -\dfrac{5}{3}x - 6$
$L_2: y = \dfrac{3}{5}x - 1$

34. $L_1: y = 3x - 1$
$L_2: y = 6 - \dfrac{x}{3}$

35. $L_1: x + y = -3$
$L_2: y - x = 1$

36. $L_1: x - 4y = 24$
$L_2: 2x - 8y = -8$

37. $L_1: 2x + 5y = 5$
$L_2: 5x + 2y = 4$

38. $L_1: x - 2y = -8$
$L_2: x + 2y = 2$

39. $L_1: 4x - 5y - 15 = 0$
$L_2: 8x - 10y + 5 = 0$

40. $L_1: x + y = 6$
$L_2: x - y = -2$

41. $L_1: 4x = 3y + 3$
$L_2: 6y = 8x + 36$

42. $L_1: 2x - 5y - 45 = 0$
$L_2: 5x + 2y - 8 = 0$

In Problems 43–54, find the equation of the line that contains the given point and is parallel to the given line. Write the equation in slope-intercept form, if possible. See Objective 2.

43. $(4, -2); y = 3x - 1$

44. $(7, -5); y = 2x + 6$

45. $(-3, 8); y = -4x + 5$

46. $(-2, 6); y = -5x - 2$

47. $(3, -7); y = 4$

48. $(-4, 5); x = -3$

49. $(-1, 10); x = 10$

50. $(4, -8); y = -1$

51. $(10, 2); 3x - 2y = 5$

52. $(6, 7); 2x + 3y = 9$

53. $(-1, -10); x + 2y = 4$

54. $(-3, -5); 2x - 5y = 6$

In Problems 55–66, find the equation of the line that contains the given point and is perpendicular to the given line. Write the equation in slope-intercept form, if possible. See Objective 4.

55. $(3, 5); y = \dfrac{1}{2}x - 2$

56. $(4, 7); y = \dfrac{1}{3}x - 3$

57. $(-4, -1); y = -4x + 1$

58. $(-2, -5); y = -2x + 5$

59. $(-2, 1); x\text{-axis}$

60. $(3, -6); y\text{-axis}$

61. $(7, 5); y\text{-axis}$

62. $(11, -6); x\text{-axis}$

63. $(0, 0); 2x + 5y = 7$

64. $(0, 0)$; $6x + 4y = 3$

65. $(-10, -3)$; $5x - 3y = 4$

66. $(-6, 10)$; $3x - 5y = 2$

Mixed Practice

In Problems 67–82, find the equation of the line that has the given properties. Write the equation in slope-intercept form, if possible. Graph each line.

67. Contains $(3, -5)$; slope $= 7$

68. Contains $(-2, 10)$; slope $= 3$

69. Contains $(2, 9)$; perpendicular to the line $y = -5x + 3$

70. Contains $(8, 3)$; parallel to the line $y = -4x + 2$

71. Contains $(6, -1)$; parallel to the line $y = -7x + 2$

72. Contains $(5, -2)$; perpendicular to the line $y = 4x + 3$

73. Contains $(-6, 2)$ and $(-1, -8)$

74. Contains $(-4, -1)$ and $(-3, -5)$

75. Slope $= 3$, y-intercept $= (0, -2)$

76. Slope $= -2$, y-intercept $= (0, 7)$

77. Contains $(5, 1)$; parallel to the line $x = -6$

78. Contains $(2, 7)$; perpendicular to the line $x = -8$

79. Contains $(3, -2)$; parallel to the line $4x + 3y = 9$

80. Contains $(-3, -2)$; perpendicular to the line $6x - 2y = 1$

81. Contains $(-1, -3)$; perpendicular to the line $x - 2y = -10$

82. Contains $(-7, 2)$; parallel to the line $x + 4y = 2$

In Problems 83–90, each line contains the given points. (a) Find the slope of each line. (b) Determine whether the lines are parallel, perpendicular, or neither.

83. L_1: $(0, -1)$ and $(-2, -7)$
L_2: $(-1, 5)$ and $(2, -4)$

84. L_1: $(-3, -14)$ and $(1, 2)$
L_2: $(0, 2)$ and $(-3, -10)$

85. L_1: $(2, 8)$ and $(7, 18)$
L_2: $(-2, -3)$ and $(6, 13)$

86. L_1: $(6, 0)$ and $(-2, 8)$
L_2: $(4, 1)$ and $(-6, -9)$

87. L_1: $(-2, -5)$ and $(4, -2)$
L_2: $(-8, -5)$ and $(0, -1)$

88. L_1: $(1, 6)$ and $(-1, -10)$
L_2: $(0, 1)$ and $(-2, 17)$

89. L_1: $(-6, -9)$ and $(3, 6)$
L_2: $(10, -8)$ and $(-5, 1)$

90. L_1: $(-8, -8)$ and $(4, 1)$
L_2: $(12, 8)$ and $(-4, -4)$

Applying the Concepts

A parallelogram is a quadrilateral in which both pairs of opposite sides are parallel. In Problems 91 and 92, plot the given points, draw the figure, and then use slope to determine whether the figure is a parallelogram.

△ **91.** $A(-1, 1)$; $B(3, 5)$; $C(6, 4)$; $D(2, 0)$

△ **92.** $A(-1, -3)$; $B(1, -1)$; $C(5, 1)$; $D(3, -2)$

A rectangle is a parallelogram that contains one right angle. That is, one pair of sides are perpendicular. In Problems 93 and 94, plot the given points, draw the figure, and then use slope to determine whether the figure is a rectangle.

△ **93.** $A(6, -1)$; $B(-3, -2)$; $C(1, -6)$; $D(2, 3)$

△ **94.** $A(1, 1)$; $B(-1, 5)$; $C(5, 8)$; $D(7, 4)$

A right triangle is a triangle that contains one right angle. In Problems 95–98, plot each point and form the triangle ABC. Verify using slope that the triangle is a right triangle.

△ **95.** $A(-2, 5)$; $B(1, 3)$; $C(3, 6)$

△ **96.** $A(-5, 3)$; $B(6, 0)$; $C(5, 5)$

△ **97.** $A(4, -3)$; $B(0, -3)$; $C(4, 2)$

△ **98.** $A(-2, 5)$; $B(12, 3)$; $C(10, -11)$

Extending the Concepts

In Problems 99 and 100, find the missing coefficient so that the lines are parallel.

99. $-3y = 6x - 12$ and $4x + By = -2$

100. $Ax + 2y = 4$ and $15x = 5y + 20$

In Problems 101 and 102, find the missing coefficient so that the lines are perpendicular.

101. $Ax + 6y = -6$ and $12 - 6y = -9x$

102. $x - By = 10$ and $3y = -6x + 9$

△ **103.** The altitude of a triangle is a line segment drawn from a vertex of the triangle perpendicular to the opposite side. Plot the following points, draw triangle ABC and the segment joining points B and D, and then determine whether $\overline{BD}$ is an altitude of the triangle. $A(-6, -3); B(-4, 7); C(-1, 2); D(-3, 0)$

△ **104.** The coordinates of the vertices of a quadrilateral are $A(2, 1), B(4, 6), C(6, 6),$ and $D(9, 2)$. Use slopes to show that the diagonals of the quadrilateral, $\overline{AC}$ and $\overline{BD}$, are perpendicular to each other.

Explaining the Concepts

105. Describe the four possible relationships of the graphs of two lines in a rectangular coordinate system.

106. You are asked to determine if the following lines are parallel, perpendicular or neither: L_1 contains the points $(1, 1)$ and $(3, 5)$ and L_2 contains the points $(-1, -1)$ and $(4, 9)$. List the steps you would follow to make this determination and describe the criteria you would use to answer the question.

Putting the Concepts Together (Sections 3.1–3.6)

We designed these problems so that you can review Sections 3.1–3.6 and show your mastery of the concepts. Take time to work these problems before proceeding with the next section. The answers are at the back of the text on page AN-13.

1. Given the linear equation $4x - 3y = 10$, determine whether the ordered pair $(1, -2)$ is a solution to the equation.

In Problems 2 and 3, graph each equation using the point-plotting method.

2. $y = \dfrac{2}{3}x - 1$

3. $-5x + 2y = 10$

4. Given the equation $-8x + 2y = 6$, determine

   (a) the x-intercept
   (b) the y-intercept

5. Graph the equation $4x + 3y = 6$ by finding the x-intercept and the y-intercept.

6. Given the equation $6x + 9y = -12$, determine

   (a) the slope
   (b) the y-intercept

7. Find the slope of the line that contains the points $(3, -5)$ and $(-6, -2)$.

8. Given the linear equation $2y = -5x - 4$, find

   (a) the slope of a line perpendicular to the given line
   (b) the slope of a line parallel to the given line

9. Determine whether the lines are parallel, perpendicular, or neither. Explain how you came to your conclusion.

$$L_1: 10x + 5y = 2 \quad L_2: y = -2x + 3$$

In Problems 10–16, write the equation of the line that satisfies the given conditions. Write the equation in slope-intercept form, if possible.

10. slope $= 3$ and y-intercept is $(0, 1)$

11. slope $= -6$ and passes through $(-1, 4)$

12. through $(4, -1)$ and $(-2, 11)$

13. through $(-8, 0)$ and perpendicular to $y = \dfrac{2}{5}x - 5$

14. through $(-8, 3)$ and parallel to $-8y + 2x = -1$

15. horizontal line through $(-6, -8)$

16. through $(2, 6)$ with undefined slope

17. Shipping Expenses The shipping department records indicate that during a week when 80 packages were shipped, the total expenses recorded for the shipping department were $1180. During a different week, 50 packages were shipped, and the expenses recorded were $850. Use the ordered pairs $(80, 1180)$ and $(50, 850)$ to determine the average rate to ship an additional package.

18. Diamonds Suppose the relation between the cost of a diamond and its weight is linear. In looking at two diamonds, we find that one of the diamonds weighs 0.7 carat and costs $3543, while the other diamond weighs 0.8 carat and costs $4378. SOURCE: *diamonds.com*

   (a) Use the ordered pairs $(0.7, 3543)$ and $(0.8, 4378)$ to find a linear equation that relates the price of a diamond to its weight.

   (b) Interpret the slope.

   (c) Predict the price of a diamond that weighs 0.76 carat.

3.7 Linear Inequalities in Two Variables

Objectives

1 Determine Whether an Ordered Pair Is a Solution to a Linear Inequality

2 Graph Linear Inequalities

3 Solve Problems Involving Linear Inequalities

Are You Ready for This Section?

Before getting started, take the following readiness quiz. If you get a problem wrong, go back to the section cited and review the material.

R1. Solve: $x - 4 > 5$ [Section 2.8, pp. 150–151]

R2. Solve: $3x + 1 \le 10$ [Section 2.8, pp. 152–154]

R3. Solve: $2(x + 1) - 6x > 18$ [Section 2.8, pp. 153–154]

▶ **1** Determine Whether an Ordered Pair Is a Solution to a Linear Inequality

In Chapter 2, we solved inequalities in one variable. In this section, we discuss linear inequalities in two variables.

> **Definition**
>
> **Linear inequalities in two variables** are inequalitties equivalent to one of the forms
>
> $$Ax + By < C \quad Ax + By > C \quad Ax + By \le C \quad Ax + By \ge C$$
>
> where A and B are not both zero. A linear inequality in two variables x and y is **satisifed** by an ordered pair (a, b) if a true statement results when x is replaced by a and y is replaced by b.

EXAMPLE 1   **Determining Whether an Ordered Pair Is a Solution to a Linear Inequality in Two Variables**

Determine which of the following ordered pairs are solutions to the linear inequality $2x + y \le 9$.

    (a) $(3, 5)$ **(b)** $(1, 3)$ **(c)** $(3, -2)$

Solution

(a) Let $x = 3$ and $y = 5$ in the inequality. If a true statement results, then $(3, 5)$ is a solution to the inequality.

$$2x + y \le 9$$

$$x = 3, y = 5: \quad 2(3) + 5 \overset{?}{\le} 9$$

$$6 + 5 \overset{?}{\le} 9$$

$$11 \le 9 \quad \text{False}$$

The statement $11 \le 9$ is false, so $(3, 5)$ is not a solution to the inequality.

(b) Let $x = 1$ and $y = 3$ in the inequality. If a true statement results, then $(1, 3)$ is a solution to the inequality.

$$2x + y \le 9$$

$$x = 1, y = 3: \quad 2(1) + 3 \overset{?}{\le} 9$$

$$2 + 3 \overset{?}{\le} 9$$

$$5 \le 9 \quad \text{True}$$

The statement $5 \le 9$ is true, so $(1, 3)$ is a solution to the inequality.

(c) Let $x = 3$ and $y = -2$ in the inequality. If a true statement results, then $(3, -2)$ is a solution to the inequality.

$$2x + y \le 9$$

$$x = 3, y = -2: \quad 2(3) + (-2) \overset{?}{\le} 9$$

$$6 - 2 \overset{?}{\le} 9$$

$$4 \le 9 \qquad \text{True}$$

The statement $4 \le 9$ is true, so $(3, -2)$ is a solution to the inequality. ●

Quick ✓

Determine which of the following ordered pairs are solutions to the given linear inequality.

(a) $(2, 1)$ **(b)** $(3, 4)$ **(c)** $(-1, 10)$

1. $2x + y > 7$ **2.** $-3x + 2y \le 8$

⏵ ② Graph Linear Inequalities

A **graph of a linear inequality in two variables** x and y consists of all points whose coordinates (x, y) satisfy the inequality.

If we replace the inequality symbol in a linear inequality of the form

$$Ax + By < C \qquad Ax + By > C \qquad Ax + By \le C \qquad Ax + By \ge C$$

with an equal sign, we obtain the equation of a line, $Ax + By = C$. This **boundary line** separates the xy-plane into two regions called **half-planes.** See Figure 59.

When graphing the boundary line, use dashes if the inequality is strict ($<$ or $>$) to indicate points on the line do not satisfy the inequality, and use a solid line if the inequality is nonstrict ($\le$ or $\ge$) to indicate points on the line do satisfy the inequality.

Let's consider the linear inequality $2x + y \le 9$ in Example 1. Figure 60 shows the graph of the equation $2x + y = 9$ and the three points $(3, 5)$, $(1, 3)$, and $(3, -2)$ examined in Example 1.

Figure 59

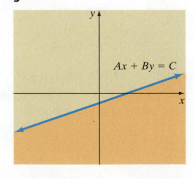

Figure 60

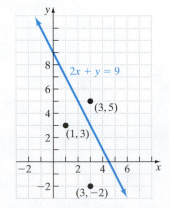

In Example 1, the two points below the line, $(1, 3)$ and $(3, -2)$, satisfy the inequality $2x + y \leq 9$, while $(3, 5)$ does not. In fact, all points below the line satisfy the inequality, and all points above the inequality do not. What can we learn from this? If a point satisfies a linear inequality in two variables, then all points in the half-plane containing that point satisfy the inequality. If a point does not satisfy a linear inequality in two variables, then all points in the opposite half-plane satisfy the inequality. For this reason, a single **test point** is all that is required to obtain the graph of a linear inequality in two variables. Even so, it is not a bad idea to check a few points to verify your algebra.

EXAMPLE 2 **How to Graph a Linear Inequality in Two Variables**

Graph the linear inequality: $y < 2x - 5$

Step-by-Step Solution

Step 1: Replace the inequality symbol with an equal sign and graph the resulting equation.

We graph the line $y = 2x - 5$ whose slope is 2 and y-intercept is $(0, -5)$ using a dashed line because the inequality is strict $(<)$. See Figure 61.

Figure 61

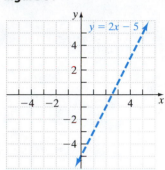

Step 2: Select any test point that is not on the line, and determine whether the test point satisfies the inequality. If the test point satisfies the inequality, we shade the half-plane containing the test point; otherwise, we shade the half-plane opposite it.

We will use $(0, 0)$ as the test point.

$$y < 2x - 5$$
$$x = 0, \quad y = 0: \quad 0 \overset{?}{<} 2(0) - 5$$
$$0 < -5 \quad \text{False}$$

Because $(0, 0)$ does not satisfy the inequality $y < 2x - 5$, we shade the half-plane opposite the half-plane containing $(0, 0)$. See Figure 62.

Figure 62
$y < 2x - 5$

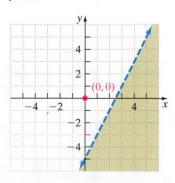

Work Smart

When the line does not contain the origin, it is usually easiest to choose the origin, $(0, 0)$, as the test point.

Work Smart

To double-check your results, choose a point, such as $(4, 0)$, in the shaded region. Does $(4, 0)$ satisfy $y < 2x - 5$? Yes!

The shaded region represents the solution to the linear inequality. Because the inequality is strict, points on the line $y = 2x - 5$ do not satisfy the inequality.

We summarize the steps for graphing a linear inequality in two variables.

> ### Graphing a Linear Inequality in Two Variables
>
> **Step 1:** Replace the inequality symbol with an equal sign, and graph the resulting equation. If the inequality is strict ($<$ or $>$), use a dashed line; if the inequality is nonstrict ($\leq$ or $\geq$), use a solid line. The graph separates the xy-plane into two half-planes.
>
> **Step 2:** Select a test point P that is not on the line.
>
>   **(a)** If the coordinates of P satisfy the inequality, then shade the half-plane containing P.
>
>   **(b)** If the coordinates of P do not satisfy the inequality, then shade the half-plane that does not contain P.
>
> **Note:** If the inequality is nonstrict, then all points on the line also satisfy the inequality. If the inequality is strict, then all points on the line *do not* satisfy the inequality.

Quick ✓

3. When drawing the boundary line for the graph of $Ax + By \geq C$, we use a _____ line. When drawing the boundary line for the graph of $Ax + By < C$, we use a _____ line.

4. The boundary line separates the xy-plane into two regions called _____.

In Problems 5 and 6, graph each linear inequality.

5. $y < -2x + 1$ **6.** $y \geq 3x + 2$

EXAMPLE 3 **Graphing a Linear Inequality in Two Variables**

Graph the linear inequality: $5x + 2y \leq 10$

Solution

Graph the equation $5x + 2y = 10$ by finding the intercepts of the equation.

Find the y-intercept:	**Find the x-intercept:**
$5x + 2y = 10$	$5x + 2y = 10$
Let $x = 0$: $5(0) + 2y = 10$	Let $y = 0$: $5x + 2(0) = 10$
$2y = 10$	$5x = 10$
Divide both sides by 2: $y = 5$	Divide both sides by 5: $x = 2$

Plot the points $(0, 5)$ and $(2, 0)$ and draw the equation of the line as a solid line because the inequality is nonstrict ($\leq$). See Figure 63(a).

Because the graph of the equation does not contain the origin, we will select the origin $(0, 0)$ as our test point.

$$5x + 2y \leq 10$$

Let $x = 0, y = 0$: $5(0) + 2(0) \overset{?}{\leq} 10$

$$0 \leq 10 \quad \text{True}$$

Work Smart

The test point $(0, 0)$ is a solution of the linear inequality, so all other points in that half-plane are also solutions. That's why we shade the half-plane that contains the origin.

Because the test point $(0, 0)$ results in a true statement, we shade the half-plane that contains $(0, 0)$. See Figure 63(b). The shaded region and all points on the line $5x + 2y = 10$ represent the solution set.

Figure 63

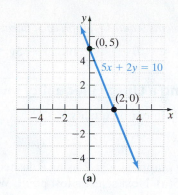
(a)

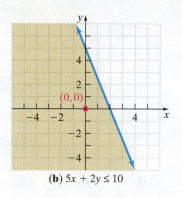
(b) $5x + 2y \leq 10$

Quick ✓

In Problems 7 and 8, graph each linear inequality.

7. $2x + 3y \leq 6$ **8.** $4x - 6y > 12$

EXAMPLE 4 **Graphing a Linear Inequality Where the Boundary Line Goes Through the Origin**

Graph the linear inequality: $-4x + 3y > 0$

Solution

Graph the equation $-4x + 3y = 0$ using a dashed line. Because this is an equation of the form $Ax + By = 0$, the graph will pass through the origin. For this reason, we write the equation in slope-intercept form.

$$-4x + 3y = 0$$

Add $4x$ to both sides: $\quad 3y = 4x$

Divide both sides by 3: $\quad y = \dfrac{4}{3}x$

The line has slope $m = \dfrac{4}{3}$ and y-intercept $(0, 0)$. See Figure 64(a).

The graph of the equation contains the origin, so we will use $(1, 3)$ as our test point.

$$-4x + 3y > 0$$

Let $x = 1$, $y = 3$: $\quad -4(1) + 3(3) \overset{?}{>} 0$

$$-4 + 9 \overset{?}{>} 0$$

$$5 > 0 \quad \text{True}$$

Work Smart

Because the boundary line contains the origin, the test point cannot be the origin.

Because the test point $(1, 3)$ results in a true statement, we shade the half-plane that contains $(1, 3)$. See Figure 64(b). The shaded region represents the solution set.

Figure 64

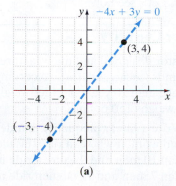
(a)

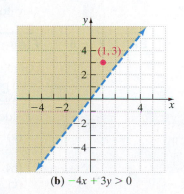
(b) $-4x + 3y > 0$

EXAMPLE 5   ### Graphing a Linear Inequality Involving a Horizontal Line

Graph the linear inequality $y < 2$.

Solution

Begin by graphing $y = 2$ using a dashed line (because the inequality is strict). See Figure 65(a). Next select a test point. We can select the origin $(0, 0)$ as our test point because the line does not contain $(0, 0)$.

$$y < 2$$

Test point $(0, 0)$: $0 < 2$ True

Because the test point satisfies the inequality, shade the half-plane that contains $(0, 0)$ as shown in Figure 65(b). The shaded region represents the solution to the linear inequality. Doesn't it seem intuitive that we shade below the line since these are the y-values that are less than 2?

Work Smart: Study Skills

Have you noticed the patterns?
If $y > mx + b$, shade above.
If $y < mx + b$, shade below.
If $y > b$, shade above.
If $y < b$, shade below.
If $x > a$, shade to the right.
If $x < a$, shade to the left.

Figure 65

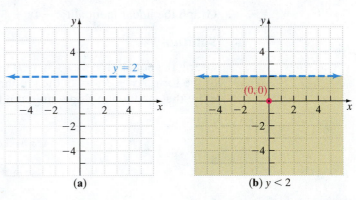

(a)   (b) $y < 2$

Graphing a linear inequality involving a vertical line is similar to graphing a linear inequality involving a horizontal line. That is, graph the vertical boundary line, choose a test point, and shade the half-plane containing the points that satisfy the inequality.

▶ ❸ ## Solve Problems Involving Linear Inequalities

Many applications of linear inequalities that involve two variables occur in areas such as nutrition, manufacturing, and sales. They are even used to describe weight loads in elevators! Let's see how.

EXAMPLE 6   ### Elevator Capacity

An elevator designed and manufactured by Otis Elevator Company has a capacity of 3000 pounds (SOURCE: *otis.com*). According to the National Center for Health Statistics, the average man 20 years or older weighs 180 pounds, and the average woman 20 years or older weighs 150 pounds.

  (a) Write a linear inequality that describes the various combinations of men and women who can ride this elevator.

  (b) Can 10 men and 9 women ride in this elevator safely?

  (c) Can 7 men and 11 women ride in this elevator safely?

Solution

(a) Use the first three steps in the problem-solving strategy given in Chapter 2 on page 124 to help develop the linear inequality.

Step 1: Identify We want to determine the number of men and women who can ride in the elevator without going over 3000 pounds.

Step 2: Name Unknowns Let m represent the number of males and f represent the number of females on the elevator.

Step 3: Translate To keep things simple, assume that all the people on the elevator are "average." If one male is on the elevator, the weight on the elevator is 180 pounds. If two males are on the elevator, the weight is $2(180) = 360$ pounds. In general, if m males are on the elevator, the weight is $180m$. Similarly, if f females are on the elevator, the weight is $150f$. Because the capacity of the elevator is 3000 pounds, we use a "less than or equal to" ($\leq$) inequality. A linear inequality that describes the weight limitations on the elevator is

$$180m + 150f \leq 3000 \quad \textcolor{red}{\text{The Model}}$$

(b) Letting $m = 10$ and $f = 9$, we obtain

$$180(10) + 150(9) \overset{?}{\leq} 3000$$
$$3150 \leq 3000 \quad \textcolor{red}{\text{False}}$$

The inequality is false, so 10 men and 9 women cannot ride the elevator safely.

(c) Letting $m = 7$ and $f = 11$, we obtain

$$180(7) + 150(11) \overset{?}{\leq} 3000$$
$$2910 \leq 3000 \quad \textcolor{red}{\text{True}}$$

The inequality is true, so 7 men and 11 women can ride the elevator safely.

Quick ✓

14. Kevin just received $2 from his grandma. He goes to a candy store where each sucker sells for $0.20 and each taffy stick sells for $0.25.

   (a) Write a linear inequality that describes the various combinations of suckers, s, and taffy sticks, t, Kevin can buy.
   (b) Can Kevin buy 6 suckers and 3 taffy sticks?
   (c) Can Kevin buy 5 suckers and 5 taffy sticks?

3.7 Exercises

 MyMathLab® Math XP PRACTICE

Exercise numbers in **green** have complete video solutions in MyMathLab.

*Problems **1–14** are the Quick ✓s that follow the **EXAMPLES**.*

Building Skills

In Problems 15–26, determine which of the following ordered pairs, if any, are solutions to the given linear inequality in two variables. See Objective 1.

15. $y > -x + 2$ $A(2, 4)$ $B(3, -6)$ $C(0, 0)$

16. $y \leq x - 5$ $A(-4, -6)$ $B(0, -8)$ $C(3, 10)$

17. $y \leq 3x - 1$ $A(-6, -15)$ $B(0, 0)$ $C(-1, -6)$

18. $y > -2x + 1$ $A(0, 0)$ $B(-2, 3)$ $C(2, -1)$

19. $3x \geq 2y$ $A(-8, -12)$ $B(3, 5)$ $C(-5, -8)$

20. $4y < 5x$ $A(-4, -6)$ $B(5, 6)$ $C(-12, -15)$

21. $2x - 3y < -6$ $A(2, -1)$ $B(4, 8)$ $C(-3, 0)$

22. $3x + 5y \geq 4$ $A(2, 0)$ $B(-3, 4)$ $C(6, -2)$

23. $x \leq 2$ $A(7, 2)$ $B(2, 5)$ $C(4, 2)$

24. $y \le -3$ $\qquad A(-3,-3)$ $B(-1,-4)$ $C(-6,-1)$

25. $y > -1$ $\qquad A(-1,1)$ $\quad B(3,-1)$ $\quad C(4,-2)$

26. $x > 10$ $\qquad A(3,12)$ $\quad B(11,-6)$ $C(10,16)$

In Problems 27–56, graph each linear inequality. See Objective 2.

27. $y > 3x - 2$ $\qquad$ **28.** $y \le 4x + 2$ $\qquad$ **29.** $y \le -x + 1$

30. $y > x - 3$ $\qquad$ **31.** $y < \dfrac{x}{2}$ $\qquad$ **32.** $y > \dfrac{x}{3}$

33. $y > 5$ $\qquad$ **34.** $y < 4$ $\qquad$ **35.** $y \le \dfrac{2}{5}x + 3$

36. $y \ge \dfrac{3}{2}x - 2$ $\qquad$ **37.** $y \ge -\dfrac{4}{3}x + 2$ **38.** $y \le -\dfrac{3}{4}x + 1$

39. $x < 2$ $\qquad$ **40.** $x > -4$ $\qquad$ **41.** $3x - 4y < 12$

42. $2x + 6y < -6$ **43.** $2x + y \ge -4$ **44.** $6x - 8y \ge 24$

45. $x + y > 0$ $\qquad$ **46.** $x - y > 0$ $\qquad$ **47.** $5x - 2y < -8$

48. $3x + 2y \le -9$ **49.** $x > -1$ $\qquad$ **50.** $y < -3$

51. $y \le 4$ $\qquad$ **52.** $x \ge 2$ $\qquad$ **53.** $\dfrac{x}{3} - \dfrac{y}{5} \ge 1$

54. $\dfrac{x}{4} + \dfrac{y}{2} < 1$ $\qquad$ **55.** $-3 \ge x - y$ **56.** $-2 > x + y$

Applying the Concepts

In Problems 57–68, translate each statement into a linear inequality. Then graph the inequality.

57. The sum of two numbers, x and y, is at least 26.

58. The ratio of a number, x, and 3 is less than -2.

59. The ratio of a number, y, and -2 is at most 4.

60. One number, x, is at least 12 more than a second number, y.

61. One number, x, is no more than 3 less than a second number, y.

62. The difference of a number, x, and half a second number, y, is positive.

63. The sum of a number, x, and 3 times a second number, y, is negative.

64. The sum of two numbers, x and y, is at least -3.

65. The difference of twice a number, x, and half a second number, y, is at least 5.

66. The product of a number, y, and 2 is at most -10.

67. The product of a number, x, and -2 is more than -1.

68. The sum of half a number, x, and twice a second number, y, is at least 12.

69. Trip to the Aquarium A kindergarten class has a maximum of $120 to spend on a trip to the aquarium. The cost of admission for students is $3, and adults must pay $5.

  (a) Write a linear inequality that describes the various combinations of the number of students s and adults a that can go on the field trip.

  (b) Is there enough money to pay for 32 students and 6 adults?

  (c) Is there enough money to pay for 29 students and 4 adults?

70. Fishing Trip Patrick's rowboat can hold a maximum of 500 pounds. In Patrick's circle of friends, the average adult weighs 160 pounds and the average child weighs 75 pounds.

  (a) Write a linear inequality that describes the various combinations of the number of adults a and children c that can go on a fishing trip in the rowboat.

  (b) Will the boat sink with 2 adults and 4 children?

  (c) Will the boat sink with 1 adult and 5 children?

71. Salesperson Leon sells two different brands of bicycles. For each ten-speed bicycle sold, he makes $50, and for each three-speed bicycle sold, he makes $40. Leon has a monthly goal of earning at least $4000.

  (a) Write a linear inequality that describes Leon's options for making his sales goal.

  (b) Will Leon make his sales goal if he sells 30 ten-speed bicycles and 70 three-speed bicycles next month?

  (c) Will Leon make his sales goal if he sells 50 ten-speed bicycles and 20 three-speed bicycles next month?

72. School Fundraiser For the Clinton Elementary School fundraiser, Guillermo earns 5 points for each magazine subscription he sells and 2 points for each novelty. It takes at least 40 points to earn the MP3 player that Guillermo wants.

  (a) Write a linear inequality that describes the various combinations of subscriptions m and novelties n that Guillermo can sell to earn enough points for the MP3 player.

  (b) Will he earn the MP3 player if he sells 8 novelties and 5 subscriptions?

  (c) Will he earn the MP3 player if he sells 20 novelties and 2 subscriptions?

Extending the Concepts

*Two inequalities joined by the word "and" or "or" form a **compound inequality**. The directions "Solve $3x - 2y > 6$ and $x + y < 2$" mean to find the ordered pairs that satisfy both inequalities at the same time. The solution of a compound inequality of this type is found by graphing the two linear inequalities in the same Cartesian coordinate plane and then shading the area* *where the graphs overlap. Graph the compound inequalities in Problems 73–78.*

73. $3x - 2y > 6$ and $x + y < 2$

74. $2x - 4y \le 4$ and $3x + 2y \ge 6$

75. $y > \dfrac{3}{4}x - 1$ and $x \ge 0$ **76.** $y > \dfrac{2}{3}x + 3$ and $y \ge 0$

77. $x < -3$ and $y \le 4$ **78.** $y < -1$ and $x \ge -2$

Explaining the Concepts

79. Describe how you can tell whether a point (a, b) is a solution to a linear inequality in two variables. How can you decide whether (a, b) lies on the boundary line?

80. Explain why $(0, 0)$ is generally a good test point. Describe how you can decide when this is a good test point to use and when you should choose a different test point.

The Graphing Calculator

Graphing calculators can be used to graph linear inequalities in two variables. The figure shows the graph of $3x + y < 7$ using a TI-84 Plus graphing calculator. To obtain the graph, we solve the inequality for y and obtain

$y < -3x + 7$. Graph the line $y = -3x + 7$ and shade below. Consult your owner's manual for specific keystrokes.

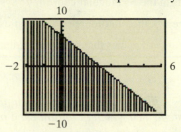

In Problems 81–92, graph the inequalities using a graphing calculator.

81. $y > 3$ **82.** $y < -2$

83. $y < 5x$ **84.** $y \ge \dfrac{2}{3}x$

85. $y > 2x + 3$ **86.** $y < -3x + 1$

87. $y \le \dfrac{1}{2}x - 5$ **88.** $y \ge -\dfrac{4}{3}x + 5$

89. $3x + y \le 4$ **90.** $-4x + y \ge -5$

91. $2x + 5y \le -10$ **92.** $3x + 4y \ge 12$

Chapter 3 Activity: Graphing Practice

Focus: Graphing and identifying the graphs of linear equations and inequalities.

Time: 15–20 minutes

Group size: 3–4

Materials needed: Blank piece of paper and graph paper for each group member.

1. On each of four pieces of paper, write one of the following equations or inequalities.

(a) $x + 3 \ge -4(y - 1)$

(b) $4(y - 2) + 5 = 5(x + 1)$

(c) $3(y - 7) + 4 = -2(x - 3) - 2$

(d) $\dfrac{2}{3}(x + 2y) + \dfrac{8}{3} \ge 0$

2. Place the four papers face down on the desk and mix them up. One by one, each group member should choose a piece of paper. Do not show your choice to the other members of the group.

3. Carefully graph the chosen equation/inequality on your graph paper. Do not label your graph.

4. When finished, place the unlabeled graphs in a pile.

5. As a group, with no help from the member who drew the graph, write each equation or inequality that was drawn.

6. As a group, discuss the outcome.

Chapter 3 Review

Section 3.1 The Rectangular Coordinate System and Equations in Two Variables

KEY TERMS

x-axis

y-axis

Origin

Rectangular or Cartesian coordinate
   system

xy-plane

Coordinate axes

Ordered pair (x, y)

Coordinates

x-coordinate

y-coordinate

Quadrants

Equation in two variables

Sides

Satisfy

You Should Be Able To...	EXAMPLE	Review Exercises
1 Plot points in the rectangular coordinate system (p. 169)	Examples 1 and 2	1–6
2 Determine whether an ordered pair satisfies an equation (p. 173)	Example 3	7, 8
3 Create a table of values that satisfy an equation (p. 175)	Examples 4 through 7	9–16

In Problems 1–4, plot the following points in the rectangular coordinate system. Tell which quadrant each point belongs to (or on which axis the point lies).

1. $A(3, -2)$ **2.** $B(-1, -3)$

3. $C(-4, 0)$ **4.** $D(0, 2)$

In Problems 5 and 6, identify the coordinates of each point labeled in the figure. Tell which quadrant each point belongs to (or on which axis the point lies).

5.

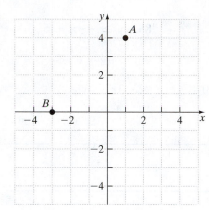

6.

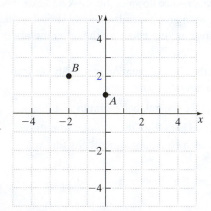

In Problems 7 and 8, determine whether the ordered pairs satisfy the given equation.

7. $y = 3x - 7$
    $A(-1, -10)$
    $B(-7, 0)$

8. $4x - 3y = 2$
    $A(2, 2)$
    $B(6, 5)$

9. Find an ordered pair that satisfies the equation $-x = 3y$ when

    (a) $x = 4$ **(b)** $y = 2$

10. Find an ordered pair that satisfies the equation $x - y = 0$ when

    (a) $x = 4$ **(b)** $y = -2$

In Problems 11–14, use the equation to complete the table. Use the table to list some of the ordered pairs that satisfy the equation.

11. $3x - 2y = 10$

x	y	(x, y)
−2		
0		
4		

12. $y = -x + 2$

x	y	(x, y)
−3		
2		
4		

13. $y = -\dfrac{1}{3}x - 4$

x	y	(x, y)
	2	
−6		
	−12	

14. $3x - 2y = 7$

x	y	(x, y)
	−8	
3		
	4	

15. Mail Order Shipping Martha purchases some clothes from a mail order catalogue. The shipping costs for her order will be $5.00 plus $2.00 per item purchased. The equation that calculates her total shipping cost, C, is $C = 5 + 2x$, where x is the number of items purchased. Complete the table below, and then plot the ordered pairs in a rectangular coordinate system.

x	C	(x, C)
1		
2		
3		

16. Department Store Wages Isabel works in a department store where her monthly earnings are $1000 plus 10% commission on her sales. Her earnings, E, are given by the equation $E = 1000 + 0.10x$, where x is Isabel's sales for the month. Complete the table below, and then plot the ordered pairs in a rectangular coordinate system.

x	E	(x, E)
500		
1000		
2000		

Section 3.2 Graphing Equations in Two Variables

KEY CONCEPTS

- **Linear Equation**

  A linear equation in two variables is an equation of the form $Ax + By = C$, where A, B, and C are real numbers. A and B cannot both be zero.

- **Procedure for Finding Intercepts**

  1. To find the x-intercept(s), if any, of the graph of an equation, let $y = 0$ in the equation and solve for x.

  2. To find the y-intercept(s), if any, of the graph of an equation, let $x = 0$ in the equation and solve for y.

- **Vertical Line**

  A vertical line is given by the equation $x = a$, where a is the x-intercept.

- **Horizontal Line**

  A horizontal line is given by the equation $y = b$, where b is the y-intercept.

KEY TERMS

Graph of an equation in two variables
Point-plotting method
Complete graph
Linear equation in two variables
Standard form of an equation of a line
Line
Intercepts
x-intercept
y-intercept

You Should Be Able To...	EXAMPLE	Review Exercises
❶ Graph a line by plotting points (p. 182)	Examples 1, 3, and 4	17–22
❷ Graph a line using intercepts (p. 187)	Examples 5 through 8	23–32
❸ Graph vertical and horizontal lines (p. 190)	Examples 9 and 10	33–36

In Problems 17–20, graph each linear equation using the point-plotting method.

17. $y = -2x$　　　　**18.** $y = x$

19. $4x + y = -2$　　**20.** $3x - y = -1$

21. Printing Cost The cost to print pamphlets for a new healthcare clinic is a \$40 set-up fee plus \$2.00 per pamphlet that will be printed. The equation that calculates the total cost for printing, C, is $C = 40 + 2p$, where p is the number of pamphlets to be printed. Complete the table below, and then graph the equation.

p	C	(p, C)
20		
50		
80		

22. Performance Fees The Crickets are the newest band in town and have a gig performing at a local concert. They agree that their total fee will be \$500 plus an additional \$3.00 per person who attends the concert. The equation that calculates the total fee for performing, F, is

$F = 500 + 3p$, where p is the number of people in attendance. Complete the table below, and then graph the equation.

p	F	(p, F)
100		
200		
500		

In Problems 23 and 24, find the intercepts of each graph.

23.

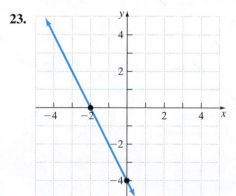

24.

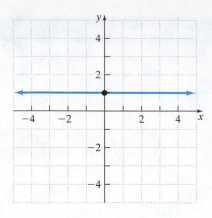

In Problems 25–28, find the x-intercept and y-intercept of each equation.

25. $-3x + y = 9$

26. $y = 2x - 6$

27. $x = 3$

28. $2x - 5y = 2$

In Problems 29–32, graph each linear equation by finding its intercepts.

29. $y - 3x = 3$

30. $2x + 5y = 0$

31. $\dfrac{x}{3} + \dfrac{y}{2} = 1$

32. $y = -\dfrac{3}{4}x + 3$

In Problems 33–36, graph each vertical or horizontal line.

33. $x = -2$

34. $y = 3$

35. $y = -4$

36. $x = 1$

Section 3.3 Slope

KEY CONCEPT

KEY TERMS

- **Slope**

  If $x_1 \neq x_2$, the slope m of the line containing (x_1, y_1) and (x_2, y_2)

  is defined by the formula $m = \dfrac{y_2 - y_1}{x_2 - x_1}$.

  The slope of a vertical line is undefined.

  The slope of a horizontal line is 0.

Run
Rise
Slope
Average rate of change

You Should Be Able To...	EXAMPLE	Review Exercises
❶ Find the slope of a line given two points (p. 195)	Examples 1 and 2	37–42
❷ Find the slope of vertical and horizontal lines (p. 199)	Examples 3 and 4	43–46
❸ Graph a line using its slope and a point on the line (p. 200)	Examples 5 and 6	47–50
❹ Work with applications of slope (p. 201)	Examples 7 and 8	51, 52

In Problems 37 and 38, find the slope of the line whose graph is shown.

37.

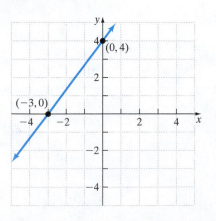

38.

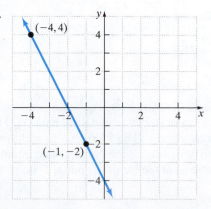

In Problems 39–46, find the slope of the line containing the given points.

39. $(-4, 6)$ and $(-3, -2)$

40. $(4, 1)$ and $(0, -7)$

41. $(-2, 3)$ and $(2, 13)$

42. $\left(-\frac{1}{2}, \frac{2}{3}\right)$ and $\left(\frac{3}{2}, \frac{1}{3}\right)$

43. $(-3, -6)$ and $(-3, -10)$

44. $(-5, -1)$ and $(-1, -1)$

45. $\left(\frac{3}{4}, \frac{1}{2}\right)$ and $\left(-\frac{1}{4}, \frac{1}{2}\right)$

46. $\left(\frac{1}{3}, -\frac{3}{5}\right)$ and $\left(\frac{3}{9}, -\frac{1}{5}\right)$

In Problems 47–50, graph the line that contains the and has the given slope.

47. $(-2, -3); m = 4$

48. $(1, -3); m = $

49. $(0, 1); m = -\dfrac{2}{3}$

50. $(2, 3); m = 0$

51. Production Cost The total cost to produce 20 bicy is \$1400, and the total cost to produce 50 bicycles is \$2750. Find and interpret the slope of the line containing the points $(20, 1400)$ and $(50, 2750)$.

52. Road Grade Scott is driving to the river on a road that falls 5 feet for every 100 feet of horizontal distance. What is the grade of this road? Express your answer as a percent.

Section 3.4 Slope-Intercept Form of a Line

KEY CONCEPT

- **Slope-intercept form of an equation of a line**

  An equation of a line with slope m and y-intercept b is $y = mx + b$.

You Should Be Able To...	EXAMPLE	Review Exercises
1 Use the slope-intercept form to identify the slope and y-intercept of a line (p. 205)	Examples 1 and 2	53–56
2 Graph a line whose equation is in slope-intercept form (p. 206)	Examples 3 and 4	57–62
3 Graph a line whose equation is in the form $Ax + By = C$ (p. 208)	Example 5	63, 64
4 Find the equation of a line given its slope and y-intercept (p. 208)	Example 6	65–70
5 Work with linear models in slope-intercept form. (p. 209)	Examples 7 and 8	71, 72

In Problems 53–56, find the slope and the y-intercept of the line whose equation is given.

53. $y = -x + \dfrac{1}{2}$

54. $y = x - \dfrac{3}{2}$

55. $3x - 4y = -4$

56. $2x + 5y = 8$

In Problems 57–64, use the slope and y-intercept to graph each line whose equation is given.

57. $y = \dfrac{1}{3}x + 1$

58. $y = -\dfrac{1}{2}x - 1$

59. $y = -\dfrac{2}{3}x - 2$

60. $y = \dfrac{3}{4}x + 3$

61. $y = x$

62. $y = -2x$

63. $2x - y = -4$

64. $-4x + 2y = 2$

In Problems 65–70, find the equation of the line with the given slope and intercept.

65. slope is -3; y-intercept is $(0, 5)$

66. slope is $\dfrac{1}{5}$; y-intercept is $(0, 10)$

67. slope is undefined; x-intercept is $(-12, 0)$

68. slope is 0; y-intercept is $(0, -4)$

69. slope is 1; y-intercept is $(0, -20)$

70. slope is -1; y-intercept is $(0, -8)$

71. Car Rentals Hot Rod Car Rentals rents spor for \$120 plus \$80 per day. The equation tha the total cost C to rent a sports car is $C = $ where d is the number of days that the c

(a) How much will it cost to rent the c

(b) If the bill came to $680, how many days was the sports car rented out?

(c) Graph the equation in a rectangular coordinate system.

72. Computer Rentals Anthony plans to rent a computer to finish his master's thesis. There is no fixed fee, only a charge for each day the computer is checked out. If he keeps the computer for 22 days, he will pay $418. If his thesis advisor asks him to redo a section and he finds that he must keep the computer for 35 days, it will cost him $665.

(a) Write two ordered pairs (d, C), where d represents the number of days the computer is rented and C represents the cost. Calculate the slope of the line that contains these two points.

(b) Interpret the meaning of the slope of this line.

(c) Write an equation that represents the total cost for Anthony to rent a computer.

(d) Calculate the cost to rent a computer for 8 days.

Section 3.5 Point-Slope Form of a Line

KEY CONCEPT

- **Point-Slope Form of an Equation of a Line**
  An equation of a nonvertical line with slope m that contains the point (x_1, y_1) is $y - y_1 = m(x - x_1)$.

You Should Be Able To...	EXAMPLE	Review Exercises
❶ Find the equation of a line given a point and a slope (p. 214)	Examples 1 through 3	73–78
❷ Find the equation of a line given two points (p. 216)	Examples 4 and 5	79–84
❸ Build linear models using the point-slope form of a line (p. 218)	Example 6	85, 86

In Problems 73–78, find the equation of the line that contains the given point and the given slope. Write the equation in slope-intercept form, if possible.

73. $(0, -3)$; slope $= 6$

74. $(4, 0)$; slope $= -2$

75. $(3, -1)$; slope $= -\dfrac{1}{2}$

76. $(-1, -3)$; slope $= \dfrac{2}{3}$

77. $\left(-\dfrac{4}{3}, -\dfrac{1}{2}\right)$; slope $= 0$

78. $\left(-\dfrac{4}{7}, \dfrac{8}{5}\right)$; slope is undefined

Problems 79–84, find the equation of the line that contains the given points. Write the equation of the line slope-intercept form, if possible.

$(-7, 0)$ and $(0, 8)$

80. $(0, -6)$ and $(4, 0)$

81. $(3, 5)$ and $(-2, -10)$

82. $(-15, 1)$ and $(-5, -3)$

83. $(-2, 7)$ and $(5, 7)$

84. $(4, -3)$ and $(4, 0)$

85. Harvesting Hay Farmer Myers noted that at the end of 3 days of harvesting, 12 acres of hay remained in his field and at the end of 5 days, 4 acres remained. Use the ordered pairs $(3, 12)$ and $(5, 4)$ to write an equation that calculates how many acres of hay, A, are left to be harvested after d days.

86. Fish Population Pat works at Scripps Institute of Oceanography and is responsible for monitoring the fish population in various waters around the world. In one bay, the sunfish population was declining at a constant rate. His initial sample netted 15 sunfish and six months later, the same location netted 13 sunfish. Use the ordered pairs $(0, 15)$ and $(6, 13)$ to write an equation that will predict how many sunfish, F, the sample will yield after m months.

Section 3.6 Parallel and Perpendicular Lines

KEY TERMS

Parallel
Perpendicular

You Should Be Able To...	EXAMPLE	Review Exercises
1 Determine whether two lines are parallel (p. 222)	Examples 1 and 2	87, 88
2 Find the equation of a line parallel to a given line (p. 223)	Examples 3 and 4	89–94
3 Determine whether two lines are perpendicular (p. 224)	Examples 5 through 7	95–98
4 Find the equation of a line perpendicular to a given line (p. 226)	Examples 8 and 9	99–102

In Problems 87 and 88, determine whether the two lines are parallel.

87.
$$y = -\frac{1}{3}x + 2$$
$$x - 3y = 3$$

88.
$$y = \frac{1}{2}x - 4$$
$$x - 2y = 6$$

In Problems 89–94, find the equation of the line that contains the given point and is parallel to the given line. Write the equation in slope-intercept form, if possible.

89. $(3, -1); y = -x + 5$

90. $(-2, 4); y = 2x - 1$

91. $(-1, 10); 3x + y = -7$

92. $(4, -5); 6x + 4y = 8$

93. $(5, 19);$ y-axis

94. $(-1, -12);$ x-axis

In Problems 95 and 96, determine the slope of the line perpendicular to the given line.

95. $x - 2y = 5$

96. $4x - 9y = 1$

In Problems 97 and 98, determine whether the two lines are perpendicular.

97. $x + 3y = 3$
$\quad\quad y = 3x + 1$

98. $5x - 2y = 2$
$$y = \frac{2}{5}x + 12$$

In Problems 99–102, write the equation of the line that contains the given point and is perpendicular to the given line. Write the equation in slope-intercept form, if possible.

99. $(-3, 4); y = -3x + 1$

100. $(4, -1); y = 2x - 1$

101. $(1, -3); 2x - 3y = 6$

102. $\left(-\frac{3}{5}, \frac{2}{5}\right); x + y = -7$

Section 3.7 Linear Inequalities in Two Variables

KEY TERMS

Satisfied
Graph of a linear inequality in two variables

Half-planes
Test point

You Should Be Able To...	EXAMPLE	Review Exercises
1 Determine whether an ordered pair is a solution to a linear inequality (p. 231)	Example 1	103, 104
2 Graph linear inequalities (p. 232)	Examples 2 through 5	105–112
3 Solve problems involving linear inequalities (p. 236)	Example 6	113, 114

In Problems 103 and 104, determine which of the following points, if any, are solutions to the given linear inequality in two variables.

103. $y \leq 3x + 4$ $A(2, 0)$ $B(-4, -8)$ $C(7, 26)$

104. $y > \dfrac{1}{3}x + 4$ $A(6, -2)$ $B(0, 4)$ $C(-18, -1)$

In Problems 105–112, graph each linear inequality.

105. $y < -\dfrac{1}{4}x + 2$

106. $y > 2x - 1$

107. $3x + 2y \geq -6$

108. $-2x + y \geq 4$

109. $x - 3y \leq 0$

110. $x - 4y \geq 4$

111. $x < -3$

112. $y > 2$

113. Coin Jar A jar holding only quarters and dimes contains at least \$12. If there are x quarters and y dimes in the jar, write a linear inequality in two variables that describes how many of each type of coin are in the jar.

114. Number The difference between twice a number, x, and half of a second number, y, is at most 10. Write a linear inequality in two variables that describes these two numbers.

Chapter 3 Test

Remember to use your Chapter Test Prep Video to see fully worked-out solutions to any of these problems you would like to review.

1. Determine whether the ordered pair $(-3, -2)$ is a solution to the equation $3x - 4y = -17$.

2. Given the equation $3x - 9y = 12$, determine

  (a) the x-intercept

  (b) the y-intercept

3. Given the equation $4x - 3y = -24$, determine

  (a) the slope

  (b) the y-intercept

In Problems 4 and 5, graph each linear equation.

4. $y = -\dfrac{3}{4}x + 2$

5. $3x - 6y = -12$

6. Find the slope of the line that contains the points $(2, -2)$ and $(-4, -1)$.

7. Given the linear equation $3y = 2x - 1$, find

  (a) the slope of a line perpendicular to the given line.

  (b) the slope of a line parallel to the given line.

8. Determine whether the lines are parallel, perpendicular, or neither. Explain how you came to your conclusion.

$$L_1: 3x - 7y = 2$$
$$L_2: y = \frac{7}{3}x + 4$$

In Problems 9–15, write the equation of the line that satisfies the given conditions. Write the equation in slope-intercept form, if possible.

9. slope $= -4$ and y-intercept is $(0, -15)$

10. slope $= 2$ and contains $(-3, 8)$

11. contains $(-3, -2)$ and $(-4, 1)$

12. contains $(4, 0)$ and parallel to $y = \dfrac{1}{2}x + 2$

13. contains $(4, 2)$ and perpendicular to $4x - 6y = 5$

14. horizontal line through $(3, 5)$

15. contains $(-2, -1)$ with undefined slope

16. Shipping Packages The shipping department records indicate that during a week when 20 packages were shipped, the total expenses recorded for the shipping department were \$560. During a different week when 30 packages were shipped, the expenses recorded were \$640. Use the ordered pairs $(20, 560)$ and $(30, 640)$ to determine the average rate to ship a package.

In Problems 17–19, graph each linear inequality.

17. $y \geq x - 3$ **18.** $-2x - 4y < 8$ **19.** $x \leq -4$

Cumulative Review Chapters 1–3

In Problems 1–3, evaluate each expression.

1. $200 \div 25 \cdot (-2)$

2. $\dfrac{3}{4} + \dfrac{1}{6} - \dfrac{2}{3}$

3. $\dfrac{8 - 3(5 - 3^2)}{7 - 2 \cdot 6}$

4. Evaluate $x^3 + 3x^2 - 5x - 7$ for $x = -3$.

5. Simplify: $8m - 5m^2 - 3 + 9m^2 - 3m - 6$

In Problems 6 and 7, solve each equation.

6. $8(n + 2) - 7 = 6n - 5$

7. $\dfrac{2}{5}x + \dfrac{1}{6} = -\dfrac{2}{3}$

8. Solve $A = \dfrac{1}{2}h(b + B)$ for B.

In Problems 9 and 10, solve each linear inequality. Express the solution using set-builder notation and interval notation. Graph the solution set.

9. $6x - 7 > -31$ 10. $5(x - 3) \geq 7(x - 4) + 3$

11. Plot the following ordered pairs in the same Cartesian plane.

$$A(-3, 0) \quad B(4, -2) \quad C(1, 5)$$
$$D(0, 3) \quad E(-4, -5) \quad F(-5, 2)$$

In Problems 12 and 13, graph the linear equation using the method you prefer.

12. $y = -\dfrac{1}{2}x + 4$ 13. $4x - 5y = 15$

In Problems 14 and 15, find the equation of the line with the given properties. Express your answer in slope-intercept form.

14. Through the points $(3, -2)$ and $(-6, 10)$

15. Parallel to $y = -3x + 10$ and through the point $(-5, 7)$

16. Graph $x - 3y > 12$.

17. **Computing Grades** Shawn really wants an A in his geometry class. His four exam scores are 94, 95, 90, and 97. The final exam is worth two exam scores. To have an A, his average must be at least 93. What scores can Shawn score on the final exam to earn an A in the course?

18. **Body Mass Index** The body mass index (BMI) of a person 62 inches tall and weighing x pounds is given by $0.2x - 2$. A person with a BMI of 30 or more is considered to be obese. For what weights would a person 62 inches tall be considered obese?

19. **Supplementary Angles** Two angles are supplementary. The measure of the larger angle is 15 degrees more than twice the measure of the smaller angle. Find the angle measures.

20. **Cylinders** Max has 100 square inches of aluminum with which to make a closed cylinder. If the radius of the cylinder must be 2 inches, how tall will the cylinder be? (Round to the nearest hundredth of an inch.)

21. **Consecutive Integers** Find three consecutive even integers such that the sum of the first two is 22 more than the third.

4 Systems of Linear Equations and Inequalities

Can you eat a healthy meal at a fast-food restaurant? The answer is "yes" if you know how to count your carbohydrates! See Problems 69 and 70 in Section 4.3.

The Big Picture: Putting It Together

In Chapter 2, we solved linear equations and inequalities in one variable. Recall that linear equations in one variable can have no solution (a contradiction), one solution (a conditional equation), or infinitely many solutions (an identity). In Chapter 3, we learned how to graph both linear equations and linear inequalities in two variables. Remember, the graph of a linear equation in two variables represents the set of all points whose ordered pairs satisfy the equation.

In this chapter, we will discuss methods for finding solutions that simultaneously satisfy two linear equations involving two variables. We are going to learn a graphical method for finding the solution and two algebraic methods. These systems can have no solution, one solution, or infinitely many solutions, just like linear equations in one variable. We conclude the chapter by looking at systems of linear inequalities. These systems require us to determine the region that satisfies two or more linear inequalities simultaneously.

Outline

4.1 Solving Systems of Linear Equations by Graphing

Objectives

1 Determine Whether an Ordered Pair Is a Solution of a System of Linear Equations

2 Solve a System of Linear Equations by Graphing

3 Classify Systems of Linear Equations as Consistent or Inconsistent

4 Solve Applied Problems Involving Systems of Linear Equations

Are You Ready for This Section?

Before getting started, take the following readiness quiz. If you get a problem wrong, go back to the section cited and review the material.

R1. Graph: $y = 2x - 3$ [Section 3.4, pp. 206–208]

R2. Graph: $3x + 4y = 12$ [Section 3.2, pp. 187–190]

R3. Determine whether $2x + 6y = 12$ is parallel to $-3x - 9y = 18$.   [Section 3.6, pp. 222–223]

In Section 3.2, we learned that an equation in two variables is linear provided that it can be written in the form $Ax + By = C$, where A and B cannot both be zero. We now learn methods for finding ordered pairs that satisfy two linear equations at the same time.

> **Definition**
>
> A **system of linear equations** is a grouping of two or more linear equations where each equation contains one or more variables.

Example 1 gives some examples of systems of linear equations containing two equations in two variables.

Ready?...Answers

R1.

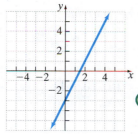

R2.

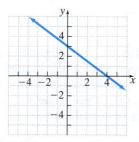

R3. Parallel

EXAMPLE 1 **Examples of Systems of Linear Equations**

(a) $\begin{cases} 2x + y = 5 \\ x - 5y = -10 \end{cases}$ Two linear equations containing two variables, x and y

(b) $\begin{cases} a + b = 5 \\ 2a - 3b = -3 \end{cases}$ Two linear equations containing two variables, a and b

We use a brace, as shown in the systems in Example 1, to remind us that we are dealing with a system of equations.

▶ **1** Determine Whether an Ordered Pair Is a Solution of a System of Linear Equations

> **Definition**
>
> A **solution** of a system of equations consists of values of the variables that satisfy each equation in the system. We represent the solution of two linear equations containing two unknowns as an ordered pair, (x, y).

To determine whether an ordered pair is a solution of a system of equations, we replace each variable with its value in each equation. If *both* equations are satisfied, the ordered pair is a solution.

EXAMPLE 2 **Determining Whether an Ordered Pair Is a Solution of a System of Linear Equations**

Determine whether each ordered pair is a solution of the system of equations.

$$\begin{cases} x + 3y = 9 \\ 4x - 2y = 8 \end{cases}$$

(a) $(-3, 4)$ **(b)** $(3, 2)$

Solution

(a) Let $x = -3$ and $y = 4$ in both equations. If both equations are true, then $(-3, 4)$ is a solution.

First equation: $\qquad\qquad x + 3y = 9$ $\qquad$ Second equation: $\qquad\qquad 4x - 2y = 8$

$x = -3, y = 4$: $\qquad -3 + 3(4) \overset{?}{=} 9$ $\qquad\qquad x = -3, y = 4$: $\quad 4(-3) - 2(4) \overset{?}{=} 8$

$\qquad\qquad\qquad\qquad -3 + 12 \overset{?}{=} 9$ $\qquad\qquad\qquad\qquad\qquad\qquad -12 - 8 \overset{?}{=} 8$

$\qquad\qquad\qquad\qquad\qquad 9 = 9$ $\quad$ True $\qquad\qquad\qquad\qquad\qquad\qquad -20 = 8$ $\quad$ False

Although $(-3, 4)$ satisfies the first equation, it does not satisfy the second equation. Because the values $x = -3$ and $y = 4$ do not satisfy both equations, $(-3, 4)$ is not a solution of the system.

(b) Let $x = 3$ and $y = 2$ in both equations. If both equations are true, then $(3, 2)$ is a solution.

First equation: $\qquad\qquad x + 3y = 9$ $\qquad$ Second equation: $\qquad\qquad 4x - 2y = 8$

$x = 3, y = 2$: $\qquad\quad 3 + 3(2) \overset{?}{=} 9$ $\qquad\qquad x = 3, y = 2$: $\quad 4(3) - 2(2) \overset{?}{=} 8$

$\qquad\qquad\qquad\qquad\quad 3 + 6 \overset{?}{=} 9$ $\qquad\qquad\qquad\qquad\qquad\qquad\quad 12 - 4 \overset{?}{=} 8$

$\qquad\qquad\qquad\qquad\qquad 9 = 9$ $\quad$ True $\qquad\qquad\qquad\qquad\qquad\qquad\quad 8 = 8$ $\quad$ True

Because $(3, 2)$ satisfies both equations, it is a solution of the system. $\qquad\qquad\qquad\qquad$ ●

Quick ✓

1. A _____ __ ____ _____ is a grouping of two or more linear equations, each of which contains one or more variables.

2. A _____ of a system of equations consists of values of the variables that satisfy each equation of the system.

In Problems 3 and 4, determine whether the given set of ordered pairs is a solution of the system of equations.

3. $\begin{cases} 2x + 3y = 7 \\ 3x + y = -7 \end{cases}$ $\qquad\qquad$ **4.** $\begin{cases} 3x - 6y = 6 \\ -2x + 4y = -4 \end{cases}$

$\quad$ **(a)** $(3, 1)$ $\quad$ **(b)** $(-4, 5)$ $\quad$ **(c)** $(-2, -1)$ $\qquad$ **(a)** $(2, 0)$ $\qquad$ **(b)** $(0, -1)$ $\qquad$ **(c)** $(4, 1)$

▶ ② Solve a System of Linear Equations by Graphing

The graph of each equation in a system of linear equations in two variables is a line. So a system of two linear equations containing two variables represents a pair of lines. We know from Chapter 3 that the graph of an equation represents the set of all ordered pairs that make the equation a true statement. Therefore, **the point, or points, at which two lines intersect, if any, represents the solution of the system of equations because both equations are satisfied at this point.**

Let's look at an example to learn how to solve a system of linear equations by graphing.

EXAMPLE 3 $\quad$ **How to Solve a System of Linear Equations by Graphing**

Solve the following system by graphing: $\begin{cases} 3x + y = 7 \\ -2x + 3y = -12 \end{cases}$

Step-by-Step Solution

Step 1: Graph the first equation in the system.

We graph the equation $3x + y = 7$ by putting the equation in slope-intercept form, $y = mx + b$.

$$3x + y = 7$$
$$y = -3x + 7$$

We graph the line whose slope is -3 and y-intercept is $(0, 7)$. See Figure 1(a).

Figure 1

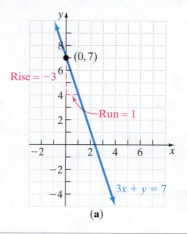

(a)

Step 2: Graph the second equation in the system in the same rectangular coordinate system as the equation in Step 1.

We graph the line $-2x + 3y = -12$ using intercepts because the equation is in standard form and we can find the intercepts easily.

x-intercept: Let $y = 0$ and solve for x.

$$-2x + 3y = -12$$
$$-2x + 3(0) = -12$$
$$-2x = -12$$
$$x = 6$$

y-intercept: Let $x = 0$ and solve for y.

$$-2x + 3y = -12$$
$$-2(0) + 3y = -12$$
$$3y = -12$$
$$y = -4$$

On the same coordinate system, plot the points $(6, 0)$ and $(0, -4)$ and draw a line through them. See Figure 1(b).

Figure 1

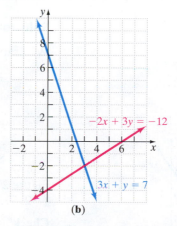

(b)

Step 3: Determine the point of intersection of the two lines, if any.

From the graphs in Figure 1(b), the lines appear to intersect at $(3, -2)$. We label this point in Figure 1(c) on the next page.

(continued)

Figure 1(continued)

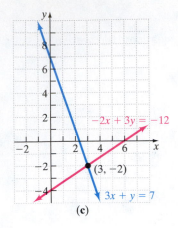

(c)

Step 4: Verify that the point of intersection determined in Step 3 is a solution of the system.

Let $x = 3$ and $y = -2$ in both equations.

First equation: $3x + y = 7$

$3(3) + (-2) \stackrel{?}{=} 7$

$9 - 2 \stackrel{?}{=} 7$

$7 = 7$ True

Second equation: $-2x + 3y = -12$

$-2(3) + 3(-2) \stackrel{?}{=} -12$

$-6 - 6 \stackrel{?}{=} -12$

$-12 = -12$ True

The solution of the system is $(3, -2)$.

Finding the Solution of A System of Linear Equations by Graphing

Step 1: Graph the first equation in the system.

Step 2: Graph the second equation in the system in the same rectangular coordinate system.

Step 3: Determine the point of intersection, if any. The point of intersection is the solution of the system.

Step 4: Verify that the point of intersection found in Step 3 is the solution of the system by substituting the ordered pair in both equations.

EXAMPLE 4   **Solving a System of Linear Equations Graphically**

Solve the following system by graphing: $\begin{cases} 3x - 2y = -16 \\ 5x + 2y = 0 \end{cases}$

Solution

We graph both equations by putting each equation in slope-intercept form.

First equation: $3x - 2y = -16$

Subtract 3x from both sides: $-2y = -3x - 16$

Divide both sides by -2: $y = \dfrac{3}{2}x + 8$

Slope: $\dfrac{3}{2}$; y-intercept: $(0, 8)$

Second equation: $5x + 2y = 0$

Subtract 5x from both sides: $2y = -5x$

Divide both sides by 2: $y = -\dfrac{5}{2}x$

Slope: $-\dfrac{5}{2}$; y-intercept: $(0, 0)$

Figure 2 shows the graph of both equations. The point of intersection appears to be $(-2, 5)$. Now, we check to see whether the solution is $x = -2, y = 5$.

Check

First equation: $3x - 2y = -16$

$x = -2, y = 5$: $3(-2) - 2(5) \stackrel{?}{=} -16$

$-6 - 10 \stackrel{?}{=} -16$

$-16 = -16$ True

Second equation: $5x + 2y = 0$

$x = -2, y = 5$: $5(-2) + 2(5) \stackrel{?}{=} 0$

$-10 + 10 \stackrel{?}{=} 0$

$0 = 0$ True

The point $(-2, 5)$ satisfies both equations, so the solution is $(-2, 5)$.

Figure 2

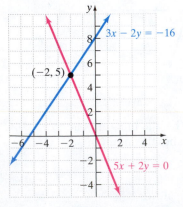

5. $\begin{cases} y = -2x + 9 \\ y = 3x - 11 \end{cases}$

6. $\begin{cases} 4x + y = -3 \\ 5x - 2y = -20 \end{cases}$

Sometimes the lines that make up a system of linear equations do not intersect at all.

▶ **EXAMPLE 5** **Solving a System of Linear Equations That Has No Point of Intersection**

Solve the system by graphing: $\begin{cases} 3x - y = -4 \\ -6x + 2y = 2 \end{cases}$

Solution

We graph both equations by putting each equation in slope-intercept form.

First equation: $3x - y = -4$ Second equation: $-6x + 2y = 2$

Subtract 3x from both sides: $-y = -3x - 4$ Add 6x to both sides: $2y = 6x + 2$

Divide both sides by -1: $y = 3x + 4$ Divide both sides by 2: $y = 3x + 1$

Slope: 3; y-intercept: $(0, 4)$ Slope: 3; y-intercept: $(0, 1)$

Figure 3

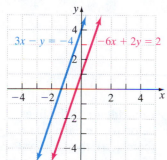

$3x - y = -4$ $-6x + 2y = 2$

Figure 3 shows the graph of both equations. The lines are parallel and therefore do not intersect. The system of equations has no solution. We could also say the solution set is the empty set, $\{\ \}$ or $\varnothing$.

7. $\begin{cases} y = 2x + 4 \\ 2x - y = 1 \end{cases}$

Another special case with systems of linear equations in two variables occurs when the graph of each equation in the system is the same line.

EXAMPLE 6 **Solving a System of Linear Equations Where the Graph of Each Equation Is the Same Line**

Solve the system by graphing: $\begin{cases} 5x + 2y = -4 \\ 10x + 4y = -8 \end{cases}$

Solution

We graph both equations in the system by putting each equation in slope-intercept form.

First equation: $5x + 2y = -4$ Second equation: $10x + 4y = -8$

Subtract 5x from both sides: $2y = -5x - 4$ Subtract 10x from both sides: $4y = -10x - 8$

Divide both sides by 2: $y = -\dfrac{5}{2}x - 2$ Divide both sides by 4: $y = -\dfrac{5}{2}x - 2$

Slope: $-\dfrac{5}{2}$; y-intercept: $(0, -2)$ Slope: $-\dfrac{5}{2}$; y-intercept: $(0, -2)$

Figure 4

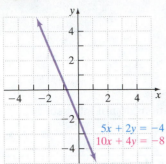

Notice that the slopes and the *y*-intercepts of the two lines are the same. Figure 4 shows the graph of each equation. The equations are the same, so the lines coincide. Any point on the graph of $5x + 2y = -4$ is also a point on the graph of $10x + 4y = -8$. For example, the ordered pairs $(-2, 3)$, $(0, -2)$, $(2, -7)$, and $(4, -12)$ all satisfy both equations in the system. In fact, the system of equations has infinitely many solutions, which are represented by all points on the line $5x + 2y = -4$. ●

Quick ✔

In Problem 8, solve the system by graphing.

8. $\begin{cases} y = 3x + 2 \\ -6x + 2y = 4 \end{cases}$

▶ ❸ Classify Systems of Linear Equations as Consistent or Inconsistent

Examples 3–6 show that systems of linear equations can have one solution, no solution, or infinitely many solutions. In Examples 3 and 4, the system had exactly one solution. In Example 5, the system had no solution. In Example 6, the system had infinitely many solutions. We classify systems of equations based on the number of solutions they have.

> **Definitions**
>
> A **consistent system** of equations has at least one solution.
>
> An **inconsistent system** of equations has no solution.

We now summarize these possibilities.

> **Classifying A System of Equations Graphically**
>
> **Intersect:** If the lines intersect, then the system of equations has one solution given by the point of intersection. We say that the system is **consistent** and the equations are **independent**. See Figure 5(a).
>
> **Parallel:** If the lines are parallel, then the system of equations has no solution because the lines do not intersect. We say that the system is **inconsistent.** See Figure 5(b).
>
> **Coincident:** If the lines lie on top of each other (coincide), then the system of equations has infinitely many solutions. The solution set is the set of all points on the line. The system is **consistent** and the equations are **dependent.** See Figure 5(c).
>
> **Figure 5**
>
>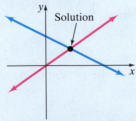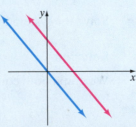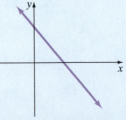
>
> (a) Intersecting lines; system has exactly one solution
>
> (b) Parallel lines; system has no solution
>
> (c) Coincident lines; system has infinitely many solutions

In addition to graphing the lines in a system to classify the system, we can also classify a system algebraically.

> **Classifying A System of Equations Algebraically**
>
> **Step 1:** Write each equation in the system in slope-intercept form.
>
> **Step 2: (a)** If the equations of the lines in the system have different slopes, then the lines will intersect. The point of intersection represents the solution. The system is consistent and the equations are independent. See Examples 3 and 4.
>
> **(b)** If the equations of the lines have the same slope but different y-intercepts, then the lines are parallel. The system has no solution. The system is inconsistent. See Example 5.
>
> **(c)** If the equations of the lines have the same slope and the same y-intercept, then the lines coincide. The system has infinitely many solutions. The system is consistent and the equations are dependent. See Example 6.

EXAMPLE 7   **Determining the Number of Solutions in a System**

Without graphing, determine the number of solutions of the system:

$$\begin{cases} 12x + 4y = -16 \\ -9x - 3y = 3 \end{cases}$$

State whether the system is consistent or inconsistent. If the system is consistent, state whether the equations are dependent or independent.

Solution

Write the equations in slope-intercept form. Then determine the slope and y-intercept to determine the number of solutions.

First equation: $12x + 4y = -16$

Subtract 12x from both sides: $4y = -12x - 16$

Divide both sides by 4: $y = -3x - 4$

Second equation: $-9x - 3y = 3$

Add 9x to both sides: $-3y = 9x + 3$

Divide both sides by -3: $y = -3x - 1$

The slope of both equations is -3. The y-intercept of the first equation is $(0, -4)$ and the y-intercept of the second equation is $(0, -1)$. The equations have the same slope but different y-intercepts, so the lines are parallel. Therefore, the system has no solution and is inconsistent. ●

Quick ✓

9. If a system of linear equations in two variables has at least one solution, the system is _____.

10. If a system of linear equations in two variables has exactly one solution, the equations are _____.

11. A system of linear equations in two variables that has no solution is _____.

12. *True or False* The visual representation of a consistent system consisting of dependent equations is one line.

In Problems 13–15, without graphing, determine the number of solutions of each system. State whether the system is consistent or inconsistent. For those systems that are consistent, state whether the equations are dependent or independent.

13. $\begin{cases} 7x - 2y = 4 \\ 2x + 7y = 7 \end{cases}$
    14. $\begin{cases} 6x + 4y = 4 \\ -12x - 8y = -8 \end{cases}$
    15. $\begin{cases} 3x - 4y = 8 \\ -6x + 8y = 8 \end{cases}$

▶ ④ Solve Applied Problems Involving Systems of Linear Equations

Many times we can use a system of two linear equations with two variables to answer a question such as "Which long-distance telephone plan should I choose?"

EXAMPLE 8   **Phone Plans**

A company offers two domestic long-distance phone plans. The "Just Call" plan charges $3.00 per month plus $0.03 per minute. The "Value Plus Flat Rate" plan has no monthly service fee and charges $0.05 per minute. Determine the number of minutes of long-distance phone calls that can be used each month in order for the two plans to charge the same fee. What is the fee for this number of minutes?

Solution

Step 1: Identify We want to know the number of minutes of long-distance phone calls so that the two plans have the same fee.

Step 2: Name Let m represent the number of minutes used and C represent the monthly cost.

Step 3: Translate For the "Just Call" plan, talking for 0 minutes costs $3.00; talking for 1 minute costs $3.00 plus 1($0.03), or $3.03; talking for 2 minutes costs of $3.00 plus 2($0.03), or $3.06. To generalize, the monthly cost for the "Just Call" plan is $3.00 plus $0.03 times the number of minutes used. This leads to the equation

$$C = 0.03m + 3.00 \quad \text{Equation (1)}$$

For the "Value Plus Flat Rate" plan, talking for 0 minutes costs $0; talking for 1 minute costs $0.05; talking for 2 minutes costs $0.10. To generalize, the monthly cost for the "Value Plus Flat Rate" plan is $0.05 times the number of minutes used. This leads to the equation

$$C = 0.05m \quad \text{Equation (2)}$$

We use equations (1) and (2) to form the following system:

$$\begin{cases} C = 0.03m + 3.00 \\ C = 0.05m \end{cases} \quad \text{The Model}$$

Step 4: Solve To draw the graph of each equation in the system, we create Table 1, which shows the monthly fee for various minutes used for each plan. Figure 6 shows the graph of the two equations in the system with the horizontal axis labeled m and the vertical axis labeled C.

Table 1

m	Cost for "Just Call" Plan	Cost for "Value Plus Flat Rate" Plan
0	$3.00	$ 0
100	$6.00	$ 5.00
200	$9.00	$10.00

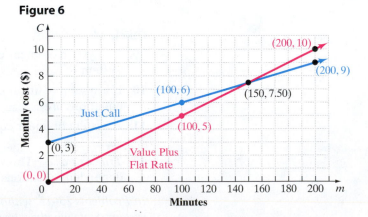

Figure 6

The graph shows that the monthly charge will be the same for 150 minutes of calls in a month. We also see that $C = 7.50$ when $m = 150$. So 150 minutes cost $7.50.

Step 5: Check For the "Just Call" plan, the bill for 150 minutes is $3.00 plus $0.03 times 150, or $3.00 plus $4.50, for a total bill of $7.50. For the "Value Plus Flat Rate" plan, the bill for 150 minutes is $0.05 times 150, or $7.50. Both bills are the same. Our answer checks.

Step 6: Answer The two plans result in the same monthly bill, $7.50, when 150 minutes are used.

> **Quick ✓**
>
> **16.** John needs to rent a 12-foot moving truck for a day. He calls two rental companies to determine their fee structure. EZ-Rental charges $20 plus $0.40 per mile, and U-Move-It charges $35 plus $0.25 per mile. After how many miles will the cost of renting the truck be the same? What is the cost at this mileage?

4.1 Exercises MyMathLab® MathXP PRACTICE

Exercise numbers in **green** have complete video solutions in MyMathLab.

*Problems **1–16** are the* **Quick ✓***s that follow the* **EXAMPLES***.*

Building Skills

In Problems 17–22, determine whether the ordered pair is a solution of the system of equations. See Objective 1.

17. $\begin{cases} x - y = -4 \\ 3x + y = -4 \end{cases}$

(a) $(2, 6)$
(b) $(-2, 2)$
(c) $(2, -2)$

18. $\begin{cases} 2x + 5y = 0 \\ x - 3y = 11 \end{cases}$

(a) $(5, -2)$
(b) $(-5, 2)$
(c) $(2, -3)$

19. $\begin{cases} 3x - y = 2 \\ -15x + 5y = -10 \end{cases}$

(a) $(1, -1)$
(b) $(-2, -8)$
(c) $(0, -2)$

20. $\begin{cases} x - 4y = 2 \\ 5x - 8y = -6 \end{cases}$

(a) $(-8, -2)$
(b) $(2, 2)$
(c) $\left(-\dfrac{10}{3}, -\dfrac{4}{3}\right)$

21. $\begin{cases} 6x - 2y = 1 \\ y = 3x + 2 \end{cases}$

(a) $\left(0, -\dfrac{1}{2}\right)$
(b) $(-2, -4)$
(c) $(0, 2)$

22. $\begin{cases} y = -4x - 1 \\ 2y + 8x = 2 \end{cases}$

(a) $\left(-\dfrac{1}{2}, 0\right)$
(b) $(0, 1)$
(c) $(-1, 0)$

In Problems 23–38, solve each system of equations by graphing. See Objective 2.

23. $\begin{cases} y = x + 5 \\ y = -\dfrac{1}{5}x - 1 \end{cases}$

24. $\begin{cases} y = -\dfrac{2}{3}x - 2 \\ y = \dfrac{1}{2}x + 5 \end{cases}$

25. $\begin{cases} y = \dfrac{3}{4}x - 4 \\ y = -\dfrac{1}{2}x + 1 \end{cases}$

26. $\begin{cases} y = -4x + 6 \\ y = -2x \end{cases}$

27. $\begin{cases} 2x - y = -1 \\ 3x + 2y = -5 \end{cases}$

28. $\begin{cases} x + 3y = 0 \\ x - y = 8 \end{cases}$

29. $\begin{cases} x - 4 = 0 \\ 3x - 5y = 22 \end{cases}$

30. $\begin{cases} y + 1 = 0 \\ x + 4y = -6 \end{cases}$

31. $\begin{cases} 3x - y = -1 \\ -6x + 2y = -4 \end{cases}$

32. $\begin{cases} -2x + 5y = -20 \\ 4x - 10y = 10 \end{cases}$

33. $\begin{cases} x + y = -2 \\ 3x - 4y = 8 \end{cases}$

34. $\begin{cases} 2x + 5y = 2 \\ -x + 2y = -1 \end{cases}$

35. $\begin{cases} 2y = 6 - 4x \\ 6x = 9 - 3y \end{cases}$

36. $\begin{cases} 3y = -4x - 4 \\ 8x + 2 = -6y - 6 \end{cases}$

37. $\begin{cases} y = -x + 3 \\ 3y = 2x + 9 \end{cases}$

38. $\begin{cases} 3y = 2x + 1 \\ y = 5x - 4 \end{cases}$

In Problems 39–46, determine the number of solutions of each system. State whether the system is consistent or inconsistent. For those systems that are consistent, state whether the equations are dependent or independent. State the solution of each system. See Objective 3.

39.

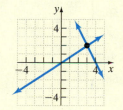

40.

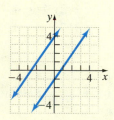

41.

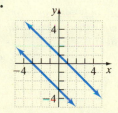

42.

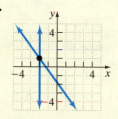

43.

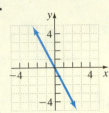

44.

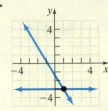

45.

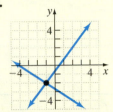

46.

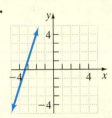

61. $\begin{cases} x - 2y = 6 \\ 2x - 4y = 0 \end{cases}$

62. $\begin{cases} 2x + 3y = -1 \\ 4x + 6y = -6 \end{cases}$

63. $\begin{cases} 3x - 2y = -2 \\ 2x + y = 8 \end{cases}$

64. $\begin{cases} 3x - y = -11 \\ -2x + y = 9 \end{cases}$

65. $\begin{cases} y = x + 2 \\ 3x - 3y = -6 \end{cases}$

66. $\begin{cases} -2x + 6y = -6 \\ 3x - 9y = 9 \end{cases}$

67. $\begin{cases} -5x - 2y = -2 \\ 10x + 4y = 4 \end{cases}$

68. $\begin{cases} 3x + y = -2 \\ 6x + 2y = -4 \end{cases}$

69. $\begin{cases} 3x = 4y - 12 \\ 2x = -4y - 8 \end{cases}$

70. $\begin{cases} 2x = -5y - 25 \\ -3x = 10y + 50 \end{cases}$

In Problems 47–58, without graphing, determine the number of solutions of each system of equations. State whether the system is consistent or inconsistent. For those systems that are consistent, state whether the equations are dependent or independent. See Objective 3.

47. $\begin{cases} y = 2x + 3 \\ y = -2x + 3 \end{cases}$

48. $\begin{cases} y = -x + 5 \\ y = x - 5 \end{cases}$

49. $\begin{cases} 6x + 2y = 12 \\ 3x + y = 12 \end{cases}$

50. $\begin{cases} 2x + y = -3 \\ -2x - y = 6 \end{cases}$

51. $\begin{cases} x + 2y = 2 \\ 2x + 4y = 4 \end{cases}$

52. $\begin{cases} -x + y = 5 \\ x - y = -5 \end{cases}$

53. $\begin{cases} y = 4 \\ x = 4 \end{cases}$

54. $\begin{cases} y - 3 = 0 \\ x - 3 = 0 \end{cases}$

55. $\begin{cases} x - 2y = -4 \\ -x + 2y = -4 \end{cases}$

56. $\begin{cases} 5x + y = -1 \\ 10x + 2y = -2 \end{cases}$

57. $\begin{cases} x + 2y = 4 \\ x + 1 = 5 - 2y \end{cases}$

58. $\begin{cases} 2y + 6 = 2x \\ x + 4 = y - 7 \end{cases}$

Mixed Practice

In Problems 59–70, solve each system of equations by graphing. Based on the graph, state whether the system is consistent or inconsistent. For those systems that are consistent, state whether the equations are dependent or independent.

59. $\begin{cases} y = 2x + 9 \\ y = -3x - 6 \end{cases}$

60. $\begin{cases} y = 3x - 5 \\ y = -x + 11 \end{cases}$

Applying the Concepts

*In business, a company's **break-even point** is the point at which the costs of production of a product will be equal to the amount of revenue from the sale of the product. In Problems 71–74, solve the system of equations by graphing to find the number of units that need to be produced for the company to break even.*

71. Producing T-Shirts The San Diego T-Shirt Company designs and sells T-shirts for promoters to sell at music concerts. The cost y to produce x T-shirts is given by the equation $y = 15x + 1000$, and the revenue y from the sale of these T-shirts is $y = 25x$. Find the break-even point: the point where the line that represents the cost, $y = 15x + 1000$, intersects the revenue line, $y = 25x$. Interpret the x-coordinate and the y-coordinate of the break-even point.

72. School Newspaper School Printers, Inc. prints and sells the school newspaper. The cost y to produce x newspapers is given by the equation $y = 0.05x + 8$, and the revenue y from the sale of these newspapers is given by the equation $y = 0.25x$. Find the break-even point: the point where the line that represents the cost, $y = 0.05x + 8$, intersects the revenue line, $y = 0.25x$. Interpret the x-coordinate and the y-coordinate of the break-even point.

73. Hockey Skates The Ice Blades Manufacturing Company produces and sells hockey skates to discount stores. The skates come in boxes of 10 pairs of skates. The cost y (in thousands of dollars) to produce x boxes of skates is given by the equation $y = 0.8x + 3.5$, and the revenue y (in thousands of dollars) from the sale of these skates is given by the equation $y = 1.5x$. Find the break-even point.

74. Patio Chairs Maumee Lumber Company produces wood patio chairs and sells them to local hardware and home improvement stores. The cost y (in hundreds of dollars) to produce x patio chairs is given by the equation $y = 0.5x + 6$. The revenue y (in hundreds of dollars) from the sale of x patio chairs is given by the equation $y = 1.5x$. Find the break-even point.

In Problems 75–78, set up a system of linear equations in two variables that models the problem. Then solve the system of linear equations.

75. Car Rental The Wheels-to-Go car rental agency rents cars for $50 daily plus $0.20 per mile. Acme Rental agency will rent the same car for $62 daily plus $0.12 per mile. On Liza's trip to Houston, she decides to rent a car. Determine the number of miles for which the cost of the car rental will be the same for both companies. If Liza plans to drive the car for 200 miles, which company should she use?

76. Car Rental The Speedy Car Rental agency charges $60 per day plus $0.14 per mile while the Slow-but-Cheap Car Rental agency charges $30 per day plus $0.20 per mile. Determine the number of miles for which the cost of the car rental will be the same for both companies. If you plan to drive the car for 400 miles, which company should you use?

77. Phone Charges A long-distance phone service provider has two different long-distance phone plans. Plan A charges a monthly fee of $8.95 plus $0.05 per minute. Plan B charges a monthly fee of $5.95 plus $0.07 per minute. Determine the number of minutes for which the cost of each plan will be the same. If you typically use 100 long-distance minutes each month, which plan should you choose?

78. Political Flyers A politician running for city council has found that PrintQuick can print pamphlets for her campaign for $0.04 per copy plus a one-time setup fee of $10. She has also learned that Print-A-Lot will print the same pamphlet for a flat fee of $20. Determine the number of political pamphlets that can be printed for the cost at PrintQuick to be the same as at Print-A-Lot.

Extending the Concepts

In Problems 79–82, determine the value of c so that the given system is consistent, but the equations are dependent.

79. $\begin{cases} 3x - y = -4 \\ \quad\ y = cx + 4 \end{cases}$

80. $\begin{cases} 2(x + 4) = 4y + 4 \\ \qquad\quad y = cx + 1 \end{cases}$

81. $\begin{cases} x + 3 = 3(x - y) \\ 2y + 3 = 2cx - y \end{cases}$

82. $\begin{cases} 5x + 2y = 3 \\ \qquad\ 4y = -10x + c \end{cases}$

The Graphing Calculator

A graphing calculator can be used to approximate the point of intersection of two equations by using its INTERSECT command. We illustrate this feature of a graphing calculator by solving the system $\begin{cases} x + y = -1 \\ -2x + y = -7 \end{cases}$.

Start by graphing each equation in the system as shown in Figure 7(a). Then use the INTERSECT command to find that the lines intersect at $x = 2$, $y = -3$. See Figure 7(b). The solution is the ordered pair $(2, -3)$.

Figure 7

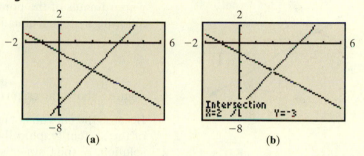

(a)   (b)

In Problems 83–88, use a graphing calculator to solve each system of equations.

83. $\begin{cases} y = 3x - 1 \\ y = -2x + 5 \end{cases}$

84. $\begin{cases} y = \dfrac{3}{2}x - 4 \\ y = -\dfrac{1}{4}x + 3 \end{cases}$

85. $\begin{cases} 3x - y = -1 \\ -4x + y = -3 \end{cases}$

86. $\begin{cases} -6x - 2y = 4 \\ 5x + 3y = -2 \end{cases}$

87. $\begin{cases} 4x - 3y = 1 \\ -8x + 6y = -2 \end{cases}$

88. $\begin{cases} -2x + 5y = -2 \\ 4x - 10y = -1 \end{cases}$

Explaining the Concepts

89. Describe graphically the three possibilities for a solution of a system of two linear equations containing two variables.

90. When solving a system of linear equations using the graphing method, you find that the solution appears to be $(3, 4)$. How do you verify that $(3, 4)$ is the solution? Graphically, what does this solution represent?

4.2 Solving Systems of Linear Equations Using Substitution

Objectives

1 Solve a System of Linear Equations Using the Substitution Method

2 Solve Applied Problems Involving Systems of Linear Equations

Are You Ready for This Section?

Before getting started, take the following readiness quiz. If you get a problem wrong, go back to the section cited and review the material.

R1. Solve $3x - y = 2$ for y.   [Section 2.4, pp. 113–115]

R2. Solve $2x + 5y = 8$ for x.   [Section 2.4, pp. 113–115]

R3. Solve: $3x - 2(5x + 1) = 12$ [Section 2.2, pp. 92–94]

Figure 8

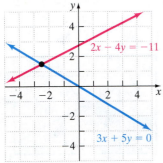

Ready?...Answers   **R1.** $y = 3x - 2$

R2. $x = -\dfrac{5}{2}y + 4$ **R3.** $\{-2\}$

1 Solve a System of Linear Equations Using the Substitution Method

Obtaining an exact result using the graphing method can be difficult if the x- and y-coordinates of the point of intersection of two lines are not integers. Consider the following system of equations:

$$\begin{cases} 3x + 5y = 0 \\ 2x - 4y = -11 \end{cases}$$

Figure 8 shows the graph of the equations in the system. The lines intersect at $\left(-\dfrac{5}{2}, \dfrac{3}{2}\right)$.

Because the lines do not intersect at integer values, it is difficult to determine the solution of the system graphically. Therefore, rather than using graphical methods to obtain solutions of some systems, we prefer to use algebraic methods. One algebraic method is the *method of substitution*.

EXAMPLE 1   **How to Solve a System of Two Equations in Two Variables by Substitution**

Solve the following system by substitution: $\begin{cases} y = -2x + 10 \\ 3x + 5y = 8 \end{cases}$

Step-by-Step Solution

First, label the equations (1) and (2) as shown.

$$\begin{cases} y = -2x + 10 & (1) \\ 3x + 5y = 8 & (2) \end{cases}$$

Step 1: Solve one of the equations for one of the unknowns.   Equation (1) is already solved for y: $y = -2x + 10$

Step 2: Substitute $y = -2x + 10$ into equation (2).   Equation (2): $3x + 5y = 8$

$3x + 5(-2x + 10) = 8$

Step 3: Solve the equation for x.

Distribute the 5: $3x - 10x + 50 = 8$

Combine like terms: $-7x + 50 = 8$

Subtract 50 from both sides: $-7x = -42$

Divide both sides by -7: $x = 6$

The x-coordinate of the solution is 6.

Step 4: Let $x = 6$ in equation (1) to find the value of y.

Equation (1): $y = -2x + 10$

$x = 6$: $y = -2(6) + 10$

$y = -12 + 10$

$y = -2$

The y-coordinate of the solution is -2.

Step 5: Verify that $x = 6$ and $y = -2$ is the solution.

Equation (1):
$$y = -2x + 10$$
$$-2 \overset{?}{=} -2(6) + 10$$
$$-2 \overset{?}{=} -12 + 10$$
$$-2 = -2 \quad \text{True}$$

Equation (2):
$$3x + 5y = 8$$
$$3(6) + 5(-2) \overset{?}{=} 8$$
$$18 - 10 \overset{?}{=} 8$$
$$8 = 8 \quad \text{True}$$

Figure 9

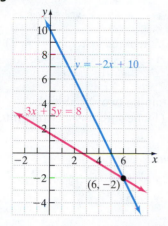

Both equations are satisfied, so the solution of the system is the ordered pair $(6, -2)$. ●

Figure 9 shows the graph of the equations in the system from Example 1. The point of intersection is $(6, -2)$, as we would expect.

Below we summarize the steps for solving a system using substitution.

> **Solving A System of Two Linear Equations Containing Two Variables by Substitution**
>
> **Step 1:** Solve one of the equations for one of the unknowns.
>
> **Step 2:** Substitute the expression solved for in Step 1 into the *other* equation. The result will be a single linear equation in one unknown.
>
> **Step 3:** Solve the linear equation in one unknown found in Step 2.
>
> **Step 4:** Substitute the value of the variable found in Step 3 into one of the *original* equations to find the value of the other variable.
>
> **Step 5:** Check your answer by substituting the ordered pair into both of the original equations.

Work Smart

Name each equation in the system as we did in Example 1 before going through the steps for solving the system.

EXAMPLE 2 | **Solving a System of Equations by Substitution**

Solve the following system by substitution: $\begin{cases} 2x - y = -15 & (1) \\ 4x + 3y = 5 & (2) \end{cases}$

Solution

When using substitution, solve for the variable whose coefficient is 1 or -1, if possible, to simplify the algebra. We will solve equation (1) for y.

Equation (1):	$2x - y = -15$
Subtract $2x$ from both sides:	$-y = -2x - 15$
Divide both sides by -1:	$y = 2x + 15$

Substitute $2x + 15$ for y in equation (2).

Equation (2):	$4x + 3y = 5$
Let $y = 2x + 15$:	$4x + 3(2x + 15) = 5$

Solve for x.

Distribute to remove parentheses:	$4x + 6x + 45 = 5$
Combine like terms:	$10x + 45 = 5$
Subtract 45 from both sides:	$10x = -40$
Divide both sides by 10:	$x = -4$

Let $x = -4$ in equation (1) and solve for y.

Equation (1):	$2x - y = -15$
Let $x = -4$:	$2(-4) - y = -15$
	$-8 - y = -15$
Add 8 to both sides:	$-y = -7$
Multiply both sides by -1:	$y = 7$

Work Smart

If you solve for a variable in equation (1), be sure to substitute into equation (2). If you solve for a variable in equation (2), be sure to substitute into equation (1).

The proposed solution is $(-4, 7)$.

Check

Equation (1):	$2x - y = -15$	Equation (2):	$4x + 3y = 5$

Let $x = -4, y = 7$: $2(-4) - 7 \overset{?}{=} -15$ Let $x = -4, y = 7$: $4(-4) + 3(7) \overset{?}{=} 5$

$-8 - 7 \overset{?}{=} -15$ $-16 + 21 \overset{?}{=} 5$

$-15 = -15$ True $5 = 5$ True

The ordered pair $(-4, 7)$ satisfies both equations. The solution of the system is $(-4, 7)$. ●

> **Quick ✓**
>
> 1. When solving a system of two linear equations containing two variables by substitution, if you solve for a variable in equation (1), be sure to substitute into equation ___.
>
> *In Problems 2 and 3, solve the system using substitution.*
>
> 2. $\begin{cases} y = 3x - 2 \\ 2x - 3y = -8 \end{cases}$ 3. $\begin{cases} 2x + y = -1 \\ 4x + 3y = 3 \end{cases}$

EXAMPLE 3   **Solving a System of Equations by Substitution**

Solve the following system by substitution: $\begin{cases} 2x + 3y = 9 & (1) \\ x + 2y = \dfrac{13}{2} & (2) \end{cases}$

Solution

We will solve equation (2) for x since the coefficient of x in equation (2) is 1. If we try to solve for x or y in equation (1) or to solve for y in equation (2), we will end up with an equation with a fractional coefficient. We'd like to avoid that situation.

Equation (2): $x + 2y = \dfrac{13}{2}$

Subtract $2y$ from both sides: $x = -2y + \dfrac{13}{2}$

We substitute $-2y + \dfrac{13}{2}$ for x in equation (1).

Equation (1): $2x + 3y = 9$

Let $x = -2y + \dfrac{13}{2}$: $2\left(-2y + \dfrac{13}{2}\right) + 3y = 9$

Now we solve this equation for y.

Distribute the 2: $-4y + 13 + 3y = 9$

Combine like terms: $-y + 13 = 9$

Subtract 13 from both sides: $-y = -4$

Multiply both sides by -1: $y = 4$

Let $y = 4$ in equation (1) and solve for x.

Equation (1): $2x + 3y = 9$

Let $y = 4$: $2x + 3(4) = 9$

$2x + 12 = 9$

Subtract 12 from both sides: $2x = -3$

Divide both sides by 2: $\dfrac{2x}{2} = -\dfrac{3}{2}$

$x = -\dfrac{3}{2}$

The proposed solution is $\left(-\dfrac{3}{2}, 4\right)$.

Check

Equation (1): $2x + 3y = 9$

Let $x = -\dfrac{3}{2}, y = 4$: $2\left(-\dfrac{3}{2}\right) + 3(4) \overset{?}{=} 9$

$-3 + 12 \overset{?}{=} 9$

$9 = 9$ True

Equation (2): $x + 2y = \dfrac{13}{2}$

Let $x = -\dfrac{3}{2}, y = 4$: $-\dfrac{3}{2} + 2(4) \overset{?}{=} \dfrac{13}{2}$

$-\dfrac{3}{2} + 8 \overset{?}{=} \dfrac{13}{2}$

$\dfrac{13}{2} = \dfrac{13}{2}$ True

Because $\left(-\dfrac{3}{2}, 4\right)$ satisfies both equations, the solution of the system is $\left(-\dfrac{3}{2}, 4\right)$.   ●

Work Smart

Solving for x or for y in equation (1) in Example 3 is possible, but it will lead to fractions, making substitution more difficult. If we solve equation (1) for y, we obtain

Equation (1): $2x + 3y = 9$

$3y = -2x + 9$

$y = -\dfrac{2}{3}x + 3$

Letting $y = -\dfrac{2}{3}x + 3$ in equation (2), $x + 2y = \dfrac{13}{2}$, we obtain

$$x + 2\left(-\dfrac{2}{3}x + 3\right) = \dfrac{13}{2}$$

$$x - \dfrac{4}{3}x + 6 = \dfrac{13}{2}$$

$$-\dfrac{1}{3}x + 6 = \dfrac{13}{2}$$

$$-\dfrac{1}{3}x = \dfrac{1}{2}$$

$$x = -\dfrac{3}{2}$$

Whew! And we still need to substitute $x = -\dfrac{3}{2}$ into equation (1) or (2) to find y!

Quick ✓

In Problems 4 and 5, solve the system using substitution.

4. $\begin{cases} -4x + y = 1 \\ 8x - y = 5 \end{cases}$

5. $\begin{cases} 3x + 2y = -3 \\ -x + y = \dfrac{11}{6} \end{cases}$

▶ Recall that a system of two linear equations containing two unknowns can be inconsistent. This means that the two lines in the system are parallel. Let's see what happens when we use the substitution method on an inconsistent system.

EXAMPLE 4   **Solving an Inconsistent System Using Substitution**

Solve the following system by substitution: $\begin{cases} x - 3y = 5 & (1) \\ -2x + 6y = 3 & (2) \end{cases}$

Solution

The coefficient of x in equation (1) is 1, so we solve that equation for x.

Equation (1):	$x - 3y = 5$
Add $3y$ to both sides:	$x = 3y + 5$
In equation (2), replace x by $3y + 5$:	$-2(3y + 5) + 6y = 3$
Distribute to remove parentheses:	$-6y - 10 + 6y = 3$
Combine like terms:	$-10 = 3$

Figure 10

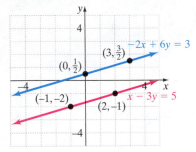

Notice that the variable, y, was eliminated and that the last equation, $-10 = 3$, is a false statement. We conclude that the system is inconsistent. The solution is $\{\ \}$ or $\varnothing$. Figure 10 shows the graph of the equations in the system. Note that the lines are parallel. ●

So if the algebraic solution ends with a false statement such as $-10 = 3$ or $-13 = 0$, the system is inconsistent.

EXAMPLE 5   ### Solving a System with Infinitely Many Solutions Using Substitution

Solve the following system by substitution: $\begin{cases} 6x - 2y = -4 & (1) \\ -3x + y = 2 & (2) \end{cases}$

Solution

It is easiest to solve equation (2) for y.

Equation (2):	$-3x + y = 2$
Add $3x$ to both sides:	$y = 3x + 2$
Let $y = 3x + 2$ in equation (1):	$6x - 2(3x + 2) = -4$
Distribute to remove parentheses:	$6x - 6x - 4 = -4$
Combine like terms:	$-4 = -4$

Figure 11

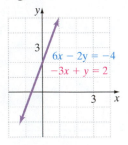

Notice that the variable x has been eliminated and that the last equation, $-4 = -4$, is true. We conclude that the system is consistent but the equations are dependent. The system has infinitely many solutions.

Figure 11 shows the graph of the equations in the system. Both equations have the same graph, indicating the system is consistent and the equations are dependent. ●

Work Smart

If a true statement with no variables occurs after the substitution (such as $-4 = -4$), then the system has infinitely many solutions. If a false statement with no variables occurs after the substitution (such as $-10 = 3$), the system has no solution.

Quick ✓

6. In the process of solving a system of equations by substitution, the variables have been eliminated, and a *false* statement such as $-3 = 0$ results. This means that the solution of the system is _____.

7. In the process of solving a system of equations by substitution, the variables have been eliminated, and a *true* statement such as $0 = 0$ results. This means that the system has _____ ____ _____.

8. *True or False* When solving a system of equations by substitution, you solve equation (1) for x, and then you substitute this expression back into equation (1), resulting in the statement $5 = 5$. This means that there are infinitely many solutions of the system.

In Problems 9–11, solve the system using substitution.

9. $\begin{cases} y = 5x + 2 \\ -10x + 2y = 4 \end{cases}$ 10. $\begin{cases} 3x - 2y = 0 \\ -9x + 6y = 5 \end{cases}$ 11. $\begin{cases} 2x - 6y = 2 \\ -3x + 9y = 4 \end{cases}$

▶ ❷ Solve Applied Problems Involving Systems of Linear Equations

EXAMPLE 6 **Supply and Demand**

The number of hot dogs h that a street vendor is willing to sell each day is given by the equation $h = 300p - 100$, where p is the price of the hot dog. The number of hot dogs h that individuals are willing to purchase each day is given by the equation $h = -150p + 1250$, where p is the price of the hot dog. The price at which supply equals demand is the equilibrium price. To find the equilibrium price of the hot dogs, solve the system of equations

$$\begin{cases} h = 300p - 100 \\ h = -150p + 1250 \end{cases}$$

How many hot dogs will the vendor sell at this price?

Solution

Name the equations (1) and (2).

$$\begin{cases} h = 300p - 100 & \text{(1)} \\ h = -150p + 1250 & \text{(2)} \end{cases}$$

We will solve this system of equations by substituting $300p - 100$ for h (from equation (1)) into equation (2). This leads to the following equation.

$$300p - 100 = -150p + 1250$$

Add 150p to both sides: $450p - 100 = 1250$

Add 100 to both sides: $450p = 1350$

Divide both sides by 450: $p = 3$

The equilibrium price of the hot dogs is \$3. The vendor will sell $h = -150(3) + 1250 = 800$ hot dogs. ●

Quick ✓

12. The number of sodas s that a street vendor is willing to sell each day is given by the equation $s = 200p + 800$, where p is the price of the soda. The number of sodas s that individuals are willing to purchase each day is given by the equation $s = -600p + 2400$, where p is the price of the soda. The price at which supply equals demand is the equilibrium price. To find the equilibrium price of the soda, solve the system of equations

$$\begin{cases} s = 200p + 800 \\ s = -600p + 2400 \end{cases}$$

How many hot dogs will the vendor sell at this price?

4.2 Exercises MyMathLab® Math XL PRACTICE

Exercise numbers in **green** have complete video solutions in MyMathLab.

Problems 1–12 are the Quick ✓s that follow the EXAMPLES.

Building Skills

In Problems 13–26, solve each system of equations using substitution. See Objective 1.

13. $\begin{cases} x + 2y = 2 \\ y = 2x - 9 \end{cases}$

14. $\begin{cases} y = 3x - 11 \\ -x + 4y = -11 \end{cases}$

15. $\begin{cases} -2x + 5y = 7 \\ x = 3y - 4 \end{cases}$

16. $\begin{cases} -4x - y = -3 \\ x = y + 7 \end{cases}$

17. $\begin{cases} x + y = -7 \\ 2x - y = -2 \end{cases}$

18. $\begin{cases} 5x + 2y = -5 \\ 3x - y = -14 \end{cases}$

19. $\begin{cases} y = \dfrac{1}{2}x - 5 \\ y = -\dfrac{3}{4}x - 10 \end{cases}$

20. $\begin{cases} y = \dfrac{2}{3}x + 1 \\ y = -\dfrac{3}{2}x + 40 \end{cases}$

21. $\begin{cases} y = 3x + 4 \\ y = -\dfrac{1}{2}x + \dfrac{5}{3} \end{cases}$

22. $\begin{cases} y = 5x - 3 \\ y = 2x - \dfrac{21}{5} \end{cases}$

23. $\begin{cases} x = -6y \\ x - 3y = 3 \end{cases}$

24. $\begin{cases} y = 4x \\ 2x - 3y = 5 \end{cases}$

25. $\begin{cases} 2x - 3y = 0 \\ 8x + 6y = 3 \end{cases}$

26. $\begin{cases} 2x + 3y = -1 \\ 2x - 9y = -9 \end{cases}$

In Problems 27–34, solve each system of equations using substitution. State whether the system is inconsistent, or consistent and with dependent equations. See Objective 1.

27. $\begin{cases} x + 3y = -12 \\ x + 3y = 6 \end{cases}$

28. $\begin{cases} 2x - y = 3 \\ y - 2x = 3 \end{cases}$

29. $\begin{cases} y = 4x - 1 \\ 8x - 2y = 2 \end{cases}$

30. $\begin{cases} y = -x + 1 \\ x + y = 1 \end{cases}$

31. $\begin{cases} x + 2y = 6 \\ x = 3 - 2y \end{cases}$

32. $\begin{cases} 4 + y = 3x \\ 6x - 2y = -2 \end{cases}$

33. $\begin{cases} 5x + 6 = 2 - y \\ y = -5x - 4 \end{cases}$

34. $\begin{cases} x + y = 2 \\ 2x + 2y = 2 \end{cases}$

Mixed Practice

In Problems 35–52, solve each system of equations using substitution.

35. $\begin{cases} x = 2y \\ x - 6y = 4 \end{cases}$

36. $\begin{cases} 2x + y = 5 \\ y = 3x \end{cases}$

37. $\begin{cases} y = \dfrac{1}{2}x \\ x = 2(y + 1) \end{cases}$

38. $\begin{cases} y = \dfrac{3}{4}x \\ x = 4(y - 1) \end{cases}$

39. $\begin{cases} y = 2x - 8 \\ x - \dfrac{1}{2}y = 4 \end{cases}$

40. $\begin{cases} x = 3y + 6 \\ y - \dfrac{1}{3}x = -2 \end{cases}$

41. $\begin{cases} 3x + 2y = 1 \\ -2x - y = 1 \end{cases}$

42. $\begin{cases} 2x + 5y = 7 \\ -x + 6y = -12 \end{cases}$

43. $\begin{cases} x - 5y = 3 \\ -2x + 10y = 8 \end{cases}$

44. $\begin{cases} -4x + y = 3 \\ 8x - 2y = 1 \end{cases}$

45. $\begin{cases} 3x - y = 1 \\ -6x + 2y = -2 \end{cases}$

46. $\begin{cases} -x + 3y = 4 \\ 2x - 6y = -8 \end{cases}$

47. $\begin{cases} 4x + 8y = -9 \\ 2x + y = \dfrac{3}{4} \end{cases}$

48. $\begin{cases} 18x - 6y = -7 \\ x + 2y = 0 \end{cases}$

49. $\begin{cases} \dfrac{3}{2}x - y = 1 \\ 3x - 2y = 2 \end{cases}$

50. $\begin{cases} \dfrac{2}{3}x + \dfrac{1}{3}y = 1 \\ \dfrac{1}{2}x + \dfrac{1}{4}y = \dfrac{3}{4} \end{cases}$

51. $\begin{cases} \dfrac{x}{2} + \dfrac{y}{3} = \dfrac{1}{12} \\ \dfrac{2x}{3} + \dfrac{y}{3} = -\dfrac{1}{3} \end{cases}$

52. $\begin{cases} \dfrac{x}{4} + \dfrac{y}{2} = \dfrac{3}{8} \\ x - \dfrac{y}{3} = \dfrac{1}{3} \end{cases}$

Applying the Concepts

In Problems 53–60, use substitution to solve each system of linear equations in two variables.

△ **53. Dimensions of a Garden** The perimeter of a rectangular garden is 34 feet. The length of the garden is 3 feet more than the width. Determine the dimensions of the garden by solving the following system of equations, where *l* and *w* represent the length and width of the garden.

$$\begin{cases} 2l + 2w = 34 & (1) \\ l = w + 3 & (2) \end{cases}$$

△ **54. Dimensions of a Rectangle** The perimeter of a rectangle is 52 inches. The length of the rectangle is 4 inches more than the width. Determine the dimensions of the rectangle by solving the following system of equations, where *l* and *w* represent the length and width of the rectangle.

$$\begin{cases} 2l + 2w = 52 & (1) \\ l = w + 4 & (2) \end{cases}$$

55. Investment Paul wants to invest part of his Ohio Lottery winnings in a safe money market fund that earns 2.5% annual interest and the rest in a risky international fund that is expected to yield 9% annual interest. The amount of money invested in the money market fund, *x*, is to be exactly twice the amount invested in the international fund, *y*. Use the system of equations to determine the amount to be invested in each fund if a total of $560 is to be earned at the end of one year.

$$\begin{cases} x = 2y & (1) \\ 0.025x + 0.09y = 560 & (2) \end{cases}$$

56. Investment Elaine wants to invest part of her $24,000 Virginia Lottery winnings in an international stock fund that yields 4% annual interest and the remainder in a domestic growth fund that yields 6.5% annually. Use the system of equations to determine the amount Elaine should invest in each account to earn $1360 interest at the end of one year, where x represents the amount invested in the international stock fund and y represents the amount invested in the domestic growth fund.

$$\begin{cases} x + \quad\;\; y = 24{,}000 & (1) \\ 0.04x + 0.065y = 1360 & (2) \end{cases}$$

57. Salary Suppose that you are offered a sales position for a pharmaceutical company. They offer you two salary options. Option A would pay you an annual base salary of $15,000 plus a commission of 2% on sales. The equation $y = 15{,}000 + 0.02x$ models salary Option A, where x represents the annual sales amount and y represents the annual salary. Option B would pay you an annual base salary of $25,000 plus a commission of 1% on sales. The equation $y = 25{,}000 + 0.01x$ models salary Option B. Determine the annual sales required for the options to result in the same annual salary.

58. Salary Melanie has been offered a sales position for a major textbook company. Melanie was offered two options. Option A would pay her an annual base salary of $20,000 plus a commission of 4% on sales. The equation $y = 20{,}000 + 0.04x$ models salary Option A, where x represents the annual sales amount and y represents the annual salary. Option B would pay Melanie an annual base salary of $25,000 plus a commission of 2% on sales. The equation $y = 25{,}000 + 0.02x$ models salary Option B. Determine the annual sales required for the two options to result in the same annual salary. What is that salary?

59. Fun with Numbers The sum of two numbers is 17, and their difference is 7. Determine the numbers by solving the following system of equations, using x and y to represent the unknown numbers.

$$\begin{cases} x + y = 17 & (1) \\ x - y = 7 & (2) \end{cases}$$

60. Fun with Numbers The sum of two numbers is 25, and their difference is 3. Determine the numbers by solving the following system of equations, using x and y to represent the unknown numbers.

$$\begin{cases} x + y = 25 & (1) \\ x - y = 3 & (2) \end{cases}$$

Extending the Concepts

61. For the system $\begin{cases} Ax + 3By = 2 \\ -3Ax + By = -11 \end{cases}$, find A and B such that $x = 3$, $y = 1$ is a solution.

62. For the system $\begin{cases} y = x - 3a \\ x + y = 7a \end{cases}$, consider x and y as the variables of the system of equations. Use substitution to solve for x and y in terms of a.

63. Write a system of equations that has $(3, 5)$ as a solution.

64. Write a system of equations that has $(-1, 4)$ as a solution.

65. Write a system of equations that has infinitely many solutions.

66. Write a system of equations that has no solution.

Explaining the Concepts

67. A test question asks you to solve the following system of equations by substitution. What would be a reasonable first step? Explain which variable you would solve for and why.

$$\begin{cases} \dfrac{1}{3}x - \dfrac{1}{6}y = \dfrac{2}{3} \\ \dfrac{3}{2}x + \dfrac{1}{2}y = -\dfrac{5}{4} \end{cases}$$

68. List two advantages of solving a system of linear equations by substitution rather than by the graphing method.

4.3 Solving Systems of Linear Equations Using Elimination

Objectives

1. Solve a System of Linear Equations Using the Elimination Method
2. Solve Applied Problems Involving Systems of Linear Equations

Are You Ready for This Section?

Before getting started, take the following readiness quiz. If you get a problem wrong, go back to the section cited and review the material.

R1. What is the additive inverse of 5?   [Section 1.4, pp. 30–31]

R2. What is the additive inverse of -8?   [Section 1.4, pp. 30–31]

R3. Distribute: $\dfrac{2}{3}(3x - 9y)$ [Section 1.8, pp. 67–68]

R4. Solve: $2y - 5y = 12$ [Section 2.2, pp. 92–93]

R5. Find the LCM of 4 and 5   [Section 1.2, pp. 9–10]

▶ ❶ Solve a System of Linear Equations Using the Elimination Method

Work Smart

Use the method of substitution if it is easy to solve for one of the variables in the system. Use elimination if substitution will lead to fractions or if solving for a variable is not easy.

We have used two methods to solve a system of linear equations containing two unknowns. The graphing method lets us visualize the solutions, but finding the exact coordinates of the point of intersection can be difficult. The substitution method lets us find an exact solution because it is an algebraic method, but it may require that we use fractions. We would rather avoid fractions, so here we introduce a second algebraic method that can be used to solve a system of equations. It is called the *elimination method*. **The elimination method is usually preferred over the substitution method when substitution leads to fractions or solving for a variable is not straight forward.**

Remember that the Addition Property of Equality states that if we add the same quantity to both sides of an equation, we obtain an equivalent equation. The elimination method takes this principle a step further.

EXAMPLE 1 Solving a System of Linear Equations Using Elimination

Solve the following system using elimination: $\begin{cases} x + y = 13 & \text{(1)} \\ 2x - y = -7 & \text{(2)} \end{cases}$

Solution

Equation (2) says that $2x - y$ equals -7. Thus, if we add $2x - y$ to the left side of equation (1) and -7 to the right side of equation (1), we are adding the same value to each side of equation (1). We will perform this addition vertically.

$$\begin{array}{r} x + y = 13 \quad \text{(1)} \\ \underline{2x - y = -7} \quad \text{(2)} \\ 3x + 0 = 6 \end{array}$$

Notice that the variable y has been eliminated, and we can solve for x.

$$3x + 0 = 6$$
$$3x = 6$$

Divide each side by 3: $x = 2$

Work Smart

This method is called "elimination" because one of the variables is eliminated through the process of addition. The elimination method is sometimes called the **addition method**.

Now substitute $x = 2$ into either equation (1) or equation (2) to find the value of y. Let's substitute $x = 2$ into equation (1).

$$x + y = 13$$

Substitute 2 for x: $2 + y = 13$

Subtract 2 from each side: $y = 11$

Ready?...Answers **R1.** -5 **R2.** 8
R3. $2x - 6y$ **R4.** $\{-4\}$ **R5.** 20

We have $x = 2$ and $y = 11$. Let's check this solution in both equations.

Check

Equation (1): $x + y = 13$ Equation (2): $2x - y = -7$

$x = 2, y = 11:$ $2 + 11 \stackrel{?}{=} 13$ $x = 2, y = 11:$ $2(2) - 11 \stackrel{?}{=} -7$

                $13 = 13$ True $4 - 11 \stackrel{?}{=} -7$

                                                        $-7 = -7$ True

The solution of the system is $(2, 11)$.   ●

The idea in using the elimination method is to get the coefficients of one of the variables to be additive inverses (or opposites). In Example 1, the coefficients of the variable y are opposites: 1 and -1. Having linear equations in which one of the variables has opposite coefficients is uncommon, so we need a strategy for solving systems in which there is no variable that has opposite coefficients.

EXAMPLE 2   **How to Solve a System of Linear Equations by Elimination—Multiply One Equation by a Number to Create Additive Inverses**

Solve the following system by elimination: $\begin{cases} 2x + 3y = -6 & (1) \\ -4x + 5y = -21 & (2) \end{cases}$

Step-by-Step Solution

Step 1: Write each equation in standard form, $Ax + By = C$.

Both equations are already in standard form.

Step 2: We need to get the coefficients of one of the variables to be additive inverses. Because the coefficient of x in equation (1) is 2, and the coefficient of x in equation (2) is -4, this can be accomplished by multiplying both sides of equation (1) by 2.

$$\begin{cases} 2x + 3y = -6 & (1) \\ -4x + 5y = -21 & (2) \end{cases}$$

Multiply equation (1) by 2: $\begin{cases} 2(2x + 3y) = 2(-6) & (1) \\ -4x + 5y = -21 & (2) \end{cases}$

Distribute the 2 in equation (1): $\begin{cases} 4x + 6y = -12 & (1) \\ -4x + 5y = -21 & (2) \end{cases}$

Step 3: Notice that the coefficients of the variable x are additive inverses. Add equations (1) and (2) to eliminate x and then solve for y.

$$\begin{cases} 4x + 6y = -12 & (1) \\ -4x + 5y = -21 & (2) \end{cases}$$

Add equations (1) and (2): $11y = -33$

Divide both sides by 11: $y = -3$

We have the y-value of the solution.

Step 4: Let $y = -3$ in either equation (1) or (2). Since equation (1) looks a little easier to work with, we will substitute $y = -3$ into equation (1) and solve for x.

Equation (1): $2x + 3y = -6$

Let $y = -3$: $2x + 3(-3) = -6$

                        $2x - 9 = -6$

Add 9 to both sides: $2x = 3$

Divide both sides by 2: $x = \dfrac{3}{2}$

We have $x = \dfrac{3}{2}, y = -3$.

(continued)

Step 5: Check Equation (1): $2x + 3y = -6$

$x = \dfrac{3}{2}, y = -3$: $2\left(\dfrac{3}{2}\right) + 3(-3) \overset{?}{=} -6$

$3 + (-9) \overset{?}{=} -6$

$-6 = -6$ True

Equation (2): $-4x + 5y = -21$

$x = \dfrac{3}{2}, y = -3$: $-4\left(\dfrac{3}{2}\right) + 5(-3) \overset{?}{=} -21$

$-6 + (-15) \overset{?}{=} -21$

$-21 = -21$ True

Both equations check, so the solution of the system is $\left(\dfrac{3}{2}, -3\right)$.   ●

Solving A System of Linear Equations in Two Variables by Elimination

Step 1: Write each equation in standard form, $Ax + By = C$.

Step 2: If necessary, multiply or divide both sides of one equation (or both equations) by a nonzero constant so that the coefficients of one of the variables are opposites (additive inverses).

Step 3: Add the equations to eliminate the variable whose coefficients are now additive inverses. Solve the resulting equation for the remaining unknown.

Step 4: Substitute the value of the variable found in Step 3 into one of the *original* equations to find the value of the remaining variable.

Step 5: Check your answer by substituting the ordered pair in both of the original equations.

Quick ✓

1. The basic idea in using the elimination method is to get the coefficients of one of the variables to be _____ _____, such as 3 and −3.

In Problems 2 and 3, solve the system using elimination.

2. $\begin{cases} x - 3y = 2 \\ 2x + 3y = -14 \end{cases}$

3. $\begin{cases} x - 2y = 2 \\ -2x + 5y = -1 \end{cases}$

EXAMPLE 3

Solve a System of Linear Equations by Elimination—Multiply Both Equations by a Number to Create Additive Inverses

Solve the following system by elimination: $\begin{cases} 3y = -2x + 3 & (1) \\ 3x + 5y = 7 & (2) \end{cases}$

Solution

Write equation (1), $3y = -2x + 3$, in standard form, $Ax + By = C$, by adding $2x$ to each side of the equation and obtain $2x + 3y = 3$.

We want the coefficients of one of the variables to be additive inverses. We cannot do this by multiplying a single equation by a nonzero constant without introducing fractions, so we will multiply *both* equations by a nonzero constant. For example, we can multiply equation (1) by 3 and equation (2) by −2 so that the coefficients of x are additive inverses.

Work Smart

There is no single right way to multiply the equations in a system by nonzero constants to get coefficients to be additive inverses. For example, we could also multiply equation (1) by 5 and equation (2) by −3. Do you see why this also works?

$\begin{cases} 2x + 3y = 3 & (1) \\ 3x + 5y = 7 & (2) \end{cases}$

Multiply equation (1) by 3:
Multiply equation (2) by −2: $\begin{cases} 3(2x + 3y) = 3 \cdot 3 & (1) \\ -2(3x + 5y) = -2 \cdot 7 & (2) \end{cases}$

$$\text{Distribute:} \quad \begin{cases} 6x + 9y = 9 & (1) \\ -6x - 10y = -14 & (2) \end{cases}$$

$$\text{Add equations (1) and (2):} \qquad -y = -5$$

$$\text{Multiply both sides by } -1: \qquad y = 5$$

Now let $y = 5$ in equation (1) to determine the value of x.

Equation (1):	$3y = -2x + 3$
Let $y = 5$ in equation (1) and find x:	$3(5) = -2x + 3$
Simplify:	$15 = -2x + 3$
Subtract 3 from both sides:	$12 = -2x$
Divide both sides by -2:	$-6 = x$

We have that $x = -6$ and $y = 5$. Now we check our answer.

Check

Equation (1):	$3y = -2x + 3$	Equation (2):	$3x + 5y = 7$
$x = -6, y = 5$:	$3(5) \overset{?}{=} -2(-6) + 3$	$x = -6, y = 5$:	$3(-6) + 5(5) \overset{?}{=} 7$
	$15 \overset{?}{=} 12 + 3$		$-18 + 25 \overset{?}{=} 7$
	$15 = 15$ True		$7 = 7$ True

The solution to the system is $(-6, 5)$. ●

Quick ✓

In Problems 4 and 5, solve the system using elimination.

4. $\begin{cases} 5x + 4y = 10 \\ -2x + 3y = -27 \end{cases}$ **5.** $\begin{cases} 4y = -6x + 30 \\ 7x + 10y = 35 \end{cases}$

▶ Now let's look at the results we get when we have systems that have no solution or infinitely many solutions.

EXAMPLE 4 **Solving a System of Equations with No Solution Using Elimination**

Solve the following system by elimination: $\begin{cases} 4x - 6y = -5 & (1) \\ 6x - 9y = 2 & (2) \end{cases}$

Solution

$$\begin{cases} 4x - 6y = -5 & (1) \\ 6x - 9y = 2 & (2) \end{cases}$$

$$\begin{aligned} \text{To eliminate } x \\ \text{multiply equation (1) by } -3: \\ \text{multiply equation (2) by 2:} \end{aligned} \begin{cases} -3(4x - 6y) = -3 \cdot (-5) & (1) \\ 2(6x - 9y) = 2 \cdot 2 & (2) \end{cases}$$

$$\text{Distribute:} \quad \begin{cases} -12x + 18y = 15 & (1) \\ 12x - 18y = 4 & (2) \end{cases}$$

$$\text{Add (1) and (2):} \qquad 0 = 19$$

The statement $0 = 19$ is false. Therefore, the system is inconsistent and has no solution. The solution set is { } or $\varnothing$. If graphed, the lines would be parallel. ●

EXAMPLE 5 Solving a System of Equations with Infinitely Many Solutions Using Elimination

Solve the following system by elimination:
$$\begin{cases} \dfrac{3}{2}x + \dfrac{2}{3}y = -4 & \text{(1)} \\ -\dfrac{9}{8}x - \dfrac{1}{2}y = 3 & \text{(2)} \end{cases}$$

Solution

We begin by multiplying equation (1) by 6, the LCD of $\dfrac{3}{2}$ and $-\dfrac{2}{3}$, and multiplying equation (2) by 8, the LCD of $-\dfrac{9}{8}$ and $-\dfrac{1}{2}$, to clear each equation of fractions.

$$\begin{cases} 6\left(\dfrac{3}{2}x + \dfrac{2}{3}y\right) = 6(-4) & \text{(1)} \\ 8\left(-\dfrac{9}{8}x - \dfrac{1}{2}y\right) = 8(3) & \text{(2)} \end{cases}$$

Distribute:
$$\begin{cases} 9x + 4y = -24 & \text{(1)} \\ -9x - 4y = 24 & \text{(2)} \end{cases}$$

Add equations (1) and (2):
$$0 = 0$$

Notice that both variables were eliminated and that the statement $0 = 0$ is true. Therefore, the system is consistent, with dependent equations. The system has infinitely many solutions. A graph of the equations would show one line in the coordinate plane. ●

Quick ✓

6. When using the elimination method to solve a system of equations, you add equation (1) and equation (2), resulting in the statement $-50 = -50$. This means that the equations are _____ and that the system has _____ many solutions.

In Problems 7–9, use elimination to determine whether the system has no solution or infinitely many solutions.

7. $\begin{cases} 2x - 6y = 10 \\ 5x - 15y = 4 \end{cases}$

8. $\begin{cases} -x + 3y = 2 \\ 3x - 9y = -6 \end{cases}$

9. $\begin{cases} -\dfrac{1}{4}x + \dfrac{1}{2}y = 1 \\ \dfrac{1}{2}x - y = -2 \end{cases}$

Summary Which Method Should I Use?

Ask yourself the following questions in order to determine the most appropriate method for solving a system of equations.

Question	Method to Use	Example	Advantages/Disadvantages
Would it be beneficial to see the solutions visually?	Graphical	Find the break-even point of the cost equation $y = 20 + 4x$ and the revenue equation $y = 9x$, where x represents the number of calculators produced and sold.	Allows us to "see" the answer, but if the coordinates of the intersection point are not integers, it can be difficult to determine the answer.
Is one of the coefficients of the variables 1 or -1?	Substitution	$\begin{cases} -x + 2y = 7 \\ 2x - 3y = -15 \end{cases}$	Gives exact solutions. The algebra can be easy if the coefficient of one of the variables is 1 or -1. If none of the coefficients is 1 or -1, the algebra can be messy.
Are the coefficients of a variable additive inverses? If not, is it easy to get the coefficients to be additive inverses?	Elimination	$\begin{cases} 3x - 2y = -5 \\ -3x + 4y = 8 \end{cases}$	Gives exact solutions. The algebra is easy when neither variable has a coefficient of 1 or -1.

▶ **2 Solve Applied Problems Involving Systems of Linear Equations**

We saw in the last section that applied problems are easily solved using substitution when at least one of the variables is isolated (that is, by itself). But many mathematical models that involve two equations containing two unknowns do not have one of the equations solved for one of the unknowns. If this occurs, it is better to use elimination to solve the problem.

EXAMPLE 6 **Hot Dogs and Soda at the Game**

Bill and Roger attend a baseball game with their kids. They get to the game early to watch batting practice and have dinner. Roger says that he will buy dinner. He gets 7 hot dogs and 5 Pepsis for $38.25. After the sixth inning everyone is hungry again, so Bill buys 5 hot dogs and 4 Pepsis for $28.50. Find the price of a hot dog and the price of a Pepsi by solving the system of equations

$$\begin{cases} 7h + 5p = 38.25 \\ 5h + 4p = 28.50 \end{cases}$$

where h represents the price of a hot dog and p represents the price of a Pepsi.

Solution

First, name the equations.

$$\begin{cases} 7h + 5p = 38.25 \quad (1) \\ 5h + 4p = 28.50 \quad (2) \end{cases}$$

Let's eliminate p.

Multiply equation (1) by -4: $\begin{cases} -4(7h + 5p) = -4 \cdot 38.25 \quad (1) \\ 5(5h + 4p) = 5 \cdot 28.50 \quad (2) \end{cases}$
Multiply equation (2) by 5:

Distribute: $\begin{cases} -28h - 20p = -153 \quad (1) \\ 25h + 20p = 142.50 \quad (2) \end{cases}$

Add equations (1) and (2): $-3h \qquad\qquad = -10.50$

Divide both sides by -3: $h \qquad\qquad = 3.50$

Let $h = 3.50$ in (2): $5(3.50) + 4p = 28.50$

$17.50 + 4p = 28.50$

Subtract 17.50 from both sides: $4p = 11$

Divide both sides by 4: $p = 2.75$

We leave it to you to verify the answer. A hot dog costs $3.50, and a Pepsi costs $2.75. ●

Quick ✓

10. At a fast-food joint, 5 cheeseburgers and 3 shakes cost $15.50. At the same fast-food joint, 3 cheeseburgers and 2 shakes cost $9.75. We can determine the price of a cheeseburger and the price of a shake by solving the system of equations

$$\begin{cases} 5c + 3s = 15.50 \\ 3c + 2s = 9.75 \end{cases}$$

where c represents the price of a cheeseburger and s represents the price of a shake. How much does a cheeseburger cost at the fast-food joint? How much does a shake cost?

4.3 Exercises MyMathLab® Math XP PRACTICE

Exercise numbers in green have complete video solutions in MyMathLab.

Problems 1–10 are the Quick ✓s that follow the EXAMPLES.

Building Skills

In Problems 11–18, solve each system of equations using elimination. See Objective 1.

11. $\begin{cases} 2x + y = 3 \\ 5x - y = 11 \end{cases}$

12. $\begin{cases} x + y = -20 \\ x - y = 10 \end{cases}$

13. $\begin{cases} 3x - 2y = 10 \\ -3x + 12y = 30 \end{cases}$

14. $\begin{cases} 2x + 6y = 6 \\ -2x + y = 8 \end{cases}$

15. $\begin{cases} 2x + 3y = -4 \\ -2x + y = 6 \end{cases}$

16. $\begin{cases} 3x + 2y = 7 \\ -3x + 4y = 2 \end{cases}$

17. $\begin{cases} 6x - 2y = 0 \\ -9x - 4y = 21 \end{cases}$

18. $\begin{cases} 2x + 5y = -2 \\ -3x + 10y = -32 \end{cases}$

In Problems 19–26, solve each system using elimination. State whether the system is inconsistent, or consistent with dependent equations. See Objective 1.

19. $\begin{cases} 2x + 2y = 1 \\ -2x - 2y = 1 \end{cases}$

20. $\begin{cases} 3x - 2y = 4 \\ -3x + 2y = 4 \end{cases}$

21. $\begin{cases} 3x + y = -1 \\ 6x + 2y = -2 \end{cases}$

22. $\begin{cases} x - y = -4 \\ -2x + 2y = -8 \end{cases}$

23. $\begin{cases} 2x - 3y = 10 \\ -4x + 6y = -20 \end{cases}$

24. $\begin{cases} 2x + 5y = 15 \\ -6x - 15y = -45 \end{cases}$

25. $\begin{cases} -4x + 8y = 1 \\ 3x - 6y = 1 \end{cases}$

26. $\begin{cases} 4x + 6y = -10 \\ 9x + \dfrac{27}{2}y = 3 \end{cases}$

In Problems 27–46, solve each system of equations using elimination. See Objective 1.

27. $\begin{cases} 2x + 3y = 14 \\ -3x + y = 23 \end{cases}$

28. $\begin{cases} 2x + y = -4 \\ 3x + 5y = 29 \end{cases}$

29. $\begin{cases} 2x + 4y = 0 \\ 5x + 2y = 6 \end{cases}$

30. $\begin{cases} -2x - 2y = 3 \\ x - 2y = 1 \end{cases}$

31. $\begin{cases} x - 3y = 4 \\ -2x + 6y = 3 \end{cases}$

32. $\begin{cases} 5x - y = 3 \\ -10x + 2y = 2 \end{cases}$

33. $\begin{cases} 2x + 3y = -3 \\ 3x + 5y = -9 \end{cases}$

34. $\begin{cases} 2x + 3y = 2 \\ 5x + 7y = 0 \end{cases}$

35. $\begin{cases} 10y = 4x - 2 \\ 2x - 5y = 1 \end{cases}$

36. $\begin{cases} x + 3y = 6 \\ 9y = -3x + 18 \end{cases}$

37. $\begin{cases} 4x + 3y = 0 \\ 3x - 5y = 2 \end{cases}$

38. $\begin{cases} 5x + 7y = 6 \\ 2x - 3y = 11 \end{cases}$

39. $\begin{cases} 4x - 3y = -10 \\ -\dfrac{2}{3}x + y = \dfrac{11}{3} \end{cases}$

40. $\begin{cases} 12x + 15y = 55 \\ \dfrac{1}{2}x + 3y = \dfrac{3}{2} \end{cases}$

41. $\begin{cases} 1.5x + 0.5y = -0.45 \\ -0.3x - 0.4y = -0.54 \end{cases}$

42. $\begin{cases} -2.4x - 0.4y = 0.32 \\ 4.2x + 0.6y = -0.54 \end{cases}$

43. $\begin{cases} \dfrac{1}{2}x + \dfrac{2}{3}y = -5 \\ \dfrac{5}{2}x + \dfrac{5}{6}y = -10 \end{cases}$

44. $\begin{cases} x + \dfrac{5}{3}y = -1 \\ \dfrac{1}{2}x + \dfrac{1}{4}y = \dfrac{1}{4} \end{cases}$

45. $\begin{cases} 0.05x + 0.10y = 5.50 \\ x + y = 80 \end{cases}$

46. $\begin{cases} x + y = 1000 \\ 0.05x + 0.02y = 32 \end{cases}$

Mixed Practice

In Problems 47–68, solve by any method: graphing, substitution, or elimination.

47. $\begin{cases} x - y = -4 \\ 3x + y = 8 \end{cases}$

48. $\begin{cases} x - 2y = 0 \\ 3x + 5y = -11 \end{cases}$

49. $\begin{cases} 3x - 10y = -5 \\ 6x - 8y = 14 \end{cases}$

50. $\begin{cases} 3x - 5y = 14 \\ -2x + 6y = -16 \end{cases}$

51. $\begin{cases} -x + 3y = 6 \\ 4x + 5y = 7 \end{cases}$

52. $\begin{cases} 2x - 3y = -10 \\ -3x + y = 1 \end{cases}$

53. $\begin{cases} 0.3x - 0.7y = 1.2 \\ 1.2x + 2.1y = 2 \end{cases}$

54. $\begin{cases} 0.25x + 0.10y = 3.70 \\ x + y = 25 \end{cases}$

55. $\begin{cases} 3x - 2y = 6 \\ \dfrac{3}{2}x - y = 3 \end{cases}$

56. $\begin{cases} 4x + 3y = 9 \\ \dfrac{4}{3}x + y = 3 \end{cases}$

57. $\begin{cases} y = -\dfrac{2}{3}x - \dfrac{7}{3} \\ y = \dfrac{3}{4}x - \dfrac{15}{4} \end{cases}$ **58.** $\begin{cases} y = \dfrac{1}{7}x - 4 \\ y = -\dfrac{3}{2}x + 19 \end{cases}$

59. $\begin{cases} \dfrac{x}{2} + \dfrac{y}{4} = -2 \\ \dfrac{3x}{2} + \dfrac{y}{5} = -6 \end{cases}$ **60.** $\begin{cases} \dfrac{x}{3} + \dfrac{y}{5} = 2 \\ \dfrac{x}{3} - \dfrac{2y}{5} = -1 \end{cases}$

61. $\begin{cases} x - 2y = -7 \\ 3x + 4y = 6 \end{cases}$ **62.** $\begin{cases} -5x + y = 3 \\ 10x + 3y = 14 \end{cases}$

63. $\begin{cases} 6x - 5y = 1 \\ 8x - 2y = -22 \end{cases}$ **64.** $\begin{cases} -3x - 2y = -19 \\ 4x + 5y = 30 \end{cases}$

65. $\begin{cases} y = 2x - 4y \\ 4x + 1 = 10y + 3 \end{cases}$ **66.** $\begin{cases} 12y = 8x + 3 \\ -2 + 4(3y - 2x) = 5 \end{cases}$

67. $\begin{cases} \dfrac{x}{2} + \dfrac{3y}{4} = \dfrac{1}{2} \\ -\dfrac{3x}{5} + \dfrac{3y}{4} = -\dfrac{1}{20} \end{cases}$ **68.** $\begin{cases} \dfrac{x}{2} - y = 1 \\ \dfrac{x}{5} + \dfrac{5y}{6} = \dfrac{14}{15} \end{cases}$

Applying the Concepts

69. Counting Calories Suppose that Kristin ate two hamburgers and drank one medium Coke, for a total of 770 calories. Kristin's friend Jack ate three hamburgers and drank two medium Cokes (Jack takes advantage of free refills) for a total of 1260 calories. How many calories are in a hamburger? How many calories are in a medium Coke? To find the answers, solve the system

$$\begin{cases} 2h + c = 770 & (1) \\ 3h + 2c = 1260 & (2) \end{cases}$$

where h represents the number of calories in a hamburger and c represents the number of calories in a medium Coke.

70. Carbs Yvette and José go to McDonald's for breakfast. Yvette orders two sausage biscuits and one 16-ounce orange juice. The entire meal had 98 grams of carbohydrates. José orders three sausage biscuits and two 16-ounce orange juices, and his meal had 168 grams of carbohydrates. How many grams of carbohydrates are in a sausage biscuit? How many grams of carbohydrates are in a 16-ounce orange juice? To find the answers, solve the system

$$\begin{cases} 2b + u = 98 & (1) \\ 3b + 2u = 168 & (2) \end{cases}$$

where b represents the number of grams of carbohydrates in a sausage biscuit and u represents the number of grams of carbohydrates in an orange juice.

71. Planting Crops Farmer Green runs an organic farm and is planting his fields. He remembers from previous years that he planted 2 acres of tomatoes and 3 acres of zucchini in 65 hours. A different year he planted 3 acres of tomatoes and 4 acres of zucchini in 90 hours. If t represents the number of hours it takes to plant an acre of tomatoes, and z represents the number of hours it takes to plant one acre of zucchini, determine how long it takes Farmer Green to plant an acre of each crop by solving the following system of equations.

$$\begin{cases} 2t + 3z = 65 & (1) \\ 3t + 4z = 90 & (2) \end{cases}$$

If he wants to plant 5 acres of tomatoes and 2 acres of zucchini, will he get the crops in if the meteorologist predicts rain in 72 hours?

72. Farmer's Daughter Farmer Green has talked his daughter, Haydee, into helping with some of the planting in the scenario in Problem 71. Haydee's specialty is planting corn and lettuce, which she has done for the past several years. One year it took her 48 hours to plant 3 acres of corn and 2 acres of lettuce. Another year it took her 68 hours to plant 4 acres of corn and 3 acres of lettuce. If c represents the number of hours to plant one acre of corn, and l represents the number of hours required to plant one acre of lettuce, determine how long it takes Haydee to plant an acre of each crop by solving the following system of equations.

$$\begin{cases} 3c + 2l = 48 & (1) \\ 4c + 3l = 68 & (2) \end{cases}$$

If Farmer Green wants to have 4 acres of corn planted before the rains begin in 72 hours, how many acres of lettuce can Haydee expect to plant?

73. Making Coffee Suppose that you want to blend two coffees in order to obtain a new blend. The blend will be made with Arabica beans that sell for $9.00 per pound and select African Robustas that sell for $11.50 per pound to obtain 100 pounds of the new blend that will sell for $10.00 per pound. How many pounds of the Arabica beans and of the Robusta beans are required? To determine the answer, solve the system

$$\begin{cases} a + \quad r = 100 & (1) \\ 9a + 11.50r = 1000 & (2) \end{cases}$$

where a represents the number of pounds of Arabica beans and r represents the number of pounds of African Robusta beans.

74. Candy A candy store sells chocolate-covered almonds for $6.50 per pound and chocolate-covered peanuts for $4.00 per pound. The manager decides to make a bridge mix that combines the almonds with the peanuts. She wants the bridge mix to sell for $6.00 per pound. How many pounds of chocolate-covered almonds and chocolate-covered peanuts are required to create 50 pounds of bridge mix? To determine the answer, solve the system

$$\begin{cases} a + \quad p = 50 & (1) \\ 6.50a + 4.00p = 300 & (2) \end{cases}$$

where a represents the number of pounds of chocolate-covered almonds and p represents the number of pounds of chocolate-covered peanuts.

△ **75. Complementary Angles** Two angles are complementary if the sum of their measures is 90°. The measure of one angle is 10° more than three times the measure of its complement. If A and B represent the measures of the two complementary angles, determine the measure of each of the angles by solving the following system of equations.

$$\begin{cases} A + B = 90 & (1) \\ A = 10 + 3B & (2) \end{cases}$$

△ **76. Supplementary Angles** Two angles are supplementary if the sum of their measures is 180°. One-half the measure of one angle is 45° more than the measure of its supplement. If A and B represent the measures of the two supplementary angles, determine the measure

of each of the angles by solving the following system of equations.

$$\begin{cases} A + B = 180 & (1) \\ \dfrac{A}{2} = 45 + B & (2) \end{cases}$$

Extending the Concepts

In Problems 77 and 78, solve the system of equations for x and y using the elimination method.

77. $\begin{cases} ax + 4y = 1 \\ -2ax - 3y = 3 \end{cases}$

78. $\begin{cases} 4x + 2by = 4 \\ 5x + 3by = 7 \end{cases}$

In Problems 79 and 80, solve the system of equations for x and y using the elimination method.

79. $\begin{cases} -3x + 2y = 6a \\ x - 2y = 2b \end{cases}$

80. $\begin{cases} -7x - 3y = 4b \\ 3x + y = -2a \end{cases}$

Explaining the Concepts

81. Suppose you are given the system of equations

$\begin{cases} x - 3y = 6 \\ 2x + y = 5 \end{cases}$. Which variable would you choose to eliminate? Why? List the steps you would use to solve this system.

82. In the process of solving a system of linear equations by elimination, what tips you off that the system is consistent with dependent equations? What tips you off that the system is inconsistent?

83. The system of equations $\begin{cases} 6x + 3y = 3 & (1) \\ -4x - y = 1 & (2) \end{cases}$ appeared on a test. You solve the system by multiplying equation (2) by 3 and adding, to eliminate the variable y. Your friend multiplies the first equation by 2 and the second equation by 3 to eliminate the variable x. Assuming that neither of you makes any algebraic errors, which of you will obtain the correct answer? Explain your response. In addition to these two strategies, are there other strategies that may be used to solve the system?

84. List the three methods presented in this chapter for solving a system of linear equations in two variables. Explain the advantages of each one.

Putting the Concepts Together (Sections 4.1–4.3)

We designed these problems so that you can review Sections 4.1–4.3 and show your mastery of the concepts. Take time to work these problems before proceeding with the next section. The answers are located at the back of the text on page AN-18.

1. Determine whether the given ordered pairs are solutions of the given system of equations.

$$\begin{cases} 4x + y = -20 \\ y = -\dfrac{1}{6}x + 3 \end{cases}$$

(a) $\left(3, \dfrac{5}{2}\right)$ **(b)** $(-6, 4)$ **(c)** $(-4, -4)$

2. Suppose that you begin to solve a system of linear equations where each equation is written in slope-intercept form. You notice that the slopes and y-intercepts of the two lines are equal.

(a) How many solutions exist?

(b) State whether the system is consistent or inconsistent.

(c) If the system is consistent, state whether the equations in the system are dependent or independent.

3. Suppose that you begin to solve a system of linear equations where each equation is written in slope-intercept form. The slopes of the two lines are not equal, but the y-intercepts are the same.

(a) How many solutions exist?

(b) State whether the system is consistent or inconsistent.

(c) If the system is consistent, state whether the equations in the system are dependent or independent.

In Problems 4–13, solve each system of equations using any method: graphing, substitution, or elimination.

4. $\begin{cases} 4x + y = 5 \\ -x + y = 0 \end{cases}$ **5.** $\begin{cases} y = -\dfrac{2}{5}x + 1 \\ y = -x + 4 \end{cases}$

6. $\begin{cases} x = 2y + 11 \\ 3x - y = 8 \end{cases}$ **7.** $\begin{cases} 4x + 3y = -4 \\ x + 5y = -1 \end{cases}$

8. $\begin{cases} y = -2x - 3 \\ y = \dfrac{1}{2}x + 7 \end{cases}$ **9.** $\begin{cases} -2x + 4y = 2 \\ 3x + 5y = -14 \end{cases}$

10. $\begin{cases} -\dfrac{3}{4}x + \dfrac{2}{3}y = \dfrac{9}{4} \\ 3x - \dfrac{1}{2}y = -\dfrac{5}{2} \end{cases}$

11. $\begin{cases} 3(y + 3) = 1 + 4(3x - 1) \\ x = \dfrac{y}{4} + 1 \end{cases}$

12. $\begin{cases} 0.4x - 2.5y = -6.5 \\ x + y = 5.5 \end{cases}$

13. $\begin{cases} 2(y + 1) = 3x + 4 \\ x = \dfrac{2}{3}y - 3 \end{cases}$

4.4 Solving Direct Translation, Geometry, and Uniform Motion Problems Using Systems of Linear Equations

Objectives

1 Model and Solve Direct Translation Problems

2 Model and Solve Geometry Problems

3 Model and Solve Uniform Motion Problems

Are You Ready For This Section?

Before getting started, take the following readiness quiz. If you get a problem wrong, go back to the section cited and review the material.

R1. If you travel at an average speed of 45 miles per hour for 3 hours, how far will you travel? [Section 2.7, pp. 142–144]

R2. Suppose that you have $3600 in a savings account. The bank pays 1.5% annual simple interest. What is the interest paid after 1 month? [Section 2.4, pp. 109–110]

In Sections 2.5–2.7 we modeled and solved problems using a linear equation with a single variable. In this section, we learn how to model problems with two unknowns using a system of equations.

Ready?...Answers **R1.** 135 miles
R2. $4.50

If you haven't done so already, review the problem-solving strategy given on page 124 in Section 2.5.

▶ ❶ Model and Solve Direct Translation Problems

Remember, direct translation problems are problems where the words describing the problem are translated to the language of mathematics.

EXAMPLE 1   **Fun with Numbers**

The sum of two numbers is 45. Twice the first number minus the second number is 27. Find the numbers.

Solution

Step 1: Identify We are looking for two unknown numbers.

Step 2: Name Let x represent the first number and y represent the second number.

Step 3: Translate The sum of the two numbers is 45, so we know that

$$x + y = 45 \quad \text{Equation (1)}$$

Twice the first number minus the second number is 27. So,

$$2x - y = 27 \quad \text{Equation (2)}$$

Combine equations (1) and (2) to form the following system:

$$\begin{cases} x + y = 45 & (1) \\ 2x - y = 27 & (2) \end{cases} \quad \text{The Model}$$

Step 4: Solve We will use the method of elimination since the coefficients of y are opposites.

$$\begin{cases} x + y = 45 & (1) \\ 2x - y = 27 & (2) \end{cases}$$

Add equations (1) and (2): $\quad 3x \quad\quad = 72$

Divide both sides by 3: $\quad\quad x \quad\quad = 24$

Let $x = 24$ in equation (1) and solve for y.

Equation (1): $\quad\quad x + y = 45$

Let $x = 24$: $\quad\quad 24 + y = 45$

Subtract 24 from both sides: $\quad\quad y = 21$

Step 5: Check The sum of 24 and 21 is 45. Twice 24 minus 21 is 48 minus 21, which equals 27.

Step 6: Answer The two numbers are 24 and 21.   ●

Quick ✓

1. The sum of two numbers is 104. The second number is 25 less than twice the first number. Find the numbers.

▶ ❷ Model and Solve Geometry Problems

Formulas from geometry are needed to solve certain types of problems. Remember that a geometry formula is a model that describes relationships and shapes.

EXAMPLE 2   **Enclosing a Yard with a Fence**

The Freese family owns a wooded lot that is on a lake. They want to enclose the lot with a fence but do not want to put up a fence along the lake. Dave Freese determined that he will need 240 feet of fence. He also knows that the width of the lot is 30 feet less than the length. See Figure 13. What are the dimensions of the lot?

Solution

Step 1: Identify We are looking for the length and the width of the lot.

Step 2: Name Let l represent the length and w represent the width of the lot.

Step 3: Translate The perimeter P of a rectangle, excluding the waterfront, is $P = l + 2w$, where l is the length and w is the width. So we know that

$$l + 2w = 240 \qquad \text{Equation (1)}$$

In addition, the width is 30 feet less than the length.

$$w = l - 30 \quad \text{Equation (2)}$$

equations (1) and (2) form the following system:

$$\begin{cases} l + 2w = 240 & (1) \\ \quad\quad w = l - 30 & (2) \end{cases} \quad \text{The Model}$$

Figure 13

Step 4: Solve Since equation (2) is already solved for w, we will use the method of substitution and let $w = l - 30$ in equation (1).

Equation (1):	$l + 2w = 240$
Let $w = l - 30$ in Equation (1):	$l + 2(l - 30) = 240$
Distribute:	$l + 2l - 60 = 240$
Combine like terms:	$3l - 60 = 240$
Add 60 to both sides:	$3l = 300$
Divide both sides by 3:	$l = 100$

Let $l = 100$ in equation (2) and solve for w.

Equation (2):	$w = l - 30$
Let $l = 100$:	$w = 100 - 30$
Simplify:	$w = 70$

Step 5: Check With $l = 100$ and $w = 70$, the perimeter (excluding the waterfront) would be $100 + 2(70) = 240$ feet. The width (70 feet) is 30 feet less than the length (100 feet).

Step 6: Answer The length of the wooded lot is 100 feet and the width is 70 feet. ●

Quick ✓

2. A rectangular field has a perimeter of 400 yards. The length of the field is three times the width. What is the length of the field? What is the width of the field?

▶   Recall from Section 2.7 that complementary angles are two angles whose measures sum to 90°. Each angle is called the *complement* of the other. For example, the angles shown in Figure 14(a) on the following page are complements because their measures sum to 90°. Supplementary angles are two angles whose measures sum to 180°. Each angle is called the *supplement* of the other. The angles shown in Figure 14(b) are supplementary.

Figure 14

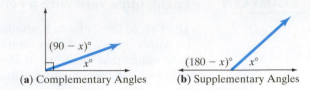

(a) Complementary Angles   (b) Supplementary Angles

The next example was first presented in Section 2.7. In Section 2.7, we solved the problem by developing a model that involved only one unknown. We now show how the same problem can be solved by developing a model of two equations containing two unknowns.

EXAMPLE 3 **Solve a Complementary Angle Problem**

Find the measure of two complementary angles such that the measure of the larger angle is 6° greater than twice the measure of the smaller angle.

Solution

Step 1: Identify This is a complementary angle problem. We are looking for the measure of the two angles whose sum is 90°.

Step 2: Name Let x represent the measure of the smaller angle and y represent the measure of the larger angle.

Step 3: Translate Because these are complementary angles, we know that the sum of the measures of the angles must be 90°. So we have

$$x + y = 90 \quad \text{Equation (1)}$$

The measure of the larger angle y is 6° more than twice the measure of the smaller angle, x. This leads to the following equation.

$$y = 2x + 6 \quad \text{Equation (2)}$$

Equations (1) and (2) form the following system.

$$\begin{cases} x + y = 90 & (1) \\ \quad\;\; y = 2x + 6 & (2) \end{cases} \quad \text{The Model}$$

Step 4: Solve Equation (2) is already solved for y, so let $y = 2x + 6$ in equation (1).

Equation (1):	$x + y = 90$
Let $y = 2x + 6$:	$x + (2x + 6) = 90$
Combine like terms:	$3x + 6 = 90$
Subtract 6 from each side of the equation:	$3x = 84$
Divide each side by 3:	$x = 28$

Now we let $x = 28$ in equation (2) to find the measure of the larger angle, y.

Equation (2):	$y = 2x + 6$
$x = 28$:	$y = 2(28) + 6$
	$y = 62$

Step 5: Check The measure of the smaller angle, x, is 28°. The measure of the larger angle, y, is 62°. The sum of the measures of these angles is 90° since 28° + 62° = 90°. The measure of the larger angle is 6° more than twice the measure of the smaller angle since 2(28°) + 6° = 62°. The answers check!

Step 6: Answer The two complementary angles measure 28° and 62°.

3. *True or False* Complementary angles are angles whose measures sum to 90°.

4. Supplementary angles are angles whose measures sum to ____.

In Problems 5 and 6, solve each problem for the unknown angle measures.

5. Find two complementary angles such that the measure of the larger angle is 18° more than the measure of the smaller angle.

6. Find two supplementary angles such that the measure of the larger angle is 16° less than three times the measure of the smaller angle.

▶ ❸ Model and Solve Uniform Motion Problems

Let's now look at a problem involving uniform motion. Remember, these problems use the fact that distance equals rate times time $(d = rt)$.

EXAMPLE 4 **Uniform Motion—Flying a Piper Aircraft**

The airspeed of a plane is its speed through the air. This speed is different from the plane's groundspeed—its speed relative to the ground. The groundspeed of an airplane is affected by the wind. Suppose that a Piper aircraft flying west a distance of 500 miles takes 5 hours. The return trip takes 4 hours. Find the airspeed of the plane and the effect that wind resistance has on the plane.

Solution

Step 1: Identify This is a uniform motion problem. We want to determine the airspeed of the plane and the effect of wind resistance.

Step 2: Name There are two unknowns in the problem—the airspeed of the plane and the effect of wind resistance on the plane. We will let a represent the airspeed of the plane and w represent the effect of wind resistance.

Step 3: Translate Going west, the plane is flying into the jet stream, so the plane is slowed down by the wind. Therefore, the groundspeed of the plane will be $a - w$. Going east, the wind (jet stream) is helping the plane, so the groundspeed of the plane will be $a + w$. We set up Table 3.

Table 3

	Distance	Rate	Time
With Wind (East)	500	$a + w$	4
Against Wind (West)	500	$a - w$	5

Going with the wind, we use distance = rate · time and obtain the equation

$$4(a + w) = 500 \quad \text{or} \quad 4a - 4w = 500 \quad \text{Equation (1)}$$

Going against the wind, we have the equation

$$5(a - w) = 500 \quad \text{or} \quad 5a - 5w = 500 \quad \text{Equation (2)}$$

Equations (1) and (2) form a system of two linear equations containing two unknowns.

$$\begin{cases} 4a + 4w = 500 & (1) \\ 5a - 5w = 500 & (2) \end{cases} \quad \text{The Model}$$

Work Smart

Instead of distributing 4 into the equation $4(a + w) = 500$, we could divide both sides by 4. The same approach can be taken with the equation $5(a - w) = 500$—divide both sides of the equation by 5. Try solving the system this way. Is it easier?

Step 4: Solve We will use the elimination method by multiplying equation (1) by 5 and equation (2) by 4 and then adding equations (1) and (2).

$$\begin{cases} 5(4a + 4w) = 5(500) & (1) \\ 4(5a - 5w) = 4(500) & (2) \end{cases}$$

Distribute:
$$\begin{cases} 20a + 20w = 2500 & (1) \\ 20a - 20w = 2000 & (2) \end{cases}$$

Add: $\quad\quad 40a \quad\quad = 4500$

Divide both sides by 40: $\quad a \quad\quad = 112.5$

We use equation (1) with $a = 112.5$ to find the effect of wind resistance.

Equation (1): $\quad 4a + 4w = 500$

$a = 112.5:\ 4(112.5) + 4w = 500$

$\quad\quad\quad\quad\quad\quad\quad 450 + 4w = 500$

Subtract 450 from both sides: $\quad 4w = 50$

Divide both sides by 4: $\quad\quad w = 12.5$

Step 5: Check Flying west, the groundspeed of the plane is $112.5 - 12.5 = 100$ miles per hour, which agrees with the plane flying 500 miles in 5 hours for an average speed of 100 miles per hour. Flying east, the groundspeed of the plane is $112.5 + 12.5 = 125$ miles per hour, which agrees with the plane flying 500 miles in 4 hours for an average speed of 125 miles per hour. Everything checks!

Step 6: Answer The airspeed of the plane is 112.5 miles per hour. The impact of wind resistance on the plane is 12.5 miles per hour.

Quick ✔

7. *True or False* In uniform motion problems we use the equation $d = rt$, where d is distance, r is rate, and t is time.

8. Suppose that a plane flying 1200 miles west requires 4 hours and flying 1200 miles east requires 3 hours. Find the airspeed of the plane and the effect that wind resistance has on the plane.

4.4 Exercises   MyMathLab®   Math XL PRACTICE

Exercise numbers in green have complete video solutions in MyMathLab.

Problems 1–8 are the Quick ✔s that follow the EXAMPLES.

Building Skills

In Problems 9 and 10, complete the system of equations. Do not solve the system. See Objective 1.

9. The sum of two numbers is 56. Two times the smaller number is 12 more than one-half the larger number. Let x represent the smaller number and y represent the larger number.

$$\begin{cases} x + y = 56 \\ \underline{\quad\quad} = \underline{\quad\quad} \end{cases}$$

10. The sum of two numbers is 90. If 20 is added to 3 times the smaller number, the result exceeds twice

the larger number by 50. Let a represent the smaller number and b represent the larger number.

$$\begin{cases} a + b = 90 \\ \underline{\quad\quad} = \underline{\quad\quad} \end{cases}$$

In Problems 11 and 12, complete the system of equations. Do not solve the system. See Objective 2.

△ 11. The perimeter of a rectangle is 59 inches. The length is 5 inches less than twice the width. Let l represent the length of the rectangle and w represent the width of the rectangle.

$$\begin{cases} l = 2w - 5 \\ \underline{\quad\quad} = 59 \end{cases}$$

△ 12. The perimeter of a rectangle is 212 centimeters. The length is 8 centimeters less than three times the width. Let *l* represent the length of the rectangle and *w* represent the width of the rectangle.

$$\begin{cases} 2w + 2l = 212 \\ \underline{\hspace{2cm}} = \underline{\hspace{1.5cm}} \end{cases}$$

In Problems 13 and 14, complete the system of equations. Do not solve the system. See Objective 3.

13. A boat is rowed down a river a distance of 16 miles in 2 hours, and it is rowed upstream the same distance in 8 hours. Let *r* represent the rate of the boat in still water in miles per hour and *c* represent the rate of the stream in miles per hour.

$$\begin{cases} 2(r + c) = 16 \\ \underline{\hspace{2cm}} = \underline{\hspace{0.5cm}} \end{cases}$$

14. A plane flew with the wind for 3 hours, covering 1200 miles. It then returned over the same route to the airport against the wind in 4 hours. Let *a* represent the airspeed of the plane and *w* represent the effect of wind resistance.

$$\begin{cases} 3(a + w) = 1200 \\ \underline{\hspace{2cm}} = \underline{\hspace{0.5cm}} \end{cases}$$

Applying the Concepts

15. Fun with Numbers Find two numbers whose sum is 82 and whose difference is 16.

16. Fun with Numbers Find two numbers whose sum is 55 and whose difference is 17.

17. Fun with Numbers Two numbers sum to 51. Twice the first subtracted from the second is 9. Find the numbers.

18. Fun with Numbers Two numbers sum to 32. Twice the larger subtracted from the smaller is −22. Find the numbers.

19. Oakland Baseball The attendance at the games on two successive nights of Oakland A's baseball was 77,000. The attendance on Thursday's game was 7000 more than two-thirds of the attendance at Friday night's game. How many people attended the baseball game each night?

20. Winning Baseball The number of games the Oakland A's are expected to win this year is 8 fewer than two-thirds of the number that they are expected to lose. If there are 162 games in a season, how many games are the A's expected to win?

21. Investments Suppose that you received an unexpected inheritance of $21,000. You have decided to invest the money by placing some of the money in stocks and the remainder in bonds. To diversify, you decide that four times the amount invested in bonds should equal three times the amount invested in stocks. How much should be invested in stocks? How much should be invested in bonds?

22. Investments Marge and Homer have $40,000 to invest. Their financial advisor has recommended that they diversify by placing some of the money in stocks and the remainder in bonds. Based upon current market conditions, he has recommended that two times the amount in bonds should equal three times the amount invested in stocks. How much should be invested in stocks? How much should be invested in bonds?

△ **23. Fencing a Garden** Melody Jackson wishes to enclose a rectangular garden with fencing, using the side of her garage as one side of the rectangle. A neighbor gave her 30 feet of fencing, and Melody wants the length of the garden along the garage to be 3 feet more than the width. What are the dimensions of the garden?

△ **24. Perimeter of a Parking Lot** A rectangular parking lot has a perimeter of 125 feet. The length of the parking lot is 10 feet more than the width. What is the length of the parking lot? What is the width?

△ **25. Perimeter** The perimeter of a rectangle is 70 meters. If the width is 40% of the length, find the dimensions of the rectangle.

△ **26. Window Dimensions** The perimeter of a rectangular window is 162 inches. If the height of the window is 80% of the width, find the dimensions of the window.

△ **27. Working with Complements** The measure of one angle is 15° more than half the measure of its complement. Find the measures of the two angles.

△ **28. Working with Complements** The measure of one angle is 10° less than the measure of three times its complement. Find the measures of the two angles.

△ **29. Finding Supplements** The measure of one angle is 30° less than one-third the measure of its supplement. Find the measures of the two angles.

△ **30. Finding Supplements** The measure of one angle is 20° more than two-thirds the measure of its supplement. Find the measures of the two angles.

31. Kayaking Michael is kayaking on the Kankakee River. His overall speed is 3.5 mph against the current and 4.3 mph with the current. Find the speed of the current and the speed Michael can paddle in still water.

32. Southwest Airlines Plane A Southwest Airlines plane can fly 455 mph against the wind and 515 mph when it flies with the wind. Find the effect of the wind and the groundspeed of the airplane.

33. Biking Suppose that José bikes into the wind for 60 miles and it takes him 6 hours. After a long rest, he returns (with the wind at his back) in 5 hours. Determine the speed at which José can ride his bike in still air and the effect that the wind had on his speed.

34. Rowing On Monday afternoon, Andrew rowed his boat with the current for 4.5 hours and covered 27 miles, stopping in the evening at a campground. On Tuesday morning he returned to his starting point against the current in 6.75 hours. Find the speed of the current and the rate at which Andrew rowed in still water.

35. Outbound from Chicago Two trains leave Chicago going opposite directions, one going north and the other going south. The northbound train is traveling 12 mph slower than the southbound train. After 4 hours the trains are 528 miles apart. Find the speed of each train.

36. Horseback Riding Monica and Gabriella enjoy riding horses at a dude ranch in Colorado. They decide to go down different trails, which travel in opposite direction, agreeing to meet back at the ranch later in the day. Monica's horse is going 4 mph faster than the Gabriella's, and after 2.5 hours, they are 20 miles apart. Find the speed of each horse.

37. Riding Bikes Vanessa and Richie are riding their bikes down a trail to the next campground. Vanessa rides at 10 mph, and Richie rides at 8 mph. Since Vanessa is a little speedier, she stays behind and cleans up camp for 30 minutes before leaving. How long has Richie been riding when Vanessa catches up to Richie?

38. Running a Marathon Rafael and Edith are running in a marathon to raise money for breast cancer. Rafael can run at 12 mph, and Edith runs at 10 mph. Unfortunately, Rafael lost his car keys and went back to look for them while Edith started down the course. If Rafael was back at the starting line 15 minutes after Edith left, how long will it take him to catch up to her?

39. Computing Wind Speed With a tailwind, a small Piper aircraft can fly 600 miles in 3 hours. Against this same wind, the Piper can fly the same distance in 4 hours. Find the effect of the wind and the average airspeed of the Piper.

40. Computing Wind Speed The average airspeed of a single-engine aircraft is 150 miles per hour. If the aircraft flew the same distance in 2 hours with the wind as it flew in 3 hours against the wind, what was the effect of the wind on the plane?

Extending the Concepts

Problems 41 and 42 are a popular type of problem that appeared in mathematics textbooks in the 1970s and 1980s. Can you find the answers?

41. Digits The sum of the digits of a two-digit number is 6. If the digits are reversed, the difference between the new number and the original number is 18. Find the original number.

42. Digits The sum of the digits of a two-digit number is 7. If the digits are reversed, the difference between the new number and the original number is 27. Find the original number.

4.5 Solving Mixture Problems Using Systems of Linear Equations

Objectives

1 Draw Up a Plan for Solving Mixture Problems

2 Set Up and Solve Money Mixture Problems

3 Set Up and Solve Dry Mixture and Percent Mixture Problems

Are You Ready for This Section?

Before getting started, take the following readiness quiz. If you get a problem wrong, go back to the section cited and review the material.

R1. Suppose that Roberta has a credit card balance of $1200. Each month, the credit card charges 14% annual simple interest on any outstanding balances. What is the interest that Roberta will be charged on this loan after one month? What is Roberta's credit card balance after one month? [Section 2.4, pp. 109–110]

R2. Solve: $0.25x = 80$ [Section 2.3, pp. 101–102]

▶ ❶ Draw Up a Plan for Solving Mixture Problems

Problems that involve mixing two or more substances are called **mixture problems.** They can be solved using the formula *number of units of the same kind · rate = amount*. The *rates* that we use in mixture problems include cost per person, interest rates, and cost per pound. A table like the one below can help organize the information given in the problem.

Work Smart

Use a chart to organize your thoughts and keep track of information.

	Number of Units	·	Rate	=	Amount
Item 1					
Item 2					
Total					

Because each problem will be slightly different, we will adjust the titles of the categories, but the formula *number of units of the same kind · rate = amount* will remain the same. Let's begin by practicing filling in the chart.

EXAMPLE 1   **Set Up a Table for a Mixture Problem**

A class of schoolchildren and their adult chaperones took a field trip to the zoo. The total cost of admission was $230, and 40 children and adults went on the field trip. If the children paid $5 each and the adults paid $8 each, how many adults and how many children went on the trip? Fill in a table that summarizes the information in the problem. Do not solve the problem.

Solution

Both the number of children and the number of adults are unknown. We let a represent the number of adults on the field trip and c represent the number of children. We can now fill in Table 4 with the quantities that we know.

Table 4

	Number	·	Cost per Person	=	Amount
Adults	a		8		$8a$
Children	c		5		$5c$
Total	40				230

Work Smart

Not every entry in the chart must be filled.

Quick ✓

1. The formula we use to solve mixture problems is number of units of the same kind · ____ = _____.

2. A one-day admission ticket to Cedar Point Amusement Park costs $42.95 for adults and $15.95 for children. Two families purchased nine tickets and spent $332.55 for the tickets. How many adult tickets did the families purchase? Fill in a chart that summarizes the information in the problem. Do not solve the problem.

▶ ❷ **Set Up and Solve Money Mixture Problems**

To solve money mixture problems, we will set up a table to organize the given information and then use the table to develop a system of equations (the model). We continue to use the six-step procedure that was introduced in Section 2.5.

(EXAMPLE 2) **Solve a Coin Problem**

A third-grade class contributed $9.20 in dimes and quarters to the Red Cross. In all there were 56 coins. Find the number of dimes and quarters that they contributed.

Solution

Step 1: Identify This is a money mixture problem. We want to know the number of dimes and quarters the children contributed, and we know that $9.20 was given.

Step 2: Name Let q represent the number of quarters and d represent the number of dimes.

Step 3: Translate We fill in Table 5 with the information that we know.

Work Smart

It's always a good idea to name your variable so that it reminds you of what it represents, as in q for the number of quarters.

Table 5

	Number of Coins	·	Value per Coin in Dollars	=	Total Value
Quarters	q		0.25		$0.25q$
Dimes	d		0.10		$0.10d$
Total	56				9.20

The total value of the coins is $9.20. Based on the information in Table 5, we have that

$$\overbrace{0.25q}^{\text{value of quarters}} + \overbrace{0.10d}^{\text{value of dimes}} = \overbrace{9.20}^{\text{total value of coins}} \qquad \text{Equation (1)}$$

There are a total of 56 coins, so

$$q + d = 56 \qquad \text{Equation (2)}$$

We use equations (1) and (2) to form a system of equations.

$$\begin{cases} 0.25q + 0.1d = 9.20 & (1) \\ q + d = 56 & (2) \end{cases} \quad \text{The Model}$$

Step 4: Solve We will use the elimination method.

$$\text{Multiply equation (2) by } -0.1: \begin{cases} 0.25q + 0.1d = 9.20 & (1) \\ -0.1(q + d) = -0.1(56) & (2) \end{cases}$$

$$\text{Distribute } -0.1 \text{ in equation (2):} \begin{cases} 0.25q + 0.1d = 9.20 & (1) \\ -0.1q - 0.1d = -5.6 & (2) \end{cases}$$

$$\text{Add equations (1) and (2):} \quad 0.15q = 3.6$$

$$\text{Divide both sides by 0.15:} \quad q = 24$$

To determine the value of d, let $q = 24$ in equation (2) and solve for d.

$$\text{Equation (2):} \quad q + d = 56$$

$$q = 24: \quad 24 + d = 56$$

$$\text{Subtract 24 from both sides:} \quad d = 32$$

Step 5: Check Because 24 quarters and 32 dimes adds up to 56 coins and has a value of $24(\$0.25) + 32(\$0.10) = \$6.00 + \$3.20 = \$9.20$, our answer is correct.

Step 6: Answer The children collected 24 quarters and 32 dimes for the Red Cross. ●

Quick ✓

3. You have a piggy bank containing a total of 85 coins in dimes and quarters. If the piggy bank contains $14.50, how many dimes are there in the piggy bank?

Recall that in Section 2.4 we introduced the simple interest formula $I = Prt$, where I is interest, P is principal (an amount borrowed or deposited), r is the annual interest rate (in decimal form), and t is time (in years). When solving money mixture problems involving interest, time is 1 year, so the simple interest formula reduces to $interest = principal \cdot rate \cdot 1$, or $interest = principal \cdot rate$.

EXAMPLE 3 **Solve a Money Mixture Problem Involving Interest**

You have $12,000 invested in a money market account that paid 2% annual interest and a stock fund that paid 8% annual interest. Suppose that you earned $480 in interest at the end of one year. How much was invested in each account?

Solution

Step 1: Identify This is a mixture problem involving simple interest. We need to know how much was invested in the money market account and how much was invested in the stock fund to earn $480 in interest.

Step 2: Name Let m represent the amount invested in the money market account and s represent the amount invested in stocks.

Step 3: Translate We organize the given information in Table 6.

Table 6

	Principal $ ·	Rate % =	Interest $
Money Market	m	0.02	0.02m
Stock Fund	s	0.08	0.08s
Total	12,000		480

The total interest is the sum of the interest from both investments:

Interest earned from money market   Interest earned from stock fund   Total interest earned

$$0.02m \quad + \quad 0.08s \quad = \quad 480 \qquad \text{Equation (1)}$$

The total investment was $12,000, so that the amount invested in the money market account plus the amount invested in the stock fund equals $12,000:

$$m + s = 12,000 \quad \text{Equation (2)}$$

We use equations (1) and (2) to form a system of equations.

$$\begin{cases} 0.02m + 0.08s = 480 & (1) \\ m + s = 12,000 & (2) \end{cases} \quad \text{The Model}$$

Step 4: Solve We will use the elimination method.

Multiply equation (2) by −0.02:
$$\begin{cases} 0.02m + 0.08s = 480 & (1) \\ -0.02(m + s) = -0.02(12,000) & (2) \end{cases}$$

Distribute −0.02 in equation (2):
$$\begin{cases} 0.02m + 0.08s = 480 & (1) \\ -0.02m - 0.02s = -240 & (2) \end{cases}$$

Add equations (1) and (2): $\quad 0.06s = 240$

Divide both sides by 0.025: $\quad s = 4000$

To determine the value of m, let $s = 4000$ in equation (2) and solve for m.

Equation (2): $\quad m + s = 12,000$

$s = 4000$: $\quad m + 4000 = 12,000$

Subtract 4000 from both sides: $\quad m = 8000$

Step 5: Check The simple interest earned each year on the money market account is ($8000)(0.02)(1) = $160 and on the stock fund is ($4000)(0.08)(1) = 320. The total interest earned is $160 + $320 = $480. Further, the total amount invested is $8000 + $4000 = $12,000. The solution checks.

Step 6: Answer You invested $8000 in the money market account and $4000 in the stock fund. ●

Quick ✓

4. The simple interest formula states that interest = _____ · ___ · ___.

5. *True or False* When we solve money mixture problems involving interest, we assume that $t = 1$.

6. Faye has recently retired and requires an extra $5500 per year in income. She has $90,000 to invest and can invest in either an Aa-rated bond that pays 5% per annum or a B-rated bond paying 7% annually. How much should be placed in each investment for Faye to achieve her goal exactly?

▶ ❸ Set Up and Solve Dry Mixture and Percent Mixture Problems

Mixture of Two Substances—Dry Mixture Problems

Often, new blends are created by mixing two quantities. For example, a chef might mix buckwheat flour with wheat flour to make buckwheat pancakes. Or a coffee shop might mix two different types of coffee to create a new coffee blend.

EXAMPLE 4 **Solve a Dry Mixture Problem Using the Mixture Model**

A store manager combined nuts and M&M's to make a trail mix. She created 10 pounds of the mix and sold it for $5.10 per pound. If the price of the nuts was $3 per pound and the price of the M&M's was $6 per pound, how many pounds of each type did she use?

Solution

Step 1: Identify This is a mixture problem. We want to know the number of pounds of nuts and the number of pounds of M&M's that were required in the trail mix.

Step 2: Name Let n represent the number of pounds of nuts and m represent the number of pounds of M&M's in the mix.

Step 3: Translate We set up Table 7.

Table 7

	Number of Pounds	·	Price $/Pound	=	Revenue
Nuts	n		3		$3n$
M&M's	m		6		$6m$
Blend	10		5.10		$5.10(10) = 51$

If the mixture contains n pounds of nuts, the revenue is $\$3n$. If the mixture contains m pounds of M&M's, the revenue is $\$6m$. Ten pounds of the blend selling for $\$5.10$ per pound yields $\$5.10(10) = \$51$ in revenue.

$$\begin{pmatrix} \text{Price per pound} \\ \text{of nuts} \\ \$3 \end{pmatrix} \begin{pmatrix} \text{Pounds of} \\ \text{nuts} \\ n \end{pmatrix} + \begin{pmatrix} \text{Price per pound} \\ \text{of M\&M's} \\ \$6 \end{pmatrix} \begin{pmatrix} \text{Pounds of} \\ \text{M\&M's} \\ m \end{pmatrix} = \begin{pmatrix} \text{Price per pound} \\ \text{of blend} \\ \$5.10 \end{pmatrix} \begin{pmatrix} \text{Pounds of} \\ \text{blend} \\ 10 \end{pmatrix}$$

We get the equation

$$3n + 6m = 51 \quad \text{Equation (1)}$$

Work Smart

Remember that mixtures can include interest (money), solids (nuts), liquids (chocolate milk), and even gases (Earth's atmosphere).

The number of pounds of nuts plus the number of M&M's equals 10 pounds:

$$n + m = 10 \quad \text{Equation (2)}$$

We use equations (1) and (2) to form a system of equations.

$$\begin{cases} 3n + 6m = 51 & (1) \\ n + m = 10 & (2) \end{cases} \quad \text{The Model}$$

Work Smart

The system $\begin{cases} 3n + 6m = 51 \\ n + m = 10 \end{cases}$ could also have been solved by elimination. Which method do you prefer?

Step 4: Solve We solve the system by substitution by solving equation (2) for n.

Equation (2):	$n + m = 10$
Subtract m from both sides:	$n = 10 - m$
Let $n = 10 - m$ in equation (1):	$3(10 - m) + 6m = 51$
Distribute the 3:	$30 - 3m + 6m = 51$
Combine like terms:	$30 + 3m = 51$
Subtract 30 from both sides:	$3m = 21$
Divide both sides by 3:	$m = 7$

If $m = 7$ pounds, then $n = 10 - m = 10 - 7 = 3$ pounds.

Step 5: Check A mixture of 3 pounds of nuts and 7 pounds of M&M's would sell for $\$3(3) + \$6(7) = \$9 + \$42 = \$51$, which equals the revenue obtained from selling the blend. In addition, the blend weighs 7 pounds + 3 pounds = 10 pounds. The solution checks.

Step 6: Answer The blend has 3 pounds of nuts and 7 pounds of M&M's.

Quick ✓

7. A coffeehouse has Brazilian coffee that sells for $\$6$ per pound and Colombian coffee that sells for $\$10$ per pound. How many pounds of each coffee should be mixed to obtain 20 pounds of a blend that costs $\$9$ per pound?

▶ **Mixture of Two Substances—Percent Mixture Problems**

The last example dealt with mixing two dry substances. We now turn our attention to mixing two liquids. These problems require using percents.

EXAMPLE 5 **Solve a Percent Mixture Problem**

You work in the chemistry stockroom and your instructor asks you to prepare 4 liters of 15% hydrochloric acid (HCl). The supply room has a bottle of 12% HCl and another of 20% HCl. How much of each should you mix so that your instructor has the required solution?

Solution

Step 1: Identify This is a percent mixture problem. We want to know the number of liters of 12% HCl that must be mixed with a 20% HCl solution to prepare 4 liters of a 15% HCl solution.

Step 2: Name We let x represent the number of liters of the 12% HCl solution and y represent the number of liters of 20% HCl.

Step 3: Translate We fill in Table 8 with the information that we know.

Table 8

	Number of Liters $\cdot$	Concentration (part of solution that is pure HCl per liter) $=$	Amount of Pure HCl
12% HCl Solution	x	0.12	$0.12x$
20% HCl Solution	y	0.20	$0.20y$
Total	4	0.15	$(0.15)(4)$

The amount of pure HCl from the 12% solution plus the amount of pure HCl from the 20% solution should yield the amount of pure HCl in 4 liters of 15% solution.

$$\underbrace{0.12x}_{\text{Pure HCl from 12\%}} + \underbrace{0.20y}_{\text{Pure HCl from 20\%}} = \underbrace{(0.15)(4)}_{\text{Part of the total that is pure HCl}} \quad \text{Equation (1)}$$

The amount of HCl solution should equal 4 liters.

$$x + y = 4 \quad \text{Equation (2)}$$

Use equations (1) and (2) to form a system of equations.

$$\begin{cases} 0.12x + 0.20y = 0.6 & (1) \\ x + y = 4 & (2) \end{cases} \quad \text{The Model}$$

Step 4: Solve Solve the system by substitution by solving equation (2) for y.

Equation (2):	$x + y = 4$
Subtract x from both sides:	$y = 4 - x$
Substitute $4 - x$ for y in equation (1):	$0.12x + 0.20(4 - x) = 0.6$
Use the Distributive Property:	$0.12x + 0.8 - 0.20x = 0.6$
Combine like terms:	$-0.08x + 0.8 = 0.6$
Subtract 0.8 from each side of the equation:	$-0.08x = -0.2$
Divide each side by -0.08:	$x = 2.5$

Because x represents the number of liters of 12% HCl solution, you need 2.5 liters of 12% HCl solution. You need $4 - x = 4 - 2.5 = 1.5$ liters of 20% HCl solution.

Step 5: Check The amount of HCl in 2.5 liters of 12% HCl is $0.12(2.5) = 0.3$ liter. The amount of HCl in 1.5 liters of 20% HCl is $0.20(1.5) = 0.3$ liter. Combined, these solutions give us 0.6 liter of pure HCl. The amount of HCl in 4 liters of 15% HCl is $0.15(4) = 0.6$ liter. Our answer is correct.

Step 6: Answer 2.5 liters of 12% HCl solution must be mixed with 1.5 liters of 20% HCl solution to form the 4 liters of 15% HCl solution.

Quick ✔

8. You are an ice cream maker and decide that ice cream with 9% butterfat is the best to sell. How many gallons of ice cream with 5% butterfat should be mixed with ice cream that is 15% butterfat to make 200 gallons of the desired blend?

4.5 Exercises

MyMathLab®    PRACTICE

Exercise numbers in green have complete video solutions in MyMathLab.

Problems 1–8 are the Quick ✔s that follow the **EXAMPLES**.

Building Skills

In Problems 9–14, fill in the table from the information given. Then write the system that models the problem. Do not solve the system. See Objective 1.

9. The PTA had an ice cream social and sold adult tickets for $4 and student tickets for $1.50. The PTA treasurer found that 215 tickets had been sold and the receipts were $580.

	Number ·	Cost per Person =	Total Value
Adult Tickets	?	?	?
Student Tickets	?	?	?
Total	?		?

10. John Murphy sells jewelry at art shows. He sells bracelets for $10 and necklaces for $15. At the end of one day, John found that he had sold 69 pieces of jewelry and had receipts of $895.

	Number ·	Cost per Item =	Total Value
Bracelets	?	?	?
Necklaces	?	?	?
Total	?		?

11. Maurice has a savings account that earns 5% simple interest per year and a money market account that earns 3% simple interest. At the end of one year, Maurice received $50 in interest on a total investment of $1600 in the two accounts.

	Principal ·	Rate =	Interest
Savings Account	?	?	?
Money Market	?	?	?
Total	?		?

12. Sherry has a savings account that earns 2.75% simple interest per year and a certificate of deposit (CD) that earns 2% simple interest annually. At the end of one year, Sherry received $37.75 in interest on a total investment of $1700 in the two accounts.

	Principal ·	Rate =	Interest
Savings Account	?	?	?
Certificate of Deposit	?	?	?
Total	?		?

13. A coffee shop wishes to blend two types of coffee to create a breakfast blend. It mixes a mild coffee that sells for $7.50 per pound with a robust coffee that sells for $10.00 per pound. The owner wants to make 12 pounds of breakfast blend that will sell for $8.75 per pound.

	Number of Pounds ·	Price per Pound =	Total Value
Mild Coffee	?	?	?
Robust Coffee	?	?	?
Total	?	?	?

14. A merchant wishes to mix peanuts worth $5 per pound and trail mix worth $2 per pound to yield 40 pounds of a nutty mixture that will sell for $3 per pound.

Number of Pounds	· Price per Pound	= Total Value
Peanuts	?	?	?
Trail Mix	?	?	?
Total	?	?	?

In Problems 15–20, complete the system of linear equations to solve the problem. Do not solve. See Objective 1.

15. A family of 11 decides to take a trip to Water World Water Park. The total cost of admission for the family is $296. If adult tickets cost $32 and children's tickets cost $24, how many adults and how many children went to Water World? Let a represent the number of adult tickets purchased, and let c represent the number of children's tickets purchased.

$$\begin{cases} a + \quad c = 11 \\ \underline{\quad} + \underline{\quad} = 296 \end{cases}$$

16. On a school field trip, 22 people attended a dress rehearsal of the Broadway play "Avenue Q." They paid $274 for the tickets, which cost $15 for each adult and $7 for each child. How many adults and how many children attended "Avenue Q"? Let a represent the number of adult tickets purchased, and let c represent the number of children's tickets purchased.

$$\begin{cases} a + \ c = 22 \\ \underline{\quad} + \underline{\quad} = 274 \end{cases}$$

17. You have a total of $2250 to invest. Account A pays 10% annual interest, and account B pays 7% annual interest. How much should you invest in each account if you would like the investment to earn $195 at the end of one year? Let A represent the amount of money invested in the account that earns 10% annual interest, and let B represent the amount of money invested in the account that earns 7% annual interest.

$$\begin{cases} A + \quad B = 2250 \\ \underline{\quad} + \underline{\quad} = 195 \end{cases}$$

18. You have a total of $2650 to invest. Account A pays 5% annual interest and account B pays 6.5% annual interest. How much should you invest in each account if you would like the investment to earn $155 at the end of one year? Let A represent the amount of money invested in the account that earns 5% annual interest, and let B represent the amount of money invested in the account that earns 6.5% annual interest.

$$\begin{cases} A + \ B = \underline{\quad} \\ \underline{\quad} + \underline{\quad} = 155 \end{cases}$$

19. Flowers on High sells flower bouquets at different prices depending on color: red for $5.85 per bouquet or yellow for $4.20 per bouquet. One day the number of red bouquets sold was 3 more than twice the number of yellow bouquets sold, and the revenue from selling the bouquets was $128.85. How many of each color were sold? Let r represent the number of red bouquets sold, and let y represent the number of yellow bouquets sold.

$$\begin{cases} r = \ 3 \ + 2y \\ \underline{\quad} + \underline{\quad} = \underline{\quad} \end{cases}$$

20. The Latte Shoppe sells Bold Breakfast coffee for $8.60 per pound and Wake-Up coffee for $5.75 per pound. One day, the amount of Bold Breakfast coffee, sold was 2 pounds less than twice the amount of Wake-Up coffee, and the revenue received from selling both types of coffee was $143.45. How many pounds of each type of coffee were sold that day? Let B represent the number of pounds of Bold Breakfast coffee sold, and let W represent the number of pounds of Wake-Up coffee sold. Complete the system of equations:

$$\begin{cases} 8.60B + 5.75W = \underline{\quad} \\ B = \underline{\quad} - \underline{\quad} \end{cases}$$

Applying the Concepts

21. Theater Tickets to a student theater production cost $8 for students and $10 for nonstudents. The receipts for opening night came to $3270 from selling 390 tickets. How many student tickets were sold?

22. Ticket Pricing A ticket on the roller coaster is priced differently for adults and children. One day there were 5 adults and 8 children in a group and the cost of their tickets was $48.50. Another group of 4 adults and 12 children paid $57. What is the price of each type of ticket?

23. Bedding Plants A girls' softball team is raising money for uniforms and travel expenses. They are selling flats of bedding plants for $13.00 and hanging baskets for $18.00. They hope to sell twice as many flats of bedding plants as hanging baskets. If they do, their total revenue will be $8800. How many flats of bedding plants do they hope to sell?

24. Amusement Park An adult discount ticket to King's Island Amusement Park costs $26, and a child's discount ticket to King's Island Amusement Park costs $24.50. A group of 13 friends purchased adult and children's tickets and paid $330.50. How many discount tickets for children were purchased?

25. Tip Jar The tip jar next to the cash register has 150 coins, all in nickels and dimes. If the total value of the tips is $12, how many of each coin are in the jar?

26. Bigger Tips Christa is a waitress and collects her tips at the table. At the end of the shift she has 68 bills in her tip wallet, all ones and fives. If the total value of her tips is $172, how many of each bill does she have?

27. Buying Stamps One day Rosemarie bought 20 first-class stamps and 10 postcard stamps for $11.10. Another day she spent $47.10 on 80 first-class stamps and 50 postcard stamps. Find the cost of each type of stamp.

28. TV Commercials Advertising on television can be expensive, depending on the expected number of viewers. The Fox network sells 30-second spots for $175,000 and one-minute spots for $250,000 during the basketball playoffs. If Fox sold 13 commercial spots and earned $2,650,000, how many of each type did it sell?

29. Investments Harry has $10,000 to invest. He invests in two different accounts, one expected to return 5% and the other expected to return 8%. If he wants to earn $575 for the year, how much should he invest at each rate?

30. Investments Ann has $5000 to invest. She invests in two different accounts, one expected to return 4.5% and the other expected to return 9%. In order to earn $382.50 for the year, how much should she invest at each rate?

31. Stock Return Esmeralda invested $5000 in two stock plans. On the risky plan she earned 12% annual interest, and on the safer plan she earned 8% annual interest. If Esmeralda saw a return on her investment of $528 last year, how much did she invest in each of the stock plans?

32. Real Estate Return Victor, Esmeralda's wealthy brother, also invested his money. He thought he would do better if he invested his money in real estate partnerships. He invested $12,000 in two groups. One of the groups yielded a 13% annual return on a downtown mall and the other a 10.5% annual return on a suburban office center. If Victor saw a profit of $1347.50 on his investment last year, how much did he invest with each group?

33. Olive Blend Juana works at a delicatessen, and it is her turn to make the zesty olive blend. If arbequina olives are $9 per pound and green olives are $4 per pound, how much of each kind of olive should she mix to get five pounds of zesty olive blend that will sell for $6 a pound?

34. Churrascaria Platter Churrascaria is a name that comes from the Brazilian gauchos of the 1800s. To celebrate, the cowboys would take a large variety of different meats and barbeque them. Today, some specialty restaurants still serve Brazilian barbeque in the Churrascaria style. Elis da Silva's Restaurant wants to offer a one-pound Churrascaria platter for a profit of $12. If the restaurant makes a profit of $8 per pound on grilled pork and $14 per pound on barbecue flank steak, how many pounds of each meat can be served on the platter to earn $12?

35. Blending Coffee A coffee manufacturer wants to market a new blend of coffee that will sell for $3.90 per pound by mixing two coffees that sell for $2.75 per pound and $5 per pound, respectively. What amounts of each coffee should be blended to obtain 100 pounds of the desired mixture?

36. Blending Nuts The Nutty Professor store wishes to mix peanuts that sell for $2 per pound with cashews that sell for $6 per pound. The store plans to use five pounds more of peanuts than of cashews in the blend and sell the mixture for a total of $26. How many pounds of each nut should be used in the mixture?

37. Grass Seed A nursery decides to make a grass-seed blend by mixing rye seed that sells for $4.20 per pound with bluegrass seed that sells for $3.75 per pound. The final mix is 180 pounds and sells for $3.95 per pound. How much of each type was used?

38. Stock Account A stock portfolio is currently valued at $5872.44. HiTech is currently selling at $82.01 per share, and IntRNet is currently selling at $26.52 per share. If the number of HiTech shares is 25 less than the number of IntRNet shares, find the number of shares of each.

39. Saline Solutions A lab technician needs 60 ml of a 50% saline solution. How many ml of 30% saline solution should she add to a 60% saline solution to obtain the required mixture?

40. Alcohol A laboratory assistant is asked to mix a 30% alcohol solution with 21 liters of an 80% alcohol solution to make a 60% alcohol solution. How many liters of the 30% alcohol solution should be used?

41. Silver Alloy How many liters of 10% silver must be added to 70 liters of 50% silver to make an alloy that is 20% silver?

42. Paint Four gallons of paint contain 2.5% pigment. How many gallons of paint that is 6% pigment must be added to make a paint that is 4% pigment?

Extending the Concepts

43. Antifreeze A radiator holds 3 gallons. How much of the 25% antifreeze solution should be drained and replaced with pure water to reduce the solution to 15% antifreeze?

44. Antifreeze A radiator holds 6 liters. How much of the 20% antifreeze solution should be drained and replaced with pure antifreeze to bring the solution up to 50% antifreeze?

45. Finance Jim Davidson invested $10,000 in two businesses. One business earned a profit of 5% for the year, while the other lost 7.5%. Find the amount he invested in each business if he earned $25 for the year.

46. Finance Marco invested money in two stocks. The stock that made a profit of 12% had $5000 less invested than the stock that lost 5.5%. Find the amount he invested in each if there was a net loss of $80 on his portfolio.

Explaining the Concepts

47. Without solving, explain what is wrong with the following mixture problem: How many liters of 25% ethanol should be added to 20 liters of 48% ethanol to obtain a solution of 58% ethanol?

48. Explain why the system $\begin{cases} a + b = 700 \\ 0.05a + 0.1b = 10,000 \end{cases}$ does not correctly model the following problem.

Molly invests $10,000 in two accounts, one paying 5% annual interest and the other paying 10% annual interest, and she earns $700 in interest at the end of the year on the two accounts. Let a represent the amount of money she invests in the account that pays 5% annual interest, and let b represent the amount of money she invests in the account that pays 10%. How much is invested in each account?

4.6 Systems of Linear Inequalities

Objectives

1. Determine Whether an Ordered Pair Is a Solution of a System of Linear Inequalities
2. Graph a System of Linear Inequalities
3. Solve Applied Problems Involving Systems of Linear Inequalities

Are you Ready for This Section?

Before getting started, take the following readiness quiz. If you get a problem wrong, go back to the section cited and review the material.

R1. Solve: $3x - 2 \geq 7$ [Section 2.8, pp. 150–156]

R2. Solve: $4(x - 1) < 6x + 4$ [Section 2.8, pp. 150–156]

R3. Graph: $y > 2x - 5$ [Section 3.7, pp. 232–236]

R4. Graph: $2x + 3y \leq 9$ [Section 3.7, pp. 232–236]

In Section 3.7, we graphed a single linear inequality in two variables. In this section, we discuss how to graph a system of linear inequalities in two variables.

Ready?...Answers **R1.** $\{x \mid x \geq 3\}$
R2. $\{x \mid x > -4\}$

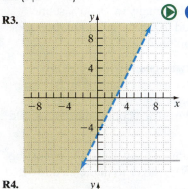

R3.

R4.

① Determine Whether an Ordered Pair Is a Solution of a System of Linear Inequalities

An ordered pair **satisfies** a system of linear inequalities if it makes each inequality in the system a true statement.

EXAMPLE 1 **Determining Whether an Ordered Pair Is a Solution of a System of Linear Inequalities**

Which of the following points, if any, satisfies the system of linear inequalities?

$$\begin{cases} 2x + y \leq 7 \\ 7x - 2y \geq 4 \end{cases}$$

(a) $(2, 1)$ **(b)** $(-2, 5)$

Solution

(a) Let $x = 2$ and $y = 1$ in each inequality in the system. If each statement is true, then $(2, 1)$ is a solution of the system.

$$2x + y \leq 7 \qquad\qquad 7x - 2y \geq 4$$

$x = 2, y = 1:$ $2(2) + 1 \overset{?}{\leq} 7 \qquad\qquad 7(2) - 2(1) \overset{?}{\geq} 4$

$$4 + 1 \overset{?}{\leq} 7 \qquad\qquad 14 - 2 \overset{?}{\geq} 4$$
$$5 \leq 7 \quad \text{True} \qquad\qquad 12 \geq 4 \quad \text{True}$$

Both inequalities are true when $x = 2$ and $y = 1$, so $(2, 1)$ is a solution of the system of inequalities.

(b) Let $x = -2$ and $y = 5$ in each inequality in the system. If each statement is true, then $(-2, 5)$ is a solution of the system.

$$2x + y \leq 7 \qquad\qquad 7x - 2y \geq 4$$
$$x = -2, y = 5: \qquad 2(-2) + 5 \overset{?}{\leq} 7 \qquad\qquad 7(-2) - 2(5) \overset{?}{\geq} 4$$
$$-4 + 5 \overset{?}{\leq} 7 \qquad\qquad -14 - 10 \overset{?}{\geq} 4$$
$$1 \leq 7 \quad \text{True} \qquad\qquad -24 \geq 4 \quad \text{False}$$

The inequality $7x - 2y \geq 4$ is not true when $x = -2$ and $y = 5$, so $(-2, 5)$ is not a solution of the system of inequalities. ●

Quick ✓

1. An ordered pair is a _____ of a system of linear inequalities if it makes each inequality in the system a true statement.

2. Determine which of the following points, if any, is a solution of the system of linear inequalities.

$$\begin{cases} 4x + y \leq 6 \\ 2x - 5y < 10 \end{cases}$$

(a) $(1, 2)$ **(b)** $(-1, -3)$

▶ ② Graph a System of Linear Inequalities

Work Smart

Don't forget that we use a solid line when the inequality is nonstrict ($\leq$ or $\geq$) and a dashed line when the inequality is strict ($<$ or $>$).

The graph of a system of inequalities in two variables x and y is the set of all points (x, y) that simultaneously satisfy *each* of the inequalities in the system. To graph a system of linear inequalities, graph each linear inequality individually. Then determine where, if at all, they intersect. The ONLY way we show the solution of a system of linear inequalities is by its graphical representation.

EXAMPLE 2   **How to Graph a System of Linear Inequalities**

Graph the system: $\begin{cases} y \geq 3x - 5 \\ y \leq -2x + 5 \end{cases}$

Step-by-Step Solution

Step 1: Graph the first inequality in the system, $y \geq 3x - 5$.

Graph $y \geq 3x - 5$ by graphing the line $y = 3x - 5$ [slope $= 3$, y-intercept $= (0, -5)$] with a solid line because the inequality is nonstrict ($\geq$). The test point $(0, 0)$ makes the inequality true ($0 \geq 3(0) - 5$), so we shade the half-plane containing $(0, 0)$. See Figure 15.

Figure 15
$y \geq 3x - 5$

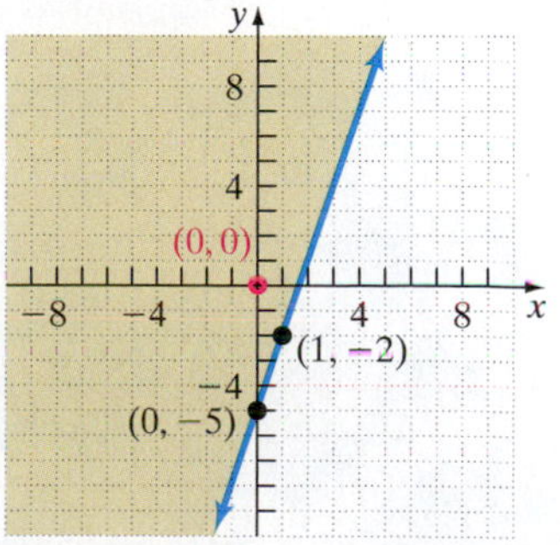

(continued)

Step 2: Graph the second inequality in the system, $y \leq -2x + 5$.

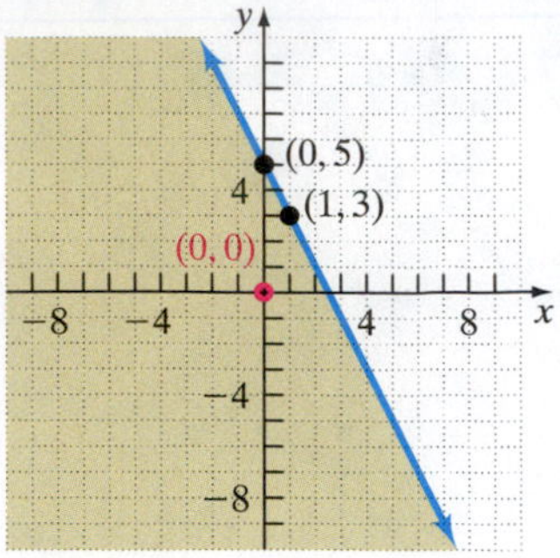

Work Smart

We can also graph inequalities by solving the inequality for y. Then, if the inequality is of the form $y >$ or $y \geq$, shade above the line. If the inequality is of the form $y <$ or $y \leq$, shade below the line.

Graph $y \leq -2x + 5$ by graphing the line $y = -2x + 5$ [slope $= -2$, y-intercept $= (0, 5)$] with a solid line because the inequality is nonstrict ($\leq$). The test point $(0, 0)$ makes the inequality true ($0 \leq -2(0) + 5$), so we shade the half-plane containing $(0, 0)$. See Figure 16.

Figure 16

$y < -2x + 5$

Step 3: Combine the graphs in Steps 1 and 2. The overlapping shaded region is the solution of the system of linear inequalities.

See Figure 17 for the graph of the system of linear inequalities.

Note: The ordered pair $(2, 1)$ represents the solution to the system

$$\begin{cases} y = 3x - 5 \\ y = -2x + 5 \end{cases}$$

Figure 17

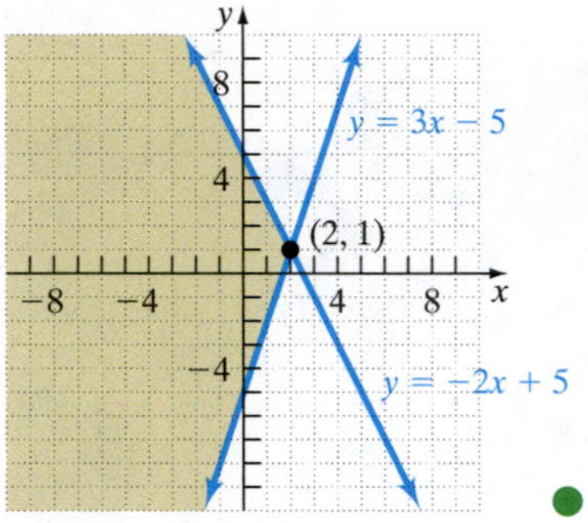

$\boxed{\text{EXAMPLE 3}}$ **Graphing a System of Linear Inequalities**

Graph the system: $\begin{cases} 2x + y < 7 \\ 7x - 2y > 4 \end{cases}$

Solution

We graph the inequality $2x + y < 7$ ($y < -2x + 7$) in Figure 18(a). We then graph the inequality $7x - 2y > 4$ $\left(y < \dfrac{7}{2}x - 2 \right)$ in Figure 18(b). Don't forget to use dashed lines!

Combine the graphs from Figures 18(a) and (b). The overlapping shaded region is the solution to the system of linear inequalities. See Figure 18(c). The point at which the two boundary lines intersect is not part of the solution of the system since the inequalities are strict.

Figure 18

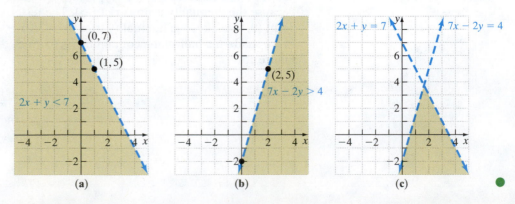

(a) (b) (c)

Quick ✓

3. When graphing linear inequalities, we use a _____ line when graphing strict inequalities ($>$ or $<$) and a ____ line when graphing nonstrict inequalities ($\geq$ or $\leq$).

In Problems 4 and 5, graph each system of linear inequalities.

4. $\begin{cases} y \geq -3x + 8 \\ y \geq 2x - 7 \end{cases}$

5. $\begin{cases} 4x + 2y < -9 \\ x + 3y < -1 \end{cases}$

Rather than using multiple graphs to find the solution of a system of linear inequalities, we can use a single graph.

EXAMPLE 4 **Graphing a System of Linear Inequalities**

Graph the system: $\begin{cases} 2x + y \geq 3 \\ 3x - 2y < 8 \end{cases}$

Figure 19

The points in this region satisfy both inequalities.

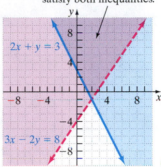

Solution

We graph the inequality $2x + y \geq 3$ ($y \geq -2x + 3$) in blue.

On the same rectangular coordinate system, we graph the inequality $3x - 2y < 8 \left(y > \dfrac{3}{2}x - 4 \right)$ in pink.

In Figure 19, the overlapping shaded region in purple represents the solution. The intersection point of the two boundary lines is not a solution of the system because the inequality $3x - 2y < 8$ is strict.

Work Smart

$$3x - 2y < 8$$
$$-2y < -3x + 8$$
$$y > \frac{3}{2}x - 4$$

Remember to reverse the inequality symbol! Now graph $y = \dfrac{3}{2}x - 4$ using a dashed line and shade above.

Quick ✓

In Problems 6 and 7, graph the system of linear inequalities.

6. $\begin{cases} x + y \leq 4 \\ -x + y > -4 \end{cases}$

7. $\begin{cases} 3x + y > -5 \\ x + 2y \leq 0 \end{cases}$

▶ ❸ **Solve Applied Problems Involving Systems of Linear Inequalities**

Now let's look at some problems that lead to systems of linear inequalities.

EXAMPLE 5 **Financial Planning**

Maurice can invest up to $50,000. His financial advisor recommends that he place at least $10,000 in Treasury notes and no more than $35,000 in corporate bonds. A system of linear inequalities that models this situation is

$$\begin{cases} c + t \leq 50{,}000 \\ c \leq 35{,}000 \\ t \geq 10{,}000 \end{cases}$$

where c represents the amount in corporate bonds and t represents the amount in Treasury notes.

(a) Graph the system.

(b) Can Maurice put $20,000 in corporate bonds and $30,000 in Treasury notes?

(c) Can Maurice put $40,000 in corporate bonds and $10,000 in Treasury notes?

Solution

Figure 20

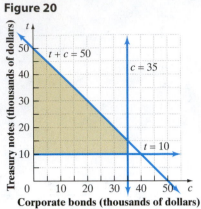

(a) Draw a rectangular coordinate system with the horizontal axis labeled c (for the amount in corporate bonds) and the vertical axis labeled t (for the amount in Treasury notes). Each axis will be in thousands, so we draw the line $t + c = 50$ and shade below. We draw the line $t = 10$ and shade above. We draw the line $c = 35$ and shade to the left. Figure 20 shows the graph of the system of linear inequalities.

(b) Yes, Maurice can put $20,000 in corporate bonds and $30,000 in Treasury notes, because these values lie within the shaded region. In other words, $c = 20$ and $t = 30$ satisfies all three inequalities.

(c) No, Maurice cannot put $40,000 in corporate bonds and $10,000 in Treasury notes because these values do not lie within the shaded region. Put another way, $c = 40$ and $t = 10$ does not satisfy the inequality $c \leq 35,000$.

Quick ✓

8. Jack and Mary can invest up to $75,000. Their financial advisor recommends that they place no more than $50,000 in corporate bonds and at least $25,000 in Treasury notes. A system of linear inequalities that models this situation is

$$\begin{cases} c + t \leq 75{,}000 \\ c \quad\;\; \leq 50{,}000 \\ \quad\; t \geq 25{,}000 \end{cases}$$

where c represents the amount in corporate bonds and t represents the amount in Treasury notes.

(a) Graph the system.

(b) Can Jack and Mary invest $30,000 in corporate bonds and $35,000 in Treasury notes?

(c) Can Jack and Mary invest $60,000 in corporate bonds and $15,000 in Treasury notes?

4.6 Exercises   MyMathLab® MathⓍP PRACTICE

Exercise numbers in green have complete video solutions in MyMathLab.

*Problems **1–8** are the Quick ✓ s that follow the **EXAMPLES**.*

Building Skills

In Problems 9–12, determine which point(s), if any, is a solution of the system of linear inequalities. See Objective 1.

9. $\begin{cases} 3x + y > 10 \\ -x + 5y \leq -10 \end{cases}$

(a) $(5, -2)$

(b) $(4, -1)$

(c) $(6, -1)$

10. $\begin{cases} 2x - 3y < 3 \\ 2x + y < -5 \end{cases}$

(a) $(-4, 1)$

(b) $\left(-\dfrac{3}{2}, -2\right)$

(c) $(-1, -2)$

11. $\begin{cases} 2x + y > -4 \\ x - y \leq 1 \end{cases}$

(a) $(-2, 1)$

(b) $(-1, -2)$

(c) $(2, -3)$

12. $\begin{cases} x - 2y > 2 \\ -3x - 2y \leq 6 \end{cases}$

(a) $\left(-1, -\dfrac{3}{2}\right)$

(b) $(0, -4)$

(c) $(4, -1)$

In Problems 13–38, graph each system of linear inequalities. See Objective 2.

13. $\begin{cases} x > 2 \\ y \leq -1 \end{cases}$

14. $\begin{cases} y > 4 \\ x < -3 \end{cases}$

15. $\begin{cases} y > -2 \\ x > -3 \end{cases}$ **16.** $\begin{cases} x \le 3 \\ y < 1 \end{cases}$

17. $\begin{cases} x + y < 3 \\ x - y > 5 \end{cases}$ **18.** $\begin{cases} x - y > 2 \\ x + y \ge -2 \end{cases}$

19. $\begin{cases} x + y > 3 \\ 2x - y > 4 \end{cases}$ **20.** $\begin{cases} x - y \ge -1 \\ x + 2y > 4 \end{cases}$

21. $\begin{cases} y \ge -\frac{1}{2}x + 1 \\ y \le \frac{1}{2}x + 3 \end{cases}$ **22.** $\begin{cases} y \ge -\frac{1}{3}x + 2 \\ y \ge \frac{2}{3}x - 1 \end{cases}$

23. $\begin{cases} y \ge -2 \\ y < 2x + 3 \end{cases}$ **24.** $\begin{cases} y > 2 \\ 2x - y \le 1 \end{cases}$

25. $\begin{cases} x > 0 \\ y \le \frac{2}{5}x - 1 \end{cases}$ **26.** $\begin{cases} y < -2 \\ y > 2x - 1 \end{cases}$

27. $\begin{cases} y \ge -x \\ 3x - y \ge -5 \end{cases}$ **28.** $\begin{cases} x + y < 2 \\ 3x - 5y \ge 0 \end{cases}$

29. $\begin{cases} x + y \le -2 \\ y \ge x + 3 \end{cases}$ **30.** $\begin{cases} 3x + y > 3 \\ y > 4x - 2 \end{cases}$

31. $\begin{cases} x + 3y \ge 0 \\ 2y < x + 1 \end{cases}$ **32.** $\begin{cases} -y < \frac{2}{3}x + 1 \\ -3x + y \le 2 \end{cases}$

33. $\begin{cases} x + y \ge 0 \\ x < 2y + 4 \end{cases}$ **34.** $\begin{cases} 2x - 3y \le 9 \\ y < -2x + 3 \end{cases}$

35. $\begin{cases} x + 3y > 6 \\ 2x - y \le 4 \end{cases}$ **36.** $\begin{cases} x - y \le 3 \\ 2x + 3y \le -9 \end{cases}$

37. $\begin{cases} -y \le 3x - 4 \\ 2x + 3y \ge -3 \end{cases}$ **38.** $\begin{cases} x + 4y \le 4 \\ 2x + 3y \ge 6 \end{cases}$

Applying the Concepts

39. House Blend The Coffee Cup coffee shop is experimenting with blending the "house" coffee. It will be made up of two varieties of coffee, French Roast and Hazelnut. The managers have decided that they will make at most 30 pounds of "house" coffee in a day. The tasters have determined that the blend should be mixed so that the amount of French Roast coffee is at least twice the amount of Hazelnut coffee. A system of linear inequalities that models this situation is

$$\begin{cases} f + h \le 30 \\ f \ge 2h \\ f \ge 0 \\ h \ge 0 \end{cases}$$

where f represents the number of pounds of French Roast coffee and h represents the number of pounds of Hazelnut coffee.

(a) Graph the system of linear inequalities.

(b) Is it possible to use 18 pounds of French Roast coffee and 11 pounds of Hazelnut coffee in the house blend?

(c) Is it possible to use 8 pounds of French Roast coffee and 4 pounds of Hazelnut coffee in the house blend?

40. Party Food Steven and Christopher are planning a party. They plan to buy bratwurst for $4.00 per pound and hamburger patties that cost $3.00 per pound. They can spend at most $70 and think they should have no more than 20 pounds of bratwurst and hamburger patties. A system of inequalities that models the situation is

$$\begin{cases} 4b + 3h \le 70 \\ b + h \le 20 \\ b \ge 0 \\ h \ge 0 \end{cases}$$

where b represents the number of pounds of bratwurst and h represents the number of pounds of hamburger patties.

(a) Graph the system of linear inequalities.

(b) Is it possible to purchase 10 pounds of bratwurst and 10 pounds of hamburger patties?

(c) Is it possible to purchase 15 pounds of bratwurst and 5 pounds of hamburger patties?

41. Auto Manufacturing A manufacturing plant has 360 hours of machine time budgeted for the production of cars and trucks on a given day. It takes 12 hours to build a car and 18 hours to build a truck. Based on experience, the plant manager knows that the number of trucks is fewer than 20 less than twice the number of cars produced. Due to limitations in the machinery, the maximum number of cars that can be produced in a day is 24. A system of linear inequalities that models this situation is

$$\begin{cases} 12c + 18t \le 360 \\ t < 2c - 20 \\ c \le 24 \end{cases}$$

where c represents the number of cars produced and t represents the number of trucks built.

(a) Graph the system of linear inequalities.

(b) Is it possible to build 17 cars and 9 trucks at this plant?

(c) Is it possible to build 12 cars and 11 trucks at this plant?

42. Breakfast at Burger King Aman decides to eat breakfast at Burger King. He is a big fan of their French toast sticks, and he likes orange juice. He wants to eat no more than 500 calories and to consume no more than 425 mg of sodium. Each French toast stick (with syrup) has 90 calories and

100 mg of sodium. Each small orange juice has 140 calories and 25 mg of sodium. The system of linear inequalities that represents the possible combination of French toast sticks and orange juice that Aman can consume is

$$\begin{cases} 90x + 140y \le 500 \\ 100x + 25y \le 425 \\ x \ge 0 \\ y \ge 0 \end{cases}$$

where x represents the number of Burger King French toast sticks and y represents the number of containers of orange juice.

(a) Graph the system of linear inequalities.

(b) Is it possible for Aman to consume 3 French toast sticks and 1 container of orange juice and stay within the allowance for calories and sodium?

(c) Is it possible for Aman to consume 2 French toast sticks and 2 containers of orange juice and stay within the allowance for calories and sodium?

Extending the Concepts

In Problems 43–46, the shaded region for the solution to each system of linear inequalities below differs from what we have seen previously. Graph the solution of each system.

43. $\begin{cases} \dfrac{y}{2} - \dfrac{x}{6} \ge 1 \\ \dfrac{x}{3} - \dfrac{y}{1} \ge 1 \end{cases}$

44. $\begin{cases} \dfrac{x}{2} + \dfrac{y}{4} > 1 \\ -\dfrac{x}{4} - \dfrac{y}{8} > 1 \end{cases}$

45. $\begin{cases} x < \dfrac{3}{2}y + \dfrac{9}{2} \\ -2x < -3(y + 2) \end{cases}$

46. $\begin{cases} x > \dfrac{5}{4}y - \dfrac{1}{2} \\ 4x < 5(y + 3) \end{cases}$

In Problems 47 and 48, write a system of linear inequalities that will produce each shaded region as its solution.

47.

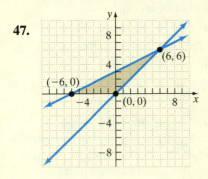

48.

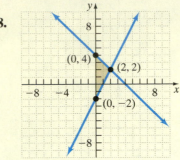

Explaining the Concepts

49. List the steps used to solve a system of linear inequalities in two variables.

50. If one inequality in a system of linear inequalities contains a strict inequality and the other inequality contains a nonstrict inequality, is the intersection point of the boundary lines a solution of the system? Explain why or why not.

51. Each of the graphs below uses two of the following inequalities:

$l_1: y \ge -2x \quad l_2: y \le -2x \quad l_3: y \ge x \quad l_4: y \le x$

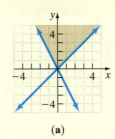

(a)

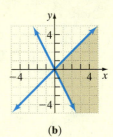

(b)

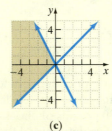

(c)

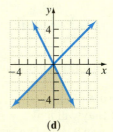

(d)

Write the system of linear inequalities for each graph and explain how you came to your conclusions.

52. When solving application problems that require solving a system of linear inequalities, typically the graph appears only in the first quadrant. Explain why this is so. Can you think of a situation when it would not be the case?

Chapter 4 Activity: Find the Numbers

Focus: Solving systems of equations
Time: 15 minutes
Group size: 2

Consider the following dialogue between two students:

Ryan: Think of two numbers between 1 and 10, and don't tell me what they are.

Melissa: OK, I've thought of two numbers.

Ryan: Now tell me the sum of the two numbers and the difference of the two numbers, and I'll tell you what your two numbers are.

Melissa: Their sum is 14 and their difference is 6.

Ryan: Your numbers are 10 and 4.

Melissa: That's right! How did you do that?

Ryan: I set up a system of equations using the sum and difference that you gave me.

1. Each group member should set up and solve the system of equations described by Ryan. Discuss your results, and be sure that you both arrive at the solutions 10 and 4.

2. Now each of you will think of two new numbers, and the other will try to find the numbers by solving a system of equations. But this time the system will be a bit trickier! Give each other the following information about your two numbers, and then figure out each other's numbers:

   six more than three times the sum of the numbers
   two less than four times the difference of the numbers

3. Would the systems in this activity work if negative numbers were used? Try it and see!

Chapter 4 Review

Section 4.1   Solving Systems of Linear Equations by Graphing

KEY CONCEPT

- **Recognizing Solutions of Systems of Two Linear Equations with Two Unknowns**
  - If the lines in a system of two linear equations containing two unknowns intersect, then the point of intersection is the solution and the system is consistent; the equations are independent.
  - If the lines in a system of two linear equations containing two unknowns are parallel, then the system has no solution and the system is inconsistent.
  - If the lines in a system of two linear equations containing two unknowns lie on top of each other (coincident), then the system has infinitely many solutions. The solution set is the set of all points on the line and the system is consistent; the equations are dependent.

KEY TERMS

System of linear equations
Solution
Consistent and independent
Inconsistent
Consistent and dependent

You Should Be Able To...	EXAMPLE	Review Exercises
1 Determine whether an ordered pair is a solution of a system of linear equations (p. 249)	Example 2	1–4
2 Solve a system of linear equations by graphing (p. 250)	Examples 3 through 6	5–12
3 Classify systems of linear equations as consistent or inconsistent (p. 254)	Example 7	13–18
4 Solve applied problems involving systems of linear equations (p. 256)	Example 8	19, 20

In Problems 1–4, determine whether the ordered pair is a solution to the system of equations.

1. $\begin{cases} x + 2y = 6 \\ 3x - y = -10 \end{cases}$ **(a)** $(3, -1)$ **(b)** $(-2, 4)$ **(c)** $(4, 1)$

2. $\begin{cases} y = 3x - 5 \\ 3y = 6x - 5 \end{cases}$ **(a)** $(2, 1)$ **(b)** $\left(0, -\dfrac{5}{3}\right)$ **(c)** $\left(\dfrac{10}{3}, 5\right)$

3. $\begin{cases} 3x - 4y = 2 \\ 20y = 15x - 10 \end{cases}$ **(a)** $\left(\dfrac{1}{2}, -\dfrac{1}{8}\right)$ **(b)** $(6, 4)$ **(c)** $(0.4, -0.2)$

4. $\begin{cases} x = -4y + 2 \\ 2x + 8y = 12 \end{cases}$

  (a) $(10, -1)$ (b) $\left(-\dfrac{1}{2}, \dfrac{1}{2}\right)$ (c) $(2, 1)$

In Problems 5–12, solve each system of equations by graphing.

5. $\begin{cases} 2x - 4y = 8 \\ x + y = 7 \end{cases}$

6. $\begin{cases} x - y = -3 \\ 3x + 2y = 6 \end{cases}$

7. $\begin{cases} y = -\dfrac{x}{2} + 2 \\ y = x + 8 \end{cases}$

8. $\begin{cases} y = -x - 5 \\ y = \dfrac{3x}{4} + 2 \end{cases}$

9. $\begin{cases} 4x - 8 = 0 \\ 3y + 9 = 0 \end{cases}$

10. $\begin{cases} x = y \\ x + y = 0 \end{cases}$

11. $\begin{cases} 0.6x + 0.5y = 2 \\ 10y = -12x + 20 \end{cases}$

12. $\begin{cases} \dfrac{1}{4}x - \dfrac{1}{2}y = 1 \\ 3x - 6y = 12 \end{cases}$

In Problems 13–18, without graphing, determine the number of solutions to each system of equations. State whether the system is consistent or inconsistent. For those systems that are consistent, state whether the equations are dependent or independent.

13. $\begin{cases} y = 3x - 4 \\ y = 3x + 4 \end{cases}$

14. $\begin{cases} -2x + 4y = -8 \\ x - 2y = -4 \end{cases}$

15. $\begin{cases} -3x + 3y = -3 \\ \dfrac{1}{2}x - \dfrac{1}{2}y = 0.5 \end{cases}$

16. $\begin{cases} x - 2 = -\dfrac{2}{3}y - \dfrac{2}{3} \\ 3x = 4 - 2y \end{cases}$

17. $\begin{cases} 2x + y = -3 \\ y = 2x + 3 \end{cases}$

18. $\begin{cases} \dfrac{y}{2} = \dfrac{x}{4} + 2 \\ \dfrac{x}{8} + \dfrac{y}{4} = -1 \end{cases}$

19. **Printing Costs** Monique is creating a flier to give to local businesses to advertise her new vintage clothing store. She is trying to decide between two quotes for the printing. Printer A has given her a quote of \$70 setup fee plus \$0.10 per flier printed. Printer B has given her a quote of \$100 plus \$0.04 per flier.

  (a) Write a system of linear equations that models the problem.

  (b) Graph the system of equations to determine how many fliers she needs to have printed in order for the cost to be the same at both printers.

  (c) If 400 fliers are printed, which printer should she choose to have the lower cost?

20. **Flooring Installation** To install carpeting in Juan's bedroom, it costs \$50 for installation plus \$40 per square yard of carpet. In the same room, Juan can install ceramic tile for a cost of \$350 for installation plus \$10 per square yard of tile.

  (a) Write a system of linear equations that models the problem.

  (b) Graph the system of equations to determine how many square yards of flooring Juan needs to have installed for the cost of carpeting and the cost of tile to be the same.

  (c) If the area of Juan's bedroom is 15 square yards, which material would be cheaper?

Section 4.2   Solving Systems of Linear Equations Using Substitution

KEY CONCEPT

- **Solving Systems of Linear Equations Using Substitution**

  Step 1: Solve one of the equations for one of the unknowns. For example, we might solve equation (1) for y in terms of x.

  Step 2: Substitute the expression solved for in Step 1 into the *other* equation. The result will be a single linear equation in one unknown. For example, if we solved equation (1) for y in terms of x in Step 1, then we would replace y in equation (2) with the expression in x.

  Step 3: Solve the linear equation in one unknown found in Step 2.

  Step 4: Substitute the value of the variable found in Step 3 into one of the *original* equations to find the value of the other variable.

  Step 5: Check your answer by substituting the ordered pair in both of the original equations.

You Should Be Able To...	EXAMPLE	Review Exercises
❶ Solve a system of linear equations using the substitution method (p. 260)	Examples 1 through 5	21–34
❷ Solve applied problems involving systems of linear equations (p. 265)	Example 6	35, 36

In Problems 21–34, solve each system of equations using substitution.

21. $\begin{cases} x + 4y = 6 \\ y = 2x - 3 \end{cases}$

22. $\begin{cases} 7x - 3y = 10 \\ y = 3x - 4 \end{cases}$

23. $\begin{cases} 2x + 5y = 4 \\ x = 3 - 2y \end{cases}$

24. $\begin{cases} 3x + y = 10 \\ x = 8 + 2y \end{cases}$

25. $\begin{cases} y = \dfrac{2}{3}x - 1 \\ y = \dfrac{1}{2}x + 2 \end{cases}$

26. $\begin{cases} y = -\dfrac{5}{6}x + 3 \\ y = -\dfrac{4}{3}x \end{cases}$

27. $\begin{cases} 2x - y = 6 \\ 4x + 3y = 2 \end{cases}$

28. $\begin{cases} 5x + 2y = 13 \\ x + 4y = -1 \end{cases}$

29. $\begin{cases} 6x + 3y = 12 \\ y = -2x + 4 \end{cases}$

30. $\begin{cases} x = 4y - 1 \\ 8y - 2x = 4 \end{cases}$

31. $\begin{cases} -6 - 2(3x - 6y) = 0 \\ 6 - 12(x - 2y) = 0 \end{cases}$

32. $\begin{cases} 6 - 2(3y + 4x) = 0 \\ 9(y - 1) + 12x = 0 \end{cases}$

33. $\begin{cases} \dfrac{1}{2}x - \dfrac{1}{4}y = \dfrac{1}{2} \\ \dfrac{1}{3}x - \dfrac{3}{4}y = -\dfrac{1}{4} \end{cases}$

34. $\begin{cases} -\dfrac{5x}{4} + \dfrac{y}{6} = -\dfrac{7}{12} \\ \dfrac{3x}{2} - \dfrac{y}{10} = \dfrac{3}{5} \end{cases}$

△ **35. Walk around the Park** A rectangular park is surrounded by a walking path. The park is 75 meters longer than it is wide. If a person walking completely around the park covers a distance of 650 meters, find the length and width of the park by solving the following system of equations, where l is the length and w is the width.

$$\begin{cases} 2l + 2w = 650 \quad \text{(1)} \\ \qquad\quad l = w + 75 \quad \text{(2)} \end{cases}$$

36. Fun with Numbers The sum of two numbers is 12. If twice the smaller number is subtracted from the larger number, the difference is 21. Determine the smaller of the two numbers by solving the following system of equations, where x and y represent the unknown numbers.

$$\begin{cases} x + y = 12 \quad \text{(1)} \\ x - 2y = 21 \quad \text{(2)} \end{cases}$$

Section 4.3 Solving Systems of Linear Equations Using Elimination

KEY CONCEPT

- **Solving Systems of Linear Equations Using Elimination**

  Step 1: Write each equation in standard form, $Ax + By = C$.

  Step 2: If necessary, multiply or divide both sides of one (or both) equations by a nonzero constant so that the coefficients of one of the variables are opposites (additive inverses).

  Step 3: Add the equations to eliminate the variable whose coefficients are now additive inverses. Solve the resulting equation for the remaining unknown.

  Step 4: Substitute the value of the variable found in Step 3 into one of the original equations to find the value of the remaining variable.

  Step 5: Check your answer by substituting the ordered pair in both of the original equations.

You Should Be Able To...	EXAMPLE	Review Exercises
1 Solve a system of linear equations using the elimination method (p. 268)	Examples 1 through 5	37–52
2 Solve applied problems involving systems of linear equations (p. 273)	Example 6	53, 54

In Problems 37–44, solve each system of equations using elimination.

37. $\begin{cases} 4x - y = 12 \\ 2x + y = -12 \end{cases}$

38. $\begin{cases} -2x + 3y = 27 \\ 2x - 5y = -41 \end{cases}$

39. $\begin{cases} -3x + 4y = 25 \\ x - 5y = -23 \end{cases}$

40. $\begin{cases} 5x + 8y = -15 \\ -2x + y = 6 \end{cases}$

41. $\begin{cases} 4x - 3y = -1 \\ 2x - \dfrac{3}{2}y = 3 \end{cases}$

42. $\begin{cases} 2x + 5y = 3 \\ -4x - 10y = -6 \end{cases}$

43. $\begin{cases} 1.3x - 0.2y = -3 \\ -0.1x + 0.5y = 1.2 \end{cases}$ **44.** $\begin{cases} 2.5x + 0.5y = 6.25 \\ -0.5x - 1.2y = 1.5 \end{cases}$

In Problems 45–52, solve each system of equations using graphing, substitution, or elimination.

45. $\begin{cases} 2x + y = -1 \\ -6x - 8y = 13 \end{cases}$ **46.** $\begin{cases} \dfrac{1}{6}x + y = \dfrac{3}{4} \\ \dfrac{2y - 8}{3} = 4x + 4 \end{cases}$

47. $\begin{cases} y + 5 = \dfrac{2}{3}x + 3 \\ \dfrac{1}{3}x - \dfrac{1}{2}y = 1 \end{cases}$ **48.** $\begin{cases} \dfrac{1}{14} - \dfrac{x}{2} = -\dfrac{y}{7} \\ y = \dfrac{1}{2} + \dfrac{7x}{3} \end{cases}$

49. $\begin{cases} -x + y = 7 \\ -3x + 4y = 8 \end{cases}$ **50.** $\begin{cases} 4x + y = 2 \\ 9y - 3x = 5 \end{cases}$

51. $\begin{cases} 4x + 6 = 3y + 5 \\ 4(-2x - 4) = 6(-y - 3) \end{cases}$

52. $\begin{cases} 3y + 2x = 16x + 2 \\ x = -\dfrac{3}{14}y - \dfrac{1}{7} \end{cases}$

53. Valentine Gift The specialty bakery around the corner is putting together a Valentine's Day gift box to sell. The gift box is going to contain two items, heart-shaped cookies and chocolate candies. If the cookies sell for $1.50 per pound and the chocolates sell for $7.75 per pound, how many pounds of each type should be included in a 4-pound box that sells for $4.00 per pound? To find the answers, solve the system

$$\begin{cases} h + c = 4 & (1) \\ 1.50h + 7.75c = 16 & (2) \end{cases}$$

where h represents the number of pounds of heart-shaped cookies and c represents the number of pounds of candies.

54. Raffle Tickets The parent booster club is selling 50–50 raffle tickets to raise money for the upcoming hockey banquet. Raffle tickets are sold in two ways: Individual chances are sold for $1.00 each, or a ticket with a block of 6 chances can be purchased for $5.00. On one particular night, the booster club brought in $2025 from the sale of 725 raffle tickets. How many of each type of raffle ticket was sold? To find the answers. Solve the system

$$\begin{cases} t + b = 725 & (1) \\ t + 5b = 2025 & (2) \end{cases}$$

where t is the number of individual tickets and b is the number of block tickets.

Section 4.4 Solving Direct Translation, Geometry, and Uniform Motion Problems Using Systems of Linear Equations

You Should Be Able To...	EXAMPLE	Review Exercises
1 Model and solve direct translation problems (p. 278)	Example 1	55–58
2 Model and solve geometry problems (p. 278)	Examples 2 and 3	59–62
3 Model and solve uniform motion problems (p. 281)	Example 4	63–66

55. Fun with Numbers Find two numbers whose difference is $\dfrac{1}{24}$ and whose sum is $\dfrac{17}{24}$.

56. Fun with Numbers The sum of two numbers is 58. If twice the smaller number is subtracted from the larger number, the difference is -20. Find the numbers.

57. Investments Aaron has $50,000 to invest. His financial advisor recommends he invest in stocks and bonds, with the amount invested in stocks equaling $4000 less than twice the amount invested in bonds. How much should be invested in stocks? How much should be invested in bonds?

58. Bookstore Sales A register at the bookstore sold only notebooks and scientific calculators. The cost of a notebook is $\dfrac{1}{3}$ of the cost of the calculator. If the calculator sells for $7.50 and the register tape shows that 24 items resulted in revenue of $110, how many of each were sold?

△ **59. Geometry** The measure of one angle is 25° less than the measure of its supplement. Find the measures of the two angles.

△ **60. Geometry** The measure of one angle is 15° more than half the measure of its complement. Find the measures of the two angles.

△ **61. Horse Corral** The O'Connells are planning to build a rectangular corral for their horses, using the river on their property as one of the sides. On the other three sides they are installing 52 meters of fencing. The longer side of the corral is opposite the water, and its length is 8 meters less than twice the shorter side. Find the dimensions of the corral.

△ **62. Geometry** In a triangle, the measure of the first angle is 10° less than three times the measure of the second. If the measure of the third angle is 30°, find the measure of the two unknown angles of the triangle.

63. Wind Resistance A plane can fly 2000 miles with a tailwind in 4 hours. It can make the return trip in 5 hours. Find the speed of the plane in still air and the effect of the wind.

64. Fishing Spot Dayle is paddling his canoe upstream, against the current, to a fishing spot 10 miles away. If he

paddles upstream for 2.5 hours and his return trip takes 1.25 hours, find the speed of the current and his paddling speed in still water.

65. Cycling in the Wind A cyclist can go 36 miles with the wind blowing at her back in 3 hours. On the return trip, after 4 hours, the cyclist still has 4 miles remaining to return to the starting point. Find the speed of the cyclist and the effect of the wind.

66. Out for a Run One speedy jogger can run 2 mph faster than his running buddy. In the last race they entered, the faster runner covered 12 miles in the same time that it took for the slower jogger to cover 9 miles. Find the speed of each of the runners.

Section 4.5 Solving Mixture Problems Using Systems of Linear Equations

You Should Be Able To...	EXAMPLE	Review Exercises
❶ Draw up a plan for solving mixture problems (p. 285)	Example 1	67, 68
❷ Set up and solve money mixture problems (p. 286)	Examples 2 and 3	69–72
❸ Set up and solve dry mixture and percent mixture problems (p. 288)	Examples 4 and 5	73–76

In Problems 67 and 68, fill in the table from the information given. DO NOT SOLVE.

67. Brendan has 35 coins, all nickels and dimes. He has a total of $2.25 in change.

	Number of Coins	·	Value of Each Coin	=	Total Value
Dimes					
Nickels					
Total					

68. Belinda invested part of her $15,000 inheritance in a savings account with an annual rate of 6.5% simple interest and the rest of her inheritance in a mutual fund with a rate of 8% simple interest. Her total yearly interest income from both accounts was $1050.

	Principal	Rate	Interest
Savings Account			
Mutual Fund			
Total			

69. Movie Tickets One day, the Beechwold Movie Theater had sales of $9929 from selling $8 regular-priced tickets and $5.50 matinee tickets to a movie. The theater sold a total of 1498 tickets. How many regular-priced tickets were sold that day?

70. Coins Sharona has $1.70 in change consisting of three more dimes than quarters. Find the number of quarters she has.

71. Investment Carlos Singer has some money invested at 5% and $5000 more than that invested at 9%. His total annual interest income is $1430. Find the amount Carlos has invested at each rate.

72. Investment Hilda invested part of her $25,000 advance in savings bonds at 7% annual simple interest and the rest in a stock portfolio yielding 8% annual simple interest. If her total yearly interest income is $1900, find the amount Hilda has invested at each rate.

73. Sugar A baker wants to mix a 60% sugar solution with a 30% sugar solution to obtain 10 quarts of a 51% sugar solution. How much of the 30% sugar solution will the baker use?

74. Peroxide How many pints of a 25% peroxide solution should be added to 10 pints of a 60%

peroxide solution to obtain a 30% peroxide solution?

75. Almond-Peanut Butter Joni works at a health food store and is planning to grind a special blend of butter by mixing peanuts, which sell for $4.00 per pound, and almonds, which sell for $6.50 per pound. How many pounds of each type should she use if she needs

5 pounds of almond-peanut butter, which sells for $5.00 per pound?

76. Mixing Chemicals Harold needs 20 liters of 55% acid. If he has 35% acid and 60% acid that he can mix together, how many liters of each should he use to obtain the desired mixture?

Section 4.6 Solving Systems of Linear Inequalities

KEY TERM

Satisfies

You Should Be Able To...	EXAMPLE	Review Exercises
1 Determine whether an ordered pair is a solution of a system of linear inequalities (p. 294)	Example 1	77–82
2 Graph a system of linear inequalities (p. 295)	Examples 2 through 4	83–94
3 Solve applied problems involving systems of linear inequalities (p. 297)	Example 5	95

In Problems 77–82, determine which point(s), if any, is a solution to the system of linear inequalities.

77. $\begin{cases} x + y \le 2 \\ 3x - 2y > 6 \end{cases}$

   (a) $(-1, -5)$

   (b) $(3, -1)$

   (c) $(4, 1)$

78. $\begin{cases} y \ge 3x + 5 \\ y \ge -2x \end{cases}$

   (a) $(-1, 0)$

   (b) $(-2, 4)$

   (c) $(1, 8)$

79. $\begin{cases} x > 5 \\ y < -2 \end{cases}$

   (a) $(10, -10)$

   (b) $(5, -3)$

   (c) $(7, -2)$

80. $\begin{cases} y > x \\ 2x - y \le 3 \end{cases}$

   (a) $(4, 3)$

   (b) $(-3, -2)$

   (c) $(-1, 4)$

81. $\begin{cases} x + 2y < 6 \\ 4y - 2x > 16 \end{cases}$

   (a) $(-4, 2)$

   (b) $(-8, 6)$

   (c) $(-2, -3)$

82. $\begin{cases} 2x - y \ge 3 \\ y \le 2x + 1 \end{cases}$

   (a) $(1, 2)$

   (b) $(3, -2)$

   (c) $(-1, 4)$

In Problems 83–94, graph the solution set of each system of linear inequalities.

83. $\begin{cases} x > -2 \\ y > 1 \end{cases}$

84. $\begin{cases} x \le 3 \\ y > -1 \end{cases}$

85. $\begin{cases} x + y \ge -2 \\ 2x - y \le -4 \end{cases}$

86. $\begin{cases} 3x + 2y < -6 \\ x - y < 2 \end{cases}$

87. $\begin{cases} x > 0 \\ y \le \dfrac{3}{4}x + 1 \end{cases}$

88. $\begin{cases} y \le 0 \\ y \le -\dfrac{1}{2}x - 3 \end{cases}$

89. $\begin{cases} -y \ge x \\ 4x - 3y \ge -12 \end{cases}$

90. $\begin{cases} y \ge -x + 2 \\ x - y \ge -4 \end{cases}$

91. $\begin{cases} 2x + 3y \ge -3 \\ y > \dfrac{3}{2}x + 1 \end{cases}$

92. $\begin{cases} x + 4y \le -4 \\ y \ge \dfrac{1}{4}x + 3 \end{cases}$

93. $\begin{cases} y > 2x - 5 \\ y - 2x \le 0 \end{cases}$

94. $\begin{cases} y < x + 2 \\ y > -\dfrac{5}{2}x + 2 \end{cases}$

95. Party Planning Alexis and Sarah are planning a barbeque for their friends. They plan to serve grilled fish and carne asada and want to spend at most $40 on the meal. The fish sells for $8 per pound, and the carne asada sells for $5 per pound. They plan to buy at least twice as much fish as carne asada. A system of linear inequalities that models this situation is

$$\begin{cases} 8x + 5y \le 40 \\ x \ge 2y \\ x \ge 0 \\ y \ge 0 \end{cases}$$

where x represents the number of pounds of fish and y represents the number of pounds of carne asada. Graph the system of linear inequalities that shows how many pounds of each they can buy and stay within their budget.

Chapter 4 Test

Step-by-step test solutions are found on the Chapter Test Prep Videos available in **MyMathLab®** *or on* **YouTube**.

In Problems 1 and 2, determine which of the following is a solution to the system of linear equations or inequalities.

1. $\begin{cases} 3x - y = -5 \\ y = \dfrac{2}{3}x - 2 \end{cases}$

  (a) $(-1, 2)$
  (b) $(-9, -8)$
  (c) $(-3, -4)$

2. $\begin{cases} 2x + y \geq 10 \\ y < \dfrac{x}{2} + 1 \end{cases}$

  (a) $(5, 3)$
  (b) $(-2, 14)$
  (c) $(-3, -1)$

3. The slopes of the two lines are negative reciprocals, and the y-intercepts are the same.

  (a) Tell how many solutions exist.
  (b) State whether the system is consistent or inconsistent.
  (c) If the system is consistent, state whether the equations are dependent or independent.

4. The slopes of the two lines are the same, and the y-intercepts are different.

  (a) Tell how many solutions exist.
  (b) State whether the system is consistent or inconsistent.
  (c) If the system is consistent, state whether the equations are dependent or independent.

In Problems 5 and 6, solve each system of equations by graphing.

5. $\begin{cases} 2x + 3y = 0 \\ x + 4y = 5 \end{cases}$

6. $\begin{cases} y = 2x - 6 \\ y = -\dfrac{1}{4}x + 3 \end{cases}$

In Problems 7 and 8, solve each system of equations using substitution.

7. $\begin{cases} 3x - y = 3 \\ 4x + 5y = -15 \end{cases}$

8. $\begin{cases} y = \dfrac{2}{3}x + 7 \\ y = \dfrac{1}{4}x + 2 \end{cases}$

In Problems 9 and 10, solve each system of equations using elimination.

9. $\begin{cases} 3x + 2y = -3 \\ 5x - y = -18 \end{cases}$

10. $\begin{cases} 4x - 5y = 12 \\ 3x + 4y = -22 \end{cases}$

In Problems 11–13, solve by graphing, substitution, or elimination.

11. $\begin{cases} \dfrac{2}{3}x - \dfrac{1}{6}y = \dfrac{25}{12} \\ -\dfrac{1}{4}x = \dfrac{3}{2}y \end{cases}$

12. $\begin{cases} 0.4x - 2.5y = -6.5 \\ x + y = 5.5 \end{cases}$

13. $\begin{cases} 4 + 3(y - 3) = -2(x + 1) \\ x + \dfrac{3y}{2} = \dfrac{3}{2} \end{cases}$

14. Plane Trip A small plane can fly 1575 miles in 7 hours with a tailwind. On the return trip, with a headwind, the same trip takes 9 hours. Find the speed of the plane in still air and the effect of the wind.

15. Ice Cream Manny is blending a peach swirl ice cream. He mixes together vanilla ice cream, which sells for $6 per container, and peach ice cream, which sells for $11.50 per container. How many containers of each should he use if he wants to have 11 containers of swirl ice cream that he can sell for $8 each?

△ **16. Supplementary Angles** The measure of one angle is 30° less than three times the measure of its supplement. Find the measures of the two angles.

17. Playground Equipment Moises is distributing new playground equipment to several elementary schools. Each school will receive some basketballs and some volleyballs. The cost of a basketball is $25 and the cost of a volleyball is $15. If each school needs 40 balls and he can spend $750 per school, how many of each type will a school receive?

In Problems 18–20, graph the solution set of each system of inequalities.

18. $\begin{cases} 4x - 2y < 8 \\ x + 3y < 6 \end{cases}$

19. $\begin{cases} x \leq -2 \\ -2x - 4y \leq 8 \end{cases}$

20. $\begin{cases} y \leq \dfrac{2}{3}x + 4 \\ x + 4y < 0 \end{cases}$

5 Exponents and Polynomials

It is very important to understand the distance that it takes a car to stop in various driving conditions, such as wet or dry pavement. This information is used to develop speed limits in school zones and busy highways with lots of traffic lights. The polynomial $0.04v^2 + 0.7v$ models the stopping distance of a car traveling on dry pavement at a speed of v miles per hour. See Problems 121 and 122 in Section 5.1 on page 317.

The Big Picture: Putting It Together

In the first two chapters of the text, we reviewed and developed fundamental skills with real numbers, algebraic expressions, and equations. We will use these skills throughout the course. In particular, we will describe how new topics and skills relate to those that we already know.

This chapter introduces polynomials. Polynomial expressions, such as the one given in the chapter opener, are used to describe many situations in business, consumer affairs, the physical and biological sciences, and even the entertainment industry. In this chapter, we will learn how to add, subtract, multiply, and divide polynomial expressions. Pay attention to these operations and how they are related to the techniques for adding, subtracting, multiplying, and dividing real numbers. Why? Algebra is easier to understand if we think of it as an extension of arithmetic.

Outline

5.1 Adding and Subtracting Polynomials

Objectives

1 Define Monomial and Determine the Degree of a Monomial

2 Define Polynomial and Determine the Degree of a Polynomial

3 Simplify Polynomials by Combining Like Terms

4 Evaluate Polynomials

Are You Ready for This Section?

Before getting started, take the following readiness quiz. If you get a problem wrong, go back to the section cited and review the material.

R1. What is the coefficient of $-4x^5$? [Section 1.8, p. 66]

R2. Combine like terms: $-3x + 2 - 2x - 6x - 7$ [Section 1.8, pp. 68–69]

R3. Use the Distributive Property to remove the parentheses: $-4(x - 3)$ [Section 1.8, pp. 67–68]

R4. Evaluate the expression $5 - 3x$ for $x = -2$. [Section 1.8, pp. 64–65]

Recall from Section 1.8 that a term is a number or the product of a number and one or more variables raised to a power. The numerical factor of a term is the coefficient. A constant is a single number, such as 2 or $-\dfrac{3}{2}$. Consider Table 1, where some algebraic expressions are given along with their terms and coefficients.

Table 1

Algebraic Expression	Terms	Coefficients
$3x + 2$	$3x, 2$	$3, 2$
$4x^2 - x + 5 = 4x^2 + (-x) + 5$	$4x^2, -x, 5$	$4, -1, 5$
$9x^2 + y^3$	$9x^2, y^3$	$9, 1$

Notice that we rewrite $4x^2 - x + 5$ given in Table 1 as a sum, $4x^2 + (-x) + 5$, to identify the terms and coefficients.

▶ 1 **Define Monomial and Determine the Degree of a Monomial**

Work Smart

The prefix "mono" means "one." For example, a monorail has one rail. So a monomial in one variable is either a constant or a single term with a variable in it.

In this chapter, we study *polynomials*. Polynomials are made up of terms that are *monomials*.

Definition

A **monomial** in one variable is the product of a constant and a variable raised to a whole number (0, 1, 2, …) power. A monomial in one variable is of the form

$$ax^k$$

where a is a constant that is any real number, x is a variable, and k is a whole number. The constant a is called the **coefficient** of the monomial. If $a \neq 0$, then k is the **degree** of the monomial.

The degree of a nonzero constant, such as 5 or -7, is zero. Because $0 = 0x = 0x^2 = 0x^3 = \cdots$, we cannot assign a degree to 0. Therefore, we say 0 has no degree.

EXAMPLE 1 **Monomials**

Monomial	Coefficient	Degree
(a) $2x^4$	2	4
(b) $-\dfrac{7}{4}x^2$	$-\dfrac{7}{4}$	2
(c) $-x = -1 \cdot x^1$	-1	1
(d) $x^3 = 1 \cdot x^3$	1	3
(e) 8	8	0
(f) 0	0	No degree

Ready?...Answers **R1.** -4
R2. $-11x - 5$ **R3.** $-4x + 12$
R4. 11

EXAMPLE 2 **Examples of Expressions That Are Not Monomials**

(a) $5x^{\frac{1}{2}}$ is not a monomial because the exponent of the variable is $\frac{1}{2}$, and $\frac{1}{2}$ is not a whole number.

(b) $2x^{-4}$ is not a monomial because the exponent of the variable is -4, and -4 is not a whole number.

In Words

The degree of a monomial can be thought of as the number of times the variable occurs as a factor. For example, because $2x^4 = 2 \cdot x \cdot x \cdot x \cdot x$, the degree of $2x^4$ is 4.

Quick ✔

1. A _____ in one variable is the product of a number and a variable raised to a whole number power.

2. The coefficient of a monomial such as x^2 or z is _.

In Problems 3–6, determine whether the expression is a monomial (yes or no). For those that are monomials, determine the coefficient and degree.

3. $12x^6$

4. $3x^{-3}$

5. 10

6. $n^{\frac{1}{3}}$

A monomial may contain more than one variable factor, such as $ax^m y^n$, where a is the coefficient, x and y are variables, and m and n are whole numbers. The **degree of the monomial** $ax^m y^n$ is the sum of the exponents, $m + n$.

EXAMPLE 3 **Monomials in More Than One Variable**

(a) $-4x^3 y^4$ is a monomial in x and y of degree $3 + 4 = 7$. The coefficient is -4.

(b) $10ab^5$ is a monomial in a and b of degree $1 + 5 = 6$. The coefficient is 10.

Quick ✔

7. The degree of a monomial in the form $ax^m y^n$ is _____.

In Problems 8–10, determine whether the expression is a monomial (yes or no). For those that are monomials, determine the coefficient and degree.

8. $3x^5 y^2$

9. $4ab^{\frac{1}{2}}$

10. $-x^2 y$

▶ ❷ **Define Polynomial and Determine the Degree of a Polynomial**

Definition

A **polynomial** is a monomial or the sum of monomials.

Certain polynomials have special names. A polynomial with exactly one term is a monomial; a polynomial that has two different monomials is called a **binomial;** and a polynomial that contains three different monomials is called a **trinomial.** So

Work Smart

The prefix "bi" means "two," as in bicycle. The prefix "tri" means "three," as in tricycle. The prefix "poly" means "many."

$-14x$ is a polynomial	but more specifically	$-14x$ is a monomial
$2x^3 - 5x$ is a polynomial	but more specifically	$2x^3 - 5x$ is a binomial
$-x^3 - 4x + 11$ is a polynomial	but more specifically	$-x^3 - 4x + 11$ is a trinomial
$3x^2 + 6xy - 2y^2$ is a polynomial	but more specifically	$3x^2 + 6xy - 2y^2$ is a trinomial

A polynomial is in **standard form** if it is written with the terms in descending order according to degree. The **degree of a polynomial** is the highest degree of all the terms of the polynomial. Remember, the degree of a nonzero constant is 0, and the number 0 has no degree.

EXAMPLE 4   **Examples of Polynomials**

Polynomial	Degree
(a) $7x^3 - 2x^2 + 6x + 4$	3
(b) $3 - 8x + x^2 = x^2 - 8x + 3$	2
(c) $-7x^4 + 24$	4
(d) $x^3y^4 - 3x^3y^2 + 2x^3y$	7
(e) $p^2q - 8p^3q^2 + 3 = -8p^3q^2 + p^2q + 3$	5
(f) 6	0
(g) 0	No degree

●

We can use letters other than x to represent the variable in a polynomial.

$$7t^4 + 14t^2 + 8 \text{ is a polynomial (in } t\text{) of degree 4}$$

$$-5z^2 + 3z - 6 \text{ is a polynomial (in } z\text{) of degree 2}$$

$$8y^3 - y^2 + 2y - 12 \text{ is a polynomial (in } y\text{) of degree 3}$$

Because a polynomial is the sum of monomials, if any term in an algebraic expression is not a monomial, then the expression is not a polynomial.

EXAMPLE 5   **Algebraic Expressions That Are Not Polynomials**

(a) $4x^{-2} - 5x + 1$ is not a polynomial because the exponent on the first term, -2, is not a whole number.

(b) $\dfrac{4}{x^3}$ is not a polynomial because a variable expression is in the denominator.

(c) $4z^2 + 9z - 3z^{\frac{1}{2}}$ is not a polynomial because the exponent on the third term, $\dfrac{1}{2}$, is not a whole number.

●

Quick ✓

11. *True or False* The degree of the polynomial $3ab^2 - 4a^2b^3 + \dfrac{6}{5}a^2b^2$ is 5.

12. A polynomial is said to be written in _____ _____ if it is written with the terms in descending order according to degree.

In Problems 13–16, determine whether the algebraic expression is a polynomial (yes or no). For those that are polynomials, determine the degree.

13. $-4x^3 + 2x^2 - 5x + 3$

14. $2m^{-1} + 7$

15. $\dfrac{-1}{x^2 + 1}$

16. $5p^3q - 8pq^2 + pq$

▶ ❸ Simplify Polynomials by Combining Like Terms

To simplify a polynomial means to perform all indicated operations and combine like terms.

EXAMPLE 6 **Simplifying Polynomials: Addition**

Find the sum: $(-5x^3 + 6x^2 + 2x - 7) + (3x^3 + 4x + 1)$

Solution

We can find the sum using horizontal or vertical addition.

Horizontal Addition: The idea here is to combine like terms.

$$(-5x^3 + 6x^2 + 2x - 7) + (3x^3 + 4x + 1) = -5x^3 + 6x^2 + 2x - 7 + 3x^3 + 4x + 1$$
$$\text{Rearrange terms:} = -5x^3 + 3x^3 + 6x^2 + 2x + 4x - 7 + 1$$
$$\text{Simplify:} = -2x^3 + 6x^2 + 6x - 6$$

Vertical Addition: Here, line up like terms in each polynomial vertically and then add the coefficients.

$$\begin{array}{r} -5x^3 + 6x^2 + 2x - 7 \\ 3x^3 \qquad\quad + 4x + 1 \\ \hline -2x^3 + 6x^2 + 6x - 6 \end{array}$$

> **Work Smart**
>
> Remember, like terms have the same variable and the same exponent on the variable.

Quick ✓

17. *True or False* $4x^3 + 7x^3 = 11x^6$

In Problems 18 and 19, add the polynomials using either horizontal or vertical addition.

18. $(9x^2 - x + 5) + (3x^2 + 4x - 2)$

19. $(4z^4 - 2z^3 + z - 5) + (-2z^4 + 6z^3 - z^2 + 4)$

Adding polynomials in two variables is handled the same way as adding polynomials in a single variable—that is, by adding like terms.

EXAMPLE 7 **Simplifying Polynomials in Two Variables: Addition**

Find the sum: $(6a^2b + 11ab^2 - 4ab) + (a^2b - 3ab^2 + 9ab)$

Solution

We present only the horizontal format.

$$(6a^2b + 11ab^2 - 4ab) + (a^2b - 3ab^2 + 9ab) = 6a^2b + 11ab^2 - 4ab + a^2b - 3ab^2 + 9ab$$
$$\text{Rearrange terms:} = 6a^2b + a^2b + 11ab^2 - 3ab^2 - 4ab + 9ab$$
$$\text{Simplify:} = 7a^2b + 8ab^2 + 5ab$$

Quick ✓

In Problem 20, simplify by adding the polynomials.

20. $(7x^2y + x^2y^2 - 5xy^2) + (-2x^2y + 5x^2y^2 + 4xy^2)$

▶ We can subtract polynomials using either the horizontal or the vertical approach as well. However, the first step in subtracting polynomials uses the method for subtracting two real numbers. Remember,

$$a - b = a + (-b)$$

So, to subtract one polynomial from another, we add the opposite of each term in the polynomial following the subtraction sign and then combine like terms.

> **Work Smart**
>
> Taking the opposite of each term following the subtraction symbol is the same as multiplying each term of the polynomial by -1.

EXAMPLE 8 **Simplifying Polynomials: Subtraction**

Find the difference: $(6z^3 + 2z^2 - 5) - (-3z^3 + 9z^2 - z + 1)$

Solution

Horizontal Subtraction: First, write the problem as an addition problem by determining the opposite of each term following the subtraction sign.

$$(6z^3 + 2z^2 - 5) - (-3z^3 + 9z^2 - z + 1) = (6z^3 + 2z^2 - 5) + (3z^3 - 9z^2 + z - 1)$$

$$\text{Remove parentheses:} \quad = 6z^3 + 2z^2 - 5 + 3z^3 - 9z^2 + z - 1$$

$$\text{Rearrange terms:} \quad = 6z^3 + 3z^3 + 2z^2 - 9z^2 + z - 5 - 1$$

$$\text{Simplify:} \quad = 9z^3 - 7z^2 + z - 6$$

Vertical Subtraction: First, line up like terms vertically.

$$6z^3 + 2z^2 \qquad - 5$$
$$-(-3z^3 + 9z^2 - z + 1)$$

Then change the sign of each coefficient of the second polynomial, and add.

$$6z^3 + 2z^2 \qquad - 5$$
$$+3z^3 - 9z^2 + z - 1$$
$$\overline{9z^3 - 7z^2 + z - 6}$$

●

> **Quick ✓**
>
> *In Problems 21 and 22, simplify by subtracting the polynomials.*
>
> **21.** $(7x^3 - 3x^2 + 2x + 8) - (2x^3 + 12x^2 - x + 1)$
>
> **22.** $(6y^3 - 3y^2 + 2y + 4) - (-2y^3 + 5y + 10)$

EXAMPLE 9 **Simplifying Polynomials in Two Variables: Subtraction**

Find the difference: $(2p^2q + 5pq^2 + pq) - (3p^2q + 9pq^2 - 6pq)$

Solution

$$(2p^2q + 5pq^2 + pq) - (3p^2q + 9pq^2 - 6pq) = (2p^2q + 5pq^2 + pq) + (-3p^2q - 9pq^2 + 6pq)$$

$$\text{Remove parentheses:} \quad = 2p^2q + 5pq^2 + pq - 3p^2q - 9pq^2 + 6pq$$

$$\text{Rearrange terms:} \quad = 2p^2q - 3p^2q + 5pq^2 - 9pq^2 + pq + 6pq$$

$$\text{Simplify:} \quad = -p^2q - 4pq^2 + 7pq$$

●

> **Quick ✓**
>
> *In Problems 23 and 24, subtract the polynomials.*
>
> **23.** $(9x^2y + 6x^2y^2 - 3xy^2) - (-2x^2y + 4x^2y^2 + 6xy^2)$
>
> **24.** $(4a^2b - 3a^2b^3 + 2ab^2) - (3a^2b - 4a^2b^3 + 5ab)$

EXAMPLE 10 **Simplifying Polynomials**

Perform the indicated operations: $(3x^2 - 2xy + y^2) - (7x^2 - y^2) + (3xy - 5y^2)$

Solution

$$(3x^2 - 2xy + y^2) - (7x^2 - y^2) + (3xy - 5y^2) = 3x^2 - 2xy + y^2 - 7x^2 + y^2 + 3xy - 5y^2$$

$$\text{Rearrange terms:} \quad = 3x^2 - 7x^2 - 2xy + 3xy + y^2 + y^2 - 5y^2$$

$$\text{Simplify:} \quad = -4x^2 + xy - 3y^2$$

●

▶ ④ Evaluate Polynomials

To **evaluate** a polynomial, we substitute the given number for the value of the variable and simplify, just as we did in Section 1.8.

EXAMPLE 11 **Evaluating a Polynomial**

Evaluate the polynomial $2x^3 - 5x^2 + x - 3$ for

(a) $x = 2$ **(b)** $x = -3$

Solution

(a) $2x^3 - 5x^2 + x - 3 = 2(2)^3 - 5(2)^2 + 2 - 3$

$= 16 - 20 + 2 - 3$

$= -5$

(b) $2x^3 - 5x^2 + x - 3 = 2(-3)^3 - 5(-3)^2 + (-3) - 3$

$= -54 - 45 - 3 - 3$

$= -105$ ●

Work Smart

Remember the order of operations: evaluate the exponent before you multiply.

EXAMPLE 12 **Evaluating a Polynomial in Two Variables**

Evaluate the polynomial $2a^2b + 3a^2b^2 - 4ab$ for $a = -3$ and $b = -1$.

Solution

Let $a = -3$ and $b = -1$ in $2a^2b + 3a^2b^2 - 4ab$.

$2a^2b + 3a^2b^2 - 4ab = 2(-3)^2(-1) + 3(-3)^2(-1)^2 - 4(-3)(-1)$

$= 2(9)(-1) + 3(9)(1) - 4(-3)(-1)$

$= -18 + 27 - 12$

$= -3$ ●

EXAMPLE 13 **How Much Revenue?**

The monthly revenue (in dollars) from selling x clocks is given by the polynomial $-0.3x^2 + 90x$. Evaluate the polynomial for $x = 200$ and explain the result.

Solution

The variable x represents the number of clocks sold in a month. When we evaluate the polynomial $-0.3x^2 + 90x$ for $x = 200$, we are finding the revenue (in dollars) when 200 clocks are sold in a month. To find the revenue, we replace x by 200 in the expression $-0.3x^2 + 90x$.

$$-0.3x^2 + 90x = -0.3(200)^2 + 90(200)$$
$$= -0.3(40,000) + 18,000$$
$$= -12,000 + 18,000$$
$$= 6000$$

The revenue from selling 200 clocks in a month is $6000.

Quick ✓

Evaluate the polynomial for the given value.

28. The polynomial $40x - 0.2x^2$ represents the monthly revenue (in dollars) realized by selling x wristwatches. Find the monthly revenue for selling 75 wristwatches by evaluating the polynomial $40x - 0.2x^2$ for $x = 75$.

5.1 Exercises MyMathLab® PRACTICE

Exercise numbers in green have complete video solutions in MyMathLab

Problems 1–28 are the Quick ✓s that follow the EXAMPLES.

Building Skills

In Problems 29–40, determine whether the given expression is a monomial (Yes or No). For those that are monomials, state the coefficient and degree. See Objective 1.

29. $8y^3$

30. $-3x^2$

31. $\dfrac{x^2}{7}$

32. $4m^{101}$

33. z^{-6}

34. $\dfrac{1}{y^7}$

35. $12mn^4$

36. $-x^6y$

37. $\dfrac{3}{n^2}$

38. y^{-1}

39. 4

40. $\dfrac{2}{3}$

In Problems 41–56, determine whether the algebraic expression is a polynomial (Yes or No). If it is a polynomial, write the polynomial in standard form, determine its degree, and state whether it is a monomial, a binomial, or a trinomial. If it is a polynomial with more than three terms, say the expression is a polynomial. See Objective 2.

41. $6x^2 - 10$

42. $4x + 1$

43. $\dfrac{-20}{n}$

44. $\dfrac{1}{x}$

45. $3y^{\frac{1}{3}} + 2$

46. $8m - 4m^{\frac{1}{2}}$

47. $\dfrac{1}{8}$

48. 32

49. $5z^3 - 10z^2 + z + 12$

50. $p^5 - 3p^4 + 7p + 8$

51. $7x^{-1} + 4$

52. $4y^{-2} + 6y - 1$

53. $3t^2 - \dfrac{1}{2}t^4 + 6t$

54. $-3x^3 + 4x^7 - 1$

55. $3x^2y^2 + 2xy^4 + 4$

56. $4mn^3 - 2m^2n^3 + mn^8$

In Problems 57–70, add the polynomials. Express your answer in standard form. See Objective 3.

57. $(4x - 3) + (3x - 7)$

58. $(-13z + 4) + (9z - 10)$

59. $(-4m^2 + 2m - 1) + (2m^2 - 2m + 6)$

60. $(x^2 - 2) + (6x^2 - x - 1)$

61. $(p - p^3 + 2) + (6 - 2p^2 + p^3)$

62. $(4r^4 + 3r - 1) + (2r^4 - 7r^3 + r^2 - 10r)$

63. $(2y - 10) + (-3y^2 - 4y + 6)$

64. $(3 - 12w^2) + (2w^2 - 5 + 6w)$

65. $\left(\dfrac{1}{2}p^2 - \dfrac{2}{3}p + 2\right) + \left(\dfrac{3}{4}p^2 + \dfrac{5}{6}p - 5\right)$

66. $\left(\dfrac{3}{8}b^2 - \dfrac{3}{5}b + 1\right) + \left(\dfrac{5}{6}b^2 + \dfrac{2}{15}b - 1\right)$

67. $(5m^2 - 6mn + 2n^2) + (m^2 + 2mn - 3n^2)$

68. $(4a^2 + ab - 9b^2) + (-6a^2 - 4ab + b^2)$

69.
$$\begin{array}{r} 4n^2 - 2n + 1 \\ +(-6n + 4) \\ \hline \end{array}$$

70.
$$\begin{array}{r} 8x^3 \quad\quad + 2x + 2 \\ +(-3x^2 - 4x - 2) \\ \hline \end{array}$$

In Problems 71–84, subtract the polynomials. Express your answer in standard form. See Objective 3.

71. $(7x - 10) - (4x + 6)$

72. $(7t + 8) - (2t + 4)$

73. $(12x^2 - 2x - 4) - (-2x^2 + x + 1)$

74. $(3x^2 + x - 3) - (x^2 - 2x + 4)$

75. $(y^3 - 2y + 1) - (-3y^3 + y + 5)$

76. $(m^4 - 3m^2 + 5) - (3m^4 - 5m^2 - 2)$

77. $(3y^3 - 2y) - (2y + y^2 + y^3)$

78. $(2x - 4x^3) - (-3 - 2x + x^3)$

79. $\left(\dfrac{5}{3}q^2 - \dfrac{5}{2}q + 4\right) - \left(\dfrac{1}{9}q^2 + \dfrac{3}{8}q + 2\right)$

80. $\left(\dfrac{7}{4}x^2 - \dfrac{5}{8}x - 1\right) - \left(\dfrac{7}{6}x^2 + \dfrac{5}{12}x + 5\right)$

81. $(-4m^2n^2 - 2mn + 3) - (4m^2n^2 + 2mn + 10)$

82. $(4m^2n - 2mn - 4) - (10m^2n - 6mn - 3)$

83.
$$\begin{array}{r} 6x - 3 \\ -(10x + 2) \\ \hline \end{array}$$

84.
$$\begin{array}{r} 6n^2 - 2n - 3 \\ -(4n + 7) \\ \hline \end{array}$$

In Problems 85–92, evaluate the polynomial for each of the given value(s). See Objective 4.

85. $2x^2 - x + 3$
 (a) $x = 0$
 (b) $x = 5$
 (c) $x = -2$

86. $-x^2 + 10$
 (a) $x = 0$
 (b) $x = -1$
 (c) $x = 1$

87. $7 - x^2$
 (a) $x = 3$
 (b) $x = -\dfrac{5}{2}$
 (c) $x = -1.5$

88. $2 + \dfrac{1}{2}n^2$
 (a) $n = 4$
 (b) $n = 0.5$
 (c) $n = -\dfrac{1}{4}$

89. $-x^2y + 2xy^2 - 3$ for $x = 2$ and $y = -3$

90. $-2ab^2 - 2a^2b - b^3$ for $a = 1$ and $b = -2$

91. $st + 2s^2t + 3st^2 - t^4$ for $s = -2$ and $t = 4$

92. $m^2n^2 - mn^2 + 3m^2 - 2$ for $m = \dfrac{1}{2}$ and $n = -1$

Mixed Practice

In Problems 93–110, perform the indicated operations. Express your answer in standard form.

93. $4t - (7t - 3)$

94. $6x^2 - (18x^2 + 2)$

95. $(5x^2 + x - 4) + (-2x^2 - 4x + 1)$

96. $(4m^2 - m + 6) + (3m^2 - 4m - 10)$

97. $(2xy^2 - 3) + (7xy^2 + 4)$

98. $(-14xy + 3) - (-xy + 10)$

99. $(4 + 8y - 2y^2) - (3 - 7y - y^2)$

100. $(9 + 2z - 6z^2) - (-2 + z + 5z^2)$

101. $\left(\dfrac{5}{6}q^2 - \dfrac{1}{3}\right) + \left(\dfrac{3}{2}q^2 + 2\right)$

102. $\left(\dfrac{7}{10}t^2 - \dfrac{5}{12}t\right) + \left(\dfrac{3}{15}t^2 + \dfrac{3}{20}t - 3\right)$

103. $14d^2 - (2d - 10) - (d^2 - 3d)$

104. $3x - (5x + 1) - 4$

105. $(4a^2 - 1) + (a^2 + 5a + 2) - (-a^2 + 4)$

106. $(2b^2 + 3b - 5) - (b^2 - 4b + 1) + (b^2 + 1)$

107. $(p^2 + 25) - (3p^2 - p + 4) - (-2p^2 - 9p + 21)$

108. $(n^2 + 4n - 5) - (2n^2 - 5n + 1) + (n^2 - 8n + 6)$

109. $(x^2 - 2xy - y^2) - (3x^2 + xy - y^2) + (xy + y^2)$

110. $(2st^4 - 3s^2t^2 + t^4) + (7s^2t^2 - 3t^4 + 8st^4)$

111. Find the sum of $3x + 10$ and $-8x + 2$.

112. Find the sum of $4x^2 - 2x - 3$ and $-5x^2 - 2x + 7$.

113. Find the difference of $-x^2 + 2x + 3$ and $-4x^2 - 2x + 6$.

114. Find the difference of $4x - 3$ and $8x - 7$.

115. Subtract $14x^2 - 2x + 3$ from $2x - 10$.

116. Subtract $-4n - 8$ from $3n^2 + 8n - 9$.

117. What polynomial should be added to $3x - 5$ so that the sum is zero?

118. What polynomial should be added to $2a + 7b$ so that the sum is $a - b$?

Applying the Concepts

119. Height of a Ball The height above the ground (in feet) of a ball dropped from the top of a 30-foot-tall building after t seconds is given by the polynomial $-16t^2 + 30$. What is the height of the ball after $\dfrac{1}{2}$ second?

120. Height of a Ball The height above the ground (in feet) of a ball tossed upward after t seconds is given by the polynomial $-16t^2 + 32t + 100$. What is the height of the ball after 2.5 seconds?

121. Stopping Distance on Dry Pavement The polynomial $0.04v^2 + 0.7v$ models the number of feet it takes for a car traveling v miles per hour to stop on dry, level concrete.

  (a) How many feet will it take a car traveling 50 miles per hour to stop on dry, level concrete?

  (b) How many feet will it take a car traveling 20 miles per hour through a school zone to stop on dry, level concrete?

122. Stopping Distance on Wet Pavement The polynomial $0.07v^2 + 0.7v$ models the number of feet it takes for a car traveling v miles per hour to stop on wet, level concrete.

  (a) How many feet will it take a car traveling 50 miles per hour to stop on wet, level concrete?

  (b) How many feet will it take a car traveling 20 miles per hour through a school zone to stop on wet, level concrete?

123. Manufacturing Calculators The revenue (in dollars) from manufacturing and selling x calculators in a day is given by the polynomial $-2x^2 + 120x$. The cost of manufacturing and selling x calculators in a day is given by the polynomial $0.125x^2 + 15x$.

  (a) Profit is equal to revenue minus costs. Write a polynomial that represents the profit from manufacturing and selling x calculators in a day.

  (b) Find the profit if 20 calculators are produced and sold each day.

124. Manufacturing DVDs Profit is equal to revenue minus cost. The revenue (in dollars) from manufacturing and selling x DVDs each day is given by the polynomial $-0.001x^2 + 6x$. The cost of manufacturing and selling x DVDs each day is given by the polynomial $0.8x + 3000$.

  (a) What a polynomial that expresses the profit from manufacturing and selling x DVDs each day.

  (b) Find the profit if 1600 DVDs are produced and sold in a day.

125. Lawn Service Marissa started a lawn-care business in Tampa, Florida, in which she fertilizes lawns. Her costs are $5 per lawn for fertilizer and $10 per lawn for labor.

  (a) If she charges $25 per lawn, write a polynomial that represents her weekly profit for fertilizing x lawns.

  (b) If Marissa fertilizes 50 lawns in a week, what is her weekly profit?

  (c) If Marissa fertilizes 50 lawns each week for 30 weeks, what is her profit?

126. Skateboard Business Kyle owns a skateboard shop. Kyle charges $40 for each skateboard he sells. The cost for manufacturing each skateboard is $12. In addition, Kyle has weekly costs of $504, regardless of how many skateboards he manufactures.

  (a) If Kyle manufactures and sells x skateboards in a week, write a polynomial that represents his weekly profit.

  (b) What is Kyle's weekly profit if he sells 45 skateboards per week?

  (c) How many skateboards does Kyle need to sell in order to break even (profit = $0)?

△ *In Problems 127–130, write the polynomial that represents the perimeter of each figure.*

127.

128.

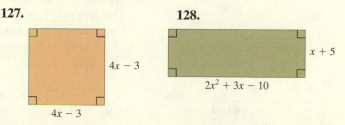

129.

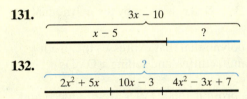

10

2x + 1

3x − 4

130.

x² − 2

4

In Problems 131 and 132, write the polynomial that represents the unknown length.

131.

3x − 10

x − 5   ?

132.

?

2x² + 5x   10x − 3   4x² − 3x + 7

Explaining the Concepts

133. When adding polynomials, you may use either a horizontal or a vertical format. Explain which format you prefer and what you view as its advantages.

134. Use the two polynomials $3x - 5$ and $2x + 3$ to make up three different problems. In the directions to the first problem, use the word "evaluate"; in the second problem use the direction "solve"; and in the third use "simplify."

135. Explain how you find the degree of a polynomial in one variable. Explain how you find the degree of a polynomial in more than one variable. How do these processes differ?

136. Suppose you are adding two polynomials of equal degree. Must the degree of the sum equal the degree of each polynomial? Explain.

137. What is the degree of a polynomial that is linear?

138. Give a definition of "polynomial" using your own words. Provide examples of polynomials that are monomials, that are binomials, and that are trinomials.

5.2 Multiplying Monomials: The Product and Power Rules

Objectives

1 Simplify Exponential Expressions Using the Product Rule

2 Simplify Exponential Expressions Using the Power Rule

3 Simplify Exponential Expressions Containing Products

4 Multiply a Monomial by a Monomial

Work Smart
The natural numbers are 1, 2, 3, ….

Are You Ready for This Section?
Before getting started, take the following readiness quiz. If you get a problem wrong, go back to the section cited and review the material.

R1. Evaluate: 4^3 [Section 1.7, pp. 56–57]

R2. Use the Distributive Property to simplify: $3(4 - 5x)$ [Section 1.8, pp. 67–68]

Recall from Section 1.7, that if a is a real number and n is a natural number, then the symbol a^n means to use a as a factor n times:

$$a^n = \underbrace{a \cdot a \cdot \ldots \cdot a}_{n \text{ factors}}$$

exponent
base

For example,

$$4^3 = \underbrace{4 \cdot 4 \cdot 4}_{3 \text{ factors}} \quad \text{or} \quad y^4 = \underbrace{y \cdot y \cdot y \cdot y}_{4 \text{ factors}}$$

In the notation a^n, a is the base and n is the power or exponent. We read a^n as "a raised to the power n" or "a raised to the nth power." We read a^2 as "a squared" and a^3 as "a cubed." The expression 4^3 is in **exponential form** and the expression $4 \cdot 4 \cdot 4$ is in **expanded form.**

▶ 1 Simplify Exponential Expressions Using the Product Rule

We can discover several rules for simplifying expressions with natural number exponents. The first rule is used when multiplying two exponential expressions with the same base. Consider the following:

$$x^2 \cdot x^4 = \underbrace{(x \cdot x)}_{2 \text{ factors}} \underbrace{(x \cdot x \cdot x \cdot x)}_{4 \text{ factors}} = \underbrace{x \cdot x \cdot x \cdot x \cdot x \cdot x}_{6 \text{ factors}} = x^6$$

Same base   Same base

Sum of powers
2 and 4

The following rule generalizes this result.

In Words

When multiplying two exponential expressions with the same base, add the exponents. Then write the common base to the power of this sum.

> **Product Rule for Exponents**
>
> If a is a real number, and m and n are natural numbers, then
> $$a^m \cdot a^n = a^{m+n}$$

EXAMPLE 1 | **Using the Product Rule to Evaluate Exponential Expressions**

Evaluate each expression:

(a) $2^2 \cdot 2^4$

(b) $(-3)^2(-3)$

Solution

(a) $2^2 \cdot 2^4 = 2^{2+4}$
$= 2^6$
$= 64$

(b) $(-3)^2(-3) = (-3)^2(-3)^1$
$= (-3)^{2+1}$
$= (-3)^3$
$= -27$

EXAMPLE 2 | **Using the Product Rule to Simplify Exponential Expressions**

Simplify the following expressions:

(a) $t^4 \cdot t^7$

(b) $a^6 \cdot a \cdot a^4$

Solution

(a) $t^4 \cdot t^7 = t^{4+7}$
$= t^{11}$

(b) $a^6 \cdot a \cdot a^4 = (a^6 \cdot a^1) \cdot a^4$
$= a^{6+1} \cdot a^4$
$= a^7 \cdot a^4$
$= a^{7+4}$
$= a^{11}$

EXAMPLE 3 | **Using the Product Rule to Simplify Exponential Expressions**

Simplify the expression: $m^3 \cdot m^5 \cdot n^9$

Solution

To use the Product Rule for Exponents, the bases must be the same.
$$m^3 \cdot m^5 \cdot n^9 = m^{3+5} \cdot n^9$$
$$= m^8 n^9$$

The expression $m^8 n^9$ is in simplest form because the bases are different.

Quick ✓

1. The expression 12^3 is written in _____ form.

2. $a^m \cdot a^n = a^{—}$.

3. *True or False* $3^4 \cdot 3^2 = 9^6$

In Problems 4–8, evaluate or simplify each expression.

4. $3^2 \cdot 3$

5. $(-5)^2(-5)^3$

6. $c^6 \cdot c^2$

7. $y^3 \cdot y \cdot y^5$

8. $a^4 \cdot a^5 \cdot b^6$

▶ ❷ **Simplify Exponential Expressions Using the Power Rule**

Let's look at an exponential expression containing a power raised to a power.

$$(3^2)^4 = \underbrace{3^2 \cdot 3^2 \cdot 3^2 \cdot 3^2}_{\text{4 factors}} = \underbrace{(3 \cdot 3)}_{\text{2 factors}} \cdot \underbrace{(3 \cdot 3)}_{\text{2 factors}} \cdot \underbrace{(3 \cdot 3)}_{\text{2 factors}} \cdot \underbrace{(3 \cdot 3)}_{\text{2 factors}} = 3^8$$

$$2 \cdot 4 = 8 \text{ factors}$$

We have the following result:

In Words

If an exponential expression contains a power raised to a power, keep the base and multiply the powers.

Power Rule For Exponents

If a is a real number, and m and n are natural numbers, then

$$(a^m)^n = a^{m \cdot n}$$

EXAMPLE 4 **Using the Power Rule to Simplify Exponential Expressions**

Simplify each expression. Write the answer in exponential form.

(a) $(2^3)^2$ **(b)** $[(-4)^2]^5$

Solution

(a) $(2^3)^2 = 2^{3 \cdot 2}$
$$= 2^6$$

(b) $[(-4)^2]^5 = (-4)^{2 \cdot 5}$
$$= (-4)^{10}$$

EXAMPLE 5 **Using the Power Rule to Simplify Exponential Expressions**

Simplify each expression. Write the answer in exponential form.

(a) $(z^4)^3$ **(b)** $[(-n)^3]^5$

Solution

(a) $(z^4)^3 = z^{4 \cdot 3}$
$$= z^{12}$$

(b) $[(-n)^3]^5 = (-n)^{3 \cdot 5}$
$$= (-n)^{15}$$

Quick ✓

9. If a is a real number, and m and n are natural numbers, then $(a^m)^n = a$___.

In Problems 10–13, simplify each expression. Write the answer in exponential form.

10. $(2^2)^4$ **11.** $[(-3)^3]^2$ **12.** $(b^2)^5$ **13.** $[(-y)^2]^4$

▶ ❸ **Simplify Exponential Expressions Containing Products**

What happens when we raise a product to a power?

$$(x \cdot y)^3 = (x \cdot y) \cdot (x \cdot y) \cdot (x \cdot y) = (x \cdot x \cdot x) \cdot (y \cdot y \cdot y) = x^3 \cdot y^3$$

The following rule generalizes this result.

In Words

When we raise a product to a power, we raise each factor to the power. Remember, each of the numbers or variables in a multiplication problem is a factor.

Product to a Power Rule for Exponents

If a, b are real numbers and n is a natural number, then

$$(a \cdot b)^n = a^n \cdot b^n$$

EXAMPLE 6 **Using the Product to a Power Rule to Simplify Exponential Expressions**

Simplify each expression:

(a) $(3b)^4$ (b) $(-4m^3)^2$ (c) $(-2a^2b)^3$

Solution

Each expression contains the product of factors, so we use the Product to a Power Rule, $(a \cdot b)^n = a^n \cdot b^n$.

(a)
$$(3b)^4 = (3)^4(b)^4$$
Evaluate 3^4: $= 81b^4$

(b)
$$(-4m^3)^2 = (-4)^2(m^3)^2$$
Evaluate $(-4)^2$; $(a^m)^n = a^{m \cdot n}$: $= 16m^{3 \cdot 2}$
$= 16m^6$

(c)
$$(-2a^2b)^3 = (-2)^3(a^2b)^3$$
$$= (-2)^3(a^2)^3(b)^3$$
Evaluate $(-2)^3$; $(a^m)^n = a^{m \cdot n}$: $= -8a^{2 \cdot 3}b^3$
$= -8a^6b^3$

Quick ✓

14. *True or False* $(ab^3)^2 = ab^6$

In Problems 15–17, simplify each expression.

15. $(2n)^3$ **16.** $(-5x^4)^3$ **17.** $(-7a^3b)^2$

Let's summarize the three rules for exponents.

Work Smart: Study Skills

Are you having trouble remembering the rules for multiplying exponents? Try making flash cards with the property on the front and an example or two on the back. Also, write in words what each rule means and when it should be applied.

Rules for Exponents

If a and b are real numbers and n is a natural number, then

- **Product Rule for Exponents** $a^m \cdot a^n = a^{m+n}$
- **Power Rule for Exponents** $(a^m)^n = a^{m \cdot n}$
- **Product to a Power Rule for Exponents** $(a \cdot b)^n = a^n \cdot b^n$

▶ ❹ **Multiply a Monomial by a Monomial**

To multiply two monomials with the same variable base, we multiply the coefficients and use the Product Rule for Exponents to multiply the variable expressions.

EXAMPLE 7 **Multiplying a Monomial by a Monomial: One Variable**

Multiply and simplify:

(a) $(5x^2)(6x^4)$ (b) $(2p^3)(-5p^2)$

Solution

In each problem, we use the Commutative Property to rearrange factors.

(a) $(5x^2)(6x^4) = 5 \cdot 6 \cdot x^2 \cdot x^4$
$a^m \cdot a^n = a^{m+n}$: $= 30x^{2+4}$
$= 30x^6$

(b) $(2p^3)(-5p^2) = 2 \cdot (-5) \cdot p^3 \cdot p^2$
$a^m \cdot a^n = a^{m+n}$: $= -10p^{3+2}$
$= -10p^5$

EXAMPLE 8 **Multiplying a Monomial by a Monomial: Two Variables**

Multiply and simplify:

(a) $(-3x^2y)(-7x^5y^3)$

(b) $\left(\frac{4}{9}ab^3\right)(-a^2b)\left(\frac{27}{2}ab^2\right)$

Solution

(a) $(-3x^2y)(-7x^5y^3) = -3 \cdot (-7) \cdot x^2 \cdot x^5 \cdot y \cdot y^3$

$a^m \cdot a^n = a^{m+n}: \quad = 21 \cdot x^{2+5} \cdot y^{1+3}$

$= 21x^7y^4$

(b) $\left(\frac{4}{9}ab^3\right)(-a^2b)\left(\frac{27}{2}ab^2\right) = \frac{4}{9} \cdot (-1) \cdot \frac{27}{2} \cdot a \cdot a^2 \cdot a \cdot b^3 \cdot b \cdot b^2$

$a^m \cdot a^n = a^{m+n}: \quad = -6 \cdot a^{1+2+1} \cdot b^{3+1+2}$

$= -6a^4b^6$ ●

Quick ✓

In Problems 18–20, multiply and simplify.

18. $(2a^6)(4a^5)$

19. $(3m^2n^4)(-6mn^5)$

20. $\left(\frac{8}{3}xy^2\right)\left(\frac{1}{2}x^2y\right)(-12xy)$

5.2 Exercises MyMathLab® PRACTICE

Exercise numbers in green have complete video solutions in MyMathLab

Problems **1–20** *are the* **Quick ✓** *s that follow the* **EXAMPLES.**

Building Skills

In Problems 21–34, simplify each expression. See Objective 1.

21. $4^2 \cdot 4^3$

22. $3 \cdot 3^3$

23. $(-2)^3(-2)^4$

24. $(-3)^2(-3)^3$

25. $m^4 \cdot m^5$

26. $a^3 \cdot a^7$

27. $b^9 \cdot b^{11}$

28. $z^8 \cdot z^{23}$

29. $x^7 \cdot x$

30. $y^{13} \cdot y$

31. $p \cdot p^2 \cdot p^6$

32. $b^5 \cdot b \cdot b^7$

33. $(-n)^3(-n)^4$

34. $(-z)(-z)^5$

In Problems 35–42, simplify each expression. See Objective 2.

35. $(2^3)^2$

36. $(3^2)^2$

37. $[(-2)^2]^3$

38. $[(-2)^3]^3$

39. $(m^2)^7$

40. $(k^8)^3$

41. $[(-b)^4]^5$

42. $[(-a)^6]^3$

In Problems 43–54, simplify each expression. See Objective 3.

43. $(3x^2)^3$

44. $(4y^3)^2$

45. $(-5z^2)^2$

46. $(-6x^3)^2$

47. $\left(\frac{1}{2}a\right)^2$

48. $\left(\frac{3}{4}n^2\right)^3$

49. $(-3p^7q^2)^4$

50. $(-2m^2n^3)^4$

51. $(-2m^2n)^3$

52. $(-4ab^2)^3$

53. $(-5xy^2z^3)^2$

54. $(-3a^6bc^4)^4$

In Problems 55–68, multiply the monomials. See Objective 4.

55. $(4x^2)(3x^3)$

56. $(5y^6)(3y^2)$

57. $(10a^3)(-4a^7)$

58. $(7b^5)(-2b^4)$

59. $(-m^3)(7m)$

60. $(-n^6)(5n)$

61. $\left(\frac{4}{5}x^4\right)\left(\frac{15}{2}x^3\right)$

62. $\left(\frac{3}{8}y^5\right)\left(\frac{4}{9}y^6\right)$

63. $(2x^2y^3)(3x^4y)$

64. $(6a^2b^3)(2a^5b)$

65. $\left(\frac{1}{4}mn^3\right)(-20mn)$

66. $\left(\frac{2}{3}s^2t^3\right)(-21st)$

67. $(4x^2y)(-5xy^3)(2x^2y^2)$

68. $\left(\frac{1}{2}b\right)(-20a^2b)\left(-\frac{2}{3}a\right)$

Mixed Practice

In Problems 69–94, simplify each expression.

69. $x^2 \cdot 4x$

70. $z^3 \cdot 5z$

71. $(-x)^2(5x^2) - (2x^2)^2$

72. $(3a^3)(-a) + (3a^2)^2$

73. $(b^3)^2 + (3b^2)^3$

74. $(m^2)^4 - (2m^4)^2$

75. $(-3b)^4 + (4b)^2$

76. $(-4a)^2 - (5a)^3$

77. $(-6p)^2\left(\frac{1}{4}p^3\right)$

78. $(-8w)^2\left(\frac{3}{16}w^5\right)$

79. $(-xy)(x^2y)(-3x)^3$

80. $(-4a)^3(bc^2)(-b)$

81. $\left(\frac{3}{5}\right)(5c)(-10d^2)$

82. $\left(-\frac{7}{3}n\right)(9n^2)\left(\frac{5}{42}n^5\right)$

83. $(5x)^2(x^2)^3$

84. $(4y)^2(y^3)^5$

85. $(x^2y)^3(-2xy^3)$

86. $(s^2t^3)^2(3st)$

87. $(-3x)^2(2x^4)^3 + (3x^{12})^2$

88. $(-2p^3)^2(3p)^3 - (2p^6)^3$

89. $\left(\frac{4}{5}q\right)^2(-5q)^2\left(\frac{1}{4}q^2\right)$

90. $\left(\frac{2}{3}m\right)^2(-3m)^3\left(\frac{3}{4}m^3\right)$

91. $(3x)^2(-2x)^4\left(\frac{1}{3}x^4\right)^2$

92. $(5y)^3(-3y)^2\left(\frac{1}{5}y^4\right)^2$

93. $-3(-5mn^3)^2\left(\frac{2}{5}m^5n\right)^3$

94. $-2(-4a^3b^2)^2\left(\frac{3}{2}ab^5\right)^3$

Applying the Concepts

△ **95. Cubes** Suppose the length of a side of a cube is x^2. Find the volume of the cube in terms of x.

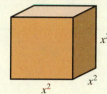

△ **96. Squares** Suppose that the length of a side of a square is $3s$. Find the area of the square in terms of s.

△ **97. Rectangle** Suppose that the width of a rectangle is $4x$ and its length is $12x$. Write an algebraic expression for the area of the rectangle in terms of x.

△ **98. Rectangle** Suppose that the width of a rectangle is $6a$ and its length is $8a$. Write an algebraic expression for the area of the rectangle in terms of a.

Explaining the Concepts

99. Provide a justification for the Product Rule for Exponents.

100. Provide a justification for the Power Rule for Exponents.

101. Provide a justification for the Product to a Power Rule for Exponents.

102. You simplified the expression $(-3a^2b^3)^4$ as $(-3a^2b^3)^4 = -3^4(a^2)^4(b^3)^4 = -81a^8b^{12}$. Your instructor marked your answer wrong. Find and explain the error, and then rework the problem correctly.

103. Explain the difference between simplifying the expression $4x^2 + 3x^2$ and simplifying the expression $(4x^2)(3x^2)$.

104. Explain why it is incorrect to simplify the product $\left(\frac{1}{2}x^2y\right)\left(\frac{4}{3}xy^4\right)$ by multiplying by the least common denominator.

5.3 Multiplying Polynomials

Objectives

❶ Multiply a Polynomial by a Monomial

❷ Multiply Two Binomials Using the Distributive Property

❸ Multiply Two Binomials Using the FOIL Method

❹ Multiply the Sum and Difference of Two Terms

❺ Square a Binomial

❻ Multiply a Polynomial by a Polynomial

Are You Ready for This Section?

Before getting started, take the following readiness quiz. If you get a problem wrong, go back to the section cited and review the material.

R1. Use the Distributive Property to simplify: $2(4x - 3)$ [Section 1.8, pp. 67–68]

R2. Find the product: $(3x^2)(-5x^3)$ [Section 5.2, pp. 321–322]

R3. Simplify: $(9a)^2$ [Section 5.2, pp. 320–321]

▶ ❶ **Multiply a Polynomial by a Monomial**

If we were asked to simplify $3(2x + 1)$, we would use the Distributive Property to obtain $3(2x) + 3(1) = 6x + 3$. In general, when we multiply a polynomial by a monomial, we use the following property.

Extended Form of the Distributive Property
$$a(b + c + \cdots + z) = a \cdot b + a \cdot c + \cdots + a \cdot z$$
where $a, b, c, \ldots, z$ are real numbers.

Ready?...Answers   **R1.** $8x - 6$

R2. $-15x^5$ **R3.** $81a^2$

EXAMPLE 1   **Multiplying a Monomial and a Trinomial**

Multiply and simplify: $2x^2(x^2 + 3x + 5)$

Solution

We are multiplying a monomial by a trinomial, so we use the Extended Form of the Distributive Property and multiply each term in parentheses by $2x^2$.

$$2x^2(x^2 + 3x + 5) = 2x^2(x^2) + 2x^2(3x) + 2x^2(5)$$
$$= 2x^4 + 6x^3 + 10x^2$$

EXAMPLE 2   **Multiplying a Monomial and a Trinomial**

Multiply and simplify: $-2xy(3x^2 + 5xy + 2y^2)$

Solution

Use the Extended Form of the Distributive Property.

$$-2xy(3x^2 + 5xy + 2y^2) = -2xy(3x^2) + (-2xy)(5xy) + (-2xy)(2y^2)$$
$$= -6x^3y - 10x^2y^2 - 4xy^3$$

EXAMPLE 3   **Multiplying a Trinomial and a Monomial**

Multiply and simplify: $\left(\dfrac{4}{3}z^2 + 8z + \dfrac{1}{4}\right)\left(\dfrac{1}{2}z^3\right)$

Solution

$$\left(\frac{4}{3}z^2 + 8z + \frac{1}{4}\right)\left(\frac{1}{2}z^3\right) = \frac{4}{3}z^2\left(\frac{1}{2}z^3\right) + 8z\left(\frac{1}{2}z^3\right) + \frac{1}{4}\left(\frac{1}{2}z^3\right)$$
$$= \frac{2}{3}z^5 + 4z^4 + \frac{1}{8}z^3$$

Work Smart

Example 3 is an algebraic **expression,** so you do not clear fractions.

Quick ✓

In Problems 1–3, multiply and simplify.

1. $3x(x^2 - 2x + 4)$

2. $-a^2b(2a^2b^2 - 4ab^2 + 3ab)$

3. $\left(5n^3 - \dfrac{15}{8}n^2 - \dfrac{10}{7}n\right)\left(\dfrac{3}{5}n^2\right)$

▶ ❷ **Multiply Two Binomials Using the Distributive Property**

To understand how to multiply two binomials, we review multiplication of two-digit numbers.

$$
\begin{array}{r}
32 \\
\times 14 \\
\end{array}
$$

$4 \times 3 \rightarrow 128 \leftarrow 4 \times 2$ ←This row represents $32 \cdot 4 = 128$.
$1 \times 3 \rightarrow \underline{32} \ \leftarrow 1 \times 2$ ←This row represents $32 \cdot 1 = 32$.
$\overline{448}$ ←Add vertically.

Now suppose we want to multiply $(3x + 2)$ and $(x + 4)$.

We proceed in exactly the same way as we did when multiplying two-digit numbers.

$$
\begin{array}{r}
3x + 2 \\
\times\ x + 4 \\
\hline
\end{array}
$$

$$4 \cdot 3x \longrightarrow \quad 12x + 8 \leftarrow 4 \cdot 2 \leftarrow \text{This row represents } 4(3x + 2)$$

$$x \cdot 3x \rightarrow 3x^2 + \ 2x \longleftarrow x \cdot 2 \leftarrow \text{This row represents } x(3x + 2)$$

$$\overline{3x^2 + 14x + 8} \leftarrow \text{Add vertically}$$

The multiplication that we just went through is known as *vertical multiplication*. We can also use *horizontal multiplication*, which requires use of the Distributive Property.

$$(3x + 2)(x + 4) = (3x + 2)x + (3x + 2) \cdot 4$$

Distribute x, distribute 4:
$$= 3x \cdot x + 2 \cdot x + 3x \cdot 4 + 2 \cdot 4$$
$$= 3x^2 + 2x + 12x + 8$$
$$= 3x^2 + 14x + 8$$

This is the same result that we obtained using vertical multiplication. Let's do a couple more examples using both horizontal and vertical multiplication.

EXAMPLE 4 **Multiplying Two Binomials**

Find the product: $(x + 3)(x - 8)$

Solution

Vertical Multiplication

$$
\begin{array}{r}
x + 3 \\
\times\ x - 8 \\
\hline
-8x - 24 \longleftarrow -8(x + 3)\\
x^2 + 3x \longleftarrow x(x + 3)\\
\hline
x^2 - 5x - 24
\end{array}
$$

Horizontal Multiplication

We distribute $x + 3$ to each term in $x - 8$.

$$(x + 3)(x - 8) = (x + 3)x + (x + 3)(-8)$$
$$= x^2 + 3x - 8x - 24$$

Combine like terms: $= x^2 - 5x - 24$

In either case, $(x + 3)(x - 8) = x^2 - 5x - 24$.

EXAMPLE 5 **Multiplying Two Binomials**

Find the product: $(3x + 4)(2x - 5)$

Solution

Vertical Multiplication

$$
\begin{array}{r}
3x + 4 \\
\times\ 2x - 5 \\
\hline
-15x - 20 \longleftarrow -5(3x + 4)\\
6x^2 + 8x \longleftarrow 2x(3x + 4)\\
\hline
6x^2 - 7x - 20
\end{array}
$$

Horizontal Multiplication

We distribute $3x + 4$ to each term in $2x - 5$.

$$(3x + 4)(2x - 5) = (3x + 4)(2x) + (3x + 4)(-5)$$
$$= 3x \cdot 2x + 4 \cdot 2x + 3x \cdot (-5) + 4 \cdot (-5)$$
$$= 6x^2 + 8x - 15x - 20$$

Combine like terms: $= 6x^2 - 7x - 20$

In either case, $(3x + 4)(2x - 5) = 6x^2 - 7x - 20$.

Quick ✔

In Problems 4–6, find the product.

4. $(x + 7)(x + 2)$ **5.** $(2n + 1)(n - 3)$ **6.** $(3p - 2)(2p - 1)$

⏵ ❸ Multiply Two Binomials Using the FOIL Method

Work Smart

The FOIL method is a form of the Distributive Property, and FOIL can be used **only** when we multiply binomials.

Another way to multiply two binomials is known as the **FOIL method.** FOIL stands for First, Outer, Inner, Last. "First" means to multiply the *first* terms in each binomial, "Outer" means to multiply the *outside* terms of each binomial, "Inner" means to multiply the *innermost* terms of each binomial, and "Last" means to multiply the *last* terms of each binomial. See below. The horizontal method is shown for comparison.

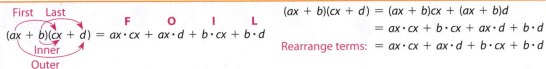

The FOIL Method	Horizontal Multiplication Using the Distributive Property
First Last $(ax + b)(cx + d) = ax \cdot cx + ax \cdot d + b \cdot cx + b \cdot d$ **F O I L** Inner Outer	$(ax + b)(cx + d) = (ax + b)cx + (ax + b)d$ $= ax \cdot cx + b \cdot cx + ax \cdot d + b \cdot d$ Rearrange terms: $= ax \cdot cx + ax \cdot d + b \cdot cx + b \cdot d$

EXAMPLE 6   **Using the FOIL Method to Multiply Two Binomials**

Find the product:

(a) $(y + 3)(y + 5)$ **(b)** $(a + 1)(a - 3)$

Solution

(a) First   Last   **F   O   I   L**
$$(y + 3)(y + 5) = (y)(y) + (y)(5) + (3)(y) + (3)(5)$$
$$= y^2 + 5y + 3y + 15$$
$$= y^2 + 8y + 15$$

(b) First   Last   **F   O   I   L**
$$(a + 1)(a - 3) = (a)(a) + (a)(-3) + (1)(a) + (1)(-3)$$
$$= a^2 - 3a + a - 3$$
$$= a^2 - 2a - 3$$

●

> **Quick ✓**
>
> **7.** FOIL stands for ____, ____, ____, ____.
>
> *In Problems 8–10, find the product using the FOIL method.*
>
> **8.** $(x + 3)(x + 4)$ **9.** $(y + 5)(y - 3)$ **10.** $(a - 1)(a - 5)$

EXAMPLE 7   **Using the FOIL Method to Multiply Two Binomials**

Find the product:

(a) $(3a + 1)(a - 4)$ **(b)** $(3a - 4b)(5a - 2b)$

Solution

(a) First   Last   **F   O   I   L**
$$(3a + 1)(a - 4) = (3a)(a) + (3a)(-4) + (1)(a) + (1)(-4)$$
$$= 3a^2 - 12a + a - 4$$
$$= 3a^2 - 11a - 4$$

(b)

$$FOIL$$

$$(3a - 4b)\,(5a - 2b) = (3a)\,(5a)\ +\ (3a)\,(-2b)\ -\ (4b)\,(5a)\ -\ (4b)\,(-2b)$$
$$= 15a^2 - 6ab - 20ab + 8b^2$$
$$= 15a^2 - 26ab + 8b^2$$

Quick ✔

In Problems 11–14, find the product using the FOIL method.

11. $(2x + 3)\,(x + 4)$ **12.** $(3y + 5)\,(2y - 3)$

13. $(2a - 1)\,(3a - 4)$ **14.** $(5x + 2y)\,(3x - 4y)$

▶ ④ Multiply the Sum and Difference of Two Terms

Certain binomials have products that result in patterns. For this reason, we call these products **special products.** We will look at several of them now.

EXAMPLE 8   **Finding a Product of the Form $(A - B)(A + B)$**

Find the product: $(x - 5)\,(x + 5)$

Solution

$$FOIL$$

$$(x - 5)\,(x + 5) = (x)\,(x)\ +\ (x)\,(5)\ -\ (5)\,(x)\ -\ (5)\,(5)$$
$$= x^2 + 5x - 5x - 25$$
$$= x^2 - 25$$

We call products of the form $(A - B)\,(A + B)$ "the sum and difference of two terms." Do you see why?

In Example 8, did you notice that the outer product, $5x$, and the inner product, $-5x$, were opposites? The sum of the outer and inner products of two binomials in the form $(A - B)\,(A + B)$ is *always* zero, so the product $(A - B)\,(A + B)$ is the *difference* $A^2 - B^2$.

Product of the Sum and Difference of Two Terms

$$(A - B)\,(A + B) = A^2 - B^2$$

EXAMPLE 9   **Finding the Product of the Sum and Difference of Two Terms**

Find each product:

    (a) $(2x + 5)\,(2x - 5)$ **(b)** $(4x - 3y)\,(4x + 3y)$

Solution

  (a) $(2x + 5)\,(2x - 5) = (2x)^2 - 5^2$ **(b)** $(4x - 3y)\,(4x + 3y) = (4x)^2 - (3y)^2$

$$= 4x^2 - 25 = 16x^2 - 9y^2$$

⊙ ❺ **Square a Binomial**

Another special product is called the square of a binomial.

$$\overset{\text{F \quad O \quad I \quad L}}{(x + 3)^2 = (x + 3)(x + 3) = x^2 + 3x + 3x + 9 = x^2 + 6x + 9}$$

Did you notice that the outer product and the inner product are the same, namely $3x$? Study Table 2 to discover the pattern.

Table 2

$(n + 5)^2 =$	$(n + 5)(n + 5) =$	$n^2 + 5n + 5n + 25 =$	$n^2 + 2(5n) + 25 =$	$n^2 + 10n + 25$
$(2a - 3)^2 =$	$(2a - 3)(2a - 3) =$	$4a^2 - 6a - 6a + 9 =$	$4a^2 - 2(6a) + 9 =$	$4a^2 - 12a + 9$
$(n + 5p)^2 =$		$n^2 + 5np + 5np + 25p^2 =$	$n^2 + 2(5np) + 25p^2 =$	$n^2 + 10np + 25p^2$
$(4b - 9)^2 =$			$16b^2 - 2(4b)(9) + 81 =$	$16b^2 - 72b + 81$

The results of Table 2 lead to the following.

$$(A + B)^2 = \underbrace{A^2}_{\text{(first term)}^2} + \underbrace{2AB}_{2\text{(product of terms)}} + \underbrace{B^2}_{\text{(second term)}^2}$$

$$(A - B)^2 = \underbrace{A^2}_{\text{(first term)}^2} - \underbrace{2AB}_{2\text{(product of terms)}} + \underbrace{B^2}_{\text{(second term)}^2}$$

Work Smart

$(x + y)^2 \neq x^2 + y^2$
$(x - y)^2 \neq x^2 - y^2$

Whenever you feel the urge to perform an operation that you're not quite sure about, try it with actual numbers. For example, does

$(3 + 2)^2 = 3^2 + 2^2$?

NO! So

$(x + y)^2 \neq x^2 + y^2$

Squares of Binomials

$$(A + B)^2 = A^2 + 2AB + B^2$$
$$(A - B)^2 = A^2 - 2AB + B^2$$

We call $A^2 + 2AB + B^2$ and $A^2 - 2AB + B^2$ **perfect square trinomials.**

EXAMPLE 10 **How to Find a Product of the Form $(A + B)^2$**

Find the product: $(y + 7)^2$

Step-by-Step Solution

Step 1: Because $(y + 7)^2$ is in the form $(A + B)^2$, we use $(A + B)^2 = A^2 + 2AB + B^2$ with $A = y$ and $B = 7$.

$$(A + B)^2 = A^2 + 2AB + B^2$$
$$(y + 7)^2 = (y)^2 + 2(y)(7) + 7^2$$

Step 2: Simplify.

$$= y^2 + 14y + 49$$ ●

EXAMPLE 11 **Finding a Product of the Form $(A - B)^2$**

Find each product: **(a)** $(3 - r)^2$ **(b)** $(9p - 4)^2$ **(c)** $(2x + 5y)^2$

Solution

(a) $(A - B)^2 = A^2 - 2AB + B^2$

$(3 - r)^2 = 3^2 - 2(3)(r) + r^2$

$= 9 - 6r + r^2$

$= r^2 - 6r + 9$

Work Smart

If you can't remember the formulas for a perfect square, don't panic! Simply use the fact that $(x + a)^2 = (x + a)(x + a)$ and then FOIL. The same logic applies to perfect squares of the form $(x - a)^2$.

(b) $(9p - 4)^2 = (9p)^2 - 2(9p)4 + 4^2$

$= 81p^2 - 72p + 16$

(c) $(2x + 5y)^2 = (2x)^2 + 2(2x)(5y) + (5y)^2$

$= 4x^2 + 20xy + 25y^2$

●

Quick ✓

19. $(A - B)^2 = $ _____; $(A + B)^2 = $ _____.

20. $x^2 + 2xy + y^2$ is referred to as a _____ _____ _____.

21. *True or False* The product of a binomial and a binomial is always a trinomial.

In Problems 22–27, find each product.

22. $(z - 9)^2$ **23.** $(p + 1)^2$

24. $(4 - a)^2$ **25.** $(3z - 4)^2$

26. $(5p + 1)^2$ **27.** $(2w + 7y)^2$

For convenience, we present a summary of binomial products.

Summary of Binomial Products

- When multiplying two binomials, we can use the Distributive Property.

$(3x + 5)(4x - 1) = (3x + 5)(4x) + (3x + 5)(-1)$

$= 12x^2 + 20x - 3x - 5$

$= 12x^2 + 17x - 5$

- When multiplying two binomials, we can use the FOIL pattern.

$$\qquad\qquad\qquad\qquad\text{F}\qquad\text{O}\qquad\text{I}\qquad\text{L}$$

$(3x + 5)(4x - 1) = 3x(4x) + 3x(-1) + 5(4x) + 5(-1)$

$= 12x^2 - 3x + 20x - 5$

$= 12x^2 + 17x - 5$

- The product of the sum and difference of two terms is $(A - B)(A + B) = A^2 - B^2$.

$(7a + 2b)(7a - 2b) = (7a)^2 - (2b)^2$

$= 49a^2 - 4b^2$

- The square of a binomial is

$(A + B)^2 = A^2 + 2AB + B^2$ or

$(A - B)^2 = A^2 - 2AB + B^2$.

The product is called a perfect square trinomial.

$(3x + 5)^2 = (3x)^2 + 2(3x)(5) + (5)^2$

$= 9x^2 + 30x + 25$

$(4y - 3)^2 = (4y)^2 - 2(4y)(3) + (3)^2$

$= 16y^2 - 24y + 9$

▶ ⑥ **Multiply a Polynomial by a Polynomial**

To find the product of two polynomials, we make repeated use of the Extended Form of the Distributive Property. We can use either a horizontal or a vertical format.

EXAMPLE 12 **Multiplying Polynomials Using Horizontal Multiplication**

Find the product $(2x + 3)(x^2 + 5x - 1)$ using

(a) horizontal multiplication and (b) vertical multiplication.

Solution

(a) Distribute $2x + 3$ to each term in the trinomial.

$$(2x + 3)(x^2 + 5x - 1) = (2x + 3) \cdot x^2 + (2x + 3) \cdot 5x + (2x + 3) \cdot (-1)$$
$$= 2x(x^2) + 3(x^2) + 2x(5x) + 3(5x) + 2x(-1) + 3(-1)$$

Simplify: $= 2x^3 + 3x^2 + 10x^2 + 15x - 2x - 3$

Combine like terms: $= 2x^3 + 13x^2 + 13x - 3$

(b) Place the polynomial with more terms on top, align terms of the same degree, and then multiply. Make sure both polynomials are written in standard form.

Now multiply the trinomial by the 3 in the binomial, and then multiply the trinomial by the $2x$ in the binomial.

$$
\begin{array}{r}
x^2 + 5x - 1 \\
\times \qquad 2x + 3 \\
\hline
\end{array}
$$

$3(x^2 + 5x - 1)$: $3x^2 + 15x - 3$

$2x(x^2 + 5x - 1)$: $2x^3 + 10x^2 - 2x$

Add vertically: $2x^3 + 13x^2 + 13x - 3$ ●

Quick ✓

28. *True or False* The product $(x - y)(x^2 + 2xy + y^2)$ can be found using the FOIL method.

In Problems 29 and 30, find the product using horizontal multiplication.

29. $(x - 2)(x^2 + 2x + 4)$ **30.** $(3y - 2)(y^2 + 2y + 4)$

In Problems 31 and 32, find the product using vertical multiplication.

31. $(x - 2)(x^2 + 2x + 4)$ **32.** $(3y - 1)(y^2 + 2y - 5)$

EXAMPLE 13 **Multiplying Three Polynomials**

Find the product: $2x(x - 4)(3x - 5)$

Solution

To find the product of three polynomials, multiply any two factors, and then multiply that product by the remaining factor. We'll start by multiplying $2x$ and $(x - 4)$.

$$2x(x - 4)(3x - 5) = (2x^2 - 8x)(3x - 5)$$

FOIL: $= 2x^2 \cdot 3x + 2x^2(-5) - 8x \cdot 3x - 8x \cdot (-5)$

$= 6x^3 - 10x^2 - 24x^2 + 40x$

Combine like terms: $= 6x^3 - 34x^2 + 40x$ ●

Quick ✓

In Problems 33 and 34, find the product.

33. $-2a(4a - 1)(3a + 5)$

34. $(x + 2)(x - 1)(x + 3)$

5.3 Exercises MathⓍP PRACTICE

Problems **1–34** are the **Quick ✓**s that follow the **EXAMPLES**.

Building Skills

In Problems 35–42, use the Distributive Property to find each product. See Objective 1.

35. $2x(3x - 5)$

36. $3m(2m - 7)$

37. $\frac{1}{2}n(4n - 6)$

38. $\frac{3}{5}b(15b - 5)$

39. $3n^2(4n^2 + 2n - 5)$

40. $4w(2w^2 + 3w - 5)$

41. $(4x^2y - 3xy^2)(x^2y)$

42. $(7r + 3s^2)(2r^2s)$

In Problems 43–48, use the Distributive Property to find each product. See Objective 2.

43. $(x + 5)(x + 7)$

44. $(x + 4)(x + 10)$

45. $(y - 5)(y + 7)$

46. $(n - 7)(n + 4)$

47. $(3m - 2y)(2m + 5y)$

48. $(5n - 2y)(2n - 3y)$

In Problems 49–62, find the product using the FOIL method. See Objective 3.

49. $(x + 2)(x + 3)$

50. $(x + 3)(x + 7)$

51. $(q - 6)(q - 7)$

52. $(n - 4)(n - 5)$

53. $(2x + 3)(3x - 1)$

54. $(3z - 2)(4z + 1)$

55. $(x^2 + 3)(x^2 + 1)$

56. $(x^2 - 5)(x^2 - 2)$

57. $(7 - x)(6 - x)$

58. $(2 - y)(4 - y)$

59. $(5u + 6v)(2u + v)$

60. $(3a - 2b)(a - 3b)$

61. $(2a - b)(5a + 2b)$

62. $(3r + 5s)(6r + 7s)$

In Problems 63–72, find the product of the sum and difference of two terms. See Objective 4.

63. $(x - 3)(x + 3)$

64. $(y + 7)(y - 7)$

65. $(2z + 5)(2z - 5)$

66. $(6r - 1)(6r + 1)$

67. $(4x^2 + 1)(4x^2 - 1)$

68. $(3a^2 + 2)(3a^2 - 2)$

69. $(2x - 3y)(2x + 3y)$

70. $(8a - 5b)(8a + 5b)$

71. $\left(x - \frac{1}{3}\right)\left(x + \frac{1}{3}\right)$

72. $\left(y + \frac{2}{9}\right)\left(y - \frac{2}{9}\right)$

In Problems 73–82, find the product. See Objective 5.

73. $(x - 2)^2$

74. $(x + 4)^2$

75. $(5k - 3)^2$

76. $(6b - 5)^2$

77. $(x + 2y)^2$

78. $(3x - 2y)^2$

79. $(2a - 3b)^2$

80. $(5x + 2y)^2$

81. $\left(x + \frac{1}{2}\right)^2$

82. $\left(y - \frac{1}{3}\right)^2$

In Problems 83–96, find the product. See Objective 6.

83. $(x - 2)(x^2 + 3x + 1)$

84. $(3a - 1)(2a^2 - 5a - 3)$

85. $(2y^2 - 6y + 1)(y - 3)$

86. $(2m^2 - m + 2)(2m + 1)$

87. $(2x - 3)(x^2 - 2x - 1)$

88. $(4y + 1)(2y^2 - y - 3)$

89. $2b(b - 3)(b + 4)$

90. $5a(a + 6)(a - 1)$

91. $-\frac{1}{2}x(2x + 6)(x - 3)$

92. $-\frac{4}{3}k(k + 7)(3k - 9)$

93. $(5y^3 - y^2 + 2)(2y^2 + y + 1)$

94. $(2m^2 - m + 4)(-m^3 - 2m - 1)$

95. $(b + 1)(b - 2)(b + 3)$

96. $(2a - 1)(a + 4)(a + 1)$

Mixed Practice

In Problems 97–128, perform the indicated operation. Express your answer as a polynomial in standard form.

97. $(x^2 - 5)(x^2 + 5)$

98. $(b^3 - 2)(b^3 + 2)$

99. $(z^2 + 9)(z + 3)(z - 3)$

100. $(y^2 + 4)(y + 2)(y - 2)$

101. $4x\left(\frac{1}{2}x + 3\right)\left(\frac{1}{2}x - 3\right) - x(x + 1)^2$

102. $25p\left(\frac{2}{5}p - 1\right)\left(\frac{2}{5}p + 1\right) - 4p(p + 2)^2$

103. $(2x^3 + 3)^2$

104. $(3a^4 - 2)^2$

105. $(x^2 + 1)(x^4 - 3)$

106. $(4y^2 - 5)(y^3 + 2)$

107. $(x + 2)(2x^2 - 3x - 1)$

108. $(x - 3)(3x^2 - x - 4)$

109. $-\frac{1}{2}x^2(10x^5 - 6x^4 + 12x^3) + (x^3)^2$

110. $-\dfrac{1}{3}(27y^2 - 9y + 6) - (3y)^2$

111. $7x^2(x + 3) - 2x(x^2 - 1)$

112. $2(3a^4 + 2b^4) - 3(b^4 - 2a^4)$

113. $-3w(w - 4)(w + 3)$ **114.** $5y(y + 5)(y - 3)$

115. $3a(a + 4)^2$ **116.** $2m(m - 3)^2$

117. $(n + 3)(n - 3) + (n + 3)^2$

118. $(s + 6)(s - 6) + (s - 6)^2$

119. $(a + 6b)^2 - (a - 6b)^2$

120. $(2a + 5b)^2 - (2a - 5b)^2$

121. $(x + 1)^2 - (2x + 1)(x - 1)$

122. $(a + 3)^2 - (a + 4)(3a - 1)$

123. Square $2x + 1$.   **124.** Square $3x - 2y$.

125. Find the cube of $x - 1$.

126. Find the cube of $2a + b$.

127. Subtract $x - 6$ from the product of $x + 3$ and $2x - 5$.

128. Add $2x + 3$ to the product of $x + 3$ and $2x - 3$.

Applying the Concepts

In Problems 129–132, find an algebraic expression that represents the area of the shaded region.

△ **129.**

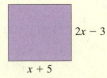

$2x - 3$

$x + 5$

△ **130.**

$3x + 5$

$x + 9$

△ **131.**

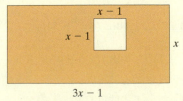

$x - 1$

$x - 1$

x

$3x - 1$

△ **132.**

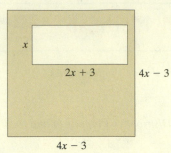

x

$2x + 3$ $4x - 3$

$4x - 3$

In Problems 133 and 134, find an algebraic expression that represents the volume of the figure.

△ **133.**

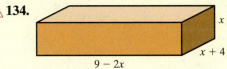

x

$x + 1$

$4x - 3$

△ **134.**

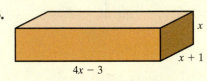

x

$x + 4$

$9 - 2x$

In Problems 135 and 136, find an algebraic expression for the area of the rectangle by finding the sum of the four interior rectangles. Then find the area of the rectangle by multiplying the width and length. Compare the two expressions. How is multiplying a binomial related to finding the area of the rectangle?

△ **135.**

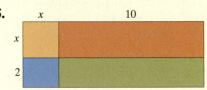

x 10

x

2

△ **136.**

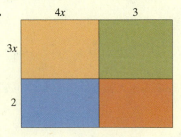

$4x$ 3

$3x$

2

137. Consecutive Integers If x represents the first of three consecutive integers, write a polynomial that represents the product of the next two consecutive integers.

138. Consecutive Integers If x represents the first of three consecutive odd integers, write a polynomial that represents the product of the first and the third integers.

△ **139. Area of a Triangle** The base of a triangle is 2 inches shorter than twice the length of its altitude, x. Write a polynomial that represents the area of the triangle.

△ **140. Area of a Triangle** The base of a triangle is 3 feet greater in length than its altitude, x. Write a polynomial that represents the area of the triangle.

△ **141. Area of a Circle** Write a polynomial for the area of a circle with radius $(x + 2)$ feet.

△ **142. Area of a Circle** Write a polynomial for the area of a circle with radius $(2y - 3)$ meters.

△ **143. Perfect Square** Why is the expression $(a + b)^2$ called a perfect square? Consider the figure below.

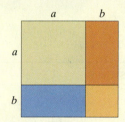

(a) Find the area of each of the four quadrilaterals.
(b) Use the result from part (a) to find the area of the entire region.
(c) Find the length and width of the entire region in terms of a and b. Use this result to find the area of the entire region. What do you notice?

△ **144. Picture Frame** A photo is 13″ by 15″ and is bordered by a matting that is x inches wide. Write the polynomial that calculates the area of the entire framed picture.

Extending the Concepts

In Problems 145–152, find the product. Write each result as a polynomial in standard form.

145. $(x - 2)(x + 2)^2$ **146.** $(x + 3)^2(x - 3)$

147. $[3 - (x + y)][3 + (x + y)]$

148. $[(x - y) + 3][(x - y) - 3]$

149. $(x + 2)^3$ **150.** $(y - 3)^3$

151. $(z + 3)^4$ **152.** $(m - 2)^4$

Explaining the Concepts

153. Explain the steps that should be followed to find the product $(3a - 5)^2$ to someone who missed the class session on squaring a binomial.

154. Explain how the product of the sum and difference of two terms can be used to calculate $19 \cdot 21$.

155. Explain when the FOIL method can be used to multiply polynomials.

156. Find and explain the error in this student's work: $2a(a^2 + 3a) + 5a = 2a(a^2) + 2a(3a) + 2a(5a)$.

5.4 Dividing Monomials: The Quotient Rule and Integer Exponents

Objectives

❶ Simplify Exponential Expressions Using the Quotient Rule

❷ Simplify Exponential Expressions Using the Quotient to a Power Rule

❸ Simplify Exponential Expressions Using Zero as an Exponent

❹ Simplify Exponential Expressions Involving Negative Exponents

❺ Simplify Exponential Expressions Using the Laws of Exponents

Are You Ready for This Section?

Before getting started, take the following readiness quiz. If you get a problem wrong, go back to the section cited and review the material.

R1. Evaluate: $(3a^2)^3$ [Section 5.2, pp. 320–321]

R2. Evaluate: $\left(\dfrac{2}{3}\right)^2$ [Section 1.7, pp. 56–57]

R3. Find the reciprocal: **(a)** 5 **(b)** $-\dfrac{6}{7}$ [Section 1.5, p. 38]

Ready?...Answers   **R1.** $27a^6$

R2. $\dfrac{4}{9}$ **R3. (a)** $\dfrac{1}{5}$ **(b)** $-\dfrac{7}{6}$

The next two sections deal with division. We begin with dividing monomials.

▶ ① Simplify Exponential Expressions Using the Quotient Rule

Consider the following:

$$\frac{y^6}{y^2} = \frac{\overbrace{y \cdot y \cdot y \cdot y \cdot y \cdot y}^{6 \text{ factors}}}{\underbrace{y \cdot y}_{2 \text{ factors}}} = \underbrace{y \cdot y \cdot y \cdot y}_{4 \text{ factors}} = y^4$$

This result suggests that $\dfrac{y^6}{y^2} = y^{6-2} = y^4$, which is true in general.

In Words

When dividing two exponential expressions with a common base, subtract the exponent in the denominator from the exponent in the numerator. Then write the base to that power.

> **Quotient Rule for Exponents**
>
> If a is a real number, and if m and n are natural numbers, then
>
> $$\frac{a^m}{a^n} = a^{m-n} \quad \text{if } a \neq 0$$

EXAMPLE 1 **Using the Quotient Rule to Simplify Expressions**

Simplify each expression.

(a) $\dfrac{6^5}{6^3}$ **(b)** $\dfrac{n^6}{n^2}$ **(c)** $\dfrac{(-7)^8}{(-7)^6}$

Solution

(a)
$$\frac{6^5}{6^3} = 6^{5-3}$$
$$= 6^2$$
Evaluate 6^2: $= 36$

(b)
$$\frac{n^6}{n^2} = n^{6-2}$$
$$= n^4$$

(c)
$$\frac{(-7)^8}{(-7)^6} = (-7)^{8-6}$$
$$= (-7)^2$$
$$= 49 \quad ●$$

EXAMPLE 2 **Using the Quotient Rule to Simplify Expressions**

Simplify each expression. All variables are nonzero.

(a) $\dfrac{25b^9}{15b^6}$ **(b)** $\dfrac{-24x^6y^4}{10x^4y}$

Solution

(a)
$$\frac{25b^9}{15b^6} = \frac{25}{15} \cdot \frac{b^9}{b^6}$$

Divide out like factors; $\dfrac{a^m}{a^n} = a^{m-n}$:
$$= \frac{5 \cdot 5}{5 \cdot 3} \cdot b^{9-6}$$
$$= \frac{5}{3}b^3$$

(b)
$$\frac{-24x^6y^4}{10x^4y} = \frac{-24}{10} \cdot \frac{x^6}{x^4} \cdot \frac{y^4}{y}$$

Divide out like factors; use $\dfrac{a^m}{a^n} = a^{m-n}$:
$$= \frac{-12 \cdot 2}{5 \cdot 2} \cdot x^{6-4} \cdot y^{4-1}$$
$$= -\frac{12}{5}x^2y^3 \quad ●$$

Quick ✓

1. $\dfrac{a^m}{a^n} = $ _____ provided that $a \neq 0$.   **2.** *True or False* $\dfrac{6^{10}}{6^4} = 1^6$.

In Problems 3–5, simplify each expression. All variables are nonzero.

3. $\dfrac{3^7}{3^5}$ **4.** $\dfrac{14c^6}{10c^5}$ **5.** $\dfrac{-21w^4z^8}{14w^3z}$

▶ ❷ Simplify Exponential Expressions Using the Quotient to a Power Rule

Now let's look at a quotient raised to a power:

$$\left(\frac{3}{2}\right)^4 = \left(\frac{3}{2}\right)\cdot\left(\frac{3}{2}\right)\cdot\left(\frac{3}{2}\right)\cdot\left(\frac{3}{2}\right) = \overbrace{\frac{3\cdot3\cdot3\cdot3}{\underbrace{2\cdot2\cdot2\cdot2}_{4\text{ factors}}}}^{4\text{ factors}} = \frac{3^4}{2^4}$$

We are led to the following result:

> **In Words**
> When a quotient is raised to a power, both the numerator and the denominator are raised to the indicated power.

Quotient to a Power Rule for Exponents

If a and b are real numbers and n is a natural number, then

$$\left(\frac{a}{b}\right)^n = \frac{a^n}{b^n} \quad \text{if } b \neq 0$$

EXAMPLE 3   **Using the Quotient to a Power Rule to Simplify an Expression**

Simplify the expression: $\left(\dfrac{z}{3}\right)^2$

Solution

This expression is a quotient, so we apply the Quotient to a Power Rule, raising both the numerator and the denominator to the indicated power.

$$\left(\frac{z}{3}\right)^2 = \frac{z^2}{3^2}$$

Evaluate 3^2: $= \dfrac{z^2}{9}$

EXAMPLE 4   **Using the Quotient to a Power Rule to Simplify Expressions**

Simplify each expression. All variables are nonzero.

(a) $\left(-\dfrac{2y}{z}\right)^3$ **(b)** $\left(\dfrac{3a^2}{b^3}\right)^4$

Solution

(a) $\left(-\dfrac{2y}{z}\right)^3 = \left(\dfrac{-2y}{z}\right)^3$

Use $\left(\dfrac{a}{b}\right)^n = \dfrac{a^n}{b^n}$: $= \dfrac{(-2y)^3}{z^3}$

Use $(ab)^n = a^nb^n$: $= \dfrac{(-2)^3(y)^3}{z^3}$

Evaluate $(-2)^3$: $= -\dfrac{8y^3}{z^3}$

(b) $\left(\dfrac{3a^2}{b^3}\right)^4 = \dfrac{(3a^2)^4}{(b^3)^4}$

Use $(ab)^n = a^nb^n$: $= \dfrac{3^4(a^2)^4}{(b^3)^4}$

Evaluate 3^4; use $(a^m)^n = a^{m\cdot n}$: $= \dfrac{81a^{2\cdot4}}{b^{3\cdot4}}$

   $= \dfrac{81a^8}{b^{12}}$

Quick ✓

6. $\left(\dfrac{a}{b}\right)^n = \underline{\quad}$ provided that $\underline{\quad\quad}$.

In Problems 7–9, simplify each expression.

7. $\left(\dfrac{p}{2}\right)^4$　　　　**8.** $\left(-\dfrac{2a^2}{b^4}\right)^3$　　　　**9.** $\left(\dfrac{-3m}{n^3}\right)^4$

▶ ❸ Simplify Exponential Expressions Using Zero as an Exponent

We will now extend the definition of exponential expressions to integer exponents. We begin with raising a real number to the 0 power.

> **Definition of Zero as an Exponent**
>
> If a is a nonzero real number (that is, $a \neq 0$), we define
> $$a^0 = 1$$

The reason why $a^0 = 1$ is based on the Quotient Rule for Exponents. That is,

$$\dfrac{a^n}{a^n} = a^{n-n} = a^0. \text{ In addition, } \dfrac{a^n}{a^n} = \dfrac{\overbrace{a \cdot a \cdot \,\ldots\, \cdot a}^{n \text{ factors}}}{\underbrace{a \cdot a \cdot \,\ldots\, \cdot a}_{n \text{ factors}}} = 1. \text{ Therefore, } a^0 = 1.$$

EXAMPLE 5　　**Using Zero as an Exponent**

In the following examples, all variables are nonzero.

(a) $3^0 = 1$ 　　　　　**(b)** $x^0 = 1$ 　　　　　**(c)** $(4y)^0 = 1$

(d) $8w^0 = 8 \cdot 1$ 　　　**(e)** $-5^0 = -1 \cdot 5^0$ 　　**(f)** $(-4y)^0 = 1$

　　　　$= 8$ 　　　　　　　　$= -1$ 　　　　　　　　　　　　　●

Work Smart

In Example 5(d), the exponent 0 applies *only* to the factor w.

Quick ✓

10. If a is a nonzero real number, then $a^0 = \underline{\quad}$.

In Problems 11 and 12, simplify each expression. When they occur, assume variables are nonzero.

11. (a) $10^0 = \underline{\quad}$ 　　**(b)** $-10^0 = \underline{\quad}$ 　　**(c)** $(-10)^0 = \underline{\quad}$

12. (a) $(2b)^0 = \underline{\quad}$ 　　**(b)** $2b^0 = \underline{\quad}$ 　　**(c)** $(-2b)^0 = \underline{\quad}$

▶ ❹ Simplify Exponential Expressions Involving Negative Exponents

Now let's look at exponents that are negative integers. Suppose that we wanted to simplify $\dfrac{z^3}{z^5}$. If we use the Quotient Rule for Exponents, we obtain

$$\dfrac{z^3}{z^5} = z^{3-5} = z^{-2}$$

We could also simplify this expression directly by dividing like factors:

$$\dfrac{z^3}{z^5} = \dfrac{\cancel{z} \cdot \cancel{z} \cdot \cancel{z}}{\cancel{z} \cdot \cancel{z} \cdot \cancel{z} \cdot z \cdot z} = \dfrac{1}{z^2}$$

This implies that $z^{-2} = \dfrac{1}{z^2}$, which leads to the following result:

> **Definition of Negative Exponent**
>
> If n is a positive integer and if a is a nonzero real number (that is, $a \neq 0$), then we define
>
> $$a^{-n} = \frac{1}{a^n}$$

Remember, the reciprocal of a nonzero number a is $\frac{1}{a}$. For example, the reciprocal of 7 is $\frac{1}{7}$ and the reciprocal of $-\frac{5}{8}$ is $-\frac{8}{5}$. Whenever we encounter a negative exponent, we should think, "take the reciprocal of the base."

EXAMPLE 6 **Simplifying Exponential Expressions Containing Negative Integer Exponents**

(a) $2^{-3} = \dfrac{1}{2^3}$

$= \dfrac{1}{8}$

(b) $(-5)^{-2} = \dfrac{1}{(-5)^2}$

$= \dfrac{1}{25}$

(c) $-4^{-2} = -1 \cdot 4^{-2}$

$= \dfrac{-1}{4^2}$

$= \dfrac{-1}{16}$

$= -\dfrac{1}{16}$

EXAMPLE 7 **Simplifying the Sum of Exponential Expressions Containing Negative Integer Exponents**

Simplify: $2^{-2} + 4^{-1}$

Solution

$$2^{-2} + 4^{-1} = \frac{1}{2^2} + \frac{1}{4^1}$$

$$\text{Evaluate:} \quad = \frac{1}{4} + \frac{1}{4}$$

$$\text{Find the sum:} \quad = \frac{2}{4}$$

$$\text{Write in lowest terms:} \quad = \frac{1}{2}$$

Quick ✓

13. If n is a positive integer and if a is a nonzero real number, then $a^{-n} = \underline{}$.

In Problems 14 and 15, simplify each expression.

14. (a) 2^{-4} (b) $(-2)^{-4}$ (c) -2^{-4} 15. $4^{-1} - 2^{-3}$

EXAMPLE 8 **Simplifying Exponential Expressions Containing a Negative Integer Exponent**

Simplify each expression so that all exponents are positive. All variables are nonzero.

(a) x^{-4} (b) $(-x)^{-4}$ (c) $-x^{-4}$

Solution

(a) $x^{-4} = \dfrac{1}{x^4}$

(b) $(-x)^{-4} = \dfrac{1}{(-x)^4}$

$= \dfrac{1}{(-1 \cdot x)^4}$

$(ab)^n = a^n b^n: \quad = \dfrac{1}{(-1)^4 x^4}$

$(-1)^4 = 1: \quad = \dfrac{1}{x^4}$

(c) $-x^{-4} = -1 \cdot x^{-4}$

$= -1 \cdot \dfrac{1}{x^4}$

$= -\dfrac{1}{x^4}$

Quick ✔

In Problem 16, simplify each expression so that all exponents are positive. All variables are nonzero.

16. (a) y^{-8} (b) $(-y)^{-8}$ (c) $-y^{-8}$

EXAMPLE 9   **Simplifying Expressions Containing a Negative Integer Exponent**

Simplify:

(a) $2a^{-3}$ (b) $(2a)^{-3}$ (c) $(-2a)^{-3}$

In each problem, $a \neq 0$.

Solution

(a) $2a^{-3} = 2 \cdot \dfrac{1}{a^3}$

$= \dfrac{2}{a^3}$

(b) $(2a)^{-3} = \dfrac{1}{(2a)^3}$

$(ab)^n = a^n \cdot b^n: \quad = \dfrac{1}{2^3 \cdot a^3}$

$= \dfrac{1}{8a^3}$

(c) $(-2a)^{-3} = \dfrac{1}{(-2a)^3}$

$(ab)^n = a^n \cdot b^n: \quad = \dfrac{1}{(-2)^3 \cdot a^3}$

$= \dfrac{1}{-8a^3}$

$\dfrac{a}{-b} = -\dfrac{a}{b}: \quad = -\dfrac{1}{8a^3}$

Quick ✔

In Problem 17, simplify each expression so that all exponents are positive. All variables are nonzero.

17. (a) $2m^{-5}$ (b) $(2m)^{-5}$ (c) $(-2m)^{-5}$

▶ Because $\dfrac{1}{a^{-n}} = \dfrac{1}{\dfrac{1}{a^n}} = 1 \cdot \dfrac{a^n}{1} = a^n$, we have the following result.

If a is a real number and n is an integer, then

$$\frac{1}{a^{-n}} = a^n \quad \text{if } a \neq 0$$

EXAMPLE 10 **Simplifying Quotients Containing Negative Integer Exponents**

Simplify each expression so that all exponents are positive integers.

(a) $\dfrac{1}{2^{-3}}$ (b) $\dfrac{7}{n^{-2}}$ (c) $\dfrac{8}{3n^{-4}}$

Solution

(a) $\dfrac{1}{2^{-3}} = 2^3 = 8$ (b) $\dfrac{7}{n^{-2}} = \dfrac{7n^2}{1} = 7n^2$ (c) $\dfrac{8}{3n^{-4}} = \dfrac{8n^4}{3} = \dfrac{8}{3}n^4$

Work Smart

$\dfrac{6}{2^{-3}} = \dfrac{6}{\frac{1}{8}}$

$= 6 \div \dfrac{1}{8}$

$= 6 \cdot 8$

$= 48$

Quick ✓

In Problems 18–20, simplify each numerical expression. Assume all variables are nonzero.

18. $\dfrac{1}{3^{-2}}$ **19.** $\dfrac{1}{-10^{-2}}$ **20.** $\dfrac{5}{2z^{-2}}$

EXAMPLE 11 **Simplifying Quotients Containing Negative Integer Exponents**

Simplify: $\left(\dfrac{3}{2}\right)^{-3}$

Solution

$$\left(\dfrac{3}{2}\right)^{-3} = \dfrac{1}{\left(\dfrac{3}{2}\right)^3}$$

Use $\left(\dfrac{a}{b}\right)^n = \dfrac{a^n}{b^n}$: $= \dfrac{1}{\dfrac{3^3}{2^3}}$

$= \dfrac{1}{\dfrac{27}{8}}$

$= 1 \cdot \dfrac{8}{27}$

$= \dfrac{8}{27}$

The following shortcut is based on the result of Example 11:

In Words

This shortcut says, "To simplify a quotient to a negative exponent, first take the reciprocal, and then raise that expression to the positive exponent."

Quotient to a Negative Power

If a and b are real numbers and n is an integer, then

$$\left(\dfrac{a}{b}\right)^{-n} = \left(\dfrac{b}{a}\right)^n \quad \text{if } a \neq 0, b \neq 0$$

Using this rule on the problem in Example 11, we obtain $\left(\dfrac{3}{2}\right)^{-3} = \left(\dfrac{2}{3}\right)^3 = \dfrac{8}{27}$.

EXAMPLE 12 **Simplifying Quotients Containing Negative Exponents**

Simplify $\left(-\dfrac{x^3}{4}\right)^{-2}, x \neq 0.$

Solution

$$\left(-\dfrac{x^3}{4}\right)^{-2} = \left(-\dfrac{4}{x^3}\right)^2$$

Use $\left(\dfrac{a}{b}\right)^n = \dfrac{a^n}{b^n}$: $= \dfrac{(-4)^2}{(x^3)^2}$

Use $(a^m)^n = a^{m \cdot n}$: $= \dfrac{16}{x^6}$

21. If a and b are real numbers and n is an integer, then $\left(\dfrac{a}{b}\right)^{-n} =$ ____.

In Problems 22–24, simplify each expression. All variables are nonzero.

22. $\left(\dfrac{7}{8}\right)^{-1}$ **23.** $\left(\dfrac{3a}{5}\right)^{-3}$ **24.** $\left(-\dfrac{2}{3n^4}\right)^{-2}$

⑤ Simplify Exponential Expressions Using the Laws of Exponents

We now summarize the Laws of Exponents where the exponents are integers.

The Laws of Exponents

If a and b are real numbers and if m and n are integers, then, assuming the expression is defined, the following rules apply.

Rule		Examples
Product Rule	$a^m \cdot a^n = a^{m+n}$	$7^3 \cdot 7 = 7^4; x^2 \cdot x^5 = x^7$
Power Rule	$(a^m)^n = a^{m \cdot n}$	$(3^4)^2 = 3^8; (x^3)^2 = x^6$
Product to Power Rule	$(a \cdot b)^n = a^n \cdot b^n$	$(x^3 y)^4 = x^{12} y^4$
Quotient Rule	$\dfrac{a^m}{a^n} = a^{m-n}$, if $a \neq 0$	$\dfrac{9^5}{9^3} = 9^2; \dfrac{y^{11}}{y^8} = y^3$
Quotient to Power Rule	$\left(\dfrac{a}{b}\right)^n = \dfrac{a^n}{b^n}$, if $b \neq 0$	$\left(\dfrac{3}{4}\right)^2 = \dfrac{9}{16}; \left(\dfrac{2x}{y^2}\right)^3 = \dfrac{8x^3}{y^6}$
Zero Exponent Rule	$a^0 = 1$, if $a \neq 0$	$3^0 = 1; \quad (5m)^0 = 1$ $5m^0 = 5 \cdot 1 = 5$
Negative Exponent Rules	$a^{-n} = \dfrac{1}{a^n}$, if $a \neq 0$ $\dfrac{1}{a^{-n}} = a^n$, if $a \neq 0$	$2^{-3} = \dfrac{1}{8}; \quad b^{-4} = \dfrac{1}{b^4}$ $\dfrac{1}{t^{-5}} = t^5$
Quotient to a Negative Power Rule	$\left(\dfrac{a}{b}\right)^{-n} = \left(\dfrac{b}{a}\right)^n$, if $a \neq 0, b \neq 0$	$\left(\dfrac{5}{k}\right)^{-2} = \left(\dfrac{k}{5}\right)^2 = \dfrac{k^2}{25}$

Let's do some examples where we use one or more of the rules listed above. To evaluate or simplify these exponential expressions, ask yourself the following questions.

Using the Laws of Exponents

- Does the exponential expression contain numerical expressions? If so, *evaluate* them. ("Evaluate" means "find the value.")
- Is the expression a product of monomials? If so, use the Product Rule:
$$a^m \cdot a^n = a^{m+n}$$
- Is the expression a quotient (division problem)? If so, use the Quotient Rule:
$$\frac{a^m}{a^n} = a^{m-n}$$
- Do you see a quantity raised to a power? If so, use one of the Power Rules:
$$(a^m)^n = a^{m \cdot n}, (a \cdot b)^n = a^n \cdot b^n, \text{ or } \left(\frac{a}{b}\right)^n = \frac{a^n}{b^n}.$$
- Is there a negative exponent in the expression? A negative exponent means "take the reciprocal of the base." An expression is not simplified if it contains negative exponents.
- Is a quantity raised to the zero power? Remember, $a^0 = 1$ for $a \neq 0$.

EXAMPLE 13   **How to Simplify Exponential Expressions Using Exponent Rules**

Simplify $\left(\dfrac{3}{2}a^3b^{-2}\right)\left(-12a^{-4}b^5\right)$. Write the answer with only positive exponents.

All variables are nonzero.

Step-by-Step Solution

Step 1: Rearrange factors.

$$\left(\dfrac{3}{2}a^3b^{-2}\right)\left(-12a^{-4}b^5\right) = \left(\dfrac{3}{2}\cdot(-12)\right)\left(a^3\cdot a^{-4}\right)\left(b^{-2}\cdot b^5\right)$$

Step 2: Find each product.

Evaluate $\left(\dfrac{3}{2}\right)\cdot(-12)$; use $a^m\cdot a^n = a^{m+n}$: $= -18a^{3+(-4)}b^{-2+5}$

$$= -18a^{-1}b^3$$

Step 3: Simplify. Write the product so that the exponents are positive.

Use $a^{-n} = \dfrac{1}{a^n}$: $= -18\cdot\dfrac{1}{a}\cdot b^3$

$$= -\dfrac{18b^3}{a}$$

EXAMPLE 14   **How to Simplify Expressions Using the Quotient Rule**
and the Negative Exponent Rule

Simplify $-\dfrac{27a^7b^{-6}}{18a^{-2}b^2}$. Write the answer with only positive exponents. All variables are nonzero.

Step-by-Step Solution

Step 1: Write the quotient as the product of factors.

Use $-\dfrac{a}{b} = \dfrac{-a}{b}$: $-\dfrac{27a^7b^{-6}}{18a^{-2}b^2} = \dfrac{-27}{18}\cdot\dfrac{a^7}{a^{-2}}\cdot\dfrac{b^{-6}}{b^2}$

Step 2: Find each quotient.

Simplify $\dfrac{-27}{18}$; use $\dfrac{a^m}{a^n} = a^{m-n}$: $= \dfrac{-3}{2}\cdot a^{7-(-2)}\cdot b^{-6-2}$

$$= -\dfrac{3}{2}\cdot a^9\cdot b^{-8}$$

Step 3: Simplify. Write the quotient so that the exponents are positive.

Use $a^{-n} = \dfrac{1}{a^n}$: $= -\dfrac{3}{2}\cdot a^9\cdot\dfrac{1}{b^8}$

$$= -\dfrac{3a^9}{2b^8}$$

Quick ✓

In Problems 25–28, simplify each expression. Write the answers with only positive exponents. All variables are nonzero.

25. $(-4a^{-3})(5a)$

26. $\left(-\dfrac{2}{5}m^{-2}n^{-1}\right)\left(-\dfrac{15}{2}mn^0\right)$

27. $-\dfrac{16a^4b^{-1}}{12ab^{-4}}$

28. $\dfrac{45x^{-2}y^{-2}}{35x^{-4}y}$

▶ (EXAMPLE 15) **Simplifying Exponential Expressions Using Exponent Rules**

Simplify each expression. Write the answer with only positive exponents. All variables are nonzero.

(a) $\left(\dfrac{25a^{-2}b}{10ab^{-1}}\right)^{-2}$

(b) $(6c^{-2}d)\left(\dfrac{3c^{-3}d^4}{2d}\right)^2$

Solution

(a) Because there are like factors inside the parentheses, we simplify within parentheses before we apply the Power Rule.

$$\left(\frac{25a^{-2}b}{10ab^{-1}}\right)^{-2} = \left(\frac{25 \cdot b \cdot b}{10 \cdot a^2 \cdot a}\right)^{-2}$$

$\dfrac{25}{10} = \dfrac{5 \cdot 5}{5 \cdot 2} = \dfrac{5}{2}$; use $a^m a^n = a^{m+n}$: $= \left(\dfrac{5b^2}{2a^3}\right)^{-2}$

Use $\left(\dfrac{a}{b}\right)^{-n} = \left(\dfrac{b}{a}\right)^n$: $= \left(\dfrac{2a^3}{5b^2}\right)^2$

Use $\left(\dfrac{a}{b}\right)^n = \dfrac{a^n}{b^n}$: $= \dfrac{(2a^3)^2}{(5b^2)^2}$

Use $(ab)^n = a^n b^n$: $= \dfrac{2^2(a^3)^2}{5^2(b^2)^2}$

Use $(a^m)^n = a^{m \cdot n}$: $= \dfrac{4a^{3 \cdot 2}}{25b^{2 \cdot 2}}$

$= \dfrac{4a^6}{25b^4}$

Work Smart

Please note that there are many ways to simplify exponential expressions. For each problem you may use a different approach from the one we illustrate. However, be sure you are using the rules in the correct manner.

(b) Since the quotient in the parentheses is messy, we will simplify it first.

$$(6c^{-2}d)\left(\frac{3c^{-3}d^4}{2d}\right)^2 = \frac{6c^{-2}d}{1}\left(\frac{3c^{-3}d^{4-1}}{2}\right)^2$$

Use $a^{-n} = \dfrac{1}{a^n}$: $= \dfrac{6d}{c^2}\left(\dfrac{3d^3}{2c^3}\right)^2$

Use $\left(\dfrac{a}{b}\right)^n = \dfrac{a^n}{b^n}$ and $(ab)^n = a^n b^n$: $= \dfrac{6d}{c^2}\left(\dfrac{3^2(d^3)^2}{2^2(c^3)^2}\right)$

Use $(a^m)^n = a^{m \cdot n}$: $= \dfrac{6d}{c^2}\left(\dfrac{9d^6}{4c^6}\right)$

Rearrange factors: $= 6 \cdot \dfrac{9}{4} \cdot \dfrac{d \cdot d^6}{c^2 \cdot c^6}$

$6 \cdot \dfrac{9}{4} = 6 \cdot \dfrac{9}{4} = \dfrac{27}{2}$; use $a^m a^n = a^{m+n}$: $= \dfrac{27}{2} \cdot \dfrac{d^{1+6}}{c^{2+6}}$

$= \dfrac{27d^7}{2c^8}$

●

Quick ✓

In Problems 29–32, simplify each expression. Write the answers with only positive exponents. All variables are nonzero.

29. $(3y^2z^{-3})^{-2}$

30. $\left(\dfrac{2wz^{-3}}{7w^{-1}}\right)^2$

31. $\left(\dfrac{-6p^{-2}}{p}\right)(3p^8)^{-1}$

32. $(-25k^5r^{-2})\left(\dfrac{2}{5}k^{-3}r\right)^2$

5.4 Exercises MyMathLab® MathXL PRACTICE

Exercise numbers in green have complete video solutions in MyMathLab

Problems **1–32** are the Quick ✓s that follow the **EXAMPLES**.

Building Skills

In Problems 33–42, use the Quotient Rule to simplify. All variables are nonzero. See Objective 1.

33. $\dfrac{2^{23}}{2^{19}}$

34. $\dfrac{10^5}{10^2}$

35. $\dfrac{x^{15}}{x^6}$

36. $\dfrac{x^{20}}{x^{14}}$

37. $\dfrac{16y^4}{4y}$

38. $\dfrac{9a^4}{27a}$

39. $\dfrac{-16m^{10}}{24m^3}$

40. $\dfrac{-36x^2y^5}{24xy^4}$

41. $\dfrac{-12m^9n^3}{-6mn}$

42. $\dfrac{-15r^9s^2}{-5r^8s}$

In Problems 43–50, use the Quotient to a Power Rule to simplify. All variables are nonzero. See Objective 2.

43. $\left(\dfrac{3}{2}\right)^3$

44. $\left(\dfrac{4}{9}\right)^2$

45. $\left(\dfrac{x^5}{3}\right)^3$

46. $\left(\dfrac{7}{y^2}\right)^2$

47. $\left(-\dfrac{x^5}{y^7}\right)^4$

48. $\left(-\dfrac{a^3}{b^{10}}\right)^5$

49. $\left(\dfrac{7a^2b}{c^3}\right)^2$

50. $\left(\dfrac{2mn^2}{q^3}\right)^4$

In Problems 51–62, use the Zero Exponent Rule to simplify. All variables are nonzero. See Objective 3.

51. 3^0

52. -100^0

53. $-\left(\dfrac{1}{2}\right)^0$

54. $\left(\dfrac{2}{5}\right)^0$

55. $18 \cdot 2^0$

56. $14^0 \cdot 3^0 \cdot 10$

57. $(-10)^0$

58. $(-8)^0$

59. $(24ab)^0$

60. $(-11xy)^0$

61. $24ab^0$

62. $-11xy^0$

In Problems 63–86, use the Negative Exponent Rules to simplify. Write answers with only positive exponents. All variables are nonzero. See Objective 4.

63. 10^{-3}

64. 5^{-1}

65. m^{-2}

66. k^{-2}

67. $-a^{-2}$

68. $-b^{-3}$

69. $-4y^{-3}$

70. $-7z^{-5}$

71. $2^{-1} + 3^{-2}$

72. $4^{-2} - 2^{-3}$

73. $\left(\dfrac{2}{5}\right)^{-2}$

74. $\left(\dfrac{5}{4}\right)^{-3}$

75. $\left(\dfrac{3}{z^2}\right)^{-1}$

76. $\left(\dfrac{4}{p^2}\right)^{-2}$

77. $\left(-\dfrac{2n}{m^2}\right)^{-3}$

78. $\left(-\dfrac{5}{3b^2}\right)^{-3}$

79. $\dfrac{1}{4^{-2}}$

80. $\dfrac{1}{6^{-2}}$

81. $\dfrac{6}{x^{-4}}$

82. $\dfrac{4}{b^{-3}}$

83. $\dfrac{5}{2m^{-3}}$

84. $\dfrac{9}{4t^{-1}}$

85. $\dfrac{5}{(2m)^{-3}}$

86. $\dfrac{9}{(4t)^{-1}}$

In Problems 87–96, use the Laws of Exponents to simplify. Write answers with positive exponents. All variables are nonzero. See Objective 5.

87. $\left(\dfrac{4}{3}y^{-2}z\right)\left(\dfrac{5}{8}y^{-2}z^4\right)$

88. $\left(\dfrac{3}{5}ab^{-3}\right)\left(\dfrac{5}{3}a^{-1}b^3\right)$

89. $\dfrac{-7m^7n^6}{3m^{-3}n^0}$

90. $\dfrac{13r^8t^3}{-5r^0t^{-5}}$

91. $\dfrac{21y^2z^{-3}}{3y^{-2}z^{-1}}$

92. $\dfrac{30ab^{-4}}{15a^{-1}b^{-2}}$

93. $(4x^2y^{-2})^{-2}$

94. $(5x^{-4}y^3)^{-2}$

95. $(3a^2b^{-1})\left(\dfrac{4a^{-1}b^2}{b^3}\right)^2$

96. $(5a^3b^{-4})\left(\dfrac{2ab^{-1}}{b^{-3}}\right)^2$

Mixed Practice

In Problems 97–130, simplify. Write answers with only positive exponents. All variables are nonzero.

97. $2^5 \cdot 2^{-3}$

98. $3^8 \cdot 3^{-6}$

99. $2^{-7} \cdot 2^4$

100. $10^{-4} \cdot 10^3$

101. $\dfrac{3}{3^{-3}}$

102. $\dfrac{5^4}{5^{-1}}$

103. $\dfrac{x^6}{x^{15}}$

104. $\dfrac{b^{24}}{b^{42}}$

105. $\dfrac{8x^2}{2x^{-1}} + x^2(3x + 5x^{-1})$

106. $\dfrac{24m^2}{4m^{-3}} - m^3(2m^2 - 5m^{-1})$

107. $\dfrac{-27xy^3z^4}{18x^4y^3z}$

108. $\dfrac{8a^{15}b^2}{-18b^{20}}$

109. $(3x^2y^{-3})(12^{-1}x^{-5}y^{-6})$

110. $(-14c^2d^4)(3c^{-6}d^{-4})$

111. $(-a^4)^{-3}$

112. $(-x^{-2})^5$

113. $(3m^{-2})^3$

114. $(2z^{-1})^3$

115. $(-3x^{-2}y^{-3})^{-2}$

116. $(-4a^{-3}b^{-2}c^{-1})^{-1}$

117. $2p^{-4} \cdot p^{-3} \cdot 3p^0$

118. $7a^{-2} \cdot 2a^0 \cdot a^{-3}$

119. $(-16a^3)(-3a^4)\left(\dfrac{1}{4}a^{-7}\right)$

120. $(-3x^{-4})(2x^{-3})\left(\dfrac{1}{36}x^{10}\right)$

121. $\dfrac{8x^2 \cdot x^5}{12x^{-3} \cdot x^4} + \dfrac{(-3x)^{-1}(2x^3)}{(x^{-2})^2}$

122. $\dfrac{32x^{-3} \cdot x^2}{24x^2 \cdot x^{-5}} + \dfrac{(-3x^2)^{-1}}{(2x^2)^{-2}}$

123. $(2x^{-3}y^{-2})^4(3x^2y^{-3})^{-3}$

124. $(4a^{-2}b)^3(2a^2b^{-4})^{-2}$

125. $\left(\dfrac{y}{2z^2}\right)^{-3} \cdot \left(\dfrac{y^2}{4z^3}\right)^2 + \left(\dfrac{2}{y}\right)^{-1}$

126. $\left(\dfrac{s^2}{4t^2}\right)^3 \cdot \left(\dfrac{s^2}{2t^4}\right)^{-3} + \left(\dfrac{2}{t^3}\right)^{-2}$

127. $(4x^2y)^3 \cdot \left(\dfrac{2x}{3y}\right)^{-3}$

128. $(2a^5b^2)^3 \cdot \left(\dfrac{4a^{-2}b^3}{3a}\right)^{-2}$

129. $\dfrac{(5a^{-3}b^2)^2}{a^{-4}b^{-4}}(15a^{-3}b)^{-1}$

130. $\left(\dfrac{2x^{-4}y^{-3}}{4xy^3}\right)^{-2}(4x^2y^{-1})^{-2}$

Applying the Concepts

△ **131. Volume of a Box** The volume of a rectangular solid is given by the equation $V = l \cdot w \cdot h$. Find the volume of a shipping box whose dimensions are $l = x$ m, $w = x$ m, and $h = 3x$ m.

△ **132. Volume of a Box** The volume of a rectangular solid is given by the equation $V = l \cdot w \cdot h$. Find the volume of a box whose dimensions are $l = 3n$ yards, $w = 2n$ yards, and $h = 6n$ yards.

△ **133. Volume of a Cylinder** The volume of a right circular cylinder is given by the equation $V = \pi r^2 h$, where r is the radius of the circular base of the cylinder and h is its height. Rewrite this equation in terms of the diameter, d, of the base of the cylinder.

△ **134. Volume of a Cylinder** The volume of a right circular cylinder is given by the equation $V = \pi r^2 h$, where r is the radius of the circular base of the cylinder and h is its height. Write an algebraic expression for the volume of a cylinder in which the radius of the circular base is equal to the height of the cylinder.

△ **135. Making Buttons** How much fabric is required to cover x buttons in the shape of a circle whose radius is $3x$?

△ **136. Making a Tablecloth** How much fabric is required to make a round tablecloth if the diameter of the tablecloth is $\left(\dfrac{1}{2}x\right)$ meters?

Extending the Concepts

In Problems 137–144, simplify. Write answers with only positive exponents.

137. $\dfrac{x^{2n}}{x^{3n}}$

138. $\dfrac{(x^{a-2})^3}{(x^{2a})^5}$

139. $\left(\dfrac{x^n y^m}{x^{4n-1}y^{m+1}}\right)^{-2}$

140. $\left(\dfrac{a^n b^{2m}}{ab^2}\right)^{-3}$

141. $(x^{2a}y^b z^{-c})^{3a}$

142. $(x^{-a}y^{-2b}z^{4c})^{2a}$

143. $\dfrac{(3a^n)^2}{(2a^{4n})^3}$

144. $\dfrac{y^a}{y^{4a}}$

Explaining the Concepts

145. Explain why $\dfrac{11^6}{11^4} \neq 1^2$.

146. Use the Quotient Rule and the expression $\left(\dfrac{a^n}{a^n}\right)$, $a \neq 0$, to explain why $a^0 = 1$, $a \neq 0$.

147. Explain the difference in simplifying the expressions $-12x^0$ and $(-12x)^0$.

148. Explain two different approaches to simplify $\left(\dfrac{x^8}{x^2}\right)^3$. Which do you prefer?

149. A friend of yours has a homework problem in which he must simplify $(x^3)^4$. He tells you that he thinks the answer is x^7. Is he right? If not, explain where he went wrong.

150. Explain why $-4x^{-3}$ is equal to $-\dfrac{4}{x^3}$. That is, why doesn't the factor -4 "move" to the denominator?

Putting the Concepts Together (Sections 5.1–5.4)

We designed these problems so that you can review Sections 5.1–5.4 and show your mastery of the concepts. Take time to work these problems before proceeding with the next section. The answers are located at the back of the text on page AN-21.

1. Determine whether the following algebraic expression is a polynomial (Yes or No). If it is a polynomial, state the degree and then state whether it is a monomial, a binomial, or a trinomial: $6x^2y^4 - 8x^5 + 3$.

2. Evaluate the polynomial $-x^2 + 3x$ for the given values: **(a)** 0 **(b)** -1 **(c)** 2.

In Problems 3–11, perform the indicated operation.

3. $(6x^4 - 2x^2 + 7) - (-2x^4 - 7 + 2x^2)$

4. $(2x^2y - xy + 3y^2) + (4xy - y^2 + 3x^2y)$

5. $-2mn(3m^2n - mn^3)$ **6.** $(5x + 3)(x - 4)$

7. $(2x - 3y)(4x - 7y)$ **8.** $(5x + 8)(5x - 8)$

9. $(2x + 3y)^2$ **10.** $2a(3a - 4)(a + 5)$

11. $(4m + 3)(4m^3 - 2m^2 + 4m - 8)$

In Problems 12–20, simplify each expression. Write answers with only positive exponents. All variables are nonzero.

12. $(-2x^7y^0z)(4xz^8)$

13. $(5m^3n^{-2})(-3m^{-4}n)$

14. $\dfrac{-18a^8b^3}{6a^5b}$ **15.** $\dfrac{16x^7y}{32x^9y^3}$

16. $\dfrac{7}{2ab^{-2}}$ **17.** $\dfrac{q^{-6}rt^5}{qr^{-4}t^7}$

18. $\left(\dfrac{3}{2}r^2\right)^{-3}$ **19.** $(4y^{-2}z^3)^{-2}$

20. $\left(\dfrac{3x^3y^{-3}}{2x^{-1}y^0}\right)^{-4} \cdot (2x^{-4}y^{-2})^2$

5.5 Dividing Polynomials

Objectives

1 Divide a Polynomial by a Monomial

2 Divide a Polynomial by a Binomial

Are You Ready for This Section?

Before getting started, take the following readiness quiz. If you get a problem wrong, go back to the section cited and review the material.

R1. Find the quotient: $\dfrac{24a^3}{6a}$ [Section 5.4, pp. 334–335]

R2. Find the quotient: $\dfrac{-49x^3}{21x^4}$ [Section 5.4, pp. 340–341]

R3. Find the product: $3x(7x - 2)$ [Section 5.3, pp. 323–324]

R4. State the degree of $4x^2 - 2x + 1$.   [Section 5.1, pp. 310–311]

Work Smart

A polynomial is a monomial or the sum of monomials.

We begin polynomial division by dividing a polynomial by a monomial. Recall that in long division, such as $15)\overline{345}$ with quotient 23, the number 15 is called the *divisor*, the number 345 is called the *dividend*, and the number 23 is called the *quotient*. We use the same language with polynomial division.

▶ **1** Divide a Polynomial by a Monomial

Dividing a polynomial by a monomial requires the Quotient Rule for Exponents, which states that if a is a real number and m and n are integers, then $\dfrac{a^m}{a^n} = a^{m-n}$, where $a \neq 0$.

Ready?...Answers **R1.** $4a^2$

R2. $-\dfrac{7}{3x}$ **R3.** $21x^2 - 6x$ **R4.** 2

In Words

To divide a polynomial by a monomial, divide each of the terms of the polynomial numerator (dividend) by the monomial denominator (divisor).

Remember that for us to add two rational numbers, the denominators must be the same. When we have the same denominator, we add the numerators and write the result over the common denominator. For example, $\dfrac{3}{11} + \dfrac{4}{11} = \dfrac{3+4}{11} = \dfrac{7}{11}$. When dividing a polynomial by a monomial, we reverse this process. So if a, b, and c are monomials, we can write $\dfrac{a+b}{c}$ as $\dfrac{a}{c} + \dfrac{b}{c}$. We can extend this result to polynomials with three or more terms.

EXAMPLE 1   **Dividing a Binomial by a Monomial**

Divide and simplify: $\dfrac{9x^3 - 21x^2}{3x}$

Solution

We divide each term in the numerator by the denominator.

$$\frac{9x^3 - 21x^2}{3x} = \frac{9x^3}{3x} - \frac{21x^2}{3x}$$

$$\frac{a^m}{a^n} = a^{m-n}: \quad = \frac{9}{3}x^{3-1} - \frac{21}{3}x^{2-1}$$

$$\text{Simplify:} \quad = 3x^2 - 7x \quad \bullet$$

EXAMPLE 2   **Dividing a Trinomial by a Monomial**

Divide and simplify: $\dfrac{12p^4 + 24p^3 + 4p^2}{4p^2}$

Work Smart

Do not make the following common mistake:

$$\frac{12p^4 + 24p^3 + 4p^2}{4p^2}$$

$$\ne \frac{12p^4 + 24p^3 + \cancel{4p^2}}{\cancel{4p^2}}$$

$$\ne 12p^4 + 24p^3 + 1$$

You must divide *each* term in the numerator by the monomial in the denominator.

Solution

We divide each term in the numerator by the denominator.

$$\frac{12p^4 + 24p^3 + 4p^2}{4p^2} = \frac{12p^4}{4p^2} + \frac{24p^3}{4p^2} + \frac{4p^2}{4p^2}$$

$$\frac{a^m}{a^n} = a^{m-n}: \quad = \frac{12}{4}p^{4-2} + \frac{24}{4}p^{3-2} + \frac{4}{4} \cdot \frac{p^2}{p^2}$$

$$\text{Simplify:} \quad = 3p^2 + 6p + 1 \quad \bullet$$

EXAMPLE 3   **Dividing a Trinomial by a Monomial**

Divide and simplify: $\dfrac{8a^2b^2 - 6a^2b + 5ab^2}{2a^2b^2}$

Solution

$$\frac{8a^2b^2 - 6a^2b + 5ab^2}{2a^2b^2} = \frac{8a^2b^2}{2a^2b^2} - \frac{6a^2b}{2a^2b^2} + \frac{5ab^2}{2a^2b^2}$$

$$\frac{a^m}{a^n} = a^{m-n} \text{ and simplify:} \quad = 4a^0b^0 - 3a^0b^{-1} + \frac{5}{2}a^{-1}b^0$$

$$a^0 = 1; \ a^{-n} = \frac{1}{a^n}: \quad = 4 \cdot 1 \cdot 1 - 3 \cdot 1 \cdot \frac{1}{b} + \frac{5}{2} \cdot \frac{1}{a} \cdot 1$$

$$= 4 - \frac{3}{b} + \frac{5}{2a} \quad \bullet$$

Quick ✓

1. The first step to simplify $\dfrac{4x^4 + 8x^2}{2x}$ is to rewrite $\dfrac{4x^4 + 8x^2}{2x}$ as $\dfrac{\overline{}}{2x} + \dfrac{\overline{}}{2x}$.

In Problems 2–4, find the quotient.

2. $\dfrac{10n^4 - 20n^3 + 5n^2}{5n^2}$

3. $\dfrac{12k^4 - 18k^2 + 5}{2k^2}$

4. $\dfrac{x^4y^4 + 8x^2y^2 - 4xy}{4x^3y}$

❷ Divide a Polynomial by a Binomial

Dividing a polynomial by a binomial is like dividing two integers. Although this procedure should be familiar to you, we review it next.

▶ EXAMPLE 4 Dividing an Integer by an Integer Using Long Division

Divide 579 by 16.

Solution

$$
\begin{array}{r}
36 \longleftarrow \text{quotient} \\
\text{divisor} \longrightarrow 16\overline{)579} \longleftarrow \text{dividend} \\
48 \longleftarrow 3 \cdot 16 = 48 \\
57 - 48 = 9 \rightarrow \quad 99 \longleftarrow \text{bring down the 9} \\
96 \longleftarrow 6 \cdot 16 = 96 \\
99 - 96 = 3 \rightarrow \quad 3 \longleftarrow \text{remainder}
\end{array}
$$

So 579 divided by 16 equals 36 with a remainder of 3. We can write this as $\dfrac{579}{16} = 36\dfrac{3}{16}$.

We can always check a long-division problem by multiplying the quotient by the divisor and adding this product to the remainder. The result should be the dividend. That is,

$$(\text{Quotient})\,(\text{Divisor}) + \text{Remainder} = \text{Dividend}$$

From Example 4, we have:

$$(36)\,(16) + 3 = 576 + 3 = 579$$

In Example 4, we wrote the solution as $\dfrac{579}{16} = 36\dfrac{3}{16}$. Remember, the mixed number $36\dfrac{3}{16}$ means $36 + \dfrac{3}{16}$. So the answer is written in the form

$$\text{Quotient} + \dfrac{\text{Remainder}}{\text{Divisor}}$$

Now we are ready to divide a polynomial by a binomial using long division.

▶ (**EXAMPLE 5**) **How to Divide a Polynomial by a Binomial Using Long Division**

Find the quotient when $x^2 + 10x + 21$ is divided by $x + 7$.

Step-by-Step Solution

To divide two polynomials, first write each polynomial in standard form (descending order of degree). The dividend is $x^2 + 10x + 21$ and the divisor is $x + 7$.

Step 1: Divide the highest-degree term of the dividend, x^2, by the highest-degree term of the divisor, x. Enter the result over the term x^2.

$$x \longleftarrow \frac{x^2}{x} = x$$
$$x + 7 \overline{)x^2 + 10x + 21}$$

Step 2: Multiply x by $x + 7$. Vertically align like terms.

$$x$$
$$x + 7 \overline{)x^2 + 10x + 21}$$
$$\underline{x^2 + 7x} \longleftarrow x(x + 7) = x^2 + 7x$$

Step 3: Subtract $x^2 + 7x$ from $x^2 + 10x + 21$.

$$x$$
$$x + 7 \overline{)x^2 + 10x + 21}$$
$$\underline{-(x^2 + 7x)}$$
$$3x + 21 \longleftarrow (x^2 + 10x + 21) - (x^2 + 7x) = 3x + 21$$

Step 4: Repeat Steps 1–3, treating $3x + 21$ as the dividend.

$$x + 3 \longleftarrow \frac{3x}{x} = 3$$
$$x + 7 \overline{)x^2 + 10x + 21}$$
$$\underline{-(x^2 + 7x)}$$
$$3x + 21$$
$$\underline{-(3x + 21)} \longleftarrow 3(x + 7) = 3x + 21$$
$$0 \longleftarrow (3x + 21) - (3x + 21) = 0$$

The quotient is $x + 3$ and the remainder is 0.

Step 5: Check your result by showing that (Quotient)(Divisor) + Remainder = Dividend.

$$(x + 3)(x + 7) + 0 = x^2 + 10x + 21$$

The product checks, so $x^2 + 10x + 21$ divided by $x + 7$ is $x + 3$. We can express this division using fractions, as follows:

$$\frac{x^2 + 10x + 21}{x + 7} = x + 3$$

●

▶ (**EXAMPLE 6**) **How to Divide a Polynomial by a Binomial Using Long Division**

Find the quotient: $\dfrac{6x^2 + 9x - 10}{2x - 1}$

Solution

Each polynomial is in standard form. The dividend is $6x^2 + 9x - 10$ and the divisor is $2x - 1$.

$$3x \longleftarrow \text{Step 1: } \frac{6x^2}{2x} = 3x$$
$$2x - 1 \overline{)6x^2 + 9x - 10}$$
$$\underline{-(6x^2 - 3x)} \longleftarrow \text{Step 2: } 3x(2x - 1) = 6x^2 - 3x$$
$$12x - 10 \longleftarrow \text{Step 3: } 6x^2 + 9x - 10 - (6x^2 - 3x) = 12x - 10$$

$$\frac{3x + 6 \longleftarrow \quad \frac{12x}{2x} = 6}{2x - 1 \overline{\smash{)}6x^2 + 9x - 10}}$$
$$\underline{- (6x^2 - 3x)}$$
$$12x - 10$$
$$\underline{- (12x - 6)} \longleftarrow 6(2x - 1) = 12x - 6$$
$$-4 \longleftarrow (12x - 10) - (12x - 6) = -4$$

Work Smart

You know that you're finished dividing when the degree of the remainder is less than the degree of the divisor.

Because the degree of -4 is less than the degree of the divisor, $2x - 1$, the process ends. The quotient is $3x + 6$ and the remainder is -4.

Check (Quotient)(Divisor) + Remainder = Dividend.

$$(3x + 6)(2x - 1) + (-4) = 6x^2 - 3x + 12x - 6 + (-4)$$

Combine like terms: $= 6x^2 + 9x - 10$

The product checks, so $\dfrac{6x^2 + 9x - 10}{2x - 1} = 3x + 6 - \dfrac{4}{2x - 1}$.

Quick ✓

5. To begin a polynomial division problem, write the divisor and the dividend in _____ form.

6. To check the result of long division, multiply the _____ and the divisor and add this result to the _____. If correct, this result will be equal to the _____.

In Problems 7–9, find the quotient by performing long division.

7. $\dfrac{x^2 - 3x - 40}{x + 5}$ **8.** $\dfrac{2x^2 - 5x - 12}{2x + 3}$

9. $\dfrac{4x^2 + 17x + 21}{x + 3}$

▶ EXAMPLE 7 **Dividing Two Polynomials Using Long Division**

Simplify by performing long division: $\dfrac{8 - 9x + 2x^2 + 12x^3 + 5x^5}{x^2 + 3}$

Solution

Do you notice that the dividend is not written in standard form? Also, when the dividend is in standard form, do you notice that there is no x^4 term? When a term is missing, its coefficient is 0, so we rewrite the division problem as follows:

$$\frac{5x^5 + 0 \cdot x^4 + 12x^3 + 2x^2 - 9x + 8}{x^2 + 3}$$

$$\begin{array}{r} 5x^3 \qquad\quad - 3x + 2 \\ x^2 + 3 \overline{\smash{)}5x^5 + 0x^4 + 12x^3 + 2x^2 - 9x + 8} \end{array}$$
$$\underline{- (5x^5 \qquad\quad + 15x^3)} \longleftarrow 5x^3(x^2 + 3)$$
$$-3x^3 + 2x^2 - 9x + 8 \longleftarrow 5x^5 + 12x^3 + 2x^2 - 9x + 8 - (5x^5 + 15x^3) = -3x^3 + 2x^2 - 9x + 8$$
$$\underline{- (-3x^3 \qquad\quad - 9x)} \longleftarrow -3x(x^2 + 3)$$
$$2x^2 \qquad\quad + 8 \longleftarrow (-3x^3 + 2x^2 - 9x + 8) - (-3x^3 - 9x) = 2x^2 + 8$$
$$\underline{- (2x^2 \qquad\quad + 6)} \longleftarrow 2(x^2 + 3)$$
$$2 \longleftarrow \text{Remainder}$$

The quotient is $5x^3 - 3x + 2$ and the remainder is 2. We now check our work.

Check (Quotient)(Divisor) + Remainder = Dividend

$$(5x^3 - 3x + 2)(x^2 + 3) + 2 = 5x^5 + 15x^3 - 3x^3 - 9x + 2x^2 + 6 + 2$$
$$= 5x^5 + 12x^3 + 2x^2 - 9x + 8$$

Our answer checks, so $\dfrac{8 - 9x + 2x^2 + 12x^3 + 5x^5}{x^2 + 3} = 5x^3 - 3x + 2 + \dfrac{2}{x^2 + 3}$.

Quick ✔

In Problems 10–12, simplify by performing long division.

10. $\dfrac{x + 1 - 3x^2 + 4x^3}{x + 2}$

11. $\dfrac{2x^3 + 3x^2 + 10}{2x - 5}$

12. $\dfrac{4x^3 - 3x^2 + x + 1}{x^2 + 2}$

5.5 Exercises MyMathLab® PRACTICE

Exercise numbers in green have complete video solutions in MyMathLab

*Problems **1–12** are the **Quick ✔**s that follow the **EXAMPLES**.*

Building Skills

In Problems 13–30, divide and simplify. See Objective 1.

13. $\dfrac{4x^2 - 2x}{2x}$

14. $\dfrac{3x^3 - 6x^2}{3x^2}$

15. $\dfrac{9a^3 + 27a^2 - 3}{3a^2}$

16. $\dfrac{16m^3 + 8m^2 - 4}{8m^2}$

17. $\dfrac{5n^5 - 10n^3 - 25n}{25n}$

18. $\dfrac{5x^3 - 15x^2 + 10x}{5x^2}$

19. $\dfrac{15r^5 - 27r^3}{9r^3}$

20. $\dfrac{16a^5 - 12a^4 + 8a^3}{20a^3}$

21. $\dfrac{3x^7 - 9x^6 + 27x^3}{-3x^5}$

22. $\dfrac{7p^4 + 21p^3 - 3p^2}{-6p^4}$

23. $\dfrac{3z + 4z^3 - 2z^2}{8z}$

24. $\dfrac{-5y^2 + 15y^4 - 16y^5}{5y^2}$

25. $\dfrac{14xy - 10y}{-2y}$

26. $\dfrac{35xy + 20y}{-5y}$

27. $\dfrac{12y - 30x}{-2x}$

28. $\dfrac{21y^2 + 35x^2}{-7x^2}$

29. $\dfrac{25a^3b^2c + 10a^2bc^3}{-5a^4b^2c}$

30. $\dfrac{16m^2n^3 - 24m^4n^3}{-8m^3n^4}$

In Problems 31–54, find the quotient using long division. See Objective 2.

31. $\dfrac{x^2 - 4x - 21}{x + 3}$

32. $\dfrac{x^2 + 18x + 72}{x + 6}$

33. $\dfrac{x^2 - 9x + 20}{x - 4}$

34. $\dfrac{x^2 + 4x - 32}{x - 4}$

35. $\dfrac{x^3 + 4x^2 - 15x + 6}{x - 2}$

36. $\dfrac{x^3 - x^2 - 40x + 12}{x + 6}$

37. $\dfrac{x^4 - x^3 + 10x - 4}{x + 2}$

38. $\dfrac{x^4 - 2x^3 + x^2 + x - 1}{x - 1}$

39. $\dfrac{x^3 - x^2 + x + 8}{x + 1}$

40. $\dfrac{x^3 - 7x^2 + 15x - 11}{x - 3}$

41. $\dfrac{2x^2 - 7x - 15}{x - 5}$

42. $\dfrac{2x^2 + 5x - 42}{x + 6}$

43. $\dfrac{x^3 + 4x^2 - 5x + 2}{x - 2}$

44. $\dfrac{x^3 - 2x^2 + x + 6}{x + 1}$

45. $\dfrac{2x^4 - 3x^3 - 11x^2 - 40x - 1}{x - 4}$

46. $\dfrac{3x^4 + 7x^3 - 5x^2 + 8x + 12}{x + 3}$

47. $\dfrac{2x^3 + 7x^2 - 10x + 5}{2x - 1}$

48. $\dfrac{12a^3 + 11a^2 + 18a + 9}{4a + 1}$

49. $\dfrac{-24 + x^2 + x}{5 + x}$

50. $\dfrac{-16x + 70 + x^2}{-9 + x}$

51. $\dfrac{4x^2 + 5}{1 + 2x}$

52. $\dfrac{9x^2 - 14}{2 + 3x}$

53. $\dfrac{x^4 + 2x^2 - 8}{x^2 - 2}$

54. $\dfrac{64x^6 - 27}{4x^2 - 3}$

Mixed Practice

In Problems 55–80, perform the indicated operation.

55. $(a - 5)(a + 6)$

56. $(7n + 3m^2 + 4m) + (-6m^2 - 7n + 4m)$

57. $(2x - 8) - (3x + x^2 - 2)$

58. $3x^2(7x^2 + 2x - 3)$

59. $(2ab + b^2 - a^2) + (b^2 - 4ab + a^2)$

60. $\dfrac{3a^4b^{-2}}{6a^{-4}b^{-3}}$

61. $\dfrac{4 + 7x^2 - 3x^4 + 6x^3}{2x^2}$

62. $(b - 3)(b + 2)$

63. $\dfrac{18x^3y^{-4}z^6}{27x^{-4}y^{-12}z^{-6}}$

64. $(4ab + 6ab^2) - (12a^2b - 2ab - 3ab^2)$

65. $\dfrac{6x^2 - 28x + 30}{3x - 5}$

66. $\dfrac{12x^2 - 25x - 50}{3x - 10}$

67. $(n - 3)^2$

68. $\dfrac{-9mn^2 - 8m^2n + 12mn}{3mn}$

69. $(x^3 + x - 4x^4)(-10x^2)$

70. $(2x^3 + 6x) + (12x - x^2 - x^3)$

71. $(2pq - q^2) + (4pq - p^2 - q^2)$

72. $(4x^2 - 6x + 3) - (x - 7)$

73. $(x^2 + x - 1)(x + 5)$

74. $(x^2 - 2x + 3)(x - 4)$

75. $(7rs^2 - 2r^2s) - (2r^2s - 8rs^2)$

76. $(x + 6)^2$

77. $(x^4 - 2x^2 + x)(-3x)$

78. $2x^4(x^2 - 2x + 3)$

79. $\dfrac{3 + 6x^2 - 11x}{2x - 3} + (x + 1)$

80. $\dfrac{7x - 5 + 6x^2}{3x + 5} - (2x + 1)$

Applying the Concepts

81. Find the quotient of $(3x^4 - 6x + 12x^2)$ and $-3x^3$.

82. Divide $x - 3$ by $x + 2$.

83. Divide the sum of $x^2 + 3x - 1$ and $x - 1$ by $-2x^3$.

84. Divide the square of the difference of $x - 9$ and $x + 3$ by x^2.

△**85. Volume of a Box** The volume of a rectangular solid is $(x^3 - 5x^2 + 6x)$ cubic feet. One side measures $(x - 2)$ feet and another measures $(x - 3)$ feet. What is the measure of the third side?

△ 86. **Area of a Rectangle** A rectangle has area $(x^2 + 2x - 48)$ square inches. If one side measures $(x + 8)$ inches, what is the measurement of the other side?

△ 87. **Area of a Rectangle** If the area of a rectangle is $(z^2 + 6z + 9)$ square inches and the length is $(z + 3)$ inches, what is the width?

△ 88. **Area of a Triangle** If the area of a triangle is $(6a^2 - 5a - 6)$ square yards and the length of the base is $(2a - 3)$ yards, what is the height?

△ 89. **Area of a Triangle** If the area of a triangle is $(6x^3 - 2x^2 - 8x)$ square feet and the height is $(3x^2 - 4x)$ feet, what is the length of the base?

△ 90. **Volume of a Box** The volume of a rectangular solid is $(x^3 + 2x^2 - x - 2)$ cubic feet. One side measures $(x + 2)$ feet and another measures $(x - 1)$ feet. What is the measure of the third side?

91. **Average Cost** The average cost of manufacturing x computers per day is given by

$$\frac{0.004x^3 - 0.8x^2 + 180x + 5000}{x}.$$

(a) Simplify the quotient by dividing each term in the numerator by the denominator.

(b) Use this result to determine the average cost of manufacturing $x = 140$ computers in one day.

92. **Average Cost** The average cost of manufacturing x digital cameras per day is given by

$$\frac{0.0024x^3 - 0.4x^2 + 46x + 4000}{x}.$$

(a) Simplify the quotient by dividing each term in the numerator by the denominator.

(b) Use this result to determine the average cost of manufacturing $x = 90$ cameras in one day.

Extending the Concepts

In Problems 93 and 94, determine the value of the missing term so that the remainder is zero.

93. $\dfrac{6x^2 - 13x + ?}{2x - 3}$

94. $\dfrac{3x^2 + ? - 5}{x - 1}$

Explaining the Concepts

95. Explain how to divide polynomials when the divisor is a monomial and then when the divisor is a binomial. Which procedures are the same and which are different?

96. The first steps of a division problem are written below. Describe what has occurred. Are there any potential errors with this presentation? What would you recommend this student do to improve his or her chances of obtaining the correct answer?

$$\begin{array}{r} 2x^2 \\ x^2 - 3 \overline{) 2x^4 - 3x^3 + 2x + 1} \\ \underline{-(2x^4 - 6x^2)} \end{array}$$

5.6 Applying Exponent Rules: Scientific Notation

Objectives

1 Convert Decimal Notation to Scientific Notation

2 Convert Scientific Notation to Decimal Notation

3 Use Scientific Notation to Multiply and Divide

Are You Ready for This Section?

Before getting started, take the following readiness quiz. If you get a problem wrong, go back to the section cited and review the material.

R1. Find the product: $(3a^6)(4.5a^4)$ [Section 5.2, pp. 321–322]

R2. Find the product: $(7n^3)(2n^{-2})$ [Section 5.4, pp. 340–342]

R3. Find the quotient: $\dfrac{3.6b^9}{0.9b^{-2}}$ [Section 5.4, pp. 340–342]

Did you know that the mass of the Sun is 1,989,000,000,000,000,000,000,000,000,000 kg? Did you know that the mass of a dust particle is 0.000000000753 kg? These numbers are difficult to write and difficult to read, so we use exponents to rewrite them.

Ready? ...Answers
R1. $13.5a^{10}$ **R2.** $14n$ **R3.** $4b^{11}$

▶ ❶ Convert Decimal Notation to Scientific Notation

Decimal notation is the notation we commonly see in newspapers or magazines. The numbers 1,989,000,000,000,000,000,000,000,000,000 kg and 0.000000000753 kg are written in decimal notation. Scientific notation expresses a number as the product of two factors: One factor is a number between 1 and 10, including 1 but not including 10, and the other factor is an integer power of 10.

Definition

A number written as the product of a number x, where $1 \leq x < 10$, and a power of 10 is said to be written in **scientific notation.** That is, a number is written in scientific notation when it is in the form

$$x \times 10^N$$

where

$$1 \leq x < 10 \text{ and } N \text{ is an integer}$$

Notice in the definition, $x < 10$. That's because when $x = 10$, we have 10^1, a power of 10. For example, in scientific notation,

$$\text{Mass of Sun} = 1.989 \times 10^{30} \text{ kilograms}$$

$$\text{Mass of a dust particle} = 7.53 \times 10^{-10} \text{ kilograms}$$

Converting from Decimal Notation to Scientific Notation

To change a positive number to scientific notation:

Step 1: Count the number N of decimal places that the decimal point must be moved in order to arrive at a number x, where $1 \leq x < 10$.

Step 2: If the original number is greater than or equal to 1, the scientific notation is $x \times 10^N$. If the original number is between 0 and 1, the scientific notation is $x \times 10^{-N}$.

In Words

For a number greater than or equal to 1, use $10^{positive\ exponent}$. For a number between 0 and 1, use $10^{negative\ exponent}$.

> **EXAMPLE 1**   **How to Convert from Decimal Notation to Scientific Notation**

Write 5283 in scientific notation.

Step-by-Step Solution

For a number to be in scientific notation, the decimal must be moved so that there is a single nonzero digit to the left of the decimal point. All remaining digits must appear to the right of the decimal point.

Step 1: The "understood" decimal point in 5283 follows the 3. Therefore, we move the decimal to the left by $N = 3$ places until it is between the 5 and the 2. Do you see why?

$$5\,2\,8\,3.$$

Step 2: The original number is greater than 1, so we write 5283 in scientific notation as

$$5.283 \times 10^3$$

EXAMPLE 2 **How to Convert from Decimal Notation to Scientific Notation**

Write 0.054 in scientific notation.

Step-by-Step Solution

Step 1: Because 0.054 is less than 1, we move the decimal point to the right $N = 2$ places until it is between the 5 and the 4.

$$0.054$$

Step 2: The original number is between 0 and 1, so we write 0.054 in scientific notation as

$$5.4 \times 10^{-2}$$

Quick ✓

1. A number written as 3.2×10^{-6} is said to be written in _____ notation.

2 When 47,000,000 is written in scientific notation, the power of 10 will be _____ (positive or negative).

3. *True or False* When a number is expressed in scientific notation, it is expressed as the product of a number $x, 0 \leq x < 1$, and a power of 10.

In Problems 4–9, write each number in scientific notation.

4. 432	**5.** 10,302	**6.** 5,432,000
7. 0.093	**8.** 0.0000459	**9.** 0.00000008

▶ **②** **Convert Scientific Notation to Decimal Notation**

Now we are going to convert a number from scientific notation to decimal notation. Study Table 3 to discover the pattern.

Table 3

Scientific Notation	Product	Decimal Notation	Location of Decimal Point
3.69×10^2	3.69×100	369	moved 2 places to the right
3.69×10^1	3.69×10	36.9	moved 1 place to the right
3.69×10^0	3.69×1	3.69	didn't move
3.69×10^{-1}	3.69×0.1	0.369	moved 1 place to the left
3.69×10^{-2}	3.69×0.01	0.0369	moved 2 places to the left

The pattern in Table 3 leads to the following steps for converting a number from scientific notation to decimal notation.

Converting a Number from Scientific Notation to Decimal Notation

Step 1: Determine the exponent, N, on the number 10.

Step 2: If the exponent is positive, then move the decimal point N decimal places to the right. If the exponent is negative, then move the decimal point $|N|$ decimal places to the left. Add zeros, as needed.

EXAMPLE 3 **How to Convert from Scientific Notation to Decimal Notation**

Write 2.3×10^3 in decimal notation.

Step-by-Step Solution

Step 1: Determine the exponent on the number 10.   The exponent on the 10 is 3.

Step 2: Since the exponent is positive, we move the decimal point three places to the $2.3\,0\,0.$
right. Notice we add zeros to the right of 3, as needed.

$$\text{So } 2.3 \times 10^3 = 2300.$$

EXAMPLE 4 **How to Convert from Scientific Notation to Decimal Notation**

Write 4.57×10^{-5} in decimal notation.

Step-by-Step Solution

Step 1: Determine the exponent on the number 10.   The exponent on the 10 is -5.

Step 2: Since the exponent is negative, we move the   Add zeros to the left of $0.0\,0\,0\,0\,4.5\,7$
decimal point five places to the left.   the original decimal point.

$$\text{So } 4.57 \times 10^{-5} = 0.0000457.$$

> **Quick ✓**
>
> **10.** *True or False* To write 3.2×10^{-6} in decimal notation, move the decimal point in 3.2 six places to the left.
>
> **11.** *True or False* To convert 2.4×10^3 to decimal notation, move the decimal point three places to the right.
>
> *In Problems 12–16, write each number in decimal notation.*
>
> **12.** 3.1×10^2 **13.** 9.01×10^{-1} **14.** 1.7×10^5
>
> **15.** 7×10^0 **16.** 8.9×10^{-4}

▶ ❸ Use Scientific Notation to Multiply and Divide

To multiply and divide numbers written in scientific notation, we use two Laws of Exponents: the Product Rule, $a^m \cdot a^n = a^{m+n}$, and the Quotient Rule, $\dfrac{a^m}{a^n} = a^{m-n}$. We will use these laws where the base is 10 as follows:

$$10^m \cdot 10^n = 10^{m+n} \quad \text{and} \quad \frac{10^m}{10^n} = 10^{m-n}$$

EXAMPLE 5 **Multiplying Using Scientific Notation**

Perform the indicated operation. Express the answer in scientific notation.
$$(3 \times 10^2) \cdot (2.5 \times 10^5)$$

Solution

$$(3 \times 10^2) \cdot (2.5 \times 10^5) = (3 \cdot 2.5) \times (10^2 \cdot 10^5)$$

Multiply $3 \cdot 2.5$; use $a^m \cdot a^n = a^{m+n}$: $= 7.5 \times 10^7$

EXAMPLE 6 **Multiplying Using Scientific Notation**

Perform the indicated operation. Express the answer in scientific notation.

(a) $(4 \times 10^{-2}) \cdot (6 \times 10^8)$ (b) $(3.2 \times 10^{-3}) \cdot (4.8 \times 10^{-4})$

Solution

(a) $(4 \times 10^{-2}) \cdot (6 \times 10^8) = (4 \cdot 6) \times (10^{-2} \cdot 10^8)$

Multiply $4 \cdot 6$; use $a^m \cdot a^n = a^{m+n}$: $= 24 \times 10^6$

Convert 24 to scientific notation: $= (2.4 \times 10^1) \times 10^6$

Use $a^m \cdot a^n = a^{m+n}$: $= 2.4 \times 10^7$

(b) $(3.2 \times 10^{-3}) \cdot (4.8 \times 10^{-4}) = (3.2 \cdot 4.8) \times (10^{-3} \cdot 10^{-4})$

Multiply $3.2 \cdot 4.8$; use $a^m \cdot a^n = a^{m+n}$: $= 15.36 \times 10^{-7}$

Convert 15.36 to scientific notation: $= (1.536 \times 10^1) \times 10^{-7}$

Use $a^m \cdot a^n = a^{m+n}$: $= 1.536 \times 10^{-6}$

●

Quick ✓

In Problems 17–20, perform the indicated operation. Express the answer in scientific notation.

17. $(3 \times 10^4) \cdot (2 \times 10^3)$ **18.** $(2 \times 10^{-2}) \cdot (4 \times 10^{-1})$

19. $(5 \times 10^{-4}) \cdot (3 \times 10^7)$ **20.** $(8 \times 10^{-4}) \cdot (3.5 \times 10^{-2})$

EXAMPLE 7 **Dividing Using Scientific Notation**

Perform the indicated operation. Express the answer in scientific notation.

(a) $\dfrac{6 \times 10^6}{2 \times 10^2}$ (b) $\dfrac{2.4 \times 10^4}{3 \times 10^{-2}}$

Solution

(a) $\dfrac{6 \times 10^6}{2 \times 10^2} = \dfrac{6}{2} \times \dfrac{10^6}{10^2}$

Divide $\dfrac{6}{2}$; use $\dfrac{a^m}{a^n} = a^{m-n}$: $= 3 \times 10^4$

(b) $\dfrac{2.4 \times 10^4}{3 \times 10^{-2}} = \dfrac{2.4}{3} \times \dfrac{10^4}{10^{-2}}$

Divide $\dfrac{2.4}{3}$; use $\dfrac{a^m}{a^n} = a^{m-n}$: $= 0.8 \times 10^{4-(-2)}$

Convert 0.8 to scientific notation: $= (8 \times 10^{-1}) \times 10^6$

Use $a^m \cdot a^n = a^{m+n}$: $= 8 \times 10^5$

●

Quick ✓

In Problems 21–24, perform the indicated operation. Express the answer in scientific notation.

21. $\dfrac{8 \times 10^6}{2 \times 10^1}$ **22.** $\dfrac{2.8 \times 10^{-7}}{1.4 \times 10^{-3}}$

23. $\dfrac{3.6 \times 10^3}{7.2 \times 10^{-1}}$ **24.** $\dfrac{5 \times 10^{-2}}{8 \times 10^2}$

EXAMPLE 8 **Visits to Facebook**

In 2012, Facebook had 1.75×10^6 visitors each day. How many visitors to Facebook were there in April 2012? Express the answer in scientific and decimal notation. (SOURCE: *Facebook*)

Solution

To find the number of visitors to Facebook for April 2012, we multiply the number of visitors each day by the number of days, 30.

In scientific notation, $30 = 3 \times 10^1$. So, we have

$$
\begin{aligned}
(1.75 \times 10^6)(3 \times 10^1) &= (1.75 \cdot 3) \times (10^6 \cdot 10^1) \\
&= 5.25 \times 10^7 \quad \text{scientific notation} \\
&= 52{,}500{,}000 \text{ visitors} \quad \text{decimal notation}
\end{aligned}
$$

In April 2012, there were 52,500,000 visitors to Facebook.

Quick ✓

25. The United States consumes 3.49×10^8 gallons of gasoline each day. How many gallons of gasoline does the United States consume in a 30-day month? Express the answer in scientific and decimal notation.

26. McDonald's has 3.6×10^6 customers each day. How many customers does McDonald's get in a 365-day year?

5.6 Exercises

Exercise numbers in green have complete video solutions in MyMathLab

Problems 1–26 *are the* **Quick ✓** *s that follow the* **EXAMPLES**.

Building Skills

In Problems 27–42, write each number in scientific notation. See Objective 1.

27. 300,000   **28.** 421,000,000

29. 64,000,000   **30.** 8,000,000,000

31. 0.00051   **32.** 0.0000001

33. 0.000000001   **34.** 0.0000283

35. 8,007,000,000   **36.** 401,000,000

37. 0.0000309   **38.** 0.000201

39. 620   **40.** 8

41. 4   **42.** 120

In Problems 43–54, a number is given in decimal notation. Write the number in scientific notation. See Objective 1.

43. World Population According to the U.S. Census Bureau, the population of the world on March 7, 2012, was approximately 6,999,000,000 persons.

44. United States Population According to the U.S. Census Bureau, the population of the United States on March 12, 2012, was approximately 313,000,000 persons.

45. Federal Debt According to the United States Treasury, the federal debt of the United States as of March 2012 was $15,500,000,000,000.

46. Interest Payments on Debt In 2011, total interest payments on the federal debt (Problem 45) totaled $197,000,000,000.

47. Distance to the Sun The average distance from Earth to the Sun is 93,000,000 miles.

48. Distance to the Moon The average distance from Earth to the moon (of Earth) is 384,000 kilometers.

49. Smallpox Virus The diameter of a smallpox virus is 0.00003 mm.

50. Human Blood Cell The diameter of a human blood cell is 0.0075 mm.

51. Dust Mites Dust mites are microscopic bugs that are a major cause of allergies and asthma. They are approximately 0.00000025 m in length.

52. DNA The length of a DNA molecule can exceed 0.025 inch in some organisms.

53. Mass of a Penny The mass of the United States' Lincoln penny is approximately 0.00311 kg.

54. Physics The radius of a typical atom is approximately 0.0000000001 m.

In Problems 55–70, write each number in decimal notation. See Objective 2.

55. 4.2×10^5

56. 3.75×10^2

57. 1×10^8

58. 6×10^6

59. 3.9×10^{-3}

60. 6.1×10^{-6}

61. 4×10^{-1}

62. 5×10^{-4}

63. 3.76×10^3

64. 4.9×10^{-1}

65. 8.2×10^{-3}

66. 5.4×10^5

67. 6×10^{-5}

68. 5.123×10^{-3}

69. 7.05×10^6

70. 7×10^8

In Problems 71–76, a number is given in scientific notation. Write the number in decimal notation. See Objective 2.

71. Time A femtosecond is equal to 1×10^{-15} second.

72. Dust Particle The mass of a dust particle is 7.53×10^{-10} kg.

73. Vitamin A One-A-Day vitamin pill contains 2.25×10^{-3} gram of zinc.

74. Water Molecule The diameter of a water molecule is 3.85×10^{-7} m.

75. Coffee Drinkers In 2012, there were 1.83×10^6 coffee drinkers in the United States.

76. Fatal Accidents In 2009, there were 4.8×10^4 drivers in fatal auto accidents in the United States. (SOURCE: *U.S. Department of Transportation*)

In Problems 77–88, perform the indicated operations. Express your answer in scientific notation. See Objective 3.

77. $(2 \times 10^6)(1.5 \times 10^3)$

78. $(3 \times 10^{-4})(8 \times 10^{-5})$

79. $(1.2 \times 10^0)(7 \times 10^{-3})$

80. $(4 \times 10^7)(2.5 \times 10^{-4})$

81. $\dfrac{9 \times 10^4}{3 \times 10^{-4}}$

82. $\dfrac{6 \times 10^3}{1.2 \times 10^5}$

83. $\dfrac{2 \times 10^{-3}}{8 \times 10^{-5}}$

84. $\dfrac{4.8 \times 10^7}{1.2 \times 10^2}$

85. $\dfrac{56,000}{0.00007}$

86. $\dfrac{0.000275}{2500}$

87. $\dfrac{300,000 \times 15,000,000}{0.0005}$

88. $\dfrac{24,000,000,000}{0.00006 \times 2000}$

Applying the Concepts

89. Speed of Light Light travels at the rate of 1.86×10^5 miles per second. How far does light travel in 1 minute $(6.0 \times 10^1$ seconds$)$?

90. Speed of Sound Sound travels at the rate of 1.127×10^3 feet per second. How far does sound travel in 1 minute $(6.0 \times 10^1$ seconds$)$?

91. Hair Growth Human hair grows at a rate of 1×10^{-8} mile per hour. How many miles will human hair grow after 100 hours? After one week? Express each answer in decimal notation.

92. Disneyland The average number of visitors to Disneyland each day is 4×10^4. How many visitors visit Disneyland in a 30-day month? Express your answer in decimal notation.

93. Ice Cream Consumption Total U.S. consumption of ice cream in 2012 amounted to about 3.9 billion pounds. (SOURCE: *U.S. Department of Agriculture*)
   (a) Write 3.9 billion pounds in scientific notation.
   (b) The population of the United States in 2012 was approximately 310,000,000 persons. Express this number in scientific notation.
   (c) Use your answers to parts (a) and (b) to find, to the nearest tenth of a pound, the average number of pounds of ice cream that were consumed per person in the United States in 2012.

94. M&M'S® Candy Over 400 million M&M'S® candies are produced in the United States every day.
   (SOURCE: *Mars.com*)
   (a) Write 400 million in scientific notation.
   (b) Assuming a 30-day month, how many M&M'S® candies are produced in a month in the United States? Write this answer in scientific notation.
   (c) The population of the United States is approximately 310,000,000 persons. Express this number in scientific notation.
   (d) Use your answers to parts (b) and (c) to find, to the nearest whole number, the average number of M&M'S candies eaten per person per month in the United States.

Extending the Concepts

Scientists often need to measure very small things, such as cells. They use the following units of measure:

millimeter $(mm) = 1 \times 10^{-3}$ meter micron $(\mu m) = 1 \times 10^{-6}$ meter

nanometer $(nm) = 1 \times 10^{-9}$ meter picometer $(pm) = 1 \times 10^{-12}$ meter

Write the following measurements in meters using scientific notation:

95. $250 \ \mu m$ **96.** 60.4 nm

97. 800 pm **98.** 40 mm

99. 71.5 nm **100.** $200 \ \mu m$

Assume a cell is in the shape of a sphere. Given that the volume of a sphere is $V = \dfrac{4}{3}\pi r^3$, find the volume of each cell whose radius is given. Express the answer as a multiple of π in cubic meters.

101. $21 \ \mu m$ **102.** 0.75 nm

103. 6 nm **104.** $108 \ \mu m$

Explaining the Concepts

105. Explain how to convert a number written in decimal notation to scientific notation.

106. Explain how to convert a number written in scientific notation to decimal notation.

107. Dana thinks that the number 34.5×10^4 is correctly written in scientific notation. Is Dana correct? If so, explain why. If not, explain why the answer is wrong and write the correct answer.

108. Explain why scientific notation is used to perform calculations that involve multiplying and dividing but not adding and subtracting.

Chapter 5 Activity: What Is the Question?

Focus: Using exponent rules, using scientific notation, and performing operations with polynomials

Time: 15–20 minutes

Group size: 2 or 4

In this activity, you will work as a team to solve eight multiple-choice questions. However, these questions are different from most multiple-choice questions. You are given the answer to a problem and must determine which of the multiple-choice options has the correct question for the given answer.

Before beginning the activity, decide how you will approach this task as a team. For example:

- If there are two members on your team, one member will always examine choices (a) and (b), and the other will always examine choices (c) and (d).

- If there are four members on your team, one member will always examine choice (a), another member will always examine choice (b), and so on.

1. The answer is $-3x^2 - 10x$. What is the question?

(a) Simplify: $2x - 3x(x^2 + 4)$

(b) Find the quotient: $\dfrac{6x^3 - 20x^2}{-2x}$

(c) Simplify: $-3x(x^2 + 3) + 1$

(d) Find the quotient: $\dfrac{-6x^4 - 20x^3}{2x^2}$

2. The answer is $-10x^4y^8$. What is the question?

(a) Simplify: $(-5x^2y^4)^2$

(b) Simplify: $(5x^3y^3)(-2xy^5)$

(c) Simplify: $\dfrac{-30x^{-2}y^6}{3x^2y^{-2}}$

(d) Simplify: $(5x^2y^4)(-2x^2y^2)$

3. The answer is $12x^2 - 16x - 3$. What is the question?

(a) Multiply: $(6x - 1)(2x + 3)$

(b) Divide: $\dfrac{24x^3 - 32x^2 + 6x}{2x}$

(c) Multiply: $(6x + 1)(2x - 3)$

(d) Simplify: $(14x^2 - x + 2) - (2x^2 + 16x + 5)$

4. The answer is $x^2 + 5x + 6$. What is the question?

(a) Find the product: $(x + 6)(x - 1)$

(b) Simplify: $2x^2 + 7x + 9 - (x^2 - 2x - 3)$

(c) Find the product: $(x + 2)(x + 3)$

(d) Simplify: $(x + 6)^2$

5. The answer is 3. What is the question?

(a) What is the name of the variable in $16z^2 + 3z - 5$?

(b) What is the degree of the polynomial $2mn + 6m - 3$?

(c) How many terms are in the polynomial $2mn + 6m - 3$?

(d) What is the coefficient of b in the polynomial $3a^2b - 9a + 5b$?

6. The answer is 43. What is the question?

(a) Evaluate: $4x^2 - 2x + 1$ for $x = -3$

(b) Evaluate: $2x^2 + 3x - 4$ for $x = -2$

(c) Evaluate: $-x^2 - 5x + 1$ for $x = -1$

(d) Evaluate: $x^2 + 4x - 8$ for $x = -5$

7. The answer is $\dfrac{2x^5}{y^5}$. What is the question?

   (a) Simplify: $(6x^{-4}y)^0\left(\dfrac{12x^{-1}y^{-2}}{x^{-4}y^3}\right)$

   (b) Simplify: $(6x^{-3}y^0)^{-1}\left(\dfrac{12xy^{-3}}{x^{-2}y^2}\right)$

   (c) Simplify: $(6x^2y^0)^{-1}\left(\dfrac{12x^{-2}y^3}{xy^{-4}}\right)$

   (d) Simplify: $(6x^{-2}y^0)^{-1}\left(\dfrac{12x^0y^{-2}}{x^{-3}y^3}\right)$

8. The answer is 2.5×10^{-8}. What is the question?

   (a) Find the quotient: $\dfrac{2 \times 10^3}{8 \times 10^{-4}}$

   (b) Find the product: $(5 \times 10^{-4}) \cdot (5 \times 10^{-4})$

   (c) Find the quotient: $\dfrac{2 \times 10^{-3}}{8 \times 10^4}$

   (d) Find the product: $(5 \times 10^{12}) \cdot (5 \times 10^{-4})$

Chapter 5 Review

Section 5.1   Adding and Subtracting Polynomials

KEY CONCEPTS

- In a monomial in the form ax^k, k is the degree of the monomial and k is a whole number.
- The degree of a polynomial is the highest degree of all the terms of the polynomial.

KEY TERMS

Monomial coefficient
Degree of a monomial
Polynomial
Binomial
Trinomial
Standard form
Degree of a polynomial

You Should Be Able To...	EXAMPLE	Review Exercises
1 Define monomial and determine the degree of a monomial (p. 309)	Examples 1 through 3	1–4
2 Define polynomial and determine the degree of a polynomial (p. 310)	Examples 4 and 5	5–10
3 Simplify polynomials by combining like terms (p. 311)	Examples 6 through 10	11–16
4 Evaluate polynomials (p. 314)	Examples 11 through 13	17–20

In Problems 1–4, determine whether the given expression is a monomial (Yes or No). For those that are monomials, state the coefficient and the degree.

1. $4x^3$ **2.** $6x^{-3}$

3. $m^{\frac{1}{2}}$ **4.** mn^2

In Problems 5–10, determine whether the algebraic expression is a polynomial (Yes or No). If it is a polynomial, state the degree and then state whether it is a monomial, a binomial, or a trinomial.

5. $4x^6 - 4x^{\frac{1}{2}}$ **6.** $\dfrac{3}{x} - \dfrac{1}{x^2}$

7. 6 **8.** $3x^3 - 4xy^4$

9. $-2x^5y - 7x^4y + 7$ **10.** $\dfrac{1}{2}x^3 + 2x^{10} - 5$

In Problems 11–14, perform the indicated operation.

11. $(6x^2 - 2x + 1) + (3x^2 + 10x - 3)$

12. $(-7m^3 - 2mn) + (8m^3 - 5m + 3mn)$

13. $(4x^2y + 10x) - (5x^2y - 2x)$

14. $(3y^2 - yz + 3z^2) - (10y^2 + 5yz - 6z^2)$

15. Find the sum of $-6x^2 + 5$ and $4x^2 - 7$.

16. Subtract $20y^2 - 10y + 5$ from $-18y + 10$.

In Problems 17–20, evaluate the polynomial for the given value(s).

17. $3x^2 - 5x$
   (a) $x = 0$
   (b) $x = -1$
   (c) $x = 2$

18. $-x^2 + 3$
   (a) $x = 0$
   (b) $x = -1$
   (c) $x = \dfrac{1}{2}$

19. $x^2y + 2xy^2$ for $x = -2$ and $y = 1$

20. $4a^2b^2 - 3ab + 2$ for $a = -1$ and $b = -3$

Section 5.2 Multiplying Monomials: The Product and Power Rules

KEY CONCEPTS

- **Product Rule for Exponents**

  If a is a real number, and m and n are natural numbers, then $a^m \cdot a^n = a^{m+n}$.

- **Power Rule for Exponents**

  If a is a real number, and m and n are natural numbers, then $(a^m)^n = a^{m \cdot n}$.

- **Product to a Power Rule for Exponents**

  If a and b are real numbers and n is a natural number, then $(a \cdot b)^n = a^n \cdot b^n$.

KEY TERMS

Base
Power
Exponent

You Should Be Able To...	EXAMPLE	Review Exercises
1 Simplify exponential expressions using the Product Rule (p. 318)	Examples 1 through 3	21, 22, 25, 26
2 Simplify exponential expressions using the Power Rule (p. 320)	Examples 4 and 5	23, 24, 27, 28
3 Simplify exponential expressions containing products (p. 320)	Example 6	29–32
4 Multiply a monomial by a monomial (p. 321)	Examples 7 and 8	33–40

In Problems 21–32, simplify each expression.

21. $6^2 \cdot 6^5$

22. $\left(-\dfrac{1}{3}\right)^2 \left(-\dfrac{1}{3}\right)^3$

23. $(4^2)^6$

24. $[(-1)^4]^3$

25. $x^4 \cdot x^8 \cdot x$

26. $m^4 \cdot m^2$

27. $(r^3)^4$

28. $(m^8)^3$

29. $(4x)^3(4x)^2$

30. $(-2n)^3(-2n)^3$

31. $(-3x^2y)^4$

32. $(2x^3y^4)^2$

In Problems 33–40, multiply.

33. $3x^2 \cdot 5x^4$

34. $-4a \cdot 9a^3$

35. $-8y^4 \cdot (-2y)$

36. $12p \cdot (-p^5)$

37. $\dfrac{8}{3}w^3 \cdot \dfrac{9}{2}w$

38. $\dfrac{1}{3}z^2 \cdot \left(-\dfrac{9}{4}z\right)$

39. $(3x^2)^3 \cdot (2x)^2$

40. $(-4a)^2 \cdot (5a^4)$

Section 5.3 Multiplying Polynomials

KEY CONCEPTS

- **Extended form of the Distributive Property**

  $a(b + c + \cdots + z) = a \cdot b + a \cdot c + \cdots + a \cdot z$ where $a, b, c, \ldots, z$ are real numbers.

- **FOIL Method for multiplying two binomials**

  $$\begin{array}{cccc} \text{F} & \text{O} & \text{I} & \text{L} \end{array}$$
  $(ax + b)(cx + d) = ax \cdot cx + ax \cdot d + b \cdot cx + b \cdot d$

- **Product of the Sum and Difference of Two Terms**

  $(A - B)(A + B) = A^2 - B^2$

- **Squares of Binomials**

  $(A + B)^2 = A^2 + 2AB + B^2$
  $(A - B)^2 = A^2 - 2AB + B^2$

KEY TERMS

FOIL method
Special products
Sum and difference of two terms
Squares of binomials
Perfect square trinomial

You Should be Able To...	EXAMPLE	Review Exercises
❶ Multiply a polynomial by a monomial (p. 323)	Examples 1 through 3	41, 42
❷ Multiply two binomials using the Distributive Property (p. 324)	Examples 4 and 5	43–46
❸ Multiply two binomials using the FOIL method (p. 326)	Examples 6 and 7	49–54
❹ Multiply the sum and difference of two terms (p. 327)	Examples 8 and 9	55, 56, 59, 60, 63, 64
❺ Square a binomial (p. 328)	Examples 10 and 11	57, 58, 61, 62, 65, 66
❻ Multiply a polynomial by a polynomial (p. 329)	Examples 12 and 13	47, 48

In Problems 41–48, multiply using the Distributive Property.

41. $-2x^3(4x^2 - 3x + 1)$ **42.** $\frac{1}{2}x^4(4x^3 + 8x^2 - 2)$

43. $(3x - 5)(2x + 1)$ **44.** $(4x + 3)(x - 2)$

45. $(x + 5)(x - 8)$ **46.** $(w - 1)(w + 10)$

47. $(4m - 3)(6m^2 - m + 1)$

48. $(2y + 3)(4y^4 + 2y^2 - 3)$

In Problems 49–54, use the FOIL method to find each product.

49. $(x + 5)(x + 3)$ **50.** $(2x - 1)(x - 8)$

51. $(2m + 7)(3m - 2)$ **52.** $(6m - 4)(8m + 1)$

53. $(3x + 2y)(7x - 3y)$ **54.** $(4x - y)(5x + 3y)$

In Problems 55–66, find the special products.

55. $(x - 4)(x + 4)$ **56.** $(2x + 5)(2x - 5)$

57. $(2x + 3)^2$ **58.** $(7x - 2)^2$

59. $(3x + 4y)(3x - 4y)$ **60.** $(8m - 6n)(8m + 6n)$

61. $(5x - 2y)^2$ **62.** $(2a + 3b)^2$

63. $(x - 0.5)(x + 0.5)$ **64.** $(r + 0.25)(r - 0.25)$

65. $\left(y + \frac{2}{3}\right)^2$

66. $2x(x + 3)(x - 3) - x(x + 2)^2$

Section 5.4 Dividing Monomials: The Quotient Rule and Integer Exponents

KEY CONCEPTS

- **Quotient Rule for Exponents**

  If a is a nonzero real number, and if m and n are integers, then $\frac{a^m}{a^n} = a^{m-n}$.

- **Definition of Zero as an Exponent**

  If a is a nonzero real number, we define $a^0 = 1$.

- **Quotient to a Power Rule for Exponents**

  If a and b are real numbers and n is an integer, then $\left(\frac{a}{b}\right)^n = \frac{a^n}{b^n}$ if $b \neq 0$.

  If n is negative or 0, then a cannot be 0.

- **Definition of a Negative Exponent**

  If n is a positive integer and if a is a nonzero real number, then we define

  $a^{-n} = \frac{1}{a^n}$ and $\frac{1}{a^{-n}} = a^n$.

- **Quotient to a Negative Power**

  If a and b are real numbers and n is an integer, then $\left(\frac{a}{b}\right)^{-n} = \left(\frac{b}{a}\right)^n$ if $a \neq 0, b \neq 0$.

You Should be Able To...	EXAMPLE	Review Exercises
1 Simplify exponential expressions using the Quotient Rule (p. 334)	Examples 1 and 2	67–70, 75, 76
2 Simplify exponential expressions using the Quotient to a Power Rule (p. 335)	Examples 3 and 4	77–80
3 Simplify exponential expressions using zero as an exponent (p. 336)	Example 5	71–74
4 Simplify exponential expressions involving negative exponents (p. 336)	Examples 6 through 12	81–86
5 Simplify exponential expressions using the Laws of Exponents (p. 340)	Examples 13 through 15	87–92

In Problems 67–92, simplify. Write answers with only positive exponents. All variables are nonzero.

67. $\dfrac{6^5}{6^3}$

68. $\dfrac{7}{7^4}$

69. $\dfrac{x^{16}}{x^{12}}$

70. $\dfrac{x^3}{x^{11}}$

71. 5^0

72. -5^0

73. $m^0, m \neq 0$

74. $-m^0, m \neq 0$

75. $\dfrac{25x^3y^7}{10xy^{10}}$

76. $\dfrac{3x^4y^2}{9x^2y^{10}}$

77. $\left(\dfrac{x^3}{y^2}\right)^5$

78. $\left(\dfrac{7}{x^2}\right)^3$

79. $\left(\dfrac{2m^2n}{p^4}\right)^3$

80. $\left(\dfrac{3mn^2}{p^5}\right)^4$

81. -5^{-2}

82. $\dfrac{1}{4^{-3}}$

83. $\left(\dfrac{2}{3}\right)^{-4}$

84. $\left(\dfrac{1}{3}\right)^{-3}$

85. $2^{-2} + 3^{-1}$

86. $4^{-1} - 2^{-3}$

87. $\dfrac{16x^{-3}y^4}{24x^{-6}y^{-1}}$

88. $\dfrac{15x^0y^{-6}}{35xy^4}$

89. $(2m^{-3}n)^{-4}(3m^{-4}n^2)^2$

90. $(4m^{-6}n^0)^3(3m^{-6}n^3)^{-2}$

91. $\left(\dfrac{3rs^{-1}}{4s^2}\right)^{-2} \cdot (2r^{-6}t^0)^{-1}$

92. $\dfrac{9a}{3a^{-2}} + a(4a^2 + 7a^{-1})$

Section 5.5   Dividing Polynomials

KEY CONCEPTS

- If a, b, and c are monomials, then $\dfrac{a+b}{c} = \dfrac{a}{c} + \dfrac{b}{c}$.
- (Quotient)(Divisor) + Remainder = Dividend

KEY TERMS

Divisor
Dividend
Quotient
Remainder

You Should Be Able To...	EXAMPLE	Review Exercises
1 Divide a polynomial by a monomial (p. 345)	Examples 1 through 3	93–98
2 Divide a polynomial by a binomial (p. 347)	Examples 4 through 7	99–104

In Problems 93–104, divide each of the following.

93. $\dfrac{36x^7 - 24x^6 + 30x^2}{6x^2}$

94. $\dfrac{15x^5 + 25x^3 - 30x^2}{5x}$

95. $\dfrac{16n^8 + 4n^5 - 10n}{4n^5}$

96. $\dfrac{30n^6 - 20n^5 - 16n^3}{5n^5}$

97. $\dfrac{2p^8 + 4p^5 - 8p^3}{-16p^5}$

98. $\dfrac{3p^4 - 6p^2 + 9}{-6p^2}$

101. $\dfrac{6x^2 + x^3 - 2x + 1}{x - 1}$

102. $\dfrac{-6x + 2x^3 - 7x^2 + 8}{x - 2}$

99. $\dfrac{8x^2 - 2x - 21}{2x + 3}$

100. $\dfrac{3x^2 + 17x - 6}{3x - 1}$

103. $\dfrac{x^3 + 8}{x + 2}$

104. $\dfrac{3x^3 + 2x - 7}{x - 5}$

Section 5.6 Applying Exponent Rules: Scientific Notation

KEY CONCEPTS

- **Definition of Scientific Notation**
  A number is written in scientific notation when it is in the form
  $x \times 10^N$, where $1 \le x < 10$ and N is an integer.

- **Convert from Decimal Notation to Scientific Notation**
  To change a positive number into scientific notation:

  Step 1: Count the number N of decimal places that the decimal point must be moved to arrive at a number x, where $1 \le x < 10$.

  Step 2: If the original number is greater than or equal to 1, the scientific notation is $x \times 10^N$. If the original number is between 0 and 1, the scientific notation is $x \times 10^{-N}$.

- **Convert from Scientific Notation to Decimal Notation**

  Step 1: Determine the exponent, N, on the number 10.

  Step 2: If the exponent is positive, then move the decimal point N decimal places to the right.
  If the exponent is negative, then move the decimal point $|N|$ decimal places to the left. Add zeros, as needed.

KEY TERMS

Decimal notation
Scientific notation

You Should Be Able To...	EXAMPLE	Review Exercises
1 Convert decimal notation to scientific notation (p. 353)	Examples 1 and 2	105–110
2 Convert scientific notation to decimal notation (p. 354)	Examples 3 and 4	111–116
3 Use scientific notation to multiply and divide (p. 355)	Examples 5 through 8	117–122

In Problems 105–110, write in scientific notation.

105. 27,000,000

106. 1,230,000,000

107. 0.00006

108. 0.00000305

109. 3

110. 8

In Problems 111–116, write in decimal notation.

111. 6×10^{-4}

112. 1.25×10^{-3}

113. 6.13×10^5

114. 8×10^4

115. 3.7×10^{-1}

116. 5.4×10^7

In Problems 117–122, perform the indicated operations. Express your answer in scientific notation.

117. $(1.2 \times 10^{-5})(5 \times 10^8)$

118. $(1.4 \times 10^{-10})(3 \times 10^2)$

119. $\dfrac{2.4 \times 10^{-6}}{1.2 \times 10^{-8}}$

120. $\dfrac{5 \times 10^6}{25 \times 10^{-3}}$

121. $\dfrac{200,000 \times 4,000,000}{0.0002}$

122. $\dfrac{1,200,000}{0.003 \times 2,000,000}$

Chapter 5 Test

Step-by-step test solutions are found on the Chapter Test Prep Videos available in MyMathLab® *or on* YouTube.

1. Determine whether the algebraic expression $6x^5 - 2x^4$ is a polynomial (Yes or No). If it is a polynomial, state the degree and then state whether it is a monomial, a binomial, or a trinomial.

2. Evaluate the polynomial $3x^2 - 2x + 5$ for the given values:
   (a) $x = 0$ (b) $x = -2$ (c) $x = 3$

In Problems 3–11, perform the indicated operation.

3. $(3x^2y^2 - 2x + 3y) + (-4x - 6y + 4x^2y^2)$

4. $(8m^3 + 6m^2 - 4) - (5m^2 - 2m^3 + 2)$

5. $-3x^3(2x^2 - 6x + 5)$ 6. $(x - 5)(2x + 7)$

7. $(2x - 7)^2$ 8. $(4x - 3y)(4x + 3y)$

9. $(3x - 1)(2x^2 + x - 8)$ 10. $\dfrac{6x^4 - 8x^3 + 9}{3x^3}$

11. $\dfrac{3x^3 - 2x^2 + 5}{x + 3}$

In Problems 12–16, simplify each expression. Write answers with only positive exponents. All variables are nonzero.

12. $(4x^3y^2)(-3xy^4)$ 13. $\dfrac{18m^5n}{27m^2n^6}$

14. $\left(\dfrac{m^{-2}n^0}{m^{-7}n^4}\right)^{-6}$ 15. $(4x^{-3}y)^{-2}(2x^4y^{-3})^4$

16. $(2m^{-4}n^2)^{-1} \cdot \left(\dfrac{16m^0n^{-3}}{m^{-3}n^2}\right)$

17. Write 0.000012 in scientific notation.

18. Write 2.101×10^5 in decimal notation.

In Problems 19 and 20, perform the indicated operation. Express your answer in scientific notation.

19. $(2.1 \times 10^{-6}) \cdot (1.7 \times 10^{10})$

20. $\dfrac{3 \times 10^{-4}}{15 \times 10^2}$

Cumulative Review Chapters 1–5

1. Use the set $\left\{-6, -\dfrac{4}{2}, 0, 1.4, \sqrt{7}, \sqrt{25}\right\}$. List all of the elements that are
   (a) natural   (b) whole   (c) integers
   (d) rational   (e) irrational   (f) real

In Problems 2 and 3, evaluate each expression.

2. $-\dfrac{1}{2} + \dfrac{2}{3} \div 4 \cdot \dfrac{1}{3}$

3. $2 + 3[3 + 10(-1)]$

In Problems 4 and 5, simplify each algebraic expression.

4. $6x^3 - (-2x^2 + 3x) + 3x^2$

5. $-4(6x - 1) + 2(3x + 2)$

6. Solve: $-2(3x - 4) + 6 = 4x - 6x + 10$

7. Translate the following statement into an equation. DO NOT SOLVE. Four times the difference of a number and 5 is equal to 10 more than twice the number.

8. **Paycheck** Kathy's monthly paycheck from her part-time job working at an electronics store totaled $659.20. This amount included a 3% raise over her previous month's earnings. What were Kathy's monthly earnings before the 3% raise?

9. **Driving** Cheyenne and Amber live 306 miles apart. They start driving toward each other and meet in 3 hours. If Cheyenne drives 12 miles per hour faster than Amber, find Amber's driving speed.

10. Solve and graph the following inequality: $-5x + 2 > 17$

11. Evaluate $\dfrac{x^2 - y^2}{z}$ when $x = 3$, $y = -2$, and $z = -10$.

12. Find the slope of the line through $(-3, 8)$ and $(1, 2)$.

13. Graph the line $2x + 3y = 24$ by finding the intercepts.

14. Graph the line $y = -3x + 8$.

15. Solve the system of linear equations:
$$\begin{cases} 2x - 3y = 27 \\ -4x + 2y = -26 \end{cases}$$

16. Solve the system of linear equations:
$$\begin{cases} 3x - 2y = 8 \\ -6x + 4y = 8 \end{cases}$$

In Problems 17–23, perform the indicated operation.

17. $(4x^2 + 6x) - (-x + 5x^2) + (6x^3 - 2x^2)$

18. $(4m - 3)(7m + 2)$

19. $(3m - 2n)(3m + 2n)$

20. $(7x + y)^2$

21. $(2m + 5)(2m^2 - 5m + 3)$

22. $\dfrac{14xy^2 + 7x^2y}{7x^2y^2}$

23. $\dfrac{x^3 + 27}{x + 3}$

In Problems 24–27, simplify the expression. Write your answers with only positive exponents. All variables are nonzero.

24. $(4m^0n^3)(-6n)$

25. $\dfrac{25m^{-6}n^{-2}}{-10m^{-4}n^{-10}}$

26. $\left(\dfrac{2xy^4}{z^{-2}}\right)^{-6}$

27. $(x^4y^{-2})^{-4} \cdot \left(\dfrac{6x^{-4}y^3}{3y^{-8}}\right)^{-1}$

28. Write 0.0000605 in scientific notation.

29. Write 2.175×10^6 in decimal notation.

30. Perform the indicated operation. Write your answer in scientific notation.

$$(3.4 \times 10^8)(2.1 \times 10^{-3})$$

6 Factoring Polynomials

David is installing new brickwork around his fireplace, but he wants an opening above the fireplace to install a plasma television. The TV he purchases is 53 inches on the diagonal, excluding the casing on the sides of the TV screen. The length of the TV is 17 inches more than the width. What should the dimensions of the opening in the brickwork be if the casing on the outside of the television is 1.5 inches? See Problem 37 in Section 6.7.

The Big Picture: Putting It Together

One of the big ideas in mathematics is to be able to "undo" any operation. In Chapter 5, we learned how to multiply polynomials. In this chapter, we write a polynomial as a product of two or more polynomials. This process is called *factoring*, and it "undoes" multiplication.

In Chapter 2, we solved linear (first-degree) equations such as $2x + 5 = 8$. In this chapter, we discuss how factoring can be used to solve equations such as $2x^2 + 7x + 3 = 0$. The approach requires that we rewrite $2x^2 + 7x + 3$ as the product of two polynomials of degree 1. This leads us to solving two linear equations, which we already know how to do! This is one of the goals of algebra: Simplify a problem until it becomes a problem you already know how to solve.

Factoring is important for solving equations, and it plays a major role throughout the rest of the course, so be sure to work hard to learn the factoring techniques presented in this chapter.

Outline

6.1 Greatest Common Factor and Factoring by Grouping

Objectives

1. Find the Greatest Common Factor of Two or More Expressions
2. Factor Out the Greatest Common Factor in Polynomials
3. Factor Polynomials by Grouping

Are You Ready for This Section?

Before getting started, take the following readiness quiz. If you get a problem wrong, go back to the section cited and review the material.

R1. Write 48 as the product of prime numbers.   [Section 1.2, pp. 8–9]

R2. Distribute: $2(5x - 3)$ [Section 1.8, pp. 67–68]

R3. Find the product: $(2x + 5)(x - 3)$ [Section 5.3, pp. 324–327]

▶ Consider the following products:

$$5 \cdot 3 = 15$$
$$5(y + 5) = 5y + 25$$
$$(3x - 1)(x + 5) = 3x^2 + 14x - 5$$

The expressions on the left side are called **factors** of the expression on the right side, so $3x - 1$ and $x + 5$ are factors of $3x^2 + 14x - 5$.

In the last chapter, we learned how to multiply expressions such as $3x - 1$ and $x + 5$ to obtain the polynomial $3x^2 + 14x - 5$. In this chapter, we learn how to obtain the factors of a polynomial. That is, we learn how to write $3x^2 + 14x - 5$ as $(3x - 1)(x + 5)$.

> **In Words**
>
> Factoring is "undoing" multiplication.

Definition

To **factor** a polynomial means to write the polynomial as a product of two or more polynomials.

The process of factoring reverses the process of multiplying, as shown below.

$$\text{Factored form} \rightarrow 3x(x - 7) \overset{\text{Multiplication}}{\underset{\text{Factoring}}{=}} 3x^2 - 21x \leftarrow \text{Product}$$

▶ **1 Find the Greatest Common Factor of Two or More Expressions**

In the illustration above, notice that $3x$ is the largest number that divides evenly into both $3x^2$ and $21x$ in the expression $3x^2 - 21x$. Thus, $3x$ is the *greatest common factor* of $3x^2 - 21x$.

Definition

The **greatest common factor (GCF)** of a list of polynomials is the largest expression that divides evenly into all the polynomials.

Ready? ...Answers   R1. $3 \cdot 2^4$
R2. $10x - 6$ **R3.** $2x^2 - x - 15$

Writing $3x^2 - 21x$ as $3x(x - 7)$ is referred to as factoring out the *greatest common factor*. But how can we find the GCF? The following example shows us.

EXAMPLE 1   **How to Find the GCF of a List of Numbers**

Find the GCF of 12 and 18.

Step-by-Step Solution

Step 1: Write each number as the product of prime factors.

$$12 = 2 \cdot 2 \cdot 3$$
$$18 = 2 \quad \cdot 3 \cdot 3$$

Step 2: Determine the common prime factors.

The common factors are 2 and 3.

Step 3: Find the product of the common factors from Step 2. This number is the GCF.

The GCF is $2 \cdot 3 = 6$.

EXAMPLE 2 **Finding the GCF of a List of Numbers**

Find the GCF of 24, 40, and 72.

Solution

We write each number as the product of prime factors.

$$24 = 4 \cdot 6 = 2 \cdot 2 \cdot 2 \cdot 3$$
$$40 = 4 \cdot 10 = 2 \cdot 2 \cdot 2 \cdot 5$$
$$72 = 8 \cdot 9 = 2 \cdot 2 \cdot 2 \cdot 3 \cdot 3$$

All three numbers have three factors of 2, so the GCF is $2 \cdot 2 \cdot 2 = 8$.

Work Smart

Remember, a prime number is a number greater than 1 that has no factors other than itself and 1. For example, 3, 7, and 13 are prime numbers, whereas $4 (= 2 \cdot 2)$, $12 (= 2 \cdot 2 \cdot 3)$, and $35 (= 5 \cdot 7)$ are not prime.

Also, it is helpful to align the common factors vertically.

The greatest common factor in Example 2 could be written as 2^3. The exponent 3 shows the number of times the factor 2 appears in the factorization of each number.

> **Quick ✓**
>
> 1. The largest expression that divides evenly into a list of polynomials is called the
>   _____ _____ ____.
>
> 2. In the product $(3x - 2)(x + 4) = 3x^2 + 10x - 8$, the polynomials $(3x - 2)$ and $(x + 4)$ are called _____ of the polynomial $3x^2 + 10x - 8$.
>
> *In Problems 3–5, find the greatest common factor of each list of numbers.*
>
> **3.** 32, 40   **4.** 12, 45   **5.** 21, 35, 84

⊳ The approach to finding the greatest common factor of two or more expressions that contain variables is the same as it is for numbers. For example, x^3, x^5, and x^6 can be written as the product of factors of x as follows:

$$x^3 = x \cdot x \cdot x$$
$$x^5 = x \cdot x \cdot x \cdot x \cdot x$$
$$x^6 = x \cdot x \cdot x \cdot x \cdot x \cdot x$$

Work Smart

The GCF of a variable factor is the lowest power of that variable.

Each of the terms contains three factors of x, so the greatest common factor is x^3. It is no coincidence that the exponent of the GCF is 3, the smallest exponent of x^3, x^5, and x^6. This approach to finding the GCF for variable expressions will work in general.

EXAMPLE 3 **Finding the Greatest Common Factor**

Find the greatest common factor (GCF) of $8y^4$ and $12y^2$.

Solution

Step 1: Find the GCF of the coefficients, 8 and 12.

$$8 = 2 \cdot 2 \cdot 2$$
$$12 = 2 \cdot 2 \cdot 3$$

The GCF of the coefficients is $2 \cdot 2 = 4$.

Step 2: The variable factors are y^4 and y^2. For each variable, determine the smallest exponent that each variable is raised to. The GCF of y^4 and y^2 is y^2.

Step 3: Multiply the common factors from Steps 1 and 2 to get the GCF, $4y^2$.

> **Finding the Greatest Common Factor**
>
> **Step 1:** Find the GCF of the coefficients of each variable expression.
>
> **Step 2:** For each variable factor common to all the expressions, determine the smallest exponent that the variable factor is raised to.
>
> **Step 3:** Find the product of the common factors found in Steps 1 and 2. This expression is the GCF.

EXAMPLE 4 **Finding the Greatest Common Factor**

Find the GCF:

(a) $3x^3, 9x^2, 21x$ (b) $10x^5y^4, 15x^2y^3, 25x^3y^5$

Solution

Find the GCF of the coefficients, and then find the variable factor with the smallest exponent. The product of these two factors is the GCF.

(a) The coefficients, 3, 9, and 21, written as products of prime numbers, are

Factor the coefficients as products of primes:
$$3 = 3$$
$$9 = 3 \cdot 3$$
$$21 = 3 \cdot 7$$

The GCF of the coefficients is 3.
The smallest exponent of x^3, x^2, and x is 1, so the GCF of the variable factors is x. Therefore, the GCF of $3x^3, 9x^2$, and $21x$ is $3x$.

(b) The coefficients are 10, 15, and 25. We write these coefficients as products of prime numbers.

Factor the coefficients as products of primes:
$$10 = 5 \cdot 2$$
$$15 = 5 \cdot 3$$
$$25 = 5 \cdot 5$$

The GCF of the coefficients is 5.
The GCF of x^5, x^2, and x^3 is x^2. The GCF of y^4, y^3, and y^5 is y^3. Therefore, the GCF of $10x^5y^4, 15x^2y^3$, and $25x^3y^5$ is $5x^2y^3$.

Quick ✔

In Problems 6–8, find the greatest common factor (GCF).

6. $14y^3, 35y^2$ **7.** $6z^3, 8z^2, 12z$ **8.** $4x^3y^5, 8x^2y^3, 24xy^4$

The greatest common factor can be a binomial, as illustrated in the following example.

EXAMPLE 5 **The GCF as a Binomial**

Find the greatest common factor of each pair of expressions.

(a) $3(x - 1)$ and $8(x - 1)$ (b) $2(z + 3)(z + 5)$ and $4(z + 5)^2$

Solution

(a) The coefficients, 3 and 8, have no common factor. However, each expression has $x - 1$ as a factor, so the GCF of $3(x - 1)$ and $8(x - 1)$ is $x - 1$.

(b) The GCF of 2 and 4 is 2. The GCF of $(z + 3)(z + 5)$ and $(z + 5)^2$ is $z + 5$. The GCF of $2(z + 3)(z + 5)$ and $4(z + 5)^2$ is $2(z + 5)$.

Quick ✔

In Problems 9 and 10, find the greatest common factor (GCF).

9. $7(2x + 3)$ and $-4(2x + 3)$ **10.** $9(k + 8)(3k - 2)$ and $12(k - 1)(k + 8)^2$

⊙ ❷ **Factor Out the Greatest Common Factor in Polynomials**

The first step in factoring any polynomial is to look for the greatest common factor of the terms of the polynomial. Then use the Distributive Property "in reverse" to factor the polynomial as shown below.

$$ab + ac = a(b + c) \quad \text{or} \quad ab - ac = a(b - c)$$

This process is called "factoring out" the greatest common factor.

EXAMPLE 6 **How to Factor Out the Greatest Common Factor in a Polynomial**

Factor $2x - 10$ by factoring out the greatest common factor.

Step-by-Step Solution

Step 1: Find the GCF of the terms of the polynomial.   The GCF of the terms, $2x$ and -10, is 2.

Step 2: Rewrite each term as the product of the GCF and the remaining factor. $\qquad 2x - 10 = 2(x) - 2(5)$

Step 3: Factor out the GCF. $\qquad\qquad\qquad\qquad\qquad = 2(x - 5)$

Step 4: Check $\qquad\qquad\qquad\qquad\qquad 2(x - 5) = 2(x) - 2(5)$
$$= 2x - 10$$

Therefore, $2x - 10 = 2(x - 5)$.

Work Smart

The Distributive Property comes in handy again for factoring out the GCF. Note how it is used in the check step, too.

> **Factoring a Polynomial Using the Greatest Common Factor**
>
> **Step 1:** Identify the greatest common factor (GCF) of the terms of the polynomial.
>
> **Step 2:** Rewrite each term as the product of the GCF and the remaining factor.
>
> **Step 3:** Use the Distributive Property "in reverse" to factor out the GCF.
>
> **Step 4:** Check using the Distributive Property.

EXAMPLE 7 **Factoring Out the Greatest Common Factor in a Binomial**

Factor the binomial $9z^3 + 36z^2$ by factoring out the greatest common factor.

Solution

The greatest common factor of 9 and 36 is 9. The greatest common factor of z^3 and z^2 is z^2. Therefore, the GCF is $9z^2$.

Rewrite each term as the product of the GCF and the remaining factor: $\quad 9z^3 + 36z^2 = 9z^2(z) + 9z^2(4)$
$$= 9z^2(z + 4)$$

Check $\qquad 9z^2(z + 4) = 9z^2(z) + 9z^2(4)$
$$= 9z^3 + 36z^2$$

So $9z^3 + 36z^2 = 9z^2(z + 4)$.

EXAMPLE 8 **Factoring Out the Greatest Common Factor in a Trinomial**

Factor the trinomial $6a^2b^2 - 8ab^3 + 18a^3b^4$ by factoring out the greatest common factor.

Solution

The GCF of $6a^2b^2 - 8ab^3 + 18a^3b^4$ is $2ab^2$. Now rewrite each term as the product of the GCF and the remaining factor.

$$6a^2b^2 - 8ab^3 + 18a^3b^4 = 2ab^2(3a) - 2ab^2(4b) + 2ab^2(9a^2b^2)$$

Factor out the GCF: $= 2ab^2(3a - 4b + 9a^2b^2)$

Check $\quad 2ab^2(3a - 4b + 9a^2b^2) = 2ab^2(3a) - 2ab^2(4b) + 2ab^2(9a^2b^2)$

$$= 6a^2b^2 - 8ab^3 + 18a^3b^4$$

Therefore, $6a^2b^2 - 8ab^3 + 18a^3b^4 = 2ab^2(3a - 4b + 9a^2b^2)$. •

> **Quick ✓**
>
> **11.** To _____ a polynomial means to write the polynomial as a product of two or more polynomials.
> **12.** When we factor a polynomial using the GCF, we use the _____ Property "in reverse."
>
> *In Problems 13–16, factor each polynomial by factoring out the greatest common factor.*
>
> **13.** $5z^2 + 30z$ **14.** $12p^2 - 12p$
> **15.** $16y^3 - 12y^2 + 4y$ **16.** $6m^4n^2 + 18m^3n^4 - 22m^2n^5$

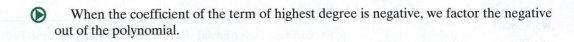 When the coefficient of the term of highest degree is negative, we factor the negative out of the polynomial.

EXAMPLE 9 **Factoring Out a Negative Greatest Common Factor**

Factor $-7a^3 + 14a$ by factoring out the greatest common factor.

Solution

This binomial is in standard form. Since the coefficient of the highest-degree term, $-7a^3$, is negative, we factor the negative out. Thus $-7a$ is the GCF.

$$-7a^3 + 14a = -7a(a^2) + (-7a)(-2)$$

Factor out the GCF: $= -7a(a^2 - 2)$

Check $\quad -7a(a^2 - 2) = -7a(a^2) + (-7a)(-2)$

$$= -7a^3 + 14a$$

Therefore, $-7a^3 + 14a = -7a(a^2 - 2)$. •

> **Quick ✓**
>
> *In Problems 17 and 18, factor out the greatest common factor.*
>
> **17.** $-4y^2 + 8y$ **18.** $-6a^3 + 12a^2 - 3a$

From Example 7, we know, the greatest common factor may be a binomial.

EXAMPLE 10 **Factoring Out a Binomial as the Greatest Common Factor**

Factor out the greatest common factor: $5x(x - 2) + 3(x - 2)$

Solution

Do you see that $x - 2$ is common to both terms? The GCF is the binomial $x - 2$.

$$5x(x - 2) + 3(x - 2) = 5x(x - 2) + 3(x - 2)$$

Factor out $x - 2$: $= (x - 2)(5x + 3)$

Check $(x - 2)(5x + 3) = (x - 2)5x + (x - 2)3$
$$= 5x(x - 2) + 3(x - 2)$$

Therefore, $5x(x - 2) + 3(x - 2) = (x - 2)(5x + 3)$. ●

Quick ✓

19. *True or False* We factor $(2x + 1)(x - 3) + (2x + 1)(2x + 7)$ as $(2x + 1)^2(3x + 4)$.

In Problems 20 and 21, factor out the greatest common factor.

20. $2a(a - 5) + 3(a - 5)$ **21.** $7z(z + 5) - 4(z + 5)$

Work Smart

Try factoring by grouping when a polynomial contains four terms.

▶ ❸ Factor Polynomials by Grouping

Sometimes a common factor does not occur in every term of the polynomial. If a polynomial contains four terms, it may be possible to find a GCF of the first two terms and a different GCF of the second two terms and have the same remaining factor from each pair. When this happens, the common factor can be factored out of each group of terms. This technique is called **factoring by grouping.**

EXAMPLE 11 **How to Factor by Grouping**

Factor by grouping: $3x - 3y + ax - ay$

Step-by-Step Solution

Step 1: Group terms with common factors. $3x - 3y + ax - ay = (3x - 3y) + (ax - ay)$

Step 2: In each grouping, factor out the GCF. $= 3(x - y) + a(x - y)$

Step 3: Factor out the common factor that remains.   Factor out $x - y$: $= (x - y)(3 + a)$

Step 4: Check   Multiply: $(x - y)(3 + a) = 3x + ax - 3y - ay$
Rearrange terms: $= 3x - 3y + ax - ay$

Therefore, $3x - 3y + ax - ay = (x - y)(3 + a)$. ●

Work Smart

We could have written the answer to Example 11 as $(3 + a)(x - y)$. Do you know why?

Based upon Example 11, we have the following steps for factoring by grouping.

Factoring a Polynomial By Grouping

Step 1: Group the terms with common factors.

Step 2: In each grouping, factor out the greatest common factor (GCF).

Step 3: If the remaining factor in each grouping is the same, factor it out.

Step 4: Check your work by finding the product of the factors.

Quick ✓

In Problem 22, factor by grouping.

22. $4x + 4y + bx + by$

▶ EXAMPLE 12   Factoring by Grouping

Work Smart

Be careful when the third term is negative (as in Example 12).

Factor by grouping: $5x - 5y - 4bx + 4by$

Solution

First, group terms with common factors. Be careful when grouping the third and fourth terms because of the minus sign with $-4bx$.

Work Smart

Notice we factor out $-4b$ in the second grouping. If we factor out $4b$ instead, we do not end up with the common factor, $x - y$.

$$5x - 5y - 4bx + 4by = 5x - 5y + (-4bx) + 4by$$
$$= (5x - 5y) + (-4bx + 4by)$$

Factor out the common factor in each group: $= 5(x - y) + (-4b)(x - y)$

Factor out the common factor that remains: $= (x - y)(5 - 4b)$

Check   Multiply: $(x - y)(5 - 4b) = 5x - 4bx - 5y + 4by$

Rearrange terms: $= 5x - 5y - 4bx + 4by$

Therefore, $5x - 5y - 4bx + 4by = (x - y)(5 - 4b)$.

In Example 12, we could have used the Commutative Property of Addition to rewrite $5x - 5y - 4bx + 4by$ as $5x - 4bx - 5y + 4by$ and factored as follows:

$$5x - 4bx - 5y + 4by = x(5 - 4b) - y(5 - 4b)$$
$$= (5 - 4b)(x - y)$$

Because $(5 - 4b)(x - y) = (x - y)(5 - 4b)$ by the Commutative Property of Multiplication, the answer is equivalent to the result of Example 12.

Work Smart

Sometimes we need to rearrange terms before factoring by grouping.

Quick ✓

In Problems 23 and 24, factor by grouping.

23. $6az - 2a - 9bz + 3b$ **24.** $8b + 4 - 10ab - 5a$

In any factoring problem, always look for a common factor first.

EXAMPLE 13   Factoring by Grouping

Factor: $3x^3 + 12x^2 - 6x - 24$

Solution

Do all four terms contain a common factor? Yes!! There is a GCF of 3, so we factor it out.

$$3x^3 + 12x^2 - 6x - 24 = 3(x^3 + 4x^2 - 2x - 8)$$

Work Smart

Whenever factoring, always look for a GCF first.

Group terms with common factors: $= 3[(x^3 + 4x^2) + (-2x - 8)]$

Factor out the common factor in each group: $= 3[x^2(x + 4) + (-2)(x + 4)]$

Factor out the common factor that remains: $= 3(x + 4)(x^2 - 2)$

Check $3(x + 4)(x^2 - 2) = 3(x^3 - 2x + 4x^2 - 8)$

Distribute the 3: $= 3x^3 - 6x + 12x^2 - 24$

Rearrange terms: $= 3x^3 + 12x^2 - 6x - 24$

Therefore, $3x^3 + 12x^2 - 6x - 24 = 3(x + 4)(x^2 - 2)$.

Quick ✔

In Problems 25 and 26, factor by grouping.

25. $3z^3 + 12z^2 + 6z + 24$ **26.** $2n^4 + 2n^3 - 4n^2 - 4n$

6.1 Exercises

 MyMathLab®   MathXL PRACTICE

Exercise numbers in green have complete video solutions in MyMathLab

Problems **1–26** are the **Quick ✔**s that follow the **EXAMPLES**.

Building Skills

In Problems 27–46, find the greatest common factor, GCF, of each group of expressions. See Objective 1.

27. 8, 12   **28.** 49, 35

29. 15, 14   **30.** 6, 55

31. 12, 36, 54   **32.** 16, 40, 64

33. x^{10}, x^2, x^8 **34.** y^3, y^5, y

35. $7x, 14x^3$ **36.** $8a^4, 20a^2$

37. $45a^2b^3, 75ab^2c$ **38.** $26xy^2, 39x^2y$

39. $4a^2bc^3, 6ab^2c^2, 8a^2b^2c^4$ **40.** $2x^2yz, xyz^2, 5x^3yz^2$

41. $3(x-1)$ and $10(x-1)$ **42.** $8(x+y)$ and $9(x+y)$

43. $2(x-4)^2$ and $4(x-4)^3$ **44.** $6(a-b)$ and $15(a-b)^3$

45. $12(x+2)(x-3)^2$ and $18(x-3)^2(x-2)$

46. $15(2a-1)^2(2a+1)$ and $18(2a-1)^3(2a+3)^2$

In Problems 47–70, factor the GCF from the polynomial. See Objective 2.

47. $12x - 18$ **48.** $3a + 6$

49. $x^2 + 12x$ **50.** $b^2 - 6b$

51. $5x^2y - 15x^3y^2$ **52.** $8a^3b^2 + 12a^5b^2$

53. $3x^3 + 6x^2 - 3x$ **54.** $5x^4 + 10x^3 - 25x^2$

55. $-5x^3 + 10x^2 - 15x$ **56.** $-2y^2 + 10y - 14$

57. $9m^5 - 18m^3 - 12m^2 + 81$ **58.** $5z^2 + 10z^4 - 15z^3 - 45z^5$

59. $-12z^3 + 16z^2 - 8z$ **60.** $-22n^4 + 18n^2 + 14n$

61. $10 - 5b - 15b^3$ **62.** $14m^2 - 16m^3 - 24m^4$

63. $15a^2b^4 - 60ab^3 + 45a^3b^2$ **64.** $12r^3s^2 + 3rs - 6rs^4$

65. $x(x-3) - 5(x-3)$ **66.** $a(a-5) + 6(a-5)$

67. $x^2(x-1) + y^2(x-1)$ **68.** $(b+2)a^2 - (b+2)b$

69. $x^2(4x+1) + 2x(4x+1) + 5(4x+1)$

70. $s^2(s^2+1) + 4s(s^2+1) + 7(s^2+1)$

In Problems 71–80, factor by grouping. See Objective 3.

71. $xy + 3y + 4x + 12$ **72.** $x^2 + ax + 2a + 2x$

73. $yz + z - y - 1$ **74.** $mn - 3n + 2m - 6$

75. $x^3 - x^2 + 2x - 2$ **76.** $z^3 + 4z^2 + 3z + 12$

77. $2t^3 - t^2 - 4t + 2$ **78.** $x^3 - x^2 - 5x + 5$

79. $2t^4 - t^3 - 6t + 3$ **80.** $6yz - 8y - 9z + 12$

Mixed Practice

In Problems 81–102, factor each polynomial.

81. $4y - 20$ **82.** $3z - 21$

83. $28m^3 + 7m^2 + 63m$ **84.** $10x^3 - 15x^2 - 5x$

85. $12m^3n^2p - 18m^2n$ **86.** $4s^2t^3 - 24st$

87. $(2p-1)(p+3) + (7p+4)(p+3)$

88. $(3x-2)(4x+1) + (3x-2)(x-10)$

89. $18ax - 9ay - 12bx + 6by$

90. $30xm + 15xn - 20ym - 10yn$

91. $(x-2)(x-3) + (x-2)$

92. $(a+2)(2b-1) - (2b-1)$

93. $15x^4 - 6x^3 + 30x^2 - 12x$

94. $2a^3 - 4a^2 + 8a - 16$ **95.** $-3x^3 + 6x^2 - 9x$

96. $-8z^4 - 12z^3 + 28z^2$ **97.** $-12b + 16b^2$

98. $-8a + 20a^2$ **99.** $12xy + 9x - 8y - 6$

100. $-12mxz - 3xz + 24mx + 6x$

101. $\dfrac{1}{3}x^3 - \dfrac{2}{9}x^2$ **102.** $\dfrac{3}{4}p^4 - \dfrac{1}{4}p^3$

Applying the Concepts

103. Height of a Toy Rocket The height of a toy rocket after t seconds, when it is fired straight up from the ground with an initial speed of 150 feet per second, is given by the polynomial $-16t^2 + 150t$. Write the polynomial $-16t^2 + 150t$ in factored form.

104. Height of a Ball The height of a ball after t seconds, when it is thrown straight up from a height of 48 feet with an initial speed of 80 feet per second, is given by the polynomial $-16t^2 + 80t + 48$. Write the polynomial $-16t^2 + 80t + 48$ in factored form.

105. Selling Calculators A manufacturer of calculators found that the number of calculators sold at a price of p dollars is given by the polynomial $21{,}000 - 150p$. Write $21{,}000 - 150p$ in factored form.

106. Revenue A manufacturer of a gas clothes dryer has found that the revenue (in dollars) from selling clothes dryers at a price of p dollars is given by the expression $-4p^2 + 4000p$. Write $-4p^2 + 4000p$ in factored form.

△ **107. Area of a Rectangle** A rectangle has an area of $(8x^5 - 28x^3)$ square feet. Write $8x^5 - 28x^3$ in factored form.

△ **108. Area of a Parallelogram** A parallelogram has an area of $(18n^4 - 15n^3 + 6n)$ square cm. Write $18n^4 - 15n^3 + 6n$ in factored form.

△ **109. Surface Area** The surface area of a cylindrical can whose radius is r inches and height is 4 inches is given by $S = (2\pi r^2 + 8\pi r)$ square inches. Express the surface area in factored form.

△ **110. Surface Area** The surface area of a cylindrical can whose radius is r inches and height is 8 inches is given by $S = (2\pi r^2 + 16\pi r)$ square inches. Express the surface area in factored form.

In Problems 111 and 112, the area of the polygon is given. Write a polynomial that represents the missing length.

△ **111.**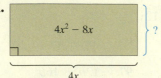

△ **112.**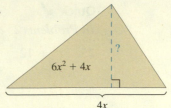

Extending the Concepts

In Problems 113–120, find the missing factor.

113. $8x^{3n} + 10x^n = 2x^n \cdot ?$

114. $3x^{2n+2} + 6x^{6n} = 3x^{2n} \cdot ?$

115. $3 - 4x^{-1} + 2x^{-3} = x^{-3} \cdot ?$

116. $1 - 3x^{-1} + 2x^{-2} = x^{-2} \cdot ?$

117. $x^2 - 3x^{-1} + 2x^{-2} = x^{-2} \cdot ?$

118. $2x^2 + x^{-3} - x^{-4} = x^{-4} \cdot ?$

119. $\dfrac{6}{35}x^4 - \dfrac{1}{7}x^2 + \dfrac{2}{7}x = \dfrac{2}{7}x \cdot ?$

120. $\dfrac{2}{15}x^3 - \dfrac{1}{9}x^2 + \dfrac{1}{3}x = \dfrac{1}{3}x \cdot ?$

Explaining the Concepts

121. Write a list of steps for finding the greatest common factor, and then write a second list of steps for factoring the GCF from a polynomial.

122. Describe how to factor a negative from a polynomial. What types of errors might happen during this process?

123. Explain the error in the following student's work:
$$3a(x + y) - 4b(x + y) = (3a - 4b)(x + y)^2$$

124. On a test, your answer for factoring
$$5z(x - 4) - 6(x - 4)$$
was $(x - 4)(5z - 6)$. Your friend Jack's answer was $(5z - 6)(x - 4)$. Explain why both answers are correct.

6.2 Factoring Trinomials of the Form x^2 bx c

Objectives

1 Factor Trinomials of the Form $x^2 + bx + c$

2 Factor Out the GCF, Then Factor $x^2 + bx + c$

Are you Ready for This Section?

Before getting started, take the following readiness quiz. If you get a problem wrong, go back to the section cited and review the material.

R1. Find two factors of 18 whose sum is 11.   [Section 1.2, pp. 8–9]

R2. Find two factors of -24 whose sum is -2.   [Section 1.4, pp. 30–33]

R3. Find two factors of -12 whose sum is 1.   [Section 1.4, pp. 30–33]

R4. Find two factors of 35 whose sum is -12.   [Section 1.4, pp. 30–33]

R5. Determine the coefficients of $3x^2 - x - 4$.   [Section 1.8, p. 66]

R6. Find the product: $-5p(p + 4)$ [Section 5.3, pp. 323–324]

R7. Find the product: $(z - 1)(z + 4)$ [Section 5.3, pp. 324–327]

R8. Find the degree: **(a)** $3x^2$ **(b)** $-z$ [Section 5.1, pp. 309–310]

In this section, we factor trinomials of degree 2. Because the word **quadratic** means "relating to a square," these trinomials are called *quadratic trinomials*.

> **Definition**
>
> A **quadratic trinomial** is a polynomial of the form $ax^2 + bx + c, a \neq 0$, where a represents the coefficient of the squared (second-degree) term, b represents the coefficient of the linear (first-degree) term, and c represents the constant.

In the trinomial $ax^2 + bx + c$, a is called the **leading coefficient.** We begin by looking at quadratic trinomials where the leading coefficient, a, is 1. Examples of quadratic trinomials whose leading coefficient is 1 are

$$x^2 + 4x + 3 \qquad a^2 + 4a - 21 \qquad p^2 - 11p + 18$$

1 Factor Trinomials of the Form $x^2 + bx + c$

The goal in factoring a second-degree trinomial is to write it as the product of two first-degree polynomials.

For example,

$$\text{Multiplication} \rightarrow$$
$$\text{Factored form} \rightarrow (x + 3)(x - 7) = x^2 - 4x - 21 \leftarrow \text{Product}$$
$$\leftarrow \text{Factoring}$$

The factors of $x^2 - 4x - 21$ are $x + 3$ and $x - 7$. Notice the following:

$$x^2 - 4x - 21 = (x + 3)(x - 7)$$

The sum of -7 and 3 is -4.   The product of -7 and 3 is -21.

In general, if $x^2 + bx + c = (x + m)(x + n)$, then $mn = c$ and $m + n = b$.

EXAMPLE 1   **How to Factor a Trinomial of the Form $x^2 + bx + c$, Where c Is Positive**

Factor: $x^2 + 7x + 12$

Step-by-Step Solution

Step 1: When we compare $x^2 + 7x + 12$ with $x^2 + bx + c$, we see that $b = 7$ and $c = 12$. Find factors of $c = 12$ whose sum is $b = 7$. Begin by listing all factors of 12 and computing the sum of these factors.

Factors of 12	1, 12	2, 6	3, 4	$-1, -12$	$-2, -6$	$-3, -4$
Sum	13	8	7	-13	-8	-7

We can see that $3 \cdot 4 = 12$ and $3 + 4 = 7$, so $m = 3$ and $n = 4$.

(continued)

Step 2: We write the trinomial in the form $(x + m)(x + n)$.

$$x^2 + 7x + 12 = (x + 3)(x + 4)$$

Step 3: Check

$$\underset{\text{F}}{(x + 3)}\underset{\text{O I L}}{(x + 4)} = x^2 + 4x + 3x + 3(4)$$
$$= x^2 + 7x + 12$$

Therefore, $x^2 + 7x + 12 = (x + 3)(x + 4)$. ●

Work Smart

Because multiplication is commutative, the order in which we list the factors does not matter.

> **Factoring a Trinomial of the Form $x^2 + bx + c$**
>
> **Step 1:** Find the pair of integers whose product is c and whose sum is b. That is, determine m and n such that $mn = c$ and $m + n = b$.
>
> **Step 2:** Write $x^2 + bx + c = (x + m)(x + n)$.
>
> **Step 3:** Check your work by multiplying the binomials.

In Example 1, notice that the coefficient of the middle term is positive and the constant is positive. **If the coefficient of the middle term and the constant are both positive, then both factors of the constant must be positive.**

EXAMPLE 2 — **Factoring a Trinomial of the Form $x^2 + bx + c$, Where c Is Positive**

Factor: $p^2 - 11p + 24$

Solution

Since $b = -11$ and $c = 24$, we want factors of 24 whose sum is -11. We begin by listing all factors of 24 and computing the sum of these factors.

Factors of 24	1, 24	2, 12	3, 8	4, 6	−1, −24	−2, −12	−3, −8	−4, −6
Sum	25	14	11	10	−25	−14	−11	−10

Because $-3 \cdot (-8) = 24$ and $-3 + (-8) = -11$, $m = -3$ and $n = -8$. Now write the trinomial in the form $(p + m)(p + n)$.

$$p^2 - 11p + 24 = (p + (-3))(p + (-8))$$
$$= (p - 3)(p - 8)$$

Check

$$(p - 3)(p - 8) = p^2 - 8p - 3p + (-3)(-8)$$
$$= p^2 - 11p + 24 \quad ●$$

Work Smart

The sum of two even numbers is even, the sum of two odd numbers is even, the sum of an odd and an even number is odd. In Example 2, the coefficient of the middle term is odd (-11), so one of the factors of 24 will be odd, and the other will be even.

Use this result on the remaining examples to reduce the list of possible factors.

In Example 2, notice that the coefficient of the middle term is negative and the constant is positive. **If the coefficient of the middle term is negative and the constant is positive, then both factors of the constant must be negative.**

Quick ✔

1. A _____ _____ is a polynomial of the form $ax^2 + bx + c$, $a \neq 0$.

2. When factoring $x^2 - 10x + 24$, look for two numbers whose product is ___ and whose sum is ___.

3. *True or False* $4 + 4x + x^2$ has a leading coefficient of 4.

In Problems 4 and 5, factor each trinomial.

4. $y^2 + 9y + 20$

5. $z^2 - 9z + 14$

EXAMPLE 3 **Factoring a Trinomial of the Form $x^2 + bx + c$, Where c Is Negative**

Factor: $z^2 + 3z - 28$

Solution

Since $b = 3$ and $c = -28$, look for factors of -28 whose sum is 3. Because c is negative, we know that one of the factors must be positive and the other negative. And because b is odd, one factor must be odd and the other even.

Factors of -28	$-1, 28$	$-4, 7$	$1, -28$	$4, -7$
Sum	27	3	-27	-3

We can see that $-4 \cdot 7 = -28$ and $-4 + 7 = 3$, so $m = -4$ and $n = 7$. Write the trinomial in the form $(z + m)(z + n)$.

$$z^2 + 3z - 28 = (z + (-4))(z + 7)$$
$$= (z - 4)(z + 7)$$

Check $\quad (z - 4)(z + 7) = z^2 + 7z - 4z + (-4)(7)$
$$= z^2 + 3z - 28$$

In Example 3, notice that the coefficient of the middle term is positive and the constant is negative. **If the constant is negative, then the factors of the constant must have opposite signs.** In addition, **if the coefficient of the middle term is positive, then the factor with the larger absolute value must be positive.**

EXAMPLE 4 **Factoring a Trinomial of the Form $x^2 + bx + c$, Where c Is Negative**

Factor: $a^2 - 7a - 18$

Solution

Find factors of $c = -18$ whose sum is $b = -7$. We begin by listing all factors of -18 and computing their sum.

Factors of -18	$-1, 18$	$-2, 9$	$-3, 6$	$1, -18$	$2, -9$	$3, -6$
Sum	17	7	3	-17	-7	-3

Because $2 \cdot (-9) = -18$ and $2 + (-9) = -7$, $m = 2$ and $n = -9$. Write the trinomial in the form $(a + m)(a + n)$.

$$a^2 - 7a - 18 = (a + 2)(a + (-9))$$
$$= (a + 2)(a - 9)$$

Check $\quad (a + 2)(a - 9) = a^2 - 9a + 2a + 2(-9)$
$$= a^2 - 7a - 18$$

In Example 4, notice that the coefficient of the middle term is negative and the constant is also negative. **If the coefficient of the middle term is negative and the constant is negative, then the factors of the constant must have opposite signs, and the factor with the larger absolute value must be negative.**

Quick ✓

6. *True or False* When factoring $x^2 - 10x - 24$, the factors of 24 must have opposite signs.

In Problems 7 and 8, factor each trinomial.

7. $y^2 - 2y - 15$

8. $w^2 + w - 12$

Table 1 summarizes the four possibilities for factoring a quadratic trinomial in the form $x^2 + bx + c = (x + m)(x + n)$.

Table 1 Factoring $x^2 + bx + c$

Signs of b and c	Signs of m and n	Example
b and c are both positive	Both factors are positive.	$x^2 + 3x + 2 = (x + 2)(x + 1)$
b is negative and c is positive	Both factors are negative.	$a^2 - 7a + 12 = (a - 4)(a - 3)$
b is positive and c is negative	Factors have opposite signs; the factor with the larger absolute value is positive.	$y^2 + 2y - 24 = (y + 6)(y - 4)$
b is negative and c is negative	Factors have opposite signs; the factor with the larger absolute value is negative.	$b^2 - 4b - 21 = (b - 7)(b + 3)$

Definition

A polynomial that cannot be written as the product of two other polynomials (other than 1 or -1) is a **prime polynomial.**

EXAMPLE 5 **Identifying a Prime Trinomial**

Show that $y^2 + 10y + 12$ is prime.

Solution

Look for factors of $c = 12$ whose sum is $b = 10$. Because both b and c are positive, the factors of 12 must both be positive, so we list only positive factors of 12 and the sum of these factors.

Factors of 12	1, 12	2, 6	3, 4
Sum	13	8	7

There are no factors of 12 whose sum is 10. Therefore, $y^2 + 10y + 12$ is prime. ●

Quick ✓

9. A polynomial that cannot be written as the product of two other polynomials (other than 1 or -1) is a _____ polynomial.

In Problems 10 and 11, factor the trinomial. If the trinomial cannot be factored, state that it is prime.

10. $z^2 - 5z + 8$ 11. $q^2 + 4q - 45$

If a trinomial has more than one variable, we take the same approach that we used for trinomials in one variable, so trinomials of the form

$$x^2 + bxy + cy^2$$

factor as

$$(x + my)(x + ny)$$

where

$$mn = c \quad \text{and} \quad m + n = b$$

EXAMPLE 6 **Factoring Trinomials in Two Variables**

Factor: $p^2 + 4pq - 21q^2$

Solution

The trinomial $p^2 + 4pq - 21q^2$ factors as $(p + mq)(p + nq)$, where $mn = -21$ and $m + n = 4$. Put another way, we want factors of -21 whose sum is 4. Also, because c is negative and b is positive, the factors of -21 have opposite signs, and the factor with the larger absolute value is positive.

Factors of -21	$-1, 21$	$-3, 7$
Sum	20	4

Because $-3 \cdot 7 = -21$ and $-3 + 7 = 4$, $m = -3$ and $n = 7$. Write the trinomial in the form $(p + mq)(p + nq)$.

$$p^2 + 4pq - 21q^2 = (p + (-3)q)(p + 7q)$$
$$= (p - 3q)(p + 7q)$$

Check $(p - 3q)(p + 7q) = p^2 + 7pq - 3pq - 21q^2$
$$= p^2 + 4pq - 21q^2$$

●

Quick ✓
In Problems 12 and 13, factor each trinomial.

12. $x^2 + 9xy + 20y^2$ **13.** $m^2 + mn - 42n^2$

EXAMPLE 7 **Factoring a Trinomial Not Written in Standard Form**

Factor: $2w + w^2 - 8$

Solution

First write the trinomial in standard form (descending order of degree) and then factor.

$$2w + w^2 - 8 = w^2 + 2w - 8$$

Look for factors of $c = -8$ whose sum is $b = 2$. Since the coefficient of the middle term is positive and the constant term is negative, the factor with the larger absolute value is positive.

Factors of -8	$-1, 8$	$-2, 4$
Sum	7	2

Because $-2 \cdot 4 = -8$ and $-2 + 4 = 2$, $m = -2$ and $n = 4$. Therefore,

$$2w + w^2 - 8 = w^2 + 2w - 8$$
$$= (w - 2)(w + 4)$$

Check $(w - 2)(w + 4) = w^2 + 4w - 2w + (-2)(4)$
$$= w^2 + 2w - 8$$

●

Quick ✓

14. Write each trinomal in standard form.

(a) $-12 - x + x^2$ (b) $9n + n^2 - 10$

In Problems 14 and 15, factor each polynomial.

15. $-56 + n^2 + n$ **16.** $y^2 + 35 - 12y$

▶ ❷ Factor Out the GCF, Then Factor $x^2 + bx + c$

Some algebraic expressions can be factored as trinomials in the form $x^2 + bx + c$ after we factor out a greatest common factor.

EXAMPLE 8 **Factoring Trinomials with a Common Factor**

Factor: $3u^3 - 21u^2 - 90u$

Solution

Did you notice that the trinomial has a GCF of $3u$? Factor the $3u$ out:

$$3u^3 - 21u^2 - 90u = 3u(u^2 - 7u - 30)$$

Now factor the trinomial in parentheses, $u^2 - 7u - 30$. Look for factors of $c = -30$ whose sum is $b = -7$. Because c is negative and b is negative, the factor of -30 with the larger absolute value is negative.

Factors of -30	$1, -30$	$2, -15$	$3, -10$	$5, -6$
Sum	-29	-13	-7	-1

Because $3(-10) = -30$ and $3 + (-10) = -7$, $m = 3$ and $n = -10$. Write the trinomial in the form $(u + m)(u + n)$.

$$u^2 - 7u - 30 = (u + 3)(u - 10)$$

Thus we have

$$3u^3 - 21u^2 - 90u = 3u(u^2 - 7u - 30)$$
$$= 3u(u + 3)(u - 10)$$

Work Smart

We could have checked the result of Example 8 as follows:

$3u(u + 3)(u - 10)$
$= (3u^2 + 9u)(u - 10)$
$= 3u^3 - 30u^2 + 9u^2 - 90u$
$= 3u^3 - 21u^2 - 90u$

Check $\overset{\text{F}\quad\text{O}\quad\text{I}\quad\text{L}}{3u(u + 3)(u - 10)} = 3u(u^2 - 10u + 3u - 30)$

Combine like terms: $= 3u(u^2 - 7u - 30)$

Distribute: $= 3u^3 - 21u^2 - 90u$ ●

We say a polynomial is **factored completely** if each factor in the final factorization is prime. For example, $3x^2 - 6x - 45 = (x - 5)(3x + 9)$ is not factored completely because $3x + 9$ has a common factor of 3. However, $3x^2 - 6x - 45 = 3(x - 5)(x + 3)$ is factored completely.

Quick ✓

17. *True or False* The polynomial $(x - 3)(3x + 6)$ is factored completely.

In Problems 18–20, completely factor each trinomial.

18. $4m^2 - 16m - 84$ **19.** $3z^3 + 12z^2 - 15z$

20. $3a^2b^2 - 24ab^2 + 36b^2$

Sometimes the leading coefficient is negative. If this is the case, factor out the negative to make factoring easier.

EXAMPLE 9 **Factoring Trinomials with a Negative Leading Coefficient**

Factor completely: $-w^2 - 5w + 24$

Solution

Notice that the leading coefficient is -1. It is easier to factor a trinomial when the leading coefficient is positive, so use -1 as the GCF and rewrite $-w^2 - 5w + 24$ as

$$-w^2 - 5w + 24 = -1(w^2 + 5w - 24)$$

Work Smart

It's easier to factor a trinomial in standard form when the leading coefficient is positive.

To factor $w^2 + 5w - 24$, look for two integers whose product is -24 and whose sum is 5. We know that one factor will be positive and the other negative. Because the coefficient of the middle term is odd, one of the factors of -24 is odd and the other is even. Lastly, the factor of -24 with the larger absolute value is positive.

Factors of -24	$-1, 24$	$-3, 8$
Sum	23	5

Factor $w^2 + 5w - 24$ as $(w - 3)(w + 8)$. But remember that we already factored out the greatest common factor, -1, so

$$\begin{aligned} -w^2 - 5w + 24 &= -1(w^2 + 5w - 24) \\ &= -1(w - 3)(w + 8) \\ &= -(w - 3)(w + 8) \end{aligned}$$

We leave the check to you.

Quick ✓

In Problems 21 and 22, factor each trinomial completely.

21. $-w^2 - 3w + 10$

22. $-2a^3 - 8a^2 + 24a$

6.2 Exercises MyMathLab® Math XL PRACTICE

Exercise numbers in **green** have complete video solutions in MyMathLab

Problems 1–22 are the Quick ✓s that follow the EXAMPLES.

Building Skills

In Problems 23–44, factor each trinomial completely. If the trinomial cannot be factored, say it is prime. See Objective 1.

23. $x^2 + 5x + 6$

24. $p^2 + 7p + 6$

25. $m^2 + 9m + 18$

26. $n^2 + 12n + 20$

27. $x^2 - 15x + 36$

28. $z^2 - 13z + 36$

29. $p^2 - 8p + 12$

30. $z^2 - 7z + 12$

31. $-x - 12 + x^2$

32. $-8y - 9 + y^2$

33. $x^2 + 6x - 20$

34. $t^2 + 2t - 38$

35. $z^2 + 12z - 45$

36. $y^2 + 6y - 40$

37. $x^2 - 5xy + 6y^2$

38. $x^2 - 14xy + 24y^2$

39. $r^2 + rs - 6s^2$

40. $p^2 + 5pq - 14q^2$

41. $x^2 - 3xy - 4y^2$

42. $x^2 - 9xy - 36y^2$

43. $z^2 + 7zy + 8y^2$

44. $m^2 + 16mn + 18n^2$

In Problems 45–58, factor each trinomial completely by factoring out the GCF first and then factoring the resulting trinomial. See Objective 2.

45. $3x^2 + 3x - 6$

46. $5x^2 + 30x + 40$

47. $3n^3 - 24n^2 + 45n$

48. $4p^4 - 4p^3 - 8p^2$

49. $5x^2z - 20xz - 160z$

50. $8x^2z^2 - 56xz^2 + 80z^2$

51. $-3x^2 + x^3 - 18x$

52. $30x - 2x^2 - 100$

53. $-2y^2 + 8y - 8$

54. $-3x^2 - 24x - 48$

55. $4x^3 - 32x^2 + x^4$

56. $-75x + x^3 + 10x^2$

57. $2x^2 + x^3 - 15x$

58. $x^3 - 20x - 8x^2$

Mixed Practice

In Problems 59–92, factor each polynomial completely. If the polynomial cannot be factored, say it is prime.

59. $x^2 - 3xy - 28y^2$

60. $x^2 - 9xy - 36y^2$

61. $x^2 + x + 6$

62. $t^2 - 8t - 7$

63. $k^2 - k - 20$

64. $x^2 - 2x - 35$

65. $2w^3 - 8w^2 - 12w$

66. $4d^4 + 28d^3 + 32d^2$

67. $s^2t^2 - 8st + 15$

68. $x^2y^2 + 3xy + 2$

69. $-3p^3 + 3p^2 + 6p$

70. $-2z^3 - 2z^2 + 24z$

71. $-x^2 - 6x - 9$

72. $-r^2 - 12r - 36$

73. $g^2 - 4g + 21$

74. $x^2 - x + 6$

75. $n^4 - 30n^2 - n^3$

76. $-16x + x^3 - 6x^2$

77. $2x^3 - 10x^2 + 6x - 30$

78. $3a^4 - 12a^3 - 6a^2 + 24a$

79. $35 + 12s + s^2$

80. $25 + 10x + x^2$

81. $n^2 - 9n - 45$

82. $x^2 - 6x - 42$

83. $m^2n^2 - 8mn + 12$

84. $x^2y^2 - 3xy - 18$

85. $-x^3 + 12x^2 + 28x$

86. $-3r^3 + 6r^2 - 3r$

87. $y^2 - 12y + 36$

88. $x^2 - 14x + 49$

89. $-36x + 20x^2 - 2x^3$

90. $-20mn^2 + 30m^2n - 5m^3$

91. $-21x^3y - 14xy^2$

92. $-12x^2y + 8xy^3$

Applying the Concepts

93. Punkin Chunkin In a recent Punkin Chunkin contest, in which a pumpkin is shot in the air with a homemade cannon, the height of a pumpkin after t seconds was given by the trinomial $(-16t^2 + 64t + 80)$ feet. Write this polynomial in factored form.

94. Punkin Chunkin In a recent Punkin Chunkin contest, in which a pumpkin is shot into the air with a catapult, the height of a pumpkin after t seconds was given by the trinomial $(-16t^2 + 16t + 32)$ feet. Write this polynomial in factored form.

△ **95. A Rectangular Field** The trinomial $(x^2 + 9x + 18)$ square meters represents the area of a rectangular field. Find two binomials that might represent the length and width of the field.

△ **96. Another Rectangular Field** The trinomial $(x^2 + 6x + 8)$ square yards represents the area of a rectangular field. Find two binomials that might represent the length and width of the field.

△ **97. Area of a Triangle** The area of a triangle is given by the trinomial $\left(\frac{1}{2}x^2 + x - \frac{15}{2}\right)$ square inches. Find algebraic expressions that might represent the base and height of the triangle. (*Hint:* Factor out $\frac{1}{2}$ as a common factor.)

△ **98. Area of a Triangle** The area of a triangle is given by the trinomial $\left(\frac{1}{2}x^2 + 5x + 12\right)$ square kilometers. Find algebraic expressions that might represent the base and height of the triangle. (*Hint:* Factor out $\frac{1}{2}$ as a common factor.)

Extending the Concepts

In Problems 99–102, find all possible values of b so that the trinomial is factorable.

99. $x^2 + bx + 6$

100. $x^2 + bx - 10$

101. $x^2 + bx - 21$

102. $x^2 + bx + 12$

In Problems 103–106, find all possible positive values of c so that the trinomial is factorable.

103. $x^2 - 2x + c$

104. $x^2 - 3x + c$

105. $x^2 - 7x + c$

106. $x^2 + 6x + c$

Explaining the Concepts

107. The answer key to a homework assignment in which you were asked to factor $10 - 3x - x^2$ states the factored form is $-(x + 5)(x - 2)$, but you found the factored form to be $(5 + x)(2 - x)$. Is your answer correct? Explain.

108. The answer key to your algebra exam said the factored form of $2 - 3x + x^2$ is $(x - 1)(x - 2)$.

You have $(1 - x)(2 - x)$ on your paper. Is your answer correct or incorrect? Explain your reasoning.

109. In the trinomial $x^2 + bx + c$, both b and c are negative. Explain how to factor the trinomial. Make up a trinomial and then use your rules to factor it.

6.3 Factoring Trinomials of the Form $ax^2 + bx + c, a \neq 1$

Objectives

❶ Factor $ax^2 + bx + c, a \neq 1$, Using Grouping

❷ Factor $ax^2 + bx + c, a \neq 1$, Using Trial and Error

Are You Ready For This Section?

Before getting started, take the following readiness quiz. If you get a problem wrong, go back to the section cited and, review the material.

R1. List the prime factorization of 24. [Section 1.2, pp. 8–9]

R2. Determine the coefficients of $5x^2 - 3x + 7$. [Section 1.8, pp. 65–66]

R3. Find the product: $(2x + 7)(3x - 1)$ [Section 5.3, pp. 324–327]

In this section, we factor trinomials of the form $ax^2 + bx + c$ in which the leading coefficient, a, is not 1. Examples of trinomials of the form $ax^2 + bx + c, a \neq 1$, are

$$2x^2 + 3x + 1 \qquad 5y^2 - y - 4 \qquad 10z^2 - 7z + 6$$

Let's begin by reviewing multiplication of binomials using FOIL.

Factored Form	F O I L	Polynomial
$(2x + 3)(x + 4) =$	$2x^2 + 8x + 3x + 12 =$	$2x^2 + 11x + 12$
$(3x - 4)(x + 7) =$	$3x^2 + 21x - 4x - 28 =$	$3x^2 + 17x - 28$
$(5m - 2n)(3m - n) =$	$15m^2 - 5mn - 6mn + 2n^2 =$	$15m^2 - 11mn + 2n^2$

To factor a trinomial, we reverse the FOIL multiplication process. For example, a trinomial such as $2x^2 + 11x + 12$ will be written as $(2x + 3)(x + 4)$. To factor trinomials in this form, we have two methods that can be used.

1. Factoring by grouping

2. Trial and error

Each method has pros and cons, which we will point out as we proceed.

❶ Factor $ax^2 + bx + c, a \neq 1$, Using Grouping

Ready?...Answers **R1.** $2^3 \cdot 3$
R2. $5, -3, 7$ **R3.** $6x^2 + 19x - 7$

We begin by showing how to factor trinomials of the form $ax^2 + bx + c, a \neq 1$, by grouping.

EXAMPLE 1 **How to Factor $ax^2 + bx + c, a \neq 1$, by Grouping**

Factor: $3x^2 + 14x + 15$

Step-by-Step Solution

First, notice that $3x^2 + 14x + 15$ has no common factors. In this trinomial, $a = 3$, $b = 14$, and $c = 15$.

(continued)

Step 1: Find the value of ac.

The value of $a \cdot c = 3 \cdot 15 = 45$.

Step 2: Find the pair of integers, m and n, whose product is ac and whose sum is b.

We want to find integer factors of 45 whose sum is 14. Because both 14 and 45 are positive, we list only the positive factors of 45.

Factors of 45	1, 45	3, 15	5, 9
Sum	46	18	14

Step 3: Write $ax^2 + bx + c$ as $ax^2 + mx + nx + c$.

Write $3x^2 + 14x + 15$ as $3x^2 + 9x + 5x + 15$

$14x = 9x + 5x$

Step 4: Factor the expression in Step 3 by grouping.

$$3x^2 + 9x + 5x + 15 = (3x^2 + 9x) + (5x + 15)$$

Common factor in 1st group: $3x$;
common factor in 2nd group: 5: $= 3x(x + 3) + 5(x + 3)$

Factor out $x + 3$: $= (x + 3)(3x + 5)$

Step 5: Check Multiply the factored form.

$$(x + 3)(3x + 5) = 3x^2 + 5x + 9x + 15$$
$$= 3x^2 + 14x + 15$$

Therefore, $3x^2 + 14x + 15 = (x + 3)(3x + 5)$.   ●

We summarize below the steps used in Example 1.

Work Smart

Notice the title in the steps to factor $ax^2 + bx + c$, $a \neq 1$, by grouping. It specifies *no common factors*. If a polynomial has a common factor, the first step is to factor it out!

> **Factoring $ax^2 + bx + c, a \neq 1$, By Grouping, Where a, b, and c Have No Common Factors**
>
> **Step 1:** Find the value of ac.
> **Step 2:** Find the pair of integers, m and n, whose product is ac and whose sum is b.
> **Step 3:** Write $ax^2 + bx + c = ax^2 + mx + nx + c$.
> **Step 4:** Factor the expression in Step 3 by grouping.
> **Step 5:** Check by multiplying the factors.

EXAMPLE 2   **Factoring $ax^2 + bx + c, a \neq 1$, by Grouping**

Factor: $12x^2 - x - 1$

Solution

There is no common factor in $12x^2 - x - 1$ and $a = 12$, $b = -1$, and $c = -1$.

The value of $a \cdot c$ is $12(-1) = -12$. We want to find factors of -12 whose sum is -1. Because the product $a \cdot c$ is -12, one factor must be positive and the other negative. Since $b = -1$, we know the factor of -12 with the larger absolute value will be negative.

Factors of -12	1, -12	2, -6	3, -4
Sum	-11	-4	-1

The factors of -12 whose sum is -1 are 3 and -4.

Write $12x^2 - x - 1$ as $12x^2 + 3x - 4x - 1$, and factor by grouping.

$$-x = 3x - 4x$$

$$12x^2 + 3x - 4x - 1 = (12x^2 + 3x) + (-4x - 1)$$

Common factor in 1st group: $3x$;
common factor in 2nd group: -1: $= 3x(4x + 1) - 1(4x + 1)$
Factor out $4x + 1$: $= (4x + 1)(3x - 1)$

Check $(4x + 1)(3x - 1) = 12x^2 - 4x + 3x - 1$
$= 12x^2 - x - 1$

Work Smart

In Example 2, we could have written
$12x^2 - x - 1$ as $12x^2 - 4x + 3x - 1$
and obtained the same result:
$12x^2 - 4x + 3x - 1$
$= (12x^2 - 4x) + (3x - 1)$
$= 4x(3x - 1) + 1(3x - 1)$
$= (3x - 1)(4x + 1)$

Quick ✓

1. When factoring $6x^2 + x - 1$ using grouping, $ac = $ ___ and $b = $ __.

2. To factor $2x^2 - 13x + 6$ using grouping, begin by finding factors whose product is __ and sum is __.

In Problems 3 and 4, factor each trinomial completely by grouping.

3. $3x^2 - 2x - 8$ 4. $10z^2 + 21z + 9$

The advantage of factoring trinomials of the form $ax^2 + bx + c$, $a \neq 1$, by grouping is that it is algorithmic (that is, step by step). However, if the product $a \cdot c$ is large, then there are a lot of factors of ac whose sum must be determined. This can get overwhelming. Under these circumstances, it may be better to employ trial and error, which we discuss in Objective 2 of this section.

EXAMPLE 3   **Factoring a Trinomial with a Negative Leading Coefficient by Grouping**

Factor: $-18x^2 + 33x + 30$

Solution

First we ask, "Is there a greatest common factor in the expression?" Remember, if the leading coefficient is negative, factor the negative out as part of the GCF. Thus the GCF is -3, which we factor out.

$$-18x^2 + 33x + 30 = -3(6x^2 - 11x - 10)$$

Now factor $6x^2 - 11x - 10$ by grouping. We see that $a = 6$, $b = -11$, and $c = -10$. The value of $a \cdot c = 6(-10) = -60$.

What factors of -60 have a sum of -11?

Factors of -60	$1, -60$	$2, -30$	$3, -20$	$4, -15$	$5, -12$	$6, -10$
Sum	-59	-28	-17	-11	-7	-4

The factors of -60 whose sum is -11 are 4 and -15.
Write $6x^2 - 11x - 10$ as $6x^2 + 4x - 15x - 10$.

$$-11x = 4x - 15x$$

Now, factor by grouping.

$$6x^2 + 4x - 15x - 10 = (6x^2 + 4x) + (-15x - 10)$$

Common factor in 1st group: $2x$;
common factor in 2nd group: -5: $= 2x(3x + 2) - 5(3x + 2)$
Factor out $3x + 2$: $= (3x + 2)(2x - 5)$

Check Don't forget, we factored out a GCF of -3.

$$-3(3x + 2)(2x - 5) = -3(6x^2 - 15x + 4x - 10)$$
$$= -3(6x^2 - 11x - 10)$$
$$= -18x^2 + 33x + 30$$

So, $-18x^2 + 33x + 30 = -3(3x + 2)(2x - 5)$.

Quick ✓

In Problems 5 and 6, factor the trinomial completely by grouping.

5. $24x^2 + 6x - 9$ **6.** $-10n^2 + 17n - 3$

EXAMPLE 4   **Factoring Trinomials with Two Variables by Grouping**

Factor: $12x^2 + xy - 6y^2$

Solution

There is no common factor in this trinomial. Factor $12x^2 + xy - 6y^2$ by grouping with $a = 12$, $b = 1$, and $c = -6$. The value of $a \cdot c = 12(-6) = -72$.
    What factors of -72 have a sum of 1?

Factors of -72	$72, -1$	$36, -2$	$24, -3$	$18, -4$	$12, -6$	$9, -8$
Sum	71	34	21	14	6	1

The factors of -72 whose sum is 1 are 9 and -8. Write

$$12x^2 + xy - 6y^2 \qquad \text{as} \qquad 12x^2 + 9xy - 8xy - 6y^2$$

$$xy = 9xy - 8xy$$

Now factor by grouping.

$$12x^2 + 9xy - 8xy - 6y^2 = (12x^2 + 9xy) + (-8xy - 6y^2)$$

Common factor in 1st group: $3x$;
Common factor in 2nd group: $-2y$: $= 3x(4x + 3y) - 2y(4x + 3y)$
Factor out $(4x + 3y)$: $= (4x + 3y)(3x - 2y)$

Check $(4x + 3y)(3x - 2y) = 12x^2 - 8xy + 9xy - 6y^2$
$$= 12x^2 + xy - 6y^2$$

Quick ✓

In Problems 7 and 8, factor the trinomial completely by grouping.

7. $6x^2 + 23xy + 21y^2$ **8.** $-36a^2 + 21ab + 30b^2$

▶ ❷ **Factor $ax^2 + bx + c$, $a \neq 1$, Using Trial and Error**

In the trial and error method, you list various binomials, and multiply to find their product until you find the binomials whose product is the original trinomial. This method may sound haphazard, but experience and logic help to minimize the number of possibilities you must try before the factored form is found.

EXAMPLE 5 **How to Factor $ax^2 + bx + c, a \neq 1$, Using Trial and Error**

Factor: $2x^2 + 7x + 5$

Step-by-Step Solution

Step 1: List the possibilities for the first terms of each binomial whose product is ax^2.

There is only one way to represent the first term, $2x^2$, since 2 is a prime number.

$$(2x + \underline{})(x + \underline{})$$

Step 2: List the possibilities for the last terms of each binomial whose product is c.

The last term, 5, is also prime. It has the factors $(1)(5)$ and $(-1)(-5)$. Notice that the coefficient of x, 7, is positive. Thus, to produce a positive sum, $7x$, we must have two positive factors. Therefore, we exclude the factors $(-1)(-5)$ in our trials.

Step 3: Write out all the combinations of factors from Steps 1 and 2. Multiply the binomials until you find a product that equals the trinomial.

Possible Factorization of $2x^2 + 7x + 5$	Product
$(2x + 1)(x + 5)$	$2x^2 + 11x + 5$
$(2x + 5)(x + 1)$	$2x^2 + 7x + 5$

The second row is the factorization that works, so $2x^2 + 7x + 5 = (2x + 5)(x + 1)$. ●

We summarize the steps used in Example 5 below.

> **Factoring $ax^2 + bx + c, a \neq 1$, Using Trial and Error, Where a, b, and c Have No Common Factors**
>
> **Step 1:** List the possibilities for the first terms of each binomial whose product is ax^2.
> $$(\square x + \underline{})(\square x + \underline{}) = ax^2 + bx + c$$
> **Step 2:** List the possibilities for the last terms of each binomial whose product is c.
> $$(\underline{}x + \square)(\underline{}x + \square) = ax^2 + bx + c$$
> **Step 3:** Write out all the combinations of factors found in Steps 1 and 2. Multiply the binomials until a product is found that equals the trinomial.

> **Quick ✓**
>
> In Problems 9 and 10, factor each trinomial by trial and error.
>
> **9.** $3x^2 + 5x + 2$ **10.** $7y^2 + 22y + 3$

EXAMPLE 6 **Factoring $ax^2 + bx + c, a \neq 1$, Using Trial and Error**

Factor completely: $7x^2 - 18x + 8$

Solution

We list possible ways of representing the first term, $7x^2$. Since 7 is a prime number, we have

$$(7x + \underline{})(x + \underline{})$$

Let's look at the last term, 8, and list its factors:

Factors of 8			
1, 8	2, 4	$-1, -8$	$-2, -4$

Now let's concentrate on the middle term, $-18x$. To produce a negative sum, $-18x$, from a positive product, 8, we must have two *negative* factors. Therefore, we do not

Work Smart

When factoring using trial and error, it is only necessary to find the sum of the outer and inner products to see which factorization works.

include the factors $1 \cdot 8$ or $2 \cdot 4$ in our list of possible factors. We also highlight the sum of the "outer" and "inner" products.

Possible Factorization	Product
$(7x - 1)(x - 8)$	$7x^2 - 57x + 8$
$(7x - 8)(x - 1)$	$7x^2 - 15x + 8$
$(7x - 2)(x - 4)$	$7x^2 - 30x + 8$
$(7x - 4)(x - 2)$	$7x^2 - 18x + 8$

The highlighted row shows that $7x^2 - 18x + 8 = (7x - 4)(x - 2)$.

Quick ✔

In Problems 11 and 12, factor each trinomial completely by trial and error.

11. $3x^2 - 13x + 12$ **12.** $5p^2 - 21p + 4$

EXAMPLE 7 **Factoring $ax^2 + bx + c$, $a \neq 1$, Using Trial and Error**

Factor completely: $10x^2 - 13x - 3$

Solution

List possible ways of representing the first term, $10x^2$.

$$(10x + \underline{\hphantom{x}})(x + \underline{\hphantom{x}})$$
$$(5x + \underline{\hphantom{x}})(2x + \underline{\hphantom{x}})$$

The last term, -3, has factors $-1 \cdot 3$ or $1 \cdot -3$.

List the possible combinations of factors and show the sum of the "outer" and "inner" products in blue.

Possible Factorization	Product
$(10x - 1)(x + 3)$	$10x^2 + 29x - 3$
$(10x + 3)(x - 1)$	$10x^2 - 7x - 3$
$(10x - 3)(x + 1)$	$10x^2 + 7x - 3$
$(10x + 1)(x - 3)$	$10x^2 - 29x - 3$
$(5x - 1)(2x + 3)$	$10x^2 + 13x - 3$
$(5x + 3)(2x - 1)$	$10x^2 + x - 3$
$(5x - 3)(2x + 1)$	$10x^2 - x - 3$
$(5x + 1)(2x - 3)$	$10x^2 - 13x - 3$

The highlighted row shows that $10x^2 - 13x - 3 = (5x + 1)(2x - 3)$.

Quick ✔

In Problems 13 and 14, factor each trinomial completely by trial and error.

13. $2n^2 - 17n - 9$ **14.** $4w^2 - 5w - 6$

Factoring trinomials of the form $ax^2 + bx + c$, $a \neq 1$ by trial and error can at first seem overwhelming. There are so many possibilities! Plus, the technique seems haphazard.

Calling the technique "trial and error," however, is not quite truth in advertising because some thought is necessary to reduce the list of possible factors. To keep the list of possibilities small, ask yourself the following questions before you begin to factor.

Hints for Using Trial and Error to Factor $ax^2 + bx + c$, $a \neq 1$

- Are there any common factors? If so, then factor out the GCF. Is the leading coefficient negative? If so, then factor it out as part of the GCF.
- Is the constant c positive? If so, then the factors of the constant, c, must be the same sign as the coefficient of the middle term, b. For example,

$$2a^2 + 11a + 5 = (2a + 1)(a + 5)$$
$$2a^2 - 11a + 5 = (2a - 1)(a - 5)$$

- Is the constant c negative? If so, then the factors of c must have opposite signs. For example,

$$10b^2 + 19b - 15 = (5b - 3)(2b + 5)$$
$$10b^2 - 19b - 15 = (5b + 3)(2b - 5)$$

- If $ax^2 + bx + c$ has no common factor, then the binomials in the factored form cannot have common factors either.
- Is the value of b small? If so, then choose factors of a and factors of c that are close to each other. If the value of b is large, then choose factors of a and factors of c that are far from each other.
- Is the value of the middle term correct, but it has the wrong sign? Then switch the signs in the binomial factors.

EXAMPLE 8

Factoring $ax^2 + bx + c$, $a \neq 1$, Using Trial and Error

Factor completely: $18x^2 + 3x - 10$

Solution

Remember, the first step in any factoring problem is to look for common factors. There are no common factors in $18x^2 + 3x - 10$. Now, list all possible ways of representing $18x^2$.

$$(18x + \underline{\quad})(x + \underline{\quad})$$
$$(9x + \underline{\quad})(2x + \underline{\quad})$$
$$(6x + \underline{\quad})(3x + \underline{\quad})$$

Let's look at the last term, -10, and list its factors:

Factors of -10	$-10, 1$	$10, -1$	$-5, 2$	$5, -2$

Before listing the possible combinations of factors, ask some questions. Is the coefficient of the middle term of the polynomial $18x^2 + 3x - 10$ small? Yes, it is $+3$. Because the coefficient of the middle term is positive and small, the binomial factors we list should have outer and inner products that sum to a positive, small number. Therefore, we will start with $(6x + \underline{\quad})(3x + \underline{\quad})$ and the factors $(2)(-5)$ and $(-2)(5)$. We do not use $6x + 2$ or $6x - 2$ as possible factors because these binomials have a common factor of 2 but $18x^2 + 3x - 10$ has no common factors.

Let's try $(6x - 5)(3x + 2)$.

$$(6x - 5)(3x + 2) = 18x^2 + 12x - 15x - 10$$
$$= 18x^2 - 3x - 10$$

Close! The only problem is that the middle term has the sign opposite to the one we want. Therefore, switch the signs of -5 and 2 in the binomials.

$$(6x + 5)(3x - 2) = 18x^2 - 12x + 15x - 10$$
$$= 18x^2 + 3x - 10$$

It works! Therefore, $18x^2 + 3x - 10 = (6x + 5)(3x - 2)$. ●

The moral of the story in Example 8 is that the name *trial and error* is a bit misleading. With some thought, you won't have to choose binomial factors haphazardly or for very long, provided that you use the helpful hints given and stay alert.

> **Quick ✓**
>
> In Problems 15 and 16, factor each trinomial completely by trial and error.
>
> **15.** $12x^2 + 17x + 6$ **16.** $12y^2 + 32y - 35$

EXAMPLE 9 **Factoring Trinomials with Two Variables Using Trial and Error**

Factor completely: $48x^2 + 4xy - 30y^2$

Solution

Did you notice that there is a greatest common factor of 2? We first factor out this GCF and obtain the polynomial $2(24x^2 + 2xy - 15y^2)$. The trinomial in parentheses factors in the form $24x^2 + 2xy - 15y^2 = (\_x + \_y)(\_x + \_y)$.

List all possible ways of representing $24x^2$.

$$(24x + \_y)(x + \_y)$$
$$(12x + \_y)(2x + \_y)$$
$$(8x + \_y)(3x + \_y)$$
$$(6x + \_y)(4x + \_y)$$

Now list the factors of the coefficient of the last term, -15.

Factors of -15	$15, -1$	$-15, 1$	$5, -3$	$-5, 3$

Do not use $24x + 3y$, $12x + 3y$, $3x + 3y$, or $6x + 3y$ as possible factors because there is a common factor in these binomials. Also, since the middle term has a small coefficient, we will start with $(6x + \_y)(4x + \_y)$ and the factors $(3)(-5)$ and $(-3)(5)$.

Try $(6x - 5y)(4x + 3y)$.

$$(6x - 5y)(4x + 3y) = 24x^2 - 2xy - 15y^2$$

Close! The only problem is that the middle term has the sign opposite to the one we want. Therefore, switch the signs of -5 and 3 in the binomials.

$$(6x + 5y)(4x - 3y) = 24x^2 - 18xy + 20xy - 15y^2$$
$$= 24x^2 + 2xy - 15y^2$$

Thus $48x^2 + 4xy - 30y^2 = 2(6x + 5y)(4x - 3y)$. ●

Work smart

Don't forget to include the GCF in the factored form!

> **Quick ✓**
>
> **17.** What is the first step in factoring any polynomial?
>
> **18.** *True or False* The trinomial $12x^2 + 22x + 6$ is completely factored as $(4x + 6)(3x + 1)$.
>
> In Problems 19 and 20, factor each trinomial completely by trial and error.
>
> **19.** $-8x^2 + 28xy + 60y^2$ **20.** $90x^2 + 21xy - 6y^2$

EXAMPLE 10

Factoring a Trinomial with a Negative Leading Coefficient

Factor: $-14x^2 + 29x + 15$

Solution

There are no common factors in $-14x^2 + 29x + 15$, but notice that the coefficient of the squared term is negative. Factor -1 out of the trinomial to obtain

$$-14x^2 + 29x + 15 = -1(14x^2 - 29x - 15)$$

Work Smart

It's easier to factor a trinomial when the leading coefficient is positive.

Now factor the expression in parentheses and obtain

$$-14x^2 + 29x + 15 = -1(14x^2 - 29x - 15)$$
$$= -1(7x + 3)(2x - 5)$$
$$= -(7x + 3)(2x - 5)$$

Quick ✔

In Problems 21 and 22, factor each trinomial completely by trial and error.

21. $-6y^2 + 23y + 4$

22. $-6x^2 - 3x + 45$

Work Smart

Let's compare the two methods presented in this section, trial and error and factoring by grouping, by factoring $3x^2 + 10x + 8$. The first question to ask is: Is there a GCF? No, there's not, so let's continue.

Trial and Error	**Grouping**
Step 1: The coefficient of x^2, 3, is prime, so we list the possibilities for the binomial factors: $(3x + \_\_)(x + \_\_)$.	**Step 1:** For the polynomial $3x^2 + 10x + 8$, $a = 3$ and $c = 8$; the value of ac is $3 \cdot 8 = 24$.
Step 2: The last term, 8, is not prime. Its factors are $(1)(8)$ or $(2)(4)$ or $(-1)(-8)$ or $(-2)(-4)$. Since $c = 8$ is positive and the coefficient of the middle term is also positive, we'll consider only $(1)(8)$ and $(2)(4)$.	**Step 2:** The two integers whose product is $ac = 24$ and whose sum is $b = 10$ are 6 and 4.
Step 3: The coefficient of the middle term is not large, so let's start by trying $(3x + 4)(x + 2)$. $(3x + 4)(x + 2) = 3x^2 + 10x + 8$ It works! Therefore, $3x^2 + 10x + 8 = (3x + 4)(x + 2)$.	**Step 3:** Rewrite $3x^2 + 10x + 8$ as $3x^2 + 10x + 8 = 3x^2 + 6x + 4x + 8$.
	Step 4: Factor the expression $3x^2 + 6x + 4x + 8$ by grouping. $3x^2 + 6x + 4x + 8$ $= 3x(x + 2) + 4(x + 2)$ $= (x + 2)(3x + 4)$ So $3x^2 + 10x + 8 = (x + 2)(3x + 4)$.

Because $(3x + 4)(x + 2) = (x + 2)(3x + 4)$, we see that both methods give the same result. Which method do you prefer?

6.3 Exercises

  MyMathLab® Math XL PRACTICE

Exercise numbers in green have complete video solutions in MyMathLab

Problems 1–22 are the Quick ✔s that follow the EXAMPLES.

Building Skills

In Problems 23–46, factor each polynomial completely using the grouping method. Hint: None of the polynomials are prime. See Objective 1.

23. $2x^2 + 13x + 15$

24. $3x^2 + 22x + 7$

25. $5w^2 + 13w - 6$

26. $7n^2 - 27n - 4$

27. $4w^2 - 8w - 5$

28. $4x^2 + 4x - 3$

29. $27z^2 + 3z - 2$

30. $25t^2 + 5t - 2$

31. $6y^2 - 5y - 6$

32. $20t^2 + 21t + 4$

33. $4m^2 + 8m - 5$

34. $6m^2 - 5m - 4$

35. $12n^2 + 19n + 5$

36. $12p^2 - 23p + 5$

37. $-5 - 9x + 18x^2$

38. $-4 - 3x + 10x^2$

39. $12x^2 + 2xy - 4y^2$

40. $18x^2 + 6xy - 4y^2$

41. $8x^2 + 28x + 12$

42. $30m^2 - 85m + 60$

43. $7x - 5 + 6x^2$

44. $15 + 8x - 12x^2$

45. $-8p^2 + 6p + 9$

46. $-10y^2 + 47y + 15$

In Problems 47–70, factor each polynomial completely using the trial and error method. Hint: None of the polynomials is prime. See Objective 2.

47. $2x^2 + 5x + 3$

48. $3x^2 + 16x + 5$

49. $5n^2 + 7n + 2$

50. $7z^2 + 22z + 3$

51. $5y^2 + 2y - 3$

52. $11z^2 + 32z - 3$

53. $-4p^2 + 11p + 3$

54. $-6w^2 + 11w + 2$

55. $5w^2 + 13w - 6$

56. $3x^2 + 16x - 12$

57. $7t^2 + 37t + 10$

58. $5x^2 + 16x + 3$

59. $6n^2 - 17n + 10$

60. $11p^2 - 46p + 8$

61. $2 - 11x + 5x^2$

62. $4 - 17x + 4x^2$

63. $2x^2 + 3xy + y^2$

64. $3x^2 + 7xy + 2y^2$

65. $2m^2 - 3mn - 2n^2$

66. $3y^2 + 7yz - 6z^2$

67. $6x^2 + 2xy - 4y^2$

68. $6x^2 - 14xy - 12y^2$

69. $-2x^2 - 7x + 15$

70. $-6x^2 - 3x + 45$

Mixed Practice

In Problems 71–96, factor completely. If a polynomial cannot be factored, say it is prime.

71. $15x^2 - 23x + 4$

72. $12x^3 - 11x^2 - 15x$

73. $-13y + 12 - 4y^2$

74. $2x + 12x^2 - 24$

75. $10x^2 - 8xy - 24y^2$

76. $12n^2 + 7n - 10$

77. $9x^2 - 18x + 10$

78. $15x^2 - 4x + 8$

79. $-12x + 9 - 24x^2$

80. $-24z^2 + 18z + 2$

81. $4x^3y^2 - 8x^2y^3 - 4x^2y^2$

82. $9x^3y + 6x^2y + 3xy$

83. $4m^2 + 13mn + 3n^2$

84. $4m^2 - 19mn + 12n^2$

85. $6x^2 - 17x - 12$

86. $8x^2 + 14x - 7$

87. $48xy + 24x^2 - 30y^2$

88. $-20y^2 + 6x^2 + 7xy$

89. $18m^2 + 39mn - 24n^2$

90. $63y^3 + 60xy^3 + 12x^2y^3$

91. $-6x^3 + 10x^2 - 4x^4$

92. $21n^2 - 18n^3 + 9n$

93. $30x + 22x^2 - 24x^3$

94. $48x - 74x^2 + 28x^3$

95. $6x^2(x^2 + 1) - 25x(x^2 + 1) + 14(x^2 + 1)$

96. $10x^2(x - 1) - x(x - 1) - 2(x - 1)$

Applying the Concepts

△ **97. Area of a Triangle** A triangle has area described by the polynomial $\left(3x^2 + \dfrac{13}{2}x - 14\right)$ square meters. Find the base and height of the triangle. *Hint:* Factor out $\dfrac{1}{2}$ as a common factor.

△ **98. Area of a Rectangle** A rectangle has area described by the polynomial $6x^2 + x - 1$ square centimeters. Find the length and width of the rectangle.

99. Suppose we know that one factor of $6x^2 - 11x - 10$ is $3x + 2$. What is the other factor?

100. Suppose we know that one factor of $8x^2 + 22x - 21$ is $2x + 7$. What is the other factor?

Extending the Concepts

In Problems 101–104, factor completely.

101. $27z^4 + 42z^2 + 16$

102. $15n^6 + 7n^3 - 2$

103. $3x^{2n} + 19x^n + 6$

104. $2x^{2n} - 3x^n - 5$

In Problems 105 and 106, find all possible integer values of b so that the polynomial is factorable.

105. $3x^2 + bx - 5$

106. $6x^2 + bx + 7$

Explaining the Concepts

107. Describe when you would use trial and error and when you would use grouping to factor a trinomial. Make up two examples that demonstrate your reasoning.

108. How can you tell if a trinomial is not factorable? Make up an example to demonstrate your reasoning.

6.4 Factoring Special Products

Objectives

1 Factor Perfect Square Trinomials

2 Factor the Difference of Two Squares

3 Factor the Sum or Difference of Two Cubes

Are You Ready For This Section?

Before getting started, take the following readiness quiz. If you get a problem wrong, go back to the section cited and review the material.

R1. Evaluate: 5^2 [Section 1.7, pp. 56–57]

R2. Evaluate: $(-2)^3$ [Section 1.7, pp. 56–57]

R3. Find the product: $(5p^2)^3$ [Section 5.2, pp. 320–321]

R4. Find the product: $(3z + 2)^2$ [Section 5.3, pp. 328–329]

R5. Find the product: $(4m + 5)(4m - 5)$ [Section 5.3, pp. 327–328]

Let's review two "special products" that we saw in Section 5.3.

Product	Example
$(A + B)^2 = A^2 + 2AB + B^2$	$(3a + 5)^2 = (3a)^2 + 2(3a)(5) + 5^2$ $= 9a^2 + 30a + 25$
$(A - B)^2 = A^2 - 2AB + B^2$	$(2p - 3)^2 = (2p)^2 - 2(2p)(3) + 3^2$ $= 4p^2 - 12p + 9$
$(A + B)(A - B) = A^2 - B^2$	$(4z + 7)(4z - 7) = (4z)^2 - 7^2$ $= 16z^2 - 49$

Recall that $A^2 + 2AB + B^2$ and $A^2 - 2AB + B^2$ are called perfect square trinomials. We call $A^2 - B^2$ the **difference of two squares.**

In this section, we factor polynomials that are special products: perfect square trinomials and the difference of two squares. By recognizing these polynomials, you'll be able to factor them quickly without using the grouping method or trial and error.

Ready?...Answers **R1.** 25   **R2.** -8
 R3. $125p^6$ **R4.** $9z^2 + 12z + 4$
 R5. $16m^2 - 25$

▶ **1** Factor Perfect Square Trinomials

When we use the formulas $(A + B)^2 = A^2 + 2AB + B^2$ and $(A - B)^2 = A^2 - 2AB + B^2$ in reverse, we obtain a method for factoring perfect square trinomials.

In Words

A perfect square trinomial is a trinomial in which the first term and the third term are perfect squares and the second term is either 2 times or -2 times the product of the expressions being squared in the first and third terms.

Perfect Square Trinomials

$$A^2 + 2AB + B^2 = (A + B)^2$$
$$A^2 - 2AB + B^2 = (A - B)^2$$

For a polynomial to be a perfect square trinomial, two conditions must be satisfied.

1. The first and last terms must be perfect squares. Any variable raised to an even exponent is a perfect square. Thus x^2, $x^4 = (x^2)^2$, and $x^6 = (x^3)^2$ are all perfect squares. Other examples of perfect squares are $49 = 7^2$, $9x^2 = (3x)^2$, and $25a^4 = (5a^2)^2$.

2. The middle term must equal 2 times or -2 times the product of the expressions being squared in the first and last term.

Work Smart

The first five perfect squares are 1, 4, 9, 16, and 25.

EXAMPLE 1   **How to Factor Perfect Square Trinomials**

Factor completely: $z^2 + 6z + 9$

Step-by-Step Solution

Step 1: Determine whether the first term and the third term are perfect squares.

The first term, z^2, is the square of z and the third term, 9, is the square of 3.

(continued)

Step 2: Determine whether the middle term is 2 times or -2 times the product of the expressions being squared in the first and last term.

The middle term, $6z$, is 2 times the product of z and 3.

Step 3: Use $A^2 + 2AB + B^2 = (A + B)^2$ to factor the expression.

$$A^2 + 2 \cdot A \cdot B + B^2$$
$$\downarrow \quad\quad \downarrow \downarrow \quad \downarrow$$
$$z^2 + 6z + 9 = z^2 + 2 \cdot z \cdot 3 + 3^2$$

Factor as $(A + B)^2$ with $A = z$ and $B = 3$: $= (z + 3)^2$

Step 4: Check

$$(z + 3)^2 = (z + 3)(z + 3)$$
$$= z^2 + 3z + 3z + 9$$
$$= z^2 + 6z + 9$$

So $z^2 + 6z + 9 = (z + 3)^2$.

EXAMPLE 2   **Factoring Perfect Square Trinomials**

Factor completely:

(a) $4x^2 - 20x + 25$ **(b)** $9x^2 + 42xy + 49y^2$

Solution

(a) The first term, $4x^2$, equals $(2x)^2$, and the third term, 25, equals 5^2. The middle term, $-20x$, equals $-2 \cdot (2x \cdot 5)$.

Factor using $A^2 \quad -2 \cdot A \cdot B + B^2$
$$\downarrow \quad\quad \downarrow \downarrow \quad \downarrow$$
$$4x^2 - 20x + 25 = (2x)^2 - 2 \cdot 2x \cdot 5 + 5^2$$
$$= (2x - 5)^2$$

Therefore, $4x^2 - 20x + 25 = (2x - 5)^2$. We leave the check to you.

(b) The first term is $9x^2 = (3x)^2$. The third term is $49y^2 = (7y)^2$. The middle term, $42xy$, equals $2 \cdot (3x \cdot 7y)$.

Factor using $A^2 \quad +2 \cdot A \cdot B + B^2$
$$\downarrow \quad\quad \downarrow \downarrow \quad \downarrow$$
$$9x^2 + 42xy + 49y^2 = (3x)^2 + 2 \cdot 3x \cdot 7y + (7y)^2$$
$$= (3x + 7y)^2$$

Therefore, $9x^2 + 42xy + 49y^2 = (3x + 7y)^2$. We leave the check to you.

Quick ✓

1. The expression $A^2 + 2AB + B^2$ is called a _____ _____ _____.

2. $A^2 - 2AB + B^2 =$ _____.

In Problems 3–5, factor each trinomial completely.

3. $x^2 - 12x + 36$ **4.** $16x^2 + 40x + 25$ **5.** $9a^2 - 60ab + 100b^2$

EXAMPLE 3   **Factoring a Trinomial**

Factor completely, if possible: $b^2 - 10b + 36$

Solution

The first term, b^2, is the square of b. The third term, 36, equals 6^2. Does the middle term, $-10b$, equal $-2(b \cdot 6)$?

$$-2(b \cdot 6) = -12b \neq -10b$$

Therefore, $b^2 - 10b + 36$ is not a perfect square trinomial. Can we factor $b^2 - 10b + 36$ using another strategy? Since $b^2 - 10b + 36$ is in the form $x^2 + bx + c$, we need two factors of 36 whose sum is -10. We choose negative factors of 36, since the coefficient of the middle term is negative. The possibilities are $(-1)(-36)$, $(-2)(-18)$, $(-3)(-12)$, or $(-9)(-4)$. None of these factors sum to -10. We conclude that $b^2 - 10b + 36$ is prime.

Quick ✓

In Problems 6–8, factor each trinomial completely, if possible.

6. $z^2 - 8z + 16$ **7.** $4n^2 + 12n + 9$ **8.** $y^2 + 10y + 36$

EXAMPLE 4 **Factoring a Perfect Square Trinomial Having a GCF**

Factor completely: $32m^4 - 48m^2 + 18$

Solution

Do you remember the first step in any factoring problem? It is to look for the GCF. There is a common factor of 2 in the trinomial, so factor it out.

Work Smart

m^4 is a square because $m^4 = (m^2)^2$.

$$32m^4 - 48m^2 + 18 = 2(16m^4 - 24m^2 + 9)$$

Now attempt to factor the trinomial $16m^4 - 24m^2 + 9$. The first term is $16m^4 = (4m^2)^2$. The third term is $9 = 3^2$. The middle term, $-24m^2$, equals $-2 \cdot (4m^2 \cdot 3)$.

$$
\begin{array}{c}
\quad\; A^2 \quad - \quad 2 \cdot A \cdot B + B^2 \\
\quad\; \downarrow \qquad\quad \downarrow \;\; \downarrow \;\; \downarrow \;\; \downarrow \\
16m^4 - 24m^2 + 9 = (4m^2)^2 - 2 \cdot 4m^2 \cdot 3 + 3^2
\end{array}
$$

Factor as $(A - B)^2$: $= (4m^2 - 3)^2$

Work Smart

Don't forget the GCF that was factored out in the first step.

Therefore,

$$
\begin{aligned}
32m^4 - 48m^2 + 18 &= 2(16m^4 - 24m^2 + 9) \\
&= 2(4m^2 - 3)^2
\end{aligned}
$$

We leave the check to you.

Quick ✓

In Problems 9 and 10, factor each trinomial completely.

9. $4z^2 + 24z + 36$ **10.** $50a^3 + 80a^2 + 32a$

Perfect square trinomials can also be factored using trial and error or grouping, but if you recognize the perfect square pattern, you won't have to use either of these approaches.

▶ ❷ **Factor the Difference of Two Squares**

In Chapter 5, we found products such as $(x + 2y)(x - 2y) = x^2 - (2y)^2 = x^2 - 4y^2$. In general,

$$(A - B)(A + B) = A^2 - B^2$$

We will now reverse the multiplication process and find the factors of the difference of two squares.

In Words

The difference of two squares is just that! A perfect square minus a perfect square. This pattern is easy to recognize: Just look for two perfect squares that have been subtracted.

Difference of Two Squares

$$A^2 - B^2 = (A - B)(A + B)$$

EXAMPLE 5 **Factoring the Difference of Two Squares**

Factor completely:

(a) $n^2 - 81$ (b) $16x^2 - 9y^2$

Solution

(a) Notice that $n^2 - 81$ is the difference of two squares, n^2 and $81 = 9^2$. Thus

$$n^2 - 81 = (n)^2 - (9)^2$$
$$A^2 - B^2 = (A - B)(A + B): \quad = (n - 9)(n + 9)$$

Check You can check the answer to any of these differences of two squares by using FOIL.

$$(n - 9)(n + 9) = n^2 + 9n - 9n - 81$$
$$= n^2 - 81 \quad \text{Our answer checks.}$$

(b) Notice that $16x^2 - 9y^2$ is the difference of two squares, $16x^2 = (4x)^2$ and $9y^2 = (3y)^2$.

$$16x^2 - 9y^2 = (4x)^2 - (3y)^2$$
$$A^2 - B^2 = (A - B)(A + B): \quad = (4x - 3y)(4x + 3y)$$

Check $(4x - 3y)(4x + 3y) = 16x^2 + 12xy - 12xy - 9y^2$
$$= 16x^2 - 9y^2$$

Work Smart

Remember that the Commutative Property of Multiplication tells us that the order of the factors in the answer doesn't matter:
$(a - b)(a + b) = (a + b)(a - b)$

Quick ✓

11. $4m^2 - 81n^2$ is called the _____ of ____ _____ and factors into two binomials.

12. $P^2 - Q^2 = (\_ - \_)(\_ + \_)$

In Problems 13–15, factor completely.

13. $z^2 - 25$ 14. $81m^2 - 16n^2$ 15. $16a^2 - \dfrac{4}{9}b^2$

For the remainder of the section, we will leave the check of the factorization to you.

EXAMPLE 6 **Factoring the Difference of Two Squares**

Factor completely:

(a) $49k^4 - 100$ (b) $p^4 - 1$

Solution

(a) Notice that $49k^4 = (7k^2)^2$ and $100 = 10^2$ are both perfect squares. Thus

$$49k^4 - 100 = (7k^2)^2 - 10^2$$
$$A^2 - B^2 = (A - B)(A + B): \quad = (7k^2 - 10)(7k^2 + 10)$$

(b) The expressions p^4 and 1 are both perfect squares; $p^4 = (p^2)^2$ and $1 = 1^2$. Thus

$$p^4 - 1 = (p^2)^2 - 1^2$$
$$A^2 - B^2 = (A - B)(A + B): \quad = (p^2 - 1)(p^2 + 1)$$

But $p^2 - 1$ is a difference of two squares! So we factor again:

$$= (p - 1)(p + 1)(p^2 + 1)$$

Therefore, $p^4 - 1 = (p - 1)(p + 1)(p^2 + 1)$. ●

You may be asking, "What about the sum of two squares—how does it factor?" If a and b are real numbers, **the sum of two squares, $a^2 + b^2$, is prime and does not factor.** Thus, binomials such as $a^2 + 1$ and $4y^2 + 81$ are prime.

Quick ✓

16. *True or False* The sum of two squares, such as $x^2 + 9$, is prime.

In Problems 17 and 18, factor each polynomial completely.

17. $100k^4 - 81w^2$ **18.** $x^4 - 16$

EXAMPLE 7 **Factoring the Difference of Two Squares Containing a GCF**

Factor completely: $50x^2 - 72y^2$

Solution

What do we look for first when we factor? A greatest common factor! Factor out 2, the GCF of 50 and 72, first.

$$50x^2 - 72y^2 = 2(25x^2 - 36y^2)$$

Difference of two squares with $A = 5x$ and $B = 6y$: $\quad = 2[(5x)^2 - (6y)^2]$

Factor as $(A - B)(A + B)$: $\quad = 2(5x - 6y)(5x + 6y)$ ●

Quick ✓

19. *True or False* $4x^2 - 16y^2$ factors completely as $(2x - 4y)(2x + 4y)$.

In Problems 20 and 21, factor each polynomial completely.

20. $147x^2 - 48$ **21.** $-2x^4y + 8y^7$

▶ ③ **Factor the Sum or Difference of Two Cubes**

Consider the following products:

$$(A + B)(A^2 - AB + B^2) = A^3 - A^2B + AB^2 + A^2B - AB^2 + B^3$$
$$= A^3 + B^3$$
$$(A - B)(A^2 + AB + B^2) = A^3 + A^2B + AB^2 - A^2B - AB^2 - B^3$$
$$= A^3 - B^3$$

These products show that we can factor the sum or difference of two cubes as follows:

Work Smart

To remember the formulas for the sum and the difference of two cubes, think of "SOAP," which stands for S̲ame (as the operation in the original polynomial), O̲pposite, A̲lways P̲ositive.

$A^3 + B^3 = (A + B)(A^2 - AB + B^2)$
$A^3 - B^3 = (A - B)(A^2 + AB + B^2)$

The Sum of Two Cubes

$$A^3 + B^3 = (A + B)(A^2 - AB + B^2)$$

The Difference of Two Cubes

$$A^3 - B^3 = (A - B)(A^2 + AB + B^2)$$

Notice in the formulas that the sign between the cubes matches the sign in the binomial factor and is the opposite of the sign of the middle term of the trinomial factor. Remember that the perfect cubes are $1^3 = 1, 2^3 = 8, 3^3 = 27$, and so on. Any variable raised to a multiple of 3 is a perfect cube. So $x^3, x^6 = (x^2)^3, x^9 = (x^3)^3$, and so on are all perfect cubes.

EXAMPLE 8 **Factoring the Sum or Difference of Two Cubes**

Factor completely:

(a) $x^3 - 8$ (b) $8m^3 + 125n^6$

Solution

(a) We have the difference of two cubes, x^3 and $8 = 2^3$. Let $A = x$ and $B = 2$.

$$A^3 - B^3 = (A - B)(A^2 + A\ B + B^2)$$

$$x^3 - 8 = x^3 - 2^3 = (x - 2)(x^2 + x(2) + 2^2)$$
$$= (x - 2)(x^2 + 2x + 4)$$

(b) Because $8m^3 = (2m)^3$ and $125n^6 = (5n^2)^3$, the expression $8m^3 + 125n^6$ is the sum of two cubes. Let $A = 2m$ and $B = 5n^2$.

$$8m^3 + 125n^6 = (2m)^3 + (5n^2)^3$$

$A^3 + B^3 = (A + B)(A^2 - AB + B^2)$:
$$= (2m + 5n^2)[(2m)^2 - (2m)(5n^2) + (5n^2)^2]$$
$$= (2m + 5n^2)(4m^2 - 10mn^2 + 25n^4)$$ ●

Quick ✓

22. The binomial $27x^3 + 64y^3$ is called the ___ of ___ ___.

23. $A^3 - B^3 = (\_ - \_)(\_ + \_ + \_)$

24. *True or False* $(x - 1)(x^2 - x + 1)$ can be factored further.

In Problems 25 and 26, factor each polynomial completely.

25. $z^3 + 125$ 26. $8p^3 - 27q^6$

EXAMPLE 9 **Factoring the Sum or Difference of Two Cubes Having a GCF**

Factor completely: $250x^4 + 54x$

Solution

Always look for a common factor first! Here, we have a common factor of $2x$.

$$250x^4 + 54x = 2x(125x^3 + 27)$$

$(5x)^3 = 125x^3$ and $3^3 = 27$:
$$= 2x[(5x)^3 + 3^3]$$
$$= 2x(5x + 3)(25x^2 - 15x + 9)$$ ●

Quick ✓

In Problems 27 and 28, factor each polynomial completely.

27. $-16a^4 - 54a$ 28. $-375b^3 + 3$

6.4 Exercises

*Problems **1–28** are the Quick ✔s that follow the **EXAMPLES**.*

Building Skills

In Problems 29–38, factor each perfect square trinomial completely. See Objective 1.

29. $x^2 + 10x + 25$

30. $m^2 + 12m + 36$

31. $4p^2 - 4p + 1$

32. $9a^2 - 12a + 4$

33. $16x^2 + 24x + 9$

34. $16y^2 - 72y + 81$

35. $x^2 - 4xy + 4y^2$

36. $4a^2 + 20ab + 25b^2$

37. $4z^2 - 12z + 9$

38. $25k^2 - 70k + 49$

In Problems 39–48, factor each difference of two squares completely. See Objective 2.

39. $x^2 - 49$

40. $m^2 - 121$

41. $4x^2 - 25$

42. $25m^2 - 9$

43. $100n^8 - 81p^4$

44. $36s^6 - 49t^4$

45. $k^8 - 256$

46. $a^4 - 16$

47. $25p^4 - 49q^2$

48. $36b^4 - 121a^2$

In Problems 49–58, factor each sum or difference of two cubes completely. See Objective 3.

49. $27 + x^3$

50. $8v^3 + 1$

51. $8x^3 - 27y^3$

52. $64r^3 - 125s^3$

53. $x^6 - 8y^3$

54. $m^9 - 27n^6$

55. $27c^3 + 64d^9$

56. $125y^3 + 27z^6$

57. $16x^3 + 250y^3$

58. $40x^3 + 135y^6$

Mixed Practice

In Problems 59–94, factor completely. If the polynomial is prime, state so.

59. $16m^2 + 40mn + 25n^2$

60. $25p^2q^2 - 80pq + 64$

61. $18 - 12x + 2x^2$

62. $50 + 20x + 2x^2$

63. $-6a^2 - a + 40$

64. $-15a^2 - 4a + 4$

65. $2x^3 + 10x^2 + 16x$

66. $-5y^3 + 20y^2 - 80y$

67. $x^4y^2 - x^2y^4$

68. $16a^2b^2 - 25a^4b^2$

69. $x^8 - 25y^{10}$

70. $32a^3 + 4b^6$

71. $b^2 + 13b + 36$

72. $p^2 + 4p - 96$

73. $3s^7 + 24s$

74. $4x - 4x^3$

75. $2x^5 - 162x$

76. $48z^4 - 3$

77. $6x^2 + 19x + 10$

78. $8x^2 + 30x - 27$

79. $4x^2 - 12x - 72$

80. $8x^2 + 16x - 64$

81. $3n^2 + 14n + 36$

82. $12m^2 - 14m + 21$

83. $2x^2 - 8x + 8$

84. $3x^2 + 18x + 27$

85. $9x^2 + y^2$

86. $16a^2 + 49b^2$

87. $x^4y^3 + 216xy^3$

88. $54b^5 - 16b^2$

89. $48n^4 - 24n^3 + 3n^2$

90. $18x^2 - 24x^3 + 8x^4$

91. $2x(x^2 - 4) + 5(x^2 - 4)$

92. $4z(z^2 - 9) + 3(z^2 - 9)$

93. $2y^3 + 5y^2 - 32y - 80$

94. $m^3 + 2m^2 - 25m - 50$

Applying the Concepts

△ **95. Area of a Square** The area of a square is given by the polynomial $(4x^2 + 20x + 25)$ square meters. What is an algebraic expression for the length of one side of the square

△ **96. Area of a Square** The area of a square is given by the polynomial $(9x^2 - 6x + 1)$ square feet. What is an algebraic expression for the length of one side of the square

Extending the Concepts

In Problems 97–106, factor completely and then simplify, if possible.

97. $(x - 2)^2 - (x + 1)^2$

98. $(m - p)^2 - (m + p)^2$

99. $(x - y)^3 + y^3$

100. $(z + 1)^3 - z^6$

101. $(x + 1)^2 - 9$

102. $(x + 3)^2 - 25$

103. $2a^2(x + 1) - 17a(x + 1) + 30(x + 1)$

104. $25a^2(x - 1)^2 - 5a(x - 1)^2 - 2(x - 1)^2$

105. $5(x + 2)^2 - 7(x + 2) - 6$

106. $9(2a + b)^2 + 6(2a + b) - 8$

Explaining the Concepts

107. Create a list of steps for factoring the sum or difference of two cubes.

108. Explain why the polynomial $4x^2 - 9$ is factorable but the polynomial $4x^2 + 9$ is prime.

109. Explain the error in factoring the polynomial $x^3 + y^3$ as $(x + y)(x^2 - xy + y^2) = (x + y)(x - y)^2$.

110. Factor $x^6 - 1$ first as a difference of two squares and then as a difference of two cubes. Are the factors the same? Explain your results.

6.5 Summary of Factoring Techniques

Objective

1 Factor Polynomials Completely

▶ **1** Factor Polynomials Completely

The objective of this section is to put together all the various factoring techniques we have discussed. Recall that a polynomial is factored completely if each factor in the final factorization is prime.

Steps for Factoring

Step 1: Is there a greatest common factor (GCF)? If so, factor it out.

Step 2: Count the number of terms.

Step 3: **(a)** Two terms (binomials)
- Is it the difference of two squares? If so,
$$A^2 - B^2 = (A - B)(A + B)$$
- Is it the sum of two squares? If so, stop! The expression is prime.
- Is it the difference of two cubes? If so,
$$A^3 - B^3 = (A - B)(A^2 + AB + B^2)$$
- Is it the sum of two cubes? If so,
$$A^3 + B^3 = (A + B)(A^2 - AB + B^2)$$

(b) Three terms (trinomials)
- Is it a perfect square trinomial? If so,
$$A^2 + 2AB + B^2 = (A + B)^2 \quad \text{or} \quad A^2 - 2AB + B^2 = (A - B)^2$$
- Is the coefficient of the squared term 1? If so,
$$x^2 + bx + c = (x + m)(x + n), \text{ where } mn = c \text{ and } m + n = b$$
- Is the coefficient of the squared term different from 1? If so, factor using grouping or trial and error.

(c) Four terms
- Use factoring by grouping.

Step 4: Check your work by multiplying the factors.

Work Smart

Be careful with the difference of two squares. One of the binomial factors may factor further. See Example 5.

► **EXAMPLE 1**   **How to Factor Completely**

Factor completely: $2x^2 - 4x - 48$

Step-by-Step Solution

Step 1: Is there a GCF?

Yes. We factor out the GCF, 2.
$$2x^2 - 4x - 48 = 2(x^2 - 2x - 24)$$

Step 2: Count the number of terms in the polynomial in parentheses.

The polynomial in parentheses has three terms.

Step 3: The trinomial in parentheses, $x^2 - 2x - 24$, is not a perfect square trinomial. The leading coefficient is 1, so try $(x + m)(x + n)$, where $mn = c$ and $m + n = b$.

Find two factors of -24 whose sum is -2. Because $-6(4) = -24$ and $-6 + 4 = -2$, $m = -6$ and $n = 4$.
$$2(x^2 - 2x - 24) = 2(x - 6)(x + 4)$$

Step 4: Check

$$
\begin{aligned}
2(x - 6)(x + 4) &= 2(x^2 + 4x - 6x - 24) \\
&= 2(x^2 - 2x - 24) \\
\text{Distribute: } &= 2x^2 - 4x - 48
\end{aligned}
$$

Therefore, $2x^2 - 4x - 48 = 2(x - 6)(x + 4)$.

> **Quick ✓**
>
> **1.** The first step in any factoring problem is to look for the _____ _____ ____.
>
> *In Problems 2 and 3, factor each polynomial completely.*
>
> **2.** $2p^2 + 8p - 90$
>
> **3.** $-45x^3 + 3x^2 + 6x$

► **EXAMPLE 2**   **How to Factor Completely**

Factor completely: $12p^2 + 5p - 3$

Step-by-Step Solution

Step 1: Is there a GCF?

There is no GCF.

Step 2: Count the number of terms.

This polynomial has three terms.

Step 3: The trinomial is not a perfect square trinomial because the first and last terms are not perfect squares. We will factor using trial and error.

Notice that the coefficient of the middle term is fairly small. For this reason, the factors of the first term are probably close in value. Thus we start by trying to factor $12p^2 + 5p - 3$ as
$$(4p + \_\_)(3p + \_\_)$$

The last term, -3, has only the factors 1, -3 or -1, 3. The factor 3 (or -3) cannot be in the binomial factor $(3p + \_\_)$ because this would result in a common factor. Since the middle term is $+5p$, we try the factor 3 with $(4p + \_\_)$ as follows:
$$(4p + 3)(3p - 1)$$

Step 4: Check

$$
\begin{aligned}
(4p + 3)(3p - 1) &= 12p^2 - 4p + 9p - 3 \\
&= 12p^2 + 5p - 3
\end{aligned}
$$

Therefore, $12p^2 + 5p - 3 = (4p + 3)(3p - 1)$.

EXAMPLE 3   **How to Factor Completely**

Factor completely: $12x^2 + 36xy + 27y^2$

Step-by-Step Solution

Step 1: Is there a GCF?

We notice that the coefficients 12, 36, and 27 are all multiples of 3, so we factor out the GCF of 3.

$$12x^2 + 36xy + 27y^2 = 3(4x^2 + 12xy + 9y^2)$$

Step 2: Count the number of terms in the polynomial in parentheses.

The polynomial in parentheses has three terms.

Step 3: Is the trinomial in parentheses a perfect square trinomial?

Yes. The first and third terms are perfect squares, $4x^2 = (2x)^2$ and $9y^2 = (3y)^2$. And, the middle term equals $2(2x \cdot 3y)$.

$$3(4x^2 + 12xy + 9y^2) = 3[(2x)^2 + 2(2x)(3y) + (3y)^2]$$

$A^2 + 2AB + B^2 = (A + B)^2$: $= 3(2x + 3y)^2$

Step 4: Check

Multiply: $3(2x + 3y)^2 = 3(2x + 3y)(2x + 3y)$

$$= 3(4x^2 + 6xy + 6xy + 9y^2)$$

$$= 3(4x^2 + 12xy + 9y^2)$$

$$= 12x^2 + 36xy + 27y^2$$

So, $12x^2 + 36xy + 27y^2 = 3(2x + 3y)^2$. ●

▶ **EXAMPLE 4**   **Factoring Completely**

Factor completely: $16m^3 + 54$

Solution

Factor out the greatest common factor of 2. The polynomial in the parentheses has two terms.

$$16m^3 + 54 = 2(8m^3 + 27)$$

Sum of two cubes with $A = 2m$ and $B = 3$: $= 2[(2m)^3 + 3^3]$

$A^3 + B^3 = (A + B)(A^2 - AB + B^2)$: $= 2[(2m + 3)((2m)^2 - (2m)(3) + (3)^2)]$

$$= 2(2m + 3)(4m^2 - 6m + 9)$$

Check $2(2m + 3)(4m^2 - 6m + 9) = 2[8m^3 - 12m^2 + 18m$

$$+ 12m^2 - 18m + 27]$$

$$= 2[8m^3 + 27]$$

$$= 16m^3 + 54$$ ●

Quick ✔

In Problems 8 and 9, factor each polynomial completely.

8. $125y^3 - 64$

9. $-24a^3 + 3b^3$

▶ **EXAMPLE 5**   **Factoring Completely**

Factor completely: $-5x^5 + 80x$

Solution

Because the leading coefficient is negative, we factor out a negative as part of the GCF, $-5x$.

$$-5x^5 + 80x = -5x(x^4 - 16)$$

The factor in parentheses has two terms. The terms $x^4 = (x^2)^2$ and $16 = 4^2$ are perfect squares. Thus $x^4 - 16$ is a difference of two squares with $A = x^2$ and $B = 4$.

Work Smart

Check each factor to determine if any factor can be factored again.

$$-5x(x^4 - 16) = -5x(x^2 - 4)(x^2 + 4).$$

Difference of two squares with $A = x$ and $B = 2$:　$= -5x(x - 2)(x + 2)(x^2 + 4)$

Check　$-5x(x - 2)(x + 2)(x^2 + 4) = -5x(x^2 - 4)(x^2 + 4)$
$$= -5x(x^4 - 16)$$
$$= -5x^5 + 80x$$ ●

Quick ✔

10. *True or False* The polynomial $x^4 - 81 = (x^2 - 9)(x^2 + 9)$ is factored completely.

In Problems 11 and 12, factor each polynomial completely.

11. $x^4 - 1$

12. $-36x^2y + 16y$

EXAMPLE 6   **Factoring Completely**

Factor completely: $4x^3 - 6x^2 - 36x + 54$

Solution

The GCF is 2 among these four terms. Since there are four terms, we attempt to factor by grouping.

Factor out the GCF of 2:　$4x^3 - 6x^2 - 36x + 54 = 2(2x^3 - 3x^2 - 18x + 27)$

Factor by grouping:　$= 2[(2x^3 - 3x^2) + (-18x + 27)]$

Factor out the common factor in each group:　$= 2[x^2(2x - 3) - 9(2x - 3)]$

Factor out $2x - 3$:　$= 2(2x - 3)(x^2 - 9)$

$x^2 - 9$ is the difference of two squares:　$= 2(2x - 3)(x - 3)(x + 3)$

Work Smart: Study Skills

You know how to begin factoring a polynomial—look for the GCF. But do you know when a polynomial is factored completely? That is, do you know when to stop? Which of the following is not completely factored?

(a) $(x + 5)(2x + 6)$
(b) $(z - 3)(z + 3)(z^2 + 9)$
(c) $(2n - 5)(5n + 7)$
(d) $(4y - 3yz)(2 + 7z)$

Did you recognize that (a) and (d) are not completely factored because each has a common factor in one of the binomials?

Check　$2(2x - 3)(x - 3)(x + 3) = 2(2x - 3)(x^2 - 9)$
$$= 2(2x^3 - 18x - 3x^2 + 27)$$
$$= 4x^3 - 6x^2 - 36x + 54$$ ●

Quick ✔

In Problems 13 and 14, factor each polynomial completely.

13. $2x^3 + 3x^2 + 4x + 6$

14. $6x^3 + 9x^2 - 6x - 9$

EXAMPLE 7 **Factoring Completely**

Factor completely: $-4xy^2 + 4xy + 12x$

Solution

Each term has a common factor of $-4x$, so we factor it out.

$$-4xy^2 + 4xy + 12x = -4x(y^2 - y - 3)$$

The polynomial in parentheses has three terms, $y^2 - y - 3$. It is not a perfect square trinomial (because 3 is not a perfect square). We look for two factors of -3 whose sum is -1. There are no such factors, so $y^2 - y - 3$ is prime.

Check $\quad -4x(y^2 - y - 3) = -4xy^2 + 4xy + 12x$

Therefore, $-4xy^2 + 4xy + 12x = -4x(y^2 - y - 3)$.

Quick ✓

In Problems 15 and 16, factor each polynomial completely.

15. $-3z^2 + 9z - 21$ **16.** $6xy^2 + 15x^3$

6.5 Exercises   MyMathLab®

Exercise numbers in green have complete video solutions in MyMathLab

Problems 1–16 are the Quick ✓s that follow the EXAMPLES.

Mixed Practice

In Problems 17–90, factor completely. If a polynomial cannot be factored, say it is prime.

17. $x^2 - 100$

18. $x^2 - 256$

19. $t^2 + t - 6$

20. $n^2 + 5n + 6$

21. $x + y + 2ax + 2ay$

22. $xy - 2ay + 3bx - 6ab$

23. $a^3 - 8$

24. $1 - y^9$

25. $a^2 - ab - 6b^2$

26. $x^2 - xy - 30y^2$

27. $2x^2 - 5x - 7$

28. $12n^2 - 23n + 5$

29. $2x^2 - 6xy - 20y^2$

30. $2x^2 + 4xy - 6y^2$

31. $2m^2 + 7m + 4$

32. $3y^2 - 8y + 3$

33. $9 - a^2$

34. $25 - m^2$

35. $u^2 - 14u + 33$

36. $x^2 + 5x - 6$

37. $xy - ay - bx + ab$

38. $2x^3 - x^2 - 18x + 9$

39. $w^2 + 6w + 8$

40. $x^2 - 8x + 15$

41. $36a^2 - 49b^4$

42. $100x^2 - 25y^2$

43. $x^2 + 2xm - 8m^2$

44. $s^2 - 11st + 24t^2$

45. $6x^2y^2 - 13xy + 6$

46. $6s^2t^2 + st - 1$

47. $x^3 + x^2 + x + 1$

48. $x^3 - 3x^2 + 2x - 6$

49. $12z^2 + 12z + 18$

50. $3y^2 + 6y + 3$

51. $14c^2 + 19c - 3$

52. $24x^2 + 66x + 45$

53. $27m^3 + 64n^6$

54. $8x^6 + 125y^3$

55. $2j^6 - 2j^2$

56. $48m - 3m^9$

57. $8a^2 + 18ab - 5b^2$

58. $10p^2 - 15q^2 + 19pq$

59. $2a^3 + 8a$

60. $4t^4 + 64t^2$

61. $12z^2 - 3$

62. $8n^3 - 18n$

63. $x^2 - x + 6$

64. $n^2 + 2n + 8$

65. $16a^4 + 2ab^3$

66. $24p^3q + 81q^4$

67. $p^2q^2 + 6pq - 7$

68. $x^2y^2 + 20xy + 96$

69. $s^2(s + 2) - 4(s + 2)$

70. $x^2(x + y) - 16(x + y)$

71. $-12x^3 + 2x^2 + 2x$

72. $-3x^3 - 13x^2 + 10x$

73. $10v^2 - 2 - v$

74. $27h - 5 + 18h^2$

75. $4n^2 - n^4 + 3n^3$

76. $4x - 2x^2 - 2x^3$

77. $-4a^3b + 2a^2b - 2ab$

78. $-4x^3y + 4x^2y - 4xy$

79. $12p - p^3 + p^2$

80. $-9x - 3x^3 - 12x^2$

81. $-32x^3 + 72xy^2$

82. $-48p^4 + 75p^2$

83. $2n^3 - 10n^2 - 6n + 30$

84. $6x^4 + 3x^3 - 24x^2 - 12x$

85. $16x^2 + 4x - 12$

86. $36x^2 + 30x - 24$

87. $14x^2 + 3x^4 + 8$

88. $14 + 53r^3 + 14r^6$

89. $-2x^3y + x^2y^2 + 3xy^3$

90. $6a^3 - 5a^2b + ab^2$

Applying the Concepts

91. Profit on Newspapers The revenue for selling x newspapers is given by the expression $x^3 + 8x$. The cost to produce x newspapers is $8x^2 - 7x$. If profit is calculated as revenue minus costs, write an expression, in factored form, that calculates the profit from selling x newspapers.

92. Profit on T-shirts Candy and her boyfriend produce silk-screened T-shirts. The cost to produce n T-shirts is given by the expression $12n - 2n^2$, and the revenue from selling the same number of T-shirts is $n^3 + 2n^2$. Write an expression, in factored form, that calculates the profit from selling n T-shirts.

△ **93. Volume of a Box** The volume of a box is given by the formula $V = lwh$. The volume of the box is represented by the expression $(4x^3 - 10x^2 - 6x)$ cubic feet. Factor this expression to determine algebraic expressions that may represent the dimensions of the box.

△ **94. Volume of a Box** The volume of a box is given by the formula $V = lwh$. The volume of the box is represented by the expression $(18x^3 + 3x^2 - 6x)$ cubic centimeters. Factor this expression to determine algebraic expressions that may represent the dimensions of the box.

Extending the Concepts

Sometimes it is possible to factor a complicated-looking polynomial by substituting one variable for another. This approach is called **factoring by substitution.**

Example: Factoring by Substitution

Factor: $2n^6 - n^3 - 15$

Solution

Notice that $2n^6 - n^3 - 15$ can be written $2(n^3)^2 - n^3 - 15$ so that the trinomial is in the form $au^2 + bu + c$, where $u = n^3$.

$$2n^6 - n^3 - 15 = 2(n^3)^2 - n^3 - 15$$
$$\text{Let } n^3 = u: = 2u^2 - u - 15$$
$$\text{Factor: } = (2u + 5)(u - 3)$$
$$\text{Let } u = n^3: = (2n^3 + 5)(n^3 - 3)$$

In Problems 95–102, factor each trinomial by substitution.

95. $y^4 - 2y^2 - 24$

96. $x^4 + 3x^2 + 2$

97. $4z^6 - 13z^3 + 10$

98. $2m^6 + 11m^3 + 12$

99. $(3r - 1)^2 - 9(3r - 1) + 20$

100. $(5z - 3)^2 - 12(5z - 3) + 32$

101. $2(y - 3)^2 + 13(y - 3) + 15$

102. $3(z + 3)^2 + 14(z + 3) + 8$

In Problems 103–108, expressions that occur in calculus are given. Factor completely each expression.

103. $2(3x + 4)^2 + (2x + 3) \cdot 2(3x + 4) \cdot 3$

104. $5(2x + 1)^2 + (5x - 6) \cdot 2(2x + 1) \cdot 2$

105. $2x(2x + 5) + x^2 \cdot 2$

106. $3x^2(8x - 3) + x^3 \cdot 8$

107. $(4x - 3)^2 + x \cdot 2(4x - 3) \cdot 4$

108. $3x^2(3x + 4)^2 + x^3 \cdot 2(3x + 4) \cdot 3$

Although the most common pattern for factoring a polynomial with four terms is to group the first two terms and group the second two terms, it is not the only possibility. In the following examples, the polynomial has been written as three terms in the first group and a single term in the second group. Study the example and then try to find a grouping that will factor each polynomial.

$$x^2 + 2xy + y^2 - z^2 = (x^2 + 2xy + y^2) - z^2$$
$$= (x + y)^2 - z^2$$
$$= (x + y + z)(x + y - z)$$

In Problems 109–112, factor completely.

109. $4m^2 - 4mn + n^2 - p^2$

110. $b^2 + 2bc + c^2 - a^2$

111. $x^2 - y^2 + 2yz - z^2$

112. $16x^2 - y^2 - 8y - 16$

Explaining the Concepts

113. You are grading the paper of a student who factors the polynomial $x^2 + 4x - 3x - 12$ by grouping as follows:

$$(x^2 + 4x)(-3x - 12) = x(x + 4) - 3(x - 4)$$
$$= (x - 3)(x + 4)$$

Did the student correctly factor the polynomial? What comments would you write on the student's paper?

114. Describe the method of factoring you find most difficult. Write a polynomial and factor it using this method.

Putting the Concepts Together (Sections 6.1–6.5)

We designed these problems so that you can review Sections 6.1 to 6.5 and show your mastery of the concepts. Take time to work these problems before proceeding with the next section. The answers are located at the back of the text on page AN-24.

1. Find the GCF of $10x^3y^4z$, $15x^5y$, and $25x^2y^7z^3$.

In Problems 2–16, factor each polynomial completely. If the polynomial cannot be factored, say it is prime.

2. $x^2 - 3x - 4$ 3. $x^6 - 27$

4. $6x(2x + 1) + 5z(2x + 1)$

5. $x^2 + 5xy - 6y^2$ 6. $x^3 + 64$

7. $4x^2 + 49y^2$ 8. $3x^2 + 12xy - 36y^2$

9. $12z^5 - 44z^3 - 24z^2$ 10. $x^2 + 6x - 5$

11. $4m^4 + 5m^3 - 6m^2$ 12. $5p^2 - 17p + 6$

13. $10m^2 + 25m - 6m - 15$ 14. $36m^2 + 6m - 6$

15. $4m^2 - 20m + 25$ 16. $5x^2 - xy - 4y^2$

17. **Surface Area** The surface area of a right circular cylinder is given by the formula $S = 2\pi rh + 2\pi r^2$. Write the right side of this formula in factored form.

18. **Rocket Height** A toy rocket is launched upward from the ground with an initial velocity of 48 feet per second. Its height, h, in feet, after t seconds, is given by the formula $h = 48t - 16t^2$. Write the right side of this formula in factored form.

6.6 Solving Polynomial Equations by Factoring

Objectives

1 Solve Quadratic Equations Using the Zero-Product Property

2 Solve Polynomial Equations of Degree 3 or Higher Using the Zero-Product Property

Are You Ready for This Section?

Before getting started, take the following readiness quiz. If you get a problem wrong, go back to the section cited and review the material.

R1. Solve: $x + 5 = 0$ [Section 2.1, pp. 83–85]

R2. Solve: $2(x - 4) - 10 = 0$ [Section 2.2, pp. 93–94]

R3. Evaluate $2x^2 + 3x - 4$ when
(a) $x = 2$ (b) $x = -1$.   [Section 1.8, pp. 64–65]

▶ Questions that you may have been asking yourself are "Why do I care about factoring? What good is it?" It turns out that there are many uses of factoring. For example, factoring is essential for solving *polynomial equations*.

Work Smart

Remember, the degree of a polynomial in one variable is the value of the **largest exponent** on the variable. For example, the degree of $4x^3 - 9x^2 + 1$ is 3.

Definitions

A **polynomial equation** is any equation that contains only polynomial expressions. The **degree of a polynomial equation** is the degree of the polynomial expression in the equation.

Some examples of polynomial equations are

$$4x + 5 = 17 \qquad 2x^2 - 5x - 3 = 0 \qquad y^3 + 4y^2 = 3y + 18$$

Polynomial equation of degree 1   Polynomial equation of degree 2   Polynomial equation of degree 3

We solved polynomial equations of degree 1 back in Chapter 2. We now learn how to solve polynomial equations of degree 2.

▶ **1** Solve Quadratic Equations Using the Zero-Product Property

We use the following property in this section.

The Zero-Product Property

If the product of two factors is zero, then at least one of the factors is 0. That is,

if $ab = 0$, then $a = 0$ or $b = 0$ or both a and b are 0.

EXAMPLE 1 **Using the Zero-Product Property**

Solve: $(x + 4)(2x - 5) = 0$

Solution

The product of two factors, $x + 4$ and $2x - 5$, is equal to 0. According to the Zero-Product Property, at least one of the factors must equal 0. Therefore, set each factor equal to 0 and solve each equation separately.

$$x + 4 = 0 \quad \text{or} \quad 2x - 5 = 0$$

Subtract 4 from both sides: $x = -4$ Add 5 to both sides: $2x = 5$

Divide both sides by 2: $\dfrac{2x}{2} = \dfrac{5}{2}$

$$x = \frac{5}{2}$$

Check $(x + 4)(2x - 5) = 0$

$x = -4$: $(-4 + 4)(2(-4) - 5) \overset{?}{=} 0$

$0(-13) \overset{?}{=} 0$

$0 = 0$ True

$(x + 4)(2x - 5) = 0$

$x = \dfrac{5}{2}$: $\left(\dfrac{5}{2} + 4\right)\left(2 \cdot \left(\dfrac{5}{2}\right) - 5\right) \overset{?}{=} 0$

$\left(\dfrac{13}{2}\right)(5 - 5) \overset{?}{=} 0$

$\dfrac{13}{2} \cdot 0 \overset{?}{=} 0$

$0 = 0$ True

The solution set is $\left\{-4, \dfrac{5}{2}\right\}$.

●

Quick ✔

1. The equation $5x^2 + 3x - 7 = 0$ is a polynomial equation of degree __.

2. The Zero-Product Property states that if $ab = 0$, then either ____ or ____.

In Problems 3 and 4, use the Zero-Product Property to solve the equation.

3. $x(x + 3) = 0$ 4. $(x - 2)(4x + 5) = 0$

▶ The Zero-Product Property comes in handy when we need to solve *quadratic equations*.

Work Smart

In a quadratic equation, why can't *a* equal 0? Because if *a* were equal to zero, the equation would be a linear equation.

Definition

A **quadratic equation** is a polynomial equation that can be written in the form

$$ax^2 + bx + c = 0$$

where a, b, and c are real numbers and $a \neq 0$.

The following are examples of quadratic equations.

$$2x^2 + 5x - 3 = 0 \qquad -6z^2 + 12z = 0 \qquad y^2 - 25 = 0 \qquad p^2 + 12p = -36$$

Notice that $y^2 - 25 = 0$ is a quadratic equation even though it is missing the "y" term and that $-6z^2 + 12z = 0$ is a quadratic equation even though it is missing a constant term.

The term "quadratic" means "of, or relating to, a square." There are many real-world situations that are modeled by quadratic equations, such as problems that involve revenue for selling x units of a good or describing the height of a projectile over time.

Sometimes a quadratic equation is called a **second-degree equation** because the polynomial in the equation is of degree 2.

> **In Words**
>
> A quadratic equation is in standard form if the polynomial is written in descending order of exponents and is set equal to zero.

> **Definition**
>
> A quadratic equation is said to be in **standard form** if it is written in the form $ax^2 + bx + c = 0$.

For example, $2x^2 + x - 5 = 0$ is in standard form, while $p^2 + 12p = -36$ is not. To write $p^2 + 12p = -36$ in standard form, add 36 to both sides of the equation to obtain $p^2 + 12p + 36 = 0$.

When a quadratic equation is written in standard form, $ax^2 + bx + c = 0$, it may be possible to factor the expression $ax^2 + bx + c$ as the product of two first-degree polynomials. If it is possible to factor the trinomial, we can use the Zero-Product Property to solve the quadratic equation.

EXAMPLE 2 **How to Solve a Quadratic Equation by Factoring**

Solve: $x^2 - 4x = 21$

Step-by-Step Solution

Step 1: Write the equation in standard form.
$$x^2 - 4x = 21$$
Subtract 21 from both sides: $x^2 - 4x - 21 = 0$

Step 2: Factor the expression on the left side of the equation.

Two integers whose product is -21 and whose sum is -4 are -7 and 3. $(x + 3)(x - 7) = 0$

Step 3: Set each factor equal to 0.
$$x + 3 = 0 \quad \text{or} \quad x - 7 = 0$$

Step 4: Solve each first-degree equation.
$$x = -3 \quad \text{or} \quad x = 7$$

Step 5: Check

$$x^2 - 4x - 21 = 0 \qquad\qquad x^2 - 4x - 21 = 0$$
$$x = -3: \; (-3)^2 - 4(-3) - 21 \stackrel{?}{=} 0 \qquad x = 7: \; 7^2 - 4(7) - 21 \stackrel{?}{=} 0$$
$$9 + 12 - 21 \stackrel{?}{=} 0 \qquad\qquad 49 - 28 - 21 \stackrel{?}{=} 0$$
$$0 = 0 \quad \text{True} \qquad\qquad 0 = 0 \quad \text{True}$$

The solution set is $\{-3, 7\}$. ●

Look at Step 3 in the solution to Example 2. To solve a quadratic equation, we wish to put the equation into a form that we already know how to solve. That is, we "transform" the quadratic equation $x^2 - 4x - 21 = 0$ into two linear equations, $x + 3 = 0$ and $x - 7 = 0$. This is an important idea to understand about math. We use procedures to reduce a problem to a simpler problem. In this case, we reduce a quadratic equation to two linear equations.

We will present methods for solving $ax^2 + bx + c = 0$ when we cannot factor the expression $ax^2 + bx + c$ later in the text.

> **Solving a Quadratic Equation By Factoring**
>
> **Step 1:** Write the quadratic equation in standard form, $ax^2 + bx + c = 0$.
>
> **Step 2:** Factor the polynomial on the left side of the equation.
>
> **Step 3:** Set each factor found in Step 2 equal to zero using the Zero-Product Property.
>
> **Step 4:** Solve each first-degree equation for the variable.
>
> **Step 5:** Check your answers by substituting the values of the variable into the *original* equation.

EXAMPLE 3 **Solving a Quadratic Equation**

Solve: $3x^2 + 5x = 14x$

Solution

First write the equation in standard form, $ax^2 + bx + c = 0$.

$$3x^2 + 5x = 14x$$

Subtract 14x from both sides of the equation: $\qquad 3x^2 - 9x = 0$

Factor: $\quad 3x(x - 3) = 0$

Set each factor equal to 0: $\qquad 3x = 0 \quad \text{or} \quad x - 3 = 0$

Solve each first-degree equation: $\qquad x = 0 \quad \text{or} \qquad x = 3$

Check Substitute $x = 0$ and $x = 3$ into the original equation.

$$3x^2 + 5x = 14x \qquad\qquad\qquad 3x^2 + 5x = 14x$$

$$x = 0: \quad 3(0)^2 + 5(0) \stackrel{?}{=} 14(0) \qquad x = 3: \quad 3(3)^2 + 5(3) \stackrel{?}{=} 14(3)$$

$$0 = 0 \quad \text{True} \qquad\qquad\qquad 27 + 15 \stackrel{?}{=} 42$$

$$42 = 42 \quad \text{True}$$

The solution set is $\{0, 3\}$.

> **Quick ✓**
>
> **5.** A _____ equation is an equation that can be written in the form $ax^2 + bx + c = 0$, where a, b, and c are real numbers and $a \neq 0$.
>
> **6.** Quadratic equations are also known as _____-degree equations.
>
> **7.** *True or False* $3x + x^2 = 6$ is written in standard form.
>
> *In Problems 8–11, use the Zero-Product Property to solve the equation.*
>
> **8.** $p^2 - 6p + 8 = 0$ $\qquad\qquad$ **9.** $2t^2 - 5t = 3$
>
> **10.** $2x^2 + 3x = 5$ $\qquad\qquad$ **11.** $z^2 + 20 = -9z$

EXAMPLE 4 **Solving a Quadratic Equation**

Solve: $3m^2 - 3m = 2 - 2m$

Solution

First write the quadratic equation in standard form, $ax^2 + bx + c = 0$.

Add 2m to both sides and $\qquad\qquad\qquad 3m^2 - 3m = 2 - 2m$

subtract 2 from both sides: $\quad 3m^2 - 3m + 2m - 2 = 2 - 2 - 2m + 2m$

$$3m^2 - m - 2 = 0$$

Factor $3m^2 - m - 2$: $\quad (3m + 2)(m - 1) = 0$

Set each factor equal to 0: $\qquad 3m + 2 = 0 \quad$ or $\quad m - 1 = 0$

Solve each first-degree equation: $\qquad 3m = -2 \quad$ or $\qquad m = 1$

$$m = -\frac{2}{3}$$

Check Substitute $m = -\frac{2}{3}$ and $m = 1$ in the original equation to check the solutions. We leave this to you.

The solution set is $\left\{ -\frac{2}{3}, 1 \right\}$.

Quick ✓

In Problems 12 and 13, solve each quadratic equation by factoring.

12. $5k^2 + 3k - 1 = 3 - 5k$ **13.** $3x^2 + 9x = 4 - 2x$

▶ **EXAMPLE 5**   **Solving a Quadratic Equation**

Solve: $(2x + 5)(x - 3) = 6x$

Solution

$$(2x + 5)(x - 3) = 6x$$

Multiply using FOIL: $\qquad 2x^2 - x - 15 = 6x$

Write in standard form: $\qquad 2x^2 - 7x - 15 = 0$

Factor the polynomial: $(2x + 3)(x - 5) = 0$

Set each factor equal to 0: $\qquad 2x + 3 = 0 \quad$ or $\quad x - 5 = 0$

Solve each first-degree equation: $\qquad 2x = -3 \quad$ or $\qquad x = 5$

$$x = -\frac{3}{2}$$

Work Smart

Do not attempt to solve $(2x + 5)(x - 3) = 6x$ by setting each factor equal to $6x$ as in $2x + 5 = 6x$ and $x - 3 = 6x$. The Zero-Product Property can be applied only when the product equals zero.

Check We leave it to you to substitute $x = -\frac{3}{2}$ and $x = 5$ into the original equation to verify the answer.

The solution set is $\left\{ -\frac{3}{2}, 5 \right\}$.

Quick ✓

14. *True or False* $x(x - 3) = 4$ means that $x = 4$ or $x - 3 = 4$.

In Problems 15 and 16, solve each quadratic equation by factoring.

15. $(x - 3)(x + 5) = 9$ **16.** $(x + 3)(2x - 1) = 7x - 3x^2$

EXAMPLE 6   **Solving a Quadratic Equation**

Solve: $4k^2 + 9 = -12k$

Solution

$$4k^2 + 9 = -12k$$

Write in standard form: $\qquad 4k^2 + 12k + 9 = 0$

Factor: $(2k + 3)(2k + 3) = 0$

Set each factor equal to 0: $\qquad 2k + 3 = 0 \quad$ or $\quad 2k + 3 = 0$

Solve each first-degree equation: $\qquad 2k = -3 \qquad\qquad 2k = -3$

$$k = -\frac{3}{2} \qquad\qquad k = -\frac{3}{2}$$

Check Substitute $k = -\dfrac{3}{2}$ into the original equation to check the solution.

The solution set is $\left\{ -\dfrac{3}{2} \right\}$.

Notice that in Example 6 the solution $k = -\dfrac{3}{2}$ occurred twice. When this occurs, the solution is called a **double root**.

> **Quick ✓**
>
> *In Problems 17 and 18, solve the quadratic equation by factoring.*
>
> **17.** $9p^2 + 16 = 24p$ **18.** $x^2 + 11x + 24 = x - 1$

EXAMPLE 7 **Solving a Quadratic Equation Containing a GCF**

Work Smart

Another technique that we could have used in Example 7 is to use the Multiplication Property of Equality and divide both sides of the equation by -3. Then we would solve the quadratic equation $x^2 - 2x - 24 = 0$ to find the solution $x = -4$ and $x = 6$.

Solve: $-3x^2 + 6x + 72 = 0$

Solution

$$-3x^2 + 6x + 72 = 0$$

Factor out the GCF, -3: $-3(x^2 - 2x - 24) = 0$

Factor the trinomial: $-3(x - 6)(x + 4) = 0$

Set each factor equal to 0: $-3 = 0$ or $x - 6 = 0$ or $x + 4 = 0$

Solve each first-degree equation: $x = 6$ $x = -4$

The statement $-3 = 0$ is false, so the solutions are 6 and -4.

Check Substitute $x = 6$ and $x = -4$ into the original equation to check. The solution set is $\{-4, 6\}$.

> **Quick ✓**
>
> *In Problems 19 and 20, solve the quadratic equation by factoring.*
>
> **19.** $4x^2 + 12x - 72 = 0$ **20.** $-2x^2 + 2x = -12$

EXAMPLE 8 **Throwing a Ball from the Top of a Building**

A ball is thrown vertically upward from the top of a building 96 feet tall with an initial velocity of 80 feet per second. Solve the equation $-16t^2 + 80t + 96 = 192$ to find the time t (in seconds) at which the ball will be 192 feet from the ground. See Figure 1.

Figure 1

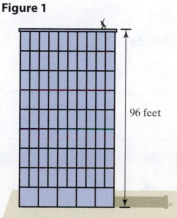

96 feet

Solution

$$-16t^2 + 80t + 96 = 192$$

Write the quadratic equation in standard form: $-16t^2 + 80t + 96 - 192 = 192 - 192$

$$-16t^2 + 80t - 96 = 0$$

Factor out the GCF, -16: $-16(t^2 - 5t + 6) = 0$

Factor the trinomial: $-16(t - 3)(t - 2) = 0$

Set each factor equal to 0: $-16 = 0$ or $t - 3 = 0$ or $t - 2 = 0$

Solve each first-degree equation: $t = 3$ or $t = 2$

The statement $-16 = 0$ is false. The solution is $t = 3$ or $t = 2$. After 2 seconds and after 3 seconds, the ball will be 192 feet from the ground. Do you see why?

Quick ✔

21. A toy rocket is shot directly up from the ground with an initial velocity of 80 feet per second. Solve the equation $-16t^2 + 80t = 64$ to find the time (in seconds) at which the toy rocket is 64 feet from the ground.

▶ ❷ Solve Polynomial Equations of Degree 3 or Higher Using the Zero-Product Property

The Zero-Product Property is also used to solve higher-degree polynomial equations.

EXAMPLE 9 How to Solve a Polynomial Equation of Degree 3 or Higher

Solve: $2x^3 + 3x^2 = 18x + 27$

Step-by-Step Solution

Step 1: Write the equation in standard form.

$$2x^3 + 3x^2 = 18x + 27$$
$$2x^3 + 3x^2 - 18x - 27 = 0$$

Step 2: Factor the polynomial.

$$(2x^3 + 3x^2) + (-18x - 27) = 0$$
$$x^2(2x + 3) - 9(2x + 3) = 0$$
$$(2x + 3)(x^2 - 9) = 0$$
$$(2x + 3)(x + 3)(x - 3) = 0$$

Step 3: Set each factor equal to 0.

$2x + 3 = 0$ or $x + 3 = 0$ or $x - 3 = 0$

Step 4: Solve each first-degree equation.

$2x = -3$ $x = -3$ $x = 3$

$x = -\dfrac{3}{2}$

Step 5: Check each solution.

We leave the check to you.

The solution set is $\left\{ -\dfrac{3}{2}, -3, 3 \right\}$

Quick ✔

22. *True or False* $x^3 - 4x^2 - 12x = 0$ can be solved using the Zero-Product Property.

In Problems 23 and 24, solve each polynomial equation.

23. $3x^3 + 9x^2 + 6x = 0$ **24.** $3x^3 - 4x^2 - 3x + 4 = 0$

6.6 Exercises MathXL PRACTICE

*Problems **1–24** are the Quick ✔s that follow the* **EXAMPLES**.

Building Skills

In Problems 25–30, solve each equation using the Zero-Product Property. See Objective 1.

25. $2x(x + 4) = 0$ **26.** $3x(x + 9) = 0$

27. $(n + 3)(n - 9) = 0$ **28.** $(a + 8)(a - 4) = 0$

29. $(3p + 1)(p - 5) = 0$ **30.** $(4z - 3)(z + 4) = 0$

In Problems 31–34, identify each equation as a linear equation or a quadratic equation. See Objective 1.

31. $3(x + 4) - 1 = 5x + 2$

32. $2x + 1 - (x + 7) = 3x + 1$

33. $x^2 - 2x = 8$

34. $(x + 2)(x - 2) = 14$

In Problems 35–58, solve each quadratic equation by factoring. See Objective 1.

35. $x^2 - 3x - 4 = 0$ **36.** $x^2 + 2x - 63 = 0$

37. $n^2 + 9n + 14 = 0$ **38.** $p^2 - 5p - 24 = 0$

39. $4x^2 + 2x = 0$ **40.** $14x - 49x^2 = 0$

41. $2x^2 - 3x - 2 = 0$ **42.** $3x^2 + x - 14 = 0$

43. $a^2 - 6a + 9 = 0$ **44.** $k^2 + 12k + 36 = 0$

45. $6x^2 = 36x$ **46.** $2x^2 = 5x$

47. $n^2 - n = 6$ **48.** $a^2 - 6a = 16$

49. $x^2 + 5x + 4 = x$ **50.** $x^2 - 5x + 6 = x - 3$

51. $1 - 5m = -4m^2$ **52.** $4p - 3 = -4p^2$

53. $n(n - 2) = 24$ **54.** $p(p + 1) = 2$

55. $(x - 2)(x - 3) = 56$ **56.** $(x + 5)(x - 3) = 9$

57. $(c + 2)^2 = 9$ **58.** $(2a - 1)^2 = 16$

In Problems 59–64, solve each polynomial equation by factoring. See Objective 2.

59. $2x^3 + 2x^2 - 12x = 0$ **60.** $3x^3 + x^2 - 14x = 0$

61. $y^3 + 3y^2 - 4y - 12 = 0$

62. $m^3 + 2m^2 - 9m - 18 = 0$

63. $2x^3 + 3x^2 = 8x + 12$ **64.** $-2x + 3 = 3x^2 - 2x^3$

Mixed Practice

In Problems 65–90, solve each equation. Be careful; the problems represent a mix of linear, quadratic, and third-degree polynomial equations.

65. $(5x + 3)(x - 4) = 0$ **66.** $(3x - 2)(x + 5) = 0$

67. $p^2 - p - 20 = 0$ **68.** $z^2 - 13z + 40 = 0$

69. $4w + 3 = 2w - 7$ **70.** $7y + 3 = 2y - 12$

71. $4a^2 - 25a = 21$ **72.** $5m^2 = 18m + 8$

73. $2a(a + 1) = a^2 + 8$ **74.** $2y(y + 5) = y^2 + 11$

75. $2x^3 + x^2 = 32x + 16$

76. $2n^3 + 4 = n^2 + 8n$

77. $4(b - 3) - 3b = 8$ **78.** $2(p - 3) = p + 1$

79. $y^2 + 5y = 5(y + 20)$ **80.** $3z^2 + 7z = 7(z + 21)$

81. $(x - 2)(x + 7) = x - 2$ **82.** $(x + 3)(x + 1) = x + 3$

83. $(2k - 3)(2k^2 - 9k - 5) = 0$

84. $(7m - 11)(3m^2 - m - 2) = 0$

85. $(w - 3)^2 = 9 + 2w$

86. $3(k + 2)^2 = 5k + 8$

87. $\frac{1}{2}x^2 + \frac{5}{4}x = 3$ **88.** $z^2 + \frac{29}{4}z = 6$

89. $8x^2 + 44x = 24$ **90.** $9q^2 = 3q + 6$

Applying the Concepts

91. Tossing a Ball A ball is thrown vertically upward from the top of a building 80 feet tall with an initial velocity of 64 feet per second. Solve the equation $-16t^2 + 64t + 80 = 128$ to find the time t (in seconds) at which the ball is 128 feet from the ground.

92. Tossing a Ball A ball is thrown vertically upward from the ground with an initial velocity of 64 feet per second. Solve the equation $-16t^2 + 64t = 48$ to find the time t (in seconds) at which the ball is 48 feet from the ground.

93. Convex Polygon A **convex polygon** is a polygon whose interior angles are between $0°$ and $180°$. The number of diagonals D in a convex polygon with n sides is given by the formula $D = \frac{1}{2}n(n - 3)$. Determine the number of sides n in a convex polygon that has 27 diagonals by solving the equation $27 = \frac{1}{2}n(n - 3)$.

94. Consecutive Integers The sum S of the consecutive integers $1, 2, 3, \ldots, n$ is given by the formula $S = \frac{1}{2}n(n+1)$.

That is, $1 + 2 + 3 + \cdots + n = \frac{1}{2}n(n+1)$. Determine the number of consecutive integers that must be added to obtain a sum of 55 by solving the equation

$55 = \frac{1}{2}n(n+1)$.

95. Consecutive Integers The product of two consecutive integers is 12. Find the integers.

96. Consecutive Odd Integers The product of two consecutive odd integers is 143. Find the integers.

97. Consecutive Even Integers Find three consecutive even integers such that the product of the first and the third is 96.

98. Consecutive Odd Integers Find three consecutive odd integers such that the product of the second and the third is 99.

△ **99. Rectangle** The length and width of two sides of a rectangle are consecutive odd integers. The area of the rectangle is 255 square units. Find the dimensions of the rectangle.

△ **100. Garden Area** The State University Landscape Club wants to establish a horticultural garden near the administration building. The length and width of the space that is available are consecutive even integers, and the area of the garden is 440 square feet. Find the dimensions of the garden.

The equation $N = \frac{t^2 - t}{2}$ models the number of soccer games that must be scheduled in a league with t teams, when each team plays every other team exactly once. Use this equation to solve Problems 101 and 102.

101. If a league has 28 games scheduled, how many teams are in the league?

102. If a league has 36 games scheduled, how many teams are in the league?

Extending the Concepts

103. Write a polynomial equation in factored form with integer coefficients that has $x = 3$ and $x = -5$ as solutions. What is the degree of this polynomial equation?

104. Write a polynomial equation in factored form with integer coefficients that has $x = 0$ and $x = -8$ as solutions. What is the degree of this polynomial equation?

105. Write a polynomial equation in factored form with integer coefficients that has $z = 6$ as a double root. What is the degree of this polynomial equation?

106. Write a polynomial equation in factored form with integer coefficients that has $x = -2$ as a double root. What is the degree of this polynomial equation?

107. Write a polynomial equation in factored form with integer coefficients that has $x = -3$, $x = 1$ and $x = 5$ as solutions. What is the degree of this polynomial equation?

108. Write a polynomial equation in factored form with integer coefficients that has $a = \frac{1}{2}$, $a = \frac{2}{3}$, and $a = 1$ as solutions. What is the degree of this polynomial equation?

In Problems 109–112, solve for x in each equation.

109. $x^2 - ax + bx - ab = 0$

110. $x^2 + ax - 6a^2 = 0$

111. $2x^3 - 4ax^2 = 0$

112. $4x^2 - 6ax + 10bx - 15ab = 0$

Explaining the Concepts

113. A student solved a quadratic equation using the following procedure. Explain the student's error, and then work the problem correctly.

$$15x^2 = 5x$$
$$\frac{15x^2}{x} = \frac{5x}{x}$$
$$15x = 5$$
$$x = \frac{5}{15} = \frac{1}{3}$$

114. A student solved a quadratic equation using the following procedure. Explain the student's error, and then work the problem correctly.

$$(x - 4)(x + 3) = -6$$
$$x - 4 = -6 \quad \text{or} \quad x + 3 = -6$$
$$x = -2 \qquad\qquad x = -9$$

115. When solving polynomial equations, we always begin by writing the equation in standard form. Explain why this is important.

116. Explain the difference between a quadratic polynomial and a quadratic equation.

6.7 Modeling and Solving Problems with Quadratic Equations

Objectives

1. Model and Solve Problems Involving Quadratic Equations
2. Model and Solve Problems Using the Pythagorean Theorem

Are You Ready For This Section?

Before getting started, take the following readiness quiz. If you get a problem wrong, go back to the section cited and review the material.

R1. Evaluate: 15^2 [Section 1.7, pp. 56–57]

R2. Solve: $x^2 - 5x - 14 = 0$ [Section 6.6, pp. 408–411]

The solutions to many applied problems require solving polynomial equations by factoring.

 ① Model and Solve Problems Involving Quadratic Equations

Let's begin by solving a quadratic equation to determine the time at which a projectile is at a certain height.

EXAMPLE 1   **Projectile Motion**

A child throws a ball upward off a cliff from a height of 240 feet above sea level. See Figure 2. The height h of the ball above the water (in feet) at any time t (in seconds) can be modeled by the equation

$$h = -16t^2 + 32t + 240$$

  (a) When will the ball be 240 feet above sea level?

  (b) When will the ball strike the water?

Solution

  (a) To determine when the ball will be 240 feet above sea level, we let $h = 240$ and solve the resulting equation.

$$-16t^2 + 32t + 240 = 240$$

Write the equation in standard form:	$-16t^2 + 32t = 0$
Factor out $-16t$:	$-16t(t - 2) = 0$
Set each factor equal to 0:	$-16t = 0$ or $t - 2 = 0$
Solve each first degree equation:	$t = 0$ or $t = 2$

The ball will be at a height of 240 feet the instant it leaves the child's hand and after 2 seconds of flight.

Figure 2

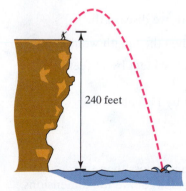

240 feet

  (b) The ball strikes the water when its height is 0, so we let $h = 0$ and solve the resulting equation.

$$-16t^2 + 32t + 240 = 0$$

Factor out -16:	$-16(t^2 - 2t - 15) = 0$
Factor the trinomial:	$-16(t - 5)(t + 3) = 0$
Set each factor equal to 0:	$-16 = 0$ or $t - 5 = 0$ or $t + 3 = 0$
Solve each first-degree equation:	$t = 5$ or $t = -3$

The equation $-16 = 0$ is false, and since t represents time, we discard the solution $t = -3$. Therefore, the ball strikes the water after 5 seconds. ●

Quick ✓

1. A model rocket is fired straight up from the ground. The height h of the rocket (in feet) at any time t (in seconds) can be modeled by the equation $h = -16t^2 + 160t$.

  (a) When will the rocket be 384 feet above the ground?

  (b) When will the rocket strike the ground?

Ready?...Answers   **R1.** 225   **R2.** $\{-2, 7\}$

The next two examples use the problem-solving strategy first presented in Section 2.5.

EXAMPLE 2 **Geometry: Area of a Rectangle**

A carpet installer finds that the length of a rectangular hallway is 3 feet more than twice the width. If the area of the hallway is 44 square feet, what are the dimensions of the hallway? See Figure 3.

Solution

Step 1: Identify This is a geometry problem involving the area of a rectangle.

Step 2: Name Because we know less about the width of the hallway, we let w represent the width. The length of the hallway is 3 feet more than twice the width, so we let $2w + 3$ represent the length.

Figure 3

Step 3: Translate The area of the hallway is 44 square feet. Because the area of a rectangle = (length)(width), we have

$$\text{area} = (\text{length})(\text{width})$$
$$44 = (2w + 3)(w) \qquad \color{red}{\text{The Model}}$$

Step 4: Solve Now solve the equation. Do you see that this equation is a quadratic equation? The first step is to put it in standard form, $ax^2 + bx + c = 0$.

$$w(2w + 3) = 44$$

$$\color{red}{\text{Distribute:}} \qquad 2w^2 + 3w = 44$$

$$\color{red}{\text{Subtract 44 from both sides:}} \qquad 2w^2 + 3w - 44 = 0$$

$$\color{red}{\text{Factor:}} \quad (2w + 11)(w - 4) = 0$$

$$\color{red}{\text{Set each factor equal to 0:}} \; 2w + 11 = 0 \quad \text{or} \quad w - 4 = 0$$

$$\color{red}{\text{Solve each first-degree equation:}} \qquad 2w = -11 \quad \text{or} \quad w = 4$$

$$\frac{2w}{2} = -\frac{11}{2}$$

$$w = -\frac{11}{2}$$

Step 5: Check Since w represents the width of the hallway, we discard the solution $w = -\frac{11}{2}$. If the width of the hallway is 4 feet, then the length would be $2w + 3 = 2(4) + 3 = 11$ feet. The area of a hallway that is 4 feet by 11 feet is $4(11) = 44$ square feet. We have the right answer!

Step 6: Answer The dimensions of the hallway are 4 feet by 11 feet.

Quick ✓

2. A rectangular plot of land has length that is 3 kilometers less than twice its width. If the area of the land is 104 square kilometers, what are the dimensions of the land?

EXAMPLE 3 **Geometry: Area of a Triangle**

The height of a triangle is 5 inches less than the length of the base, and the area of the triangle is 42 square inches. Find the height of the triangle.

Solution

Step 1: Identify This is a geometry problem involving the area of a triangle.

Step 2: Name The height of the triangle is 5 inches less than its base. We will let b represent the base and $b - 5$ represent the height. See Figure 4.

Figure 4

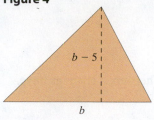

Step 3: Translate We know that the area of a triangle is given by the formula $\text{Area} = \frac{1}{2}(\text{base})(\text{height})$. We also know that the area is 42 square feet, so

$$\text{Area} = \frac{1}{2}(\text{base})(\text{height})$$

$$42 = \frac{1}{2}(b)(b-5) \quad \textcolor{red}{\text{The Model}}$$

Step 4: Solve Our model is a quadratic equation. Put the equation in standard form, $ax^2 + bx + c = 0$.

$$42 = \frac{1}{2}(b)(b-5)$$

Work Smart

Multiply only the $\frac{1}{2}$ by 2—don't multiply the factors b and $(b-5)$ by 2 also.

Multiply by 2 to clear fractions: $\quad 2(42) = 2\left[\frac{1}{2}(b)(b-5)\right]$

$$84 = b(b-5)$$

Distribute: $\quad 84 = b^2 - 5b$

Subtract 84 from both sides: $\quad 0 = b^2 - 5b - 84$

Factor: $\quad 0 = (b-12)(b+7)$

Set each factor equal to 0: $\quad b - 12 = 0 \quad \text{or} \quad b + 7 = 0$

Solve each first-degree equation: $\quad b = 12 \quad \text{or} \quad b = -7$

Step 5: Check Since b represents the base of the triangle, we discard the solution $b = -7$. The base of the triangle is 12 inches, so the height is $b - 5 = 12 - 5 = 7$ inches. The area of a triangle with a 12-inch base and a 7-inch height is $\frac{1}{2} \cdot 12 \cdot 7 = 42$ square inches. We have the right answer!

Step 6: Answer The height of the triangle is 7 inches. ●

> **Quick ✓**
>
> **3.** The base of a triangular garden is 4 yards longer than the height, and the area of the garden is 48 square yards. Find the dimensions of the triangle.

▶ ❷ **Model and Solve Problems Using the Pythagorean Theorem**

The Pythagorean Theorem is a statement about *right triangles*.

> **Definitions**
>
> A **right triangle** is one that contains a **right angle**—that is, an angle that measures 90°. The side of the triangle opposite the 90° angle is the **hypotenuse;** the remaining two sides are the **legs.**

Figure 5

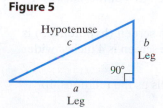

In Figure 5, c represents the length of the hypotenuse and a and b represent the lengths of the legs. Notice the use of the symbol ⌐ to show the 90° angle.

> **The Pythagorean Theorem**
>
> In a right triangle, the square of the length of the hypotenuse equals the sum of the squares of the lengths of the legs. That is, in the right triangle shown in Figure 5,
> $$a^2 + b^2 = c^2 \quad \text{or} \quad \text{leg}^2 + \text{leg}^2 = \text{hypotenuse}^2$$

EXAMPLE 4 Using the Pythagorean Theorem

Find the lengths of the sides of the right triangle in Figure 6.

Solution

Figure 6 shows a right triangle, so we use the Pythagorean Theorem. The legs have lengths x and $x + 7$, and the hypotenuse has length 13.

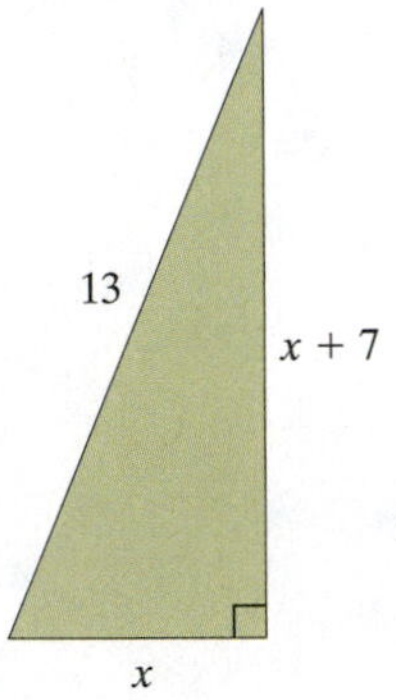

Pythagorean Theorem: $\qquad\qquad a^2 + b^2 = c^2$

Substitute $a = x, b = x + 7, c = 13$: $\qquad x^2 + (x + 7)^2 = 13^2$

$(A + B)^2 = A^2 + 2AB + B^2$: $\quad x^2 + x^2 + 14x + 49 = 169$

Combine like terms: $\qquad\qquad 2x^2 + 14x + 49 = 169$

Write the equation in standard form: $\qquad 2x^2 + 14x - 120 = 0$

Factor out GCF 2: $\qquad\qquad 2(x^2 + 7x - 60) = 0$

Factor: $\qquad\qquad 2(x + 12)(x - 5) = 0$

Set each factor equal to 0: $\quad 2 = 0 \quad$ or $\quad x + 12 = 0 \quad$ or $\quad x - 5 = 0$

Solve each first-degree equation: $\qquad\qquad\qquad x = -12 \quad$ or $\quad x = 5$

The equation $2 = 0$ is false, and the solution $x = -12$ makes no sense because x is a length. Thus $x = 5$ is the length of one leg of the triangle. The other leg is $x + 7 = 5 + 7 = 12$.

To check our solution, replace a by 5 and b by 12 in the Pythagorean Theorem. Because $5^2 + 12^2 = 13^2$, or $25 + 144 = 169$, our answers are correct. ●

Figure 6

13
$x + 7$

x

Quick ✓

4. Find the length of each leg of the right triangle pictured below.

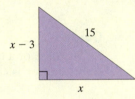

15
$x - 3$
x

5. One leg of a right triangle is 17 inches longer than the other leg. The hypotenuse has a length of 25 inches. Find the lengths of the two legs of the triangle.

EXAMPLE 5 Will the Television Fit in the Media Cabinet?

A rectangular 20-inch television screen (measured diagonally) is 4 inches wider than it is tall. Will this TV fit in your new media cabinet that is 20 inches wide?

Solution

Step 1: Identify This is a geometry problem involving the lengths of the sides of a right triangle.

Step 2: Name We know that the diagonal of the television screen (hypotenuse) is 20 inches long. Let x represent the height of the screen. The screen is 4 inches wider than it is tall, so $x + 4$ represents the width.

Step 3: Translate Use the Pythagorean Theorem, which tells us the relationship between the lengths of the sides of a right triangle.

Pythagorean Theorem: $\qquad\qquad a^2 + b^2 = c^2$

Substitute $a = x, b = x + 4, c = 20$: $\quad x^2 + (x + 4)^2 = 20^2$

Work Smart

Remember

$(x + 4)^2 \neq x^2 + 16$

$(x + 4)^2 = x^2 + 8x + 16$

Step 4: Solve

$$x^2 + (x + 4)^2 = 20^2$$

$(A + B)^2 = A^2 + 2AB + B^2$: $x^2 + x^2 + 8x + 16 = 400$

Combine like terms: $2x^2 + 8x + 16 = 400$

Subtract 400 from both sides: $2x^2 + 8x - 384 = 0$

Factor: $2(x^2 + 4x - 192) = 0$

Factor: $2(x + 16)(x - 12) = 0$

Set each factor equal to 0: $2 = 0$ or $x + 16 = 0$ or $x - 12 = 0$

Solve: $x = -16$ or $x = 12$

Work Smart

Rather than setting 2 equal to 0, we could have divided both sides of the equation by 2.

Step 5: Check The statement $2 = 0$ is false. Since x represents a length, we discard the solution $x = -16$. If the shorter leg of the triangle (height of the TV) is 12 inches, then the longer leg (width of the television) is $x + 4 = 12 + 4 = 16$ inches. Because $12^2 + 16^2 = 20^2$, our answer is correct!

Step 6: Answer The television screen is 16 inches wide, so it will fit in the new cabinet.

> **Quick ✓**
>
> **6.** A rectangular 10-inch television screen (measured diagonally) is 2 inches wider than it is tall. What are the dimensions of the TV screen?

6.7 Exercises   MyMathLab®   MathXL PRACTICE

Exercise numbers in green have complete video solutions in MyMathLab

*Problems **1–6** are the Quick ✓s that follow the EXAMPLES.*

Building Skills

In Problems 7–10, use the given area to find the missing sides of the rectangle. See Objective 1.

△ **7.**

$A = 56$

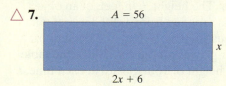

x

$2x + 6$

△ **8.** $A = 250$

△ **9.** $A = 18$

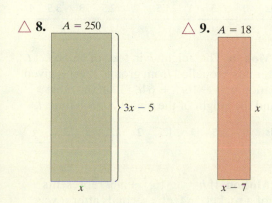

$3x - 5$

x

x

$x - 7$

△ **10.** $A = 364$

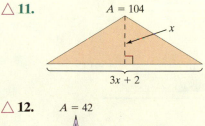

x

$x + 12$

In Problems 11–14, use the given area to find the height and base of the triangle. See Objective 1.

△ **11.** $A = 104$

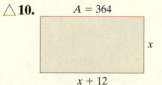

x

$3x + 2$

△ **12.** $A = 42$

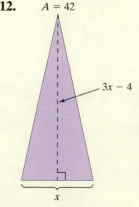

$3x - 4$

x

△ **13.**

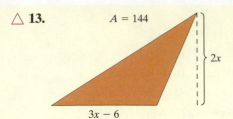

$A = 144$
$2x$
$3x - 6$

△ **14.**

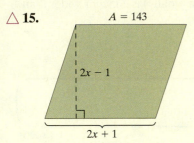

$A = 84$
$x + 4$
$2x - 8$

In Problems 15–18, use the given area to find the dimensions of the quadrilateral. See Objective 1.

△ **15.**

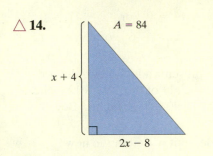

$A = 143$
$2x - 1$
$2x + 1$

△ **16.**

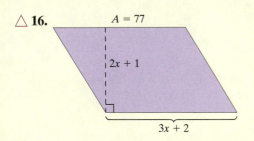

$A = 77$
$2x + 1$
$3x + 2$

△ **17.**

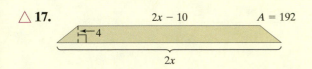

$2x - 10$ $A = 192$
4
$2x$

△ **18.**

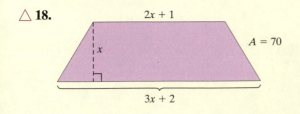

$2x + 1$
x
$A = 70$
$3x + 2$

In Problems 19–22, use the Pythagorean Theorem to find the lengths of the sides of the triangle. See Objective 2.

△ **19.**

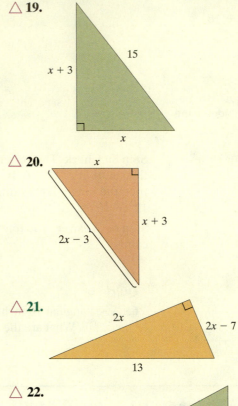

15
$x + 3$
x

△ **20.**

x
$x + 3$
$2x - 3$

△ **21.**

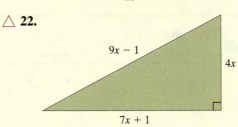

$2x$ $2x - 7$
13

△ **22.**

$9x - 1$
$4x$
$7x + 1$

Applying the Concepts

23. Projectile Motion The height, h in feet, of an object t seconds after it is dropped from a cliff 256 feet tall is given by the equation $h = -16t^2 + 256$. Suppose you dropped your glasses off the cliff. Fill in the table below to find the height of your glasses at each time, t.

Time, in Seconds	0	0.5	1	1.5	2	2.5	3	3.5	4
Height, in Feet									

24. Projectile Motion The height, h in feet, of an object t seconds after it is propelled from ground level is given by the equation $h = -16t^2 + 64t$. Fill in the table below to find the height of the object at each time, t.

Time, in Seconds	0	0.5	1	1.5	2	2.5	3	3.5	4
Height, in Feet									

25. Projectile Motion If $h = -16t^2 + 96t$ represents the height of a rocket, in feet, t seconds after it was fired, when will the rocket hit the ground? (*Hint:* The rocket is on the ground when $h = 0$.)

26. Projectile Motion If $h = -16t^2 + 96t$ represents the height of a rocket, in feet, t seconds after it was fired, when will the rocket be 144 feet high?

△ **27. Rectangular Room** The length of a rectangular room is 8 meters more than the width. If the area of the room is 48 square meters, find the dimensions of the room.

△ **28. Rectangular Room** The width of a rectangular room is 4 feet less than the length. If the area of the room is 21 square feet, find the dimensions of the room.

△ **29. Sailboat** The sail on a sailboat is in the shape of a triangle. If the height of the sail is 3 times the length of the base, and the area is 54 square feet, find the dimensions of the sail.

△ **30. Triangle** The base of a triangle is 2 meters more than the height. If the area of the triangle is 24 square meters, find the base and height of the triangle.

△ **31. Big-Screen TV** Your big-screen TV measures 50 inches on the diagonal. If the front of the TV measures 40 inches across the bottom, find the height of the TV.

△ **32. Big-Screen TV** Hannah owns a 29-inch TV (that is, it measures 29 inches on the diagonal). If the television is 20 inches tall, find the distance across the bottom.

△ **33. Dimensions of a Rectangle** The length of a rectangle is 1 mm more than twice the width. If the area is 300 square mm, find the dimensions of the rectangle.

△ **34. Playing Field Dimensions** The length of a rectangular playing field is 3 yards less than twice the width. If the area of the field is 104 square yards, what are its dimensions?

△ **35. Rectangle and Square** A rectangle and a square have the same area. The width of the rectangle is 6 cm less than the side of the square, and the length of the rectangle is 5 cm more than twice the side of the square. What are the dimensions of the rectangle?

△ **36. Square and Rectangle** A rectangle and a square have the same area. The width of the rectangle is 2 in. less than the side of the square, and the length of the rectangle is 3 in. less than twice the side of the square. What are the dimensions of the rectangle?

△ **37. Hanging a Plasma TV** David is installing new brickwork around his fireplace, but he wants an opening above the fireplace to install a plasma television. The TV he purchases is 53 inches on the

diagonal, excluding the casing around the TV screen. The length of the TV is 17 inches more than the width.

(a) What are the dimensions of the television screen?

(b) If the casing around the entire TV is 1.5 inches, what should the dimensions of the opening in the brickwork be?

△ **38. Jasper's Big-Screen TV** Jasper purchased a new big-screen TV. The TV screen is 10 inches wider than it is tall and is surrounded by a casing that is 2 inches wide.

(a) Jasper lost his tape measure but sees on the box that the TV measures 50 inches on the diagonal, including the casing. What are the dimensions of the TV screen?

(b) Jasper knows that the size of the opening where he wants the TV installed is 29 by 42 inches. Will this TV fit into his space?

△ **39. How Tall Is the Pole?** The owners of A & L Auto Sales want to hang decorative banners to draw attention to their used car offerings. They have 10 feet of wire to attach banners and know that, according to zoning regulations, the distance from the pole to the point where the wire attaches to the ground must be 2 feet greater than the height of the pole. How tall can the pole be?

△ **40. Sailing on Lake Erie** A sail on a sailboat is in the shape of a right triangle. The longest side of the sail is 13 feet long, and one side of the sail is 7 feet longer than the other. Find the dimensions of the sail.

Extending the Concepts

△ **41. Gardening** Beth has 28 feet of fence to enclose a small garden. The length of the garden lies along her house, so only three sides require fencing. Find the dimensions of the garden if she encloses 98 square feet of the garden with the fence.

△ **42. Volume of a Box** A rectangular solid has a square base and is 8 meters high. What are the dimensions of the base if the volume of the solid is 128 cubic meters?

Explaining the Concepts

43. In a projectile motion problem such as Quick Check 1, two positive numbers satisfy the problem. Explain the meaning of each of these numbers, and then give a case where you would have only one positive solution.

44. Can the Pythagorean Theorem be used to find the length of a side of any triangle? Explain your answer.

Chapter 6 Activity: Which One Does Not Belong?

Focus: Factoring polynomials.

Time: 20–30 minutes

Group size: 3–4

Each group member should decide which polynomial from each row does not belong. Answers will vary. As a group, discuss each

member's results. Each group member should be prepared to explain WHY that particular polynomial does not belong with the other three. Be creative!

	A	B	C	D
1.	$15x^2 - 10ax - 15xy + 10ay$	$xr - 3xs - ry + 3sy$	$2ax - 14bx - 2ay + 14by$	$3ab - 3ay - 3bx + 3xy$
2.	$-x^2 - x + 6$	$3x^2 + 39x + 120$	$2x^2 - 12xy - 54y^2$	$7x^2 + 7x - 140$
3.	$3x^2 - 16x + 21$	$6x^2 - 3x + 15$	$3x^2 - 13x - 56$	$3x^2 + x + 1$
4.	$9a^2 + 30a + 25$	$16x^2 - 24x + 9$	$4x^2 - 12xy + 9y^2$	$18s^2 - 60s + 9$
5.	$4x^2 + 13x + 10$	$16x^2 + 40xy + 25y^2$	$16x^2 - 25y^2$	$8x^2 - 2xy - 15y^2$

Chapter 6 Review

Section 6.1 Greatest Common Factor and Factoring by Grouping

KEY CONCEPTS

- **To find the greatest common factor of two or more expressions,**

  Step 1: Find the GCF of the coefficients of each variable expression.

  Step 2: For each variable, determine the smallest exponent that the variable is raised to.

  Step 3: Find the product of the common factors found in Steps 1 and 2. This expression is the GCF.

- **To factor a polynomial using the greatest common factor,**

  Step 1: Identify the greatest common factor (GCF) of the terms that make up the polynomial.

  Step 2: Rewrite each term as the product of the GCF and the remaining factor.

  Step 3: Use the Distributive Property "in reverse" to factor out the GCF.

  Step 4: Check using the Distributive Property.

KEY TERMS

Factors
Greatest common factor
Factoring by grouping

You Should Be Able To...	EXAMPLE	Review Exercises
1 Find the greatest common factor (GCF) of two or more expressions (p. 368)	Examples 1 through 5	1–10
2 Factor out the greatest common factor in polynomials (p. 371)	Examples 6 through 10	11–16
3 Factor polynomials by grouping (p. 373)	Examples 11 through 13	17–20

In Problems 1–10, find the greatest common factor (GCF) of each group of expressions.

1. 24, 36

2. 27, 54

3. 10, 20, 30

4. 8, 16, 28

5. x^4, x^2, x^8

6. m^3, m, m^5

7. $30a^2b^4, 45ab^2$

8. $18x^4y^2z^3, 24x^3y^5z$

9. $4(2a + 1)^2$ and $6(2a + 1)^3$

10. $9(x - y)$ and $18(x - y)$

In Problems 11–16, factor the GCF from the polynomial.

11. $-18a^3 - 24a^2$

12. $-9x^2 + 12x$

13. $15y^2z + 5y^7z + 20y^3z$

14. $7x^3y - 21x^2y^2 + 14xy^3$

15. $x(5 - y) + 2(5 - y)$ **16.** $z(a + b) + y(a + b)$

18. $2xy + y^2 + 2x^2 + xy$

19. $8x + 16 - xy - 2y$

In Problems 17–20, factor by grouping.

20. $xy^2 + x - 3y^2 - 3$

17. $5m^2 + 2mn + 15mn + 6n^2$

Section 6.2 Factoring Trinomials of the Form $x^2 + bx + c$

KEY CONCEPT

- **To factor a trinomial of the form $x^2 + bx + c$, we use the following steps.**

  Step 1: Find the pair of integers whose product is c and whose sum is b. That is, determine m and n such that $mn = c$ and $m + n = b$.

  Step 2: Write $x^2 + bx + c = (x + m)(x + n)$.

  Step 3: Check your work by multiplying the binomials.

KEY TERMS

Quadratic trinomial
Leading coefficient
Prime polynomial

You Should Be Able To...	EXAMPLE	Review Exercises
❶ Factor trinomials of the form $x^2 + bx + c$ (p. 377)	Examples 1 through 7	21–28
❷ Factor out the GCF, then factor $x^2 + bx + c$ (p. 382)	Examples 8 and 9	29–34

In Problems 21–34, factor completely. If the polynomial cannot be factored, say it is prime.

21. $x^2 + 5x + 6$ **22.** $x^2 + 6x + 8$

23. $x^2 - 21 - 4x$ **24.** $3x + x^2 - 10$

25. $m^2 + m + 20$ **26.** $m^2 - 6m - 5$

27. $x^2 - 8xy + 15y^2$ **28.** $m^2 + 4mn - 5n^2$

29. $-p^2 - 11p - 30$ **30.** $-y^2 + 2y + 15$

31. $3x^3 + 33x^2 + 36x$ **32.** $4x^2 + 36x + 32$

33. $2x^2 - 2xy - 84y^2$ **34.** $4y^3 + 12y^2 - 40y$

Section 6.3 Factoring Trinomials of the Form $ax^2 + bx + c$, $a \neq 1$

KEY CONCEPTS

- **Factoring $ax^2 + bx + c$, $a \neq 1$, by grouping, where a, b, and c have no common factors**

  Step 1: Find the value of ac.

  Step 2: Find the pair of integers, m and n, whose product is ac and whose sum is b.

  Step 3: Write $ax^2 + bx + c = ax^2 + mx + nx + c$.

  Step 4: Factor the expression in Step 3 by grouping.

  Step 5: Check by multiplying the factors.

- **Factoring $ax^2 + bx + c$, $a \neq 1$, using trial and error, where a, b, and c have no common factors**

  Step 1: List the possibilities for the first terms of each binomial whose product is ax^2.

  $$(\Box x + \_\_)(\Box x + \_\_) = ax^2 + bx + c$$

  Step 2: List the possibilities for the last terms of each binomial whose product is c.

  $$(\_\_x + \Box)(\_\_x + \Box) = ax^2 + bx + c$$

  Step 3: Write out all the combinations of factors found in Steps 1 and 2. Multiply the binomials out until a product is found that equals the trinomial.

You Should Be Able To...	EXAMPLE	Review Exercises
❶ Factor $ax^2 + bx + c$, $a \neq 1$, using grouping (p. 385)	Examples 1 through 4	35–44
❷ Factor $ax^2 + bx + c$, $a \neq 1$, using trial and error (p. 388)	Examples 5 through 10	35–44

In Problems 35–44, factor completely using any method you wish. If a polynomial cannot be factored, say it is prime.

35. $5y^2 + 14y - 24$

36. $6y^2 - 41y - 7$

37. $-5x + 2x^2 + 3$

38. $23x + 6x^2 + 7$

39. $2x^2 - 7x - 6$

40. $8m^2 + 18m + 9$

41. $9m^3 + 30m^2n + 21mn^2$

42. $14m^2 + 16mn + 2n^2$

43. $-15x^3 - x^2 + 2x$

44. $-12p^4 - 2p^3 + 2p^2$

Section 6.4   Factoring Special Products

KEY CONCEPTS

- **Perfect Square Trinomials**
  $A^2 + 2AB + B^2 = (A + B)^2$; $A^2 - 2AB + B^2 = (A - B)^2$

- **Difference of Two Squares**
  $A^2 - B^2 = (A - B)(A + B)$

- **Sum or Difference of Two Cubes**
  $A^3 + B^3 = (A + B)(A^2 - AB + B^2)$; $A^3 - B^3 = (A - B)(A^2 + AB + B^2)$

KEY TERMS

Perfect square trinomial
Difference of two squares
Sum of two cubes
Difference of two cubes

You Should Be Able To...	EXAMPLE	Review Exercises
1 Factor perfect square trinomials (p. 395)	Examples 1 through 4	45–50
2 Factor the difference of two squares (p. 397)	Examples 5 through 7	51–56
3 Factor the sum or difference of two cubes (p. 399)	Examples 8 and 9	57–62

In Problems 45–62, factor completely. If a polynomial cannot be factored, say it is prime.

45. $4x^2 - 12x + 9$

46. $x^2 - 10x + 25$

47. $x^2 + 6xy + 9y^2$

48. $9x^2 + 24xy + 4y^2$

49. $8m^2 + 8m + 2$

50. $2m^2 - 24m + 72$

51. $4x^2 - 25y^2$

52. $49x^2 - 36y^2$

53. $x^2 + 25$

54. $x^2 + 100$

55. $x^4 - 81$

56. $x^4 - 625$

57. $m^3 + 27$

58. $m^3 + 125$

59. $27p^3 - 8$

60. $64p^3 - 1$

61. $y^9 + 64z^6$

62. $8y^3 + 27z^6$

Section 6.5   Summary of Factoring Techniques

Steps for Factoring

Step 1: Is there a greatest common factor (GCF)? If so, factor out the GCF.

Step 2: Count the number of terms.

Step 3:

(a) Two terms (binomials)

- Is it the difference of two squares? If so,
  $$A^2 - B^2 = (A - B)(A + B)$$

- Is it the sum of two squares? If so, Stop! The expression is prime.

- Is it the difference of two cubes? If so,
  $$A^3 - B^3 = (A - B)(A^2 + AB + B^2)$$

- Is it the sum of two cubes? If so,
  $$A^3 + B^3 = (A + B)(A^2 - AB + B^2)$$

(b) Three terms (trinomials)

- Is it a perfect square trinomial? If so,
  $$A^2 + 2AB + B^2 = (A + B)^2$$
  or $A^2 - 2AB + B^2 = (A - B)^2$

- Is the leading coefficient 1? If so,
  $$x^2 + bx + c = (x + m)(x + n)$$
  where $mn = c$ and $m + n = b$

- Is the coefficient of the squared term different from 1? If so, factor using trial and error or grouping.

(c) Four terms

- Use factoring by grouping

Step 4: Check your work by multiplying out the factors.

You Should Be Able To...	EXAMPLE	Review Exercises
1 Factor polynomials completely (p. 402)	Examples 1 through 7	63–78

In Problems 63–78, factor completely. If a polynomial cannot be factored, say it is prime.

63. $15a^3 - 6a^2b - 25ab^2 + 10b^3$

64. $12a^2 - 9ab + 4ab - 3b^2$

65. $x^2 - xy - 48y^2$ **66.** $x^2 - 10xy - 24y^2$

67. $x^3 - x^2 - 42x$ **68.** $3x^6 - 30x^5 + 63x^4$

69. $6x^2 + 11x + 3$ **70.** $10z^2 + 9z - 9$

71. $27x^3 + 8$ **72.** $8z^3 - 1$

73. $4y^2 + 18y - 10$ **74.** $5x^3y^2 - 8x^2y^2 + 3xy^2$

75. $25k^2 - 81m^2$ **76.** $x^4 - 9$

77. $16m^2 + 1$ **78.** $m^4 + 25$

Section 6.6 Solving Polynomial Equations by Factoring

KEY CONCEPT

- **The Zero-Product Property**

  If the product of two factors is zero, then at least one of the factors is 0. That is, if $ab = 0$, then $a = 0$ or $b = 0$ or both a and b are 0.

KEY TERMS

Polynomial equation
Degree of a polynomial
  equation
Quadratic equation
Second-degree equation
Standard form

You Should Be Able To...	EXAMPLE	Review Exercises
1 Solve quadratic equations using the Zero-Product Property (p. 408)	Examples 1 through 8	79–88
2 Solve polynomial equations of degree 3 or higher using the Zero-Product Property (p. 414)	Example 9	89–92

In Problems 79–92, solve each equation by factoring.

79. $(x - 4)(2x - 3) = 0$

80. $(2x + 1)(x + 7) = 0$

81. $x^2 - 12x - 45 = 0$

82. $x^2 - 7x + 10 = 0$

83. $3x^2 + 6x = 0$

84. $4x^2 + 18x = 0$

85. $3x(x + 1) = 2x^2 + 5x + 3$

86. $2x^2 + 6x = (3x + 1)(x + 3)$

87. $5x^2 + 7x + 16 = 3x^2 - 5x$

88. $8x^2 - 10x = -2x + 6$

89. $x^3 = -11x^2 + 42x$

90. $-3x^2 = -x^3 + 18x$

91. $3x^3 - 4x^2 - 27x + 36 = 0$

92. $2x^3 + 5x^2 - 8x = 20$

Section 6.7 Modeling and Solving Problems with Quadratic Equations

KEY CONCEPT

- **The Pythagorean Theorem**

  In a right triangle, the square of the length of the hypotenuse is equal to the sum of the squares of the lengths of the legs. That is, if a and b are the lengths of the legs and c is the length of the hypotenuse, then $a^2 + b^2 = c^2$ or $\text{leg}^2 + \text{leg}^2 = \text{hypotenuse}^2$.

KEY TERMS

Right triangle
Right angle
Hypotenuse
Legs

You Should Be Able To...	EXAMPLE	Review Exercies
1 Model and solve problems involving quadratic equations (p. 417)	Examples 1 through 3	93–96
2 Model and solve problems using the Pythagorean Theorem (p. 419)	Examples 4 and 5	97, 98

93. Geysers If $h = -16t^2 + 80t$ represents the height of a jet of water from a geyser t seconds after the geyser erupts, when will the water hit the ground?

94. More Geysers If $h = -16t^2 + 80t$ represents the height of the jet of water from a geyser at time t, when will the jet of water be 96 feet high?

△ **95. Tabletop** The length of a rectangular tabletop is 3 feet shorter than twice its width. If the area of the tabletop is 54 square feet, what are the dimensions of the tabletop?

△ **96. Tarp** The length of a rectangular tarp is 1 yard shorter than twice its width. If the area of the tarp is 15 square yards, what are the dimensions of the tarp?

△ **97. Right Triangle** The shorter leg of a right triangle is 2 feet shorter than the longer leg. The hypotenuse is 10 feet. How long is each leg?

△ **98. Right Triangle** The shorter leg of a right triangle is 14 feet shorter than the longer leg. The hypotenuse is 26 feet. How long is each leg?

Chapter 6 Test

Remember to use your Chapter Test Prep Video to see fully worked-out solutions to any of these problems you would like to review.

1. Find the GCF of $16x^5y^2$, $20x^4y^6$, and $24x^6y^8$.

In Problems 2–16, factor each polynomial completely. If the polynomial cannot be factored, say it is prime.

2. $x^4 - 256$

3. $18x^3 - 9x^2 - 27x$

4. $xy - 7y - 4x + 28$

5. $27x^3 + 125$

6. $y^2 - 8y - 48$

7. $6m^2 - m - 5$

8. $4x^2 + 25$

9. $4(x - 5) + y(x - 5)$

10. $3x^2y - 15xy - 42y$

11. $x^2 + 4x + 12$

12. $2x^6 - 54y^3$

13. $9x^3 + 39x^2 + 12x$

14. $6m^2 + 7m + 2$

15. $4m^2 - 6mn + 4$

16. $25x^2 + 70xy + 49y^2$

In Problems 17 and 18, solve the equation by factoring.

17. $5x^2 = -16x - 3$

18. $5x^3 - 20x^2 + 20x = 0$

19. The length of a rectangle is 8 inches shorter than three times its width. If the area of the rectangle is 35 square inches, find the dimensions of the rectangle.

20. The hypotenuse of a right triangle is 1 inch longer than the longer leg. The shorter leg is 7 inches shorter than the longer leg. Find the length of all three sides of the triangle.

7 Rational Expressions and Equations

Environmental scientists often use cost-benefit models to estimate the cost of removing a certain percentage of a pollutant from the environment. An equation such as $C = \dfrac{25x}{100 - x}$ may be used by environmental protection authorities to determine the cost C, in millions of dollars, to remove x percent of the pollutants from a lake or stream. See Problem 91 in Section 7.7.

The Big Picture: Putting It Together

In Section 1.5, we discussed how to simplify, add, subtract, multiply, and divide rational numbers expressed as fractions. Recall that a rational number is a quotient of two integers, where the denominator is not zero. In this chapter, we discuss *rational expressions*. A *rational expression* is the quotient of two polynomials where the denominator is not zero.

The skills we discussed in Section 1.5 apply to simplifying, adding, subtracting, multiplying, and dividing rational expressions. Performing these operations also requires the skills learned in Chapter 5 for polynomial operations plus the factoring skills acquired in Chapter 6. Being able to factor a polynomial plays a *huge* role in being able to simplify a rational expression, so make sure you are good at factoring. Because the methods we use to perform operations on rational expressions are identical to those we use on rational numbers, we really aren't learning any new methods here—just new ways to apply skills we already have!

7.1 Simplifying Rational Expressions

Objectives

1. Evaluate a Rational Expression
2. Determine Values for Which a Rational Expression Is Undefined
3. Simplify Rational Expressions

Are You Ready for This Section?

R1. Evaluate $\dfrac{3x + y}{2}$ for $x = -1$ and $y = 7$. [Section 1.8, pp. 64–65]

R2. Factor: $2x^2 + x - 3$ [Section 6.3, pp. 385–393]

R3. Solve: $3x^2 - 5x - 2 = 0$ [Section 6.6, pp. 408–414]

R4. Write $\dfrac{21}{70}$ in lowest terms. [Section 1.2, pp. 12–13]

R5. Divide: $\dfrac{x^3 y^4}{xy^2}$ [Section 5.4, pp. 334–335]

Recall that a rational number is the quotient of two integers, where the denominator is not zero.

Work Smart

The numbers $\dfrac{4}{3}, -\dfrac{5}{2}$, and 18 are examples of rational numbers.

> ### Definition
>
> A **rational expression** is the quotient of two polynomials. That is, a rational expression is written in the form $\dfrac{p}{q}$, where p and q are polynomials and $q \neq 0$.

Some examples of rational expressions are

(a) $\dfrac{x - 3}{4x + 1}$ **(b)** $\dfrac{x^2 - 4x - 12}{x^2 - 5}$ **(c)** $\dfrac{3a^2 + 7b + 2b^2}{a^2 - 2ab + 8b^2}$

Expressions (a) and (b) are rational expressions in one variable, x, and expression (c) is a rational expression in two variables, a and b.

▶ 1 Evaluate a Rational Expression

To **evaluate** a rational expression, we replace the variable with its assigned numerical value and perform the arithmetic, using the order of operations.

EXAMPLE 1 **Evaluating a Rational Expression**

Evaluate $\dfrac{-2}{x + 5}$ for **(a)** $x = -3$ **(b)** $x = 11$.

Solution

(a) We substitute -3 for x and simplify.

$$\frac{-2}{x + 5} = \frac{-2}{-3 + 5}$$
$$= \frac{-2}{2} = -1$$

(b) Substitute 11 for x and simplify.

$$\frac{-2}{x + 5} = \frac{-2}{11 + 5}$$
$$= \frac{-2}{16}$$

Divide out common factors: $\qquad = \dfrac{-1 \cdot 2}{8 \cdot 2}$

$$= -\frac{1}{8}$$

Ready? ...Answers

R1. 2   **R2.** $(2x + 3)(x - 1)$

R3. $\left\{-\dfrac{1}{3}, 2\right\}$ **R4.** $\dfrac{3}{10}$ **R5.** $x^2 y^2$

EXAMPLE 2 **Evaluating a Rational Expression of a Higher Degree**

Evaluate $\dfrac{p^2 - 9}{2p^2 + p - 10}$ for **(a)** $p = 1$ **(b)** $p = -2$.

Solution

(a) Substitute 1 for p and simplify.

$$\frac{p^2 - 9}{2p^2 + p - 10} = \frac{(1)^2 - 9}{2(1)^2 + 1 - 10}$$

$$= \frac{1 - 9}{2(1) + 1 - 10}$$

$$= \frac{-8}{2 + 1 - 10}$$

$$= \frac{-8}{-7}$$

$$= \frac{8}{7}$$

Work Smart

Remember to use parentheses when substituting a number for a variable with an exponent (as in Example 2(b)).

(b) Substitute -2 for p and simplify.

$$\frac{p^2 - 9}{2p^2 + p - 10} = \frac{(-2)^2 - 9}{2(-2)^2 + (-2) - 10}$$

$$= \frac{4 - 9}{2(4) - 2 - 10}$$

$$= \frac{-5}{8 - 2 - 10}$$

$$= \frac{-5}{-4}$$

$$= \frac{5}{4}$$

●

Quick ✓

1. The quotient of two polynomials is called a _____ _____.

In Problems 2 and 3, evaluate the rational expression for (a) $x = -5$ (b) $x = 3$.

2. $\dfrac{3}{5x + 1}$

3. $\dfrac{x^2 + 6x + 9}{x + 1}$

EXAMPLE 3 **Evaluating a Rational Expression with More Than One Variable**

Evaluate: **(a)** $\dfrac{2a + 4b}{3c - 1}$ for $a = 1, b = -2, c = 3$ **(b)** $\dfrac{w - 3v}{v + 3w}$ for $w = 2$ and $v = -4$

Solution

(a) We substitute 1 for a, -2 for b, and 3 for c.

$$\frac{2a + 4b}{3c - 1} = \frac{2(1) + 4(-2)}{3(3) - 1}$$

$$= \frac{2 + (-8)}{9 - 1}$$

$$= \frac{-6}{8}$$

Divide out common factors: $$= \frac{2 \cdot -3}{2 \cdot 4}$$

$$= -\frac{3}{4}$$

(b) We substitute 2 for w and -4 for v.

$$\frac{w - 3v}{v + 3w} = \frac{2 - 3(-4)}{-4 + 3(2)}$$

$$= \frac{2 + 12}{-4 + 6}$$

$$= \frac{14}{2}$$

$$= 7$$

Quick ✓

In Problems 4 and 5, evaluate the rational expression for the given values of each variable.

4. $\dfrac{3m - 5n}{p - 4}$ for $m = 2, n = -1$, and $p = 6$ **5.** $\dfrac{5y + 2}{2y - z}$ for $y = 2, z = 1$

▶ ❷ Determine Values for Which a Rational Expression Is Undefined

Work Smart

A rational expression is undefined if the *denominator* equals 0. It is okay for the *numerator* to equal 0.

A rational expression is **undefined** for values of the variable(s) that make the denominator zero. **To find the values that make a rational expression undefined, set the denominator equal to zero and solve the resulting equation.** Finding the values that make a rational expression undefined is important because it will play a role in solving rational equations, presented later in this chapter.

EXAMPLE 4   **Determining the Values for Which a Rational Expression Is Undefined**

Find the value of x for which the rational expression $\dfrac{2}{x + 3}$ is undefined.

Solution

Find all values of x in the rational expression $\dfrac{2}{x + 3}$ that make the denominator, $x + 3$, zero.

Set the denominator equal to 0: $x + 3 = 0$
Subtract 3 from both sides: $x = -3$

Since -3 makes the denominator, $x + 3$, equal 0, $\dfrac{2}{x + 3}$ is undefined for $x = -3$.

EXAMPLE 5   **Determining the Values for Which a Rational Expression Is Undefined**

Find the values of x for which the rational expression $\dfrac{8x}{x^2 - 2x - 3}$ is undefined.

Solution

Find all values of x in the rational expression $\dfrac{8x}{x^2 - 2x - 3}$ that make the denominator, $x^2 - 2x - 3$, zero.

Set the denominator equal to 0: $x^2 - 2x - 3 = 0$
Use the Zero-Product Property: $(x - 3)(x + 1) = 0$
Set each factor equal to 0: $x - 3 = 0$ or $x + 1 = 0$
Solve each equation for x: $x = 3$ or $x = -1$

Work Smart

Do not divide out like factors when finding values of the variable for which the rational expression is undefined.

The rational expression $\dfrac{8x}{x^2 - 2x - 3}$ is undefined for $x = 3$ or $x = -1$.

Work Smart

In Example 5, the expression $\dfrac{8x}{x^2 - 2x - 3}$ is *not* undefined at $x = 0$ because 0 causes the *numerator* to equal 0, which is okay.

Check Let's verify our answer by evaluating $\dfrac{8x}{x^2 - 2x - 3}$ for each of these values.

If $x = 3$: $\dfrac{8(3)}{(3)^2 - 2(3) - 3} = \dfrac{24}{9 - 6 - 3} = \dfrac{24}{0}$ undefined

If $x = -1$: $\dfrac{8(-1)}{(-1)^2 - 2(-1) - 3} = \dfrac{-8}{1 + 2 - 3} = \dfrac{-8}{0}$ undefined

Quick ✓

6 A rational expression is _____ for those values of the variable(s) that make the denominator zero.

In Problems 7–9, find the value(s) for which the rational expression is undefined.

7. $\dfrac{3}{x + 7}$

8. $\dfrac{n - 3}{3n + 5}$

9. $\dfrac{8x}{x^2 + 2x - 3}$

▶ ❸ Simplify Rational Expressions

Remember that we write fractions in lowest terms by using the fact that $\dfrac{a \cdot c}{b \cdot c} = \dfrac{a}{b}$. Thus, to write $\dfrac{18}{45}$ in lowest terms, we factor both 18 and 45 and divide out common factors.

$$\frac{18}{45} = \frac{2 \cdot 3 \cdot 3}{3 \cdot 3 \cdot 5} = \frac{2}{5}$$

To **simplify** a rational expression means to write the rational expression in the form $\dfrac{p}{q}$, where p and q are polynomials that have no common factors. We use the same ideas to simplify rational expressions as we use to write fractions in lowest terms.

In Words

A rational expression is simplified if the numerator and the denominator share no common factor other than 1.

Simplifying Rational Expressions

If p, q, and r are polynomials, then

$$\frac{p \cdot r}{q \cdot r} = \frac{p}{q} \quad \text{if } q \neq 0 \text{ and } r \neq 0$$

To simplify a rational expression, factor the numerator, factor the denominator, and divide out the common factors.

EXAMPLE 6 **How to Simplify a Rational Expression**

Simplify: $\dfrac{7x^2 + 14x}{49x}$

Step-by-Step Solution

Step 1: Completely factor the numerator and the denominator.

$$\frac{7x^2 + 14x}{49x} = \frac{7x(x + 2)}{7 \cdot 7 \cdot x}$$

Step 2: Divide out common factors.

$$= \frac{\cancel{7}\cancel{x}(x + 2)}{\cancel{7} \cdot 7 \cdot \cancel{x}}$$

$$= \frac{x + 2}{7}$$

When we simplify a rational expression by dividing out common factors, we change the replacement values allowed for the variable. In Example 6, we found that $\dfrac{7x^2 + 14x}{49x}$ equals $\dfrac{x + 2}{7}$. This is not quite true since x cannot equal 0 in

$\dfrac{7x^2 + 14x}{49x}$, but x can equal 0 in $\dfrac{x + 2}{7}$. Thus we should write the values of the variable that make the original expression undefined. For example, we should write $\dfrac{7x^2 + 14x}{49x} = \dfrac{x + 2}{7}$, $x \neq 0$. For the remainder of the text, we assume a variable cannot take on any values that result in division by zero to maintain equality.

Simplifying a Rational Expression

Step 1: Completely factor the numerator and denominator of the rational expression.

Step 2: Divide out common factors.

EXAMPLE 7 **Simplifying a Rational Expression**

Simplify: (a) $\dfrac{2x - 10}{4x^2 - 20x}$ (b) $\dfrac{x^2 - 9}{2x^2 - 3x - 9}$

Solution

(a) First, we factor the numerator and denominator.

$$\frac{2x - 10}{4x^2 - 20x} = \frac{2(x - 5)}{4x(x - 5)}$$

Divide out common factors:
$$= \frac{2\,\cancel{(x - 5)}}{2 \cdot 2x\,\cancel{(x - 5)}}$$

$$= \frac{1}{2x}$$

Work Smart

Notice in Example 7(a) that when all the factors in the numerator divide out, we are left with a factor of 1, not 0!

(b) Factor the numerator and denominator:
$$\frac{x^2 - 9}{2x^2 - 3x - 9} = \frac{(x + 3)(x - 3)}{(2x + 3)(x - 3)}$$

Divide out common factors:
$$= \frac{(x + 3)\,\cancel{(x - 3)}}{(2x + 3)\,\cancel{(x - 3)}}$$

$$= \frac{x + 3}{2x + 3}$$

Work Smart

When simplifying, we can divide out only common factors, not common terms.

WRONG! $\dfrac{x + 3}{x} = \dfrac{\cancel{x} + 3}{\cancel{x}} = 3$ **WRONG!** $\dfrac{x^2 - 4x + 3}{x + 3} = \dfrac{x^2 - 4\cancel{x} + \cancel{3}}{\cancel{x} + \cancel{3}} = x^2 - 4$

If you aren't sure what you can divide out, try the computation with numbers to see whether it works. For example, does $\dfrac{2 + 3}{2} = \dfrac{\cancel{2} + 3}{\cancel{2}} = 3$? No!

Quick ✓

10. To _____ a rational expression means to write the rational expression in the form $\dfrac{p}{q}$, where p and q are polynomials that have no common factors.

11. *True or False* $\dfrac{2n + 3}{2n + 6} = \dfrac{\cancel{2n} + 3}{\cancel{2n} + 6} = \dfrac{3}{6} = \dfrac{1}{2}$

In Problems 12–14, simplify the rational expression.

12. $\dfrac{3n^2 + 12n}{6n}$ 13. $\dfrac{2z^2 + 6z + 4}{-4z - 8}$ 14. $\dfrac{a^2 + 3a - 28}{2a^2 - a - 28}$

EXAMPLE 8 **Simplifying a Rational Expression**

Simplify: $\dfrac{a^3 + 3a^2 + 2a + 6}{a^2 + 6a + 9}$

Solution

Factor the numerator and the denominator:
$$\frac{a^3 + 3a^2 + 2a + 6}{a^2 + 6a + 9} = \frac{a^2(a+3) + 2(a+3)}{(a+3)^2}$$

$$= \frac{(a+3)(a^2+2)}{(a+3)^2}$$

Divide out common factors:
$$= \frac{\cancel{(a+3)}(a^2+2)}{\cancel{(a+3)}(a+3)}$$

$$= \frac{a^2+2}{a+3}$$

Work Smart

Remember: You can simplify only after the numerator and denominator have been completely factored!

> **Quick ✓**
>
> *In Problems 15 and 16, simplify the rational expression.*
>
> **15.** $\dfrac{z^3 + z^2 - 3z - 3}{z^2 - z - 2}$ **16.** $\dfrac{4k^2 + 4k + 1}{4k^2 - 1}$

Notice what happens when a factor in the numerator is the *opposite* of the factor in the denominator.

$$3 \text{ and } -3 \text{ are opposites, so } \frac{3}{-3} = -1$$

$$\frac{5-x}{x-5} = \frac{-1(-5+x)}{x-5} = \frac{-1(x-5)}{x-5} = -1$$

$$\frac{4x-7}{7-4x} = \frac{4x-7}{-1(-7+4x)} = \frac{4x-7}{-1(4x-7)} = -1$$

⊙ EXAMPLE 9 **Simplifying a Rational Expression Containing Opposite Factors**

Simplify: $\dfrac{4 - x^2}{2x^2 - x - 6}$

Solution

Factor the numerator and denominator:
$$\frac{4-x^2}{2x^2-x-6} = \frac{(2+x)(2-x)}{(2x+3)(x-2)}$$

Factor -1 from $2-x$:
$$= \frac{(2+x)(-1)(-2+x)}{(2x+3)(x-2)}$$

Divide out common factors:
$$= \frac{(x+2)(-1)\cancel{(x-2)}}{(2x+3)\cancel{(x-2)}}$$

$$= \frac{-(x+2)}{2x+3}$$

$$= -\frac{x+2}{2x+3}$$

Quick ✓

17. *True or False* $\dfrac{a+b}{a-b} = -1$

In Problems 18–20, simplify the rational expression.

18. $\dfrac{7a - 7b}{b - a}$ **19.** $\dfrac{12 - 4x}{4x^2 - 13x + 3}$ **20.** $\dfrac{25z^2 - 1}{3 - 15z}$

7.1 Exercises   MyMathLab®

Exercise numbers in green have complete video solutions in MyMathLab

*Problems **1–20** are the **Quick ✓**s that follow the **EXAMPLES**.*

Building Skills

In Problems 21–28, evaluate each expression for the given values. See Objective 1.

21. $\dfrac{x}{x-5}$
 (a) $x = 10$
 (b) $x = -5$
 (c) $x = 0$

22. $\dfrac{x}{x+4}$
 (a) $x = 8$
 (b) $x = -4$
 (c) $x = 0$

23. $\dfrac{2a-3}{a}$
 (a) $a = 3$
 (b) $a = -3$
 (c) $a = 9$

24. $\dfrac{2m-1}{m}$
 (a) $m = 4$
 (b) $m = 1$
 (c) $m = -1$

25. $\dfrac{x^2 - 2x}{x - 4}$
 (a) $x = 3$
 (b) $x = 2$
 (c) $x = -3$

26. $\dfrac{a^2 - 2a}{a - 4}$
 (a) $a = 5$
 (b) $a = -1$
 (c) $a = -4$

27. $\dfrac{x^2 - y^2}{2x - y}$
 (a) $x = 2, y = 2$
 (b) $x = 3, y = 4$
 (c) $x = 1, y = -2$

28. $\dfrac{b^2 - a^2}{(a - b)^2}$
 (a) $a = 3, b = 2$
 (b) $a = 5, b = 4$
 (c) $a = -2, b = 2$

In Problems 29–40, find the value(s) of the variable for which the rational expression is undefined. See Objective 2.

29. $\dfrac{2 - 4x}{3x}$ **30.** $\dfrac{-x + 1}{5x}$

31. $\dfrac{3p}{p - 5}$ **32.** $\dfrac{5m^3}{m + 8}$

33. $\dfrac{8}{3 - 2x}$ **34.** $\dfrac{12}{4a - 3}$

35. $\dfrac{6z}{z^2 - 36}$ **36.** $\dfrac{5x}{25 - x^2}$

37. $\dfrac{x}{x^2 - 7x + 10}$ **38.** $\dfrac{2x^2}{x^2 + x - 2}$

39. $\dfrac{12x + 5}{x^3 - x^2 - 6x}$ **40.** $\dfrac{3h + 2}{h^3 + 5h^2 + 4h}$

In Problems 41–52, simplify each rational expression. Assume that no variable has a value that results in a denominator with a value of zero. See Objective 3.

41. $\dfrac{5x - 10}{15}$ **42.** $\dfrac{3n + 9}{3}$

43. $\dfrac{3z^2 + 6z}{3z}$ **44.** $\dfrac{6p^2 + 3p}{3p}$

45. $\dfrac{p - 3}{p^2 - p - 6}$ **46.** $\dfrac{x - 3}{x^2 - 4x + 3}$

47. $\dfrac{2 - x}{x - 2}$ **48.** $\dfrac{4 - z}{z - 4}$

49. $\dfrac{2k^2 - 14k}{7 - k}$ **50.** $\dfrac{2v^2 - 6v}{3 - v}$

51. $\dfrac{x^2 - 1}{2x^2 + 5x + 3}$ **52.** $\dfrac{x^2 - 9}{x^2 + 5x + 6}$

Mixed Practice

In Problems 53–76, simplify each rational expression. Assume that no variable has a value that results in a denominator with a value of zero.

53. $\dfrac{6x + 30}{x^2 - 25}$ **54.** $\dfrac{n^2 - 49}{2n + 14}$

55. $\dfrac{-3x - 3y}{x + y}$ **56.** $\dfrac{-2a - 2b}{a + b}$

57. $\dfrac{b^2 - 25}{4b + 20}$ **58.** $\dfrac{3p - 12}{p^2 - 16}$

59. $\dfrac{x^2 - 2x - 15}{x^2 - 8x + 15}$ **60.** $\dfrac{x^2 + x - 2}{x^2 + 5x + 6}$

61. $\dfrac{3x^3 + 8x^2 - 16x}{x^3 - x^2 - 20x}$

62. $\dfrac{4x^3 + 5x^2 - 6x}{x^3 - 2x^2 - 6x}$

63. $\dfrac{x^2 - y^2}{y - x}$

64. $\dfrac{a - b}{b^2 - a^2}$

65. $\dfrac{x^3 + 4x^2 + x + 4}{x^2 + 5x + 4}$

66. $\dfrac{z^3 + 3z^2 + 2z + 6}{z^2 + 5z + 6}$

67. $\dfrac{16 - c^2}{(c - 4)^2}$

68. $\dfrac{9 - x^2}{(x - 3)^2}$

69. $\dfrac{4x^2 - 20x + 24}{6x^2 - 48x + 90}$

70. $\dfrac{2x^2 + 5x - 3}{4x^2 - 8x + 3}$

71. $\dfrac{6 + x - x^2}{6x^2 + 7x - 10}$

72. $\dfrac{5 + 4x - x^2}{4x^2 - 17x - 15}$

73. $\dfrac{2t^2 - 18}{t^4 - 81}$

74. $\dfrac{x^3 + 4x}{x^4 - 16}$

75. $\dfrac{12w - 3w^2}{w^3 - 5w^2 + 4w}$

76. $\dfrac{7a - a^2}{a^3 - 5a^2 - 14a}$

Applying the Concepts

77. Drug Concentration The concentration, C, in mg/mL, of a drug in a patient's bloodstream t minutes after an injection is given by the formula

$$C = \frac{50t}{t^2 + 25}.$$

(a) Find the concentration in a patient 5 minutes after an injection.

(b) Find the concentration in a patient 10 minutes after an injection.

78. Drug Concentration The concentration, D, in mg/mL, of a drug in a patient's bloodstream t minutes after an injection is given by the formula

$$D = \frac{t}{2t^2 + 1}.$$

(a) Find the concentration in a patient 30 minutes after an injection.

(b) Find the concentration in a patient 60 minutes after an injection.

79. Body Mass Index Body Mass Index, or BMI, is used to determine if a person's weight is in a healthy range. The formula used to determine Body Mass Index is given by BMI $= \dfrac{k}{m^2}$, where k is the person's weight in kilograms and m is the person's height in meters. If a BMI range of 19 to 24.9 is considered healthy and 25 to 29.9 is considered overweight, should a person 2 meters tall weighing 110 kilograms start looking for a weight-loss program?

80. Body Mass Index A formula to calculate the BMI when the weight, w, is given in pounds and the height, h, is given in inches is BMI $= \dfrac{705w}{h^2}$. What is the BMI of a person weighing 120 pounds who is 5 feet tall?

81. Cost of a Car The average cost, in thousands of dollars, to produce Chevy Cobalts is given by the rational expression $\dfrac{0.2x^3 - 2.3x^2 + 14.3x + 10.2}{x}$, where x is the number of cars produced. If two Cobalts are produced, what is the average cost per car?

82. Cost of a Car Use the rational expression from Problem 81 to find the average cost of producing 10 Cobalts. Do you believe that the cost will continue to decrease as more cars are made?

Extending the Concepts

In Problems 83–88, simplify each rational expression.

83. $\dfrac{c^8 - 1}{(c^4 - 1)(c^6 + 1)}$

84. $\dfrac{(1 - x)(x^6 + 1)}{x^4 - 1}$

85. $\dfrac{x^4 - x^2 - 12}{x^4 + 2x^3 - 9x - 18}$

86. $\dfrac{n^3 + 3n^2 - 16n - 48}{n^3 - 4n^2 + 3n - 12}$

87. $\dfrac{(t + 2)^3(t^4 - 16)}{(t^3 + 8)(t + 2)(t^2 - 4)}$

88. $\dfrac{24x^4 - 16x^3 + 27x - 18}{18 - 27x + 16x^2 - 24x^3}$

Explaining the Concepts

89. Explain the error in simplifying each of the following:

(a) $\dfrac{5x^2 - 5}{x^2 - 1} = \dfrac{5\cancel{x^2} - 5}{\cancel{x^2} - 1} = \dfrac{5 - 5}{-1} = \dfrac{0}{-1} = 0$

(b) $\dfrac{x - 2}{3x - 6} = \dfrac{x - 2}{3(x - 2)} = \dfrac{\cancel{x - 2}}{3(\cancel{x - 2})} = 3$

90. Explain why we can divide out only common *factors*, not common *terms*.

91. What does it mean for a rational expression to be undefined? Explain why the expression $\dfrac{x^2 - 9}{3x - 6}$ is undefined for $x = 2$.

92. Compare the process for simplifying a rational number such as $-\dfrac{21}{35}$ with simplifying a rational algebraic expression such as $\dfrac{x^2 - 2x - 3}{1 - x^2}$.

93. Why is $\dfrac{x - 7}{7 - x} = -1$ whereas $\dfrac{x - 7}{x + 7}$ is in simplest form?

94. Rational expressions that represent the same quantity are said to be *equivalent*. Are the rational expressions $\dfrac{x - y}{y - x}$ and $\dfrac{y - x}{x - y}$ equivalent? Explain how to determine whether two rational expressions are equivalent.

7.2 Multiplying and Dividing Rational Expressions

Objectives

1. Multiply Rational Expressions
2. Divide Rational Expressions

Are You Ready For This Section?

Before getting started, take the following readiness quiz. If you get a problem wrong, go back to the section cited and review the material.

R1. Find the product: $\dfrac{3}{14} \cdot \dfrac{28}{9}$ [Section 1.5, pp. 37–38]

R2. Find the reciprocal of $\dfrac{5}{8}$ [Section 1.4, pp. 33–34]

R3. Find the quotient: $\dfrac{12}{25} \div \dfrac{12}{5}$ [Section 1.5, pp. 38–39]

R4. Factor: $3x^3 - 27x$ [Section 6.4, pp. 397–399]

R5. Simplify the rational expression: $\dfrac{3x + 12}{5x^2 + 20x}$ [Section 7.1, pp. 433–436]

▶ ① Multiply Rational Expressions

Multiplying rational expressions is similar to multiplying rational numbers. Recall that,

$$\frac{2}{3} \cdot \frac{12}{5} = \frac{2 \cdot 12}{3 \cdot 5} = \frac{2 \cdot 3 \cdot 4}{3 \cdot 5} = \frac{2 \cdot 4}{5} = \frac{8}{5}$$

EXAMPLE 1 **How to Multiply Rational Expressions**

Multiply: $\dfrac{x - 3}{5} \cdot \dfrac{5x + 35}{x^2 - 9}$

Step-by-Step Solution

Step 1: Completely factor the polynomials in each numerator and denominator.

$$\frac{x - 3}{5} \cdot \frac{5x + 35}{x^2 - 9} = \frac{x - 3}{5} \cdot \frac{5(x + 7)}{(x + 3)(x - 3)}$$

Step 2: Multiply.

$$= \frac{5(x - 3)(x + 7)}{5(x + 3)(x - 3)}$$

Step 3: Divide out common factors in the numerator and denominator.

The common factors are $x - 3$ and 5:

$$= \frac{\cancel{5}\,(\cancel{x - 3})\,(x + 7)}{\cancel{5}\,(x + 3)\,(\cancel{x - 3})}$$

$$= \frac{x + 7}{x + 3}$$

Remember that $\dfrac{x + 7}{x + 3}$ cannot be simplified further. Divide out *factors*, not *terms*.

In Words

To multiply two rational expressions, factor each polynomial, write the expression as a single fraction in factored form, and then divide out common factors.

Multiplying Rational Expressions

Step 1: Factor the polynomials in each numerator and denominator.

Step 2: Use the fact that if $\dfrac{a}{b}$ and $\dfrac{c}{d}$, where $b \neq 0$ and $d \neq 0$, are two rational expressions, then $\dfrac{a}{b} \cdot \dfrac{c}{d} = \dfrac{ac}{bd}$.

Step 3: Divide out common factors in the numerator and denominator. Leave the remaining factors in factored form.

Ready?...Answers **R1.** $\dfrac{2}{3}$ **R2.** $\dfrac{8}{5}$
R3. $\dfrac{1}{5}$ **R4.** $3x(x + 3)(x - 3)$ **R5.** $\dfrac{3}{5x}$

EXAMPLE 2 **Multiplying Rational Expressions**

Multiply: $\dfrac{7y + 28}{4y} \cdot \dfrac{8}{y^2 + 2y - 8}$

Solution

We begin by factoring each polynomial in the numerator and denominator.

$$\frac{7y + 28}{4y} \cdot \frac{8}{y^2 + 2y - 8} = \frac{7(y + 4)}{4y} \cdot \frac{4 \cdot 2}{(y + 4)(y - 2)}$$

$$\text{Multiply:} \quad = \frac{7(y + 4) \cdot 4 \cdot 2}{4y(y + 4)(y - 2)}$$

$$\text{Divide out common factors, 4 and } y + 4: \quad = \frac{7\cancel{(y + 4)} \cdot \cancel{4} \cdot 2}{\cancel{4} \cdot y \cancel{(y + 4)}(y - 2)}$$

$$= \frac{7 \cdot 2}{y(y - 2)}$$

$$\text{Multiply:} \quad = \frac{14}{y(y - 2)}$$

EXAMPLE 3 **Multiplying Rational Expressions**

Multiply: $\dfrac{p^2 - 9}{p^2 + 5p + 6} \cdot \dfrac{p + 2}{6 - 2p}$.

Solution

$$\text{Factor each polynomial:} \quad \frac{p^2 - 9}{p^2 + 5p + 6} \cdot \frac{p + 2}{6 - 2p} = \frac{(p + 3)(p - 3)}{(p + 3)(p + 2)} \cdot \frac{p + 2}{(-2)(p - 3)}$$

Work Smart

$6 - 2p = 2(3 - p) = -2(p - 3)$

$$\text{Multiply:} \quad = \frac{(p + 3)(p - 3)(p + 2)}{(p + 3)(p + 2)(-2)(p - 3)}$$

$$\text{Divide out common factors, } p + 3, p - 3, \text{ and } p + 2: \quad = \frac{\cancel{(p + 3)}\,\cancel{(p - 3)}\,\cancel{(p + 2)}}{\cancel{(p + 3)}\,\cancel{(p + 2)}(-2)\cancel{(p - 3)}}$$

Work Smart

When all the factors in the numerator divide out, we are left with a factor of 1, *not* 0.

$$= \frac{1}{-2}$$

$$\frac{a}{-b} = -\frac{a}{b}: \quad = -\frac{1}{2}$$

Quick ✓

1. *True or False* $\dfrac{x + 4}{24} \cdot \dfrac{24}{(x + 4)(x - 4)} = \dfrac{\cancel{(x + 4)} \cdot 24}{24 \cdot \cancel{(x + 4)}(x - 4)} = x - 4$

2. *True or False* $\dfrac{8}{x - 7} \cdot \dfrac{7 - x}{16x} = \dfrac{1}{2x}$

In Problems 3–6, multiply and simplify the rational expressions.

3. $\dfrac{p^2 - 9}{5} \cdot \dfrac{25p}{2p - 6}$

4. $\dfrac{2x + 8}{x^2 + 3x - 4} \cdot \dfrac{7x - 7}{6x + 30}$

5. $\dfrac{x - 7}{x^2 - 9} \cdot \dfrac{x + 3}{7 - x}$

6. $\dfrac{15 - 3a}{7} \cdot \dfrac{3 + 2a}{2a^2 - 7a - 15}$

EXAMPLE 4 **Multiplying Rational Expressions with Two Variables**

Multiply: $\dfrac{m^2 - n^2}{10m^2 - 10mn} \cdot \dfrac{10m + 5n}{2m^2 + 3mn + n^2}$

Solution

Factor each polynomial: $\dfrac{m^2 - n^2}{10m^2 - 10mn} \cdot \dfrac{10m + 5n}{2m^2 + 3mn + n^2} = \dfrac{(m + n)(m - n)}{10m(m - n)} \cdot \dfrac{5(2m + n)}{(2m + n)(m + n)}$

Multiply: $= \dfrac{(m + n)(m - n) \cdot 5(2m + n)}{5 \cdot 2m(m - n)(2m + n)(m + n)}$

Divide out common factors, $m + n, m - n, 2m + n,$ and 5: $= \dfrac{\cancel{(m + n)}\ \cancel{(m - n)} \cdot \cancel{5}\ \cancel{(2m + n)}}{\cancel{5} \cdot 2m \cancel{(m - n)}\ \cancel{(2m + n)}\ \cancel{(m + n)}}$

$= \dfrac{1}{2m}$ ●

> **Quick ✔**
>
> In Problem 7, multiply the rational expressions.
>
> 7. $\dfrac{a^2 + 2ab + b^2}{3a^2 + 3ab} \cdot \dfrac{a - b}{a^2 - b^2}$

▶ ❷ **Divide Rational Expressions**

Recall that in the expression $\dfrac{20x^5}{3y} \div \dfrac{4x^2}{15y^5}, \dfrac{20x^5}{3y}$ is called the *dividend* and $\dfrac{4x^2}{15y^5}$ is called the *divisor*. The answer is called the *quotient*.

Finding the quotient of rational expressions is similar to finding the quotient of rational numbers. For example:

reciprocal of divisor
↓
$$\dfrac{2}{3} \div \dfrac{7}{9} = \dfrac{2}{3} \cdot \dfrac{9}{7} = \dfrac{2 \cdot 3 \cdot 3}{3 \cdot 7} = \dfrac{2 \cdot 3}{7} = \dfrac{6}{7}$$
↑
dividend

EXAMPLE 5 **How to Divide Rational Expressions**

Divide: $\dfrac{x + 3}{2x - 8} \div \dfrac{9}{4x}$

Step-by-Step Solution

Step 1: Multiply the dividend by the reciprocal of the divisor.

The reciprocal of $\dfrac{9}{4x}$ is $\dfrac{4x}{9}$
↓
$\dfrac{x + 3}{2x - 8} \div \dfrac{9}{4x} = \dfrac{x + 3}{2x - 8} \cdot \dfrac{4x}{9}$

Step 2: Completely factor the polynomials in each numerator and denominator.

$= \dfrac{x + 3}{2(x - 4)} \cdot \dfrac{2 \cdot 2 \cdot x}{9}$

(continued)

Step 3: Multiply.

$$= \frac{(x+3)\cdot 2 \cdot 2 \cdot x}{2(x-4)\cdot 9}$$

Step 4: Divide out common factors in the numerator and denominator.

Simplify:
$$= \frac{(x+3)\cdot \cancel{2} \cdot 2 \cdot x}{\cancel{2}(x-4)\cdot 9}$$

$$= \frac{2x(x+3)}{9(x-4)}$$

Work Smart

In Step 2 of Example 5, we factored 4 as $2 \cdot 2$, but we did not factor 9. Why? Because we anticipated dividing out a common factor of 2. We could see there were no factors of 9 in the numerator, so we did not factor 9.

> **Dividing Rational Expressions**
>
> **Step 1:** Multiply the dividend by the reciprocal of the divisor. That is, $\frac{a}{b} \div \frac{c}{d} = \frac{a}{b} \cdot \frac{d}{c}$.
>
> **Step 2:** Factor each polynomial in the numerator and denominator.
>
> **Step 3:** Multiply.
>
> **Step 4:** Divide out common factors in the numerator and denominator. Leave the remaining factors in factored form.

EXAMPLE 6 **Dividing Rational Expressions**

Find the quotient: $\dfrac{y^2-9}{2y^2-y-15} \div \dfrac{3y^2+10y+3}{2y^2+y-10}$

Solution

Multiply the dividend by the reciprocal of the divisor.

$$\frac{y^2-9}{2y^2-y-15} \div \frac{3y^2+10y+3}{2y^2+y-10} = \frac{y^2-9}{2y^2-y-15} \cdot \frac{2y^2+y-10}{3y^2+10y+3}$$

Factor each polynomial in the numerator and denominator:
$$= \frac{(y-3)(y+3)}{(2y+5)(y-3)} \cdot \frac{(2y+5)(y-2)}{(3y+1)(y+3)}$$

Multiply:
$$= \frac{(y-3)(y+3)(2y+5)(y-2)}{(2y+5)(y-3)(3y+1)(y+3)}$$

Divide out common factors:
$$= \frac{\cancel{(y-3)}\,\cancel{(y+3)}\,\cancel{(2y+5)}\,(y-2)}{\cancel{(2y+5)}\,\cancel{(y-3)}\,(3y+1)\,\cancel{(y+3)}}$$

$$= \frac{y-2}{3y+1}$$

> **Quick** ✓
>
> **8.** $\dfrac{8}{9} \div \dfrac{2}{3} = \dfrac{8}{9} \cdot \dfrac{}{} = \dfrac{}{}$
>
> *In Problems 9 and 10, find the quotient.*
>
> **9.** $\dfrac{12}{x^2-x} \div \dfrac{4x-2}{x^2-1}$
>
> **10.** $\dfrac{x^2-9}{x^2-16} \div \dfrac{x^2-x-12}{x^2+x-12}$

EXAMPLE 7 **Dividing Rational Expressions**

Find the quotient: $\dfrac{a^2+3a+2}{a^2-9} \div (a+2)$

Solution

We write the divisor $a + 2$ as $\dfrac{a + 2}{1}$ and proceed as we did in Example 6.

$$\frac{a^2 + 3a + 2}{a^2 - 9} \div (a + 2) = \frac{a^2 + 3a + 2}{a^2 - 9} \div \frac{a + 2}{1}$$

$$\text{Find the reciprocal of } \frac{a + 2}{1} \text{ and multiply:} \quad = \frac{a^2 + 3a + 2}{a^2 - 9} \cdot \frac{1}{a + 2}$$

$$\text{Factor:} \quad = \frac{(a + 2)(a + 1)}{(a + 3)(a - 3)} \cdot \frac{1}{a + 2}$$

$$\text{Multiply:} \quad = \frac{(a + 2)(a + 1)}{(a + 3)(a - 3)(a + 2)}$$

$$\text{Divide out common factors:} \quad = \frac{\cancel{(a + 2)}(a + 1)}{(a + 3)(a - 3)\cancel{(a + 2)}}$$

$$= \frac{a + 1}{(a + 3)(a - 3)} \quad \bullet$$

> **Quick ✔**
>
> *In Problem 11, find the quotient.*
>
> **11.** $\dfrac{q^2 - 6q - 7}{q^2 - 25} \div (q - 7)$

EXAMPLE 8 **Dividing Rational Expressions Written Vertically**

Find the quotient: $\dfrac{\dfrac{12x}{5x + 20}}{\dfrac{4x^2}{x^2 - 16}}$

Solution

$$\text{Multiply by the reciprocal of the divisor}$$

$$\frac{\dfrac{12x}{5x + 20}}{\dfrac{4x^2}{x^2 - 16}} = \frac{12x}{5x + 20} \cdot \frac{x^2 - 16}{4x^2}$$

$$\text{Factor each polynomial in the numerator and denominator:} \quad = \frac{4 \cdot 3 \cdot x}{5(x + 4)} \cdot \frac{(x - 4)(x + 4)}{4 \cdot x \cdot x}$$

$$\text{Multiply:} \quad = \frac{4 \cdot 3 \cdot x(x - 4)(x + 4)}{5(x + 4) \cdot 4 \cdot x \cdot x}$$

$$\text{Divide out common factors:} \quad = \frac{\cancel{4} \cdot 3 \cdot \cancel{x}(x - 4)\cancel{(x + 4)}}{5\cancel{(x + 4)} \cdot \cancel{4} \cdot \cancel{x} \cdot x}$$

$$= \frac{3(x - 4)}{5x} \quad \bullet$$

Work Smart

$$\frac{\dfrac{12x}{5x + 2}}{\dfrac{4x^2}{x^2 - 16}} = \frac{12x}{5x + 2} \div \frac{4x^2}{x^2 - 16}$$

> **Quick ✔**
>
> *In Problems 12 and 13, find the quotient.*
>
> **12.** $\dfrac{\dfrac{2}{x + 1}}{\dfrac{8}{x^2 - 1}}$ **13.** $\dfrac{\dfrac{x + 3}{x^2 - 4}}{\dfrac{4x + 12}{7x^2 + 14x}}$

EXAMPLE 9   **Dividing Rational Expressions with More Than One Variable**

Find the quotient: $\dfrac{5a - 5b}{7c^2} \div \dfrac{10b - 10a}{21c}$

Solution

Multiply by the reciprocal of the divisor

$$\dfrac{5a - 5b}{7c^2} \div \dfrac{10b - 10a}{21c} \overset{\downarrow}{=} \dfrac{5a - 5b}{7c^2} \cdot \dfrac{21c}{10b - 10a}$$

Factor each polynomial in the numerator and denominator:
$$= \dfrac{5(a - b)}{7 \cdot c \cdot c} \cdot \dfrac{7 \cdot 3 \cdot c}{5 \cdot 2 \cdot (-1)(a - b)}$$

Multiply:
$$= \dfrac{5(a - b) \cdot 7 \cdot 3 \cdot c}{7 \cdot c \cdot c \cdot 5 \cdot 2 \cdot (-1)(a - b)}$$

Divide out common factors:
$$= \dfrac{\cancel{5}\,\cancel{(a - b)} \cdot \cancel{7} \cdot 3 \cdot \cancel{c}}{\cancel{7} \cdot c \cdot \cancel{c} \cdot \cancel{5} \cdot 2 \cdot (-1)\,\cancel{(a - b)}}$$

$$= \dfrac{3}{c \cdot 2 \cdot (-1)}$$

Multiply:
$$= -\dfrac{3}{2c}$$

Quick ✓

In Problem 14, find the quotient and simplify.

14. $\dfrac{3m - 6n}{5n} \div \dfrac{m^2 - 4n^2}{10mn}$

7.2 Exercises

MyMathLab®   Math XP   PRACTICE   Exercise numbers in green have complete video solutions in MyMathLab

*Problems **1–14** are the Quick ✓s that follow the **EXAMPLES**.*

Building Skills

In Problems 15–22, multiply. See Objective 1.

15. $\dfrac{x + 5}{7} \cdot \dfrac{14x}{x^2 - 25}$

16. $\dfrac{x - 4}{4} \cdot \dfrac{12x}{x^2 - 16}$

17. $\dfrac{x^2 - x}{x^2 - x - 2} \cdot \dfrac{x - 2}{x^2 - 1}$

18. $\dfrac{8n - 8}{n^2 - 3n + 2} \cdot \dfrac{n + 2}{12}$

19. $\dfrac{p^2 - 1}{2p - 3} \cdot \dfrac{2p^2 + p - 6}{p^2 + 3p + 2}$

20. $\dfrac{z^2 - 4}{3z - 2} \cdot \dfrac{3z^2 + 7z - 6}{z^2 + z - 6}$

21. $(x + 1) \cdot \dfrac{x - 6}{x^2 - 5x - 6}$

22. $(x - 3) \cdot \dfrac{x - 2}{x^2 - 5x + 6}$

In Problems 23–30, divide. See Objective 2.

23. $\dfrac{x - 4}{2x - 8} \div \dfrac{3x}{2}$

24. $\dfrac{p + 7}{3p + 21} \div \dfrac{p}{6}$

25. $\dfrac{m^2 - 16}{6m} \div \dfrac{m^2 + 8m + 16}{12m}$

26. $\dfrac{z^2 - 25}{10z} \div \dfrac{z^2 - 10z + 25}{5z}$

27. $\dfrac{x^2 - x}{x^2 - 1} \div \dfrac{x + 2}{x^2 + 3x + 2}$

28. $\dfrac{3y^2 + 6y}{8} \div \dfrac{y + 2}{12y - 12}$

29. $\dfrac{\dfrac{x^2 + 4x + 4}{x^2 - 4}}{\dfrac{x^2 - x - 6}{x^2 - 5x + 6}}$

30. $\dfrac{\dfrac{4z^2 + 12z + 9}{8z + 16}}{\dfrac{4z^2 + 12z + 9}{12z + 24}}$

Mixed Practice

In Problems 31–66, perform the indicated operation.

31. $\dfrac{14}{9} \cdot \dfrac{15}{7}$

32. $\dfrac{7}{52} \cdot \dfrac{77}{13}$

33. $-\dfrac{20}{16} \div \left(-\dfrac{30}{24}\right)$

34. $-\dfrac{60}{35} \div \dfrac{15}{77}$

35. $\dfrac{8}{11} \div (-2)$

36. $\dfrac{16}{3} \div (-4)$

37. $\dfrac{3y}{y^2 - y - 6} \cdot \dfrac{4y + 8}{9y^2}$

38. $\dfrac{x - 4}{12x - 18} \cdot \dfrac{6}{x^2 - 16}$

39. $\dfrac{4a + 8b}{a^2 + 2ab} \cdot \dfrac{a}{12}$

40. $\dfrac{m^2 - n^2}{3m - 3n} \cdot \dfrac{6}{2m + 2n}$

41. $\dfrac{3x^2 - 6x}{x^2 - 2x - 8} \div \dfrac{x - 2}{x + 2}$

42. $\dfrac{x + 5}{3} \div \dfrac{30x}{4x + 20}$

43. $\dfrac{\dfrac{2w^2 - 16w + 32}{4w^2 - 17w + 4}}{\dfrac{2w^2 - 11w + 12}{8w^2 - 10w - 3}}$

44. $\dfrac{\dfrac{-3w^2 + 12w - 12}{5w^2 - 7w - 6}}{\dfrac{2w^2 - 7w + 6}{10w^2 - 9w - 9}}$

45. $\dfrac{3xy - 2y^2 - x^2}{x + y} \cdot \dfrac{x^2 - y^2}{x^2 - 2xy}$

46. $\dfrac{2r^2 + rs - 3s^2}{r^2 - s^2} \cdot \dfrac{r^2 - 2rs - 3s^2}{2r + 3s}$

47. $\dfrac{\dfrac{2c - 4}{8}}{\dfrac{2 - c}{2}}$

48. $\dfrac{\dfrac{2x - 3}{3}}{\dfrac{}{6x - 9}}$

49. $\dfrac{8n^2 - 6n - 9}{6n + 18} \cdot \dfrac{9n^2 - 81}{2n^2 + 5n - 12}$

50. $\dfrac{2x^2 - 5x + 3}{x^2 - 1} \cdot \dfrac{x^2 + 1}{2x^2 - x - 3}$

51. $\dfrac{x^2y}{2x^2 - 5xy + 2y^2} \div \dfrac{(2xy^2)^2}{2x^2y - xy^2}$

52. $\dfrac{(x - 3)^2}{8xy^2} \div \dfrac{x^2 - 9}{(4x^2y)^2}$

53. $\dfrac{2a^2 + 3ab - 2b^2}{a^2 - b^2} \cdot \dfrac{a^2 - ab}{2a^3 + 4a^2b}$

54. $\dfrac{3y^2 - 3x^2}{2x^2 + xy - y^2} \cdot \dfrac{3x - 6y}{6x - 6y}$

55. $\dfrac{6x^2 + 17x + 5}{2x^2 - 2x - 24} \cdot \dfrac{4x^2 - 13x - 12}{12x^2 + 13x + 3}$

56. $\dfrac{6x^2 + 21x - 45}{3x^2 - 11x - 20} \cdot \dfrac{6x^2 + 11x + 4}{4x^2 - 4x - 3}$

57. $\dfrac{3x^2 + 12x + 12}{x^2 - 4} \div \dfrac{-x^2 + x + 6}{x^2 - 5x + 6}$

58. $\dfrac{4z^2 + 12z + 9}{8z + 16} \div \dfrac{(2z + 3)^3}{12z + 24}$

59. $\dfrac{9 - x^2}{x^2 + 5x + 4} \div \dfrac{x^2 - 2x - 3}{x^2 + 4x}$

60. $\dfrac{1}{b^2 + b - 12} \div \dfrac{1}{b^2 - 5b - 36}$

61. $\dfrac{\dfrac{a^2 - b^2}{a^2 + b^2}}{\dfrac{4a - 4b}{2a^2 + 2b^2}}$

62. $\dfrac{\dfrac{t^2}{t^2 - 16}}{\dfrac{t^2 - 3t}{t^2 - t - 12}}$

63. $\dfrac{\dfrac{x^3 - 1}{x^4 - 1}}{\dfrac{3x^2 + 3x + 3}{x^3 + x^2 + x + 1}}$

64. $\dfrac{\dfrac{p^3 - 8q^3}{p^2 - 4q^2}}{\dfrac{p^2 + 4pq + 4q^2}{(p + 2q)^2}}$

65. $\dfrac{xy - ay + xb - ab}{xy + ay - xb - ab} \cdot \dfrac{2xy - 2ay - 2xb + 2ab}{4b + 4y}$

66. $\dfrac{x^3 - x^2 + x - 1}{x + 1} \cdot \dfrac{x^2 + 2x + 1}{1 - x^2}$

Applying the Concepts

67. Find the product of $\dfrac{x^2 + 3xy + 2y^2}{x^2 - y^2}$ and $\dfrac{3x - 3y}{9x^2 + 9xy - 18y^2}$.

68. Find the product of $\dfrac{x}{6y^2}$ and $\dfrac{21x^2y}{(7x)^2}$.

69. Find the quotient of $\dfrac{x}{2y}$ and $\dfrac{(2xy)^2}{9xy^3}$.

70. Find the quotient of $\dfrac{x - 3}{2x + 6}$ and $\dfrac{x - 9}{4x^2 - 36}$.

71. Find $\dfrac{x - y}{x + y}$ squared divided by $x^2 - y^2$.

72. Find $(a - b)$ squared divided by $a^2 - b^2$.

73. What is $\dfrac{3x - 9}{2x + 4}$ divided into $\dfrac{6x}{x^2 - 4}$?

74. What is $\dfrac{2x^2}{3}$ divided into $\dfrac{x^3 - 3x}{6x}$?

△ **75. Area of a Rectangle** Write an algebraic expression for the area of the rectangle.

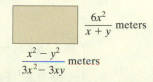

$\dfrac{3x + 9}{27x^2}$ feet

$\dfrac{9x}{x + 3}$ feet

△ **76. Area of a Rectangle** Write an algebraic expression for the area of the rectangle.

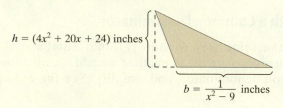

$\dfrac{6x^2}{x + y}$ meters

$\dfrac{x^2 - y^2}{3x^2 - 3xy}$ meters

△ **77. Area of a Triangle** Write an algebraic expression for the area of the triangle.

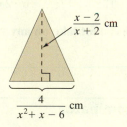

$h = (4x^2 + 20x + 24)$ inches

$b = \dfrac{1}{x^2 - 9}$ inches

△ **78. Area of a Triangle** Write an algebraic expression for the area of the triangle.

$\dfrac{x - 2}{x + 2}$ cm

$\dfrac{4}{x^2 + x - 6}$ cm

Extending the Concepts

In Problems 79–86, perform the indicated operation.

79. $\dfrac{x^2 + x - 12}{x^2 - 2x - 35} \div \dfrac{x + 4}{x^2 + 4x - 5} \div \dfrac{12 - 4x}{x - 7}$

80. $\dfrac{x^2 - 6xy + 9y^2}{x^2 - 4y^2} \div \dfrac{x - 3y}{x^2 - 5xy + 6y^2} \div \dfrac{x^2 - 9y^2}{x^2 - xy - 6y^2}$

81. $\dfrac{a^2 - 2ab}{2b - 3a} \div \dfrac{3a^2 - 4ab - 4b^2}{16a^2b^2 - 36a^4} \div (6a)$

82. $\dfrac{(2v - 1)^6}{(1 - 2v)^5} \cdot \dfrac{1 - 8v^3}{4v^2 - 4v + 1} \div (2v - 1)$

83. $\dfrac{x^2 + xy - 3x - 3y}{x^3 + y^3} \cdot \dfrac{x^2 + 2xy + y^2}{x^2 - x - 6}$

84. $\dfrac{a^2 + 4a - 12}{a^2 + 6a - 27} \cdot \dfrac{a^2 - 9}{2a^2 + 12a} \div (4a^2)$

85. $\dfrac{p^3 - 27q^3}{9pq} \cdot \dfrac{(3p^2q)^3}{p^2 - 9q^2} \div \dfrac{1}{p^2 + 2pq - 3q^2}$

86. $\dfrac{x^2 - 1}{x^4 - 81} \div \dfrac{x^2 + 1}{(x - 3)^2} \cdot \dfrac{x^3 + x}{x^2 - 2x - 3}$

In Problems 87 and 88, find the missing expression.

87. $\dfrac{2x}{x^3 - 3x^2} \cdot \dfrac{x^2 - x - 6}{?} = \dfrac{1}{3x}$

88. $\dfrac{x^2 + 12x + 36}{?} \div \dfrac{x^2 + 11x + 30}{x^2 + 3x - 10} = \dfrac{x + 6}{x - 2}$

Explaining the Concepts

89. List the steps for multiplying two rational algebraic expressions.

90. List the steps for dividing two rational algebraic expressions.

91. Find and describe the error in the following simplification, and then work the problem correctly.

$$\dfrac{n^2 - 2n - 24}{6n - n^2} = \dfrac{(n - 6)(n + 4)}{n(6 - n)} = \dfrac{n + 4}{n}$$

92. A unit fraction is any representation of 1. For instance, both $\dfrac{x}{x}$ and $\dfrac{1 \text{ foot}}{12 \text{ inches}}$ are unit fractions. Since multiplying by 1 does not change the value of any real number, use this concept to explain how to convert 56 inches into feet.

7.3 Adding and Subtracting Rational Expressions with a Common Denominator

Objectives

1 Add Rational Expressions with a Common Denominator

2 Subtract Rational Expressions with a Common Denominator

3 Add or Subtract Rational Expressions with Opposite Denominators

Are You Ready for This Section?

Before getting started, take the following readiness quiz. If you get a problem wrong, go back to the section cited and review the material.

R1. Write $\dfrac{12}{15}$ in lowest terms.   [Section 1.2, pp. 12–13]

R2. Find each sum and write in lowest terms:

(a) $\dfrac{7}{5} + \dfrac{2}{5}$ (b) $\dfrac{5}{6} + \dfrac{11}{6}$ [Section 1.5, pp. 39–40]

R3. Find each difference and write in lowest terms:

(a) $\dfrac{7}{9} - \dfrac{5}{9}$ (b) $\dfrac{7}{8} - \dfrac{5}{8}$ [Section 1.5, pp. 39–40]

R4. Determine the additive inverse of 5.   [Section 1.4, pp. 30–31]

R5. Simplify: $-(x - 2)$ [Section 1.8, pp. 67–68]

▶ 1 Add Rational Expressions with a Common Denominator

We add rational expressions the same way we add rational numbers. If the denominators of two rational expressions are the same, then we add the numerators, write the result over the common denominator, and simplify. See the examples below.

Ready?...Answers R1. $\dfrac{4}{5}$

R2. (a) $\dfrac{9}{5}$ (b) $\dfrac{8}{3}$ R3. (a) $\dfrac{2}{9}$ (b) $\dfrac{1}{4}$

R4. -5 R5. $2 - x$

ADDING RATIONAL NUMBERS

$$\frac{7}{12} + \frac{11}{12} = \frac{18}{12} = \frac{6 \cdot 3}{6 \cdot 2} = \frac{3}{2}$$

ADDING RATIONAL EXPRESSIONS

$$\frac{7}{12z} + \frac{11}{12z} = \frac{18}{12z} = \frac{6 \cdot 3}{6 \cdot 2 \cdot z} = \frac{3}{2z}, z \neq 0$$

Throughout the section, we assume that a variable cannot take on any values that result in division by zero.

EXAMPLE 1 **How to Add Rational Expressions with a Common Denominator**

Find the sum: $\dfrac{5}{x + 1} + \dfrac{3}{x + 1}$

Step-by-Step Solution

The rational expressions in the sum $\dfrac{5}{x + 1} + \dfrac{3}{x + 1}$ have a common denominator, $x + 1$.

Step 1: Add the numerators and write the result over the common denominator.

Use $\dfrac{a}{c} + \dfrac{b}{c} = \dfrac{a + b}{c}$: $\dfrac{5}{x + 1} + \dfrac{3}{x + 1} = \dfrac{5 + 3}{x + 1}$

Combine like terms in the numerator: $= \dfrac{8}{x + 1}$

Step 2: Simplify the rational expression by dividing out like factors.

The sum is already simplified.

So, $\dfrac{5}{x + 1} + \dfrac{3}{x + 1} = \dfrac{8}{x + 1}$.

In Words

To add rational expressions with a common denominator, add the numerators and write the result over the common denominator. Then simplify.

> **Adding Rational Expressions with a Common Denominator**
>
> **Step 1:** Use the fact that if $\dfrac{a}{c}$ and $\dfrac{b}{c}$, $c \neq 0$, are two rational expressions, then
>
> $$\frac{a}{c} + \frac{b}{c} = \frac{a+b}{c}$$
>
> **Step 2:** Simplify the sum by dividing out like factors.

EXAMPLE 2 **Adding Rational Expressions with a Common Denominator**

Find the sum:

(a) $\dfrac{x^2}{x+7} + \dfrac{7x}{x+7}$

(b) $\dfrac{2x^2+x}{x^2-4} + \dfrac{x-x^2}{x^2-4}$

Solution

(a) Because the denominators are the same, we add the numerators and write the sum over the common denominator.

$$\frac{x^2}{x+7} + \frac{7x}{x+7} = \frac{x^2+7x}{x+7}$$

Factor the numerator: $= \dfrac{x(x+7)}{x+7}$

Divide out common factors: $= \dfrac{x\cancel{(x+7)}}{\cancel{x+7}}$

$= x$

(b)

Use $\dfrac{a}{c} + \dfrac{b}{c} = \dfrac{a+b}{c}$: $\dfrac{2x^2+x}{x^2-4} + \dfrac{x-x^2}{x^2-4} = \dfrac{2x^2+x+x-x^2}{x^2-4}$

Combine like terms in the numerator: $= \dfrac{x^2+2x}{x^2-4}$

Factor the numerator and denominator: $= \dfrac{x(x+2)}{(x+2)(x-2)}$

Divide out common factors: $= \dfrac{x\cancel{(x+2)}}{\cancel{(x+2)}(x-2)}$

$= \dfrac{x}{x-2}$ ●

Quick ✔

1. If $\dfrac{a}{c}$ and $\dfrac{b}{c}$, $c \neq 0$, are two rational expressions, then $\dfrac{a}{c} + \dfrac{b}{c} = $ _____.

2. *True or False* $\dfrac{2x}{a} + \dfrac{4x}{a} = \dfrac{6x}{2a} = \dfrac{3x}{a}$

In Problems 3–6, find the sum.

3. $\dfrac{1}{x-2} + \dfrac{3}{x-2}$

4. $\dfrac{2x+1}{x+1} + \dfrac{x^2}{x+1}$

5. $\dfrac{9x}{6x-5} + \dfrac{2x-3}{6x-5}$

6. $\dfrac{2x-2}{2x^2-7x-15} + \dfrac{5}{2x^2-7x-15}$

▶ ❷ Subtract Rational Expressions with a Common Denominator

The steps for subtracting rational expressions are the same as the steps for subtracting rational numbers. If the denominators of two rational expressions are the same, then we subtract the numerators and write the result over the common denominator.

SUBTRACTING RATIONAL NUMBERS

$$\frac{3}{10} - \frac{7}{10} = \frac{3-7}{10} = \frac{-4}{10} = \frac{-2 \cdot 2}{5 \cdot 2} = -\frac{2}{5}$$

SUBTRACTING RATIONAL EXPRESSIONS

$$\frac{3}{10b} - \frac{7}{10b} = \frac{3-7}{10b} = \frac{-4}{10b} = \frac{-2 \cdot 2}{5 \cdot 2 \cdot b} = -\frac{2}{5b}$$

Once again, we exclude values of the variable that result in division by zero.

EXAMPLE 3 How to Subtract Rational Expressions with a Common Denominator

Find the difference: $\dfrac{2n^2}{n^2-1} - \dfrac{2n}{n^2-1}$

Step-by-Step Solution

Step 1: Subtract the numerators and write the result over the common denominator.

Use $\dfrac{a}{c} - \dfrac{b}{c} = \dfrac{a-b}{c}$: $\dfrac{2n^2}{n^2-1} - \dfrac{2n}{n^2-1} = \dfrac{2n^2 - 2n}{n^2-1}$

Step 2: Simplify the rational expression by dividing out common factors.

Factor the numerator and denominator: $= \dfrac{2n(n-1)}{(n-1)(n+1)}$

Divide out common factors: $= \dfrac{2n(n-1)}{(n-1)(n+1)}$

$= \dfrac{2n}{n+1}$ ●

In Words

To subtract rational expressions with a common denominator, subtract the numerators and write the result over the common denominator. Then simplify.

Subtracting Rational Expressions with a Common Denominator

Step 1: Use the fact that if $\dfrac{a}{c}$ and $\dfrac{b}{c}$, $c \neq 0$, are two rational expressions, then

$$\frac{a}{c} - \frac{b}{c} = \frac{a-b}{c}$$

Step 2: Simplify the difference by dividing out common factors.

Quick ✓

In Problems 7 and 8, find the difference.

7. $\dfrac{8y}{2y-5} - \dfrac{6}{2y-5}$

8. $\dfrac{10+3z}{6z} - \dfrac{7}{6z}$

If the numerator of the rational expression being subtracted contains more than one term, enclose the terms in parentheses. This will remind you to distribute the minus sign across all the terms.

EXAMPLE 4 **Subtracting Rational Expressions in Which a Numerator Contains More Than One Term**

Find the difference: $\dfrac{9x+1}{x+1} - \dfrac{6x-2}{x+1}$

Solution

The rational expressions have the same denominator, $x + 1$, so we subtract the numerators and write the result over the common denominator, $x + 1$.

Work Smart

Enclose the numerator terms that follow the subtraction sign in parentheses because this will remind you to distribute the minus sign across all the terms.

Notice the use of parentheses: $\dfrac{9x + 1}{x + 1} - \dfrac{6x - 2}{x + 1} = \dfrac{9x + 1 - (6x - 2)}{x + 1}$

Distribute the minus sign across the parentheses: $= \dfrac{9x + 1 - 6x + 2}{x + 1}$

Combine like terms: $= \dfrac{3x + 3}{x + 1}$

Factor the numerator: $= \dfrac{3(x + 1)}{x + 1}$

Divide out common factors: $= \dfrac{3\cancel{(x + 1)}}{\cancel{(x + 1)}}$

$= 3$

Quick ✓

9. $\dfrac{3x + 1}{x - 4} - \dfrac{x + 3}{x - 4} = \dfrac{3x + 1 - (\underline{\quad\quad})}{x - 4}$

In Problems 10 and 11, find the difference.

10. $\dfrac{2x^2 - 5x}{3x} - \dfrac{x^2 - 13x}{3x}$

11. $\dfrac{3x^2 + 8x - 1}{x^2 - 3x - 28} - \dfrac{2x^2 + 2x - 9}{x^2 - 3x - 28}$

▶ ❸ Add or Subtract Rational Expressions with Opposite Denominators

Suppose you were asked to find the sum $\dfrac{w^2 - 3w}{w - 4} + \dfrac{4}{4 - w}$. Notice the denominators are additive inverses (opposites) of each other. Recall that $4 - w = -w + 4 = -1(w - 4)$. We use this idea to rewrite the sum so that each rational expression has the same denominator.

EXAMPLE 5 **Adding Rational Expressions with Opposite Denominators**

Find the sum: $\dfrac{w^2 - 3w}{w - 4} + \dfrac{4}{4 - w}$

Solution

To rewrite the denominators so they are the same, factor -1 from $4 - w$.

$\dfrac{w^2 - 3w}{w - 4} + \dfrac{4}{4 - w} = \dfrac{w^2 - 3w}{w - 4} + \dfrac{4}{\underline{\quad\quad}}$

Use $\dfrac{a}{-b} = \dfrac{-a}{b}$: $= \dfrac{w^2 - 3w}{w - 4} + \dfrac{-4}{w - 4}$

Use $\dfrac{a}{c} + \dfrac{b}{c} = \dfrac{a + b}{c}$: $= \dfrac{w^2 - 3w - 4}{w - 4}$

Factor: $= \dfrac{(w - 4)(w + 1)}{w - 4}$

Divide out common factors: $= \dfrac{\cancel{(w - 4)}(w + 1)}{\cancel{(w - 4)}}$

$= w + 1$

Quick ✓

12. *True or False* $8 - x = -1(x - 8)$

In Problems 13 and 14, find the sum.

13. $\dfrac{3x}{x - 5} + \dfrac{1}{5 - x}$

14. $\dfrac{a^2 + 2a}{a - 7} + \dfrac{a^2 + 14}{7 - a}$

EXAMPLE 6 **Subtracting Rational Expressions with Opposite Denominators**

Find the difference: $\dfrac{b^2 - 11}{b^2 - 25} - \dfrac{3b + 1}{25 - b^2}$

Solution

To rewrite the denominators so they are the same, factor -1 from $25 - b^2$.

$$\dfrac{b^2 - 11}{b^2 - 25} - \dfrac{3b + 1}{25 - b^2} = \dfrac{b^2 - 11}{b^2 - 25} - \dfrac{3b + 1}{-(b^2 - 25)}$$

$-\dfrac{a}{-b} = +\dfrac{a}{b}:\quad = \dfrac{b^2 - 11}{b^2 - 25} + \dfrac{3b + 1}{b^2 - 25}$

Use $\dfrac{a}{c} + \dfrac{b}{c} = \dfrac{a + b}{c}:\quad = \dfrac{b^2 - 11 + 3b + 1}{b^2 - 25}$

Combine like terms: $\quad = \dfrac{b^2 + 3b - 10}{b^2 - 25}$

Factor: $\quad = \dfrac{(b + 5)(b - 2)}{(b + 5)(b - 5)}$

Divide out common factors: $\quad = \dfrac{b - 2}{b - 5}$

Quick ✓

In Problems 15 and 16, find the difference.

15. $\dfrac{2n}{n^2 - 9} - \dfrac{6}{9 - n^2}$

16. $\dfrac{2k}{6k - 6} - \dfrac{9 + 4k}{6 - 6k}$

7.3 Exercises MyMathLab® | Math⊗XL PRACTICE

Exercise numbers in green have complete video solutions in MyMathLab

Problems 1–16 are the Quick ✓s that follow the EXAMPLES.

Building Skills

In Problems 17–32, add the rational expressions. See Objective 1.

17. $\dfrac{3p}{8} + \dfrac{11p}{8}$

18. $\dfrac{2m}{9} + \dfrac{4m}{9}$

19. $\dfrac{n}{2} + \dfrac{3n}{2}$

20. $\dfrac{3}{x} + \dfrac{y}{x}$

21. $\dfrac{4a - 1}{3a} + \dfrac{2a - 2}{3a}$

22. $\dfrac{4n + 1}{8} + \dfrac{12n - 1}{8}$

23. $\dfrac{8c - 3}{c - 1} + \dfrac{2c - 1}{c - 1}$

24. $\dfrac{4n}{6n + 27} + \dfrac{18}{6n + 27}$

25. $\dfrac{2x}{x + y} + \dfrac{2y}{x + y}$

26. $\dfrac{4}{2x + 1} + \dfrac{8x}{2x + 1}$

27. $\dfrac{14x}{x+2} + \dfrac{7x^2}{x+2}$

28. $\dfrac{4p-1}{p-1} + \dfrac{p+1}{p-1}$

49. $\dfrac{2p^2-1}{p-1} - \dfrac{2p+2}{1-p}$

50. $\dfrac{x^2+3}{x-1} - \dfrac{x+3}{1-x}$

29. $\dfrac{12a-1}{2a+6} + \dfrac{13-8a}{2a+6}$

30. $\dfrac{3x^2+4}{3x-6} + \dfrac{3x^2-28}{3x-6}$

51. $\dfrac{2a}{a-b} + \dfrac{2a-4b}{b-a}$

52. $\dfrac{4k^2}{2k-1} + \dfrac{2k}{1-2k}$

31. $\dfrac{a}{a^2-3a-10} + \dfrac{2}{a^2-3a-10}$

53. $\dfrac{3}{p^2-3} + \dfrac{p^2}{3-p^2}$

54. $\dfrac{x^2+6x}{x^2-1} + \dfrac{4x+3}{1-x^2}$

32. $\dfrac{2x}{2x^2-x-15} + \dfrac{5}{2x^2-x-15}$

55. $\dfrac{2x}{x^2-y^2} - \dfrac{2y}{y^2-x^2}$

56. $\dfrac{m^2+2mn}{n^2-m^2} - \dfrac{n^2}{m^2-n^2}$

In Problems 33–44, subtract the rational expressions. See Objective 2.

33. $\dfrac{x^2}{x^2-9} - \dfrac{3x}{x^2-9}$

34. $\dfrac{4a^2}{3a-1} - \dfrac{a^2+a}{3a-1}$

35. $\dfrac{x^2-x}{2x} - \dfrac{2x^2-x}{2x}$

36. $\dfrac{3x+3x^2}{x} - \dfrac{x^2-x}{x}$

37. $\dfrac{2c-3}{4c} - \dfrac{6c+9}{4c}$

38. $\dfrac{7x+13}{3x} - \dfrac{x+1}{3x}$

39. $\dfrac{x^2-6}{x-2} - \dfrac{x^2-3x}{x-2}$

40. $\dfrac{2x-3}{x-1} - \dfrac{x+1}{x-1}$

41. $\dfrac{2}{c^2-4} - \dfrac{c^2-2}{c^2-4}$

42. $\dfrac{2z^2+7z}{z^2-1} - \dfrac{z^2-6}{z^2-1}$

43. $\dfrac{2x^2+5x}{2x+3} - \dfrac{4x+3}{2x+3}$

44. $\dfrac{2n^2+n}{n^2-n-2} - \dfrac{n^2+6}{n^2-n-2}$

In Problems 45–56, add or subtract the rational expressions. See Objective 3.

45. $\dfrac{n}{n-3} + \dfrac{3}{3-n}$

46. $\dfrac{4}{x-1} + \dfrac{8x}{1-x}$

47. $\dfrac{12q}{2p-2q} - \dfrac{8p}{2q-2p}$

48. $\dfrac{2a}{a-b} - \dfrac{4b}{b-a}$

Mixed Practice

In Problems 57–76, perform the indicated operation.

57. $\dfrac{4}{7x} - \dfrac{11}{7x}$

58. $\dfrac{3}{2x} - \left(\dfrac{-5}{2x}\right)$

59. $\dfrac{5}{2x-5} - \dfrac{2x}{2x-5}$

60. $\dfrac{2b+1}{b+1} - \dfrac{b}{b+1}$

61. $\dfrac{2n^3}{n-1} - \dfrac{2n^3}{n-1}$

62. $\dfrac{3n^2}{2n-1} - \dfrac{3n^2}{2n-1}$

63. $\dfrac{n^2-3}{2n+3} + \dfrac{n^2+n}{2n+3}$

64. $\dfrac{5x}{x^2+1} + \dfrac{x^2-3x}{x^2+1}$

65. $\dfrac{3v-1}{v^2-9} - \dfrac{8}{v^2-9}$

66. $\dfrac{x^2-x}{x^2-1} - \dfrac{2}{x^2-1}$

67. $\dfrac{x^2}{x^2-1} + \dfrac{2x+1}{x^2-1}$

68. $\dfrac{x^2}{x^2-9} + \dfrac{6x+9}{x^2-9}$

69. $\dfrac{2x^2+3}{x^2-3x} - \dfrac{x^2}{x^2-3x}$

70. $\dfrac{4x}{x^2-4} + \dfrac{8}{x^2-4}$

71. $\dfrac{3x^2}{x+1} - \dfrac{x^2+2}{x+1}$

72. $\dfrac{x^2}{x^2-1} - \dfrac{3x+2}{x^2-1}$

73. $\dfrac{3x-1}{x-y} + \dfrac{1-3y}{y-x}$

74. $\dfrac{2s-3t}{s-t} + \dfrac{6s-4t}{t-s}$

75. $\dfrac{2p - 3q}{3p^2 - 18q^2} - \dfrac{9q + p}{18q^2 - 3p^2}$

76. $\dfrac{n + 3}{2n^2 - n - 3} - \dfrac{-n^2 - 3n}{2n^2 - n - 3}$

Applying the Concepts

77. Find the sum of $\dfrac{7a^2}{a}$ and $\dfrac{5a^2}{a}$.

78. Find the sum of $\dfrac{3n}{3n + 2}$ and $\dfrac{n^2 + 2}{3n + 2}$.

79. Find the difference of $\dfrac{2n}{n + 1}$ and $\dfrac{n + 3}{n + 1}$.

80. Find the difference of $\dfrac{p^2 - 4}{p - 4}$ and $\dfrac{4p}{p - 4}$.

81. Subtract $\dfrac{3}{2n}$ from $\dfrac{11}{2n}$.

82. Subtract $\dfrac{3x + 4}{4}$ from $\dfrac{x - 4}{4}$.

83. Find a rational expression that, when subtracted from $\dfrac{x^2}{x^2 - 9}$, gives a difference of $\dfrac{3x}{x^2 - 9}$.

84. Find a rational expression that, when subtracted from $\dfrac{6x}{x - 3}$, gives a difference of 1.

85. Find a rational expression that, when added to $\dfrac{-3x + 4}{x + 2}$, gives a sum of 1.

86. Find a rational expression that, when added to $\dfrac{x - 3}{x + 1}$, gives a sum of $\dfrac{2x - 5}{x + 1}$.

△ **87. Perimeter of a Rectangle** Find the perimeter of the rectangle.

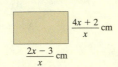

$\dfrac{4x + 2}{x}$ cm

$\dfrac{2x - 3}{x}$ cm

△ **88. Perimeter of a Rectangle** Find the perimeter of the rectangle.

$\dfrac{2n + 3}{2n + 1}$ yd

$\dfrac{n - 3}{2n + 1}$ yd

Extending the Concepts

In Problems 89–94, perform the indicated operations.

89. $\dfrac{7x - 3y}{x^2 - y^2} - \left(\dfrac{2x - 3y}{x^2 - y^2} - \dfrac{x + 7y}{x^2 - y^2} \right)$

90. $\dfrac{2g}{g - 3} - \left(\dfrac{g + 3}{g - 3} - \dfrac{g - 4}{g - 3} \right)$

91. $\dfrac{x^2}{3x^2 + 5x - 2} - \dfrac{x}{3x - 1} \cdot \dfrac{2x - 1}{x + 2}$

92. $\dfrac{4}{x - 2} \cdot \dfrac{x - 2}{x + 2} + \dfrac{3x - 1}{x^2 - 4}$

93. $\dfrac{5x}{x - 2} + \dfrac{2(x + 3)}{x - 2} - \dfrac{6(2x - 1)}{x - 2}$

94. $\dfrac{2a - b}{a - b} - \dfrac{6(a - 2b)}{a - b} + \dfrac{3(2b + a)}{b - a}$

In Problems 95 and 96, find the missing expression.

95. $\dfrac{2n + 1}{n - 3} - \dfrac{?}{n - 3} = \dfrac{6n + 7}{n - 3}$

96. $\dfrac{?}{n^2 - 1} - \dfrac{n - 3}{n^2 - 1} = \dfrac{1}{n + 1}$

Explaining the Concepts

97. Is the following student work correct or incorrect? Explain your reasoning.

$$\dfrac{x - 2}{x} - \dfrac{x + 4}{x} = \dfrac{x - 2 - x + 4}{x} = \dfrac{2}{x}$$

98. The denominators in the expression $\dfrac{4x}{x - 2} - \dfrac{2x}{2 - x}$

are opposites in sign. Explain how to rewrite this difference so that the terms have a common denominator, and then describe the steps to simplify the result.

99. On a multiple-choice quiz, you see the following problem.

Find the sum and simplify: $\dfrac{5}{x} + \dfrac{2}{x}$

The possible answers are **(a)** $\dfrac{7}{2x}$ **(b)** $\dfrac{7}{x^2}$ **(c)** $\dfrac{7}{x}$

Determine the correct answer, and explain why the others are incorrect.

100. Find and explain the error in the following problem. Then work the problem correctly.

$$\dfrac{a}{a^2 - b^2} - \dfrac{b}{a^2 - b^2} = \dfrac{a - b}{a^2 - b^2} = \dfrac{a - b}{a^2 - b^2} = \dfrac{1}{a - b}$$

7.4 Finding the Least Common Denominator and Forming Equivalent Rational Expressions

Objectives

① Find the Least Common Denominator of Two or More Rational Expressions

② Write a Rational Expression Equivalent to a Given Rational Expression

③ Use the LCD to Write Equivalent Rational Expressions

Are You Ready for This Section?

Before getting started, take the following readiness quiz. If you get a problem wrong, go back to the section cited and review the material.

R1. Write $\frac{5}{12}$ as a fraction with 24 as its denominator. [Section 1.2, pp. 10–12]

R2. Find the least common denominator (LCD) of $\frac{4}{15}$ and $\frac{7}{25}$. [Section 1.2, p. 11]

▶ ① Find the Least Common Denominator of Two or More Rational Expressions

Work Smart

Don't confuse the LCD (least common denominator) with the GCF (greatest common factor)!

For us to add or subtract rational expressions the denominators must be the same. What if the denominators are different? We use the same idea we used for adding or subtracting rational numbers with unlike denominators: Rewrite each rational expression as an equivalent rational expression over the least common denominator (LCD).

Let's review how to find the LCD of rational numbers.

EXAMPLE 1   **How to Find the Least Common Denominator of Rational Numbers**

Find the least common denominator (LCD) of $\frac{5}{6}$ and $\frac{11}{45}$.

Step-by-Step Solution

Step 1: Write each denominator as the product of prime factors.

$$6 = 2 \cdot 3$$
$$45 = 3 \cdot 3 \cdot 5$$

Step 2: Find the product of each prime factor the greatest number of times it appears in any factorization.

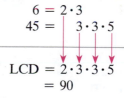

$$LCD = 2 \cdot 3 \cdot 3 \cdot 5$$
$$= 90$$

In Example 1, we could have written the factorization as follows:

$$6 = 2 \cdot 3$$
$$45 = 3^2 \cdot 5$$
$$LCD = 2 \cdot 3^2 \cdot 5$$
$$= 90$$

We write $3 \cdot 3$ as 3^2 because the greatest number of times that 3 appears in any factorization is two.

We will follow the same approach for finding the LCD of rational expressions.

Definition

The **least common denominator (LCD)** of two or more rational expressions is the polynomial of least degree that is a multiple of each denominator in the rational expressions.

EXAMPLE 2 **How to Find the Least Common Denominator of Rational Expressions**

Find the least common denominator of the rational expressions $\dfrac{3}{4x^3}$ and $\dfrac{5}{6x}$.

Step-by-Step Solution

Step 1: Write each denominator as the product of prime factors.

$$4x^3 = 2^2 \quad \cdot x^3$$
$$6x = 2 \cdot 3 \cdot x$$

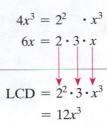

Step 2: Find the product of each prime factor the greatest number of times it appears in any factorization.

$$\text{LCD} = 2^2 \cdot 3 \cdot x^3$$
$$= 12x^3$$

Finding the Least Common Denominator of Rational Expressions

Step 1: Factor each denominator completely. Write the factored form using powers. For example, write $x^2 + 4x + 4$ as $(x + 2)^2$.

Step 2: The LCD is the product of each prime factor the greatest number of times it appears in any factorization.

EXAMPLE 3 **Finding the Least Common Denominator of Rational Expressions with Monomial Denominators**

Find the least common denominator of the rational expressions $\dfrac{1}{15xy^2}$ and $\dfrac{7}{18x^3y}$.

Solution

Write each denominator as the product of prime factors:

$$15xy^2 = \quad 3 \cdot 5 \cdot x \cdot y^2$$
$$18x^3y = 2 \cdot 3^2 \quad \cdot x^3 \cdot y$$

Work Smart

List the common factor with the highest power in the LCD.

Find the product of each factor the greatest number of times it appears in any factorization:

$$\text{LCD} = 2 \cdot 3^2 \cdot 5 \cdot x^3 \cdot y^2$$
$$= 90\, x^3y^2$$

Quick ✓

1. The ____ _____ _____ of two or more rational expressions is the smallest polynomial that is a multiple of each denominator in the rational expressions to be added or subtracted.

2. *True or False* The least common denominator of the expressions $\dfrac{3}{4a^2b}$ and $\dfrac{5}{4ab}$ is $16a^3b^2$.

In Problems 3 and 4, find the least common denominator of the rational expressions.

3. $\dfrac{5}{8x^2y}$ and $\dfrac{1}{12xy^3}$

4. $\dfrac{1}{6a^3b^2}$ and $\dfrac{5}{21ab^3}$

EXAMPLE 4 **Finding the Least Common Denominator of Rational Expressions with Polynomial Denominators**

Find the LCD of the rational expressions $\dfrac{7}{3a}$ and $\dfrac{5}{6a + 6}$.

Solution

Factor the denominators: $\quad 3a \qquad\qquad = \quad 3 \cdot a$

$$6a + 6 = 6(a + 1) = 2 \cdot 3 \cdot \;\Big|\;\; (a + 1)$$

Write the LCD: $\quad \text{LCD} = 2 \cdot 3 \cdot a \cdot (a + 1)$

$$= 6a(a + 1)$$

Work Smart

The a in the denominator $3a$ is a factor; the a in $(a + 1)$ is a *term*.

EXAMPLE 5 **Finding the Least Common Denominator of Rational Expressions with Polynomial Denominators**

Find the LCD of the rational expressions $\dfrac{7}{x^2 - x - 2}$ and $\dfrac{3}{x^2 + 2x + 1}$.

Solution

Factor the denominators: $\quad x^2 - x - 2 = (x - 2)\,(x + 1)$

$$x^2 + 2x + 1 = \qquad\qquad (x + 1)^2$$

Write the LCD: $\quad \text{LCD} = (x - 2)\,(x + 1)^2$

Quick ✓

5. *True or False* The least common denominator of two rational expressions is unique.

In Problems 6 and 7, find the least common denominator of the rational expressions.

6. $\dfrac{2}{15z}$ and $\dfrac{7}{5z^2 + 5z}$
 $\qquad$ **7.** $\dfrac{3}{x^2 + 4x - 5}$ and $\dfrac{1}{x^2 + 10x + 25}$

EXAMPLE 6 **Finding the Least Common Denominator with Opposite Factors**

Find the LCD of the rational expressions $\dfrac{1}{x^2 - 25}$ and $\dfrac{9}{10 - 2x}$.

Solution

Factor the denominators: $\quad x^2 - 25 = (x + 5)\,(x - 5)$

$$10 - 2x = 2(5 - x)$$

Work Smart

In Example 6 we could also have factored $10 - 2x$ as $-2(-5 + x) = -2(x - 5)$.

It looks as though there are no common factors. But you should notice that the factors $x - 5$ and $5 - x$ are opposites. So we can write the factorization of $10 - 2x$ as follows:

$$10 - 2x = 2(5 - x)$$

Factor out -1: $\quad = 2(-1)\,(x - 5)$

$$= -2(x - 5)$$

Now we have

$$x^2 - 25 = \quad (x - 5)\,(x + 5)$$

$$10 - 2x = -2(x - 5)$$

Write the LCD: $\quad \text{LCD} = -2(x - 5)\,(x + 5)$

Quick ✓

In Problem 8, find the least common denominator of the rational expressions.

8. $\dfrac{3}{21 - 3x}$ and $\dfrac{8}{x^2 - 49}$

▶ ❷ Write a Rational Expression Equivalent to a Given Rational Expression

Rational expressions that represent the same quantity are *equivalent*. If we multiply the numerator and denominator of a rational expression by the same quantity, we get an equivalent rational expression.

$$\frac{p}{q} = \frac{p}{q} \cdot 1 = \frac{p}{q} \cdot \frac{r}{r} = \frac{pr}{qr}$$

In other words, we are multiplying the rational expression by 1, the multiplicative identity. Let's review how to do this using rational numbers.

EXAMPLE 7 **Forming Equivalent Rational Numbers**

Write $\dfrac{2}{5}$ as an equivalent fraction with a denominator of 20.

Solution

We want to change from a denominator of 5 to a denominator of 20:

$$\frac{2}{5} \cdot \frac{?}{?} = \frac{\Box}{20}$$

Since $5 \cdot 4 = 20$, form the factor of $1 = \dfrac{4}{4}$.

$$\frac{2}{5} \cdot \frac{4}{4} = \frac{8}{20}$$

The equivalent fraction is $\dfrac{8}{20}$.

The approach used to form equivalent fractions is the same as the approach we use to form equivalent rational expressions.

EXAMPLE 8 **How to Form an Equivalent Rational Expression**

Write the rational expression $\dfrac{4}{5x^2 + x}$ as an equivalent rational expression with denominator $10x^2 + 2x$.

Step-by-Step Solution

Step 1: Write each denominator in factored form.

Original denominator: $\quad 5x^2 + x = \quad x(5x + 1)$

New denominator: $\quad 10x^2 + 2x = 2 \cdot x(5x + 1)$

Step 2: Determine the missing factor(s).

The original denominator is missing a factor of 2.

Step 3: Multiply the original rational expression by $1 = \dfrac{2}{2}$.

$$\frac{4}{5x^2 + x} = \frac{4}{x(5x + 1)} \cdot \frac{2}{2}$$

Step 4: Find the product. Leave the denominator in factored form.

$$= \frac{8}{2x(5x + 1)}$$

So $\dfrac{4}{5x^2 + x} = \dfrac{8}{2x(5x + 1)}$. Notice that $2x(5x + 1) = 10x^2 + 2x$, as required.

> **Forming Equivalent Rational Expressions**
>
> **Step 1:** Write each denominator in factored form.
>
> **Step 2:** Determine the factor(s) the new denominator has that is (are) missing from the original denominator.
>
> **Step 3:** Multiply the original rational expression by $1 = \dfrac{\text{missing factor(s)}}{\text{missing factor(s)}}$.
>
> **Step 4:** Find the product. Leave the denominator in factored form.

Quick ✔

9. If we want to rewrite the rational expression $\dfrac{2x+1}{x-1}$ with a denominator of $(x-1)(x+3)$, we multiply the numerator and denominator of $\dfrac{2x+1}{x-1}$ by ____

In Problem 10, write the rational expression as an equivalent rational expression with the given denominator.

10. $\dfrac{3}{4p^2 - 8p}$ with denominator $16p^3(p-2)$.

▶ ❸ Use the LCD to Write Equivalent Rational Expressions

There is one more skill we need before we add or subtract rational expressions with unlike denominators. It is to find the LCD of two or more rational expressions and then write them as equivalent rational expressions.

EXAMPLE 9 **Using the LCD to Write Equivalent Rational Expressions**

Find the LCD of the rational expressions $\dfrac{4}{x^2 + 3x + 2}$ and $\dfrac{9}{x^2 - 4}$. Then rewrite each rational expression with the LCD.

Solution

Factor each denominator to find the LCD.

$$x^2 + 3x + 2 = (x+1)(x+2)$$
$$x^2 - 4 = \qquad (x+2)(x-2)$$
$$\text{LCD} = (x+1)(x+2)(x-2)$$

Rewrite each rational expression with a denominator of $(x+1)(x+2)(x-2)$. Multiply the numerators, but leave the denominator in factored form.

$$\frac{4}{x^2 + 3x + 2} = \frac{4}{(x+2)(x+1)} \cdot \frac{(x-2)}{(x-2)}$$
$$= \frac{4x - 8}{(x+2)(x+1)(x-2)}$$
$$\frac{9}{x^2 - 4} = \frac{9}{(x-2)(x+2)} \cdot \frac{(x+1)}{(x+1)}$$
$$= \frac{9x + 9}{(x-2)(x+2)(x+1)}$$

Quick ✔

In Problems 11 and 12, find the least common denominator of the rational expressions. Then rewrite each rational expression with the LCD.

11. $\dfrac{3}{6a^2b}$ and $\dfrac{-5}{20ab^3}$

12. $\dfrac{5}{x^2 - 4x - 5}$ and $\dfrac{-3}{x^2 - 7x + 10}$

7.4 Exercises MyMathLab® MathXL PRACTICE

Exercise numbers in green have complete video solutions in MyMathLab

*Problems **1–12** are the Quick ✓s that follow the EXAMPLES.*

Building Skills

In Problems 13–32, find the LCD of the given rational expressions. See Objective 1.

13. $\dfrac{14}{5x^2}; \dfrac{4}{25x}$

14. $\dfrac{3}{7y^2}; \dfrac{4}{49y}$

15. $\dfrac{7}{12xy^2}; \dfrac{4}{15x^3y}$

16. $\dfrac{5}{36xy^2}; \dfrac{1}{24x^2y}$

17. $\dfrac{3}{x}; \dfrac{4}{x+1}$

18. $\dfrac{2x-1}{2x+1}; \dfrac{3x}{2}$

19. $\dfrac{4}{2x+1}; 2$

20. $\dfrac{7}{y-1}; 3$

21. $\dfrac{7}{2b-6}; \dfrac{3b}{4b-12}$

22. $\dfrac{2}{6x-18}; \dfrac{3}{8x-24}$

23. $\dfrac{6}{p^2+p}; \dfrac{7}{p^2-p-2}$

24. $\dfrac{5}{2x^2-12x+18}; \dfrac{4}{4x^2-36}$

25. $\dfrac{3}{r^2+4r+4}; \dfrac{4}{r^2-r-2}$

26. $\dfrac{8}{x^2-1}; \dfrac{3}{x^2-2x+1}$

27. $\dfrac{2}{x-4}; \dfrac{3}{4-x}$

28. $\dfrac{7}{3-a}; \dfrac{1}{a-3}$

29. $\dfrac{8}{x^2-9}; \dfrac{2}{6-2x}$

30. $\dfrac{-1}{c^2-49}; \dfrac{5}{21-3c}$

31. $\dfrac{1}{(x-1)(x+2)}; \dfrac{11x}{(1-x)(2+x)}$

32. $\dfrac{3z}{(z+5)(z-4)}; \dfrac{-z}{(4-z)(5+z)}$

In Problems 33–42, write an equivalent rational expression with the given denominator. See Objective 2.

33. $\dfrac{4}{x}$ with denominator $3x^3$

34. $\dfrac{7}{x}$ with denominator $2x^2$

35. $\dfrac{3+c}{a^2b^2c}$ with denominator $a^2b^2c^2$

36. $\dfrac{7a+1}{abc}$ with denominator a^2b^2c

37. $\dfrac{x+2}{x+4}$ with denominator x^2-16

38. $\dfrac{x-5}{x+2}$ with denominator x^2+5x+6

39. $\dfrac{3n}{2n+2}$ with denominator $6n^2-6$

40. $\dfrac{7a}{3a^2+9a+6}$ with denominator $6a^2+18a+12$

41. $4t$ with denominator $t-1$

42. 7 with denominator t^2+1

In Problems 43–60, find the LCD of the rational expressions. Then rewrite each as an equivalent rational expression with the LCD. See Objective 3.

43. $\dfrac{2x}{3y}; \dfrac{4}{9y^2}$

44. $\dfrac{4}{5n^2}; \dfrac{3}{7n}$

45. $\dfrac{3a+1}{2a^3}; \dfrac{4a-1}{4a}$

46. $\dfrac{2p^2-1}{6p}; \dfrac{3p^3+2}{8p^2}$

47. $\dfrac{2}{m}; \dfrac{3}{m+1}$

48. $\dfrac{x+2}{x}; \dfrac{x}{x+2}$

49. $\dfrac{y-2}{4y}; \dfrac{y}{8y-4}$

50. $\dfrac{3b}{7a}; \dfrac{2a}{7a+14b}$

51. $\dfrac{1}{x-1}; \dfrac{2x}{1-x^2}$

52. $\dfrac{2a}{a-2}; \dfrac{a}{4-a^2}$

53. $\dfrac{4x}{x^2 - 4}; \dfrac{2}{x + 2}$

54. $\dfrac{4a}{a^2 - 1}; \dfrac{7}{a + 1}$

55. $\dfrac{3x}{x^2 - 6x - 7}; \dfrac{5}{x^2 - 4x - 21}$

56. $\dfrac{4}{m^2 - 5m - 6}; \dfrac{2m}{m^2 - 12m + 36}$

57. $\dfrac{x + 1}{x^2 - 9}; \dfrac{x + 2}{x^2 - 6x + 9}$

58. $\dfrac{4n}{n^2 - n - 6}; \dfrac{2}{n^2 + 4n + 4}$

59. $\dfrac{3}{x + 4}; \dfrac{2x - 1}{2x^2 + 7x - 4}$

60. $\dfrac{3}{2n^2 - 7n + 3}; \dfrac{2n}{n - 3}$

Applying the Concepts

61. Painting a Room It takes an experienced painter x hours to paint a room alone; assuming that he works at a constant rate, he completes $\dfrac{1}{x}$ of the job per hour. It takes an apprentice 3 times as long to paint the same room as it takes the experienced painter, so the apprentice completes $\dfrac{1}{3x}$ of the job per hour. Find the LCD of the two rational expressions that represent the rate of work of the experienced painter and the apprentice, and then write each as an equivalent rational expression using this denominator.

62. Mowing the Lawn It takes Mario 2 hours longer to mow the lawn than it takes Marco. If Marco takes x hours, he completes $\dfrac{1}{x}$ of the job per hour. Since Mario requires 2 hours more, he completes $\dfrac{1}{x + 2}$ of the job per hour. Find the LCD of the two rational expressions, and then write each as an equivalent rational expression using this denominator.

63. Traveling on a River The time to complete a journey can be calculated by the formula $t = \dfrac{d}{r}$, where t is the time, d is the distance traveled, and r is rate. Suppose the time for a boat to travel a distance of 12 miles upstream on a river whose current is 4 miles per hour is given by $\dfrac{12}{r - 4}$. The time for the boat to travel 12 miles downstream on the same river is $\dfrac{12}{r + 4}$. Find the LCD of the two rational expressions, and then write each as an equivalent rational expression using this denominator.

64. Traveling by Plane The time to complete a journey can be calculated by the formula $t = \dfrac{d}{r}$, where t is the time, d is the distance traveled, and r is rate. Suppose a plane flies 500 miles west into a headwind of 50 miles per hour in a time of $\dfrac{500}{r - 50}$ hours, where r is the speed of the plane in still air. The plane flies 500 miles east with a tailwind of 50 miles per hour in a time of $\dfrac{500}{r + 50}$ hours. Find the LCD of the two rational expressions, and then write each as an equivalent rational expression using this denominator.

Extending the Concepts

In Problems 65–70, identify the LCD of the rational expressions.

65. $\dfrac{4}{x^3 + 8}; \dfrac{1}{x^2 - 4}; \dfrac{5}{x^3 - 8}$

66. $\dfrac{x}{2x^3 - 2}; \dfrac{1}{3x^2 - 3}; \dfrac{2x + 1}{4x - 4}$

67. $\dfrac{11}{p^2 - p}; \dfrac{-2}{p^3 - p^2}; \dfrac{8}{p^2 - 4p + 3}$

68. $\dfrac{7}{4n^3 - 2n^2}; \dfrac{9}{8n^2 - 4n}; \dfrac{6}{4n^2 - 1}$

69. $\dfrac{a}{2a^4 - 2a^2b^2}; \dfrac{b}{4ab^2 + 4b^3}; \dfrac{ab}{a^3b - b^4}$

70. $\dfrac{5}{3x^2 + xy - 2y^2}; \dfrac{2}{x^2 - xy - 2y^2}$

Explaining the Concepts

71. If you were teaching the class how to write an equivalent rational expression with a given denominator, what steps would you list on the board for the class?

72. A member of your class says that $\dfrac{1}{(x + 3)(x + 2)}$ and $\dfrac{4}{(x + 2)}$ have an LCD of $(x + 3)(x + 2)^2$. This class member then rewrites the rational expressions as $\dfrac{x + 2}{(x + 3)(x + 2)^2}$ and $\dfrac{4(x + 3)(x + 2)}{(x + 3)(x + 2)^2}$. Use this result to explain why $(x + 3)(x + 2)^2$ is not the least common denominator.

7.5 Adding and Subtracting Rational Expressions with Unlike Denominators

Objective

1 Add and Subtract Rational Expressions with Unlike Denominators

Are You Ready for This Section?

R1. Find the least common denominator (LCD) of 15 and 24. [Section 1.2, p. 12]

R2. Factor: $2x^2 + 3x + 1$ [Section 6.3, pp. 385–392]

R3. Simplify: $\dfrac{3x + 9}{6x}$ [Section 7.1, pp. 433–434]

R4. Simplify: $\dfrac{1 - x^2}{2x - 2}$ [Section 7.1, pp. 433–434]

R5. Find the sum: $\dfrac{5}{2x} + \dfrac{7}{2x}$ [Section 7.3, pp. 446–447]

Ready?...Answers **R1.** 120
R2. $(2x + 1)(x + 1)$
R3. $\dfrac{x + 3}{2x}$ **R4.** $-\dfrac{1 + x}{2}$ **R5.** $\dfrac{6}{x}$

▶ **1** Add and Subtract Rational Expressions with Unlike Denominators

Before we add and subtract rational expressions with unlike denominators, let's review how to add rational numbers with unlike denominators.

EXAMPLE 1 **How to Add Rational Numbers with Unlike Denominators**

Evaluate $\dfrac{1}{6} + \dfrac{9}{14}$. Be sure to simplify the result.

Step-by-Step Solution

Step 1: Find the least common denominator.

$$6 = 2 \cdot 3$$
$$14 = 2 \cdot 7$$
$$\text{LCD} = 2 \cdot 3 \cdot 7 = 42$$

Step 2: Write each fraction as an equivalent fraction, with the least common denominator.

Because $6 \cdot 7 = 42$, use $1 = \dfrac{7}{7}$ to rewrite $\dfrac{1}{6}$ using the LCD.

Since $14 \cdot 3 = 42$, use $1 = \dfrac{3}{3}$ to rewrite $\dfrac{9}{14}$ using the LCD:

$$\dfrac{1}{6} \cdot \dfrac{7}{7} + \dfrac{9}{14} \cdot \dfrac{3}{3} = \dfrac{7}{42} + \dfrac{27}{42}$$

Step 3: Find the sum of the numerators, written over the common denominator.

$$= \dfrac{7 + 27}{42}$$
$$= \dfrac{34}{42}$$

Step 4: Simplify.

Factor 34 and 42: $= \dfrac{2 \cdot 17}{2 \cdot 21}$

Divide out common factors: $= \dfrac{\cancel{2} \cdot 17}{\cancel{2} \cdot 21}$

$$= \dfrac{17}{21}$$

Now that you've practiced adding and subtracting rational numbers, let's explore adding and subtracting rational expressions.

EXAMPLE 2 **How to Add Rational Expressions with Unlike Monomial Denominators**

Find the sum: $\dfrac{5}{6x^2} + \dfrac{4}{15x}$

Step-by-Step Solution

Step 1: Find the least common denominator.

$$6x^2 = 2 \cdot 3 \cdot x^2$$
$$15x = 3 \cdot 5 \cdot x$$
$$\text{LCD} = 2 \cdot 3 \cdot 5 \cdot x^2 = 30x^2$$

Step 2: Rewrite each rational expression with the *common* denominator.

We need the denominator of $\dfrac{5}{6x^2}$ to be $30x^2$, so we multiply $\dfrac{5}{6x^2}$ by $1 = \dfrac{5}{5}$. Similarly, multiply $\dfrac{4}{15x}$ by $1 = \dfrac{2x}{2x}$. Do you see why?

$$\frac{5}{6x^2} = \frac{5}{6x^2} \cdot \frac{5}{5} = \frac{25}{30x^2} \qquad\qquad \frac{4}{15x} = \frac{4}{15x} \cdot \frac{2x}{2x} = \frac{8x}{30x^2}$$

Step 3: Add the rational expressions found in Step 2.

$$\frac{5}{6x^2} + \frac{4}{15x} = \frac{25}{30x^2} + \frac{8x}{30x^2}$$

Add using $\dfrac{a}{c} + \dfrac{b}{c} = \dfrac{a+b}{c}$:

$$= \frac{25 + 8x}{30x^2}$$

Step 4: Simplify.

The rational expression is already simplified.

Therefore, $\dfrac{5}{6x^2} + \dfrac{4}{15x} = \dfrac{25 + 8x}{30x^2}$.

Adding or Subtracting Rational Expressions with Unlike Denominators

Step 1: Find the least common denominator.

Step 2: Rewrite each rational expression as an equivalent rational expression with the common denominator.

Step 3: Add or subtract the rational expressions found in Step 2.

Step 4: Simplify the result.

EXAMPLE 3 **Adding Rational Expressions with Unlike Polynomial Denominators**

Find the sum: $\dfrac{-1}{x+3} + \dfrac{3}{x+2}$

Solution

The least common denominator is $(x+3)(x+2)$ because the two denominators have no common factors. To rewrite each rational expression as an equivalent rational expression with the LCD, we multiply $\dfrac{-1}{x+3}$ by $1 = \dfrac{x+2}{x+2}$ and we multiply $\dfrac{3}{x+2}$ by $1 = \dfrac{x+3}{x+3}$.

$$\frac{-1}{x+3} + \frac{3}{x+2} = \frac{-1}{x+3} \cdot \frac{x+2}{x+2} + \frac{3}{x+2} \cdot \frac{x+3}{x+3}$$

Multiply the numerators; leave the denominators in factored form:
$$= \frac{-x-2}{(x+3)(x+2)} + \frac{3x+9}{(x+2)(x+3)}$$

Add using $\dfrac{a}{c} + \dfrac{b}{c} = \dfrac{a+b}{c}$:
$$= \frac{-x-2+3x+9}{(x+3)(x+2)}$$

Combine like terms:
$$= \frac{2x+7}{(x+3)(x+2)}$$

EXAMPLE 4 **Adding Rational Expressions with Unlike Polynomial Denominators**

Find the sum: $\dfrac{4}{x+1} + \dfrac{16}{x^2-2x-3}$

Solution

The first step is to find the least common denominator. We begin by factoring each denominator.

$$x+1 = \qquad x+1$$
$$x^2-2x-3 = (x-3)(x+1)$$

The LCD is $(x - 3)(x + 1)$. Now rewrite each rational expression as an equivalent rational expression with the least common denominator.

$$\frac{4}{x + 1} + \frac{16}{x^2 - 2x - 3} = \frac{4}{x + 1} + \frac{16}{(x - 3)(x + 1)}$$

Rewrite each rational expression with the common denominator:
$$= \frac{4}{x + 1} \cdot \frac{x - 3}{x - 3} + \frac{16}{(x + 1)(x - 3)}$$

Don't simplify $\dfrac{4(x - 3)}{(x + 1)(x - 3)}$ or you'll be right back where you started!
$$= \frac{4(x - 3)}{(x + 1)(x - 3)} + \frac{16}{(x + 1)(x - 3)}$$

Distribute the 4:
$$= \frac{4x - 12}{(x + 1)(x - 3)} + \frac{16}{(x + 1)(x - 3)}$$

Add using $\dfrac{a}{c} + \dfrac{b}{c} = \dfrac{a + b}{c}$:
$$= \frac{4x - 12 + 16}{(x + 1)(x - 3)}$$

Add:
$$= \frac{4x + 4}{(x + 1)(x - 3)}$$

Factor the numerator:
$$= \frac{4(x + 1)}{(x + 1)(x - 3)}$$

Divide out like factors:
$$= \frac{4}{x - 3}$$

Quick ✓

In Problems 10 and 11, find each sum.

10. $\dfrac{-z + 1}{z^2 + 7z + 10} + \dfrac{2}{z + 5}$

11. $\dfrac{1}{x^2 + 5x} + \dfrac{1}{x^2 - 5x}$

▶ **EXAMPLE 5**   **How to Subtract Rational Expressions with Unlike Denominators**

Find the difference: $\dfrac{8}{x} - \dfrac{2}{x + 1}$

Step-by-Step Solution

Step 1: Find the least common denominator.

The LCD is $x(x + 1)$.

Step 2: Rewrite each rational expression as an equivalent rational expression with the common denominator. Multiply the numerators, but leave the denominators in factored form.

Multiply $\dfrac{8}{x}$ by $1 = \dfrac{x + 1}{x + 1}$; multiply $\dfrac{2}{x + 1}$ by $1 = \dfrac{x}{x}$:
$$\frac{8}{x} - \frac{2}{x + 1} = \frac{8}{x} \cdot \frac{x + 1}{x + 1} - \frac{2}{x + 1} \cdot \frac{x}{x}$$

$$= \frac{8(x + 1)}{x(x + 1)} - \frac{2 \cdot x}{x(x + 1)}$$

$$= \frac{8x + 8}{x(x + 1)} - \frac{2x}{x(x + 1)}$$

Step 3: Subtract the rational expressions found in Step 2.

Subtract using $\dfrac{a}{c} - \dfrac{b}{c} = \dfrac{a - b}{c}$:
$$= \frac{8x + 8 - 2x}{x(x + 1)}$$

Combine like terms:
$$= \frac{6x + 8}{x(x + 1)}$$

(continued)

Step 4: Simplify the result.

$$\text{Factor the numerator:} \quad = \frac{2(3x + 4)}{x(x + 1)}$$

Since there are no common factors, the expression is fully simplified.

Quick ✓

In Problems 12 and 13, find each difference.

12. $\dfrac{-4}{5ab^2} - \dfrac{3}{4a^2b^3}$

13. $\dfrac{5}{x} - \dfrac{3}{x - 4}$

EXAMPLE 6 **Adding Rational Expressions with Unlike Denominators—Both Denominators Factor**

Perform the indicated operation: $\dfrac{-2}{a^2 - 4a} + \dfrac{2}{4a - 16}$

Solution

Factor each denominator to find the LCD.

$$a^2 - 4a = \quad a \cdot (a - 4)$$
$$4a - 16 = 4 \cdot \quad (a - 4)$$

The LCD is $4a(a - 4)$. Rewrite each rational expression as an equivalent expression using the LCD $4a(a - 4)$. Multiply $\dfrac{-2}{a(a - 4)}$ by $1 = \dfrac{4}{4}$ and multiply $\dfrac{2}{4(a - 4)}$ by $1 = \dfrac{a}{a}$.

$$\frac{-2}{a(a - 4)} + \frac{2}{4(a - 4)} = \frac{-2}{a(a - 4)} \cdot \frac{4}{4} + \frac{2}{4(a - 4)} \cdot \frac{a}{a}$$

$$\text{Multiply the numerators:} \quad = \frac{-8}{4a(a - 4)} + \frac{2a}{4a(a - 4)}$$

$$\text{Add using } \frac{a}{c} + \frac{b}{c} = \frac{a + b}{c}: \quad = \frac{2a - 8}{4a(a - 4)}$$

$$\text{Factor the numerator and denominator:} \quad = \frac{2(a - 4)}{2 \cdot 2a(a - 4)}$$

$$\text{Divide out common factors:} \quad = \frac{1}{2a}$$

EXAMPLE 7 **Subtracting Rational Expressions with Unlike Denominators—Both Denominators Factor**

Perform the indicated operation: $\dfrac{a - 2}{a^2 + 5a + 6} - \dfrac{2a - 3}{3a^2 + 9a}$

Solution

First, factor each denominator.

$$a^2 + 5a + 6 = \quad (a + 3)(a + 2)$$
$$3a^2 + 9a = 3a(a + 3)$$

The LCD is $3a(a + 3)(a + 2)$. Now rewrite each expression as an equivalent rational expression using the LCD $3a(a + 3)(a + 2)$.

$$\frac{a - 2}{a^2 + 5a + 6} - \frac{2a - 3}{3a^2 + 9a} = \frac{a - 2}{(a + 3)(a + 2)} \cdot \frac{3a}{3a} - \frac{2a - 3}{3a(a + 3)} \cdot \frac{a + 2}{a + 2}$$

Multiply the numerators:
$$= \frac{(a - 2) \cdot 3a}{3a(a + 3)(a + 2)} - \frac{(2a - 3)(a + 2)}{3a(a + 3)(a + 2)}$$

$$= \frac{3a^2 - 6a}{3a(a + 3)(a + 2)} - \frac{2a^2 + a - 6}{3a(a + 3)(a + 2)}$$

Subtract using $\frac{a}{c} - \frac{b}{c} = \frac{a - b}{c}$; don't forget the parentheses:
$$= \frac{3a^2 - 6a - (2a^2 + a - 6)}{3a(a + 3)(a + 2)}$$

Distribute the minus sign:
$$= \frac{3a^2 - 6a - 2a^2 - a + 6}{3a(a + 3)(a + 2)}$$

Combine like terms:
$$= \frac{a^2 - 7a + 6}{3a(a + 3)(a + 2)}$$

Factor the numerator:
$$= \frac{(a - 6)(a - 1)}{3a(a + 3)(a + 2)}$$

●

Quick ✓

In Problems 14 and 15, perform the indicated operation.

14. $\dfrac{3}{(x - 5)(x + 4)} - \dfrac{2}{(x - 5)(x - 1)}$

15. $\dfrac{x - 2}{x^2 - 3x} + \dfrac{x + 3}{4x - 12}$

EXAMPLE 8   **Adding Rational Expressions Containing Opposite Factors**

Find the sum: $\dfrac{1}{1 - n} + \dfrac{2n}{n^2 - 1}$

Solution

First, factor each denominator:

$$1 - n = 1 - n$$
$$n^2 - 1 = (n + 1)(n - 1)$$

Do you see that the factors $n - 1$ and $1 - n$ are opposites? Factor -1 from $1 - n$ so that $1 - n = -1(-1 + n) = -1(n - 1)$. Rewrite the sum as

$$\frac{1}{1 - n} + \frac{2n}{n^2 - 1} = \frac{1}{-1(n - 1)} + \frac{2n}{(n + 1)(n - 1)}$$

Use $\dfrac{a}{-b} = \dfrac{-a}{b}$:
$$= \frac{-1}{n - 1} + \frac{2n}{(n + 1)(n - 1)}$$

In this form, the LCD is $(n - 1)(n + 1)$. Now rewrite each expression as an equivalent expression with the common denominator.

$$\text{Multiply } \frac{-1}{n-1} \text{ by } 1 = \frac{n+1}{n+1}: \quad = \frac{-1}{n-1} \cdot \frac{n+1}{n+1} + \frac{2n}{(n+1)(n-1)}$$

$$\text{Multiply the numerators:} \quad = \frac{-1(n+1)}{(n+1)(n-1)} + \frac{2n}{(n+1)(n-1)}$$

$$\text{Distribute the } -1: \quad = \frac{-n-1}{(n+1)(n-1)} + \frac{2n}{(n+1)(n-1)}$$

$$\text{Add using } \frac{a}{c} + \frac{b}{c} = \frac{a+b}{c}: \quad = \frac{-n-1+2n}{(n+1)(n-1)}$$

$$\text{Combine like terms:} \quad = \frac{n-1}{(n+1)(n-1)}$$

$$\text{Divide out common factors:} \quad = \frac{1}{n+1}$$

Work Smart

Remember that when like factors divide, they form a quotient of 1. That's why

$$\frac{n-1}{(n+1)(n-1)}$$

simplifies to

$$\frac{\cancel{n-1}}{(n+1)\cancel{(n-1)}} = \frac{1}{n+1}$$

Quick ✓

16. *True or False* $\dfrac{3}{x} + \dfrac{6}{-x} = \dfrac{3-6}{x}$

In Problems 17 and 18, perform the indicated operation.

17. $\dfrac{2p^2 - 8}{p^2 + p - 2} + \dfrac{p + 1}{1 - p}$

18. $\dfrac{y^2 - 13}{y^2 - 2y - 3} - \dfrac{1}{3 - y}$

EXAMPLE 9   **Adding an Integer and a Rational Expression**

Find the sum: $5 + \dfrac{2}{3x - 4}$

Solution

Since $5 = \dfrac{5}{1}$, rewrite the expression as $\dfrac{5}{1} + \dfrac{2}{3x - 4}$. The LCD is $3x - 4$.

$$\frac{5}{1} + \frac{2}{3x-4} = \frac{5}{1} \cdot \frac{3x-4}{3x-4} + \frac{2}{3x-4}$$

$$\text{Multiply the numerators:} \quad = \frac{5(3x-4)}{3x-4} + \frac{2}{3x-4}$$

$$= \frac{15x-20}{3x-4} + \frac{2}{3x-4}$$

$$\text{Add using } \frac{a}{c} + \frac{b}{c} = \frac{a+b}{c}: \quad = \frac{15x-20+2}{3x-4}$$

$$\text{Combine like terms:} \quad = \frac{15x-18}{3x-4}$$

$$\text{Factor the numerator:} \quad = \frac{3(5x-6)}{3x-4}$$

Quick ✓

In Problems 19 and 20, perform the indicated operation.

19. $2 + \dfrac{3}{x-1}$

20. $\dfrac{6}{x+3} - 3$

EXAMPLE 10 **Adding and Subtracting Rational Expressions**

Perform the indicated operations.

$$\frac{5}{n-2} + \frac{5}{n+2} - \frac{20}{n^2-4}$$

Solution

Factor each denominator to find the LCD. Since $n^2 - 4 = (n-2)(n+2)$, the LCD is $(n-2)(n+2)$.

$$\frac{5}{n-2} + \frac{5}{n+2} - \frac{20}{n^2-4} = \frac{5}{n-2} + \frac{5}{n+2} - \frac{20}{(n-2)(n+2)}$$

Rewrite each expression with the common denominator:
$$= \frac{5}{n-2} \cdot \frac{n+2}{n+2} + \frac{5}{n+2} \cdot \frac{n-2}{n-2} - \frac{20}{(n-2)(n+2)}$$

Multiply the numerators:
$$= \frac{5n+10}{(n-2)(n+2)} + \frac{5n-10}{(n-2)(n+2)} - \frac{20}{(n-2)(n+2)}$$

Add using $\dfrac{a}{c} + \dfrac{b}{c} = \dfrac{a+b}{c}$:
$$= \frac{5n+10+5n-10-20}{(n-2)(n+2)}$$

Combine like terms:
$$= \frac{10n-20}{(n-2)(n+2)}$$

Factor the numerator:
$$= \frac{10(n-2)}{(n-2)(n+2)}$$

Divide out common factors:
$$= \frac{10}{n+2}$$

Quick ✓

In Problems 21 and 22, perform the indicated operations.

21. $\dfrac{1}{x+1} - \dfrac{2}{x^2-1} + \dfrac{3}{x-1}$

22. $\dfrac{3}{x} - \left(\dfrac{1}{x-2} - \dfrac{6}{x^2-2x}\right)$

7.5 Exercises **MyMathLab®** PRACTICE

Exercise numbers in **green** have complete video solutions in MyMathLab

Problems **1–22** are the Quick ✓s that follow the **EXAMPLES**.

Building Skills

In Problems 23–34, find each sum and simplify. See Objective 1.

23. $-\dfrac{4}{3} + \dfrac{1}{2}$

24. $\dfrac{7}{12} + \dfrac{3}{4}$

25. $\dfrac{2}{3x} + \dfrac{1}{x}$

26. $\dfrac{5}{2x} + \dfrac{6}{5}$

27. $\dfrac{a}{2a-1} + \dfrac{3}{2a+1}$

28. $\dfrac{2}{x-1} + \dfrac{x-1}{x+1}$

29. $\dfrac{3}{x-4} + \dfrac{5}{4-x}$

30. $\dfrac{7}{n-4} + \dfrac{8}{4-n}$

31. $\dfrac{a+5}{5a-a^2} + \dfrac{a+3}{4a-20}$

32. $\dfrac{2x-6}{x^2-x-6} + \dfrac{x+4}{x+2}$

33. $\dfrac{2x + 4}{x^2 + 2x} + \dfrac{3}{x}$

34. $\dfrac{3}{x - 4} + \dfrac{x + 4}{x^2 - 16}$

61. $\dfrac{3x - 1}{x} - \dfrac{9}{x^2 - 9x}$

62. $\dfrac{2x}{8x - 12} - \dfrac{3}{2x + 2}$

In Problems 35–44, find each difference and simplify. See Objective 1.

35. $\dfrac{7}{15} - \dfrac{9}{25}$

36. $\dfrac{8}{21} - \dfrac{6}{35}$

63. $\dfrac{3}{x^2 - 4x - 5} - \dfrac{2}{x^2 - 6x + 5}$

37. $m - \dfrac{16}{m}$

38. $\dfrac{9}{x} - x$

64. $\dfrac{3}{x^2 + 7x + 10} - \dfrac{4}{x^2 + 6x + 5}$

39. $\dfrac{x}{x - 3} - \dfrac{x - 2}{x + 3}$

40. $\dfrac{x - 2}{x + 2} - \dfrac{x + 2}{x - 2}$

65. $\dfrac{5}{2a - a^2} - \dfrac{3}{2a^2 - 4a}$

41. $\dfrac{x + 2}{x + 3} - \dfrac{x^2 + 3x}{x^2 + 6x + 9}$

42. $\dfrac{x}{x + 4} - \dfrac{-4}{x^2 + 8x + 16}$

66. $\dfrac{-1}{3n - n^2} + \dfrac{1}{3n^2 - 9n}$

43. $\dfrac{-3x - 9}{x^2 + x - 6} - \dfrac{x + 3}{2 - x}$

44. $\dfrac{p + 1}{p^2 - 2p} - \dfrac{2p + 3}{2 - p}$

67. $\dfrac{2n + 1}{n^2 - 4} + \dfrac{3n}{6 - n^2 - n}$

68. $\dfrac{a - 5}{2a^2 - 6a} - \dfrac{6}{12a^2 - 4a^3}$

Mixed Practice

69. $\dfrac{m + 3}{m^2 + 2m - 8} - \dfrac{m + 2}{m^2 - 4}$

In Problems 45–78, perform the indicated operation.

45. $\dfrac{5}{3y^2} - \dfrac{3}{4y}$

46. $\dfrac{5}{4n} - \dfrac{3}{n^2}$

70. $\dfrac{n + 3}{n^2 - 8n + 15} + \dfrac{n + 3}{n^2 - 9}$

47. $\dfrac{9}{5x} - \dfrac{6}{10x}$

48. $\dfrac{7}{4b^2} - \dfrac{3b^2}{2b}$

71. $\dfrac{2n + 1}{n + 3} + \dfrac{7 - 2n^2}{n^2 + n - 6}$

49. $\dfrac{2x}{2x + 3} - 1$

50. $a + \dfrac{a}{a - 2}$

72. $\dfrac{4}{x^2 + 2x - 15} + \dfrac{3}{x^2 - x - 6}$

51. $\dfrac{n}{n - 2} + \dfrac{n + 2}{n}$

52. $\dfrac{6}{x + 2} + \dfrac{4}{x - 3}$

73. $\dfrac{-3x - 9}{x^2 + x - 6} - \dfrac{x + 3}{2 - x}$

53. $\dfrac{2x}{x - 3} - \dfrac{5}{x}$

54. $\dfrac{x}{x - 2} - \dfrac{2}{x - 1}$

74. $\dfrac{2x - 1}{1 - x} - \dfrac{-x^2 - 2x}{x^2 + x - 2}$

55. $\dfrac{y + 8}{y - 3} - \dfrac{10y + 30}{y^2 - 9}$

56. $\dfrac{y + 3}{y + 2} - \dfrac{4y + 4}{y^2 - 4}$

75. $\dfrac{4}{x + 2} + \dfrac{-5x - 2}{x^2 + 2x} - \dfrac{3 - x}{x}$

57. $\dfrac{4}{2n + 1} + 1$

58. $\dfrac{x - 2}{x + 3} + 2$

76. $\dfrac{x + 1}{x - 3} + \dfrac{x + 2}{x - 2} - \dfrac{x^2 + 3}{x^2 - 5x + 6}$

77. $\dfrac{2}{m + 2} - \dfrac{3}{m} + \dfrac{m + 10}{m^2 - 4}$

59. $\dfrac{-12}{3x - 6} + \dfrac{4x - 1}{x - 2}$

60. $\dfrac{2}{4x + 4} + \dfrac{8}{3x + 3}$

78. $\dfrac{7}{w - 3} - \dfrac{5}{w} - \dfrac{2w + 6}{w^2 - 9}$

Applying the Concepts

In Problems 79–86, perform the indicated operation.

79. Find the quotient of $\dfrac{x-3}{x+2}$ and $\dfrac{x^2-9}{x^2+4}$.

80. Find the quotient of $\dfrac{x}{x+1}$ and $\dfrac{5}{3x+3}$.

81. Find the difference of $\dfrac{4x+6}{2x^2+x-3}$ and $\dfrac{x-1}{x^2-1}$.

82. Find the difference of $\dfrac{x+4}{2x^2-8}$ and $\dfrac{3}{4x-8}$.

83. Find the sum of $\dfrac{x+2}{x^2+x-6}$ and $\dfrac{x-3}{x^2+5x+6}$.

84. Find the sum of $\dfrac{x^2}{x-4}$ and $\dfrac{16}{4-x}$.

85. Find the product of $\dfrac{2x-3}{x+6}$ and $\dfrac{x-1}{x-7}$.

86. Find the product of $\dfrac{x^2-3x+2}{x^2+2x-3}$ and $\dfrac{x^2+x-6}{x^2-4}$.

△ **87. Perimeter of a Rectangle** Find the perimeter of the rectangle:

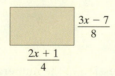

$\dfrac{3x-7}{8}$

$\dfrac{2x+1}{4}$

△ **88. Area of a Rectangle** Find the area of the rectangle:

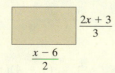

$\dfrac{2x+3}{3}$

$\dfrac{x-6}{2}$

Extending the Concepts

In Problems 89–98, perform the indicated operation.

89. $\dfrac{2}{a} - \left(\dfrac{2}{a-1} - \dfrac{3}{(a-1)^2} \right)$

90. $\dfrac{2}{b} - \left(\dfrac{2}{b+2} - \dfrac{2}{(b+2)^2} \right)$

91. $\dfrac{1}{x} - \dfrac{2}{x^2+x} + \dfrac{3}{x^3-x^2}$

92. $\dfrac{x}{(x-1)^2} + \dfrac{2}{x} - \dfrac{x+1}{x^3-x^2}$

93. $\dfrac{2a+b}{a-b} \cdot \dfrac{2}{a+b} - \dfrac{3a+3b}{a^2-b^2}$

94. $\dfrac{2a+b}{a-b} \div \dfrac{a+b}{2} + \dfrac{3}{a+b}$

95. $\dfrac{x-3}{x-4} + \dfrac{x+2}{x-4} \cdot \dfrac{4}{x+1}$

96. $\dfrac{3}{m+n} - \dfrac{m-2n}{m-n} \cdot \dfrac{2}{m+n}$

97. $\dfrac{x+2}{x-2} - \dfrac{x-2}{x+2} \div \dfrac{1}{x^2-4}$

98. $\dfrac{a+3}{a-3} - \dfrac{a+3}{a-3} \cdot \dfrac{a^2-4a+3}{a^2+5a+6}$

Explaining the Concepts

99. Explain the steps that should be used to add or subtract rational expressions with unlike denominators.

100. There are two ways that you could add $\dfrac{1}{x-7} + \dfrac{3}{7-x}$. Describe these two ways. Which method do you prefer?

101. Suppose you are asked to simplify $\dfrac{3x+1}{(2x+5)(x-1)} - \dfrac{x-4}{(2x+5)(x-1)}$. Explain the steps you should use to simplify the expression.

102. Explain the error in the following problem, and then rework the problem correctly.

$$\dfrac{2}{x-1} - \dfrac{x+2}{x} = \dfrac{2}{x-1} \cdot \dfrac{x}{x} - \dfrac{x+2}{x} \cdot \dfrac{x-1}{x-1}$$

$$= \dfrac{2x}{x(x-1)} - \dfrac{(x+2)(x-1)}{x(x-1)}$$

$$= \dfrac{2x-(x+2)(x-1)}{x(x-1)}$$

$$= \dfrac{2x-(x+2)(x-1)}{x(x-1)}$$

$$= \dfrac{2x-x-2}{x}$$

$$= \dfrac{x-2}{x}$$

7.6 Complex Rational Expressions

Objectives

① Simplify a Complex Rational Expression by Simplifying the Numerator and Denominator Separately (Method I)

② Simplify a Complex Rational Expression Using the Least Common Denominator (Method II)

Are You Ready For This Section?

Before getting started, take the following readiness quiz. If you get a problem wrong, go back to the section cited and review the material.

R1. Factor: $6y^2 - 5y - 6$ [Section 6.3, pp. 385–393]

R2. Find the quotient: $\dfrac{x+3}{12} \div \dfrac{x^2-9}{15}$ [Section 7.2, pp. 440–443]

Definition

A **complex rational expression** is a fraction in which the numerator and/or the denominator contains the sum or difference of two or more rational expressions.

The following are examples of complex rational expressions.

$$\frac{3}{\dfrac{4}{x}-\dfrac{x}{3}} \qquad \frac{\dfrac{x+1}{x+4}-1}{16} \qquad \frac{3-\dfrac{1}{x}}{1+\dfrac{1}{x}}$$

To **simplify** a complex rational expression means to write the rational expression in the form $\dfrac{p}{q}$, where p and q are polynomials that have no common factors.

In this section, we use the division techniques introduced in Section 7.2, which we review in Example 1.

▶ EXAMPLE 1 Dividing Rational Expressions

Find the quotient: $\dfrac{\dfrac{3}{x+4}}{\dfrac{9}{x^2-16}}$

Solution

To find the quotient, we multiply the rational expression in the numerator by the reciprocal of the rational expression in the denominator.

$$\frac{\dfrac{3}{x+4}}{\dfrac{9}{x^2-16}} = \frac{3}{x+4} \cdot \frac{x^2-16}{9}$$

Factor the numerator and denominator: $= \dfrac{3}{x+4} \cdot \dfrac{(x+4)(x-4)}{3 \cdot 3}$

Multiply: $= \dfrac{3(x+4)(x-4)}{(x+4) \cdot 3 \cdot 3}$

Divide out common factors: $= \dfrac{\cancel{3}\,\cancel{(x+4)}(x-4)}{\cancel{(x+4)} \cdot \cancel{3} \cdot 3}$

$= \dfrac{x-4}{3}$

Ready? ...Answers

R1. $(3y+2)(2y-3)$ **R2.** $\dfrac{5}{4(x-3)}$

●

Quick ✓

1. An expression such as $\dfrac{\dfrac{x}{2} + \dfrac{5}{x}}{\dfrac{2x - 1}{3}}$ is called a _____ _____ _____.

2. To _____ a complex rational expression means to write the rational expression in the form $\dfrac{p}{q}$, where p and q are polynomials that have no common factors.

In Problems 3 and 4, simplify each expression.

3. $\dfrac{\dfrac{2}{k + 3}}{\dfrac{4}{k^2 + 4k + 3}}$

4. $\dfrac{\dfrac{2}{n + 3}}{\dfrac{8n}{2n + 6}}$

▶ ❶ Simplify a Complex Rational Expression by Simplifying the Numerator and Denominator Separately (Method I)

There are two ways to simplify a complex rational expression. We'll start with Method I, simplifying the numerator and denominator separately.

EXAMPLE 2 **How to Simplify a Complex Rational Expression (Method I)**

Simplify: $\dfrac{\dfrac{1}{5} + \dfrac{1}{x}}{\dfrac{x + 5}{2}}$

Step-by-Step Solution

Step 1: Write the numerator of the complex fraction as a single rational expression.

The numerator is $\dfrac{1}{5} + \dfrac{1}{x}$.

$\text{LCD} = 5x:\quad \dfrac{1}{5} + \dfrac{1}{x} = \dfrac{1}{5} \cdot \dfrac{x}{x} + \dfrac{1}{x} \cdot \dfrac{5}{5}$

$= \dfrac{x}{5x} + \dfrac{5}{5x}$

$\text{Use } \dfrac{a}{c} + \dfrac{b}{c} = \dfrac{a + b}{c}:\quad = \dfrac{x + 5}{5x}$

Step 2: Write the denominator of the complex rational expression as a single rational expression.

The denominator is $\dfrac{x + 5}{2}$. It is already written as a single rational expression.

Step 3: Rewrite the complex rational expression using the rational expressions found in Steps 1 and 2.

$\dfrac{\dfrac{1}{5} + \dfrac{1}{x}}{\dfrac{x + 5}{2}} = \dfrac{\dfrac{x + 5}{5x}}{\dfrac{x + 5}{2}}$

Step 4: Simplify the rational expression using the techniques for dividing rational expressions from Section 7.2.

Multiply the numerator by the reciprocal of the denominator: $= \dfrac{x + 5}{5x} \cdot \dfrac{2}{x + 5}$

$= \dfrac{(x + 5) \cdot 2}{5x (x + 5)}$

Multiply; divide out common factors: $= \dfrac{2}{5x}$

We summarize the steps of Method I next.

Simplifying a Complex Rational Expression by Simplifying the Numerator and Denominator Separately (Method I)

Step 1: Write the numerator of the complex rational expression as a single rational expression.

Step 2: Write the denominator of the complex rational expression as a single rational expression.

Step 3: Rewrite the complex rational expression using the rational expressions found in Steps 1 and 2.

Step 4: Simplify the rational expression using the techniques for dividing rational expressions from Section 7.2.

▶ **EXAMPLE 3** **Simplifying a Complex Rational Expression (Method I)**

Simplify: $\dfrac{\dfrac{2}{y} - \dfrac{8}{y^3}}{\dfrac{2}{y^2} + \dfrac{1}{y}}$

Solution

Write the numerator and the denominator of the complex rational expression as single rational expressions.

Work Smart

First simplify the numerator and the denominator into a single rational expression before you take the reciprocal and multiply.

NUMERATOR:

LCD is y^3: $\dfrac{2}{y} - \dfrac{8}{y^3} = \dfrac{2}{y} \cdot \dfrac{y^2}{y^2} - \dfrac{8}{y^3}$

$= \dfrac{2y^2 - 8}{y^3}$

$= \dfrac{2(y^2 - 4)}{y^3}$

$= \dfrac{2(y + 2)(y - 2)}{y^3}$

DENOMINATOR:

LCD is y^2: $\dfrac{2}{y^2} + \dfrac{1}{y} = \dfrac{2}{y^2} + \dfrac{1}{y} \cdot \dfrac{y}{y}$

$= \dfrac{2}{y^2} + \dfrac{y}{y^2}$

$= \dfrac{2 + y}{y^2}$

$= \dfrac{y + 2}{y^2}$

Now rewrite the complex rational expression with the new numerator and denominator.

$$\dfrac{\dfrac{2}{y} - \dfrac{8}{y^3}}{\dfrac{2}{y^2} + \dfrac{1}{y}} = \dfrac{\dfrac{2(y + 2)(y - 2)}{y^3}}{\dfrac{y + 2}{y^2}}$$

Multiply the numerator by the reciprocal of the denominator:

$$= \dfrac{2(y + 2)(y - 2)}{y^3} \cdot \dfrac{y^2}{y + 2}$$

Multiply; divide out common factors:

$$= \dfrac{2(y + 2)(y - 2) \cdot y^2}{y^2 \cdot y(y + 2)}$$

$$= \dfrac{2(y - 2)}{y}$$

●

EXAMPLE 4   **Simplifying a Complex Rational Expression (Method I)**

Simplify: $\dfrac{\dfrac{1}{x + 2} + 1}{x - \dfrac{3}{x + 2}}$

Solution

Write the numerator and denominator of the complex rational expression as a single rational expression.

NUMERATOR:

$$\frac{1}{x + 2} + 1 = \frac{1}{x + 2} + 1 \cdot \frac{x + 2}{x + 2}$$

$$= \frac{1}{x + 2} + \frac{x + 2}{x + 2}$$

$\dfrac{a}{c} + \dfrac{b}{c} = \dfrac{a + b}{c}$: $= \dfrac{1 + x + 2}{x + 2}$

Combine like terms: $= \dfrac{x + 3}{x + 2}$

DENOMINATOR:

$$x - \frac{3}{x + 2} = \frac{x}{1} \cdot \frac{x + 2}{x + 2} - \frac{3}{x + 2}$$

Distribute: $= \dfrac{x^2 + 2x}{x + 2} - \dfrac{3}{x + 2}$

$\dfrac{a}{c} - \dfrac{b}{c} = \dfrac{a - b}{c}$: $= \dfrac{x^2 + 2x - 3}{x + 2}$

Factor: $= \dfrac{(x + 3)(x - 1)}{x + 2}$

Rewrite the complex rational expression using the numerator and denominator just found, and then simplify.

$$\frac{\dfrac{1}{x + 2} + 1}{x - \dfrac{3}{x + 2}} = \frac{\dfrac{x + 3}{x + 2}}{\dfrac{(x + 3)(x - 1)}{x + 2}}$$

Multiply the numerator by the reciprocal of the denominator: $= \dfrac{x + 3}{x + 2} \cdot \dfrac{x + 2}{(x + 3)(x - 1)}$

Multiply; divide out common factors: $= \dfrac{\cancel{(x + 3)}\ \cancel{(x + 2)}}{\cancel{(x + 2)}\ \cancel{(x + 3)}\ (x - 1)}$

$$= \frac{1}{x - 1}$$

▶ ❷ **Simplify a Complex Rational Expression Using the Least Common Denominator (Method II)**

We now introduce a second method for simplifying complex rational expressions.

EXAMPLE 5 **How to Simplify a Complex Rational Expression Using the Least Common Denominator (Method II)**

Simplify: $\dfrac{\dfrac{2}{y} - \dfrac{8}{y^3}}{\dfrac{2}{y^2} + \dfrac{1}{y}}$

Step-by-Step Solution

Step 1: Find the least common denominator among all the denominators in the complex rational expression.

The denominators of the complex rational expression are y, y^3, and y^2. The least common denominator is y^3.

Step 2: Multiply both the numerator and the denominator of the complex rational expression by the least common denominator found in Step 1.

Multiply by $1 = \dfrac{y^3}{y^3}$: $\left(\dfrac{\dfrac{2}{y} - \dfrac{8}{y^3}}{\dfrac{2}{y^2} + \dfrac{1}{y}}\right) \cdot \left(\dfrac{y^3}{y^3}\right) = \dfrac{\dfrac{2}{y} \cdot y^3 - \dfrac{8}{y^3} \cdot y^3}{\dfrac{2}{y^2} \cdot y^3 + \dfrac{1}{y} \cdot y^3}$

Divide out common factors: $= \dfrac{\dfrac{2}{y} \cdot y^3 y^2 - \dfrac{8}{y^3} \cdot y^3 1}{\dfrac{2}{y^2} \cdot y^3 y + \dfrac{1}{y} \cdot y^3 y^2}$

Step 3: Simplify.

Multiply: $= \dfrac{2y^2 - 8}{2y + y^2}$

Factor: $= \dfrac{2(y^2 - 4)}{y(2 + y)}$

$= \dfrac{2(y + 2)(y - 2)}{y(y + 2)}$

Divide out common factors: $= \dfrac{2(y - 2)}{y}$ ●

Compare the approach used in Example 5 with that used in Example 3.

> **Simplifying a Complex Rational Expression Using the Least Common Denominator (Method II)**
>
> **Step 1:** Find the least common denominator among all the denominators in the complex rational expression.
>
> **Step 2:** Multiply both the numerator and the denominator of the complex rational expression by the least common denominator found in Step 1.
>
> **Step 3:** Simplify the rational expression.

▶ **EXAMPLE 6** **Simplifying a Complex Rational Expression Using the Least Common Denominator (Method II)**

Simplify: $\dfrac{\dfrac{1}{x + 2} + 1}{x - \dfrac{3}{x + 2}}$

Solution

The least common denominator among all denominators is $x + 2$, so we multiply by $1 = \dfrac{x + 2}{x + 2}$.

$$\dfrac{\dfrac{1}{x + 2} + 1}{x - \dfrac{3}{x + 2}} = \dfrac{\dfrac{1}{x + 2} + 1}{x - \dfrac{3}{x + 2}} \cdot \dfrac{x + 2}{x + 2}$$

Distribute the LCD to each term: $= \dfrac{\dfrac{1}{x + 2} \cdot (x + 2) + 1 \cdot (x + 2)}{x \cdot (x + 2) - \dfrac{3}{x + 2} \cdot (x + 2)}$

Divide out common factors: $= \dfrac{\dfrac{1}{\cancel{x + 2}} \cdot \cancel{(x + 2)} + x + 2}{x \cdot (x + 2) - \dfrac{3}{\cancel{x + 2}} \cdot \cancel{(x + 2)}}$

$= \dfrac{1 + x + 2}{x^2 + 2x - 3}$

Combine like terms; factor: $= \dfrac{x + 3}{(x + 3)(x - 1)}$

Divide out common factors: $= \dfrac{1}{x - 1}$

Work Smart

$\dfrac{1}{x + 2} \cdot (x + 2) = \dfrac{1}{x + 2} \cdot \dfrac{x + 2}{1} = 1$

Compare the approach used in Example 6 with that used in Example 4.

> **Quick ✓**
>
> In Problems 9 and 10, simplify the complex rational expression using Method II.
>
> **9.** $\dfrac{\dfrac{3}{2} + \dfrac{2}{3}}{\dfrac{1}{2} - \dfrac{1}{3}}$
>
> **10.** $\dfrac{\dfrac{1}{x} + \dfrac{2}{y}}{\dfrac{2}{x} - \dfrac{1}{y}}$

7.6 Exercises MyMathLab®

Exercise numbers in **green** have complete video solutions in MyMathLab

*Problems **1–10** are the Quick ✓s that follow the **EXAMPLES**.*

In Problems 11–24, simplify the complex rational expression using Method I. See Objective 1.

11. $\dfrac{1 - \dfrac{3}{4}}{\dfrac{1}{8} + 2}$

12. $\dfrac{\dfrac{2}{3} - \dfrac{3}{4}}{\dfrac{1}{6} + \dfrac{1}{2}}$

13. $\dfrac{\dfrac{x^2}{12} - \dfrac{1}{3}}{\dfrac{x + 2}{18}}$

14. $\dfrac{\dfrac{4}{t^2} - 1}{\dfrac{t + 2}{t^3}}$

15. $\dfrac{\dfrac{x + 3}{x^2}}{\dfrac{x^2}{9} - 1}$

16. $\dfrac{\dfrac{n + 2}{4}}{\dfrac{4}{n^2} - 1}$

17. $\dfrac{\dfrac{m}{2} + n}{\dfrac{m}{n}}$

18. $\dfrac{\dfrac{1}{x} - \dfrac{1}{y}}{\dfrac{2}{xy}}$

19. $\dfrac{\dfrac{5}{a} + \dfrac{4}{b^2}}{\dfrac{5b + 4}{b^2}}$

20. $\dfrac{\dfrac{2}{x} + \dfrac{3}{x^2}}{\dfrac{2x + 3}{x}}$

21. $\dfrac{\dfrac{8}{y + 3} - 2}{y - \dfrac{4}{y + 3}}$

22. $\dfrac{\dfrac{5}{n - 1} + 3}{n - \dfrac{2}{n - 1}}$

23. $\dfrac{\dfrac{1}{4} - \dfrac{6}{y}}{\dfrac{5}{6y} - y}$

24. $\dfrac{\dfrac{2}{a + b}}{\dfrac{1}{a} + \dfrac{1}{b}}$

In Problems 25–38, simplify the complex rational expression using Method II. See Objective 2.

25. $\dfrac{\dfrac{3}{2} - \dfrac{1}{4}}{\dfrac{5}{6} + \dfrac{1}{2}}$

26. $\dfrac{\dfrac{5}{4} - \dfrac{1}{8}}{\dfrac{3}{8} + \dfrac{9}{2}}$

27. $\dfrac{\dfrac{3}{m} + \dfrac{2}{m^2}}{\dfrac{6}{m} + \dfrac{4}{m^2}}$

28. $\dfrac{\dfrac{3c}{4} + \dfrac{3d}{10}}{\dfrac{3c}{2} - \dfrac{6d}{5}}$

29. $\dfrac{1 - \dfrac{49}{b^2}}{1 + \dfrac{7}{b}}$

30. $\dfrac{\dfrac{a}{2} + 4}{2 - \dfrac{a}{2}}$

31. $\dfrac{1 + \dfrac{5}{x}}{1 + \dfrac{1}{x+4}}$

32. $\dfrac{\dfrac{x}{x+1}}{1 + \dfrac{1}{x-1}}$

33. $\dfrac{\dfrac{1}{x^2} - \dfrac{1}{y^2}}{x - y}$

34. $\dfrac{6x + \dfrac{3}{y}}{\dfrac{9x+3}{y}}$

35. $\dfrac{a + \dfrac{1}{b}}{a + \dfrac{2}{b}}$

36. $\dfrac{4 + \dfrac{1}{x}}{8 + \dfrac{2}{x}}$

37. $\dfrac{12}{\dfrac{4}{n} - \dfrac{2}{3n}}$

38. $\dfrac{7}{\dfrac{2}{a} + \dfrac{3}{4a}}$

Mixed Practice

In Problems 39–52, simplify the complex rational expression using either Method I or Method II.

39. $\dfrac{\dfrac{x}{x-y} + \dfrac{y}{x+y}}{\dfrac{xy}{x^2-y^2}}$

40. $\dfrac{\dfrac{b}{b+1} - 1}{\dfrac{b+3}{b} - 2}$

41. $\dfrac{\dfrac{2}{x+4}}{\dfrac{2}{x+4} - 4}$

42. $\dfrac{\dfrac{4}{x+1} - 1}{\dfrac{4}{x+1} - 3}$

43. $\dfrac{\dfrac{b^2}{b^2-16} - \dfrac{b}{b+4}}{\dfrac{b}{b^2-16} - \dfrac{1}{b-4}}$

44. $\dfrac{\dfrac{-6}{y^2+5y+6}}{\dfrac{2}{y+3} - \dfrac{3}{y+2}}$

45. $\dfrac{1 - \dfrac{2}{x} - \dfrac{3}{x^2}}{1 - \dfrac{9}{x^2}}$

46. $\dfrac{2 - \dfrac{3}{x} - \dfrac{2}{x^2}}{1 - \dfrac{5}{x} + \dfrac{6}{x^2}}$

47. $\dfrac{1 - \dfrac{a^2}{4b^2}}{1 + \dfrac{a}{2b}}$

48. $\dfrac{1 - \dfrac{n^2}{25m^2}}{1 - \dfrac{n}{5m}}$

49. $\dfrac{\dfrac{x+2}{x-3} - 5}{\dfrac{x+2}{x-3} + 1}$

50. $\dfrac{\dfrac{x+4}{2x-5} - 1}{\dfrac{x+4}{2x-5} - 2}$

51. $\dfrac{\dfrac{3}{n-2} + 1}{5 + \dfrac{1}{n-2}}$

52. $\dfrac{\dfrac{5}{2b+3} + \dfrac{1}{2b-3}}{\dfrac{6b}{8b^2-18}}$

Applying the Concepts

53. **Finding the Mean** The arithmetic mean of a set of numbers is found by adding the numbers and then dividing by the number of entries on the list. Write a complex rational expression to find the arithmetic mean of the expressions $\dfrac{n}{6}$, $\dfrac{n+3}{2}$, and $\dfrac{2n-1}{8}$, and then simplify the complex rational expression.

54. **Finding the Mean** See Problem 53. Write a complex rational expression to find the arithmetic mean of the expressions $\dfrac{z}{2}$, $\dfrac{z-3}{4}$, and $\dfrac{2z-1}{6}$, and then simplify the complex rational expression.

△ 55. **Area of a Rectangle** In a rectangle, the length can be found by dividing the area by the width. If the area of a rectangle is $\left(\dfrac{2x+3}{x^2} - \dfrac{3}{x}\right)$ ft^2 and the width is $\dfrac{x^2-9}{x^5}$ feet, write a complex rational expression to find the length and then simplify the complex rational expression.

△ 56. **Area of a Rectangle** In a rectangle, the length can be found by dividing the area by the width. If the area of a rectangle is $\left(\dfrac{x-4}{x^3} - \dfrac{2}{x^2}\right)$ ft^2 and the width is $\dfrac{x^2-16}{x}$ feet, write a complex rational expression to find the length and then simplify the complex rational expression.

57. Electric Circuits An electric circuit contains two resistors connected in parallel, as shown in the figure. If they have resistance R_1 and R_2 ohms, respectively, then their combined resistance R is given by the formula

$$R = \frac{1}{\dfrac{1}{R_1} + \dfrac{1}{R_2}}$$

(a) Express R as a simplified rational expression.
(b) Evaluate the rational expression if $R_1 = 6$ ohms and $R_2 = 10$ ohms.

58. Electric Circuits An electric circuit contains three resistors connected in parallel. If the circuits have resistance R_1, R_2, and R_3 ohms, respectively, then their combined resistance is given by the formula

$$R = \frac{1}{\dfrac{1}{R_1} + \dfrac{1}{R_2} + \dfrac{1}{R_3}}$$

(a) Express R as a simplified rational expression.
(b) Evaluate the rational expression if $R_1 = 4$ ohms, $R_2 = 6$ ohms, and $R_3 = 10$ ohms.

Extending the Concepts

In Problems 59–62, use the definition of a negative exponent to simplify each complex rational expression.

59. $\dfrac{x^{-1} - 3}{x^{-2} - 9}$

60. $\dfrac{1 - 16x^{-2}}{1 - 3x^{-1} - 4x^{-2}}$

61. $\dfrac{2x^{-1} + 5}{4x^{-2} - 25}$

62. $\dfrac{x^{-2} - 3x^{-1}}{x^{-2} - 9}$

In Problems 63–66, simplify each expression.

63. $1 + \dfrac{1}{1 + \dfrac{1}{x}}$

64. $1 - \dfrac{1}{1 - \dfrac{1}{x - 2}}$

65. $\dfrac{1}{1 - \dfrac{1}{2 - \dfrac{1}{3 - x}}}$

66. $1 + \dfrac{2}{1 + \dfrac{2}{x + \dfrac{2}{x}}}$

Explaining the Concepts

67. Is the expression $\dfrac{\dfrac{2}{x + 1}}{\dfrac{1}{x - 5}}$ in simplified form? Explain why or why not.

68. Describe the steps to simplify a complex rational expression using Method I.

69. State which method you prefer to use when simplifying complex rational expressions. State your reasons for choosing this method, and then explain how to use this method to simplify $\dfrac{\dfrac{1}{x} + \dfrac{1}{y}}{\dfrac{1}{x^2} - \dfrac{1}{y^2}}$.

70. Explain the property of real numbers that is being applied in Method II. Make up a complex rational expression, and explain how to use Method II to simplify the expression.

Putting the Concepts Together (Sections 7.1–7.6)

We designed these problems so that you can review Sections 7.1–7.6 and show your mastery of the concepts. Take time to work these problems before proceeding with the next section. The answers are located at the back of the text on page AN-27.

1. Evaluate $\dfrac{a^3 - b^3}{a + b}$ when $a = 2$ and $b = -3$.

2. Find the values for which the following rational expressions are undefined:

(a) $\dfrac{-3a}{a - 6}$ 　　(b) $\dfrac{y + 2}{y^2 + 4y}$

3. Simplify:

(a) $\dfrac{ax + ay - 4bx - 4by}{2x + 2y}$ 　　(b) $\dfrac{x^2 + x - 2}{1 - x^2}$

4. Find the least common denominator (LCD) of the rational expressions:

$$\frac{5}{2a + 4b}, \frac{-1}{4a + 8b}, \frac{3ab}{8a - 32b}$$

5. Write $\dfrac{7}{3x^2 - x}$ as an equivalent rational expression with denominator $5x^2(3x - 1)$.

In Problems 6–13, perform the indicated operations.

6. $\dfrac{y^2 - y}{3y} \cdot \dfrac{6y^2}{1 - y^2}$

7. $\dfrac{m^2 + m - 2}{m^3 - 6m^2} \cdot \dfrac{2m^2 - 14m + 12}{m + 2}$

8. $\dfrac{4y + 12}{5y - 5} \div \dfrac{2y^2 - 18}{y^2 - 2y + 1}$

9. $\dfrac{4x^2 + x}{x + 3} + \dfrac{12x + 3}{x + 3}$

10. $\dfrac{x^2}{x^3 + 1} - \dfrac{x - 1}{x^3 + 1}$

11. $\dfrac{-8}{2x - 1} - \dfrac{9}{1 - 2x}$

12. $\dfrac{4}{m + 3} + \dfrac{3}{3m + 2}$

13. $\dfrac{3m}{m^2 + 7m + 10} - \dfrac{2m}{m^2 + 6m + 8}$

14. Simplify the complex rational expression:

$$\dfrac{\dfrac{-1}{m + 1} - 1}{m - \dfrac{2}{m + 1}}$$

15. Simplify the complex rational expression: $\dfrac{\dfrac{4}{a} + \dfrac{1}{6}}{\dfrac{3}{a^2} + \dfrac{1}{2}}$

7.7 Rational Equations

Objectives

1 Solve Equations Containing Rational Expressions

2 Solve for a Variable in a Rational Equation

Are You Ready for This Section?

Before getting started, take the following readiness quiz. If you get a problem wrong, go back to the section cited and review the material.

R1. Solve: $3k - 2(k + 1) = 6$ [Section 2.2, pp. 92–94]

R2. Factor: $3p^2 - 7p - 6$ [Section 6.3, pp. 385–393]

R3. Solve: $8z^2 - 10z - 3 = 0$ [Section 6.6, pp. 408–414]

R4. Find the values for which the expression

$\dfrac{x + 4}{x^2 - 2x - 24}$ is undefined.   [Section 7.1, pp. 432–433]

R5. Solve for y: $4x - 2y = 10$ [Section 2.4, pp. 114–115]

▶ **1 Solve Equations Containing Rational Expressions**

So far, we have solved linear equations (Sections 2.1–2.3), quadratic equations (Section 6.6), and equations that contain polynomial expressions that can be factored (Section 6.6). Now, we look at another type of equation.

> **Definition**
>
> A **rational equation** is an equation that contains a rational expression.

Ready?...Answers **R1.** {8}

R2. $(3p + 2)(p - 3)$

R3. $\left\{ -\dfrac{1}{4}, \dfrac{3}{2} \right\}$ **R4.** $x = -4$ or $x = 6$

R5. $y = 2x - 5$

Examples of rational equations are

$$\frac{3}{x + 4} = \frac{1}{x + 2} \quad \text{and} \quad \frac{4}{3} + \frac{7}{x - 4} = \frac{x - 1}{3x - 12}$$

The idea in solving a rational equation is to use algebraic techniques to rewrite it as an equation you already know how to solve, such as a linear or quadratic equation. In Chapter 2, you learned how to solve equations such as $\dfrac{3x}{4} - \dfrac{x}{2} = \dfrac{1}{2}$. Remember that one approach to solving an equation whose coefficients are fractions is to multiply each side of the equation by the least common denominator of the fractions, which, when simplified, will result in an equivalent equation without fractions.

To solve an equation such as $\dfrac{8}{p} + \dfrac{1}{4p} = \dfrac{11}{8}$ we use the same process. Let's compare the approaches.

EXAMPLE 1 **Solving a Rational Equation**

Solve:

(a) $\dfrac{3x}{4} - \dfrac{x}{2} = \dfrac{1}{2}$ **(b)** $\dfrac{8}{p} + \dfrac{1}{4p} = \dfrac{11}{8}, \quad p \neq 0$

Solution

Notice that Example 1(a) is a linear equation with fractional coefficients. Example 1(b) is a rational equation because it contains a rational expression. Notice that the steps for solving the two equations are identical and are the same steps that we used in Chapter 2.

(a)
$$\frac{3x}{4} - \frac{x}{2} = \frac{1}{2}$$

Find the LCD: The LCD is 4.

Multiply each side of the equation by the LCD:
$$4\left(\frac{3x}{4} - \frac{x}{2}\right) = 4\left(\frac{1}{2}\right)$$

Distribute and divide out common factors:
$$4\left(\frac{3x}{4}\right) - 4\left(\frac{x}{2}\right) = 4\left(\frac{1}{2}\right)$$
$$3x - 2x = 2$$
$$x = 2$$

We leave the check to you. The solution set is {2}.

(b)
$$\frac{8}{p} + \frac{1}{4p} = \frac{11}{8}, \quad p \neq 0$$

Find the LCD: The LCD is $8p$.

Multiply each side of the equation by the LCD:
$$8p\left(\frac{8}{p} + \frac{1}{4p}\right) = 8p\left(\frac{11}{8}\right)$$

Distribute and divide out common factors:
$$8p\left(\frac{8}{p}\right) + 8p\left(\frac{1}{4p}\right) = 8p\left(\frac{11}{8}\right)$$
$$64 \quad + \quad 2 \quad = \quad 11p$$

Solve the linear equation:
$$66 = 11p$$
$$6 = p$$

We leave the check to you. The solution set is {6}.   ●

Did you notice the restriction that $p \neq 0$ for the equation $\dfrac{8}{p} + \dfrac{1}{4p} = \dfrac{11}{8}$ in

Example 1(b)? We include that condition because if p is replaced by 0, both $\dfrac{8}{p}$ and $\dfrac{1}{4p}$ are

undefined. When solving a rational equation, always find the value(s) of the variable that will cause the expression to be undefined.

Also notice in Example 1(b) that by multiplying both sides by the LCD, we obtain a linear equation—an equation we already know how to solve. Again, this is a key to mathematics: Solve a problem by rewriting it as one you already know how to solve.

EXAMPLE 2 **How to Solve a Rational Equation**

Solve: $\dfrac{3}{2b - 1} = \dfrac{4}{3b}$

Step-by-Step Solution

Step 1: Determine the value(s) of the variable that cause the rational expression(s) in the rational equation to be undefined.

We need to find the values of the variable that cause either denominator to equal 0. If $2b - 1 = 0$, then $b = \dfrac{1}{2}$; if $3b = 0$, then $b = 0$. Therefore, $b \neq 0$, $b \neq \dfrac{1}{2}$.

Step 2: Determine the least common denominator (LCD) of all the denominators.

The LCD is $3b(2b - 1)$.

Step 3: Multiply both sides of the equation by the LCD, and simplify the expressions on each side.

$$\frac{3}{2b - 1} = \frac{4}{3b}$$

Multiply both sides by $3b(2b - 1)$: $\ 3b(2b - 1)\left(\dfrac{3}{2b - 1}\right) = 3b(2b - 1)\left(\dfrac{4}{3b}\right)$

Divide out common factors: $\ 3b\,\cancel{(2b - 1)} \cdot \dfrac{3}{\cancel{2b - 1}} = \cancel{3b}(2b - 1) \cdot \dfrac{4}{\cancel{3b}}$

Step 4: Solve the resulting equation.

$$3b(3) = (2b - 1)4$$

Multiply and distribute: $\qquad 9b = 8b - 4$

Subtract $8b$ from both sides: $\qquad b = -4$

Step 5: Check Verify your solution using the original equation.

Let $b = -4$: $\quad \dfrac{3}{2(-4) - 1} \overset{?}{=} \dfrac{4}{3(-4)}$

$$\frac{3}{-8 - 1} \overset{?}{=} \frac{4}{-12}$$

$$\frac{3}{-9} \overset{?}{=} \frac{4}{-12}$$

$$-\frac{1}{3} = -\frac{1}{3} \qquad \text{True}$$

The solution checks, so the solution set is $\{-4\}$.

Solving a Rational Equation

Step 1: Determine the value(s) of the variable that result in an undefined rational expression in the rational equation.

Step 2: Determine the least common denominator (LCD) of all the denominators.

Step 3: Multiply both sides of the equation by the LCD, and simplify the expressions on each side of the equation.

Step 4: Solve the resulting equation.

Step 5: Verify your solution using the original equation.

Quick ✓

1. A _____ _____ is an equation that contains a rational expression.

In Problems 2–4, solve the rational equation.

2. $\dfrac{5}{2} + \dfrac{1}{z} = 4$

3. $\dfrac{8}{x+4} = \dfrac{12}{x-3}$

4. $\dfrac{4}{3b} + \dfrac{1}{6b} = \dfrac{7}{2b} + \dfrac{1}{3}$

EXAMPLE 3   **Solving a Rational Equation**

Solve: $\dfrac{4}{3} + \dfrac{7}{x-4} = \dfrac{x-1}{3x-12}$

Solution

Work Smart

Writing the denominator in factored form makes it easier to find the LCD and the values of the variable that make the equation undefined.

First factor the denominator $3x - 12$. Factoring the denominator helps us find the undefined values of the variable and the LCD. Because $3x - 12 = 3(x - 4)$, we write the equation as

$$\frac{4}{3} + \frac{7}{x-4} = \frac{x-1}{3(x-4)}$$

Now we solve $x - 4 = 0$ to find the values of x that result in an undefined rational expression.

$$x - 4 = 0$$
$$x = 4$$

The rational expression is undefined when $x = 4$, so $x \neq 4$. The factored form of the denominators also shows that the LCD is $3(x - 4)$.

$$\frac{4}{3} + \frac{7}{x-4} = \frac{x-1}{3(x-4)}$$

Multiply both sides of the equation by the LCD, $3(x-4)$: $3(x-4)\left(\dfrac{4}{3} + \dfrac{7}{x-4}\right) = 3(x-4) \cdot \dfrac{x-1}{3(x-4)}$

Distribute: $3(x-4) \cdot \dfrac{4}{3} + 3(x-4) \cdot \dfrac{7}{x-4} = 3(x-4) \cdot \dfrac{x-1}{3(x-4)}$

Divide out common factors: $3(x-4) \cdot \dfrac{4}{3} + 3(x-4) \cdot \dfrac{7}{x-4} = 3(x-4) \cdot \dfrac{x-1}{3(x-4)}$

$$4(x-4) + 3(7) = x - 1$$

Distribute the 4; multiply: $4x - 16 + 21 = x - 1$

Combine like terms: $4x + 5 = x - 1$

Subtract x from both sides: $3x + 5 = -1$

Subtract 5 from both sides: $3x = -6$

Divide both sides by 3: $x = -2$

Check Let $x = -2$ in the *original* equation.

$$\frac{4}{3} + \frac{7}{x-4} = \frac{x-1}{3x-12}$$

Let $x = -2$: $\dfrac{4}{3} + \dfrac{7}{-2-4} \stackrel{?}{=} \dfrac{-2-1}{3(-2)-12}$

$$\frac{4}{3} + \frac{7}{-6} \stackrel{?}{=} \frac{-3}{-6-12}$$

$$\frac{4}{3} - \frac{7}{6} \stackrel{?}{=} \frac{-3}{-18}$$

(continued)

$$\frac{8}{6} - \frac{7}{6} \overset{?}{=} \frac{1}{6}$$

$$\frac{1}{6} = \frac{1}{6} \quad \text{True}$$

The solution checks, so the solution set is $\{-2\}$.

Quick ✔

In Problem 5, solve the rational equation.

5. $\dfrac{3}{2} + \dfrac{5}{x - 3} = \dfrac{x + 9}{2x - 6}$

EXAMPLE 4 **Solving a Rational Equation in Which the LCD Is the Product of the Denominators**

Solve: $\dfrac{5}{x + 2} + \dfrac{3}{x - 2} = 1$

Solution

The rational expression $\dfrac{5}{x + 2} + \dfrac{3}{x - 2}$ is undefined when $x = -2$ and $x = 2$. The LCD is $(x + 2)(x - 2)$.

$$\frac{5}{x + 2} + \frac{3}{x - 2} = 1$$

Multiply both sides of the equation by the LCD:
$$(x + 2)(x - 2)\left(\frac{5}{x + 2} + \frac{3}{x - 2}\right) = 1(x + 2)(x - 2)$$

Distribute:
$$(x + 2)(x - 2)\left(\frac{5}{x + 2}\right) + (x + 2)(x - 2)\left(\frac{3}{x - 2}\right) = (x + 2)(x - 2) \cdot 1$$

Divide out common factors:
$$\cancel{(x + 2)}(x - 2)\left(\frac{5}{\cancel{x + 2}}\right) + (x + 2)\cancel{(x - 2)}\left(\frac{3}{\cancel{x - 2}}\right) = (x + 2)(x - 2) \cdot 1$$

$$(x - 2)5 + (x + 2)(3) = (x + 2)(x - 2)$$

Multiply: $5x - 10 + 3x + 6 = x^2 - 4$

Combine like terms: $8x - 4 = x^2 - 4$

Write the quadratic equation in standard form: $0 = x^2 - 8x$

Factor the polynomial: $0 = x(x - 8)$

Use the Zero-Product Property: $x = 0 \quad \text{or} \quad x - 8 = 0$

Solve for x: $x = 8$

Work Smart

Recall that a quadratic equation is a second-degree equation. It is solved by writing the equation in standard form, $ax^2 + bx + c = 0$, factoring, and then using the Zero-Product Property.

Check Substitute $x = 0$ and $x = 8$ in the *original* equation. We leave it to you to confirm that the solution set is $\{0, 8\}$.

Quick ✔

In Problem 6, solve the rational equation.

6. $\dfrac{4}{x - 3} - \dfrac{3}{x + 3} = 1$

▶ Did you wonder why we determine the values of the variable that make a rational expression in the rational equation undefined? Well, in some instances the solution process gives results that do not satisfy the original equation.

Definition

An **extraneous solution** is a solution found through the solving process that does not satisfy the original equation.

> **EXAMPLE 5** **Solving a Rational Equation with an Extraneous Solution**

Solve: $\dfrac{x^2 + 8x + 6}{x^2 + 3x - 4} = \dfrac{3}{x - 1} - \dfrac{2}{x + 4}$

Solution

Rewrite the equation so that each denominator is in factored form. Because $x^2 + 3x - 4 = (x - 1)(x + 4)$, write the equation as

$$\frac{x^2 + 8x + 6}{(x - 1)(x + 4)} = \frac{3}{x - 1} - \frac{2}{x + 4}$$

Notice that $x \neq 1$ and $x \neq -4$ since these values cause the denominator to equal 0. We also see that the LCD is $(x - 1)(x + 4)$.

Now multiply both sides of the equation by the LCD and divide out common factors.

$$(x - 1)(x + 4)\left[\frac{x^2 + 8x + 6}{(x - 1)(x + 4)}\right] = \left(\frac{3}{x - 1} - \frac{2}{x + 4}\right)(x - 1)(x + 4)$$

$$(x - 1)(x + 4) \cdot \frac{x^2 + 8x + 6}{(x - 1)(x + 4)} = (x - 1)(x + 4)\left(\frac{3}{x - 1}\right) - (x - 1)(x + 4)\left(\frac{2}{x + 4}\right)$$

$$\cancel{(x - 1)}\cancel{(x + 4)}\left(\frac{x^2 + 8x + 6}{\cancel{(x - 1)}\cancel{(x + 4)}}\right) = \cancel{(x - 1)}(x + 4)\left(\frac{3}{\cancel{x - 1}}\right) - (x - 1)\cancel{(x + 4)}\left(\frac{2}{\cancel{x + 4}}\right)$$

$$x^2 + 8x + 6 = 3(x + 4) - 2(x - 1)$$

Distribute: $\quad x^2 + 8x + 6 = 3x + 12 - 2x + 2$

Combine like terms: $\quad x^2 + 8x + 6 = x + 14$

Write the quadratic equation in standard form: $\quad x^2 + 7x - 8 = 0$

Factor the polynomial: $(x + 8)(x - 1) = 0$

Use the Zero-Product Property: $\quad x + 8 = 0 \quad$ or $\quad x - 1 = 0$

Solve each equation for x: $\quad x = -8 \quad$ or $\quad x = 1$

Since $x = 1$ is one of the values of x that results in division by zero, we reject it as a solution. We leave it to you to verify that the solution set is $\{-8\}$.

Quick ✓

7. Solutions obtained through the solving process that do not satisfy the original equation are called _____ _____.

In Problem 8, solve the rational equation.

8. $\dfrac{5}{z - 4} + \dfrac{3}{z - 2} = \dfrac{z^2 - z - 2}{z^2 - 6z + 8}$

▶ (**EXAMPLE 6**) **Solving a Rational Equation with No Solution**

Solve: $\dfrac{6}{z^2 - 1} = \dfrac{5}{z - 1} - \dfrac{3}{z + 1}$

Solution

$$\frac{6}{z^2 - 1} = \frac{5}{z - 1} - \frac{3}{z + 1}$$

Factor $z^2 - 1$: $\dfrac{6}{(z - 1)(z + 1)} = \dfrac{5}{z - 1} - \dfrac{3}{z + 1}$

Notice that $z \neq 1$ and $z \neq -1$. Also, the LCD is $(z - 1)(z + 1)$. Multiply both sides of the equation by the LCD and divide out common factors.

$$(z - 1)(z + 1) \cdot \frac{6}{(z - 1)(z + 1)} = (z - 1)(z + 1)\left(\frac{5}{z - 1} - \frac{3}{z + 1}\right)$$

$$\cancel{(z - 1)}\cancel{(z + 1)} \cdot \frac{6}{\cancel{(z - 1)}\cancel{(z + 1)}} = \cancel{(z - 1)}(z + 1)\left(\frac{5}{\cancel{z - 1}}\right) - (z - 1)\cancel{(z + 1)}\left(\frac{3}{\cancel{z + 1}}\right)$$

$$6 = 5(z + 1) - 3(z - 1)$$

Distribute: $6 = 5z + 5 - 3z + 3$

Combine like terms: $6 = 2z + 8$

Subtract 8 from both sides: $-2 = 2z$

Divide both sides by 2: $-1 = z$

Since z cannot equal -1 (because it results in division by 0), we reject $z = -1$ as a solution. Therefore, the equation has no solution. The solution set is { } or $\varnothing$. ●

Work Smart: Study Skills

Showing your work makes it easier for you to see how you get the extraneous solution. Don't skip steps to get the answer. Double-check every step as you proceed through the problem to catch possible errors.

Work Smart

The solution set { } is not the same as the solution set {0}.

> **Quick** ✔
>
> **9.** *True or False* A rational equation can have no solution.
>
> *In Problem 10, solve the rational equation.*
>
> **10.** $\dfrac{4}{y + 1} = \dfrac{7}{y - 1} - \dfrac{8}{y^2 - 1}$

Work Smart

When we solve a rational equation, we multiply both sides of the equation by the LCD, but when we add or subtract rational expressions, we retain the LCD. Notice the difference in the directions: We *simplify* expressions and *solve* equations.

Simplify: $\dfrac{2}{x} - \dfrac{1}{6} + \dfrac{5}{2x} - \dfrac{1}{3}$

$\dfrac{2}{x} - \dfrac{1}{6} + \dfrac{5}{2x} - \dfrac{1}{3} = \dfrac{2}{x} \cdot \dfrac{6}{6} - \dfrac{1}{6} \cdot \dfrac{x}{x} + \dfrac{5}{2x} \cdot \dfrac{3}{3} - \dfrac{1}{3} \cdot \dfrac{2x}{2x}$

$ = \dfrac{12}{6x} - \dfrac{x}{6x} + \dfrac{15}{6x} - \dfrac{2x}{6x}$

$ = \dfrac{12 - x + 15 - 2x}{6x}$

$ = \dfrac{27 - 3x}{6x}$

$ = \dfrac{9 - x}{2x}$

Solve: $\dfrac{2}{x} - \dfrac{1}{6} = \dfrac{5}{2x} - \dfrac{1}{3}$

$6x\left(\dfrac{2}{x} - \dfrac{1}{6}\right) = 6x\left(\dfrac{5}{2x} - \dfrac{1}{3}\right)$

$6x\left(\dfrac{2}{x}\right) - 6x\left(\dfrac{1}{6}\right) = 6x\left(\dfrac{5}{2x}\right) - 6x\left(\dfrac{1}{3}\right)$

$6(2) - x = 3(5) - 2x$

$12 - x = 15 - 2x$

$12 + x = 15$

$x = 3$

EXAMPLE 7 **An Application of Rational Equations: Average Daily Cost**

Suppose that the average daily cost $\overline{C}$, in dollars, of manufacturing x bicycles is given by the equation

$$\overline{C} = \frac{x^2 + 75x + 5000}{x}$$

Determine the number of bicycles for which the average daily cost will be $225.

Solution

First, notice that x must be greater than 0. Do you know why? Since we want to know the level of production to obtain an average daily cost of $225, we solve the equation $\overline{C} = 225$.

$$\overline{C} = \frac{x^2 + 75x + 5000}{x}$$

$$225 = \frac{x^2 + 75x + 5000}{x}$$

Multiply both sides by x: $\quad x \cdot 225 = \frac{x^2 + 75x + 5000}{x} \cdot x$

$$225x = x^2 + 75x + 5000$$

Write the quadratic equation in standard form: $\quad 0 = x^2 - 150x + 5000$

Factor the polynomial: $\quad 0 = (x - 50)(x - 100)$

Use the Zero-Product Property: $\quad x - 50 = 0 \quad \text{or} \quad x - 100 = 0$

Solve each equation: $\quad x = 50 \quad \text{or} \quad x = 100$

The average daily cost is $225 when the level of production is at 50 or 100 bicycles. ●

Quick ✓

11. The concentration C in milligrams per liter of a certain drug in a patient's bloodstream t hours after injection is given by $C = \dfrac{50t}{t^2 + 25}$. When will the concentration of the drug be 4 milligrams per liter?

 2 **Solve for a Variable in a Rational Equation**

Work Smart

The steps that we follow when solving formulas for a variable are identical to those that we follow when solving rational equations.

Recall from Section 2.4 that "solve for the variable" means getting the variable by itself on one side of the equation, with all other variable and constant terms, if any, on the other side. We use the same steps to solve formulas for a variable that we use to solve rational equations.

EXAMPLE 8 **How to Solve for a Variable in a Rational Equation**

Solve $\dfrac{x}{1 + y} = z$ for y.

Step-by-Step Solution

We need to isolate the variable y. That is, we need to get y by itself on one side of the equation and the other terms on the other side.

Step 1: Determine the value(s) of the variable that cause any rational expression in the rational equation to be undefined.

Since $y = -1$ results in division by zero, $y \neq -1$.

Step 2: Find the least common denominator (LCD) of all the denominators.

The LCD is $1 + y$.

Step 3: Multiply both sides of the equation by the LCD, and simplify the expressions on each side of the equation.

$$\frac{x}{1 + y} = z$$

$$(1 + y)\left(\frac{x}{1 + y}\right) = z(1 + y)$$

$$x = z(1 + y)$$

Step 4: Solve the resulting equation for y.

Divide both sides by z:

$$\frac{x}{z} = \frac{z(1 + y)}{z}$$

$$\frac{x}{z} = 1 + y$$

Subtract 1 from both sides:

$$\frac{x}{z} - 1 = 1 - 1 + y$$

$$\frac{x}{z} - 1 = y$$

Look back at Step 4, where we divided both sides of the equation by z. Did you think about using the Distributive Property to simplify $x = z(1 + y)$? If you did, that is perfectly correct! We show this approach in solving $\frac{x}{1 + y} = z$ for y below:

$$x = z(1 + y)$$

Use the Distributive Property:

$$x = z + zy$$

Isolate the expression containing the variable y:

$$x - z = z - z + zy$$

$$x - z = zy$$

Divide both sides by z:

$$\frac{x - z}{z} = \frac{zy}{z}$$

$$\frac{x - z}{z} = y$$

The result that we get using this method, $y = \frac{x - z}{z}$, is equivalent to the equation $y = \frac{x}{z} - 1$, since $\frac{x - z}{z} = \frac{x}{z} - \frac{z}{z} = \frac{x}{z} - 1$.

Quick ✔

In Problems 12 and 13, solve the equation for the indicated variable.

12. Solve $R = \dfrac{4g}{x}$ for x

13. Solve $S = \dfrac{a}{1 - r}$ for r

EXAMPLE 9 **Solving for a Variable in a Rational Equation**

Solve $\dfrac{1}{a} + \dfrac{1}{b} = \dfrac{1}{c}$ for b.

We can see that neither a, b, nor c can be equal to zero. Further, the LCD is abc.

$$\frac{1}{a} + \frac{1}{b} = \frac{1}{c}$$

Multiply both sides by the LCD: $\quad (abc)\left(\frac{1}{a} + \frac{1}{b}\right) = (abc)\left(\frac{1}{c}\right)$

Distribute: $\quad (abc)\left(\frac{1}{a}\right) + (abc)\left(\frac{1}{b}\right) = (abc)\left(\frac{1}{c}\right)$

Divide out common factors: $\quad (\cancel{a}bc)\left(\frac{1}{\cancel{a}}\right) + (a\cancel{b}c)\left(\frac{1}{\cancel{b}}\right) = (ab\cancel{c})\left(\frac{1}{\cancel{c}}\right)$

$$bc + ac = ab$$

Work Smart

To solve for a variable that occurs on both sides of the equation, get all terms containing that variable on the same side of the equation. Then factor out the variable you wish to solve for.

To solve for b, we need to get both terms containing b on the same side of the equation.

Subtract bc from both sides: $\quad ac = ab - bc$

Factor out b as a common factor: $\quad ac = b(a - c)$

Divide each side by $a - c$: $\quad \dfrac{ac}{a - c} = \dfrac{b(a - c)}{a - c}$

$$\frac{ac}{a - c} = b$$

When we solve $\dfrac{1}{a} + \dfrac{1}{b} = \dfrac{1}{c}$ for b, we obtain $b = \dfrac{ac}{a - c}$.

●

> **Quick ✓**
>
> *In Problem 14, solve the equation for the indicated variable.*
>
> **14.** $\dfrac{1}{f} = \dfrac{1}{p} + \dfrac{1}{q}$ for p

7.7 Exercises   MyMathLab® PRACTICE

Problems 1–14 are the Quick ✓s that follow the EXAMPLES.

Building Skills

In Problems 15–46, solve each equation and state the solution set. Remember to identify the values of the variable for which the expressions in each rational equation are undefined. See Objective 1.

15. $\dfrac{5}{3y} - \dfrac{1}{2} = \dfrac{5}{6y} - \dfrac{1}{12}$

16. $\dfrac{4}{x} - \dfrac{11}{5} = \dfrac{3}{2x} - \dfrac{6}{5}$

17. $\dfrac{6}{x} + \dfrac{2}{3} = \dfrac{4}{2x} - \dfrac{14}{3}$

18. $\dfrac{4}{p} - \dfrac{5}{4} = \dfrac{5}{2p} + \dfrac{3}{8}$

19. $\dfrac{4}{x - 1} = \dfrac{3}{x + 1}$

20. $\dfrac{4}{x - 4} = \dfrac{5}{x + 4}$

21. $\dfrac{2}{x + 2} + 2 = \dfrac{7}{x + 2}$

22. $\dfrac{4}{x + 1} + 2 = \dfrac{3}{x + 1}$

23. $\dfrac{r - 4}{3r} + \dfrac{2}{5r} = \dfrac{1}{5}$

24. $\dfrac{x - 2}{4x} - \dfrac{x + 2}{3x} = \dfrac{1}{2x} - \dfrac{1}{2}$

25. $\dfrac{2}{a - 1} + \dfrac{3}{a + 1} = \dfrac{-6}{a^2 - 1}$

26. $\dfrac{6}{t} - \dfrac{2}{t - 1} = \dfrac{2 - 4t}{t^2 - t}$

27. $\dfrac{1}{4 - x} + \dfrac{2}{x^2 - 16} = \dfrac{1}{x - 4}$

28. $\dfrac{4}{x - 3} - \dfrac{3}{x - 2} = \dfrac{2x + 1}{x^2 - 5x + 6}$

29. $\dfrac{3}{2t - 2} + 4 = \dfrac{2t}{3t - 3}$

30. $\dfrac{2x + 3}{x - 1} - 2 = \dfrac{3x - 1}{4x - 4}$

31. $\dfrac{2}{5} + \dfrac{3 - 2a}{10a - 20} = \dfrac{2a + 1}{a - 2}$

32. $\dfrac{1}{x - 2} + \dfrac{2}{2 - x} = \dfrac{5}{x + 1}$

33. $\dfrac{6}{j^2 - 1} - \dfrac{4j}{j^2 - 5j + 4} = -\dfrac{4}{j - 1}$

34. $\dfrac{3}{2x + 2} - \dfrac{5}{4x - 4} = \dfrac{2x}{x^2 - 1}$

35. $\dfrac{x}{x - 2} = \dfrac{3}{x + 8}$

36. $\dfrac{x + 5}{x - 7} = \dfrac{x - 3}{x + 7}$

37. $\dfrac{x}{x + 3} = \dfrac{6}{x - 3} + 1$

38. $\dfrac{3x^2}{x + 1} = 2 + \dfrac{3x}{x + 1}$

39. $x = \dfrac{2 - x}{6x}$

40. $x = \dfrac{6 - 5x}{6x}$

41. $\dfrac{2x + 3}{x - 1} = \dfrac{x - 2}{x + 1} + \dfrac{6x}{x^2 - 1}$

42. $\dfrac{2x}{x + 4} = \dfrac{x + 1}{x + 2} - \dfrac{7x + 12}{x^2 + 6x + 8}$

43. $\dfrac{2}{b + 2} - \dfrac{5b + 6}{b^2 - b - 6} = \dfrac{-b}{b - 3}$

44. $\dfrac{2x}{x + 3} - \dfrac{2x^2 + 2}{x^2 - 9} = \dfrac{-6}{x - 3} + 1$

45. $\dfrac{1}{n - 3} = \dfrac{3n - 1}{9 - n^2}$

46. $\dfrac{5}{x - 2} - \dfrac{2}{2 - x} = \dfrac{4}{x + 1}$

In Problems 47–64, solve the equation for the indicated variable. See Objective 2.

47. $x = \dfrac{2}{y}$ for y

48. $4 = \dfrac{k}{m}$ for m

49. $I = \dfrac{E}{R}$ for R

50. $T = \dfrac{D}{R}$ for R

51. $h = \dfrac{2A}{B + b}$ for b

52. $m = \dfrac{y - z}{x - w}$ for y

53. $\dfrac{x}{3 + y} = z$ for y

54. $\dfrac{a}{b + 2} = c$ for b

55. $\dfrac{1}{R} = \dfrac{1}{S} + \dfrac{1}{T}$ for S

56. $\dfrac{3}{i} - \dfrac{4}{j} = \dfrac{8}{k}$ for j

57. $m = \dfrac{n}{y} - \dfrac{p}{ay}$ for y

58. $\dfrac{1}{a} = \dfrac{1}{b} + \dfrac{1}{c}$ for c

59. $A = \dfrac{xy}{x + y}$ for x

60. $B = \dfrac{k}{x} - \dfrac{m}{cx}$ for x

61. $\dfrac{2}{x} - \dfrac{1}{y} = \dfrac{6}{z}$ for y

62. $y = \dfrac{a}{a + b}$ for b

63. $y = \dfrac{x}{x - c}$ for x

64. $X = \dfrac{ab}{a - b}$ for a

Mixed Practice

In Problems 65–88, simplify the expression or solve the equation.

65. $\dfrac{1}{x} + \dfrac{3}{x + 5}$

66. $\dfrac{2}{x - 5} + \dfrac{5}{x - 5}$

67. $x - \dfrac{6}{x} = 1$

68. $z + \dfrac{3}{z} = 4$

69. $\dfrac{3}{x - 1} \cdot \dfrac{x^2 - 1}{6} + 3$

70. $5 - \dfrac{n^2 - 3n - 4}{n^2 - 4}$

71. $2b - \dfrac{5}{3} = \dfrac{1}{3b}$

72. $3a + \dfrac{7}{3} = \dfrac{2}{3a}$

73. $\dfrac{5x}{2x - 3} = \dfrac{3x}{x - 1} - \dfrac{5}{2x^2 - 5x + 3}$

74. $\dfrac{2x + 1}{x^2 + 2x - 3} = \dfrac{x - 1}{x^2 + 5x + 6} + \dfrac{x + 1}{x^2 + x - 2}$

75. $\dfrac{x^2}{x^2 - 4} - \dfrac{1}{x}$

76. $\dfrac{x}{x + 1} + \dfrac{2x - 3}{x - 1}$

77. $\dfrac{x^2 - 9}{x + 1} \div \dfrac{3x^2 + 9x}{x^2 - 1}$

78. $\dfrac{2t + 1}{t^2 - 16} \cdot \dfrac{t^2 + t - 12}{4t + 2}$

79. $\dfrac{\dfrac{1}{a} + 3}{\dfrac{3a + 1}{5}}$

80. $\dfrac{3n + 4}{5n} - \dfrac{2}{n^2 + 2n}$

81. $\dfrac{a}{2a - 2} - \dfrac{2}{3a + 3} = \dfrac{5a^2 - 2a + 9}{12a^2 - 12}$

82. $\dfrac{x}{12} + \dfrac{1}{2} = \dfrac{1}{3x} + \dfrac{2}{x^2}$

83. $\dfrac{\dfrac{3}{b+c} + \dfrac{5}{b-c}}{\dfrac{4b+c}{b^2-c^2}}$

84. $\dfrac{\dfrac{1}{2a} + \dfrac{3}{a^2}}{\dfrac{1}{3a} - \dfrac{4}{a^2}}$

85. $\dfrac{2x^2 + 5x + 2}{x^2 + 6} = 2$

86. $\dfrac{x^2 + x - 12}{x^2 - 4} = 1$

87. $\dfrac{x^3 + 12}{x^2 - 4} = x$

88. $\dfrac{x^3 + 8}{x^2 + 2} = x$

Applying the Concepts

89. Drug Concentration The concentration C of a drug in a patient's bloodstream, in milligrams per liter, t hours after ingestion is modeled by $C = \dfrac{40t}{t^2 + 9}$. When will the concentration of the drug be 4 milligrams per liter?

90. Drug Concentration The concentration C of a drug in a patient's bloodstream, in milligrams per liter, t hours after ingestion is modeled by $C = \dfrac{40t}{t^2 + 3}$. When will the concentration of the drug be 10 milligrams per liter?

91. Cost-Benefit Model Environmental scientists often use cost-benefit models to estimate the cost of removing a certain percentage of a pollutant from the environment. Suppose a cost-benefit model for the cost C (in millions of dollars) of removing x percent of the pollutants from Maple Lake is given by $C = \dfrac{25x}{100 - x}$. If the federal government budgets $100 million to clean up the lake, what percent of the pollutants can be removed?

92. Average Cost Suppose that the average daily cost $\overline{C}$ in dollars of manufacturing x bicycles is given by the equation $\overline{C} = \dfrac{x^2 + 75x + 5000}{x}$. Determine the level of production for which the average daily cost will be $240.

Extending the Concepts

93. For what value of k will the solution set of $\dfrac{4x+3}{k} = \dfrac{x-1}{3}$ be $\{2\}$?

94. For what value of k will the solution set of $\dfrac{1}{2} + \dfrac{3x}{k} = 1 + \dfrac{x}{3}$ be $\{3\}$?

Explaining the Concepts

95. On a test, a student wrote the following steps to solve the equation $\dfrac{2}{x-1} - \dfrac{4}{x^2-1} = -2$ but then got stuck and quit. Find and explain the error in the following *incorrect* solution.

$$\dfrac{2}{x-1} - \dfrac{4}{x^2-1} = -2$$

$$\dfrac{2}{x-1} - \dfrac{4}{(x-1)(x+1)} = -2$$

$$[(x-1)(x+1)]\left[\dfrac{2}{x-1} - \dfrac{4}{(x-1)(x+1)}\right] = -2[(x-1)(x+1)]$$

$$2(x+1) - 4 = -2(x-1)(x+1)$$

$$2x + 2 - 4 = (-2x+2)(-2x-2)$$

$$2x - 2 = 4x^2 - 4$$

$$0 = 4x^2 + 2x - 2$$

96. Explain why it is important to find the values of the variable that make the denominator zero when solving a rational equation.

97. Explain the difference between the directions *simplify* and *solve*. Make up a problem using each of these directions with the rational expressions $\dfrac{5}{x+1}$ and $\dfrac{6}{x-1}$.

98. Explain how using the LCD to solve an equation that contains rational expressions is different from using the LCD in a problem that requires adding and subtracting rational expressions with unlike denominators.

7.8 Models Involving Rational Equations

Objectives

1 Model and Solve Ratio and Proportion Problems

2 Model and Solve Problems with Similar Figures

3 Model and Solve Work Problems

4 Model and Solve Uniform Motion Problems

Are You Ready for This Section?

Before getting started, take the following readiness quiz. If you get the problem wrong, go back to the section cited and review the material.

R1. Solve: $\dfrac{150}{r} = \dfrac{250}{r + 20}$ [Section 7.7, pp. 478–480]

▶ **1** Model and Solve Ratio and Proportion Problems

We begin with a definition.

> **Definition**
>
> A **ratio** is the quotient of two numbers or two quantities. The ratio of two numbers a and b can be written as
>
> a to b or $a:b$ or $\dfrac{a}{b}$

In algebraic problems, we write ratios as $\dfrac{a}{b}$. For example, the probability of winning the "Pick Three" Instant Ohio Lottery is 1 in 1000, which is written as the ratio $\dfrac{1}{1000}$. Ratios are written as fractions reduced to lowest terms. If 8 oz of a cleaner is needed for every 4 gallons of water, the ratio is $\dfrac{8 \text{ oz}}{4 \text{ gal}} = \dfrac{2 \text{ oz}}{1 \text{ gal}}$.

A rational equation that involves two ratios is called a *proportion*.

> **Definition**
>
> A **proportion** is an equation of the form $\dfrac{a}{b} = \dfrac{c}{d}$, where $b \neq 0$ and $d \neq 0$. The **terms** of the proportion are a, b, c, and d. The **means** of the proportion are b and c, and the **extremes** of the proportion are a and d.

We solve a proportion the same way we solve a rational equation.

EXAMPLE 1 **Solving a Proportion**

Solve the proportion: $\dfrac{x}{5} = \dfrac{63}{105}$

Solution

Because $105 = 21 \cdot 5$, the LCD is 105.

$$\frac{x}{5} = \frac{63}{105}$$

Multiply both sides of the equation by 105: $105\left(\dfrac{x}{5}\right) = 105\left(\dfrac{63}{105}\right)$

Divide out common factors: $21x = 63$

Divide both sides by 21: $x = 3$

The solution set is $\{3\}$.

Work Smart

The equation $\dfrac{3}{8} = \dfrac{7}{x} + 1$ is NOT a proportion. Do you see why?

Another method used to solve a proportion is **cross multiplication**, which is based on the *Means-Extremes Theorem*.

> **Means-Extremes Theorem**
>
> If $\dfrac{a}{b} = \dfrac{c}{d}$, where $b \neq 0$ and $d \neq 0$, then $a \cdot d = b \cdot c$.

We can use cross multiplication to solve a proportion because the LCD for the proportion $\dfrac{a}{b} = \dfrac{c}{d}$, is bd. If we multiply both sides of the equation $\dfrac{a}{b} = \dfrac{c}{d}$ by the LCD, we obtain

$$\cancel{bd}\left(\dfrac{a}{\cancel{b}}\right) = \left(\dfrac{c}{\cancel{d}}\right)\cancel{bd}$$

$$ad = bc$$

If we solve the proportion in Example 1, $\dfrac{x}{5} = \dfrac{63}{105}$, using cross multiplication, we obtain

$$\dfrac{x}{5} = \dfrac{63}{105}$$

Cross multiply: $\quad 105x = 5 \cdot 63$

$$105x = 315$$

Divide both sides by 105: $\quad x = 3$

Work Smart

Use this visual for cross multiplication.

$$\dfrac{x}{5} \;\diagtimes\; \dfrac{63}{105}$$

Thus

$$105x = 5 \cdot 63$$

> **Quick ✓**
>
> 1. A _____ is an equation of the form $\dfrac{a}{b} = \dfrac{c}{d}$, where $b \neq 0$ and $d \neq 0$.
>
> 2. If $\dfrac{a}{b} = \dfrac{c}{d}$, then $a \cdot \_\_ = b \cdot \_\_$.
>
> *In Problems 3 and 4, solve the proportion for the indicated variable.*
>
> 3. $\dfrac{2p + 1}{4} = \dfrac{p}{8}$
>
> 4. $\dfrac{6}{x^2} = \dfrac{2}{x}$

EXAMPLE 2 **Solving an Exchange Rate Problem**

Last summer Laura took a Study Abroad trip to Italy. In Venice she spent 159 euros (€) on a digital camera. If \$1 US is equal to 0.75013 euro (€), then to the nearest cent, how much did the camera cost in U.S. dollars?

Solution

Step 1: Identify We want to know the cost of the camera in U.S. dollars.

Step 2: Name We let c represent the cost of the camera in U.S. dollars.

Step 3: Translate We want to know how many dollars are equivalent to 159 euros (€). Since c represents the cost of the camera (in dollars), and we know 1 dollar equals 0.75013 euros, we set up the proportion using $\dfrac{\text{dollars}}{\text{euros}}$.

$$\dfrac{\$1}{0.75013\ \text{€}} = \dfrac{c}{159\ \text{€}} \qquad \text{\color{red}The Model}$$

Work Smart

You could also set up the model using $\dfrac{\text{euros}}{\text{dollars}}$.

$$\dfrac{0.75013\text{€}}{\$1} = \dfrac{159\text{€}}{c}$$

Notice we include the units of measure in the model. This is done to keep track of the units. Our answer should be in dollars.

Step 4: Solve We'll now solve the equation.

Cross-multiply: $\$1 \cdot 159€ = (0.75013€)c$

Divide both sides by 0.75013 €: $\dfrac{\$1 \cdot 159€}{0.75013€} = \dfrac{(0.75013€)c}{0.75013€}$

Simplify using a calculator: $\$211.96 \approx c$

Step 5: Check Is this answer reasonable? First, our answer is in dollars. Next, the ratio \$1 to 0.75013€ should equal the ratio of \$211.97 to 159 €. That is, $\dfrac{1}{0.75013}$ should equal $\dfrac{211.96}{159}$. Because $\dfrac{1}{0.75013} \approx 1.3331$ and $\dfrac{211.96}{159} \approx 1.3331$, our answer checks.

Step 6: Answer The camera that Laura purchased in Italy cost \$211.96 U.S. ●

> **Quick ✓**
>
> **5.** An automobile manufacturer is advertising special financing on its autos. A buyer will have a monthly payment of \$16.67 for every \$1000 borrowed. If Clem borrows \$14,000, find his monthly payment.

Sometimes a proportion can be used as a general model for economic behavior.

EXAMPLE 3 **Model and Solve a Problem from Business**

A real estate agent knows that an 1800-square-foot house in a particular neighborhood sold for \$150,000. For how much should the agent appraise a 2100-square-foot house in the same neighborhood?

Solution

Step 1: Identify We want to know the price of a 2100-square-foot house in the same neighborhood as an 1800-square-foot house that sold for \$150,000.

Step 2: Name Let p represent the price of the 2100-square-foot house.

Step 3: Translate The selling price of an 1800-square-foot house is \$150,000. We need the selling price p of a 2100-square foot house. We use the ratio $\dfrac{\text{square feet}}{\text{selling price}}$ to set up the proportion

$$\frac{1800 \text{ ft}^2}{\$150{,}000} = \frac{2100 \text{ ft}^2}{p} \qquad \text{The Model}$$

Step 4: Solve

Work Smart

You could also cross multiply to solve the equation in Example 3.

Multiply by the LCD 150,000p: $\$150{,}000p\left(\dfrac{1800 \text{ ft}^2}{\$150{,}000}\right) = \$150{,}000p\left(\dfrac{2100 \text{ ft}^2}{p}\right)$

Divide out common factors: $(1800 \text{ ft}^2)p = (\$150{,}000)(2100 \text{ ft}^2)$

Divide both sides by 1800 ft²: $p = \dfrac{(\$150{,}000)(2100 \text{ ft}^2)}{1800 \text{ ft}^2}$

$p = \$175{,}000$

Step 5: Check The answer is in dollars, and the bigger house is appraised at a higher price. Plus, the ratio $\dfrac{1800}{150,000} = 0.012$ equals the ratio $\dfrac{2100}{175,000} = 0.012$, so our answer checks.

Step 6: Answer The appraisal should be $175,000.

> **Quick** ✔
>
> 6. On a map, $\dfrac{1}{4}$ inch represents a distance of 15 miles. According to the map, Springfield and Brookhaven are $3\dfrac{1}{2} = \dfrac{7}{2}$ inches apart. Find the number of miles between the two cities.

▶ ❷ **Model and Solve Problems with Similar Figures**

You may be familiar with using proportions to solve problems involving *similar figures* from geometry.

Figure 1

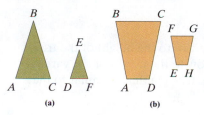

(a)   (b)

> **Definition**
>
> Two figures are **similar** if their corresponding angle measures are equal and their corresponding sides are in the same ratio.

Figure 1 shows examples of similar figures.

In Figure 1(a), $\triangle ABC$ is similar to $\triangle DEF$, and in Figure 1(b), quadrilateral $ABCD$ is similar to quadrilateral $EFGH$. Do you see that similar figures have the same shape but differ in size? Because $\triangle ABC$ is similar to $\triangle DEF$, we know that the ratio of AB to DE equals the ratio of AC to DF. That is,

$$\frac{AB}{DE} = \frac{AC}{DF}$$

The principle that the ratios of corresponding sides are equal can be used to find unknown lengths in similar figures.

EXAMPLE 4 | **Solving a Problem with Similar Triangles**

Find the length of side DE in triangle $\triangle DEF$, given that $\triangle ABC$ is similar to $\triangle DEF$, as shown in Figure 2. All measurements are in inches.

Solution

We know that $\triangle ABC$ is similar to $\triangle DEF$. Also, the corresponding sides have the same ratio, so $\dfrac{AB}{DE} = \dfrac{AC}{DF}$. We substitute the known values and solve for the unknown value, letting $x = DE$.

Figure 2

$$\frac{AB}{DE} = \frac{AC}{DF}$$

$AB = 45, AC = 10, DF = 4$: $\quad \dfrac{45}{x} = \dfrac{10}{4} \qquad$ The Model

Cross-multiply: $\quad 10x = 45 \cdot 4$

Multiply: $\quad 10x = 180$

Divide both sides by 10: $\quad x = 18$

The length of side DE is 18 inches.

In Example 4, we could have used the proportion $\dfrac{AB}{AC} = \dfrac{DE}{DF}$ and solved the proportion $\dfrac{45}{10} = \dfrac{x}{4}$.

Quick ✔

7. In geometry, two figures are _____ if their corresponding angle measures are equal and their corresponding sides have the same ratio.

In Problem 8, find the length of side XY given that $\triangle MNP$ is similar to $\triangle XYZ$.

8.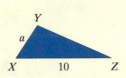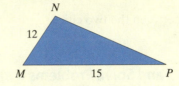

> **EXAMPLE 5**   **Finding the Height of a Tree**

A fifth-grade student is conducting an experiment to find the height of a tree in the schoolyard. The student measures the length of the tree's shadow and then immediately measures the length of the shadow that a yardstick forms. The tree's shadow is 24 feet long, and the yardstick's shadow is 4 feet long. Find the height of the tree.

Solution

See Figure 3. The ratio of the length of the tree's shadow to the height of the tree, h, equals the ratio of the length of the yardstick's shadow to the height of the yardstick, 3 feet. This is because the tree and its shadow form a triangle that is similar to that formed by the yardstick and its shadow.

We set up the proportion problem as

$$\frac{24}{4} = \frac{h}{3} \qquad \text{The Model}$$

Cross multiply: $\quad 4h = 24 \cdot 3$

Multiply: $\quad 4h = 72$

Divide both sides by 4: $\quad h = 18$

The height of the tree is 18 feet.

Figure 3

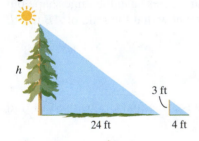

Work Smart

We could have also solved the proportion $\dfrac{4}{3} = \dfrac{24}{h}$. Do you see why?

Quick ✔

9. A tree casts a shadow 25 feet long. At the same time of day, a 6-foot-tall man casts a shadow 2.5 feet long. How tall is the tree?

▶ ❸ Model and Solve Work Problems

We now look at work problems. These problems assume that jobs are performed at a **constant rate,** which means an individual works at the same pace throughout the entire job, no matter how many other people are working on the same job. Although this assumption is reasonable for machines, it is not reasonable for people because of the old phrase "too many chefs spoil the broth." Think of it this way—if you continually add more people to paint a room, the time to complete the job may decrease initially, but eventually the painters get in each other's way, and the time to completion actually

Work Smart

Remember, when we model we make simplifying assumptions to make the math easier to deal with.

increases. While we could take this into account when modeling situations such as this, we will make the "constant-rate" assumption for humans to keep the mathematics manageable.

The constant-rate assumption states that if it takes t units of time to complete a job, then $\frac{1}{t}$ of the job is done in 1 unit of time. For example, if it takes 5 hours to paint a room, then $\frac{1}{5}$ of the room should be painted in 1 hour.

EXAMPLE 6   **Planting Seedlings**

Dan, an experienced horticulturist, can plant 1000 seedlings in 2 hours. Hala, a student assistant, takes 5 hours to plant 1000 seedlings. To the nearest tenth of an hour, how long would it take Dan and Hala working together to plant 1000 seedlings?

Solution

Step 1: Identify We want to know how long it will take for Dan and Hala, working together, to plant 1000 seedlings.

Step 2: Name We let t represent the time (in hours) it takes them to plant the seedlings working together. Then in 1 hour they will complete $\frac{1}{t}$ of the job.

Step 3: Translate Since Dan can finish the job in 2 hours, Dan will finish $\frac{1}{2}$ of the job in 1 hour. Since Hala can finish the job in 5 hours, Hala will finish $\frac{1}{5}$ of the job in 1 hour. Using this information, we set up Table 1.

Table 1

	Number of Hours to Complete the Job	Part of the Job Completed per Hour
Dan	2	$\frac{1}{2}$
Hala	5	$\frac{1}{5}$
Together	t	$\frac{1}{t}$

Work Smart

The part of the job completed per hour is the reciprocal of the number of hours to complete the job.

We set up the model using the following logic:

$$\left(\begin{array}{c}\text{Part done by Dan} \\ \text{in 1 hour}\end{array}\right) + \left(\begin{array}{c}\text{Part done by Hala} \\ \text{in 1 hour}\end{array}\right) = \left(\begin{array}{c}\text{Part done together} \\ \text{in 1 hour}\end{array}\right)$$

$$\frac{1}{2} \qquad + \qquad \frac{1}{5} \qquad = \qquad \frac{1}{t} \qquad \text{The Model}$$

Step 4: Solve

$$\frac{1}{2} + \frac{1}{5} = \frac{1}{t}$$

Multiply both sides by the LCD, 10t:

$$10t \cdot \left(\frac{1}{2} + \frac{1}{5}\right) = 10t \cdot \frac{1}{t}$$

Distribute:

$$10t \cdot \frac{1}{2} + 10t \cdot \frac{1}{5} = 10t \cdot \frac{1}{t}$$

Divide out common factors:	$5t + 2t = 10$
Combine like terms:	$7t = 10$
Divide both sides by 7:	$t = \dfrac{10}{7}$
	$t \approx 1.4$

Work Smart

We convert 0.4 hour to minutes by multiplying 0.4 by 60 minutes and obtain 24 minutes.

Step 5: Check We check only to make sure the answer is reasonable. We expect our answer to be greater than 0 but less than 2 (because the job takes Dan 2 hours working by himself). Our answer of 1.4 hours, or 1 hour 24 minutes, seems reasonable.

Step 6: Answer It will take Dan and Hala about 1 hour 24 minutes to plant 1000 seedlings.

Quick ✓

10. When solving work problems, we assume jobs are performed at a _____ rate. Thus, if a job can be completed in 4 hours, ___ of the job is completed in 1 hour.

11. It takes Molly 3 hours to shovel her driveway after a snowstorm working by herself. It takes her elderly neighbor 6 hours to shovel his driveway working by himself. How long would it take Molly and her neighbor to shovel the neighbor's driveway working together, assuming both driveways have the same dimensions?

EXAMPLE 7 **Mowing the Lawn**

Sara can cut the grass in 3 hours working by herself. When Sara cuts the grass with her younger brother Brian, it takes 2 hours. How long would it take Brian to cut the grass if he worked by himself?

Solution

Step 1: Identify We want to know how long it will take Brian to cut the grass by himself.

Step 2: Name Let t represent the time (in hours) that it takes Brian to cut the grass by himself. Then, in 1 hour he will complete $\dfrac{1}{t}$ of the job.

Step 3: Translate Since Sara can finish the job in 3 hours, she will finish $\dfrac{1}{3}$ of the job in 1 hour. Together the job is completed in 2 hours, so $\dfrac{1}{2}$ of the job is completed in 1 hour. We set up the model using the following logic:

$$\begin{pmatrix} \text{Part done by} \\ \text{Sara in 1 hour} \end{pmatrix} + \begin{pmatrix} \text{Part done by} \\ \text{Brian in 1 hour} \end{pmatrix} = \begin{pmatrix} \text{Part done} \\ \text{together in 1 hour} \end{pmatrix}$$

$$\dfrac{1}{3} \qquad + \qquad \dfrac{1}{t} \qquad = \qquad \dfrac{1}{2} \qquad \text{The Model}$$

Step 4: Solve

$$\dfrac{1}{3} + \dfrac{1}{t} = \dfrac{1}{2}$$

Multiply both sides by the LCD, 6t:
$$6t \cdot \left(\dfrac{1}{3} + \dfrac{1}{t} \right) = 6t \cdot \dfrac{1}{2}$$

Distribute:
$$6t \cdot \dfrac{1}{3} + 6t \cdot \dfrac{1}{t} = 6t \cdot \dfrac{1}{2}$$

| Divide out common factors: | $2t + 6 = 3t$ |
| Subtract $2t$ from each side: | $6 = t$ |

Step 5: Check Our answer of 6 hours is reasonable since this time is greater than the time required when working together. The mathematics is correct since
$$\frac{2}{6} + \frac{1}{6} = \frac{3}{6} = \frac{1}{2}.$$

Step 6: Answer It will take Brian 6 hours to cut the grass, working by himself.

> **Quick ✓**
>
> 12. It takes Leon 5 hours to seal his driveway working by himself. If Michael helps Leon, they can seal the driveway in 2 hours. How long would it take Michael to seal the driveway alone?

▶ ④ Model and Solve Uniform Motion Problems

We introduced uniform motion problems in Section 2.7. Recall that these problems use the fact that distance equals rate times time—that is, $d = rt$. When modeling uniform motion problems that lead to rational equations, we often use the model, $t = \dfrac{d}{r}$. These problems use the idea of constant rate, just as work problems do. For example, a bicyclist may not ride for several hours at exactly the same speed. To make the mathematics easier, however, we assume that speed is constant throughout the time period.

EXAMPLE 8 **A Boat Trip on the Mighty Olentangy**

The Olentangy River has a current of 2 miles per hour. A motorboat goes 48 miles downstream in the same time it takes to go 36 miles upstream. What is the speed of the boat in still water?

Solution

Step 1: Identify This is a uniform motion problem. We want to know the speed of the boat in still water.

Step 2: Name Let r represent the speed of the boat in still water.

Step 3: Translate The current pushes the boat when it is going downstream, so the total speed of the boat is the speed of the boat in still water plus the speed of the current. We let $r + 2$ represent the speed of the boat downstream. Going upstream, the current slows the boat, so the total speed of the boat upstream will be $r - 2$. We set up Table 2.

Work Smart

Remember, $d = r \cdot t$, so $\dfrac{d}{r} = t$.

Table 2

	Distance (miles)	Rate (miles per hour)	Time $= \dfrac{\text{Distance}}{\text{Rate}}$ (hours)
Downstream	48	$r + 2$	$\dfrac{48}{r + 2}$
Upstream	36	$r - 2$	$\dfrac{36}{r - 2}$

The amount of time traveling downstream is the same as the amount of time spent traveling upstream, so

$$\frac{48}{r+2} = \frac{36}{r-2} \quad \text{The Model}$$

Step 4: Solve We wish to solve for r:

$$\frac{48}{r+2} = \frac{36}{r-2}$$

Cross-multiply: $48(r-2) = 36(r+2)$

Distribute: $48r - 96 = 36r + 72$

Subtract $36r$ from both sides: $12r - 96 = 72$

Add 96 to both sides: $12r = 168$

Divide each side by 12: $\dfrac{12r}{12} = \dfrac{168}{12}$

$$r = 14$$

Step 5: Check Are the downstream and upstream travel times the same? Going downstream, the boat travels at $14 + 2 = 16$ miles per hour, so it takes $\dfrac{48}{16} = 3$ hours. Going upstream, the boat travels at $14 - 2 = 12$ miles per hour, and it takes $\dfrac{36}{12} = 3$ hours. Our answer checks!

Step 6: Answer The boat travels 14 miles per hour in still water.

Quick ✓

13. A small airplane can travel 120 mph in still air. The plane can fly 700 miles with the wind in the same time that it can travel 500 miles against the wind. Find the speed of the wind.

EXAMPLE 9　**The Ride Back Through the Forest**

Every weekend, you ride your bicycle on a forest preserve path. The path is 12 miles long and ends at a waterfall, where you relax before making the trip back to the starting point. One weekend, you notice that because the ride to the waterfall is partially uphill, your cycling speed returning is 2 miles per hour faster than the speed riding to the waterfall. If the round trip takes 5 hours (excluding resting time), find the average speed cycling back from the waterfall.

Solution

Step 1: Identify This is a uniform motion problem. We wish to know the average speed returning from the waterfall.

Step 2: Name Let r represent your cycling speed when you are cycling to the waterfall. Then $r + 2$ represents the speed cycling back from the waterfall.

Step 3: Translate The distance to the waterfall is 12 miles. We set up Table 3.

Table 3

	Distance (miles)	Rate (miles per hour)	Time $= \dfrac{\text{Distance}}{\text{Rate}}$ (hours)
Going to the Waterfall	12	r	$\dfrac{12}{r}$
Returning Home	12	$r + 2$	$\dfrac{12}{r + 2}$

We know that the total time spent cycling was 5 hours, so we set up the equation

$$\left(\begin{array}{c}\text{Time cycling to} \\ \text{waterfall}\end{array}\right) + \left(\begin{array}{c}\text{Time cycling to} \\ \text{starting point}\end{array}\right) = 5 \text{ hours}$$

$$\frac{12}{r} + \frac{12}{r + 2} = 5 \quad \text{The Model}$$

Step 4: Solve Solve for r.

$$\frac{12}{r} + \frac{12}{r + 2} = 5$$

Multiply both sides by the LCD, $r(r + 2)$:

$$r(r + 2)\left(\frac{12}{r} + \frac{12}{r + 2}\right) = (5)\, r(r + 2)$$

Distribute: $\quad r(r + 2)\left(\dfrac{12}{r}\right) + r(r + 2)\left(\dfrac{12}{r + 2}\right) = (5)r(r + 2)$

Divide out common factors: $\quad (r + 2)12 + r(12) = 5r(r + 2)$

Multiply and distribute: $\quad 12r + 24 + 12r = 5r^2 + 10r$

Combine like terms: $\quad 24r + 24 = 5r^2 + 10r$

Write the equation in standard form: $\quad 0 = 5r^2 - 14r - 24$

Factor the polynomial: $\quad 0 = (5r + 6)(r - 4)$

Use the Zero-Product Property: $\quad 5r + 6 = 0 \quad \text{or} \quad r - 4 = 0$

Solve each equation: $\quad 5r = -6 \qquad\qquad r = 4$

$$r = -\frac{6}{5}$$

We discard the solution $r = -\dfrac{6}{5}$ because the cycling speed, r, cannot be negative. Thus $r = 4$ miles per hour is the speed cycling to the waterfall. The speed cycling back to the starting point is $r + 2 = 4 + 2 = 6$ mph.

Step 5: Check We should find that the time cycling to the waterfall and back is 5 hours. Because $\dfrac{12 \text{ mi}}{4 \text{ mph}} + \dfrac{12 \text{ mi}}{6 \text{ mph}} = 3 \text{ hr} + 2 \text{ hr} = 5 \text{ hr}$, our answer is correct.

Step 6: Answer Your average speed cycling from the waterfall is 6 mph.

Quick ✓

14. To prepare for a marathon run, Sue ran 18 miles before she blistered her heel, and then she walked 1 additional mile. Her running speed was 12 times as fast as her walking speed. She was running and walking for 5 hours. Find Sue's running speed.

7.8 Exercises Exercise numbers in green have complete video solutions in MyMathLab

Problems 1–14 are the Quick ✔s that follow the EXAMPLES.

Building Skills

In Problems 15–34, solve the proportion. See Objective 1.

15. $\dfrac{9}{x} = \dfrac{3}{4}$

16. $\dfrac{x}{5} = \dfrac{12}{4}$

17. $\dfrac{4}{7} = \dfrac{2x}{9}$

18. $\dfrac{6}{5} = \dfrac{8}{3x}$

19. $\dfrac{6}{5} = \dfrac{x+2}{15}$

20. $\dfrac{5}{9} = \dfrac{2x+3}{6}$

21. $\dfrac{b}{b+6} = \dfrac{4}{9}$

22. $\dfrac{k}{k+3} = \dfrac{6}{15}$

23. $\dfrac{y}{y-10} = \dfrac{2}{27}$

24. $\dfrac{y}{y-3} = \dfrac{3}{10}$

25. $\dfrac{p+2}{4} = \dfrac{2p+4}{5}$

26. $\dfrac{n+6}{3} = \dfrac{n+4}{5}$

27. $\dfrac{2z-1}{z} = \dfrac{3}{5}$

28. $\dfrac{2k-3}{k} = \dfrac{4}{3}$

29. $\dfrac{2}{v^2-v} = \dfrac{1}{3-v}$

30. $\dfrac{n-2}{4n+7} = \dfrac{-2}{n+1}$

31. $\dfrac{10-x}{4x} = \dfrac{1}{x-1}$

32. $\dfrac{4}{x^2} = \dfrac{1}{x+3}$

33. $\dfrac{2p-3}{p^2+12p+32} = \dfrac{1}{p+4}$

34. $\dfrac{1}{z+3} = \dfrac{2z+1}{z^2+10z+21}$

In Problems 35–38, △ABC is similar to △XYZ. See Objective 2.

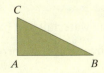

△ **35.** If $AB = 6$, $XY = 9$, $AC = 8$, find XZ.

△ **36.** If $XY = 8$, $YZ = 7$, $AB = 24$, find BC.

△ **37.** If $XY = n$, $ZY = 2n - 1$, $BC = 5$, and $AB = 3$, find n and ZY.

△ **38.** If $AB = n$, $BC = n + 2$, $XY = 5$, and $YZ = 15$, find n and BC.

In Problems 39 and 40, rectangle ABCD is similar to rectangle EFGH. See Objective 2.

△ **39.** If $AB = x - 2$, $EF = 4$, $BC = 2x + 3$, and $FG = 9$, find x and AB.

△ **40.** If $DC = 6$, $HG = x - 1$, $DA = 8$, and $HE = 2x - 3$, find x and HE.

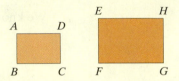

In Problems 41–44, write an algebraic expression that represents each phrase. See Objective 3.

41. If Mariko worked x hours and Natalie worked twice as many hours as Mariko, write an algebraic expression that represents the number of hours Natalie worked.

42. If Brian worked for n days and Kellen worked for half as many days, write an algebraic expression that represents the number of days Kellen worked.

43. If David painted for 3 days less than Pierre and Pierre painted for t days, write an algebraic expression that represents the number of days David painted.

44. If Valerie danced for p minutes and Melody danced for 45 minutes less than Valerie, write an algebraic expression that represents the number of minutes Melody danced.

In Problems 45–48, write an equation that could be used to model each of the following. SET UP BUT DO NOT SOLVE THE EQUATION. See Objective 3.

45. Painting Chairs Christina can paint a chair in 5 hours, and Victoria can paint a chair in 3 hours. How many hours will it take to paint the chair when the two girls work together?

46. Weeding the Yard Bill can weed the backyard in 6 hours. When he worked with Tamra, it took only 2 hours. How long would it take Tamra to weed the backyard if she worked alone?

47. Filling a Tank There are two different inlet pipes that can be used to fill a 3000-gallon tank. Pipe A takes 4 hours longer than Pipe B to fill the tank. With both pipes open, it takes 7 hours to fill the tank. How long would it take Pipe B alone to fill the tank?

48. Canning Peaches A factory that was producing canned peaches had an old canning machine. They decided to add a newer machine that could work twice as fast. With both machines on line, the plant could produce 10,000 cans in 6 hours. How long would it take to produce the same number of cans using only the newer machine?

In Problems 49–52, write an algebraic expression that represents each phrase. See Objective 4.

49. If the rate of the current of a stream is 2 mph and Joe can swim r mph in still water, write an algebraic expression that represents Joe's rate when he swims upstream.

50. If the wind is blowing at 25 mph and a plane flies at r mph in still air, write an algebraic expression that represents the rate the plane is traveling when it flies with the wind.

51. The wind is blowing at 48 mph, and a plane flies at r mph in still air. Write an algebraic expression that represents the rate the plane is traveling when it flies against the wind.

52. If the rate of the current in a stream is 3 mph and Betsy can paddle her canoe at a rate of r mph in still water, write an algebraic expression that represents Betsy's rate when she paddles downstream.

In Problems 53–56, write an equation that could be used to model each of the following. SET UP BUT DO NOT SOLVE THE EQUATION. See Objective 4.

53. Bob's River Trip Bob's boat travels at 14 mph in still water. Find the speed of the current if he can go 4 miles upstream in the same time that it takes to go 7 miles downstream.

54. Marty's River Trip A stream has a current of 3 mph. Find the speed of Marty's boat in still water if she can go 10 miles downstream in the same time it takes to go 4 miles upstream.

55. Assen's River Trip Assen can travel 12 miles downstream in his motorboat in the same time he travels 8 miles upstream. If the current of the river is 3 mph, what is the boat's rate in still water?

56. Airplane Travel A small airplane takes the same amount of time to fly 200 miles against a 30-mph wind as it does to fly 300 miles with a 30-mph wind. Find the speed of the airplane in still air.

Applying the Concepts

57. Yasmine's Map On Yasmine's map, $\frac{1}{4}$ inch represents 10 miles. According to the map, Yellow Springs and Hillsboro are $3\frac{5}{8}$ inches apart. Find the number of miles between the two towns.

58. Beth's Map On Beth's map, 0.5 cm represents 60 miles. If it is 1840 miles from San Diego to New Orleans, how far is this distance on her map?

59. Making Bread If it takes 5 lb of flour to make 3 loaves of bread, how much flour is needed to make 5 loaves of bread?

60. Buying Candy If 2 lb of candy costs $3.50, how much candy can be purchased for $8.75?

61. Comparing Rubles and Dollars. If $5 U.S. is worth approximately 143.25 Russian rubles, how many rubles can Hortencia purchase for $1300 U.S.?

62. Comparing Pesos and Pounds If 50 Mexican pesos are worth approximately 2.5 British pounds and if Sean takes 80 pounds as spending money in Mexico City, how many pesos will he have?

63. Tree Shadow If a 6-ft man casts a shadow that is 2.5 ft long, how tall is a tree that has a shadow 10 ft long?

64. Telephone Pole Shadow A bush that is 4 m tall casts a shadow that is 1.75 m long. How tall is a telephone pole if its shadow is 8.75 m?

65. Cleaning the Math Building Josh can clean the math building on his campus in 3 hours. Ken takes 5 hours to clean the same building. If they work together, how long will it take them to clean this building?

66. Pruning Trees Martin can prune his fruit trees in 4 hours. His neighbor can prune the same trees for him in 7 hours. If they work together on this job, how long, to the nearest tenth of an hour, will it take to prune the trees?

67. Retrieving Volleyballs After hitting practice for the Long Beach State volleyball team, Dyanne can retrieve all of the balls in the gym in 8 minutes. It takes Makini 6 minutes to retrieve all the balls. If they work together, how long, to the nearest tenth of a minute, will it take these two players to return the volleyballs and be ready to start the next round of hitting practice?

68. Stuffing Envelopes It took Kirsten 6 hours to stuff 3000 envelopes. When she worked with Brittany, it took only 4 hours to stuff the next 3000 envelopes. How long would it take Brittany working alone to stuff the 3000 envelopes?

69. Painting City Hall Three people were given the job of painting city hall. They divided the job into equal portions and decided they would each go home after completing their assigned portion. It took José 9 hours to paint his part, and then he went home. It took Joaquín 12 hours to complete his part. The third person played around, never got started, and was promptly fired from the job. If José and Joaquín went back the second day and worked together to paint the undone portion, how long, to the nearest tenth of an hour, did it take them?

70. Mowing the Grass It takes a mower 18 minutes to cut the grass in the outfield of the stadium. A new mower was purchased, and with both mowers working, the same task takes 8 minutes. How long would it take the new mower to cut the outfield grass?

71. Replacing Pipes It takes an apprentice twice as long as the experienced plumber to replace the pipes under an old house. If it takes them 5 hours when they work together, how long would it take the apprentice alone?

72. Making Care Packages It takes Rocco 3 times as long as Traci to make 100 Red Cross care packages for refugees. If it takes them 10 hours when they work together to make 100 care packages, how long would it take Rocco working alone to make the same number?

73. Filling the Pool Using a single hose, Janet can fill a pool in 6 hours. The same pool can be drained in 8 hours by opening a drainpipe. If Janet forgets to close the drainpipe, how long will it take her to fill the pool?

74. Jill's Bucket Jill has a bucket with a hole in it. When the hole did not exist, Jill could fill the bucket in 30 seconds. With the hole, the bucket now empties in 210 seconds (3.5 minutes). How long will it take Jill to fill the bucket now that it has a hole in it?

75. Travel by Boat A boat can travel 12 km down the river in the same time it can go 4 km up the river. If the current in the river is 2 km per hour, how fast can the boat travel in still water?

76. Travel by Plane A small plane can travel 1000 miles with the wind in the same time it can go 600 miles against the wind. If the speed of the plane in still air is 180 mph, what is the speed of the wind?

77. Iron Man Training While training for an iron man competition, Tony bikes for 60 miles and runs for 15 miles. If his biking speed is 8 times his running speed and it takes 5 hours to complete the training, how long did he spend on his bike?

78. Robin's Workout Robin can run twice as fast as she walks. If she runs for 12 miles and walks for 8 miles, the total time to complete the trip is 5 hours. To the nearest tenth of an hour, how many hours did she run?

79. Driving During a Snowstorm Claire and Chris were driving to their home in Milwaukee when they came upon a snowstorm. They drove at an average rate of 20 miles per hour slower for the last 60 miles during the snowstorm than they drove for the first 90 miles. Find their average rate of speed during the last 60 miles if the trip took 3 hours.

80. Driving During Road Construction David and Lisa drove 150 miles through road construction at a certain average rate. By increasing their speed by 20 miles per hour, they traveled the next 250 miles without road construction in the same time they spent on the 150-mile leg of their trip. Find their average rate during the part of the trip that had road construction.

81. Tough Commute You have a 20-mile commute into work. Since you leave very early, the trip going to work is easier than the trip home. You can travel to work in the same time that it takes you to travel 16 miles on the trip back home. Your average speed coming home is 7 miles per hour slower than your average speed going to work. What is your average speed going to work?

82. A Bike Trip A bicyclist rides his bicycle 12 miles up a hill and then 16 miles on level terrain. His speed on level ground is 3 miles per hour faster than his speed going uphill. The cyclist rides for the same amount of time going uphill and peddling on level ground. Find his speed going uphill.

Extending the Concepts

In geometry, when a line is parallel to one side of a triangle, the intersection of the parallel line with the other two sides of the triangle will form a new triangle that is similar to the original triangle. In the figure below, $\overline{XY}$ is parallel to $\overline{BC}$, so $\triangle XAY$ is similar to $\triangle BAC$.

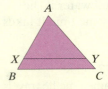

In Problems 83–88, solve for x.

△ **83.** $XA = 4$, $BA = 12$, $CA = 18$, and $YA = x$.

△ **84.** $XA = 4$, $BA = 10$, $CA = 30$, and $YA = x$.

△ **85.** $XA = 5$, $XB = 9$, $XY = 7$, and $BC = x$.

△ **86.** $XA = 3$, $XB = 15$, $XY = 4$, and $BC = x$.

△ **87.** $AY = 5$, $YC = 12$, $AX = 6$, and $XB = x$.

△ **88.** $AY = 4$, $YC = x$, $AX = x$, and $XB = 9$.

Explaining the Concepts

89. A student reads the following sentence from a word problem: *Barbara rows her boat in a stream that has a current of 1 mph. She travels 10 miles upstream and returns to her starting point 5 hours later*. She makes the accompanying chart. Is the information correctly placed for this part of the problem?

d	r	t
10	$r + 1$	5
10	$r - 1$	5

90. When modeling work problems, we assume that individuals work at a constant rate. When modeling uniform motion problems, we assume that individuals travel at a constant rate. Explain what these assumptions mean.

91. You are teaching the class how to solve the following problem: *Marcos can write a budget analysis in 8 hours, and Alonso can complete the same analysis in 6 hours. If they work together, how long will it take to finish the analysis?* Write an equation that could be used to find the answer to the question, and then explain your justification for setting up the equation your way. If you can think of another way to do the problem, explain how to set it up differently, and then explain why this logic will also solve the problem correctly.

92. Explain why the following proportions all yield the same result.

(a) $\dfrac{x}{2} = \dfrac{4}{9}$ **(b)** $\dfrac{2}{x} = \dfrac{9}{4}$

(c) $\dfrac{x}{4} = \dfrac{2}{9}$ **(d)** $\dfrac{9}{2} = \dfrac{4}{x}$

(e) $\dfrac{4}{9} = \dfrac{x}{2}$

7.9 Variation

Objectives

1 Model and Solve Direct Variation Problems

2 Model and Solve Inverse Variation Problems

Are You Ready For This Section?

Before getting started, take the following readiness quiz. If you get a problem wrong, go back to the section cited and review the material.

R1. Solve $10 = 2k$ for k. [Section 2.1, pp. 85–89]

R2. Solve $3 = \dfrac{k}{5}$ for k. [Section 2.1, pp. 85–89]

▶ 1 Model and Solve Direct Variation Problems

Often two variables are related as proportions. For example, we say, "Revenue is proportional to sales" or "Force is proportional to acceleration." When we say that one variable is proportional to another variable, we are talking about *variation*. **Variation** describes how one quantity changes in relation to another quantity. In this text, we will discuss two types of variation: *direct* and *inverse*. We will discuss direct variation first.

> **Definition**
>
> If x and y represent two quantities, we say that y **varies directly** with x, or y is **directly proportional to** x, if there is a nonzero number k such that
>
> $$y = kx$$
>
> The number k is called the **constant of proportionality,** or the **constant of variation.**

Figure 4

The graph in Figure 4 shows the relationship between y and x if y varies directly with x and $k > 0$, $x \geq 0$. The constant of proportionality k is the slope of the line.

If we know that two quantities vary directly, then knowing the value of each quantity in one instance enables us to write a formula that is true in all cases. Although the definition for direct variation uses y and x as the variables, any variables can be used.

EXAMPLE 1 **Direct Variation**

Suppose that y varies directly with x for $x \geq 0$. Find an equation that relates y and x if $y = 20$ when $x = 4$. Graph the equation.

Solution

Because y varies directly with x, we know that $y = kx$. Now we can use the given information to find k, the constant of proportionality.

$$y = kx$$
$$y = 20, x = 4: \quad 20 = k(4)$$
Divide both sides by 4: $\quad 5 = k$
Symmetric Property: $\quad k = 5$

Since $k = 5$, we know that $y = 5x$. Because $x \geq 0$, we only need to graph the equation in quadrant I. See Figure 5.

Figure 5

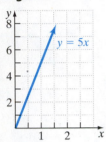

Work Smart

The constant of variation, k, is the slope of the line.

> **Quick ✓**
>
> 1. The statement "t varies directly with s" is given by the equation _____.
>
> 2. In the equation $y = kx$, k is called the _____ _ _____ and represents the rate of change between the variables x and y.
>
> 3. Suppose that y varies directly with x for $x \geq 0$. Find an equation that relates y and x if it is known that $y = 15$ when $x = 5$. Graph the equation.
>
> 4. Suppose that q varies directly with w for $w \geq 0$.
>
>   (a) Find an equation that relates q and w if it is known that $q = 10$ when $w = 40$.
>
>   (b) Use the equation found in part (a) to determine q when $w = 60$.
>   (c) Graph the equation found in part (a).

EXAMPLE 2 **Car Payments**

Brandon just bought a used car for $12,000. He decides to put $2000 down on the car and borrow the remaining $10,000. The bank lends Brandon $10,000 at 5.9% interest for 48 months. His payments are $234.39. The monthly payment p on a car varies directly with the amount borrowed b. If Brandon puts $3000 down on the car instead, what will his monthly payment be?

Solution

Step 1: Identify We want to know the monthly payment if Brandon puts $3000 down. The model will involve variables that vary directly.

Step 2: Name The variables have been named already: p is the monthly payment and b is the amount borrowed.

Step 3: Translate The monthly payment p varies directly with the amount borrowed b:

$$p = kb$$

$p = 234.39$ when $b = 10,000$: $\quad 234.39 = k(10,000)$

Divide both sides by 10,000 to obtain $k = 0.023439$ and we have the model:

$$p = 0.023439b \quad \text{The Model}$$

Work Smart

If possible, find the exact value of k. If you cannot avoid approximate values, do not round the value of k to fewer than 5 decimal places.

Step 4: Solve If Brandon puts $3000 down, he will need to borrow $9000. We let $b = 9000$ in the model to determine the monthly payment.

$$p = 0.023439b$$

Let $b = 9000$: $p = 0.023439(9000)$

Evaluate: $p \approx 210.95$

Step 5: Check We can check the reasonableness of our answer. Brandon's payments are $234.39 when he borrows $10,000. It seems reasonable that his payments will decrease a little when he puts more money down. We found that with $3000 down, his payment will be $210.95. This seems reasonable.

Step 6: Answer When he puts $3000 down, Brandon's monthly payment will be $210.95.

> **Quick ✓**
>
> 5. The cost of gas C varies directly with the number of gallons pumped, g. Suppose that the cost of pumping 8 gallons of gas is $34. If 6.8 gallons are pumped into your car, what will the cost be?

▶ ❷ Model and Solve Inverse Variation Problems

We now discuss another type of variation, which describes relations between two variables when an increase in one variable results in a decrease in a second variable. For example, as the price for a product increases, the quantity demanded of the product decreases.

Figure 6 $y = \dfrac{k}{x}, k > 0, x > 0$

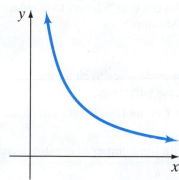

> **Definition**
>
> If x and y represent two quantities, we say that y **varies inversely** with x, or y is **inversely proportional** to x, if there is a nonzero number k such that
>
> $$y = \frac{k}{x}$$
>
> where k is the **constant of variation.**

The graph in Figure 6 illustrates the relationship between y and x if y varies inversely with x with $k > 0$ and $x > 0$. Notice from the graph that as x increases, the values of y decrease.

(**EXAMPLE 3**)   **Inverse Variation**

Suppose that y varies inversely with x for $x > 0$.

   (a) Find an equation that relates y and x if it is known that $y = 20$ when $x = 3$.

   (b) Use the equation found in part (a) to determine y when $x = 5$.

Solution

   (a) Because y varies inversely with x, we know that $y = \dfrac{k}{x}$. Now we can use the given information to find k.

$$y = \frac{k}{x}$$

$y = 20, x = 3$: $20 = \dfrac{k}{3}$

Multiply both sides by 3: $60 = k$

Symmetric Property: $k = 60$

Since $k = 60$, we know the equation for this inverse variation is $y = \dfrac{60}{x}$.

(b) We can let $x = 5$ in the equation $y = \dfrac{60}{x}$ to find y. We obtain

$$y = \frac{60}{5}$$
$$= 12$$

When $x = 5$, $y = 12$. ●

Quick ✓

6. The statement "f varies inversely with d" is given by the equation _____.

7. *True or False* If your employer reimburses you 52¢ per mile for driving your car on business trips, then the amount you are reimbursed is inversely proportional to the miles you drive on business.

8. Suppose that y varies inversely with x for $x > 0$.
   (a) Find an equation that relates y and x if it is known that $y = 6$ when $x = 2$.
   (b) Use the equation found in part (a) to determine y when $x = 4$.

9. Suppose that p varies inversely with q for $q > 0$.
   (a) Find an equation that relates p and q if it is known that $p = 10$ when $q = 4$.
   (b) Use the equation found in part (a) to determine p when $q = 8$.

EXAMPLE 4   **Cleaning a Stadium**

The amount of time T it takes to clean a stadium after a game varies inversely with the number of employees cleaning, n. Suppose 20 employees take 5 hours to clean a stadium. How long will 16 employees take to clean the stadium?

Solution

Step 1: Identify We want to find the time it will take 16 employees to clean the stadium. We also know that the model will involve inverse variation.

Step 2: Name The variables have been named already: T is the time it takes to clean the stadium, and n is the number of employees.

Step 3: Translate The time to clean T varies inversely with the number of employees n:

$$T = \frac{k}{n}$$

$T = 5$ when $n = 20$: $5 = \dfrac{k}{20}$

Multiply both sides by 20: $100 = k$

A model for the time T to clean the stadium with n employees is given by

$$T = \frac{100}{n} \quad \text{The Model}$$

Step 4: Solve We let $n = 16$ in the model to predict the time to clean the stadium.

$$T = \frac{100}{n}$$

Let $n = 16$: $T = \dfrac{100}{16}$

Evaluate: $T = 6.25$

Step 5: Check We can check the reasonableness of our answer. Because it takes 20 employees 5 hours to clean the stadium, it seems reasonable that 16 employees will take more time (but not that much more). We are predicting with 16 employees, it will take 6.25 hours. This seems reasonable.

Step 6: Answer It takes 6.25 hours for 16 employees to clean the stadium. ●

Quick ✓

10. The rate of vibration (in oscillations per second) V of a string under constant tension varies inversely with the length l. If a string is 30 inches long and vibrates 20 times per second, determine the rate of vibration of a string that is 24 inches long.

7.9 Exercises

 MyMathLab® Math XP PRACTICE

Exercise numbers in **green** have complete video solutions in MyMathLab

*Problems **1–10** are the **Quick ✓**s that follow the **EXAMPLES**.*

Building Skills

In Problems 11–16, suppose y varies directly with x. Find an equation that relates x and y given the following. See Objective 1.

11. $y = 3$ when $x = 6$ **12.** $y = 15$ when $x = 45$

13. $x = 12$ when $y = 24$ **14.** $x = 2$ when $y = 8$

15. $y = -6$ when $x = 8$ **16.** $y = -10$ when $x = 24$

In Problems 17–20, use the direct variation model to find each of the following. See Objective 1.

17. Suppose p varies directly with g.

  (a) Find an equation that relates p and g if it is known that $p = 12$ when $g = 36$.

  (b) Use this equation to determine p if g is 9.

18. Suppose d varies directly with t.

  (a) Find an equation that relates d and t if it is known that $d = 320$ when $t = 8$.

  (b) Use this equation to determine d if t is 5.

19. Suppose y varies directly with x.

  (a) Find an equation that relates y and x if it is known that $x = 12$ when $y = -9$.

  (b) Use this equation to determine x if y is $\dfrac{5}{4}$.

20. Suppose n varies directly with m.

  (a) Find an equation that relates n and m if it is known that $m = 12$ when $n = 9$.

  (b) Use this equation to determine m if n is $-\dfrac{8}{5}$.

In Problems 21–26, suppose y varies inversely with x. Find an equation that relates x and y given the following. See Objective 2.

21. $x = 2$ when $y = 3$ **22.** $x = 7$ when $y = 10$

23. $y = 3$ when $x = 4$ **24.** $y = 2$ when $x = 8$

25. $x = \dfrac{1}{2}$ when $y = \dfrac{4}{7}$ **26.** $x = \dfrac{3}{4}$ when $y = -\dfrac{2}{9}$

In Problems 27–30, use the inverse variation model to find each of the following. See Objective 2.

27. Suppose e varies inversely with n.

  (a) Find an equation that relates e and n if it is known that $e = 4$ when $n = 2$.

  (b) Use this equation to determine e if n is 16.

28. Suppose y varies inversely with x.

  (a) Find an equation that relates y and x if it is known that $y = 4$ when $x = 3$.

  (b) Use this equation to determine y if x is 18.

29. Suppose b varies inversely with a.

  (a) Find an equation that relates b and a if it is known that $b = \dfrac{3}{2}$ when $a = \dfrac{1}{2}$.

  (b) Use this equation to determine a if b is $\dfrac{9}{10}$.

30. Suppose f varies inversely with d.

  (a) Find an equation that relates f and d if it is known that $f = \dfrac{2}{9}$ when $d = \dfrac{15}{2}$.

  (b) Use this equation to determine d if f is $\dfrac{15}{4}$.

Mixed Practice

In Problems 31–38, indicate whether the equation represents direct variation, inverse variation, or neither. If it is a variation equation, identify the constant of proportionality.

31. $y = \dfrac{2x}{3}$ **32.** $x = 4$

33. $y = \dfrac{1}{2}$ **34.** $4x = y$

35. $xy = 9$ **36.** $5y = 2x$

37. $6x = 3y$ **38.** $xy = 4$

In Problems 39–46, find the quantity indicated.

39. x varies inversely with y. If $x = 6$ when $y = 3$, find x when $y = 4$.

40. p varies inversely with v. If $p = 50$ when $v = 24$, find p when $v = 75$.

41. J is directly proportional to P. If $J = 80$ when $P = 120$, find P when $J = 50$.

42. A is directly proportional to B. If $A = 360$ when $B = 72$, find B when $A = 400$.

43. s varies directly with t. If $t = 18$ when $s = 21$, find t when $s = 49$.

44. m varies directly with r. If $r = 24$ when $m = 9$, find r when $m = 24$.

45. b is inversely proportional to a. If $a = 4$ when $b = 6$, find b when $a = \dfrac{4}{3}$.

46. x is inversely proportional to y. If $y = 14$ when $x = 4$, find x when $y = \dfrac{2}{3}$.

Applying the Concepts

47. House of Representatives Apportionment The number of representatives that each state has in the U.S. House of Representatives is directly proportional to the state's population. According to the 2010 Census, Ohio's 16 delegates represent a state whose population is 11.537 million. Find the number of representatives from Massachusetts if its population is 6.548 million.

48. House of Representatives Apportionment The number of representatives that each state has in the U.S. House of Representatives is directly proportional to the state's population. According to the 2010 Census, California's 53 delegates represent a state whose population is 37.254 million. Find the number of representatives from Florida if its population is 18.801 million.

49. Buying Lumber The cost of a certain type of lumber at Beechwold Lumber varies directly with the number of board feet purchased. If 70 board feet of this type of lumber cost Christine $717.50, find the number of board feet in Elizabeth's order of $1230.

50. Buying Gasoline The cost to purchase a tank of gasoline varies directly with the number of gallons purchased. You notice that the person in front of you spent $61.50 on 15 gallons of gas. If your SUV needs 35 gallons of gas, how much will you spend?

51. Measuring Pressure A fixed amount of gas is placed in a collapsible cylinder. The pressure in the cylinder, P, varies inversely with the volume V. If the pressure is measured at 5 atmospheres when the volume is 2.5 liters, what will the pressure be when the cylinder is collapsed to 2 liters?

52. Measuring Sound The frequency of sound varies inversely with the wavelength. If a radio station broadcasts at a frequency of 90 megahertz, the wavelength of the sound is approximately 3 meters. Find the wavelength of a radio broadcast at 20 megahertz.

53. Demand Suppose that the demand D for candy at the movie theater is inversely related to the price p. When the price of candy is $2.50 per bag, the theater sells 180 bags of candy. Determine the number of bags of candy that will be sold if the price is raised to $3 a bag.

54. Driving to School The time t that it takes to drive to school varies inversely with your average speed s. It takes you 24 minutes to drive to school when your average speed is 35 miles per hour. Suppose that your average speed to school yesterday was 30 miles per hour. How long did it take you to get to school?

Extending the Concepts

Joint variation is a model in which a variable is proportional to the product of two or more variables. For example, we may say that y varies jointly as x and z, or y = kxz. Use joint variation to solve Problems 55–58.

55. m varies jointly as r and s. If $m = 12$ when $r = 0.5$ and $s = 4$, find m when $s = 9$ and $r = \dfrac{5}{3}$.

56. m varies jointly as r and s. If $m = 9$ when $r = 6$ and $s = 2$, find m when $s = 3$ and $r = 16$.

57. p varies jointly as q and the square of r. If $p = 162$ when $q = 9$ and $r = 6$, find q when $p = 300$ and $r = 5$.

58. p varies jointly as q and the square of r. If $p = 6$ when $q = \dfrac{1}{8}$ and $r = 4$, find q when $p = 40$ and $r = \dfrac{2}{3}$.

The energy produced by a photon of light, e (in joules), is inversely proportional to the wavelength of the light, λ (in meters). The constant of variation is the product of Planck's Constant (6.63×10^{-34}) and the speed of light (3×10^8).

59. (a) Write an equation to calculate the energy produced by a photon with a wavelength, λ.

(b) Calculate the energy produced when the wavelength is 500×10^{-9} meter.

60. Use the equation from Problem 59(a) to calculate the energy produced when the wavelength is 600 nanometer. (1 nanometer $= 10^{-9}$ meter)

Explaining the Concepts

61. List three everyday occurrences of direct variation. Make up an example for each, and explain how to use this equation to acquire information about the situation.

62. Describe the conditions that are necessary for two variables to be directly proportional. Also describe the conditions necessary for two variables to be inversely proportional.

Chapter 7 Activity: Correct the Quiz

Focus: Performing operations with rational expressions and solving rational equations
Time: 15–20 minutes
Group size: 2

In this activity you will work as a team to grade the student quiz shown below. If an answer is correct, mark it correct. If an answer is wrong, mark it wrong and show the correct answer.

Once all of the quiz questions are graded, compute the final score for the quiz. Be prepared to discuss your results with the rest of the class.

Student Quiz

Name: *Ima Student*   Quiz Score: _____

(1) Multiply: $\dfrac{6x - 12}{4x + 8} \cdot \dfrac{3x + 6}{2x - 4}$

Answer: $\dfrac{9}{4}$

(2) Subtract: $\dfrac{xy}{x^2 - y^2} - \dfrac{y}{x + y}$

Answer: $\dfrac{y^2}{x^2 - y^2}$

(3) Solve: $\dfrac{5x}{x + 1} = 2 + \dfrac{2x}{x + 1}$

Answer: {2}

(4) Divide: $\dfrac{x^2 + x - 6}{5x^2 - 7x - 6} \div \dfrac{3x^2 + 13x + 12}{6x^2 + 17x + 12}$

Answer: $\dfrac{2}{5}$

(5) Solve: Joe can mow his lawn in 2 hr. Mike can mow the same lawn in 3 hr. If they work together, how long will it take them to mow the lawn?

Answer: *1 hr, 12 min*

(6) Simplify: $\dfrac{\dfrac{1}{x + 5} - \dfrac{2}{x - 7}}{\dfrac{4}{x - 7} + \dfrac{1}{x + 5}}$

Answer: $\dfrac{x + 17}{5x + 13}$

(7) Add: $\dfrac{x}{x^2 + 10x + 25} + \dfrac{4}{x^2 + 6x + 5}$

Answer: $\dfrac{x^2 + 5x + 20}{(x + 5)(x + 1)}$

Chapter 7 Review

Section 7.1   Simplifying Rational Expressions

KEY CONCEPTS

- To determine undefined values of a rational expression, set the denominator equal to zero, and solve for the variable.

- **Simplifying a Rational Expression**

  First, completely factor the numerator and the denominator of the rational expression; then divide out common factors using the fact that if p, q, and r are polynomials, then $\dfrac{p \cdot r}{q \cdot r} = \dfrac{p}{q}$ if $q \neq 0$ and $r \neq 0$.

KEY TERMS

Rational expression
Evaluate
Undefined
Simplify

You Should Be Able To...	EXAMPLE	Review Exercises
❶ Evaluate a rational expression (p. 430)	Examples 1 through 3	1–4
❷ Determine values for which a rational expression is undefined (p. 432)	Examples 4 through 5	5–10
❸ Simplify rational expressions (p. 433)	Examples 6 through 9	11–16

In Problems 1–4, evaluate each expression for the given values.

1. $\dfrac{2x}{x + 3}$
   (a) $x = 1$
   (b) $x = -2$
   (c) $x = 3$

2. $\dfrac{3x}{x - 4}$
   (a) $x = 0$
   (b) $x = 1$
   (c) $x = -2$

3. $\dfrac{x^2 + 2xy + y^2}{x - y}$
   (a) $x = 1$
       $y = -1$
   (b) $x = 2$
       $y = 1$
   (c) $x = -3$
       $y = -4$

4. $\dfrac{2x^2 - x - 3}{x + z}$
   (a) $x = 1$
       $z = 1$
   (b) $x = 1$
       $z = 2$
   (c) $x = 5$
       $z = -4$

In Problems 5–10, find the value(s) of the variable for which the rational expression is undefined.

5. $\dfrac{3x}{3x - 7}$

6. $\dfrac{x + 1}{4x - 2}$

7. $\dfrac{5}{x^2 + 25}$

8. $\dfrac{17}{4x^2 + 49}$

9. $\dfrac{5x + 2}{x^2 + 12x + 20}$

10. $\dfrac{x + 1}{x^2 - 3x - 4}$

In Problems 11–16, simplify each rational expression, if possible. Assume that no variable has a value that results in a denominator with a value of zero.

11. $\dfrac{5y^2 + 10y}{25y}$

12. $\dfrac{2x^3 - 8x^2}{10x}$

13. $\dfrac{3k - 21}{k^2 - 5k - 14}$

14. $\dfrac{-2x - 2}{x^2 - 2x - 3}$

15. $\dfrac{x^2 + 8x + 15}{2x^2 + 5x - 3}$

16. $\dfrac{3x^2 + 5x - 2}{2x^3 + 16}$

Section 7.2 Multiplying and Dividing Rational Expressions

KEY CONCEPTS

- **Multiplying Rational Expressions**

  If $\dfrac{a}{b}$ and $\dfrac{c}{d}$, $b \neq 0, d \neq 0$, are two rational expressions, then $\dfrac{a}{b} \cdot \dfrac{c}{d} = \dfrac{ac}{bd}$.

- **Dividing Rational Expressions**

  If $\dfrac{a}{b}$ and $\dfrac{c}{d}$, $b \neq 0, c \neq 0, d \neq 0$, are two rational expressions, then $\dfrac{\frac{a}{b}}{\frac{c}{d}} = \dfrac{a}{b} \div \dfrac{c}{d} = \dfrac{a}{b} \cdot \dfrac{d}{c} = \dfrac{ad}{bc}$.

You Should Be Able To...	EXAMPLE	Review Exercises
❶ Multiply rational expressions (p. 438)	Examples 1 through 4	17, 18, 21, 22, 25, 26
❷ Divide rational expressions (p. 440)	Examples 5 through 9	19, 20, 23, 24, 27, 28

In Problems 17–28, perform the indicated operations.

17. $\dfrac{12m^4n^3}{7} \cdot \dfrac{21}{18m^2n^5}$

18. $\dfrac{3x^2y^4}{4} \cdot \dfrac{2}{9x^3y}$

19. $\dfrac{10m^2n^4}{9m^3n} \div \dfrac{15mn^6}{21m^2n}$

20. $\dfrac{5ab^3}{3b^4} \div \dfrac{10a^2b^8}{6b^2}$

21. $\dfrac{5x-15}{x^2-x-12} \cdot \dfrac{x^2-6x+8}{3-x}$

22. $\dfrac{4x-24}{x^2-18x+81} \cdot \dfrac{x^2-9}{6-x}$

23. $\dfrac{\dfrac{x^2-4}{x^2-8x+15}}{\dfrac{12x+24}{3x-15}}$

24. $\dfrac{\dfrac{5x^3+10x^2}{3x}}{\dfrac{2x+4}{18x^2}}$

25. $\dfrac{3x^2+14x-5}{x^2+x-30} \cdot \dfrac{x^2-2x-15}{3x^2+8x-3}$

26. $\dfrac{y^2-5y-14}{y^2-2y-35} \cdot \dfrac{y^2+6y+5}{y^2-y-6}$

27. $\dfrac{x^2-9x}{x^2+3x+2} \div \dfrac{x^2-81}{x^2+2x}$

28. $\dfrac{y^2-9}{2y^2-y-15} \div \dfrac{3y^2+10y+3}{2y^2+y-10}$

Section 7.3 Adding and Subtracting Rational Expressions with a Common Denominator

KEY CONCEPT

- **Adding/Subtracting Rational Expressions**

  If $\dfrac{a}{c}$ and $\dfrac{b}{c}$, $c \neq 0$, then $\dfrac{a}{c} + \dfrac{b}{c} = \dfrac{a+b}{c}$ and $\dfrac{a}{c} - \dfrac{b}{c} = \dfrac{a-b}{c}$. Simplify the sum or difference by dividing out like factors.

You Should Be Able To...	EXAMPLE	Review Exercies
1 Add rational expressions with a common denominator (p. 446)	Examples 1 and 2	29–34
2 Subtract rational expressions with a common denominator (p. 448)	Examples 3 and 4	35–38
3 Add or subtract rational expressions with opposite denominators (p. 449)	Examples 5 and 6	39–42

In Problems 29–42, perform the indicated operation.

29. $\dfrac{4}{x-3} + \dfrac{5}{x-3}$

30. $\dfrac{3}{x+4} + \dfrac{7}{x+4}$

31. $\dfrac{m^2}{m+3} + \dfrac{3m}{m+3}$

32. $\dfrac{1}{6m} + \dfrac{5}{6m}$

33. $\dfrac{-m+1}{m^2-4} + \dfrac{2m+1}{m^2-4}$

34. $\dfrac{2m^2}{m+1} + \dfrac{5m+3}{m+1}$

35. $\dfrac{11}{15m} - \dfrac{6}{15m}$

36. $\dfrac{15b}{2b^2} - \dfrac{11b}{2b^2}$

37. $\dfrac{2y^2}{y-7} - \dfrac{y^2+49}{y-7}$

38. $\dfrac{6m+5}{m^2-36} - \dfrac{5m-1}{m^2-36}$

39. $\dfrac{3}{x-4} + \dfrac{10}{4-x}$

40. $\dfrac{7}{a-b} + \dfrac{3}{b-a}$

41. $\dfrac{2x}{x^2-25} - \dfrac{x-5}{25-x^2}$

42. $\dfrac{x+5}{2x-6} - \dfrac{x+3}{6-2x}$

Section 7.4 Finding the Least Common Denominator and Forming Equivalent Rational Expressions

KEY CONCEPTS

- The steps to find the LCD of rational expressions are given on page 454.
- The steps to form equivalent rational expressions are given on page 457.

KEY TERMS

Least common denominator (LCD)
Equivalent rational expressions

You Should Be Able To...	EXAMPLE	Review Exercises
1 Find the least common denominator (LCD) of two or more rational expressions (p. 453)	Examples 1 through 6	43–48
2 Write a rational expression equivalent to a given rational expression (p. 456)	Examples 7 and 8	49–52
3 Use the LCD to write equivalent rational expressions (p. 457)	Example 9	53–58

In Problems 43–48, identify the LCD of the given rational expressions.

43. $\dfrac{6}{4x^2y^7}; \dfrac{8}{6x^4y}$

44. $\dfrac{3}{20a^3bc^4}; \dfrac{7}{30ab^4c^7}$

45. $\dfrac{11}{4a}; \dfrac{7a}{8a + 16}$

46. $\dfrac{6}{5a}; \dfrac{17}{a^2 + 6a}$

47. $\dfrac{x + 1}{4x - 12}; \dfrac{3x}{x^2 - 2x - 3}$

48. $\dfrac{11}{x^2 - 7x}; \dfrac{x + 2}{x^2 - 49}$

In Problems 49–52, write an equivalent rational expression with the given denominator.

49. $\dfrac{6}{x^3y}$ with denominator x^4y^7

50. $\dfrac{11}{a^3b^2}$ with denominator a^7b^5

51. $\dfrac{x - 1}{x - 2}$ with denominator $x^2 - 4$

52. $\dfrac{m + 2}{m + 7}$ with denominator $m^2 + 5m - 14$

In Problems 53–58, identify the LCD and then write each as an equivalent rational expression with that denominator.

53. $\dfrac{5y}{6x^3}; \dfrac{7}{8x^5}$

54. $\dfrac{6}{5a^3}; \dfrac{11b}{10a}$

55. $\dfrac{4x}{x - 2}; \dfrac{6}{4 - x^2}$

56. $\dfrac{3}{m - 5}; \dfrac{-2m}{15 + 2m - m^2}$

57. $\dfrac{2}{m^2 + 5m - 14}; \dfrac{m + 1}{m^2 + 9m + 14}$

58. $\dfrac{n - 2}{n^2 - 5n}; \dfrac{n}{n^2 - 25}$

Section 7.5 Adding and Subtracting Rational Expressions with Unlike Denominators

KEY CONCEPT

- The steps for adding or subtracting rational expressions with unlike denominators are given on page 461.

You Should Be Able To...	EXAMPLE	Review Exercises
1 Add and subtract rational expressions with unlike denominators (p. 460)	Examples 2 through 10	59–72

In Problems 59–72, perform the indicated operations.

59. $\dfrac{4}{3x^2} + \dfrac{8}{9x}$

60. $\dfrac{x}{2x^3y} + \dfrac{y}{10xy^3}$

61. $\dfrac{x}{x + 7} + \dfrac{2}{x - 7}$

62. $\dfrac{4}{2x + 3} + \dfrac{x + 1}{2x - 3}$

63. $\dfrac{x + 5}{x} - \dfrac{x + 7}{x - 2}$

64. $\dfrac{p + 1}{p + 3} - \dfrac{p + 17}{p^2 - p - 12}$

65. $\dfrac{3x - 4}{4x + 1} + \dfrac{3x + 6}{4x^2 + 9x + 2}$

66. $\dfrac{7}{3x^2 + x - 4} + \dfrac{9x + 2}{3x^2 - 2x - 8}$

67. $\dfrac{m}{m - 3} - \dfrac{4m + 12}{m^2 - 9}$

68. $\dfrac{3}{y^2 + 7y + 10} - \dfrac{4}{y^2 + 6y + 5}$

69. $\dfrac{3}{m - 2} + \dfrac{12}{4 - m^2}$

70. $\dfrac{m}{m - n} - \dfrac{n}{m + n}$

71. $4 + \dfrac{x}{x + 3}$

72. $7 - \dfrac{1}{x + 2}$

Section 7.6 Complex Rational Expressions

KEY CONCEPT

- There are two methods that can be used to simplify a complex rational expression. The steps for Method I are presented on page 472, and the steps or Method II are presented on page 474.

KEY TERMS

Complex rational expression
Simplify

You Should Be Able To...	EXAMPLE	Review Exercises
1 Simplify a complex rational expression by simplifying the numerator and denominator separately (Method I) (p. 471)	Examples 2 through 4	73–80
2 Simplify a complex rational expression using the least common denominator (Method II) (p. 474)	Examples 5 and 6	73–80

In Problems 73–80, simplify the complex rational expressions.

73. $\dfrac{\frac{1}{2} - \frac{2}{3}}{\frac{4}{9} + \frac{5}{6}}$

74. $\dfrac{\frac{1}{4} + \frac{1}{2}}{\frac{5}{8} - \frac{1}{6}}$

77. $\dfrac{\frac{x}{4} - \frac{1}{2}}{\frac{3x}{2} - 3}$

78. $\dfrac{\frac{8}{y + 4} + 2}{\frac{12}{y + 4} - 2}$

75. $\dfrac{\frac{1}{5} - \frac{1}{m}}{\frac{1}{10} + \frac{1}{m^2}}$

76. $\dfrac{\frac{1}{m^2} + \frac{2}{3}}{\frac{1}{m} - \frac{5}{6}}$

79. $\dfrac{\frac{m - 4}{m + 4} + \frac{m - 4}{m - 2}}{1 - \frac{2}{m-2}}$

80. $\dfrac{\frac{a^2}{a^2 - 36} - \frac{a}{a + 6}}{\frac{a}{a^2 - 36} - \frac{1}{a - 6}}$

Section 7.7 Rational Equations

KEY CONCEPTS

- The steps for solving a rational equation are given on page 480.
- To solve for a variable in a formula, get the variable by itself on one side of the equation with all other variable and constant terms, if any, on the other side.

KEY TERMS

Rational equation
Extraneous solution

You Should Be Able To...	EXAMPLE	Review Exercises
1 Solve equations containing rational expressions (p. 478)	Examples 1 through 7	83–92
2 Solve for a variable in a rational equation (p. 485)	Examples 8 and 9	93–96

In Problems 81 and 82, state the values of the variable for which the expressions in the rational equation are undefined.

81. $\dfrac{4}{x + 5} + \dfrac{1}{x} = 3$

82. $\dfrac{2}{x^2 - 36} = \dfrac{1}{x + 6}$

In Problems 83–92, solve each equation and state the solution set. Be sure to check for extraneous solutions.

83. $\dfrac{2}{x} - \dfrac{3}{4} = \dfrac{5}{x}$

84. $\dfrac{4}{x} + \dfrac{3}{4} = \dfrac{2}{3x} + \dfrac{23}{4}$

85. $\dfrac{4}{m} - \dfrac{3}{2m} = \dfrac{1}{2}$

86. $\dfrac{2}{m} + \dfrac{5}{m + 2} = -3$

87. $\dfrac{m + 4}{m - 3} = \dfrac{m + 10}{m + 2}$

88. $\dfrac{m - 6}{m + 5} = \dfrac{m - 3}{m + 1}$

89. $\dfrac{2x}{x - 1} - 5 = \dfrac{2}{x - 1}$

90. $\dfrac{1}{x + 3} + \dfrac{1}{x - 3} = \dfrac{-5}{x^2 - 9}$

91. $\dfrac{2x}{x - 2} - 1 = \dfrac{8x - 24}{x^2 - 5x + 6}$

92. $\dfrac{3}{5 - x} - \dfrac{11}{x^2 - 2x - 15} = \dfrac{4}{x + 3}$

In Problems 93–96, solve the equation for the indicated variable.

93. $y = \dfrac{4}{k}$ for k

94. $6 = \dfrac{x}{y}$ for y

95. $\dfrac{1}{x} + \dfrac{1}{y} = \dfrac{1}{z}$ for y

96. $\dfrac{1}{x} + \dfrac{1}{y} = \dfrac{1}{z}$ for z

Section 7.8 Models Involving Rational Equations

KEY CONCEPT

- **Means-Extremes Theorem**

  If $\dfrac{a}{b} = \dfrac{c}{d}$, then $a \cdot d = b \cdot c$, b and $d \neq 0$.

KEY TERMS

Ratio	Extremes
Proportion	Cross multiplication
Terms	Similar figures
Means	Constant rate

You Should Be Able To...	EXAMPLE	Review Exercises
1 Model and solve ratio and proportion problems (p. 490)	Examples 1 through 3	97–100, 103, 104
2 Model and solve problems with similar figures (p. 493)	Examples 4 and 5	101, 102
3 Model and solve work problems (p. 494)	Examples 6 and 7	105–108
4 Model and solve uniform motion problems (p. 497)	Examples 8 and 9	109–112

In Problems 97–100, solve the proportion.

97. $\dfrac{6}{4y + 5} = \dfrac{2}{7}$

98. $\dfrac{2}{y - 3} = \dfrac{5}{y}$

99. $\dfrac{y + 1}{8} = \dfrac{1}{4}$

100. $\dfrac{6y + 7}{10} = \dfrac{2y + 9}{6}$

In Problems 101 and 102, use the similar triangles to solve for x.

101.

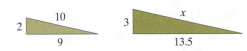

102.

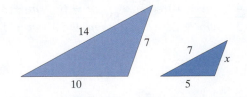

103. Cement A cement mixer uses 4 tanks of water to mix 15 bags of cement. How many tanks of water are needed to mix 30 bags of cement?

104. Pizza If 4 small pizzas cost $15.00, find the cost of 7 small pizzas.

105. Washing Lucille can wash the walls in 3 hours working alone, and Teresa can wash the walls in 2 hours. To the nearest tenth of an hour, how long would it take them to wash the walls working together?

106. Carpet Fred can install the carpet in a room in 3 hours, but Barney needs 5 hours to install the carpet. To the nearest tenth of an hour, how long would it take them to complete the carpet installation if they worked together?

107. Dishes Working alone, it takes Jake 5 minutes longer to wash the dishes than it takes Adrienne when she washes the dishes alone. Washing the dishes together, Jake and Adrienne can finish the job in 6 minutes. How long does it take Jake to wash the dishes by himself?

108. Painting Working together, Donovan and Ben can paint a room in 4 hours. Working alone, it takes Donovan 7 hours to paint the room. To the nearest tenth of an hour, how long would it take Ben to paint the room working alone?

109. Paddling Paul can paddle a kayak 15 miles upstream in the same amount of time it takes him to paddle a kayak 27 miles downstream. If the current is 2 mph, what is Paul's speed as he paddles downstream?

110. Cruising A cruise ship traveled for 275 miles with the current in the same amount of time it traveled 175 miles against the current. The speed of the current was 10 mph. What was the speed of the cruise ship as it traveled with the current?

111. Vacation On their vacation, a family traveled 135 miles by train and then traveled 855 miles by plane. The speed of the plane was three times the speed of the train. If the total time of the trip was 6 hours, what was the speed of the train?

112. Driving Tamika drove for 90 miles in the city. When she got on the highway, she increased her speed by 20 mph and drove for 130 miles. If Tamika drove a total of 4 hours, how fast did she drive in the city?

Section 7.9 Variation

KEY TERMS

Variation
Varies directly
Directly proportional to
Constant of proportionality

Constant of variation
Varies inversely
Inversely proportional

You Should Be Able To...	EXAMPLE	Review Exercises
1 Model and solve direct variation problems (p. 503)	Examples 1 and 2	113–118
2 Model and solve inverse variation problems (p. 505)	Examples 3 and 4	119–124

In Problems 113 and 114, assume that y varies directly with x. Find an equation that relates x and y given the following:

113. $x = 12$ when $y = 3$

114. $x = 3$ when $y = 18$

115. Suppose that f varies directly with g. If $f = 12$ when $g = 18$, find f when $g = 24$.

116. Suppose that p varies directly with q. If $p = 25$ when $q = 20$, find p when $q = 88$.

117. CD Sales The income from CD sales varies directly with the number of customers that enter the store. If 20 customers produce CD sales of $290, how much in sales will 35 customers produce?

118. Map Distances The distance between two points on a map varies directly with the actual distance between the cities. If two cities are 3 cm apart on a map and their actual distance is 60 miles, find the actual distance for two cities that measure 8 cm apart on the same map.

In Problems 119 and 120, suppose that y varies inversely with x. Find an equation that relates x and y given the following:

119. $x = 8$ when $y = 6$

120. $x = 6$ when $y = 9$

121. Suppose that r varies inversely with s. If $r = 18$ when $s = 2$, find r when $s = 4$.

122. Suppose that s varies inversely with u. If $s = 3$ when $u = 2$, find s when $u = 24$.

123. Road Trip The time that it takes to drive from Los Angeles to San Francisco is inversely proportional to the speed the car is driven. If the trip takes 6.5 hours at 60 mph, what speed is necessary to make the trip in 5 hours?

124. Measuring Pressure The volume of a gas varies inversely with the pressure exerted on it. If the volume is 8 liters when the pressure is 100 grams per liter, what is the volume of this gas when the pressure is 32 grams per liter?

Chapter 7 Test

Step-by-step test solutions are found on the Chapter Test Prep Videos available in MyMathLab® or on YouTube.

1. Evaluate $\dfrac{3x - 2y^2}{6z}$ when $x = 2, y = -3$, and $z = -1$.

2. Find the values for which the following rational expression is undefined: $\dfrac{x+5}{x^2 - 3x - 10}$

3. Simplify: $\dfrac{x^2 - 4x - 21}{14 - 2x}$

In Problems 4–11, perform the indicated operation.

4. $\dfrac{\dfrac{35x^6}{9x^4}}{\dfrac{25x^5}{18x}}$

5. $\dfrac{5x - 15}{3x + 9} \cdot \dfrac{5x + 15}{3x - 9}$

6. $\dfrac{\dfrac{2x^2 - 5xy - 12y^2}{x^2 + xy - 20y^2}}{\dfrac{4x^2 - 9y^2}{x^2 + 4xy - 5y^2}}$

7. $\dfrac{y^2}{y + 3} + \dfrac{3y}{y + 3}$

8. $\dfrac{x^2}{x^2 - 9} - \dfrac{8x - 15}{x^2 - 9}$

9. $\dfrac{6}{y - z} + \dfrac{7}{z - y}$

10. $\dfrac{x}{x - 2} + \dfrac{3}{2x + 1}$

11. $\dfrac{2x}{x^2 + 5x + 6} - \dfrac{x + 1}{x^2 + 2x - 3}$

12. Simplify the following complex rational expression:

$$\dfrac{\dfrac{1}{9} - \dfrac{1}{y^2}}{\dfrac{1}{3} + \dfrac{1}{y}}$$

In Problems 13 and 14, solve the rational equations. Check for extraneous solutions.

13. $\dfrac{m}{5} + \dfrac{5}{m} = \dfrac{m + 3}{4}$

14. $\dfrac{4}{x + 3} + \dfrac{5}{x - 6} = \dfrac{4x + 1}{x^2 - 3x - 18}$

15. Solve $\dfrac{1}{x} + \dfrac{1}{y} = \dfrac{1}{z}$ for y.

16. Solve the proportion: $\dfrac{2}{y + 1} = \dfrac{1}{y - 2}$

17. Use the similar triangles to solve for x.

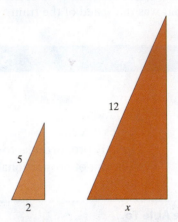

18. Barrettes If 8 hair barrettes cost $22.00, how much would 12 hair barrettes cost?

19. Car Washing It takes Frank 18 more minutes than Juan to wash a car. If they can wash the car together in 12 minutes, how long does it take Frank to wash the car by himself?

20. Excursion On a vacation excursion, some tourists walked 7 miles on a nature path and then hiked 12 miles up a mountainside. The tourists walked 3 mph faster than they hiked. The total time of the excursion was 4 hours. At what rate did the tourists hike?

21. m varies directly with n. If $m = 12$ when $n = 8$, find m when $n = 20$.

Cumulative Review Chapters 1–7

In Problems 1 and 2, simplify each expression.

1. $-6^2 + 4(-5 + 2)^3$

2. $3(4x - 2) - (3x + 5)$

In Problems 3–5, solve each equation.

3. $-3(x - 5) + 2x = 5x - 4$

4. $3(2x - 1) + 5 = 6x + 2$

5. $0.25x + 0.10(x - 3) = 0.05(22)$

6. Movie Poster In the entranceway of the Sawgrass Cinema, there is a large rectangular poster advertising a movie. If the poster's length is 3 feet less than twice its width, and its perimeter is 24 feet, what is the length of the poster?

7. Integers The sum of three consecutive even integers is 138. Find the integers.

8. Solve and graph the following inequality:
$$2(x - 3) - 5 \le 3(x + 2) - 18$$

9. Multiply: $(3x - 2y)^2$

In Problems 10 and 11, simplify each expression. Write your answers with positive exponents only.

10. $\left(\dfrac{2a^5b}{4ab^{-2}}\right)^{-4}$

11. $(3x^0y^{-4}z^3)^3$

12. Factor completely: $8a^2b + 34ab - 84b$

13. Solve the equation using the Zero-Product Property:
$$6x^3 - 31x^2 = -5x$$

14. Rectangle The length of a rectangle is 2 cm longer than twice its width. If the area of the rectangle is 40 square centimeters, what is the length of the rectangle?

15. Find the value of $\dfrac{x + 5}{x^2 + 25}$ when $x = -5$.

In Problems 16–19, perform the indicated operation and simplify, if possible.

16. $\dfrac{3x + 3}{5x - 5x^2} \cdot \dfrac{2x^2 + x - 3}{4x^2 - 9}$

17. $\dfrac{x^2 - x - 2}{10} \div \dfrac{2x + 4}{5}$

18. $\dfrac{2x + 3}{x^2 - x - 30} - \dfrac{x - 2}{x^2 - x - 30}$

19. $\dfrac{15}{2x - 4} + \dfrac{x}{x^2 - 4}$

20. Simplify the following complex rational expression:

$$\dfrac{\dfrac{2}{x^2} - \dfrac{3}{5x}}{\dfrac{4}{x} + \dfrac{1}{4x}}$$

In Problems 21 and 22, solve each equation.

21. $\dfrac{3}{x - 4} = \dfrac{5x + 4}{x^2 - 16} - \dfrac{4}{x + 4}$

22. $\dfrac{x - 5}{3} = \dfrac{x + 2}{2}$

23. Map On a city map, 4 inches represents 50 miles. How many inches would represent 125 miles?

24. Inventory It takes Trent 9 hours longer than Sharona to do a store's inventory. If they can finish the inventory in 6 hours working together, how long does it take Sharona to do the inventory working alone?

25. Racing Francisco ran for 35 miles and then walked for 6 miles. Francisco runs 4 mph faster than he walks. If it took him 7 hours to finish the race, how fast was he running?

26. Find the slope of the line joining the points $(-1, 3)$ and $(5, 11)$ in the Cartesian plane.

27. Find the equation of the line that has slope $-\dfrac{3}{2}$ and contains the point $(4, -1)$.

28. Graph the line $2x - 3y = -12$ by finding the intercepts of the graph.

29. Find the slope of the line parallel to $3x + 5y = 15$. What is the slope of the line perpendicular to $3x + 5y = 15$?

30. Solve: $\begin{cases} 2x + 3y = 1 \\ -3x + 2y = 18 \end{cases}$

8 Roots and Radicals

Did you ever think that dropping a water balloon from the top of a building could have a mathematical context? The time t that it takes for an object to fall h feet is modeled by the formula $t = \sqrt{\dfrac{h}{16}}$. See Problems 99 and 100 in Section 8.6.

The Big Picture: Putting It Together

In Chapter 5, you multiplied polynomial expressions. In Chapter 6, you learned that polynomial multiplication can be "undone" by factoring. Throughout your mathematics career, you will often learn a process and a method for "undoing" the process.

In this chapter, you will learn to "undo" integer exponents. Sections 8.1 and 8.2 are about "undoing" the squaring operation. This technique involves a symbol called a radical. In Sections 8.3 through 8.5, you will add, subtract, multiply, and divide radical expressions. We turn our attention to solving equations involving radicals in Section 8.6. Finally, Section 8.7 deals with "undoing" positive integer exponents that are 3 or greater and raising a real number to a rational exponent.

8.1 Introduction to Square Roots

Objectives

1. Evaluate Square Roots
2. Determine Whether a Square Root Is Rational, Irrational, or Not a Real Number
3. Find Square Roots of Variable Expressions

Are You Ready for This Section?

Before getting started, take the following readiness quiz. If you get a problem wrong, go to the section cited and review the material.

In Problems 1–3, use the set $\left\{-4, \dfrac{5}{3}, 0, \sqrt{2}, 6.95, 13, \pi\right\}$.

R1. Which of the numbers are integers? [Section 1.3, pp. 19–22]

R2. Which of the numbers are rational numbers? [Section 1.3, pp. 19–22]

R3. Which of the numbers are irrational numbers? [Section 1.3, pp. 19–22]

R4. Evaluate: **(a)** $\left(\dfrac{3}{2}\right)^2$ **(b)** $(0.4)^2$ [Section 1.7, pp. 56–57]

In Section 1.7, we introduced exponents. Exponents indicate repeated multiplication. For example, 4^2 means $4 \cdot 4$, so $4^2 = 16$; $(-6)^2 = (-6) \cdot (-6) = 36$. In this section, we will "undo" the process of raising a number to the second power and ask questions such as, "What number, or numbers, when squared, give me 16?"

1 Evaluate Square Roots

▶ You know that both 5^2 and $(-5)^2$ equal 25. Thus, if you were asked, "What number, when squared, equals 25?", you would respond, "Either -5 or 5."

> **In Words**
> Taking the square root of a number "undoes" raising a number to the second power.

> **Definition**
> For any real numbers a and b, b is a **square root** of a if $b^2 = a$.

Put another way, the square roots of a are the numbers that, when squared, give a. For example, the square roots of 25 are -5 and 5 because $(-5)^2 = 25$ and $5^2 = 25$.

EXAMPLE 1 | **Finding the Square Roots of Numbers**

Find the square roots: **(a)** 36 **(b)** 0

Solution

(a) 6 is a square root of 36 because $6^2 = 36$.
-6 is also a square root of 36 because $(-6)^2 = 36$.

(b) 0 is the only square root of 0 because only $0^2 = 0$. ●

EXAMPLE 2 | **Finding the Square Roots of Numbers**

Work Smart

When finding square roots of fractions, look at the numerator and denominator separately.

Find the square roots: **(a)** $\dfrac{16}{49}$ **(b)** 0.09

Solution

(a) $-\dfrac{4}{7}$ and $\dfrac{4}{7}$ are the square roots of $\dfrac{16}{49}$ because $\left(-\dfrac{4}{7}\right)^2 = \dfrac{16}{49}$ and $\left(\dfrac{4}{7}\right)^2 = \dfrac{16}{49}$.

(b) -0.3 and 0.3 are the square roots of 0.09 because $(-0.3)^2 = 0.09$ and $(0.3)^2 = 0.09$. ●

Ready?...Answers **R1.** $-4, 0, 13$

R2. $-4, \dfrac{5}{3}, 0, 6.95, 13$ **R3.** $\sqrt{2}, \pi$

R4. (a) $\dfrac{9}{4}$ **(b)** 0.16

Notice in Examples 1 and 2 that every positive number has a positive and a negative square root. We use the symbol $\sqrt{}$, called a **radical,** to denote the **principal square root,** or nonnegative (zero or positive) square root. For example, if we want the positive square root of 25, we write $\sqrt{25} = 5$. We read $\sqrt{25} = 5$ as "The positive square root of 25 is 5." But what if we want the negative square root of a real number? In that case, we use the expression $-\sqrt{25} = -5$.

Work Smart

Symbols and notation in math enable us to express our thoughts using shorthand. The symbol $\sqrt{100}$ means "the positive number whose square is 100." If you want the negative number whose square is 100, write $-\sqrt{100}$.

> **Properties of Square Roots**
>
> - Every positive real number has two square roots, one positive and one negative.
> - The square root of 0 is 0. In symbols, $\sqrt{0} = 0$.
> - We use the symbol $\sqrt{}$, called a radical, to denote the nonnegative square root of a real number. The nonnegative square root is called the principal square root.
> - The number under the radical is called the **radicand.** For example, the radicand in $\sqrt{25}$ is 25.

EXAMPLE 3 **Evaluating Square Roots**

Evaluate: **(a)** $\sqrt{64}$ **(b)** $-\sqrt{81}$

Solution

(a) The notation $\sqrt{64}$ means "the positive number whose square is 64." Because $8^2 = 64$,

$$\sqrt{64} = 8$$

(b) The notation $-\sqrt{81}$ means "the opposite of the positive number whose square is 81." Thus

$$-\sqrt{81} = -1 \cdot \sqrt{81} = -1 \cdot 9 = -9$$ ●

EXAMPLE 4 **Evaluating Square Roots**

Evaluate: **(a)** $\sqrt{\dfrac{1}{4}}$ **(b)** $-\sqrt{0.01}$

Solution

(a) $\sqrt{\dfrac{1}{4}} = \dfrac{1}{2}$ because $\left(\dfrac{1}{2}\right)^2 = \dfrac{1}{4}$.

(b) $-\sqrt{0.01} = -0.1$ because $0.1^2 = 0.01$. ●

Figure 1

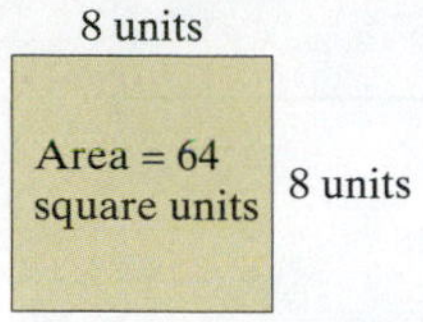

8 units

Area = 64 square units   8 units

A rational number is a **perfect square** if it is the square of a rational number. Examples 3 and 4 show that 64, 81, $\frac{1}{4}$, and 0.01 are all perfect squares because their square roots are rational numbers. We can think of perfect squares geometrically as in Figure 1, which shows a square whose area is 64 square units. The square root of the area, $\sqrt{64}$, gives us the length of each side of the square, 8 units.

▶ EXAMPLE 5   Evaluating an Expression Containing Square Roots

Evaluate: $-4\sqrt{25}$

Solution

The expression $-4\sqrt{25}$ means "-4 times the positive square root of 25". First, find the positive square root of 25, and then multiply this result by -4.

$$-4\sqrt{25} = -4 \cdot 5$$
$$= -20$$

EXAMPLE 6   Evaluating an Expression Containing Square Roots

Evaluate:

(a) $\sqrt{9} + \sqrt{16}$ **(b)** $\sqrt{9 + 16}$ **(c)** $\sqrt{8^2 - 4 \cdot 7 \cdot 1}$

Solution

(a) $\sqrt{9} + \sqrt{16} = 3 + 4$ **(b)** $\sqrt{9 + 16} = \sqrt{25}$
$\qquad\qquad\qquad = 7$ $\qquad\qquad\qquad\quad = 5$

(c) $\sqrt{8^2 - 4 \cdot 7 \cdot 1} = \sqrt{64 - 28}$
$\qquad\qquad\qquad\quad = \sqrt{36}$
$\qquad\qquad\qquad\quad = 6$

Work Smart

In Examples 6(a) and (b), notice that $\sqrt{9} + \sqrt{16} \neq \sqrt{9 + 16}$. In general,

$$\sqrt{a} + \sqrt{b} \neq \sqrt{a + b}$$

The radical acts like a grouping symbol, so always simplify the radicand before taking the square root.

Quick ✓

In Problems 11–14, evaluate each expression.

11. $2\sqrt{36}$ **12.** $\sqrt{25 + 144}$ **13.** $\sqrt{25} + \sqrt{144}$ **14.** $\sqrt{5^2 - 4 \cdot 3 \cdot (-2)}$

▶ ❷ Determine Whether a Square Root Is Rational, Irrational, or Not a Real Number

Because there is no rational number whose square is 5, $\sqrt{5}$ is not a rational number. In fact, $\sqrt{5}$ is an *irrational* number. Recall that an irrational number is a number that cannot be written as the quotient of two integers.

What if we wanted to evaluate $\sqrt{-16}$? Because any positive real number squared is positive, any negative real number squared is also positive, and 0 squared is 0, there is no real number whose square is -16. We conclude: **Negative real numbers do not have square roots that are real numbers!**

The following comments regarding square roots are important.

Work Smart

The perfect squares of positive integers are

$1^2 = 1 \qquad 5^2 = 25$
$2^2 = 4 \qquad 6^2 = 36$
$3^2 = 9 \qquad 7^2 = 49$
$4^2 = 16 \qquad 8^2 = 64$

and so on.

Work Smart

The square roots of negative real numbers are not real.

> **More Properties of Square Roots**
>
> - The square root of a perfect square number is a rational number.
> - The square root of a positive rational number that is not a perfect square is an irrational number. For example, $\sqrt{20}$ is an irrational number because 20 is not a perfect square.
> - The square root of a negative real number is not a real number. For example, $\sqrt{-2}$ is not a real number.

If needed, find decimal approximations for irrational square roots using a calculator or the table in Appendix A.

EXAMPLE 7 **Approximating Square Roots**

Approximate $\sqrt{5}$ by writing it rounded to two decimal places.

Solution

Because $\sqrt{4} = 2$ and $\sqrt{9} = 3$, we expect $\sqrt{5}$ to be between 2 and 3 since 5 is between 4 and 9. Use a calculator or Appendix A and find $\sqrt{5} \approx 2.23607$. Rounded to two decimal places, $\sqrt{5} \approx 2.24$.

Work Smart

The notation $\approx$ means "is approximately equal to."

> **Quick ✓**
>
> In Problems 15 and 16, express each square root as a decimal rounded to two decimal places.
>
> **15.** $\sqrt{35}$ **16.** $-\sqrt{6}$

EXAMPLE 8 **Determining Whether a Square Root of an Integer Is Rational, Irrational, or Not a Real Number**

Determine whether each square root is rational, irrational, or not a real number. Evaluate each real square root, if possible. Express any irrational square root as a decimal rounded to two decimal places.

  (a) $\sqrt{51}$ **(b)** $\sqrt{169}$ **(c)** $\sqrt{-81}$

Solution

  (a) $\sqrt{51}$ is irrational because 51 is not a perfect square. There is no rational number whose square is 51. Using a calculator or Appendix A, we find $\sqrt{51} \approx 7.14$.

  (b) $\sqrt{169}$ is a rational number because $13^2 = 169$. So $\sqrt{169} = 13$.

  (c) $\sqrt{-81}$ is not a real number because there is no real number whose square is -81.

Work Smart

$\sqrt{-81}$ is not a real number, but $-\sqrt{81}$ is because $-\sqrt{81} = -9$. Note the placement of the negative sign!

> **Quick ✓**
>
> **17.** *True or False* If a positive number is not a perfect square, then its square root is an irrational number.
>
> In Problems 18–21, determine whether each square root is rational, irrational, or not a real number. Evaluate each square root, if possible. For each square root that is irrational, round your answer to two decimal places.
>
> **18.** $\sqrt{49}$ **19.** $\sqrt{71}$ **20.** $\sqrt{-25}$ **21.** $-\sqrt{16}$

▶ **③ Find Square Roots of Variable Expressions**

What is $\sqrt{4^2}$? Because $4^2 = 16$, we have that $\sqrt{4^2} = \sqrt{16} = 4$. From this result, we might conclude that $\sqrt{a^2} = a$ for any real number a. But consider $\sqrt{(-4)^2}$. Our "formula" says that $\sqrt{a^2} = a$, so $\sqrt{(-4)^2} = -4$, right? Wrong! $\sqrt{(-4)^2} = \sqrt{16} = 4$. So $\sqrt{4^2} = 4$ and $\sqrt{(-4)^2} = 4$. How can we fix our "formula"? In Section 1.3, we learned that the absolute value of a, $|a|$, is a positive number if a is nonzero. From this, we have the following result:

In Words

The square root of a nonzero number squared is the absolute value of that number, which is always positive.

> For any **real number** a,
>
> $$\sqrt{a^2} = |a|$$

For example, $\sqrt{(-4)^2} = |-4| = 4$. We use this result when evaluating square roots that have variable expressions because it guarantees the result will be nonnegative regardless of whether the variable is negative, positive, or zero.

EXAMPLE 9 **Finding Square Roots of Variable Expressions**

(a) $\sqrt{(2x)^2} = |2x| = 2|x|$ for any real number x.

(b) $\sqrt{(a-4)^2} = |a - 4|$ for any real number a.

Quick ✓

22. For any real number a, $\sqrt{a^2} = $ __ .

In Problems 23–25, simplify each square root. The variable represents any real number.

23. $\sqrt{b^2}$

24. $\sqrt{(4p)^2}$

25. $\sqrt{(w+3)^2}$

If it is stated that $a \geq 0$, then the absolute value bars are not needed because $|a| = a$ when $a \geq 0$. Thus $\sqrt{a^2} = a$ when $a \geq 0$.

EXAMPLE 10 **Finding Square Roots of Variable Expressions**

Simplify each square root.

(a) $\sqrt{(3y)^2}, y \geq 0$

(b) $\sqrt{(b+3)^2}, b + 3 \geq 0$

Solution

(a) $\sqrt{(3y)^2} = 3y$ because $y \geq 0$.

(b) $\sqrt{(b+3)^2} = b + 3$ because $b + 3 \geq 0$.

Quick ✓

In Problems 26–28, simplify each square root.

26. $\sqrt{(5b)^2}, b \geq 0$

27. $\sqrt{(z-6)^2}, z - 6 \geq 0$

28. $\sqrt{(2h-5)^2}, 2h - 5 \geq 0$

8.1 Exercises

MyMathLab® PRACTICE

Exercise numbers in green have complete video solutions in MyMathLab.

*Problems **1–28** are the Quick ✓s that follow the **EXAMPLES**.*

Building Skills

In Problems 29–36, find the value(s) of each expression. See Objective 1.

29. the square roots of 1

30. the square roots of 4

31. the square roots of $\frac{1}{9}$

32. the square roots of $\frac{9}{4}$

33. the square roots of 169

34. the square roots of 49

35. the square roots of 0.25

36. the square roots of 0.16

In Problems 37–60, find the exact value of each square root without a calculator. See Objective 1.

37. $\sqrt{144}$

38. $\sqrt{4}$

39. $-\sqrt{9}$

40. $-\sqrt{1}$

41. $\sqrt{225}$

42. $\sqrt{169}$

43. $\sqrt{\frac{1}{121}}$

44. $\sqrt{\frac{1}{49}}$

45. $\sqrt{0.04}$

46. $\sqrt{0.09}$

47. $-6\sqrt{9}$

48. $-4\sqrt{4}$

49. $\sqrt{\dfrac{16}{81}}$

50. $\sqrt{\dfrac{25}{9}}$

51. $5\sqrt{49}$

52. $4\sqrt{16}$

53. $\sqrt{36 + 64}$

54. $\sqrt{81 + 144}$

55. $\sqrt{36} + \sqrt{64}$

56. $\sqrt{81} + \sqrt{144}$

57. $\sqrt{4 - (4)(1)(-15)}$

58. $\sqrt{9 - (4)(1)(-10)}$

59. $\sqrt{7^2 - (4)(15)(-2)}$

60. $\sqrt{10^2 - (4)(8)(3)}$

In Problems 61–66, use a calculator or Appendix A to find the approximate value of the square root, rounded to the indicated place. See Objective 2.

61. $\sqrt{8}$ to 3 decimal places

62. $\sqrt{12}$ to 3 decimal places

63. $\sqrt{30}$ to the nearest hundredth

64. $\sqrt{18}$ to the nearest hundredth

65. $\sqrt{57}$ to the nearest tenth

66. $\sqrt{42}$ to the nearest tenth

In Problems 67–76, tell if the square root is rational, irrational, or not a real number. If the square root is rational, find the exact value; if the square root is irrational, write the approximate value rounded to two decimal places. See Objective 2.

67. $\sqrt{-4}$

68. $\sqrt{-100}$

69. $\sqrt{400}$

70. $\sqrt{900}$

71. $\sqrt{\dfrac{1}{4}}$

72. $\sqrt{\dfrac{49}{64}}$

73. $\sqrt{54}$

74. $\sqrt{24}$

75. $\sqrt{50}$

76. $\sqrt{12}$

In Problems 77–84, simplify each square root. See Objective 3.

77. $\sqrt{d^2},\, d \geq 0$

78. $\sqrt{k^2},\, k \geq 0$

79. $\sqrt{(5y)^2}$

80. $\sqrt{(7n)^2}$

81. $\sqrt{(x - 9)^2}$

82. $\sqrt{(y - 16)^2},\, y - 16 \geq 0$

83. $\sqrt{(m - 3)^2},\, m - 3 \geq 0$ **84.** $\sqrt{(p + 9)^2}$

Mixed Practice

In Problems 85–100, evaluate each square root if possible. If the square root is irrational, write the approximate value rounded to two decimal places. If the square root is not a real number, so state.

85. $\sqrt{3}$

86. $\sqrt{2}$

87. $\sqrt{0.4}$

88. $\sqrt{\dfrac{4}{81}}$

89. $\sqrt{-2}$

90. $\sqrt{-10}$

91. $3\sqrt{4}$

92. $-10\sqrt{9}$

93. $3\sqrt{\dfrac{25}{9}} - \sqrt{169}$

94. $2\sqrt{\dfrac{9}{4}} - \sqrt{4}$

95. $\sqrt{4^2 - 24}$

96. $\sqrt{(-3)^2 - 20}$

97. $\sqrt{7^2 - (4)(-2)(-3)}$

98. $\sqrt{13^2 - (4)(-6)(-6)}$

99. $\sqrt{(11 - 8)^2 + (11 - 5)^2}$

100. $\sqrt{(-3 - 5)^2 + (-5 - (-1))^2}$

Applying the Concepts

The area, A, of a square whose side has length s is given by $A = s^2$. We can calculate the length, s, of the side of a square as the positive square root of the area, A, using $s = \sqrt{A}$. In Problems 101–104, find the length of the side of the square whose area is given.

△ **101.** 625 square feet

△ **102.** 256 square meters

△ **103.** 256 square kilometers

△ **104.** 400 square inches

The area, A, of a circle whose radius is r is given by the formula $A = \pi r^2$. We can calculate the radius when the area is given using $r = \sqrt{\dfrac{A}{\pi}}$. In Problems 105–108, find the radius of the circle with the following areas.

△ **105.** 49π square meters

△ **106.** 25π square feet

△ **107.** 196π square inches

△ **108.** 289π square centimeters

△ **109. Great Pyramid at Giza** The volume V of a pyramid with a square base s and height h is $V = \dfrac{1}{3}s^2 h$. If we solve this formula for s, we obtain $s = \sqrt{\dfrac{3V}{h}}$. The Great Pyramid at Giza, built around 2500 B.C., has a volume of approximately 7,700,000 cubic meters. Find the length, to the nearest meter, of the side of the Great Pyramid if you know that the height is approximately 146 meters.

△ **110. Sun Pyramid** Use the formula given in Problem 109 to the find the length, to the nearest foot, of the side of the Sun Pyramid if the volume of the pyramid is approximately 114,000,000 cubic feet and the height is 210 feet. The Sun Pyramid was built in the ancient city of Teotihuacán, around 200 A.D.

Extending the Concepts

In Problems 111–114, simplify each expression.

111. $\sqrt{\sqrt{16}}$

112. $\sqrt{\sqrt{81}}$

113. $\sqrt{5 \cdot \sqrt{25}}$

114. $\sqrt{2 \cdot \sqrt{4}}$

Explaining the Concepts

115. Explain why $-\sqrt{9}$ is a real number but $\sqrt{-9}$ is not a real number.

116. You don't have a calculator handy but need an estimate for $\sqrt{20}$. Explain how you might find an approximate value for this number without a calculator.

8.2 Simplifying Square Roots

Objectives

1. Use the Product Rule to Simplify Square Roots of Constants
2. Use the Product Rule to Simplify Square Roots of Variable Expressions
3. Use the Quotient Rule to Simplify Square Roots

Work Smart

The perfect square integers are

$1^2 = 1$	$5^2 = 25$
$2^2 = 4$	$6^2 = 36$
$3^2 = 9$	$7^2 = 49$
$4^2 = 16$	$8^2 = 64$

and so on.

Ready?...Answers **R1.** $2^1 = 2, 2^2 = 4,$
$2^3 = 8, 2^4 = 16, 2^5 = 32, 2^6 = 64$
R2. $3^1 = 3, 3^2 = 9, 3^3 = 27, 3^4 = 81$
R3. 7

In Words
$\sqrt{ab} = \sqrt{a} \cdot \sqrt{b}$ can be
stated as follows: "The square
root of a product equals the
product of the square roots."

Are You Ready for This Section?

Before getting started, take the following readiness quiz. If you get a problem wrong, go to the section cited and review the material.

R1. List the powers of 2 that are less than 100.

R2. List the powers of 3 that are less than 100.

R3. Simplify: $\sqrt{49}$ [Section 8.1, p. 520]

▶ ❶ Use the Product Rule to Simplify Square Roots of Constants

Recall that a number that is the square of a rational number is called a perfect square. For example, $1^2 = 1, 2^2 = 4, 3^2 = 9, \left(\dfrac{2}{3}\right)^2 = \dfrac{4}{9}$ are all perfect squares.

Definition

A square root expression is **simplified** if the radicand does not contain any factors that are perfect squares.

For example, $\sqrt{50}$ is not simplified because 25 is a factor of 50, and 25 is a perfect square. But how do we simplify a square root expression? Consider the following:

$$\sqrt{25 \cdot 4} = \sqrt{100} = 10 \quad \text{and} \quad \sqrt{25} \cdot \sqrt{4} = 5 \cdot 2 = 10$$

Because both expressions equal 10, we might conclude that

$$\sqrt{25 \cdot 4} = \sqrt{25} \cdot \sqrt{4}$$

which suggests the following result:

Product Rule for Square Roots

If $\sqrt{a}$ and $\sqrt{b}$ are nonnegative real numbers, then

$$\sqrt{ab} = \sqrt{a} \cdot \sqrt{b}$$

EXAMPLE 1 How to Use the Product Property to Simplify Square Roots

Simplify: $\sqrt{12}$

Step-by-Step Solution

Step 1: Write the radicand as the product of factors, one of which is a perfect square.

Write 12 as $4 \cdot 3$ since 4 is a factor of 12, and 4 is a perfect square: $\sqrt{12} = \sqrt{4 \cdot 3}$

(continued)

Step 2: Write the radicand as the product of two or more radicals.

Use $\sqrt{ab} = \sqrt{a} \cdot \sqrt{b}$: $\quad = \sqrt{4} \cdot \sqrt{3}$

Step 3: Take the square root of any perfect square.

$\sqrt{4} = 2$: $\quad = 2\sqrt{3}$ ●

Below, we summarize the steps used to simplify a square root expression.

Simplifying a Square Root Expression

Step 1: Write the radicand as the product of factors, one of which is a perfect square. Look for the largest factor of the radicand that is a perfect square.

Step 2: Use the Product Rule of Square Roots to write the radicand as the product of two or more radicals.

Step 3: Take the square root of the perfect square.

EXAMPLE 2 · **Using the Product Rule to Simplify Square Roots**

Simplify each of the following:

(a) $\sqrt{45}$ (b) $\sqrt{48}$

Solution

(a) Since 9 is a factor of 45, and 9 is a perfect square, we write 45 as $9 \cdot 5$.

$$\sqrt{45} = \sqrt{9 \cdot 5}$$
$$\sqrt{ab} = \sqrt{a} \cdot \sqrt{b}: \quad = \sqrt{9} \cdot \sqrt{5}$$
$$\sqrt{9} = 3: \quad = 3\sqrt{5}$$

(b) Since 16 is a factor of 48, and 16 is a perfect square, we write 48 as $16 \cdot 3$.

$$\sqrt{48} = \sqrt{16 \cdot 3}$$
$$\sqrt{ab} = \sqrt{a} \cdot \sqrt{b}: \quad = \sqrt{16} \cdot \sqrt{3}$$
$$\sqrt{16} = 4: \quad = 4\sqrt{3}$$ ●

Work Smart

Sometimes there is more than one way to factor a number. Use the product that produces the **largest** perfect square.

Did you notice in Example 2(b) that 48 can be factored in more than one way? $48 = 6 \cdot 8$ and $48 = 4 \cdot 12$. Either one of these factorizations of 48 would also have produced $4\sqrt{3}$. For example,

$$\sqrt{48} = \sqrt{4 \cdot 12} = \sqrt{4}\sqrt{12} = 2 \cdot \sqrt{3 \cdot 4} = 2 \cdot 2\sqrt{3} = 4\sqrt{3}$$

Choosing the *largest* factor of a radicand that is a *perfect square* makes your mathematical life easier. Be sure to Work Smart!

Quick ✓

1. If $\sqrt{a}$ and $\sqrt{b}$ are real numbers, then $\sqrt{ab} = $ _____.
2. To simplify $\sqrt{75}$, write it as $\sqrt{\underline{} \cdot 3}$.

In Problems 3–5, simplify the square root.

3. $\sqrt{20}$ 4. $\sqrt{24}$ 5. $\sqrt{72}$

Watch out! The direction "simplify" does not mean find a decimal approximation. The simplified form of a radical is its exact value, not its decimal approximation.

EXAMPLE 3 **Simplifying an Expression Involving a Square Root**

Simplify: $\dfrac{4 - \sqrt{20}}{2}$

Solution

We start by recognizing that 4 is the largest perfect square factor of 20.

Work Smart

$\dfrac{4 - \sqrt{20}}{2} \neq 2 - \sqrt{20}$

$$\frac{4 - \sqrt{20}}{2} = \frac{4 - \sqrt{4 \cdot 5}}{2}$$

Use $\sqrt{a \cdot b} = \sqrt{a} \cdot \sqrt{b}$:
$$= \frac{4 - \sqrt{4} \cdot \sqrt{5}}{2}$$

$$= \frac{4 - 2 \cdot \sqrt{5}}{2}$$

At this point, we have two options to finish simplifying the expression.

Factor out the 2 in the numerator: $= \dfrac{2(2 - \sqrt{5})}{2}$ | Use $\dfrac{a - b}{c} = \dfrac{a}{c} - \dfrac{b}{c}$: $= \dfrac{4}{2} - \dfrac{2 \cdot \sqrt{5}}{2}$

Divide out the common factor: $= 2 - \sqrt{5}$ | Divide out the common factors: $= 2 - \sqrt{5}$ ●

> **Quick ✓**
>
> In Problems 6 and 7, simplify the expression.
>
> 6. $\dfrac{6 + \sqrt{45}}{3}$ 7. $\dfrac{-2 + \sqrt{32}}{4}$

▶ ❷ **Use the Product Rule to Simplify Square Roots of Variable Expressions**

In Section 8.1, we learned that $\sqrt{a^2} = a$ if $a \geq 0$. Consider the following:

$$(a^1)^2 = a^2 \quad (a^2)^2 = a^4 \quad (a^3)^2 = a^6 \quad (a^4)^2 = a^8$$

and so on. Based on this, we conclude that if a is a nonnegative real number, then

$$\sqrt{a^2} = a \quad \sqrt{a^4} = \sqrt{(a^2)^2} = a^2 \quad \sqrt{a^6} = \sqrt{(a^3)^2} = a^3 \quad \sqrt{a^8} = \sqrt{(a^4)^2} = a^4$$

Work Smart

Even powers of a variable expression are perfect squares.

In other words, even powers of a variable expression are perfect squares. **Throughout this section, if a variable appears in the radicand of a radical expression, we assume that any variable appearing in a radicand represents a nonnegative real number.**

EXAMPLE 4 **Simplifying a Square Root Containing Variables to Even Powers**

Simplify each expression. Assume the variables are nonnegative real numbers.

(a) $\sqrt{49x^6}$ (b) $\sqrt{54m^4n^6}$

Solution

(a) $\sqrt{a \cdot b} = \sqrt{a} \cdot \sqrt{b}$: $\sqrt{49x^6} = \sqrt{49} \cdot \sqrt{x^6}$

$$= \sqrt{49} \cdot \sqrt{(x^3)^2}$$

Simplify using $\sqrt{a^2} = a$ for $a \geq 0$: $= 7x^3$

(b) 9 is a factor of 54 and 9 is a perfect square: $\sqrt{54m^4n^6} = \sqrt{9 \cdot 6 \cdot m^4n^6}$

$9 = 3^2$; $m^4 = (m^2)^2$ and $n^6 = (n^3)^2$ are perfect squares: $= \sqrt{9m^4n^6 \cdot 6}$

$\sqrt{ab} = \sqrt{a} \cdot \sqrt{b}$: $= \sqrt{9m^4n^6} \cdot \sqrt{6}$

$\sqrt{9} = 3$; $\sqrt{m^4} = m^2$; $\sqrt{n^6} = n^3$: $= 3m^2n^3\sqrt{6}$ ●

Quick ✓

8. *True or False* To simplify a square root expression, write the radicand as a product of any two factors.

In Problems 9–11, simplify the expression. Assume all variables represent nonnegative real numbers.

9. $\sqrt{36y^8}$ **10.** $\sqrt{45x^{12}}$ **11.** $\sqrt{28a^2b^{10}}$

What if the variable expression in the radicand does not have an even exponent? We factor the variable expression so that one of the factors is a perfect square, just as we did for constants.

EXAMPLE 5   **Simplify a Square Root in Which the Variable Is Not a Perfect Square**

Simplify each of the following. Assume the variables represent nonnegative real numbers.

(a) $\sqrt{36x^5}$ (b) $\sqrt{300m^4n^9}$

Solution

(a) x^4 is a perfect square: $\sqrt{36x^5} = \sqrt{36x^4 \cdot x}$

$\sqrt{ab} = \sqrt{a} \cdot \sqrt{b}$: $= \sqrt{36x^4} \cdot \sqrt{x}$

$\sqrt{36} = 6; \sqrt{x^4} = x^2$: $= 6x^2\sqrt{x}$

Work Smart

To simplify square roots that contain variable expressions, we can use the following "trick of the trade":

$$\sqrt{p^{15}} = p^7\sqrt{p} \text{ if } p \geq 0$$

because $15 \div 2 = 7$ with a remainder of 1.

(b) 100 is a perfect square: m^4 and n^8
are perfect squares: $\sqrt{300m^4n^9} = \sqrt{100m^4n^8 \cdot 3n}$

$\sqrt{ab} = \sqrt{a} \cdot \sqrt{b}$: $= \sqrt{100m^4n^8} \cdot \sqrt{3n}$

$\sqrt{100} = 10; \sqrt{m^4} = m^2; \sqrt{n^8} = n^4$: $= 10m^2n^4\sqrt{3n}$ ●

Quick ✓

In Problems 12–15, simplify the expression. Assume the variables represent nonnegative real numbers.

12. $\sqrt{25y^{11}}$ **13.** $\sqrt{121a^5}$ **14.** $\sqrt{32ab^9}$ **15.** $\sqrt{12x^7y^{13}}$

▶ ❸ **Use the Quotient Rule to Simplify Square Roots**

Now consider the following:

$$\sqrt{\frac{64}{4}} = \sqrt{16} = 4 \text{ and } \sqrt{\frac{64}{4}} = \frac{\sqrt{16}}{\sqrt{4}} = \frac{8}{2} = 4$$

Because both expressions equal 4, we might conclude that

$$\sqrt{\frac{64}{4}} = \frac{\sqrt{64}}{\sqrt{4}}$$

which suggests the following result:

In Words

$\sqrt{\dfrac{a}{b}} = \dfrac{\sqrt{a}}{\sqrt{b}}$ means the square root of the quotient equals the quotient of the square roots.

Quotient Rule for Square Roots

If $\sqrt{a}$ and $\sqrt{b}$ are nonnegative real numbers and $b \neq 0$, then

$$\sqrt{\frac{a}{b}} = \frac{\sqrt{a}}{\sqrt{b}}$$

We know that a square root expression is simplified if the radicand does not contain any factors that are perfect squares. There is a second requirement that must be satisfied for a square root to be simplified.

A square root is simplified when the radicand does not contain any fractions.

Thus $\sqrt{\dfrac{4}{81}}$ and $\sqrt{\dfrac{7}{36}}$ are not simplified, but we can simplify them using the Quotient Rule for Square Roots.

EXAMPLE 6   **Simplifying Square Roots of Quotients Using the Quotient Rule**

Simplify:

(a) $\sqrt{\dfrac{4}{81}}$

(b) $\sqrt{\dfrac{7}{36}}$

Solution

(a) $\sqrt{\dfrac{a}{b}} = \dfrac{\sqrt{a}}{\sqrt{b}}$: $\sqrt{\dfrac{4}{81}} = \dfrac{\sqrt{4}}{\sqrt{81}}$

Simplify: $= \dfrac{2}{9}$

(b) $\sqrt{\dfrac{a}{b}} = \dfrac{\sqrt{a}}{\sqrt{b}}$: $\sqrt{\dfrac{7}{36}} = \dfrac{\sqrt{7}}{\sqrt{36}}$

Simplify: $= \dfrac{\sqrt{7}}{6}$

EXAMPLE 7   **Simplifying Square Roots of Quotients Involving Variables Using the Quotient Rule**

Simplify: $\sqrt{\dfrac{75x^5}{y^6}}, x > 0, y > 0$

Solution

$$\sqrt{\dfrac{a}{b}} = \dfrac{\sqrt{a}}{\sqrt{b}}: \quad \sqrt{\dfrac{75x^5}{y^6}} = \dfrac{\sqrt{75x^5}}{\sqrt{y^6}}$$

x^4 is a perfect square: $= \dfrac{\sqrt{25x^4 \cdot 3x}}{\sqrt{y^6}}$

$\sqrt{ab} = \sqrt{a} \cdot \sqrt{b}$: $= \dfrac{\sqrt{25x^4} \cdot \sqrt{3x}}{\sqrt{y^6}}$

Simplify each radical: $= \dfrac{5x^2\sqrt{3x}}{y^3}$

Quick ✔

16. If $\sqrt{a}$ and $\sqrt{b}$ are nonnegative real numbers and $b \neq 0$, then $\sqrt{\dfrac{a}{b}} = $ ___.

In Problems 17 and 18, find the quotient and simplify. Assume that variables represent nonnegative real numbers and that the variable in the denominator is not zero.

17. $\sqrt{\dfrac{49a^3}{100}}$

18. $\sqrt{\dfrac{99}{16m^2}}$

Sometimes we perform the division in the radicand first and then simplify the radical. This approach is useful when dividing first results in a radicand that is a perfect square.

EXAMPLE 8 **Simplifying Square Roots of Quotients**

Simplify: $\sqrt{\dfrac{28a^5}{7a^2}}, a > 0$

Solution

Look carefully at the radicand. Do you see that if we divide 28 by 7 we get 4, a perfect square? For this reason, we will divide first.

Divide radicand first: $\sqrt{\dfrac{28a^5}{7a^2}} = \sqrt{4a^3}$

a^2 is a perfect square: $\qquad = \sqrt{4a^2 \cdot a}$

$\sqrt{ab} = \sqrt{a} \cdot \sqrt{b}$: $\qquad = \sqrt{4a^2} \cdot \sqrt{a}$

Simplify: $\qquad = 2a\sqrt{a}$ •

> **Quick ✓**
>
> *In Problems 19 and 20, find the quotient and simplify. Assume the variable represents a nonnegative real number.*
>
> **19.** $\sqrt{\dfrac{125}{5}}$ **20.** $\sqrt{\dfrac{144b^5}{16}}$

As Examples 7 and 8 show, **to simplify square roots, if the radicand has a numerator or denominator containing a perfect square, simplify it first. (See Example 7.) If dividing in the radicand first results in a perfect square, then divide first. (See Example 8.)**

8.2 Exercises

Exercise numbers in green have complete video solutions in MyMathLab.

Problems 1–20 are the Quick ✓s that follow the EXAMPLES.

Building Skills

In Problems 21–42, simplify each square root. See Objective 1.

21. $\sqrt{8}$ **22.** $\sqrt{27}$

23. $\sqrt{40}$ **24.** $\sqrt{32}$

25. $\sqrt{18}$ **26.** $\sqrt{52}$

27. $\sqrt{33}$ **28.** $\sqrt{21}$

29. $\sqrt{45}$ **30.** $\sqrt{50}$

31. $\sqrt{125}$ **32.** $\sqrt{200}$

33. $\sqrt{42}$ **34.** $\sqrt{30}$

35. $\sqrt{98}$ **36.** $\sqrt{80}$

37. $\sqrt{8^2 - 4(-2)(8)}$ **38.** $\sqrt{(-4)^2 - 4(-5)(2)}$

39. $\dfrac{4 + \sqrt{36}}{2}$ **40.** $\dfrac{5 - \sqrt{100}}{5}$

41. $\dfrac{9 + \sqrt{18}}{3}$ **42.** $\dfrac{10 - \sqrt{75}}{5}$

In Problems 43–58, simplify each square root. Assume all variables represent nonnegative real numbers. See Objective 2.

43. $\sqrt{x^{10}}$ **44.** $\sqrt{y^{12}}$

45. $\sqrt{n^{144}}$ **46.** $\sqrt{m^{100}}$

47. $\sqrt{9y^4}$ **48.** $\sqrt{25x^8}$

49. $\sqrt{24z^{14}}$ **50.** $\sqrt{32p^{16}}$

51. $\sqrt{49x^7}$ **52.** $\sqrt{81y^3}$

53. $\sqrt{48b^5}$ **54.** $\sqrt{60n^7}$

55. $\sqrt{18m^3n^4}$ **56.** $\sqrt{50a^6b^5}$

57. $\sqrt{75c^4d^2}$ **58.** $\sqrt{125s^2t^6}$

In Problems 59–76, simplify each square root. Assume all variables represent positive real numbers. See Objective 3.

59. $\sqrt{\dfrac{25}{36}}$ **60.** $\sqrt{\dfrac{9}{16}}$

61. $\sqrt{\dfrac{11}{9}}$

62. $\sqrt{\dfrac{13}{4}}$

63. $\sqrt{\dfrac{y^{16}}{121}}$

64. $\sqrt{\dfrac{x^{10}}{36}}$

65. $\sqrt{\dfrac{3}{a^4}}$

66. $\sqrt{\dfrac{5}{b^2}}$

67. $\sqrt{\dfrac{x^5}{y^8}}$

68. $\sqrt{\dfrac{a^7}{b^6}}$

69. $\sqrt{\dfrac{20x^2y^5}{z^4}}$

70. $\sqrt{\dfrac{24a^3b^6}{c^2}}$

71. $\sqrt{\dfrac{105x^2}{7x^6}}$

72. $\sqrt{\dfrac{30y}{3y^5}}$

73. $\sqrt{\dfrac{100a^3}{16a}}$

74. $\sqrt{\dfrac{36b^{10}}{144b^4}}$

75. $\sqrt{\dfrac{64n^9}{9n^4}}$

76. $\sqrt{\dfrac{81m^{11}}{4m^6}}$

Mixed Practice

In Problems 77–96, simplify each expression. Assume all variables represent positive real numbers.

77. $\sqrt{63}$

78. $\sqrt{54}$

79. $\sqrt{\dfrac{12}{25}}$

80. $\sqrt{\dfrac{18}{49}}$

81. $\sqrt{144a^6}$

82. $\sqrt{225b^4}$

83. $\sqrt{\dfrac{10x^5}{90}}$

84. $\sqrt{\dfrac{6}{24y^8}}$

85. $\sqrt{12a^4b^2c^6}$

86. $\sqrt{18x^2y^8z^8}$

87. $\sqrt{\dfrac{4x^7}{y^{12}}}$

88. $\sqrt{\dfrac{81x^3}{y^4}}$

89. $\sqrt{27a^{25}}$

90. $\sqrt{32x^{49}}$

91. $\sqrt{4^2 + 6^2}$

92. $\sqrt{2^2 + 8^2}$

93. $\dfrac{-4 - \sqrt{162}}{6}$

94. $\dfrac{-6 + \sqrt{48}}{8}$

95. $\dfrac{7 - \sqrt{98}}{14}$

96. $\dfrac{-6 + \sqrt{108}}{6}$

Applying the Concepts

△ *In Problems 97–104, use the following information. According to the Pythagorean Theorem, the length of the hypotenuse of a right triangle is $c = \sqrt{a^2 + b^2}$, where c is the length of the hypotenuse and a and b are the lengths of the legs. The formula $c = \sqrt{a^2 + b^2}$ can also be used to find the diagonal of a square or a rectangle. Find the missing length, x, and simplify the result.*

97.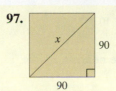

98.

99.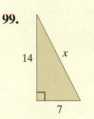

100.

101.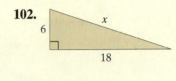

102. (image)

103.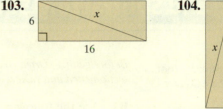

104.

In Problems 105 and 106, use the following information. The frequency of a wave is the number of full oscillations it completes per unit of time. For instance, if you stood on a pier in the Pacific Ocean and counted the number of times that a wave went past you within a certain amount of time, you could calculate the wave frequency. Frequency is measured in oscillations per second. One oscillation per second is equal to 1 hertz, Hz. The frequency of an object attached to a spring can be calculated using the formula

$$f = \dfrac{1}{2\pi}\sqrt{\dfrac{k}{m}}$$ *cycles per second. In this formula, k is a constant that depends on the material that is oscillating, and m stands for the mass of the object.*

105. Suppose you place a 5-kg weight at the end of a spring.
(a) If the constant k is 120, what is the frequency of the vibration of the spring? Write your result as a simplified radical, leaving π in your answer.

(b) Approximate this value to the nearest hundredth.

106. Suppose you place a 16-kg weight at the end of a spring.
(a) If the constant k is 200, what is the frequency of the vibration of the spring? Write your result as

a simplified radical, leaving π in your answer.

 (b) Approximate this value to the nearest hundredth.

Extending the Concepts

In Problems 107–110, find the missing expression. Assume all variables represent positive real numbers.

107. $\sqrt{?} = x^{12}$

108. $\sqrt{?} = a^{20}$

109. $\sqrt{?} = 4s^{9n}$

110. $\sqrt{?} = 9p^{16m}$

Explaining the Concepts

111. Why do we state that $y > 0$ when simplifying $\sqrt{\dfrac{x^4}{y}}$?

112. Explain how you know when a radical expression is completely simplified.

113. Is the following simplified? If not, complete the problem by simplifying. Is there an alternative method for simplifying the quotient? If so, explain the steps you would use to simplify.

$$\sqrt{\frac{315}{5}} = \frac{\sqrt{315}}{\sqrt{5}} = \frac{\sqrt{9 \cdot 35}}{\sqrt{5}} = \frac{3\sqrt{35}}{\sqrt{5}}$$

114. Example 7 and Example 8 show two different approaches to simplifying a radical that contains a quotient. Explain what to look for in the radicand when deciding which method to use.

8.3 Adding and Subtracting Square Roots

Objectives

1 Add and Subtract Square Root Expressions with Like Square Roots

2 Add and Subtract Square Root Expressions with Unlike Square Roots

Are You Ready for This Section?

Before getting started, take the following readiness quiz. If you get a problem wrong, go to the section cited and review the material.

R1. Are the following like terms? [Section 1.8, pp. 66–67]
  (a) $3a$ and $5a$ **(b)** $4m$ and $-6n$

R2. Add: $(3x^2 - 2x + 4) + (x^2 - 4x - 1)$ [Section 5.1, pp. 311–312]

R3. Subtract: $(3 - 5k + k^2) - (6 - 3k - 4k^2)$ [Section 5.1, pp. 311–314]

R4. Simplify: $3x(x - 7)$ [Section 5.3, pp. 323–324]

▶ **1** Add and Subtract Square Root Expressions with Like Square Roots

> **In Words**
>
> Remember, like terms are terms that have the same variables and the same exponent on each variable. Like square roots have the same radicand.

Definition

Square root expressions are **like square roots** if each square root has the same radicand.

For example, $5\sqrt{7}$ and $3\sqrt{7}$ are like square roots because they have the same radicand, 7. Additionally, $3\sqrt{2x}$ and $-5\sqrt{2x}$ are like square roots. Do you see why? However, $\sqrt{3}$ and $\sqrt{11}$ are not like square roots.

To add or subtract square root expressions, we combine like square roots using the Distributive Property "in reverse." Compare the approach for adding square root expressions to the approach for combining like terms from Section 1.8.

Combining Like Terms	Adding Like Square Roots
$6x + 8x = (6 + 8)x$	$6\sqrt{3} + 8\sqrt{3} = (6 + 8)\sqrt{3}$
$= 14x$	$= 14\sqrt{3}$

So all we are doing here is using an old technique in a new way.

EXAMPLE 1 **Adding and Subtracting Square Root Expressions with Like Square Roots**

Add or subtract, as indicated.

(a) $5\sqrt{3} + 9\sqrt{3}$ (b) $2\sqrt{7} - 3\sqrt{7} + 9\sqrt{5}$

Solution

(a) These terms are like square roots because they both have the radicand 3.

Apply the Distributive Property "in reverse": $5\sqrt{3} + 9\sqrt{3} = (5 + 9)\sqrt{3}$

$$= 14\sqrt{3}$$

Work Smart

Remember, that to add or subtract square roots, the square roots must have the same radicand.

(b) The first two terms are like square roots, so we find their difference using the Distributive Property.

$$2\sqrt{7} - 3\sqrt{7} + 9\sqrt{5} = (2 - 3)\sqrt{7} + 9\sqrt{5}$$
$$= (-1)\sqrt{7} + 9\sqrt{5}$$
$$= -\sqrt{7} + 9\sqrt{5}$$

EXAMPLE 2 **Adding and Subtracting Square Root Expressions with Radicands That Contain Variables**

Perform the indicated operations: $4\sqrt{11p} + 9\sqrt{11p} - 5\sqrt{11p}$

Solution

Apply the Distributive Property "in reverse": $4\sqrt{11p} + 9\sqrt{11p} - 5\sqrt{11p} = (4 + 9 - 5)\sqrt{11p}$

$$= 8\sqrt{11p}$$

Quick ✓

1. Square root expressions are like square roots if each square root has the same _____.

In Problems 2–5, add or subtract, as indicated.

2. $6\sqrt{3} + 8\sqrt{3}$ 3. $7\sqrt{11} - 8\sqrt{11} + 2\sqrt{10}$
4. $3\sqrt{13} + 4\sqrt{6} - 11\sqrt{13} - 3\sqrt{6}$ 5. $4\sqrt{5x} - 5\sqrt{5x} + 13\sqrt{5x}$

▶ ❷ **Add and Subtract Square Root Expressions with Unlike Square Roots**

What if the square root expressions to be added or subtracted aren't like square roots? In this case, we try to simplify the square roots so that they are like. Then we add or subtract the like square roots.

EXAMPLE 3 **Adding Unlike Square Roots**

Add: $\sqrt{12} + \sqrt{75}$

Solution

The radicands are different. Simplify each square root. Remember, to simplify, begin by writing each square root as the product of factors, with one of the factors a perfect square.

Work Smart

$\sqrt{a} + \sqrt{b} \neq \sqrt{a + b}$

For example,

$\sqrt{12} + \sqrt{75} \neq \sqrt{87}$

4 and 25 are perfect squares: $\sqrt{12} + \sqrt{75} = \sqrt{4 \cdot 3} + \sqrt{25 \cdot 3}$

Simplify each radical: $= 2\sqrt{3} + 5\sqrt{3}$

Add like radicals: $= 7\sqrt{3}$

EXAMPLE 4   **Subtracting Unlike Square Roots**

Subtract: $\dfrac{1}{2}\sqrt{32} - 2\sqrt{72}$

Solution

The radicands are different, so simplify the square roots to obtain like square roots.

16 and 36 are perfect squares: $\dfrac{1}{2}\sqrt{32} - 2\sqrt{72} = \dfrac{1}{2}\sqrt{16\cdot 2} - 2\sqrt{36\cdot 2}$

Simplify each radical: $= \dfrac{1}{2}\cdot 4\sqrt{2} - 2\cdot 6\sqrt{2}$

$= 2\sqrt{2} - 12\sqrt{2}$

Subtract like radicals: $= -10\sqrt{2}$ ●

Quick ✓

6. *True or False* $\sqrt{5} + \sqrt{6} = \sqrt{5 + 6}$

7. *True or False* $\sqrt{8}$ and $\sqrt{18}$ cannot be added because the radicands are different.

In Problems 8–10, add or subtract, as indicated.

8. $\sqrt{50} + \sqrt{32}$ **9.** $\sqrt{48} - \sqrt{108}$ **10.** $\sqrt{45} + \sqrt{20} - \dfrac{1}{2}\sqrt{72}$

EXAMPLE 5   **Subtracting Unlike Square Roots That Contain Variables**

Find the difference: $x\sqrt{18} - 2\sqrt{128x^2}$, $x > 0$

Solution

The radicands are different, so simplify each square root and then combine like radicals.

9 and 64 are perfect squares: $x\sqrt{18} - 2\sqrt{128x^2} = x\sqrt{9\cdot 2} - 2\sqrt{64x^2\cdot 2}$

Simplify each radical: $= 3x\sqrt{2} - 2\cdot 8x\sqrt{2}$

$= 3x\sqrt{2} - 16x\sqrt{2}$

Distributive Property "in reverse": $= (3x - 16x)\sqrt{2}$

Combine like radicals: $= -13x\sqrt{2}$ ●

Quick ✓

In Problems 11 and 12, add or subtract, as indicated. All variables are positive real numbers.

11. $\sqrt{48z} + \sqrt{108z} - \sqrt{12z}$ **12.** $\dfrac{1}{3}\sqrt{108a^3} - 2a\sqrt{75a}$

8.3 Exercises

Exercise numbers in green have complete video solutions in MyMathLab.

*Problems **1–12** are the Quick ✓s that follow the **EXAMPLES**.*

Building Skills

In Problems 13–26, add or subtract the square root expressions.
See Objective 1.

13. $\sqrt{3} + 5\sqrt{3}$ **14.** $\sqrt{2} + 7\sqrt{2}$

15. $a\sqrt{2a} + a\sqrt{2a}$ **16.** $k\sqrt{11k} + k\sqrt{11k}$

17. $\sqrt{15} + \sqrt{7}$

18. $\sqrt{3} + \sqrt{10}$

19. $5\sqrt{19} - 5\sqrt{19}$

20. $\sqrt{34} - \sqrt{34}$

21. $4x\sqrt{2} - 5x\sqrt{2}$

22. $8y\sqrt{3} - 9y\sqrt{3}$

23. $-3\sqrt{6} - 6\sqrt{11}$

24. $-11\sqrt{2} - 13\sqrt{7}$

25. $-\sqrt{22} - (-3\sqrt{22})$ **26.** $-14\sqrt{30} - (-\sqrt{30})$

In Problems 27–44, add or subtract the square root expressions. Assume that variables represent positive real numbers. See Objective 2.

27. $\sqrt{12} - \sqrt{27}$ **28.** $\sqrt{8} - \sqrt{32}$

29. $-\sqrt{75} + 6\sqrt{12}$ **30.** $-\sqrt{8} + 3\sqrt{18}$

31. $7\sqrt{5xy} + 3\sqrt{45xy}$ **32.** $-16\sqrt{2ab} + 7\sqrt{18ab}$

33. $2\sqrt{3} - \frac{1}{2}\sqrt{300}$ **34.** $7\sqrt{6} - \frac{1}{3}\sqrt{54}$

35. $6\sqrt{60} + 3\sqrt{135}$ **36.** $2\sqrt{27} + 5\sqrt{48}$

37. $4\sqrt{20} + \sqrt{45} - 2\sqrt{24}$ **38.** $-3\sqrt{63} + 2\sqrt{28} - 4\sqrt{49}$

39. $\sqrt{54x^3} - \sqrt{96x^3}$ **40.** $\sqrt{80y^5} - \sqrt{20y^5}$

41. $\frac{1}{3}\sqrt{108a^3} - 2a\sqrt{75a}$ **42.** $\frac{5}{2}\sqrt{300a^5} - a^2\sqrt{27a}$

43. $\frac{2}{5}\sqrt{20} - \frac{4}{3}\sqrt{45}$ **44.** $\frac{3}{5}\sqrt{27} - \frac{2}{3}\sqrt{12}$

Mixed Practice

In Problems 45–80, add or subtract, as indicated. Assume all variables represent positive real numbers.

45. $15\sqrt{14} - 8\sqrt{14}$ **46.** $18\sqrt{6} - 20\sqrt{6}$

47. $3\sqrt{2} - 3\sqrt{2}$ **48.** $4\sqrt{5} - 4\sqrt{5}$

49. $y^2\sqrt{y} - 6y^2\sqrt{y}$ **50.** $9x\sqrt{7} - 12x\sqrt{7}$

51. $15\sqrt{6} - 14\sqrt{6} + 9\sqrt{5}$ **52.** $8\sqrt{x} + \sqrt{x} + \sqrt{x^2}$

53. $\sqrt{2x} + 8\sqrt{2x} - 5\sqrt{2}$ **54.** $13\sqrt{3p} - 6\sqrt{3p} - 4\sqrt{3}$

55. $2\sqrt{5n^3} - 4n\sqrt{5n}$ **56.** $6\sqrt{2a^3} - 15a\sqrt{2a}$

57. $3\sqrt{20} - 3\sqrt{125} + 4\sqrt{45}$

58. $-2\sqrt{96} + 4\sqrt{24} - 3\sqrt{150}$

59. $2\sqrt{75} + 6\sqrt{12} - \sqrt{48}$

60. $-2\sqrt{243} + 8\sqrt{48} - \sqrt{108}$

61. $5\sqrt{72} - 4\sqrt{243} - \sqrt{288}$

62. $4\sqrt{300} + 3\sqrt{108} + 2\sqrt{242}$

63. $5\sqrt{32y^4} - 4\sqrt{50y^4}$

64. $3\sqrt{12x^2} - 2\sqrt{27x^2}$

65. $\sqrt{45x^2y} + x\sqrt{20y} - \sqrt{125x^2}$

66. $\sqrt{63xy^2} + y\sqrt{28x} - \sqrt{20xy^2}$

67. $2a^2\sqrt{40a^5} - a\sqrt{90a^7} + 3\sqrt{12a^3}$

68. $3n^2\sqrt{54n^4} - 2n\sqrt{150n^6} + 2\sqrt{24n^9}$

69. $3 + 2\sqrt{8} + 4 + 3\sqrt{18}$

70. $\sqrt{5} - 5 + 4 - 3\sqrt{20}$

71. $-1 - 2\sqrt{27} + 4\sqrt{48} - \sqrt{75}$

72. $2 + 3\sqrt{8} - \sqrt{32} - 2\sqrt{18}$

73. $\frac{1}{2}\sqrt{3} + \frac{2}{3}\sqrt{3}$ **74.** $\frac{1}{4}\sqrt{5} + \frac{5}{6}\sqrt{5}$

75. $\sqrt{\frac{3}{25}} + \sqrt{\frac{3}{16}}$ **76.** $\sqrt{\frac{2}{9}} + \sqrt{\frac{2}{49}}$

77. $4\sqrt{\frac{15}{9}} - 2\sqrt{\frac{15}{16}}$ **78.** $3\sqrt{\frac{3}{4}} - \frac{1}{2}\sqrt{\frac{3}{16}}$

79. $\frac{2}{3}\sqrt{\frac{18}{3}} + \frac{3}{2}\sqrt{\frac{120}{5}}$

80. $\frac{4}{3}\sqrt{\frac{18}{6}} + \frac{1}{2}\sqrt{\frac{72}{6}}$

Applying the Concepts

△ **81.** Find the perimeter of a square that measures $3\sqrt{12}$ cm on each side.

△ **82.** Find the perimeter of a square that measures $4\sqrt{18}$ ft on each side.

△ **83.** Find the perimeter of a rectangle that measures $3\sqrt{18}$ inches on one side and $4\sqrt{48}$ inches on the other.

△ **84.** Find the perimeter of a rectangle that measures $2\sqrt{45}$ meters on one side and $9\sqrt{20}$ meters on the other.

△ **85. Building a Shed** Alex is planning to strengthen the side of a wooden shed, which measures 10 feet long and 6 feet tall, by purchasing two metal bars and constructing a support in the shape of an X. If he buys one metal bar and cuts it into 2 pieces to make the X, what length of metal, to the nearest tenth of a foot, does he need to buy? (*Hint:* Use $c = \sqrt{a^2 + b^2}$, where $a = 10$ feet and $b = 6$ feet.)

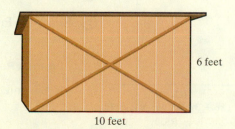

6 feet

10 feet

△ **86. Strengthening a Gate** Barry sees his neighbor, Alex (from Problem 85), working on his shed and thinks that he should use the same plan to strengthen his gate. If his gate measures 4 feet by 6 feet, how long, to the nearest tenth of a foot, is the metal bar that he needs to buy?

In Problems 87 and 88, find the perimeter of each figure.

△ **87.**

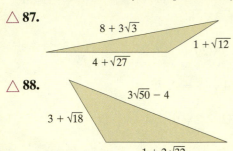

$8 + 3\sqrt{3}$

$1 + \sqrt{12}$

$4 + \sqrt{27}$

△ **88.**

$3\sqrt{50} - 4$

$3 + \sqrt{18}$

$1 + 2\sqrt{32}$

Explaining the Concepts

89. Look at the problem below. Tell which terms are like terms and which terms are unlike terms. Then, tell whether the problem has been correctly simplified. If not, explain how you would simplify the problem.

$$6 - 2\sqrt{x} - 10 + 5\sqrt{x} = 4\sqrt{x} - 5\sqrt{x} = -\sqrt{x}$$

90. Give an example of two square roots that are like and two that are unlike. Make up a problem

that requires adding or subtracting these four expressions. Explain the solution to your problem.

91. Explain why $\sqrt{a \cdot b} = \sqrt{a} \cdot \sqrt{b}$, but $\sqrt{a + b} \neq \sqrt{a} + \sqrt{b}$.

The symbol indicating the square root of a number, $\sqrt{\ }$, did not become commonplace until the seventeenth century. In ancient times, Arab scholars thought of numbers that were perfect squares as growing out of a root in the ground. Later translations of Arabic work used the word radix, *meaning root, when dealing with these numbers. Today, mathematicians still use the phrase "extracting roots" when simplifying radicals.*

Over the centuries, many different symbols have been used to indicate the square root of a number. Scholars writing in Latin used R, as it was derived from the word radix. *During the sixteenth century, all of the following were used to indicate the square root of a number: $\sqrt{\ } \sqrt{\ } \sqrt{\ } \sqrt{\ }$. The $\sqrt{\ }$ symbol appeared in the work of Christoff Rudolff in 1525 and is the basis of the square root symbol used today, although it took until the seventeenth century to become the standard notation.*

92. Math history has many interesting stories to tell. Go to the library or the Internet and research a person, notation, or concept. Topics of interest might include the Pythagorean Theorem and the Pythagorean Society, ancient numeration systems used by the Egyptians or Mayans, early algebra, prime numbers, and the origins of zero. Write a brief report on your findings.

8.4 Multiplying Expressions with Square Roots

Objectives

❶ Find the Product of Square Roots Containing One Term

❷ Find the Product of Square Roots Using the Distributive Property

❸ Find the Product of Square Roots Using FOIL

❹ Find the Product of Square Roots Using Special Products: $(A + B)^2$, $(A - B)^2$, and $(A + B)(A - B)$

Are You Ready for This Section?

Before getting started, take the following readiness quiz. If you get a problem wrong, go to the section cited and review the material.

R1. Find the product: $(3x)^2$ [Section 5.2, pp. 320–321]

R2. Find the product: $3(2x^2 - 4x + 1)$ [Section 5.3, pp. 323–324]

R3. Find the product: $(2a + 3)(3a - 4)$ [Section 5.3, pp. 324–327]

R4. Find the product: $(5b + 2)^2$ [Section 5.3, pp. 328–329]

R5. Find the product: $(3y - 4)(3y + 4)$ [Section 5.3, pp. 327–328]

In this section, we multiply square roots using the same techniques we used to find polynomial products in Section 5.3. **Throughout the section, assume that all variables represent nonnegative real numbers.**

❶ Find the Product of Square Roots Containing One Term

▶ Recall the Product Rule for Square Roots from Section 8.2:

Product Rule for Square Roots

If $\sqrt{a}$ and $\sqrt{b}$ are nonnegative real numbers, then

$$\sqrt{a} \cdot \sqrt{b} = \sqrt{ab}$$

EXAMPLE 1 **Using the Product Rule to Multiply Square Roots**

Find each product.

(a) $\sqrt{5} \cdot \sqrt{7}$

(b) $\sqrt{3} \cdot \sqrt{2a}$

Solution

(a) $\sqrt{5} \cdot \sqrt{7} = \sqrt{5 \cdot 7}$
$= \sqrt{35}$

(b) $\sqrt{3} \cdot \sqrt{2a} = \sqrt{3 \cdot 2a}$
$= \sqrt{6a}$

> **Quick ✔**
>
> **1.** If $\sqrt{a}$ and $\sqrt{b}$ are nonnegative real numbers, then $\sqrt{a} \cdot \sqrt{b} = $ ___.
>
> *In Problems 2 and 3, find the product.*
>
> **2.** $\sqrt{3} \cdot \sqrt{5}$ **3.** $\sqrt{2z} \cdot \sqrt{7}$

Sometimes after two radicands are multiplied, the resulting square root can be simplified. Remember, to simplify square roots, look for factors of the radicand that are perfect squares.

EXAMPLE 2 **Multiplying and Simplifying Square Root Expressions**

Find the product and simplify, if possible.

(a) $\sqrt{3} \cdot \sqrt{12}$

(b) $\sqrt{6} \cdot \sqrt{15}$

Solution

(a) $\sqrt{a} \cdot \sqrt{b} = \sqrt{ab}$: $\sqrt{3} \cdot \sqrt{12} = \sqrt{3 \cdot 12}$
$= \sqrt{36}$
$= 6$

(b) $\sqrt{a} \cdot \sqrt{b} = \sqrt{ab}$: $\sqrt{6} \cdot \sqrt{15} = \sqrt{6 \cdot 15}$
$= \sqrt{90}$

9 is the largest perfect square factor of 90: $= \sqrt{9 \cdot 10}$

$\sqrt{a \cdot b} = \sqrt{a} \cdot \sqrt{b}$: $= \sqrt{9} \cdot \sqrt{10}$
$= 3\sqrt{10}$

EXAMPLE 3 **Multiplying and Simplifying Square Root Expressions Containing Variables**

Find the product and simplify, if possible. Assume all variables represent nonnegative real numbers.

(a) $\sqrt{6p} \cdot \sqrt{2p^3}$

(b) $\sqrt{10m^2} \cdot \sqrt{5m^3}$

Solution

(a) $\sqrt{a} \cdot \sqrt{b} = \sqrt{ab}$: $\sqrt{6p} \cdot \sqrt{2p^3} = \sqrt{6p \cdot 2p^3}$
$= \sqrt{12p^4}$

4 is a perfect square factor of 12: $= \sqrt{4p^4 \cdot 3}$

$\sqrt{a \cdot b} = \sqrt{a} \cdot \sqrt{b}$: $= \sqrt{4p^4} \cdot \sqrt{3}$

$\sqrt{4p^4} = 2p^2$: $= 2p^2\sqrt{3}$

(b) $\sqrt{a} \cdot \sqrt{b} = \sqrt{ab}$: $\sqrt{10m^2} \cdot \sqrt{5m^3} = \sqrt{50m^5}$

25 is the largest perfect square factor of 50: $= \sqrt{25m^4 \cdot 2m}$

$\sqrt{a \cdot b} = \sqrt{a} \cdot \sqrt{b}$: $= \sqrt{25m^4} \cdot \sqrt{2m}$

$\sqrt{25m^4} = 5m^2$: $= 5m^2\sqrt{2m}$

▶ **EXAMPLE 4** **Multiplying and Simplifying Expressions with Square Roots**

Multiply and simplify: $\sqrt{2w^3} \cdot \sqrt{12w^6}$. Assume all variables represent nonnegative real numbers.

Solution

Sometimes more than one approach will lead to the correct solution.

METHOD 1:

$\sqrt{a} \cdot \sqrt{b} = \sqrt{ab}$: $\sqrt{2w^3} \cdot \sqrt{12w^6} = \sqrt{24w^9}$

$\phantom{\sqrt{a} \cdot \sqrt{b} = \sqrt{ab}: \sqrt{2w^3} \cdot \sqrt{12w^6}} = \sqrt{4w^8 \cdot 6w}$

$\sqrt{a \cdot b} = \sqrt{a} \cdot \sqrt{b}$: $= \sqrt{4w^8} \cdot \sqrt{6w}$

$\sqrt{4w^8} = 2w^4$: $= 2w^4\sqrt{6w}$

METHOD 2:

Simplify first: $\sqrt{2w^3} \cdot \sqrt{12w^6} = \sqrt{w^2 \cdot 2w} \cdot \sqrt{4w^6 \cdot 3}$

$\sqrt{a \cdot b} = \sqrt{a} \cdot \sqrt{b}$: $= \sqrt{w^2} \cdot \sqrt{2w} \cdot \sqrt{4w^6} \cdot \sqrt{3}$

$\sqrt{w^2} = w; \sqrt{4w^6} = 2w^3$: $= w\sqrt{2w} \cdot 2w^3\sqrt{3}$

Use Commutative Property of

Multiplication: $= w \cdot 2w^3 \cdot \sqrt{2w} \cdot \sqrt{3}$

Multiply: $= 2w^4\sqrt{6w}$

Which approach do you like better?

Work Smart

Remember, for a radical to be in simplified form, the exponent on the variable in the radicand must be less than 2.

> **Quick** ✓
>
> *In Problems 4–6, multiply and simplify. Assume all variables represent nonnegative real numbers.*
>
> **4.** $\sqrt{15b} \cdot \sqrt{5b}$ 　　　**5.** $\sqrt{24k} \cdot \sqrt{3k^3}$ 　　　**6.** $\sqrt{2z^3} \cdot \sqrt{10z^4}$

What happens when we square an expression containing a square root? Let's see.

EXAMPLE 5 **Squaring a Square Root**

Simplify:

 (a) $(\sqrt{3})^2$ 　　　　　　　　　　　　**(b)** $(-\sqrt{7})^2$

Solution

 (a) $(\sqrt{3})^2 = \sqrt{3} \cdot \sqrt{3}$ 　　　　　**(b)** $(-\sqrt{7})^2 = (-\sqrt{7}) \cdot (-\sqrt{7})$

$\phantom{(a)(\sqrt{3})^2} = \sqrt{9}$ 　　　　　　　　　　　$\phantom{(b)(-\sqrt{7})^2} = \sqrt{49}$

$\phantom{(a)(\sqrt{3})^2} = 3$ 　　　　　　　　　　　　$\phantom{(b)(-\sqrt{7})^2} = 7$

Work Smart

$(a)^2 = a \cdot a$ so $(\sqrt{a})^2 = \sqrt{a} \cdot \sqrt{a}$.

We can generalize the results of Example 5 as follows:

> **Squaring a Square Root**
>
> $(\sqrt{a})^2 = a$ and $(-\sqrt{a})^2 = a$ for any real number $a \geq 0$.

> **Quick** ✓
>
> **7.** If $a \geq 0$, then $(\sqrt{a})^2 = \_\_$ and $(-\sqrt{a})^2 = \_\_$.
>
> *In Problems 8 and 9, simplify the expression.*
>
> **8.** $(\sqrt{11})^2$ 　　　　　　　**9.** $(-\sqrt{31})^2$

EXAMPLE 6 **Multiplying Square Root Expressions**

Find each product:

    (a) $(3\sqrt{5})(2\sqrt{5})$ **(b)** $(7\sqrt{2})(-4\sqrt{12})$

Solution

For each problem, first use the Commutative Property of Multiplication to rearrange factors.

$$\textbf{(a)}\quad (3\sqrt{5})(2\sqrt{5}) = 3\cdot 2\cdot \sqrt{5}\cdot \sqrt{5}$$
$$= 6\cdot (\sqrt{5})^2$$
$$(\sqrt{a})^2 = a:\quad = 6\cdot 5$$
$$= 30$$

$$\textbf{(b)}\qquad\qquad (7\sqrt{2})(-4\sqrt{12}) = 7\cdot(-4)\cdot\sqrt{2}\cdot\sqrt{12}$$
$$\sqrt{a}\cdot\sqrt{b} = \sqrt{a\cdot b}:\quad = -28\cdot\sqrt{24}$$
$$\text{4 is the largest perfect square factor of 24:}\quad = -28\sqrt{4\cdot 6}$$
$$\sqrt{4\cdot 6} = \sqrt{4}\cdot\sqrt{6} = 2\sqrt{6}:\quad = -28\cdot 2\sqrt{6}$$
$$= -56\sqrt{6}$$

> **Quick** ✔
>
> *In Problems 10–12, find the product.*
>
> **10.** $(7\sqrt{3})(\sqrt{3})$ **11.** $(2\sqrt{5})(-3\sqrt{10})$ **12.** $(-2\sqrt{11})^2$

▶ ❷ Find the Product of Square Roots Using the Distributive Property

Now we multiply square root expressions using the Distributive Property.

EXAMPLE 7 **Multiplying Square Roots Using the Distributive Property**

Multiply and simplify:

    (a) $2(4 + \sqrt{3})$ **(b)** $-\sqrt{2}(4 - 3\sqrt{2})$

Solution

    (a) Use the Distributive Property and multiply each term in the parentheses by 2.

$$2(4 + \sqrt{3}) = 2\cdot 4 + 2\cdot \sqrt{3}$$
$$\text{Multiply:} = 8 + 2\sqrt{3}$$

    (b) Use the Distributive Property.

$$-\sqrt{2}(4 - 3\sqrt{2}) = -\sqrt{2}\cdot 4 - (-\sqrt{2})(3\sqrt{2})$$
$$= -4\sqrt{2} + 3\cdot\sqrt{2}\cdot\sqrt{2}$$
$$\text{Multiply radicals:} = -4\sqrt{2} + 3\cdot 2$$
$$= -4\sqrt{2} + 6$$

> **Quick** ✔
>
> *In Problems 13–16, find the product.*
>
> **13.** $-3(5 - \sqrt{7})$ **14.** $\sqrt{5}(3 - \sqrt{5})$ **15.** $\sqrt{2}(-4 + \sqrt{10})$ **16.** $-\sqrt{8}(\sqrt{2} - \sqrt{6})$

▶ ❸ **Find the Product of Square Roots Using FOIL**

When we multiplied polynomials such as $(2x + 1)(x - 4)$, we used the **F**irst, **O**uter, **I**nner, **L**ast (FOIL) method as follows:

$$(2x + 1)(x - 4) = 2x \cdot x + 2x \cdot (-4) + 1 \cdot x + 1 \cdot (-4)$$
$$= 2x^2 - 8x + x - 4$$
$$= 2x^2 - 7x - 4$$

We'll use the same process to multiply square root expressions such as $(7 + \sqrt{2})(6 + \sqrt{2})$.

EXAMPLE 8 | **Multiplying Square Roots Using FOIL**

Find each product. Assume all variables represent nonnegative real numbers.

(a) $(7 + \sqrt{2})(6 + \sqrt{2})$ (b) $(3 - \sqrt{2a})(5 - \sqrt{2a})$

Solution

(a) We use the FOIL pattern to multiply.

$$(7 + \sqrt{2})(6 + \sqrt{2}) = 7 \cdot 6 + 7\sqrt{2} + 6\sqrt{2} + \sqrt{2} \cdot \sqrt{2}$$
Multiply: $= 42 + 7\sqrt{2} + 6\sqrt{2} + 2$
Combine like terms: $= 44 + 13\sqrt{2}$

(b)
$$(3 - \sqrt{2a})(5 - \sqrt{2a}) = 3 \cdot 5 - 3(\sqrt{2a}) - \sqrt{2a} \cdot 5 - \sqrt{2a}(-\sqrt{2a})$$
Multiply: $= 15 - 3\sqrt{2a} - 5\sqrt{2a} + \sqrt{4a^2}$
Combine like terms; simplify: $= 15 - 8\sqrt{2a} + 2a$ ●

Quick ✓

In Problems 17 and 18, find the product.

17. $(3 + \sqrt{7})(4 + \sqrt{7})$ **18.** $(7 - \sqrt{5n})(6 + \sqrt{5n})$

EXAMPLE 9 | **Multiplying Square Roots Using FOIL**

Find the product: $(9 + 5\sqrt{10})(1 - 3\sqrt{10})$

Solution

$$(9 + 5\sqrt{10})(1 - 3\sqrt{10}) = 9 \cdot 1 - 9(3\sqrt{10}) + 5\sqrt{10} \cdot 1 + (5\sqrt{10})(-3\sqrt{10})$$
Multiply: $= 9 - 27\sqrt{10} + 5\sqrt{10} - 15(\sqrt{10})^2$
Combine like terms; simplify: $= 9 - 22\sqrt{10} - 15 \cdot 10$
Multiply: $= 9 - 22\sqrt{10} - 150$
Combine like terms: $= -141 - 22\sqrt{10}$ ●

Quick ✓

In Problems 19 and 20, find the product.

19. $(4 - 2\sqrt{3})(3 + 5\sqrt{3})$ **20.** $(2 - 3\sqrt{7})(5 + 2\sqrt{7})$

▶ ④ **Find the Product of Square Roots Using Special Products:**
$(A + B)^2, (A - B)^2,$ **and** $(A + B)(A - B)$

We can use our special products formulas from Section 5.3 to multiply radicals. First, we use the formulas for perfect square trinomials,

$$(A + B)^2 = A^2 + 2AB + B^2 \quad \text{and} \quad (A - B)^2 = A^2 - 2AB + B^2$$

EXAMPLE 10 **Multiplying Square Roots Using Perfect Square Trinomials**

Find each product. Assume all variables represent nonnegative real numbers.

(a) $(3 + \sqrt{5})^2$ (b) $(4 - 2\sqrt{3y})^2$

Solution

(a) Use $(A + B)^2 = A^2 + 2AB + B^2$ to find the product.

$$(A + B)^2 \quad = A^2 + 2 \cdot A \cdot \quad B + B^2$$
$$(3 + \sqrt{5})^2 = 3^2 + 2 \cdot 3 \cdot \sqrt{5} + (\sqrt{5})^2$$
$$= 9 + 6\sqrt{5} + 5$$

Combine like terms: $\quad = 14 + 6\sqrt{5}$

(b) Use $(A - B)^2 = A^2 - 2AB + B^2$ to find the product.

$$(A - B)^2 \quad = A^2 - \quad 2 \cdot A \cdot B \quad + \quad B^2$$
$$(4 - 2\sqrt{3y})^2 = 4^2 - 2 \cdot 4 \cdot 2\sqrt{3y} + (2\sqrt{3y})^2$$

$(2\sqrt{3y})^2 = 2^2(\sqrt{3y})^2 = 4 \cdot 3y = 12y:$ $\quad = 16 - 16\sqrt{3y} + 12y$ ●

Quick ✓

21. *True or False* $(\sqrt{a} + \sqrt{b})^2 = (\sqrt{a})^2 + (\sqrt{b})^2$

In Problems 22–24, find the product.

22. $(8 + \sqrt{3})^2$ **23.** $(2 - \sqrt{7z})^2$ **24.** $(-2 + 3\sqrt{5})^2$

Now let's look at products in the form of the Difference of Two Squares.

$$(A + B)(A - B) = A^2 - B^2$$

EXAMPLE 11 **Multiplying Square Roots Using the Difference of Two Squares**

Find each product:

(a) $(5 + \sqrt{6})(5 - \sqrt{6})$ (b) $(2 - 3\sqrt{5})(2 + 3\sqrt{5})$

Solution

(a)
$$(A + \quad B)(A - B) \quad = A^2 - \quad B^2$$
$$(5 + \sqrt{6})(5 - \sqrt{6}) = 5^2 - (\sqrt{6})^2$$
$$= 25 - 6$$
$$= 19$$

(b)
$$(A - \quad B)(A + B) \quad = A^2 - \quad B^2$$
$$(2 - 3\sqrt{5})(2 + 3\sqrt{5}) = 2^2 - (3\sqrt{5})^2$$

$(3\sqrt{5})^2 = 3^2 \cdot (\sqrt{5})^2 = 9 \cdot 5 = 45:$ $\quad = 4 - 45$
$$= -41$$ ●

Notice that the products found in Example 11 are integers. That is, there are no square roots in the product. Square root expressions such as $5 + \sqrt{6}$ and $5 - \sqrt{6}$ are called **conjugates.** When we multiply square root expressions that are conjugates, the result never contains a square root. This result plays an important role in the next section.

Quick ✓

25. Square root expressions such as $-3 + \sqrt{11}$ and $-3 - \sqrt{11}$ are called _____.

In Problems 26–28, find the product. Assume all variables represent nonnegative real numbers.

26. $(6 + \sqrt{3})(6 - \sqrt{3})$ **27.** $(\sqrt{2} - 7)(\sqrt{2} + 7)$ **28.** $(1 + 4\sqrt{5n})(1 - 4\sqrt{5n})$

8.4 Exercises

 MyMathLab® MathXL PRACTICE

Exercise numbers in **green** have complete video solutions in MyMathLab.

*Problems **1–28** are the Quick ✓s that follow the **EXAMPLES**.*

Building Skills

In Problems 29–54, find the product and simplify, if possible. Assume all variables represent nonnegative real numbers. See Objective 1.

29. $\sqrt{3} \cdot \sqrt{3}$ **30.** $\sqrt{2} \cdot \sqrt{2}$

31. $\sqrt{7x} \cdot \sqrt{7x}$ **32.** $\sqrt{5n} \cdot \sqrt{5n}$

33. $\sqrt{3} \cdot \sqrt{13}$ **34.** $\sqrt{2} \cdot \sqrt{5}$

35. $\sqrt{2} \cdot \sqrt{7a}$ **36.** $\sqrt{5x} \cdot \sqrt{3}$

37. $-\sqrt{10} \cdot \sqrt{6}$ **38.** $-\sqrt{10} \cdot \sqrt{15}$

39. $\sqrt{24} \cdot \sqrt{72}$ **40.** $\sqrt{90} \cdot \sqrt{54}$

41. $-\sqrt{3x} \cdot \sqrt{27x^3}$ **42.** $-\sqrt{2a^2} \cdot \sqrt{50a^6}$

43. $3\sqrt{12} \cdot 4\sqrt{18}$ **44.** $\sqrt{6q^3} \cdot \sqrt{8q^4}$

45. $5\sqrt{12} \cdot 9\sqrt{18}$ **46.** $8\sqrt{15} \cdot 3\sqrt{20}$

47. $-\sqrt{45} \cdot 3\sqrt{30}$ **48.** $-4\sqrt{24} \cdot \sqrt{40}$

49. $2\sqrt{6x} \cdot 4\sqrt{8x^4}$ **50.** $10\sqrt{6n^2} \cdot 5\sqrt{18n}$

51. $(\sqrt{37})^2$ **52.** $(-\sqrt{42})^2$

53. $(-\sqrt{7x})^2$ **54.** $(\sqrt{6y})^2$

In Problems 55–62, find the product and simplify. Assume all variables represent nonnegative real numbers. See Objective 2.

55. $4(6 + 2\sqrt{5})$ **56.** $3(7\sqrt{2} - 5)$

57. $\sqrt{6}(2\sqrt{6} - 3)$ **58.** $\sqrt{8}(2 + 3\sqrt{8})$

59. $\sqrt{3}(7\sqrt{15} + 4)$ **60.** $\sqrt{5}(5 + 3\sqrt{15})$

61. $\sqrt{a}(\sqrt{2a} + 2)$ **62.** $\sqrt{b}(\sqrt{3b} - 3)$

In Problems 63–74, find the product using the FOIL method, and simplify. Assume all variables represent nonnegative real numbers. See Objective 3.

63. $(6 + \sqrt{3})(5 + \sqrt{3})$ **64.** $(8 - \sqrt{2})(7 - \sqrt{2})$

65. $(2 + \sqrt{5})(3 - \sqrt{5})$ **66.** $(4 + \sqrt{11})(2 - \sqrt{11})$

67. $(5 - \sqrt{2})(4 - \sqrt{8})$ **68.** $(8 + \sqrt{12})(1 + \sqrt{3})$

69. $(3 - 2\sqrt{x})(1 - 4\sqrt{x})$ **70.** $(7 + \sqrt{y})(2 + 5\sqrt{y})$

71. $(5 + 2\sqrt{3})(4 - 3\sqrt{2})$ **72.** $(7 + 3\sqrt{3})(2 - 3\sqrt{2})$

73. $(2\sqrt{3x} + \sqrt{x})(5\sqrt{x} - 2\sqrt{3x})$

74. $(7\sqrt{x} - 3\sqrt{2x})(2\sqrt{x} + 3\sqrt{2x})$

In Problems 75–86, find the product using special products: $(A + B)^2$, $(A - B)^2$, and $(A - B)(A + B)$, and simplify. Assume all variables represent nonnegative real numbers. See Objective 4.

75. $(4 + \sqrt{3})^2$ **76.** $(3 - \sqrt{2})^2$

77. $(3\sqrt{x} - 2)^2$ **78.** $(2\sqrt{x} + 3)^2$

79. $(3 + 2\sqrt{2})(3 - 2\sqrt{2})$ **80.** $(4 - 3\sqrt{5})(4 + 3\sqrt{5})$

81. $(5 + 2\sqrt{2})^2$ **82.** $(-1 - 3\sqrt{7})^2$

83. $(4 - 9\sqrt{5})(4 + 9\sqrt{5})$ **84.** $(2\sqrt{3} - 8)(2\sqrt{3} + 8)$

85. $(3\sqrt{5} - \sqrt{11})(3\sqrt{5} + \sqrt{11})$

86. $(5\sqrt{3} - \sqrt{7})(5\sqrt{3} + \sqrt{7})$

Mixed Practice

In Problems 87–112, perform the indicated operation and simplify. Assume all variables represent nonnegative real numbers.

87. $\sqrt{15m^5} \cdot \sqrt{3m}$ **88.** $\sqrt{18n^2} \cdot \sqrt{6n^2}$

89. $3\sqrt{2} - 4\sqrt{8}$ **90.** $7\sqrt{18} + 6\sqrt{3}$

91. $3(2\sqrt{2n} + \sqrt{8n})$ **92.** $5(\sqrt{27x} - 3\sqrt{3x})$

93. $(4\sqrt{2})^2$ **94.** $(3\sqrt{5})^2$

95. $(\sqrt{a} + \sqrt{b})^2 - \sqrt{ab}$ **96.** $(\sqrt{p} - \sqrt{q})^2 + 3\sqrt{pq}$

97. $(3\sqrt{5} - 2\sqrt{6})(5\sqrt{5} + 4\sqrt{6}) + 3\sqrt{10} \cdot \sqrt{3}$

98. $(6\sqrt{15} + 2\sqrt{5})(3\sqrt{15} - 7\sqrt{5}) + 20\sqrt{12}$

99. $(2 - \sqrt{x + 1})^2$ **100.** $(4 - \sqrt{x + 3})^2$

101. $(3\sqrt{s} - 2)^2 + 4\sqrt{9s}$ **102.** $(5 + 3\sqrt{t})^2 - 6\sqrt{25t}$

103. $-3\sqrt{5} \cdot 4\sqrt{20}$ **104.** $-6\sqrt{28} \cdot 3\sqrt{63}$

105. $\sqrt{12a} \cdot \sqrt{3a^4} + 4a^2\sqrt{a}$

106. $\sqrt{8b^2} \cdot \sqrt{2b^3} - 6b^2\sqrt{b}$

107. $\dfrac{\sqrt{27}}{2} - 3\sqrt{12}$ **108.** $\sqrt{54} - \dfrac{\sqrt{24}}{3}$

109. $(3 - \sqrt{5})(3 + \sqrt{5})$ **110.** $(2 + \sqrt{10})(2 - \sqrt{10})$

111. $\sqrt{3}(\sqrt{3} + \sqrt{12})$ **112.** $\sqrt{2}(\sqrt{18} + \sqrt{2})$

Applying the Concepts

The area A of a trapezoid is calculated with the formula $A = \dfrac{1}{2}h(B + b)$, where h is the height, B is the length of one base, and b is the length of the other base (see the figure to the right). In Problems 113–118, find the area of each trapezoid.

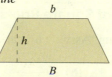

△ **113.** The height is $3\sqrt{6}$ and the bases have lengths $3\sqrt{32}$ and $10\sqrt{18}$.

△ **114.** The height is $2\sqrt{10}$ and the bases have lengths $3\sqrt{45}$ and $2\sqrt{20}$.

△ **115.** The height is $2\sqrt{14}$ and the bases have lengths $4\sqrt{128}$ and $6\sqrt{32}$.

△ **116.** The height is $5\sqrt{6}$ and the bases have lengths $4\sqrt{75}$ and $8\sqrt{48}$.

△ **117.** B is $\sqrt{27}$, h is 4, and b is $6 + 3\sqrt{12}$.

△ **118.** B is $\sqrt{98}$, h is 5, and b is $3 + 2\sqrt{8}$.

In Problems 119–122, use the formula $A = \pi r^2$ to find the exact area of each circle. Write your answer as an expression containing π.

△ **119.** Find the area of a circle if its radius is $7\sqrt{3}$ meters.

△ **120.** Find the area of a circle if its radius is $12\sqrt{6}$ yards.

△ **121.** Find the area of a circle if its diameter is $3\sqrt{56}$ inches.

△ **122.** Find the area of a circle if its diameter is $2\sqrt{84}$ centimeters.

Extending the Concepts

Heron's Formula, $A = \sqrt{s(s - a)(s - b)(s - c)}$, can be used to find the area of a triangle when the lengths of the three sides are known. See the figure below. In this formula, a, b, and c are the lengths of the three sides of the triangle, and s represents the semiperimeter $\left(s = \dfrac{1}{2}(a + b + c)\right)$. In Problems 123–126, use Heron's Formula to find the exact area of each triangle. (This formula has always been credited to Heron of Alexandria, but it is possible that it is actually the work of Archimedes.)

△ **123.** Find the area of a triangle with sides of lengths 11, 12, and 17.

△ **124.** Find the area of a triangle with sides of lengths 12, 21, and 31.

△ **125.** Find the area of a triangle with sides of lengths 15, 25, and 30.

△ **126.** Find the area of a triangle with sides of lengths 34, 42, and 68.

Explaining the Concepts

127. Is $(\sqrt{5} + \sqrt{7})^2 = (\sqrt{5})^2 + (\sqrt{7})^2$? Explain your answer.

128. Explain why the product of two conjugates never contains a square root.

129. A student saw the following question on a quiz: Find the product and simplify: $\sqrt{12x^2} \cdot \sqrt{2x^3}$. The student wrote the answer $\sqrt{24x^5}$. If this question was worth 5 points on the quiz and you were the instructor, how many points would you award the student and why?

130. Find the number to fill in the blanks that will make the statement true. Explain how you arrived at your answer.

$$(7 + \sqrt{\underline{}})(7 - \sqrt{\underline{}}) = 44$$

8.5 Dividing Expressions with Square Roots

Objectives

1 Find the Quotient of Two Square Roots

2 Rationalize a Denominator Containing One Term

3 Rationalize a Denominator Containing Two Terms

Are You Ready for This Section?

Before getting started, take the following readiness quiz. If you get a problem wrong, go to the section cited and review the material.

R1. Find the quotient: $\dfrac{55x^8}{11x}$ [Section 5.4, pp. 334–335]

R2. Find the quotient: $\sqrt{\dfrac{45x^7}{5x^3}}$ [Section 8.2, pp. 528–530]

R3. Find the quotient: $\sqrt{\dfrac{50x^6}{4x^4}}, x > 0$ [Section 8.2, pp. 528–530]

Because division by zero is not defined, throughout this section, we assume that all variables represent positive real numbers.

▶ **1** Find the Quotient of Two Square Roots

Recall from Section 8.2 that we have two requirements for a square root to be simplified:

1. The radicand cannot contain any factors that are perfect squares.

2. The radicand cannot contain any fractions.

We now introduce a third (and final) requirement for a square root to be simplified.

3. No square roots may appear as denominators in a fraction.

When square roots appear in denominators $\left(\text{as in } \dfrac{\sqrt{18}}{\sqrt{50}} \text{ and } \dfrac{\sqrt{69x^3}}{\sqrt{12x}}\right)$, we can use the Quotient Rule to write the quotient of two square roots as the square root of a quotient, as in $\dfrac{\sqrt{a}}{\sqrt{b}} = \sqrt{\dfrac{a}{b}}$.

EXAMPLE 1   **Simplifying Quotients of Square Roots**

Simplify each of the following. Assume all variables are positive real numbers.

(a) $\dfrac{\sqrt{18}}{\sqrt{50}}$ **(b)** $\dfrac{\sqrt{69x^3}}{\sqrt{12x}}$

Work Smart

The Quotient Rule states that
$$\sqrt{\dfrac{a}{b}} = \dfrac{\sqrt{a}}{\sqrt{b}}.$$

Solution

In each case, the denominator contains a square root. We will use the Quotient Rule to simplify the expression.

(a) $\dfrac{\sqrt{a}}{\sqrt{b}} = \sqrt{\dfrac{a}{b}}:$ $\dfrac{\sqrt{18}}{\sqrt{50}} = \sqrt{\dfrac{18}{50}}$

$$= \sqrt{\dfrac{9 \cdot 2}{25 \cdot 2}}$$

$$= \sqrt{\dfrac{9}{25}}$$

$$= \dfrac{3}{5}$$

(b) $\dfrac{\sqrt{69x^3}}{\sqrt{12x}} = \sqrt{\dfrac{69x^3}{12x}}$

$= \sqrt{\dfrac{23 \cdot 3 \cdot x \cdot x^2}{4 \cdot 3 \cdot x}}$

$= \sqrt{\dfrac{23\,x^2}{4}}$

$= \dfrac{\sqrt{23}\,\sqrt{x^2}}{\sqrt{4}}$

$= \dfrac{\sqrt{23}}{2}\,x$

Quick ✓

1. $\dfrac{\sqrt{a}}{\sqrt{b}} = $ _____

In Problems 2–4, simplify the quotient.

2. $\dfrac{\sqrt{108}}{\sqrt{12}}$

3. $\dfrac{\sqrt{35y}}{\sqrt{5y}}$

4. $\dfrac{\sqrt{32z^5}}{\sqrt{50z}}$

EXAMPLE 2 **Dividing Square Root Expressions with More Than One Term in the Numerator**

Find the quotient: $\dfrac{\sqrt{8} - \sqrt{12}}{\sqrt{2}}$

Solution

There are two terms in the numerator and one term in the denominator, so use the fact that $\dfrac{a + b}{c} = \dfrac{a}{c} + \dfrac{b}{c}$ to simplify.

$\dfrac{a + b}{c} = \dfrac{a}{c} + \dfrac{b}{c}:$ $\quad \dfrac{\sqrt{8} - \sqrt{12}}{\sqrt{2}} = \dfrac{\sqrt{8}}{\sqrt{2}} - \dfrac{\sqrt{12}}{\sqrt{2}}$

$\dfrac{\sqrt{a}}{\sqrt{b}} = \sqrt{\dfrac{a}{b}}:$ $\quad = \sqrt{\dfrac{8}{2}} - \sqrt{\dfrac{12}{2}}$

Divide: $\quad = \sqrt{4} - \sqrt{6}$

Simplify: $\quad = 2 - \sqrt{6}$

Quick ✓

In Problem 5, find the quotient.

5. $\dfrac{\sqrt{30} - \sqrt{75}}{\sqrt{5}}$

▶ **② Rationalize a Denominator Containing One Term**

Work Smart

Recall that we can simplify a square root to a rational number only if the radicand is a perfect square. Thus

$\sqrt{4}\ (= 2)$ and $\sqrt{\dfrac{25}{36}}\ \left(= \dfrac{5}{6}\right)$

are rational numbers, but $\sqrt{11}$ is not.

We stated earlier in this section that no square roots may appear as denominators in a fraction. In Example 1(a) of this section, we simplified $\dfrac{\sqrt{18}}{\sqrt{50}}$ using the Quotient Rule and obtained $\dfrac{3}{5}$. However, the method from Example 1(a) does not always work. For example, if we were asked to simplify $\dfrac{3}{\sqrt{2}}$, we would not be able to use the approach of Example 1 to remove the square root from the denominator. We now present a method that can be used to make denominators rational numbers.

> **Definition**
>
> Rewriting a quotient in which the denominator contains a square root that is irrational as an equivalent quotient in which the denominator is rational is called **rationalizing the denominator.**

Work Smart

Multiply the denominator and numerator by a square root that makes the radicand in the denominator a perfect square.

To rationalize the denominator, we need to make the radicand in the denominator a perfect square. For example, to simplify $\dfrac{3}{\sqrt{2}}$ we must make the denominator, $\sqrt{2}$, a rational number. We can do this by multiplying $\sqrt{2}$ by $\sqrt{2}$. The result is $\sqrt{4}$, which equals 2, a rational number! But be careful: If you multiply the denominator by $\sqrt{2}$, you must also multiply the numerator by $\sqrt{2}$.

In general, to rationalize a denominator that is a single term containing a square root, multiply the denominator and numerator by a square root that makes the radicand in the denominator a perfect square. Since we are multiplying both the numerator and the denominator by the same value, we are multiplying the radical expression by 1.

EXAMPLE 3 **How to Rationalize a Denominator Containing a Single Square Root**

Rationalize the denominator of the expression $\dfrac{1}{\sqrt{5}}$.

Step-by-Step Solution

Step 1: Multiply the numerator and denominator by a square root that makes the radicand in the denominator a perfect square.

Because $5 \cdot 5 = 25$, a perfect square, multiply the numerator and denominator of $\dfrac{1}{\sqrt{5}}$ by $\sqrt{5}$.

$$\text{Multiply } \frac{1}{\sqrt{5}} \text{ by } 1 = \frac{\sqrt{5}}{\sqrt{5}}: \quad \frac{1}{\sqrt{5}} = \frac{1}{\sqrt{5}} \cdot \frac{\sqrt{5}}{\sqrt{5}}$$

Step 2: Multiply numerators; multiply denominators.

$$= \frac{\sqrt{5}}{\sqrt{25}}$$

Step 3: Simplify.

$$= \frac{\sqrt{5}}{5}$$

EXAMPLE 4 **Rationalizing a Denominator Containing a Single Square Root**

Rationalize the denominator of each expression:

(a) $-\dfrac{9}{2\sqrt{3}}$ **(b)** $\sqrt{\dfrac{3}{32}}$

Work Smart

The process of rationalizing the denominator is the same method we use to write equivalent fractions. For example, $\dfrac{2}{3} \cdot \dfrac{3}{3} = \dfrac{6}{9}$.

Solution

(a) The denominator of the expression $-\dfrac{9}{2\sqrt{3}}$ has a radicand of 3. If we multiply 3 by 3, we get 9, a perfect square. Since the factor 2 in the denominator is already a rational number, it does not play a role in rationalizing the denominator.

$$\text{Multiply by } 1 = \frac{\sqrt{3}}{\sqrt{3}}: \quad -\frac{9}{2\sqrt{3}} = -\frac{9}{2\sqrt{3}} \cdot \frac{\sqrt{3}}{\sqrt{3}}$$

$$\text{Multiply numerators;} \atop \text{multiply denominators:} \quad = -\frac{9\sqrt{3}}{2\sqrt{9}}$$

$$= -\frac{9\sqrt{3}}{2 \cdot 3}$$

$$\text{Simplify:} \quad = -\frac{3\sqrt{3}}{2}$$

(b) First, we write the square root as the quotient of two square roots:

$\sqrt{\dfrac{3}{32}} = \dfrac{\sqrt{3}}{\sqrt{32}}$. This denominator has a radicand of 32. We multiply $\dfrac{\sqrt{3}}{\sqrt{32}}$ by

$1 = \dfrac{\sqrt{2}}{\sqrt{2}}$ because $32 \cdot 2 = 64$, a perfect square.

$$\dfrac{\sqrt{3}}{\sqrt{32}} = \dfrac{\sqrt{3}}{\sqrt{32}} \cdot \dfrac{\sqrt{2}}{\sqrt{2}}$$

Multiply numerators;
multiply denominators: $= \dfrac{\sqrt{6}}{\sqrt{64}}$

Simplify: $= \dfrac{\sqrt{6}}{8}$

Work Smart

In Example 4(b) we could also simplify $\sqrt{32}$ first so that it does not contain any factors that are perfect squares.

$$\sqrt{\dfrac{3}{32}} = \dfrac{\sqrt{3}}{\sqrt{32}} = \dfrac{\sqrt{3}}{\sqrt{16 \cdot 2}} = \dfrac{\sqrt{3}}{4\sqrt{2}} = \dfrac{\sqrt{3}}{4\sqrt{2}} \cdot \dfrac{\sqrt{2}}{\sqrt{2}} = \dfrac{\sqrt{6}}{4\sqrt{4}} = \dfrac{\sqrt{6}}{4 \cdot 2} = \dfrac{\sqrt{6}}{8}$$

Also, note that we did not multiply $\dfrac{\sqrt{3}}{\sqrt{32}}$ by $\dfrac{\sqrt{32}}{\sqrt{32}}$. Although this approach will work, it is not the most efficient approach. Try this approach to see why.

Quick ✓

6. To _____ ___ _____ means to rewrite a quotient in an equivalent form that has a rational number in the denominator.

7. *True or False* $\dfrac{9}{\sqrt{7}}$ requires multiplying by $1 = \dfrac{\sqrt{7}}{\sqrt{7}}$ to rationalize the denominator.

In Problems 8–10, rationalize the denominator of each expression.

8. $\dfrac{1}{\sqrt{2}}$　　　　　9. $-\dfrac{\sqrt{2}}{5\sqrt{3}}$　　　　　10. $\dfrac{5}{\sqrt{20}}$

▶ ❸ Rationalize a Denominator Containing Two Terms

Recall from Section 8.4 that expressions like $5 + \sqrt{6}$ and $5 - \sqrt{6}$ are called conjugates.

EXAMPLE 5　　**Determining the Conjugate of a Square Root Expression**

   (a) The conjugate of $3 + \sqrt{5}$ is $3 - \sqrt{5}$.

   (b) The conjugate of $\sqrt{7} - 5$ is $\sqrt{7} + 5$.

   (c) The conjugate of $-2 + 3\sqrt{11}$ is $-2 - 3\sqrt{11}$.

   (d) The conjugate of $2\sqrt{7} + \sqrt{6}$ is $2\sqrt{7} - \sqrt{6}$.

In Example 11(a) of Section 8.4 we found the product $(5 + \sqrt{6})(5 - \sqrt{6})$ to be 19, a rational number. The product of two conjugates containing square roots is always a rational number. To rationalize a denominator containing two terms, multiply both the numerator and denominator by the conjugate of the denominator. For example, to rationalize $\dfrac{3}{2 + \sqrt{3}}$, multiply both the numerator and the denominator by the conjugate of $2 + \sqrt{3}$, which is $2 - \sqrt{3}$. Since we are multiplying both the numerator and the denominator by $2 - \sqrt{3}$, we are multiplying $\dfrac{3}{2 + \sqrt{3}}$ by 1, where $1 = \dfrac{2 - \sqrt{3}}{2 - \sqrt{3}}$.

EXAMPLE 6 **How to Rationalize a Denominator Containing Two Terms**

Rationalize the denominator: $\dfrac{4}{2 - \sqrt{6}}$

Step-by-Step Solution

Step 1: Multiply the numerator and denominator of the quotient by the conjugate of the denominator.

Multiply $\dfrac{4}{2 - \sqrt{6}}$ by $1 = \dfrac{2 + \sqrt{6}}{2 + \sqrt{6}}$: $\dfrac{4}{2 - \sqrt{6}} = \dfrac{4}{2 - \sqrt{6}} \cdot \dfrac{2 + \sqrt{6}}{2 + \sqrt{6}}$

Step 2: Multiply numerators; multiply denominators.

$$= \dfrac{4(2 + \sqrt{6})}{(2 - \sqrt{6})(2 + \sqrt{6})}$$

$(A + B)(A - B) = A^2 - B^2$ in the denominator:
$$= \dfrac{4(2 + \sqrt{6})}{(2)^2 - (\sqrt{6})^2}$$

$$= \dfrac{4(2 + \sqrt{6})}{4 - 6}$$

$$= \dfrac{4(2 + \sqrt{6})}{-2}$$

Step 3: Simplify.

Find the quotient $\dfrac{4}{-2} = -2$: $= \dfrac{\overset{-2}{\cancel{4}}(2 + \sqrt{6})}{\underset{1}{\cancel{-2}}}$

$$= -2(2 + \sqrt{6})$$

Distribute -2: $= -4 - 2\sqrt{6}$ ●

> **Work Smart**
>
> In Example 6 we didn't use the Distributive Property in the numerator until after we had simplified the denominator. We kept the numerator in factored form so that we could easily simplify the expression.

> **Quick** ✔
>
> **11.** *True or False* The conjugate of $-5 - 2\sqrt{7}$ is $5 + 2\sqrt{7}$.
>
> *In Problems 12 and 13, rationalize the denominator.*
>
> **12.** $\dfrac{2}{2 + \sqrt{3}}$ **13.** $\dfrac{12}{3 - \sqrt{5}}$

EXAMPLE 7 **Rationalizing a Denominator Containing Two Terms**

Rationalize the denominator: $\dfrac{\sqrt{2}}{\sqrt{x} + \sqrt{5}}$

Solution

Multiply the numerator and denominator by the conjugate of $\sqrt{x} + \sqrt{5}$, which is $\sqrt{x} - \sqrt{5}$.

Work Smart: Study Skills

Can you distinguish between the different types of problems involving division presented in this chapter? Make a list of examples, such as

$\sqrt{\dfrac{100x^6}{4x^4}}, \dfrac{\sqrt{18}}{\sqrt{50}}, \dfrac{4 + \sqrt{20}}{2},$

$\dfrac{\sqrt{8} - \sqrt{12}}{\sqrt{2}}, \dfrac{1}{\sqrt{5}},$ and $\dfrac{4}{2 - \sqrt{6}}.$

Can you simplify each one?

Multiply by $1 = \dfrac{\sqrt{x} - \sqrt{5}}{\sqrt{x} - \sqrt{5}}$: $\dfrac{\sqrt{2}}{\sqrt{x} + \sqrt{5}} = \dfrac{\sqrt{2}}{\sqrt{x} + \sqrt{5}} \cdot \dfrac{\sqrt{x} - \sqrt{5}}{\sqrt{x} - \sqrt{5}}$

Multiply the numerators and denominators: $= \dfrac{\sqrt{2}(\sqrt{x} - \sqrt{5})}{(\sqrt{x})^2 - (\sqrt{5})^2}$

$(A + B)(A - B) = A^2 - B^2$ in the denominator: $= \dfrac{\sqrt{2}(\sqrt{x} - \sqrt{5})}{x - 5}$

Distribute $\sqrt{2}$: $= \dfrac{\sqrt{2x} - \sqrt{10}}{x - 5}$ ●

Quick ✔

In Problems 14–16, rationalize the denominator.

14. $\dfrac{\sqrt{3}}{4 + \sqrt{7}}$ 15. $\dfrac{\sqrt{3}}{3 - \sqrt{x}}$ 16. $\dfrac{-\sqrt{2}}{\sqrt{3} + \sqrt{2}}$

Summary

A square root expression is simplified if

- no radicand contains factors that are perfect squares.
- no radicand contains fractions.
- the denominator of a fraction contains no square roots.

8.5 Exercises

MyMathLab® PRACTICE

Exercise numbers in green have complete video solutions in MyMathLab.

*Problems **1–16** are the **Quick ✔**s that follow the **EXAMPLES**.*

Building Skills

In Problems 17–28, simplify the quotient. Assume all variables represent positive real numbers. See Objective 1.

17. $\dfrac{\sqrt{96}}{\sqrt{6}}$ 18. $\dfrac{\sqrt{108}}{\sqrt{3}}$ 19. $-\dfrac{\sqrt{5}}{\sqrt{125}}$

20. $-\dfrac{\sqrt{10}}{\sqrt{90}}$ 21. $\dfrac{\sqrt{54}}{\sqrt{3}}$ 22. $\dfrac{\sqrt{144}}{\sqrt{3}}$

23. $\dfrac{\sqrt{196}}{\sqrt{7}}$ 24. $\dfrac{\sqrt{300}}{\sqrt{2}}$ 25. $\dfrac{-\sqrt{75y^2}}{\sqrt{3y}}$

26. $\dfrac{-\sqrt{18x^3}}{\sqrt{2x}}$ 27. $\sqrt{\dfrac{128a^2b^5}{2a^2b^5}}$ 28. $\sqrt{\dfrac{147a^3b^4}{3ab^4}}$

In Problems 29–48, rationalize the denominator of each expression. Assume all variables represent positive real numbers. See Objective 2.

29. $\dfrac{3}{\sqrt{5}}$ 30. $\dfrac{9}{\sqrt{11}}$

31. $\dfrac{20}{\sqrt{6}}$ 32. $\dfrac{15}{\sqrt{21}}$

33. $-\dfrac{16}{3\sqrt{2}}$ 34. $-\dfrac{15}{2\sqrt{3}}$

35. $\dfrac{2}{-\sqrt{x}}$ 36. $\dfrac{5}{-\sqrt{y}}$

37. $\dfrac{9y}{\sqrt{2y}}$ 38. $\dfrac{3x}{\sqrt{5x}}$

39. $\dfrac{6}{\sqrt{48}}$ 40. $\dfrac{10}{\sqrt{28}}$

41. $\dfrac{6n}{\sqrt{8n^3}}$ 42. $\dfrac{4p}{\sqrt{12p^5}}$

43. $\dfrac{4\sqrt{2}}{3\sqrt{6}}$ 44. $\dfrac{5\sqrt{3}}{2\sqrt{10}}$

45. $\dfrac{2\sqrt{a}}{7\sqrt{6b}}$ 46. $-\dfrac{5\sqrt{a}}{2\sqrt{10b}}$

47. $-\dfrac{7\sqrt{5}}{2\sqrt{18}}$ 48. $\dfrac{3\sqrt{3}}{2\sqrt{20}}$

In Problems 49–56, determine the conjugate of each expression. Then find the product of the expression and its conjugate. See Objective 3.

49. $2 + \sqrt{3}$ 50. $4 - \sqrt{5}$

51. $\sqrt{6} - 1$ 52. $\sqrt{10} + 3$

53. $-1 - 3\sqrt{14}$ 54. $-8 + 5\sqrt{21}$

55. $\sqrt{p} + 3\sqrt{q}$ 56. $\sqrt{x} - 2\sqrt{y}$

In Problems 57–76, rationalize the denominator of each expression. Assume all variables represent positive real numbers. See Objective 3.

57. $\dfrac{1}{3 + \sqrt{5}}$ 58. $\dfrac{1}{6 + \sqrt{2}}$

59. $\dfrac{9}{9 - \sqrt{3}}$ 60. $\dfrac{4}{4 - \sqrt{10}}$

61. $\dfrac{4}{\sqrt{7} + 6}$ 62. $\dfrac{12}{\sqrt{3} + 2}$

63. $\dfrac{3}{\sqrt{y} - 6}$ 64. $\dfrac{8}{\sqrt{x} - 2}$

65. $\dfrac{16}{-2\sqrt{2} + 3}$ 66. $\dfrac{9}{-2\sqrt{3} + 15}$

67. $\dfrac{\sqrt{2}}{6 - \sqrt{21}}$ 68. $\dfrac{\sqrt{3}}{4 - \sqrt{10}}$

69. $\dfrac{n}{\sqrt{n} + 1}$ 70. $\dfrac{z}{\sqrt{z} + 2}$

71. $\dfrac{\sqrt{x}}{2 + 3\sqrt{x}}$ 72. $\dfrac{\sqrt{y}}{1 + 2\sqrt{y}}$

73. $\dfrac{\sqrt{5}}{\sqrt{11} - \sqrt{6}}$ 74. $\dfrac{\sqrt{22}}{\sqrt{3} - \sqrt{5}}$

75. $\dfrac{\sqrt{2}}{\sqrt{6} - \sqrt{2}}$ 76. $\dfrac{\sqrt{3}}{\sqrt{6} - \sqrt{3}}$

Mixed Practice

In Problems 77–104, simplify the expression. Assume all variables represent positive real numbers.

77. $\dfrac{\sqrt{6} - \sqrt{12}}{\sqrt{2}}$

78. $\dfrac{\sqrt{27} + \sqrt{24}}{\sqrt{3}}$

79. $\sqrt{\dfrac{1}{2}}$

80. $\sqrt{\dfrac{1}{3}}$

81. $\dfrac{\sqrt{18} - \sqrt{20}}{2\sqrt{10}}$

82. $\dfrac{\sqrt{8} + \sqrt{15}}{3\sqrt{6}}$

83. $\dfrac{2}{3}\sqrt{18} + \dfrac{5}{2}\sqrt{8}$

84. $\dfrac{3}{5}\sqrt{50} + \dfrac{3}{4}\sqrt{32}$

85. $-\sqrt{\dfrac{10}{15}}$

86. $-\sqrt{\dfrac{6}{14}}$

87. $\dfrac{5}{14\sqrt{7}}$

88. $\dfrac{3}{20\sqrt{5}}$

89. $\dfrac{4x}{\sqrt{2x}}$

90. $\dfrac{10y}{\sqrt{5y}}$

91. $\sqrt{\dfrac{4}{3}}$

92. $\sqrt{\dfrac{2}{5}}$

93. $-4\sqrt{8} + 3\sqrt{20}$

94. $-3\sqrt{12} + 2\sqrt{27}$

95. $\sqrt{\dfrac{3}{4}} \cdot \sqrt{\dfrac{6}{15}}$

96. $\sqrt{\dfrac{5}{8}} \cdot \sqrt{\dfrac{20}{3}}$

97. $\dfrac{2}{2 + \sqrt{3}}$

98. $\dfrac{4}{4 - \sqrt{5}}$

99. $\dfrac{1}{\sqrt{50}} + \sqrt{\dfrac{8}{25}}$

100. $\dfrac{1}{\sqrt{27}} + \sqrt{\dfrac{12}{81}}$

101. $\sqrt{\dfrac{5x}{y}} \cdot \sqrt{\dfrac{1}{35x^4}}$

102. $\sqrt{\dfrac{3x}{y}} \cdot \sqrt{\dfrac{y^2}{2x^4}}$

103. $(3 - \sqrt{x+2})^2$

104. $(4 + \sqrt{y-3})^2$

Applying the Concepts

In Problems 105–112, use the following information. An answer that contains a square root is said to be "the exact answer" when it contains a radical in simplified form. Irrational numbers, such as some of the square roots we have seen in this chapter, do not have an exact decimal representation. Any decimal form of an answer containing irrational numbers is only an approximation. For each expression, do the following:

 (a) *Use your calculator to find the approximate value of the expression, to four decimal places.*

   (b) *Rationalize the denominator to find the exact value of this expression.*

 (c) *Use your calculator to find the approximate value of (b).*

   (d) *Compare your results.*

105. $\dfrac{1}{2 + \sqrt{3}}$

106. $\dfrac{1}{3 - \sqrt{5}}$

107. $\dfrac{5}{2\sqrt{3} - 1}$

108. $\dfrac{2}{3\sqrt{5} + 2}$

109. $\dfrac{\sqrt{8}}{\sqrt{2} - 3}$

110. $\dfrac{\sqrt{12}}{\sqrt{3} - 4}$

111. $\dfrac{\sqrt{6}}{3 + 5\sqrt{3}}$

112. $\dfrac{\sqrt{5}}{5 + 2\sqrt{2}}$

Extending the Concepts

In Problems 113–118, rationalize the denominator of each expression.

113. $\dfrac{1 + \sqrt{3}}{1 - \sqrt{3}}$

114. $\dfrac{2 + \sqrt{2}}{2 - \sqrt{2}}$

115. $\dfrac{\sqrt{6} - \sqrt{3}}{\sqrt{2} - \sqrt{3}}$

116. $\dfrac{\sqrt{5} - \sqrt{2}}{\sqrt{3} - \sqrt{2}}$

117. $\dfrac{2\sqrt{x} + \sqrt{y}}{\sqrt{x} + 4\sqrt{y}}$

118. $\dfrac{\sqrt{x} + 6\sqrt{y}}{3\sqrt{x} + \sqrt{y}}$

Explaining the Concepts

119. Explain why x cannot be divided out in the expression $\dfrac{3\sqrt{xy}}{x}$.

120. Explain why $\dfrac{1}{\sqrt{2}}$ can be simplified by multiplying by $1 = \dfrac{\sqrt{2}}{\sqrt{2}}$, but $\dfrac{1}{1 - \sqrt{2}}$ cannot be simplified by multiplying by $1 = \dfrac{\sqrt{2}}{\sqrt{2}}$.

121. The following question appeared on an exam: Rationalize the denominator of $\dfrac{3}{\sqrt{2} + \sqrt{3}}$. One student wrote $\dfrac{3}{\sqrt{2} + \sqrt{3}} \cdot \dfrac{\sqrt{2} + \sqrt{3}}{\sqrt{2} + \sqrt{3}}$. A second student wrote $\dfrac{3}{\sqrt{2} + \sqrt{3}} \cdot \dfrac{\sqrt{2} + \sqrt{3}}{\sqrt{2} - \sqrt{3}}$. Is either approach correct? If not, what is the correct approach?

122. There was another question on this exam that asked the student to rationalize the denominator of $\dfrac{3 + \sqrt{6}}{6 + \sqrt{2}}$. The paper you are grading has the following steps. Is this procedure correct or incorrect? Explain how you would rationalize the denominator of $\dfrac{3 + \sqrt{6}}{6 + \sqrt{2}}$.

$$\dfrac{3 + \sqrt{6}}{6 + \sqrt{2}} = \dfrac{3}{6} + \dfrac{\sqrt{6}}{\sqrt{2}} = \dfrac{1}{2} + \sqrt{3}$$

Putting The Concepts Together (Sections 8.1–8.5)

We designed these problems so that you can review Sections 8.1 through 8.5 and show your mastery of the concepts. Take time to work these problems before proceeding with the next section. The answers are located at the back of the text on page AN-30.

1. Determine whether each square root is rational or irrational. Then evaluate the square root. If the square root is irrational, express the square root as a decimal rounded to two decimal places.

   (a) $\sqrt{98}$ **(b)** $\sqrt{361}$

In Problems 2–6, simplify each of the following. Assume all variables represent positive real numbers.

2. $-2\sqrt{18}$ **3.** $\sqrt{32x^3y^6}$

4. $\sqrt{\dfrac{100x^3}{16x^5}}$ **5.** $\sqrt{121 - 4(6)(3)}$

6. $\dfrac{\sqrt{72m^6}}{\sqrt{8}}$

In Problems 7–16, perform the indicated operation and simplify. Assume all variables represent positive real numbers.

7. $5\sqrt{3} - \sqrt{12}$ **8.** $3\sqrt{50} - 6\sqrt{8}$

9. $\sqrt{12} \cdot \sqrt{18}$ **10.** $(2\sqrt{7})(-3\sqrt{7})$

11. $6\sqrt{5}(2\sqrt{10} + 3)$ **12.** $(6 - \sqrt{3})(2 - 4\sqrt{3})$

13. $(3 - 2\sqrt{5})(3 + 2\sqrt{5})$ **14.** $(\sqrt{7} - 2)^2$

15. $\dfrac{-4 + \sqrt{28}}{2}$ **16.** $\dfrac{\sqrt{90x^2}}{\sqrt{5x^6}}$

In Problems 17 and 18, rationalize the denominator.

17. $\dfrac{8}{\sqrt{24}}$ **18.** $\dfrac{12}{\sqrt{3} + 3}$

8.6 Solving Equations Containing Square Roots

Objectives

1 Determine Whether a Number Is a Solution of a Radical Equation

2 Solve Equations Containing One Square Root

3 Solve Equations Containing Two Square Roots

4 Solve Problems Modeled by Radical Equations

Are You Ready for This Section?

Before getting started, take the following readiness quiz. If you get a problem wrong, go to the section cited and review the material.

R1. Solve: $4x - 5 = 0$ [Section 2.2, pp. 91–92]

R2. Solve: $2p^2 + 4p - 16 = 0$ [Section 6.6, pp. 408–414]

R3. Find the product: $(2x - 3)^2$ [Section 5.3, pp. 328–329]

R4. Find the product: $(\sqrt{x} + 2)^2$ [Section 8.4, p. 541]

When the variable in an equation occurs in a square root, the equation is called a **radical equation.** Examples of radical equations are

$$\sqrt{3x + 1} = 5 \qquad \sqrt{x - 2} - \sqrt{2x + 5} = 2$$

▶ **1 Determine Whether a Number Is a Solution of a Radical Equation**

A number is a **solution** of a radical equation if, when we substitute the number for the variable, a true statement results.

  (**EXAMPLE 1**)   **Determining Whether a Number Is a Solution of a Radical Equation**

      (a) Determine whether $x = 5$ is a solution of the equation $\sqrt{3x + 1} = 4$.

      (b) Is $x = 1$ a solution of $\sqrt{5x - 1} = -2$?

Ready? ...Answers **R1.** $\left\{\dfrac{5}{4}\right\}$

R2. $\{-4, 2\}$ **R3.** $4x^2 - 12x + 9$
R4. $x + 4\sqrt{x} + 4$

Solution

(a) To determine whether $x = 5$ is a solution of $\sqrt{3x + 1} = 4$, substitute 5 for x.

$$\sqrt{3x + 1} = 4$$

Substitute 5 for x: $\quad \sqrt{3(5) + 1} \stackrel{?}{=} 4$

Multiply: $\quad \sqrt{15 + 1} \stackrel{?}{=} 4$

$$\sqrt{16} \stackrel{?}{=} 4$$

$$4 = 4 \quad \text{True}$$

Since $x = 5$ produces a true statement, $x = 5$ is a solution of $\sqrt{3x + 1} = 4$.

(b) Substitute 1 for x in the equation $\sqrt{5x - 1} = -2$.

$$\sqrt{5x - 1} = -2$$

Substitute 1 for x: $\quad \sqrt{5(1) - 1} \stackrel{?}{=} -2$

Multiply: $\quad \sqrt{5 - 1} \stackrel{?}{=} -2$

$$\sqrt{4} \stackrel{?}{=} -2$$

$$2 = -2 \quad \text{False}$$

Because $x = 1$ gives a false statement, $x = 1$ is not a solution of $\sqrt{5x - 1} = -2$.

> **Quick ✓**
>
> 1. An equation such as $\sqrt{x + 3} = 5$ is called a(n) _____ _____.
>
> *In Problems 2–4, determine whether the number is a solution to the given equation.*
>
> 2. $\sqrt{6x + 4} = 8$; $x = 10$
>
> 3. $\sqrt{n + 2} = n$; $n = -1$
>
> 4. $\sqrt{2y + 11} = 3$; $y = -1$

❷ Solve Equations Containing One Square Root

▶ Remember when we solved quadratic equations? We used factoring to rewrite the quadratic equation as two linear equations. When we solved rational equations, we multiplied both sides of the equation by the least common denominator and rewrote the rational equation as a linear or quadratic equation. In short, each time a problem came up, we used techniques to rewrite it as a problem we already knew how to solve.

To solve radical equations, we also rewrite the equation as one we already know how to solve. To turn a radical equation into one we already know how to solve, we use the fact that

$$(\sqrt{a})^2 = a$$

provided that $a > 0$. Examples 2 and 3 illustrate how to solve a radical equation.

(**EXAMPLE 2**)   **How to Solve a Radical Equation Containing One Radical**

Solve: $\sqrt{x - 3} = 2$

Step-by-Step Solution

Step 1: Isolate the radical.

The radical is isolated: $\quad \sqrt{x - 3} = 2$

Step 2: Square both sides of the equation.

$$(\sqrt{x - 3})^2 = 2^2$$
$$x - 3 = 4$$

Step 3: Solve the equation that results.

Add 3 to both sides: $\quad x - 3 + 3 = 4 + 3$

$$x = 7$$

Step 4: Check

$$\sqrt{x-3} = 2$$

Let $x = 7$ in the original equation: $\sqrt{7-3} \overset{?}{=} 2$

$$\sqrt{4} \overset{?}{=} 2$$

$$2 = 2 \quad \text{True}$$

The solution set is $\{7\}$.

EXAMPLE 3 **How to Solve a Radical Equation Containing One Radical**

Solve: $3\sqrt{4x+1} - 2 = 13$

Step-by-Step Solution

Step 1: Isolate the radical.

$$3\sqrt{4x+1} - 2 = 13$$

Add 2 to each side of the equation: $3\sqrt{4x+1} - 2 + 2 = 13 + 2$

$$3\sqrt{4x+1} = 15$$

Divide both sides of the equation by 3: $\sqrt{4x+1} = 5$

Step 2: Square both sides of the equation.

$$(\sqrt{4x+1})^2 = 5^2$$

$$4x + 1 = 25$$

Step 3: Solve the equation that results.

Subtract 1 from both sides: $4x + 1 - 1 = 25 - 1$

$$4x = 24$$

Divide both sides of the equation by 4: $x = 6$

Step 4: Check

$$3\sqrt{4x+1} - 2 = 13$$

Let $x = 6$ in the original equation: $3\sqrt{4(6)+1} - 2 \overset{?}{=} 13$

$$3\sqrt{25} - 2 \overset{?}{=} 13$$

$$3(5) - 2 \overset{?}{=} 13$$

$$13 = 13 \quad \text{True}$$

The solution set is $\{6\}$.

Solving a Radical Equation Containing One Radical

Step 1: Isolate the radical. That is, get the radical by itself on one side of the equation.

Step 2: Square both sides of the equation. This will eliminate the radical from the equation.

Step 3: Solve the equation that results.

Step 4: Check your answer in the original equation.

Quick ✓

5. *True or False* To solve an equation such as $\sqrt{x} + 1 = 16$, the first step should be to isolate the radical.

In Problems 6–9, solve the equation and check the solution.

6. $\sqrt{y-5} = 3$ **7.** $\sqrt{x+17} + 4 = 7$

8. $2\sqrt{p} = 12$ **9.** $4\sqrt{t-3} - 2 = 10$

Work Smart

Extraneous solutions are likely to occur when one side of the equation contains a square root expression and the other side contains a variable expression.

▶ In Section 7.7, we found that when we solve a rational equation, some apparent solutions to the equation do not satisfy the original equation. Recall that we call these apparent solutions **extraneous solutions.** Equations containing square roots may also have extraneous solutions.

We can identify extraneous solutions by substituting apparent solutions into the original equation. If the solution does not satisfy the equation, it is extraneous.

EXAMPLE 4 **Solving a Radical Equation with an Extraneous Solution**

Solve: $\sqrt{x+6} = x$

Solution

The radical is already isolated, so we square both sides of the equation:

$$\sqrt{x+6} = x$$
$$(\sqrt{x+6})^2 = x^2$$
$$x+6 = x^2$$

The resulting equation, $x + 6 = x^2$, is a quadratic equation (do you see why?), so we write it in standard form, $ax^2 + bx + c = 0$, and solve.

Subtract x and 6 from each side:	$x - x + 6 - 6 = x^2 - x - 6$
Write in standard form:	$0 = x^2 - x - 6$
Factor the polynomial:	$0 = (x-3)(x+2)$
Use the Zero-Product Property:	$x - 3 = 0$ or $x + 2 = 0$
Solve the resulting equations:	$x = 3$ or $x = -2$

Check

Let $x = 3$ in the original equation:
$x = 3$: $\sqrt{x+6} = x$
$\sqrt{3+6} \overset{?}{=} 3$
$\sqrt{9} \overset{?}{=} 3$
$3 = 3$ True

$x = 3$ checks, so $x = 3$ is a solution of the equation.

Let $x = -2$ in the original equation:
$x = -2$: $\sqrt{x+6} = x$
$\sqrt{-2+6} \overset{?}{=} -2$
$\sqrt{4} \overset{?}{=} -2$
$2 = -2$ False

$x = -2$ does not check, so $x = -2$ is an extraneous solution.

The solution set is $\{3\}$. ●

> **Quick ✓**
>
> **10.** Apparent solutions that do not satisfy the original equation are called _____ _____.
>
> *In Problems 11 and 12, solve the equation and check the solution.*
>
> **11.** $\sqrt{2a+3} = a$ **12.** $\sqrt{2p+8} = p$

EXAMPLE 5 **Using the Square of a Binomial to Solve a Radical Equation**

Solve: $\sqrt{w} + 6 = w$

Solution

$$\sqrt{w} + 6 = w$$

Isolate the radical expression:	$\sqrt{w} + 6 - 6 = w - 6$
	$\sqrt{w} = w - 6$
Square both sides of the equation:	$(\sqrt{w})^2 = (w-6)^2$
$(A-B)^2 = A^2 - 2AB + B^2$:	$w = w^2 - 12w + 36$

Work Smart

$(A-B)^2 \neq A^2 - B^2$
$(A-B)^2 = A^2 - 2AB + B^2$

Write the quadratic equation in standard form and solve.

Subtract w from each side:	$w - w = w^2 - 12w - w + 36$
Write in standard form:	$0 = w^2 - 13w + 36$
Factor the polynomial:	$0 = (w-9)(w-4)$
Use the Zero-Product Property:	$w - 9 = 0$ or $w - 4 = 0$
Solve the resulting equations:	$w = 9$ or $w = 4$

Let's check these two apparent solutions.

Check

$w = 9$: $\sqrt{w} + 6 = w$
Let $w = 9$ in the
original equation: $\sqrt{9} + 6 \overset{?}{=} 9$

$3 + 6 \overset{?}{=} 9$

$9 = 9$ True

$w = 9$ checks, so $w = 9$ is a solution of the equation.

$w = 4$: $\sqrt{w} + 6 = w$
Let $w = 4$ in the
original equation: $\sqrt{4} + 6 \overset{?}{=} 4$

$2 + 6 \overset{?}{=} 4$

$8 = 4$ False

$w = 4$ does not check, so $w = 4$ is an extraneous solution.

The solution set is $\{9\}$.

Quick ✓

13. *True or False* The first step in solving $x + \sqrt{x + 3} = 5$ is to square both sides of the equation to remove the square root.

In Problems 14 and 15, solve the equation and check the solution.

14. $\sqrt{m + 20} = m + 8$

15. $\sqrt{17 - 2x} + 1 = x$

Is it possible for a radical equation to have no solution? Yes!

EXAMPLE 6 **Solving a Radical Equation with No Solution**

Solve: $\sqrt{5p - 3} + 7 = 3$

Solution

$$\sqrt{5p - 3} + 7 = 3$$

Isolate the radical expression: $\sqrt{5p - 3} + 7 - 7 = 3 - 7$

$$\sqrt{5p - 3} = -4$$

Square both sides: $(\sqrt{5p - 3})^2 = (-4)^2$

$$5p - 3 = 16$$

$$5p = 19$$

$$p = \frac{19}{5}$$

Work Smart

Look closely at the equation

$\sqrt{5p - 3} = -4$.

Because the principal square root of a number cannot be negative, we know that there is no real solution to the equation $\sqrt{5p - 3} + 7 = 3$. Thus the solution set is $\varnothing$ or $\{\ \}$.

Check

$$\sqrt{5p - 3} + 7 = 3$$

$p = \frac{19}{5}$: $\sqrt{5\left(\dfrac{19}{5}\right) - 3} + 7 \overset{?}{=} 3$

$$\sqrt{19 - 3} + 7 \overset{?}{=} 3$$

$$\sqrt{16} + 7 \overset{?}{=} 3$$

$$11 = 3 \quad \text{False}$$

Because $p = \dfrac{19}{5}$ does not satisfy the original equation, it is extraneous. The solution set is $\{\ \}$ or $\varnothing$.

▶ **③ Solve Equations Containing Two Square Roots**

If a radical equation has two square roots, place one radical on one side of the equation and the other radical on the other side. Then square both sides of the equation and solve the resulting equation.

EXAMPLE 7 | **Solving a Radical Equation with Two Square Roots**

Solve: $\sqrt{x + 9} = \sqrt{2x + 5}$

Solution

Notice that each side of the equation contains one radical expression. Square both sides of the equation and solve the resulting equation.

$$\sqrt{x + 9} = \sqrt{2x + 5}$$

Square both sides of the equation: $(\sqrt{x + 9})^2 = (\sqrt{2x + 5})^2$

$$x + 9 = 2x + 5$$

Subtract x from both sides: $x - x + 9 = 2x - x + 5$

$$9 = x + 5$$

Subtract 5 from both sides: $4 = x$

Check

$$\sqrt{x + 9} = \sqrt{2x + 5}$$

$x = 4$: $\sqrt{4 + 9} \overset{?}{=} \sqrt{2(4) + 5}$

$$\sqrt{13} \overset{?}{=} \sqrt{8 + 5}$$

$$\sqrt{13} = \sqrt{13} \qquad \text{True}$$

Because $x = 4$ satisfies the equation, the solution set is $\{4\}$. ●

EXAMPLE 8 | **Solving a Radical Equation with Two Square Roots**

Solve: $\sqrt{y + 6} - 2\sqrt{y} = 0$

Solution

Notice that both radicals are on the same side of the equation. First, add $2\sqrt{y}$ to both sides to get one radical expression on each side of the equation.

$$\sqrt{y + 6} - 2\sqrt{y} = 0$$

$$\sqrt{y + 6} - 2\sqrt{y} + 2\sqrt{y} = 0 + 2\sqrt{y}$$

$$\sqrt{y + 6} = 2\sqrt{y}$$

Square both sides of the equation: $(\sqrt{y + 6})^2 = (2\sqrt{y})^2$

$(2\sqrt{y})^2 = 2^2 \cdot (\sqrt{y})^2 = 4y$: $y + 6 = 4y$

Subtract y from both sides: $y - y + 6 = 4y - y$

$$6 = 3y$$

Divide both sides by 3: $\dfrac{6}{3} = \dfrac{3y}{3}$

$$2 = y$$

Check

$$\sqrt{y+6} - 2\sqrt{y} = 0$$
$$y = 2: \quad \sqrt{2+6} - 2\sqrt{2} \overset{?}{=} 0$$
$$\sqrt{8} - 2\sqrt{2} \overset{?}{=} 0$$
$$2\sqrt{2} - 2\sqrt{2} \overset{?}{=} 0$$
$$0 = 0 \quad \text{True}$$

Since $y = 2$ checks, the solution set is $\{2\}$.

Quick ✓

In Problems 17–19, solve each equation and check the solution.

17. $\sqrt{3x+1} = \sqrt{2x+7}$ **18.** $\sqrt{2w^2 - 3w - 4} = \sqrt{w^2 + 6w + 6}$

19. $2\sqrt{k+5} - \sqrt{8k+4} = 0$

We now discuss one more type of equation with two radicals. In these equations, it's necessary to square both sides of the equation twice to eliminate radicals. After we square the first time, a radical remains. We then follow the same steps for solving an equation with one radical.

EXAMPLE 9   **Solving a Radical Equation—Squaring Twice**

Solve: $\sqrt{x} + 1 = \sqrt{x+5}$

Solution

There is a radical expression on each side of the equation, so we square both sides.

$$\sqrt{x} + 1 = \sqrt{x+5}$$
Square both sides: $\quad (\sqrt{x}+1)^2 = (\sqrt{x+5})^2$

$(\sqrt{x}+1)^2 = (\sqrt{x})^2 + 2(\sqrt{x})(1) + 1^2$
$= x + 2\sqrt{x} + 1$

$$x + 2\sqrt{x} + 1 = x + 5$$

Isolate the remaining radical: $\quad x - x + 2\sqrt{x} + 1 - 1 = x - x + 5 - 1$
$$2\sqrt{x} = 4$$
Square both sides: $\quad (2\sqrt{x})^2 = 4^2$
$$4x = 16$$
Divide both sides of the equation by 4: $\quad x = 4$

Work Smart

Another strategy that may be used to solve Example 9 at the Step $2\sqrt{x} = 4$ is

$$2\sqrt{x} = 4$$
$$\frac{2\sqrt{x}}{2} = \frac{4}{2}$$
$$\sqrt{x} = 2$$
$$(\sqrt{x})^2 = (2)^2$$
$$x = 4$$

Check

$$\sqrt{x} + 1 = \sqrt{x+5}$$
$$x = 4: \quad \sqrt{4} + 1 \overset{?}{=} \sqrt{4+5}$$
$$2 + 1 \overset{?}{=} 3$$
$$3 = 3 \quad \text{True}$$

Since $x = 4$ results in a true statement, the solution set is $\{4\}$.

Quick ✓

20. *True or False* To solve $\sqrt{x} + 3 = \sqrt{2x+1}$, the first step is $(\sqrt{x})^2 + (3)^2 = (\sqrt{2x+1})^2$.

In Problem 21, solve the equation and check the solution.

21. $\sqrt{x} = \sqrt{x+9} - 1$

▶ ❹ **Solve Problems Modeled by Radical Equations**

Let's see how square roots are used in real-world applications.

EXAMPLE 10 **How Much Money?**

The annual interest rate r (expressed as a decimal) required in order for a depositor to have A dollars after 2 years after depositing P dollars is given by the equation

$r = \sqrt{\dfrac{A}{P}} - 1$. Suppose you deposit \$1000 in an account that pays 5% annual interest. How much money will you have after 2 years?

Solution

We want to find A, the amount of money that will be in the account after 2 years. We know that $P = 1000$ and that $r = 5\% = 0.05$.

$$r = \sqrt{\frac{A}{P}} - 1$$

Substitute $r = 0.05, P = 1000$: $\quad 0.05 = \sqrt{\dfrac{A}{1000}} - 1$

Isolate the square root by adding
1 to both sides of the equation: $\quad 0.05 + 1 = \sqrt{\dfrac{A}{1000}} - 1 + 1$

$$1.05 = \sqrt{\frac{A}{1000}}$$

Square both sides: $\quad (1.05)^2 = \left(\sqrt{\dfrac{A}{1000}}\right)^2$

$$1.1025 = \frac{A}{1000}$$

Multiply both sides by 1000: $\quad 1000\,(1.1025) = \left(\dfrac{A}{1000}\right)1000$

$$1102.50 = A$$

There will be \$1102.50 in the account after 2 years. ●

Quick ✓

22. A method to "curve" grades on an exam uses the model $N = 10\sqrt{O}$, where N is the new grade and O is the original grade. (This method is sometimes referred to as a "square root curve.") Use this model to find the original score on a test that has a new score of 70.

8.6 Exercises   MathXL® PRACTICE

Exercise numbers in green have complete video solutions in MyMathLab.

Problems 1–22 are the Quick ✓s that follow the EXAMPLES.

Building Skills

In Problems 23–30, determine whether the number is a solution to the equation. See Objective 1.

23. $\sqrt{6x + 1} = 5; x = 4$

24. $\sqrt{4n - 7} = 3; n = 4$

25. $\sqrt{5n - 6} + 6 = 0; n = -6$

26. $\sqrt{3y - 4} + 5 = 0; y = -7$

27. $\sqrt{p + 6} = p + 4; p = -2$

28. $\sqrt{p + 6} = p + 4; p = -5$

29. $\sqrt{2y + 3} = \sqrt{y + 5} - 1; y = -1$

30. $\sqrt{3y + 15} = \sqrt{y + 6} + 1; y = -2$

In Problems 31–56, solve the equation and check the solution.
See Objective 2.

31. $\sqrt{a} = 4$

32. $\sqrt{x} = 9$

33. $\sqrt{5 - x} = 3$

34. $\sqrt{3 - y} = 2$

35. $5 = \sqrt{4x - 3}$

36. $4 = \sqrt{3x + 10}$

37. $3\sqrt{p} = 6$

38. $5\sqrt{a - 2} = 10$

39. $2\sqrt{x} + 2 = 8$

40. $3\sqrt{n} + 3 = 6$

41. $3 = 6 + \sqrt{n}$

42. $10 = 23 + \sqrt{p}$

43. $\sqrt{2y + 11} - 4 = -1$

44. $\sqrt{3y + 25} - 3 = 1$

45. $\sqrt{9z - 18} = z$

46. $\sqrt{7v - 10} = v$

47. $y = \sqrt{4 - 3y}$

48. $y = \sqrt{24 - 5y}$

49. $n + 1 = \sqrt{n^2 - 3}$

50. $p + 3 = \sqrt{p^2 - 15}$

51. $\sqrt{15 - 7x} = x + 9$

52. $\sqrt{8 - 2x} = x + 8$

53. $\sqrt{31x + 10} - 8 = x$

54. $\sqrt{17x + 13} - 5 = x$

55. $6 + \sqrt{x} - x = 0$

56. $4 + \sqrt{2x} - x = 0$

In Problems 57–76, solve the equation and check the solution.
See Objective 3.

57. $\sqrt{3x - 4} = \sqrt{x + 8}$

58. $\sqrt{5x + 1} = \sqrt{3x - 15}$

59. $\sqrt{x^2 + 3x} = \sqrt{x^2 + 6x - 3}$

60. $\sqrt{x^2 + 4} = \sqrt{x^2 + 8x}$

61. $2\sqrt{a + 5} - \sqrt{7a + 20} = 0$

62. $\sqrt{a + 3} - 2\sqrt{a - 3} = 0$

63. $\sqrt{2x^2 - 10} - \sqrt{x^2 - 3x} = 0$

64. $\sqrt{2x^2 + 8} - \sqrt{x^2 - 3x + 6} = 0$

65. $2\sqrt{n} = \sqrt{n^2 - 5}$

66. $\sqrt{y^2 - 36} = 4\sqrt{y}$

67. $\sqrt{2x + 2} = \sqrt{2x - 12}$

68. $\sqrt{x + 3} = \sqrt{x + 15}$

69. $\sqrt{z + 10} = 2 + \sqrt{5z + 6}$

70. $2\sqrt{t + 9} = \sqrt{4t + 3} + 3$

71. $\sqrt{p + 13} = \sqrt{p - 8} - 7$

72. $2\sqrt{s - 4} = 2 + \sqrt{4s}$

73. $\sqrt{x + 2} + 2 = \sqrt{2x + 3}$

74. $\sqrt{x - 1} = \sqrt{2x - 7}$

75. $\sqrt{3x + 4} - \sqrt{4x + 1} = -1$

76. $\sqrt{4x - 7} - \sqrt{2x} = 1$

Mixed Practice

In Problems 77–98, solve the equation and check the solution.

77. $-2 = \sqrt{5x + 6} - 3$

78. $-4 = \sqrt{2x + 11} - 7$

79. $2\sqrt{x} - 8 = 0$

80. $3\sqrt{x} - 27 = 0$

81. $3x - 2 = \sqrt{9x^2 - 20}$

82. $2x - 1 = \sqrt{4x^2 - 11}$

83. $3 = \sqrt{w + 12} - 2w$

84. $1 = \sqrt{3 - 3c} - 2c$

85. $x(x - 5) = 24$

86. $\dfrac{3}{2}x + 5 = \dfrac{1}{4}x + 4$

87. $\sqrt{a^2 - 3a + 5} = \sqrt{2a^2 - 6a - 23}$

88. $\sqrt{2n^2 - 5n - 20} = \sqrt{n^2 - 3n + 15}$

89. $\sqrt{2y + 1} - 2 = 1$

90. $\sqrt{4z - 3} - 6 = -3$

91. $x - \dfrac{2}{x} = 1$

92. $2x^2 - 12 = 7x + 3$

93. $\sqrt{p + 4} + 1 = \sqrt{p - 5}$

94. $\sqrt{p + 8} - 1 = \sqrt{p + 3}$

95. $2x + 3(x - 4) = 7x - 18$

96. $\dfrac{3}{x - 2} - \dfrac{12}{x^2 - 4} = 1$

97. $\sqrt{11x - 18} = -x$

98. $\sqrt{10x - 21} = -x$

Applying the Concepts

Problems 99 and 100 use the following information. The time t that it takes an object to fall h feet is modeled by the formula

$$t = \sqrt{\frac{h}{16}}$$

99. (a) If a ball is dropped off a cliff that is 1024 feet tall, how long will it take to hit the ground?

(b) If a water balloon is dropped from the top of a building that is 300 feet tall, exactly how long will it take to hit the ground?

100. (a) If a ball is dropped off a cliff that is 64 feet tall, how long will it take to hit the ground?

(b) If a water balloon is dropped from the top of a building that is 400 feet tall, exactly how long will it take to hit the ground?

Problems 101 and 102 use the following information. When one is driving on dry asphalt, the distance in feet, d, to stop a car going S miles per hour is modeled by the formula

$$S = \sqrt{22.5d}$$

101. Braking Distance Mark's car is going down a street made of asphalt. If he is driving at 25 mph and slams on the brakes, how far, to the nearest foot, will the car skid before it comes to a stop?

102. Braking Distance Dale is driving his Hummer down a county highway made of asphalt in the desert. If he is going 60 mph and slams on the brakes, how far will the car skid before it comes to a stop?

Problems 103 and 104 use the following information. When one is driving on gravel, the distance in feet, d, to stop a car going S miles per hour is modeled by the formula

$$S = \sqrt{15d}$$

103. Braking Distance Mark gets off the asphalt street and drives onto the gravel road leading to his house, still going 25 mph, when he hits the brakes to avoid a jackrabbit.

(a) To the nearest foot, how far will the car skid before it comes to a stop?

(b) If Mark spotted the jackrabbit 40 feet away when he hit his brakes, would Mark hit the rabbit?

104. Braking Distance Dale decides to go off-road driving in his Hummer and turns off the pavement onto a gravel road at 60 mph. After driving a short time, he sees a huge ditch in the road and hits his brakes.

(a) How far will the Hummer skid before it comes to a stop?

(b) Dale spotted the ditch 241 feet away when he hit the brakes. Did he land in the ditch or manage to avoid a big crash?

105. Fun with Numbers The square root of twice the difference of a number and 3 is 4. Find the number.

106. Fun with Numbers The square root of 5 less than twice a number is 7. Find the number.

107. Fun with Numbers Three times the square root of 1 more than an unknown number is the same as twice the square root of 3 more than double the unknown number. Find the number.

108. Fun with Numbers Twice the square root of 1 more than the square of an unknown number is the same as the square root of the sum of the unknown number and 4. Find the number.

Extending the Concepts

In Problems 109–116, solve for the indicated variable.

109. $\sqrt{\dfrac{A}{\pi}} = r$; solve for A

110. $v = \sqrt{2gh}$; solve for g

111. $S = \sqrt{\dfrac{3V}{h}}$; solve for h

112. $r = \sqrt{\dfrac{3V}{\pi h}}$; solve for h

113. $b = \sqrt{2a + b}$; solve for a

114. $m = \sqrt{m - 5n}$; solve for n

115. $2\sqrt{m - 2n} = \sqrt{4n - 2m}$; solve for n

116. $2\sqrt{3p + 5q} = \sqrt{5q - 9p}$; solve for p

Explaining the Concepts

117. An exam asked students to solve the radical equation $\sqrt{x - 1} = 3 - \sqrt{x + 6}$. The following steps appear on a student's paper. Is the student's work correct or incorrect? What property did the student apply to get the second line? If there is an error, explain the correct steps to solve the equation.

$$\sqrt{x - 1} = 3 - \sqrt{x + 6}$$
$$x - 1 = 9 - (x + 6)$$

118. Explain how we know that the solution set of $\sqrt{2x - 5} = -3$ is $\varnothing$ without solving the equation.

8.7 Higher Roots and Rational Exponents

Objectives

1. Evaluate Higher Roots
2. Use Product and Quotient Rules to Simplify Higher Roots
3. Define and Evaluate Expressions of the Form $a^{\frac{1}{n}}$
4. Define and Evaluate Expressions of the Form $a^{\frac{m}{n}}$
5. Use Laws of Exponents to Simplify Expressions with Rational Exponents

Are You Ready for This Section?

Before getting started, take the following readiness quiz. If you get a problem wrong, go to the section cited and review the material.

R1. Evaluate: **(a)** 2^4 **(b)** $(-4)^3$ [Section 1.7, pp. 56–57]

R2. Simplify: $x^3 \cdot x^5$ [Section 5.2, pp. 318–319]

R3. Simplify: $\dfrac{y^5}{y}$ [Section 5.4, pp. 334–335]

R4. Simplify: $(b^3)^4$ [Section 5.2, p. 320]

1 Evaluate Higher Roots

▶ Finding the square root of a number "reverses" the squaring process. We can also reverse the process of raising a number to other positive integer powers.

> **In Words**
>
> When you see the notation $\sqrt[n]{a} = b$, think to yourself, "Find a number b that, raised to the nth power, gives me a."

Definition

The **principal nth root of a number a,** symbolized by $\sqrt[n]{a}$, where $n \geq 2$ is an integer, is defined as follows:

$$\sqrt[n]{a} = b \quad \text{means} \quad a = b^n$$

- If $n \geq 2$ and even, then a and b must be greater than or equal to 0.
- If $n \geq 3$ and odd, then a and b can be any real number.

In the notation $\sqrt[n]{a}$, the integer n, $n \geq 2$, is called the **index.** A radical written without the index is understood to mean the square root, so $\sqrt{a}$ represents the square root of a. If the index is 3, we call $\sqrt[3]{a}$ the **cube root** of a. If the index is even, then the radicand must be greater than or equal to 0. If the index is odd, then the radicand can be any real number. Do you know why? Since $\sqrt[n]{a} = b$ means $b^n = a$, if the index n is even, then $b^n \geq 0$, so $a \geq 0$. If we have an odd index n, then b^n can be any real number, so a can be any real number.

Before we evaluate nth roots, we list some perfect powers (squares, cubes, and so on) in the table below. This list will help in finding roots in the examples that follow. Notice that perfect cubes can be negative, but perfect fourths cannot. Do you know why?

Perfect Squares	Perfect Cubes		Perfect Fourths	Perfect Fifths	
$1^2 = 1$	$1^3 = 1$	$(-1)^3 = -1$	$1^4 = 1$	$1^5 = 1$	$(-1)^5 = -1$
$2^2 = 4$	$2^3 = 8$	$(-2)^3 = -8$	$2^4 = 16$	$2^5 = 32$	$(-2)^5 = -32$
$3^2 = 9$	$3^3 = 27$	$(-3)^3 = -27$	$3^4 = 81$	$3^5 = 243$	$(-3)^5 = -243$
$4^2 = 16$	$4^3 = 64$	$(-4)^3 = -64$	$4^4 = 256$	$4^5 = 1024$	$(-4)^5 = -1024$
$5^2 = 25$	$5^3 = 125$	$(-5)^3 = -125$	$5^4 = 625$	$5^5 = 3125$	$(-5)^5 = -3125$
and so on	and so on		and so on	and so on	

EXAMPLE 1   **Evaluating Higher Roots**

(a) $\sqrt[3]{-1} = -1$ because $(-1)^3 = -1$ **(b)** $\sqrt[4]{16} = 2$ because $2^4 = 16$

(c) $\sqrt[5]{-32} = -2$ because $(-2)^5 = -32$ **(d)** $\sqrt[3]{\dfrac{1}{64}} = \dfrac{1}{4}$ because $\left(\dfrac{1}{4}\right)^3 = \dfrac{1}{64}$

(e) $\sqrt[4]{-256}$ is not a real number. There is no real number b such that $b^4 = -256$.

Work Smart

$\sqrt[3]{-8} = -2$ because $(-2)^3 = -8$. But $\sqrt[4]{-16}$ is not a real number because there's no real number b such that $b^4 = -16$. Remember, we can take odd roots of any real number, but we can take only even roots of nonnegative real numbers.

Quick ✓

1. In the expression $\sqrt[n]{a}$, a is called the _____ and n is called the _____.

2. Because $(-5)^3 = -125$, $\sqrt[3]{-125} = $ ___.

3. *True or False* The expression $\sqrt[n]{a}$ is not a real number when the index n is even and the radicand a is negative.

In Problems 4–7, evaluate each root, if possible.

4. $\sqrt[5]{32}$ 5. $\sqrt[3]{-27}$

6. $\sqrt[4]{-16}$ 7. $\sqrt[6]{64}$

Recall from Section 8.1 that $\sqrt{a^2} = a$ if $a \geq 0$. A similar rule applies to nth roots. For example, what is $\sqrt[3]{4^3}$? Here we are asking, what number, when raised to the 3rd power, gives me 4^3? The answer is 4, so $\sqrt[3]{4^3} = 4$. What is $\sqrt[4]{10^4}$? We are asking, what number, when raised to the 4th power, gives me 10^4? The answer is 10, so $\sqrt[4]{10^4} = 10$. We have the following general result.

In Words

Odd roots of positive numbers are positive and odd roots of negative numbers are negative. If the radicand is nonnegative, the even root is nonnegative.

Simplifying $\sqrt[n]{a^n}$

If $n \geq 2$ is an integer, then

$$\sqrt[n]{a^n} = |a| \quad \text{if } n \text{ is even}$$
$$\sqrt[n]{a^n} = a \quad \text{if } n \text{ is odd}$$

If $a \geq 0$, then $\sqrt[n]{a^n} = a$ for $n \geq 2$ is an integer.

EXAMPLE 2   **Simplifying Expressions of the Form $\sqrt[n]{a^n}$**

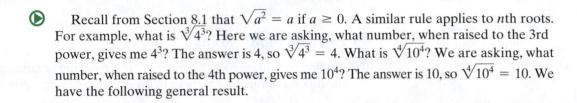

(a) $\sqrt[4]{5^4} = 5$ (b) $\sqrt[5]{x^5} = x$ (c) $\sqrt[6]{y^6} = y$ for $y \geq 0$ (d) $\sqrt[3]{(6z)^3} = 6z$ ●

Quick ✓

In Problems 8–10, simplify the radical.

8. $\sqrt[6]{7^6}$ 9. $\sqrt[4]{w^4}$ for $w \geq 0$ 10. $\sqrt[5]{(12p)^5}$

❷ Use Product and Quotient Rules to Simplify Higher Roots

Recall that we simplified square roots such as $\sqrt{18}$ by looking for a factor of 18 that is a perfect square. In the same way, we can simplify radicals such as $\sqrt[3]{16}$ and $\sqrt[4]{32}$. To do this, we use the Product Rule for Radicals.

In Words

$\sqrt[n]{ab} = \sqrt[n]{a} \cdot \sqrt[n]{b}$ means "The product of the roots equals the root of the product."

Product Rule for Radicals

If $\sqrt[n]{a}$ and $\sqrt[n]{b}$ are real numbers and $n \geq 2$ is an integer, then
$$\sqrt[n]{ab} = \sqrt[n]{a} \cdot \sqrt[n]{b}$$

To simplify higher roots, write the radicand as the product of two or more factors, at least one of which is a perfect power of the index, and then take the root. For example, to simplify a cube root, factor the radicand so that one of the factors is a perfect cube.

EXAMPLE 3 **Simplifying Higher Roots Using the Product Rule**

Simplify each of the following:

(a) $\sqrt[3]{54}$ **(b)** $\sqrt[4]{32x^4}$, $x \geq 0$

Solution

(a)

27 is a factor of 54, and 27 is a perfect cube, so we write 54 as $27 \cdot 2$: $\sqrt[3]{54} = \sqrt[3]{27 \cdot 2}$

$\sqrt[n]{ab} = \sqrt[n]{a} \cdot \sqrt[n]{b}$: $= \sqrt[3]{27} \cdot \sqrt[3]{2}$

$27 = 3^3$: $= \sqrt[3]{3^3} \cdot \sqrt[3]{2}$

$\sqrt[n]{a^n} = a$: $= 3\sqrt[3]{2}$

(b)

16 is a factor of 32, and 16 is a fourth power $(16 = 2^4)$, so we write 32 as $16 \cdot 2$: $\sqrt[4]{32x^4} = \sqrt[4]{16x^4 \cdot 2}$

$\sqrt[n]{ab} = \sqrt[n]{a} \cdot \sqrt[n]{b}$: $= \sqrt[4]{16x^4} \cdot \sqrt[4]{2}$

$16x^4 = 2^4x^4 = (2x)^4$: $= \sqrt[4]{(2x)^4} \cdot \sqrt[4]{2}$

$\sqrt[n]{a^n} = a$: $= 2x\sqrt[4]{2}$

●

Quick ✓

11. If $\sqrt[n]{a}$ and $\sqrt[n]{b}$ are real numbers and $n \geq 2$ is an integer, then $\sqrt[n]{ab} =$ _____.

12. *True or False* $\sqrt[4]{32} = \sqrt[4]{16 \cdot 2} = 4\sqrt[4]{2}$

In Problems 13–15, simplify the radical.

13. $\sqrt[3]{-32}$ **14.** $\sqrt[3]{250n^3}$ **15.** $\sqrt[4]{162b^{12}}$, $b \geq 0$

The Quotient Rule for Radicals can also be generalized to any integer index $n \geq 2$.

In Words

$$\sqrt[n]{\frac{a}{b}} = \frac{\sqrt[n]{a}}{\sqrt[n]{b}}$$

means "The root of the quotient equals the quotient of the roots."

Quotient Rule for Radicals

If $\sqrt[n]{a}$ and $\sqrt[n]{b}$ are real numbers, $b \neq 0$, and $n \geq 2$ is an integer, then

$$\sqrt[n]{\frac{a}{b}} = \frac{\sqrt[n]{a}}{\sqrt[n]{b}}$$

EXAMPLE 4 **Simplifying Higher Roots Using the Quotient Rule**

Simplify:

(a) $\dfrac{\sqrt[3]{-48}}{\sqrt[3]{3}}$ **(b)** $\dfrac{\sqrt[4]{64z^9}}{\sqrt[4]{2z}}$, $z > 0$

Solution

(a)

$\dfrac{\sqrt[n]{a}}{\sqrt[n]{b}} = \sqrt[n]{\dfrac{a}{n}}$: $\dfrac{\sqrt[3]{-48}}{\sqrt[3]{3}} = \sqrt[3]{\dfrac{-48}{3}}$

$= \sqrt[3]{-16}$

-8 is a factor of -16, and -8 is a perfect cube $(-8 = (-2)^3)$, so we write -16 as $-8 \cdot 2$: $= \sqrt[3]{-8 \cdot 2}$

$\sqrt[n]{ab} = \sqrt[n]{a} \cdot \sqrt[n]{b}$: $= \sqrt[3]{-8} \cdot \sqrt[3]{2}$

Simplify each radical: $= -2\sqrt[3]{2}$

(b)

$$\frac{\sqrt[n]{a}}{\sqrt[n]{b}} = \sqrt[n]{\frac{a}{b}}: \qquad \frac{\sqrt[4]{64z^9}}{\sqrt[4]{2z}} = \sqrt[4]{\frac{64z^9}{2z}}$$

$$= \sqrt[4]{32z^8}$$

16 is a factor of 32, and $16 = 2^4$: $= \sqrt[4]{16z^8 \cdot 2}$

$\sqrt[n]{ab} = \sqrt[n]{a} \cdot \sqrt[n]{a}$: $= \sqrt[4]{16z^8} \cdot \sqrt[4]{2}$

$\sqrt[4]{16} = 2; \sqrt[4]{z^8} = z^2$: $= 2z^2\sqrt[4]{2}$

> **Quick** ✔
>
> *In Problems 16 and 17, find the quotient and simplify.*
>
> **16.** $\dfrac{\sqrt[4]{324}}{\sqrt[4]{4}}$ **17.** $\dfrac{\sqrt[3]{512m^{10}}}{\sqrt[3]{4m}}, m \neq 0$

▶ ❸ Define and Evaluate Expressions of the Form $a^{\frac{1}{n}}$

Now we extend the rules for integer exponents to rational exponents.

Work Smart

A rational number is a number that can be written as the quotient of two integers where the denominator is not zero.

We start by defining "a raised to the power $\dfrac{1}{n}$," where a is a real number and n is a positive integer. Remember, raising an expression to an integer exponent means repeated multiplication. We use this to help determine what $a^{\frac{1}{n}}$ means.

$$a^2 = a \cdot a \qquad (5^{\frac{1}{2}})^2 = 5^{\frac{1}{2}} \cdot 5^{\frac{1}{2}}$$

$$a^m \cdot a^n = a^{m+n}: \qquad = 5^{\frac{1}{2}+\frac{1}{2}}$$

$$= 5^1$$

$$= 5$$

We also know that $(\sqrt{5})^2 = 5$, so it is reasonable to conclude that

$$5^{\frac{1}{2}} = \sqrt{5}$$

This suggests the following definition:

> **Definition of $a^{\frac{1}{n}}$**
>
> If a is a real number and $n \geq 2$ is an integer, then
>
> $$a^{\frac{1}{n}} = \sqrt[n]{a}$$
>
> provided that $\sqrt[n]{a}$ exists.

EXAMPLE 5   **Evaluating Expressions of the Form $a^{\frac{1}{n}}$**

Write each expression as a radical and evaluate, if possible.

(a) $16^{\frac{1}{2}}$ **(b)** $(-27)^{\frac{1}{3}}$ **(c)** $-81^{\frac{1}{2}}$ **(d)** $(-81)^{\frac{1}{2}}$

Solution

(a) $a^{\frac{1}{2}} = \sqrt{a}$: $16^{\frac{1}{2}} = \sqrt{16}$ **(b)** $a^{\frac{1}{3}} = \sqrt[3]{a}$: $(-27)^{\frac{1}{3}} = \sqrt[3]{-27}$

$\qquad\qquad\qquad\quad = 4$ $= -3$

(c) $-81^{\frac{1}{2}} = -1 \cdot 81^{\frac{1}{2}}$ **(d)** $(-81)^{\frac{1}{2}} = \sqrt{-81}$ is not a

$a^{\frac{1}{2}} = \sqrt{a}$: $= -\sqrt{81}$ real number because there

$\qquad\qquad\qquad\quad = -9$ is no real number whose

                                                  square is -81.

EXAMPLE 6 **Writing an Expression of the Form $a^{\frac{1}{n}}$ as a Radical**

Write each expression as a radical:

(a) $z^{\frac{1}{3}}$ (b) $3a^{\frac{1}{2}}$ (c) $(3a)^{\frac{1}{2}}$

Solution

(a) $a^{\frac{1}{3}} = \sqrt[3]{a}$: $z^{\frac{1}{3}} = \sqrt[3]{z}$

(b) $\qquad\quad 3a^{\frac{1}{2}} = 3 \cdot a^{\frac{1}{2}}$

$\quad a^{\frac{1}{2}} = \sqrt{a}$: $\qquad = 3 \cdot \sqrt{a}$

$\qquad\qquad\qquad = 3\sqrt{a}$

(c) $a^{\frac{1}{2}} = \sqrt{a}$: $(3a)^{\frac{1}{2}} = \sqrt{3a}$

Quick ✔

18. If a is a nonnegative real number and $n \geq 2$ is an integer, then $a^{\frac{1}{n}} =$___ .

In Problem 19, write each expression as a radical and evaluate.

19. (a) $49^{\frac{1}{2}}$ **(b)** $(-64)^{\frac{1}{3}}$ **(c)** $-100^{\frac{1}{2}}$ **(d)** $(-100)^{\frac{1}{2}}$

In Problem 20, write each expression as a radical.

20. (a) $p^{\frac{1}{5}}$ **(b)** $(4n)^{\frac{1}{3}}$ **(c)** $4n^{\frac{1}{3}}$

If an expression is written as a radical, we can rewrite it with rational exponents, using $\sqrt[n]{a} = a^{\frac{1}{n}}$. **For the remainder of the section, we will assume that each variable represents a nonnegative real number.**

EXAMPLE 7 **Writing Radicals with Rational Exponents**

Write each expression with a rational exponent.

(a) $\sqrt[3]{w}$ (b) $6\sqrt[5]{z}$ (c) $\sqrt{11b}$

Solution

(a) $\sqrt[n]{a} = a^{\frac{1}{n}}$: $\sqrt[3]{w} = w^{\frac{1}{3}}$

(b) $\qquad\quad 6\sqrt[5]{z} = 6 \cdot \sqrt[5]{z}$

$\quad \sqrt[n]{a} = a^{\frac{1}{n}}$: $\qquad = 6 \cdot z^{\frac{1}{5}}$

$\qquad\qquad\qquad = 6z^{\frac{1}{5}}$

(c) $\sqrt[n]{a} = a^{\frac{1}{n}}$: $\sqrt{11b} = (11b)^{\frac{1}{2}}$

Quick ✔

In Problem 21, rewrite each radical expression with a rational exponent. Assume that each variable is a nonnegative real number.

21. (a) $\sqrt{b}$ **(b)** $5\sqrt[3]{p}$ **(c)** $\sqrt[4]{10p}$

▶ ④ **Define and Evaluate Expressions of the Form $a^{\frac{m}{n}}$**

We now define $a^{\frac{m}{n}}$, where m and n are integers.

Work Smart

The expression $a^{\frac{m}{n}}$ is not a real number when a is a negative real number and n is an even integer.

Definition of $a^{\frac{m}{n}}$

If a is a real number and $\dfrac{m}{n}$ is a rational number in lowest terms with $n \geq 2$, then

$$a^{\frac{m}{n}} = \sqrt[n]{a^m} = (\sqrt[n]{a})^m$$

provided that $\sqrt[n]{a}$ exists.

In the expression $a^{\frac{m}{n}}$, $\dfrac{m}{n}$ is reduced to lowest terms and $n \geq 2$. The definition obeys all the laws of exponents presented earlier. For example,

$$a^{\frac{m}{n}} = a^{m \cdot \frac{1}{n}} = (a^m)^{\frac{1}{n}} = \sqrt[n]{a^m}$$

and

$$a^{\frac{m}{n}} = a^{\frac{1}{n} \cdot m} = (a^{\frac{1}{n}})^m = (\sqrt[n]{a})^m$$

When evaluating or simplifying $a^{\frac{m}{n}}$, you may use either $\sqrt[n]{a^m}$ or $(\sqrt[n]{a})^m$, whichever makes simplifying the expression easier. Generally, taking the root first, as in $(\sqrt[n]{a})^m$, is easier.

EXAMPLE 8 | **Evaluating Expressions of the Form $a^{\frac{m}{n}}$**

Write each expression as a radical and evaluate, if possible.

(a) $16^{\frac{3}{2}}$ **(b)** $27^{\frac{2}{3}}$

Work Smart

When simplifying expressions of the form $a^{\frac{m}{n}}$, it is typically easier to evaluate the radical first. In Example 8(a), $16^{\frac{3}{2}} = (\sqrt{16})^3 = 4^3 = 64$ is easier than $16^{\frac{3}{2}} = \sqrt{16^3} = \sqrt{4096} = 64$.

Solution

(a) $16^{\frac{3}{2}} = (\sqrt{16})^3$
$\qquad = 4^3$
$\qquad = 64$

(b) $27^{\frac{2}{3}} = (\sqrt[3]{27})^2$
$\qquad = 3^2$
$\qquad = 9$

EXAMPLE 9 | **Evaluating Expressions of the Form $a^{\frac{m}{n}}$**

Write each expression as a radical and evaluate, if possible.

(a) $-9^{\frac{3}{2}}$ **(b)** $(-64)^{\frac{4}{3}}$ **(c)** $(-25)^{\frac{7}{2}}$

Solution

(a) $-9^{\frac{3}{2}} = -1 \cdot 9^{\frac{3}{2}}$
$\qquad = -1 \cdot (\sqrt{9})^3$
$\qquad = -1 \cdot 3^3$
$\qquad = -1 \cdot 27$
$\qquad = -27$

(b) $(-64)^{\frac{4}{3}} = (\sqrt[3]{-64})^4$
$\qquad = (-4)^4$
$\qquad = 256$

(c) $(-25)^{\frac{7}{2}}$ is not a real number because $(-25)^{\frac{7}{2}} = (\sqrt{-25})^7$, and $\sqrt{-25}$ is not a real number.

Quick ✓

22. If a is a real number and m/n is a rational number in lowest terms with $n \geq 2$, then $a^{\frac{m}{n}} = $ _____ or _____.

23. *True or False* To evaluate $8^{\frac{2}{3}}$, it is best to begin by squaring 8.

In Problem 24, write each expression as a radical and evaluate.

24. (a) $25^{\frac{3}{2}}$ **(b)** $64^{\frac{2}{3}}$ **(c)** $-16^{\frac{3}{4}}$

(d) $(-8)^{\frac{4}{3}}$ **(e)** $(-49)^{\frac{5}{2}}$

If an expression is written as a radical, we can rewrite it with rational exponents, using the definition of a rational exponent.

EXAMPLE 10   **Writing a Radical Expression with Rational Exponents**

Write each radical with a rational exponent:

(a) $\sqrt[3]{z^2}$ (b) $\sqrt[4]{p^7}$ (c) $(\sqrt[4]{8p^3})^5$

Solution

(a) $\sqrt[3]{z^2} = z^{\frac{2}{3}}$ (b) $\sqrt[4]{p^7} = p^{\frac{7}{4}}$ (c) $(\sqrt[4]{8p^3})^5 = (8p^3)^{\frac{5}{4}}$ ●

Quick ✔

In Problem 25, write each radical with a rational exponent.

25. (a) $\sqrt[4]{d^3}$ **(b)** $\sqrt[5]{y^8}$ **(c)** $(\sqrt[4]{11a^2b})^3$

Likewise, if an expression is written with rational exponents, we can rewrite it in radical form, using the definition of a rational exponent.

EXAMPLE 11   **Writing an Expression with Rational Exponents as a Radical**

Write each exponential expression as a radical:

(a) $b^{\frac{2}{3}}$ (b) $7x^{\frac{3}{5}}$ (c) $(7x)^{\frac{3}{5}}$

Solution

(a) $b^{\frac{2}{3}} = \sqrt[3]{b^2}$ (b) $7x^{\frac{3}{5}} = 7\sqrt[5]{x^3}$

$\sqrt[n]{a^m} = (\sqrt[n]{a})^m$: $= (\sqrt[3]{b})^2$ $\sqrt[n]{a^m} = (\sqrt[n]{a})^m$: $= 7(\sqrt[5]{x})^3$

(c) $(7x)^{\frac{3}{5}} = \sqrt[5]{(7x)^3}$

$\sqrt[n]{a^m} = (\sqrt[n]{a})^m$: $= (\sqrt[5]{7x})^3$ ●

Quick ✔

In Problem 26, write each expression in radical form.

26. (a) $z^{\frac{4}{5}}$ **(b)** $2n^{\frac{2}{3}}$ **(c)** $(2n)^{\frac{2}{3}}$

We can extend the definition of a negative integer exponent, given in Section 5.4 on page 337, to rational exponents.

Definition Negative-Exponent Rule

If $\dfrac{m}{n}$ is a rational number and $a \neq 0$, then

$$a^{-\frac{m}{n}} = \frac{1}{a^{\frac{m}{n}}} \quad \text{or} \quad \frac{1}{a^{-\frac{m}{n}}} = a^{\frac{m}{n}}$$

EXAMPLE 12   **Evaluating Expressions with Negative Rational Exponents**

Rewrite each of the following with a positive exponent and evaluate, if possible.

(a) $25^{-\frac{1}{2}}$ (b) $27^{-\frac{2}{3}}$ (c) $-4^{-\frac{3}{2}}$

Solution

(a) $a^{-\frac{m}{n}} = \frac{1}{a^{\frac{m}{n}}}$: $\quad 25^{-\frac{1}{2}} = \frac{1}{25^{\frac{1}{2}}}$

$a^{\frac{1}{2}} = \sqrt{a}$: $\qquad = \frac{1}{\sqrt{25}}$

$\qquad\qquad\qquad = \frac{1}{5}$

(b) $a^{-\frac{m}{n}} = \frac{1}{a^{\frac{m}{n}}}$: $\quad 27^{-\frac{2}{3}} = \frac{1}{27^{\frac{2}{3}}}$

$a^{\frac{m}{n}} = (\sqrt[n]{a})^{m}$: $\qquad = \frac{1}{(\sqrt[3]{27})^{2}}$

$\qquad\qquad\qquad = \frac{1}{3^{2}}$

$\qquad\qquad\qquad = \frac{1}{9}$

Work Smart

A negative exponent doesn't mean a negative answer.

(c) $\qquad\qquad -4^{-\frac{3}{2}} = -1 \cdot 4^{-\frac{3}{2}}$

$a^{-\frac{m}{n}} = \frac{1}{a^{\frac{m}{n}}}$: $\qquad = \frac{-1}{4^{\frac{3}{2}}}$

$a^{\frac{m}{n}} = (\sqrt[n]{a})^{m}$: $\qquad = \frac{-1}{(\sqrt{4})^{3}}$

$\qquad\qquad\qquad = \frac{-1}{2^{3}}$

$\qquad\qquad\qquad = \frac{-1}{8} = -\frac{1}{8}$

Quick ✓

In Problem 27, rewrite each of the following with positive exponents and evaluate, if possible.

27. (a) $36^{-\frac{1}{2}}$ **(b)** $8^{-\frac{4}{3}}$ **(c)** $-25^{-\frac{3}{2}}$

▶ **❺ Use Laws of Exponents to Simplify Expressions with Rational Exponents**

The Laws of Exponents for integer exponents (Chapter 5) also apply to rational exponents.

Laws of Exponents

If a and b are real numbers and if r and s are rational numbers, then, assuming the expression is defined,

Product Rule: $\qquad\qquad a^{r} \cdot a^{s} = a^{r+s}$

Quotient Rule: $\qquad\qquad \dfrac{a^{r}}{a^{s}} = a^{r-s} = \dfrac{1}{a^{s-r}}$ if $a \neq 0$

Power Rule: $\qquad\qquad (a^{r})^{s} = a^{r \cdot s}$

Product-to-a-Power Rule: $(ab)^{r} = a^{r} \cdot b^{r}$

Negative-Exponent Rule: $\qquad a^{-r} = \dfrac{1}{a^{r}}$ $\qquad\qquad$ if $a \neq 0$

The direction **simplify** shall mean the following:

- All the exponents are positive.
- The base only occurs once.
- There are no parentheses in the expression.
- There are no powers written to powers.

EXAMPLE 13 **Using Laws of Exponents to Simplify Expressions Involving Rational Exponents**

Simplify each of the following:

(a) $3^{\frac{1}{2}} \cdot 3^{\frac{3}{2}}$ (b) $\dfrac{7^{\frac{2}{3}}}{7^{\frac{5}{3}}}$ (c) $(x^4)^{\frac{1}{2}}$

Solution

(a) $a^r \cdot a^s = a^{r+s}$: $3^{\frac{1}{2}} \cdot 3^{\frac{3}{2}} = 3^{\frac{1}{2}+\frac{3}{2}}$

$= 3^{\frac{4}{2}}$

$= 3^2$

$= 9$

(b) $\dfrac{a^r}{a^s} = a^{r-s}$: $\dfrac{7^{\frac{2}{3}}}{7^{\frac{5}{3}}} = 7^{\frac{2}{3}-\frac{5}{3}}$

$= 7^{-\frac{3}{3}}$

$= 7^{-1}$

$= \dfrac{1}{7}$

(c) $(a^r)^s = a^{r \cdot s}$: $(x^4)^{\frac{1}{2}} = x^{4 \cdot \frac{1}{2}}$

$= x^2$

EXAMPLE 14 **Using Laws of Exponents to Simplify Expressions Involving Rational Exponents**

Simplify each of the following:

(a) $\left(16x^{\frac{2}{3}}\right)^{\frac{3}{4}}$ (b) $\dfrac{n^{\frac{4}{3}} \cdot n^{-\frac{5}{3}}}{n^{-\frac{2}{3}}}$

Solution

(a) $(ab)^r = a^r \cdot b^r$: $\left(16x^{\frac{2}{3}}\right)^{\frac{3}{4}} = 16^{\frac{3}{4}} \cdot \left(x^{\frac{2}{3}}\right)^{\frac{3}{4}}$

$(a^r)^s = a^{r \cdot s}$: $= 16^{\frac{3}{4}} \cdot \left(x^{\frac{2 \cdot 3}{3 \cdot 4}}\right)$

Simplify the exponent: $= 16^{\frac{3}{4}} \cdot \left(x^{\frac{1}{2}}\right)$

Write $16^{\frac{3}{4}}$ as a radical: $= \left(\sqrt[4]{16}\right)^3 \cdot x^{\frac{1}{2}}$

Simplify: $= 2^3 \cdot x^{\frac{1}{2}}$

$= 8x^{\frac{1}{2}}$

(b) $a^r \cdot a^s = a^{r+s}$: $\dfrac{n^{\frac{4}{3}} \cdot n^{-\frac{5}{3}}}{n^{-\frac{2}{3}}} = \dfrac{n^{\frac{4}{3}+\left(-\frac{5}{3}\right)}}{n^{-\frac{2}{3}}}$

$= \dfrac{n^{-\frac{1}{3}}}{n^{-\frac{2}{3}}}$

$\dfrac{a^r}{a^s} = a^{r-s}$: $= n^{-\frac{1}{3}-\left(-\frac{2}{3}\right)}$

$= n^{-\frac{1}{3}+\left(\frac{2}{3}\right)}$

$= n^{\frac{1}{3}}$

Quick ✓

In Problem 28, use Laws of Exponents to simplify each expression.

28. (a) $7^{-\frac{5}{12}} \cdot 7^{\frac{7}{12}}$ **(b)** $\dfrac{11^{\frac{6}{5}}}{11^{\frac{3}{5}}}$ **(c)** $\left(2^{\frac{3}{8}}\right)^{\frac{16}{3}}$ **(d)** $\dfrac{a^{\frac{3}{2}} \cdot a^{\frac{5}{2}}}{a^{-\frac{3}{2}}}$

8.7 Exercises MyMathLab®   Math XL PRACTICE   Exercise numbers in green have complete video solutions in MyMathLab.

*Problems **1–28** are the Quick ✓s that follow the **EXAMPLES**.*

Building Skills

In Problems 29–48, evaluate each expression. See Objective 1.

29. the cube root of 8   **30.** the cube root of -1

31. the fourth root of 16   **32.** the fourth root of -1

33. the fourth root of 81   **34.** the cube root of 27

35. the fifth root of -32 **36.** the cube root of -125

37. $\sqrt[3]{27}$ **38.** $\sqrt[3]{64}$

39. $\sqrt[4]{-1}$ **40.** $\sqrt[4]{-81}$

41. $\sqrt[3]{-125}$ **42.** $\sqrt[3]{-1000}$

43. $\sqrt[4]{625}$ **44.** $\sqrt[4]{81}$

45. $\sqrt[6]{0}$ **46.** $\sqrt[6]{1}$

47. $\sqrt[5]{243}$ **48.** $\sqrt[5]{32}$

In Problems 49–56, simplify the radical. See Objective 1.

49. $\sqrt[6]{5^6}$ **50.** $\sqrt[8]{7^8}$

51. $\sqrt[4]{n^4}$, $n \geq 0$ **52.** $\sqrt[12]{z^{12}}$, $z \geq 0$

53. $\sqrt[5]{r^5}$ **54.** $\sqrt[3]{(16p)^3}$

55. $\sqrt[7]{(3a)^7}$ **56.** $\sqrt[3]{w^3}$

In Problems 57–74, simplify each expression. All variables represent positive real numbers. See Objective 2.

57. $\sqrt[3]{16}$ **58.** $\sqrt[3]{24}$

59. $\sqrt[4]{-64}$ **60.** $\sqrt[4]{-162}$

61. $\sqrt[3]{-40}$ **62.** $\sqrt[3]{-48}$

63. $\dfrac{\sqrt[4]{243}}{\sqrt[4]{3}}$ **64.** $\dfrac{\sqrt[4]{64}}{\sqrt[4]{4}}$

65. $-\sqrt[4]{32n^8}$ **66.** $-\sqrt[4]{48n^{12}}$

67. $-\sqrt[3]{-128}$ **68.** $-\sqrt[3]{-250}$

69. $\sqrt[3]{81x^3}$ **70.** $\sqrt[3]{54x^3}$

71. $\dfrac{\sqrt[3]{56b^5}}{\sqrt[3]{7b^2}}$ **72.** $\dfrac{\sqrt[3]{320w^7}}{\sqrt[3]{5w}}$

73. $\dfrac{\sqrt[3]{-108z^8}}{\sqrt[3]{2z^2}}$ **74.** $\dfrac{\sqrt[4]{96a^{13}}}{\sqrt[4]{2a}}$

In Problems 75–82, evaluate each expression. See Objective 3.

75. $-16^{\frac{1}{4}}$ **76.** $-81^{\frac{1}{4}}$

77. $-8^{\frac{1}{3}}$ **78.** $-216^{\frac{1}{3}}$

79. $(-216)^{\frac{1}{3}}$ **80.** $(-8)^{\frac{1}{3}}$

81. $(-36)^{\frac{1}{2}}$ **82.** $(-25)^{\frac{1}{2}}$

In Problems 83–88, write the expression as a radical and simplify. See Objective 3.

83. $(2a)^{\frac{1}{2}}$ **84.** $3x^{\frac{1}{3}}$

85. $7z^{\frac{1}{4}}$ **86.** $(125p)^{\frac{1}{3}}$

87. $(-27r)^{\frac{1}{3}}$ **88.** $9b^{\frac{1}{5}}$

In Problems 89–94, write each radical as a rational expression. Assume that each variable is a nonnegative real number. See Objective 3.

89. $\sqrt{c}$ **90.** $\sqrt[3]{x}$

91. $\sqrt{2y}$ **92.** $\sqrt[3]{5z}$

93. $7\sqrt{a}$ **94.** $-11\sqrt[3]{x}$

In Problems 95–106, write each expression as a radical and evaluate. See Objective 4.

95. $64^{\frac{3}{2}}$ **96.** $8^{\frac{2}{3}}$

97. $81^{-\frac{1}{2}}$ **98.** $4^{-\frac{1}{2}}$

99. $64^{\frac{2}{3}}$ **100.** $27^{\frac{4}{3}}$

101. $-9^{-\frac{3}{2}}$ **102.** $(-16)^{\frac{5}{4}}$

103. $(-81)^{\frac{3}{4}}$ **104.** $-16^{-\frac{3}{2}}$

105. $49^{-\frac{3}{2}}$ **106.** $9^{-\frac{5}{2}}$

In Problems 107–112, write each exponential expression as a radical. See Objective 4.

107. $u^{\frac{2}{3}}$ **108.** $v^{\frac{3}{5}}$

109. $4x^{\frac{2}{3}}$ **110.** $(4x)^{\frac{3}{4}}$

111. $(5n)^{\frac{2}{3}}$ **112.** $-7y^{\frac{4}{5}}$

In Problems 113–118, write each radical as an expression with a rational exponent. See Objective 4.

113. $\sqrt[4]{p^5}$ **114.** $\sqrt[4]{n^3}$

115. $5\sqrt[4]{t^3}$ **116.** $-4\sqrt[3]{z^2}$

117. $\sqrt[3]{(5a)^2}$ **118.** $\sqrt[4]{(7y)^3}$

In Problems 119–130, use Laws of Exponents to simplify each of the following. All variables represent positive real numbers. See Objective 5.

119. $4^{\frac{2}{3}} \cdot 4^{\frac{4}{3}}$

120. $6^{\frac{12}{5}} \cdot 6^{-\frac{2}{5}}$

121. $\dfrac{6^{\frac{3}{2}}}{6^{\frac{7}{2}}}$

122. $\dfrac{5^{\frac{1}{3}}}{5^{\frac{4}{3}}}$

123. $(x^9)^{\frac{2}{3}}$

124. $(y^4)^{\frac{5}{2}}$

125. $(8n^{\frac{1}{2}})^{\frac{2}{3}}$

126. $(16y^{\frac{1}{3}})^{\frac{3}{4}}$

127. $\dfrac{x^{-\frac{1}{4}} \cdot x^{\frac{3}{4}}}{x^{\frac{1}{4}}}$

128. $\dfrac{x^{-\frac{3}{5}} \cdot x^{\frac{9}{5}}}{x^{\frac{2}{5}}}$

129. $\dfrac{(2x)^{-\frac{3}{2}}}{(2x)^{-\frac{1}{2}}}$

130. $\dfrac{(3x)^{\frac{5}{4}}}{(3x)^{-\frac{3}{4}}}$

Mixed Practice

In Problems 131–144, simplify each expression. Assume that all variables represent positive real numbers.

131. $-25^{\frac{3}{2}}$

132. $-16^{\frac{3}{2}}$

133. $\dfrac{6^{\frac{5}{2}}}{6^{\frac{7}{2}}}$

134. $\dfrac{5^{\frac{1}{3}}}{5^{\frac{7}{3}}}$

135. $-125^{-\frac{2}{3}}$

136. $-64^{-\frac{4}{3}}$

137. $4^{\frac{1}{2}} + 25^{\frac{3}{2}}$

138. $100^{\frac{1}{2}} - 4^{\frac{3}{2}}$

139. $(\sqrt[4]{25})^2$

140. $(\sqrt[6]{27})^2$

141. $(25^{\frac{3}{4}} \cdot 4^{-\frac{3}{4}})^2$

142. $(36^{-\frac{1}{4}} \cdot 9^{\frac{1}{4}})^{-2}$

143. $\dfrac{y^{-\frac{1}{3}} \cdot y^{-\frac{1}{3}}}{y^{\frac{4}{3}}}$

144. $\dfrac{n^{-\frac{5}{4}} \cdot n^{-\frac{1}{4}}}{n^{\frac{1}{2}}}$

Applying the Concepts

145. Evaluate $2x^{-\frac{3}{2}}$ for $x = 4$.

146. Evaluate $10x^{-\frac{2}{3}}$ for $x = 8$.

147. Evaluate $(27x)^{\frac{1}{3}}$ for $x = 8$.

148. Evaluate $(36x)^{\frac{1}{2}}$ for $x = 4$.

Extending the Concepts

In Problems 149–154, simplify each expression. Express your answer in radical form. All variables represent positive real numbers.

149. $x^{\frac{1}{2}} \cdot x^{-\frac{3}{4}}$

150. $x^{-\frac{3}{2}} \cdot x^{\frac{7}{6}}$

151. $\dfrac{(-8a^4)^{\frac{1}{3}}}{a^{\frac{5}{6}}}$

152. $\dfrac{(100a^3)^{\frac{1}{2}}}{(16a)^{\frac{3}{4}}}$

153. $\dfrac{(3x^{\frac{2}{3}})^4}{x^{\frac{5}{6}}}$

154. $\dfrac{(2x^{\frac{2}{5}})^3}{x^{\frac{7}{10}}}$

Explaining the Concepts

155. Explain why $a^{\frac{1}{n}}$ is undefined when a is negative and n is even. Give an example to support your response.

156. Explain the difference between $-9^{\frac{1}{2}}$ and $(-9)^{\frac{1}{2}}$.

Chapter 8 Activity: Working Together with Radicals

Focus: As a group, perform operations on radicals.

Time: 20 minutes

Group size: 3–4

As a group, answer the following questions:

1. Assuming that $q = 81x^9y^2$, find: $\sqrt{q}, \sqrt[3]{q}, \sqrt[4]{q}$

2. Assuming that $x = a\sqrt[3]{125a^3b^4c^2}$

   $y = 10b\sqrt[3]{a^6bc^2}$, find: $x + y, x - y, x \cdot y, x \div y$

3. Assuming that $a = 2x^{\frac{5}{4}}y^{\frac{1}{3}}$

   $b = 6x^{\frac{2}{3}}y^{\frac{5}{2}}$

   $c = 4x^{\frac{1}{5}}y$, find: $a + c, a \cdot b, b \div c, \dfrac{ab}{c}$

Chapter 8 Review

Section 8.1 Introduction to Square Roots

KEY CONCEPTS

- The principal square root of any positive number is positive.
- The principal square root of 0 is 0 because $0^2 = 0$. That is, $\sqrt{0} = 0$.
- The square root of a perfect square is a rational number.
- The square root of a positive rational number that is not a perfect square is an irrational number.
- The square root of a negative real number is not a real number.
- For any real number a, $\sqrt{a^2} = |a|$. If $a \geq 0$, $\sqrt{a^2} = a$.

KEY TERMS

Square root
Radical
Principal square root
Radicand
Perfect square

You Should Be Able To...	EXAMPLE	Review Exercises
1 Evaluate square roots (p. 519)	Examples 1 through 6	1–12
2 Determine whether a square root is rational, irrational, or not a real number (p. 521)	Examples 7 and 8	13–16
3 Find square roots of variable expressions (p. 522)	Examples 9 and 10	17, 18

In Problems 1 and 2, find the value of each expression.

1. the square roots of 4

2. the square roots of 81

In Problems 3–12, find the exact value of each expression.

3. $-\sqrt{1}$

4. $-\sqrt{25}$

5. $\sqrt{0.16}$

6. $\sqrt{0.04}$

7. $\dfrac{3}{2}\sqrt{\dfrac{25}{36}}$

8. $\dfrac{4}{3}\sqrt{\dfrac{81}{4}}$

9. $\sqrt{25 - 9}$

10. $\sqrt{169 - 25}$

11. $\sqrt{9^2 - (4)(5)(-18)}$

12. $\sqrt{13^2 - (4)(-3)(-4)}$

In Problems 13–16, determine whether each square root is rational, irrational, or not a real number. Then evaluate the square root. For each square root that is irrational, use a calculator or Appendix A to round your answer to two decimal places.

13. $-\sqrt{9}$

14. $-\dfrac{1}{2}\sqrt{48}$

15. $\sqrt{14}$

16. $\sqrt{-2}$

In Problems 17 and 18, simplify each square root. Assume that the variable can be any real number.

17. $\sqrt{(4x - 9)^2}$

18. $\sqrt{(16m - 25)^2}$

Section 8.2 Simplifying Square Roots

KEY CONCEPTS

- **Product Rule for Square Roots**
  If $\sqrt{a}$ and $\sqrt{b}$ are nonnegative real numbers, then $\sqrt{ab} = \sqrt{a} \cdot \sqrt{b}$.
- **Quotient Rule for Square Roots**
  If $\sqrt{a}$ and $\sqrt{b}$ are nonnegative real numbers and $b \neq 0$, then $\sqrt{\dfrac{a}{b}} = \dfrac{\sqrt{a}}{\sqrt{b}}$.

You Should Be Able To...	EXAMPLE	Review Exercises
1 Use the Product Rule to simplify square roots of constants (p. 525)	Examples 1 through 3	19–24
2 Use the Product Rule to simplify square roots of variable expressions (p. 527)	Examples 4 and 5	25–30
3 Use the Quotient Rule to simplify square roots (p. 528)	Examples 6 through 8	31–34

In Problems 19–24, simplify each of the square roots.

19. $\sqrt{28}$ **20.** $\sqrt{45}$

21. $\sqrt{200}$ **22.** $\sqrt{150}$

23. $\dfrac{-2 + \sqrt{8}}{2}$ **24.** $\dfrac{-3 + \sqrt{27}}{6}$

In Problems 25–30, simplify each of the following. Assume that the variables represent nonnegative real numbers.

25. $\sqrt{a^{36}}$ **26.** $\sqrt{x^{16}}$

27. $\sqrt{16x^{10}}$ **28.** $\sqrt{49a^{12}}$

29. $\sqrt{18n^9}$ **30.** $\sqrt{8y^{25}}$

In Problems 31–34, find the quotient and simplify. Assume that variables represent positive real numbers.

31. $\sqrt{\dfrac{27}{3x^8}}$ **32.** $\sqrt{\dfrac{8}{2x^4}}$

33. $\sqrt{\dfrac{81y^5}{25y}}$ **34.** $\sqrt{\dfrac{36n}{121n^9}}$

Section 8.3   Adding and Subtracting Square Roots

KEY CONCEPT

- To add or subtract square roots, the square roots must have the same radicand.

KEY TERM

Like square roots

You Should Be Able To...	EXAMPLE	Review Exercises
1 Add and subtract square root expressions with like square roots (p. 532)	Examples 1 and 2	35–40
2 Add and subtract square root expressions with unlike square roots (p. 533)	Examples 3 through 5	41–50

In Problems 35–40, add or subtract as indicated. Assume all variables represent nonnegative real numbers.

35. $\sqrt{7} - 3\sqrt{7}$ **36.** $4\sqrt{11} - \sqrt{11}$

37. $4\sqrt{x} - 3\sqrt{x}$ **38.** $5\sqrt{n} - 6\sqrt{n}$

39. $20a\sqrt{ab} + 5a\sqrt{ba}$ **40.** $2x\sqrt{3xy} + x\sqrt{3yx}$

In Problems 41–50, add or subtract as indicated. Assume all variables represent nonnegative real numbers.

41. $-2\sqrt{12} + 3\sqrt{27}$ **42.** $-4\sqrt{18} + 5\sqrt{32}$

43. $4 + 2\sqrt{20} - \sqrt{45}$ **44.** $6 - 2\sqrt{27} + \sqrt{48}$

45. $2n\sqrt{8n^3} + 5\sqrt{18n^5}$ **46.** $3\sqrt{56a^5} + a^2\sqrt{126a}$

47. $\dfrac{3}{4}\sqrt{32} + \dfrac{2}{3}\sqrt{27} - \dfrac{1}{2}\sqrt{8}$

48. $\dfrac{5}{2}\sqrt{24} + \dfrac{2}{9}\sqrt{27} - \dfrac{1}{6}\sqrt{54}$

49. $3\sqrt{\dfrac{3}{16}} - \dfrac{1}{2}\sqrt{\dfrac{12}{25}}$

50. $2\sqrt{\dfrac{2}{25}} - 3\sqrt{\dfrac{8}{9}}$

Section 8.4   Multiplying Expressions with Square Roots

KEY CONCEPTS

- If $\sqrt{a}$ and $\sqrt{b}$ are nonnegative real numbers, then $\sqrt{a} \cdot \sqrt{b} = \sqrt{ab}$.
- $(\sqrt{a})^2 = a$ and $(-\sqrt{a})^2 = a$ for any real number a, $a \geq 0$.

KEY TERM

Conjugates

You Should Be Able To...	EXAMPLE	Review Exercises
❶ Find the product of square roots containing one term (p. 536)	Examples 1 through 6	51–60
❷ Find the product of square roots using the Distributive Property (p. 539)	Example 7	61, 62
❸ Find the product of square roots using FOIL (p. 540)	Examples 8 and 9	63–66
❹ Find the product of square roots using special products: $(A + B)^2$, $(A - B)^2$, and $(A + B)(A - B)$ (p. 541)	Examples 10 and 11	67–72

In Problems 51–60, find the product and simplify. Assume all variables represent nonnegative real numbers.

51. $\sqrt{12} \cdot \sqrt{8}$ **52.** $\sqrt{24} \cdot \sqrt{10}$

53. $\sqrt{14x^3} \cdot \sqrt{21x^2}$ **54.** $\sqrt{6y} \cdot \sqrt{30y^4}$

55. $-4\sqrt{20} \cdot 8\sqrt{8}$ **56.** $-2\sqrt{10} \cdot 5\sqrt{40}$

57. $(2\sqrt{2a})^2$ **58.** $(3\sqrt{5n})^2$

59. $4\sqrt{2} \cdot (-3\sqrt{2})$ **60.** $7\sqrt{11} \cdot (-2\sqrt{11})$

In Problems 61 and 62, find the product and simplify. Assume all variables represent nonnegative real numbers.

61. $\sqrt{3}(\sqrt{3a} + \sqrt{15a})$ **62.** $\sqrt{x}(\sqrt{4x} + \sqrt{8x})$

In Problems 63–66, find the product and simplify. Assume all variables represent nonnegative real numbers.

63. $(3 + 2\sqrt{6})(2 + 4\sqrt{3})$ **64.** $(5 - 3\sqrt{3})(3 + 2\sqrt{6})$

65. $(4 - 2\sqrt{2y})(\sqrt{y} + 3)$ **66.** $(1 + 3\sqrt{3x})(\sqrt{x} + 9)$

In Problems 67–72, find the product and simplify. Assume all variables represent nonnegative real numbers.

67. $(3 - \sqrt{2})^2$ **68.** $(4 - \sqrt{5})^2$

69. $(\sqrt{2x} + 3\sqrt{x})^2$ **70.** $(\sqrt{n} + 2\sqrt{3n})^2$

71. $(3 + 5\sqrt{x})(3 - 5\sqrt{x})$ **72.** $(2\sqrt{x} - 4)(2\sqrt{x} + 4)$

Section 8.5 Dividing Expressions with Square Roots

KEY CONCEPTS

- To rationalize the denominator when the denominator contains a single square root, multiply by a factor of 1, such that the product forms a radicand in the denominator that is a perfect square.

- To rationalize a denominator containing two terms, multiply the numerator and the denominator by the conjugate of the denominator.

KEY TERM

Rationalizing the denominator

You Should Be Able To...	EXAMPLE	Review Exercises
❶ Find the quotient of two square roots (p. 544)	Examples 1 and 2	73–78
❷ Rationalize a denominator containing one term (p. 545)	Examples 3 and 4	79–82
❸ Rationalize a denominator containing two terms (p. 547)	Examples 5 through 7	85–88

In Problems 73–78, simplify the quotient. Assume all variables represent nonnegative real numbers.

73. $\dfrac{\sqrt{108}}{\sqrt{2}}$ **74.** $\dfrac{\sqrt{135}}{\sqrt{3}}$

75. $\sqrt{\dfrac{6x}{4x^3}}$ **76.** $\sqrt{\dfrac{15x^4}{36x^2}}$

77. $\dfrac{\sqrt{10} + \sqrt{12}}{\sqrt{2}}$ **78.** $\dfrac{\sqrt{30} - \sqrt{15}}{\sqrt{5}}$

In Problems 79–82, rationalize the denominator of each expression.

79. $\dfrac{14}{\sqrt{7}}$ **80.** $\dfrac{25}{\sqrt{5}}$

81. $\dfrac{3}{2\sqrt{6}}$ **82.** $\dfrac{2}{5\sqrt{10}}$

In Problems 83 and 84, determine the conjugate of each expression. Then find the product of the expression and its conjugate.

83. $-2 + 3\sqrt{5}$

84. $-3 - 2\sqrt{7}$

In Problems 85–88, rationalize the denominator.

85. $\dfrac{4}{2 - \sqrt{2}}$

86. $\dfrac{3}{6 + \sqrt{3}}$

87. $\dfrac{\sqrt{12}}{\sqrt{3} + \sqrt{6}}$

88. $\dfrac{\sqrt{8}}{\sqrt{6} - \sqrt{2}}$

Section 8.6   Solving Equations Containing Square Roots

KEY CONCEPT

- Steps to solve a radical equation containing one radical (see p. 553)

KEY TERMS

Radical equation

Solution to a radical equation

Extraneous solution

You Should Be Able To...	EXAMPLE	Review Exercises
1 Determine whether a number is a solution of a radical equation (p. 551)	Example 1	89, 90
2 Solve equations containing one square root (p. 552)	Examples 2 through 6	91–98
3 Solve equations containing two square roots (p. 556)	Examples 7 through 9	99–102
4 Solve problems modeled by radical equations (p. 558)	Example 10	103, 104

In Problems 89 and 90, determine whether the number is a solution to the equation.

89. Is $x = 8$ a solution of $\sqrt{3x + 1} = 5$?

90. Is $x = 4$ a solution of $3\sqrt{x} - 9 = -3$?

In Problems 91–98, solve the equation and check the solution.

91. $\sqrt{3 - x} = 4$

92. $5 = \sqrt{2x - 1}$

93. $3\sqrt{2x} + 4 = 10$

94. $5\sqrt{x} - 3 = 12$

95. $x = \sqrt{8x - 15}$

96. $\sqrt{3x + 18} = x$

97. $n + 2 = \sqrt{n^2 + 5}$

98. $n + 3 = \sqrt{n^2 - 9}$

In Problems 99–102, solve the equation and check the solution.

99. $\sqrt{2x + 7} = \sqrt{5x - 2}$

100. $3\sqrt{a} - \sqrt{4a + 10} = 0$

101. $\sqrt{2x - 8} = 3 + \sqrt{2x + 1}$

102. $\sqrt{x - 6} = 2 - \sqrt{x - 10}$

△ **103.** Use the formula $r = \sqrt{\dfrac{A}{\pi}}$, where r represents the radius and A represents the area of a circle, to find the exact area of a circle whose radius is $3\sqrt{2}$ cm.

△ **104.** The formula $r = \sqrt{\dfrac{3V}{\pi h}}$ represents the relation among the radius of a cone, its volume, and its height. If the radius of the base of the cone, r, is 7 inches and the volume, V, is 98π cubic inches, find the height, h, of the cone.

Section 8.7   Higher Roots and Rational Exponents

KEY CONCEPTS

- The principal nth root of a number a, symbolized by $\sqrt[n]{a}$, where $n \geq 2$ is an integer, is defined as follows: $\sqrt[n]{a} = b$ means $a = b^n$.
  - If $n \geq 2$ and even, then a and b must be greater than or equal to 0.
  - If $n \geq 3$ and odd, then a and b can be any real number.
- If $n \geq 2$ is an integer, then

  $\sqrt[n]{a^n} = a$ if n is odd $\sqrt[n]{a^n} = |a|$ if n is even

- If $\sqrt[n]{a}$ and $\sqrt[n]{b}$ are real numbers and $n \geq 2$ is an integer, then $\sqrt[n]{ab} = \sqrt[n]{a} \cdot \sqrt[n]{b}$.

KEY TERMS

Cube root

Principal nth root

Index

(continued)

- If $\sqrt[n]{a}$ and $\sqrt[n]{b}$ are real numbers, $b \neq 0$, and $n \geq 2$ is an integer, then
$$\sqrt[n]{\frac{a}{b}} = \frac{\sqrt[n]{a}}{\sqrt[n]{b}}.$$

- If a is a real number and $n \geq 2$ is an integer, then $a^{\frac{1}{n}} = \sqrt[n]{a}$ provided that $\sqrt[n]{a}$ exists.

- If a is a real number and $\dfrac{m}{n}$ is a rational number in lowest terms with $n \geq 2$, then $a^{\frac{m}{n}} = \sqrt[n]{a^m} = (\sqrt[n]{a})^m$ provided that $\sqrt[n]{a}$ exists.

- If $\dfrac{m}{n}$ is a rational number and if a is a nonzero real number, then $a^{-\frac{m}{n}} = \dfrac{1}{a^{\frac{m}{n}}}$, or
$$\frac{1}{a^{-\frac{m}{n}}} = a^{\frac{m}{n}}.$$

- If a and b are real numbers and if r and s are rational numbers, then, assuming the expression is defined,

  Product Rule: $a^r \cdot a^s = a^{r+s}$

  Power Rule: $(a^r)^s = a^{r \cdot s}$

  Negative-Exponent Rule: $a^{-r} = \dfrac{1}{a^r}$ if $a \neq 0$

  Quotient Rule: $\dfrac{a^r}{a^s} = a^{r-s} = \dfrac{1}{a^{s-r}}$ if $a \neq 0$

  Power-of-a-Product Rule: $(ab)^r = a^r b^r$

You Should Be Able To...	EXAMPLE	Review Exercises
❶ Evaluate higher roots (p. 561)	Examples 1 and 2	105, 106
❷ Use Product and Quotient Rules to simplify higher roots (p. 562)	Examples 3 and 4	107–110
❸ Define and evaluate expressions of the form $a^{\frac{1}{n}}$ (p. 564)	Examples 5 through 7	111, 112
❹ Define and evaluate expressions of the form $a^{\frac{m}{n}}$ (p. 565)	Examples 8 through 12	113–120
❺ Use Laws of Exponents to simplify expressions with rational exponents (p. 568)	Examples 13 and 14	121–124

In Problems 105 and 106, evaluate each expression.

105. $\sqrt[3]{-8}$ **106.** $\sqrt[4]{-16}$

In Problems 107–110, simplify each expression. Assume all variables represent positive real numbers.

107. $\sqrt[4]{48x^4}$ **108.** $\sqrt[3]{625x^3}$

109. $\dfrac{\sqrt[3]{48n^6}}{\sqrt[3]{2n^2}}$ **110.** $\dfrac{\sqrt[4]{324z^7}}{\sqrt[4]{2z}}$

In Problems 111 and 112, evaluate each expression.

111. $-16^{\frac{1}{4}}$ **112.** $(-16)^{\frac{1}{4}}$

In Problems 113 and 114, write each radical with a rational exponent.

113. $\sqrt[4]{x^3}$ **114.** $\sqrt[3]{z}$

In Problems 115 and 116, write each expression as a radical.

115. $2x^{\frac{3}{2}}$ **116.** $(3v)^{\frac{1}{3}}$

In Problems 117–120, evaluate each expression, if possible.

117. $27^{\frac{2}{3}}$ **118.** $16^{\frac{5}{4}}$

119. $-16^{-\frac{3}{2}}$ **120.** $32^{-\frac{3}{5}}$

In Problems 121–124, simplify each of the following. Assume all variables represent positive real numbers.

121. $2^{\frac{2}{3}} \cdot 2^{\frac{4}{3}}$ **122.** $6^{\frac{1}{2}} \cdot 6^{\frac{3}{2}}$

123. $\dfrac{x^{\frac{3}{4}} \cdot x^{-\frac{1}{4}}}{x^{\frac{5}{4}}}$ **124.** $\left(27a^{-\frac{1}{2}}\right)^{-\frac{2}{3}}$

Chapter 8 Test

Step-by-step test solutions are found on the Chapter Test Prep Videos available in MyMathLab® *or on* YouTube™.

In Problems 1–4, simplify each of the following. Assume all variables represent positive real numbers.

1. $-3\sqrt{12}$

2. $\sqrt{50x^2y^3}$

3. $\sqrt{\dfrac{16x^4}{9x^2}}$

4. $\sqrt{25 - 4(3)(-2)}$

In Problems 5–12, perform the indicated operation and simplify. Assume all variables represent positive real numbers.

5. $3\sqrt{2} - \sqrt{8}$

6. $3\sqrt{27} + 2\sqrt{48}$

7. $\sqrt{45} \cdot \sqrt{15}$

8. $2\sqrt{2}(3\sqrt{6} - 2)$

9. $(4 - \sqrt{3})^2$

10. $\dfrac{\sqrt{36x^4}}{\sqrt{3x}}$

11. $(2 - 3\sqrt{2})(4 + \sqrt{2})$

12. $\dfrac{-3 + \sqrt{162}}{6}$

In Problems 13 and 14, rationalize each denominator.

13. $\dfrac{12}{\sqrt{20}}$

14. $\dfrac{6}{\sqrt{2} + 2}$

In Problems 15–17, solve each equation.

15. $3\sqrt{2x - 1} = 9$

16. $2\sqrt{x + 11} - 3 = x$

17. $\sqrt{x^2 - 3x} = \sqrt{x + 21}$

In Problems 18 and 19, simplify each expression.

18. $\sqrt[3]{108x^5}$

19. $27^{-\frac{2}{3}}$

20. Two legs of a right triangle measure 3 feet and 6 feet. Find the exact length of the hypotenuse, and then find the decimal approximation to the nearest tenth of a foot. (*Hint:* $c = \sqrt{a^2 + b^2}$, where c is the length of the hypotenuse and a and b are the lengths of the legs.)

9 Quadratic Equations

Did you know that an equation from economics may be used to predict your annual income on the basis of your age? At what age can you expect an annual income of \$50,000? See Quick Check Question 7 on page 614.

The Big Picture: Putting It Together

In Chapter 2, we solved linear (first-degree) equations. In Chapter 6, we introduced quadratic equations, which are second-degree equations of the form $ax^2 + bx + c = 0$. We solved quadratic equations in Chapter 6 by factoring the expression $ax^2 + bx + c$ and setting each factor equal to zero. But not all quadratic equations can be solved using factoring.

In Chapter 9, we use our knowledge of radicals from Chapter 8 to develop additional techniques for solving any quadratic equation.

9.1 Solving Quadratic Equations Using the Square Root Property

Objectives

1. Solve Quadratic Equations Using the Square Root Property
2. Solve Problems Using the Pythagorean Theorem

Are You Ready for This Section?

Before getting started, take the following readiness quiz. If you get a problem wrong, go back to the section cited and review the material.

R1. Simplify: **(a)** $\sqrt{25}$ **(b)** $\sqrt{48}$ [Section 8.1, pp. 519–521; Section 8.2, pp. 525–526]

R2. Simplify: $\dfrac{3 - \sqrt{72}}{6}$ [Section 8.2, p. 527]

R3. Factor: $z^2 - 8z + 16$ [Section 6.4, pp. 395–397]

In Chapter 6, you learned to solve quadratic equations by factoring. A quadratic equation can be written in the form $ax^2 + bx + c = 0$, where a, b, and c are real numbers and $a \neq 0$. Consider how we solve the quadratic equation $x^2 - 2x - 8 = 0$:

$$x^2 - 2x - 8 = 0$$

Factor the polynomial: $(x - 4)(x + 2) = 0$

Use the Zero-Product Property: $x - 4 = 0$ or $x + 2 = 0$

Solve the resulting equations: $x = 4$ or $x = -2$

The solution set is $\{-2, 4\}$.

However, some quadratic equations of the form $ax^2 + bx + c = 0$ cannot be solved by factoring, so we need other methods. The next three sections contain three more methods for solving quadratic equations. This section presents the *Square Root Property*.

1 Solve Quadratic Equations Using the Square Root Property

▶ Suppose we want to solve the quadratic equation

$$x^2 = p$$

where p is any real number. This equation is saying, "Give me all numbers whose square is p." For example, the equation $x^2 = 16$ means we want "all numbers whose square is 16." The two numbers whose square is 16 are -4 and 4, so the solution set to $x^2 = 16$ is $\{-4, 4\}$.

The following method is used to solve equations of the form $x^2 = p$.

Work Smart

The Square Root Property is useful for solving equations of the form "some unknown squared equals a real number." To solve this equation, we take the square root of both sides of the equation. Because any positive real number has two square roots, don't forget the $\pm$ symbol!

The Square Root Property

If $x^2 = p$, where $p \geq 0$, then $x = \sqrt{p}$ or $x = -\sqrt{p}$.

When using the Square Root Property to solve an equation such as $x^2 = p$, we usually write the solutions as $x = \pm\sqrt{p}$, which is read "x equals plus or minus the square root of p." For example, the two solutions of the equation

$$x^2 = 16$$

are

$$x = \pm\sqrt{16}$$
$$= \pm 4$$

Ready? …Answers **R1. (a)** 5 **(b)** $4\sqrt{3}$
R2. $\dfrac{1 - 2\sqrt{2}}{2}$ or $\dfrac{1}{2} - \sqrt{2}$
R3. $(z - 4)^2$

EXAMPLE 1 **How to Solve a Quadratic Equation Using the Square Root Property**

Solve the quadratic equation $x^2 - 32 = 0$.

Step-by-Step Solution

Step 1: Isolate the expression containing the squared term.

$$x^2 - 32 = 0$$

Add 32 to each side: $\quad x^2 = 32$

Step 2: Use the Square Root Property. Don't forget the $\pm$ symbol.

Use the Square Root Property: $\quad x = \pm\sqrt{32}$

Simplify the radical: $\quad x = \pm\sqrt{16 \cdot 2}$

$$x = \pm 4\sqrt{2}$$

Step 3: Isolate the variable, if necessary.

The variable is already isolated.

Step 4: Verify your solutions.

$x = -4\sqrt{2}$: $\quad \left(-4\sqrt{2}\right)^2 - 32 \overset{?}{=} 0$ $\qquad\qquad$ $x = 4\sqrt{2}$: $\quad \left(4\sqrt{2}\right)^2 - 32 \overset{?}{=} 0$

$(ab)^2 = a^2b^2$: $\quad (-4)^2\left(\sqrt{2}\right)^2 - 32 \overset{?}{=} 0$ $\qquad$ $(ab)^2 = a^2b^2$: $\quad (4)^2\left(\sqrt{2}\right)^2 - 32 \overset{?}{=} 0$

$$16 \cdot 2 - 32 \overset{?}{=} 0 \qquad\qquad\qquad\qquad\qquad 16 \cdot 2 - 32 \overset{?}{=} 0$$

$$0 = 0 \text{ True} \qquad\qquad\qquad\qquad\qquad\qquad 0 = 0 \text{ True}$$

The solution set is $\left\{-4\sqrt{2}, 4\sqrt{2}\right\}$.

You can solve a quadratic equation in the form $x^2 = p$, where p is a perfect square, using the Square Root Property or by factoring.

SQUARE ROOT PROPERTY $\qquad\qquad\qquad\qquad\qquad\qquad\qquad$ **FACTORING**

$$y^2 = 25 \qquad\qquad\qquad\qquad\qquad\qquad\qquad\qquad\qquad\qquad y^2 = 25$$

Use the Square Root Property: $\quad y = \pm\sqrt{25}$ $\qquad$ Subtract 25 from both sides: $\qquad y^2 - 25 = 0$

$$y = \pm 5 \qquad \text{Factor the difference of two squares:} \qquad (y - 5)(y + 5) = 0$$

$$y = 5 \text{ or } y = -5 \qquad \text{Zero-Product Property:} \quad y - 5 = 0 \quad \text{or} \quad y + 5 = 0$$

$$y = 5 \quad \text{or} \qquad y = -5$$

Which approach do you prefer?

    Below, we summarize the steps that are taken to solve quadratic equations using the Square Root Property.

Work Smart

To use the Square Root Property, the equation must be in the form "some unknown squared equals a number."

Solving a Quadratic Equation Using the Square Root Property

Step 1: Isolate the expression containing the squared term.

Step 2: Use the Square Root Property: If $x^2 = p$, then $x = \pm\sqrt{p}$.

Step 3: Isolate the variable, if necessary.

Step 4: Verify your solution(s).

EXAMPLE 2

Solving a Quadratic Equation in the Form $ax^2 = p$

Solve each quadratic equation:

(a) $3z^2 = 36$

(b) $\dfrac{9}{2}y^2 + 4 = 15$

Solution

(a)

$$3z^2 = 36$$

Divide both sides by 3: $\dfrac{3z^2}{3} = \dfrac{36}{3}$

$$z^2 = 12$$

Use the Square Root Property: $z = \pm\sqrt{12}$

Simplify the radical expression: $z = \pm\sqrt{4 \cdot 3}$

$$z = \pm 2\sqrt{3}$$

We leave it to you to check the solutions. The solution set is $\left\{-2\sqrt{3}, 2\sqrt{3}\right\}$.

(b)

$$\dfrac{9}{2}y^2 + 4 = 15$$

Subtract 4 from both sides: $\dfrac{9}{2}y^2 + 4 - 4 = 15 - 4$

$$\dfrac{9}{2}y^2 = 11$$

Multiply both sides by $\dfrac{2}{9}$: $\dfrac{2}{9} \cdot \dfrac{9}{2}y^2 = \dfrac{2}{9} \cdot 11$

$$y^2 = \dfrac{22}{9}$$

Use the Square Root Property;

don't forget the $\pm$ symbol: $y = \pm\sqrt{\dfrac{22}{9}}$

$\sqrt{\dfrac{a}{b}} = \dfrac{\sqrt{a}}{\sqrt{b}}$: $y = \pm\dfrac{\sqrt{22}}{\sqrt{9}}$

$$y = \pm\dfrac{\sqrt{22}}{3}$$

We leave it to you to check the solutions. The solution set is $\left\{-\dfrac{\sqrt{22}}{3}, \dfrac{\sqrt{22}}{3}\right\}$. ●

EXAMPLE 3 **Solving a Quadratic Equation in the Form $ax^2 = p$**

Solve the quadratic equation $3x^2 - 25 = 0$.

Solution

$$3x^2 - 25 = 0$$

Add 25 to both sides: $\qquad 3x^2 = 25$

Divide both sides by 3: $\qquad x^2 = \dfrac{25}{3}$

Use the Square Root Property: $\qquad x = \pm\sqrt{\dfrac{25}{3}}$

$\sqrt{\dfrac{a}{b}} = \dfrac{\sqrt{a}}{\sqrt{b}}:$ $\qquad x = \pm\dfrac{\sqrt{25}}{\sqrt{3}}$

$\qquad x = \pm\dfrac{5}{\sqrt{3}}$

Rationalize the denominator: $\qquad x = \pm\dfrac{5}{\sqrt{3}} \cdot \dfrac{\sqrt{3}}{\sqrt{3}}$

$\qquad x = \pm\dfrac{5\sqrt{3}}{3}$

We leave it to you to check the solutions. The solution set is $\left\{-\dfrac{5\sqrt{3}}{3}, \dfrac{5\sqrt{3}}{3}\right\}$.

Quick ✔

In Problems 7 and 8, solve the quadratic equation using the Square Root Property.

7. $x^2 = \dfrac{1}{2}$

8. $5x^2 - 16 = 0$

Solving Quadratic Equations of the Form $(ax + b)^2 = p$

▶ When solving a quadratic equation such as $(3y - 6)^2 - 2 = 25$, we isolate the expression containing the squared term and then use the Square Root Property.

EXAMPLE 4 **How to Solve a Quadratic Equation in the Form $(ax + b)^2 = p$**

Solve: $(3y - 6)^2 - 2 = 25$

Step-by-Step Solution

Step 1: Isolate the expression containing the squared term.

$$(3y - 6)^2 - 2 = 25$$

Add 2 to each side: $\quad (3y - 6)^2 - 2 + 2 = 25 + 2$

$$(3y - 6)^2 = 27$$

Step 2: Use the Square Root Property. Don't forget the $\pm$ symbol.

Use the Square Root Property: $\quad 3y - 6 = \pm\sqrt{27}$

Simplify the radical expression: $\quad 3y - 6 = \pm\sqrt{9 \cdot 3}$

$$3y - 6 = \pm 3\sqrt{3}$$

Step 3: Isolate the variable, if necessary.

Add 6 to each side: $3y - 6 + 6 = \pm 3\sqrt{3} + 6$

Apply the Commutative Property of Addition to rearrange terms: $3y = 6 \pm 3\sqrt{3}$

Divide each side by 3: $\dfrac{3y}{3} = \dfrac{6 \pm 3\sqrt{3}}{3}$

$y = \dfrac{6 \pm 3\sqrt{3}}{3}$

Factor the numerator: $y = \dfrac{3(2 \pm \sqrt{3})}{3}$

Divide out common factors: $y = 2 \pm \sqrt{3}$

Step 4: Verify your solutions.

$y = 2 - \sqrt{3}$:

$$(3y - 6)^2 - 2 = 25$$
$$\left[3(2 - \sqrt{3}) - 6\right]^2 - 2 \stackrel{?}{=} 25$$
$$\left[6 - 3\sqrt{3} - 6\right]^2 - 2 \stackrel{?}{=} 25$$
$$\left[-3\sqrt{3}\right]^2 - 2 \stackrel{?}{=} 25$$
$(ab)^2 = a^2b^2 \quad 9 \cdot 3 - 2 \stackrel{?}{=} 25$
$$27 - 2 \stackrel{?}{=} 25$$
$$25 = 25 \quad \text{True}$$

$y = 2 + \sqrt{3}$:

$$(3y - 6)^2 - 2 = 25$$
$$\left[3(2 + \sqrt{3}) - 6\right]^2 - 2 \stackrel{?}{=} 25$$
$$\left[6 + 3\sqrt{3} - 6\right]^2 - 2 \stackrel{?}{=} 25$$
$$\left[3\sqrt{3}\right]^2 - 2 \stackrel{?}{=} 25$$
$(ab)^2 = a^2b^2 \quad 9 \cdot 3 - 2 \stackrel{?}{=} 25$
$$27 - 2 \stackrel{?}{=} 25$$
$$25 = 25 \quad \text{True}$$

The solution set is $\left\{2 - \sqrt{3}, 2 + \sqrt{3}\right\}$. ●

In Step 3 of Example 4, we wrote $3y = \pm 3\sqrt{3} + 6$ as $3y = 6 \pm 3\sqrt{3}$. It is common practice to put rational numbers first, followed by square roots. After dividing both sides by 3, we factored the numerator and divided out common factors to simplify $\dfrac{6 \pm 3\sqrt{3}}{3}$. Another way to simplify $\dfrac{6 \pm 3\sqrt{3}}{3}$ is to use $\dfrac{a + b}{c} = \dfrac{a}{c} + \dfrac{b}{c}$, and write $y = \dfrac{6}{3} \pm \dfrac{3\sqrt{3}}{3} = 2 \pm \sqrt{3}$.

Quick ✓

9. When using the Square Root Property to solve an equation in the form $(ax + b)^2 = p$, where p is a positive real number, you will have (how many) ___ unique solutions.

In Problems 10 and 11, solve the quadratic equation using the Square Root Property.

10. $(y - 3)^2 = 121$

11. $(5k + 1)^2 - 2 = 26$

▶ (EXAMPLE 5) **Solving a Quadratic Equation Containing a Perfect Square Trinomial**

Solve: $x^2 + 6x + 9 = 7$

Solution

Did you notice that the left side of the equation is a perfect square trinomial? Use the fact that $A^2 + 2AB + B^2 = (A + B)^2$ along with the Square Root

Property to solve the equation.

$$x^2 + 6x + 9 = 7$$

Factor the trinomial $x^2 + 6x + 9$: $\qquad (x + 3)^2 = 7$

Use the Square Root Property: $\qquad x + 3 = \pm \sqrt{7}$

Subtract 3 from both sides: $\qquad x = -3 \pm \sqrt{7}$

The solution set is $\left\{ -3 - \sqrt{7}, -3 + \sqrt{7} \right\}$.

Quick ✓

In Problems 12 and 13, solve the quadratic equation using the Square Root Property.

12. $y^2 + 8y + 16 = 9$ $\qquad$ **13.** $4n^2 + 16n + 16 = 32$

EXAMPLE 6 **Solving a Quadratic Equation of the Form $(ax + b)^2 = p$, Where p Is Negative**

Solve: $(2x + 1)^2 = -10$

Solution

The equation states that the square of some value is negative 10. But there is no real number whose square is negative, so there is no real solution to this equation.

Quick ✓

In Problems 14 and 15, solve the quadratic equation.

14. $(n + 3)^2 = -4$ $\qquad$ **15.** $(2x + 5)^2 + 8 = 7$

▶ ❷ **Solve Problems Using the Pythagorean Theorem**

Recall from Section 6.7 that the Pythagorean Theorem is a statement about *right triangles*. A right triangle is one that contains a right angle—that is, an angle of 90°. The side of the triangle opposite the 90° angle is the hypotenuse; the remaining two sides are the legs. In Figure 1, we use c to represent the length of the hypotenuse and a and b to represent the lengths of the legs. Notice the use of the symbol ⌐ to show the 90° angle.

Figure 1

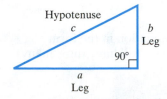

The Pythagorean Theorem

In a right triangle, the square of the length of the hypotenuse equals the sum of the squares of the lengths of the legs. That is, in the right triangle shown in Figure 1,

$$a^2 + b^2 = c^2$$

EXAMPLE 7 **Finding the Hypotenuse of a Right Triangle**

In a right triangle, one leg is 4 inches long and the other is 6 inches long. Find the exact length and the approximate length (to the nearest tenth of an inch) of the hypotenuse.

Solution

Figure 2 shows the right triangle with the legs and the hypotenuse labeled. Because the triangle is a right triangle, we use the Pythagorean Theorem with $a = 4$ and $b = 6$ to find the length c of the hypotenuse.

Figure 2

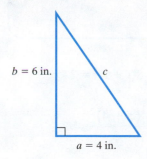

$b = 6$ in. c

$a = 4$ in.

Work Smart

Notice that the **exact** answer contains a radical, and the **approximate** answer is a decimal.

$$a^2 + b^2 = c^2$$
$$4^2 + 6^2 = c^2$$
$$16 + 36 = c^2$$
$$52 = c^2$$
$$c^2 = 52$$

Use the Square Root Property: $c = \pm\sqrt{52}$

$c = \pm\sqrt{4 \cdot 13}$

$c = \pm 2\sqrt{13}$

Use only the positive square root because c represents the length of a side of a triangle and a negative length does not make sense. The exact length of the hypotenuse is $2\sqrt{13}$ inches. Use a calculator or Appendix A to approximate the hypotenuse at 7.2 inches.   ●

Quick ✓

16. *True or False* In a right triangle, the length of the hypotenuse is equal to the sum of the squares of the lengths of the two legs.

In Problems 17–19, the lengths of the legs of a right triangle are given. Find the exact length and the approximate length (to the nearest tenth) of the hypotenuse.

17. $a = 5, b = 12$ 18. $a = 5, b = 15$ 19. $a = 6, b = 6$

EXAMPLE 8   **Spring Cleaning: Washing Windows**

Bob wants to wash the windows of his house. He has a 25-foot ladder and places the base of the ladder 10 feet from the wall of his house. To the nearest tenth of a foot, how far up the wall will the ladder reach?

Solution

We use the problem-solving approach from Chapter 2.

Step 1: Identify We wish to know how far up the wall the ladder reaches.

Step 2: Name Let a represent the height the ladder reaches on the side of the house.

Step 3: Translate To visualize the problem, make a sketch to represent the information given. See Figure 3. The base of the triangle is 10 feet—the distance to the bottom of the ladder is from the house. The hypotenuse is 25 feet-the length of the ladder. Express the relation among the three sides of the right triangle using the Pythagorean Theorem.

$$a^2 + b^2 = c^2$$

Substitute $b = 10$ and $c = 25$: $a^2 + 10^2 = 25^2$

Figure 3

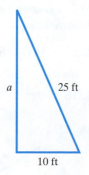

a 25 ft

10 ft

Step 4: Solve

$$a^2 + 100 = 625$$
$$a^2 = 525$$

Use the Square Root Property: $a = \pm\sqrt{525}$

Because a represents the height on the building, use the positive square root, $a = \sqrt{525} = 5\sqrt{21}$ ft. Using a calculator, we find $a \approx 22.9$ feet.

Step 5: Check Does $\left(\sqrt{525}\right)^2 + 10^2 = 25^2$? Because $525 + 100 = 625$, our answer checks.

Step 6: Answer The ladder reaches approximately 22.9 feet up the wall.   ●

Work Smart

It doesn't matter which leg we name a or b. But the hypotenuse is always labeled c.

Quick ✓

20. Stefan needs to clean his gutters. He has a 30-foot ladder. The manufacturer recommends that the base of the ladder be 8 feet from the wall. If the gutters are 25 feet above the ground, can Stefan use this ladder to clean his gutters?

*Problems **1–20** are the Quick ✓s that follow the EXAMPLES.*

Building Skills

In Problems 21–70, solve each quadratic equation using the Square Root Property. Express radicals in simplest form. See Objective 1.

21. $x^2 = 144$ **22.** $x^2 = 81$

23. $u^2 = 0$ **24.** $y^2 = 1$

25. $12 = x^2$ **26.** $48 = t^2$

27. $s^2 = \dfrac{4}{9}$ **28.** $x^2 = \dfrac{25}{4}$

29. $x^2 = \dfrac{4}{3}$ **30.** $d^2 = \dfrac{36}{5}$

31. $\dfrac{1}{2}r^2 = 16$ **32.** $\dfrac{1}{3}w^2 = 9$

33. $x^2 - 169 = 0$ **34.** $r^2 - 36 = 0$

35. $x^2 - 50 = 0$ **36.** $x^2 - 20 = 0$

37. $p^2 + 16 = 0$ **38.** $0 = h^2 + 4$

39. $27x^2 = 3$ **40.** $64x^2 = 4$

41. $\dfrac{1}{16}n^2 - 4 = 0$ **42.** $\dfrac{1}{4}m^2 - 25 = 0$

43. $65 = 2n^2 - 7$ **44.** $76 = 3v^2 + 1$

45. $3x^2 + 20 = 45$ **46.** $7x^2 + 8 = 24$

47. $2k^2 + 12 = 10$ **48.** $4 = 3z^2 + 10$

49. $2x^2 - 15 = 49$ **50.** $100 = 2x^2 + 10$

51. $4 = (x + 4)^2$ **52.** $16 = (x + 5)^2$

53. $27(x - 1)^2 = 3$ **54.** $8(v + 1)^2 = 2$

55. $\dfrac{(n + 5)^2}{3} = 2$ **56.** $\dfrac{(z + 4)^2}{5} = 3$

57. $\dfrac{2}{3}(x - 6)^2 = \dfrac{16}{3}$ **58.** $\dfrac{2}{15}(p - 3)^2 = \dfrac{8}{5}$

59. $\left(p + \dfrac{1}{3}\right)^2 = \dfrac{16}{9}$ **60.** $\left(y - \dfrac{1}{2}\right)^2 = \dfrac{9}{4}$

61. $\left(x - \dfrac{5}{2}\right)^2 = \dfrac{15}{4}$ **62.** $\left(x + \dfrac{2}{5}\right)^2 = \dfrac{21}{25}$

63. $(x - 1)^2 - 7 = 9$ **64.** $(x + 3)^2 - 12 = 24$

65. $(8k + 3)^2 - 5 = -4$ **66.** $(8g - 3)^2 - 2 = -1$

67. $30 = (9x + 2)^2 - 24$ **68.** $20 = (3y + 2)^2 + 2$

69. $(2w + 8)^2 - 12 = -24$ **70.** $(3x - 6)^2 - 40 = -48$

In Problems 71–74, solve each quadratic equation by first factoring the perfect square trinomial on one side of the equation and then using the Square Root Property. See Objective 1.

71. $x^2 + 8x + 16 = 25$ **72.** $x^2 + 16x + 64 = 36$

73. $49 = 4w^2 - 4w + 1$ **74.** $16 = 9z^2 - 12z + 4$

In Problems 75–80, use the right triangle shown to the right and find the missing length. Give exact answers and decimal approximations rounded to two decimal places. See Objective 2.

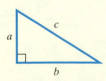

75. $a = 6, b = 8$ **76.** $a = 5, b = 12$

77. $a = 3, b = 3$ **78.** $a = 4, b = 2$

79. $b = 4, c = 12$ **80.** $b = 6, c = 10$

Mixed Practice

In Problems 81–104, solve each quadratic equation by either factoring or applying the Square Root Property. For each problem, choose the method that is most efficient.

81. $x^2 - 13x + 36 = 0$

82. $x^2 - 14x + 48 = 0$

83. $2n^2 = 16$

84. $3q^2 = 36$

85. $3x^2 - 9 = 36$

86. $81 = 2x^2 + 9$

87. $2m^2 = 1 - m$

88. $3x^2 = 2x + 1$

89. $0 = 15x^2 - 9$

90. $0 = 6t^2 - 14$

91. $2 = r^2 + 6$

92. $x^2 + 17 = 8$

93. $12 = x^2 + x$

94. $15 = n^2 + 2n$

95. $d^2 - 27 = 0$

96. $n^2 - 2 = 0$

97. $x^2 - 10x + 25 = 0$

98. $x^2 + 14x + 49 = 0$

99. $2(x + 2)^2 + 3 = 8$

100. $3(z - 1)^2 + 4 = 6$

101. $3x^2 + 4 = 6$

102. $8 = 2z^2 + 5$

103. $2x^2 - 5x - 12 = 0$

104. $2y^2 - 5y - 3 = 0$

In Problems 105–108, solve each equation.

105. $\sqrt{x^2 - 15} = 5$

106. $\sqrt{p^2 + 4} = 4$

107. $\sqrt{q^2 + 13} + 3 = 6$

108. $\sqrt{z^2 - 2} + 9 = 4$

Applying the Concepts

In Problems 109 and 110, find the length of the diagonal in each figure.

109.

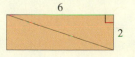

110.

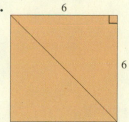

111. Firefighter's Ladder A firefighter has a 13-foot ladder. Suppose he places the ladder 7 feet from a building.

    (a) Exactly how far up the building will his ladder reach?

    (b) Approximate this height to the nearest tenth of a foot.

112. Picture Frame A rectangular picture frame measures 17 inches on the diagonal and is 8 inches high. Exactly how wide is this frame?

113. Golf A golfer hits an errant tee shot that lands in the rough. The golfer finds that the ball is exactly 20 yards to the right of the 100-yard marker that indicates the distance to the center of the green, as shown in the figure. To the nearest yard, how far is the ball from the center of the green?

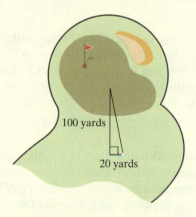

100 yards

20 yards

114. Baseball Justin Upton plays right field for the Arizona Diamondbacks. He catches a fly ball 30 feet from the right-field foul line, as indicated in the figure. How far is it to home plate? Round your answer to the nearest foot.

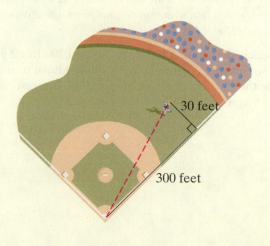

30 feet

300 feet

△ **115. Football Goal Post** The groundskeeper at the football stadium suspects that the crossbar of the goalpost is beginning to lean slightly. He decides to determine whether it is square by measuring the distance from the end of the crossbar to the point where the pole goes in the ground.

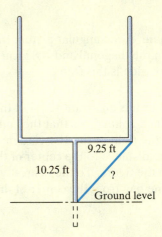

9.25 ft

10.25 ft

?

Ground level

(a) How long should this distance be if the distance from the center of the crossbar to the end is 9.25 feet and the goalpost is 10.25 feet high? Find the distance, to the nearest tenth of a foot.

(b) What is this distance in feet and inches?

△ **116. Ham Radio Antenna** Bob is installing a new ham radio antenna in his back yard. The antenna is 16 feet high, and he wants to place a guy wire 7 feet from the antenna for added support during windy weather.

(a) Find the length of the guy wire to the nearest tenth of a foot.

(b) What is this distance in inches and feet?

The formula for calculating the amount of money in an interest-earning account after 2 years is given by the formula $A = P(1 + r)^2$, where A is the new amount, P is the principal, and r is the interest rate, written in decimal form. Use this formula in Problems 117 and 118.

117. Leticia deposited $500 and had $583.20 after 2 years. Use this information in the formula above to solve for the rate of interest, r. Express the interest rate as a percent.

118. James deposited $1200 and had $1323 after 2 years. Use this information in the formula $A = P(1 + r)^2$ to solve for the rate of interest, r. Express the interest rate as a percent.

Extending the Concepts

In Problems 119–124, find a quadratic equation that has the indicated solutions. Write the equation in standard form, $ax^2 + bx + c = 0, a \neq 0$.

119. The solutions to the equation are $x = 4$ and $x = -4$ $(x = \pm 4)$.

120. The solutions to the equation are $x = 9$ and $x = -9$ $(x = \pm 9)$.

121. The solutions to the equation are $x = \pm\sqrt{6}$.

122. The solutions to the equation are $x = \pm\sqrt{15}$.

123. The solutions to the equation are $x = \pm 3\sqrt{5}$.

124. The solutions to the equation are $x = \pm 2\sqrt{7}$.

Explaining the Concepts

125. Find and explain the error in the solution of the quadratic equation. Then solve the equation correctly.

$$(x - 5)^2 - 4 = 0$$
$$(x - 5)^2 = 4$$
$$x - 5 = 2$$
$$x = 7$$

126. Consider the two quadratic equations $x^2 + 36 = 0$ and $x^2 - 36 = 0$. Describe how to solve each equation. Explain the difference between the two solutions.

9.2 Solving Quadratic Equations by Completing the Square

Objectives

1 Complete the Square in One Variable

2 Solve Quadratic Equations by Completing the Square

Are You Ready for This Section?

Before getting started, take the following readiness quiz. If you get a problem wrong, go back to the section cited and review the material.

R1. Factor: $a^2 + 10a + 25$ [Section 6.4, pp. 395–397]

R2. Solve: $(x + 3)^2 = 16$ [Section 9.1, pp. 582–583]

R3. Find the quotient: $\dfrac{3x^2 - 5x + 4}{3}$ [Section 5.5, pp. 345–347]

You know two methods for solving quadratic equations: factoring and using the Square Root Property. A third method is called the method of *completing the square*. Before we use this method to solve a quadratic equation, we first discuss how to complete the square.

▶ **1** Complete the Square in One Variable

The idea behind **completing the square** is to "adjust" the left side of a quadratic equation of the form $ax^2 + bx + c = 0$ to make it a perfect square trinomial. Recall that perfect square trinomials are of the form

$$A^2 + 2AB + B^2 = (A + B)^2 \quad \text{or} \quad A^2 - 2AB + B^2 = (A - B)^2$$

For example, $x^2 + 8x + 16$ is a perfect square trinomial because $x^2 + 8x + 16 = (x + 4)^2$.

We "adjust" the equation by adding a number to the left side of the equation to make it a perfect square trinomial. We also have to add this number to the right side. For example, to make $x^2 + 10x$ a perfect square, we add 25. Why did we choose 25? Look at the coefficient of the first-degree term, 10. If we divide 10 by 2 and then square the result, we obtain 25. This approach works in general.

Obtaining a Perfect Square Trinomial by Completing the Square

Step 1: Identify b, the coefficient of the first-degree term.

Step 2: Multiply b by $\dfrac{1}{2}$ and square the result. That is, compute $\left(\dfrac{1}{2}b\right)^2$.

Step 3: Add $\left(\dfrac{1}{2}b\right)^2$ to $x^2 + bx$ to obtain a perfect square trinomial.

EXAMPLE 1   **Obtaining a Perfect Square Trinomial**

Determine the number that must be added to each expression to make it a perfect square trinomial. Then factor the expression as $A^2 + 2AB + B^2 = (A + B)^2$ or $A^2 - 2AB + B^2 = (A - B)^2$.

Start	Add	Result	Factored Form
$y^2 + 8y$	$\left(\dfrac{1}{2}\cdot 8\right)^2 = 4^2 = 16$	$y^2 + 8y + 16$	$(y + 4)^2$
$z^2 - 20z$	$\left(\dfrac{1}{2}\cdot(-20)\right)^2 = (-10)^2 = 100$	$z^2 - 20z + 100$	$(z - 10)^2$
$p^2 - 5p$	$\left(\dfrac{1}{2}\cdot(-5)\right)^2 = \left(-\dfrac{5}{2}\right)^2 = \dfrac{25}{4}$	$p^2 - 5p + \dfrac{25}{4}$	$\left(p - \dfrac{5}{2}\right)^2$
$n^2 + \dfrac{3}{2}n$	$\left(\dfrac{1}{2}\cdot\dfrac{3}{2}\right)^2 = \left(\dfrac{3}{4}\right)^2 = \dfrac{9}{16}$	$n^2 + \dfrac{3}{2}n + \dfrac{9}{16}$	$\left(n + \dfrac{3}{4}\right)^2$

Ready?...Answers   **R1.** $(a + 5)^2$

R2. $\{-7, 1\}$ **R3.** $x^2 - \dfrac{5}{3}x + \dfrac{4}{3}$

Did you notice in the factored form that the perfect square trinomial always factors so that

$$x^2 + bx + \left(\frac{b}{2}\right)^2 = \left(x + \frac{b}{2}\right)^2 \quad \text{or} \quad x^2 - bx + \left(\frac{b}{2}\right)^2 = \left(x - \frac{b}{2}\right)^2 ?$$

The $\dfrac{b}{2}$ represents half the value of the coefficient of the first-degree term.

Quick ✔

1. The polynomial $x^2 - 20x +$ ___ is a perfect square trinomial and factors as _____.

2. In general $x^2 - bx + \left(\dfrac{b}{2}\right)^2 = \left(\underline{}\right)^2$

In Problems 3–5, determine the number that must be added to the expression to make it a perfect square trinomial. Then factor the expression.

3. $p^2 + 6p$ 4. $b^2 - 18b$ 5. $w^2 + 3w$

Figure 4

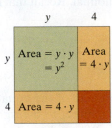

Are you wondering why we call this process "completing the square"? Consider the expression $y^2 + 8y$ given in Example 1, which we have geometrically represented in Figure 4. The green area is y^2 and each of the orange areas is $4y$ (for a total orange area of $8y$). But what is the area of the red region in order to make the square complete? The dimensions of the red region must be 4 by 4, so its area is 16. The area of the entire square region, $(y + 4)^2$, equals the sum of the areas of the four regions that make up the square: $y^2 + 4y + 4y + 16 = y^2 + 8y + 16$.

② Solve Quadratic Equations by Completing the Square

▶ We're now ready to learn how to use the method of completing the square to solve quadratic equations in the form $x^2 + bx + c = 0$.

EXAMPLE 2 How to Solve a Quadratic Equation by Completing the Square

Solve: $p^2 + 10p + 21 = 0$

Step-by-Step Solution

Step 1: Rewrite $x^2 + bx + c = 0$ as $x^2 + bx = -c$.

$$p^2 + 10p + 21 = 0$$

Subtract 21 from both sides: $p^2 + 10p = -21$

Step 2: Complete the square on the expression $x^2 + bx$. Remember, whatever is added to one side of the equation must also be added to the other side!

$\left(\dfrac{1}{2} \cdot 10\right)^2 = 5^2 = 25$, so add 25 to both sides of the equation:

$$p^2 + 10p + 25 = -21 + 25$$
$$p^2 + 10p + 25 = 4$$

Step 3: Factor the perfect square trinomial on the left side of the equation.

$$(p + 5)^2 = 4$$

Step 4: Solve the equation using the Square Root Property.

Use the Square Root Property: $p + 5 = \pm\sqrt{4}$

$$p + 5 = \pm 2$$

$$p + 5 = -2 \quad \text{or} \quad p + 5 = 2$$

Subtract 5 from both sides: $p = -7 \quad \text{or} \quad p = -3$

Step 5: Verify your solution(s).

$p = -7:$ $p^2 + 10p + 21 = 0$ $p = -3:$ $p^2 + 10p + 21 = 0$

$(-7)^2 + 10(-7) + 21 \overset{?}{=} 0$ $(-3)^2 + 10(-3) + 21 \overset{?}{=} 0$

$49 - 70 + 21 \overset{?}{=} 0$ $9 - 30 + 21 \overset{?}{=} 0$

$0 = 0$ True $0 = 0$ True

The solution set is $\{-7, -3\}$. ●

> **Solving a Quadratic Equation in the Form $x^2 + bx + c = 0$ by Completing the Square**
>
> **Step 1:** Rewrite $x^2 + bx + c = 0$ as $x^2 + bx = -c$ by subtracting (or adding) the constant from (or to) both sides of the equation.
>
> **Step 2:** Complete the square on the expression $x^2 + bx$ to make it a perfect square trinomial. Don't forget, whatever is added to the left side of the equation must also be added to the right side.
>
> **Step 3:** Factor the perfect square trinomial on the left side of the equation.
>
> **Step 4:** Solve the equation using the Square Root Property.
>
> **Step 5:** Verify your solutions.

Quick ✓

6. *True or False* To solve $x^2 + 18x = 6$ by completing the square, add 9 to both sides of the equation.

In Problems 7 and 8, solve the quadratic equation by completing the square.

7. $n^2 + 8n + 7 = 0$ **8.** $y^2 + 4y - 32 = 0$

The solutions in Example 2 were rational numbers. We know from solving quadratic equations using the Square Root Property that some quadratic equations have irrational solutions.

(**EXAMPLE 3**) **Solving a Quadratic Equation Having Irrational Solutions**

Solve: $z^2 + 8z = -4$

Solution

The equation is already written in the form $x^2 + bx = -c$. Now complete the square on the left-hand side of the equation:

$$z^2 + 8z = -4$$

Add 16 to both sides since $\left(\dfrac{1}{2} \cdot 8\right)^2 = 4^2 = 16$: $z^2 + 8z + 16 = -4 + 16$

Factor: $(z + 4)^2 = 12$

Use the Square Root Property: $z + 4 = \pm\sqrt{12}$

Simplify the radical: $z + 4 = \pm 2\sqrt{3}$

$z + 4 = -2\sqrt{3} \quad \text{or} \quad z + 4 = 2\sqrt{3}$

Subtract 4 from both sides: $z = -4 - 2\sqrt{3} \quad \text{or} \quad z = -4 + 2\sqrt{3}$

Check $z = -4 - 2\sqrt{3}$: $z^2 + 8z = -4$ $z = -4 + 2\sqrt{3}$: $z^2 + 8z = -4$

$(-4 - 2\sqrt{3})^2 + 8(-4 - 2\sqrt{3}) \overset{?}{=} -4$ $(-4 + 2\sqrt{3})^2 + 8(-4 + 2\sqrt{3}) \overset{?}{=} -4$

$16 + 16\sqrt{3} + 12 - 32 - 16\sqrt{3} \overset{?}{=} -4$ $16 - 16\sqrt{3} + 12 - 32 + 16\sqrt{3} \overset{?}{=} -4$

$-4 = -4$ True $-4 = -4$ True

The solution set is $\{-4 - 2\sqrt{3}, -4 + 2\sqrt{3}\}$.

Quick ✓

In Problems 9 and 10, solve the quadratic equation by completing the square.

9. $z^2 + 8z = -9$ **10.** $t^2 - 10t = -19$

EXAMPLE 4   **Solving a Quadratic Equation Not in Standard Form**

Solve: $(x - 1)(x + 4) = 9$

Solution

What is the first thing that you notice about this equation? It's not in the form $x^2 + bx = -c$. Our first step will be to find the product $(x - 1)(x + 4)$.

$$(x - 1)(x + 4) = 9$$

Multiply: $x^2 + 3x - 4 = 9$

Add 4 to both sides: $x^2 + 3x - 4 + 4 = 9 + 4$

$$x^2 + 3x = 13$$

Work Smart

Do NOT solve $(x - 1)(x + 4) = 9$ by setting each factor equal to 9 as in

$x - 1 = 9$ or $x + 4 = 9$

Do you know why?

Now complete the square.

Add $\dfrac{9}{4}$ to both sides since $\left(\dfrac{1}{2} \cdot 3\right)^2 = \left(\dfrac{3}{2}\right)^2 = \dfrac{9}{4}$: $x^2 + 3x + \dfrac{9}{4} = 13 + \dfrac{9}{4}$

$$x^2 + 3x + \dfrac{9}{4} = \dfrac{52}{4} + \dfrac{9}{4}$$

$$x^2 + 3x + \dfrac{9}{4} = \dfrac{61}{4}$$

Factor: $\left(x + \dfrac{3}{2}\right)^2 = \dfrac{61}{4}$

Use the Square Root Property: $x + \dfrac{3}{2} = \pm\sqrt{\dfrac{61}{4}}$

Subtract $\dfrac{3}{2}$ from both sides; use $\sqrt{\dfrac{a}{b}} = \dfrac{\sqrt{a}}{\sqrt{b}}$: $x = -\dfrac{3}{2} \pm \dfrac{\sqrt{61}}{\sqrt{4}}$

$$x = -\dfrac{3}{2} \pm \dfrac{\sqrt{61}}{2}$$

$$x = \dfrac{-3 - \sqrt{61}}{2} \quad \text{or} \quad x = \dfrac{-3 + \sqrt{61}}{2}$$

We leave the check to you. The solution set is $\left\{\dfrac{-3 - \sqrt{61}}{2}, \dfrac{-3 + \sqrt{61}}{2}\right\}$.

Quick ✓

11. *True or False* The equation $(x - 1)(x + 5) = 10$ can be solved by solving

$x - 1 = 10$ and $x + 5 = 10$.

In Problems 12 and 13, solve the quadratic equation by completing the square.

12. $(n + 5)(n + 3) = 24$ **13.** $(w + 6)(w - 1) = 4$

▶ So far, the quadratic equations we have solved by completing the square were in the form $x^2 + bx = -c$. In other words, the leading coefficient was 1. What happens if the leading coefficient is not 1?

EXAMPLE 5   **Solving a Quadratic Equation Whose Leading Coefficient Is Not 1**

Solve: $2x^2 - 5x - 12 = 0$

Solution

First, we must get the coefficient of the squared term to be 1, so divide each side of the equation by 2.

Work Smart

For a quadratic equation to be solved by completing the square, the coefficient of the squared term must be 1.

$$2x^2 - 5x - 12 = 0$$

Divide each side by 2: $\quad \dfrac{2x^2 - 5x - 12}{2} = \dfrac{0}{2}$

Divide 2 into each term in the numerator: $\quad x^2 - \dfrac{5}{2}x - 6 = 0$

Add 6 to both sides: $\quad x^2 - \dfrac{5}{2}x = 6$

Because $\left(\dfrac{1}{2} \cdot -\dfrac{5}{2}\right)^2 = \left(-\dfrac{5}{4}\right)^2 = \dfrac{25}{16}$, add $\dfrac{25}{16}$ to both sides: $\quad x^2 - \dfrac{5}{2}x + \dfrac{25}{16} = 6 + \dfrac{25}{16}$

Factor: $\quad \left(x - \dfrac{5}{4}\right)^2 = \dfrac{96}{16} + \dfrac{25}{16}$

$$\left(x - \dfrac{5}{4}\right)^2 = \dfrac{121}{16}$$

Use the Square Root Property: $\quad x - \dfrac{5}{4} = \pm\sqrt{\dfrac{121}{16}}$

$$x - \dfrac{5}{4} = \pm\dfrac{11}{4}$$

$$x - \dfrac{5}{4} = -\dfrac{11}{4} \quad \text{or} \quad x - \dfrac{5}{4} = \dfrac{11}{4}$$

Add $\dfrac{5}{4}$ to both sides: $\quad x = \dfrac{5}{4} - \dfrac{11}{4} \quad \text{or} \quad x = \dfrac{5}{4} + \dfrac{11}{4}$

$$x = \dfrac{-6}{4} \quad \text{or} \quad x = \dfrac{16}{4}$$

$$x = -\dfrac{3}{2} \quad \text{or} \quad x = 4$$

We leave it to you to verify the solutions. The solution set is $\left\{-\dfrac{3}{2}, 4\right\}$.

Quick ✓

14. To solve $2x^2 + 4x = 9$ by completing the square, the first step is to divide both sides of the equation by __.

In Problems 15 and 16, solve the quadratic equation by completing the square.

15. $2m^2 = m + 10$ **16.** $3y^2 - 4y - 2 = 0$

Do you recall from Section 9.1 that quadratic equations of the form $(ax + b)^2 = p$, where p is negative, have no real solution? We illustrate this point again in the next example.

EXAMPLE 6   **Solving a Quadratic Equation Using Completing the Square**

Solve: $x(x + 6) + 28 = 10$

Solution

What's the first step? Write the equation in the form $x^2 + bx = -c$.

$$x(x + 6) + 28 = 10$$

Use the Distributive Property: $\quad x^2 + 6x + 28 = 10$

Subtract 28 from each side: $\quad x^2 + 6x + 28 - 28 = 10 - 28$

$$x^2 + 6x = -18$$

Add 9 to both sides because $\left(\frac{1}{2} \cdot 6\right)^2 = 3^2 = 9$: $\quad x^2 + 6x + 9 = -18 + 9$

$$x^2 + 6x + 9 = -9$$

Factor: $\quad (x + 3)^2 = -9$

Work Smart

No real solution does not mean *no solution*. We will learn how to find the solution to equations such as the one in Example 6 in Section 9.5.

No real number has a square of -9, so this equation has no real solution. ●

Quick ✓

In Problems 17 and 18, solve the quadratic equation by completing the square.

17. $n(n - 4) = -12$ **18.** $-x^2 + 6x = 7$

9.2 Exercises MyMathLab®   PRACTICE Exercise numbers in **green** have complete video solutions in MyMathLab

*Problems **1–18** are the Quick ✓s that follow the **EXAMPLES**.*

Building Skills

In Problems 19–30, determine the number that must be added to the expression to make it a perfect square trinomial. Then factor the expression. See Objective 1.

19. $x^2 + 16x$ **20.** $x^2 + 24x$

21. $z^2 - 12z$ **22.** $p^2 - 36p$

23. $k^2 + \frac{4}{3}k$ **24.** $y^2 + \frac{6}{7}y$

25. $y^2 - 9y$ **26.** $z^2 - 5z$

27. $t^2 + t$ **28.** $w^2 - w$

29. $s^2 - \frac{s}{2}$ **30.** $r^2 + \frac{r}{4}$

In Problems 31–64, solve the quadratic equation using completing the square. See Objective 2.

31. $x^2 - 4x = 21$ **32.** $x^2 + 6x = 40$

33. $x^2 + 4x = 7$ **34.** $x^2 + 2x = 5$

35. $m^2 - 3m - 18 = 0$ **36.** $z^2 - 5z + 6 = 0$

37. $r^2 - 8r + 15 = 0$ **38.** $p^2 - 6p + 8 = 0$

39. $x^2 + 2x + 24 = 0$

40. $x^2 - 8x + 18 = 0$

41. $x^2 = 3x + 4$

42. $r^2 = 7r - 6$

43. $t^2 = 8t + 12$

44. $n^2 = 6n + 18$

45. $-10 = x^2 + 7x$

46. $-4 = x^2 + 5x$

47. $n^2 + 7n - 50 = 0$

48. $p^2 + 7p - 48 = 0$

49. $10x = -16 - x^2$

50. $12x = -32 - x^2$

51. $3x^2 + 6x - 9 = 0$

52. $2x^2 - 6x + 4 = 0$

53. $4x^2 - 4x - 3 = 0$

54. $5x^2 + 8x - 4 = 0$

55. $3y^2 + 4y = 2$

56. $3x^2 = 4 - 2x$

57. $4x^2 = 5 - 4x$

58. $4n^2 + 12n + 1 = 0$

59. $4m^2 + 5m + 7 = 0$

60. $-9z = 2z^2 + 44$

61. $4p^2 - 12p = -7$

62. $2r^2 - 5r + 1 = 0$

63. $4r^2 - 24r - 1 = 0$

64. $5 = 4n^2 - 32n$

Mixed Practice

In Problems 65–92, solve the quadratic equation using factoring, the Square Root Property, or completing the square.

65. $(x + 2)(x - 4) = 16$

66. $(x - 8)(x - 1) = 1$

67. $4x^2 - 4x + 1 = 0$

68. $4x^2 + 12x + 9 = 0$

69. $n^2 + 6n - 3 = 0$

70. $n^2 + 10n - 15 = 0$

71. $3z^2 + 12 = 72$

72. $2y^2 + 15 = 51$

73. $3t^2 = 2 - 6t$

74. $2p^2 = 4p - 1$

75. $5x = 18 - x^2$

76. $x^2 + 3x = 6$

77. $h^2 = 9$

78. $k^2 = 36$

79. $(x + 2)^2 - 16 = 0$

80. $(x - 9)^2 - 4 = 0$

81. $n^2 + 10n = 24$

82. $n^2 - 13n = 30$

83. $y^2 - 18 = 0$

84. $z^2 - 48 = 0$

85. $2x^2 - 24 = 0$

86. $3x^2 - 24 = 0$

87. $12u + 2u^2 - 27 = 0$

88. $3v = 3v^2 - 10$

89. $r(3r - 11) = -6$

90. $t(2t + 9) = -4$

91. $(x + 6)(x - 4) = 9$

92. $(x - 2)(x - 3) = 8$

Applying the Concepts

△ **93. The Pythagorean Theorem** In a right triangle, the lengths of the two legs are equal. If the length of the hypotenuse is 30, find the length of the legs.

△ **94. The Pythagorean Theorem** In a right triangle, the lengths of the two legs are equal. If the length of the hypotenuse is 18, find the length of the legs.

△ **95. The Pythagorean Theorem** In a right triangle, the length of the hypotenuse is twice the length of the shorter leg. If the length of the longer leg is $\sqrt{27}$, find the lengths of the other two sides of the triangle.

△ **96. The Pythagorean Theorem** In a right triangle, the length of the hypotenuse is twice the length of the shorter leg. If the length of the longer leg is $\sqrt{48}$, find the lengths of the other two sides of the triangle.

97. Fun with Numbers The product of a number and one less than half of the number is 12. Find the two numbers that satisfy this condition.

98. Fun with Numbers The product of a number and one more than one-quarter of this number is 3. Find the two numbers that satisfy this condition.

99. Fun with Numbers The square of one more than one-third of a number is the same as twice the number. Find the two numbers that satisfy this condition.

100. Fun with Numbers The square of one less than half of a number is the same as three more than twice the number. Find the two numbers that satisfy this condition.

Extending the Concepts

When a quadratic equation is written in the form $x^2 + bx + c = 0$, the sum of the two solutions of the equation is $-b$, the opposite of the coefficient of the linear term. In Problems 101–104, use this fact to (a) identify the sum of the solutions for each quadratic equation, (b) find the solutions of the quadratic equation by completing the square, and (c) verify that the sum of the solutions is $-b$. Hint: In Problems 103 and 104, first write the equation in the form $x^2 + bx + c = 0$.

101. $x^2 - 9x + 18 = 0$ **102.** $x^2 + 5x - 24 = 0$

103. $2x^2 - 4x - 1 = 0$ **104.** $2x^2 + 4x - 4 = 0$

When a quadratic equation is written in the form $x^2 + bx + c = 0$, the product of the two solutions is c, the constant term. In Problems 105–108, use this fact to (a) identify the product of the solutions for each quadratic equation, (b) find the solutions of the quadratic equation by completing the square, and (c) verify that the product of the solutions is c. Hint: First write the equation in the form $x^2 + bx + c = 0$.

105. $4x^2 + 4x - 3 = 0$ **106.** $3x^2 - 8x + 4 = 0$

107. $2x^2 = 4x + 14$ **108.** $2x^2 - 4x = 13$

An equation of the form $(x - h)^2 + (y - k)^2 = r^2$ is the equation of a circle whose radius is r and center is (h, k). In Problems 109–112, use completing the square twice (once in x, once in y) to write the equations in this form. State the center and radius of each circle.

109. $x^2 + y^2 - 10x + 6y + 30 = 0$

110. $x^2 + y^2 + 8x - 4y + 19 = 0$

111. $x^2 + y^2 - 20x - 16y + 22 = 0$

112. $x^2 + y^2 + 14x + 6y - 12 = 0$

Explaining the Concepts

113. Describe how to find the number that must be added to $x^2 - 9x$ to make a perfect square trinomial. Then express the perfect square trinomial in factored form.

114. We now have three techniques to solve quadratic equations: factoring, using the Square Root Property, and completing the square. Which method do you believe is easiest to use when solving $x^2 + 3x - 208 = 0$ and why?

9.3 Solving Quadratic Equations Using the Quadratic Formula

Objectives

1 Solve Quadratic Equations Using the Quadratic Formula

2 Use the Discriminant to Determine Which Method to Use When Solving a Quadratic Equation

Ready? …Answers **R1.** $-4 + 2\sqrt{2}$
R2. $\dfrac{3 + \sqrt{15}}{3}$ or $1 + \dfrac{\sqrt{15}}{3}$
R3. (a) $x^2 - 4x + 12 = 0$
 (b) $n^2 - n - 12 = 0$ **R4.** {3}

Are You Ready for This Section?

Before getting started, take the following readiness quiz. If you get a problem wrong, go back to the section cited and review the material.

R1. Evaluate $-4 + \sqrt{16 - 4(2)(1)}$. Give the exact value. [Section 8.2, pp. 525–527]

R2. Evaluate $\dfrac{6 + \sqrt{6^2 - 4(3)(-2)}}{6}$. Give the exact value. [Section 8.2, pp. 525–527]

R3. Write each quadratic equation in standard form:
   (a) $4x - x^2 = 12$ **(b)** $(n + 2)(n - 3) = 6$ [Section 6.6, p. 410]

R4. Solve: $\dfrac{3}{x} + 4 = \dfrac{15}{x}$ [Section 7.7, pp. 478–484]

Currently, we have three methods for solving quadratic equations: (1) factoring, (2) using the Square Root Property, and (3) completing the square. Why do we need three methods? Because each method provides the quickest route to the solution, when used appropriately. For example, if the quadratic expression is easy to factor, factoring will get you to the solution most easily. If the equation is in the form $x^2 = p$ or $(ax \pm b)^2 = p$, using the Square Root Property is easiest. To solve the remaining quadratic equations, use completing the square. But completing the square is tedious, so you may be asking yourself, "Is there an alternative to completing the square?" Yes!

❶ Solve Quadratic Equations Using the Quadratic Formula

We can find a general formula for solving the quadratic equation $ax^2 + bx + c = 0$, $a \neq 0$, using the method of completing the square. To complete the square, we first must get the constant c on the right-hand side of the equation.

$$ax^2 + bx = -c, \qquad a \neq 0$$

Divide both sides of the equation by a: $\quad x^2 + \dfrac{b}{a}x = -\dfrac{c}{a}$

Now we can complete the square on the left side by adding the square of half of the coefficient of x to both sides of the equation. That is, we add

$$\left(\frac{1}{2} \cdot \frac{b}{a}\right)^2 = \frac{b^2}{4a^2}$$

to both sides of the equation. We now have

$$x^2 + \frac{b}{a}x + \frac{b^2}{4a^2} = -\frac{c}{a} + \frac{b^2}{4a^2}$$

or

$$x^2 + \frac{b}{a}x + \frac{b^2}{4a^2} = \frac{b^2}{4a^2} - \frac{c}{a}$$

Combine terms on the right-hand side. The least common denominator (LCD) on the right-hand side is $4a^2$, so we multiply $-\dfrac{c}{a}$ by $1 = \dfrac{4a}{4a}$:

$$x^2 + \frac{b}{a}x + \frac{b^2}{4a^2} = \frac{b^2}{4a^2} - \frac{c}{a} \cdot \frac{4a}{4a}$$

$$x^2 + \frac{b}{a}x + \frac{b^2}{4a^2} = \frac{b^2}{4a^2} - \frac{4ac}{4a^2}$$

$$x^2 + \frac{b}{a}x + \frac{b^2}{4a^2} = \frac{b^2 - 4ac}{4a^2}$$

Factor the left-hand side and obtain

$$\left(x + \frac{b}{2a}\right)^2 = \frac{b^2 - 4ac}{4a^2}$$

Assume that $a > 0$ (you'll see why in a little while). This assumption does not compromise the results because if $a < 0$, we could multiply both sides of the equation $ax^2 + bx + c = 0$ by -1 to make a positive. With this assumption, we use the Square Root Property and get

$$x + \frac{b}{2a} = \pm\sqrt{\frac{b^2 - 4ac}{4a^2}}$$

$$\sqrt{\frac{a}{b}} = \frac{\sqrt{a}}{\sqrt{b}}: \qquad x + \frac{b}{2a} = \pm\frac{\sqrt{b^2 - 4ac}}{\sqrt{4a^2}}$$

Work Smart

To factor any perfect square trinomial of the form $x^2 + bx + c$, we write

$$\left(x + \frac{b}{2}\right)^2.$$

Work Smart

This step shows why $a > 0$. Without this assumption, $\sqrt{4a^2} = 2|a|$.

$\sqrt{4a^2} = 2a$ since $a > 0$: $\quad x + \dfrac{b}{2a} = \pm \dfrac{\sqrt{b^2 - 4ac}}{2a}$

Subtract $\dfrac{b}{2a}$ from both sides: $\quad x = -\dfrac{b}{2a} \pm \dfrac{\sqrt{b^2 - 4ac}}{2a}$

Write over a common denominator: $\quad x = \dfrac{-b \pm \sqrt{b^2 - 4ac}}{2a}$

This gives us the *quadratic formula*.

In Words

The quadratic formula says that the solution of the equation $ax^2 + bx + c = 0$ is "the opposite of *b* plus or minus the square root of *b* squared minus 4*ac* all over 2*a*."

The Quadratic Formula

The solution(s) to the quadratic equation $ax^2 + bx + c = 0$, $a \neq 0$, are given by the **quadratic formula**

$$x = \dfrac{-b \pm \sqrt{b^2 - 4ac}}{2a}$$

▶ **EXAMPLE 1** **How to Solve a Quadratic Equation Using the Quadratic Formula**

Solve: $2x^2 + 11x + 15 = 0$

Step-by-Step Solution

Step 1: Write the equation in standard form $ax^2 + bx + c = 0$ and identify the values of *a*, *b*, and *c*.

The equation is already in standard form.

$a = 2, b = 11, c = 15$: $\quad 2x^2 + 11x + 15 = 0$

Step 2: Substitute the values of *a*, *b*, and *c* into the quadratic formula.

Write the quadratic formula: $\quad x = \dfrac{-b \pm \sqrt{b^2 - 4ac}}{2a}$

Substitute 2 for *a*, 11 for *b*, and 15 for *c*: $\quad x = \dfrac{-11 \pm \sqrt{11^2 - 4(2)(15)}}{2(2)}$

Step 3: Simplify the expression found in Step 2.

$$= \dfrac{-11 \pm \sqrt{121 - 120}}{2(2)}$$

$$= \dfrac{-11 \pm \sqrt{1}}{4}$$

Work Smart: Study Skills

Notice that in Step 2 we wrote out the quadratic formula. When solving homework problems, always write the quadratic formula as part of the solution to help you memorize the formula.

$$= \dfrac{-11 \pm 1}{4}$$

$$x = \dfrac{-11 - 1}{4} \quad \text{or} \quad x = \dfrac{-11 + 1}{4}$$

$$x = \dfrac{-12}{4} \quad \text{or} \quad x = \dfrac{-10}{4}$$

Simplify: $\quad x = -3 \quad \text{or} \quad x = -\dfrac{5}{2}$

Step 4: Check Verify your solutions.

$x = -3$: $\quad 2x^2 + 11x + 15 = 0$ $\qquad x = -\dfrac{5}{2}$: $\quad 2x^2 + 11x + 15 = 0$

$2(-3)^2 + 11(-3) + 15 \stackrel{?}{=} 0$ $\qquad 2\left(-\dfrac{5}{2}\right)^2 + 11\left(-\dfrac{5}{2}\right) + 15 \stackrel{?}{=} 0$

$$2 \cdot 9 - 33 + 15 \overset{?}{=} 0 \qquad 2 \cdot \frac{25}{4} - \frac{55}{2} + 15 \overset{?}{=} 0$$

$$18 - 33 + 15 \overset{?}{=} 0 \qquad \frac{25}{2} - \frac{55}{2} + 15 \overset{?}{=} 0$$

$$0 = 0 \quad \text{True} \qquad -\frac{30}{2} + 15 \overset{?}{=} 0$$

$$0 = 0 \quad \text{True}$$

The solution set is $\left\{ -3, -\dfrac{5}{2} \right\}$.

Solving a Quadratic Equation Using the Quadratic Formula

Step 1: Write the equation in standard form $ax^2 + bx + c = 0$, and identify the values of a, b, and c.

Step 2: Substitute the values of a, b, and c into the quadratic formula.

Step 3: Simplify the expression found in Step 2.

Step 4: Verify your solution(s).

Quick ✔

1. The solutions to the quadratic equation $ax^2 + bx + c = 0$, $a \neq 0$, are given by the quadratic formula $x = $ _____.

2. For each of the following quadratic equations, identify the values of a, b, and c.

   (a) $x^2 + 5x - 3 = 0$ **(b)** $4x^2 = 3x - 1$

In Problems 3 and 4, solve each equation using the quadratic formula.

3. $3x^2 + 7x + 4 = 0$ **4.** $2x^2 + 13x + 6 = 0$

EXAMPLE 2　**Solving a Quadratic Equation Using the Quadratic Formula**

Solve: $6x^2 + 7x = 20$

Solution

The quadratic equation must be in standard form, $ax^2 + bx + c = 0$.

$$6x^2 + 7x = 20$$

$$6x^2 + 7x - 20 = 20 - 20$$

$$6x^2 + 7x - 20 = 0$$

$$\underset{a}{\underbrace{}} \quad \underset{b}{\underbrace{}} \quad \underset{c}{\underbrace{}}$$

Write the quadratic formula: $\quad x = \dfrac{-b \pm \sqrt{b^2 - 4ac}}{2a}$

Substitute $a = 6, b = 7, c = -20$: $\quad x = \dfrac{-7 \pm \sqrt{7^2 - 4(6)(-20)}}{2(6)}$

$$= \dfrac{-7 \pm \sqrt{49 + 480}}{12}$$

$$= \dfrac{-7 \pm \sqrt{529}}{12}$$

Work Smart

A common error is to accidentally write the quadratic formula as

$$x = -b \pm \frac{\sqrt{b^2 - 4ac}}{2a} \quad \text{WRONG!}$$

Be sure to extend the fraction bar under the entire numerator, as in

$$x = \frac{-b \pm \sqrt{b^2 - 4ac}}{2a}$$

Work Smart

The quadratic equation solved in Example 2 could also be solved by factoring. Try it for yourself. Which method do you prefer? Why?

Evaluate the square root:
$$= \frac{-7 \pm 23}{12}$$

$$x = \frac{-7 - 23}{12} \quad \text{or} \quad x = \frac{-7 + 23}{12}$$

$$= \frac{-30}{12} \quad \text{or} \quad = \frac{16}{12}$$

$$= -\frac{5}{2} \quad \text{or} \quad = \frac{4}{3}$$

We leave it to you to verify the solutions. The solution set is $\left\{ -\frac{5}{2}, \frac{4}{3} \right\}$.

Quick ✔

In Problems 5 and 6, solve each equation using the quadratic formula.

5. $6x^2 + 6 = -13x$ **6.** $8x^2 + 6x = 5$

Not every quadratic equation has rational solutions.

▶ **EXAMPLE 3**   **Solving a Quadratic Equation Having Irrational Solutions**

Solve: $4z^2 + 1 = 8z$

Solution

First write the equation in standard form by setting the equation equal to 0.

$$4z^2 + 1 = 8z$$

Subtract 8z from both sides: $4z^2 - 8z + 1 = 8z - 8z$

$$4z^2 - 8z + 1 = 0$$

We see that $a = 4$, $b = -8$, and $c = 1$. Now use the quadratic formula.

Work Smart

Notice that we write the quadratic formula as $z =$ because the variable in the equation is z.

Write the quadratic formula: $z = \dfrac{-b \pm \sqrt{b^2 - 4ac}}{2a}$

Substitute $a = 4, b = -8, c = 1$: $z = \dfrac{-(-8) \pm \sqrt{(-8)^2 - 4(4)(1)}}{2(4)}$

$$= \frac{8 \pm \sqrt{64 - 16}}{8}$$

$$= \frac{8 \pm \sqrt{48}}{8}$$

Simplify the square root: $= \dfrac{8 \pm \sqrt{16 \cdot 3}}{8}$

$$= \frac{8 \pm 4\sqrt{3}}{8}$$

Work Smart

We could also simplify $z = \dfrac{8 \pm 4\sqrt{3}}{8}$ as follows:

$$z = \frac{8 \pm 4\sqrt{3}}{8}$$
$$= \frac{8}{8} \pm \frac{4\sqrt{3}}{8}$$
$$= 1 \pm \frac{\sqrt{3}}{2}$$

These answers are equivalent to the ones given in Example 3.

Factor the numerator: $= \dfrac{\overset{1}{4}(2 \pm \sqrt{3})}{\underset{2}{8}}$

Divide out common factors: $= \dfrac{2 \pm \sqrt{3}}{2}$

We leave it to you to check the solutions. The solution set is $\left\{ \dfrac{2 - \sqrt{3}}{2}, \dfrac{2 + \sqrt{3}}{2} \right\}$.

Quick ✓

7. *True or False* To solve $2x^2 + 5x + 3 = 0$ using the quadratic formula, we write

$$x = -5 \pm \frac{\sqrt{5^2 - 4(2)(3)}}{2(2)}.$$

In Problems 8 and 9, solve each equation using the quadratic formula.

8. $y^2 - 2 = -4y$

9. $16x^2 - 24x + 7 = 0$

When fractions appear as coefficients in a quadratic equation, multiply both sides of the equation by the least common denominator of the coefficients to obtain an equivalent equation without fractions. Then use the quadratic formula without fractional values for a, b, and c.

EXAMPLE 4 **Solving a Quadratic Equation with Fractional Coefficients**

Solve: $\dfrac{3}{4} + \dfrac{5}{8}x - \dfrac{1}{2}x^2 = 0$

Solution

Because the coefficients are fractions, multiply both sides of the equation by 8, the least common denominator of the coefficients.

$$\frac{3}{4} + \frac{5}{8}x - \frac{1}{2}x^2 = 0$$

Multiply both sides by 8:
$$8\left(\frac{3}{4} + \frac{5}{8}x - \frac{1}{2}x^2\right) = 8 \cdot 0$$

Use the Distributive Property:
$$8 \cdot \frac{3}{4} + 8 \cdot \frac{5}{8}x - 8 \cdot \frac{1}{2}x^2 = 8 \cdot 0$$

$$6 + 5x - 4x^2 = 0$$

Rearrange terms:
$$-4x^2 + 5x + 6 = 0$$

Work Smart

Although not necessary, writing a quadratic equation in standard form with a leading coefficient that is positive is a good idea, because it makes computation easier later on.

Multiply both sides by -1 because the coefficient of x^2 is negative.

$$-1 \cdot (-4x^2 + 5x + 6) = -1 \cdot 0$$

Distribute:
$$4x^2 - 5x - 6 = 0$$

The equation is in standard form, with $a = 4$, $b = -5$, and $c = -6$.

Write the quadratic formula:
$$x = \frac{-b \pm \sqrt{b^2 - 4ac}}{2a}$$

Substitute $a = 4, b = -5, c = -6$:
$$x = \frac{-(-5) \pm \sqrt{(-5)^2 - 4(4)(-6)}}{2(4)}$$

$$= \frac{5 \pm \sqrt{25 - (-96)}}{8}$$

$$= \frac{5 \pm \sqrt{121}}{8}$$

Simplify the radicand:
$$= \frac{5 \pm 11}{8}$$

$$x = \frac{5 - 11}{8} \quad \text{or} \quad x = \frac{5 + 11}{8}$$

$$x = \frac{-6}{8} \quad \text{or} \quad x = \frac{16}{8}$$

Simplify: $= -\frac{3}{4} \quad \text{or} \quad = 2$

We leave the check to you. The solution set is $\left\{ -\frac{3}{4}, 2 \right\}$.

Quick ✓

10. *True or False* To solve $\frac{1}{2}x^2 + \frac{2}{3}x + 1 = 0$ using the quadratic formula, we

can use either $a = \frac{1}{2}, b = \frac{2}{3}, c = 1$ or $a = 3, b = 4, c = 6$.

In Problem 11, solve the equation using the quadratic formula.

11. $x - \frac{3}{8} = \frac{1}{4}x^2$

EXAMPLE 5 **Solving a Quadratic Equation Having a Repeated Solution**

Solve $4x(x - 3) + 9 = 0$ using the quadratic formula.

Solution

What's the first step? Write the equation in standard form!

$$4x(x - 3) + 9 = 0$$

Use the Distributive Property: $4x^2 - 12x + 9 = 0$

Write the quadratic formula: $x = \dfrac{-b \pm \sqrt{b^2 - 4ac}}{2a}$

Substitute $a = 4, b = -12, c = 9$: $x = \dfrac{-(-12) \pm \sqrt{(-12)^2 - 4(4)(9)}}{2(4)}$

$$= \dfrac{12 \pm \sqrt{144 - 144}}{8}$$

$$= \dfrac{12 \pm \sqrt{0}}{8}$$

$$= \dfrac{12 \pm 0}{8} = \dfrac{12}{8} = \dfrac{3}{2}$$

Work Smart

Did you notice that the quadratic equation solved in Example 5 can be solved by factoring? Try solving it this way. Which method do you prefer?

The solution $x = \dfrac{3}{2}$ is called a **repeated solution** or **double root** because it appears twice, once in the form $x = \dfrac{12 - 0}{8} = \dfrac{3}{2}$ and once in the form $x = \dfrac{12 + 0}{8} = \dfrac{3}{2}$. The solution set is $\left\{ \dfrac{3}{2} \right\}$.

Quick ✓

In Problem 12, solve the equation using the quadratic formula.

12. $16n^2 - 40n = -25$

Some linear equations have no real solution, and some radical equations have no real solution. The same holds for quadratic equations.

EXAMPLE 6 **Solving a Quadratic Equation Having No Real Solution**

Solve: $\frac{1}{2}m^2 + \frac{1}{2}m + 4 = 0$

Solution

The equation has fractional coefficients, so we multiply both sides of the equation by the LCD, 2.

$$\frac{1}{2}m^2 + \frac{1}{2}m + 4 = 0$$

Clear fractions: $\quad 2\left(\frac{1}{2}m^2 + \frac{1}{2}m + 4\right) = 2 \cdot 0$

Distribute and simplify: $\quad m^2 + m + 8 = 0$

Write the quadratic formula: $\quad m = \dfrac{-b \pm \sqrt{b^2 - 4ac}}{2a}$

Substitute $a = 1$, $b = 1$, $c = 8$: $\quad m = \dfrac{-1 \pm \sqrt{1^2 - 4(1)(8)}}{2(1)}$

$$= \dfrac{-1 \pm \sqrt{1 - 32}}{2}$$

$$= \dfrac{-1 \pm \sqrt{-31}}{2}$$

Because $\sqrt{-31}$ is not a real number, the equation has no real solution.

●

> **Quick ✓**
>
> *In Problem 13, solve the equation using the quadratic formula.*
>
> **13.** $4n^2 - 3n + 5 = 0$

EXAMPLE 7 **Revenue from Selling Sunglasses**

The equation $R = -0.1x^2 + 70x$ models the revenue a company receives from selling x pairs of sunglasses per week. If the company always sells at least 250 pairs of sunglasses per week, find the number of pairs it must sell to have weekly revenue of $10,000.

Solution

Because we want to find the value of x that gives a revenue of $10,000, let $R = 10,000$ in the equation $R = -0.1x^2 + 70x$ and get $10,000 = -0.1x^2 + 70x$.

$$-0.1x^2 + 70x = 10,000$$

Multiply both sides by -10 to clear decimals: $\quad -10(-0.1x^2 + 70x) = -10 \cdot 10,000$

Distribute: $\quad x^2 - 700x = -100,000$

Add 100,000 to both sides: $\quad x^2 - 700x + 100,000 = -100,000 + 100,000$

$$x^2 - 700x + 100,000 = 0$$

Write the quadratic formula: $\quad x = \dfrac{-b \pm \sqrt{b^2 - 4ac}}{2a}$

Work Smart

Multiply both sides of the equation by -10 to clear decimals and make the coefficient of x^2 positive.

Substitute $a = 1, b = -700, c = 100,000$: $x = \dfrac{-(-700) \pm \sqrt{(-700)^2 - 4(1)(100,000)}}{2(1)}$

$$= \dfrac{700 \pm \sqrt{490,000 - 400,000}}{2}$$

$$= \dfrac{700 \pm \sqrt{90,000}}{2}$$

Evaluate the square root: $= \dfrac{700 \pm 300}{2}$

$$x = \dfrac{700 - 300}{2} \quad \text{or} \quad x = \dfrac{700 + 300}{2}$$

$$= 200 \quad \text{or} \quad = 500$$

The company sells at least 250 pairs of sunglasses per week, so it must sell 500 pairs of sunglasses to have revenue of $10,000. ●

> ### Quick ✓
>
> **14.** The daily revenue received from a company selling x all-day passes to a theme park is given by the equation $R = -0.02x^2 + 24x$. Determine the number of all-day passes that must be sold for the company to have daily revenue of $4000. Past records of the company show that the park sells more than 400 all-day passes daily.

▶ ❷ Use the Discriminant to Determine Which Method to Use When Solving a Quadratic Equation

We now have four methods for solving quadratic equations:

1. Factoring
2. Square Root Property
3. Completing the square
4. The quadratic formula

You may be wondering which method to use or whether one method is better than another. The answer is "It depends." When solving a quadratic equation, ask yourself the following questions: "Is the expression in the quadratic equation factorable?" and "Is the equation of the form $x^2 = p$?" The answers will help you to decide the most efficient method of solving the equation. Another strategy to use when determining which method to use is to find the value of the *discriminant*.

In Words
The discriminant is the radicand of the quadratic formula.

> **Definition**
>
> For a quadratic equation $ax^2 + bx + c = 0$, the expression $b^2 - 4ac$ is called the **discriminant.**

Why is the discriminant important? The following supplies the answer.

> **The Discriminant and Solving a Quadratic Equation**
>
> The discriminant of a quadratic equation $ax^2 + bx + c = 0$ is $b^2 - 4ac$.
>
> 1. If $b^2 - 4ac > 0$, and
>   - $b^2 - 4ac$ is a perfect square, we can solve the quadratic equation by factoring.
>   - $b^2 - 4ac$ is not a perfect square, we cannot solve the quadratic equation by factoring. Completing the square or the quadratic formula should be used to solve the equation.
> 2. If $b^2 - 4ac = 0$, we can solve the quadratic equation by factoring.
> 3. If $b^2 - 4ac < 0$, the equation has no real solution.

EXAMPLE 8 **Using the Discriminant to Decide How to Solve a Quadratic Equation**

For each quadratic equation, determine the discriminant, determine the most efficient method to solve the quadratic equation, and solve.

(a) $x^2 - 5x + 3 = 0$ 　　　　　　　(b) $2x^2 - 5x - 12 = 0$

Solution

(a) The equation $x^2 - 5x + 3 = 0$ is in standard form with $a = 1$, $b = -5$, and $c = 3$. Substitute these values into the formula for the discriminant, $b^2 - 4ac$:

$$b^2 - 4ac = (-5)^2 - 4(1)(3)$$
$$= 25 - 12$$
$$= 13$$

Because $b^2 - 4ac = 13$ is positive but not a perfect square, we use the quadratic formula to solve the equation.

$$x = \frac{-b \pm \sqrt{b^2 - 4ac}}{2a}$$

Substitute $a = 1$, $b = -5$, and $b^2 - 4ac = 13$:

$$= \frac{-(-5) \pm \sqrt{13}}{2(1)}$$

$$= \frac{5 \pm \sqrt{13}}{2}$$

We leave the check to you. The solution set is $\left\{ \dfrac{5 - \sqrt{13}}{2}, \dfrac{5 + \sqrt{13}}{2} \right\}$.

(b) The equation is in standard form with $a = 2$, $b = -5$, and $c = -12$. The value of the discriminant, $b^2 - 4ac$, is

$$b^2 - 4ac = (-5)^2 - 4(2)(-12)$$
$$= 25 + 96$$
$$= 121$$

Work Smart

If the discriminant is zero, then the quadratic equation has a repeated solution, which is $\left\{ -\dfrac{b}{2a} \right\}$.

Since the discriminant is positive and a perfect square, solve the quadratic equation $2x^2 - 5x - 12 = 0$ by factoring.

$$2x^2 - 5x - 12 = 0$$
$$(2x + 3)(x - 4) = 0$$
$$2x + 3 = 0 \quad \text{or} \quad x - 4 = 0$$
$$2x = -3 \quad \text{or} \quad x = 4$$
$$x = -\frac{3}{2}$$

We leave the check to you. The solution set is $\left\{ -\dfrac{3}{2}, 4 \right\}$. 　　●

Quick ✓

15. The discriminant of $ax^2 + bx + c = 0$ is given by the formula _____.

16. If the discriminant of a quadratic equation has a value of zero, the most efficient way of solving the quadratic equation is _____.

17. If the value of the discriminant of a quadratic equation is a perfect square, the most efficient way of solving the quadratic equation is _____.

18. If the value of the discriminant of a quadratic equation is a positive number that is not a perfect square, the most efficient way of solving the quadratic equation is __ _____ _____.

(*continued*)

In Problems 19–21, determine the value of the discriminant. Use the discriminant to determine the most efficient method to solve the quadratic equation, and then solve.

19. $9q^2 - 6q + 1 = 0$

20. $3w^2 = 5w - 2$

21. $4z^2 - 2z = 1$

Table 1 summarizes all four methods for solving a quadratic equation, guidance on when to use each one, and an example of each.

Table 1

Method	When to Use	Example
Square Root Property	When the quadratic equation is written in the form $$x^2 = p$$ where p is any real number	$x^2 = 45$ Square Root Property: $x = \pm\sqrt{45}$ $= \pm 3\sqrt{5}$
Square Root Property	When the quadratic equation is written in the form $$ax^2 - c = 0$$	$3p^2 - 12 = 0$ Add 12 to both sides: $3p^2 = 12$ Divide both sides by 3: $p^2 = 4$ Square Root Property: $p = \pm\sqrt{4}$ $p = \pm 2$
Square Root Property	When the quadratic equation is written in the form $$(ax + b)^2 = p$$ where p is any real number	$(4n - 3)^2 = 12$ Square Root Property: $4n - 3 = \pm\sqrt{12}$ Simplify the radicand: $4n - 3 = \pm 2\sqrt{3}$ Add 3 to both sides: $4n = 3 \pm 2\sqrt{3}$ Divide both sides by 4: $n = \dfrac{3 \pm 2\sqrt{3}}{4}$
Factoring or the Quadratic Formula	When the quadratic equation is written in the form $$ax^2 + bx + c = 0$$ where $b^2 - 4ac$ is a perfect square. Use factoring if the quadratic expression is easy to factor. Otherwise, use the quadratic formula.	$2m^2 + m - 10 = 0$ $a = 2, b = 1, c = -10$ $b^2 - 4ac = 1^2 - 4(2)(-10) = 1 + 80 = 81$ 81 is a perfect square, so we can use factoring: $2m^2 + m - 10 = 0$ $(2m + 5)(m - 2) = 0$ $2m + 5 = 0$ or $m - 2 = 0$ $m = -\dfrac{5}{2}$ or $m = 2$
Quadratic Formula	When the quadratic equation is written in the form $$ax^2 + bx + c = 0$$ where $b^2 - 4ac$ is positive, but not a perfect square	$2x^2 + 4x - 1 = 0$ $a = 2, b = 4, c = -1$ $b^2 - 4ac = 4^2 - 4(2)(-1) = 16 + 8 = 24$ 24 is positive, but is not a perfect square, so we use the quadratic formula to solve: $x = \dfrac{-b \pm \sqrt{b^2 - 4ac}}{2a}$ $a = 2, b = 4, b^2 - 4ac = 24:$ $x = \dfrac{-4 \pm \sqrt{24}}{2(2)}$ $= \dfrac{-4 \pm 2\sqrt{6}}{4}$ $= \dfrac{-2 \pm \sqrt{6}}{2}$
Quadratic Formula	When the quadratic equation is written in the form $$ax^2 + bx + c = 0$$ where $b^2 - 4ac$ is negative. The equation has no real solution. We discuss this type of equation in more detail in Section 9.5.	$2n^2 + 5n + 8 = 0$ $a = 2, b = 5, c = 8$ $b^2 - 4ac = 5^2 - 4(2)(8) = 25 - 64 = -39$ The value of the discriminant is a negative number, so the quadratic equation has no real solution.

Notice that we did not recommend completing the square as one of the methods to use in solving a quadratic equation. This is because the quadratic formula is based on completing the square of $ax^2 + bx + c = 0$. Besides, completing the square may be a cumbersome task, and the quadratic formula is straightforward. Completing the square was worth your time, however, because it is needed to understand the quadratic formula. Also, completing the square is a skill that you will need in future math courses.

Quick ✔

In Problems 22–24, solve each quadratic equation using any method you wish.

22. $4n^2 - 24 = 0$

23. $2y^2 = 3y + 35$

24. $1 = 3q^2 + 4q$

9.3 Exercises MyMathLab®   PRACTICE

Exercise numbers in green have complete video solutions in MyMathLab.

Problems 1–24 are the Quick ✔s that follow the EXAMPLES.

Building Skills

In Problems 25–32, write the quadratic equation in standard form and then identify the values assigned to a, b, and c. Do not solve the equation. See Objective 1.

25. $x + 2x^2 = -4$

26. $2x^2 = 1 - x$

27. $x^2 + 4 = 0$

28. $3x^2 - 6 = 0$

29. $2 + \dfrac{6}{5}x = \dfrac{3}{2}x^2$

30. $\dfrac{3}{7}x^2 = x + \dfrac{1}{2}$

31. $0.5x^2 = x - 3$

32. $0.1x - x^2 = 2$

In Problems 33–76, solve each equation using the quadratic formula. See Objective 1.

33. $x^2 + 5x - 24 = 0$

34. $x^2 - 5x - 6 = 0$

35. $2x^2 + 11x + 12 = 0$

36. $2x^2 + 9x - 5 = 0$

37. $x^2 - x - 5 = 0$

38. $q^2 - 7q - 3 = 0$

39. $3r^2 + 4r - 1 = 0$

40. $2s^2 + 2s - 5 = 0$

41. $2t^2 - 4t + \dfrac{3}{2} = 0$

42. $5y^2 + \dfrac{1}{2}y - 1 = 0$

43. $z^2 = -1 - 7z$

44. $x^2 = 3x + 9$

45. $2x^2 + 1 = 5x$

46. $3x^2 + 3 = 7x$

47. $x^2 + 4x = -12$

48. $x^2 + 7x = -14$

49. $3z^2 - 4z = 4$

50. $2y^2 + 15 = 11y$

51. $\dfrac{3}{2}n + 1 + \dfrac{1}{4}n^2 = 0$

52. $0 = \dfrac{1}{2}v + \dfrac{1}{2} - \dfrac{1}{4}v^2$

53. $2k^2 + 1 - 4k = 0$

54. $3x^2 - 2 + 4x = 0$

55. $x(2x + 5) - 1 = 0$

56. $x(2x + 3) - 4 = 0$

57. $15m^2 - 8m = -1$

58. $4n^2 + 12n = -5$

59. $4x^2 + 18x + 9 = 6x$

60. $9x^2 - 12x + 6 = 2$

61. $\dfrac{k^2}{3} - 6 = -k$

62. $\dfrac{1}{12}p^2 = \dfrac{1}{6}p + 2$

63. $4z = \dfrac{3}{2}z^2 + 1$

64. $n^2 + n = \dfrac{3}{2}$

65. $x = \dfrac{x^2}{6} + \dfrac{3}{2}$

66. $\dfrac{x^2}{6} - \dfrac{4}{3}x + \dfrac{8}{3} = 0$

67. $(x + 3)(x - 2) = 9$

68. $(x + 4)(x - 1) = 4$

69. $t(1 - 2t) = 1$

70. $2(t - t^2) = 3$

71. $3w^2 = 36$

72. $2q^2 = 64$

73. $\dfrac{2}{9}w^2 - w - 2 = 0$

74. $v^2 + \dfrac{10}{3}v + 1 = 0$

75. $8x = 2x^2 - 2x$

76. $5x = 3x^2 - 4x$

In Problems 77–88, determine the discriminant and then use this value to determine the most efficient way of solving the quadratic equation. DO NOT SOLVE THE EQUATION. See Objective 2.

77. $0 = 2x^2 + 3x - 2$

78. $0 = 3x^2 - 8x - 3$

79. $2p^2 - p + 5 = 0$

80. $3s^2 - s + 4 = 0$

81. $4x(x + 3) = -9$

82. $2x = 1 + x^2$

83. $y = y^2 + 7y - 1$

84. $n(2 + n) = 7$

85. $0 = -20z - 4z^2 - 25$

86. $0 = 4z - 1 - 4z^2$

87. $\dfrac{x^2}{2} - \dfrac{x}{3} + \dfrac{5}{6} = 0$

88. $\dfrac{x^2}{2} - \dfrac{x}{2} + \dfrac{1}{5} = 0$

Mixed Practice

In Problems 89–106, solve each equation.

89. $x^2 - 2x + \dfrac{2}{3} = 0$

90. $x^2 - 2x - \dfrac{5}{2} = 0$

91. $(3p + 1)^2 = 3p(3p + 1)$

92. $(2m - 3)^2 = 2m(2m - 3)$

93. $(y + 1)(y + 5) + 9 = 0$

94. $(r - 3)(r - 2) + 4 = 0$

95. $x + 1 = -\dfrac{1}{4x}$

96. $x - 5 = -\dfrac{25}{4x}$

97. $\dfrac{2u}{u + 1} = \dfrac{-10}{u + 5}$

98. $\dfrac{6v}{4v + 1} = \dfrac{-3}{v - 5}$

99. $2n^2 = -12 + 6(n + 2)$

100. $3k^2 = -24 + 12(k + 2)$

101. $(2x + 7)^2 - 9 = 0$

102. $(3x - 2)^2 - 16 = 0$

103. $x + \dfrac{7}{2} - \dfrac{2}{x} = 0$

104. $\dfrac{x}{4} - 1 - \dfrac{3}{x} = 0$

105. $w^2 + 10 = 2$

106. $q^2 + 23 = 5$

Applying the Concepts

107. Self-Help Magazine The annual profit, P, from selling x subscriptions to *Enjoying Life While Studying Algebra* is given by the equation $P = 40 + 30x - 0.01x^2$.

(a) What is the annual profit from selling 50 subscriptions to this popular magazine?

(b) Approximately how many subscriptions would have to be sold if the publisher wanted an annual profit of $20,000?

108. Heavenly Spa The monthly revenue from selling x passes to Heavenly Spa is given by the equation $R = 0.02x^2 + 40x$.

(a) What is the revenue during a month that the spa sells 25 passes?

(b) How many passes does Heavenly Spa need to sell to have a monthly income of $2050?

109. Engineering Majors The number of students graduating from a local high school and choosing to enter college as engineering majors is given by the equation $N = 0.05x^2 - 1.5x + 43$, where x is the number of years after 2010.

(a) How many students chose engineering as a major in 2010?

(b) Approximately how many students chose engineering as a major in 2015?

(c) How many students will choose engineering as a major in 2020?

(d) In what year will 60 students choose engineering as a major?

110. Preschool Attendance The number of children attending Happy Days Preschool in a given year is given by the equation $N = 0.3x^2 - 3.2x + 50$, where x is the number of years after 2010.

(a) How many children attended this school in 2010?

(b) Approximately how many children attended Happy Days Preschool in 2013?

(c) Approximately how many children will attend this school in 2019?

(d) In what year will 60 students attend Happy Days Preschool?

111. Fun with Numbers Find the number or numbers such that the square of three more than the number is 18.

112. Fun with Numbers Find the number or numbers such that the square of five less than the number is 13.

113. Fun with Numbers Four times the sum of a number and two is the same as three more than the square of the number. Find the number(s).

114. Fun with Numbers The square of five more than twice a number is the same as three more than the number. Find the numbers.

Extending the Concepts

Up to this point, we have seen values of a, b, and c that were rational (that is, positive or negative whole numbers, fractions, or terminating or repeating decimals). It is possible to have quadratic equations that have irrational coefficients and constants. In Problems 115–118, use the quadratic formula to solve each equation. Find only the exact solutions expressed in simplest form.

115. $x^2 + 3\sqrt{2}x - 8 = 0$

116. $x^2 + 4\sqrt{3}x - 6 = 0$

117. $\sqrt{3}x^2 - 5x + \sqrt{12} = 0$

118. $\sqrt{2}x^2 - \sqrt{12}x - \sqrt{18} = 0$

Explaining the Concepts

119. Explain how you can use the discriminant to tell whether a quadratic equation is factorable.

120. Explain the error in the solution of the quadratic equation.

$$2x^2 - 8x + 5 = 0$$

$$x = \frac{-b \pm \sqrt{b^2 - 4ac}}{2a}$$

$$x = \frac{8 \pm \sqrt{-8^2 - 4(2)(5)}}{2(2)}$$

$$= \frac{8 \pm \sqrt{-64 - 40}}{4}$$

$$= \frac{8 \pm \sqrt{-104}}{4}$$

There is no real solution.

Putting the Concepts Together (Sections 9.1–9.3)

We designed these problems so that you can review Sections 9.1 through 9.3 and show your mastery of the concepts. Take time to work these problems before proceeding with the next section. The answers are located at the back of the text on page AN-32.

1. Solve using the Square Root Method:

(a) $w^2 = \dfrac{5}{4}$

(b) $(y - 2)^2 = 12$

(c) $\left(y + \dfrac{3}{2}\right)^2 = \dfrac{3}{4}$

2. Determine the number that must be added to the expression to make it a perfect square trinomial. Then factor the expression.

(a) $z^2 - 18z$

(b) $y^2 + 9y$

(c) $m^2 + \dfrac{4}{5}m$

3. Solve by completing the square:

(a) $x^2 + 12x + 35 = 0$

(b) $z^2 - 4z + 9 = 0$

(c) $m^2 = 8m + 2$

4. Solve using the quadratic formula:

(a) $x^2 + 8x - 4 = 0$

(b) $2y^2 = 5y - 4$

(c) $4p^2 + 36p + 81 = 0$

In Problems 5–7, solve the quadratic equation using any appropriate method.

5. $(x + 3)(x - 4) = -6$

6. $n + 1 = \dfrac{5}{2n}$

7. $\frac{1}{2}z^2 + \frac{1}{8} = -z$

8. Consider the quadratic equation $2x = -5 + 3x^2$.

   (a) Determine the value of the discriminant.

   (b) Determine which method you would choose to solve the equation.

   (c) Explain why this method is appropriate for this equation.

9. **Painter's Ladder** A painter has a ladder that can be extended to 20 feet. The painter decides to place the bottom of the ladder 5 feet from the house to be painted.

   (a) Exactly how far up the house will the ladder reach?

   (b) Approximate this height to the nearest tenth of a foot.

10. **Bookstore Sales** Rock Bottom Discount Bookstore determines that the revenue from selling x calculus texts can be given by the equation $R = -x^2 + 220x$.

    (a) What is the revenue from selling 100 calculus texts?

    (b) How many texts must the bookstore sell to have revenue of $9600?

9.4 Problem Solving Using Quadratic Equations

Objectives

1. Model and Solve Direct Translation Problems

2. Solve Problems Modeled by Quadratic Equations

Are You Ready for This Section?

Before getting started, take the following readiness quiz. If you get a problem wrong, go back to the section cited and review the material.

R1. Use the Pythagorean Theorem to find a when $b = 7$ and $c = 25$. [Section 6.7, pp. 419–421]

Many applied problems require solving quadratic equations. As always, we will use the problem-solving strategy from Section 2.5.

▶ 1 Model and Solve Direct Translation Problems

EXAMPLE 1 **Finding the Dimensions of a Rectangle from Its Area**

The area of a rectangle is 112 square inches. The length of the rectangle is 6 inches more than its width. What are the dimensions of the rectangle?

Solution

Step 1: Identify We must find the length and width of the rectangle.

Step 2: Name The length of the rectangle is expressed in terms of its width, so let w represent the width of the rectangle. The problem says the length is 6 inches more than the width, so we let $w + 6$ represent the length.

Step 3: Translate See Figure 5. The area of a rectangle is found by multiplying the length by the width: $A = l \cdot w$. Develop the model by substituting $A = 112$ and $l = w + 6$.

Figure 5

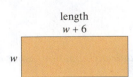

length
$w + 6$

w

$$A = l \cdot w$$
$$112 = (w + 6)w \quad \text{The Model}$$

Multiply out the right side: $112 = w^2 + 6w$

Write the equation in standard form: $0 = w^2 + 6w - 112$

Step 4: Solve Solve the equation by factoring.

$$w^2 + 6w - 112 = 0$$
$$(w + 14)(w - 8) = 0$$
$$w + 14 = 0 \quad \text{or} \quad w - 8 = 0$$
$$w = -14 \quad \text{or} \quad w = 8$$

Step 5: Check Discard the solution $w = -14$ because w represents the width of the rectangle and cannot be negative. Does a rectangle whose dimensions are 8 inches by $8 + 6 = 14$ inches have an area that is 112 square inches?

$$(8)(14) \overset{?}{=} 112$$
$$112 = 112 \quad \text{True}$$

Step 6: Answer The dimensions of the rectangle are 8 inches by 14 inches. ●

> **Quick ✓**
>
> 1. *True or False* In a rectangle, the width is 5 feet less than the length. If l represents the length, then the width is $5 - l$.
> 2. The height of a triangle is 2 inches more than the base, and the area of the triangle is 40 square inches. Find the measurements of the base and the height of the triangle.

(**EXAMPLE 2**) **Putting in a Flower Garden**

Melissa plans to plant a flower garden at the corner of her lot. The garden is in the shape of a right triangle with the hypotenuse measuring 12 feet, as shown in Figure 6. Because of the location of a dogwood tree, one leg of the triangle will be 2 feet longer than the other leg. Find the approximate lengths of the sides of the flower bed, to the nearest tenth of a foot.

Figure 6

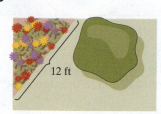

12 ft

Solution

Step 1: Identify We must find the length of each leg of the right triangle that is formed by Melissa's garden.

Step 2: Name Let x represent the length of the shorter leg of the right triangle. Because "one leg of the triangle will be 2 feet longer than the other leg," $x + 2$ represents the length of the longer leg.

Step 3: Translate Develop the model by stating the relationship among the three sides of the right triangle using the Pythagorean Theorem with $a = x$ and $b = x + 2$.

$$a^2 + b^2 = c^2$$
$$x^2 + (x + 2)^2 = 12^2$$

$(A + B)^2 = A^2 + 2AB + B^2:\quad x^2 + (x^2 + 4x + 4) = 144$

Write in standard form: $\quad 2x^2 + 4x - 140 = 0$

Divide each side by 2: $\quad \dfrac{2x^2 + 4x - 140}{2} = \dfrac{0}{2}$

$$x^2 + 2x - 70 = 0$$

Step 4: Solve Solve the quadratic equation using the quadratic formula.

$$x^2 + 2x - 70 = 0$$

$$x = \frac{-b \pm \sqrt{b^2 - 4ac}}{2a}$$

Substitute $a = 1$, $b = 2$, $c = -70$:

$$x = \frac{-2 \pm \sqrt{2^2 - 4(1)(-70)}}{2(1)}$$

$$= \frac{-2 \pm \sqrt{284}}{2}$$

$$= \frac{-2 \pm \sqrt{4 \cdot 71}}{2}$$

$$= \frac{-2 \pm 2\sqrt{71}}{2}$$

$$= \frac{2(-1 \pm \sqrt{71})}{2}$$

Divide out common factors: $$= \frac{2(-1 \pm \sqrt{71})}{\cancel{2}}$$

$$= -1 \pm \sqrt{71}$$

The two solutions are $x = -1 - \sqrt{71} \approx -9.4$ and $x = -1 + \sqrt{71} \approx 7.4$. We discard the negative solution because x represents a length. One leg is approximately 7.4 feet long. The other leg is $x + 2 \approx 7.4 + 2 \approx 9.4$ feet long.

Step 5: Check Because $\left(-1 + \sqrt{71}\right)^2 + \left(1 + \sqrt{71}\right)^2 = \left(1 - 2\sqrt{71} + 71\right) + \left(1 + 2\sqrt{71} + 71\right) = 144 = 12^2$ our answer checks.

Step 6: Answer The sides of the triangular flower bed are approximately 7.4 feet and 9.4 feet.   ●

Quick ✓

3. The Pythagorean Theorem tells the relationship between the lengths of the legs and the hypotenuse of a right triangle and is given by the formula _____.

4. A rectangular plot of land is designed so that its length is 4 meters more than its width. The diagonal of the land is 14 meters. To the nearest tenth of a meter, what are the dimensions of the land?

EXAMPLE 3 **Building a Flower Box**

Jonathan is building a 1-foot-tall flower box for his patio. The flower box is to be 2 feet longer than it is wide, and Jonathan wants to fill the planter with 6 cubic feet of potting soil. How long, to the nearest tenth of a foot, should he make the flower box?

Solution

Step 1: Identify We want to find the length of the flower box.

Figure 7

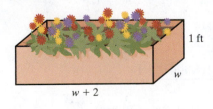

1 ft

w

$w + 2$

Step 2: Name We know that the box will be 2 feet longer than it is wide, so let w represent the width and let $w + 2$ represent the length of the box. See Figure 7.

Step 3: Translate The volume of the rectangular box is $V = l \cdot w \cdot h$. The volume is 6 cubic feet, and the height is 1 foot, so substitute $V = 6$, $l = w + 2$, and $h = 1$ into the equation to develop the model.

$$V = lwh$$
$$6 = (w + 2)(w)(1) \quad \text{The Model}$$

Use the Distributive Property: $6 = w^2 + 2w$

Write in standard form: $0 = w^2 + 2w - 6$

Step 4: Solve Solve the quadratic equation using the quadratic formula.

$$w^2 + 2w - 6 = 0$$

$$w = \frac{-b \pm \sqrt{b^2 - 4ac}}{2a}$$

Substitute $a = 1$, $b = 2$, $c = -6$:

$$w = \frac{-2 \pm \sqrt{2^2 - 4(1)(-6)}}{2(1)}$$

$$= \frac{-2 \pm \sqrt{28}}{2}$$

Simplify the radical:

$$= \frac{-2 \pm 2\sqrt{7}}{2}$$

Factor; divide out common factors:

$$= \frac{\cancel{2}(-1 \pm \sqrt{7})}{\cancel{2}}$$

$$= -1 \pm \sqrt{7}$$

The solutions are $w = -1 - \sqrt{7} \approx -3.6$ feet or $w = -1 + \sqrt{7} \approx 1.6$ feet. We discard the negative solution because w represents the width of the flower box and cannot be negative. The flower box should be approximately $1.6 + 2 = 3.6$ feet long.

Step 5: Check If we multiply length $\cdot$ width $\cdot$ height, we should obtain a volume of 6 cubic feet. Because $(w + 2) \cdot w \cdot 1 = (-1 + \sqrt{7} + 2)(-1 + \sqrt{7})(1)$ $= (1 + \sqrt{7})(-1 + \sqrt{7})(1) = -1 + 7 = 6$, our dimensions are correct.

Step 6: Answer Jonathan should construct the flower box to be approximately 1.6 feet wide and 3.6 feet long.

> ## Quick ✓
>
> 5. *True or False* The length of a rectangular box could be either $2 + \sqrt{7}$ feet or $2 - \sqrt{7}$ feet.
> 6. The compost bin that Alyssa is constructing must be 4 feet high and have a length that is 1 foot more than the width. The bin should hold 50 cubic feet of compost. To the nearest tenth of a foot, what should the length and width of the compost bin be?

▶ ❷ Solve Problems Modeled by Quadratic Equations

Numerous problems from the physical, behavioral, and social sciences can be modeled with quadratic equations.

(**EXAMPLE 4**)   **Launching a Toy Rocket**

The height of a toy rocket t seconds after it is fired straight up from an initial height of 2 feet with an initial speed of 80 feet per second can be modeled by the equation $h = -16t^2 + 80t + 2$, where t is time in seconds. After how many seconds will the rocket be 82 feet above the ground?

Solution

Substitute 82 for h in the equation $h = -16t^2 + 80t + 2$ and then solve for t.

$$-16t^2 + 80t + 2 = 82$$

Subtract 82 from each side:

$$-16t^2 + 80t - 80 = 0$$

Factor out -16:

$$-16(t^2 - 5t + 5) = 0$$

Divide both sides by -16:

$$t^2 - 5t + 5 = 0$$

With $a = 1$, $b = -5$, and $c = 5$, $b^2 - 4ac = (-5)^2 - 4(1)(5) = 25 - 20 = 5$. Use the quadratic formula to solve the equation.

Quadratic formula:

$$t = \frac{-b \pm \sqrt{b^2 - 4ac}}{2a}$$

Substitute:

$$t = \frac{-(-5) \pm \sqrt{5}}{2(1)}$$

$$= \frac{5 \pm \sqrt{5}}{2}$$

$$t = \frac{5 - \sqrt{5}}{2} \quad \text{or} \quad t = \frac{5 + \sqrt{5}}{2}$$

$$\approx 1.38 \quad \text{or} \quad \approx 3.62$$

Approximately 1.38 seconds after liftoff and approximately 3.62 seconds after liftoff, the rocket is 82 feet above the ground. Do you understand why both answers are possible? ●

Quick ✔

7. A part of a theory from economics states that a worker's income depends on his or her age. The equation $I = -55a^2 + 5119a - 54,448$ represents the relationship between average annual income I and age a. For what ages does average income I equal \$50,000? Round your answer to the nearest year.

8. The height of a cannonball shot upward from a height of 3 feet above the ground with an initial velocity of 100 feet/second is given by the equation $h = -16t^2 + 100t + 3$. After how many seconds will the cannonball be 50 feet above the ground? Round your answer to the nearest hundredth of a second.

9.4 Exercises    PRACTICE

Exercise numbers in green have complete video solutions in MyMathLab.

---

*Problems **1–8** are the Quick ✔ s that follow the EXAMPLES.*

Applying the Concepts

In Problems 9–12, find the dimensions of each rectangle.

△ **9.** The length of a rectangle is 3 meters more than twice the width. If the area of the rectangle is 90 square meters, find the dimensions.

△ **10.** The length of a rectangle is 5 feet less than twice the width. If the area of the rectangle is 150 square feet, find the dimensions.

△ **11.** The width of a rectangle is 3 inches less than half of the length. If the area of the rectangle is 36 square inches, find the dimensions.

△ **12.** The length of a rectangle is 7 centimeters more than half of the width. If the area of the rectangle is 120 square centimeters, find the dimensions.

In Problems 13–16, find the unknown values in each triangle.

△ **13.** In a triangle, the base is 3 centimeters longer than the height. If the area of the triangle is 20 square centimeters, find the base and height of the triangle.

△ **14.** In a triangle, the height is 1 inch less than twice the base. If the area of the triangle is 60 square inches, find the base and height of the triangle.

△ **15.** In a triangle, the height is 5 feet more than half of the base. If the area of the triangle is 66 square feet, find the base and height of the triangle.

△ **16.** In a triangle, the base is 7 meters less than the height. If the area of the triangle is 99 square meters, find the base and height of the triangle.

In Problems 17–20, use the Pythagorean Theorem to find the unknown values.

△ **17.** In a right triangle, one leg is 3 inches shorter than the other. If the length of the hypotenuse is $3\sqrt{2}$ inches, find the exact length of each leg. Then approximate these lengths to the nearest tenth of an inch.

△ **18.** In a right triangle, one leg is 5 kilometers less than the other. If the length of the hypotenuse is 10 kilometers, find the exact length of each leg. Then approximate these lengths to the nearest tenth of a kilometer.

△ **19.** In a right triangle, the hypotenuse is twice as long as one leg. If the length of the other leg is 9 meters, find the exact lengths of the remaining leg and the hypotenuse. Then approximate these lengths to the nearest tenth of a meter.

△ **20.** In a right triangle, the hypotenuse is 1 yard less than twice one of the legs. If the length of the other leg is 3 yards, find the exact lengths of the remaining leg and the hypotenuse. Then approximate these lengths to the nearest tenth of a yard.

In Problems 21–28, write an equation to represent the unknown number and then solve.

21. The product of two consecutive integers is 272. Find two pairs of integers that have this product.

22. The product of two consecutive integers is 182. Find two pairs of integers that have this product.

23. Consider two consecutive integers. The product of twice the smaller integer and 10 less than the larger integer is 140. Find the positive integers that satisfy these conditions.

24. Consider two consecutive integers. The product of the larger and half of the smaller is 55. Find the positive integers that satisfy these conditions.

25. Consider any real number such that its square increased by twice the number is 23. Find the exact value of the number(s).

26. Consider any real number such that its square decreased by four times the number is 14. Find the exact value of the number(s).

27. Consider any real number such that its square is the same as 13 less than 10 times the number. Find the exact value of the number(s).

28. Consider any real number such that the square of the number decreased by four times the number is negative one. Find the exact value of the number(s).

29. Model Rocket A model rocket is launched from the ground with an initial speed of 90 feet per second. The equation that models its height, h (in feet), t seconds after it was fired is $h = -16t^2 + 90t$.

(a) How high is the rocket 0.5 second after it was fired?

(b) How high is the rocket 2 seconds after it was fired?

(c) How long will it take the rocket to reach an altitude of 100 feet?

(d) How long will it take the rocket to return to Earth? (*Hint*: It is on the ground when $h = 0$.)

30. Projectile Motion Omar shot a projectile up vertically with a initial speed of 92 feet per second. If he was standing on a building 25 feet high, the equation that models the height of the projectile is $h = -16t^2 + 92t + 25$.

(a) How high was the projectile $\frac{3}{4}$ of a second after it was launched?

(b) How high is it after 3 seconds?

(c) To the nearest hundredth of a second, how long does it take for the projectile to reach an altitude of 75 feet?

(d) To the nearest hundredth of a second, how long will it take to hit the ground? (*Hint*: It is on the ground when $h = 0$.)

31. Fruit Fly Experiment Ruben is conducting an experiment with fruit flies in a Petri dish. Ruben discovers that the population of fruit flies in the Petri dish can be modeled by the equation $P = \dfrac{4000t}{4t^2 + 90}$, where t represents the number of hours after starting the experiment.

(a) How many fruit flies are in the Petri dish after 1 hour?

(b) How many fruit flies are in the Petri dish after 2 hours?

(c) How long does it take the population to first reach 80 flies?

32. Another Experiment Tuan is doing the same experiment as Ruben (Problem 31) but with a different variety of fruit flies. The number of fruit flies in his Petri dish is given by the equation $P = \dfrac{4000t}{4t^2 + 50}$.

(a) How many fruit flies are in the Petri dish after 1 hour?

(b) How many fruit flies are in the Petri dish after 2 hours?

(c) How long does it take the population to first reach 100 flies?

△ **33. Diagonals of a Polygon** In geometry, a convex polygon is a many-sided closed figure, with no sides collapsing in toward the middle. The points where the sides intersect are called vertices and the line segment joining any two vertices is called a diagonal of the polygon. For example, a rectangle has two diagonals. The number of diagonals of other polygons is given by the formula $D = \dfrac{n(n - 3)}{2}$, where n is the number of sides.

(a) Use this formula to calculate the number of diagonals in a hexagon (6 sides).

(b) How many diagonals does an octagon have? (*Hint:* An octagon has 8 sides.)

(c) If a polygon has 90 diagonals, how many sides does it have?

△ **34. Diagonals of a Polygon** Using the formula given in Problem 33, find the following.

(a) How many diagonals does a heptagon have? (*Hint:* A heptagon has 7 sides.)

(b) If a polygon has 9 diagonals, how many sides does it have?

△ **35. Supplementary Angles** Two angles are supplementary if the sum of their measures is 180°.

(a) If the measure of one angle is $x°$, what is the measure of its supplement?

(b) One-third of the product of the measure of an angle and 25° less than its supplement is 1250°. Find the measure of the angle.

△ **36. Complementary Angles** Two angles are complementary if the sum of their measures is 90°.

(a) If the measure of one angle is $y°$, what is the measure of its complement?

(b) Half of the product of the measure of an angle and its complement is 700°. Find the measure of the angle.

37. Profit from Jewelry The monthly profit, *P,* that Dorothy will earn from her jewelry business is given by the equation $P = -0.1x^2 + 50x - 150$, where x is the number of months that she has been in business.

(a) What will be her monthly profit after 6 months?

(b) What will be her monthly profit after 2 years?

(c) To the nearest month, how long does Dorothy have to stay in business before she breaks even? The break-even point occurs when $P = 0$.

(d) To the nearest month, how many months will it take Dorothy to get to the point where she is earning $500 per month?

38. Advertising Costs Joel has started a catering business, so he decides to buy advertising. He predicts that as time goes by, this expense will decrease because he will have all the business he can handle. A model that predicts the monthly advertising expenditures N (in dollars) in the xth year his business is open is given by $N = \dfrac{2000}{x^2 + 5}$.

(a) How much did Joel spend on advertising each month during the first year he was open?

(b) To the nearest dollar, what was his annual expenditure for advertising during the third year of operation?

(c) To the nearest year, after how many years will Joel be spending $25 per month on advertising?

39. A Day on the River Patti and John decide to take their speedboat to the Kankakee River. The current of the river is 3 mph. If they travel 30 miles downstream and then return in 6 hours, what is the average speed of the boat in still water? Round your answer to the nearest tenth of a mile.

40. Chicago to Vegas Samer is a pilot for a major airline. Today, he must fly from Chicago to Las Vegas (1500 miles) and back. The total flight time is 9 hours. If the average effect of the wind resistance (wind speed) is 20 miles per hour, what is the average speed of the plane in still air? Round your answer to the nearest tenth of a mile.

Extending the Concepts

Problems 41–50 use the following discussion. Two right triangles have special properties worth noting. One of these is the isosceles right triangle, sometimes called the 45-45-90° right triangle, in which two angles have a measure of 45°. The other is the 30-60-90° triangle, in which one of the angles measures 30° and the other angle measures 60°. The relationships among the sides for each triangle are given.

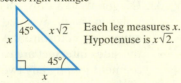

isosceles right triangle

Each leg measures x. Hypotenuse is $x\sqrt{2}$.

30-60-90 triangle

Short leg measures x. Hypotenuse is $2x$. Long leg is $x\sqrt{3}$.

△ *Use the relationships among the sides of isosceles and 30-60-90 triangles to find the missing lengths of the sides of each triangle.*

	Isosceles Right Triangle		
	Leg	Leg	Hypotenuse
	x	x	$x\sqrt{2}$
41.	1	?1	?$\sqrt{2}$
42.	?3	3	?3$\sqrt{2}$
43.	$\sqrt{6}$	?$\sqrt{6}$	?2$\sqrt{3}$
44.	?6$\sqrt{2}$	?6$\sqrt{2}$	12
45.	?2	?2	$\sqrt{8}$

	30-60-90° Right Triangle		
	Short Leg	Long Leg	Hypotenuse
	x	$x\sqrt{3}$	$2x$
46.	1	?$\sqrt{3}$	?2
47.	5	?5$\sqrt{3}$	?10
48.	?8	?8$\sqrt{3}$	16
49.	?3$\sqrt{3}$	9	?6$\sqrt{3}$
50.	?$\sqrt{2}$	$\sqrt{6}$	?2$\sqrt{2}$

9.5 The Complex Number System

Objectives

1. Evaluate the Square Root of a Negative Real Number
2. Add or Subtract Complex Numbers
3. Multiply Complex Numbers
4. Divide Complex Numbers
5. Solve Quadratic Equations with Complex Solutions

Are You Ready for This Section?

Before getting started, take the following readiness quiz. If you get a problem wrong, go back to the section cited and review the material.

R1. List the numbers in the set   [Section 1.3, pp. 19–22]

$$\left\{ 6, -\frac{5}{6}, -14, 0, \sqrt{3}, 1.\overline{65}, -\frac{18}{3}, \sqrt{-2} \right\} \text{ that are}$$

  (a) natural numbers   **(b)** whole numbers
  (c) integers   **(d)** rational numbers
  (e) irrational numbers   **(f)** real numbers

R2. Distribute: $2x(3x - 4)$ [Section 5.3, pp. 323–324]

R3. Multiply: $(m + 2)(3m - 2)$ [Section 5.3, pp. 324–327]

R4. Multiply: $(6y + 7)(6y - 7)$ [Section 5.3, pp. 327–328]

▶ Each time we encounter a situation where a number system can't handle a problem, we expand the number system. For example, if we considered only the whole numbers, we could not describe a negative balance in a checking account, so we introduce integers. But if the world had to be described by integers alone, then we could not talk about parts of a whole such as $\frac{1}{2}$ a pizza or $\frac{3}{4}$ of a dollar, so we introduce rational numbers. If we considered only rational numbers, then we wouldn't be able to find a number whose square is 2, so we introduce the irrational numbers so that $\left(\sqrt{2}\right)^2 = 2$. By combining the rational numbers with the irrational numbers, we created the real number system.

Now, suppose we wanted to determine a number whose square is -1. We know the square of any real number is never negative. We call this property of real numbers the *Nonnegativity Property*.

Nonnegativity Property of Real Numbers

For any real number a, $a^2 \geq 0$.

Because of the Nonnegativity Property, there is no real number solution to the equation

$$x^2 = -1$$

To remedy this situation, we introduce a new number.

Definition

The **imaginary unit,** denoted by i, is the number whose square is -1. That is,

$$i^2 = -1$$

By introducing the number i, we now have a new number system called the **complex number system.**

Definition

Complex numbers are numbers of the form $a + bi$, where a and b are real numbers. The real number a is called the **real part** of the number $a + bi$; the real number b is called the **imaginary part** of $a + bi$.

Ready?...Answers **R1. (a)** 6
(b) $6, 0$ **(c)** $6, -14, 0, -\dfrac{18}{3}$
(d) $6, -\dfrac{5}{6}, -14, 0, -\dfrac{18}{3}, 1.\overline{65}$
(e) $\sqrt{3}$ **(f)** $6, -\dfrac{5}{6}, -14, 0, \sqrt{3}, 1.\overline{65}, -\dfrac{18}{3}$
R2. $6x^2 - 8x$ **R3.** $3m^2 + 4m - 4$
R4. $36y^2 - 49$

In Words

The real number system is a subset of the complex number system. This means that all real numbers are, more generally, complex numbers.

For example, the complex number $3 + 4i$ has the real part 3 and the imaginary part 4. The complex number $6 - 2i = 6 + (-2)i$ has the real part 6 and the imaginary part -2.

When a complex number is written in the form $a + bi$, where a and b are real numbers, we say that it is in **standard form.** The complex number $a + 0i$ is typically written as a. This serves as a reminder that the real number system is a subset of the complex number system. The complex number $0 + bi$ is usually written as bi. Any number of the form bi is called a **pure imaginary number.** Figure 8 shows how the various number systems are related.

Figure 8
The Complex Number System

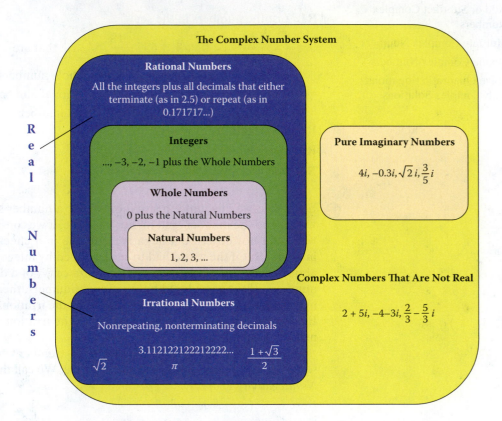

▶ ❶ Evaluate the Square Root of a Negative Real Number

First, we define the square root of a negative number.

Square Root of Negative Number

If N is a positive real number, we define the **principal square root of $-N$,** denoted by $\sqrt{-N}$, as

$$\sqrt{-N} = \sqrt{N}\, i$$

where i is the imaginary unit and $i^2 = -1$.

EXAMPLE 1   **Evaluating the Square Root of a Negative Number**

Write each of the following as a pure imaginary number.

(a) $\sqrt{-25}$ (b) $\sqrt{-12}$

Solution

(a) $\sqrt{-25} = \sqrt{25} \cdot i$
       $= 5i$

(b) $\sqrt{-12} = \sqrt{12} \cdot i$
       $= 2\sqrt{3}i$

EXAMPLE 2   **Writing Complex Numbers in Standard Form**

Write each of the following in standard form.

    (a) $3 - \sqrt{-100}$ **(b)** $5 + \sqrt{-28}$

Solution

  (a) $\begin{aligned} 3 - \sqrt{-100} &= 3 - \sqrt{100}\,i \\ &= 3 - 10i \end{aligned}$ **(b)** $\begin{aligned} 5 + \sqrt{-28} &= 5 + \sqrt{28}\cdot i \\ &= 5 + 2\sqrt{7}i \end{aligned}$ ●

EXAMPLE 3   **Writing a Complex Number in Standard Form**

Write in standard form: $\dfrac{4 - \sqrt{-72}}{10}$

Solution

$$\frac{4 - \sqrt{-72}}{10} = \frac{4 - \sqrt{72}\,i}{10}$$

$\sqrt{72} = \sqrt{36}\cdot\sqrt{2} = 6\sqrt{2}:$ $= \dfrac{4 - 6\sqrt{2}i}{10}$

Factor out 2: $= \dfrac{2\left(2 - 3\sqrt{2}i\right)}{2\cdot 5}$

Divide out the 2s: $= \dfrac{\cancel{2}\left(2 - 3\sqrt{2}i\right)}{\cancel{2}\cdot 5}$

Write as $a + bi:$ $= \dfrac{2}{5} - \dfrac{3\sqrt{2}}{5}i$ ●

Work Smart

Complex numbers must be written in the form $a + bi$. In Example 3, it is incorrect to write $\dfrac{2 - 3\sqrt{2}i}{5}$. We must write $\dfrac{2}{5} - \dfrac{3\sqrt{2}}{5}i$

▶ ❷ **Add or Subtract Complex Numbers**

Throughout your math career, whenever you were introduced to a new number system, the operations of addition, subtraction, multiplication, and division using that number system were discussed. We are now going to examine these operations on complex numbers.

    Complex numbers are added by adding the real parts and then adding the imaginary parts.

The Sum of Complex Numbers

$$(a + bi) + (c + di) = (a + c) + (b + d)i$$

To subtract two complex numbers, we use this rule:

The Difference of Complex Numbers

$$(a + bi) - (c + di) = (a - c) + (b - d)i$$

EXAMPLE 4 **Adding Complex Numbers**

Add:

(a) $(4 + 6i) + (-3 + 5i)$ **(b)** $\left(8 + \sqrt{-9}\right) + \left(2 - \sqrt{-64}\right)$

Solution

(a) $(4 + 6i) + (-3 + 5i) = [4 + (-3)] + (6 + 5)i$
$$= 1 + 11i$$

$\sqrt{-9} = 3i; \sqrt{-64} = 8i$

Work Smart

Always write a number in standard form before you add or subtract.

(b) $\left(8 + \sqrt{-9}\right) + \left(2 - \sqrt{-64}\right) = (8 + 3i) + (2 - 8i)$
$$= (8 + 2) + (3 - 8)i$$
$$= 10 + (-5)i$$
$$= 10 - 5i$$

EXAMPLE 5 **Subtracting Complex Numbers**

Subtract:

(a) $(4 - 2i) - (-3 + 7i)$ **(b)** $\left(4 - \sqrt{-4}\right) - \left(-7 + \sqrt{-9}\right)$

Solution

(a) $(4 - 2i) - (-3 + 7i) = (4 - (-3)) + (-2 - 7)i$
$$= 7 + (-9)i$$
$$= 7 - 9i$$

$\sqrt{-4} = 2i; \sqrt{-9} = 3i$

Work Smart

Notice how adding complex numbers is like combining like terms. For example,
$(4 + 6x) + (-3 + 5x)$
$= \left[4 + (-3)\right] + (6 + 5)x$
$= 1 + 11x$
and
$(4 + 6i) + (-3 + 5i)$
$= \left[4 + (-3)\right] + (6 + 5)i$
$= 1 + 11i$

(b) $\left(4 - \sqrt{-4}\right) - \left(-7 + \sqrt{-9}\right) = (4 - 2i) - (-7 + 3i)$
$$= (4 - (-7)) + (-2 - 3)i$$
$$= (4 + 7) + (-2 - 3)i$$
$$= 11 + (-5i)$$
$$= 11 - 5i$$

Quick ✓

In Problems 11–13, add or subtract as indicated.

11. $(4 - 3i) + (-2 + 5i)$ **12.** $(-3 + 7i) - (5 - 4i)$

13. $\left(3 + \sqrt{-12}\right) - \left(-2 - \sqrt{-27}\right)$

▶ ❸ Multiply Complex Numbers

Multiplying complex numbers is similar to multiplying polynomials.

EXAMPLE 6 | **Using the Distributive Property to Multiply Complex Numbers**

Multiply and write the answer in standard form: $3i(5 - 4i)$

Solution

Distribute the $3i$ to each term in the parentheses.

$$3i(5 - 4i) = 3i \cdot 5 - 3i \cdot 4i$$
$$= 15i - 12i^2$$
$$i^2 = -1: \quad = 15i - 12(-1)$$
$$= 12 + 15i$$

EXAMPLE 7 | **Using the FOIL Method to Multiply Complex Numbers**

Multiply and write the answer in standard form: $(-2 + 5i)(4 - 2i)$

Solution

$$\begin{array}{c} \quad \overset{F}{} \quad \overset{O}{} \quad \overset{I}{} \quad \overset{L}{} \\ (-2 + 5i)(4 - 2i) = -2 \cdot 4 - 2 \cdot (-2i) + 5i \cdot 4 + 5i \cdot (-2i) \end{array}$$
$$= -8 + 4i + 20i - 10i^2$$
$$\text{Combine like terms; } i^2 = -1: \quad = -8 + 24i - 10(-1)$$
$$= -8 + 24i + 10$$
$$= 2 + 24i$$

> **Quick ✓**
>
> *In Problems 14 and 15, multiply.*
>
> **14.** $4i(3 - 6i)$ **15.** $(-2 + 4i)(3 - i)$

Work Smart

$\sqrt{a} \cdot \sqrt{b} = \sqrt{a \cdot b}$ only if a and b are nonnegative (zero or positive) real numbers.

Look back at the Product Rule for Square Roots in Section 8.4. You should notice that the rule applies only when $\sqrt{a}$ and $\sqrt{b}$ are real numbers. This means that

$$\sqrt{a} \cdot \sqrt{b} \neq \sqrt{ab} \quad \text{if} \quad a < 0 \text{ or } b < 0$$

So how do we perform this multiplication? We write the radical as a complex number, using the fact that $\sqrt{-N} = \sqrt{N}i$, and then perform the multiplication.

EXAMPLE 8 | **Multiplying Square Roots of Negative Numbers**

Multiply:

(a) $\sqrt{-9} \cdot \sqrt{-36}$ **(b)** $\left(2 - \sqrt{-36}\right)\left(4 - \sqrt{-25}\right)$

Solution

Work Smart

$\sqrt{-9} \cdot \sqrt{-36} \neq \sqrt{(-9)(-36)}$ because the Product Property of Square Roots applies only when the radicand is a nonnegative real number.

(a) We cannot use the Product Rule of Radicals to multiply these radicals, because $\sqrt{-9}$ and $\sqrt{-36}$ are not real numbers. Therefore, we express the radicals as complex numbers and then multiply.

$$\sqrt{-9} \cdot \sqrt{-36} = 3i \cdot 6i$$
$$= 18i^2$$
$$i^2 = -1: \quad = 18(-1)$$
$$= -18$$

(b) First, rewrite each expression as a complex number in standard form.

$$\left(2 - \sqrt{-36}\right)\left(4 - \sqrt{-25}\right) = \left(2 - 6i\right)\left(4 - 5i\right)$$

FOIL: $= 2 \cdot 4 + 2 \cdot (-5i) - 6i \cdot 4 - 6i \cdot (-5i)$

$= 8 - 10i - 24i + 30i^2$

Combine like terms; $i^2 = -1$: $= 8 - 34i + 30(-1)$

$= 8 - 34i - 30$

$= -22 - 34i$

Quick ✔

In Problems 16 and 17, multiply.

16. $\sqrt{-25} \cdot \sqrt{-4}$

17. $\left(2 - \sqrt{-16}\right)\left(1 - \sqrt{-4}\right)$

Conjugates

We now introduce a special product that involves the *conjugate* of a complex number.

In Words

The complex conjugate of $a + bi$ is $a - bi$. The complex conjugate of $a - bi$ is $a + bi$.

> **The Conjugate of a Complex Number**
>
> If $a + bi$ is a complex number, then its **conjugate** is defined as $a - bi$.
>
> The expression $a - bi$ is called the *complex conjugate* of $a + bi$.

For example,

Complex Number	Complex Conjugate
$2 + 3i$	$2 - 3i$
$-12 - 7i$	$-12 + 7i$
$5i$	$-5i$

The next example shows the product of a complex number and its conjugate.

EXAMPLE 9   **Multiplying a Complex Number by Its Conjugate**

Find the product of $5 + 2i$ and its conjugate, $5 - 2i$.

Solution

$$(5 + 2i)(5 - 2i) = 5 \cdot 5 + 5 \cdot (-2i) + 2i \cdot 5 + 2i \cdot (-2i)$$

$$= 25 - 10i + 10i - 4i^2$$

$$= 25 - 4(-1)$$

$$= 25 + 4$$

$$= 29$$

Wow! The product of $5 + 2i$ and its conjugate $5 - 2i$ is 29, a real number. In fact, the results of Example 9 are true in general.

> **The Product of a Complex Number and Its Conjugate**
>
> The product of a complex number and its conjugate is a nonnegative real number:
>
> $$(a + bi)(a - bi) = a^2 + b^2$$

Work Smart

$(a + b)(a - b) = a^2 - b^2$

but

$(a + bi)(a - bi) = a^2 + b^2$

Notice that multiplying a complex number and its conjugate is different from multiplying the sum and difference of the same two terms:

$$(a + b)(a - b) = a^2 - b^2, \text{ whereas } (a + bi)(a - bi) = a^2 + b^2$$

Quick ✔

18. The complex conjugate of $-1 + 5i$ is _____.

In Problems 19 and 20, multiply.

19. $(2 - 7i)(2 + 7i)$

20. $(-4 + 9i)(-4 - 9i)$

▶ ④ Divide Complex Numbers

Now that we know the product of a complex number and its conjugate is a nonnegative real number, we can divide complex numbers. In Examples 10 and 11, notice how the approach is similar to rationalizing denominators. (See Examples 3 and 6 in Section 8.5.)

EXAMPLE 10 **How to Divide Complex Numbers**

Divide: $\dfrac{15}{2 + i}$

Step-by-Step Solution

Step 1: Write the numerator and denominator in standard form, $a + bi$.

The numerator and denominator are in standard form.

Step 2: Multiply the numerator and denominator by the complex conjugate of the denominator.

Multiply by $1 = \dfrac{2 - i}{2 - i}$: $\quad \dfrac{15}{2 + i} \cdot \dfrac{2 - i}{2 - i} = \dfrac{15(2 - i)}{(2 + i)(2 - i)}$

$(a + bi)(a - bi) = a^2 + b^2$: $\qquad\qquad = \dfrac{15(2 - i)}{2^2 + 1^2}$

Divide out like factors: $\qquad\qquad = \dfrac{\overset{3}{\cancel{15}}(2 - i)}{\underset{1}{\cancel{5}}}$

$\qquad\qquad\qquad\qquad\qquad = 3(2 - i)$

Step 3: Simplify by writing the quotient in standard form, $a + bi$.

$\qquad\qquad = 6 - 3i$

●

Work Smart

Dividing complex numbers is a lot like rationalizing a denominator.

$\dfrac{15}{2 + \sqrt{3}} = \dfrac{15}{2 + \sqrt{3}} \cdot \dfrac{2 - \sqrt{3}}{2 - \sqrt{3}}$

$= \dfrac{15(2 - \sqrt{3})}{4 - 2\sqrt{3} + 2\sqrt{3} - \sqrt{9}}$

$= \dfrac{15(2 - \sqrt{3})}{4 - 3}$

$= 15(2 - \sqrt{3})$

$= 30 - 15\sqrt{3}$

We now summarize the steps used to divide complex numbers.

Dividing Complex Numbers

Step 1: Write the numerator and denominator in standard form, $a + bi$.

Step 2: Multiply the numerator and denominator by the complex conjugate of the denominator.

Step 3: Simplify by writing the quotient in standard form, $a + bi$.

Quick ✔

In Problems 21 and 22, divide.

21. $\dfrac{26}{-2 + 3i}$

22. $\dfrac{-12}{4 - i}$

EXAMPLE 11 **Dividing Complex Numbers**

Divide: $\dfrac{8 + 5i}{\sqrt{-4}}$

Solution

Begin by writing the numerator and denominator in standard form. Because $\sqrt{-4} = 2i$, we have

$$\frac{8 + 5i}{\sqrt{-4}} = \frac{8 + 5i}{2i}$$

The complex conjugate of the denominator, $2i = 0 + 2i$, is $0 - 2i$. Since $0 - 2i = -2i$, multiply the numerator and denominator by $-2i$.

$$\text{Multiply by } \frac{-2i}{-2i}: \quad \frac{8 + 5i}{2i} = \frac{8 + 5i}{2i} \cdot \frac{-2i}{-2i}$$

$$= \frac{(8 + 5i)(-2i)}{(2i)(-2i)}$$

$$\text{Distribute } -2i: \quad = \frac{8(-2i) + 5i(-2i)}{-4i^2}$$

$$= \frac{-16i - 10i^2}{-4i^2}$$

$$i^2 = -1: \quad = \frac{-16i - (10)(-1)}{-4(-1)}$$

$$= \frac{-16i + 10}{4}$$

$$\text{Divide each term in the numerator by 4:} \quad = \frac{-16i}{4} + \frac{10}{4}$$

$$= -4i + \frac{5}{2}$$

$$\text{Write in standard form:} \quad = \frac{5}{2} - 4i$$

Quick ✔

In Problems 23 and 24, divide.

23. $\dfrac{8 + 3i}{3i}$

24. $\dfrac{1 + 5i}{2i}$

Work Smart

Just as all integers are rational numbers, and all rational numbers are real numbers, all real numbers are complex numbers. For example, -2 can be written $-2 + 0i$.

▶ **5** **Solve Quadratic Equations with Complex Solutions**

Equations such as $x^2 = -9$ and $(y + 2)^2 + 3 = 1$ do not have real solutions, but they do have solutions that are nonreal complex numbers. Before we solve such equations, let's see how to determine whether a given complex number is a solution of a quadratic equation.

EXAMPLE 12 **Determining Whether a Complex Number Is a Solution of a Quadratic Equation**

Is $x = 2 + i$ a solution of the equation $x^2 - 4x + 5 = 0$?

Solution

Does $x = 2 + i$ satisfy $x^2 - 4x + 5 = 0$? If so, it is a solution.

$$x^2 - 4x + 5 = 0$$

Substitute $2 + i$ for x: $\quad (2 + i)^2 - 4(2 + i) + 5 \stackrel{?}{=} 0$

$$2^2 + 2(2)(i) + (i)^2 - 4 \cdot 2 - 4 \cdot i + 5 \stackrel{?}{=} 0$$

$$4 + 4i + i^2 - 8 - 4i + 5 \stackrel{?}{=} 0$$

$$4 + 4i - 1 - 8 - 4i + 5 \stackrel{?}{=} 0$$

$$0 = 0 \quad \text{True}$$

We have a true statement, so $x = 2 + i$ is a solution of $x^2 - 4x + 5 = 0$. ●

Now let's solve quadratic equations that have nonreal complex solutions.

EXAMPLE 13 **Solving a Quadratic Equation Having Complex Solutions**

Solve the quadratic equation: $(x - 3)^2 + 9 = 5$

Solution

$$(x - 3)^2 + 9 = 5$$

Isolate the squared term: $\quad (x - 3)^2 + 9 - 9 = 5 - 9$

$$(x - 3)^2 = -4$$

Use the Square Root Property: $\quad x - 3 = \pm\sqrt{-4}$

$\sqrt{-4} = 2i$: $\quad x - 3 = \pm 2i$

Add 3 to both sides: $\quad x - 3 + 3 = \pm 2i + 3$

$$x = 3 \pm 2i$$

Work Smart

A quadratic equation may be solved by factoring, using the Square Root Property, or using the quadratic formula. Study the form of the quadratic equation to see which method is most efficient.

Check $\quad x = 3 - 2i$: $\quad (x - 3)^2 + 9 = 5 \qquad x = 3 + 2i$: $\quad (x - 3)^2 + 9 = 5$

$$[(3 - 2i) - 3]^2 + 9 \stackrel{?}{=} 5 \qquad\qquad [(3 + 2i) - 3]^2 + 9 \stackrel{?}{=} 5$$

$$[3 - 2i - 3]^2 + 9 \stackrel{?}{=} 5 \qquad\qquad [3 + 2i - 3]^2 + 9 \stackrel{?}{=} 5$$

$$[-2i]^2 + 9 \stackrel{?}{=} 5 \qquad\qquad [2i]^2 + 9 \stackrel{?}{=} 5$$

$$4i^2 + 9 \stackrel{?}{=} 5 \qquad\qquad 4i^2 + 9 \stackrel{?}{=} 5$$

$$-4 + 9 \stackrel{?}{=} 5 \qquad\qquad -4 + 9 \stackrel{?}{=} 5$$

$$5 = 5 \quad \text{True} \qquad\qquad 5 = 5 \quad \text{True}$$

Both answers check, so the solution set is $\{3 - 2i, 3 + 2i\}$. ●

EXAMPLE 14 **Solving a Quadratic Equation Having Complex Solutions**

Solve the quadratic equation: $x^2 - 2x + 7 = 0$

Solution

We'll use the quadratic formula to solve $x^2 - 2x + 7 = 0$.

Quadratic formula: $\quad x = \dfrac{-b \pm \sqrt{b^2 - 4ac}}{2a}$

Substitute $a = 1, b = -2, c = 7$: $x = \dfrac{-(-2) \pm \sqrt{(-2)^2 - 4(1)7}}{2(1)}$

$$= \dfrac{2 \pm \sqrt{-24}}{2}$$

Work Smart

The expression $\dfrac{2 \pm 2\sqrt{6}i}{2}$ can also be

simplified using $\dfrac{A + B}{C} = \dfrac{A}{C} + \dfrac{B}{C}$:

$\dfrac{2 \pm 2\sqrt{6}i}{2} = \dfrac{2}{2} \pm \dfrac{2\sqrt{6}}{2}i$

$= 1 + \sqrt{6}i$

Simplify the square root: $= \dfrac{2 \pm 2\sqrt{6}i}{2}$

Factor; divide out the 2s: $= \dfrac{2(1 \pm \sqrt{6}i)}{2}$

$$= 1 \pm \sqrt{6}i$$

We leave the check to you. The solution set is $\{1 - \sqrt{6}i, 1 + \sqrt{6}i\}$.

●

Quick ✓

In Problems 25 and 26, solve the quadratic equation.

25. $2w^2 + 5 = 0$

26. $n^2 + 6n = -17$

9.5 Exercises MyMathLab® MathXP PRACTICE

Exercise numbers in green have complete video solutions in MyMathLab.

Problems 1–26 are the Quick ✓s that follow the EXAMPLES.

Building Skills

In Problems 27–30, write each of the following as a pure imaginary number. See Objective 1.

27. $\sqrt{-625}$

28. $\sqrt{-169}$

29. $-\sqrt{-52}$

30. $-\sqrt{-54}$

In Problems 31–38, write each of the following in standard form. See Objective 1.

31. $5 + \sqrt{-4}$

32. $7 + \sqrt{-36}$

33. $12 - \sqrt{-27}$

34. $15 - \sqrt{-45}$

35. $\dfrac{-9 + \sqrt{-18}}{6}$

36. $\dfrac{-6 + \sqrt{-12}}{2}$

37. $\dfrac{2 - \sqrt{-48}}{-4}$

38. $\dfrac{10 - \sqrt{-50}}{-5}$

In Problems 39–52, add or subtract as indicated. Express your answer in standard form. See Objective 2.

39. $(3 - 2i) + (-2 + 5i)$

40. $(-5 - 2i) + (4 - 9i)$

41. $(-4 + 6i) - (6 + 6i)$

42. $(-5 + 4i) - (7 + 4i)$

43. $9i - (3 - 12i)$

44. $(4 + 2i) - i$

45. $(10 - \sqrt{-4}) - (-3 + \sqrt{-64})$

46. $(8 + \sqrt{-25}) + (-3 - \sqrt{-16})$

47. $(-6 + \sqrt{-64}) + (-2 - \sqrt{-81})$

48. $(13 - \sqrt{-121}) - (8 + \sqrt{-49})$

49. $(11 + \sqrt{-18}) + (1 - \sqrt{-8})$

50. $(-4 - \sqrt{-24}) + (-2 + \sqrt{-54})$

51. $(-1 - \sqrt{-40}) - (-1 + \sqrt{-90})$

52. $(1 + \sqrt{-18}) - (2 - \sqrt{-2})$

In Problems 53–70, multiply. Express your answer in standard form. See Objective 3.

53. $3i(4 - 5i)$

54. $5i(2 + 7i)$

55. $(1 - i)(-10i)$

56. $(5 + 3i)(-i)$

57. $(-2 + 7i)(-2 + i)$

58. $(-3 - 3i)(1 - 2i)$

59. $(16i)^2$

60. $(100i)^2$

61. $(6 + 12i)(-2 - 3i)$

62. $(6 + 4i)(3 + 2i)$

63. $(4 - 3i)^2$

64. $(-8 + 6i)^2$

65. $\sqrt{-25} \cdot \sqrt{-16}$

66. $\sqrt{-4} \cdot \sqrt{-1}$

67. $\sqrt{-6} \cdot \sqrt{-12}$

68. $\sqrt{-3} \cdot \sqrt{-15}$

69. $(-5 - \sqrt{-1})(1 + \sqrt{-36})$

70. $(-2 + \sqrt{-4})(-3 - \sqrt{-25})$

In Problems 71–74, write the complex conjugate of the given complex number, and then find the product of the complex number and its conjugate. See Objective 3.

71. $-3 + 2i$

72. $4 - 5i$

73. $6 + \sqrt{-18}$

74. $2 - \sqrt{-12}$

In Problems 75–86, divide. Express your answer in standard form. See Objective 4.

75. $\dfrac{20}{1 - 2i}$

76. $\dfrac{15}{1 - 2i}$

77. $\dfrac{-4 + 6i}{2i}$

78. $\dfrac{2 - 5i}{i}$

79. $\dfrac{3}{1 - i}$

80. $\dfrac{-6i}{1 + i}$

81. $\dfrac{-5i}{-2 + 4i}$

82. $\dfrac{9i}{-2 - 6i}$

83. $\dfrac{8 + 27i}{-4i}$

84. $\dfrac{-5 - 4i}{-5i}$

85. $\dfrac{-1}{3 + 2i}$

86. $\dfrac{-5}{-3 + i}$

In Problems 87–102, solve the quadratic equation by any method. See Objective 5.

87. $2x^2 - 4x + 5 = 0$

88. $2x^2 - 3x + 6 = 0$

89. $(n + 1)^2 + 12 = 0$

90. $(n - 2)^2 + 18 = 0$

91. $(w - 5)^2 + 6 = 2$

92. $(z + 6)^2 + 13 = 4$

93. $z^2 + 6z + 9 = 2z$

94. $r^2 + 3r + 6 = -r$

95. $2x^2 - 13x = 24$

96. $3x^2 + 14x = 5$

97. $2x^2 - 16 = 0$

98. $2x^2 - 90 = 0$

99. $3 = 2t - t^2$

100. $2p = 4p^2 + 7$

101. $13 = k(2k - 1)$

102. $x(x + 6) = 9$

Applying the Concepts

103. Fun with Numbers The square of one more than a number is the same as twice the number squared increased by six. Find the number(s).

104. Fun with Numbers Three times one less than a number is the same as the sum of the number and its square. Find the number(s).

105. Fun with Numbers The square of a number is the same as ten less than twice the number. Find the number(s).

106. Fun with Numbers Three times the reciprocal of a number is the same as four decreased by twice the number. Find the number(s).

Explaining the Concepts

107. Explain why the product of two complex conjugates is always a real number.

108. Is it possible to have a solution set to a quadratic equation that contains one real number and one complex number that is not real? Explain your response.

Complex numbers have not always enjoyed the acceptance that they have today. In 1545, Girolamo Cardano published a paper on solutions to higher-degree equations. In this work, he recognized the existence of negative roots but claimed them to be "fictitious." Complex numbers became more accepted as established mathematicians such as Descartes, Euler, and Gauss used them in their work. Descartes originated the terms "real part" and "imaginary parts" (in 1637), Euler introduced the notation $i = \sqrt{-1}$ (in 1777), and Gauss coined the term "complex" (in 1831).

109. Part of Cardano's gradual acceptance of negative roots stemmed from his investigation into the following problem: "Divide 10 into two parts such that the product of one part times the remainder is 40." Here "remainder" is not a part of a division problem but rather another way of saying that a number, x, and its remainder $(10 - x)$ sum to 10. Can you help Cardano find an answer to this puzzle?

Chapter 9 Activity: The Math Game

Focus: Solving quadratic equations

Time: 15 minutes

Group size: 2–4

- The instructor will announce when the groups may begin solving the problems to the right.
- When your group has completed all of the problems, ask the instructor to check the answers. The instructor will tell you how many answers are correct, but not which ones.
- The first group to complete all of the problems correctly will win a prize, as determined by the instructor.

Solve the following quadratic equations. Express radicals in simplest form, and express complex answers in standard form.

1. $2x^2 + 9x - 5 = 0$ **2.** $x^2 - 20 = 5$

3. $3x^2 - 2 = -4x$ **4.** $3x^2 + 6x + 6 = 0$

5. $x^2 - 3x + 9 = 0$ **6.** $3x^2 - 1 = 2x$

7. $5(x - 1)^2 + 3 = 6$ **8.** $5x = 2x^2 + 1$

Chapter 9 Review

Section 9.1   Solving Quadratic Equations Using the Square Root Property

KEY CONCEPT

- **The Square Root Property**
  If $x^2 = p$, where $p \geq 0$, then $x = \sqrt{p}$ or $x = -\sqrt{p}$.

You Should Be Able To...	EXAMPLE	Review Exercises
① Solve quadratic equations using the Square Root Property (p. 579)	Examples 1 through 6	1–14
② Solve problems using the Pythagorean Theorem (p. 584)	Examples 7 and 8	15–20

In Problems 1–14, solve each quadratic equation using the Square Root Property. Express radicals in simplest form.

1. $x^2 - 16 = 0$ **2.** $0 = x^2 - 144$

3. $4p^2 = 9$ **4.** $25t^2 = 4$

5. $3x^2 - 10 = 26$ **6.** $2x^2 + 5 = 41$

7. $(3x + 1)^2 = 16$ **8.** $(2x - 5)^2 = 49$

9. $9 = (2n + 10)^2 + 1$ **10.** $25 = (4d - 8)^2 - 7$

11. $(2x + 7)^2 + 6 = 2$ **12.** $(5x - 1)^2 - 5 = -30$

13. $x^2 - 4x + 4 = 24$ **14.** $x^2 + 8x + 16 = 28$

In Problems 15–18, use the right triangle shown to the right and find the missing length. Give exact answers and decimal approximations rounded to two decimal places.

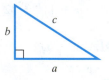

15. $a = 12, b = 8$ **16.** $a = 6, b = 12$

17. $b = 8, c = 14$ **18.** $a = 4, c = 12$

19. Picture Window A square window in the front of the Browns' house measures 6 feet on the diagonal.

(a) Exactly how wide and how tall is the window?

(b) Approximate the dimensions of the window rounded to two decimal places.

20. Painter's Ladder A painter has a ladder that can be extended to 15 feet. The painter decides to place the bottom of the ladder 5 feet from the house to be painted.

(a) Exactly how far up the house will the ladder reach?

(b) Approximate this height to the nearest tenth of a foot.

Section 9.2 Solving Quadratic Equations by Completing the Square

KEY CONCEPT

- Steps to Solve a Quadratic Equation by Completing the Square. See page 591.

KEY TERM

Complete the square

You Should Be Able To...	EXAMPLE	Review Exercises
1 Complete the square in one variable (p. 589)	Example 1	21–26
2 Solve quadratic equations by completing the square (p. 590)	Examples 2 through 6	27–40

In Problems 21–26, determine the number that must be added to the expression to make it a perfect square trinomial. Then factor the expression.

21. $x^2 - 14x$

22. $x^2 + 10x$

23. $y^2 + \dfrac{2}{3}y$

24. $k^2 - \dfrac{4}{5}k$

25. $n^2 - \dfrac{n}{2}$

26. $t^2 + \dfrac{t}{3}$

In Problems 27–38, solve the quadratic equation by completing the square.

27. $x^2 - 6x = 7$

28. $x^2 - 12x = -11$

29. $x^2 = 4 - 3x$

30. $x^2 = -14 - 9x$

31. $x^2 - 4x - 16 = 0$

32. $\dfrac{1}{2}x^2 - x - \dfrac{7}{2} = 0$

33. $2z^2 + z - 10 = 0$

34. $3n^2 + 11n + 6 = 0$

35. $(x - 5)(x + 2) = 1$

36. $(x + 3)(x - 2) = 2$

37. $x(x - 10) - 15 = 5$

38. $x(x - 8) + 2 = 10$

39. Fun with Numbers The product of a number and two more than the number is $\dfrac{5}{4}$. Find the pair(s) of numbers that satisfy this condition.

40. Fun with Numbers The sum of the square of a number and six times this number is 16. Find the two numbers that satisfy this condition.

Section 9.3 Solving Quadratic Equations Using the Quadratic Formula

KEY CONCEPTS

- The solution(s) to the quadratic equation $ax^2 + bx + c = 0, a \neq 0$, are given by

$$x = \frac{-b \pm \sqrt{b^2 - 4ac}}{2a}$$

- Steps to Solve a Quadratic Equation Using the Quadratic Formula. See page 599.

- The discriminant of a quadratic equation is the expression $b^2 - 4ac$.

KEY TERMS

Quadratic formula
Repeated solution
Discriminant

You Should Be Able To...	EXAMPLE	Review Exercises
1 Solve quadratic equations using the quadratic formula (p. 597)	Examples 1 through 7	41–54
2 Use the discriminant to determine which method to use when solving a quadratic equation (p. 604)	Example 8; Table 1	55–60

In Problems 41 and 42, write the quadratic equation in standard form and then identify the values assigned to a, b, and c. Do not solve the equation.

41. $3 = x - 2x^2$

42. $\frac{1}{2}x^2 = \frac{3}{4} - x$

In Problems 43–52, solve each equation using the quadratic formula.

43. $2x^2 - 13x + 15 = 0$ **44.** $4x^2 + 8x + 3 = 0$

45. $x^2 - 12x = -5$ **46.** $x^2 - 3x = 9$

47. $p(3p + 5) - 8 = 0$ **48.** $v(2v + 7) - 4 = 0$

49. $2x = 3 - 2x^2$ **50.** $x^2 = 2 - 4x$

51. $\frac{z}{3} + \frac{1}{12z} = 1$ **52.** $\frac{3n}{2} + \frac{1}{n} = 4$

53. Bookstore Sales Dirt Cheap Discount Bookstore determines that the revenue R from selling x algebra texts is given by the equation $R = -\frac{1}{2}x^2 + 130x$.

(a) What is the revenue from selling 150 algebra texts?

(b) How many texts must the bookstore sell to have revenue of $8000?

54. Fancy Pet Grooming Martin owns a dog-grooming business and has determined that his profit P from grooming x dogs is given by the equation $P = x^2 + 30x - 60$.

(a) What is Martin's profit when he grooms 8 dogs?

(b) How many dogs must he groom to have a profit of $615?

In Problems 55–60, determine the discriminant, and then use this value to determine the most efficient way of solving the quadratic equation. Last, solve the quadratic equation over the set of real numbers by any appropriate method.

55. $5t^2 + 2 = 8t$ **56.** $x^2 - 5x = 6$

57. $3x^2 + 7x = 10$ **58.** $n^2 - 2 = -3n$

59. $2 + 3d^2 = d$ **60.** $3(p^2 + 1) + 5p = 0$

Section 9.4   Problem Solving Using Quadratic Equations

You Should Be Able To...	EXAMPLE	Review Exercises
❶ Model and solve direct translation problems (p. 610)	Examples 1 through 3	61–70
❷ Solve problems modeled by quadratic equations (p. 613)	Example 4	71–74

61. Consecutive Integers The product of two consecutive even integers is 288. Find both pairs of integers that have this product.

62. Fun with Numbers The sum of a number and its reciprocal is $\frac{13}{6}$. Find the number and the reciprocal.

△ **63. A Rectangle** The length of a rectangle is 2 meters less than twice the width. If the area of the rectangle is 112 square meters, find the dimensions of the rectangle.

△ **64. A Rectangle** The width of a rectangle is 1 yard more than one-third of the length. If the area of the rectangle is 60 square yards, find the dimensions of the rectangle.

△ **65. A Triangle** In a right triangle, the length of one leg is double the length of the other. If the length of the hypotenuse is $4\sqrt{5}$ feet, find the length of each of the legs of the triangle.

△ **66. A Triangle** In a right triangle, the hypotenuse is 3 centimeters longer than twice the length of one leg. If the length of the other leg is 5 centimeters, find the exact lengths of the unknown sides of the triangle.

△ **67. A Triangle** In a triangle, the base is 5 meters more than the height. If the area of the triangle is 42 square meters, find the base and height of the triangle.

△ **68. A Triangle** In a triangle, the base is one and a half inches more than the height. If the area of the triangle is $\frac{9}{4}$ square inches, find the base and height of the triangle.

△ **69. Jewelry Box** Kaysie is taking a jewelry-making class and is building a box to hold her treasures. The height of her box is to be 5 inches, and the volume will be 200 cubic inches. The base of the box is a rectangle whose width is 1 inch less than half the length. Find the dimensions of the box.

△ **70. Harold's Driveway** Harold is ordering cement for his new driveway. He orders 60 cubic feet of cement that will fill his rectangular driveway 6 inches deep. If the length of the driveway is 1 foot less than twice the width, find the dimensions of his driveway.

71. Projectile Motion Ron and Ellen are shooting a toy rocket from a slingshot. The rocket leaves their hands 4 feet above the ground with an initial velocity of 80 feet per second. The equation that models the height of the rocket h feet above the ground t seconds after it was fired is $h = -16t^2 + 80t + 4$. How long does it take for the rocket to hit the ground? Find the exact answer and then approximate it, rounding to the nearest tenth of a second.

72. Model Rocket A model rocket is launched from the ground with an initial speed of 120 feet per second. The equation that models its height, h (in feet), off the ground t seconds after it was fired is $h = -16t^2 + 120t$. How long will it take for the rocket to be 80 feet off the ground? Find the exact answer and then approximate it, rounding to the nearest tenth of a second.

*The point at which a company's profit equals zero is called the company's **break-even point**. In Problems 73 and 74, let R represent a company's revenue, let C represent the company's costs, and let x represent the number of units produced and sold. Find the company's break-even point; that is, find x so that R − C = 0. Give an exact answer and, if necessary, approximate this value rounded to the nearest whole number.*

73. $R = -\dfrac{x^2}{5} + 10x$

$C = 4x + 25$

74. $R = -\dfrac{x^2}{10} + 8x$

$C = 3x + 35$

Section 9.5 The Complex Number System

KEY CONCEPTS

- For any real number a, $a^2 \geq 0$
- $i^2 = -1$
- $\sqrt{-N} = \sqrt{N}i$, where N is a positive real number.
- $(a + bi) + (c + di) = (a + c) + (b + d)i$
- $(a + bi) - (c + di) = (a - c) + (b - d)i$
- $(a + bi)(a - bi) = a^2 + b^2$

KEY TERMS

Imaginary unit
Complex number
Real part
Imaginary part
Standard form
Pure imaginary number
Principal square root of $-N$
Complex conjugate

You Should Be Able To...	EXAMPLE	Review Exercises
❶ Evaluate the square root of a negative real number (p. 618)	Examples 1 through 3	75–80
❷ Add or subtract complex numbers (p. 619)	Examples 4 and 5	81–86
❸ Multiply complex numbers (p. 621)	Examples 6 through 9	87–96
❹ Divide complex numbers (p. 623)	Examples 10 and 11	97–102
❺ Solve quadratic equations with complex solutions (p. 624)	Examples 12 through 14	103–106

In Problems 75–80, write each complex number in standard form.

75. $-\sqrt{-100}$

76. $-\sqrt{-49}$

77. $5 + \sqrt{-20}$

78. $4 - \sqrt{-60}$

79. $\dfrac{-3 - \sqrt{-27}}{3}$

80. $\dfrac{-5 + \sqrt{-75}}{10}$

In Problems 81–94, perform the indicated operation and express your answer in standard form.

81. $(4 + 10i) + (-3 - 2i)$

82. $(-3 + 5i) + (3 + 6i)$

83. $(-8 + 5i) - (8 - 5i)$

84. $(-13 + i) - (13 - 12i)$

85. $\left(2 + \sqrt{-9}\right) - \left(12 - \sqrt{-16}\right)$

86. $\left(12 - \sqrt{-25}\right) + \left(2 + \sqrt{-1}\right)$

87. $2i(6 + 5i)$

88. $-3i(1 - i)$

89. $(4 - 5i)(3 - 2i)$

90. $(12 + i)(2 - 3i)$

91. $(2 - 3i)^2$

92. $(3 + 5i)^2$

93. $3\sqrt{-12} \cdot 2\sqrt{-6}$

94. $4\sqrt{-10} \cdot 3\sqrt{-8}$

In Problems 95 and 96, write the conjugate of the given complex number, and then find the product of the complex number and its conjugate.

95. $5 + \sqrt{-4}$

96. $3 - \sqrt{-9}$

In Problems 97–102, divide and express your answer in standard form.

97. $\dfrac{9 - 3i}{9i}$

98. $\dfrac{-6 + 4i}{2i}$

99. $\dfrac{5}{3 - \sqrt{-16}}$

100. $\dfrac{6}{4 + \sqrt{-4}}$

101. $\dfrac{5i}{1 - i}$

102. $\dfrac{2i}{3 + 5i}$

In Problems 103–106, solve each quadratic equation by any appropriate method. Express your answer in standard form.

103. $2x^2 + 2x + 5 = 0$

104. $2x^2 - 6x + 9 = 0$

105. $(x - 2)^2 + 12 = 4$

106. $(x + 3)^2 + 9 = -23$

Chapter 9 Test

Step-by-step test solutions are found on the Chapter Test Prep Videos available in MyMathLab® or on YouTube.

1. Solve using the Square Root Property:
$4x^2 - 12 = 24$

2. Solve by completing the square:
$x^2 + 6x - 40 = 0$

3. Solve using the quadratic formula:
$x^2 + 6x + 1 = 0$

In Problems 4–6, solve the quadratic equation by any appropriate method.

4. $(x - 2)(x + 1) = 4$

5. $2n = 4 + \dfrac{3}{n}$

6. $\dfrac{1}{2}t^2 = \dfrac{1}{8} + \dfrac{3}{4}t$

7. Consider the quadratic equation $x = 5 - 3x^2$.

(a) Determine the value of the discriminant.

(b) Determine which method you would choose to solve the equation.

(c) Explain why the method you chose is appropriate for this equation.

8. Write the complex number in standard form:
$\dfrac{-3 - \sqrt{-81}}{6}$

In Problems 9–15, perform the indicated operation and write your answer in standard form.

9. $(-12 + 4i) + (-2 - 8i)$

10. $\left(3 - \sqrt{-4}\right) - \left(2 - \sqrt{-25}\right)$

11. $2\sqrt{-3} \cdot 3\sqrt{-12}$

12. $(1 - 4i)(3 - 2i)$

13. $(4 - 6i)^2$

14. $\dfrac{4 + 12i}{3i}$

15. $\dfrac{6}{5 + 3i}$

In Problems 16 and 17, solve the quadratic equation by any appropriate method. Express complex answers in standard form.

16. $x^2 - 4x + 5 = 0$

17. $2p^2 + 8p + 15 = 0$

18. **Sibling Rivalry** Cybil and her sister, Kara, share their bedroom. During a bout of sibling rivalry, they decide they are going to split the bedroom into two parts by installing a room divider along the diagonal of the room, creating two triangular spaces. The divider is 16 feet long, and the room is 2 feet longer than it is wide.

   (a) Find the exact dimensions of Cybil and Kara's bedroom.

   (b) Approximate these dimensions, rounding to the nearest tenth of a foot.

19. **Pirate Battle** Pirates, protecting their harbor, fire a cannonball toward an intruder's sailing ship. The cannon sits on the cliffs above the harbor, and the pirates use the equation $h = -16t^2 + 100t + 40$ to calculate how long it will take for the cannonball to strike the ship, where h is the height of the cannonball above the water in feet, and t is the time, in seconds, the ball travels.

   (a) If the cannonball strikes the ship at the water line (that is, $h = 0$), calculate exactly how long the cannonball was in the air.

   (b) Approximate this time, rounding to the nearest tenth of a second.

20. **Cornfield** Farmer Green is planting this season's cornfield. The length of the rectangular field is 100 yards less than twice the width, and it has an area of 200,000 square yards.

   (a) Find the exact dimensions of the field.

   (b) Approximate these dimensions, rounding to the nearest hundredth of a yard.

Cumulative Review Chapters 1–9

In Problems 1–3, perform the indicated operations.

1. $\dfrac{-8 + 4(-2)^2}{-1 - (-1)}$

2. $\dfrac{\dfrac{1}{2} \cdot \dfrac{8}{3} - \dfrac{3}{4} \cdot \dfrac{16}{9}}{\left(\dfrac{1}{2}\right)^2}$

3. $\dfrac{4 - 7}{-3 - (-9)}$

In Problems 4 and 5, simplify each expression. Write your answers with positive exponents only.

4. $(3x^2y^{-2})^3(-x^3y^2)^2$

5. $\left(\dfrac{2x^2y}{6xy^{-2}}\right)^{-2}$

6. Consider the polynomial $-2x^2 - 3x + 1$.

   (a) Evaluate the polynomial for $x = -2$.

   (b) What is the leading coefficient?

   (c) What is the degree of the polynomial?

   (d) Is the polynomial a monomial, a binomial, a trinomial, or none of these?

7. Consider the equation $3x - 2y = 12$.

   (a) Solve for y.

   (b) When the equation is solved for y, identify the coefficient of x.

   (c) When the equation is solved for y, identify the constant.

In Problems 8–11, solve each equation.

8. $3 - 5(x + 2) = -3(x - 7)$

9. $\dfrac{x - 1}{4} = \dfrac{2x + 1}{5}$

10. $\dfrac{x^2}{3} + \dfrac{x}{6} = 1$

11. $x(x - 2) = 3$

12. Graph the linear equation: $2x - 3y = 6$

13. Write the equation of the line that contains the points $(-2, 4)$ and $(4, -5)$.

14. Solve the system of equations: $\begin{cases} 2x - 3y = -12 \\ 3x + 2y = -5 \end{cases}$

15. Graph the inequality: $y \geq \dfrac{1}{2}x + 1$

16. Suppose that x varies directly with y. If $y = 16$ when $x = 2.5$, find y when $x = 15$.

In Problems 17–19, factor completely, if possible.

17. $6h^2 - 11h - 10$

18. $3x^3 + 24$

19. $xy + 2x - 3y - 6$

In Problems 20 and 21, perform the indicated operations. Write your answer in lowest terms, if possible.

20. $\dfrac{6x^3 + 5x^2 - x + 3}{2x + 1}$

21. $\dfrac{3x^2 + 6x - 4}{2x}$

In Problems 22 and 23, evaluate each expression.

22. $-8^{\frac{2}{3}}$

23. $9^{-\frac{3}{2}}$

24. Write in scientific notation:
34,000,000,000

25. Write in decimal notation: 3.04×10^{-4}

In Problems 26 and 27, perform the indicated operation and write your answer in standard form.

26. $\left(-3 + \sqrt{2}\right)^2$

27. $\dfrac{4 + 3i}{6i}$

28. Old Globe Tickets All 580 seats to the opening night performance of *Twelfth Night* at the Old Globe Theatre were sold. Tickets were sold at two different rates: $20 for general seating and $50 for reserved seats. If the Old Globe Theatre took in $17,300 for the night, how many of each type of ticket were sold?

29. Popcorn Demand The demand D for popcorn at the movie theater is inversely related to the price p. When the price of popcorn is $3.50 per bag, the theater sells 240 bags of popcorn at a matinee. Determine the number of bags of popcorn that will be sold at a matinee if the price is raised to $4 per bag.

30. Pole Vault Misha is practicing pole vaulting for the upcoming track meet. The height of his jump, h (in feet), can be approximated by the equation $h = -2t^2 + 9t + 6$, where t is the number of seconds after takeoff.

(a) How high is Misha 4 seconds after he plants his vaulting pole?

(b) Exactly how long does it take for him to land on the mat in the pit?

(c) Approximate the time found in part (b) to the nearest hundredth of a second.

10

Graphs of Quadratic Equations in Two Variables and an Introduction to Functions

As you may know, a car's gas mileage increases with speed, up to a point, and then decreases. For example, a car gets better gas mileage at 55 mph than it does at either 35 mph or 75 mph. Engineers can compute the number of miles per gallon a car will get at various speeds by using an equation—you can too! See Problem 77 in Section 10.1.

The Big Picture: Putting It All Together

We have reached the last chapter of the text! This chapter begins with an introduction to graphing quadratic equations in two variables. When we graphed linear equations in two variables in Chapter 3, we learned that it is usually more efficient to use the properties of linear equations (such as slope and intercepts) than to plot points. The same is true for graphing quadratic equations in two variables. We will learn how to use the properties of a quadratic equation to obtain its graph. Just as linear equations describe everyday phenomena, quadratic equations and their graphs are all around us. You can "see" quadratic equations in the path of a shot hurled by a shot-putter, in car headlights, in satellite dishes, and even in the wok in your kitchen.

The chapter ends with two sections that introduce functions. Those of you who continue your study of algebra will see functions often. It is worthwhile to start learning about functions now so the concept will not be entirely new in your next class.

10.1 Quadratic Equations in Two Variables

Objectives

1. Graph a Quadratic Equation by Plotting Points
2. Find the Intercepts of the Graph of a Quadratic Equation
3. Find the Vertex and Axis of Symmetry of the Graph of a Quadratic Equation
4. Graph a Quadratic Equation Using Its Vertex, Intercepts, and Axis of Symmetry
5. Determine the Maximum or Minimum Value of a Quadratic Equation

Are You Ready for This Chapter?

Before getting started, take the following readiness quiz. If you get a problem wrong, go back to the section cited and review the material.

R1. Graph: $y = 3x - 8$ [Section 3.4, pp. 206–208]

R2. Graph: $x = -1$ [Section 3.2, pp. 190–191]

R3. Solve: $x^2 - 6x + 8 = 0$ [Section 6.6, pp. 408–414]

R4. Solve: $2x^2 + x - 5 = 0$ [Section 9.3, pp. 596–604]

R5. Find the intercepts of $3x - 5y = 30$.   [Section 3.2, pp. 187–190]

We first graphed linear equations in two variables by plotting points and then graphed a linear equation by using its properties, such as slope and y-intercept.

We now graph a *quadratic* equation first by plotting points and then by using its properties.

> **Definition**
>
> A **quadratic equation** in two variables is an equation of the form
> $$y = ax^2 + bx + c$$
> where *a*, *b*, and *c* are real numbers and $a \neq 0$.

Just as we did for linear equations, we begin by graphing quadratic equations using the point-plotting method.

▶ 1 Graph a Quadratic Equation by Plotting Points

The point-plotting method for obtaining the graph of an equation requires that we choose certain x-values and then use the equation to find the corresponding y-values (or choose y-values and find the corresponding x-values.) We plot the ordered pairs and connect the points in a smooth curve.

EXAMPLE 1 **Graphing a Quadratic Equation in Two Variables by Plotting Points**

Graph $y = x^2 - 2x - 3$ by plotting points.

Solution

Begin by creating a table of values. Without knowing much about the graph of the equation, it is hard to tell which x-values to choose. Start with the values shown in Table 1. Then use the equation to determine the corresponding values of y shown in the second column of Table 1.

Plot the ordered pairs shown in the third column of Table 1 and connect the points with a smooth curve. See Figure 1.

Ready?...Answers

R1.
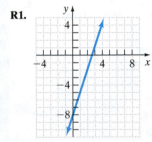

R2.
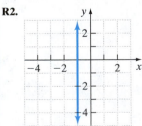

R3. $\{2, 4\}$

R4. $\left\{ \dfrac{-1 - \sqrt{41}}{4}, \dfrac{-1 + \sqrt{41}}{4} \right\}$

R5. $(10, 0), (0, -6)$

Table 1

x	y	(x, y)
−2	$y = (-2)^2 - 2(-2) - 3 = 5$	$(-2, 5)$
−1	$y = (-1)^2 - 2(-1) - 3 = 0$	$(-1, 0)$
0	$y = (0)^2 - 2(0) - 3 = -3$	$(0, -3)$
1	$y = (1)^2 - 2(1) - 3 = -4$	$(1, -4)$
2	$y = (2)^2 - 2(2) - 3 = -3$	$(2, -3)$
3	$y = (3)^2 - 2(3) - 3 = 0$	$(3, 0)$
4	$y = (4)^2 - 2(4) - 3 = 5$	$(4, 5)$

Figure 1

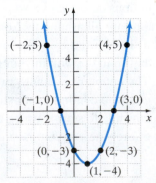

Now let's look at another example.

EXAMPLE 2 · Graphing a Quadratic Equation by Plotting Points

Graph $y = -x^2 + 6x - 8$ by plotting points.

Solution

Begin by creating a table of values. As in Example 1, it is difficult to determine which x-values to choose without knowing much about the graph. We will start with the values shown in Table 2. Then use the equation to determine the corresponding values of y shown in the second column of Table 2.

Plot the ordered pairs shown in the third column of Table 2 and connect the points with a smooth curve. See Figure 2.

Table 2

x	y	(x, y)
0	$y = -(0)^2 + 6(0) - 8 = -8$	$(0, -8)$
1	$y = -(1)^2 + 6(1) - 8 = -3$	$(1, -3)$
2	$y = -(2)^2 + 6(2) - 8 = 0$	$(2, 0)$
3	$y = -(3)^2 + 6(3) - 8 = 1$	$(3, 1)$
4	$y = -(4)^2 + 6(4) - 8 = 0$	$(4, 0)$
5	$y = -(5)^2 + 6(5) - 8 = -3$	$(5, -3)$
6	$y = -(6)^2 + 6(6) - 8 = -8$	$(6, -8)$

Figure 2
$y = -x^2 + 6x - 8$

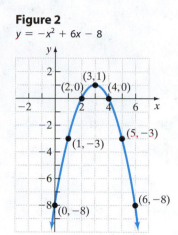

We stated in Examples 1 and 2 that choosing the x-values to use to obtain the graph of the quadratic equation is difficult. For this reason, we will supply the x-values. Your job will be to find the corresponding y-values, plot the points, and connect the points with a smooth curve.

Quick ✓

1. A _____ _____ is an equation in the form $y = ax^2 + bx + c$, where a, b, and c are real numbers and $a \neq 0$.

In Problems 2–4, graph each quadratic equation by filling in the table and plotting points.

2. $y = x^2 - 3$

x	y
-3	
-2	
-1	
0	
1	
2	
3	

3. $y = x^2 + 4x - 5$

x	y
-5	
-4	
-3	
-2	
-1	
0	
1	

4. $y = -x^2 + 8x - 15$

x	y
1	
2	
3	
4	
5	
6	
7	

▶ ❷ Find the Intercepts of the Graph of a Quadratic Equation

Figure 3

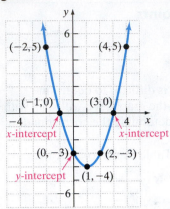

Figure 3 shows the graph of the quadratic equation $y = x^2 - 2x - 3$ from Example 1. We have labeled the x- and y-intercepts.

The y-intercept is the value of the quadratic equation $y = ax^2 + bx + c$ at $x = 0$.

The x-intercepts, if there are any, are found by letting $y = 0$ in the quadratic equation $y = ax^2 + bx + c$ and solving the equation

$$ax^2 + bx + c = 0$$

Recall that the solutions to the equation

$$ax^2 + bx + c = 0 \quad \text{are given by} \quad x = \frac{-b \pm \sqrt{b^2 - 4ac}}{2a}$$

The discriminant, $b^2 - 4ac$, is the radicand of the square root. Remember from our study of square roots that

- The square root of a positive number (a number greater than 0) is a real number.
- The square root of 0 is 0.
- The square root of a negative number is not a real number.

This leads to the following result.

> **Finding the x-Intercepts of the Graph of $y = ax^2 + bx + c$**
>
> 1. If the discriminant $b^2 - 4ac > 0$, then the graph has two different x-intercepts. The graph will cross the x-axis at the solutions to the equation $ax^2 + bx + c = 0$.
> 2. If the discriminant $b^2 - 4ac = 0$, then the graph has one x-intercept. The graph will touch the x-axis at the solution to the equation $ax^2 + bx + c = 0$.
> 3. If the discriminant $b^2 - 4ac < 0$, then the graph has no x-intercepts. The graph will not cross or touch the x-axis.

As Examples 1 and 2 illustrate, the graph of a quadratic equation can either *open up* or *open down*. Figure 4 illustrates the possibilities for x-intercepts for graphs that open up.

Figure 4

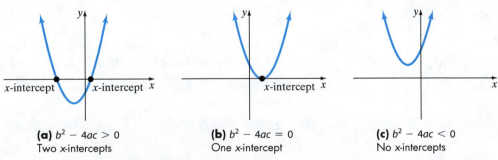

(a) $b^2 - 4ac > 0$
Two x-intercepts

(b) $b^2 - 4ac = 0$
One x-intercept

(c) $b^2 - 4ac < 0$
No x-intercepts

The intercepts of a graph represent one of the "interesting" features of the graph, so all graphs should have the intercepts labeled.

EXAMPLE 3 **Finding the Intercepts of the Graph of a Quadratic Equation**

Find the intercepts of the graph of $y = x^2 - 2x - 15$.

Solution

To find the y-intercept, let $x = 0$ in the equation $y = x^2 - 2x - 15$.

$$y = (0)^2 - 2(0) - 15$$
$$= -15$$

The y-intercept is $(0, -15)$. To find the x-intercepts, first find the value of the discriminant, $b^2 - 4ac$. Comparing $y = x^2 - 2x - 15$ to $y = ax^2 + bx + c$, we find that $a = 1, b = -2$, and $c = -15$. Thus

$$b^2 - 4ac = (-2)^2 - 4(1)(-15)$$
$$= 4 + 60$$
$$= 64$$

Because the discriminant, 64, is positive, the graph of the equation will have two x-intercepts. Find the x-intercepts by letting $y = 0$ in the equation $y = x^2 - 2x - 15$.

$$y = x^2 - 2x - 15$$

Let $y = 0$: $\quad 0 = x^2 - 2x - 15$

Factor: $\quad 0 = (x - 5)(x + 3)$

Zero-Product Property: $\quad x - 5 = 0 \quad$ or $\quad x + 3 = 0$

$$x = 5 \quad \text{or} \quad x = -3$$

The x-intercepts are $(-3, 0)$ and $(5, 0)$. The graph passes through the points $(0, -15)$, $(-3, 0)$, and $(5, 0)$. ●

EXAMPLE 4 | **Finding the Intercepts of the Graph of a Quadratic Equation**

Find the intercepts of the graph of $y = -2x^2 + x - 3$.

Solution

To find the y-intercept, let $x = 0$ in the equation $y = -2x^2 + x - 3$.

$$y = -2(0)^2 + 0 - 3$$
$$= -3$$

The y-intercept is $(0, -3)$. Find the x-intercepts by first determining the value of the discriminant, $b^2 - 4ac$. For $y = -2x^2 + x - 3$, $a = -2, b = 1$, and $c = -3$. Thus

$$b^2 - 4ac = (1)^2 - 4(-2)(-3)$$
$$= 1 - 24$$
$$= -23$$

Because the discriminant, -23, is negative, the graph of the equation has no x-intercepts. The graph passes through the point $(0, -3)$. ●

Quick ✓

5. *True or False* To find the y-intercept of the quadratic equation $y = x^2 - 2x - 3$, let $x = 0$ and solve for y.

6. *True or False* The value of the discriminant of the quadratic equation $y = 2x^2 - 2x - 5$ is 44. This means that the graph of $y = 2x^2 - 2x - 5$ has two x-intercepts.

7. For a quadratic equation $y = ax^2 + bx + c, a \neq 0$, if $b^2 - 4ac = 0$, then the graph has ____ (how many) x-intercept(s)?

In Problems 8–10, determine the intercepts of the graph of each quadratic equation.

8. $y = x^2 + 5x + 6$ 9. $y = -3x^2 + 4x - 5$ 10. $y = 4x^2 + 4x + 1$

▶ ❸ **Find the Vertex and Axis of Symmetry of the Graph of a Quadratic Equation**

The graphs we drew in Examples 1 and 2 are typical graphs of quadratic equations. The graph of a quadratic equation is a **parabola** (pronounced puh-rab-o-luh). Refer to Figure 5 on the next page, where two parabolas are shown. The shape of parabolas should be familiar—you see them in satellite dishes, car headlights, and some bridges.

Figure 5

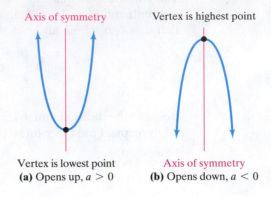

Axis of symmetry

Vertex is highest point

Vertex is lowest point
(a) Opens up, $a > 0$

Axis of symmetry
(b) Opens down, $a < 0$

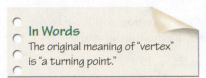

In Words
The original meaning of "vertex" is "a turning point."

Symmetry is all around us—do you see the symmetry in the butterfly?

Work Smart

The axis of symmetry is *not* part of the graph of the parabola. However, it serves as a guide in graphing the parabola.

In Words
If the leading coefficient is positive, the graph opens up, and if the leading coefficient is negative, it opens down.

The parabola in Figure 5(a) **opens up** and has a lowest point. The parabola in Figure 5(b) **opens down** and has a highest point. The lowest or highest point of a parabola is called the **vertex**. The vertical line passing through the vertex is called the **axis of symmetry**. If we were to take the portion of the parabola to the right of the vertex and fold it over the axis of symmetry, it would lie directly on top of the portion of the parabola to the left of the vertex. Therefore, we say that the parabola is symmetric about its axis of symmetry. The axis of symmetry is not part of the graph of the quadratic equation.

How do we know whether a parabola opens up or down? Examples 1 and 2 provide the answer. In Example 1, the coefficient of the squared term, a, is 1, and the parabola opens up. In Example 2, $y = -x^2 + 6x - 8$, so $a = -1$, and the parabola opens down. This leads us to the following conclusion.

The graph of the quadratic equation $y = ax^2 + bx + c$ opens up if $a > 0$ and opens down if $a < 0$.

How do we find the vertex and axis of symmetry of a parabola? Recall, the x-intercepts of the graph of $y = ax^2 + bx + c$ are found by solving the equation $ax^2 + bx + c = 0$. The solution(s) to this equation using the quadratic formula are given by

$$x = \frac{-b - \sqrt{b^2 - 4ac}}{2a} \text{ and } x = \frac{-b + \sqrt{b^2 - 4ac}}{2a}$$

The x-coordinate of the vertex of a parabola must be halfway between its x-intercepts, so the average of the x-intercepts is the x-coordinate of the vertex.

$$x = \frac{\dfrac{-b - \sqrt{b^2 - 4ac}}{2a} + \dfrac{-b + \sqrt{b^2 - 4ac}}{2a}}{2}$$

Use $\dfrac{a}{c} + \dfrac{b}{c} = \dfrac{a + b}{c}$:
$$= \frac{\dfrac{-b - \sqrt{b^2 - 4ac} - b + \sqrt{b^2 - 4ac}}{2a}}{2}$$

Combine like terms:
$$= \frac{\dfrac{-2b}{2a}}{2}$$

Take the reciprocal of the denominator and multiply:
$$= \frac{-2b}{2a} \cdot \frac{1}{2}$$

$$= -\frac{b}{2a}$$

The Vertex and Axis of Symmetry of a Parabola

The x-coordinate of the vertex of any quadratic equation $y = ax^2 + bx + c$, $a \neq 0$, is

$$-\frac{b}{2a}$$

The y-coordinate of the vertex is found by evaluating the equation at the x-coordinate of the vertex.

The axis of symmetry of a parabola is given by the equation

$$x = -\frac{b}{2a}$$

EXAMPLE 5 **Determining Whether a Parabola Opens Up or Down and Finding Its Vertex and Axis of Symmetry**

Consider the quadratic equation $y = -2x^2 + 8x - 1$.

(a) Determine whether the parabola opens up or down.

(b) Determine the vertex of the parabola.

(c) Determine the axis of symmetry of the parabola.

Solution

In this quadratic equation, $a = -2$, $b = 8$, and $c = -1$.

(a) Because a is negative, the parabola opens down.

(b) The x-coordinate of the vertex is

$$x = -\frac{b}{2a}$$

$$a = -2, b = 8: \quad = -\frac{8}{2(-2)} = -\frac{8}{-4}$$

$$= 2$$

The x-coordinate of the vertex is 2. To find the y-coordinate of the vertex, let $x = 2$ in the equation $y = -2x^2 + 8x - 1$.

$$y = -2x^2 + 8x - 1$$

$$x = 2: \quad = -2(2)^2 + 8(2) - 1$$

$$= -2(4) + 16 - 1$$

$$= 7$$

The vertex is $(2, 7)$.

(c) The axis of symmetry is $x = -\frac{b}{2a}$. Since $-\frac{b}{2a} = 2$, the axis of symmetry is $x = 2$.

Work Smart

Do not write that the axis of symmetry is 2. The axis of symmetry is a vertical line and must be written in the form $x = -\frac{b}{2a}$. Don't forget the "$x = $" part!

Quick ✓

11. When a quadratic equation is graphed, if the leading coefficient is positive, the graph opens ___, and if the leading coefficient is negative, the graph opens _____.

12. The lowest or highest point of the graph of $y = ax^2 + bx + c$ is called the _____. Its x-coordinate can be determined using the formula _____.

13. The axis of symmetry of a parabola is given by the equation _____.

(Continued)

For each quadratic equation in Problems 14 and 15, (a) determine whether the parabola opens up or down, (b) find the vertex of the parabola, and (c) find the axis of symmetry.

14. $y = 3x^2 - 6x + 2$ **15.** $y = -2x^2 - 12x - 7$

❹ Graph a Quadratic Equation Using Its Vertex, Intercepts, and Axis of Symmetry

We now know how to determine whether the graph of a quadratic equation opens up or down, how to find its intercepts, how to find its vertex, and how to find its axis of symmetry. We can use this information to draw the graph of the parabola.

Ⓟ EXAMPLE 6 How to Graph a Quadratic Equation Using Its Properties

Graph $y = x^2 + 2x - 8$ using its properties.

Step-by-Step Solution

Step 1: Determine whether the parabola opens up or down.

In the equation $y = x^2 + 2x - 8$, we see that $a = 1$, $b = 2$, and $c = -8$. The parabola opens up because a is positive.

Step 2: Determine the vertex and axis of symmetry.

The x-coordinate of the vertex is

$$x = -\frac{b}{2a} = -\frac{2}{2(1)} = -1$$

The y-coordinate of the vertex is found by letting $x = -1$ in the equation $y = x^2 + 2x - 8$.

$$\begin{aligned} y &= (-1)^2 + 2(-1) - 8 \\ &= 1 - 2 - 8 \\ &= -9 \end{aligned}$$

The vertex is $(-1, -9)$.
The axis of symmetry is the line

$$x = -\frac{b}{2a} = -1$$

Step 3: Determine the y-intercept by letting $x = 0$ in the equation.

$$\begin{aligned} y &= 0^2 + 2(0) - 8 \\ &= -8 \end{aligned}$$

Step 4: Find the discriminant, $b^2 - 4ac$, to determine the number of x-intercepts. Then find the x-intercepts, if any.

With $a = 1$, $b = 2$, and $c = -8$, $b^2 - 4ac = (2)^2 - 4(1)(-8) = 36$. Because the discriminant is positive, the parabola has two different x-intercepts. We find the x-intercepts by letting $y = 0$ and solving the resulting equation

$$x^2 + 2x - 8 = 0$$
$$(x + 4)(x - 2) = 0$$
$$x + 4 = 0 \quad \text{or} \quad x - 2 = 0$$
$$x = -4 \quad \text{or} \quad x = 2$$

The x-intercepts are $(-4, 0)$ and $(2, 0)$.

Step 5: Plot the points and draw the graph of the quadratic equation. Use the axis of symmetry to find an additional point.

See Figure 6. Notice how we use the axis of symmetry to find the additional point $(-2, -8)$. The y-intercept, $(0, -8)$, is 1 unit to the right of the axis of symmetry, so there must be a point with the same y-coordinate 1 unit to the left of the axis of symmetry, $(-2, -8)$.

Figure 6
$y = x^2 + 2x - 8$

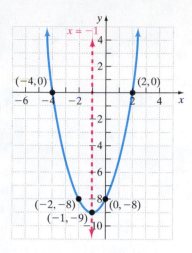

We summarize the steps used in Example 6.

> ## Graphing a Quadratic Equation $y = ax^2 + bx + c, a \neq 0,$ Using Its Properties
>
> **Step 1:** Determine whether the parabola opens up or down.
>
> **Step 2:** Determine the vertex and axis of symmetry.
>
> **Step 3:** Determine the y-intercept by evaluating the equation at $x = 0$.
>
> **Step 4:** Find the x-intercept(s) by first determining the discriminant, $b^2 - 4ac$.
>
> (a) If $b^2 - 4ac > 0$, the parabola has two x-intercepts, which are found by solving the equation $ax^2 + bx + c = 0$.
>
> (b) If $b^2 - 4ac = 0$, the vertex is the x-intercept.
>
> (c) If $b^2 - 4ac < 0$, there are no x-intercepts.
>
> **Step 5:** Plot the points, use the axis of symmetry to find an additional point, and draw the graph.

Quick ✓

In Problems 16 and 17, graph each quadratic equation using its properties.

16. $y = x^2 + 6x + 5$

17. $y = -x^2 - 2x + 3$

To find the x-intercepts of a quadratic equation, let $y = 0$ and solve for x. In Example 6, the quadratic equation was factorable. When a quadratic equation cannot be solved by factoring, we use the quadratic formula to find the x-intercepts. To graph the quadratic equation, approximate the x-intercepts rounded to two decimal places.

▶ **EXAMPLE 7** **Graphing a Quadratic Equation Using Its Properties**

Graph $y = -2x^2 + 4x + 3$ using its properties.

Solution

For $y = -2x^2 + 4x + 3$, we find that $a = -2, b = 4$, and $c = 3$. Because a is negative, the parabola opens down. The x-coordinate of the vertex is

$$x = -\frac{b}{2a}$$

$$a = -2, b = 4: \quad = -\frac{4}{2(-2)}$$

$$= 1$$

Find the y-coordinate of the vertex by evaluating $y = -2x^2 + 4x + 3$ at $x = 1$.

$$y = -2(1)^2 + 4(1) + 3$$
$$= -2 + 4 + 3$$
$$= 5$$

The vertex is at $(1, 5)$. The axis of symmetry is the line $x = 1$. The y-intercept is found by evaluating $y = -2x^2 + 4x + 3$ at $x = 0$.

$$y = -2(0)^2 + 4(0) + 3$$
$$= 3$$

The y-intercept is $(0, 3)$. The x-intercepts are found by letting $y = 0$ in the equation $y = -2x^2 + 4x + 3$ and solving for x.

$$0 = -2x^2 + 4x + 3$$

The discriminant is $b^2 - 4ac = 4^2 - 4(-2)(3) = 40$. Because the discriminant is positive, we know the parabola will have two x-intercepts. The equation cannot be solved by factoring, because the discriminant is not a perfect square. We will find the x–intercepts using the quadratic formula.

$$x = \frac{-b \pm \sqrt{b^2 - 4ac}}{2a}$$

$a = -2, b = 4, b^2 - 4ac = 40:$
$$= \frac{-4 \pm \sqrt{40}}{2(-2)}$$

$\sqrt{40} = \sqrt{4 \cdot 10} = 2\sqrt{10}:$
$$= \frac{-4 \pm 2\sqrt{10}}{-4}$$

Factor out -2:
$$= \frac{-2(2 \pm \sqrt{10})}{-4}$$

Simplify:
$$= \frac{2 \pm \sqrt{10}}{2}$$

Using a calculator, we find $\dfrac{2 - \sqrt{10}}{2} \approx -0.58$ and $\dfrac{2 + \sqrt{10}}{2} \approx 2.58$. The x-intercepts are approximately $(-0.58, 0)$ and $(2.58, 0)$. We plot the vertex, intercepts, and axis of symmetry and trace in a smooth curve as shown in Figure 7. Notice how the axis of symmetry was used to find the additional point $(2, 3)$.

Figure 7
$y = -2x^2 + 4x + 3$

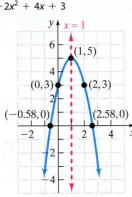

Quick ✔

In Problems 18 and 19, graph each quadratic equation using its properties.

18. $y = 2x^2 + 8x - 1$

19. $y = -3x^2 - 12x + 5$

▶ ⑤ Determine the Maximum or Minimum Value of a Quadratic Equation

Recall that the graph of a quadratic equation $y = ax^2 + bx + c$ is a parabola that opens up (if $a > 0$) or down (if $a < 0$). The vertex of the parabola will be the highest point on the graph if $a < 0$ and the lowest point on the graph if $a > 0$.

Definitions

If the vertex is the highest point on the graph of a parabola $(a < 0)$, then the y-coordinate of the vertex represents the **maximum value** of the equation. If the vertex is the lowest point on the graph of a parabola $(a > 0)$, then the y-coordinate of the vertex represents the **minimum value** of the equation.

EXAMPLE 8 | **Finding the Maximum or Minimum Value of a Quadratic Equation**

Determine whether the quadratic equation

$$y = 3x^2 + 12x - 1$$

has a maximum or a minimum value, and then find that value.

Solution

For $y = 3x^2 + 12x - 1$, we find that $a = 3$, $b = 12$, and $c = -1$. Because a is positive, the equation has a minimum value at

$$x = -\frac{b}{2a} = -\frac{12}{2(3)} = -2$$

Work Smart

The maximum or minimum value of the quadratic equation is the y-coordinate of the vertex.

The minimum value at $x = -2$ is

$$
\begin{aligned}
y &= 3(-2)^2 + 12(-2) - 1 \\
&= 3(4) - 24 - 1 \\
&= -13
\end{aligned}
$$

The minimum value of the equation is -13 and occurs at $x = -2$. ●

Quick ✓

20. If the vertex is the highest point on the graph of a parabola, the y-coordinate of the vertext represents the _____ _____ of the equation.

21. *True or False* The parabola determined by the equation $y = -3x^2 + 2x - 1$ has a maximum value.

In Problems 22 and 23, determine whether the quadratic equation has a maximum or a minimum value; then find the maximum or minimum value of the equation.

22. $y = 2x^2 - 8x + 1$ **23.** $y = -x^2 + 10x + 8$

EXAMPLE 9 | **Projectile Motion: Firing a Rocket**

Kevin just received a "slingshot rocket" for his birthday. If Kevin fires the rocket straight up from an initial height of 4 feet with an initial speed of 64 feet per second, then the height h of the rocket after t seconds can be modeled by the equation

$$h = -16t^2 + 64t + 4$$

(a) Determine the time at which the rocket is at a maximum height.

(b) Determine the maximum height of the rocket.

Solution

Notice that the graph of the quadratic equation opens down because a is negative. Therefore, the vertex will be the highest point of the graph.

(a) Let $a = -16$ and $b = 64$ in the expression $t = -\dfrac{b}{2a}$.

$$t = -\frac{64}{2(-16)} = 2$$

The rocket will be at a maximum height after 2 seconds.

(b) We evaluate the equation at $t = 2$.

$$
\begin{aligned}
h &= -16t^2 + 64t + 4 \\
\text{Let } t = 2: \quad &= -16(2)^2 + 64(2) + 4 \\
&= -64 + 128 + 4 \\
&= 68
\end{aligned}
$$

The rocket will have a maximum height of 68 feet. ●

Quick ✓

24. The height h of a ball fired straight up from an initial height of 15 feet with an initial speed of 128 feet per second after t seconds is given by the equation

$$h = -16t^2 + 128t + 15$$

(a) Determine the time at which the ball is highest by finding $t = -\dfrac{b}{2a}$.

(b) Determine the maximum height of the ball by evaluating the equation at the time found in part (a).

10.1 Exercises

 MyMathLab® Math XL PRACTICE

Exercise numbers in green have complete video solutions in MyMathLab

*Problems **1–24** are the **Quick** ✓s that follow the **EXAMPLES**.*

Building Skills

In Problems 25–32, graph each quadratic equation by filling in the table and plotting the points. See Objective 1.

25. $y = x^2$

x	y
−3	
−2	
−1	
0	
1	
2	
3	

26. $y = -x^2$

x	y
−3	
−2	
−1	
0	
1	
2	
3	

27. $y = -x^2 + 4$

x	y
−3	
−2	
−1	
0	
1	
2	
3	

28. $y = x^2 - 2$

x	y
−3	
−2	
−1	
0	
1	
2	
3	

29. $y = -x^2 + 3x + 4$

x	y
−2	
−1	
0	
1	
2	
3	
4	

30. $y = x^2 - 5x + 6$

x	y
−1	
0	
1	
2	
3	
4	
5	

31. $y = 2x^2 + 5x - 3$

x	y
−4	
−3	
−2	
−1	
0	
1	
2	

32. $y = -2x^2 - x + 3$

x	y
−3	
−2	
−1	
0	
1	
2	
3	

In Problems 33–44, determine the x- and y-intercepts of the graph of each quadratic equation. If an intercept is an irrational number, give both the exact and the approximate value. See Objective 2.

33. $y = -x^2 + 6x - 8$ **34.** $y = x^2 - 2x - 3$ **35.** $y = 6x^2 - x - 1$ **36.** $y = -3x^2 - 5x - 2$

37. $y = x^2 + 3x$ **38.** $y = x^2 + 6x$ **39.** $y = -x^2 - 5$ **40.** $y = x^2 - 2$

41. $y = -2x^2 + x + 2$ **42.** $y = 3x^2 + 5x + 1$ **43.** $y = x^2 + 4x + 2$ **44.** $y = -2x^2 - 2x + 3$

In Problems 45–52, for each quadratic equation, (a) determine whether the parabola opens up or down, (b) find the vertex of the parabola, and (c) find the axis of symmetry. See Objective 3.

45. $y = x^2 + 6$

46. $y = -x^2 + 3$

47. $y = -x^2 + 8x - 15$

48. $y = x^2 - 2x - 8$

49. $y = -2x^2 + 3x + 5$

50. $y = 2x^2 - 9x + 4$

51. $y = 3x^2 - 8x + 4$

52. $y = -6x^2 + 2x + 1$

In Problems 53–68, graph each quadratic equation by determining whether the parabola opens up or down and finding the vertex, the intercepts, and the axis of symmetry. See Objective 4.

53. $y = x^2 + 2x - 3$

54. $y = x^2 - 4x - 5$

55. $y = -x^2 - 2x$

56. $y = -x^2 + 4x$

57. $y = -x^2 - 8x - 9$

58. $y = -x^2 - 6x + 16$

59. $y = x^2 - 9$

60. $y = -x^2 + 25$

61. $y = x^2 - 7x + 10$

62. $y = -x^2 + 5x - 4$

63. $y = x^2 - 6x + 9$

64. $y = 4x^2 + 12x + 9$

65. $y = -2x^2 - 4x + 5$

66. $y = \frac{1}{2}x^2 - 4x + 6$

67. $y = x^2 - 2x - 5$

68. $y = x^2 - 4x + 1$

In Problems 69–76, determine whether the quadratic equation has a maximum or minimum value; then find the maximum or minimum value of the equation. See Objective 5.

69. $y = x^2 - 6x + 5$

70. $y = x^2 + 8x - 2$

71. $y = 2x^2 + 8x + 15$

72. $y = 4x^2 - 16x + 19$

73. $y = -2x^2 + 6x + 3$

74. $y = -2x^2 - 4x + 5$

75. $y = -x^2 - 3x - 7$

76. $y = -x^2 + 2x - 6$

Applying the Concepts

77. Maximizing Miles per Gallon An engineer has determined that the miles per gallon, m, of a Ford Taurus can be modeled by the quadratic equation

$$m = -0.02s^2 + 2s - 25$$

where s represents the speed in miles per hour.

(a) Determine the speed s that maximizes the miles per gallon by finding $s = -\dfrac{b}{2a}$.

(b) Determine the maximum miles per gallon that a Ford Taurus can achieve by evaluating the equation at the value of s found in part (a).

78. Height of a Shot Put A track and field athlete throws a shot put, and during one competition, the path of the shot put could be modeled by the quadratic equation

$$h = -0.04x^2 + 3x + 6$$

where h is the height (in feet) of the shot put at the instant it has traveled x feet horizontally.

(a) Determine the distance x at which the height of the shot put is a maximum by finding $x = -\dfrac{b}{2a}$.

(b) Determine the maximum height of the shot put by evaluating the equation at the value of x found in part (a).

△ **79. Maximizing an Enclosed Area** A farmer has 100 yards of fence and wishes to enclose a rectangular field for grazing cattle. The area A of the enclosed field is given by

$$A = -x^2 + 50x$$

where x is the length of one of the sides of the enclosed field. See the figure.

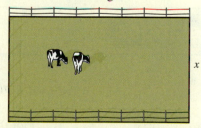

(a) Determine the length of the side x that maximizes the enclosed area by finding $x = -\dfrac{b}{2a}$.

(b) Determine the maximum enclosed area by evaluating the equation at the value of x found in part (a).

△ **80. Maximizing an Enclosed Area** Mayra is planning to provide child care in her home. She needs to enclose an area in her backyard so that the children will be safe when they play. She has 60 feet of chain link fence that she can use to make the play area. The area A of the enclosed region is given by

$$A = -x^2 + 30x$$

where x is the length of one of the sides of the enclosed region.

(a) Determine the length of the side x that maximizes the enclosed area by finding $x = -\dfrac{b}{2a}$.

(b) Determine the maximum enclosed area by evaluating the equation at the value of x found in part (a).

△ **81. Maximizing Corral Area** Kevin is adding a corral for his horses on his property. There is a river on one side, so he only has to put the fencing on three sides, enclosing the corral. If he has 200 feet of fencing, the maximum area he can enclose can be determined from the equation

$$A = -2x^2 + 200x$$

where x is the length of one of the sides of the enclosed region. See the figure.

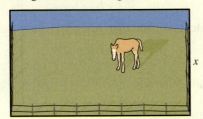

(a) Determine the length of the side x that maximizes the enclosed area by finding $x = -\dfrac{b}{2a}$.

(b) Determine the maximum enclosed area by evaluating the equation at the value of x found in part (a).

△ **82. Maximizing a Dog Pen** Martin is making a dog run in which he will keep his show dogs and decides to enclose an area of his backyard. He uses one side of his house for the pen and encloses the other three sides with fencing. If he has 50 feet of fencing, the maximum area he can enclose can be determined from the equation

$$A = -2x^2 + 50x$$

where x is the length of one of the sides of the enclosed region.

(a) Determine the length of the side x that maximizes the enclosed area by finding $x = -\dfrac{b}{2a}$.

(b) Determine the maximum enclosed area by evaluating the equation at the value of x found in part (a).

Extending the Concepts

In Problems 83–86, graph each quadratic equation using its properties.

83. $y = x^2 + 6x + 10$

84. $y = x^2 + 4x + 5$

85. $y = -2x^2 + 6x - 7$

86. $y = -3x^2 + 4x - 2$

Explaining the Concepts

87. Given a quadratic equation, without graphing, explain how you can tell the number of x-intercepts the graph has and how you can find them.

88. How many y-intercepts can the graph of a quadratic equation have? How can you find the y-intercept(s)?

89. How do you find the axis of symmetry of a parabola? How can the axis of symmetry be used to find additional points on the graph of the quadratic equation?

90. How many different parabolas have a given vertex and axis of symmetry?

10.2 Relations

Objectives

❶ Define Relations

❷ Find the Domain and the Range of a Relation

❸ Graph a Relation Defined by an Equation

Are You Ready for This Section?

Before getting started, take this readiness quiz. If you get a problem wrong, go back to the section cited and review the material.

R1. Plot the points $(1, 5)$, $(-4, 0)$, and $(3, -1)$ in the rectangular coordinate system. [Section 3.1, pp. 169–172]

R2. Graph: $y = -2x + 6$ [Section 3.4, pp. 206–208]

▶ ❶ Define Relations

Frequently, the value of one variable is somehow linked to the value of some other variable. For example, the total revenue of Brownie Troop 45 is related to the number of boxes of Girl Scout cookies they sell. And the total cholesterol of an individual is related to the amount of saturated fat consumed.

Definition

When the elements in one set are associated with the elements in a second set, we have a **relation**. If x is an element in the first set and y is an element in the second set, and if a relation exists between x and y, then we say that x **corresponds to** y or that y **depends on** x, and write $x \rightarrow y$. We may also write y depends on x as an ordered pair (x, y).

EXAMPLE 1 **Illustrating a Relation**

Figure 8

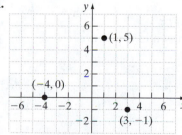

Figure 8 shows a correspondence between states and number of representatives in the U.S. House of Representatives. Figure 8 represents a relation as a **map**, in which we draw an arrow from an element from the first set (State) to an element in the second set (Number of Representatives). We could also represent the relation in Figure 8 using ordered pairs: {(Alabama, 7), (Colorado, 7), (Georgia, 14), (Massachusetts, 9), (New York, 27)}. The first value in the ordered pair is the state and the second value in the ordered pair is the number of representatives. ●

Quick ✓

1. When the elements in one set are associated with elements of a second set, we have a _____.

2. If a relation exists between the variables x and y, then we say that x _____ to y or that y _____ on x.

3. Use the map to represent the relation as a set of ordered pairs.

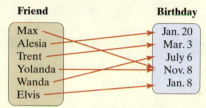

4. Use the set of ordered pairs to represent the relation as a map.

$$\{ (1, 5), (2, 4), (3, 7), (8, 12) \}$$

Ready?...Answers

R1.

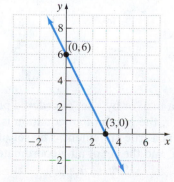

R2.

2 Find the Domain and the Range of a Relation

In a relation we say that y depends on x, and we can write the relation as a set of ordered pairs (x, y). We can think of the set of all x values as the **inputs** to the relation. The set of all y values can be thought of as the **outputs** of the relation. We use this interpretation of a relation to define *domain* and *range*.

Definitions

The **domain** of a relation is the set of all inputs to the relation. The **range** is the set of all outputs of the relation.

EXAMPLE 2 **Finding the Domain and the Range of a Relation**

Find the domain and the range of the relation presented in Figure 8.

Solution

The set of inputs—and therefore the domain—of the relation is

{Alabama, Colorado, Georgia, Massachusetts, New York}

The set of outputs—and therefore the range—of the relation is

{7, 14, 9, 27} ●

Did you notice that we did not list 7 twice in the range? The domain and the range are sets, and we never list elements in a set more than once.

Quick ✔

5. The _____ of a relation is the set of all inputs to the relation. The _____ is the set of all outputs of the relation.

6. State the domain and the range of the relation.

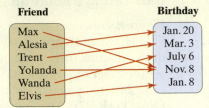

7. State the domain and the range of the relation.

$$\{(1,5), (2,4), (3,7), (8,7)\}$$

Work Smart

The order in which we list elements in the domain or the range does not matter. Never list elements in the domain or range more than once.

A relation can also be represented by plotting a set of ordered pairs. The set of all *x*-coordinates is the domain of the relation, and the set of all *y*-coordinates is the range of the relation.

EXAMPLE 3 **Finding the Domain and Range of a Relation**

Figure 9

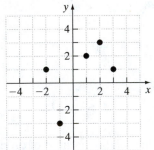

Figure 9 shows the graph of a relation. Identify the domain and the range of the relation.

Solution

The ordered pairs of the points in the graph are $(-2, 1)$, $(-1, -3)$, $(1, 2)$, $(2, 3)$, and $(3, 1)$. The domain is the set of all *x*-coordinates: $\{-2, -1, 1, 2, 3\}$. The range is the set of all *y*-coordinates: $\{-3, 1, 2, 3\}$.

Quick ✔

8. Identify the domain and range of the relation shown in the figure.

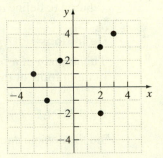

Work Smart

First write the points as ordered pairs to assist in finding the domain and range.

We have learned that a relation can be defined by a map or by a set of ordered pairs. A relation can also be defined by a graph. Remember, the graph of an equation is the set of all ordered pairs (x, y) such that the equation is a true statement. The set of all *x*-coordinates for which the graph exists is the domain, and the set of all *y*-coordinates is the range.

Figure 10

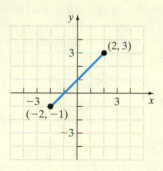

EXAMPLE 4   **Identifying the Domain and Range of a Relation from Its Graph**

Determine the domain and range of the relation shown in Figure 10.

Solution

The domain of the relation is the set of all x-coordinates of the graph. Therefore, the domain is $\{x \mid -3 \leq x \leq 8\}$.

The range of the relation is the set of all y-coordinates of the graph, $-2 \leq y \leq 2$. Therefore, the range is $\{y \mid -2 \leq y \leq 2\}$.

Quick ✓

9. *True or False* If the graph of a relation does not exist for $x = 5$, then 5 is not in the domain of the relation.

10. *True or False* The domain of the relation shown in the graph is $\{-2, -1, 0, 1, 2\}$.

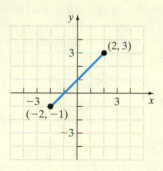

In Problems 11 and 12, identify the domain and the range of the relation from its graph.

11.
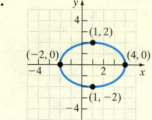

12.
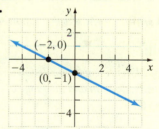

Work Smart

We can define relations four ways:

1. Maps
2. Sets of ordered pairs
3. Graphs
4. Equations

▶ ❸ **Graph a Relation Defined by an Equation**

Another way to define relations (instead of using maps, sets of ordered pairs, or graphs) is using equations such as $x + y = 4$ or $x = y^2$. If a relation is defined by an equation, we usually graph the relation to see how y depends on x. The graph of the relation also helps to identify its domain and range.

EXAMPLE 5   **Relations Defined by Equations**

Graph the relation $y = x^2 + 1$ to determine its domain and range.

Solution

We use the point-plotting method to graph the relation. Table 3 on the next page lists some of the points on the graph. Figure 11 shows a graph of the relation.

Table 3

x	$y = x^2 + 1$	(x, y)
-3	$(-3)^2 + 1 = 10$	$(-3, 10)$
-2	$(-2)^2 + 1 = 5$	$(-2, 5)$
-1	2	$(-1, 2)$
0	1	$(0, 1)$
1	2	$(1, 2)$
2	5	$(2, 5)$
3	10	$(3, 10)$

Figure 11 $y = x^2 + 1$

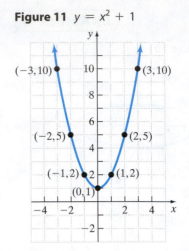

The graph extends indefinitely to the left and to the right (that is, the graph exists for all x-values), so the domain of the relation is the set of all real numbers, or $\{x \mid x \text{ is any real number}\}$. Notice from the graph that there are no y-values less than 1, but the graph exists for all y-values greater than or equal to 1. Therefore, the range of the relation is $\{y \mid y \geq 1\}$.

Quick ✔

In Problems 13 and 14, graph each relation. Use the graph to identify the domain and range of the relation.

13. $y = 2x - 5$

14. $y = x^2 - 3$

10.2 Exercises

 MyMathLab® Math XL PRACTICE

Exercise numbers in **green** have complete video solutions in MyMathLab.

*Problems **1–14** are the Quick ✔s that follow the EXAMPLES.*

Building Skills

In Problems 15–18, represent each relation as a set of ordered pairs. Then state the domain and the range of each relation. See Objectives 1 and 2.

15.

Occupation — Average Salary ($ thousands)

Physician, Teacher, Mathematician, Computer Engineer, Lawyer → 120, 37, 64, 82

16. Student — ID Number

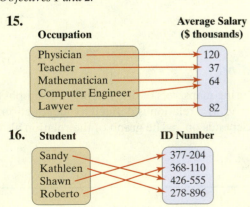

Sandy, Kathleen, Shawn, Roberto → 377-204, 368-110, 426-555, 278-896

17.

Country — Life Expectancy at Birth

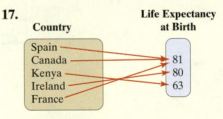

Spain, Canada, Kenya, Ireland, France → 81, 80, 63

Source: CIA World Fact Book

18.

Dieter — Number of Pounds Lost

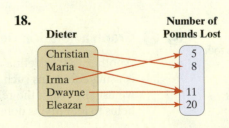

Christian, Maria, Irma, Dwayne, Eleazar → 5, 8, 11, 20

In Problems 19–24, use the set of ordered pairs to represent the relation as a map. Then state the domain and the range of each relation. See Objectives 1 and 2.

19. $\{(-1,3); (-4,3); (0,2); (1,1)\}$

20. $\{(-2,3); (-4,3); (-2,2); (1,1)\}$

21. $\{(2,-1); (2,-2); (3,-1); (-2,-2)\}$

22. $\{(3,0); (-1,-6); (-1,6); (1,0)\}$

23. $\{(a,z); (b,y); (c,x); (d,w)\}$

24. $\{(b,o); (o,k); (b,a); (g,s)\}$

In Problems 25–40, identify the domain and the range of the relation shown in the figure. See Objective 2.

25.
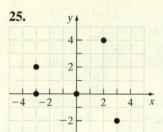

26.
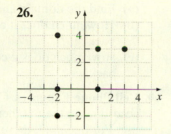

27.
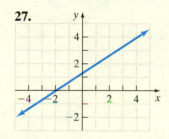

28.
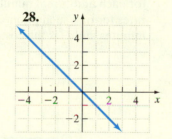

29.
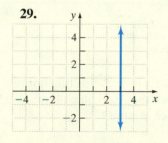

30.
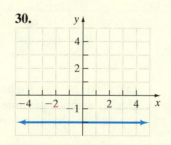

31.
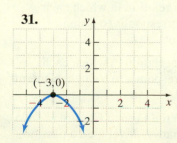

32.
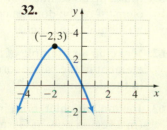

33.
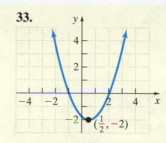

34.
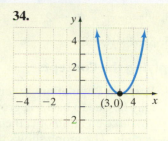

35.
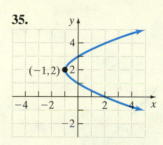

36.
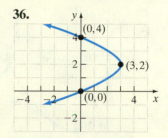

37.
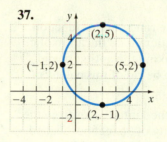

38.
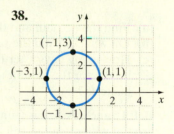

39.
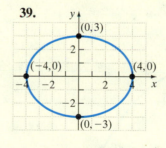

40.
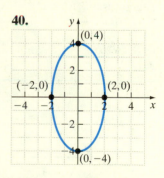

In Problems 41–56, graph each relation. Use the graph to identify the domain and the range of the relation. See Objective 3.

41. $y = -2x + 1$

42. $y = x + 4$

43. $y = -\dfrac{2}{3}x - 1$

44. $y = 3x - 2$

45. $y = 4x - 2x^2$

46. $y = -x^2 + 4x - 4$

47. $y = x^2 - 7x + 10$

48. $y = 3x^2 + 6x$

49. $y = 4x$

50. $3x + 3y = 2$

51. $y = 2x^2 - 5$

52. $y = -x^2 + 5x + 2$

53. $y = -x^2 + 4x + 2$ **54.** $y = x^2 - 5x + 1$

55. $3x + 4y = 8$ **56.** $2 = x - 2y$

Applying the Concepts

57. Bacteria Population When you are ill, your illness may be due to bacteria growing in your body. Once you take an antibiotic, the population of bacteria will decline. The graph shows the population of bacteria after t days.

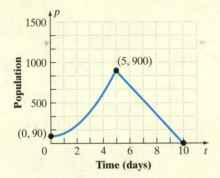

(a) Find the domain and the range of this function.
(b) On what day did the patient take the antibiotic?
(c) How many days after taking the antibiotic did it take for the patient to be rid of this colony of bacteria?

58. Ball Launch A ball is launched into the air from the ground with an initial velocity of 96 feet per second. The graph shows the height, h, of the ball after t seconds.

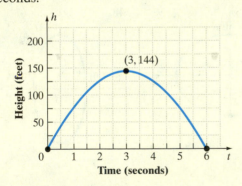

(a) Find the domain and the range of this function.
(b) How many seconds does it take for the ball to get to its highest point?
(c) What is the maximum height of the ball?

59. Mailing a Package When you send a package in the mail, you pay according to the weight of the package. All packages weighing within a certain number of ounces pay one rate. Once you go over that weight, the price jumps to the next higher rate (the open circle in the graph is the point where the jump occurs). The graph shows the cost

to send a package by priority mail, where C represents the cost and w represents the weight in pounds.

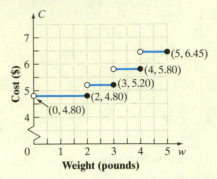

(a) Find the domain and the range of this function.
(b) How much does it cost to send a 10-ounce package?
(c) How much does it cost to send a 40-ounce package?

60. Cell Phone Expenses Cell phone users may pay according to the number of minutes of air time they use. For one plan, there is a fixed rate for a specified number of minutes, and after that the user is charged for each additional minute. The graph shows the rates for one such phone plan, where t is the time in minutes and C is the cost in dollars.

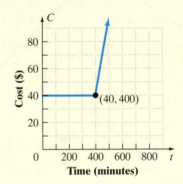

(a) Find the domain and the range of this function.
(b) How much air time does the fixed plan allow before charging for each minute?

Extending the Concepts

61. Make up an example of a relation in which each element of the domain corresponds to a single element in the range.

62. Make up an example of a relation in which each element in the range corresponds to a single element in the domain.

63. Johann Dirichlet, born in the French Empire in 1805 (in a region now in Germany), developed a passion for mathematics at an early age and spent his pocket money buying math books. One of the things he is famous for is the following mapping:

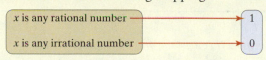

(a) Is it possible to list every ordered pair represented in Dirichlet's mapping?

(b) Show six ordered pairs that demonstrate how this relation maps all real numbers.

(c) Could you make a graph of this mapping similar to those in Problems 57–60? Why or why not?

64. Draw the graph of a relation whose domain is all real numbers but whose range is a single real number. Compare your graph with those of your classmates. How are they similar?

65. Draw the graph of a relation whose domain is a single real number but whose range is all real numbers. Compare your graph with those of your classmates. How are they similar?

66. Draw the graph of a relation that contains all ordered pairs (x, y), such that $y = x$ and x is any real number.

Explaining the Concepts

67. Explain what a relation is. Be sure to include an explanation of *domain* and *range*.

68. State the four methods presented in this section to describe a relation. When is using ordered pairs most appropriate? When is using a graph most appropriate? Support your opinion with an example.

Putting The Concepts Together (Sections 10.1 and 10.2)

We designed these problems so that you can review Sections 10.1 and 10.2 and show your mastery of the concepts. Take time to work these problems to make sure you understand these concepts before proceeding with the next section. The answers to these problems are located on page AN-36.

1. For the quadratic equation $y = x^2 - 2x - 3$,

   (a) Determine whether the parabola opens up or down.

   (b) Find the vertex of the parabola.

   (c) Find the equation of the axis of symmetry.

   (d) Find all intercepts.

In Problems 2–4, graph each quadratic equation using its properties. Label the vertex, intercepts, and axis of symmetry. Based on the graph, determine the domain and the range of the relation.

2. $y = x^2 - 6x$

3. $y = x^2 + 4x - 1$

4. $y = -x^2 + 9$

5. For the quadratic equation $y = 2x^2 + 12x - 3$,

   (a) Determine whether the equation has a maximum or a minimum value.

   (b) Find the maximum or minimum value of the equation.

In Problems 6–9, determine the domain and the range of the given relation.

6. $\{ (a, 1), (b, 2), (c, 1) \}$

7. $\{ (1, -1), (3, -3), (3, -1) \}$

8.

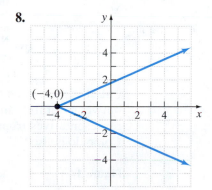

9.

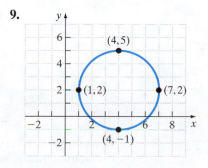

10.3 An Introduction to Functions

Objectives

1. Determine Whether a Relation Expressed as a Map or as Ordered Pairs Represents a Function
2. Determine Whether a Relation Expressed as an Equation Represents a Function
3. Identify the Graph of a Function Using the Vertical Line Test
4. Find the Value of a Function

Work Smart

Not all relations are functions!

Are You Ready for This Section?

Before getting started, take this readiness quiz. If you get the problem wrong, go back to the section cited and review the material.

R1. Evaluate the expression $2x^2 - 5x$ for   [Section 1.8, pp. 64–65]
(a) $x = 1$, **(b)** $x = 4$, and **(c)** $x = -3$.

1 Determine Whether a Relation Expressed as a Map or as Ordered Pairs Represents a Function

We now present one of the most important concepts in algebra—the *function*. A function is a special type of relation. Consider the relation in Figure 12. In this correspondence between people and their telephone numbers, notice that Carter has two phone numbers, "929-9486" and "430-5968." In other words, not every input "person" corresponds to a single output "telephone number."

Figure 12

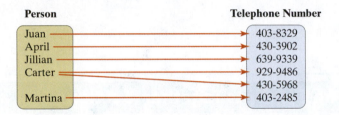

Now consider the correspondence between states and their populations, presented in Figure 13. In this relation, each input "state" corresponds to exactly one output "population."

Figure 13

Let's look at one more relation. The relation in Figure 14 shows a correspondence between "state" and "number of representatives." Notice that the number of representatives from Alabama is "7" and that the number of representatives from Colorado is also "7."

Figure 14

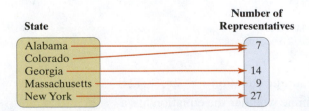

The relations in Figures 13 and 14 have something in common. Can you tell what it is? Each input corresponds to only one output. Remember that the set of all inputs is called the domain of the relation, and the set of all outputs is called the range of the relation. This leads to the definition of a *function*.

In Words

For a relation to be classified as a function, each input must have exactly one output.

Definition

A **function** is a relation in which each element in the domain of the relation corresponds to exactly one element in the range of the relation.

The idea behind functions is predictability. If an input is known, a function can be used to determine the output with 100% certainty (Figures 13 and 14). With relations that aren't functions, we don't have this predictability. In Figure 12, Carter has two telephone numbers (two outputs), so the relation is not a function.

EXAMPLE 1 **Determining Whether a Relation Defined by a Map Is a Function**

Determine whether the following relations are functions. If the relation is a function, then state its domain and range.

(a) For this relation, the domain represents the length (mm) of the right humerus, and the range represents the length (mm) of the right tibia, for each of five rats after their trip into space.

Source: NASA Life Sciences Data Archive

(b) For this relation, the domain represents the state, and the range represents the percent of the population living in poverty in 2010.

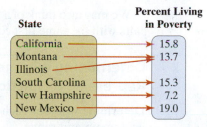

Source: United States Census Bureau

(c) For the following relation, the domain represents the average total cholesterol (mg/dL), and the range represents the age category.

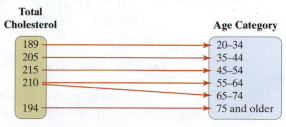

Source: Centers for Disease Control and Prevention

Solution

(a) The relation is a function because each element in the domain corresponds to exactly one element in the range. The domain is {24.80, 24.59, 24.29, 23.81, 24.87}. The range is {36.05, 35.57, 34.58, 34.20, 34.73}.

(b) The relation is a function because each element in the domain corresponds to exactly one element in the range. The domain is {California, Montana, Illinois, South Carolina, New Hampshire, New Mexico}. The range is {15.8, 13.7, 15.3, 7.2, 19.0}.

(c) The relation is not a function because one element in the domain, 210 mg/dL, corresponds to more than one element in the range. There are two age groups, 55–64 and 65–74, that have an average total cholesterol of 210 mg/dL.

Quick ✓

1. A _____ is a relation in which each element in the domain of the relation corresponds to exactly one element in the range of the relation.

2. *True or False* Every relation is a function.

In Problems 3–5, determine whether the relation represents a function. If the relation is a function, then state its domain and range.

3.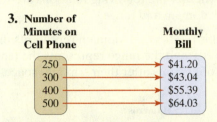

4.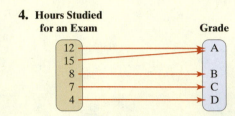

5.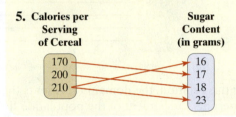

We may also think of a function as a set of ordered pairs (x, y) in which no two ordered pairs with the same first coordinate have different second coordinates.

EXAMPLE 2 **Determining Whether a Relation Written as Ordered Pairs Represents a Function**

Determine whether each relation is a function. If the relation is a function, then state its domain and range.

(a) $\{(1, 3), (-1, 4), (0, 5), (2, 7)\}$

(b) $\{(-2, 8), (-1, 4), (0, 1), (1, 6), (2, 8)\}$

(c) $\{(0, 3), (1, 4), (4, 5), (9, 5), (4, 1)\}$

Solution

(a) This relation is a function because no ordered pairs have the same first coordinate and different second coordinates. The domain is the set of all first coordinates, $\{-1, 0, 1, 2\}$; the range is the set of all second coordinates, $\{3, 4, 5, 7\}$.

(b) This relation is a function because there are no ordered pairs with the same first coordinate but different second coordinates. The domain of the function is the set of all first coordinates, $\{-2, -1, 0, 1, 2\}$; the range of the function is the set of all second coordinates, $\{1, 4, 6, 8\}$. Remember that it is okay for two different inputs to correspond to the same output.

(c) This relation is not a function because there are two ordered pairs, $(4, 5)$ and $(4, 1)$, with the same first coordinate but different second coordinates. In other words, this is not a function because an element in the domain, 4, corresponds to two different elements in the range, 1 and 5.

In Example 2(b), notice that both -2 and 2 in the domain correspond to 8 in the range. This does not violate the definition of a function—two different first coordinates can have the same second coordinate. A violation of the definition occurs when two ordered pairs have the same first coordinate and different second coordinates, as in Example 2(c).

> **Quick ✓**
>
> *In Problems 6–8, determine whether each relation represents a function. If the relation is a function, then state its domain and range.*
>
> **6.** $\{(-3, 3), (-2, 2), (-1, 1), (0, 0), (1, 1)\}$
>
> **7.** $\{(-3, 2), (-2, 5), (-1, 8), (-3, 6)\}$ **8.** $\{(3, 8), (4, 10), (5, 11), (6, 8)\}$

▶ ❷ Determine Whether a Relation Expressed as an Equation Represents a Function

To determine whether an equation, where y depends on x, is a function, it is often easiest to solve the equation for y. If a value of x corresponds to exactly one y, then the equation defines a function. Otherwise, it does not define a function.

EXAMPLE 3   **Determining Whether a Linear Equation Represents a Function**

Determine whether the equation $y = 2x + 7$ is a function.

Solution

The rule for getting from x to y is to multiply x by 2 and then add 7. Since there is only one output y that can result from performing these operations on any given input x, the equation is a function.   ●

EXAMPLE 4   **Determining Whether an Equation Represents a Function**

Determine whether the equation $y = \pm\sqrt{x}$ is a function.

Solution

Notice that for any value of x greater than 0, two values of y result. For example, if $x = 4$, then $y = \pm 2$. Because a single x corresponds to more than one y, the equation is not a function.   ●

> **Quick ✓**
>
> *In Problems 9–12, determine whether each equation is that of a function.*
>
> **9.** $y = -5x + 1$ **10.** $y = \pm 6x$
>
> **11.** $y = x^2 + 3x$ **12.** $x^2 + y^2 = 4$

Work Smart: Study Skills

Here are two student definitions of a function:

1. A **function** is a relation in which each input matches up with only one output.
2. A **function** is a relation in which every input appears only once in the correspondence.

Each of these definitions is correct. How do these definitions compare with the one presented in this section? How will they help you remember the definition of a function?

▶ ❸ Identify the Graph of a Function Using the Vertical Line Test

Remember that in a function, each number x in the domain corresponds to exactly one number y in the range. This means that the graph of an equation does *not* represent a function if two points with the same x-coordinate have different y-coordinates.

Vertical Line Test

A set of points in the coordinate plane is the graph of a function if and only if every vertical line intersects the graph in at most one point.

EXAMPLE 5 **Using the Vertical Line Test to Identify Graphs of Functions**

Which of the graphs in Figure 15 are graphs of functions?

Figure 15

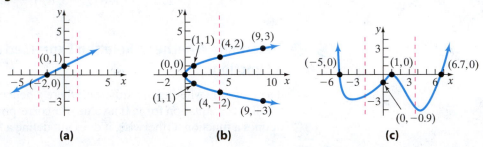

(a) (b) (c)

Solution

The graph in Figure 15(a) is a function because every vertical line intersects the graph in at most one point. The graph in Figure 15(b) is not a function because a vertical line intersects the graph in more than one point. The graph in Figure 15(c) is a function because any vertical line intersects the graph in at most one point.

Quick ✓

13. *True or False* For a graph to be a function, any vertical line can intersect the graph in at most one point.

In Problems 14 and 15, use the vertical line test to determine whether the graph is that of a function.

14.

15.

▶ ④ **Find the Value of a Function**

A relation is a function if each input corresponds to exactly one output. When a function is written as an equation, we use a special notation called *function notation* to represent the function. The following table shows the similarity between an equation in two variables that is solved for *y* and function notation.

Equation in Two Variables	Function Notation
$y = \dfrac{1}{2}x + 5$	$f(x) = \dfrac{1}{2}x + 5$
$y = -3x + 1$	$f(x) = -3x + 1$
$y = 2x^2 - 3x$	$f(x) = 2x^2 - 3x$

Observe that the expression $f(x)$ replaces the variable y. Functions are often denoted by letters such as f, F, g, G, and so on. If f is a function, then for each number x in its domain, the corresponding value in the range is denoted $f(x)$, which is read as "f of x" or as "f at x." Note that $f(x)$ does *not* mean "f times x."

We call $f(x)$ the *value of f at the number x*—that is, $f(x)$ is the number that results when the function is applied to x. Finding the value of a function is like finding the value of y in an equation in two variables.

Equation in Two Variables	**Function Notation**
Given the equation $y = \dfrac{1}{2}x + 5$, find the value of y when $x = -2$ and $x = 4$.	Given the function $f(x) = \dfrac{1}{2}x + 5$, evaluate $f(-2)$ and $f(4)$.

$$y = \frac{1}{2}(-2) + 5 \qquad y = \frac{1}{2}(4) + 5$$
$$= -1 + 5 \qquad\qquad = 2 + 5$$
$$= 4 \qquad\qquad\quad = 7$$

$$f(-2) = \frac{1}{2}(-2) + 5 \qquad f(4) = \frac{1}{2}(4) + 5$$
$$= -1 + 5 \qquad\qquad = 2 + 5$$
$$= 4 \qquad\qquad\quad = 7$$

For equations, we say, "the value of y when $x = -2$ is 4." For functions, we say, "f of -2 equals 4." In both cases, we replace x with the indicated number and evaluate.

EXAMPLE 6 **Finding Values of a Function Defined by a Linear Equation**

For the function defined by $G(x) = 3x - 8$, evaluate:

    (a) $G(4)$ **(b)** $G(-3)$

Solution

    (a) Substitute 4 for x in the equation defining the function G.

$$G(4) = 3(4) - 8$$
$$= 12 - 8$$
$$= 4$$

    (b) $G(-3) = 3(-3) - 8$
$$= -9 - 8$$
$$= -17$$

EXAMPLE 7 **Finding Values of a Function**

For the function defined by $f(x) = x^2 + 2x$, evaluate:

    (a) $f(3)$ **(b)** $f(-2)$

Solution

    (a) Substitute 3 for x in the equation defining the function f.

$$f(3) = 3^2 + 2(3)$$
$$= 9 + 6$$
$$= 15$$

    (b) $f(-2) = (-2)^2 + 2(-2)$
$$= 4 + (-4)$$
$$= 0$$

Quick ✔

16. *True or False* The notation $f(2)$ means to replace the value of x in the given function with the number 2 and evaluate.

In Problems 17–20, let $f(x) = 4x + 1$ and $g(x) = 2x^2 + 5x - 3$ to evaluate each function.

17. $f(4)$ **18.** $f(-2)$ **19.** $g(2)$ **20.** $g(-5)$

Summary **Important Facts About Functions**

1. For each x in the domain, there corresponds exactly one y in the range.

2. We use $f(x)$ to denote the function. It represents the value of the function at x.

10.3 Exercises

MyMathLab® MathXL PRACTICE

Exercise numbers in green have complete video solutions in MyMathLab.

*Problems **1–20** are the Quick ✔s that follow the EXAMPLES.*

Building Skills

In Problems 21–24, determine whether each relation represents a function. If the relation is a function, state its domain and range. See Objective 1.

21.

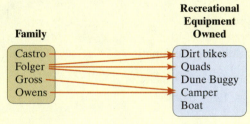

Family → Recreational Equipment Owned

22.

Textbook → Number of Pages

23.

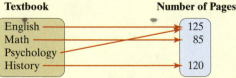

Fruit → Cost per Pound

24.

Student → Courses Enrolled

In Problems 25–30, determine whether each relation represents a function. If the relation is a function, state its domain and range. See Objective 1.

25. $\{(-1, -1), (0, 0), (1, 1), (2, 2)\}$

26. $\{(3, 1), (2, 1), (0, 1), (-1, 1)\}$

27. $\{(-1, 1), (0, 0), (1, -1), (-1, -2)\}$

28. $\{(5, 0), (3, 1), (-1, 2), (3, 2)\}$

29. $\{(a, a), (b, a), (c, b), (d, b)\}$

30. $\{(a, z), (b, z), (c, z), (d, z)\}$

In Problems 31–44, determine whether each equation represents y as a function of x. See Objective 2.

31. $3x - y = 2$ **32.** $2x + 3y = 1$

33. $x = 3$ **34.** $x = -4$

35. $y = -1$ **36.** $y = 0$

37. $y = -x^2 + 2x - 3$ **38.** $y = x^2 - 3x + 2$

39. $x = y^2$ **40.** $y = \sqrt{x}$

41. $y = \pm x$ **42.** $y = \pm\sqrt{x + 5}$

43. $x^2 + y^2 = 9$ **44.** $x^2 - y^2 = 1$

In Problems 45–58, determine whether the graph is that of a function. See Objective 3.

45.

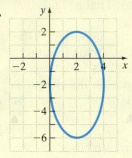

46.

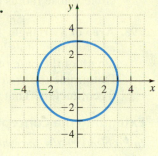

47.

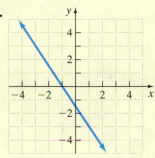

48.

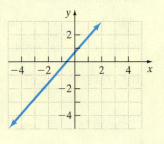

49.

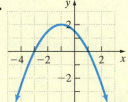

50.

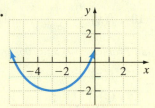

51.

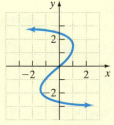

52.

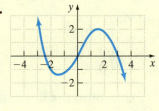

53.

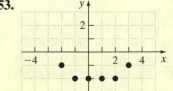

54.

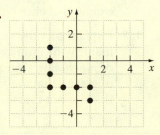

55.

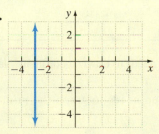

56.

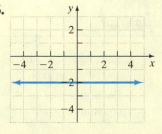

57.

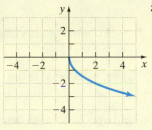

58.

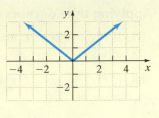

In Problems 59–72, find the following values for each function. See Objective 4.

(a) $f(0)$ **(b)** $f(3)$ **(c)** $f(-2)$

59. $f(x) = -x - 3$

60. $f(x) = 2x + 4$

61. $f(x) = 4 - 2x$

62. $f(x) = 1 - x$

63. $f(x) = 3$

64. $f(x) = -1$

65. $f(x) = \dfrac{1}{2}x - 3$

66. $f(x) = -\dfrac{2}{3}x + 4$

67. $f(x) = 3 - 5x$

68. $f(x) = -3 + 4x$

69. $f(x) = x^2 - 3x$

70. $f(x) = x^2 + 4$

71. $f(x) = -2x^2 - 5x + 1$

72. $f(x) = -2x^2 + x - 3$

In Problems 73–86, find the indicated value of each function. See Objective 4.

73. $f(x) = x^2 - 3x; f(-3)$

74. $f(x) = -x^2 + 4x; f(4)$

75. $g(x) = -2x^2 + x - 5; g(-1)$

76. $g(x) = 3x^2 - 2; g(0)$

77. $a(x) = 2 + x - x^2; a(-1)$

78. $a(x) = 3 - x + x^2; a(-3)$

79. $h(x) = \sqrt{x + 2}; h(2)$

80. $h(x) = \sqrt{2x - 1}; h(5)$

81. $h(x) = \dfrac{-2}{2x + 3}; h\left(\dfrac{1}{2}\right)$

82. $h(x) = \dfrac{-4}{5x - 1}; h(-1)$

83. $f(x) = |3x^2 - 10|; f(-1)$

84. $f(x) = |2x^2 - 8|; f(-2)$

85. $g(x) = \sqrt{16 - x^2}; g(-2)$

86. $g(x) = \sqrt{25 - x^2}; g(3)$

Applying the Concepts

87. If $f(x) = 2x + C$ and $f(3) = 11$, what is the value of C?

88. If $f(x) = 3x - C$ and $f(-2) = -10$, what is the value of C?

89. If $f(x) = -2x^2 + 5x + C$ and $f(-2) = -15$, what is the value of C?

90. If $f(x) = 3x^2 - x + C$ and $f(3) = 18$, what is the value of C?

91. If $f(x) = \dfrac{2x + 5}{x - A}$ and $f(0) = -1$, what is the value of A?

92. If $f(x) = \dfrac{-x + B}{x - 5}$ and $f(3) = -1$, what is the value of B?

93. Salary Express the gross salary, G, of Jackie, who earns $12 per hour, as a function of the number of hours worked, h. Determine the gross salary of Jackie if she works 20 hours. That is, find $G(20)$.

94. Commission Roberta is a commissioned salesperson. She earns a base weekly salary of $350 per week plus a 10% commission on her sales. Express her gross salary, G, as a function of the price, p, of the items sold. Determine the gross salary of Roberta if the value of the items sold is $1500. That is, find $G(1500)$.

95. Homicide Rate The number of homicides in the United States can be modeled by the function $h(x) = -0.00535x^2 + 0.353x + 2.357$, where x is the number of years after 1900 and h is the number of homicides per 100,000 people.

  (a) Find $h(30)$ and interpret your results.

  (b) Find $h(50)$ and interpret your results.

96. Salary The percent of workers, p, who earn a given salary, x (in thousands of dollars), can be modeled by the function $p(x) = 0.00311x^2 - 0.567x + 26.619$.

  (a) Find $p(30)$ and interpret your results.

  (b) Find $p(60)$ and interpret your results.

  (c) Find $p(90)$ and interpret your results.

△ **97. Area of a Circle** Express the area, A, of a circle as a function of its radius, r. Determine the area of a circle whose radius is 3 inches. That is, find $A(3)$.

△ **98. Area of a Triangle** Express the area, A, of a triangle as a function of its height, h, assuming that the length of the base of the triangle is 5 centimeters. Determine the area of this triangle if its height is 8 centimeters. That is, find $A(8)$.

Extending the Concepts

In Problems 99–104, the value of a function is given. Find the value of x that corresponds to the given function value.

99. $f(x) = 2x + 5$; find x if $f(x) = 9$.

100. $f(x) = 3x - 8$; find x if $f(x) = -10$.

101. $f(x) = 2 - \dfrac{1}{2}x$; find x if $f(x) = 1$.

102. $f(x) = -4 - \dfrac{2}{3}x$; find x if $f(x) = 8$.

103. $f(x) = \dfrac{12}{x + 2}$; find x if $f(x) = -6$.

104. $f(x) = \dfrac{x}{x + 4}$; find x if $f(x) = 2$.

In Problems 105 and 106, find the values for the given function.

105. $f(x) = 1 - 2x$

  (a) $f(3)$ **(b)** $f(h)$

  (c) $f(h + 3)$ **(d)** $f(h) + f(3)$

  (e) $f(h + 3) - f(3)$

106. $f(x) = 5 - 3x$

  (a) $f(2)$ **(b)** $f(h)$

  (c) $f(h + 2)$ **(d)** $f(h) + f(2)$

  (e) $f(h + 2) - f(2)$

Explaining the Concepts

107. In this section, four ways to represent a function were presented. List these four ways and give an example of each.

108. Explain why the vertical line test can be used to identify the graph of a function. Use the definition of *function* in your response. Then show two cases of graphs that are functions and two cases of graphs that are not functions.

Chapter 10 Activity: Discovering Shifting

Focus: Discovering the rules for graphing quadratic equations by shifting

Time: 20–30 minutes

Group size: 3–4

Materials needed: Graph paper

1. Each member of the group needs to draw a coordinate plane and label the axes.

2. Each member of the group needs to graph the primary equation $y = x^2$.

3. On the same coordinate plane, each group member should choose one of the following quadratic equations and graph by plotting points. Be sure your graph of the primary equation and your graph of one of the equations (a)–(d) are on the same coordinate plane.

(a) $y = x^2 + 4$ **(b)** $y = x^2 - 4$

(c) $y = (x - 4)^2$ **(d)** $y = (x + 4)^2$

4. As a group, discuss the following:

(a) What shape are the graphs?

(b) Each member of the group should share the difference between her or his graph and the primary graph.

(c) As a group, can you develop possible rules for these differences?

5. With these rules in mind, repeat the above procedure with the primary equation $y = x^2$ and one of the following equations. As a group, decide whether your rules held.

(a) $y = (x - 2)^2$

(b) $y = x^2 - 2$

(c) $y = (x + 2)^2$

(d) $y = x^2 + 2$

Chapter 10 Review

Section 10.1 Quadratic Equations in Two Variables

KEY CONCEPTS

- Any quadratic equation $y = ax^2 + bx + c, a \neq 0$, has a vertex whose x-coordinate is $-\dfrac{b}{2a}$.

- The y-coordinate of the vertex is found by evaluating the equation at the x-coordinate of the vertex.

- The axis of symmetry of a parabola is given by the equation $x = -\dfrac{b}{2a}$.

KEY TERMS

Quadratic equation in two variables
Parabola
Opens up
Opens down
Vertex
Axis of symmetry
Maximum value
Minimum value

The x-intercepts of the Graph of a Quadratic Equation

1. If the discriminant $b^2 - 4ac > 0$, then the graph of $y = ax^2 + bx + c$ has two different x-intercepts. The graph will cross the x-axis at the solutions to the equation $ax^2 + bx + c = 0$.

2. If the discriminant $b^2 - 4ac = 0$, then the graph of $y = ax^2 + bx + c$ has one x-intercept. The graph will touch the x-axis at the solution to the equation $ax^2 + bx + c = 0$.

3. If the discriminant $b^2 - 4ac < 0$, then the graph of $y = ax^2 + bx + c$ has no x-intercepts. The graph will not cross or touch the x-axis.

You Should Be Able To...	EXAMPLE	Review Exercises
1 Graph a quadratic equation by plotting points (p. 636)	Examples 1 and 2	1–6
2 Find the intercepts of the graph of a quadratic equation (p. 638)	Examples 3 and 4	7–14
3 Find the vertex and axis of symmetry of the graph of a quadratic equation (p. 639)	Example 5	15–22
4 Graph a quadratic equation using its vertex, intercepts, and axis of symmetry (p. 642)	Examples 6 and 7	23–28
5 Determine the maximum or minimum value of a quadratic equation (p. 644)	Examples 8 and 9	29–34

In Problems 1–6, graph each quadratic equation by making a table of values and plotting the points.

1. $y = -x^2$ **2.** $y = x^2$

3. $y = 2x^2 - 3$ **4.** $y = -\dfrac{1}{2}x^2 + 4$

5. $y = x^2 - 4x - 3$ **6.** $y = -2x^2 + 4x + 1$

In Problems 7–14, find the intercepts of the graph.

7. $y = x^2 + x - 6$

8. $y = x^2 + 6x + 8$

9. $y = -4x^2 + 12x - 9$

10. $y = -9x^2 - 6x - 1$

11. $y = -4x^2 + 8x - 5$

12. $y = -3x^2 + 2x - 1$

13. $y = x^2 - 4x + 2$

14. $y = -x^2 - 4x + 14$

In Problems 15–22, for each quadratic equation,
(a) determine whether the parabola opens up or down,
(b) find the vertex of the parabola, and (c) find the axis of symmetry.

15. $y = x^2 + 5$

16. $y = -x^2 - 4$

17. $y = -2x^2 + 4x - 1$

18. $y = 3x^2 - 12x + 4$

19. $y = \frac{3}{2}x^2 + 9x$

20. $y = -\frac{3}{2}x^2 - 6x$

21. $y = -x^2 + 3x - 4$

22. $y = x^2 + 5x + 5$

In Problems 23–28, graph each quadratic equation by using its properties.

23. $y = x^2 + 2x - 3$

24. $y = -x^2 + 6x - 8$

25. $y = -x^2 + 4$

26. $y = x^2 - 9$

27. $y = -x^2 + 3x + 4$

28. $y = x^2 - 5x + 4$

In Problems 29–32, determine whether the quadratic equation has a maximum or a minimum value; then find the maximum or minimum value of the equation.

29. $y = -2x^2 - 8x + 7$

30. $y = 3x^2 + 12x - 4$

31. $y = x^2 - 4x - 6$

32. $y = -x^2 - 10x - 15$

33. Throwing a Baseball A baseball is thrown from right field to home plate. The height, h, of the path of the baseball x feet from where it is thrown can be modeled by the following quadratic equation:

$$h = -0.00175x^2 + 0.325x + 5$$

(a) Determine the distance x (to the nearest tenth of a foot) that maximizes the height of the throw.

(b) Determine the maximum height (to the nearest tenth of a foot) of the path of the baseball.

(c) How far (to the nearest tenth of a foot) from the right fielder does the baseball land?

34. Enclosing a Pasture A farmer has 400 meters of fencing to enclose a rectangular pasture. The area, A, of the pasture is given by the quadratic equation

$$A = -x^2 + 200x$$

where x is the length of the pasture (in meters).

(a) Determine the length of the side x that maximizes the enclosed area.

(b) Determine the maximum area of the pasture that can be enclosed with 400 meters of fence.

Section 10.2 Relations

KEY CONCEPTS

- **Relation**
  A correspondence between two variables x and y where y depends on x. Relations can be represented through maps, sets of ordered pairs, equations, or graphs.

KEY TERMS

Relation
Corresponds to
Depends on
Map
Input
Output
Domain
Range

You Should Be Able To...	EXAMPLE	Review Exercises
1 Define relations (p. 648)	Example 1	35, 36
2 Find the domain and the range of a relation (p. 649)	Examples 2 through 4	37–50
3 Graph a relation defined by an equation (p. 651)	Example 5	51–54

In Problems 35 and 36, represent each relation as a set of ordered pairs.

35.

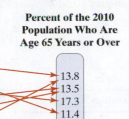

State | Percent of the 2010 Population Who Are Age 65 Years or Over

Alabama
California
Florida
Nebraska
New York
Massachusetts

13.8
13.5
17.3
11.4

36.

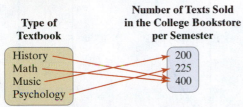

Type of Textbook | Number of Texts Sold in the College Bookstore per Semester

History
Math
Music
Psychology

200
225
400

37. State the domain and range of the relation shown in Problem 35.

38. State the domain and range of the relation shown in Problem 36.

In Problems 39–50, identify the domain and range of each relation.

39. $\{(-1, 3), (-1, 5), (0, -2), (0, -4)\}$

40. $\{(a, x^2), (b, x^3), (c, x^4), (d, x^5)\}$

41. $\{(a_1, 1), (a_2, 4), (a_3, 9), (a_4, 16)\}$

42. $\{(1, -3), (3, -3), (5, 2), (7, -2)\}$

43.

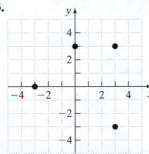

44.

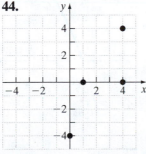

45.

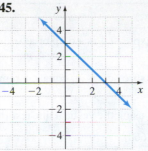

46.

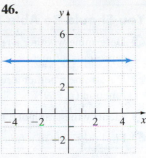

47.

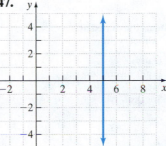

48.

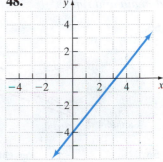

49.

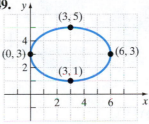

50.

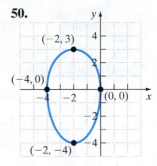

In Problems 51–54, graph each relation. Use the graph to identify the domain and the range of the relation.

51. $y = -\dfrac{3}{2}x + 4$

52. $y = \dfrac{4}{3}x - 5$

53. $y = 8x - 4x^2$

54. $y = x^2 - 4x + 3$

Section 10.3 An Introduction to Functions

KEY CONCEPTS

KEY TERMS

- **Function**

  A relation in which each element in the domain of the relation corresponds to exactly one element in the range. Functions can be represented through maps, sets of ordered pairs, equations, or graphs.

- **Vertical Line Test**

  A set of points in the *xy*-plane is the graph of a function if and only if every vertical line intersects the graph in at most one point.

Function
Vertical line test
Value of *f* at the number *x*

You Should Be Able To...	EXAMPLE	Review Exercises
1 Determine whether a relation expressed as a map or as ordered pairs represents a function (p. 656)	Examples 1 and 2	55–60
2 Determine whether a relation expressed as an equation represents a function (p. 659)	Examples 3 and 4	61–70
3 Identify the graph of a function using the vertical line test (p. 659)	Example 5	71–78
4 Find the value of a function (p. 660)	Examples 6 and 7	79–87

In Problems 55–60, determine whether each relation represents a function. If the relation is a function, state its domain and range.

55.

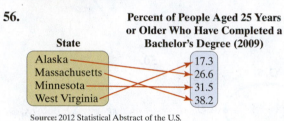

Day of Week — **Items Purchased at Grocery Store**

Monday, Sunday → Dairy, Fruit/Vegetables, Meat, Snacks

56.

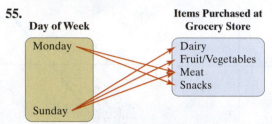

State — **Percent of People Aged 25 Years or Older Who Have Completed a Bachelor's Degree (2009)**

Alaska, Massachusetts, Minnesota, West Virginia → 17.3, 26.6, 31.5, 38.2

Source: 2012 Statistical Abstract of the U.S.

57. $\{(-4, 0), (-3, 0), (-2, 1), (-1, 1)\}$

58. $\{(1, 4), (3, 3), (1, 2), (-1, 1)\}$

59. $\{(a, 9), (b, 8), (b, 7), (a, 6)\}$

60. $\{(5, z), (3, x), (1, x), (-1, z)\}$

In Problems 61–70, determine whether each equation represents a function.

61. $x = y + 2$

62. $x = y^2$

63. $x = -3$

64. $y = 2x$

65. $y = \pm\sqrt{x}$

66. $y = 7$

67. $y = x^2 + 4x$

68. $x^2 + y^2 = 25$

69. $x^2 - y^2 = 16$

70. $x^2 = y - 1$

In Problems 71–78, use the vertical line test to determine whether the graph is that of a function.

71.

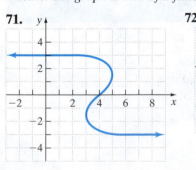

72.

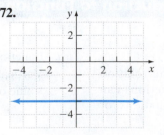

73.

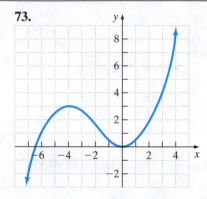

74.

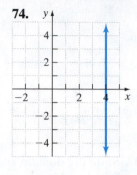

75.

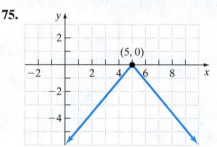

76.

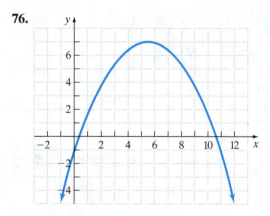

77.

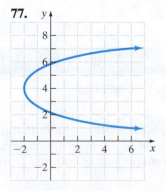

78.

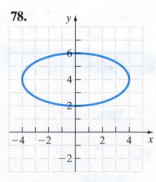

In Problems 79–86, find the indicated value for each function.

79. $f(x) = -3x - 2; f(-1)$

80. $g(x) = 6 - 4x; g(-3)$

81. $h(x) = -3; h(2)$

82. $F(x) = -10; F(-2)$

83. $G(x) = x^2 - 3x + 4; G(-2)$

84. $H(x) = -x^2 + 6x - 1; H(4)$

85. $A(x) = \sqrt{2x + x^2}; A(4)$

86. $B(x) = \left| \dfrac{3x + 2}{2} \right|; B(-4)$

87. Car Rental The cost to rent a car for a week is $175 plus $0.10 per mile driven.

   (a) Express the cost, C, to rent a car as a function of the number of miles, m, driven.

   (b) Determine the total cost to rent a car and drive 130 miles in a week.

Chapter 10 Test

Step-by-step test solutions are found on the Chapter Test Prep Videos available in MyMathLab® *or on* You Tube.

1. Consider the quadratic equation $y = x^2 + 4x - 5$.

   (a) Determine whether the parabola opens up or down.

   (b) Find the vertex of the parabola.

   (c) Find the equation of the axis of symmetry.

   (d) Find all intercepts.

In Problems 2–4, graph each quadratic equation using its properties. Label the vertex, intercepts, and axis of symmetry.

2. $y = -x^2 + 6x - 9$

3. $y = x^2 - 4$

4. $y = x^2 - 4x - 3$

5. Consider the quadratic equation $y = -2x^2 - 4x + 5$.

   (a) Determine whether the equation has a maximum or a minimum value.

   (b) Find the maximum or minimum value of the equation.

In Problems 6–9, determine the domain and the range of the given relation.

6. $\{(1, -1), (1, -2), (1, -3)\}$

7. $\{(-2, a), (-3, b), (-3, c)\}$

8.

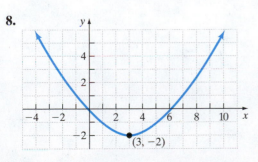

9.

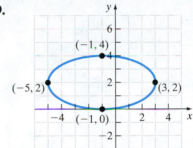

In Problems 10–15, determine whether each of the following represents a function.

10. $\{(-5, 10), (-5, 9), (-5, 8)\}$

11. $\{(4, -2), (3, -2), (2, -2)\}$

12. $x = \sqrt{y + 2}$

13. $x = 4 + y$

14.

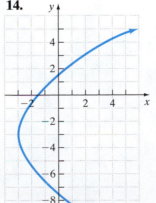

15.

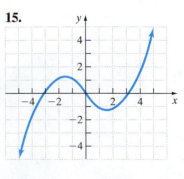

In Problems 16–18, find the indicated value for each function.

16. $f(x) = -x^2 + 3x - 1; f(-1)$

17. $g(x) = \dfrac{-2}{3x + 4}; g(2)$

18. $h(x) = 2\sqrt{x + 4}; h(12)$

19. Height of a Ball The height, h, in feet, of the path of a ball can be approximated by the quadratic equation $h(t) = -16t^2 + 32t$, where t is the number of seconds after the ball was thrown.

    (a) How long does it take for the ball to reach its maximum height?

    (b) What is the maximum height of the ball?

20. Oil Price The price per barrel of crude oil rose sharply and then came back down between 1969 and 1989. The equation that can be used to model the cost of a barrel of crude oil between 1969 and 1989 is given by the function $P(t) = -0.33t^2 + 6.5t + 3$, where P is the price of a barrel of crude oil (not adjusted for inflation) and t is the number of years after 1969.

    (a) What was the price per barrel of crude oil in 1969?

    (b) How many years, to the nearest tenth, after 1969 did the price reach its maximum?

    (c) What was the highest price per barrel, to the nearest dollar, during this time?

    (d) Find the price per barrel of crude oil, to the nearest dollar, in 1974.

Appendix A

Table of Square Roots

n	$\sqrt{n}$	n	$\sqrt{n}$	n	$\sqrt{n}$	n	$\sqrt{n}$
1	1	26	5.09902	51	7.14143	76	8.71780
2	1.41421	27	5.19615	52	7.21110	77	8.77496
3	1.73205	28	5.29150	53	7.28011	78	8.83176
4	2	29	5.38516	54	7.34847	79	8.88819
5	2.23607	30	5.47723	55	7.41620	80	8.94427
6	2.44949	31	5.56776	56	7.48331	81	9
7	2.64575	32	5.65685	57	7.54983	82	9.05539
8	2.82843	33	5.74456	58	7.61577	83	9.11043
9	3	34	5.83095	59	7.68115	84	9.16515
10	3.16228	35	5.91608	60	7.74597	85	9.21954
11	3.31662	36	6	61	7.81025	86	9.27362
12	3.46410	37	6.08276	62	7.87401	87	9.32738
13	3.60555	38	6.16441	63	7.93725	88	9.38083
14	3.74166	39	6.24500	64	8	89	9.43398
15	3.87298	40	6.32456	65	8.06226	90	9.48683
16	4	41	6.40312	66	8.12404	91	9.53939
17	4.12311	42	6.48074	67	8.18535	92	9.59166
18	4.24264	43	6.55744	68	8.24621	93	9.64365
19	4.35890	44	6.63325	69	8.30662	94	9.69536
20	4.47214	45	6.70820	70	8.36660	95	9.74679
21	4.58258	46	6.78233	71	8.42615	96	9.79796
22	4.69042	47	6.85565	72	8.48528	97	9.84886
23	4.79583	48	6.92820	73	8.54400	98	9.89949
24	4.89898	49	7	74	8.60233	99	9.94987
25	5	50	7.07107	75	8.66025	100	10

Appendix B

Geometry Review

B.1 Lines and Angles

Objectives

1. Understand the Terms Point, Line, and Plane
2. Work with Angles
3. Find the Measures of Angles Formed by Parallel Lines

Figure 1

Figure 2

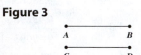

Figure 3

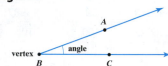

Figure 4

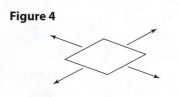

Figure 5

Work Smart

One full rotation is represented below.

The word *geometry* comes from the Greek words *geo*, meaning "earth" and *metra*, meaning "measure." The Greek scholar Euclid collected and organized the geometry known in his day into a logical system more than two thousand years ago. Euclid's system forms the basis of the geometry we study today.

❶ Understand the Terms Point, Line, and Plane

A **point** has no size, only position, and is usually designated by a capital letter as shown below.

$$\overset{\bullet}{P}$$

A **line** is a set of points extending infinitely far in opposite directions. A line has no width or height, just length, and is uniquely determined by two points. For example, the line in Figure 1 passes through the points A and B. The notation for the line shown is $\overleftrightarrow{AB}$.

A **ray** is a half-line with one **endpoint,** and extends infinitely far in one direction. See Figure 2. The notation for the ray shown is $\overrightarrow{AB}$.

A **line segment** is a portion of a line that has a beginning and an end. See Figure 3. If two line segments have the same length, they are said to be **congruent.** The notation for the congruent line segments shown are $\overline{AB}$ and $\overline{CD}$.

A **plane** is the set of points that forms a flat surface that extends indefinitely. A plane has no thickness. See Figure 4. The arrows indicate that the plane extends indefinitely in each direction.

❷ Work with Angles

Suppose we draw two rays with a common endpoint as shown in Figure 5. The amount of rotation from one ray to the second ray is called the **angle** between the rays. The common endpoint is called the **vertex.** In Figure 5, the name of the angle is $\angle ABC$, $\angle CBA$, or $\angle B$. Angles are measured in **degrees,** which is symbolized°. One full rotation represents 360°. The notation $m\angle A = 60°$ means "the measure of angle A is 60 degrees." Because 60° is $\frac{1}{6}$ of 360°, an angle whose measure is 60° is $\frac{1}{6}$ of a full rotation. Two angles with the same measure are **congruent.**

Some angles are classified by their measure.

> **Definitions**
>
> An angle that measures 90° is a **right angle.** The symbol ⌐ denotes a right angle. A right angle is $\frac{1}{4}$ of a full rotation. See Figure 6(a).
>
> An angle whose measure is between 0° and 90° is an **acute angle.** See Figure 6(b).
>
> An angle whose measure is between 90° and 180° is an **obtuse angle.** See Figure 6(c).
>
> An angle whose measure is 180° is a **straight angle.** A straight angle is $\frac{1}{2}$ of a full rotation. See Figure 6(d).

Figure 6

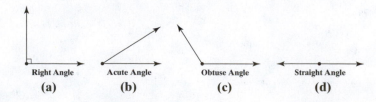

Right Angle   Acute Angle   Obtuse Angle   Straight Angle
(a)   (b)   (c)   (d)

Definitions

Two angles whose measures sum to 90° are **complementary** angles. Each angle is the **complement** of the other. Two angles whose measures sum to 180° are **supplementary** angles. Each angle is the **supplement** of the other. See Figure 7.

Figure 7

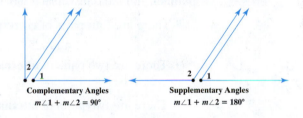

Complementary Angles
$m\angle 1 + m\angle 2 = 90°$

Supplementary Angles
$m\angle 1 + m\angle 2 = 180°$

EXAMPLE 1 **Finding the Complement of an Angle**

Find the complement of an angle whose measure is 18°.

Solution

Two angles are complementary if their sum is 90°. An angle that is complementary to an 18° angle measures $90° - 18° = 72°$. ●

EXAMPLE 2 **Finding the Supplement of an Angle**

Find the supplement of an angle whose measure is 97°.

Solution

Two angles are supplementary if their sum is 180°. An angle that is supplementary to a 97° angle measures $180° - 97° = 83°$. ●

❸ Find the Measures of Angles Formed by Parallel Lines

Lines that lie in the same plane are **coplanar.**

Definitions

Parallel lines are lines in the same plane that never meet, as shown in Figure 8(a). **Intersecting lines** meet or cross in one point. See Figure 8(b). Two lines that intersect to form right (90°) angles are called **perpendicular lines.** Figure 8(c) shows perpendicular lines.

Figure 8

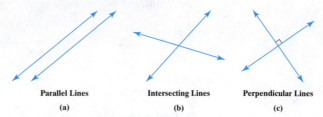

Parallel Lines   Intersecting Lines   Perpendicular Lines
(a)   (b)   (c)

Figure 9

$m\angle 1 = m\angle 3$
$m\angle 2 = m\angle 4$

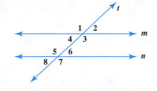

Two lines that intersect form four angles. Two angles that are opposite each other are called **vertical angles.** Vertical angles have equal measures. **Adjacent angles** have the same vertex and share a side. In Figure 9, angles 1 and 3 are vertical angles, and angles 1 and 2 are adjacent angles. Angles 1 and 2 are also supplementary angles. Other pairs of adjacent angles are angles 2 and 3, angles 3 and 4, and angles 1 and 4.

A line that cuts two parallel lines is called a **transversal.** In Figure 10, lines m and n are parallel, and the transversal is labeled t. In this figure, certain angles have special names.

- There are four pairs of **corresponding angles:**

$\angle 1$ and $\angle 5$ $\angle 2$ and $\angle 6$ $\angle 3$ and $\angle 7$ $\angle 4$ and $\angle 8$

- There are two pairs of **alternate interior angles:**

$\angle 3$ and $\angle 5$ $\angle 4$ and $\angle 6$

- There are two pairs of **alternate exterior angles:**

$\angle 1$ and $\angle 7$ $\angle 2$ and $\angle 8$

Parallel lines and these angles are related in the following way.

Figure 10

Parallel Lines Cut by a Transversal

If two parallel lines are cut by a transversal, then

- Corresponding angles are equal in measure.
- Alternate interior angles are equal in measure.
- Alternate exterior angles are equal in measure.

EXAMPLE 3   **Finding the Measure of Corresponding and Alternate Interior Angles**

Given that lines m and n are parallel, t is a transversal, and the measure of angle 1 is 85°, find the measures of angles 2, 3, 4, 5, 6, and 7.

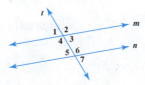

Solution

$m\angle 2 = 180° - 85° = 95°$ because $\angle 1$ and $\angle 2$ are supplementary angles.

$m\angle 3 = 85°$ because $\angle 1$ and $\angle 3$ are vertical angles.

$m\angle 4 = 180° - 85° = 95°$ because $\angle 3$ and $\angle 4$ are supplementary angles. $\angle 1$ and $\angle 4$ are also supplementary angles.

$m\angle 5 = 85°$ because $\angle 1$ and $\angle 5$ are corresponding angles (or because $\angle 3$ and $\angle 5$ are alternate interior angles).

$m\angle 6 = 95°$ because $\angle 2$ and $\angle 6$ are corresponding angles.

$m\angle 7 = 85°$ because $\angle 1$ and $\angle 7$ are alternate exterior angles.   ●

Quick ✓

11. _____ lines are lines in the same plane that never meet.

12. Suppose two lines intersect. The two angles that are opposite each other are called _____ angles.

13. *True or False* If two parallel lines are cut by a transversal, the corresponding angles are equal in measure.

14. Find the measure of angles 1–7, given that lines *m* and *n* are parallel and *t* is a transversal.

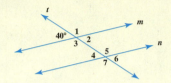

B.1 Exercises MyMathLab® Math XL PRACTICE

*Problems **1–14** are the **Quick ✓**s that follow the **EXAMPLES**.*

Building Skills

In Problems 15–22, classify each angle as right, acute, obtuse, or straight. See Objective 2.

15.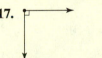

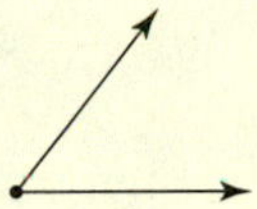

16.

17.

18.

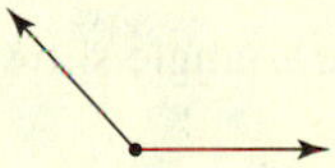

19.

20.

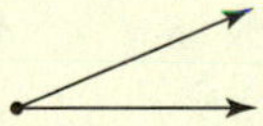

21.

22.

In Problems 23–26, find the complement of each angle. See Objective 2.

23. 32°   24. 19°

25. 73°   26. 51°

In Problems 27–30, find the supplement of each angle. See Objective 2.

27. 67°   28. 145°

29. 8°   30. 106°

In Problems 31 and 32, find the measure of angles 1–7, given that lines m and n are parallel and t is a transversal. See Objective 3.

31.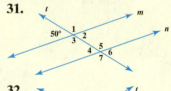

32.

B.2 Polygons

Objectives

1. Define Polygon
2. Work with Triangles
3. Identify Quadrilaterals
4. Work with Circles

❶ Define Polygon

A **polygon** is a closed figure in a plane consisting of line segments that meet at the **vertices**. A **regular polygon** is a polygon in which the sides are congruent and the angles are congruent. Figure 11 shows four regular polygons.

Figure 11
Regular polygons: All the sides are the same length; all the angles have the same measure.

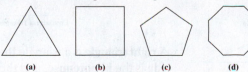

(a)   (b)   (c)   (d)

A polygon is named according to the number of sides. Table 1 summarizes the names of the polygons with 3 to 10 sides. A **triangle** is a polygon with three sides. Figure 11(a) is a regular triangle. A **quadrilateral** is a polygon with four sides. Figure 11(b) is a regular quadrilateral (a *square*). A **pentagon** is a polygon with five sides. Figure 5(c) is a regular pentagon. An **octagon** is a polygon with eight sides. Figure 5(d) shows a regular octagon.

Table 1	
Polygons	
Number of Sides	**Name of Polygon**
3	Triangle
4	Quadrilateral
5	Pentagon
6	Hexagon
7	Heptagon
8	Octagon
9	Nonagon
10	Decagon

Figure 12

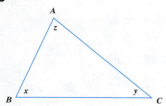

② Work with Triangles

In triangle *ABC,* shown in Figure 12, angles *A, B,* and *C* are called **interior angles.** The sum of the measures of the interior angles of a triangle is 180°. If *x, y,* and *z* represent the measures of angles *A, B,* and *C,* respectively, then

$$x + y + z = 180°$$

EXAMPLE 1 **Finding the Measure of an Interior Angle of a Triangle**

Find the measure of angle *C* in the triangle shown in Figure 13.

Figure 13

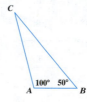

Solution

The measure of angle *A* is 100° and the measure of angle *B* is 50°. We compute the measure of angle *C* as

$$m\angle C = 180° - 50° - 100° = 30°$$

We classify triangles by the lengths of their sides. A triangle with three congruent sides is an **equilateral** triangle. A triangle with two congruent sides is an **isosceles** triangle, and a triangle with no congruent sides is a **scalene** triangle. See Figure 14.

Figure 14

Equilateral Triangle   Isosceles Triangle   Scalene Triangle

Figure 15

A **right triangle** is a triangle that contains a right (90°) angle. In a right triangle, the longest side is the **hypotenuse**, and the remaining two sides are the **legs.** See Figure 15.

EXAMPLE 2 **Finding the Measure of an Angle of a Right Triangle**

Find the measure of angle *B* in the right triangle shown in Figure 16.

Figure 16

Solution

There are 180° degrees in a triangle and the right angle measures 90°, so

$$m\angle B = 180° - 90° - 35° = 55°$$

Quick ✓

1. A triangle in which two sides are congruent is a(n) _____ triangle.

2. A ____ triangle is a triangle that contains a 90° angle.

3. The sum of the measures of the interior angles in a triangle is ___ degrees.

In Problems 4 and 5, find the measure of angle B in each triangle.

4.

5.

Congruent Triangles

Figure 17

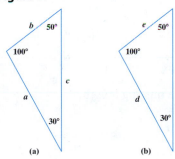

Two triangles are **congruent** if the corresponding angles have the same measure and the corresponding sides have the same length. See Figures 17(a) and 17(b). We see that the corresponding angles in triangles (a) and (b) are equal. Also, the lengths of the corresponding sides are equal: $a = d$, $b = e$, and $c = f$.

It is not necessary to verify that all three angles and all three sides are the same measure to determine whether two triangles are congruent.

Determining Congruent Triangles

1. Two triangles are congruent if two pairs of corresponding angles and the length of the corresponding side between the two angles are equal. This is called the Angle-Side-Angle (ASA) postulate. See Figure 18(a).

2. Two triangles are congruent if the lengths of the corresponding sides of the triangles are equal. This is called the Side-Side-Side (SSS) postulate. See Figure 18(b).

3. Two triangles are congruent if the lengths of two pairs of corresponding sides and the measure of the angle between the two sides are equal. This is called the Side-Angle-Side (SAS) postulate. See Figure 18(c).

Figure 18

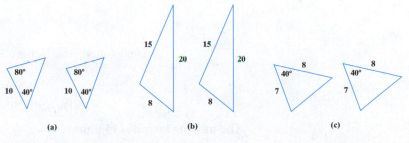

Figure 19

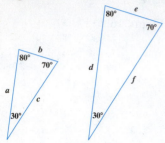

Similar Triangles

Two triangles are **similar** if the corresponding angles are equal and the lengths of the corresponding sides are proportional. That is, triangles are similar if they have the same shape. In Figure 19, the triangles are similar because the corresponding angles are equal and the corresponding sides are proportional: $\dfrac{d}{a} = \dfrac{e}{b} = \dfrac{f}{c}$. It is not necessary to verify that all the angles are congruent and all the sides are proportional to determine whether two triangles are similar.

> **Determining Similar Triangles**
>
> **1.** Two triangles are similar if two pairs of corresponding angles are equal. See Figure 20(a).
>
> **2.** Two triangles are similar if the lengths of all the corresponding sides of the triangles are proportional. See Figure 20(b).
>
> **3.** Two triangles are similar if two pairs of corresponding sides are proportional and the angles between the corresponding sides are congruent. See Figure 20(c).

Figure 20

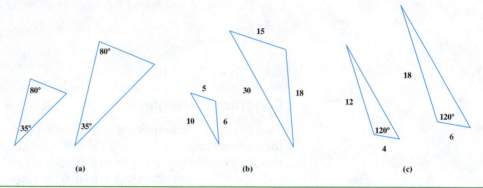

(a)   (b)   (c)

EXAMPLE 3   **Finding the Missing Length in Similar Triangles**

Given that the triangles in Figure 21 are similar, find the missing length.

Figure 21

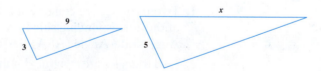

Solution

Because the triangles are similar, the corresponding sides are proportional. That is, $\dfrac{3}{5} = \dfrac{9}{x}$. Solve this equation for x.

$$\frac{3}{5} = \frac{9}{x}$$

Multiply both sides by the LCD, 5x: $5x \cdot \left(\dfrac{3}{5}\right) = 5x \cdot \left(\dfrac{9}{x}\right)$

Simplify: $3x = 45$

Divide both sides by 3: $x = 15$

The missing length is 15 units.

Quick ✓

6. Two triangles are _____ if the corresponding angles have the same measure and the corresponding sides have the same length.

7. Two triangles are _____ if corresponding angles of the triangle are equal and the lengths of the corresponding sides are proportional.

In Problem 8, given that the following triangles are similar, find the missing length.

8.

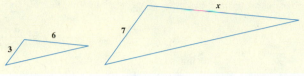

❸ Identify Quadrilaterals

A **quadrilateral** is a polygon with four sides. A **parallelogram** is a quadrilateral in which both pairs of opposite sides are parallel. A **rectangle** is a parallelogram that contains a right angle. A **square** is a rectangle with all sides of equal length. A **rhombus** is a parallelogram that has all sides equal in length. A **trapezoid** is a quadrilateral with exactly one pair of opposite sides that are parallel.

Figure	Sketch
Parallelogram	
Rectangle	
Square	
Rhombus	
Trapezoid	

❹ Work with Circles

Figure 22

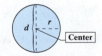

A **circle** is a figure made up of all points in the plane that are a fixed distance from a point called the **center**. The **radius** of the circle is any line segment drawn from the center of the circle to any point on the circle. The **diameter** of the circle is any line segment that has endpoints on the circle and passes through the center of the circle. See Figure 22. Notice that the length of the diameter d of a circle is twice the length of the radius r. That is, $d = 2r$.

EXAMPLE 4 **Finding the Length of the Diameter of a Circle**

Find the length of the diameter of a circle with radius 4 centimeters (cm).

Solution

The length of the diameter is twice the length of the radius.

$$d = 2 \cdot r$$
$$d = 2 \cdot 4 \text{ cm}$$
$$d = 8 \text{ cm}$$

The diameter of the circle is 8 cm.

EXAMPLE 5 **Finding the Length of the Radius of a Circle**

Find the length of the radius of a circle with diameter 18 yards.

Solution

The length of the radius is one-half the length of the diameter.

$$r = \frac{1}{2} \cdot d$$

$$r = \frac{1}{2} \cdot 18 \text{ yards}$$

$$r = 9 \text{ yards}$$

The radius of the circle is 9 yards.

Quick ✓

9. The _____ of a circle is any line segment drawn from the center of the circle to any point on the circle.

10. *True or False* The length of the diameter of a circle is exactly twice the length of the radius.

In Problems 11–14, find the length of the radius or diameter of each circle.

11. $d = 15$ inches, find r.

12. $d = 24$ feet, find r.

13. $r = 3.6$ yards, find d.

14. $r = 9$ cm, find d.

B.2 Exercises — MyMathLab® MathXL PRACTICE

*Problems **1–14** are the **Quick ✓**s that follow the **EXAMPLES**.*

Building Skills

In Problems 15–18, find the measure of the missing angle of the triangle. See Objective 2.

15.

16.

17.

18.

In Problems 19–22, determine the length of the missing side of the triangle. (These are similar triangles.) See Objective 2.

19.

20.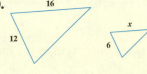

21.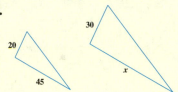

22.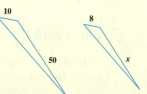

In Problems 23–26, find the length of the diameter of the circle. See Objective 4.

23. $r = 5$ in.

24. $r = 16$ feet

25. $r = 2.5$ cm

26. $r = 5.9$ in.

In Problems 27–30, find the length of the radius of the circle. See Objective 4.

27. $d = 14$ cm

28. $d = 58$ inches

29. $d = 11$ yards

30. $d = 27$ feet

B.3 Perimeter and Area of Polygons and Circles

Objectives

1 Find the Perimeter and Area of a Rectangle and a Square

2 Find the Perimeter and Area of a Parallelogram and a Trapezoid

3 Find the Perimeter and Area of a Triangle

4 Find the Circumference and Area of a Circle

The **perimeter** of a polygon is the distance around the polygon. Put another way, the perimeter of a polygon is the sum of the lengths of the sides.

1 Find the Perimeter and Area of a Rectangle and a Square

A rectangle is a polygon, so the perimeter of a rectangle is the sum of the lengths of the sides.

EXAMPLE 1 **Finding the Perimeter of a Rectangle**

Find the perimeter of the rectangle in Figure 23.

Solution

The perimeter of the rectangle is the sum of the lengths of the sides, so

$$\text{Perimeter} = 11 \text{ feet} + 8 \text{ feet} + 11 \text{ feet} + 8 \text{ feet}$$
$$= 38 \text{ feet}$$

Figure 23

Did you notice that the perimeter from Example 1 can also be written as follows?

$$\text{Perimeter} = 2 \cdot 11 \text{ feet} + 2 \cdot 8 \text{ feet}$$

In general, the perimeter of a rectangle is written $P = 2l + 2w$, where l is the length and w is the width of the rectangle.

A different measure of a polygon is its *area*. The **area** of a polygon is the amount of surface the polygon covers. Consider the rectangle shown in Figure 24. If we count the number of 1-unit by 1-unit squares within the rectangle, we see that the area of the rectangle is 6 square units. The area can also be found by multiplying the number of units of length by the number of units of width. In other words, the area of a rectangle is the product of its length and width.

Figure 24

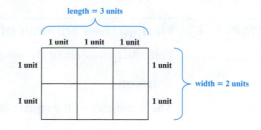

EXAMPLE 2 **Finding the Area of a Rectangle**

Find the area of the rectangle in Figure 25.

Figure 25

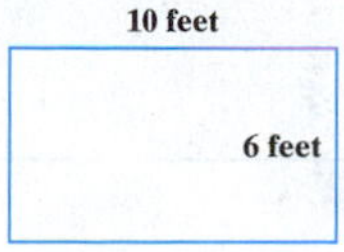

Solution

The area of the rectangle is the product of the length and the width, so

$$\text{Area} = 6 \text{ feet} \cdot 10 \text{ feet}$$
$$= 60 \text{ square feet}$$

Below is a summary of the formulas for the perimeter and area of a rectangle.

Figure	Sketch	Perimeter	Area
Rectangle	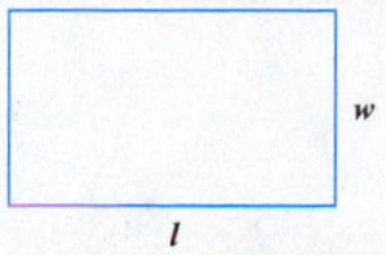	$P = 2l + 2w$	$A = lw$

EXAMPLE 3 **Finding the Perimeter and Area of a Rectangle**

Find **(a)** the perimeter and **(b)** the area of the rectangle shown in Figure 26.

Figure 26

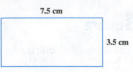

7.5 cm

3.5 cm

Solution

(a) The perimeter of the rectangle is

$$P = 2l + 2w$$
$$= 2 \cdot 7.5 \text{ cm} + 2 \cdot 3.5 \text{ cm}$$
$$= 15 \text{ cm} + 7 \text{ cm}$$
$$= 22 \text{ cm}$$

(b) The area of the rectangle is

$$A = lw$$
$$= 7.5 \text{ cm} \cdot 3.5 \text{ cm}$$
$$= 26.25 \text{ square cm}$$

●

Quick ✓

1. The _____ of a polygon is the distance around the polygon.

2. The ____ of a polygon is the amount of surface the polygon covers.

In Problems 3 and 4, find the perimeter and area of each rectangle.

3.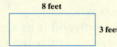

8 feet

3 feet

4.

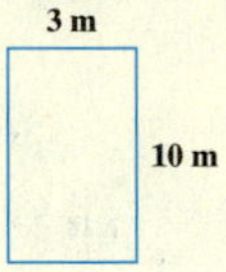

3 m

10 m

A square is a rectangle that has four congruent sides, so the perimeter of a square is

$$\text{Perimeter} = \text{side} + \text{side} + \text{side} + \text{side} = 4 \cdot \text{side} = 4s$$

where *s* is the length of a side.

EXAMPLE 4 **Finding the Perimeter of a Square**

Find the perimeter of the square in Figure 27.

Figure 27

3 cm

Solution

The perimeter of the square is

$$\text{Perimeter} = 4 \cdot s$$
$$= 4 \cdot 3 \text{ cm}$$
$$= 12 \text{ cm}$$

●

We know that the area of a rectangle is the product of the length and the width. In a square, the sides are congruent, so

$$\text{Area} = \text{side} \cdot \text{side} = \text{side}^2$$

EXAMPLE 5 **Finding the Area of a Square**

Figure 28

7 inches

Find the area of the square in Figure 28.

Solution

The area of the square is

$$\text{Area} = \text{side}^2$$
$$= (7 \text{ inches})^2$$
$$= 49 \text{ square inches}$$

●

Below is a summary of the formulas for the perimeter and area of a square.

Figure	Sketch	Perimeter	Area
Square		$P = 4s$	$A = s^2$

EXAMPLE 6   **Finding the Perimeter and Area of a Geometric Figure**

Find **(a)** the perimeter and **(b)** the area of the region shown in Figure 29.

Figure 29

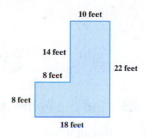

Solution

(a) The perimeter is the distance around the polygon, so

$$\text{Perimeter} = 8 \text{ feet} + 8 \text{ feet} + 14 \text{ feet} + 10 \text{ feet} + 22 \text{ feet} + 18 \text{ feet}$$
$$= 80 \text{ feet}$$

Figure 30

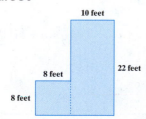

(b) The region can be divided into an 8-foot by 8-foot square plus a 10-foot by 22-foot rectangle. See Figure 30. The total area is the area of the square plus the area of the rectangle.

$$\text{Area} = \text{area of square} + \text{area of rectangle}$$
$$= (8 \text{ feet})^2 + (10 \text{ feet})(22 \text{ feet})$$
$$= 64 \text{ square feet} + 220 \text{ square feet}$$
$$= 284 \text{ square feet}$$

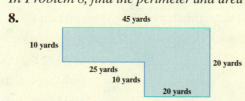

EXAMPLE 7 **Painting a Room**

You've decided to paint your rectangular bedroom. Two walls are 14 feet long and 7 feet high, and the other two walls are 10 feet long and 7 feet high.

(a) Ignoring the window and door openings in the bedroom, what is the area of the walls in your bedroom?

(b) You know that 1 gallon of paint will cover 300 square feet. How many 1-gallon cans of paint must you purchase to paint your bedroom?

Solution

(a) The bedroom has two walls that are 14 feet long and 7 feet high. The area of these two walls is

$$\text{Area} = l \cdot w \cdot 2$$
$$= 14 \text{ feet} \cdot 7 \text{ feet} \cdot 2$$
$$= 196 \text{ square feet}$$

The area of the other two walls is

$$\text{Area} = l \cdot w \cdot 2$$
$$= 10 \text{ feet} \cdot 7 \text{ feet} \cdot 2$$
$$= 140 \text{ square feet}$$

The total area to be painted is

$$\text{Area} = 196 \text{ square feet} + 140 \text{ square feet}$$
$$= 336 \text{ square feet}$$

(b) A 1-gallon can of paint covers 300 square feet. You have 336 square feet to paint, so you will need to purchase 2 gallons of paint.

② Find the Perimeter and Area of a Parallelogram and a Trapezoid

Recall that a parallelogram is a quadrilateral with parallel opposite sides. A trapezoid is a quadrilateral with exactly one pair of opposite sides that are parallel. The following table gives the formulas for the perimeter and area of parallelograms and trapezoids.

Figure	Sketch	Perimeter	Area
Parallelogram		$P = 2a + 2b$	$A = b \cdot h$
Trapezoid		$P = a + b + c + B$	$A = \frac{1}{2}h(b + B)$

EXAMPLE 8 **Finding the Perimeter and Area of a Parallelogram**

Find **(a)** the perimeter and **(b)** the area of parallelogram shown in Figure 31.

Figure 31

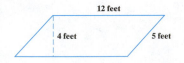

Solution

(a) The perimeter of the parallelogram is

$$P = 2l + 2w$$
$$= 2 \cdot 12 \text{ feet} + 2 \cdot 5 \text{ feet}$$
$$= 24 \text{ feet} + 10 \text{ feet}$$
$$= 34 \text{ feet}$$

(b) The area of the parallelogram is

$$A = b \cdot h$$
$$= 12 \text{ feet} \cdot 4 \text{ feet}$$
$$= 48 \text{ square feet}$$

⬤

(**EXAMPLE 9**) **Finding the Perimeter and Area of a Trapezoid**

Find **(a)** the perimeter and **(b)** the area of the trapezoid shown in Figure 32.

Figure 32

10 inches

8 inches   6 inches   7 inches

15 inches

Solution

(a) The perimeter of the trapezoid is

$$\text{Perimeter} = 7 \text{ inches} + 8 \text{ inches} + 10 \text{ inches} + 15 \text{ inches}$$
$$= 40 \text{ inches}$$

(b) The area of the trapezoid is

$$A = \frac{1}{2} h (b + B)$$
$$= \frac{1}{2} \cdot 6 \text{ inches} \cdot (15 \text{ inches} + 10 \text{ inches})$$
$$= \frac{1}{2} \cdot 6 \text{ inches} \cdot 25 \text{ inches}$$
$$= 75 \text{ square inches}$$

⬤

Quick ✓

9. To find the area of a trapezoid, we use the formula $A =$ _____ ,

where __ is the height of the trapezoid and the bases have lengths __ and __.

In Problems 10 and 11, find the perimeter and area of each figure.

10.

7 m   8 m

10 m

11.

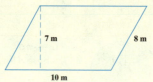

5 yards

7 yards   6 yards   9 yards

12 yards

---

③ Find the Perimeter and Area of a Triangle

Recall that a triangle is a polygon with three sides. The following table gives the formulas for the perimeter and area of a triangle.

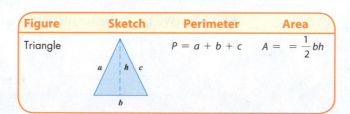

Figure	Sketch	Perimeter	Area
Triangle		$P = a + b + c$	$A = = \frac{1}{2} bh$

EXAMPLE 10 **Finding the Perimeter and Area of a Triangle**

Find **(a)** the perimeter and **(b)** the area of the triangle shown in Figure 33.

Figure 33

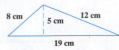

Solution

(a) To find the perimeter of the triangle, add the lengths of the three sides of the triangle.

$$\text{Perimeter} = a + b + c$$
$$= 8 \text{ cm} + 12 \text{ cm} + 19 \text{ cm}$$
$$= 39 \text{ cm}$$

(b) We use $A = \frac{1}{2} bh$ with base $b = 19$ cm and height $h = 5$ cm.

$$A = \frac{1}{2} bh = \frac{1}{2} \cdot 19 \text{ cm} \cdot 5 \text{ cm} = 47.5 \text{ square cm}$$

Quick ✓

12. *True or False* The area of a triangle is given by the formula $A = \frac{1}{2} bh$, where b is the base and h is the height.

In Problems 13 and 14, find the perimeter and area of each triangle.

13.

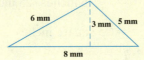

14.

❹ Find the Circumference and Area of a Circle

The **circumference** of a circle is the distance around a circle. We use the diameter or the radius of the circle to find the circumference of a circle according to the formulas given below. We also give the formula for the area of a circle.

Figure	Sketch	Perimeter	Area
Circle		$C = \pi d$ where d is the length of the diameter $C = 2\pi r$ where r is the length of the radius	$A = \pi r^2$

EXAMPLE 11 **Finding the Circumference of a Circle**

Find the circumference of the circles in Figure 34.

Figure 34

(a)

(b)

Solution

(a) We know the length of the radius is 4 cm, so we use the formula $C = 2\pi r$.

$$C = 2\pi r$$
$$r = 4 \text{ cm:} \quad = 2 \cdot \pi \cdot 4 \text{ cm}$$
$$= 8 \cdot \pi \text{ cm}$$
Use a calculator: $\quad \approx 25.13 \text{ cm}$

The exact circumference of the circle is 8π cm, and the approximate circumference is 25.13 cm, rounded to the nearest hundredth of a centimeter.

(b) The length of the diameter is 12 inches, so we use $C = \pi d$.

$$C = \pi d$$

$d = 12$ inches: $\qquad = \pi \cdot 12$ inches

$\qquad\qquad\qquad = 12\pi$ inches

Use a calculator: $\qquad \approx 37.70$ inches

The exact circumference of the circle is 12π in., and the approximate circumference is 37.70 in., rounded to the nearest hundredth of an inch. ●

EXAMPLE 12 **Finding the Area of a Circle**

Find the area of the circle in Figure 35.

Figure 35

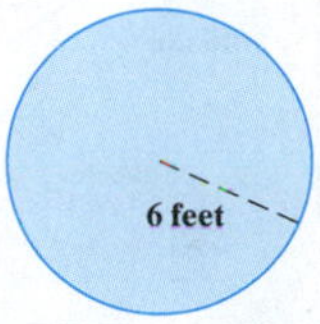

6 feet

Solution

The circle in Figure 35 has radius with length 6 feet, so we substitute $r = 6$ feet in the equation $A = \pi r^2$.

$$A = \pi r^2$$

$\qquad = \pi \cdot (6 \text{ feet})^2$

$\qquad = \pi(36) \text{ square feet}$

$\qquad = 36\pi \text{ square feet}$

Use a calculator: $\qquad \approx 113.10 \text{ square feet}$

The area of the circle is exactly 36π square feet, or approximately 113.10 square feet. ●

Quick ✓

15. The _____ of a circle is the distance around the circle.

16. *True or False* The area of a circle is given by the formula $A = \pi d^2$.

In Problems 17 and 18, find the circumference and area of each circle.

17.

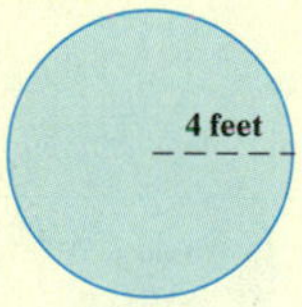

4 feet

18.

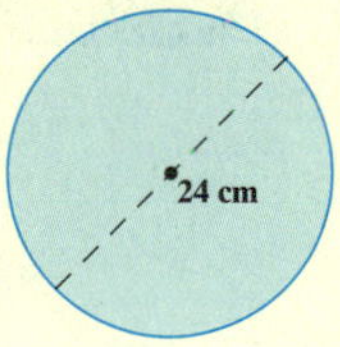

24 cm

B.3 Exercises MyMathLab® Math XL PRACTICE

Problems 1–18 are the Quick ✓ s that follow the EXAMPLES.

Building Skills

In Problems 19–22, find the perimeter and area of each rectangle. See Objective 1.

19.

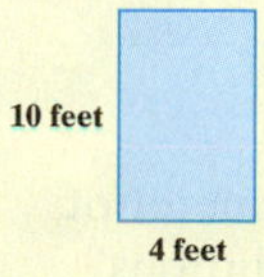

10 feet

4 feet

20.

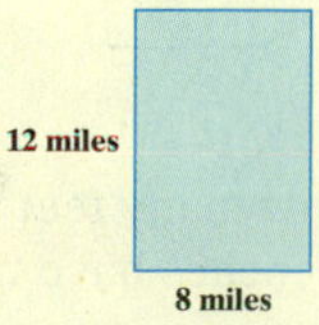

12 miles

8 miles

21.

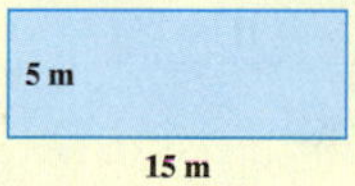

5 m

15 m

22.

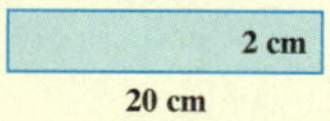

2 cm

20 cm

In Problems 23 and 24, find the perimeter and area of each square. See Objective 1.

23.

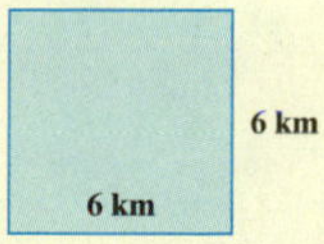

6 km

6 km

24.

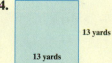

13 yards

13 yards

In Problems 25–28, find the perimeter and area of each figure. See Objective 1.

25.

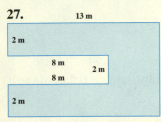

26.

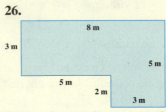

27.

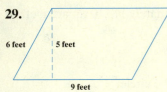

28.

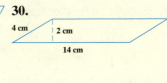

In Problems 37–40, find the perimeter and area of each triangle. See Objective 3.

37.

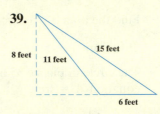

38.

39.

40.

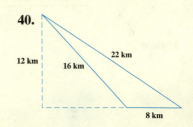

In Problems 29–36, find the perimeter and area of each quadrilateral. See Objective 2.

29.

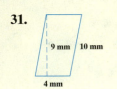

30.

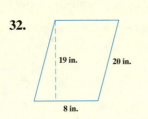

31.

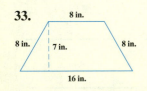

32.

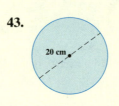

33.

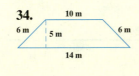

34.

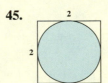

35.

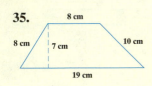

36.

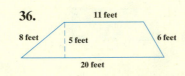

In Problems 41–44, find (a) the circumference and (b) the area of each circle. For both the circumference and the area, provide exact answers and approximate answers rounded to the nearest hundredth. See Objective 4.

41.

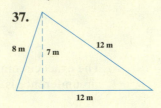

42.

43.

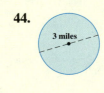

44.

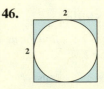

Applying the Concepts

In Problems 45 and 46, find the area of the shaded region.

45.

46.

47. How many feet will a wheel with a diameter of 20 inches have traveled after five revolutions?

48. How many feet will a wheel with a diameter of 18 inches have traveled after three revolutions?

B.4 Volume and Surface Area

Objectives

❶ Identify Solid Figures

❷ Find the Volumes and Surface Areas of Solid Figures

In Words
Polyhedra is the plural form of the Greek word *polyhedron*.

Figure 36
Hexagonal Dipyramid

Figure 37

1 inch
1 inch
1 inch

Work Smart

Volume is measured in cubic units. Surface area is measured in square units.

❶ Identify Solid Figures

A **geometric solid** is a three-dimensional region of space enclosed by planes and curved surfaces. Examples of geometric solids are cubes, pyramids, spheres, cylinders, and cones.

You see and use geometric solids every day. When you grab a box of cereal, you are holding a rectangular solid. When you reach for a can of soup, you are reaching for a circular cylinder.

We first discuss *polyhedra*. A **polyhedron** is a three-dimensional solid formed by connecting polygons. Figure 36 shows an example of a polyhedron called a hexagonal dipyramid. Notice that the top and bottom are formed by six connected triangles.

Each of the planes of a polyhedron is called a **face.** The line segment that is the intersection of any two faces of a polyhedron is called an **edge.** The point of intersection of three or more edges is called a **vertex.**

❷ Find the Volumes and Surface Areas of Solid Figures

The **volume** of a polyhedron is the measure of the number of units of space contained in the solid. Volume can be used to describe the amount of soda in a can or the amount of cereal in a box. Volume is measured in cubic units. For example, the cube in Figure 37 represents 1 cubic inch.

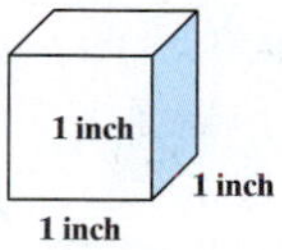

The **surface area** of a polyhedron is the sum of the areas of the faces of the polyhedron. For example, because each face in the cube shown in Figure 37 has an area of 1 square inch, and a cube has six faces, the surface area of the cube is 6 square inches. Because surface area is the sum of the areas of each polygon in the polyhedron, surface area is measured in square units.

Table 2 shows some common geometric solids, along with formulas to find their volume and surface area.

Table 2

Solids		Formulas
Cube	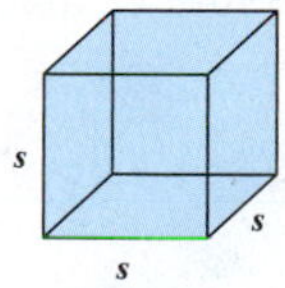	**Volume:** $V = s^3$ **Surface Area:** $S = 6s^2$
Rectangular Solid	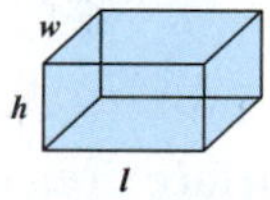	**Volume:** $V = lwh$ **Surface Area:** $S = 2lw + 2lh + 2wh$
Sphere	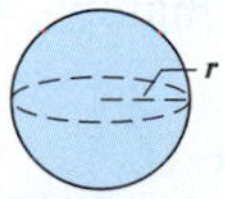	**Volume:** $V = \dfrac{4}{3}\pi r^3$ **Surface Area:** $S = 4\pi r^2$
Right Circular Cylinder		**Volume:** $V = \pi r^2 h$ **Surface Area:** $S = 2\pi r^2 + 2\pi rh$
Cone	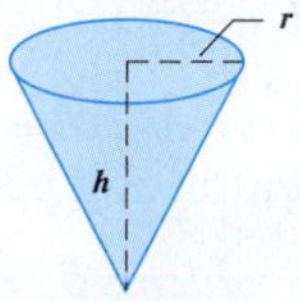	**Volume:** $V = \dfrac{1}{3}\pi r^2 h$
Square Pyramid	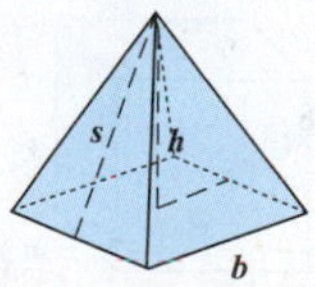	**Volume:** $V = \dfrac{1}{3}b^2 h$ **Surface Area:** $S = b^2 + 2bs$

Figure 38

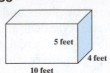

EXAMPLE 1 **Finding the Volume and Surface Area of a Rectangular Solid**

Find the volume and surface area of the rectangular solid shown in Figure 38.

Solution

From Figure 38, $l = 10$ feet, $h = 5$ feet, and $w = 4$ feet. The volume of the rectangular solid is

$$V = lwh$$
$$= (10 \text{ feet})(4 \text{ feet})(5 \text{ feet})$$
$$= 200 \text{ cubic feet}$$

The surface area of the rectangular solid is

$$S = 2lw + 2lh + 2wh$$
$$= 2(10 \text{ feet})(4 \text{ feet}) + 2(10 \text{ feet})(5 \text{ feet}) + 2(4 \text{ feet})(5 \text{ feet})$$
$$= 220 \text{ square feet}$$

Figure 39

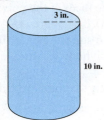

EXAMPLE 2 **Finding the Volume of a Right Circular Cylinder**

Find the volume and surface area of the right circular cylinder shown in Figure 39.

Solution

From Figure 39, $h = 10$ inches and $r = 3$ inches. The volume of the right circular cylinder is

$$V = \pi r^2 h$$
$$= \pi (3 \text{ in.})^2 (10 \text{ in.})$$
$$= 90\pi \text{ in.}^3$$

Use a calculator: $\approx 282.74 \text{ in.}^3$

The volume of the right circular cylinder is exactly 90π cubic inches and approximately 282.74 cubic inches.

The surface area of the right circular cylinder is

$$V = 2\pi r^2 + 2\pi rh$$
$$= 2\pi (3 \text{ in.})^2 + 2\pi (3 \text{ in.})(10 \text{ in.})$$
$$= 18\pi \text{ in.}^2 + 60\pi \text{ in.}^2$$
$$= 78\pi \text{ in.}^2$$

Use a calculator: $\approx 245.04 \text{ in.}^2$

The surface area of the right circular cylinder is exactly 78π square inches and approximately 245.04 square inches.

Quick ✓

1. A _____ is a three-dimensional solid formed by connecting polygons.

2. The _____ of a polyhedron is the sum of the areas of the faces of the polyhedron.

3. *True or False* The volume of a right circular cylinder is measured in square units.

4. *True or False* The volume of a rectangular solid is the product of its length, width, and height.

In Problems 5 and 6, find the volume and surface area of the following solids.

5.

6.

B.4 Exercises

*Problems **1–6** are the **Quick** ✔s that follow the **EXAMPLES**.*

Building Skills

See Objective 2.

7. Find the volume *V* and surface area *S* of a rectangular box with length 10 feet, width 5 feet, and height 12 feet.

8. Find the volume *V* and surface area *S* of a rectangular box with length 2 meters, width 6 meters, and height 9 meters.

9. Find the volume *V* and surface area *S* of a sphere with radius 6 centimeters.

10. Find the volume *V* and surface area *S* of a sphere with radius 10 inches.

11. Find the volume *V* and surface area *S* of a right circular cylinder with radius 2 inches and height 8 inches.

12. Find the volume *V* and surface area *S* of a right circular cylinder with radius 3 inches and height 6 inches.

13. Find the volume *V* of a cone with radius 10 mm and height 8 mm.

14. Find the volume *V* of a cone with radius 20 feet and height 30 feet.

15. Find the volume *V* and surface area *S* of a square pyramid with height 10 feet, slant height 12 feet, and base 8 feet.

16. Find the volume *V* and surface area *S* of a square pyramid with height 5 m, slant 8 m, and base 10 m.

Applying the Concepts

17. **Rain Gutter** A rain gutter is in the shape of a rectangular solid. How much water, in cubic inches, can the gutter hold if it is 4 inches in height, 3 inches wide, and 12 feet long?

18. **Water for the Horses** A trough for horses in the shape of a rectangular solid is 10 feet long, 2 feet wide, and 3 feet deep. How much water can the trough hold?

19. **A Can of Peaches** A can of peaches is in the shape of a right circular cylinder. The can has a 4-inch diameter and is 6 inches tall. What is the volume of the can? What is the surface area of the can? Express your answers as decimals rounded to the nearest hundredth.

20. **Coffee Can** A coffee can is in the shape of a right circular cylinder. The can has an 8-inch diameter and is 10 inches tall. What is the volume of the can? What is the surface area of the can? Express your answers as decimals rounded to the nearest hundredth.

21. **Ice Cream Cone** A waffle cone for ice cream has a diameter of 8 cm and a height of 16 cm. How much ice cream can the cone hold if the ice cream is flush with the top of the cone? Express your answer as a decimal rounded to the nearest hundredth.

22. **Water Cooler** The cups at the water cooler are cone-shaped. How much water can a cup hold if it has a 5-inch diameter and is 8 inches in height? Express your answer as a decimal rounded to the nearest hundredth.

Answers to Selected Exercises

Chapter 1 Operations on Real Numbers and Algebraic Expressions

Section 1.1 Success in Mathematics   Answers will vary.

Section 1.2 Fractions, Decimals, and Percents **1.** prime **2.** factors; product **3.** $2 \cdot 2 \cdot 3$ **4.** $2 \cdot 2 \cdot 2 \cdot 3 \cdot 5$ **5.** prime **6.** $3 \cdot 3 \cdot 13$

7. least common multiple **8.** 24 **9.** 360 **10.** 126 **11.** 180 **12.** numerator; denominator **13.** equivalent fractions **14.** $\frac{5}{10}$ **15.** $\frac{30}{48}$ **16.** least common denominator

17. $\frac{1}{4} = \frac{3}{12}; \frac{5}{6} = \frac{10}{12}$ **18.** $\frac{9}{20} = \frac{36}{80}; \frac{11}{16} = \frac{55}{80}$ **19.** lowest terms **20.** $\frac{9}{16}$ **21.** in lowest terms **22.** $\frac{2}{7}$ **23.** hundredths **24.** tenths **25.** thousands **26.** thousandths

27. 0.2 **28.** 0.93 **29.** 1.40 **30.** 690.00 **31.** 60.0 **32.** 0.4 **33.** 0.833... or $0.8\overline{3}$ **34.** 1.375 **35.** $0.\overline{428571}$ **36.** $\frac{3}{5}$ **37.** $\frac{17}{100}$ **38.** $\frac{5}{8}$ **39.** hundred; 35; 35 **40.** 0.23

41. 0.01 **42.** 0.724 **43.** 1.27 **44.** 100% **45.** 15% **46.** 80% **47.** 137.2% **48.** 0.4% **49.** $5 \cdot 5$ **51.** $2 \cdot 2 \cdot 7$ **53.** $3 \cdot 7$ **55.** $2 \cdot 2 \cdot 3 \cdot 3$ **57.** $2 \cdot 5 \cdot 5$ **59.** 53 is prime.

61. $2 \cdot 2 \cdot 3 \cdot 3 \cdot 7$ **63.** 42 **65.** 126 **67.** 210 **69.** 90 **71.** 60 **73.** 72 **75.** $\frac{8}{12}$ **77.** $\frac{18}{24}$ **79.** $\frac{21}{3}$ **81.** $\frac{4}{8}$ and $\frac{3}{8}$ **83.** $\frac{9}{15}$ and $\frac{10}{15}$ **85.** $\frac{3}{36}$ and $\frac{10}{36}$ **87.** $\frac{20}{90}$ and $\frac{35}{90}$ and $\frac{21}{90}$ **89.** $\frac{2}{3}$

91. $\frac{19}{9}$ **93.** $\frac{1}{3}$ **95.** 6 **97.** hundredths place **99.** tens place **101.** thousandths place **103.** 578.2 **105.** 354.68 **107.** 3682.010 **109.** 30 **111.** 0.625 **113.** $0.\overline{285714}$

115. 0.3125 **117.** $0.\overline{230769}$ **119.** 1.16 **121.** $\frac{3}{4}$ **123.** $\frac{1}{2}$ **125.** $\frac{491}{500}$ **127.** 0.37 **129.** 0.0602 **131.** 0.001 **133.** 20% **135.** 27.5% **137.** 200% **139.** 2.2 **141.** 2.67

143. 0.519 **145.** 84 months **147.** every 20 days **149.** $\frac{13}{20}$ **151.** 70% **153.** 77.27% **155. (a)** 33.33% **(b)** 16.67% **(c)** 25% **157.** 21.43% **159.** 2, 3, 5, 7, 11, 13, 17, 19, 23, 29, 31, 37, 41, 43, 47, 53, 59, 61, 67, 71, 73, 79, 83, 89, 97

Section 1.3 The Number Systems and the Real Number Line **1.** {1, 3, 5, 7} **2.** {Alabama, Alaska, Arkansas, Arizona} **3.** $\varnothing$ or { } **4.** True **5.** rational

6. 12, 0 **7.** 12, 0 **8.** $-5, 12, 0$ **9.** $\frac{11}{5}, -5, 12, 2.\overline{76}, 0, \frac{18}{4}$ **10.** 2.737737773... **11.** All numbers listed **12.** origin **13.** (number line)

14. inequality **15.** < **16.** < **17.** > **18.** > **19.** = **20.** < **21.** absolute value **22.** 15 **23.** $\frac{3}{4}$ **24.** -4 **25.** $A = \{0, 1, 2, 3, 4\}$ **27.** $D = \{1, 2, 3, 4\}$

29. $E = \{ \}$ or $\varnothing$ **31.** 3 **33.** $-4, 3, 0$ **35.** 2.303003000... **37.** All numbers listed **39.** π **41.** $\frac{5}{5} = 1$ **43.** (number line) **45.** True

47. True **49.** True **51.** False **53.** < **55.** > **57.** > **59.** = **61.** 12 **63.** 4 **65.** $\frac{3}{8}$ **67.** 2.1 **69. (a)** (number line)

(b) $-4.5, -1, -\frac{1}{2}, \frac{3}{5}, 1, 3.5, |-7| = 7$ **(c) (i)** $-1, 1, |-7| = 7$ **(ii)** All numbers listed

71. -100, Integers, Rational, Real **73.** -10.5, Rational, Real **75.** $\frac{75}{25}$, Natural, Whole, Integers, Rational, Real **77.** 7.56556555..., Irrational, Real

79. True **81.** False **83.** True **85.** True **87.** True **89.** Irrational numbers **91.** Real numbers **93.** 0 **95.** True **97.** True **99.** {7, 8, 9, 10, 11, 12, 13, 14, 15}

101. {10, 11, 12, 13, 14, 15} **103.** {11, 12} **105.** {2, 4, 6, 8, 10} **107. (a)** {1}, {2}, {3}, {4}, {1, 2}, {1, 3}, {1, 4}, {2, 3}, {2, 4}, {3, 4}, {1, 2, 3}, {1, 2, 4}, {1, 3, 4}, {2, 3, 4}, {1, 2, 3, 4}, $\varnothing$ **(b)** 16 **109.** A rational number is any number that can be written as the quotient of two integers, denominator not equal to zero. Natural numbers, whole numbers, and integers are rational numbers. Terminating and repeating decimals are also rational numbers.

Section 1.4 Adding, Subtracting, Multiplying, and Dividing Integers **1.** sum **2.** 14 **3.** -8 **4.** 3 **5.** -1 **6.** -4 **7.** 14 **8.** negative **9.** -4 **10.** -3

11. -24 **12.** -56 **13.** additive inverse; opposite; 0 **14.** -7 **15.** 21 **16.** $\frac{8}{5}$ **17.** -5.75 **18.** difference **19.** (-10) **20.** 80 **21.** -178 **22.** 21 **23.** 0 **24.** -58

25. 12 **26.** 550 **27.** positive **28.** -21 **29.** -52 **30.** 80 **31.** 108 **32.** 325 **33.** True **34.** 108 **35.** 360 **36.** $\frac{1}{6}$ **37.** $-\frac{1}{2}$ **38.** dividend; divisor; quotient **39.** True

40. -5 **41.** $-\frac{18}{5}$ **42.** 9 **43.** 15 **45.** 4 **47.** -4 **49.** -19 **51.** 21 **53.** -328 **55.** -11 **57.** 43 **59.** 325 **61.** -125 **63.** 11 **65.** -8 **67.** -28 **69.** 54 **71.** 0 **73.** -41

75. -31 **77.** 172 **79.** 40 **81.** -56 **83.** 0 **85.** 144 **87.** -126 **89.** -90 **91.** 210 **93.** 120 **95.** $\frac{1}{8}$ **97.** $-\frac{1}{4}$ **99.** 1 **101.** 5 **103.** 7 **105.** -15 **107.** $\frac{7}{2}$ **109.** $-\frac{10}{7}$

111. $\frac{35}{4}$ **113.** -72 **115.** 60 **117.** 171 **119.** -15 **121.** -42 **123.** $\frac{15}{4}$ **125.** 40 **127.** -238 **129.** $28 + (-21) = 7$ **131.** $-21 - 47 = -68$ **133.** $-12 \cdot 18 = -216$

135. $-36 \div (-108)$ or $\frac{-36}{-108} = \frac{1}{3}$ **137.** -3.25 dollars **139.** -6 yards **141.** $-\$48$ **143.** 14 miles **145.** \$655 **147.** No; -125 cases **149.** 25,725 feet

151. $-3, -5$ **153.** $-12, 2$ **155. (a)** 1, 2, 1.5, $1.\overline{6}$, 1.6, 1.625, 1.615, 1.619, 1.618,... **(b)** 1.618 **(c)** Answers may vary **157.** The problem $42 \div 4$ may be written equivalently as $42 \cdot \frac{1}{4}$.

Section 1.5 Adding, Subtracting, Multiplying, and Dividing Rational Numbers **1.** $\frac{27}{32}$ **2.** $-\frac{8}{3}$ **3.** $-\frac{6}{25}$ **4.** $\frac{3}{4}$ **5.** $-\frac{3}{22}$ **6.** 1 **7.** $\frac{1}{12}$ **8.** $\frac{5}{7}$ **9.** -4 **10.** $-\frac{20}{31}$

11. $\frac{50}{49}$ **12.** $-\frac{3}{8}$ **13.** $-\frac{16}{7}$ **14.** $\frac{5}{27}$ **15.** $-5; 7$ **16.** $\frac{10}{11}$ **17.** $-\frac{3}{7}$ **18.** $\frac{1}{7}$ **19.** $-\frac{6}{5}$ **20.** $\frac{29}{42}$ **21.** $\frac{5}{36}$ **22.** $-\frac{13}{4}$ **23.** $-\frac{3}{20}$ **24.** $-\frac{25}{16}$ **25.** $\frac{15}{4}$ **26.** 21.014 **27.** 64.57

28. -59.448 **29.** -71.412 **30.** 4.78 **31.** -899.5 **32.** -0.1035 **33.** 0.0135 **34.** 0.25 **35.** 8.36 **36.** -0.094 **37.** $\frac{2}{3}$ **39.** $-\frac{19}{9}$ **41.** $-\frac{1}{2}$ **43.** $\frac{12}{25}$ **45.** -25 **47.** $-\frac{2}{3}$

49. 8 **51.** $\frac{1}{6}$ **53.** $\frac{5}{3}$ **55.** $-\frac{1}{5}$ **57.** $\frac{5}{6}$ **59.** $-\frac{1}{9}$ **61.** $-\frac{2}{3}$ **63.** $\frac{1}{11}$ **65.** 32 **67.** $\frac{3}{2}$ **69.** $\frac{1}{2}$ **71.** 2 **73.** $\frac{1}{3}$ **75.** $\frac{5}{2}$ **77.** $-\frac{13}{12}$ **79.** $\frac{1}{4}$ **81.** $-\frac{1}{6}$ **83.** $\frac{33}{40}$ **85.** -6.5 **87.** 8.4 **89.** 55.92

91. 1.49 **93.** 42.55 **95.** 24.94 **97.** -9 **99.** 24.3 **101.** -490 **103.** $-\frac{11}{30}$ **105.** $-\frac{4}{3}$ **107.** $-\frac{1}{12}$ **109.** $-\frac{1}{4}$ **111.** $-\frac{129}{35}$ **113.** 1.6 **115.** $-\frac{2}{21}$ **117.** -50.526 **119.** -16

121. -15 **123.** -6.58 **125.** -7.9 **127.** $-\dfrac{25}{14}$ **129.** 58.39 **131.** -30.96 **133.** $\dfrac{1}{8}$ **135.** 21 hours **137.** 18 students **139.** \$18 **141.** \$2.40 **143.** 13.2 **145.** $\dfrac{86}{15}$

147. The expression $6 \div 2$ means to take 6 and divide it into groups of 2. How many times can we divide 6 into groups of 2? The answer is 3, so $6 \div 2 = 3$. The expression $6 \div \dfrac{1}{2}$ means to take 6 and divide it into groups of $\dfrac{1}{2}$. There are twelve $\dfrac{1}{2}$'s in 6, as shown in the figure, so $6 \div \dfrac{1}{2} = 12$.

Putting the Concepts Together (Sections 1.2–1.5) **1.** $\dfrac{7}{8} = \dfrac{35}{40}; \dfrac{9}{20} = \dfrac{18}{40}$ **2.** $\dfrac{1}{3}$ **3.** $0.\overline{285714}$ **4.** $\dfrac{3}{8}$ **5.** 0.123 **6.** 6.25% **7.** **(a)** $-12, -\dfrac{14}{7} = -2, 0, 3$

(b) $-12, -\dfrac{14}{7} = -2, -1.25, 0, 3, 11.2$ **(c)** $\sqrt{2}$ **(d)** All the numbers in the set are real numbers. **8.** $<$ **9.** -11 **10.** -65 **11.** -27 **12.** 27 **13.** -5.5 **14.** -10

15. -100 **16.** 72 **17.** -5 **18.** 16 **19.** -9 **20.** -3 **21.** $\dfrac{31}{5}$ **22.** $\dfrac{31}{36}$ **23.** $-\dfrac{17}{36}$ **24.** $\dfrac{9}{5}$ **25.** $-\dfrac{1}{28}$ **26.** 0 **27.** 10.76 **28.** 7.646 **29.** 1.46 **30.** 22.232

Section 1.6 Properties of Real Numbers **1.** multiplicative identity **2.** 8 feet **3.** 8 hours, 20 minutes **4.** 5 pounds, 8 ounces **5.** b; a **6.** a; b; b; a **7.** 22

8. $\dfrac{3}{20}$ **9.** 11.98 **10.** -13 **11.** $-\dfrac{36}{331}$ **12.** 349 **13.** 14 **14.** 14 **15.** -24.2 **16.** $\dfrac{50}{13}$ **17.** 0 **18.** undefined **19.** 156 inches **21.** 45 meters **23.** 10 gallons, 2 quarts

25. 11 pounds, 4 ounces **27.** $4\dfrac{1}{2}$ hours = 4 hours 30 min **29.** Additive Inverse Property **31.** Identity Property of Multiplication

33. Multiplicative Inverse Property **35.** Additive Inverse Property **37.** $\dfrac{0}{a} = 0$ **39.** Commutative Property of Multiplication

41. Commutative Property of Addition **43.** $\dfrac{a}{0}$ is undefined **45.** 29 **47.** 18 **49.** -65 **51.** 347 **53.** -90 **55.** undefined **57.** -34 **59.** 0 **61.** 0 **63.** $-\dfrac{20}{3}$

65. \$203.16 **67.** $-3 - (4 - 10)$ **69.** $-15 + 10 - (4 - 8)$ **71.** 44 feet per second **73.** There is no real number that equals 1 when multiplied by 0.

75. The quotient $\dfrac{0}{4} = 0$ because $4 \cdot 0 = 0$. The quotient $\dfrac{4}{0}$ is undefined because we should be able to determine a real number $\square$ such that $0 \cdot \square = 4$. But because the product of 0 and every real number is 0, there is no single replacement value for $\square$. **77.** The product of a nonzero real number and its multiplicative inverse (reciprocal) equals 1, the multiplicative identity.

Section 1.7 Exponents and the Order of Operations **1.** base; exponent; power **2.** 11^5 **3.** $(-7)^4$ **4.** 16 **5.** 49 **6.** $-\dfrac{1}{216}$ **7.** 0.81 **8.** -16 **9.** 16 **10.** 15 **11.** 24

12. 23 **13.** -19 **14.** 40 **15.** -63 **16.** 4 **17.** $-\dfrac{8}{7}$ **18.** $\dfrac{4}{7}$ **19.** $\dfrac{5}{3}$ **20.** 20 **21.** 40 **22.** -9 **23.** 48 **24.** $-\dfrac{1}{6}$ **25.** 12 **26.** -108 **27.** 10 **28.** 4 **29.** 5^2 **31.** $\left(-\dfrac{3}{5}\right)^3$ **33.** 64

35. 64 **37.** 1000 **39.** -1000 **41.** -1000 **43.** 2.25 **45.** -64 **47.** -1 **49.** 0 **51.** $\dfrac{1}{64}$ **53.** $-\dfrac{1}{27}$ **55.** 14 **57.** -3 **59.** 2500 **61.** 160 **63.** 20 **65.** 4 **67.** $\dfrac{3}{5}$ **69.** -1

71. 42 **73.** 5 **75.** -4 **77.** 115 **79.** -5 **81.** $\dfrac{169}{4}$ **83.** -24 **85.** $-\dfrac{13}{12}$ **87.** 11 **89.** 12 **91.** $-\dfrac{1}{2}$ **93.** 36 **95.** 1 **97.** 24 **99.** 5 **101.** $\dfrac{2}{5}$ **103.** 3 **105.** $\dfrac{3}{2}$ **107.** $\dfrac{64}{27}$ **109.** $-\dfrac{1}{6}$

111. $2^3 \cdot 3^2$ **113.** $2^4 \cdot 3$ **115.** $(4 \cdot 3 + 6) \cdot 2$ **117.** $(4 + 3) \cdot (4 + 2)$ **119.** $(6 - 4) + (3 - 1)$ **121.** \$514.93 **123.** 603.19 in.^{2} **125.** \$1060.90 **127.** 115.75°

129. The expression -3^2 means "take the opposite of three squared": $-(3 \cdot 3) = -9$. The base is the number 3. The expression $(-3)^2$ means use -3 as a base twice: $-3 \cdot -3 = 9$.

Section 1.8 Simplifying Algebraic Expressions **1.** variable **2.** evaluate **3.** -7 **4.** 2 **5.** \$216 **6.** True **7.** $5x^2$; $3xy$ **8.** $9ab$; $-3bc$; $5ac$; $-ac^2$

9. $\dfrac{2mn}{5}$; $-\dfrac{3n}{7}$ **10.** 2 **11.** 1 **12.** -1 **13.** 5 **14.** $-\dfrac{2}{3}$ **15.** False **16.** like **17.** like **18.** unlike **19.** unlike **20.** like **21.** b; c **22.** $6x + 12$ **23.** $-5x - 10$

24. $-2k + 14$ **25.** $6x + 9$ **26.** 4; 9 **27.** $-5x$ **28.** $-4x^2$ **29.** $-8x + 3$ **30.** $-2a + 9b - 4$ **31.** $12ac - 5a + b$ **32.** $8ab^2 - a^2b$ **33.** $2rs - \dfrac{3}{2}r^2 - 5$

34. remove all parentheses and combine like terms. **35.** $-2x - 1$ **36.** $-2m - n - 7$ **37.** $a - 11b$ **38.** $6x - 2$ **39.** 13 **41.** 17 **43.** -21 **45.** $\dfrac{1}{4}$ **47.** 81

49. 4 **51.** $2x^3, 3x^2$; $-x, 6$; $2, 3, -1, 6$ **53.** $z^2, \dfrac{2y}{3}$; $1, \dfrac{2}{3}$ **55.** unlike **57.** like **59.** like **61.** unlike **63.** $3m + 6$ **65.** $18n^2 + 12n - 6$ **67.** $-x + y$ **69.** $-4x + 3y$

71. $3x$ **73.** $6z$ **75.** $10m + 10n$ **77.** $2.2x^7$ **79.** $10y^6$ **81.** $-6w - 12y + 13z$ **83.** $-3k + 15$ **85.** $4n - 8$ **87.** $-3x + 3$ **89.** $4n - 2$ **91.** $-4n + 20$ **93.** $\dfrac{5}{6}x$

95. $-\dfrac{11}{2}$ **97.** $-3.5x - 6$ **99.** 32 **101.** 27 **103.** 0 **105.** -13 **107.** -7 **109.** -12 **111.** 44 **113.** $-\dfrac{3}{2}$ **115.** 36 **117.** \$78.70 **119.** \$4819 **121.** **(a)** $8w - 8$ **(b)** 32 yards

123. \$228.88 **125.** $-3x^2 + 7x - 3$ **127.** The sum $2x^2 + 4x^2$ is not equal to $6x^4$ because when we combine like terms, we add the coefficients of the like terms and keep the variables and exponents the same. Put another way, $2x^2 + 4x^2 = (2 + 4)x^2 = 6x^2$.

Chapter 1 Review **1.** $3 \cdot 5 \cdot 5$ **2.** $3 \cdot 29$ **3.** $3 \cdot 3 \cdot 3 \cdot 3$ **4.** prime **5.** 72 **6.** 72 **7.** $\dfrac{14}{30}$ **8.** $\dfrac{12}{4}$ **9.** $\dfrac{1}{6} = \dfrac{4}{24}; \dfrac{3}{8} = \dfrac{9}{24}$ **10.** $\dfrac{9}{16} = \dfrac{27}{48}; \dfrac{7}{24} = \dfrac{14}{48}$ **11.** $\dfrac{5}{12}$ **12.** $\dfrac{1}{2}$

13. $\dfrac{4}{5}$ **14.** 21.76 **15.** 15 **16.** $0.88\ldots = 0.\overline{8}$ **17.** 0.28125 **18.** 1.83 **19.** 2.4 **20.** $\dfrac{3}{5}$ **21.** $\dfrac{3}{8}$ **22.** $\dfrac{108}{125}$ **23.** 0.41 **24.** 7.60 **25.** 0.0903 **26.** 0.0035 **27.** 23\% **28.** 117\%

29. 4.5\% **30.** 300\% **31.** **(a)** $\dfrac{3}{5}$ **(b)** 60\% **32.** $A = \{0, 1, 2, 3, 4, 5, 6\}$ **33.** $B = \{1, 2, 3\}$ **34.** $C = \{-2, -1, 0, 1, 2, 3, 4, 5\}$ **35.** $D = \{\ \}$ or $\varnothing$

36. $\dfrac{9}{3} = 3, 11$ **37.** $0, \dfrac{9}{3} = 3, 11$ **38.** $-6, 0, \dfrac{9}{3} = 3, 11$ **39.** $-6, -3.25, 0, \dfrac{9}{3} = 3, 11, \dfrac{5}{7}$ **40.** $5.030030003\ldots$ **41.** All numbers listed

42. **43.** False **44.** True **45.** True **46.** True **47.** False **48.** $-\dfrac{1}{2}$ **49.** 7 **50.** -6 **51.** $=$ **52.** $<$ **53.** $>$ **54.** $=$ **55.** $>$

56. $<$ **57.** Rational numbers are numbers that can be written as the quotient of two integers, provided the denominator does not equal 0. Rational numbers can also be written as decimals that either terminate, or do not terminate but repeat a block of decimals. Irrational numbers can be written as decimals that neither terminate nor repeat. **58.** natural numbers **59.** 7 **60.** -4 **61.** -34 **62.** -95 **63.** -4 **64.** 47 **65.** -53 **66.** -115 **67.** -22 **68.** -7 **69.** 21 **70.** 67

71. 26 **72.** -36 **73.** 12 **74.** -40 **75.** -1118 **76.** -8037 **77.** -715 **78.** $-11,130$ **79.** 5 **80.** -12 **81.** 5 **82.** -25 **83.** -8 **84.** $-\dfrac{16}{5}$ **85.** $-\dfrac{10}{3}$ **86.** $-\dfrac{30}{7}$ **87.** -13

88. 45 **89.** $-43 + 101 = 58$ **90.** $45 + (-28) = 17$ **91.** $-10 - (-116) = 106$ **92.** $74 - 56 = 18$ **93.** $13 + (-8) = 5$ **94.** $-60 - (-10) = -50$

95. $-21 \cdot (-3) = 63$ **96.** $54 \cdot (-18) = -972$ **97.** $-34 \div (-2)$ or $\dfrac{-34}{-2} = 17$ **98.** $-49 \div 14$ or $\dfrac{-49}{14} = -\dfrac{7}{2}$ **99.** 26 yards **100.** $-3°F$ **101.** $24°F$

102. 87 points **103.** $\dfrac{5}{4}$ **104.** $-\dfrac{5}{28}$ **105.** $-\dfrac{1}{20}$ **106.** $-\dfrac{3}{2}$ **107.** $\dfrac{4}{17}$ **108.** $-\dfrac{2}{3}$ **109.** $-\dfrac{3}{10}$ **110.** -32 **111.** $\dfrac{1}{3}$ **112.** $-\dfrac{2}{5}$ **113.** $\dfrac{3}{7}$ **114.** 3 **115.** $\dfrac{7}{20}$ **116.** $\dfrac{31}{36}$ **117.** $-\dfrac{59}{245}$

118. $\dfrac{13}{12}$ **119.** $-\dfrac{19}{12}$ **120.** $-\dfrac{11}{4}$ **121.** 0 **122.** $-\dfrac{23}{24}$ **123.** 48.5 **124.** -24.66 **125.** 82.98 **126.** -53.74 **127.** 0.0804 **128.** -260.154 **129.** 18.4 **130.** -25.79 **131.** 2.3

132. -22.9 **133.** -5.418 **134.** -0.732 **135.** $-\$98.93$; yes **136.** 24 friends **137.** $\dfrac{23}{2}$ or $11\dfrac{1}{2}$ inches **138.** $\$186.81$ **139.** Associative Property of Multiplication

140. Multiplicative Inverse Property **141.** Multiplicative Inverse Property **142.** Commutative Property of Multiplication **143.** Commutative Property of Multiplication **144.** Additive Inverse Property **145.** Identity Property of Addition **146.** Identity Property of Addition **147.** Commutative Property of Addition **148.** Multiplicative Identity Property **149.** Multiplication Property of Zero **150.** Associative Property of Addition **151.** 29 **152.** 99 **153.** 18

154. 121 **155.** 3.4 **156.** 5.3 **157.** -33 **158.** 6 **159.** undefined **160.** 0 **161.** -334 **162.** 2 **163.** 0 **164.** 1 **165.** 0 **166.** 130 **167.** $-\dfrac{5}{3}$ **168.** $-\dfrac{150}{13}$ **169.** 3^4

170. $\left(\dfrac{2}{3}\right)^3$ **171.** $(-4)^2$ **172.** $(-3)^3$ **173.** 125 **174.** -125 **175.** 81 **176.** -125 **177.** -81 **178.** $\dfrac{1}{64}$ **179.** -4 **180.** -76 **181.** 206 **182.** 32 **183.** 2 **184.** $\dfrac{1}{2}$

185. $\dfrac{6}{5}$ **186.** $\dfrac{7}{5}$ **187.** 21 **188.** -18 **189.** -729 **190.** -3 **191.** $3x^2, -x, 6; 3, -1, 6$ **192.** $2x^2y^3, -\dfrac{y}{5}; 2, -\dfrac{1}{5}$ **193.** like **194.** unlike **195.** unlike **196.** like **197.** $-3x$

198. $-4x - 15$ **199.** $-4.1x^4 + 0.3x^3$ **200.** $-3x^4 + 6x^2 + 12$ **201.** $18 - x$ **202.** $4x - 18$ **203.** $9x - 4$ **204.** -1 **205.** $\$98.70$

Chapter 1 Test **1.** 42 **2.** $\dfrac{7}{22}$ **3.** 1.44 **4.** $\dfrac{17}{40}$ **5.** 0.006 **6.** 18.3% **7.** $\dfrac{1}{3}$ **8.** $\dfrac{9}{4}$ **9.** $-\dfrac{320}{3}$ **10.** -102 **11.** -12.16 **12.** -20 **13.** undefined **14.** -14 **15.** 55

16. (a) 6 **(b)** 0, 6 **(c)** $-2, 0, 6$ **(d)** $-2, -\dfrac{1}{2}, 0, 2.5, 6$ **(e)** none **(f)** All those listed. **17.** $<$ **18.** $=$ **19.** -7 **20.** 15 **21.** -102 **22.** -343 **23.** $-16x - 28$

24. $-4x^2 + 5x + 4$ **25.** $\$531.85$ **26.** $4x + 10$

Chapter 2 Equations and Inequalities in One Variable

Section 2.1 Linear Equations: The Addition and Multiplication Properties of Equality
1. solution **2.** yes **3.** no **4.** yes **5.** no **6.** True **7.** $\{32\}$

8. $\{14\}$ **9.** $\{12\}$ **10.** $\{-15\}$ **11.** $\left\{\dfrac{7}{3}\right\}$ **12.** $\left\{-\dfrac{13}{12}\right\}$ **13.** $\$12,455$ **14.** Multiplication **15.** $\{2\}$ **16.** $\{-2\}$ **17.** $\left\{\dfrac{5}{2}\right\}$ **18.** $\left\{-\dfrac{7}{3}\right\}$ **19.** $\{9\}$ **20.** $\{-9\}$ **21.** $\{-30\}$

22. $\{6\}$ **23.** $\left\{\dfrac{8}{3}\right\}$ **24.** $\left\{-\dfrac{5}{14}\right\}$ **25.** Yes **27.** No **29.** Yes **31.** Yes **33.** $\{20\}$ **35.** $\{-12\}$ **37.** $\{19\}$ **39.** $\{-13\}$ **41.** $\{2\}$ **43.** $\left\{\dfrac{1}{4}\right\}$ **45.** $\left\{\dfrac{19}{24}\right\}$ **47.** $\{-6.1\}$

49. $\{5\}$ **51.** $\{-4\}$ **53.** $\left\{\dfrac{7}{2}\right\}$ **55.** $\left\{-\dfrac{5}{2}\right\}$ **57.** $\{21\}$ **59.** $\{121\}$ **61.** $\{40\}$ **63.** $\left\{\dfrac{3}{5}\right\}$ **65.** $\left\{-\dfrac{1}{3}\right\}$ **67.** $\{9\}$ **69.** $\left\{-\dfrac{4}{9}\right\}$ **71.** $\left\{-\dfrac{5}{3}\right\}$ **73.** $\{2\}$ **75.** $\{-3\}$ **77.** $\left\{\dfrac{2}{3}\right\}$

79. $\{-6\}$ **81.** $\{19\}$ **83.** $\{283\}$ **85.** $\{50\}$ **87.** $\{41.1\}$ **89.** $\left\{\dfrac{20}{3}\right\}$ **91.** $\{-4\}$ **93.** $\{-12\}$ **95.** $\left\{\dfrac{1}{2}\right\}$ **97.** $\left\{\dfrac{3}{16}\right\}$ **99.** $\left\{-\dfrac{5}{4}\right\}$ **101.** $\$24,963.27$ **103.** $\$68$ **105.** 12

107. $r = 0.18$ or 18% **109.** $x = 48 - \lambda$ **111.** $x = \dfrac{14}{\theta}$ **113.** $\lambda = \dfrac{50}{9}$ **115.** $\theta = -\dfrac{6}{7}$ **117.** To find the solution of an equation means to find all values of the variable that satisfy the equation. **119.** An algebraic expression differs from an equation in that the expression does not contain an equal sign, and an equation does. An equation using the expression $x - 10$ is $x - 10 = 22$. Solving for x, $x = 32$. **121.** The Addition Property states that we may add (or subtract) any real number to (or from) both sides of an equation and maintain the equality. It is used to isolate the variable. To solve $x - 5 = 12$, use the Addition Property of Equality to add 5 to each side.

Section 2.2 Linear Equations: Using the Properties Together

1. True **2.** $\{3\}$ **3.** $\{18\}$ **4.** $\{2\}$ **5.** $\left\{\dfrac{3}{2}\right\}$ **6.** $\{2\}$ **7.** $\{-18\}$ **8.** $\left\{\dfrac{5}{2}\right\}$ **9.** $\{-8\}$ **10.** $\{2\}$ **11.** $\left\{\dfrac{4}{3}\right\}$ **12.** $\{13\}$ **13.** $\{6\}$ **14.** $\{2\}$ **15.** $\{6\}$ **16.** $\left\{-\dfrac{7}{2}\right\}$ **17.** False **18.** $\{1\}$

19. $\left\{-\dfrac{5}{3}\right\}$ **20.** 52 hours **21.** $\{1\}$ **23.** $\{-2\}$ **25.** $\{-3\}$ **27.** $\left\{\dfrac{3}{2}\right\}$ **29.** $\{-3\}$ **31.** $\{12\}$ **33.** $\{4\}$ **35.** $\{-2\}$ **37.** $\{-5\}$ **39.** $\{-8\}$ **41.** $\{-5\}$ **43.** $\{-21\}$ **45.** $\{-6\}$

47. $\{-8\}$ **49.** $\{3\}$ **51.** $\left\{-\dfrac{7}{2}\right\}$ **53.** $\{7\}$ **55.** $\{-18\}$ **57.** $\left\{\dfrac{5}{3}\right\}$ **59.** $\left\{-\dfrac{27}{16}\right\}$ **61.** $\{2\}$ **63.** $\left\{-\dfrac{3}{4}\right\}$ **65.** $\left\{\dfrac{1}{3}\right\}$ **67.** $\{0\}$ **69.** $\left\{\dfrac{15}{2}\right\}$ **71.** $\left\{\dfrac{1}{2}\right\}$ **73.** $\{-7\}$ **75.** $\left\{\dfrac{7}{2}\right\}$

77. McDonald's: 16 grams; Burger King: 22 grams **79.** width: $\dfrac{13}{3}$ feet or 4 feet 4 inches; length: $\dfrac{32}{3}$ feet or 10 feet 8 inches **81.** $\$12$ **83.** Yes

85. $\left\{\dfrac{51}{7}\right\}$ **87.** $\{2.47\}$ **89.** $\{-5.6\}$ **91.** $\dfrac{20}{3}$ **93.** $-\dfrac{5}{4}$ **95.** The first is an expression because it does not have an equal sign. The second is an equation because it does have an equal sign. In general, an expression contains the sum, difference, product, and/or quotient of terms. An equation may be thought of as the equality of two algebraic expressions. **97.** Adding $2x$ to both sides will lead to the correct solution, but it may be easier to combine like terms on the left side of the equation instead of adding $2x$ to each side. A series of steps leading to the solution would be (1) combining like terms on the left side of the equation; (2) isolating x on the right side by subtracting $5x$ from both sides; (3) isolating the constants by adding 5 to each side; (4) dividing both sides by 4 to obtain $x = 2$.

Section 2.3 Solving Linear Equations Involving Fractions and Decimals; Classifying Equations   **1.** least common denominator   **2.** {10}

3. $\left\{-\dfrac{3}{7}\right\}$ **4.** 70   **5.** {−11}   **6.** $\left\{-\dfrac{10}{3}\right\}$ **7.** 100   **8.** {100}   **9.** {50}   **10.** 1.25   **11.** {50}   **12.** {160}   **13.** {4}   **14.** {3000}   **15.** conditional equation   **16.** contradiction; identity

17. True   **18.** ∅ or { }   **19.** all real numbers   **20.** all real numbers   **21.** ∅ or { }   **22.** all real numbers; identity   **23.** {4}; conditional   **24.** ∅ or { }; contradiction

25. ∅ or { }; contradiction   **26.** $500   **27.** {−6}   **29.** {6}   **31.** {−2}   **33.** $\left\{\dfrac{9}{2}\right\}$ **35.** $\left\{\dfrac{2}{3}\right\}$ **37.** $\left\{-\dfrac{2}{5}\right\}$ **39.** {−20}   **41.** {1}   **43.** {30}   **45.** {−4}   **47.** {50}   **49.** {4.8}

51. {150}   **53.** {−12}   **55.** {60}   **57.** {5}   **59.** {2}   **61.** {75}   **63.** ∅ or { }; contradiction   **65.** all real numbers; identity   **67.** $\left\{-\dfrac{1}{2}\right\}$; conditional equation

69. ∅ or { }; contradiction   **71.** ∅ or { }; contradiction   **73.** all real numbers; identity   **75.** $\left\{-\dfrac{3}{4}\right\}$ **77.** all real numbers   **79.** $\left\{-\dfrac{1}{3}\right\}$ **81.** {−20}   **83.** $\left\{\dfrac{7}{2}\right\}$

85. ∅ or { }   **87.** {−2}   **89.** {3}   **91.** ∅ or { }   **93.** {−4}   **95.** {39}   **97.** {0}   **99.** $\left\{\dfrac{2}{3}\right\}$ **101.** {−1.70}   **103.** {−13.2}   **105.** $50   **107.** $18,000   **109.** $8.50   **111.** $80

113. 15 quarters   **115.** 6 units   **117.** $12,050   **119.** Answers will vary. A linear equation with one solution is $2x + 5 = 11$. A linear equation with no solution is $2x + 5 = 6 + 2x − 9$. A linear equation that is an identity is $2(x + 5) − 3 = 4x − (2x − 7)$. To form an identity or a contradiction, the variable expressions must be eliminated, leaving either a true (identity) or a false (contradiction) statement.   **121.** The student didn't multiply each term of the equation by 6. Solving using that (incorrect) method gives the second step $4x − 5 = 3x$, and solving for x gives $x = 5$. The correct method to solve the equation is to multiply ALL terms by 6. The correct second step is $4x − 30 = 3x$, producing the correct solution $x = 30$.

Section 2.4 Evaluating Formulas and Solving Formulas for a Variable   **1.** formula   **2.** 59°F   **3.** size 40   **4.** $312.50   **5.** 5936 persons

6. principal; Interest   **7.** $50 interest; $2550 total   **8.** perimeter   **9.** area   **10.** volume   **11.** radius   **12.** True   **13.** 36 square inches   **14. (a)** 187 ft² **(b)** $46.75

15. 28.27 ft²   **16.** The extra large pizza is the better buy.   **17.** $t = \dfrac{d}{r}$ **18.** $l = \dfrac{V}{wh}$ **19.** $C = \dfrac{5}{9}(F − 32)$ **20.** $h = \dfrac{S − 2\pi r^2}{2\pi r}$ **21.** $y = 2x − 6$

22. $y = \dfrac{15 − 4x}{6}$ or $y = -\dfrac{2}{3}x + \dfrac{5}{2}$ **23.** 176 miles   **25.** $104   **27.** $650   **29.** 20°C   **31.** $3   **33. (a)** 50 units **(b)** 144 square units   **35. (a)** 36.2 meters **(b)** 70 square meters   **37. (a)** 36 units **(b)** 81 square units   **39. (a)** 31.4 cm **(b)** 78.5 cm²   **41.** $\dfrac{616}{9}$ in.²   **43.** $r = \dfrac{d}{t}$ **45.** $d = \dfrac{C}{\pi}$ **47.** $t = \dfrac{I}{Pr}$ **49.** $b = \dfrac{2A}{h}$

51. $a = P − b − c$ **53.** $r = \dfrac{A − P}{Pt}$ **55.** $b = \dfrac{2A}{h} − B$ **57.** $y = -3x + 12$ **59.** $y = 2x − 5$ **61.** $y = \dfrac{-4x + 13}{3}$ or $y = -\dfrac{4}{3}x + \dfrac{13}{3}$ **63.** $y = 3x − 12$

65. (a) $C = R − P$ **(b)** $450   **67. (a)** $r = \dfrac{I}{Pt}$ **(b)** 3%   **69. (a)** $x = Z\sigma + \sigma$ **(b)** 130   **71. (a)** $m = \dfrac{y − 5}{x}$ **(b)** −2   **73. (a)** $r = \dfrac{A − P}{Pt}$ **(b)** 0.04 or 4%

75. (a) $h = \dfrac{V}{\pi r^2}$ **(b)** 5 mm   **77. (a)** $b = \dfrac{2A}{h}$ **(b)** 18 ft   **79.** 1834.05   **81. (a)** $h = \dfrac{V}{\pi r^2}$ **(b)** 10 inches   **83.** medium (12")   **85. (a)** 12 hours **(b)** $336

87. 50.5 square inches   **89.** 96π cm³ ≈ 301.59 cm³   **91. (a)** $I = 4(T − 4867.5) + 35,350$ **(b)** $75,000   **93. (a)** 62 tiles **(b)** $372 **(c)** Yes   **95. (a)** 4948 ft² **(b)** $1237   **97. (a)** 1080 in.² **(b)** 7.5 ft²   **99.** Multiply by $\dfrac{1 \text{ ft}^2}{144 \text{ in}^2}$ **101.** Both answers are correct. When solving for y, the first student left the entire numerator over 2. The second student simplified the expression to $y = \dfrac{-x + 6}{2} = \dfrac{-x}{2} + \dfrac{6}{2} = -\dfrac{1}{2}x + 3$.

Putting the Concepts Together (Sections 2.1–2.4)   **1. (a)** Yes **(b)** No   **2. (a)** No **(b)** Yes   **3.** $\left\{-\dfrac{2}{3}\right\}$ **4.** {−40}   **5.** {−6}   **6.** {3}   **7.** {−6}   **8.** $\left\{-\dfrac{7}{3}\right\}$

9. $\left\{-\dfrac{57}{2}\right\}$ **10.** {25}   **11.** {6}   **12.** $\left\{\dfrac{74}{5}\right\}$ or {14.8}   **13.** ∅ or { }   **14.** all real numbers   **15.** $5000   **16. (a)** $b = \dfrac{2A}{h} − B$ **(b)** 6 in.   **17. (a)** $h = \dfrac{V}{\pi r^2}$ **(b)** 13 in.

18. $y = -\dfrac{3}{2}x + 7$

Section 2.5 Problem Solving: Direct Translation   **1.** $5 + 17$ **2.** $7 − 4$ **3.** $\dfrac{25}{3}$ **4.** $-2 \cdot 6$ **5.** $2a − 2$ **6.** $5(m − 6)$ **7.** $z + 50$ **8.** $x − 15$ **9.** $75 − d$

10. $3l − 2$ **11.** $2q + 3$ **12.** $3b − 5$ **13.** equations   **14.** $3y = 21$ **15.** $x − 10 = \dfrac{x}{2}$ **16.** $3(n + 2) = 15$ **17.** $3n + 2 = 15$ **18.** mathematical modeling

19. $s + \dfrac{2}{3}s = 15$, where s represents the amount Sean pays; Sean pays $9 and Connor pays $6   **20.** False   **21.** $n + (n + 2) + (n + 4) = 270$; 88, 90, 92

22. $n + (n + 2) + (n + 4) + (n + 6) = 72$; 15, 17, 19, 21   **23.** $76 = x + (x + 24) + \dfrac{1}{2}(x + 24)$, where $x =$ length of smallest piece of ribbon; 16 inches, 40 inches, 20 inches   **24.** $x + 2x = 18,000$, where $x =$ amount invested in stocks; $6000 in stocks; $12,000 in bonds   **25.** 150 miles   **26.** 100 minutes

27. $5 + x$ **29.** $x\left(\dfrac{2}{3}\right)$ or $\dfrac{2}{3}x$ **31.** $\dfrac{1}{2}x$ **33.** $x − (−25)$ **35.** $\dfrac{x}{3}$ **37.** $x + \dfrac{1}{2}$ **39.** $6x + 9$ **41.** $2(13.7 + x)$ **43.** $2x + 31$ **45.** Braves: r; Clippers: $r + 5$ **47.** Bill's amount: b; Jan's amount: $b + 0.55$ **49.** Janet's share: j; Kathy's share: $200 − j$ **51.** number adults: a; number children: $1433 − a$ **53.** $x + 15 = −34$

55. $35 = 3x − 7$ **57.** $\dfrac{x}{−4} + 5 = 36$ **59.** $2(x + 6) = x + 3$ **61.** 83   **63.** −14   **65.** 54, 55, 56   **67.** Verrazano-Narrows Bridge; 4260 ft; Golden Gate Bridge: 4200 ft   **69.** $11,215   **71.** CDs: $8500; bonds: $11,500   **73.** Stocks: $20,000; bonds: $12,000   **75.** Smart Start: 2 g; Go Lean: 8 g   **77.** $29,140   **79.** 150 miles   **81.** 2000 pages   **83.** Jensen: $36,221; Maureen: $35,972   **85.** 94   **87.** Answers may vary.   **89.** Answers may vary.   **91.** 20°, 40°, 120°   **93.** The process of taking a verbal description of a problem and developing a mathematical equation that can be used to solve the problem is mathematical modeling. We often make assumptions to make the mathematics more manageable in the mathematical models.   **95.** Answers may vary. Both students are correct; however, the value for n will not be the same for both students.

Section 2.6 Problem Solving: Problems Involving Percent   **1.** 100   **2.** True   **3.** 54   **4.** 2.4   **5.** 36   **6.** 3.5   **7.** 40%   **8.** 37.5%   **9.** 110%   **10.** 50   **11.** 80   **12.** 75   **13.** 40,000,000   **14.** $7300   **15.** $60,475   **16.** $3.90 per gallon.   **17.** $659   **18.** $151,020   **19.** 80   **21.** 14   **23.** 4.8   **25.** 120   **27.** 72   **29.** 20   **31.** 40%   **33.** 7.5%

35. 200% **37.** $54 **39.** $102,000 **41.** $25,000 **43.** $2240 **45.** $490 **47.** $400 **49.** winner: 530; loser: 318 **51.** $285,700 **53.** 1,820,000 **55.** 33.3 million
57. 17.2%; 19.4% **59.** $24 **61.** 28.6% **63. (a)** $31,314.38 **(b)** 43.7% **65.** 2.1% **67.** The equation should be $x + 0.05x = 12.81$. Jack's new hourly wage is a percentage of his current hourly wage, so multiplying 5% by his original wage gives his hourly raise. His current hourly wage is $12.20.

Section 2.7 Problem Solving: Geometry and Uniform Motion
1. 90 **2.** 39° and 51° **3.** 70° and 110° **4.** 180 **5.** 35°, 40°, and 105° **6.** True **7.** False
8. width is 1.5 feet; length is 3 feet **9.** 5 feet **10.** False **11.** José's speed is 13 mph and Luis' speed is 8 mph. **12.** It takes $\frac{1}{2}$ hour to catch up to Tanya.
Each of you has traveled 20 miles. **13.** 45° and 135° **15.** 44°, 46° **17.** 42°, 44°, 94° **19.** 58°, 60°, 62° **21.** length = 26 feet; width = 18 feet
23. 42 yards by 84 yards **25.** Shorter base: 15 m; longer base: 30 m. **27. (a)** $62t$ **(b)** $68t$ **(c)** $62t + 68t$ **(d)** $62t + 68t = 585$ **29.** $528(t + 10) = 880t$
31. 26.5 inches **33.** 40 ft **35.** 40 ft; 32 ft **37. (a)** length = 11 yards; width = 19 yards **(b)** 209 yd² **39.** 13 hours **41.** fast car: 60 mph; slow car: 48 mph
43. freeway: 4.5 hours; 2-lane: 1.5 hours **45.** 6 mph **47.** 49°, 49°, 82° **49.** $x = 3$ **51.** $x = 15$ **53.** $x = 12$ **55.** Complementary angles are those whose
measures sum to 90° and supplementary angles have measures that sum to 180°. **57.** Answers may vary. The equation $65t + 40t = 115$ is describing
the sum of distances, whereas the equation $65t - 40t = 115$ represents the difference of distances.

Section 2.8 Solving Linear Inequalities in One Variable
1. True **2.**  **3.**

4. **5.** **6.** False **7.** False **8.** $[-3, \infty)$

9. $(-\infty, 12)$ **10.** $(-\infty, 2.5]$ **11.** $(125, \infty)$ **12.** solve **13.** $10 < 15$; Addition Property

14. $\{n \mid n > 3\}$; $(3, \infty)$ **15.** $\{x \mid x < 4\}$; $(-\infty, 4)$

16. $\{n \mid n \le -4\}$; $(-\infty, -4]$ **17.** $\{x \mid x > -1\}$; $(-1, \infty)$

18. $1 < 4$; Multiplication Property **19.** $-3 > -5$; Multiplication Property **20.** $\{k \mid k < 12\}$; $(-\infty, 12)$

21. $\{n \mid n \ge -3\}$; $[-3, \infty)$ **22.** $\{k \mid k < -8\}$; $(-\infty, -8)$

23. $\left\{p \mid p \ge \frac{3}{5}\right\}$; $\left[\frac{3}{5}, \infty\right)$ **24.** $\{x \mid x > 7\}$; $(7, \infty)$

25. $\{n \mid n > -3\}$; $(-3, \infty)$ **26.** $\{x \mid x > -4\}$; $(-4, \infty)$

27. $\{x \mid x \le -6\}$; $(-\infty, -6]$ **28.** $\{x \mid x > 8\}$; $(8, \infty)$

29. $\left\{x \mid x \le \frac{19}{8}\right\}$; $\left(-\infty, \frac{19}{8}\right]$ **30.** True **31.** True

32. $\{x \mid x \ge 3\}$; $[3, \infty)$ **33.** $\varnothing$ or $\{\ \}$

34. $\{x \mid x > 4\}$; $(4, \infty)$ **35.** $\{x \mid x$ is any real number$\}$; $(-\infty, \infty)$

36. at most 20 boxes **37.** $(4, \infty)$ **39.** $(-\infty, -1]$ **41.** $[-3, \infty)$

43. $(-\infty, 4)$ **45.** $(-\infty, 3)$ **47.** $\varnothing$ or $\{\ \}$ **49.** $(-\infty, \infty)$ **51.** $<$; Addition Property of Inequality

53. $>$; Multiplication Property of Inequality **55.** $\le$; Addition Property of Inequality **57.** $\le$; Multiplication Property of Inequality

59. $\{x \mid x < 4\}$; $(-\infty, 4)$ **61.** $\{x \mid x \ge 2\}$; $[2, \infty)$ **63.** $\{x \mid x \le 5\}$; $(-\infty, 5]$

65. $\{x \mid x > -7\}$; $(-7, \infty)$ **67.** $\{x \mid x > 3\}$; $(3, \infty)$ **69.** $\{x \mid x \ge 2\}$; $[2, \infty)$

71. $\{x \mid x \ge -1\}$; $[-1, \infty)$ **73.** $\{x \mid x > -7\}$; $(-7, \infty)$ **75.** $\{x \mid x \le 0\}$; $(-\infty, 0]$

77. $\{x \mid x < -20\}$; $(-\infty, -20)$ **79.** $\varnothing$ or $\{\ \}$

81. $\{n \mid n$ is any real number$\}$; $(-\infty, \infty)$ **83.** $\{n \mid n > 5\}$; $(5, \infty)$

85. $\{w \mid w$ is any real number$\}$; $(-\infty, \infty)$ **87.** $\left\{y \mid y < -\frac{3}{2}\right\}$; $\left(-\infty, -\frac{3}{2}\right)$

89. $x \geq 16{,}000$ **91.** $x \leq 20{,}000$ **93.** $x > 12{,}000$ **95.** $x > 0$ **97.** $x \leq 0$ **99.** $\{x \mid x > 4\}$; $(4, \infty)$

101. $\left\{x \mid x < \dfrac{3}{4}\right\}$; $\left(-\infty, \dfrac{3}{4}\right)$ **103.** $\{x \mid x$ is any real number$\}$; $(-\infty, \infty)$

105. $\{a \mid a < -1\}$; $(-\infty, -1)$ **107.** $\{n \mid n$ is any real number$\}$; $(-\infty, \infty)$

109. $\left\{x \mid x \geq \dfrac{4}{3}\right\}$; $\left[\dfrac{4}{3}, \infty\right)$ **111.** $\{x \mid x < 25\}$; $(-\infty, 25)$ **113.** $\varnothing$ or $\{\ \}$

115. $\{x \mid x > 5.9375\}$; $(5.9375, \infty)$ **117.** at most 1250 miles **119.** at least 32 **121.** more than 400 minutes

123. greater than \$50,361.11 **125.** at least 74 **127.** $\{x \mid -33 < x < -14\}$ **129.** $\{x \mid -4 \leq x \leq 6\}$ **131.** $\{x \mid -2 \leq x < 9\}$ **133.** $\{x \mid -8 \leq x < 4\}$

135. A left parenthesis is used to indicate that the solution is greater than a number. A left bracket is used to show that the solution is greater than or equal to a given number. **137.** In solving an inequality, when the variables are eliminated and a true statement results, the solution is all real numbers. In solving an inequality, when the variables are eliminated and a false statement results, the solution is the empty set.

Chapter 2 Review **1.** No **2.** No **3.** No **4.** Yes **5.** $\{16\}$ **6.** $\{20\}$ **7.** $\{-16\}$ **8.** $\{-7\}$ **9.** $\{-95\}$ **10.** $\{50\}$ **11.** $\{24\}$ **12.** $\{80\}$ **13.** $\{-6\}$ **14.** $\{5\}$

15. $\left\{-\dfrac{1}{3}\right\}$ **16.** $\left\{\dfrac{3}{8}\right\}$ **17.** $\{4\}$ **18.** $\{5\}$ **19.** \$20,100 **20.** \$2.55 **21.** $\{-4\}$ **22.** $\{4\}$ **23.** $\{9\}$ **24.** $\{-21\}$ **25.** $\{-4\}$ **26.** $\{-6\}$ **27.** $\{4\}$ **28.** $\{-5\}$ **29.** $\{6\}$ **30.** $\{-6\}$

31. $\left\{\dfrac{4}{3}\right\}$ **32.** $\{5\}$ **33.** $\{2\}$ **34.** $\{7\}$ **35.** 14 **36.** width = 19 yards; length = 29 yards **37.** $\left\{-\dfrac{35}{12}\right\}$ **38.** $\left\{-\dfrac{62}{3}\right\}$ **39.** $\{-2\}$ **40.** $\left\{\dfrac{6}{5}\right\}$ **41.** $\left\{\dfrac{3}{2}\right\}$ **42.** $\{3\}$

43. $\{4\}$ **44.** $\{-1\}$ **45.** $\left\{-\dfrac{7}{2}\right\}$ **46.** $\{-3\}$ **47.** $\{58\}$ **48.** $\{-10.5\}$ **49.** $\varnothing$ or $\{\ \}$; contradiction **50.** $\varnothing$ or $\{\ \}$; contradiction **51.** $\{0\}$; conditional equation

52. $\{0\}$; conditional equation **53.** all real numbers; identity **54.** all real numbers; identity **55.** \$15.75 **56.** 3 dimes **57.** 48 in.² **58.** 64 cm

59. $\dfrac{3}{2}$ yards **60.** 15 mm **61.** $H = \dfrac{V}{LW}$ **62.** $P = \dfrac{I}{rt}$ **63.** $W = \dfrac{S - 2LH}{2L + 2H}$ **64.** $M = \dfrac{\rho - mv}{V}$ **65.** $y = \dfrac{-2x + 10}{3}$ or $y = -\dfrac{2}{3}x + \dfrac{10}{3}$

66. $y = 2x - 5$ **67. (a)** $P = \dfrac{A}{(1 + r)^t}$ **(b)** \$2238.65 **68. (a)** $h = \dfrac{A - 2\pi r^2}{2\pi r}$ **(b)** 5 cm **69.** \$11.25 **70.** $\dfrac{9}{4}\pi$ ft² ≈ 7.1 ft² **71.** $x - 6$

72. $x - 8$ **73.** $-8x$ **74.** $\dfrac{x}{10}$ **75.** $2(6 + x)$ **76.** $4(5 - x)$ **77.** $6 + x = 2x + 5$ **78.** $6x - 10 = 2x + 1$ **79.** $x - 8 = \dfrac{1}{2}x$ **80.** $\dfrac{6}{x} = 10 + x$

81. $4(2x + 8) = 16$ **82.** $5(2x - 8) = -24$ **83.** Sarah's age: s; Jacob's age: $s + 7$ **84.** Consuelo's speed: c; Jose's speed: $2c$

85. Max's amount: m; Irene's amount: $m - 6$ **86.** Victor's amount: v; Larry's amount: $350 - v$ **87.** 153 pounds **88.** 12, 13, 14

89. Juan: \$11,000; Roberto: \$9000 **90.** 100 miles **91.** 5.2 **92.** 60 **93.** 13% **94.** 50 **95.** \$18.50 **96.** \$30 **97.** \$40 **98.** \$200 **99.** \$125,000

100. winner: 500 votes; loser: 400 votes **101.** 80°, 10° **102.** 100°, 80° **103.** 30°, 60°, 90° **104.** 50°, 45°, 85° **105.** length = 31 in.; width = 8 in.

106. length = 28 cm; width = 7 cm **107. (a)** length = 20 ft; width = 40 ft **(b)** 800 ft² **108.** 50 ft; 40 ft **109.** 5 hours **110.** 65 mph

111. **112.** **113.** **114.**

115. **116.** **117.** $(-\infty, -4)$ **118.** $[7, \infty)$ **119.** $[2, \infty)$ **120.** $(-\infty, 3)$

121. $\left\{x \mid x < -\dfrac{13}{2}\right\}$; $\left(-\infty, -\dfrac{13}{2}\right)$ **122.** $\left\{x \mid x \geq -\dfrac{7}{3}\right\}$; $\left[-\dfrac{7}{3}, \infty\right)$

123. $\left\{x \mid x \geq -\dfrac{4}{5}\right\}$; $\left[-\dfrac{4}{5}, \infty\right)$ **124.** $\{x \mid x > -12\}$; $(-12, \infty)$

125. $\varnothing$ or $\{\ \}$ **126.** $\{x \mid x$ is any real number$\}$; $(-\infty, \infty)$

127. $\{x \mid x > 3\}$; $(3, \infty)$ **128.** $\left\{x \mid x < -\dfrac{38}{5}\right\}$; $\left(-\infty, -\dfrac{38}{5}\right)$. **129.** at most 65 miles

130. more than 150

Chapter 2 Test **1.** $\{-17\}$ **2.** $\left\{-\dfrac{4}{9}\right\}$ **3.** $\{4\}$ **4.** $\left\{\dfrac{10}{13}\right\}$ **5.** $\left\{\dfrac{5}{8}\right\}$ **6.** $\{5\}$ **7.** $\varnothing$ or $\{\ \}$ **8.** all real numbers **9. (a)** $l = \dfrac{V}{wh}$ **(b)** 9 in.

10. (a) $y = -\dfrac{2}{3}x + 4$ **(b)** $y = -\dfrac{4}{3}$ **11.** $6(x - 8) = 2x - 5$ **12.** 60 **13.** 15, 16, 17 **14.** 10 in., 24 in., 26 in. **15.** 3.5 hours

16. shorter piece is 5 feet; longer is 16 feet **17.** \$36 **18.** $\{x \mid x \leq 6\}$; $(-\infty, 6]$ **19.** $\left\{x \mid x > \dfrac{5}{4}\right\}$; $\left(\dfrac{5}{4}, \infty\right)$

20. at most 200 minutes

Chapter 3 Introduction to Graphing and Equations of Lines

Section 3.1 The Rectangular Coordinate System and Equations in Two Variables

1. x-axis; y-axis; origin **2.** x-coordinate; y-coordinate **3.** False **4.** False

5. $\left(-\dfrac{3}{2}, \dfrac{5}{2}\right)$

(a) I (b) III (c) IV (d) x-axis (e) y-axis (f) II

6.

(a) II (b) I (c) III (d) x-axis (e) y-axis (f) IV

7. (a) $(2, 3)$ (b) $(1, -3)$ (c) $(-3, 0)$ (d) $(-2, -1)$ (e) $(0, 2)$ **8.** True **9.** (a) Yes (b) No (c) No

10. (a) No (b) Yes (c) Yes **11.** $(3, 4)$ **12.** $(-3, 1)$ **13.** $\left(\dfrac{1}{2}, -\dfrac{2}{3}\right)$

14.

x	y	(x, y)
-2	-12	$(-2, -12)$
0	-2	$(0, -2)$
1	3	$(1, 3)$

15.

x	y	(x, y)
-1	7	$(-1, 7)$
2	-2	$(2, -2)$
5	-11	$(5, -11)$

16.

x	y	(x, y)
-5	2	$(-5, 2)$
-2	-4	$(-2, -4)$
2	-12	$(2, -12)$

17.

x	y	(x, y)
-6	-6	$(-6, -6)$
-1	-4	$(-1, -4)$
2	$-\dfrac{14}{5}$	$\left(2, -\dfrac{14}{5}\right)$

18. (a)

x therms	50 therms	100 therms	150 therms
C ($)	$54.34	$96.68	$140.01

$(50, 53.34)$; $(100, 96.68)$; $(150, 140.01)$

(b)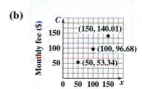

19.

Quadrant I: B
Quadrant II: A, E
Quadrant III: C
Quadrant IV: D, F

21.

Quadrant I: C, E
Quadrant III: F
Quadrant IV: B
x-axis: A, G;
y-axis: D, G

23.

Positive x-axis: A
Negative x-axis: D
Positive y-axis: C
Negative y-axis: B

25. $A(4, 0)$: x-axis; $B(-3, 2)$: quadrant II; $C(1, -4)$: quadrant IV; $D(-2, -4)$: quadrant III; $E(3, 5)$: quadrant I; $F(0, -3)$: y-axis

27. A No
B Yes
C Yes

29. A Yes
B No
C Yes

31. A Yes
B No
C Yes

33. $(4, 1)$ **35.** $(5, -1)$ **37.** $(-3, 3)$

39.

x	y	(x, y)
-3	3	$(-3, 3)$
0	0	$(0, 0)$
1	-1	$(1, -1)$

41.

x	y	(x, y)
-2	7	$(-2, 7)$
-1	4	$(-1, 4)$
4	-11	$(4, -11)$

43.

x	y	(x, y)
-1	8	$(-1, 8)$
2	2	$(2, 2)$
3	0	$(3, 0)$

45.

x	y	(x, y)
-4	6	$(-4, 6)$
1	6	$(1, 6)$
12	6	$(12, 6)$

47.

x	y	(x, y)
1	$\dfrac{7}{2}$	$\left(1, \dfrac{7}{2}\right)$
-4	1	$(-4, 1)$
-2	2	$(-2, 2)$

49.

x	y	(x, y)
4	7	$(4, 7)$
-4	3	$(-4, 3)$
-6	2	$(-6, 2)$

51.

x	y	(x, y)
0	-3	$(0, -3)$
-2	0	$(-2, 0)$
2	-6	$(2, -6)$

53. $A(2, -16)$
$B(-3, -1)$
$C\left(-\dfrac{1}{3}, -9\right)$

55. $A(2, -6)$
$B(0, 0)$
$C\left(\dfrac{1}{6}, -\dfrac{1}{2}\right)$

57. $A(4, -8)$
$B(4, -19)$
$C(4, 5)$

59. $A(3, 4)$
$B(-6, -2)$
$C\left(\dfrac{1}{2}, \dfrac{7}{3}\right)$

61. $A\left(-4, -\dfrac{4}{3}\right)$
$B(-2, -1)$
$C\left(-\dfrac{2}{3}, -\dfrac{7}{9}\right)$

63. $A(20, 23)$
$B(-4, -17)$
$C(2.6, -6)$

65. (a) $265,000 (b) $235,000 (c) After 8 years (d) After 3 years, the book value is $255,000.

67. (a) 25.6% (b) 30.01% (c) 34.42% (d) 2055 (e) Answers may vary. The result is not reasonable since it is unlikely the trend will be linear over this time frame.

69.

a	b	(a, b)
2	−8	(2, −8)
0	−4	(0, −4)
−5	6	(−5, 6)

71.

p	q	(p, q)
0	$\frac{10}{3}$	$\left(0, \frac{10}{3}\right)$
$\frac{5}{2}$	0	$\left(\frac{5}{2}, 0\right)$
−10	$\frac{50}{3}$	$\left(-10, \frac{50}{3}\right)$

73. $k = 4$ **75.** $k = 2$
77. $k = \frac{1}{2}$
79. Points may vary; line

81.

x	y	(x, y)
−2	0	(−2, 0)
−1	−3	(−1, −3)
0	−4	(0, −4)
1	−3	(1, −3)
2	0	(2, 0)

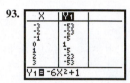

83.

x	y	(x, y)
−2	10	(−2, 10)
−1	3	(−1, 3)
0	2	(0, 2)
1	1	(1, 1)
2	−6	(2, −6)

85. The first quadrant is the upper right-hand corner of the rectangular coordinate system. The quadrants are then II, III, and IV going in a counterclockwise direction. Points in quadrant I have both the *x*- and *y*-coordinates positive; points in quadrant II have a negative *x*-coordinate and a positive *y*-coordinate; points in quadrant III have both the *x*- and *y*-coordinates negative; points in quadrant IV have a positive *x*-coordinate and a negative *y*-coordinate. A point on the *x*-axis has *y*-coordinate equal to zero. A point on the *y*-axis has *x*-coordinate that is equal to zero.

87.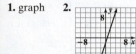

89.

91.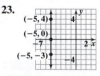

93.

Section 3.2 Graphing Equations in Two Variables

1. graph   **2.**   **3.**   **4.** linear; standard form   **5.** Linear   **6.** Not linear   **7.** Linear   **8.** line   **9.**

10.   **11. (a)** (0, 3000), (10,000, 3800), (25,000, 5000)   **(b)** 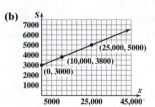   **12.** intercepts

13. Intercepts: (0, 3), (4, 0); *x*-intercept: (4, 0); *y*-intercept: (0, 3)   **14.** Intercept: (0, −2); *y*-intercept: (0, −2); no *x*-intercept   **15.** False

16.   **17.**   **18.**   **19.**    **20.**   **21.** vertical; (*a*, 0)   **23.**
22. horizontal; (0, *b*)

24.   **25.** Linear   **27.** Not linear   **29.** Not linear   **31.** Linear   **33.** $y = 2x$ **35.** $y = 4x - 2$

37. $y = -2x + 5$ **39.** $x + y = 5$ **41.** $-2x + y = 6$ **43.** $4x - 2y = -8$ **45.** $x = -4y$ **47.** $y + 7 = 0$

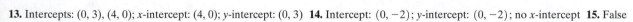

49. $y - 2 = 3(x + 1)$ **51.** $(0, -5), (5, 0)$ **53.** $(0, 4), (2, 0)$ **55.** $(0, -3)$ **57.** $(-5, 0)$ **59.** $(0, -4), (-6, 0)$

61. $(0, 0)$ **63.** $(0, -5), (5, 0)$ **65.** $(0, 8), (6, 0)$ **67.** $(4, 0)$ **69.** $(0, -2)$

71. $3x + 6y = 18$ **73.** $-x + 5y = 15$ **75.** $\frac{1}{2}x = y + 3$ **77.** $9x - 2y = 0$ **79.** $y = -\frac{1}{2}x + 3$ **81.** $\frac{1}{3}y + 2 = 2x$

               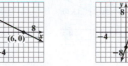

83. $\frac{x}{2} + \frac{y}{3} = 1$ **85.** $4y - 2x + 1 = 0$ **87.** $x = 5$ **89.** $y = -6$ **91.** $y - 12 = 0$ **93.** $3x - 5 = 0$

95. $y = 2x - 5$ **97.** $y = -5$ **99.** $2x + 5y = -20$ **101.** $2x = -6y + 4$ **103.** $x - 3 = 0$ **105.** $3y - 12 = 0$

      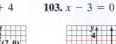   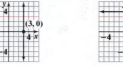   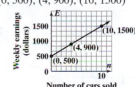

107. $y = 2$ **109.** $x = 7$ **111.** $y = 5$ **113.** $x = -2$ **115.** $y = 4$ **123. (a)** $(0, 500), (4, 900), (10, 1500)$

117. $x = -9$ **(b)**

119. $x = 2y$

121. $y = x + 2$

      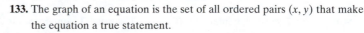

(c) If she sells 0 cars, her weekly
earnings are $500.

125. The "steepness" of the lines
is the same.

127. The lines get more steep
as the coefficient of x gets
larger.

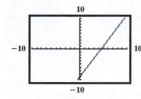

129. $(0, -6), (-2, 0), (3, 0)$

131. $(0, 14), (-3, 0), (2, 0), (5, 0)$

133. The graph of an equation is the set of all ordered pairs (x, y) that make
the equation a true statement.

135. Two points are needed to graph a line. A third point is used to verify
your results.

137. $y = 2x - 9$ **139.** $y + 2x = 13$ or $y = -2x + 13$ **141.** $y = -6x^2 + 1$

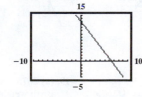

Section 3.3 Slope   **5.**   **6.**   **7.** 0; undefined

1. $\frac{3}{5}$

2. False

3. True

4. positive

8. Slope undefined; when y increases
by 1, there is no change in x.

   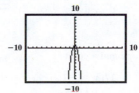

$m = 2$; y increases by 2
when x increases by 1

$m = -\frac{9}{5}$; y decreases by 9 when x increases by 5,
or y increases by 9 when x decreases by 5.

9. $m = 0$; there is no change in y when x increases by 1 unit.

10. (a) **(b)** 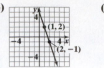 **(c)**

11. 8%   **12.** $m = 0.12$; between 10,000 and 14,000 miles driven, the average annual cost of operating a Chevy Cobalt is $0.12 per mile.

13. $-\dfrac{3}{2}$ **15.** $\dfrac{1}{2}$ **17.** $-\dfrac{2}{3}$

19. (a), (b)   **(c)** $m = \dfrac{1}{2}$; for every 2-unit increase in x, there is a 1-unit increase in y.

21. (a), (b) 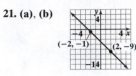   **(c)** $m = -2$; the value of y decreases by 2 when x increases by 1.

23. $m = -2$; y decreases by 2 when x increases by 1.   **25.** $m = -1$; y decreases by 1 when x increases by 1.   **27.** $m = -\dfrac{5}{3}$; y decreases by 5 when x increases by 3.   **29.** $m = \dfrac{3}{2}$; y increases by 3 when x increases by 2.   **31.** $m = \dfrac{2}{3}$; y increases by 2 when x increases by 3.   **33.** $m = \dfrac{1}{3}$; y increases by 1 when x increases by 3.   **35.** $m = 2$; y increases by 2 when x increases by 1.   **37.** m is undefined.   **39.** $m = 0$ **41.** $m = 0$; the line is horizontal, so there is no change in y when x increases by 1.   **43.** slope is undefined; the line is vertical, so there is no change in x when y increases by 1.

45.   **47.**   **49.**   **51.**   **53.**   **55.**

57.   **59.**   **61.**   **63.**   **65.**   **67.** $\dfrac{1}{3}$ **69.** 12 in. or 1 ft

71. 16%   **73.** $m = 2.325$ million; the population was increasing at an average rate of about 2.325 million people per year.

75. Points may vary. $(-2, 1)$, $(0, -5)$; $m = -3$

77. Points may vary. $(-2, -2)$, $(0, 4)$ $m = 3$

79. $m = -2$ **81.** $m = \dfrac{q}{p}$ **83.** $m = \dfrac{6}{a - 6}$ **85.** $MR = 2$; For every hot dog sold, revenue increases by $2.   **87.** The line is a vertical line. Answers will vary, but the points should be of the form (a, y_1) and (a, y_2), where a is a specific value and y_1, y_2 are any two different values. Because the line is a vertical line, the slope is undefined.

Section 3.4 Slope-Intercept Form of a Line

1. $y = mx + b$ **2.** slope: 4, y-intercept: $(0, -3)$ **3.** slope: -3, y-intercept: $(0, 7)$ **4.** slope: $-\dfrac{2}{5}$, y-intercept: $(0, 3)$ **5.** slope: 0; y-intercept: $(0, 8)$

6. slope: undefined; y-intercept: none

7.   **8.**   **9.**   **10.**   **11.**   **12.**   **13.**

14. point-plotting, using intercepts, using slope and a point

15. $y = 3x - 2$ **16.** $y = -\dfrac{1}{4}x + 3$ **17.** $y = -1$ **18.** $y = x$

19. (a) 2075 grams   **(b)** 2933 grams   **(c)** slope $= 143$. Birth weight increases by 143 grams for each additional week of pregnancy.   **(d)** A gestation period of 0 weeks does not make sense.   **(e)**

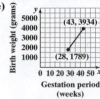

20. (a) $y = 0.38x + 50$ **(b)** $78.50 **(c)** 90 miles **(d)**

21. $m = 5$; y-intercept $= (0, 2)$ **23.** $m = 1$; y-intercept $= (0, -9)$

25. $m = -10$; y-intercept $= (0, 7)$ **27.** $m = -1$; y-intercept $= (0, -9)$

29. $m = -2$; y-intercept $= (0, 4)$ **31.** $m = -\dfrac{2}{3}$; y-intercept $= (0, 8)$

33. $m = \dfrac{5}{3}$; y-intercept $= (0, -3)$ **35.** $m = \dfrac{1}{2}$; y-intercept $= \left(0, -\dfrac{5}{2}\right)$ **37.** $m = 0$; y-intercept $= (0, -5)$ **39.** m is undefined; no y-intercept

41. 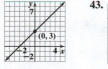 **43.**   **45.** 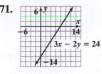 **47.**   **49.**   **51.**   **53.**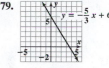

55. 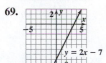 **57.** $y = -x + 8$ **59.** $y = \dfrac{6}{7}x - 6$ **61.** $y = -\dfrac{1}{3}x + \dfrac{2}{3}$ **63.** $x = -5$ **65.** $y = 3$ **67.** $y = 5x$

69. 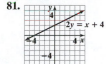 **71.**   **73.**   **75.**   **77.**   **79.**

81.   **83.**   **85.** 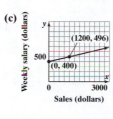 **87.** 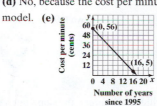 **89.**   **91.**

93. (a) $y = 0.08x + 400$ **(b)** $496 **(c)**

95. (a) 43.25 cents **(b)** 2007 **(c)** Each year, the cost per minute declines 3.1875 cents.
(d) No, because the cost per minute would eventually become negative using this model. **(e)**

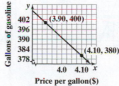

97. $B = -4$ **99.** $A = -4$ **101.** $B = -3$
103. (a) $y = 40x + 4000$ **(b)** $24,000$
(c) 375 calculators **(d)**

105. The line has a positive slope with a negative y-intercept. Possible equations are (a) and (e).

Section 3.5 Point-Slope Form of a Line

1. $y - y_1 = m(x - x_1)$ **2.** True

3. $y = 3x - 5$ **4.** $y = \dfrac{1}{3}x - 5$ **5.** $y = -4x - 3$ **6.** $y = -\dfrac{5}{2}x - 5$ **7.** $y = 3$ **8.** $y = x + 2$ **9.** $y = -3x + 1$

10. $x = 3$

11. Horizontal line: $y = b$;
Vertical line: $x = a$;
Point-slope: $y - y_1 = m(x - x_1)$;
Slope-intercept: $y = mx + b$;
Standard form: $Ax + By = C$

12. (a) $y = -100x + 790$ **(b)** 390 gallons
(c) The number of gallons of gasoline sold will decrease by 100 if the price per gallon increases by $1.

13. $y = 3x - 1$

15. $y = -2x$

17. $y = \frac{1}{4}x - 3$

19. $y = -6x + 13$

21. $y = -7$

23. $x = -4$

25. $y = \frac{2}{3}x + 2$

27. $y = -\frac{3}{4}x$

29. $x = -3$ **31.** $y = -5$ **33.** $y = -4.3$ **35.** $x = \frac{1}{2}$ **37.** $y = 2x + 4$ **39.** $y = -4x + 6$

41. $y = -\frac{3}{2}x - \frac{5}{2}$ **43.** $y = 2x - 5$ **45.** $y = -3$ **47.** $x = 2$ **49.** $y = 0.25x + 0.575$

51. $y = x - \frac{11}{4}$

53. $y = 5x - 22$

55. $y = 5$

57. $y = x + 2$

59. $y = \frac{1}{2}x + 4$

61. $x = 5$

63. $y = -7x + 2$

65. $y = -\frac{2}{3}x + 7$

67. $y = -\frac{3}{2}x$

69. $y = \frac{2}{5}x - 2$

71. (a) When 60 packages are shipped, the expenses are \$1635.

(b)

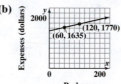

(c) $y = \frac{9}{4}x + 1500$
(d) \$1950
(e) Expenses increase by \$2.25 for each additional package.

73. (a) (600, 15); (750, 6)

(b)

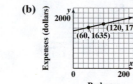

(c) $y = -0.06x + 51$ **(d)** 9% **(e)** For each 1-point increase in credit score, interest rate goes down 0.06%. **75.** $y = 3x + 14$ **77.** $y = -2x - 2$ **79.** $y = 6x - \frac{7}{2}$

81. $y = -x - 7$ **83.** $y = \frac{1}{5}x - 2$ **85.** $y = -\frac{3}{2}x + 2$

87. Yes. Although the point-slope form of the line will look different, the slope-intercept form of the line will be the same.

Section 3.6 Parallel and Perpendicular Lines

1. slopes; y-intercepts; x-intercepts

2. Not parallel **3.** Parallel **4.** Not parallel **5.** $y = 2x - 1$

6. $y = -\frac{3}{2}x$ **7.** $x = 3$ **8.** $y = 5$

9. perpendicular **10.** False **11.** $\frac{1}{4}$ **12.** $-\frac{4}{5}$ **13.** 5

14. Perpendicular **15.** Not perpendicular **16.** Perpendicular **17.** $y = -\frac{1}{2}x$ **18.** $y = \frac{3}{2}x + 2$ **19.** $y = -5$

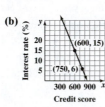

20. $x = 3$

	Slope of the Given Line	Slope of a Line Parallel to the Given Line	Slope of a Line Perpendicular to the Given Line
21.	$m = -3$	-3	$\frac{1}{3}$
23.	$m = \frac{1}{2}$	$\frac{1}{2}$	-2
25.	$m = -\frac{4}{9}$	$-\frac{4}{9}$	$\frac{9}{4}$
27.	$m = 0$	0	undefined

29. perpendicular **31.** parallel **33.** perpendicular **35.** perpendicular **37.** neither **39.** parallel **41.** parallel

43. $y = 3x - 14$ **45.** $y = -4x - 4$ **47.** $y = -7$ **49.** $x = -1$ **51.** $y = \frac{3}{2}x - 13$ **53.** $y = -\frac{1}{2}x - \frac{21}{2}$ **55.** $y = -2x + 11$

57. $y = \frac{1}{4}x$ **59.** $x = -2$ **61.** $y = 5$ **63.** $y = \frac{5}{2}x$ **65.** $y = -\frac{3}{5}x - 9$

67. $y = 7x - 26$ **69.** $y = \frac{1}{5}x + \frac{43}{5}$ **71.** $y = -7x + 41$ **73.** $y = -2x - 10$ **75.** $y = 3x - 2$ **77.** $x = 5$

79. $y = -\frac{4}{3}x + 2$ **81.** $y = -2x - 5$ **83. (a)** $m_1 = 3, m_2 = -3$ **(b)** neither **85. (a)** $m_1 = 2, m_2 = 2$ **(b)** parallel

87. (a) $m_1 = \frac{1}{2}, m_2 = \frac{1}{2}$ **(b)** parallel **89. (a)** $m_1 = \frac{5}{3}, m_2 = -\frac{3}{5}$ **(b)** perpendicular

91.   **93.**   **95.**   **97.**   **99.** $B = 2$ **101.** $A = 4$ **103.**

parallelogram rectangle right triangle right triangle not an altitude

105. L_1 could be parallel to L_2. L_1 could be perpendicular to L_2. L_1 and L_2 could intersect, but not at right angles. L_1 could be the same as, or coincident with, L_2.

Putting the Concepts Together (Sections 3.1–3.6)

1. yes. $(1, -2)$ is a solution **2.** 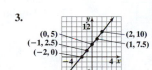 **3.**  **4. (a)** x-intercept is $\left(-\frac{3}{4}, 0\right)$ **(b)** y-intercept is $(0, 3)$

5. x-intercept is $\left(\frac{3}{2}, 0\right)$; y-intercept is $(0, 2)$ **6. (a)** slope $= -\frac{2}{3}$ **(b)** y-intercept $= \left(0, -\frac{4}{3}\right)$ **7.** slope $= -\frac{1}{3}$

8. (a) $m = \frac{2}{5}$ **(b)** $m = -\frac{5}{2}$ **9.** The lines are parallel. Answers may vary. **10.** $y = 3x + 1$ **11.** $y = -6x - 2$ **12.** $y = -2x + 7$

13. $y = -\frac{5}{2}x - 20$ **14.** $y = \frac{1}{4}x + 5$ **15.** $y = -8$ **16.** $x = 2$ **17.** $m = 11$. Every package increases expenses by \$11.

18. (a) $y = 8350x - 2302$, where x is the weight in carats, and y is the price in dollars. **(b)** For every 1-carat increase in weight, the cost increases by \$8350.
   (c) \$4044

Section 3.7 Linear Inequalities in Two Variables

1. (a) No **(b)** Yes **(c)** Yes **2. (a)** Yes **(b)** Yes **(c)** No **3.** solid; dashed **4.** half-planes

5. **6.** **7.** **8.** **9.** **10.** **11.**

12. **13.** **14. (a)** $0.2s + 0.25t \leq 2$ **(b)** Yes **(c)** No **15.** A is a solution. **17.** C is a solution.

19. A and C are solutions. **21.** B is a solution. **23.** B is a solution. **25.** A is a solution.

27. **29.** **31.** **33.** **35.** **37.**

39. **41.** **43.** **45.** **47.** **49.**

51. **53.** **55.** **57.** $x + y \geq 26$ **59.** $\dfrac{y}{-2} \leq 4$ **61.** $x \leq y - 3$

63. $x + 3y < 0$ **65.** $2x - \dfrac{1}{2}y \geq 5$ **67.** $-2x > -1$ **69. (a)** $3s + 5a \leq 120$ **(b)** No **(c)** Yes

71. (a) $50x + 40y \geq 400$ **(b)** Yes **(c)** No

73. **75.** **77.**  **79.** If (a, b) satisfies the inequality, then (a, b) is a solution to the linear inequality. If (a, b) satisfies $Ax + By = C$, then (a, b) lies on the boundary line.

81. 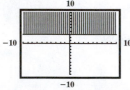 **83.** 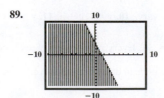 **85.**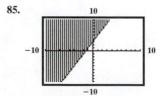

87. 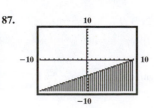 **89.** 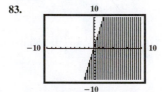 **91.**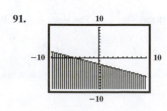

Chapter 3 Review

1–4.

5. $A(1, 4)$; quadrant I; $B(-3, 0)$; x-axis   **6.** $A(0, 1)$; y-axis; $B(-2, 2)$; quadrant II   **7.** A: Yes; B: No

8. A: Yes; B: No   **9. (a)** $\left(4, -\frac{4}{3}\right)$ **(b)** $(-6, 2)$ **10. (a)** $(4, 4)$ **(b)** $(-2, -2)$

$A(3, -2)$ quadrant IV;
$B(-1, -3)$ quadrant III;
$C(-4, 0)$ x-axis; $D(0, 2)$
y-axis

11.

x	y	(x, y)
-2	-8	(-2, -8)
0	-5	(0, -5)
4	1	(4, 1)

12.

x	y	(x, y)
-3	5	(-3, 5)
2	0	(2, 0)
4	-2	(4, -2)

13.

x	y	(x, y)
-18	2	(-18, 2)
-6	-2	(-6, -2)
24	-12	(24, -12)

14.

x	y	(x, y)
-3	-8	(-3, -8)
3	1	(3, 1)
5	4	(5, 4)

15.

x	C	(x, C)
1	7	(1, 7)
2	9	(2, 9)
3	11	(3, 11)

16.

x	E	(x, E)
500	1050	(500, 1050)
1000	1100	(1000, 1100)
2000	1200	(2000, 1200)

17.

18.

19.

20.

21.

p	C	(p, C)
20	80	(20, 80)
50	140	(50, 140)
80	200	(80, 200)

22.

p	F	(p, F)
100	800	(100, 800)
200	1100	(200, 1100)
500	2000	(500, 2000)

23. $(-2, 0), (0, -4)$ **24.** $(0, 1)$ **25.** $(0, 9), (-3, 0)$ **26.** $(0, -6), (3, 0)$ **27.** $(3, 0)$ **28.** $\left(0, -\frac{2}{5}\right), (1, 0)$

29. 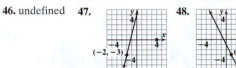   **30.**   **31.**   **32.**   **33.**   **34.**

35.   **36.**   **37.** $\frac{4}{3}$ **38.** -2 **39.** -8 **40.** 2 **41.** $\frac{5}{2}$ **42.** $-\frac{1}{6}$ **43.** undefined   **44.** 0 **45.** 0

46. undefined   **47.**   **48.**   **49.**   **50.**

51. $m = 45$; the cost to produce 1 additional bicycle is \$45.   **52.** 5%   **53.** $m = -1$; $b = \frac{1}{2}$ **54.** $m = 1$; $b = -\frac{3}{2}$

55. $m = \frac{3}{4}$; $b = 1$ **56.** $m = -\frac{2}{5}$; $b = \frac{8}{5}$

57.   **58.**   **59.**   **60.**   **61.**   **62.**

63.   **64.**   **65.** $y = -3x + 5$ **66.** $y = \frac{1}{5}x + 10$ **67.** $x = -12$ **68.** $y = -4$ **69.** $y = x - 20$

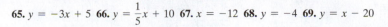

70. $y = -x - 8$ **71. (a)** \$360 **(b)** 7 days **(c)**

72. (a) $(22, 418)$, $(35, 665)$, $m = 19$ **(b)** It costs \$19 more for each additional day. **(c)** $C = 19d$ **(d)** \$152 **73.** $y = 6x - 3$ **74.** $y = -2x + 8$

75. $y = -\frac{1}{2}x + \frac{1}{2}$ **76.** $y = \frac{2}{3}x - \frac{7}{3}$ **77.** $y = -\frac{1}{2}$ **78.** $x = -\frac{4}{7}$

79. $y = \frac{8}{7}x + 8$ **80.** $y = \frac{3}{2}x - 6$ **81.** $y = 3x - 4$ **82.** $y = -\frac{2}{5}x - 5$ **83.** $y = 7$ **84.** $x = 4$ **85.** $A = -4d + 24$ **86.** $F = -\frac{1}{3}m + 15$ **87.** not parallel

88. parallel **89.** $y = -x + 2$ **90.** $y = 2x + 8$ **91.** $y = -3x + 7$ **92.** $y = -\frac{3}{2}x + 1$ **93.** $x = 5$ **94.** $y = -12$ **95.** -2 **96.** $-\frac{9}{4}$ **97.** perpendicular

98. not perpendicular **99.** $y = \frac{1}{3}x + 5$ **100.** $y = -\frac{1}{2}x + 1$ **101.** $y = -\frac{3}{2}x - \frac{3}{2}$ **102.** $y = x + 1$ **103.** A, B are solutions. **104.** C is a solution.

105.   **106.**   **107.**   **108.**   **109.**   **110.**

111.   **112.**   **113.** $0.25x + 0.1y \geq 12$ **114.** $2x - \frac{1}{2}y \leq 10$

Chapter 3 Test **1.** No **2. (a)** $(4, 0)$ **(b)** $\left(0, -\frac{4}{3}\right)$ **3. (a)** $m = \frac{4}{3}$ **(b)** $(0, 8)$

4.   **5.**   **6.** $-\frac{1}{6}$ **7. (a)** $-\frac{3}{2}$ **(b)** $\frac{2}{3}$ **8.** neither; Answers may vary. **9.** $y = -4x - 15$

10. $y = 2x + 14$ **11.** $y = -3x - 11$ **12.** $y = \frac{1}{2}x - 2$ **13.** $y = -\frac{3}{2}x + 8$

14. $y = 5$ **15.** $x = -2$ **16.** \$8 per package

17.   **18.** 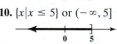 **19.**

Cumulative Reviews Chapters 1–3

1. -16 **2.** $\frac{1}{4}$ **3.** -4 **4.** 8 **5.** $4m^2 + 5m - 9$ **6.** $\{-7\}$ **7.** $\left\{-\frac{25}{12}\right\}$ **8.** $B = \frac{2A - hb}{h}$ or $B = \frac{2A}{h} - b$

9. $\{x \mid x > -4\}$ or $(-4, \infty)$

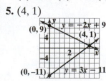

10. $\{x \mid x \leq 5\}$ or $(-\infty, 5]$

11. 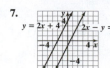 **12.** 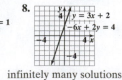 **13.**

14. $y = -\frac{4}{3}x + 2$ **15.** $y = -3x - 8$ **16.**   **17.** Shawn needs to score at least 91 on the final exam to earn an A.

18. A person 62 inches tall would be considered obese if she or he weighed 160 pounds or more. **19.** The angles measure 55° and 125°. **20.** The cylinder should be about 5.96 inches tall. **21.** The three consecutive even integers are 24, 26, and 28.

Chapter 4 Systems of Linear Equations and Inequalities

Section 4.1 Solving Systems of Linear Equations by Graphing **1.** system of linear equations **2.** solution **3. (a)** No **(b)** Yes **(c)** No
4. (a) Yes **(b)** Yes **(c)** Yes

5. $(4, 1)$ **6.** $(-2, 5)$ **7.**   **8.**   **9.** consistent **10.** independent
11. inconsistent **12.** True
no solution; $\{\,\}$ or $\varnothing$ infinitely many solutions **13.** One solution; consistent; independent
14. infinitely many solutions; consistent; dependent **15.** no solution; inconsistent

16. After 100 miles; the cost is $60.

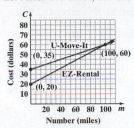

17. (a) No **(b)** Yes **(c)** No

19. (a) No **(b)** Yes **(c)** Yes

21. (a) No **(b)** No **(c)** No

23. $(-5p)$

25. $(4, -1)$

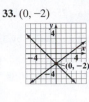

27. $(-1, 1)$

29. $(4, -2)$

31.

no solution;
$\{\}$ or $\varnothing$

33. $(0, -2)$

35.

infinitely many
solutions

37. $(0, 3)$

39. one solution; consistent; independent; $(3, 2)$
41. no solution; inconsistent **43.** infinitely many
solutions; consistent; dependent **45.** one solution;
consistent; independent; $(-1, -2)$ **47.** one solution;
consistent; independent **49.** no solution;
inconsistent **51.** infinitely many solutions; consistent;
dependent **53.** one solution; consistent; independent

55. no solution; inconsistent **57.** infinitely many solutions; consistent; dependent

59.

$(-3, 3)$; consistent;
independent

61.

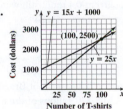

no solution $\{\}$ or $\varnothing$;
inconsistent

63.

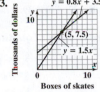

$(2, 4)$; consistent; independent

65.

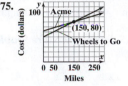

infinitely many solutions;
consistent; dependent

67.

infinitely many solutions;
consistent; dependent

69.

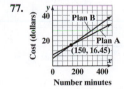

$(-4, 0)$; consistent;
independent

71.

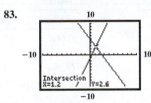

The break-even point is (100, 2500).
San Diego T-shirt Company must sell
100 T-shirts to break even at a cost/
revenue of $2500.
79. $c = 3$ **81.** $c = 1$

73.

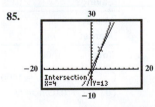

The break-even point is (5, 7.5). The
company needs to sell 5 boxes of
skates to break even at a cost/
revenue of $7500.

75.

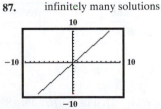

$\begin{cases} y = 62 + 0.12x \\ y = 50 + 0.2x \end{cases}$

The cost for driving 150 miles is
the same for both companies ($80).
Choose Acme to drive 200 miles.

77.

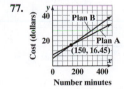

$\begin{cases} y = 8.95 + 0.05x \\ y = 5.95 + 0.07x \end{cases}$

The cost is the same
($16.45) when 150 minutes
are used. Choose Plan B for
100 long-distance minutes.

83.

85.

87. infinitely many solutions

89. The three possibilities for a solution of a system of two linear equations containing two variables are (1) lines intersecting at a single point (one solution); (2) the lines are coincident (a single line in the plane) so an infinite number of
solutions; (3) parallel lines (no solution).

Section 4.2 Solving Systems of Linear Equations Using Substitution
1. (2) **2.** $(2, 4)$ **3.** $(-3, 5)$ **4.** $\left(\dfrac{3}{2}, 7\right)$ **5.** $\left(-\dfrac{4}{3}, \dfrac{1}{2}\right)$ **6.** $\varnothing$ or $\{\}$

7. infinitely many solutions **8.** False **9.** infinitely many solutions **10.** no solution; $\{\}$ or $\varnothing$ **11.** no solution; $\{\}$ or $\varnothing$ **12.** Equilibrium price $= \$2$;

Number of sodas $= 1200$ **13.** $(4, -1)$ **15.** $(-1, 1)$ **17.** $(-3, -4)$ **19.** $(-4, -7)$ **21.** $\left(-\dfrac{2}{3}, 2\right)$ **23.** $\left(2, -\dfrac{1}{3}\right)$ **25.** $\left(\dfrac{1}{4}, \dfrac{1}{6}\right)$ **27.** no solution; $\{\}$ or $\varnothing$; incon-

sistent **29.** infinitely many solutions; consistent and dependent **31.** no solution; inconsistent **33.** infinitely many solutions; consistent and dependent
35. $(-2, -1)$ **37.** no solution; $\{\}$ or $\varnothing$ **39.** infinitely many solutions **41.** $(-3, 5)$ **43.** no solution; $\{\}$ or $\varnothing$ **45.** infinitely many solutions

47. $\left(\dfrac{5}{4}, -\dfrac{7}{4}\right)$ **49.** infinitely many solutions **51.** $\left(-\dfrac{5}{2}, 4\right)$ **53.** length 10 ft; width 7 ft **55.** $8000 in money market; $4000 in international fund

57. $1,000,000 **59.** 5 and 12 **61.** $A = \dfrac{7}{6}$; $B = -\dfrac{1}{2}$ **63.** Answers may vary. **65.** Answers may vary. **67.** A reasonable first step is to multiply the first

equation by 6 to clear fractions and to multiply the second equation by 4 to clear fractions. Then the system would be $\begin{cases} 2x - y = 4 \\ 6x + 2y = -5 \end{cases}$. Solve for y,

because its coefficient is -1. Answers may vary.

Section 4.3 Solving Systems of Linear Equations Using Elimination   **1.** additive inverses   **2.** $(-4, -2)$ **3.** $(8, 3)$ **4.** $(6, -5)$ **5.** $(5, 0)$
6. dependent; infinitely   **7.** no solution; $\{\ \}$ or $\varnothing$ **8.** infinitely many solutions   **9.** infinitely many solutions   **10.** Cheeseburger: \$1.75; shake: \$2.25
11. $(2, -1)$ **13.** $(6, 4)$ **15.** $\left(-\dfrac{11}{4}, \dfrac{1}{2}\right)$ **17.** $(-1, -3)$ **19.** no solution; $\{\ \}$ or $\varnothing$; inconsistent   **21.** infinitely many solutions; consistent, dependent
23. infinitely many solutions; consistent, dependent   **25.** no solution; $\{\ \}$ or $\varnothing$; inconsistent   **27.** $(-5, 8)$ **29.** $\left(\dfrac{3}{2}, -\dfrac{3}{4}\right)$ **31.** no solution; $\{\ \}$ or $\varnothing$
33. $(12, -9)$ **35.** infinitely many solutions   **37.** $\left(\dfrac{6}{29}, -\dfrac{8}{29}\right)$ **39.** $\left(\dfrac{1}{2}, 4\right)$ **41.** $(-1, 2.1)$ **43.** $(-2, -6)$ **45.** $(50, 30)$ **47.** $(1, 5)$ **49.** $(5, 2)$ **51.** $\left(-\dfrac{9}{17}, \dfrac{31}{17}\right)$
53. $\left(\dfrac{8}{3}, -\dfrac{4}{7}\right)$ **55.** infinitely many solutions   **57.** $(1, -3)$ **59.** $(-4, 0)$ **61.** $\left(-\dfrac{8}{5}, \dfrac{27}{10}\right)$ **63.** $(-4, -5)$ **65.** no solution; $\{\ \}$ or $\varnothing$ **67.** $\left(\dfrac{1}{2}, \dfrac{1}{3}\right)$
69. hamburger 280 cal; Coke 210 cal   **71.** tomatoes: 10 hr; zucchini: 15 hr; no   **73.** 60 lb Arabica; 40 lb Robusta   **75.** 70° and 20°   **77.** $\left(-\dfrac{3}{a}, 1\right)$
79. $\left(-3a - b, \dfrac{3}{2}a - \dfrac{3}{2}b\right)$ **81.** It is easier to eliminate y because the signs on y are opposites. Multiply the second equation by 3, to form $6x + 3y = 15$.
Then add the equations, obtaining $7x = 21$. So $x = 3$. Substitute $x = 3$ into either the first equation or the second equation and solve for y, $y = -1$.
83. Both of you will obtain the correct solution. Other strategies are: multiplying equation (1) by one-third and adding; solving equation (2) for y and
substituting this expression into equation (1); and graphing the equations and finding the point of intersection.

Putting the Concepts Together (Section 4.1–4.3)   **1. (a)** No   **(b)** Yes   **(c)** No   **2. (a)** infinitely many   **(b)** consistent   **(c)** dependent   **3. (a)** one
(b) consistent   **(c)** independent   **4.** $(1, 1)$ **5.** $(5, -1)$ **6.** $(1, -5)$ **7.** $(-1, 0)$ **8.** $(-4, 5)$ **9.** $(-3, -1)$ **10.** $\left(-\dfrac{1}{3}, 3\right)$ **11.** infinitely many solutions
12. $(2.5, 3)$ **13.** no solution; $\{\ \}$ or $\varnothing$

Section 4.4 Section Solving Direct Translation, Geometry, and Uniform Motion Problems Using Systems of Linear Equations   **1.** 43 and 61
2. length: 150 yards width: 50 yards;   **3.** True   **4.** 180°   **5.** 36° and 54°   **6.** 49° and 131°   **7.** True   **8.** airspeed: 350 mph; wind resistance: 50 mph
9. $2x$; $12 + \dfrac{1}{2}y$ **11.** $2l + 2w$ **13.** $8(r - c)$; 16   **15.** 33, 49   **17.** 14, 37   **19.** Thursday 35,000; Friday 42,000   **21.** \$12,000 in stocks; \$9000 in bonds
23. length 12 ft; width 9 ft   **25.** length 25 m; width 10 m   **27.** 40°, 50°   **29.** 22.5°, 157.5°   **31.** current 0.4 mph; still water 3.9 mph   **33.** bike 11 mph;
wind 1 mph   **35.** northbound 60 mph; southbound 72 mph   **37.** 4 hr   **39.** wind 25 mph; Piper 175 mph   **41.** 24

Section 4.5 Solving Mixture Problems Using Systems of Linear Equations   **1.** rate; amount

2.

	Number ·	Cost per Person =	Amount
Adults	a	42.95	42.95a
Children	c	15.95	15.95c
Total	9		332.55

3. 45 dimes   **4.** principal; rate; time   **5.** True   **6.** \$40,000 in Aa-rated bond;
\$50,000 in B-rated bond   **7.** 5 pounds of Brazilian coffee and 15 pounds of
Colombian coffee   **8.** Mix 120 gallons of the ice cream with 5% butterfat and
80 gallons of the ice cream with 15% butterfat.

9.

	Number ·	Cost =	Total
Adult	a	4	4a
Student	s	1.5	1.5s
Total	215		580

$\begin{cases} a + s = 215 \\ 4a + 1.5s = 580 \end{cases}$

11.

	P ·	r =	I
Savings	s	0.05	0.05s
Money Market	m	0.03	0.03m
Total	1600		50

$\begin{cases} s + m = 1600 \\ 0.005s + 0.03m = 50 \end{cases}$

13.

	lb ·	Price =	Total
Mild	m	7.5	7.5m
Robust	r	10	10r
Total	12	8.75	8.75(12)

$\begin{cases} m + r = 12 \\ 7.5m + 10r = 8.75(12) \end{cases}$

15. 32a; 24c **17.** 0.1A; 0.07B **19.** 5.85r; 4.20y; 128.85   **21.** 315 student tickets   **23.** 400 flats   **25.** 60 nickels, 90 dimes   **27.** first-class \$0.42; postcard
\$0.27   **29.** \$7500 in 5% account; \$2500 in 8% account   **31.** \$3200 in risky plan; \$1800 in safer plan   **33.** 2 lb arbequina; 3 lb green   **35.** 48.9 lb of the \$2.75
per pound coffee; 51.1 lb of the \$5 per pound coffee   **37.** 80 lb rye; 100 lb bluegrass   **39.** 20 ml 30% saline solution   **41.** 210 liters   **43.** 1.2 gal   **45.** \$6200 at
5%; \$3800 at 7.5% loss   **47.** The percentage of ethanol in the final solution is greater than the percentage of ethanol in either of the two original solutions.

Section 4.6 Systems of Linear Inequalities   **1.** solution   **2. (a)** solution   **(b)** not a solution   **3.** dashed; solid

4.   **5.**   **6.**   **7.**   **8. (a)**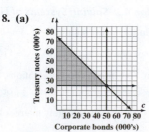

9. (a) Yes   **(b)** No   **(c)** Yes   **11. (a)** Yes   **(b)** No   **(c)** No

13.   **15.**   **17.**   **19.**   **21.** 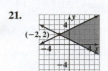   **(b)** Yes   **(c)** No

23. **25.** **27.** **29.** **31.** **33.**

35. **37.** **39. (a)** 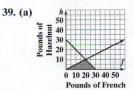 **(b)** No **(c)** Yes **41. (a)** 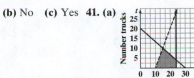 **(b)** No **(c)** No

43. **45.** **47.** $\begin{cases} y \geq 0 \\ y \geq x \\ y \leq \frac{1}{2}x + 3 \end{cases}$

49. Answers may vary. Graph each inequality in the system. The overlapping shaded region represents the solution of the system.

51. (a) $\begin{cases} y \geq x \\ y \geq -2x \end{cases}$ **(b)** $\begin{cases} y \leq x \\ y \geq -2x \end{cases}$ **(c)** $\begin{cases} y \geq x \\ y \leq -2x \end{cases}$ **(d)** $\begin{cases} y \leq x \\ y \leq -2x \end{cases}$ Answers may vary.

no solution; { } or ∅

Chapter 4 Review **1. (a)** No **(b)** Yes **(c)** No **2. (a)** No **(b)** No **(c)** Yes **3. (a)** Yes **(b)** Yes **(c)** Yes **4. (a)** No **(b)** No **(c)** No

5. **6.** **7.** **8.** 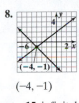 **9.** 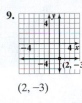 **10.** **11.**

(6, 1) (0, 3) (−4, 4) (−4, −1) (2, −3) (0, 0) no solution

12. **13.** none; inconsistent **14.** none; inconsistent **15.** infinitely many; consistent; dependent

16. infinitely many; consistent; dependent **17.** one; consistent; independent **18.** one; consistent; independent

infinitely many solutions

19. (a) $\begin{cases} y = 70 + 0.1x \\ y = 100 + 0.04x \end{cases}$ **20. (a)** $\begin{cases} y = 50 + 40x \\ y = 350 + 10x \end{cases}$ **21.** (2, 1) **22.** (1, −1) **23.** (7, −2) **24.** (4, −2) **25.** (18, 11)

(b) 500 fliers **(b)** 10 sq yd **26.** (−6, 8) **27.** (2, −2) **28.** (3, −1)

29. infinitely many solutions **30.** no solution; { } or ∅

31. no solution; { } or ∅ **32.** infinitely many solutions

33. $\left(\frac{3}{2}, 1\right)$ **34.** $\left(\frac{1}{3}, -1\right)$ **35.** width 125 m; length 200 m

36. −3 **37.** (0, −12) **38.** (−3, 7) **39.** (−3, 4)

40. (−3, 0) **41.** no solution; { } or ∅

42. infinitely many solutions **43.** (−2, 2)

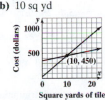

(c) Printer A **(c)** tile

44. (3, −2.5) **45.** $\left(\frac{1}{2}, -2\right)$ **46.** $\left(-\frac{3}{2}, 1\right)$ **47.** infinitely many solutions **48.** $\left(\frac{6}{7}, \frac{5}{2}\right)$ **49.** (−20, −13) **50.** $\left(\frac{1}{3}, \frac{2}{3}\right)$ **51.** no solution; { } or ∅

52. $\left(-\frac{1}{7}, 0\right)$ **53.** 2.4 lb cookies; 1.6 lb chocolates **54.** 400 individual tickets; 325 block tickets **55.** $\frac{3}{8}, \frac{1}{3}$ **56.** 26, 32

57. $32,000 in stocks, $18,000 in bonds **58.** 14 notebooks, 10 calculators **59.** 77.5°, 102.5° **60.** 40°, 50° **61.** 15 m by 22 m **62.** 110°, 40°

63. plane 450 mph, wind 50 mph **64.** current 2 mph, paddling 6 mph **65.** cyclist 10 mph, wind 2 mph **66.** faster 8 mph; slower 6 mph

67.

	Number ·	Value =	Total Value
Dimes	d	0.10	0.10d
Nickels	n	0.05	0.05n
Total	35		2.25

68.

	P ·	r =	I
Savings	s	0.065	0.065s
Mutual Fund	m	0.08	0.08m
Total	15,000		1050

69. 676
70. 4 quarters
71. $7000 at 5%, $12,000 at 9%
72. $10,000 bonds, $15,000 stocks
73. 3 quarts 30% sugar solution

74. 60 pints **75.** 3 lb peanuts; 2 lb almonds **76.** 4 liters 35% acid; 16 liters 60% acid **77. (a)** Yes **(b)** Yes **(c)** No

78. (a) No **(b)** Yes **(c)** Yes **79. (a)** Yes **(b)** No **(c)** No **80. (a)** No **(b)** Yes **(c)** Yes **81. (a)** No **(b)** Yes **(c)** No

82. (a) No **(b)** Yes **(c)** No

83. **84.** **85.** **86.** **87.**

88. **89.** **90.** **91.** **92.**

93. **94.** **95.**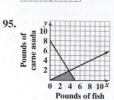

Chapter 4 Test 1. (a) No **(b)** No **(c)** Yes **2. (a)** Yes **(b)** No **(c)** No **3. (a)** one **(b)** consistent **(c)** independent
4. (a) none **(b)** inconsistent **(c)** not applicable **5.**   **6.**  **7.** $(0, -3)$ **8.** $(-12, -1)$ **9.** $(-3, 3)$

$(-3, 2)$ $\qquad$ $(4, 2)$ $\qquad$ **10.** $(-2, -4)$ **11.** $\left(3, -\dfrac{1}{2}\right)$ **12.** $(2.5, 3)$

13. infinitely many solutions
14. airplane 200 mph; wind 25 mph
15. 7 containers vanilla; 4 containers peach
16. 52.5°, 127.5°
17. 15 basketballs, 25 volleyballs

18. **19.** **20.**

Chapter 5 Exponents and Polynomials

Section 5.1 Adding and Subtracting Polynomials 1. monomial **2.** 1 **3.** Yes; Coefficient 12; Degree 6 **4.** No **5.** Yes; Coefficient 10; Degree 0 **6.** No
7. $m + n$ **8.** Yes; Coefficient 3; Degree 7 **9.** No **10.** Yes; Coefficient -1; Degree 3 **11.** True **12.** standard form **13.** Yes; degree 3 **14.** No
15. No **16.** Yes; degree 4 **17.** False **18.** $12x^2 + 3x + 3$ **19.** $2z^4 + 4z^3 - z^2 + z - 1$ **20.** $5x^2y + 6x^2y^2 - xy^2$ **21.** $5x^3 - 15x^2 + 3x + 7$
22. $8y^3 - 3y^2 - 3y - 6$ **23.** $11x^2y + 2x^2y^2 - 9xy^2$ **24.** $a^2b + a^2b^3 + 2ab^2 - 5ab$ **25.** $7a^2 - 4ab - 12b^2$ **26. (a)** 1 **(b)** -214 **(c)** 101 **27. (a)** 40 **(b)** -20
28. \$1875 **29.** Yes; coefficient 8; degree 3 **31.** Yes; coefficient $\dfrac{1}{7}$; degree 2 **33.** No **35.** Yes; coefficient 12; degree 5 **37.** No **39.** Yes; coefficient 4; degree 0
41. Yes; $6x^2 - 10$; degree 2; binomial **43.** No **45.** No **47.** Yes; $\dfrac{1}{8}$; degree 0; monomial **49.** Yes; $5z^3 - 10z^2 + z + 12$; degree 3; polynomial **51.** No
53. Yes; $-\dfrac{1}{2}t^4 + 3t^2 + 6t$; degree 4; trinomial **55.** Yes; $2xy^4 + 3x^2y^2 + 4$; degree 5; trinomial **57.** $7x - 10$ **59.** $-2m^2 + 5$ **61.** $-2p^2 + p + 8$
63. $-3y^2 - 2y - 4$ **65.** $\dfrac{5}{4}p^2 + \dfrac{1}{6}p - 3$ **67.** $6m^2 - 4mn - n^2$ **69.** $4n^2 - 8n + 5$ **71.** $3x - 16$ **73.** $14x^2 - 3x - 5$ **75.** $4y^3 - 3y - 4$ **77.** $2y^3 - y^2 - 4y$
79. $\dfrac{14}{9}q^2 - \dfrac{23}{8}q + 2$ **81.** $-8m^2n^2 - 4mn - 7$ **83.** $-4x - 5$ **85. (a)** 3 **(b)** 48 **(c)** 13 **87. (a)** -2 **(b)** $\dfrac{3}{4}$ **(c)** 4.75 **89.** 45 **91.** -328 **93.** $-3t + 3$
95. $3x^2 - 3x - 3$ **97.** $9xy^2 + 1$ **99.** $-y^2 + 15y + 1$ **101.** $\dfrac{7}{3}q^2 + \dfrac{5}{3}$ **103.** $13d^2 + d + 10$ **105.** $6a^2 + 5a - 3$ **107.** $10p$ **109.** $-2x^2 - 2xy + y^2$
111. $-5x + 12$ **113.** $3x^2 + 4x - 3$ **115.** $-14x^2 + 4x - 13$ **117.** $-3x + 5$ **119.** 26 ft **121. (a)** 135 feet **(b)** 30 feet **123. (a)** $-2.125x^2 + 105x$
(b) \$1250 **125. (a)** $10x$ **(b)** \$500 **(c)** \$15,000 **127.** $16x - 12$ **129.** $7x + 18$ **131.** $2x - 5$ **133.** Answers may vary. **135.** To find the degree of a polynomial
in one variable, use the largest exponent on any term in the polynomial. To find the degree of a polynomial in more than one variable, find the degree
of each term in the polynomial. That is, the degree of the term ax^my^n is the sum of the exponents, $m + n$. The highest degree of all of the terms is the
degree of the polynomial. **137.** 1

Section 5.2 Multiplying Monomials: The Product and Power Rules 1. exponential **2.** $m + n$ **3.** False **4.** 27 **5.** -3125 **6.** c^8 **7.** y^9 **8.** a^9b^6 **9.** $m \cdot n$ **10.** 2^8
11. $(-3)^6$ **12.** b^{10} **13.** $(-y)^8$ **14.** False **15.** $8n^3$ **16.** $-125x^{12}$ **17.** $49a^6b^2$ **18.** $8a^{11}$ **19.** $-18m^3n^9$ **20.** $-16x^4y^4$ **21.** 1024 **23.** -128 **25.** m^9 **27.** b^{20}
29. x^8 **31.** p^9 **33.** $(-n)^7$ **35.** 64 **37.** 64 **39.** m^{14} **41.** $(-b)^{20}$ **43.** $27x^6$ **45.** $25z^4$ **47.** $\dfrac{1}{4}a^2$ **49.** $81p^{28}q^8$ **51.** $-8m^6n^3$ **53.** $25x^2y^4z^6$ **55.** $12x^5$ **57.** $-40a^{10}$
59. $-7m^4$ **61.** $6x^7$ **63.** $6x^6y^4$ **65.** $-5m^2n^4$ **67.** $-40x^5y^6$ **69.** $4x^3$ **71.** x^4 **73.** $28b^6$ **75.** $81b^4 + 16b^2$ **77.** $9p^5$ **79.** $27x^6y^2$ **81.** $-30cd^2$ **83.** $25x^8$ **85.** $-2x^7y^6$
87. $72x^{14} + 9x^{24}$ **89.** $4q^6$ **91.** $16x^{14}$ **93.** $-\dfrac{24}{5}m^{17}n^9$ **95.** x^6 **97.** $48x^2$ **99.** The product rule for exponential expressions is a "shortcut" for writing out each
exponential expression in expanded form. For example, $x^2 \cdot x^3 = (x \cdot x) \cdot (x \cdot x \cdot x) = x^5$. **101.** The product to a power rule generalizes the result of
using a product as a factor several times. That is, $(a \cdot b)^n = (a \cdot b) \cdot (a \cdot b) \cdot (a \cdot b) \ldots = a \cdot a \cdot a \cdot \ldots \cdot b \cdot b \cdot b \ldots = a^n \cdot b^n$. **103.** The expression
$4x^2 + 3x^2$ is a sum of terms: Because the bases are the same, add the coefficient and retain the common base. The expression $(4x^2)(3x^2)$ is a product:
Multiply the constant terms and, since the bases are the same, add the exponents on the variable bases.

Section 5.3 Multiplying Polynomials **1.** $3x^3 - 6x^2 + 12x$ **2.** $-2a^4b^3 + 4a^3b^3 - 3a^3b^2$ **3.** $3n^5 - \frac{9}{8}n^4 - \frac{6}{7}n^3$ **4.** $x^2 + 9x + 14$ **5.** $2n^2 - 5n - 3$

6. $6p^2 - 7p + 2$ **7.** First, Outer, Inner, Last **8.** $x^2 + 7x + 12$ **9.** $y^2 + 2y - 15$ **10.** $a^2 - 6a + 5$ **11.** $2x^2 + 11x + 12$ **12.** $6y^2 + y - 15$
13. $6a^2 - 11a + 4$ **14.** $15x^2 - 14xy - 8y^2$ **15.** $A^2 - B^2$ **16.** $a^2 - 16$ **17.** $9w^2 - 49$ **18.** $x^2 - 4y^6$ **19.** $A^2 - 2AB + B^2$; $A^2 + 2AB + B^2$
20. perfect square trinomial **21.** False **22.** $z^2 - 18z + 81$ **23.** $p^2 + 2p + 1$ **24.** $16 - 8a + a^2$ **25.** $9z^2 - 24z + 16$ **26.** $25p^2 + 10p + 1$
27. $4w^2 + 28wy + 49y^2$ **28.** False **29.** $x^3 - 8$ **30.** $3y^3 + 4y^2 + 8y - 8$ **31.** $x^3 - 8$ **32.** $3y^3 + 5y^2 - 17y + 5$ **33.** $-24a^3 - 34a^2 + 10a$
34. $x^3 + 4x^2 + x - 6$ **35.** $6x^2 - 10x$ **37.** $2n^2 - 3n$ **39.** $12n^4 + 6n^3 - 15n^2$ **41.** $4x^4y^2 - 3x^3y^3$ **43.** $x^2 + 12x + 35$ **45.** $y^2 + 2y - 35$
47. $6m^2 + 11my - 10y^2$ **49.** $x^2 + 5x + 6$ **51.** $q^2 - 13q + 42$ **53.** $6x^2 + 7x - 3$ **55.** $x^4 + 4x^2 + 3$ **57.** $42 - 13x + x^2$ **59.** $10u^2 + 17uv + 6v^2$

61. $10a^2 - ab - 2b^2$ **63.** $x^2 - 9$ **65.** $4z^2 - 25$ **67.** $16x^4 - 1$ **69.** $4x^2 - 9y^2$ **71.** $x^2 - \frac{1}{9}$ **73.** $x^2 - 4x + 4$ **75.** $25k^2 - 30k + 9$ **77.** $x^2 + 4xy + 4y^2$

79. $4a^2 - 12ab + 9b^2$ **81.** $x^2 + x + \frac{1}{4}$ **83.** $x^3 + x^2 - 5x - 2$ **85.** $2y^3 - 12y^2 + 19y - 3$ **87.** $2x^3 - 7x^2 + 4x + 3$ **89.** $2b^3 + 2b^2 - 24b$ **91.** $-x^3 + 9x$

93. $10y^5 + 3y^4 + 4y^3 + 3y^2 + 2y + 2$ **95.** $b^3 + 2b^2 - 5b - 6$ **97.** $x^4 - 25$ **99.** $z^4 - 81$ **101.** $-2x^2 - 37x$ **103.** $4x^6 + 12x^3 + 9$
105. $x^6 + x^4 - 3x^2 - 3$ **107.** $2x^3 + x^2 - 7x - 2$ **109.** $-5x^7 + 4x^6 - 6x^5$ **111.** $5x^3 + 21x^2 + 2x$ **113.** $-3w^3 + 3w^2 + 36w$ **115.** $3a^3 + 24a^2 + 48a$
117. $2n^2 + 6n$ **119.** $24ab$ **121.** $-x^2 + 3x + 2$ **123.** $4x^2 + 4x + 1$ **125.** $x^3 - 3x^2 + 3x - 1$ **127.** $2x^2 - 9$ **129.** $2x^2 + 7x - 15$ **131.** $2x^2 + x - 1$
133. $4x^3 + x^2 - 3x$ **135.** $x^2 + 12x + 20$ **137.** $x^2 + 3x + 2$ **139.** $(x^2 - x)$ square inches **141.** $(\pi x^2 + 4\pi x + 4\pi)$ square feet **143. (a)** a^2, ab, ab, b^2
(b) $a^2 + 2ab + b^2$ **(c)** length: $a + b$, width: $a + b$, area: $(a + b)^2 = a^2 + 2ab + b^2$ **145.** $x^3 + 2x^2 - 4x - 8$ **147.** $9 - x^2 - 2xy - y^2$
149. $x^3 + 6x^2 + 12x + 8$ **151.** $z^4 + 12z^3 + 54z^2 + 108z + 81$ **153.** To square the binomial $(3a - 5)^2$, use the pattern $A^2 - 2AB + B^2$.
This means square the first term, take the product of -2 times the first term times the second term, and square the second term. So
$(3a - 5)^2 = (3a)^2 - 2(3a)(5) + 5^2 = 9a^2 - 30a + 25$. **155.** The FOIL method can only be used to multiply two binomials.

Section 5.4 Dividing Monomials: The Quotient Rule and Integer Exponents **1.** a^{u-n} **2.** False **3.** $3^2 = 9$ **4.** $\frac{7c}{5}$ **5.** $-\frac{3wz^7}{2}$ **6.** $\frac{a^n}{b^n}$; $b \neq 0$ **7.** $\frac{p^4}{16}$

8. $-\frac{8a^6}{b^{12}}$ **9.** $\frac{81m^4}{n^{12}}$ **10.** 1 **11. (a)** 1 **(b)** -1 **(c)** 1 **12. (a)** 1 **(b)** 2 **(c)** 1 **13.** $\frac{1}{a^n}$ **14. (a)** $\frac{1}{16}$ **(b)** $\frac{1}{16}$ **(c)** $-\frac{1}{16}$ **15.** $\frac{1}{8}$ **16. (a)** $\frac{1}{y^8}$ **(b)** $\frac{1}{y^8}$ **(c)** $-\frac{1}{y^8}$ **17. (a)** $\frac{2}{m^5}$

(b) $\frac{1}{32m^5}$ **(c)** $-\frac{1}{32m^5}$ **18.** 9 **19.** -100 **20.** $\frac{5z^2}{2}$ **21.** $\left(\frac{b}{a}\right)^n$ **22.** $\frac{8}{7}$ **23.** $\frac{125}{27a^3}$ **24.** $\frac{9n^8}{4}$ **25.** $-\frac{20}{a^2}$ **26.** $\frac{3}{mn}$ **27.** $-\frac{4a^3b^3}{3}$ **28.** $\frac{9x^2}{7y^3}$ **29.** $\frac{z^6}{9y^4}$ **30.** $\frac{4w^4}{49z^6}$ **31.** $-\frac{2}{p^{11}}$ **32.** $-\frac{4}{k}$

33. 16 **35.** x^9 **37.** $4y^3$ **39.** $-\frac{2m^7}{3}$ **41.** $2m^8n^2$ **43.** $\frac{27}{8}$ **45.** $\frac{x^{15}}{27}$ **47.** $\frac{x^{20}}{y^{28}}$ **49.** $\frac{49a^4b^2}{c^6}$ **51.** 1 **53.** -1 **55.** 18 **57.** 1 **59.** 1 **61.** $24a$ **63.** $\frac{1}{1000}$ **65.** $\frac{1}{m^2}$ **67.** $-\frac{1}{a^2}$

69. $-\frac{4}{y^3}$ **71.** $\frac{11}{18}$ **73.** $\frac{25}{4}$ **75.** $\frac{z^2}{3}$ **77.** $-\frac{m^6}{8n^3}$ **79.** 16 **81.** $6x^4$ **83.** $\frac{5m^3}{2}$ **85.** $40m^3$ **87.** $\frac{5z^5}{6y^4}$ **89.** $-\frac{7m^{10}n^6}{3}$ **91.** $\frac{7y^4}{z^2}$ **93.** $\frac{y^4}{16x^4}$ **95.** $\frac{48}{b^3}$ **97.** 4 **99.** $\frac{1}{8}$ **101.** 81 **103.** $\frac{1}{x^9}$

105. $7x^3 + 5x$ **107.** $\frac{4}{x^3}$ **107.** $-\frac{3z^3}{2x^3}$ **109.** $\frac{1}{4x^3y^9}$ **111.** $-\frac{1}{a^{12}}$ **113.** $\frac{27}{m^6}$ **115.** $\frac{x^4y^6}{9}$ **117.** $\frac{6}{p^7}$ **119.** 12 **121.** 0 **123.** $\frac{16y}{27x^{18}}$ **125.** y **127.** $216x^3y^6$

129. $\frac{5ab^7}{3}$ **131.** $3x^3$ cubic meters **133.** $V = \frac{\pi d^2h}{4}$ **135.** $9\pi x^3$ square units **137.** $\frac{1}{x^n}$ **139.** $x^{6n-2}y^2$ **141.** $\frac{x^{6a^2}y^{3ab}}{z^{3ac}}$ **143.** $\frac{9}{8a^{10n}}$ **145.** The quotient $\frac{11^6}{11^4} \neq 1^2$

because $\frac{11^6}{11^4} = \frac{11 \cdot 11 \cdot 11 \cdot 11 \cdot 11 \cdot 11}{11 \cdot 11 \cdot 11 \cdot 11} = \frac{\cancel{11} \cdot \cancel{11} \cdot \cancel{11} \cdot \cancel{11} \cdot 11 \cdot 11}{\cancel{11} \cdot \cancel{11} \cdot \cancel{11} \cdot \cancel{11}} = 11^2$. **147.** When we simplify the expression $-12x^0$, the exponent 0 applies only to
the base x, so $-12x^0 = -12 \cdot 1 = -12$. However, $(-12x)^0 = 1$ because the entire expression is raised to the 0 power. **149.** His answer is incorrect. To
raise a factor to a power, multiply exponents. Your friend added exponents.

Putting the Concepts Together (Sections 5.1–5.4) **1.** Yes; degree 6; trinomial **2. (a)** 0 **(b)** -4 **(c)** 2 **3.** $8x^4 - 4x^2 + 14$ **4.** $5x^2y + 3xy + 2y^2$
5. $-6m^3n^2 + 2m^2n^4$ **6.** $5x^2 - 17x - 12$ **7.** $8x^2 - 26xy + 21y^2$ **8.** $25x^2 - 64$ **9.** $4x^2 + 12xy + 9y^2$ **10.** $6a^3 + 22a^2 - 40a$
11. $16m^4 + 4m^3 + 10m^2 - 20m - 24$ **12.** $-8x^8z^9$ **13.** $-\frac{15}{mn}$ **14.** $-3a^3b^2$ **15.** $\frac{1}{2x^2y^2}$ **16.** $\frac{7b^2}{2a}$ **17.** $\frac{r^5}{q^7t^2}$ **18.** $\frac{8}{27r^6}$ **19.** $\frac{y^4}{16x^6}$ **20.** $\frac{64y^8}{81x^{24}}$

Section 5.5 Dividing Polynomials **1.** $4x^4$; $8x^2$ **2.** $2n^2 - 4n + 1$ **3.** $6k^2 - 9 + \frac{5}{2k^2}$ **4.** $\frac{xy^3}{4} + \frac{2y}{x} - \frac{1}{x^2}$ **5.** standard **6.** quotient; remainder; dividend

7. $x - 8$ **8.** $x - 4$ **9.** $4x + 5 + \frac{6}{x + 3}$ **10.** $4x^2 - 11x + 23 - \frac{45}{x + 2}$ **11.** $x^2 + 4x + 10 + \frac{60}{2x - 5}$ **12.** $4x - 3 + \frac{-7x + 7}{x^2 + 2}$ **13.** $2x - 1$

15. $3a + 9 - \frac{1}{a^2}$ **17.** $\frac{n^4}{5} - \frac{2n^2}{5} - 1$ **19.** $\frac{5r^2}{3} - 3$ **21.** $-x^2 + 3x - \frac{9}{x^2}$ **23.** $\frac{3}{8} + \frac{z^2}{2} - \frac{z}{4}$ **25.** $-7x + 5$ **27.** $-\frac{6y}{x} + 15$ **29.** $-\frac{5}{a} - \frac{2c^2}{a^2b}$ **31.** $x - 7$

33. $x - 5$ **35.** $x^2 + 6x - 3$ **37.** $x^3 - 3x^2 + 6x - 2$ **39.** $x^2 - 2x + 3 + \frac{5}{x + 1}$ **41.** $2x + 3$ **43.** $x^2 + 6x + 7 + \frac{16}{x - 2}$ **45.** $2x^3 + 5x^2 + 9x - 4 - \frac{17}{x - 4}$

47. $x^2 + 4x - 3 + \frac{2}{2x - 1}$ **49.** $x - 4 - \frac{4}{5 + x}$ **51.** $2x - 1 + \frac{6}{1 + 2x}$ **53.** $x + 2 + \frac{2x - 4}{x^2 - 2}$ **55.** $a^2 + a - 30$ **57.** $-x^2 - x - 6$ **59.** $-2ab + 2b^2$

61. $\frac{2}{x^2} + \frac{7}{2} - \frac{3x^2}{2} + 3x$ **63.** $\frac{2x^7y^8z^{12}}{3}$ **65.** $2x - 6$ **67.** $n^2 - 6n + 9$ **69.** $-10x^5 - 10x^3 + 40x^6$ **71.** $6pq - 2q^2 - p^2$ **73.** $x^3 + 6x^2 + 4x - 5$

75. $15rs^2 - 4r^2s$ **77.** $-3x^5 + 6x^3 - 3x^2$ **79.** $4x$ **81.** $-x + \frac{2}{x^2} - \frac{4}{x}$ **83.** $\frac{1}{2x} - \frac{2}{x^2} + \frac{1}{x^3}$ **85.** x ft **87.** $(z + 3)$ in. **89.** $(4x + 4)$ ft

91. (a) $0.004x^2 - 0.8x + 180 + \dfrac{5000}{x}$ **(b)** $182.11 per computer **93.** 6 **95.** When the divisor is a monomial, divide each term in the numerator by the denominator. When dividing by a binomial, use long division. Answers may vary.

Section 5.6 Applying Exponent Rules: Scientific Notation
1. scientific **2.** positive **3.** False **4.** 4.32×10^2 **5.** 1.0302×10^4 **6.** 5.432×10^6
7. 9.3×10^{-2} **8.** 4.59×10^{-5} **9.** 8×10^{-8} **10.** True **11.** True **12.** 310 **13.** 0.901 **14.** 170,000 **15.** 7 **16.** 0.00089 **17.** 6×10^7 **18.** 8×10^{-3}
19. 1.5×10^4 **20.** 2.8×10^{-5} **21.** 4×10^5 **22.** 2×10^{-4} **23.** 5×10^3 **24.** 6.25×10^{-5} **25.** $1.047 \times 10^{10} = 10,470,000,000$ gallons
26. $1.314 \times 10^9 = 1,314,000,000$ customers **27.** 3×10^5 **29.** 6.4×10^7 **31.** 5.1×10^{-4} **33.** 1×10^{-9} **35.** 8.007×10^9 **37.** 3.09×10^{-5} **39.** 6.2×10^2
41. 4×10^0 **43.** 6.999×10^9 **45.** 1.55×10^{13} **47.** 9.3×10^7 miles **49.** 3×10^{-5} mm **51.** 2.5×10^{-7} m **53.** 3.11×10^{-3} kg **55.** 420,000 **57.** 100,000,000
59. 0.0039 **61.** 0.4 **63.** 3760 **65.** 0.0082 **67.** 0.00006 **69.** 7,050,000 **71.** 0.000000000000001 **73.** 0.00225 **75.** 500,000 **77.** 3×10^9 **79.** 8.4×10^{-3}
81. 3×10^8 **83.** 2.5×10^1 **85.** 8×10^8 **87.** 9×10^{15} **89.** 1.116×10^7 miles **91.** 0.000001 mile; 0.00000168 mile **93. (a)** 3.9×10^9 pounds **(b)** 3.1×10^8
(c) 12.6 pounds **95.** 2.5×10^{-4} m **97.** 8×10^{-10} m **99.** 7.15×10^{-8} m **101.** $1.2348\pi \times 10^{-14}$ m^3 **103.** $2.88\pi \times 10^{-25}$ m^3 **105.** To convert a number written in decimal notation to scientific notation, count the number of decimal places, N, that the decimal point must be moved to arrive at a number x such that $1 \le x < 10$. If the original number is greater than or equal to 1, move the decimal point to the left that many places and write the number in the form $x \times 10^N$. If the original number is between 0 and 1, move the decimal point to the right that many places and write the number in the form $x \times 10^{-N}$. **107.** The number 34.5×10^4 is incorrect because the number 34.5 is not a number between 1 (inclusive) and 10. The correct answer is 3.45×10^5.

Chapter 5 Review Exercises
1. Yes; coefficient 4; degree 3 **2.** No **3.** No **4.** Yes; coefficient 1; degree 3 **5.** No **6.** No **7.** Yes; degree 0; monomial
8. Yes; degree 5; binomial **9.** Yes; degree 6; trinomial **10.** Yes; degree 10; trinomial **11.** $9x^2 + 8x - 2$ **12.** $m^3 + mn - 5m$ **13.** $-x^2y + 12x$
14. $-7y^2 - 6yz + 9z^2$ **15.** $-2x^2 - 2$ **16.** $-20y^2 - 8y + 5$ **17. (a)** 0 **(b)** 8 **(c)** 2 **18. (a)** 3 **(b)** 2 **(c)** $\dfrac{11}{4}$ **19.** 0 **20.** 29 **21.** 279,936 **22.** $-\dfrac{1}{243}$
23. 16,777,216 **24.** 1 **25.** x^{13} **26.** m^6 **27.** r^{12} **28.** m^{24} **29.** $1024x^5$ **30.** $64n^6$ **31.** $81x^8y^4$ **32.** $4x^6y^8$ **33.** $15x^6$ **34.** $-36a^4$ **35.** $16y^5$ **36.** $-12p^6$
37. $12w^4$ **38.** $-\dfrac{3}{4}z^3$ **39.** $108x^8$ **40.** $80a^6$ **41.** $-8x^5 + 6x^4 - 2x^3$ **42.** $2x^7 + 4x^6 - x^4$ **43.** $6x^2 - 7x - 5$ **44.** $4x^2 - 5x - 6$ **45.** $x^2 - 3x - 40$
46. $w^2 + 9w - 10$ **47.** $24m^3 - 22m^2 + 7m - 3$ **48.** $8y^5 + 12y^4 + 4y^3 + 6y^2 - 6y - 9$ **49.** $x^2 + 8x + 15$ **50.** $2x^2 - 17x + 8$ **51.** $6m^2 + 17m - 14$
52. $48m^2 - 26m - 4$ **53.** $21x^2 + 5xy - 6y^2$ **54.** $20x^2 + 7xy - 3y^2$ **55.** $x^2 - 16$ **56.** $4x^2 - 25$ **57.** $4x^2 + 12x + 9$ **58.** $49x^2 - 28x + 4$ **59.** $9x^2 - 16y^2$
60. $64m^2 - 36n^2$ **61.** $25x^2 - 20xy + 4y^2$ **62.** $4a^2 + 12ab + 9b^2$ **63.** $x^2 - 0.25$ **64.** $r^2 - 0.0625$ **65.** $y^2 + \dfrac{4}{3}y + \dfrac{4}{9}$ **66.** $x^3 - 4x^2 - 22x$ **67.** 36 **68.** $\dfrac{1}{343}$
69. x^4 **70.** $\dfrac{1}{x^8}$ **71.** 1 **72.** -1 **73.** 1 **74.** -1 **75.** $\dfrac{5x^2}{2y^3}$ **76.** $\dfrac{x^2}{3y^8}$ **77.** $\dfrac{x^{15}}{y^{10}}$ **78.** $\dfrac{343}{x^6}$ **79.** $\dfrac{8m^6n^3}{p^{12}}$ **80.** $\dfrac{81m^4n^8}{p^{20}}$ **81.** $-\dfrac{1}{25}$ **82.** 64 **83.** $\dfrac{81}{16}$ **84.** 27 **85.** $\dfrac{7}{12}$ **86.** $\dfrac{1}{8}$ **87.** $\dfrac{2x^3y^5}{3}$
88. $\dfrac{3}{7xy^{10}}$ **89.** $\dfrac{9m^4}{16}$ **90.** $\dfrac{64}{9m^6n^6}$ **91.** $\dfrac{8r^4s^6}{9}$ **92.** $7a^3 + 7$ **93.** $6x^5 - 4x^4 + 5$ **94.** $3x^4 + 5x^2 - 6x$ **95.** $4n^3 + 1 - \dfrac{5}{2n^4}$ **96.** $6n - 4 - \dfrac{16}{5n^2}$ **97.** $-\dfrac{p^3}{8} - \dfrac{1}{4} + \dfrac{1}{2p^2}$
98. $-\dfrac{p^2}{2} + 1 - \dfrac{3}{2p^2}$ **99.** $4x - 7$ **100.** $x + 6$ **101.** $x^2 + 7x + 5 + \dfrac{6}{x-1}$ **102.** $2x^2 - 3x - 12 - \dfrac{16}{x-2}$ **103.** $x^2 - 2x + 4$ **104.** $3x^2 + 15x + 77 + \dfrac{378}{x-5}$
105. 2.7×10^7 **106.** 1.23×10^9 **107.** 6×10^{-5} **108.** 3.05×10^{-6} **109.** 3×10^0 **110.** 8×10^0 **111.** 0.0006 **112.** 0.00125 **113.** 613,000 **114.** 80,000 **115.** 0.37
116. 54,000,000 **117.** 6×10^3 **118.** 4.2×10^{-8} **119.** 2×10^2 **120.** 2×10^8 **121.** 4×10^{15} **122.** 2×10^2

Chapter 5 Test
1. Yes; degree 5; binomial **2. (a)** 5 **(b)** 21 **(c)** 26 **3.** $7x^2y^2 - 6x - 3y$ **4.** $10m^3 + m^2 - 6$ **5.** $-6x^5 + 18x^4 - 15x^3$ **6.** $2x^2 - 3x - 35$
7. $4x^2 - 28x + 49$ **8.** $16x^2 - 9y^2$ **9.** $6x^3 + x^2 - 25x + 8$ **10.** $2x - \dfrac{8}{3} + \dfrac{3}{x^3}$ **11.** $3x^2 - 11x + 33 + \dfrac{-94}{x+3}$ **12.** $-12x^4y^6$ **13.** $\dfrac{2m^3}{3n^5}$ **14.** $\dfrac{n^{24}}{m^{30}}$ **15.** $\dfrac{x^{22}}{y^{14}}$
16. $\dfrac{8m^7}{n^7}$ **17.** 1.2×10^{-5} **18.** 210,100 **19.** 3.57×10^4 **20.** 2×10^{-7}

Cumulative Review Chapters 1–5
1. (a) $\left\{\sqrt{25}\right\}$ **(b)** $\left\{0, \sqrt{25}\right\}$ **(c)** $\left\{-6, -\dfrac{4}{2}, 0, \sqrt{25}\right\}$ **(d)** $\left\{-6, -\dfrac{4}{2}, 0, 1.4, \sqrt{25}\right\}$ **(e)** $\left\{\sqrt{7}\right\}$
(f) $\left\{-6, -\dfrac{4}{2}, 0, 1.4, \sqrt{7}, \sqrt{25}\right\}$ **2.** $-\dfrac{4}{9}$ **3.** -19 **4.** $6x^3 + 5x^2 - 3x$ **5.** $-18x + 8$ **6.** $\{1\}$ **7.** $4(x - 5) = 2x + 10$ **8.** $640 **9.** 45 mph
10. $\{x \mid x < -3\}$ or $(-\infty, -3)$; **11.** $-\dfrac{1}{2}$ **12.** $-\dfrac{3}{2}$ **13.** **14.** **15.** $(3, -7)$ **16.** No solution
17. $6x^3 - 3x^2 + 7x$ **18.** $28m^2 - 13m - 6$
19. $9m^2 - 4n^2$ **20.** $49x^2 + 14xy + y^2$
21. $4m^3 - 19m + 15$ **22.** $\dfrac{2}{x} + \dfrac{1}{y}$ **23.** $x^2 - 3x + 9$ **24.** $-24n^4$ **25.** $-\dfrac{5n^8}{2m^2}$ **26.** $\dfrac{1}{64x^6y^{24}z^{12}}$ **27.** $\dfrac{1}{2x^{12}y^3}$ **28.** 6.05×10^{-5} **29.** 2,175,000 **30.** 7.14×10^5

Chapter 6 Factoring Polynomials

Section 6.1 *Greatest Common Factor and Factoring by Grouping* **1.** greatest common factor **2.** factors **3.** 8 **4.** 3 **5.** 7 **6.** $7y^2$ **7.** $2z$ **8.** $4xy^3$
9. $2x + 3$ **10.** $3(k + 8)$ **11.** factor **12.** Distributive **13.** $5z(z + 6)$ **14.** $12p(p - 1)$ **15.** $4y(4y^2 - 3y + 1)$ **16.** $2m^2n^2(3m^2 + 9mn^2 - 11n^3)$
17. $-4y(y - 2)$ **18.** $-3a(2a^2 - 4a + 1)$ **19.** False **20.** $(a - 5)(2a + 3)$ **21.** $(z + 5)(7z - 4)$ **22.** $(x + y)(4 + b)$ **23.** $(3z - 1)(2a - 3b)$
24. $(2b + 1)(4 - 5a)$ **25.** $3(z + 4)(z^2 + 2)$ **26.** $2n(n + 1)(n^2 - 2)$ **27.** 4 **29.** 1 **31.** 6 **33.** x^2 **35.** $7x$ **37.** $15ab^2$ **39.** $2abc^2$ **41.** $x - 1$
43. $2(x - 4)^2$ **45.** $6(x - 3)^2$ **47.** $6(2x - 3)$ **49.** $x(x + 12)$ **51.** $5x^2y(1 - 3xy)$ **53.** $3x(x^2 + 2x - 1)$ **55.** $-5x(x^2 - 2x + 3)$
57. $3(3m^5 - 6m^3 - 4m^2 + 27)$ **59.** $-4z(3z^2 - 4z + 2)$ **61.** $-5(3b^3 + b - 2)$ **63.** $15ab^2(ab^2 - 4b + 3a^2)$ **65.** $(x - 3)(x - 5)$
67. $(x - 1)(x^2 + y^2)$ **69.** $(4x + 1)(x^2 + 2x + 5)$ **71.** $(x + 3)(y + 4)$ **73.** $(y + 1)(z - 1)$ **75.** $(x - 1)(x^2 + 2)$ **77.** $(2t - 1)(t^2 - 2)$
79. $(2t - 1)(t^3 - 3)$ **81.** $4(y - 5)$ **83.** $7m(4m^2 + m + 9)$ **85.** $6m^2n(2mnp - 3)$ **87.** $3(p + 3)(3p + 1)$ **89.** $3(2x - y)(3a - 2b)$

91. $(x - 2)(x - 2)$ or $(x - 2)^2$ **93.** $3x(5x - 2)(x^2 + 2)$ **95.** $-3x(x^2 - 2x + 3)$ **97.** $4b(4b - 3)$ **99.** $(4y + 3)(3x - 2)$ **101.** $\frac{1}{3}x^2\left(x - \frac{2}{3}\right)$

103. $-2t(8t - 75)$ **105.** $150(140 - p)$ **107.** $4x^3(2x^2 - 7)$ **109.** $2\pi r(r + 4)$ **111.** $x - 2$ **113.** $4x^{2n} + 5$ **115.** $3x^3 - 4x^2 + 2$ **117.** $x^4 - 3x + 2$

119. $\frac{3}{5}x^3 - \frac{1}{2}x + 1$ **121.** To find the greatest common factor: (1) Determine the GCF of the coefficients of the variable factors (the largest number that
divides evenly into a set of numbers). (2) For each variable factor common to all the terms, determine the smallest exponent that the variable factor is raised
to. (3) Find the product of these common factors. This is the GCF. To factor the GCF from a polynomial: (1) Identify the GCF of the terms that make up the
polynomial. (2) Rewrite each term as the product of the GCF and the remaining factor. (3) Use the Distributive Property "in reverse" to factor out the GCF.
(4) Check using the Distributive Property. **123.** $3a(x + y) - 4b(x + y) \neq (3a - 4b)(x + y)^2$ because when checking, the product $(3a - 4b)(x + y)^2$
is equal to $(3a - 4b)(x + y)^2 = (3a - 4b)(x^2 + 2xy + y^2)$, not to $3a(x + y) - 4b(x + y) = 3ax + 3ay - 4bx - 4by$.

Section 6.2 *Factoring Trinomials of the Form* $x^2 + bx + c$ **1.** quadratic trinomial **2.** $24; -10$ **3.** False **4.** $(y + 5)(y + 4)$ **5.** $(z - 7)(z - 2)$
6. True **7.** $(y - 5)(y + 3)$ **8.** $(w - 3)(w + 4)$ **9.** prime **10.** prime **11.** $(q + 9)(q - 5)$ **12.** $(x + 4y)(x + 5y)$ **13.** $(m + 7n)(m - 6n)$
14. (a) $x^2 - x - 12$ **(b)** $n^2 + 9n - 10$ **15.** $(n + 8)(n - 7)$ **16.** $(y - 7)(y - 5)$ **17.** False **18.** $4(m + 3)(m - 7)$ **19.** $3z(z - 1)(z + 5)$
20. $3b^2(a - 6)(a - 2)$ **21.** $-1(w + 5)(w - 2)$ **22.** $-2a(a + 6)(a - 2)$ **23.** $(x + 2)(x + 3)$ **25.** $(m + 3)(m + 6)$ **27.** $(x - 3)(x - 12)$
29. $(p - 2)(p - 6)$ **31.** $(x - 4)(x + 3)$ **33.** prime **35.** $(z - 3)(z + 15)$ **37.** $(x - 2y)(x - 3y)$ **39.** $(r - 2s)(r + 3s)$ **41.** $(x - 4y)(x + y)$
43. prime **45.** $3(x + 2)(x - 1)$ **47.** $3n(n - 3)(n - 5)$ **49.** $5z(x - 8)(x + 4)$ **51.** $x(x - 6)(x + 3)$ **53.** $-2(y - 2)^2$ **55.** $x^2(x + 8)(x - 4)$
57. $x(x + 5)(x - 3)$ **59.** $(x + 4y)(x - 7y)$ **61.** prime **63.** $(k - 5)(k + 4)$ **65.** $2w(w^2 - 4w - 6)$ **67.** $(st - 3)(st - 5)$
69. $-3p(p - 2)(p + 1)$ **71.** $-(x + 3)^2$ **73.** prime **75.** $n^2(n - 6)(n + 5)$ **77.** $2(x - 5)(x^2 + 3)$ **79.** $(s + 5)(s + 7)$ **81.** prime
83. $(mn - 6)(mn - 2)$ **85.** $-x(x - 14)(x + 2)$ **87.** $(y - 6)^2$ **89.** $-2x(x^2 - 10x + 18)$ **91.** $-7xy(3x^2 + 2y)$ **93.** $-16(t + 1)(t - 5)$
95. $(x + 6)$ meters and $(x + 3)$ meters **97.** $(x + 5)$ inches and $(x - 3)$ inches **99.** $-7, -5, 5, 7$ **101.** $-20, -4, 4, 20$ **103.** 1 **105.** 6, 10, 12
107. Yes, $(5 + x)(2 - x) = (x + 5)(-x + 2) = -1(x + 5)(x - 2)$ **109.** In the trinomial $x^2 + bx + c$, where both b and c are negative, find two
numbers m and n so that $m \cdot n = c$ and $m + n = b$. m and n will have opposite signs and the factor with the larger absolute value is negative. For
example, $x^2 - 3x - 28 = (x - 7)(x + 4)$.

Section 6.3 *Factoring Trinomials of the Form* $ax^2 + bx + c, a \neq 1$ **1.** $-6; 1$ **2.** $12; -13$ **3.** $(3x + 4)(x - 2)$ **4.** $(5z + 3)(2z + 3)$
5. $3(4x + 3)(2x - 1)$ **6.** $-(5n - 1)(2n - 3)$ **7.** $(3x + 7y)(2x + 3y)$ **8.** $-3(4a - 5b)(3a + 2b)$ **9.** $(3x + 2)(x + 1)$ **10.** $(7y + 1)(y + 3)$
11. $(3x - 4)(x - 3)$ **12.** $(5p - 1)(p - 4)$ **13.** $(2n + 1)(n - 9)$ **14.** $(4w + 3)(w - 2)$ **15.** $(3x + 2)(4x + 3)$ **16.** $(6y - 5)(2y + 7)$
17. Look for a common factor **18.** False **19.** $-4(2x + 3y)(x - 5y)$ **20.** $3(6x - y)(5x + 2y)$ **21.** $-(6y + 1)(y - 4)$ **22.** $-3(2x - 5)(x + 3)$
23. $(2x + 3)(x + 5)$ **25.** $(5w - 2)(w + 3)$ **27.** $(2w + 1)(2w - 5)$ **29.** $(9z - 2)(3z + 1)$ **31.** $(3y + 2)(2y - 3)$ **33.** $(2m - 1)(2m + 5)$
35. $(3n + 1)(4n + 5)$ **37.** $(6x - 5)(3x + 1)$ **39.** $2(3x + 2y)(2x - y)$ **41.** $4(2x + 1)(x + 3)$ **43.** $(3x + 5)(2x - 1)$ **45.** $-(4p + 3)(2p - 3)$
47. $(2x + 3)(x + 1)$ **48.** $(5n + 2)(n + 1)$ **51.** $(5y - 3)(y + 1)$ **53.** $-(4p + 1)(p - 3)$ **55.** $(5w - 2)(w + 3)$ **57.** $(7t + 2)(t + 5)$ **59.**
$(6n - 5)(n - 2)$ **61.** $(x - 2)(5x - 1)$ **63.** $(2x + y)(x + y)$ **65.** $(2m + n)(m - 2n)$ **67.** $2(3x - 2y)(x + y)$ **69.** $-(2x - 3)(x + 5)$
71. $(3x - 4)(5x - 1)$ **73.** $-(y + 4)(4y - 3)$ **75.** $2(x - 2y)(5x + 6y)$ **77.** prime **79.** $-3(8x^2 + 4x - 3)$ **81.** $4x^2y^2(x - 2y - 1)$
83. $(m + 3n)(4m + n)$ **85.** prime **87.** $6(2x + 5y)(2x - y)$ **89.** $3(3m + 8n)(2m - n)$ **91.** $-2x^2(2x + 5)(x - 1)$ **93.** $-2x(4x + 3)(3x - 5)$
95. $(x^2 + 1)(2x - 7)(3x - 2)$ **97.** $(2x + 7)$ meters and $(3x - 4)$ meters **99.** $2x - 5$ **101.** $(9z^2 + 8)(3z^2 + 2)$ **103.** $(3x^n + 1)(x^n + 6)$
105. ± 2 or ± 14 **107.** Trial and error may be used when the value of a and/or the value of c are prime numbers, or numbers with few factors.
Grouping can be used when the product of a and c is not a very large number. Examples may vary.

Section 6.4 *Factoring Special Products* **1.** perfect square trinomial **2.** $(A - B)^2$ **3.** $(x - 6)^2$ **4.** $(4x + 5)^2$ **5.** $(3a - 10b)^2$
6. $(z - 4)^2$ **7.** $(2n + 3)^2$ **8.** prime **9.** $4(z + 3)^2$ **10.** $2a(5a + 4)^2$ **11.** difference; two squares **12.** $(P - Q)(P + Q)$ **13.** $(z - 5)(z + 5)$

14. $(9m - 4n)(9m + 4n)$ **15.** $\left(4a - \frac{2}{3}b\right)\left(4a + \frac{2}{3}b\right)$ **16.** True **17.** $(10k^2 - 9w)(10k^2 + 9w)$ **18.** $(x - 2)(x + 2)(x^2 + 4)$ **19.** False

20. $3(7x - 4)(7x + 4)$ **21.** $-2y(x^2 - 2y^3)(x^2 + 2y^3)$ **22.** sum; two cubes **23.** $(A - B)(A^2 + AB + B^2)$ **24.** False **25.** $(z + 5)(z^2 - 5z + 25)$
26. $(2p - 3q^2)(4p^2 - 6pq^2 + 9q^4)$ **27.** $-2a(2a + 3)(4a^2 - 6a + 9)$ **28.** $-3(5b - 1)(25b^2 + 5b + 1)$ **29.** $(x + 5)^2$ **31.** $(2p - 1)^2$
33. $(4x + 3)^2$ **35.** $(x - 2y)^2$ **37.** $(2z - 3)^2$ **39.** $(x - 7)(x + 7)$ **41.** $(2x - 5)(2x + 5)$ **43.** $(10n^4 - 9p^2)(10n^4 + 9p^2)$
45. $(k - 2)(k + 2)(k^2 + 4)(k^4 + 16)$ **47.** $(5p^2 - 7q)(5p^2 + 7q)$ **49.** $(3 + x)(9 - 3x + x^2)$ **51.** $(2x - 3y)(4x^2 + 6xy + 9y^2)$
53. $(x^2 - 2y)(x^4 + 2x^2y + 4y^2)$ **55.** $(3c + 4d^3)(9c^2 - 12cd^3 + 16d^6)$ **57.** $2(2x + 5y)(4x^2 - 10xy + 25y^2)$ **59.** $(4m + 5n)^2$ **61.** $2(x - 3)^2$
63. $-(2a - 5)(3a + 8)$ **65.** $2x(x^2 + 5x + 8)$ **67.** $x^2y^2(x - y)(x + y)$ **69.** $(x^4 - 5y^5)(x^4 + 5y^5)$ **71.** $(b + 9)(b + 4)$ **73.** $3s(s^2 + 2)(s^4 - 2s^2 + 4)$
75. $2x(x - 3)(x + 3)(x^2 + 9)$ **77.** $(2x + 5)(3x + 2)$ **79.** $4(x - 6)(x + 3)$ **81.** prime **83.** $2(x - 2)^2$ **85.** prime **87.** $xy^3(x + 6)(x^2 - 6x + 36)$
89. $3n^2(4n - 1)^2$ **91.** $(x - 2)(x + 2)(2x + 5)$ **93.** $(2y + 5)(y - 4)(y + 4)$ **95.** $(2x + 5)$ meters **97.** $-3(2x - 1)$ **99.** $x(x^2 - 3xy + 3y^2)$
101. $(x - 2)(x + 4)$ **103.** $(x + 1)(2a - 5)(a - 6)$ **105.** $x(5x + 13)$ **107.** First identify the problem as a sum or a difference of cubes. Identify A

and B, and rewrite the problem as $A^3 + B^3$ or $A^3 - B^3$. The sum of cubes can be factored into the product of a binomial and a trinomial, where the binomial factor is $A + B$ and the trinomial factor is A^2 minus the product AB plus B^2. The difference of cubes also can be factored into the product of a binomial and a trinomial, but here the binomial factor is $A - B$ and the trinomial factor is the sum of A^2, the product AB, and B^2. **109.** The correct factorization of $x^3 + y^3$ is $(x + y)(x^2 - xy + y^2)$. The trinomial factor $(x^2 - xy + y^2)$ is not a perfect square trinomial. A perfect square trinomial is $(x - y)^2 = (x^2 - 2xy + y^2)$.

Section 6.5 Summary of Factoring Techniques
1. greatest common factor **2.** $2(p - 5)(p + 9)$ **3.** $-3x(3x + 1)(5x - 2)$
4. $(4x + 1)(2x + 5)$ **5.** $(6m + 7)(2m - 5)$ **6.** $(p - 6q)^2$ **7.** $3(5x + 3)^2$ **8.** $(5y - 4)(25y^2 + 20y + 16)$ **9.** $-3(2a - b)(4a^2 + 2ab + b^2)$
10. False **11.** $(x - 1)(x + 1)(x^2 + 1)$ **12.** $-4y(3x - 2)(3x + 2)$ **13.** $(2x + 3)(x^2 + 2)$ **14.** $3(2x + 3)(x - 1)(x + 1)$ **15.** $-3(z^2 - 3z + 7)$
16. $3x(2y^2 + 5x^2)$ **17.** $(x - 10)(x + 10)$ **19.** $(t + 3)(t - 2)$ **21.** $(x + y)(1 + 2a)$ **23.** $(a - 2)(a^2 + 2a + 4)$ **25.** $(a - 3b)(a + 2b)$
27. $(2x - 7)(x + 1)$ **29.** $2(x - 5y)(x + 2y)$ **31.** prime **33.** $(3 - a)(3 + a)$ **35.** $(u - 11)(u - 3)$ **37.** $(x - a)(y - b)$ **39.** $(w + 2)(w + 4)$
41. $(6a - 7b^2)(6a + 7b^2)$ **43.** $(x + 4m)(x - 2m)$ **45.** $(3xy - 2)(2xy - 3)$ **47.** $(x + 1)(x^2 + 1)$ **49.** $6(2z^2 + 2z + 3)$ **51.** $(7c - 1)(2c + 3)$
53. $(3m + 4n^2)(9m^2 - 12mn^2 + 16n^4)$ **55.** $2j^2(j - 1)(j + 1)(j^2 + 1)$ **57.** $(4a - b)(2a + 5b)$ **59.** $2a(a^2 + 4)$ **61.** $3(2z - 1)(2z + 1)$
63. prime **65.** $2a(2a + b)(4a^2 - 2ab + b^2)$ **67.** $(pq + 7)(pq - 1)$ **69.** $(s + 2)^2(s - 2)$ **71.** $-2x(3x + 1)(2x - 1)$ **73.** $(5v + 2)(2v - 1)$
75. $-n^2(n - 4)(n + 1)$ **77.** $-2ab(2a^2 - a + 1)$ **79.** $-p(p - 4)(p + 3)$ **81.** $-8x(2x - 3y)(2x + 3y)$ **83.** $2(n - 5)(n^2 - 3)$
85. $4(4x - 3)(x + 1)$ **87.** $(3x^2 + 2)(x^2 + 4)$ **89.** $-xy(2x - 3y)(x + y)$ **91.** $x(x - 5)(x - 3)$ **93.** $2x$ ft., $(2x + 1)$ ft., $(x - 3)$ ft.
95. $(y^2 + 4)(y^2 - 6)$ **97.** $(z^3 - 2)(4z^3 - 5)$ **99.** $3(r - 2)(3r - 5)$ **101.** $(y + 2)(2y - 3)$ **103.** $2(3x + 4)(9x + 13)$ **105.** $2x(3x + 5)$
107. $3(4x - 3)(4x - 1)$ **109.** $(2m - n + p)(2m - n - p)$ **111.** $(x + y - z)(x - y + z)$ **113.** The student initially wrote the sum of terms as a product of terms by enclosing the binomial $-3x - 12$ in parentheses. Then, in the second step, the student writes the factor $-3(x - 4)$ and concludes that the factorization of $x^2 + 4x - 3x - 12$ is $(x - 3)(x + 4)$, even though both terms do not have the common factor of $(x + 4)$. The student obtained the correct answer, but the first two steps are incorrect; the correct factorization is $x^2 + 4x - 3x - 12 = x(x + 4) - 3(x + 4) = (x + 4)(x - 3)$.

Putting the Concepts Together (Sections 6.1–6.5)
1. $5x^2y$ **2.** $(x - 4)(x + 1)$ **3.** $(x^2 - 3)(x^4 + 3x^2 + 9)$ **4.** $(2x + 1)(6x + 5z)$
5. $(x + 6y)(x - y)$ **6.** $(x + 4)(x^2 - 4x + 16)$ **7.** prime **8.** $3(x + 6y)(x - 2y)$ **9.** $4z^2(3z^3 - 11z - 6)$ **10.** prime **11.** $m^2(m + 2)(4m - 3)$
12. $(5p - 2)(p - 3)$ **13.** $(2m + 5)(5m - 3)$ **14.** $6(3m - 1)(2m + 1)$ **15.** $(2m - 5)^2$ **16.** $(5x + 4y)(x - y)$ **17.** $S = 2\pi r(h + r)$
18. $h = 16t(3 - t)$

Section 6.6 Solving Polynomial Equations by Factoring
1. 2 **2.** $a = 0$; $b = 0$ **3.** $\{-3, 0\}$ **4.** $\left\{-\dfrac{5}{4}, 2\right\}$ **5.** quadratic **6.** second
7. False **8.** $\{2, 4\}$ **9.** $\left\{-\dfrac{1}{2}, 3\right\}$ **10.** $\left\{-\dfrac{5}{2}, 1\right\}$ **11.** $\{-5, -4\}$ **12.** $\left\{-2, \dfrac{2}{5}\right\}$ **13.** $\left\{-4, \dfrac{1}{3}\right\}$ **14.** False **15.** $\{-6, 4\}$ **16.** $\left\{-\dfrac{3}{5}, 1\right\}$ **17.** $\left\{\dfrac{4}{3}\right\}$
18. $\{-5\}$ **19.** $\{-6, 3\}$ **20.** $\{-2, 3\}$ **21.** After 1 second and after 4 seconds **22.** True **23.** $\{-2, -1, 0\}$ **24.** $\left\{-1, 1, \dfrac{4}{3}\right\}$ **25.** $\{-4, 0\}$
27. $\{-3, 9\}$ **29.** $\left\{-\dfrac{1}{3}, 5\right\}$ **31.** linear **33.** quadratic **35.** $\{-1, 4\}$ **37.** $\{-7, -2\}$ **39.** $\left\{-\dfrac{1}{2}, 0\right\}$ **41.** $\left\{-\dfrac{1}{2}, 2\right\}$ **43.** $\{3\}$ **45.** $\{0, 6\}$
47. $\{-2, 3\}$ **49.** $\{-2\}$ **51.** $\left\{\dfrac{1}{4}, 1\right\}$ **53.** $\{-4, 6\}$ **55.** $\{-5, 10\}$ **57.** $\{-5, 1\}$ **59.** $\{-3, 0, 2\}$ **61.** $\{-3, -2, 2\}$ **63.** $\left\{-\dfrac{3}{2}, 2, -2\right\}$
65. $\left\{-\dfrac{3}{5}, 4\right\}$ **67.** $\{-4, 5\}$ **69.** $\{-5\}$ **71.** $\left\{-\dfrac{3}{4}, 7\right\}$ **73.** $\{-4, 2\}$ **75.** $\left\{-4, -\dfrac{1}{2}, 4\right\}$ **77.** $\{20\}$ **79.** $\{-10, 10\}$ **81.** $\{-6, 2\}$ **83.** $\left\{-\dfrac{1}{2}, \dfrac{3}{2}, 5\right\}$
85. $\{0, 8\}$ **87.** $\left\{-4, \dfrac{3}{2}\right\}$ **89.** $\left\{-6, \dfrac{1}{2}\right\}$ **91.** 1 sec, 3 sec **93.** 9 **95.** -4 and -3 or 3 and 4 **97.** $-12, -10$, and -8, or 8, 10, and 12 **99.** 15 by 17
101. 8 teams **103.** $(x - 3)(x + 5) = 0$; at least degree 2 **105.** $(z - 6)^2 = 0$; at least degree 2 **107.** $(x + 3)(x - 1)(x - 5) = 0$; at least degree 3
109. $\{a, -b\}$ **111.** $\{0, 2a\}$

113. The student divided by the variable x instead of writing the quadratic equation in standard form, factoring, and using the Zero-Product Property. The correct solution is

$$15x^2 = 5x$$
$$15x^2 - 5x = 0$$
$$5x(3x - 1) = 0$$
$$5x = 0 \quad \text{or} \quad 3x - 1 = 0$$
$$x = 0 \quad \text{or} \quad 3x = 1$$
$$x = \frac{1}{3}$$

115. We write a quadratic equation in standard form, $ax^2 + bx + c = 0$, so that when the quadratic polynomial on the left side of the equation is factored, we can apply the Zero-Product Property, which says that if $a \cdot b = 0$, then $a = 0$ or $b = 0$.

Section 6.7 Modeling and Solving Problems with Quadratic Equations
1. (a) After 4 seconds and after 6 seconds (b) After 10 seconds
2. 8 km by 13 km **3.** height = 8 yards, base = 12 yards **4.** $x = 12$, $x - 3 = 9$ **5.** 7 inches and 24 inches **6.** 6 inches high, 8 inches wide **7.** 4, 14
9. 2, 9 **11.** base = 26; height = 8 **13.** base = 18; height = 16 **15.** base = 13; height = 11 **17.** $B = 53$; $b = 43$ **19.** 9, 12 **21.** 12, 5 **23.** 256, 252,
240, 220, 192, 156, 112, 60, 0 feet **25.** 6 sec **27.** width = 4 m; length = 12 m **29.** base = 6 ft; height = 18 ft **31.** 30 inches
33. width = 12 mm; length = 25 mm **35.** width = 4 cm; length = 25 cm **37.** (a) length = 45 in.; width = 28 in. (b) 48 in. by 31 in. **39.** 6 ft
41. width = 7 ft; length = 14 feet **43.** If you have two positive numbers that satisfy the projectile motion word problem, one answer is the height of the object on its upward path, and the other represents the height of the object on its downward path. There is only one positive number that satisfies the motion problem if the object is dropped from a height above the ground or if the answer represents the maximum height of the projectile.

Chapter 6 Review **1.** 12 **2.** 27 **3.** 10 **4.** 4 **5.** x^2 **6.** m **7.** $15ab^2$ **8.** $6x^3y^2z$ **9.** $2(2a+1)^2$ **10.** $9(x-y)$ **11.** $-6a^2(3a+4)$ **12.** $-3x(3x-4)$
13. $5y^2z(3+y^5+4y)$ **14.** $7xy(x^2-3xy+2y^2)$ **15.** $(5-y)(x+2)$ **16.** $(a+b)(z+y)$ **17.** $(5m+2n)(m+3n)$ **18.** $(2x+y)(y+x)$
19. $(x+2)(8-y)$ **20.** $(y^2+1)(x-3)$ **21.** $(x+3)(x+2)$ **22.** $(x+2)(x+4)$ **23.** $(x-7)(x+3)$ **24.** $(x+5)(x-2)$ **25.** prime
26. prime **27.** $(x-3y)(x-5y)$ **28.** $(m+5n)(m-n)$ **29.** $-(p+6)(p+5)$ **30.** $-(y-5)(y+3)$ **31.** $3x(x^2+11x+12)$
32. $4(x+1)(x+8)$ **33.** $2(x-7y)(x+6y)$ **34.** $4y(y+5)(y-2)$ **35.** $(5y-6)(y+4)$ **36.** $(y-7)(6y+1)$ **37.** $(2x-3)(x-1)$
38. $(2x+7)(3x+1)$ **39.** prime **40.** $(4m+3)(2m+3)$ **41.** $3m(m+n)(3m+7n)$ **42.** $2(7m+n)(m+n)$ **43.** $-x(5x+2)(3x-1)$
44. $-2p^2(3p-1)(2p+1)$ **45.** $(2x-3)^2$ **46.** $(x-5)^2$ **47.** $(x+3y)^2$ **48.** prime **49.** $2(2m+1)^2$ **50.** $2(m-6)^2$ **51.** $(2x-5y)(2x+5y)$
52. $(7x-6y)(7x+6y)$ **53.** prime **54.** prime **55.** $(x-3)(x+3)(x^2+9)$ **56.** $(x-5)(x+5)(x^2+25)$ **57.** $(m+3)(m^2-3m+9)$
58. $(m+5)(m^2-5m+25)$ **59.** $(3p-2)(9p^2+6p+4)$ **60.** $(4p-1)(16p^2+4p+1)$ **61.** $(y^3+4z^2)(y^6-4y^3z^2+16z^4)$
62. $(2y+3z^2)(4y^2-6yz^2+9z^4)$ **63.** $(5a-2b)(3a^2-5b^2)$ **64.** $(4a-3b)(3a+b)$ **65.** prime **66.** $(x-12y)(x+2y)$ **67.** $x(x-7)(x+6)$
68. $3x^4(x-7)(x-3)$ **69.** $(3x+1)(2x+3)$ **70.** $(2z+3)(5z-3)$ **71.** $(3x+2)(9x^2-6x+4)$ **72.** $(2z-1)(4z^2+2z+1)$
73. $2(2y-1)(y+5)$ **74.** $xy^2(5x-3)(x-1)$ **75.** $(5k-9m)(5k+9m)$ **76.** $(x^2-3)(x^2+3)$ **77.** prime **78.** prime **79.** $\left\{\dfrac{3}{2}, 4\right\}$
80. $\left\{-7, -\dfrac{1}{2}\right\}$ **81.** $\{-3, 15\}$ **82.** $\{2, 5\}$ **83.** $\{0, -2\}$ **84.** $\left\{-\dfrac{9}{2}, 0\right\}$ **85.** $\{-1, 3\}$ **86.** $\{-3, -1\}$ **87.** $\{-4, -2\}$ **88.** $\left\{-\dfrac{1}{2}, \dfrac{3}{2}\right\}$
89. $\{-14, 0, 3\}$ **90.** $\{-3, 0, 6\}$ **91.** $\left\{-3, \dfrac{4}{3}, 3\right\}$ **92.** $\left\{-\dfrac{5}{2}, -2, 2\right\}$ **93.** 5 sec **94.** 2 sec and 3 sec **95.** 9 ft by 6 ft **96.** 3 yd by 5 yd
97. 8 ft, 6 ft **98.** 24 ft, 10 ft

Chapter 6 Test **1.** $4x^4y^2$ **2.** $(x-4)(x+4)(x^2+16)$ **3.** $9x(2x-3)(x+1)$ **4.** $(x-7)(y-4)$ **5.** $(3x+5)(9x^2-15x+25)$
6. $(y-12)(y+4)$ **7.** $(6m+5)(m-1)$ **8.** prime **9.** $(x-5)(4+y)$ **10.** $3y(x-7)(x+2)$ **11.** prime
12. $2(x^2-3y)(x^4+3x^2y+9y^2)$ **13.** $3x(3x+1)(x+4)$ **14.** $(3m+2)(2m+1)$ **15.** $2(2m^2-3mn+2)$ **16.** $(5x+7y)^2$
17. $\left\{-\dfrac{1}{5}, -3\right\}$ **18.** $\{0, 2\}$ **19.** length $= 7$ in.; width $= 5$ in. **20.** legs: 12 in., 5 in., hypotenuse: 13 in.

Chapter 7 Rational Expressions and Equations

Section 7.1 Simplifying Rational Expressions
1. rational expression **2.** (a) $-\dfrac{1}{8}$ (b) $\dfrac{3}{16}$ **3.** (a) -1 (b) 9 **4.** $\dfrac{11}{2}$ **5.** 4
6. undefined **7.** $x = -7$ **8.** $n = -\dfrac{5}{3}$ **9.** $x = -3$ or $x = 1$ **10.** simplify **11.** False **12.** $\dfrac{n+4}{2}$ **13.** $-\dfrac{z+1}{2}$ **14.** $\dfrac{a+7}{2a+7}$ **15.** $\dfrac{z^2-3}{z-2}$ **16.** $\dfrac{2k+1}{2k-1}$
17. False **18.** -7 **19.** $\dfrac{-4}{4x-1}$ **20.** $-\dfrac{5z+1}{3}$ **21.** (a) 2 (b) $\dfrac{1}{2}$ (c) 0 **23.** (a) 1 (b) 3 (c) $\dfrac{5}{3}$ **25.** (a) -3 (b) 0 (c) $-\dfrac{15}{7}$
27. (a) 0 (b) $-\dfrac{7}{2}$ (c) $-\dfrac{3}{4}$ **29.** 0 **31.** 5 **33.** $\dfrac{3}{2}$ **35.** $-6, 6$ **37.** 2, 5 **39.** $-2, 0, 3$ **41.** $\dfrac{x-2}{3}$ **43.** $z+2$ **45.** $\dfrac{1}{p+2}$ **47.** -1 **49.** $-2k$ **51.** $\dfrac{x-1}{2x+3}$
53. $\dfrac{6}{x-5}$ **55.** -3 **57.** $\dfrac{b-5}{4}$ **59.** $\dfrac{x+3}{x-3}$ **61.** $\dfrac{3x-4}{x-5}$ **63.** $-(x+y)$ **65.** $\dfrac{x^2+1}{x+1}$ **67.** $-\dfrac{4+c}{c-4}$ **69.** $\dfrac{2(x-2)}{3(x-5)}$ **71.** $-\dfrac{x-3}{6x-5}$ **73.** $\dfrac{2}{t^2+9}$ **75.** $\dfrac{-3}{w-1}$
77. (a) 5 mg/mL (b) 4 mg/mL **79.** Yes; BMI $= 27.5$ **81.** \$15,600 **83.** $\dfrac{c^4+1}{c^6+1}$ **85.** $\dfrac{(x-2)(x^2-3)}{x^3-9}$ **87.** $\dfrac{(t^2+4)(t+2)}{t^2-2t+4}$ **89.** (a) The x^2s
cannot be divided out because they are not factors. (b) When the factors $x-2$ are divided out, the result is $\dfrac{1}{3}$, not 3. **91.** A rational expression is
undefined if the denominator is equal to zero. Division by zero is not allowed in the real number system. The expression $\dfrac{x^2-9}{3x-6}$ is undefined for
$x = 2$ because when 2 is substituted for x, we get $\dfrac{2^2-9}{3(2)-6} = \dfrac{4-9}{6-6} = \dfrac{-5}{0}$, which is undefined. **93.** $\dfrac{x-7}{7-x} = -1$ because $x-7$ and $7-x$ are opposites:
$\dfrac{x-7}{7-x} = \dfrac{x-7}{-1(-7+x)} = \dfrac{x-7}{-1(x-7)} = \dfrac{\cancel{x-7}}{-1\cancel{(x-7)}} = \dfrac{1}{-1} = -1$. The expression $\dfrac{x-7}{x+7}$ is in simplest form because $x-7$ and $x+7$ are not opposites
and they have no common factor.

Section 7.2 Multiplying and Dividing Rational Expressions
1. False **2.** False **3.** $\dfrac{5p(p+3)}{2}$ **4.** $\dfrac{7}{3(x+5)}$ **5.** $-\dfrac{1}{x-3}$ **6.** $-\dfrac{3}{7}$ **7.** $\dfrac{1}{3a}$ **8.** $\dfrac{3}{2}; \dfrac{4}{3}$
9. $\dfrac{6(x+1)}{x(2x-1)}$ **10.** $\dfrac{(x-3)^2}{(x-4)^2}$ **11.** $\dfrac{q+1}{(q-5)(q+5)}$ **12.** $\dfrac{x-1}{4}$ **13.** $\dfrac{7x}{4(x-2)}$ **14.** $\dfrac{6m}{m+2n}$ **15.** $\dfrac{2x}{x-5}$ **17.** $\dfrac{x}{(x+1)^2}$ **19.** $p-1$ **21.** 1 **23.** $\dfrac{1}{3x}$
25. $\dfrac{2(m-4)}{(m+4)}$ **27.** x **29.** 1 **31.** $\dfrac{10}{3}$ **33.** 1 **35.** $-\dfrac{4}{11}$ **37.** $\dfrac{4}{3y(y-3)}$ **39.** $\dfrac{1}{3}$ **41.** $\dfrac{3x}{x-4}$ **43.** $\dfrac{2(4w+1)}{4w-1}$ **45.** $-\dfrac{(x-y)^2}{x}$ **47.** $-\dfrac{1}{2}$ **49.** $\dfrac{3(4n+3)(n-3)}{2(n+4)}$
51. $\dfrac{x}{4y^2(x-2y)}$ **53.** $\dfrac{2a-b}{2a(a+b)}$ **55.** $\dfrac{2x+5}{2(x+3)}$ **57.** -3 **59.** $-\dfrac{x(x+3)}{(x+1)^2}$ **61.** $\dfrac{a+b}{2}$ **63.** $\dfrac{1}{3}$ **65.** $\dfrac{(x-a)^2}{2(x+a)}$ **67.** $\dfrac{1}{3(x-y)}$ **69.** $\dfrac{9}{8}$ **71.** $\dfrac{x-y}{(x+y)^3}$
73. $\dfrac{4x}{(x-2)(x-3)}$ **75.** $\dfrac{1}{x}$ ft^2 **77.** $\dfrac{2(x+2)}{x-3}$ in.^{2} **79.** $\dfrac{1-x}{4}$ **81.** $\dfrac{2a^2}{3}$ **83.** $\dfrac{(x+y)^2}{(x^2-xy+y^2)(x+2)}$ **85.** $3p^5q^2(p^2+3pq+9q^2)(p-q)$ **87.** $6x+12$
89. To multiply two rational expressions, (1) factor the polynomials in the numerator and denominator; (2) multiply the numerators and denominators
using $\dfrac{a}{b} \cdot \dfrac{c}{d} = \dfrac{a \cdot c}{b \cdot d}$; (3) divide out common factors in the numerators and denominators. Leave the result in factored form. **91.** The error is in
incorrectly dividing out the factors $(n-6)$ and $(6-n)$. The correct method is $\dfrac{n^2-2n-24}{6n-n^2} = \dfrac{-1}{n\cancel{(6-n)}} \cdot \dfrac{\cancel{(n-6)}(n+4)}{n\cancel{(6-n)}} = \dfrac{-(n+4)}{n}.$

Section 7.3 Adding and Subtracting Rational Expressions with a Common Denominator **1.** $\dfrac{a+b}{c}$ **2.** False **3.** $\dfrac{4}{x-2}$ **4.** $x+1$ **5.** $\dfrac{11x-3}{6x-5}$

6. $\dfrac{1}{x-5}$ **7.** $\dfrac{2(4y-3)}{2y-5}$ **8.** $\dfrac{z+1}{2z}$ **9.** $x+3$ **10.** $\dfrac{x+8}{3}$ **11.** $\dfrac{x+2}{x-7}$ **12.** True **13.** $\dfrac{3x-1}{x-5}$ **14.** 2 **15.** $\dfrac{2}{n-3}$ **16.** $\dfrac{2k+3}{2(k-1)}$ **17.** $\dfrac{7p}{4}$ **19.** $2n$

21. $\dfrac{2a-1}{a}$ **23.** $\dfrac{2(5c-2)}{c-1}$ **25.** 2 **27.** $7x$ **29.** 2 **31.** $\dfrac{1}{a-5}$ **33.** $\dfrac{x}{x+3}$ **35.** $-\dfrac{x}{2}$ **37.** $-\dfrac{c+3}{c}$ **39.** 3 **41.** -1 **43.** $x-1$ **45.** 1 **47.** $\dfrac{2(2p+3q)}{p-q}$

49. $\dfrac{2p^2+2p+1}{p-1}$ **51.** $\dfrac{4b}{a-b}$ **53.** -1 **55.** $\dfrac{2}{x-y}$ **57.** $-\dfrac{1}{x}$ **59.** -1 **61.** 0 **63.** $n-1$ **65.** $\dfrac{3}{v+3}$ **67.** $\dfrac{x+1}{x-1}$ **69.** $\dfrac{x^2+3}{x(x-3)}$ **71.** $2(x-1)$

73. $\dfrac{3x+3y-2}{x-y}$ **75.** $\dfrac{p+2q}{p^2-6q^2}$ **77.** $12a$ **79.** $\dfrac{n-3}{n+1}$ **81.** $\dfrac{4}{n}$ **83.** $\dfrac{x}{x+3}$ **85.** $\dfrac{4x-2}{x+2}$ **87.** $\dfrac{2(6x-1)}{x}$ cm **89.** $\dfrac{6x+7y}{(x+y)(x-y)}$ **91.** $\dfrac{-x(x-1)}{(3x-1)(x+2)}$

93. $\dfrac{-5x+12}{x-2}$ **95.** $-4n-6$ **97.** The expression is incorrect because the subtraction symbol was not applied to both terms of the second numerator. The correct method is $\dfrac{x-2}{x}-\dfrac{x+4}{x}=\dfrac{x-2-x-4}{x}=\dfrac{-6}{x}$. **99.** The correct answer is (c). The answer (a) $\dfrac{7}{2x}$ is incorrect because the denominators were added; and the answer (b) $\dfrac{7}{x^2}$ is incorrect because the denominators were multiplied.

Section 7.4 Finding the Least Common Denominator and Forming Equivalent Rational Expressions **1.** least common denominator

2. False **3.** $24x^2y^3$ **4.** $42a^3b^3$ **5.** True **6.** $15z(z+1)$ **7.** $(x-1)(x+5)^2$ **8.** $-3(x-7)(x+7)$ **9.** $x+3$ **10.** $\dfrac{12p^2}{16p^3(p-2)}$

11. LCD $=60a^2b^3$; $\dfrac{3}{6a^2b}=\dfrac{30b^2}{60a^2b^3}$, $\dfrac{-5}{20ab^3}=\dfrac{-15a}{60a^2b^3}$ **12.** LCD $=(x-5)(x-2)(x+1)$; $\dfrac{5}{x^2-4x-5}=\dfrac{5x-10}{(x-5)(x-2)(x+1)}$,

$\dfrac{-3}{x^2-7x+10}=\dfrac{-3x-3}{(x-5)(x-2)(x+1)}$ **13.** $25x^2$ **15.** $60x^3y^2$ **17.** $x(x+1)$ **19.** $2x+1$ **21.** $4(b-3)$ **23.** $p(p+1)(p-2)$ **25.** $(r+2)^2(r+1)(r-2)$

27. $-(x-4)$ **29.** $-2(x+3)(x-3)$ **31.** $-(x-1)(x+2)$ **33.** $\dfrac{12x^2}{3x^3}$ **35.** $\dfrac{3c+c^2}{a^2b^2c^2}$ **37.** $\dfrac{x^2-2x-8}{(x-4)(x+4)}$ **39.** $\dfrac{9n^2-9n}{6(n-1)(n+1)}$ **41.** $\dfrac{4t^2-4t}{t-1}$

43. $\dfrac{6xy}{9y^2}$, $\dfrac{4}{9y^2}$ **45.** $\dfrac{6a+2}{4a^3}$, $\dfrac{4a^3-a^2}{4a^3}$ **47.** $\dfrac{2m+2}{m(m+1)}$, $\dfrac{3m}{m(m+1)}$ **49.** $\dfrac{2y^2-5y+2}{4y(2y-1)}$, $\dfrac{y^2}{4y(2y-1)}$ **51.** $\dfrac{x+1}{(x-1)(x+1)}$, $\dfrac{-2x}{(x-1)(x+1)}$

53. $\dfrac{4x}{(x+2)(x-2)}$, $\dfrac{2x-4}{(x+2)(x-2)}$ **55.** $\dfrac{3x^2+9x}{(x-7)(x+3)(x+1)}$, $\dfrac{5x+5}{(x-7)(x+3)(x+1)}$ **57.** $\dfrac{x^2-2x-3}{(x+3)(x-3)^2}$, $\dfrac{x^2+5x+6}{(x+3)(x-3)^2}$

59. $\dfrac{6x-3}{(x+4)(2x-1)}$, $\dfrac{2x-1}{(x+4)(2x-1)}$ **61.** $3x$; $\dfrac{3}{3x}$, $\dfrac{1}{3x}$ **63.** $(r-4)(r+4)$; $\dfrac{12r+48}{(r-4)(r+4)}$, $\dfrac{12r-48}{(r-4)(r+4)}$ **65.** $(x+2)(x-2)(x^2-2x+4)(x^2+2x+4)$

67. $p^2(p-1)(p-3)$ **69.** $4a^2b^2(a+b)(a-b)(a^2+ab+b^2)$ **71.** To write an equivalent rational expression with a given denominator: (1) Write each denominator in factored form. (2) Determine the "missing factor(s)" (Answer the question "What factor(s) does the new denominator have that is missing from the original denominator?") (3) Multiply the original expression by $1=\dfrac{\text{missing factor}}{\text{missing factor}}$. (4) Find the product and leave the denominator in factored form.

Section 7.5 Adding and Subtracting Rational Expressions with Unlike Denominators **1.** 72 **2.** $\dfrac{25}{36}$ **3.** $\dfrac{7}{30}$ **4.** least common denominator

5. True **6.** $\dfrac{3b+10a^2}{24a^3b^2}$ **7.** $\dfrac{6x^2+35y}{90x^3y^2}$ **8.** $\dfrac{2(4x-1)}{(x-4)(x+2)}$ **9.** $\dfrac{3n-13}{(n-3)(n+1)}$ **10.** $\dfrac{1}{z+2}$ **11.** $\dfrac{2}{(x-5)(x+5)}$ **12.** $\dfrac{-16ab-15}{20a^2b^3}$ **13.** $\dfrac{2(x-10)}{x(x-4)}$

14. $\dfrac{x-11}{(x-5)(x+4)(x-1)}$ **15.** $\dfrac{(x+8)(x-1)}{4x(x-3)}$ **16.** True **17.** $\dfrac{p-5}{p-1}$ **18.** $\dfrac{y+4}{y+1}$ **19.** $\dfrac{2x+1}{x-1}$ **20.** $\dfrac{-3(x+1)}{x+3}$ **21.** $\dfrac{4x}{(x-1)(x+1)}$ **22.** $\dfrac{2}{x-2}$

23. $-\dfrac{5}{6}$ **25.** $\dfrac{5}{3x}$ **27.** $\dfrac{2a^2+7a-3}{(2a-1)(2a+1)}$ **29.** $-\dfrac{2}{x-4}$ **31.** $\dfrac{a+4}{4a}$ **33.** $\dfrac{5}{x}$ **35.** $\dfrac{8}{75}$ **37.** $\dfrac{(m-4)(m+4)}{m}$ **39.** $\dfrac{2(4x-3)}{(x-3)(x+3)}$ **41.** $\dfrac{2}{x+3}$ **43.** $\dfrac{x}{x-2}$

45. $\dfrac{20-9y}{12y^2}$ **47.** $\dfrac{6}{5x}$ **49.** $-\dfrac{3}{2x+3}$ **51.** $\dfrac{2(n^2-2)}{n(n-2)}$ **53.** $\dfrac{2x^2-5x+15}{x(x-3)}$ **55.** $\dfrac{y-2}{y-3}$ **57.** $\dfrac{2n+5}{2n+1}$ **59.** $\dfrac{4x-5}{x-2}$ **61.** $\dfrac{3x-28}{x-9}$ **63.** $\dfrac{1}{(x+1)(x-1)}$

65. $-\dfrac{13}{2a(a-2)}$ **67.** $-\dfrac{n^2-n-3}{(n+2)(n-2)(n+3)}$ **69.** $\dfrac{-1}{(m-2)(m-4)}$ **71.** $\dfrac{-3n+5}{(n+3)(n-2)}$ **73.** $\dfrac{x}{x-2}$ **75.** $\dfrac{x-4}{x}$ **77.** $\dfrac{6}{m(m-2)}$ **79.** $\dfrac{x^2+4}{(x+2)(x+3)}$

81. $\dfrac{x+3}{(x+1)(x-1)}$ **83.** $\dfrac{2x^2-x+10}{(x+3)(x-2)(x+2)}$ **85.** $\dfrac{2x^2-5x+3}{(x+6)(x-7)}$ **87.** $\dfrac{7x-5}{4}$ units **89.** $\dfrac{a+2}{a(a-1)^2}$ **91.** $\dfrac{x^3-2x^2+4x+3}{x^2(x+1)(x-1)}$ **93.** $\dfrac{1}{a+b}$

95. $\dfrac{x^2+2x+5}{(x-4)(x+1)}$ **97.** $\dfrac{-x^3+6x^2-11x+10}{x-2}$ **99.** The steps to add or subtract rational expressions with unlike denominators are (1) find the least common denominator; (2) rewrite each rational expression with the common denominator; (3) add or subtract the rational expressions from Step 2; (4) simplify the result. **101.** To complete the subtraction problem $\dfrac{3x+1}{(2x+5)(x-1)}-\dfrac{x-4}{(2x+5)(x-1)}$, distribute the subtraction sign to both the x and the -4 in the numerator of the second term to form $\dfrac{3x+1-x+4}{(2x+5)(x-1)}=\dfrac{2x+5}{(2x+5)(x-1)}$. Then divide out the common factor $(2x+5)$ to arrive at the result $\dfrac{1}{x-1}$.

Section 7.6 Complex Rational Expressions

1. complex rational expression **2.** simplify **3.** $\dfrac{k+1}{2}$ **4.** $\dfrac{1}{2n}$ **5.** False **6.** $\dfrac{3}{2}$ **7.** $\dfrac{1}{3+y}$ **8.** $\dfrac{3}{x+5}$

9. 13 **10.** $\dfrac{y+2x}{2y-x}$ **11.** $\dfrac{2}{17}$ **13.** $\dfrac{3(x-2)}{2}$ **15.** $\dfrac{9}{x-3}$ **17.** $\dfrac{n(m+2n)}{2m}$ **19.** $\dfrac{5b^2+4a}{a(5b+4)}$ **21.** $-\dfrac{2}{y+4}$ **23.** $\dfrac{3(y-24)}{2(5-6y^2)}$ **25.** $\dfrac{15}{16}$ **27.** $\dfrac{1}{2}$ **29.** $\dfrac{b-7}{b}$

31. $\dfrac{x+4}{x}$ **33.** $-\dfrac{x+y}{x^2y^2}$ **35.** $\dfrac{ab+1}{ab+2}$ **37.** $\dfrac{18n}{5}$ **39.** $\dfrac{x^2+2xy-y^2}{xy}$ **41.** $-\dfrac{1}{2x+7}$ **43.** $-b$ **45.** $\dfrac{x+1}{x+3}$ **47.** $\dfrac{2b-a}{2b}$ **49.** $\dfrac{-4x+17}{2x-1}$ **51.** $\dfrac{n+1}{5n-9}$

53. $\dfrac{11(2n+3)}{72}$ **55.** $-\dfrac{x^3}{x+3}$ **57. (a)** $R=\dfrac{R_1R_2}{R_2+R_1}$ **(b)** $\dfrac{15}{4}$ ohms **59.** $\dfrac{x}{1+3x}$ **61.** $\dfrac{x}{2-5x}$ **63.** $\dfrac{2x+1}{x+1}$ **65.** $\dfrac{5-2x}{2-x}$ **67.** The expression $\dfrac{\dfrac{2}{x+1}}{\dfrac{1}{x-5}}$ is

not in simplified form. Simplified form is the form $\dfrac{p}{q}$, where p and q are polynomials that have no common factors. **69.** Answers may vary.

Putting the Concepts Together (Section 7.1–7.6)

1. -35 **2. (a)** The expression is undefined for $a=6$. **(b)** The expression is undefined for
$y=0$ and $y=-4$. **3. (a)** $\dfrac{a-4b}{2}$ **(b)** $\dfrac{x+2}{x+1}$ **4.** The LCD is $8(a+2b)(a-4b)$. **5.** $\dfrac{7}{3x^2-x}=\dfrac{35x}{5x^2(3x-1)}$ **6.** $\dfrac{-2y^2}{y+1}$ **7.** $\dfrac{2(m-1)^2}{m^2}$
8. $\dfrac{2(y-1)}{5(y-3)}$ **9.** $4x+1$ **10.** $\dfrac{1}{x+1}$ **11.** $\dfrac{1}{2x-1}$ **12.** $\dfrac{15m+17}{(m+3)(3m+2)}$ **13.** $\dfrac{m}{(m+5)(m+4)}$ **14.** $-\dfrac{1}{m-1}$ **15.** $\dfrac{a(a+24)}{3(a^2+6)}$

Section 7.7 Rational Equations

1. rational equation **2.** $\left\{\dfrac{2}{3}\right\}$ **3.** $\{-18\}$ **4.** $\{-6\}$ **5.** $\{4\}$ **6.** $\{-5,6\}$ **7.** extraneous solutions

8. $\{5\}$ **9.** True **10.** $\{\ \}$ or $\varnothing$ **11.** After 2.5 hours and after 10 hours **12.** $x=\dfrac{4g}{R}$ **13.** $r=1-\dfrac{a}{S}$ or $r=\dfrac{S-a}{S}$ **14.** $p=\dfrac{fq}{q-f}$

15. $\{2\}$ **17.** $\left\{-\dfrac{3}{4}\right\}$ **19.** $\{-7\}$ **21.** $\left\{\dfrac{1}{2}\right\}$ **23.** $\{7\}$

25. $\{\ \}$ or $\varnothing$ **27.** $\{-3\}$ **29.** $\left\{\dfrac{3}{4}\right\}$ **31.** $\left\{-\dfrac{5}{6}\right\}$ **33.** $\{-4\}$

35. $\{-3,-2\}$ **37.** $\{-1\}$ **39.** $\left\{-\dfrac{2}{3},\dfrac{1}{2}\right\}$ **41.** $\{\ \}$ or $\varnothing$

43. $\{-3,4\}$ **45.** $\left\{-\dfrac{1}{2}\right\}$ **47.** $y=\dfrac{2}{x}$ **49.** $R=\dfrac{E}{I}$

51. $b=\dfrac{2A}{h}-B$ or $b=\dfrac{2A-Bh}{h}$ **53.** $y=\dfrac{x}{z}-3$ or $y=\dfrac{x-3z}{z}$

55. $S=\dfrac{RT}{T-R}$ **57.** $y=\dfrac{an-p}{am}$ **59.** $x=\dfrac{Ay}{y-A}$

61. $y=\dfrac{xz}{2z-6x}$ **63.** $x=\dfrac{cy}{y-1}$ **65.** $\dfrac{4x+5}{x(x+5)}$ **67.** $\{-2,3\}$

69. $\dfrac{x+7}{2}$ **71.** $\left\{-\dfrac{1}{6},1\right\}$ **73.** $\{-1,5\}$ **75.** $\dfrac{x^3-x^2+4}{x(x+2)(x-2)}$

77. $\dfrac{(x-3)(x-1)}{3x}$ **79.** $\dfrac{5}{a}$ **81.** $\{\ \}$ or $\varnothing$ **83.** 2 **85.** $\{2\}$ **87.** $\{-3\}$

89. After 1 hour and after 9 hours **91.** 80% **93.** $k=33$ **95.** The error
occurred when the student incorrectly multiplied $-2[(x-1)(x+1)]$ on
the right side of the equation. The correct solution is

$$\dfrac{2}{x-1}-\dfrac{4}{x^2-1}=-2$$

$$\dfrac{2}{x-1}-\dfrac{4}{(x-1)(x+1)}=-2$$

$$[(x-1)(x+1)]\left(\dfrac{2}{x-1}-\dfrac{4}{(x-1)(x+1)}\right)=-2[(x-1)(x+1)]$$

$$2(x+1)-4=-2(x-1)(x+1)$$

$$2x+2-4=-2(x^2-1)$$

$$2x-2=-2x^2+2$$

$$2x^2+2x-4=0$$

$$x^2+x-2=0$$

$$(x+2)(x-1)=0$$

$$x+2=0 \quad \text{or} \quad x-1=0$$

$$x=-2 \qquad\qquad x=1$$

97. To *simplify* a rational expression means to add, subtract, multiply, or divide the rational expressions and express the result in a form in which there
are no common factors in the numerator or denominator. To *solve* an equation means to find the value(s) of the variable that satisfy the equation.
Answers will vary.

Section 7.8 Models Involving Rational Equations

1. proportion **2.** $d;c$ **3.** $\left\{-\dfrac{2}{3}\right\}$ **4.** $\{3\}$ **5.** $\$233.38$ **6.** 210 miles **7.** similar **8.** $XY=8$

9. 60 feet **10.** constant; $\dfrac{1}{4}$ **11.** 2 hours **12.** $3\dfrac{1}{3}$ hours or 3 hours, 20 minutes **13.** 20 mph **14.** 6 mph **15.** $\{12\}$ **17.** $\left\{\dfrac{18}{7}\right\}$ **19.** $\{16\}$ **21.** $\left\{\dfrac{24}{5}\right\}$

23. $\left\{-\dfrac{4}{5}\right\}$ **25.** $\{-2\}$ **27.** $\left\{\dfrac{5}{7}\right\}$ **29.** $\{-3,2\}$ **31.** $\{2,5\}$ **33.** $\{11\}$ **35.** $XZ=12$ **37.** $n=3$; $ZY=5$ **39.** $x=30$; $AB=28$ **41.** $2x$ **43.** $t-3$

45. $\dfrac{1}{5}+\dfrac{1}{3}=\dfrac{1}{t}$ **47.** $\dfrac{1}{b+4}+\dfrac{1}{b}=\dfrac{1}{7}$ **49.** $r-2$ **51.** $r-48$ **53.** $\dfrac{4}{14-c}=\dfrac{7}{14+c}$ **55.** $\dfrac{12}{r+3}=\dfrac{8}{r-3}$ **57.** 145 miles **59.** $8\dfrac{1}{3}$ lb **61.** 37, 245 rubles

63. 24 ft **65.** 1.875 hr **67.** $\dfrac{24}{7}\approx 3.4$ minutes **69.** $\dfrac{36}{7}\approx 5.1$ hr **71.** 15 hr **73.** 24 hr **75.** 4 km per hour **77.** 1 hour, 40 minutes **79.** 40 mph

81. 35 mph **83.** $x=6$ **85.** $x=\dfrac{98}{5}$ **87.** $x=\dfrac{72}{5}$ **89.** The time component is not correctly placed. The times in the table should be $\dfrac{10}{r+1}$ and $\dfrac{10}{r-1}$.
The equation will incorporate the 5 hours: time traveled downstream + time traveled upstream = 5 hours. **91.** One equation to solve the problem is
$\dfrac{1}{8}+\dfrac{1}{6}=\dfrac{1}{t}$, where t is the number of hours required to complete the job together. Answers may vary.

Section 7.9 Variation   **1.** $t = ks$ **2.** constant of proportionality   **3.** $y = 3x$ **4. (a)** $q = \dfrac{1}{4}w$ **(b)** 15   **(c)**

5. \$28.90   **6.** $f = \dfrac{k}{d}$ **7.** False   **8.(a)** $y = \dfrac{12}{x}$ **(b)** 3

9. (a) $p = \dfrac{40}{q}$ **(b)** 5   **10.** 25 times per second   **11.** $y = \dfrac{1}{2}x$ **13.** $y = 2x$ **15.** $y = -\dfrac{3}{4}x$ **17. (a)** $p = \dfrac{1}{3}g$ **(b)** $p = 3$

19. (a) $y = -\dfrac{3}{4}x$ **(b)** $x = -\dfrac{5}{3}$ **21.** $y = \dfrac{6}{x}$ **23.** $y = \dfrac{12}{x}$ **25.** $y = \dfrac{2}{7x}$ **27. (a)** $e = \dfrac{8}{n}$ **(b)** $e = \dfrac{1}{2}$ **29. (a)** $b = \dfrac{3}{4a}$ **(b)** $a = \dfrac{5}{6}$ **31.** direct variation; $\dfrac{2}{3}$

33. neither   **35.** inverse variation; 9   **37.** direct variation; 2   **39.** $\dfrac{9}{2}$ **41.** $P = 75$ **43.** $t = 42$ **45.** $b = 18$ **47.** 9 representatives   **49.** 120 board feet

51. 6.25 atmospheres   **53.** 150 bags   **55.** $m = 90$ **57.** $q = 24$ **59. (a)** $e = \dfrac{1.989 \times 10^{-25}}{\lambda}$ **(b)** 3.978×10^{-19} joule   **61.** Answers will vary.

Chapter 7 Review   **1. (a)** $\dfrac{1}{2}$ **(b)** -4 **(c)** 1   **2. (a)** 0   **(b)** -1 **(c)** 1   **3. (a)** 0   **(b)** 9   **(c)** 49   **4. (a)** -1 **(b)** $-\dfrac{2}{3}$ **(c)** 42   **5.** $\dfrac{7}{3}$ **6.** $\dfrac{1}{2}$

7. none   **8.** none   **9.** -10 or -2 **10.** -1 or 4   **11.** $\dfrac{y+2}{5}$ **12.** $\dfrac{x(x-4)}{5}$ **13.** $\dfrac{3}{k+2}$ **14.** $-\dfrac{2}{x-3}$ **15.** $\dfrac{x+5}{2x-1}$ **16.** $\dfrac{3x-1}{2(x^2-2x+4)}$ **17.** $\dfrac{2m^2}{n^2}$ **18.** $\dfrac{y^3}{6x}$

19. $\dfrac{14}{9n^2}$ **20.** $\dfrac{1}{ab^7}$ **21.** $-\dfrac{5(x-2)}{x+3}$ **22.** $-\dfrac{4(x-3)(x+3)}{(x-9)^2}$ **23.** $\dfrac{x-2}{4(x-3)}$ **24.** $15x^3$ **25.** $\dfrac{x+5}{x+6}$ **26.** $\dfrac{y+1}{y-3}$ **27.** $\dfrac{x^2}{(x+1)(x+9)}$ **28.** $\dfrac{y-2}{3y+1}$ **29.** $\dfrac{9}{x-3}$

30. $\dfrac{10}{x+4}$ **31.** m **32.** $\dfrac{1}{m}$ **33.** $\dfrac{1}{m-2}$ **34.** $2m+3$ **35.** $\dfrac{1}{3m}$ **36.** $\dfrac{2}{b}$ **37.** $y+7$ **38.** $\dfrac{1}{m-6}$ **39.** $-\dfrac{7}{x-4}$ **40.** $\dfrac{4}{a-b}$ **41.** $\dfrac{3x-5}{x^2-25}$ **42.** $\dfrac{x+4}{x-3}$

43. $12x^4y^7$ **44.** $60a^3b^4c^7$ **45.** $8a(a+2)$ **46.** $5a(a+6)$ **47.** $4(x-3)(x+1)$ **48.** $x(x-7)(x+7)$ **49.** $\dfrac{6xy^6}{x^4y^7}$ **50.** $\dfrac{11a^4b^3}{a^7b^5}$ **51.** $\dfrac{(x-1)(x+2)}{(x-2)(x+2)}$

52. $\dfrac{(m+2)(m-2)}{(m+7)(m-2)}$ **53.** $\dfrac{20x^2y}{24x^5}; \dfrac{21}{24x^5}$ **54.** $\dfrac{12}{10a^3}; \dfrac{11a^2b}{10a^3}$ **55.** $\dfrac{4x^2+8x}{(x-2)(x+2)}; \dfrac{-6}{(x-2)(x+2)}$ **56.** $\dfrac{3m+9}{(m-5)(m+3)}; \dfrac{2m}{(m-5)(m+3)}$

57. $\dfrac{2m+4}{(m+7)(m-2)(m+2)}; \dfrac{m^2-m-2}{(m+7)(m-2)(m+2)}$ **58.** $\dfrac{n^2+3n-10}{n(n-5)(n+5)}; \dfrac{n^2}{n(n-5)(n+5)}$ **59.** $\dfrac{12+8x}{9x^2}$ **60.** $\dfrac{5y+x}{10x^2y^2}$ **61.** $\dfrac{x^2-5x+14}{(x+7)(x-7)}$

62. $\dfrac{2x^2+13x-9}{(2x+3)(2x-3)}$ **63.** $\dfrac{-2(2x+5)}{x(x-2)}$ **64.** $\dfrac{p-7}{p-4}$ **65.** $\dfrac{3x-1}{4x+1}$ **66.** $\dfrac{3x-4}{(x-1)(x-2)}$ **67.** $\dfrac{m-4}{m-3}$ **68.** $-\dfrac{1}{(y+1)(y+2)}$ **69.** $\dfrac{3}{m+2}$ **70.** $\dfrac{m^2+n^2}{(m-n)(m+n)}$

71. $\dfrac{5x+12}{x+3}$ **72.** $\dfrac{7x+13}{x+2}$ **73.** $-\dfrac{3}{23}$ **74.** $\dfrac{18}{11}$ **75.** $\dfrac{2m^2-10m}{m^2+10}$ **76.** $\dfrac{6+4m^2}{6m-5m^2}$ **77.** $\dfrac{1}{6}$ **78.** $-\dfrac{y+8}{y-2}$ **79.** $\dfrac{2(m+1)}{m+4}$ **80.** $-a$ **81.** $x = -5$ or $x = 0$

82. $x = -6$ or $x = 6$ **83.** $\{-4\}$ **84.** $\left\{\dfrac{2}{3}\right\}$ **85.** $\{5\}$ **86.** $\left\{-4, -\dfrac{1}{3}\right\}$ **87.** $\{38\}$ **88.** $\left\{\dfrac{9}{7}\right\}$ **89.** $\{\,\}$ or $\varnothing$ **90.** $\left\{-\dfrac{5}{2}\right\}$ **91.** $\{6\}$ **92.** $\{0\}$ **93.** $k = \dfrac{4}{y}$

94. $y = \dfrac{x}{6}$ **95.** $y = \dfrac{xz}{x-z}$ **96.** $z = \dfrac{xy}{x+y}$ **97.** $\{4\}$ **98.** $\{5\}$ **99.** $\{1\}$ **100.** $\{3\}$ **101.** $x = 15$ **102.** $x = 3.5$ **103.** 8 tanks   **104.** \$26.25   **105.** 1.2 hours

106. $\dfrac{15}{8} \approx 1.9$ hours   **107.** 15 min   **108.** $\dfrac{28}{3} \approx 9.3$ hours   **109.** 9 mph   **110.** 55 mph   **111.** 70 mph   **112.** 45 mph   **113.** $y = \dfrac{1}{4}x$ **114.** $y = 6x$

115. $f = 16$ **116.** $p = 110$ **117.** \$507.50   **118.** 160 mi   **119.** $y = \dfrac{48}{x}$ **120.** $y = \dfrac{54}{x}$ **121.** $r = 9$ **122.** $s = \dfrac{1}{4}$ **123.** 78 mph   **124.** 25 liters

Chapter 7 Test   **1.** 2   **2.** $x = 5$ or $x = -2$ **3.** $-\dfrac{x+3}{2}$ **4.** $\dfrac{14}{5x^2}$ **5.** $\dfrac{25}{9}$ **6.** $\dfrac{x-y}{2x-3y}$ **7.** y **8.** $\dfrac{x-5}{x+3}$ **9.** $-\dfrac{1}{y-z}$ or $\dfrac{1}{z-y}$ **10.** $\dfrac{2(x+3)(x-1)}{(x-2)(2x+1)}$

11. $\dfrac{x^2-5x-2}{(x+2)(x+3)(x-1)}$ **12.** $\dfrac{y-3}{3y}$ **13.** $\{-20, 5\}$ **14.** $\{2\}$ **15.** $y = \dfrac{xz}{x-z}$ **16.** $\{5\}$ **17.** $x = \dfrac{24}{5}$ **18.** \$33   **19.** 36 min   **20.** 4 mph   **21.** $m = 30$

Cumulative Review Chapter 1–7   **1.** -144 **2.** $9x - 11$ **3.** $\left\{\dfrac{19}{6}\right\}$ **4.** all real numbers   **5.** $\{4\}$ **6.** 7 feet   **7.** 44, 46, 48   **8.** $x \geq 1$;

9. $9x^2 - 12xy + 4y^2$ **10.** $\dfrac{16}{a^{16}b^{12}}$ **11.** $\dfrac{27z^9}{y^{12}}$ **12.** $2b(4a-7)(a+6)$ **13.** $\left\{0, \dfrac{1}{6}, 5\right\}$ **14.** 10 cm   **15.** 0   **16.** $-\dfrac{3(x+1)}{5x(2x-3)}$ **17.** $\dfrac{(x-2)(x+1)}{4(x+2)}$ **18.** $\dfrac{1}{x-6}$

19. $\dfrac{17x+30}{2(x-2)(x+2)}$ **20.** $\dfrac{4(10-3x)}{85x}$ **21.** $\{\,\}$ or $\varnothing$ **22.** $\{-16\}$ **23.** 10 inches   **24.** 9 hr   **25.** 7 mph   **26.** $\dfrac{4}{3}$ **27.** $y = -\dfrac{3}{2}x + 5$ or $3x + 2y = 10$

28.    **29.** $-\dfrac{3}{5}; \dfrac{5}{3}$ **30.** $(-4, 3)$

Chapter 8 Roots and Radicals

Section 8.1 Introduction to Square Roots 1. $b^2; a$ 2. -8 and 8 3. $-\dfrac{5}{7}$ and $\dfrac{5}{7}$ 4. -0.6 and 0.6 5. principal square root 6. radicand 7. 10 8. -3

9. $\dfrac{5}{7}$ 10. 0.6 11. 12 12. 13 13. 17 14. 7 15. ≈ 5.92 16. ≈ -2.45 17. True 18. rational; 7 19. irrational; ≈ 8.43 20. not a real number

21. rational; -4 22. $|a|$ 23. $|b|$ 24. $4|p|$ 25. $|w + 3|$ 26. $5b$ 27. $z - 6$ 28. $2h - 5$ 29. $-1, 1$ 31. $-\dfrac{1}{3}, \dfrac{1}{3}$ 33. $-13, 13$ 35. $-0.5, 0.5$

37. 12 39. -3 41. 15 43. $\dfrac{1}{11}$ 45. 0.2 47. -18 49. $\dfrac{4}{9}$ 51. 35 53. 10 55. 14 57. 8 59. 13 61. 2.828 63. 5.48 65. 7.5 67. not a real number

69. rational; 20 71. rational; $\dfrac{1}{2}$ 73. irrational; 7.35 75. irrational; 7.07 77. d 79. $5|y|$ 81. $|x - 9|$ 83. $m - 3$ 85. 1.73 87. 0.63

89. not a real number 91. 6 93. -8 95. not a real number 97. 5 99. 6.71 101. 25 ft 103. 16 km 105. 7 m 107. 14 in. 109. 398 m 111. 2
113. 5 115. The number $-\sqrt{9}$ is a real number because it denotes the opposite, or negative, of the principal square root of 9, namely 3. The number $\sqrt{-9}$ is not a real number because there is no real number whose square is -9.

Section 8.2 Simplifying Square Roots 1. $\sqrt{a} \cdot \sqrt{b}$ 2. 25 3. $2\sqrt{5}$ 4. $2\sqrt{6}$ 5. $6\sqrt{2}$ 6. $2 + \sqrt{5}$ 7. $\dfrac{-1 + 2\sqrt{2}}{2}$ or $-\dfrac{1}{2} + \sqrt{2}$ 8. False 9. $6y^4$

10. $3x^6\sqrt{5}$ 11. $2ab^5\sqrt{7}$ 12. $5y^5\sqrt{y}$ 13. $11a^2\sqrt{a}$ 14. $4b^4\sqrt{2ab}$ 15. $2x^3y^6\sqrt{3xy}$ 16. $\dfrac{\sqrt{a}}{\sqrt{b}}$ 17. $\dfrac{7a\sqrt{a}}{10}$ 18. $\dfrac{3\sqrt{11}}{4m}$ 19. 5 20. $3b^2\sqrt{b}$ 21. $2\sqrt{2}$

23. $2\sqrt{10}$ 25. $3\sqrt{2}$ 27. $\sqrt{33}$ 29. $3\sqrt{5}$ 31. $5\sqrt{5}$ 33. $\sqrt{42}$ 35. $7\sqrt{2}$ 37. $8\sqrt{2}$ 39. 5 41. $3 + \sqrt{2}$ 43. x^5 45. n^{72} 47. $3y^2$ 49. $2z^7\sqrt{6}$

51. $7x^3\sqrt{x}$ 53. $4b^2\sqrt{3b}$ 55. $3mn^2\sqrt{2m}$ 57. $5c^2d\sqrt{3}$ 59. $\dfrac{5}{6}$ 61. $\dfrac{\sqrt{11}}{3}$ 63. $\dfrac{y^8}{11}$ 65. $\dfrac{\sqrt{3}}{a^2}$ 67. $\dfrac{x^2\sqrt{x}}{y^4}$ 69. $\dfrac{2xy^2\sqrt{5y}}{z^2}$ 71. $\dfrac{\sqrt{15}}{x^2}$ 73. $\dfrac{5a}{2}$

75. $\dfrac{8n^2\sqrt{n}}{3}$ 77. $3\sqrt{7}$ 79. $\dfrac{2\sqrt{3}}{5}$ 81. $12a^3$ 83. $\dfrac{x^2\sqrt{x}}{3}$ 85. $2a^2bc^3\sqrt{3}$ 87. $\dfrac{2x^3\sqrt{x}}{y^6}$ 89. $3a^{12}\sqrt{3a}$ 91. $2\sqrt{13}$ 93. $\dfrac{-4 - 9\sqrt{2}}{6}$

95. $\dfrac{1 - \sqrt{2}}{2}$ 97. $x = 90\sqrt{2}$ 99. $x = 7\sqrt{5}$ 101. $x = 2\sqrt{34}$ 103. $x = 2\sqrt{73}$ 105. (a) $\dfrac{\sqrt{6}}{\pi}$Hz (b) 0.78 107. x^{24} 109. $16s^{18n}$

111. When simplifying $\sqrt{\dfrac{x^4}{y}}$, it must be stated that $y > 0$ because if $y = 0$, the denominator would be 0; division by zero is undefined in the set of real numbers. If $y < 0$, the radicand is negative, but the square root of a negative number is not a real number. 113. The expression $\sqrt{\dfrac{315}{5}} = \dfrac{3\sqrt{35}}{\sqrt{5}}$ is not completely simplified because $\dfrac{3\sqrt{35}}{\sqrt{5}} = 3\sqrt{\dfrac{35}{5}} = 3\sqrt{7}$. Another strategy to simplify $\sqrt{\dfrac{315}{5}}$ is $\sqrt{\dfrac{315}{5}} = \sqrt{63} = \sqrt{9 \cdot 7} = 3\sqrt{7}$.

Section 8.3 Adding and Subtracting Square Roots 1. radicand 2. $14\sqrt{3}$ 3. $-\sqrt{11} + 2\sqrt{10}$ 4. $-8\sqrt{13} + \sqrt{6}$ 5. $12\sqrt{5x}$ 6. False
7. False 8. $9\sqrt{2}$ 9. $-2\sqrt{3}$ 10. $5\sqrt{5} - 3\sqrt{2}$ 11. $8\sqrt{3z}$ 12. $-8a\sqrt{3a}$ 13. $6\sqrt{3}$ 15. $2a\sqrt{2a}$ 17. $\sqrt{15} + \sqrt{7}$ 19. 0 21. $-x\sqrt{2}$ or $-\sqrt{2}x$
23. $-3\sqrt{6} - 6\sqrt{11}$ 25. $2\sqrt{22}$ 27. $-\sqrt{3}$ 29. $7\sqrt{3}$ 31. $16\sqrt{5xy}$ 33. $-3\sqrt{3}$ 35. $21\sqrt{15}$ 37. $11\sqrt{5} - 4\sqrt{6}$ 39. $-x\sqrt{6x}$ 41. $-8a\sqrt{3a}$
43. $-\dfrac{16}{5}\sqrt{5}$ 45. $7\sqrt{14}$ 47. 0 49. $-5y^2\sqrt{y}$ 51. $\sqrt{6} + 9\sqrt{5}$ 53. $9\sqrt{2x} - 5\sqrt{2}$ 55. $-2n\sqrt{5n}$ 57. $3\sqrt{5}$ 59. $18\sqrt{3}$ 61. $18\sqrt{2} - 36\sqrt{3}$
63. 0 65. $5x\sqrt{5y} - 5x\sqrt{5}$ 67. $a^4\sqrt{10a} + 6a\sqrt{3a}$ 69. $7 + 13\sqrt{2}$ 71. $-1 + 5\sqrt{3}$ 73. $\dfrac{7}{6}\sqrt{3}$ 75. $\dfrac{9}{20}\sqrt{3}$ 77. $\dfrac{5}{6}\sqrt{15}$ 79. $\dfrac{11}{3}\sqrt{6}$

81. $24\sqrt{3}$ cm 83. $\left(18\sqrt{2} + 32\sqrt{3}\right)$ inches 85. 23.3 ft 87. $13 + 8\sqrt{3}$ 89. In the expression $6 - 2\sqrt{x} - 10 + 5\sqrt{x}$, the terms 6 and -10 are like terms, and $-2\sqrt{x}$ and $5\sqrt{x}$ are like terms. The expression $6 - 2\sqrt{x} - 10 + 5\sqrt{x}$ has not been correctly simplified. The correct simplification $6 - 2\sqrt{x} - 10 + 5\sqrt{x} = 6 - 10 - 2\sqrt{x} + 5\sqrt{x} = -4 + 3\sqrt{x}$. 91. When multiplying square roots, we multiply the radicands: $\sqrt{a \cdot b} = \sqrt{a} \cdot \sqrt{b}$ is the Product Rule for square roots. However, when we add square roots, the square root symbol acts as a grouping symbol. The operation under the radical sign must be done first, and then the radicand may be simplified, if possible. Thus $\sqrt{a + b} \neq \sqrt{a} + \sqrt{b}$.

Section 8.4 Multiplying Expressions with Square Roots 1. $\sqrt{ab}$ 2. $\sqrt{15}$ 3. $\sqrt{14z}$ 4. $5b\sqrt{3}$ 5. $6k^2\sqrt{2}$ 6. $2z^3\sqrt{5z}$ 7. $a; a$ 8. 11 9. 31 10. 21
11. $-30\sqrt{2}$ 12. 44 13. $-15 + 3\sqrt{7}$ 14. $3\sqrt{5} - 5$ 15. $-4\sqrt{2} + 2\sqrt{5}$ 16. $-4 + 4\sqrt{3}$ 17. $19 + 7\sqrt{7}$ 18. $42 + \sqrt{5n} - 5n$ 19. $-18 + 14\sqrt{3}$
20. $-32 - 11\sqrt{7}$ 21. False 22. $67 + 16\sqrt{3}$ 23. $4 - 4\sqrt{7z} + 7z$ 24. $49 - 12\sqrt{5}$ 25. conjugates 26. 33 27. -47 28. $1 - 80n$ 29. 3
31. $7x$ 33. $\sqrt{39}$ 35. $\sqrt{14a}$ 37. $-2\sqrt{15}$ 39. $24\sqrt{3}$ 41. $-9x^2$ 43. $72\sqrt{6}$ 45. $270\sqrt{6}$ 47. $-45\sqrt{6}$ 49. $32x^2\sqrt{3x}$ 51. 37 53. $7x$
55. $24 + 8\sqrt{5}$ 57. $12 - 3\sqrt{6}$ 59. $21\sqrt{5} + 4\sqrt{3}$ 61. $a\sqrt{2} + 2\sqrt{a}$ 63. $33 + 11\sqrt{3}$ 65. $1 + \sqrt{5}$ 67. $24 - 14\sqrt{2}$ 69. $3 - 14\sqrt{x} + 8x$
71. $20 - 15\sqrt{2} + 8\sqrt{3} - 6\sqrt{6}$ 73. $8x\sqrt{3} - 7x$ 75. $19 + 8\sqrt{3}$ 77. $9x - 12\sqrt{x} + 4$ 79. 1 81. $33 + 20\sqrt{2}$ 83. -389 85. 34 87. $3m^3\sqrt{5}$
89. $-5\sqrt{2}$ 91. $12\sqrt{2n}$ 93. 32 95. $a + \sqrt{ab} + b$ 97. $27 + 5\sqrt{30}$ 99. $x + 5 - 4\sqrt{x + 1}$ 101. $9s + 4$ 103. -120 105. $10a^2\sqrt{a}$
107. $-\dfrac{9}{2}\sqrt{3}$ 109. 4 111. 9 113. $126\sqrt{3}$ square units 115. $112\sqrt{7}$ square units 117. $(12 + 18\sqrt{3})$ square units 119. 147π m^2 121. 126π in.^{2}
123. $12\sqrt{30}$ square units 125. $50\sqrt{14}$ square units 127. No. $(\sqrt{5} + \sqrt{7})^2 \neq (\sqrt{5})^2 + (\sqrt{7})^2$. When squaring a binomial, we use the pattern $A^2 + 2AB + B^2$. So $(\sqrt{5} + \sqrt{7})^2 = (\sqrt{5})^2 + 2(\sqrt{5})(\sqrt{7}) + (\sqrt{7})^2 = 5 + 2\sqrt{35} + 7 = 12 + 2\sqrt{35}$. 129. Answers will vary. The answer is correct, but it is not in simplest form. The correct answer is $\sqrt{12x^2} \cdot \sqrt{2x^3} = \sqrt{24x^5} = \sqrt{4x^4 \cdot 6x} = 2x^2\sqrt{6x}$.

Section 8.5 Dividing Expressions with Square Roots 1. $\sqrt{\dfrac{a}{b}}$ 2. 3 3. $\sqrt{7}$ 4. $\dfrac{4}{5}z^2$ 5. $\sqrt{6} - \sqrt{15}$ 6. rationalize the denominator 7. True 8. $\dfrac{\sqrt{2}}{2}$

9. $-\dfrac{\sqrt{6}}{15}$ 10. $\dfrac{\sqrt{5}}{2}$ 11. False 12. $4 - 2\sqrt{3}$ 13. $9 + 3\sqrt{5}$ 14. $\dfrac{4\sqrt{3} - \sqrt{21}}{9}$ 15. $\dfrac{3\sqrt{3} + \sqrt{3x}}{9 - x}$ 16. $2 - \sqrt{6}$ 17. 4 19. $-\dfrac{1}{5}$ 21. $3\sqrt{2}$

23. $2\sqrt{7}$ **25.** $-5\sqrt{y}$ **27.** 8 **29.** $\dfrac{3\sqrt{5}}{5}$ **31.** $\dfrac{10\sqrt{6}}{3}$ **33.** $-\dfrac{8\sqrt{2}}{3}$ **35.** $-\dfrac{2\sqrt{x}}{x}$ **37.** $\dfrac{9\sqrt{2y}}{2}$ **39.** $\dfrac{\sqrt{3}}{2}$ **41.** $\dfrac{3\sqrt{2n}}{2n}$ **43.** $\dfrac{4\sqrt{3}}{9}$ **45.** $\dfrac{\sqrt{6ab}}{21b}$ **47.** $-\dfrac{7\sqrt{10}}{12}$

49. $2-\sqrt{3}; 1$ **51.** $\sqrt{6}+1; 5$ **53.** $-1+3\sqrt{14}; -125$ **55.** $\sqrt{p}-3\sqrt{q}; p-9q$ **57.** $\dfrac{3-\sqrt{5}}{4}$ **59.** $\dfrac{27+3\sqrt{3}}{26}$ **61.** $\dfrac{24-4\sqrt{7}}{29}$ **63.** $\dfrac{3\sqrt{y}+18}{y-36}$

65. $32\sqrt{2}+48$ **67.** $\dfrac{6\sqrt{2}+\sqrt{42}}{15}$ **69.** $\dfrac{n\sqrt{n}-n}{n-1}$ **71.** $\dfrac{2\sqrt{x}-3x}{4-9x}$ **73.** $\dfrac{\sqrt{55}+\sqrt{30}}{5}$ **75.** $\dfrac{\sqrt{3}+1}{2}$ **77.** $\sqrt{3}-\sqrt{6}$ **79.** $\dfrac{\sqrt{2}}{2}$ **81.** $\dfrac{3\sqrt{5}}{10}-\dfrac{\sqrt{2}}{2}$

83. $7\sqrt{2}$ **85.** $-\dfrac{\sqrt{6}}{3}$ **87.** $\dfrac{5\sqrt{7}}{98}$ **89.** $2\sqrt{2x}$ **91.** $\dfrac{2\sqrt{3}}{3}$ **93.** $-8\sqrt{2}+6\sqrt{5}$ **95.** $\dfrac{\sqrt{30}}{10}$ **97.** $4-2\sqrt{3}$ **99.** $\dfrac{\sqrt{2}}{2}$ **101.** $\dfrac{\sqrt{7xy}}{7x^2y}$ **103.** $x+11-6\sqrt{x+2}$

105. (a) 0.2679 **(b)** $2-\sqrt{3}$ **(c)** 0.2679 **(d)** They are the same. **107. (a)** 2.0291 **(b)** $\dfrac{10\sqrt{3}+5}{11}$ **(c)** 2.0291 **(d)** They are the same.

109. (a) -1.7836 **(b)** $-\dfrac{4+6\sqrt{2}}{7}$ **(c)** -1.7836 **(d)** They are the same. **111. (a)** 0.2101 **(b)** $\dfrac{5\sqrt{2}-\sqrt{6}}{22}$ **(c)** 0.2101 **(d)** They are the same.

113. $-2-\sqrt{3}$ **115.** $-2\sqrt{3}-3\sqrt{2}+\sqrt{6}+3$ **117.** $\dfrac{2x-7\sqrt{xy}-4y}{x-16y}$ **119.** x cannot be canceled in the expression $\dfrac{3\sqrt{xy}}{x}$ because the x in the numerator is part of the radicand and the x in the denominator is not. The quotient rule says $\dfrac{\sqrt{a}}{\sqrt{b}}=\sqrt{\dfrac{a}{b}}$. **121.** Neither student is correct. The correct approach is

$$\dfrac{3}{\sqrt{2}+\sqrt{3}}\cdot\dfrac{\sqrt{2}-\sqrt{3}}{\sqrt{2}-\sqrt{3}}=\dfrac{3(\sqrt{2}-\sqrt{3})}{(\sqrt{2})^2-(\sqrt{3})^2}=\dfrac{3(\sqrt{2}-\sqrt{3})}{2-3}=\dfrac{3(\sqrt{2}-\sqrt{3})}{-1}=-3\sqrt{2}+3\sqrt{3}.$$

Putting the Concepts Together (Section 8.1–Section 8.5) **1. (a)** irrational; $\sqrt{98}\approx 9.90$ **(b)** rational; $\sqrt{361}=19$ **2.** $-6\sqrt{2}$ **3.** $4xy^3\sqrt{2x}$
4. $\dfrac{5}{2x}$ **5.** 7 **6.** $3m^3$ **7.** $3\sqrt{3}$ **8.** $3\sqrt{2}$ **9.** $6\sqrt{6}$ **10.** -42 **11.** $60\sqrt{2}+18\sqrt{5}$ **12.** $24-26\sqrt{3}$ **13.** -11 **14.** $11-4\sqrt{7}$ **15.** $-2+\sqrt{7}$
16. $\dfrac{3\sqrt{2}}{x^2}$ **17.** $\dfrac{2\sqrt{6}}{3}$ **18.** $-2\sqrt{3}+6$

Section 8.6 Solving Equations Containing Square Roots **1.** radical equation **2.** yes **3.** no **4.** yes **5.** True **6.** $\{14\}$ **7.** $\{-8\}$ **8.** $\{36\}$ **9.** $\{12\}$
10. extraneous solutions **11.** $\{3\}$ **12.** $\{4\}$ **13.** False **14.** $\{-4\}$ **15.** $\{4\}$ **16.** $\{\ \}$ or $\varnothing$ **17.** $\{6\}$ **18.** $\{-1,10\}$ **19.** $\{4\}$ **20.** False
21. $\{16\}$ **22.** 49 **23.** Yes **25.** No **27.** Yes **29.** Yes **31.** $\{16\}$ **33.** $\{-4\}$ **35.** $\{7\}$ **37.** $\{4\}$ **39.** $\{9\}$ **41.** $\{\ \}$ or $\varnothing$ **43.** $\{-1\}$ **45.** $\{3,6\}$
47. $\{1\}$ **49.** $\{\ \}$ or $\varnothing$ **51.** $\{-3\}$ **53.** $\{6,9\}$ **55.** $\{9\}$ **57.** $\{6\}$ **59.** $\{1\}$ **61.** $\{0\}$ **63.** $\{-5\}$ **65.** $\{5\}$ **67.** $\{\ \}$ or $\varnothing$ **69.** $\{-1\}$
71. $\{\ \}$ or $\varnothing$ **73.** $\{23\}$ **75.** $\{20\}$ **77.** $\{-1\}$ **79.** $\{16\}$ **81.** $\{2\}$ **83.** $\left\{\dfrac{1}{4}\right\}$ **85.** $\{-3,8\}$ **87.** $\{-4,7\}$ **89.** $\{4\}$ **91.** $\{-1,2\}$ **93.** $\{\ \}$ or $\varnothing$
95. $\{3\}$ **97.** $\{\ \}$ or $\varnothing$ **99. (a)** 8 seconds **(b)** $\dfrac{5\sqrt{3}}{2}$ seconds **101.** about 28 ft **103. (a)** about 42 ft **(b)** Yes **105.** 11 **107.** 3 **109.** $A=\pi r^2$
111. $h=\dfrac{3V}{S^2}$ **113.** $a=\dfrac{b^2-b}{2}$ **115.** $n=\dfrac{1}{2}m$ **117.** The student's work is incorrect. The student did not square the first step correctly to obtain the second step. Using the property $(\sqrt{a})^2=a$ for $a>0$ is correct, but the student incorrectly squared the right side of the equation. The student should have used the pattern $A^2-2AB+B^2$ to simplify the right side of the equation and then proceeded to solve for x.

Section 8.7 Higher Roots and Rational Exponents **1.** radicand; index **2.** -5 **3.** True **4.** 2 **5.** -3 **6.** not a real number **7.** 2 **8.** 7
9. w **10.** $12p$ **11.** $\sqrt[n]{a}\cdot\sqrt[n]{b}$ **12.** False **13.** $-2\sqrt[3]{4}$ **14.** $5n\sqrt[3]{2}$ **15.** $3b^3\sqrt[4]{2}$ **16.** 3 **17.** $4m^3\sqrt[3]{2}$ **18.** $\sqrt[n]{a}$ **19. (a)** $\sqrt{49}=7$ **(b)** $\sqrt[3]{-64}=-4$
(c) $-\sqrt{100}=-10$ **(d)** $\sqrt{-100}$ not a real number **20. (a)** $\sqrt[5]{p}$ **(b)** $\sqrt[3]{4n}$ **(c)** $4\sqrt[3]{n}$ **21. (a)** $b^{\frac{1}{2}}$ **(b)** $5p^{\frac{1}{3}}$ **(c)** $(10p)^{\frac{1}{4}}$ **22.** $\sqrt[n]{a^m}; (\sqrt[n]{a})^m$ **23.** False
24. (a) $(\sqrt{25})^3=125$ **(b)** $(\sqrt[3]{64})^2=16$ **(c)** $-(\sqrt[4]{16})^3=-8$ **(d)** $(\sqrt[3]{-8})^4=16$ **(e)** $(\sqrt{-49})^5$; not a real number **25. (a)** $d^{\frac{3}{4}}$ **(b)** $y^{\frac{8}{5}}$ **(c)** $\left(11a^2b\right)^{\frac{3}{4}}$
26. (a) $\sqrt[5]{z^4}=(\sqrt[5]{z})^4$ **(b)** $2\sqrt[3]{n^2}=2(\sqrt[3]{n})^2$ **(c)** $\sqrt[3]{(2n)^2}=(\sqrt[3]{2n})^2$ **27. (a)** $\dfrac{1}{36^{\frac{1}{2}}}=\dfrac{1}{6}$ **(b)** $\dfrac{1}{8^{\frac{4}{3}}}=\dfrac{1}{16}$ **(c)** $-\dfrac{1}{25^{\frac{3}{2}}}=-\dfrac{1}{125}$ **28. (a)** $7^{\frac{5}{6}}$ **(b)** $11^{\frac{1}{5}}$
(c) 4 **(d)** $a^{\frac{11}{2}}$ **29.** $\sqrt[3]{8}$ or 2 **31.** $\sqrt[4]{16}$ or 2 **33.** $\sqrt[4]{81}$ or 3 **35.** $\sqrt[5]{-32}$ or -2 **37.** 3 **39.** not a real number **41.** -5 **43.** 5 **45.** 0 **47.** 3 **49.** 5
51. n **53.** r **55.** $3a$ **57.** $2\sqrt[3]{2}$ **59.** not a real number **61.** $-2\sqrt[3]{5}$ **63.** 3 **65.** $-2n^2\sqrt[4]{2}$ **67.** $4\sqrt[3]{2}$ **69.** $3x\sqrt[3]{3}$ **71.** $2b$ **73.** $-3z^2\sqrt[3]{2}$ **75.** -2 **77.** -2
79. -6 **81.** not a real number **83.** $\sqrt{2a}$ **85.** $7\sqrt[5]{z}$ **87.** $-3\sqrt[3]{r}$ **89.** $c^{\frac{1}{2}}$ **91.** $(2y)^{\frac{1}{2}}$ **93.** $7a^{\frac{1}{2}}$ **95.** $(\sqrt{64})^3=512$ **97.** $\dfrac{1}{\sqrt{81}}=\dfrac{1}{9}$ **99.** $(\sqrt[3]{64})^2=16$
101. $-\dfrac{1}{(\sqrt{9})^3}=-\dfrac{1}{27}$ **103.** $(\sqrt[4]{-81})^3$ not a real number **105.** $\dfrac{1}{(\sqrt{49})^3}=\dfrac{1}{343}$ **107.** $\sqrt[3]{u^2}=(\sqrt[3]{u})^2$ **109.** $4\sqrt[3]{x^2}=4(\sqrt[3]{x})^2$
111. $\sqrt[3]{(5n)^2}=(\sqrt[3]{5n})^2$ **113.** $p^{\frac{5}{4}}$ **115.** $5t^{\frac{3}{4}}$ **117.** $(5a)^{\frac{2}{3}}$ **119.** 16 **121.** $\dfrac{1}{36}$ **123.** x^6 **125.** $4n^{\frac{1}{3}}$ or $4\sqrt[3]{n}$ **127.** $x^{\frac{1}{4}}$ or $\sqrt[4]{x}$ **129.** $\dfrac{1}{2x}$ **131.** -125 **133.** $\dfrac{1}{6}$
135. $-\dfrac{1}{25}$ **137.** 127 **139.** 5 **141.** $\dfrac{125}{8}$ **143.** $\dfrac{1}{y^2}$ **145.** $\dfrac{1}{4}$ **147.** 6 **149.** $\dfrac{1}{\sqrt[4]{x}}$ **151.** $-2\sqrt{a}$ **153.** $81x\sqrt[6]{x^5}$ **155.** The expression $a^{\frac{1}{n}}$ is undefined when a is negative and n is even, because an even root of a negative number is not defined. For example, $\sqrt{-16}$ is undefined because the principal square root of a negative real number is not a real number.

Chapter 8 Review **1.** $-2,2$ **2.** $-9,9$ **3.** -1 **4.** -5 **5.** 0.4 **6.** 0.2 **7.** $\dfrac{5}{4}$ **8.** 6 **9.** 4 **10.** 12 **11.** 21 **12.** 11 **13.** rational; -3 **14.** irrational; -3.46
15. irrational; 3.74 **16.** not a real number **17.** $|4x-9|$ **18.** $|16m-25|$ **19.** $2\sqrt{7}$ **20.** $3\sqrt{5}$ **21.** $10\sqrt{2}$ **22.** $5\sqrt{6}$ **23.** $-1+\sqrt{2}$
24. $\dfrac{-1+\sqrt{3}}{2}$ **25.** a^{18} **26.** x^8 **27.** $4x^5$ **28.** $7a^6$ **29.** $3n^4\sqrt{2n}$ **30.** $2y^{12}\sqrt{2y}$ **31.** $\dfrac{3}{x^4}$ **32.** $\dfrac{2}{x^2}$ **33.** $\dfrac{9y^2}{5}$ **34.** $\dfrac{6}{11n^4}$ **35.** $-2\sqrt{7}$ **36.** $3\sqrt{11}$ **37.** $\sqrt{x}$

38. $-\sqrt{n}$ **39.** $25a\sqrt{ab}$ **40.** $3x\sqrt{3xy}$ **41.** $5\sqrt{3}$ **42.** $8\sqrt{2}$ **43.** $4+\sqrt{5}$ **44.** $6-2\sqrt{3}$ **45.** $19n^2\sqrt{2n}$ **46.** $9a^2\sqrt{14a}$ **47.** $2\sqrt{2}+2\sqrt{3}$
48. $\frac{9}{2}\sqrt{6}+\frac{2}{3}\sqrt{3}$ **49.** $\frac{11}{20}\sqrt{3}$ **50.** $-\frac{8}{5}\sqrt{2}$ **51.** $4\sqrt{6}$ **52.** $4\sqrt{15}$ **53.** $7x^2\sqrt{6x}$ **54.** $6y^2\sqrt{5y}$ **55.** $-128\sqrt{10}$ **56.** -200 **57.** $8a$ **58.** $45n$ **59.** -24
60. -154 **61.** $3\sqrt{a}+3\sqrt{5a}$ **62.** $2x+2x\sqrt{2}$ **63.** $6+12\sqrt{3}+4\sqrt{6}+24\sqrt{2}$ **64.** $15+10\sqrt{6}-9\sqrt{3}-18\sqrt{2}$ **65.** $4\sqrt{y}+12-2y\sqrt{2}-6\sqrt{2y}$
66. $\sqrt{x}+9+3x\sqrt{3}+27\sqrt{3x}$ **67.** $11-6\sqrt{2}$ **68.** $21-8\sqrt{5}$ **69.** $11x+6x\sqrt{2}$ **70.** $13n+4n\sqrt{3}$ **71.** $9-25x$ **72.** $4x-16$ **73.** $3\sqrt{6}$
74. $3\sqrt{5}$ **75.** $\frac{\sqrt{6}}{2x}$ **76.** $\frac{x\sqrt{15}}{6}$ **77.** $\sqrt{5}+\sqrt{6}$ **78.** $\sqrt{6}-\sqrt{3}$ **79.** $2\sqrt{7}$ **80.** $5\sqrt{5}$ **81.** $\frac{\sqrt{6}}{4}$ **82.** $\frac{\sqrt{10}}{25}$ **83.** $-2-3\sqrt{5};-41$
84. $-3+2\sqrt{7};-19$ **85.** $4+2\sqrt{2}$ **86.** $\frac{6-\sqrt{3}}{11}$ **87.** $-2+2\sqrt{2}$ **88.** $\sqrt{3}+1$ **89.** Yes **90.** Yes **91.** $\{-13\}$ **92.** $\{13\}$ **93.** $\{2\}$ **94.** $\{9\}$
95. $\{3,5\}$ **96.** $\{6\}$ **97.** $\left\{\frac{1}{4}\right\}$ **98.** $\{-3\}$ **99.** $\{3\}$ **100.** $\{2\}$ **101.** $\{\ \}$ or $\varnothing$ **102.** $\{10\}$ **103.** 18π square cm **104.** 6 in. **105.** -2
106. not a real number **107.** $2x\sqrt[4]{3}$ **108.** $5x\sqrt[3]{5}$ **109.** $2n\sqrt[3]{3n}$ **110.** $3z\sqrt[4]{2z^2}$ **111.** -2 **112.** not a real number **113.** $x^{\frac{3}{4}}$ **114.** $z^{\frac{1}{3}}$
115. $2\sqrt{x^3}=2x\sqrt{x}$ **116.** $\sqrt[3]{3v}$ **117.** 9 **118.** 32 **119.** $-\frac{1}{64}$ **120.** $\frac{1}{8}$ **121.** 4 **122.** 36 **123.** $\frac{1}{x^{\frac{3}{4}}}$ **124.** $\frac{a^{\frac{1}{3}}}{9}$

Chapter 8 Test **1.** $-6\sqrt{3}$ **2.** $5xy\sqrt{2y}$ **3.** $\frac{4x}{3}$ **4.** 7 **5.** $\sqrt{2}$ **6.** $17\sqrt{3}$ **7.** $15\sqrt{3}$ **8.** $12\sqrt{3}-4\sqrt{2}$ **9.** $19-8\sqrt{3}$ **10.** $2x\sqrt{3x}$ **11.** $2-10\sqrt{2}$
12. $\frac{-1+3\sqrt{2}}{2}$ **13.** $\frac{6\sqrt{5}}{5}$ **14.** $-3\sqrt{2}+6$ **15.** $\{5\}$ **16.** $\{5\}$ **17.** $\{7,-3\}$ **18.** $3x\sqrt[3]{4x^2}$ **19.** $\frac{1}{9}$ **20.** $3\sqrt{5}\approx 6.7$ feet

Chapter 9 Quadratic Equations

Section 9.1 Solving Quadratic Equations Using the Square Root Property **1.** $\sqrt{p};-\sqrt{p}$ **2.** False **3.** $\{-6,6\}$ **4.** $\{-2\sqrt{5},2\sqrt{5}\}$
5. $\{-5,5\}$ **6.** $\{-3\sqrt{2},3\sqrt{2}\}$ **7.** $\left\{-\frac{\sqrt{2}}{2},\frac{\sqrt{2}}{2}\right\}$ **8.** $\left\{-\frac{4\sqrt{5}}{5},\frac{4\sqrt{5}}{5}\right\}$ **9.** two **10.** $\{-8,14\}$ **11.** $\left\{\frac{-1-2\sqrt{7}}{5},\frac{-1+2\sqrt{7}}{5}\right\}$ **12.** $\{-7,-1\}$
13. $\{-2-2\sqrt{2},-2+2\sqrt{2}\}$ **14.** no real solution **15.** no real solution **16.** False **17.** 13 **18.** $5\sqrt{10},15.8$ **19.** $6\sqrt{2},8.5$
20. Yes. The ladder reaches 28.9 feet up on the wall. **21.** $\{-12,12\}$ **23.** $\{0\}$ **25.** $\{-2\sqrt{3},2\sqrt{3}\}$ **27.** $\left\{-\frac{2}{3},\frac{2}{3}\right\}$ **29.** $\left\{-\frac{2\sqrt{3}}{3},\frac{2\sqrt{3}}{3}\right\}$
31. $\{-4\sqrt{2},4\sqrt{2}\}$ **33.** $\{-13,13\}$ **35.** $\{-5\sqrt{2},5\sqrt{2}\}$ **37.** no real solution **39.** $\left\{-\frac{1}{3},\frac{1}{3}\right\}$ **41.** $\{-8,8\}$ **43.** $\{-6,6\}$ **45.** $\left\{-\frac{5\sqrt{3}}{3},\frac{5\sqrt{3}}{3}\right\}$
47. no real solution **49.** $\{-4\sqrt{2},4\sqrt{2}\}$ **51.** $\{-6,-2\}$ **53.** $\left\{\frac{2}{3},\frac{4}{3}\right\}$ **55.** $\{-5-\sqrt{6},-5+\sqrt{6}\}$ **57.** $\{6-2\sqrt{2},6+2\sqrt{2}\}$
59. $\left\{-\frac{5}{3},1\right\}$ **61.** $\left\{\frac{5-\sqrt{15}}{2},\frac{5+\sqrt{15}}{2}\right\}$ **63.** $\{-3,5\}$ **65.** $\left\{-\frac{1}{2},-\frac{1}{4}\right\}$ **67.** $\left\{\frac{-2-3\sqrt{6}}{9},\frac{-2+3\sqrt{6}}{9}\right\}$ **69.** no real solution **71.** $\{-9,1\}$
73. $\{-3,4\}$ **75.** $c=10$ **77.** $c=3\sqrt{2}\approx 4.24$ **79.** $a=8\sqrt{2}\approx 11.31$ **81.** $\{4,9\}$ **83.** $\{-2\sqrt{2},2\sqrt{2}\}$ **85.** $\{-\sqrt{15},\sqrt{15}\}$ **87.** $\left\{-1,\frac{1}{2}\right\}$
89. $\left\{-\frac{\sqrt{15}}{5},\frac{\sqrt{15}}{5}\right\}$ **91.** no real solution **93.** $\{-4,3\}$ **95.** $\{-3\sqrt{3},3\sqrt{3}\}$ **97.** $\{5\}$ **99.** $\left\{\frac{-4-\sqrt{10}}{2},\frac{-4+\sqrt{10}}{2}\right\}$ **101.** $\left\{-\frac{\sqrt{6}}{3},\frac{\sqrt{6}}{3}\right\}$
103. $\left\{-\frac{3}{2},4\right\}$ **105.** $\{-2\sqrt{10},2\sqrt{10}\}$
107. no real solution **109.** $2\sqrt{10}$
111. (a) $2\sqrt{30}$ ft **(b)** 11.0 ft
113. $20\sqrt{26}\approx 102$ yd
115. (a) 13.8 ft **(b)** 13 ft 9.6 in.
117. (b) 8%
119. $x^2-16=0$ **121.** $x^2-6=0$ **123.** $x^2-45=0$

125. The error is in the third step, where only the positive square root was taken. The correct solution is

$$(x-5)^2-4=0$$
$$(x-5)^2=4$$
$$x-5=\pm 2$$
$$x-5=2 \text{ or } x-5=-2$$
$$x=7 \qquad x=3$$

The solution set is $\{3,7\}$

Section 9.2 Solving Quadratic Equations by Completing the Square **1.** $100;(x-10)^2$ **2.** $x-\frac{b}{2}$ **3.** $9;(p+3)^2$ **4.** $81;(b-9)^2$
5. $\frac{9}{4};\left(w+\frac{3}{2}\right)^2$ **6.** False **7.** $\{-7,-1\}$ **8.** $\{-8,4\}$ **9.** $\{-4-\sqrt{7},-4+\sqrt{7}\}$ **10.** $\{5-\sqrt{6},5+\sqrt{6}\}$ **11.** False **12.** $\{-9,1\}$
13. $\left\{\frac{-5-\sqrt{65}}{2},\frac{-5+\sqrt{65}}{2}\right\}$ **14.** 2 **15.** $\left\{-2,\frac{5}{2}\right\}$ **16.** $\left\{\frac{2-\sqrt{10}}{3},\frac{2+\sqrt{10}}{3}\right\}$ **17.** no real solution **18.** $\{3-\sqrt{2},3+\sqrt{2}\}$
19. $64;(x+8)^2$ **21.** $36;(z-6)^2$ **23.** $\frac{4}{9};\left(k+\frac{2}{3}\right)^2$ **25.** $\frac{81}{4};\left(y-\frac{9}{2}\right)^2$ **27.** $\frac{1}{4};\left(t+\frac{1}{2}\right)^2$ **29.** $\frac{1}{16};\left(s-\frac{1}{4}\right)^2$ **31.** $\{-3,7\}$
33. $\{-2-\sqrt{11},-2+\sqrt{11}\}$ **35.** $\{-3,6\}$ **37.** $\{3,5\}$ **39.** no real solution **41.** $\{-1,4\}$ **43.** $\{4-2\sqrt{7},4+2\sqrt{7}\}$ **45.** $\{-5,-2\}$

47. $\left\{\dfrac{-7 - \sqrt{249}}{2}, \dfrac{-7 + \sqrt{249}}{2}\right\}$ **49.** $\{-8, -2\}$ **51.** $\{-3, 1\}$ **53.** $\left\{-\dfrac{1}{2}, \dfrac{3}{2}\right\}$ **55.** $\left\{\dfrac{-2 - \sqrt{10}}{3}, \dfrac{-2 + \sqrt{10}}{3}\right\}$ **57.** $\left\{\dfrac{-1 - \sqrt{6}}{2}, \dfrac{-1 + \sqrt{6}}{2}\right\}$

59. no real solution   **61.** $\left\{\dfrac{3 - \sqrt{2}}{2}, \dfrac{3 + \sqrt{2}}{2}\right\}$ **63.** $\left\{\dfrac{6 - \sqrt{37}}{2}, \dfrac{6 + \sqrt{37}}{2}\right\}$ **65.** $\{-4, 6\}$ **67.** $\left\{\dfrac{1}{2}\right\}$ **69.** $\left\{-3 - 2\sqrt{3}, -3 + 2\sqrt{3}\right\}$

71. $\left\{-2\sqrt{5}, 2\sqrt{5}\right\}$ **73.** $\left\{\dfrac{-3 - \sqrt{15}}{3}, \dfrac{-3 + \sqrt{15}}{3}\right\}$ **75.** $\left\{\dfrac{-5 - \sqrt{97}}{2}, \dfrac{-5 + \sqrt{97}}{2}\right\}$ **77.** $\{-3, 3\}$ **79.** $\{-6, 2\}$ **81.** $\{-12, 2\}$ **83.** $\left\{-3\sqrt{2}, 3\sqrt{2}\right\}$

85. $\left\{-2\sqrt{3}, 2\sqrt{3}\right\}$ **87.** $\left\{\dfrac{-6 - 3\sqrt{10}}{2}, \dfrac{-6 + 3\sqrt{10}}{2}\right\}$ **89.** $\left\{\dfrac{2}{3}, 3\right\}$ **91.** $\left\{-1 - \sqrt{34}, -1 + \sqrt{34}\right\}$ **93.** $15\sqrt{2}$ **95.** 3 and 6   **97.** -4 or 6

99. $6 - 3\sqrt{3}$ or $6 + 3\sqrt{3}$ **101. (a)** 9   **(b)** $\{3, 6\}$ **(c)** $3 + 6 = 9$ **103. (a)** 2   **(b)** $\left\{1 - \dfrac{\sqrt{6}}{2}, 1 + \dfrac{\sqrt{6}}{2}\right\}$ **(c)** $1 - \dfrac{\sqrt{6}}{2} + 1 + \dfrac{\sqrt{6}}{2} = 2$

105. (a) $-\dfrac{3}{4}$;   **(b)** $\left\{-\dfrac{3}{2}, \dfrac{1}{2}\right\}$;   **(c)** $\left(-\dfrac{3}{2}\right)\left(\dfrac{1}{2}\right) = -\dfrac{3}{4}$ **107. (a)** -7;   **(b)** $\left\{1 - 2\sqrt{2}, 1 + 2\sqrt{2}\right\}$;   **(c)** $\left(1 - 2\sqrt{2}\right)\left(1 + 2\sqrt{2}\right) = -7$

109. center $= (5, -3)$, radius $= 2$ **111.** center $= (10, 8)$, radius $= \sqrt{142}$ **113.** To find the number that must be added to $x^2 - 9x$ to make a perfect square trinomial, take half of the coefficient of x, which is -9, and square that value: $\left(\dfrac{1}{2} \cdot -9\right)^2 = \left(-\dfrac{9}{2}\right)^2 = \dfrac{81}{4}$. So $x^2 - 9x + \dfrac{81}{4} = \left(x - \dfrac{9}{2}\right)^2$.

Section 9.3   Solving Quadratic Equations Using the Quadratic Formula

1. $\dfrac{-b \pm \sqrt{b^2 - 4ac}}{2a}$ **2. (a)** $a = 1; b = 5; c = -3$
(b) $a = 4; b = -3; c = 1$ **3.** $\left\{-\dfrac{4}{3}, -1\right\}$ **4.** $\left\{-6, -\dfrac{1}{2}\right\}$ **5.** $\left\{-\dfrac{2}{3}, -\dfrac{3}{2}\right\}$ **6.** $\left\{-\dfrac{5}{4}, \dfrac{1}{2}\right\}$ **7.** False   **8.** $\left\{-2 - \sqrt{6}, -2 + \sqrt{6}\right\}$

9. $\left\{\dfrac{3 - \sqrt{2}}{4}, \dfrac{3 + \sqrt{2}}{4}\right\}$ **10.** True   **11.** $\left\{\dfrac{4 - \sqrt{10}}{2}, \dfrac{4 + \sqrt{10}}{2}\right\}$ **12.** $\left\{\dfrac{5}{4}\right\}$ **13.** no real solution

14. 1000   **15.** $b^2 - 4ac$ **16.** factoring   **17.** factoring   **18.** the quadratic formula   **19.** 0; factoring; $\left\{\dfrac{1}{3}\right\}$ **20.** 1; factoring; $\left\{\dfrac{2}{3}, 1\right\}$

21. 20; quadratic formula; $\left\{\dfrac{1 - \sqrt{5}}{4}, \dfrac{1 + \sqrt{5}}{4}\right\}$ **22.** $\left\{-\sqrt{6}, \sqrt{6}\right\}$ **23.** $\left\{-\dfrac{7}{2}, 5\right\}$ **24.** $\left\{\dfrac{-2 - \sqrt{7}}{3}, \dfrac{-2 + \sqrt{7}}{3}\right\}$

25. $2x^2 + x + 4 = 0; a = 2, b = 1, c = 4$ **27.** $x^2 + 0x + 4 = 0; a = 1, b = 0, c = 4$ **29.** $\dfrac{3}{2}x^2 - \dfrac{6}{5}x - 2 = 0; a = \dfrac{3}{2}, b = -\dfrac{6}{5}, c = -2$

31. $0.5x^2 - x + 3 = 0; a = 0.5, b = -1, c = 3$ **33.** $\{-8, 3\}$ **35.** $\left\{-4, -\dfrac{3}{2}\right\}$ **37.** $\left\{\dfrac{1 - \sqrt{21}}{2}, \dfrac{1 + \sqrt{21}}{2}\right\}$ **39.** $\left\{\dfrac{-2 - \sqrt{7}}{3}, \dfrac{-2 + \sqrt{7}}{3}\right\}$

41. $\left\{\dfrac{1}{2}, \dfrac{3}{2}\right\}$ **43.** $\left\{\dfrac{-7 - 3\sqrt{5}}{2}, \dfrac{-7 + 3\sqrt{5}}{2}\right\}$ **45.** $\left\{\dfrac{5 - \sqrt{17}}{4}, \dfrac{5 + \sqrt{17}}{4}\right\}$ **47.** no real solution   **49.** $\left\{-\dfrac{2}{3}, 2\right\}$ **51.** $\left\{-3 - \sqrt{5}, -3 + \sqrt{5}\right\}$

53. $\left\{\dfrac{2 - \sqrt{2}}{2}, \dfrac{2 + \sqrt{2}}{2}\right\}$ **55.** $\left\{\dfrac{-5 - \sqrt{33}}{4}, \dfrac{-5 + \sqrt{33}}{4}\right\}$ **57.** $\left\{\dfrac{1}{5}, \dfrac{1}{3}\right\}$ **59.** $\left\{-\dfrac{3}{2}\right\}$ **61.** $\{-6, 3\}$ **63.** $\left\{\dfrac{4 - \sqrt{10}}{3}, \dfrac{4 + \sqrt{10}}{3}\right\}$ **65.** $\{3\}$

67. $\left\{\dfrac{-1 - \sqrt{61}}{2}, \dfrac{-1 + \sqrt{61}}{2}\right\}$ **69.** no real solution   **71.** $\left\{-2\sqrt{3}, 2\sqrt{3}\right\}$ **73.** $\left\{-\dfrac{3}{2}, 6\right\}$ **75.** $\{0, 5\}$ **77.** -39; no real solution   **79.** 25; factoring

81. 0; factoring   **83.** 40; quadratic formula   **85.** 0; factoring   **87.** -56; no real solution   **89.** $\left\{\dfrac{3 - \sqrt{3}}{3}, \dfrac{3 + \sqrt{3}}{3}\right\}$ **91.** $\left\{-\dfrac{1}{3}\right\}$ **93.** no real solution

95. $\left\{-\dfrac{1}{2}\right\}$ **97.** $\left\{-5 - 2\sqrt{5}, -5 + 2\sqrt{5}\right\}$ **99.** $\{0, 3\}$ **101.** $\{-5, -2\}$ **103.** $\left\{-4, \dfrac{1}{2}\right\}$ **105.** no real solution   **107. (a)** \$1515   **(b)** 996 or 2004

109. (a) 43   **(b)** 37   **(c)** 33   **(d)** 2049   **111.** $-3 - 3\sqrt{2}$ and $-3 + 3\sqrt{2}$ **113.** -1 or 5   **115.** $\left\{-4\sqrt{2}, \sqrt{2}\right\}$ **117.** $\left\{\dfrac{2\sqrt{3}}{3}, \sqrt{3}\right\}$

119. The quadratic equation is factorable if the discriminant is a perfect square or if the discriminant is equal to zero.

Putting the Concepts Together (Section 9.1–9.3)

1. (a) $\left\{-\dfrac{\sqrt{5}}{2}, \dfrac{\sqrt{5}}{2}\right\}$ **(b)** $\left\{2 - 2\sqrt{3}, 2 + 2\sqrt{3}\right\}$ **(c)** $\left\{\dfrac{-3 - \sqrt{3}}{2}, \dfrac{-3 + \sqrt{3}}{2}\right\}$

2. (a) add 81; $z^2 - 18z + 81 = (z - 9)^2$ **(b)** add $\dfrac{81}{4}$; $y^2 + 9y + \dfrac{81}{4} = \left(y + \dfrac{9}{2}\right)^2$ **(c)** add $\dfrac{4}{25}$; $m^2 + \dfrac{4}{5}m + \dfrac{4}{25} = \left(m + \dfrac{2}{5}\right)^2$ **3. (a)** $\{-7, -5\}$

(b) no real solution   **(c)** $\left\{4 - 3\sqrt{2}, 4 + 3\sqrt{2}\right\}$ **4. (a)** $\left\{-4 - 2\sqrt{5}, -4 + 2\sqrt{5}\right\}$ **(b)** no real solution   **(c)** $\left\{-\dfrac{9}{2}\right\}$ **5.** $\{-2, 3\}$

6. $\left\{\dfrac{-1 - \sqrt{11}}{2}, \dfrac{-1 + \sqrt{11}}{2}\right\}$ **7.** $\left\{\dfrac{-2 - \sqrt{3}}{2}, \dfrac{-2 + \sqrt{3}}{2}\right\}$ **8. (a)** 64   **(b)** factoring   **(c)** Answers may vary.   **9. (a)** $5\sqrt{15}$ feet   **(b)** 19.4 feet

10. (a) \$12,000   **(b)** 60 or 160 texts

Section 9.4   Problem Solving Using Quadratic Equations

1. False   **2.** base $= 8$ inches; height $= 10$ inches   **3.** $a^2 + b^2 = c^2$
4. 7.7 m by 11.7 m   **5.** False   **6.** length $= 4.1$ feet; width $= 3.1$ feet   **7.** ages 30 and 63 years   **8.** The cannonball will be 50 feet above the ground after 0.51 s and after 5.74 s.   **9.** width $= 6$ m, length $= 15$ m   **11.** width $= 3$ in.; length $= 12$ in.   **13.** height $= 5$ cm; base $= 8$ cm

15. height $= 11$ ft; base $= 12$ ft   **17.** Longer leg: $\dfrac{3 + 3\sqrt{3}}{2} \approx 4.1$ inches; shorter leg: $\dfrac{-3 + 3\sqrt{3}}{2} \approx 1.1$ inches.

19. leg $= 3\sqrt{3} \approx 5.2$ m; hypotenuse $= 6\sqrt{3} \approx 10.4$ m   **21.** $-17, -16$ and 16, 17   **23.** 14, 15   **25.** $-1 - 2\sqrt{6}$ or $-1 + 2\sqrt{6}$
27. $5 - 2\sqrt{3}$ or $5 + 2\sqrt{3}$ **29. (a)** 41 ft   **(b)** 116 ft   **(c)** 1.52 s and 4.10 s   **(d)** 5.625 s   **31. (a)** 43 fruit flies   **(b)** 75 fruit flies   **(c)** 2.18 hours

33. (a) 9 **(b)** 20 **(c)** 15 **35. (a)** $(180 - x)°$ **(b)** 30° or 125° **37. (a)** \$146.40 **(b)** \$992.40 **(c)** approximately 3 months **(d)** approximately 13 months
39. 10.8 mph **41.** leg = 1, hypotenuse = $\sqrt{2}$ **43.** leg = $\sqrt{6}$, hypotenuse = $2\sqrt{3}$ **45.** leg = 2, leg = 2 **47.** long leg = $5\sqrt{3}$; hypotenuse = 10
49. short leg = $3\sqrt{3}$; hypotenuse = $6\sqrt{3}$

Section 9.5 The Complex Number System **1.** -1 **2.** $\sqrt{N}i$ **3.** real; imaginary **4.** True **5.** $4i$ **6.** $\sqrt{2}i$ **7.** $2\sqrt{6}i$ **8.** $9 + 12i$ **9.** $-7 - 2\sqrt{5}i$

10. $2 - \sqrt{2}i$ **11.** $2 + 2i$ **12.** $-8 + 11i$ **13.** $5 + 5\sqrt{3}i$ **14.** $24 + 12i$ **15.** $-2 + 14i$ **16.** -10 **17.** $-6 - 8i$ **18.** $-1 - 5i$ **19.** 53 **20.** 97

21. $-4 - 6i$ **22.** $-\dfrac{48}{17} - \dfrac{12}{17}i$ **23.** $1 - \dfrac{8}{3}i$ **24.** $\dfrac{5}{2} - \dfrac{1}{2}i$ **25.** $\left\{-\dfrac{\sqrt{10}i}{2}, \dfrac{\sqrt{10}i}{2}\right\}$ **26.** $\left\{-3 - 2\sqrt{2}i, -3 + 2\sqrt{2}i\right\}$ **27.** $25i$ **29.** $-2\sqrt{13}i$

31. $5 + 2i$ **33.** $12 - 3\sqrt{3}i$ **35.** $-\dfrac{3}{2} + \dfrac{\sqrt{2}}{2}i$ **37.** $-\dfrac{1}{2} + \sqrt{3}i$ **39.** $1 + 3i$ **41.** -10 **43.** $-3 + 21i$ **45.** $13 - 10i$ **47.** $-8 - i$ **49.** $12 + \sqrt{2}i$

51. $-5\sqrt{10}i$ **53.** $15 + 12i$ **55.** $-10 - 10i$ **57.** $-3 - 16i$ **59.** -256 **61.** $24 - 42i$ **63.** $7 - 24i$ **65.** -20 **67.** $-6\sqrt{2}$ **69.** $1 - 31i$

71. $-3 - 2i$; 13 **73.** $6 - 3\sqrt{2}i$; 54 **75.** $4 + 8i$ **77.** $3 + 2i$ **79.** $\dfrac{3}{2} + \dfrac{3}{2}i$ **81.** $-1 + \dfrac{1}{2}i$ **83.** $-\dfrac{27}{4} + 2i$ **85.** $-\dfrac{3}{13} + \dfrac{2}{13}i$ **87.** $\left\{1 - \dfrac{\sqrt{6}}{2}i, 1 + \dfrac{\sqrt{6}}{2}i\right\}$

89. $\left\{-1 - 2\sqrt{3}i, -1 + 2\sqrt{3}i\right\}$ **91.** $\{5 - 2i, 5 + 2i\}$ **93.** $\left\{-2 - \sqrt{5}i, -2 + \sqrt{5}i\right\}$ **95.** $\left\{-\dfrac{3}{2}, 8\right\}$ **97.** $\left\{-2\sqrt{2}, 2\sqrt{2}\right\}$ **99.** $\left\{1 - \sqrt{2}i, 1 + \sqrt{2}i\right\}$

101. $\left\{\dfrac{1 - \sqrt{105}}{4}, \dfrac{1 + \sqrt{105}}{4}\right\}$ **103.** $\dfrac{1}{3} - \dfrac{\sqrt{14}}{3}i$ or $\dfrac{1}{3} + \dfrac{\sqrt{14}}{3}i$ **105.** $1 - 3i$ or $1 + 3i$ **107.** The product of two complex conjugates is always a
real number because the product of the conjugates is in the form $(a + bi)(a - bi) = (a)^2 - (abi) + (abi) - (bi)^2$. When simplified, the terms
$-(abi) + (abi) = 0$, and $(bi)^2 = (b^2)(i)^2 = (b^2)(-1) = -b^2$. Thus the product contains no imaginary numbers, only real numbers.

109. $\left\{5 - \sqrt{15}i, 5 + \sqrt{15}i\right\}$

Chapter 9 Review **1.** $\{-4, 4\}$ **2.** $\{-12, 12\}$ **3.** $\left\{-\dfrac{3}{2}, \dfrac{3}{2}\right\}$ **4.** $\left\{-\dfrac{2}{5}, \dfrac{2}{5}\right\}$ **5.** $\left\{-2\sqrt{3}, 2\sqrt{3}\right\}$ **6.** $\left\{-3\sqrt{2}, 3\sqrt{2}\right\}$ **7.** $\left\{-\dfrac{5}{3}, 1\right\}$ **8.** $\{-1, 6\}$

9. $\left\{-5 - \sqrt{2}, -5 + \sqrt{2}\right\}$ **10.** $\left\{2 - \sqrt{2}, 2 + \sqrt{2}\right\}$ **11.** no real solution **12.** no real solution **13.** $\left\{2 - 2\sqrt{6}, 2 + 2\sqrt{6}\right\}$

14. $\left\{-4 - 2\sqrt{7}, -4 + 2\sqrt{7}\right\}$ **15.** $c = 4\sqrt{13} \approx 14.42$ **16.** $c = 6\sqrt{5} \approx 13.42$ **17.** $a = 2\sqrt{33} \approx 11.49$ **18.** $b = 8\sqrt{2} \approx 11.31$

19. (a) $3\sqrt{2}$ ft by $3\sqrt{2}$ ft **(b)** 4.24 ft by 4.24 ft **20. (a)** $10\sqrt{2}$ ft **(b)** 14.1 ft **21.** 49; $(x - 7)^2$ **22.** 25; $(x + 5)^2$ **23.** $\dfrac{1}{9}; \left(y + \dfrac{1}{3}\right)^2$ **24.** $\dfrac{4}{25}; \left(k - \dfrac{2}{5}\right)^2$

25. $\dfrac{1}{16}; \left(n - \dfrac{1}{4}\right)^2$ **26.** $\dfrac{1}{36}; \left(t + \dfrac{1}{6}\right)^2$ **27.** $\{-1, 7\}$ **28.** $\{1, 11\}$ **29.** $\{-4, 1\}$ **30.** $\{-7, -2\}$ **31.** $\left\{2 - 2\sqrt{5}, 2 + 2\sqrt{5}\right\}$ **32.** $\left\{1 - 2\sqrt{2}, 1 + 2\sqrt{2}\right\}$

33. $\left\{-\dfrac{5}{2}, 2\right\}$ **34.** $\left\{-3, -\dfrac{2}{3}\right\}$ **35.** $\left\{\dfrac{3 - \sqrt{53}}{2}, \dfrac{3 + \sqrt{53}}{2}\right\}$ **36.** $\left\{\dfrac{-1 - \sqrt{33}}{2}, \dfrac{-1 + \sqrt{33}}{2}\right\}$ **37.** $\left\{5 - 3\sqrt{5}, 5 + 3\sqrt{5}\right\}$ **38.** $\left\{4 - 2\sqrt{6}, 4 + 2\sqrt{6}\right\}$

39. $-\dfrac{5}{2}$ and $-\dfrac{1}{2}$ or $\dfrac{1}{2}$ and $\dfrac{5}{2}$ **40.** -8 or 2 **41.** $2x^2 - x + 3 = 0; a = 2, b = -1, c = 3$ **42.** $\dfrac{1}{2}x^2 + x - \dfrac{3}{4} = 0; a = \dfrac{1}{2}, b = 1, c = -\dfrac{3}{4}$ **43.** $\left\{\dfrac{3}{2}, 5\right\}$

44. $\left\{-\dfrac{3}{2}, -\dfrac{1}{2}\right\}$ **45.** $\left\{6 - \sqrt{31}, 6 + \sqrt{31}\right\}$ **46.** $\left\{\dfrac{3 - 3\sqrt{5}}{2}, \dfrac{3 + 3\sqrt{5}}{2}\right\}$ **47.** $\left\{-\dfrac{8}{3}, 1\right\}$ **48.** $\left\{-4, \dfrac{1}{2}\right\}$ **49.** $\left\{\dfrac{-1 - \sqrt{7}}{2}, \dfrac{-1 + \sqrt{7}}{2}\right\}$

50. $\left\{-2 - \sqrt{6}, -2 + \sqrt{6}\right\}$ **51.** $\left\{\dfrac{3 - 2\sqrt{2}}{2}, \dfrac{3 + 2\sqrt{2}}{2}\right\}$ **52.** $\left\{\dfrac{4 - \sqrt{10}}{3}, \dfrac{4 + \sqrt{10}}{3}\right\}$ **53. (a)** \$8250 **(b)** 100 or 160 texts **54. (a)** \$244 **(b)** 15 dogs

55. $b^2 - 4ac = 24$; quadratic formula; $\left\{\dfrac{4 - \sqrt{6}}{5}, \dfrac{4 + \sqrt{6}}{5}\right\}$ **56.** $b^2 - 4ac = 49$; factoring; $\{-1, 6\}$ **57.** $b^2 - 4ac = 169$; factoring; $\left\{-\dfrac{10}{3}, 1\right\}$

58. $b^2 - 4ac = 17$; quadratic formula; $\left\{\dfrac{-3 - \sqrt{17}}{2}, \dfrac{-3 + \sqrt{17}}{2}\right\}$ **59.** $b^2 - 4ac = -23$; no real solution **60.** $b^2 - 4ac = -11$; no real solution

61. $-18, -16$ and 16, 18 **62.** $\dfrac{3}{2}; \dfrac{2}{3}$ **63.** width = 8 m, length = 14 m **64.** width = 5 yd, length = 12 yd **65.** 4 ft, 8 ft

66. leg = $\dfrac{-6 + 2\sqrt{21}}{3}$ cm, hypotenuse = $\dfrac{-3 + 4\sqrt{21}}{3}$ cm **67.** base = 12 m, height = 7 m **68.** base = 3 in., height = $\dfrac{3}{2}$ in.

69. length = 10 in., width = 4 in. **70.** length = 15 ft, width = 8 ft **71.** $\dfrac{5 + \sqrt{26}}{2} \approx 5.0$ s **72.** $\dfrac{15 - \sqrt{145}}{4} \approx 0.7$ s and $\dfrac{15 + \sqrt{145}}{4} \approx 6.8$ s

73. 5 or 25 units. **74.** $(25 - 5\sqrt{11}) \approx 8$ units or $(25 + 5\sqrt{11}) \approx 42$ units **75.** $-10i$ **76.** $-7i$ **77.** $5 + 2\sqrt{5}i$ **78.** $4 - 2\sqrt{15}i$ **79.** $-1 - \sqrt{3}i$

80. $-\dfrac{1}{2} + \dfrac{\sqrt{3}}{2}i$ **81.** $1 + 8i$ **82.** $11i$ **83.** $-16 + 10i$ **84.** $-26 + 13i$ **85.** $-10 + 7i$ **86.** $14 - 4i$ **87.** $-10 + 12i$ **88.** $-3 - 3i$ **89.** $2 - 23i$

90. $27 - 34i$ **91.** $-5 - 12i$ **92.** $-16 + 30i$ **93.** $-36\sqrt{2}$ **94.** $-48\sqrt{5}$ **95.** $5 - 2i$; 29 **96.** $3 + 3i$; 18 **97.** $-\dfrac{1}{3} - i$ **98.** $2 + 3i$ **99.** $\dfrac{3}{5} + \dfrac{4}{5}i$

100. $\dfrac{6}{5} - \dfrac{3}{5}i$ **101.** $-\dfrac{5}{2} + \dfrac{5}{2}i$ **102.** $\dfrac{5}{17} + \dfrac{3}{17}i$ **103.** $\left\{-\dfrac{1}{2} - \dfrac{3}{2}i, -\dfrac{1}{2} + \dfrac{3}{2}i\right\}$ **104.** $\left\{\dfrac{3}{2} - \dfrac{3}{2}i, \dfrac{3}{2} + \dfrac{3}{2}i\right\}$ **105.** $\left\{2 - 2\sqrt{2}i, 2 + 2\sqrt{2}i\right\}$

106. $\left\{-3 - 4\sqrt{2}i, -3 + 4\sqrt{2}i\right\}$

Chapter 9 Test **1.** $\{-3, 3\}$ **2.** $\{-10, 4\}$ **3.** $\left\{-3 - 2\sqrt{2}, -3 + 2\sqrt{2}\right\}$ **4.** $\{-2, 3\}$ **5.** $\left\{\dfrac{2 - \sqrt{10}}{2}, \dfrac{2 + \sqrt{10}}{2}\right\}$ **6.** $\left\{\dfrac{3 - \sqrt{13}}{4}, \dfrac{3 + \sqrt{13}}{4}\right\}$

7. (a) 61 **(b)** quadratic formula **(c)** Answers may vary. **8.** $-\dfrac{1}{2} - \dfrac{3}{2}i$ **9.** $-14 - 4i$ **10.** $1 + 3i$ **11.** -36 **12.** $-5 - 14i$ **13.** $-20 - 48i$

14. $4 - \dfrac{4}{3}i$ **15.** $\dfrac{15}{17} - \dfrac{9}{17}i$ **16.** $\{2 - i, 2 + i\}$ **17.** $\left\{-2 - \dfrac{\sqrt{14}}{2}i, -2 + \dfrac{\sqrt{14}}{2}i\right\}$ **18. (a)** width = $(-1 + \sqrt{127})$ ft, length = $(1 + \sqrt{127})$ ft

(b) width ≈ 10.3 ft, length ≈ 12.3 ft **19. (a)** $\dfrac{25 + \sqrt{785}}{8}$ s **(b)** 6.6 s **20. (a)** width = $(25 + 25\sqrt{161})$ yards, length = $(-50 + 50\sqrt{161})$ yards
(b) width ≈ 342.21 yards, length ≈ 584.43 yards

Cumulative Review Chapters 1–9 **1.** undefined **2.** 0 **3.** $-\dfrac{1}{2}$ **4.** $\dfrac{27x^{12}}{y^2}$ **5.** $\dfrac{9}{x^2 y^6}$ **6. (a)** -1 **(b)** -2 **(c)** 2 **(d)** trinomial

7. (a) $y = \dfrac{3}{2}x - 6$ **(b)** $\dfrac{3}{2}$ **(c)** -6 **8.** $\{-14\}$ **9.** $\{-3\}$ **10.** $\left\{-2, \dfrac{3}{2}\right\}$ **11.** $\{-1, 3\}$ **12.** 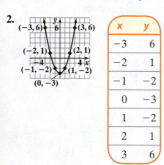 **13.** $y = -\dfrac{3}{2}x + 1$ **14.** $(-3, 2)$ **15.**
16. 96 **17.** $(2h - 5)(3h + 2)$ **18.** $3(x + 2)(x^2 - 2x + 4)$ **19.** $(y + 2)(x - 3)$
20. $3x^2 + x - 1 + \dfrac{4}{2x + 1}$ **21.** $\dfrac{3}{2}x + 3 - \dfrac{2}{x}$ **22.** -4 **23.** $\dfrac{1}{27}$ **24.** 3.4×10^{10}
25. 0.000304 **26.** $11 - 6\sqrt{2}$ **27.** $\dfrac{1}{2} - \dfrac{2}{3}i$ **28.** There were 390 general seating tickets and 190 reserved seat tickets sold.

29. 210 bags **30. (a)** 10 feet **(b)** $\dfrac{9 + \sqrt{129}}{4}$ s **(c)** 5.09 seconds

Chapter 10 Graphs of Quadratic Equations in Two Variables and an Introduction to Functions

Section 10.1 Quadratic Equations in Two Variables **1.** quadratic equation

2.

x	y
−3	6
−2	1
−1	−2
0	−3
1	−2
2	1
3	6

3.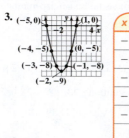

x	y
−5	0
−4	−5
−3	−8
−2	−9
−1	−8
0	−5
1	0

4.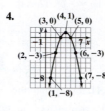

x	y
1	−8
2	−3
3	0
4	1
5	0
6	−3
7	−8

5. True **6.** True **7.** one **8.** y-intercept: $(0, 6)$, x-intercepts: $(-3, 0)$ and $(-2, 0)$ **9.** y-intercept: $(0, -5)$, no x-intercepts **10.** y-intercept: $(0, 1)$;
x-intercept: $\left(-\dfrac{1}{2}, 0\right)$ **11.** up; down **12.** vertex; $x = -\dfrac{b}{2a}$ **13.** $x = -\dfrac{b}{2a}$ **14. (a)** Opens up **(b)** $(1, -1)$ **(c)** $x = 1$

15. (a) Opens down **(b)** $(-3, 11)$ **(c)** $x = -3$

16. 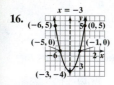 **17.**   **18.** 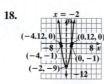 **19.** 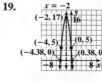 **20.** maximum value
21. True
22. Minimum; -7
23. Maximum; 33
24. (a) After 4 seconds **(b)** 271 feet
25.   **27.**   **29.** 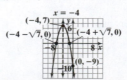 **31.**   **33.** x-intercepts: $(2, 0), (4, 0)$;
y-intercept: $(0, -8)$
35. x-intercepts: $\left(-\dfrac{1}{3}, 0\right), \left(\dfrac{1}{2}, 0\right)$;
y-intercept: $(0, -1)$

37. x-intercepts: $(-3, 0), (0, 0)$; y-intercept: $(0, 0)$ **39.** x-intercepts: none; y-intercept: $(0, -5)$
41. x-intercepts: $\left(\dfrac{1 - \sqrt{17}}{4}, 0\right) \approx (-0.78, 0), \left(\dfrac{1 + \sqrt{17}}{4}, 0\right) \approx (1.28, 0)$; y-intercept: $(0, 2)$ **43.** x-intercepts: $(-2 - 2\sqrt{2}, 0) \approx (-3.41, 0)$,
$(-2 + \sqrt{2}, 0) \approx (-0.59, 0)$; y-intercept: $(0, 2)$ **45. (a)** up **(b)** $(0, 6)$ **(c)** $x = 0$ **47. (a)** down **(b)** $(4, 1)$ **(c)** $x = 4$ **49. (a)** down **(b)** $\left(\dfrac{3}{4}, \dfrac{49}{8}\right)$
(c) $x = \dfrac{3}{4}$ **51. (a)** up **(b)** $\left(\dfrac{4}{3}, -\dfrac{4}{3}\right)$ **(c)** $x = \dfrac{4}{3}$

53. opens up; vertex: $(-1, -4)$;
x-intercepts: $(-3, 0), (1, 0)$;
y-intercept: $(0, -3)$; axis of
symmetry: $x = -1$

55. opens down; vertex: $(-1, 1)$;
x-intercepts: $(-2, 0), (0, 0)$;
y-intercept: $(0, 0)$; axis of symmetry: $x = -1$

57. opens down; vertex: $(-4, 7)$;
x-intercepts: $(-4 - \sqrt{7}, 0)$,
$(-4 + \sqrt{7}, 0)$; y-intercept:
$(0, -9)$; axis of symmetry:
$x = -4$

59. opens up; vertex: $(0, -9)$;
x-intercepts: $(-3, 0), (3, 0)$;
y-intercept: $(0, -9)$; axis of
symmetry: $x = 0$

61. opens up; vertex: $\left(\frac{7}{2}, -\frac{9}{4}\right)$;
x-intercepts: $(2, 0)$, $(5, 0)$
y-intercept: $(0, 10)$; axis of
symmetry: $x = \frac{7}{2}$

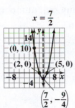

63. opens up; vertex: $(3, 0)$;
x-intercept: $(3, 0)$ y-intercept:
$(0, 9)$; axis of symmetry: $x = 3$

65. opens down; vertex: $(-1, 7)$;
x-intercepts: $\left(\frac{-2 - \sqrt{14}}{2}, 0\right), \left(\frac{-2 + \sqrt{14}}{2}, 0\right)$
y-intercept: $(0, 5)$; axis of symmetry: $x = -1$

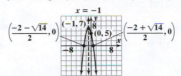

67. opens up; vertex: $(1, -6)$;
x-intercepts: $(1 - \sqrt{6}, 0)$,
$(1 + \sqrt{6}, 0)$; y-intercept: $(0, -5)$
axis of symmetry: $x = 1$

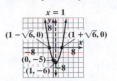

69. minimum; -4 at $x = 3$ **71.** minimum; 7 at $x = -2$ **73.** maximum; $\frac{15}{2}$ at $x = \frac{3}{2}$ **75.** maximum; $-\frac{19}{4}$ at $x = -\frac{3}{2}$ **77. (a)** 50 mph **(b)** 25 miles
per gallon **79. (a)** 25 yd **(b)** 625 yd^2 **81. (a)** 50 ft **(b)** 5000 ft^2

83. opens up; vertex $(-3, 1)$; axis of
symmetry: $x = -3$; y-intercept:
10 x-intercepts: none

85. opens down;
vertex $\left(\frac{3}{2}, -\frac{5}{2}\right)$
axis of symmetry: $x = \frac{3}{2}$
y-intercept: -7
x-intercepts: none

87. To find the number of x-intercepts the graph of a quadratic equation
has, find the discriminant, $b^2 - 4ac$. If $b^2 - 4ac$ is positive, then the parabola
has two x-intercepts, which are found by letting $y = 0$ in the equation and
solving for x. If $b^2 - 4ac = 0$, then the vertex is the x-intercept. If $b^2 - 4ac$ is
negative, then there are no x-intercepts.
89. The axis of symmetry is found using the equation $x = -\frac{b}{2a}$. The axis of
symmetry serves as a guide in graphing a parabola, because every point that
lies on the graph of a parabola contains a mirror-image point on the graph
that lies on the opposite side of the axis of symmetry.

Section 10.2 Relations **1.** relation **2.** corresponds; depends **3.** {(Max, Nov. 8), (Alesia, Jan. 20), (Trent, Mar. 3), (Yolanda, Nov. 8),
(Wanda, July 6), (Elvis, Jan. 8)}

4.

5. domain; range **6.** Domain: {Max, Alesia, Trent, Yolanda, Wanda, Elvis}; Range: {Jan. 20, Mar. 3, July 6, Nov. 8, Jan. 8}
7. Domain: {1, 2, 3, 8}; Range: {5, 4, 7} **8.** Domain: $\{-3, -2, -1, 2, 3\}$; Range: $\{-2, -1, 1, 2, 3, 4\}$ **9.** True
10. False **11.** Domain: $\{x \mid -2 \leq x \leq 4\}$; Range: $\{y \mid -2 \leq y \leq 2\}$ **12.** Domain: $\{x \mid x$ is any real number$\}$; Range:
$\{y \mid y$ is any real number$\}$

13.

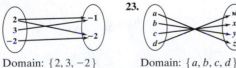

Domain: $\{x \mid x$ is any real number$\}$
Range: $\{y \mid y$ is any real number$\}$

14.

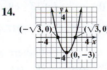

Domain: $\{x \mid x$ is any real number$\}$
Range: $\{y \mid y \geq -3\}$

15. {(Physician, 120), (Teacher, 37), (Mathematician, 64), (Computer
Engineer, 64), (Lawyer, 82)}; Domain: {(Physician, Teacher, Mathematician,
Computer Engineer, Lawyer}; Range: {120, 37, 64, 82}
17. {(Spain, 81), (Canada, 81), (Kenya, 63), (Ireland, 80), (France, 81)}; Domain:
{Spain, Canada, Kenya, Ireland, France}; Range: {81, 80, 63}

19.

Domain: $\{-1, -4, 0, 1\}$
Range: {3, 2, 1}

21.

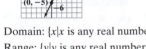

Domain: $\{2, 3, -2\}$
Range: $\{-1, -2\}$

23.

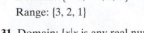

Domain: $\{a, b, c, d\}$
Range: $\{w, x, y, z\}$

25. Domain: $\{-3, 0, 2, 3\}$
Range: $\{-2, 0, 2, 4\}$
27. Domain: $\{x \mid x$ is any real number$\}$
Range: $\{y \mid y$ is any real number$\}$
29. Domain: {3}
Range: $\{y \mid y$ is any real number$\}$

31. Domain: $\{x \mid x$ is any real number$\}$
Range: $\{y \mid y \leq 0\}$
33. Domain: $\{x \mid x$ is any real number$\}$
Range: $\{y \mid y \geq -2\}$
35. Domain: $\{x \mid x \geq -1\}$
Range: $\{y \mid y$ is any real number$\}$
37. Domain: $\{x \mid -1 \leq x \leq 5\}$
Range: $\{y \mid -1 \leq y \leq 5\}$
39. Domain: $\{x \mid -4 \leq x \leq 4\}$
Range: $\{y \mid -3 \leq y \leq 3\}$

41.

Domain: $\{x \mid x$ is any real number$\}$
Range: $\{y \mid y$ is any real number$\}$

43.

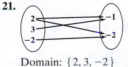

Domain: $\{x \mid x$ is any real number$\}$
Range: $\{y \mid y$ is any real number$\}$

45.

Domain: $\{x \mid x$ is any real number$\}$
Range: $\{y \mid y \leq 2\}$

47.

$\left(\frac{7}{2}, -\frac{9}{4}\right)$

Domain: $\{x|x$ is any real number$\}$

Range: $\left\{y\middle|y \geq -\frac{9}{4}\right\}$

49.

Domain: $\{x|x$ is any real number$\}$

Range: $\{y|y$ is any real number$\}$

51.

Domain: $\{x|x$ is any real number$\}$

Range: $\{y|y \geq -5\}$

53.

Domain: $\{x|x$ is any real number$\}$

Range: $\{y|y \leq 6\}$

55.

Domain: $\{x|x$ is any real number$\}$

Range: $\{y|y$ is any real number$\}$

57. (a) Domain: $\{t|0 \leq t \leq 10\}$; Range: $\{p|0 \leq p \leq 900\}$ **(b)** day 5   **(c)** 5 days

59. (a) Domain: $\{w|0 < w \leq 5\}$; Range: $\{4.80, 5.20, 5.80, 6.45\}$ **(b)** \$4.80   **(c)** \$5.20

61. Answers may vary.   **63. (a)** No   **(b)** Answers may vary.   **(c)** No; answers may vary.

65. Answers may vary. All graphs are vertical lines.   **67.** A relation is a correspondence between the elements of two sets. The domain of the relation is the set of all inputs of the relation, and the range is the set of all outputs of the relation.

Putting the Concepts Together (Sections 10.1 and 10.2)   **1. (a)** Opens up   **(b)** $(1, -4)$ **(c)** $x = 1$ **(d)** $(0, -3)$, $(-1, 0)$, $(3, 0)$

2.

Domain: $\{x|x$ is any real number$\}$

Range: $\{y|y \geq -9\}$

3.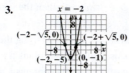

Domain: $\{x|x$ is any real number$\}$

Range: $\{y|y \geq -5\}$

4.

Domain $\{x|x$ is any real number$\}$

Range: $\{y|y \leq 9\}$

5. (a) Minimum   **(b)** -21

6. Domain: $\{a, b, c\}$; Range: $\{1, 2\}$

7. Domain: $\{1, 3\}$; Range: $\{-1, -3\}$

8. Domain: $\{x|x \geq -4\}$

Range: $\{y|y$ is any real number$\}$

9. Domain: $\{x|1 \leq x \leq 7\}$

Range: $\{y|-1 \leq y \leq 5\}$

Section 10.3 An Introduction to Functions   **1.** Function   **2.** False   **3.** Function; Domain: $\{250, 300, 400, 500\}$; Range: $\{\$41.20, \$43.04, \$55.39, \$64.03\}$

4. Function; Domain: $\{12, 15, 8, 7, 4\}$; Range: $\{A, B, C, D\}$ **5.** not a function   **6.** Function; Domain: $\{-3, -2, -1, 0, 1\}$; Range: $\{3, 2, 1, 0\}$

7. not a function   **8.** Function; Domain: $\{3, 4, 5, 6\}$; Range: $\{8, 10, 11\}$ **9.** function   **10.** not a function   **11.** function   **12.** not a function   **13.** True

14. function   **15.** not a function   **16.** True   **17.** 17   **18.** -7 **19.** 15   **20.** 22   **21.** not a function   **23.** function; Domain: $\{$Apples, Pears, Oranges, Grapes$\}$; Range: $\{\$0.99, \$1.29, \$1.99\}$ **25.** function; Domain: $\{-1, 0, 1, 2\}$; Range: $\{-1, 0, 1, 2\}$ **27.** not a function   **29.** function; Domain: $\{a, b, c, d\}$; Range: $\{a, b\}$ **31.** function   **33.** not a function   **35.** function   **37.** function   **39.** not a function   **41.** not a function   **43.** not a function

45. not a function   **47.** function   **49.** function   **51.** not a function   **53.** function   **55.** not a function   **57.** function   **59. (a)** -3 **(b)** -6 **(c)** -1

61. (a) 4   **(b)** -2 **(c)** 8   **63. (a)** 3   **(b)** 3   **(c)** 3   **65. (a)** -3 **(b)** $-\frac{3}{2}$ **(c)** -4 **67. (a)** 3   **(b)** -12 **(c)** 13   **69. (a)** 0   **(b)** 0   **(c)** 10

71. (a) 1   **(b)** -32 **(c)** 3   **73.** 18   **75.** -8 **77.** 0   **79.** 2   **81.** $-\frac{1}{2}$ **83.** 7   **85.** $2\sqrt{3}$ **87.** $C = 5$ **89.** $C = 3$ **91.** $A = 5$ **93.** $G(h) = 12h$; \$240

95. (a) $h(30) = 8.132$; in 1930, there were 8.132 homicides per 100,000 people.   **(b)** $h(50) = 6.632$; in 1950, there were 6.632 homicides per 100,000 people.   **97.** $A(r) = \pi r^2$; $A(3) = 9\pi$ in.^{2} **99.** $x = 2$ **101.** $x = 2$ **103.** $x = -4$ **105. (a)** -5 **(b)** $1 - 2h$ **(c)** $-5 - 2h$ **(d)** $-4 - 2h$ **(e)** $-2h$

107. The four ways to represent a function are maps, ordered pairs, equations, and graphs. Answers may vary.

Chapter 10 Review

1.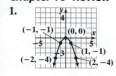

2.

3.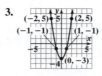

4.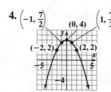

5.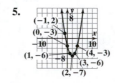

6.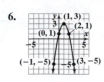

7. x-intercepts: $(-3, 0)$, $(2, 0)$;   **8.** x-intercepts: $(-2, 0)$, $(-4, 0)$;   **9.** x-intercept: $\left(\frac{3}{2}, 0\right)$;   **10.** x-intercept: $\left(-\frac{1}{3}, 0\right)$;   **11.** x-intercepts: none;

y-intercept: $(0, -6)$ y-intercept: $(0, 8)$ y-intercept: $(0, -9)$ y-intercept: $(0, -1)$ y-intercept: $(0, -5)$

12. x-intercepts: none;   **13.** x-intercepts: $(2 - \sqrt{2}, 0)$, $(2 + \sqrt{2}, 0)$;   **14.** x-intercepts: $(-2 - 3\sqrt{2}, 0)$, $(-2 + 3\sqrt{2}, 0)$;

y-intercept: $(0, -1)$ y-intercept: $(0, 2)$ y-intercept: $(0, 14)$

15. (a) up   **(b)** $(0, 5)$ **(c)** $x = 0$ **16. (a)** down   **(b)** $(0, -4)$ **(c)** $x = 0$ **17. (a)** down   **(b)** $(1, 1)$ **(c)** $x = 1$ **18. (a)** up   **(b)** $(2, -8)$ **(c)** $x = 2$

19. (a) up   **(b)** $\left(-3, -\frac{27}{2}\right)$ **(c)** $x = -3$ **20. (a)** down   **(b)** $(-2, 6)$ **(c)** $x = -2$ **21. (a)** down   **(b)** $\left(\frac{3}{2}, -\frac{7}{4}\right)$ **(c)** $x = \frac{3}{2}$

22. (a) up   **(b)** $\left(-\frac{5}{2}, -\frac{5}{4}\right)$ **(c)** $x = -\frac{5}{2}$

23. opens up; vertex $(-1, -4)$; axis of symmetry: $x = -1$ x-intercepts: $(-3, 0)$, $(1, 0)$ y-intercept: $(0, -3)$

24. opens down; vertex $(3, 1)$; axis of symmetry: $x = 3$; x-intercepts: $(2, 0)$, $(4, 0)$ y-intercept: $(0, -8)$

25. opens down; vertex $(0, 4)$; axis of symmetry: $x = 0$ x-intercepts: $(-2, 0)$, $(2, 0)$ y-intercept: $(0, 4)$

26. opens up; vertex $(0, -9)$; axis of symmetry: $x = 0$; x-intercepts: $(-3, 0)$, $(3, 0)$; y-intercept: $(0, -9)$

27. opens down; vertex $\left(\dfrac{3}{2}, \dfrac{25}{4}\right)$; axis of symmetry: $x = \dfrac{3}{2}$; x-intercepts: $(-1, 0)$, $(4, 0)$; y-intercept: $(0, 4)$

28. opens up; vertex $\left(\dfrac{5}{2}, -\dfrac{9}{4}\right)$; axis of symmetry: $x = \dfrac{5}{2}$ x-intercepts: $(1, 0)$, $(4, 0)$ y-intercept: $(0, 4)$

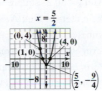

29. maximum; 15 at $x = -2$ **30.** minimum; -16 at $x = -2$
31. minimum; -10 at $x = 2$ **32.** maximum; 10 at $x = -5$
33. (a) 92.9 ft   **(b)** 20.1 ft   **(c)** 200 ft   **34. (a)** 100 m   **(b)** 10,000 m^2
35. {(Alabama, 13.8), (California, 11.4), (Florida, 17.3), (Nebraska, 13.5), (New York, 13.5), (Massachusetts, 13.8)}
36. {(History, 400), (Math, 400), (Music, 200), (Psychology, 225)}
37. Domain: {Alabama, California, Florida, Nebraska, New York, Massachusetts} Range: {13.8, 13.5, 17.3, 11.4}
38. Domain: {History, Math, Music, Psychology}   **39.** Domain: $\{-1, 0\}$
Range: {200, 225, 400}   Range: $\{3, 5, -2, -4\}$

40. Domain: $\{a, b, c, d\}$ **41.** Domain: $\{a_1, a_2, a_3, a_4\}$ **42.** Domain: $\{1, 3, 5, 7\}$ **43.** Domain: $\{-3, 0, 3\}$ **44.** Domain: $\{0, 1, 4\}$
Range: $\{x^2, x^3, x^4, x^5\}$ Range: $\{1, 4, 9, 16\}$ Range: $\{-3, 2, -2\}$ Range: $\{-3, 0, 3\}$ Range: $\{-4, 0, 4\}$

45. Domain: $\{x \mid x \text{ is any real number}\}$ **46.** Domain: $\{x \mid x \text{ is any real number}\}$ **47.** Domain: $\{5\}$
Range: $\{y \mid y \text{ is any real number}\}$ Range: $\{4\}$ Range: $\{y \mid y \text{ is any real number}\}$
48. Domain: $\{x \mid x \text{ is any real number}\}$ **49.** Domain: $\{x \mid 0 \le x \le 6\}$ **50.** Domain: $\{x \mid -4 \le x \le 0\}$
Range: $\{y \mid y \text{ is any real number}\}$ Range: $\{y \mid 1 \le y \le 5\}$ Range: $\{y \mid -4 \le y \le 3\}$

51.

Domain: $\{x \mid x \text{ is any real number}\}$
Range: $\{y \mid y \text{ is any real number}\}$

52.

Domain: $\{x \mid x \text{ is any real number}\}$
Range: $\{y \mid y \text{ is any real number}\}$

53.

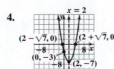

Domain: $\{x \mid x \text{ is any real number}\}$
Range: $\{y \mid y \le 4\}$

54.

Domain: $\{x \mid x \text{ is any real number}\}$
Range: $\{y \mid y \ge -1\}$

55. not a function   **56.** function; Domain: {Alaska, Massachusetts, Minnesota, West Virginia}; Range: {17.3, 26.6, 31.5, 38.2}
57. function; Domain: $\{-4, -3, -2, -1\}$; Range: $\{0, 1\}$ **58.** not a function   **59.** not a function   **60.** function; Domain: $\{5, 3, 1, -1\}$; Range: $\{x, z\}$
61. function   **62.** not a function   **63.** not a function   **64.** function   **65.** not a function   **66.** function   **67.** function   **68.** not a function
69. not a function   **70.** function   **71.** not a function   **72.** function   **73.** function   **74.** not a function   **75.** function   **76.** function   **77.** not a function
78. not a function   **79.** 1   **80.** 18   **81.** -3 **82.** -10 **83.** 14   **84.** 7   **85.** $2\sqrt{6}$ **86.** 5   **87. (a)** $C(m) = 0.10m + 175$ **(b)** $188

Chapter 10 Test   **1. (a)** up   **(b)** $(-2, -9)$ **(c)** $x = -2$ **(d)** x-intercepts: $(-5, 0)$, $(1, 0)$; y-intercept: $(0, -5)$

2.

3.

4.

5. (a) maximum   **6.** Domain: $\{1\}$
(b) 7 at $x = -1$ Range: $\{-1, -2, -3\}$
7. Domain: $\{-2, -3\}$ **8.** Domain: $\{x \mid x \text{ is any real number}\}$
Range: $\{a, b, c\}$ Range: $\{y \mid y \ge -2\}$

9. Domain: $\{x \mid -5 \le x \le 3\}$; Range: $\{y \mid 0 \le y \le 4\}$ **10.** not a function   **11.** function   **12.** function   **13.** function   **14.** not a function

15. function   **16.** -5 **17.** $-\dfrac{1}{5}$ **18.** 8   **19. (a)** 1 sec   **(b)** 16 ft   **20. (a)** $3   **(b)** 9.8 years   **(c)** $35   **(d)** $27

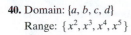

Appendix B Geometry Review

Section B.1 Lines and Angles   **1.** congruent   **2.** angle   **3.** right   **4.** acute   **5.** obtuse   **6.** straight   **7.** right   **8.** False   **9.** 75°; 165°
10. 30°; 120°   **11.** Parallel   **12.** False   **13.** True   **14.** $m\angle 1 = 140°$; $m\angle 2 = 40°$; $m\angle 3 = 140°$; $m\angle 4 = 40°$; $m\angle 5 = 140°$;
$m\angle 6 = 40°$; $m\angle 7 = 140°$ **15.** acute   **17.** right   **19.** straight   **21.** obtuse   **23.** 58°   **25.** 17°   **27.** 113°   **29.** 172°
31. $m\angle 1 = 130°$; $m\angle 2 = 50°$; $m\angle 3 = 130°$; $m\angle 4 = 50°$; $m\angle 5 = 130°$; $m\angle 6 = 50°$; $m\angle 7 = 130°$

Section B.2 Polygons   **1.** isosceles   **2.** right   **3.** 180   **4.** 70°   **5.** 42°   **6.** Congruent   **7.** Similar   **8.** 14 units   **9.** radius   **10.** True
11. $\frac{15}{2}$ inches or 7.5 inches   **12.** 12 feet   **13.** 7.2 yards   **14.** 18 cm   **15.** 55°   **17.** 48°   **19.** 4 units   **21.** 67.5 units   **23.** 10 inches   **25.** 5 cm
27. 7 cm   **29.** $\frac{11}{2}$ yards or 5.5 yards

Section B.3 Perimeter and Areas of Polygons and Circles   **1.** perimeter   **2.** area   **3.** Perimeter: 22 feet; Area: 24 square feet
4. Perimeter: 26 m; Area: 30 square m   **5.** False   **6.** Perimeter: 16 cm; Area 16 square cm   **7.** Perimeter: 6 yards; Area: 2.25 square yards
8. Perimeter: 130 yards: Area: 650 square yards   **9.** $\frac{1}{2}h(b + B)$; h; b; B **10.** Perimeter: 36 m; Area: 70 square m
11. Perimeter: 33 yards; Area: 51 square yards   **12.** True   **13.** Perimeter: 19 mm: Area: 12 square mm   **14.** Perimeter: 30 feet; Area: 30 square feet
15. Circumference   **16.** False   **17.** Circumference: 8π feet $\approx$ 25.13 feet; Area: 16π square feet $\approx$ 50.27 square feet
18. Circumference: 24π cm $\approx$ 75.40 cm; Area: 144π square cm $\approx$ 452.39 square cm   **19.** Perimeter: 28 feet; Area: 40 square feet
21. Perimeter: 40 m; Area: 75 square m   **23.** Perimeter; 24 km; Area: 36 square km   **25.** Perimeter: 72 feet: Area: 218 square feet
27. Perimeter: 54 m; Area: 62 square m   **29.** Perimeter: 30 feet: Area: 45 square feet   **31.** Perimeter: 28 mm; Area: 36 square mm
33. Perimeter: 40 in; Area: 84 square in.   **35.** Perimeter: 45 cm; Area: 94.5 square cm   **37.** Perimeter: 32 m; Area: 42 square m
39. Perimeter: 32 feet; Area: 24 square feet   **41.** Circumference; 32π in. $\approx$ 100.53 in; Area: 256π square in. $\approx$ 804.25 square in.
43. Circumference: 20π cm $\approx$ 62.83 cm; Area 100π square cm $\approx$ 314.16 square cm   **45.** π square units   **47.** about 26.18 feet

Section B. 4 Volume and Surface Area   **1.** polyhedron   **2.** surface area   **3.** False   **4.** True   **5.** Volume: 125 cubic m; Surface area: 150 square m
6. Volume: $\frac{256}{3}\pi$ cubic in. $\approx$ 268.08 cubic in.; Surface area: 64π square in. $\approx$ 201.06 square in.   **7.** Volume: 600 cubic feet; Surface Area: 460 square feet
9. Volume: 288π cubic centimeters $\approx$ 904.78 cubic centimeters; Surface Area: 144π square centimeters $\approx$ 452.39 square centimeters
11. Volume: 32π cubic inches $\approx$ 100.53 cubic inches: Surface Area: 40π square inches $\approx$ 125.66 square inches
13. Volume: $\frac{800}{3}\pi$ cubic mm $\approx$ 837.76 cubic mm   **15.** Volume $\frac{640}{3}$ cubic feet; Surface Area: 256 square feet   **17.** 1728 cubic inches
19. 75.40 cubic inches; 100.53 square inches   **21.** approximately 268.08 cubic cm

Applications Index

Subject Index

Photo Credits

Working with Square Roots (Chapter 8)

- If a is a real number, then $\sqrt{a^2} = |a|$

- **Product Property of Radicals**
  If $\sqrt{a}$ and $\sqrt{b}$ are real numbers, then $\sqrt{a} \cdot \sqrt{b} = \sqrt{ab}$

- **Quotient Property of Radicals**
  If $\sqrt{a}$ and $\sqrt{b}$ are real numbers, $b \neq 0$, then $\dfrac{\sqrt{a}}{\sqrt{b}} = \sqrt{\dfrac{a}{b}}$

- If a is a real number and $n \geq 2$ is an integer, then $a^{\frac{1}{n}} = \sqrt[n]{a}$ provided that $\sqrt[n]{a}$ exists.

- If a is a real number, and if $\dfrac{m}{n}$ is a rational number in lowest terms with $n \geq 2$, then
  $a^{\frac{m}{n}} = \sqrt[n]{a^m} = (\sqrt[n]{a})^m$ provided that $\sqrt[n]{a}$ exists.

- If $\dfrac{m}{n}$ is a rational number, and if a is a nonzero real number, then $a^{-\frac{m}{n}} = \dfrac{1}{a^{\frac{m}{n}}}$ or $\dfrac{1}{a^{-\frac{m}{n}}} = a^{\frac{m}{n}}$

Quadratic Equations (Chapter 9)

- **Square Root Property**
  If $x^2 = p$, then $x = \sqrt{p}$ or $x = -\sqrt{p}$

- **Pythagorean Theorem**
  In a right triangle, the square of the length of the hypotenuse is equal to the sum of the squares of the lengths of the legs. That is, $\text{leg}^2 + \text{leg}^2 = \text{hypotenuse}^2$.

- **The Quadratic Formula**
  The solutions to the equation $ax^2 + bx + c = 0$, $a \neq 0$, are given by $x = \dfrac{-b \pm \sqrt{b^2 - 4ac}}{2a}$

- **Discriminant**
  For the quadratic equation $ax^2 + bx + c = 0$, $a \neq 0$:
  - If $b^2 - 4ac > 0$, the equation has two unequal real solutions.
    - If $b^2 - 4ac$ is a perfect square, the equation has two rational solutions.
    - If $b^2 - 4ac$ is not a perfect square, the equation has two irrational solutions.
  - If $b^2 - 4ac = 0$, the equation has a repeated real solution.
  - If $b^2 - 4ac < 0$, the equation has two complex solutions that are not real.

Steps for Factoring (Chapter 6)

Step 1: Factor out the greatest common factor (GCF), if any exists.

Step 2: Count the number of terms.

Step 3: **(a)** Two terms

- Is it the difference of two squares? If so, $A^2 = B^2 = (A - B)(A + B)$
- Is it the sum of two squares? If so, stop! The expression is prime.
- Is it the difference of two cubes? If so, $A^3 - B^3 = (A - B)(A^2 + AB + B^2)$
- Is it the sum of two cubes? If so, $A^3 + B^3 = (A + B)(A^2 - AB + B^2)$

(b) Three terms
- Is it a perfect square trinomial? If so, $A^2 + 2AB + B^2 = (A + B)^2$ or $A^2 - 2AB + B^2 = (A - B)^2$
- Is the coefficient of the square term 1? If so, $x^2 + bx + c = (x + m)(x + n)$, where $mn = c$ and $m + n = b$
- Is the coefficient of the square term different from 1? If so,
  (a) Use factoring by grouping
  (b) Use trial and error

(c) Four terms
- Use factoring by grouping

Step 4: Check your work by multiplying out the factored form.

Formulas for Lines and Slope (Chapter 3)

Standard form of a line	$Ax + By = C$
Equation of a vertical line	$x = a$, where $(a, 0)$ is the x-intercept
Equation of a horizontal line	$y = b$, where $(0, b)$ is the y-intercept
Slope of a line	$m = \dfrac{y_2 - y_1}{x_2 - x_1}, x_1 \neq x_2$ Slope undefined if $x_1 = x_2$
Point-slope form of a line	$y - y_1 = m(x - x_1)$
Slope-intercept form of a line	$y = mx + b$

The Rules of Exponents (Chapter 5, Chapter 8)

If a and b are real numbers and if r and s are rational numbers, then assuming the expression is defined,

Zero-Exponent Rule:	$a^0 = 1$	if $a \neq 0$
Negative-Exponent Rule:	$a^{-r} = \dfrac{1}{a^r}$	if $a \neq 0$
Product Rule:	$a^r \cdot a^s = a^{r+s}$	
Quotient Rule:	$\dfrac{a^r}{a^s} = a^{r-s} = \dfrac{1}{a^{s-r}}$	if $a \neq 0$
Power Rule:	$(a^r)^s = a^{r \cdot s}$	
Product-to-a-Power Rule:	$(a \cdot b)^r = a^r \cdot b^r$	
Quotient-to-a-Power Rule:	$\left(\dfrac{a}{b}\right)^r = \dfrac{a^r}{b^r}$	if $b \neq 0$
Quotient-to-a-Negative-Power Rule:	$\left(\dfrac{a}{b}\right)^{-r} = \left(\dfrac{b}{a}\right)^r$	if $a \neq 0, b \neq 0$

Working with Rational Expressions (Chapter 7)

Multiplying Rational Expressions	$\dfrac{a}{b} \cdot \dfrac{c}{d} = \dfrac{ac}{bd}$	$b \neq 0, d \neq 0$
Adding Rational Expressions	$\dfrac{a}{c} + \dfrac{b}{c} = \dfrac{a + b}{c}$	$c \neq 0$
Subtracting Rational Expressions	$\dfrac{a}{c} - \dfrac{b}{c} = \dfrac{a - b}{c}$	$c \neq 0$
Dividing Rational Expressions	$\dfrac{a}{b} \div \dfrac{c}{d} = \dfrac{\frac{a}{b}}{\frac{c}{d}} = \dfrac{a}{b} \cdot \dfrac{d}{c} = \dfrac{ad}{bc}$	$b \neq 0, c \neq 0,$ $d \neq 0$